INFORMATION CONTROL PROBLEMS IN MANUFACTURING 2004

(INCOM 2004)

AF180358

A Proceedings volume from the 11th IFAC Symposium,
Salvador, Brazil, 5 – 7 April 2004

Edited by

P. KOPACEK
Institute for Handling Devices and Robotics,
Vienna University of Technology, Austria

C.E. PEREIRA
Department of Electrical Engineering,
Federal University of Rio Grande do Sul,
Porto Alegre, Brazil

and

G. MOREL
Research Centre for Automatic Control,
CRAN, Faculté des Sciences,
Vandoeuvre-les-Nancy, France

(In two volumes)

Volume 1

Published for the

INTERNATIONAL FEDERATION OF AUTOMATIC CONTROL

by

ELSEVIER LIMITED

ELSEVIER Ltd
The Boulevard, Langford Lane
Kidlington, Oxford OX5 1GB, UK

Elsevier Internet Homepage
http://www.elsevier.com

Consult the Elsevier Homepage for full catalogue information on all books, journals and electronic products and services.

IFAC Publications Internet Homepage
http://www.elsevier.com/locate/ifac

Consult the IFAC Publications Homepage for full details on the preparation of IFAC meeting papers, published/forthcoming IFAC books, and information about the IFAC Journals and affiliated journals.

Copyright © 2005 IFAC

All Rights Reserved. No part of this publication may be reproduced, stored in a retrieval system or transmitted in any form or by any means: electronic, electrostatic, magnetic tape, mechanical, photocopying, recording or otherwise, without permission in writing from the copyright holders.

First edition 2005

Library of Congress Cataloging in Publication Data

A catalogue record for this book is available from the Library of Congress

British Library Cataloguing in Publication Data

A catalogue record for this book is available from the British Library

ISBN 0-08-044249 8
ISSN 1474-6670

These proceedings were reproduced from manuscripts supplied by the authors, therefore the reproduction is not completely uniform but neither the format nor the language have been changed in the interests of rapid publication. Whilst every effort is made by the publishers to see that no inaccurate or misleading data, opinion or statement appears in this publication, they wish to make it clear that the data and opinions appearing in the articles herein are the sole responsibility of the contributor concerned. Accordingly, the publisher, editors and their respective employers, officers and agents accept no responsibility or liability whatsoever for the onsequences of any such inaccurate or misleading data, opinion or statement.

Printed in Great Britain

To Contact the Publisher

Elsevier welcomes enquiries concerning publishing proposals: books, journal special issues, conference proceedings, etc. All formats and media can be considered. Should you have a publishing proposal you wish to discuss, please contact, without obligation, the publisher responsible for Elsevier's industrial and control engineering publishing programme:

Christopher Greenwell
Publishing Editor
Elsevier Ltd
The Boulevard, Langford Lane
Kidlington, Oxford
OX5 1GB, UK

Phone: +44 1865 843230
Fax: +44 1865 843920
E.mail: c.greenwell@elsevier.com

General enquiries, including placing orders, should be directed to Elsevier's Regional Sales Offices – please access the Elsevier homepage for full contact details (homepage details at the top of this page).

11th IFAC SYMPOSIUM ON INFORMATION CONTROL PROBLEMS IN MANUFACTURING 2004

Sponsored by
International Federation of Automatic Control (IFAC)
IFAC Technical Committee on Manufacturing Plant Control

Co-sponsored by
IFAC Technical Committees on:
- Discrete Event Dynamic Systems
- Computers for Control
- Components and Instruments
- Cost-Oriented Automation
- Human Machine Systems
- Manufacturing Modeling for Management and Control
- Enterprise Integration and Networking
- Large Scale Complex Systems
- Social Impact of Automation

IFIP Working Group 5.7 - Integration on Product Management
IMS - NOE - Network of Excellence - Europe
SBA - Brazilian Automation Society (Brazilian IFAC NMO)
SBA - TC on Industrial Automation
MANET - Brazilian Research Network on Manufacturing Automation

Conference Chairs

Pereira, C.E. (BR)	(General Chair)
Morel, G. (F)	(IPC Chair)
Silva, J.R. (BR)	(Local Organization Chair)
Bernhardt, R. (DE)	(Industrial Chair)

International Programme Committee (IPC)
Albertos, P. (ES)
Banaszak, Z. (PL)
Bititci, U. (UK)
Boer, C.R. (I)
Boverie, S. (F)
Diedrich, C. (DE)
Dolgui, A. (F)
Erbe, H.-H. (DE)
Faure, J.-M. (F)
Florin, F. (RO)
Fonseca, J.A. (P)
Frey, G. (DE)
Garetti, M. (I)
Grabot, B. (F)
Johnson, T. (USA)
Kopacek, P. (A)
Lee, J. (USA)
McFarlane, D. (UK)
Molina, A. (MX)
Monostori, L. (HU)
Morel, G. (F)
Neagu, G. (RO)
Nof, S. (USA)
Pereira, C.E. (BR)
Sanz, R. (ES)
Silva, J.R. (BR)
Stahre, J. (S)
Valckenaers, P. (B)
Vingerhoeds, R. (F)
Zaremba, M. (CAN)
Zaytoon, J. (F)
Zuehlke, D. (DE)

Local Organization
SBA : Brazilian IFAC National Member Organization
MANET: Manufacturing Automation Network – Brazilian Network of Excellence
GCAR/UFRGS: Control, Automation, and Robotics Group - Federal University of Rio Grande do Sul
SENAI/CIMATEC
UFBA
UNIFACS
FTC

Financial Support
CNPq
FINEP
CAPES
MANET
IMS - NOE
KUKA Roboter
FANUC Robotics
FAPESB
USP/EPUSP
UFRGS/EE/DELET/GCAR

PREFACE

INCOM'04 is a tri-annual event of symposia series organized by IFAC under the sponsoring of the Technical Committee 5.1 on ***Manufacturing Plant Control***. Previous editions were held in Vienna (2001), Nancy (1998) and Beijing (1995).

This 11th edition was co-sponsored by 9 others IFAC Technical Committees covering in large theoretical, technological and applications issues of manufacturing plant automation for Shop-Floor Process-Control as well as for Enterprise-Control System Integration.

INCOM'04 main purpose was to point up international researches and developments dealing with all the applications of automation, information and communication technologies in order to control and to manage the manufacturing plant within the e-enterprise, involving all methodological and technological aspects to embed a technical "intelligence" within the components. The symposium addressed automation scientific challenges and issues raised by the IMS paradigm in order to apply MEMS, MECHATRONICS, MES, MAS, HMS and e-technologies to digitally control with more agility the entire manufacturing chain, from supply and design through manufacturing, to maintenance and service, over the whole product and processes life cycle. Information issues at the interface of the technical and business processes were discussed as well as societal and human impact (safety, dependability, usability, knowledge, and experience) of these emerging e-manufacturing technologies..

INCOM'04 Technical Program was organized around 6 technical tracks under track chairs ensuring the scientific relevance of 6 Manufacturing Plant Control Areas:

- **Production & Logistics over Manufacturing Networking**, with an invited plenary lecture on *Collaborative E-Work And E-Mfg.: The State Of The Art And Challenges For Production And Logistics Managers*, and 2 regular sessions on ***Business to Manufacturing Issues***,
- **Manufacturing Automation over Networks**, with an invited plenary lecture on ***Communication In Industrial Automation - What Is Going On?***, an invited session on ***Real-Time Distributed Embedded Systems,*** and a regular session on ***Networked Controlled Systems,***
- **Dependable Manufacturing Systems Control**, with an invited plenary lecture on *Improving Automation Software Dependability: A Role for Formal Methods?*, an invited session on ***Industrial controllers software dependability : formal methods applications***, and a regular session on ***Design and implementation of dependable systems***,
- **Discrete Event Systems in Manufacturing**, with an invited plenary lecture on *Analysis of Hybrid Systems- State of the Art*, and 2 regular sessions on ***Petri nets based approaches*** and ***Design and control synthesis of DES***,
- **E-Manufacturing Technologies and Facilities**, with an invited plenary lecture on *Infotronics Technologies and Predictive Tools for Next-Generation Maintenance*

Systems, and 2 regular sessions on ***Cooperative Manufacturing and Engineering*** and ***Methods and Technologies towards E_Maintenance***,
- **IMS Modelling and Experiments**, with 2 invited plenary lectures on *Industrial Applications of the Agent-Based Technology* and *How Information Technology Enables the Potential Flexibility and Re-configurability of Highly Automated Manufacturing Plants*, and 2 regular sessions on ***IMS Engineering trends and issues*** and ***B2M Performance Issue.***

This core technical programme is completed by 2 other areas:

- **Research Networks**, with 2 invited sessions on ***Remote Experiments for e-learning (Virtual Laboratories)***, and ***B2M Systems Interoperability: Modeling and Benchmarking Issues,***
- Posters, with 3 regular sessions on *Manufacturing Modelling Issues*, *Manufacturing Automation Issues*, and *Manufacturing Assembly and Robotics Issues,* as well as a special session on *Work-In-Progress.*

All the **163 papers submitted** to **INCOM'04,** which came from 25 different countries, have been reviewed by at least 3 reviewers among a panel of **171** relevant scientists and industrialists, leading to the presentation of **7 keynote talks** and **76 papers** in the 6 technical tracks of **INCOM'04** core technical programme and the presentation of *20 papers* in the complementary area of the complete technical programme.

We hope that all readers find the contents of this conference proceedings intellectually stimulating and offering an opportunity to address the above important challenge of defining a coherent framework of concepts, theories, models, methods, methodologies, languages and tools for *Manufacturing Plant Automation and Integration.*

Last but not least, we would like to express our gratitude to all people acting in this promising event, coming from the National Organizing Committee for their enthusiastic and unflagging contribution, and coming from the IPC, IFAC TCs, IFIP TCs, and the Manufacturing community in large for their high scientific contribution.

Salvador da Bahia, April 2004.

Gérard Morel
Carlos Eduardo Pereira
José Reinaldo Silva
Alceu Heinke Frigeri

CONTENTS

VOLUME 1

PRODUCTION AND LOGISTICS OVER MANUFACTURING NETWORKING

Real-Time Distributed Embedded Systems

DEPENDABLE MANUFACTURING SYSTEMS CONTROL

Keynote Paper III

Industrial Controllers Software Dependability: Formal Methods Applications

Design and Implementation of Dependable Systems

WIP – WORK IN PROGRESS PAPERS

Work in Progress I

Work in Progress II

Work in Progress III

Copyright © IFAC Information Control Problems in
Manufacturing, Salvador, Brazil, 2004

ELSEVIER
IFAC
PUBLICATIONS
www.elsevier.com/locate/ifac

COLLABORATIVE e-WORK AND e-MFG.: CHALLENGES FOR PRODUCTION AND LOGISTICS MANAGERS

Shimon Y. Nof

PRISM Center – Production, Robotics, and Integration Software for Mfg. & Management
Purdue University, West Lafayette, Indiana, 47907-2023, USA
nof@purdue.edu

Abstract: The foundations and scope of e-Work and e-Mfg., and investigations of fundamental design principles for their effectiveness are reviewed. The premise is that without effective e-Work, potential benefits of emerging and promising electronic work activities, such as virtual manufacturing, tele-robotic maintenance, automated inspection, e-supply and e-logistics cannot be materialized. Challenges and emerging solutions are described, clustered in four areas: 1. e-Work; 2. Integration, Coordination and Collaboration; 3. Distributed Decision Support; and 4. Active Middleware. Design principles, models and tools are described, and challenges in production and logistics management are illustrated. It is intended to stimulate and inspire fruitful future investigations, revolutionizing the way we can design beneficial e-Work.
Copyright © 2004 IFAC
Keywords: Active protocols, \middleware, autonomous agents, collaborative robotics, conflict resolution, design principles, distributed operations, integration, parallelism.

1. INTRODUCTION: SMART TEAMS IN e-WORK AND e-MFG.

Collaborative e-Work and e-Mfg. are based on smart teams. e-Work was defined by the PRISM Center as any collaborative, computer-supported and communication-enabled productive activities in highly distributed organizations of humans and/or robots or autonomous systems (Figure 1). In essence, e-Work is comprised of e-activities, namely, activities based on and executed through information technologies. In Figure 1, those e-activities include v-(virtual) Design, e-Business, e-Commerce, e-Manufacturing, v-Factories, v-Enterprises, e-Logistics, and similarly, intelligent robotics, intelligent transportation, and so on. All these e-activities rely on computer support and communication technologies, and all of them require collaboration in their inherent interactions between machines, humans, and computers. e-Work includes applications such as tele-robotics and remote services. Telework is also part of e-Work according to the general definition above. Increasingly, however, it is recognized that human work and work organizations have been

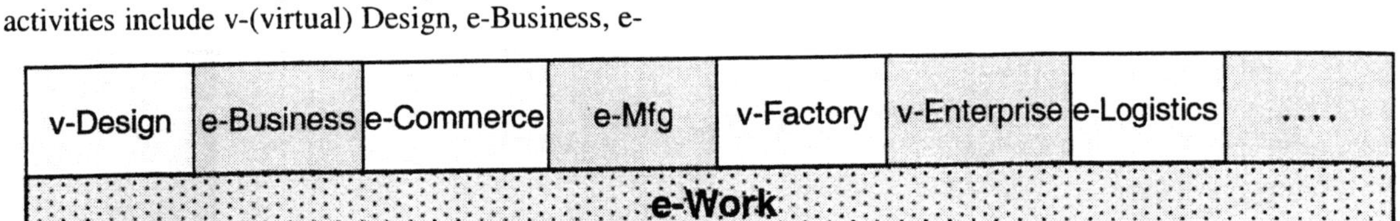

Fig. 1. The scope of e-Work (Nof, 1999)

and still are being fundamentally transformed, as it has been throughout history, by the combination of social organization and technology advancements.

The principles of the work system have been transformed, and the focus of concern in the design of work systems has changed. While advancement in technology over time has increased our ability to work and produce, the challenges and complexities have also increased, and seem to be increasing at an exponential rate. The present overall challenge with e-Work systems is to reduce and possibly, optimize both the information overload and the task overload.

The objective of this article is to describe briefly the foundations and scope of e-Work and e-Mfg., review fundamental design principles for e-Work effectiveness, which lead to e-Mfg and e-Logistics effectiveness, and outline challenges to the management of such emerging systems. The premise is that without effective e-Work, the potential advantages of emerging and promising electronic work activities, such as e-manufacturing, e-logistics and e-business, cannot be fully materialized. A typical and recent example is the frustration of workers over supply chains/networks with complex ERP and other information systems, originally designed to simplify and improve their performance. Even if the human-computer interface for such systems is highly effective, the information overload and expectations from the human operators can be overwhelming. As a result, managers of production and logistics are challenged.

The view of the "e" in e-Work as the foundation of effective e-activities implies that there are functions that can be significantly better performed by enabling technologies, e.g.,: Better connectivity and ability for farther, global reach; enhanced communication and coordination; acceleration of knowledge sharing and distribution; better interactivity, flexibility, customization-ability; higher velocity of work tasks and exchanges; reduced communication costs; and lower agency cost by replacing humans with software agents.

Potential advantages of e-activities impact on the four primary factors of "what really works" (Nohria, 2003) for the success of winning companies, as found in a survey of 160 companies over the 1990s: "Devise and maintain a clearly stated, focused strategy; Develop and maintain flawless operational execution; Develop and maintain a performance-oriented culture; Build and maintain a fast, flexible, flat organization." Cost-effective maintenance of these four factors in any modern company would not be possible without effective e-Work.

Despite the advantages entailed with e-Work and e-Mfg., there are common challenges that emerge with collaboration: (1) Greater work complexity; (2) Limitations caused by higher inter-dependence; (3) Issues of integrity and trust; (4) Greater need for coordination, cooperation, synchronization; (5) Communication challenges and failures, complicating systems' requirements; (6) Mismatch problems, e.g., inconsistent versions, cultural differences, etc.; (7) New users training requirements and associated costs. Failing to overcome these common challenges may explain why often, e-activities do not fulfill the expectation promised by technology.

2. MAJOR CHALLENGES AND EMERGING e-WORK SOLUTIONS

How is e-Work different from e- Business and e-Commerce? The same way work is different from business and commerce. They are highly related, but not the same. Figure 2 depicts some of the scope differences. One way to define the distinction between these three related areas is:

e-Commerce: Commercial e-exchange among customers and providers (suppliers). This exchange is mostly between separate, external entities.

e-Business: Same e-exchanges as in e-Commerce, except the emphasis is on internal e-operations and e-services within the scope of a given enterprise.

e-Work: e-operations and e-support to enable the above two, as well as many non-commercial and non-business work activities, from creative artistic projects, education and learning, to medical help, charity activities, police work, scientific exploration, and many more productive work endeavors.

The main reason to distinguish the focus on e-Work is the realization that there are certain common principles in the design of effective e-Work, without which manufacturing, production, logistics, and other e-activities cannot succeed (Nof, 2001). Modern production and logistics systems depend on communication, including telecommunication. Table 1 shows examples of Internet-based and non-Internet based e-Work, and e-Work's impacts on e-Mfg. and e-Logistics.

Table 1 e-Work with communication (not only Internet), and impacts on e-Mfg. e-Logistics

e-Work Enabled by Communication	Impact on e-Mfg.	Impact on e-Logistics
Distributed Computing and Information Exchanges		
1. Teleconference	✓	✓
2. EDI (data interchange)	✓	✓
3. EFT (fund transfer)		✓
4. Virtual reality for training	✓	✓
5. Virtual reality for design	✓	✓
6. GPS-based monitoring		✓
Collaborative CNC/Robotics/Human-Robot Teams		
1. Tele-robotic facility repair	✓	✓
2. Networked CNC, test, inspection	✓	✓
3. Tele-assembly in clean room	✓	
4. Diagnostics by sensor web	✓	✓
5. Robotic laser drilling	✓	
6. Robotic load/unload devices	✓	✓

With the *Computer Age*, new concerns emerged, increasingly, for better **Interaction** and for **Integration** ("from islands of automation to total systems, from local optimization to global optimization"). These concerns reflected the challenges of new work abilities to interface more complex systems, involving more information transfers, many more transactions, and the dawn of computer and robot servitude: Beginning the delegation of control and cognitive tasks from humans to computers, machines, and robots. One of the main new concerns has been **Information Overload.** It means that humans, each with a single brain, have to interact and contend with computer-based systems and machines that may have certain superior skills not only in physical tasks, but also in information handling and computing. The objective has been to reduce the information overload (only for the humans) by better design of interfaces and of computer systems, and by more training for the human users/participants.

Another natural solution has been the increasing reliance on **Teamwork**, not only in physical tasks, but increasingly, in design and engineering, planning, control, supply, and management tasks. Concurrent design, concurrent engineering, total information systems, and total quality management are examples of the urge to team and collaborate to solve an increasingly complex world of production and service. Further teaming has been evident in the trend of outsourcing. **Outsourcing** certain tasks of production and services is, practically, teaming with external enterprises. The models of **supply chains and supply networks** reflect further teaming among enterprises. Obviously, the enabling technologies of computer communication increased the human ability to share and exchange information:

- Among team members;
- Among distributed machines and robots, including tele-operations and sensors; and
- Among producers, suppliers, distributors and customers.

But at the same time, computer communication has increased the complexity and challenges for human workers and for the designers of work-systems.

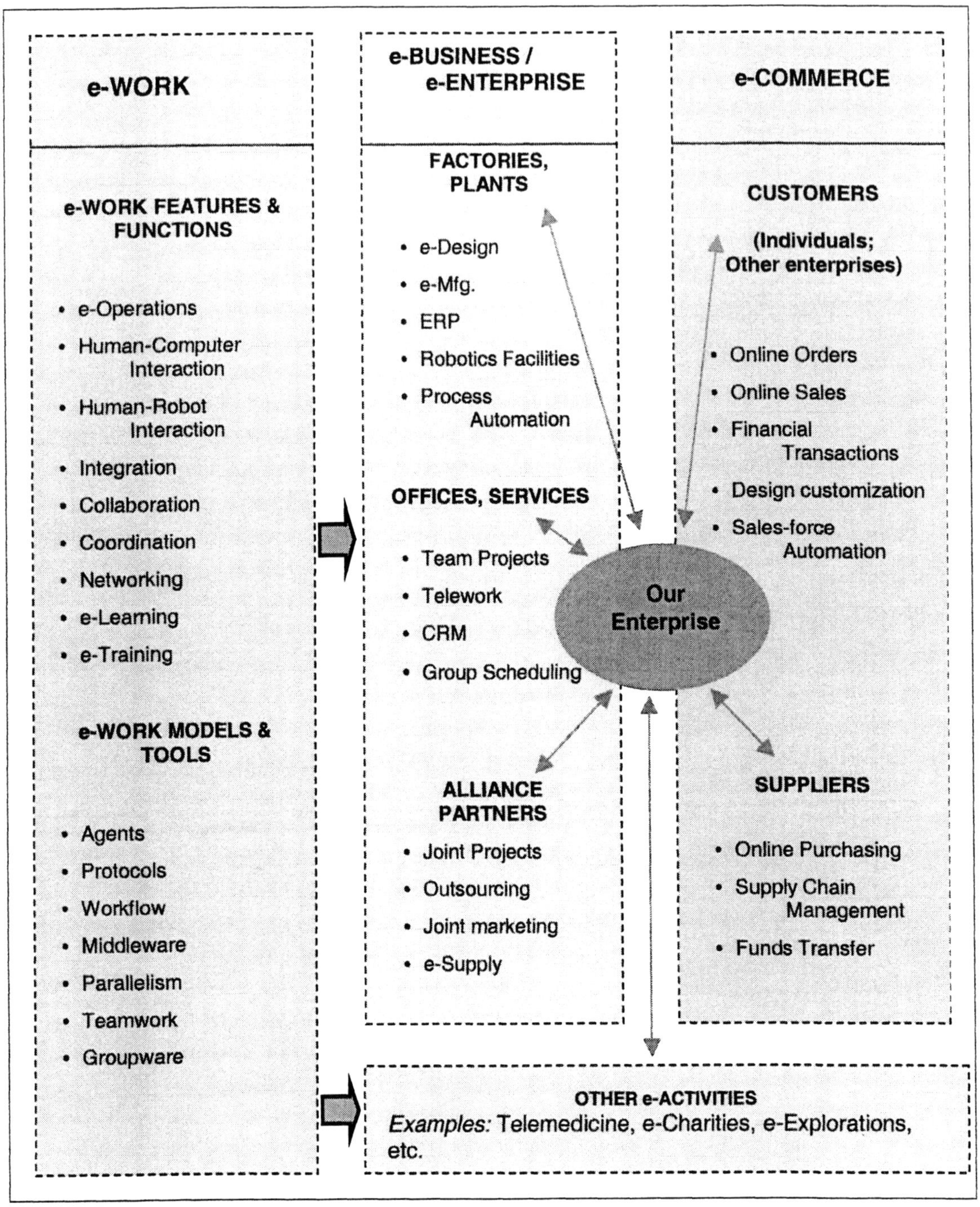

Fig. 2. e-Work is fundamental to e-Business, e-Commerce, e-Mfg., e-Logistics, and other e-activities (just as work is fundamental to business, commerce, and other activities) (Nof, 2003).

The advent of the *Information Age* has enabled new levels of work augmentation, including telecommunication, computer internetworking, and mobile computing. With these new powerful dimensions of work abilities, the field of e-Work has emerged. Telecommunication and the Internet significantly change the cost and time of e-activities, in certain ways potentially increasing them, in other ways potentially reducing them. For example: The cost of agency and the needed layers of agency are reduced. On the other hand, the lack of agency services requires more efforts of coordination and synchronization by the remaining layers of work. Outsourcing becomes attractive by delegating tasks to those who specialize in the outsourced work and can reduce costs and delivery time. At the same time, the cost and time of performance may actually increase because of the added needs for coordination and communication among distributed peers.

The significant reductions in information transfer-cost and response-time in the *Information Age* explain the proliferation of distributed networks of producers, suppliers, consumers and customers. At the same time, there is further growth, to the point of explosion, in **Information Overload and Task Overload** across larger markets and wider distributed networks of operations. The development and emergence of solutions to these challenges are described in the next section.

3. THE FOUR CIRCLES OF THE "e-" IN e-WORK, AND THE 15 e-DIMENSIONS

An overview of recent, related research about the "e-" in e-Work is shown in Figure 3. The four circles reflect four clusters of knowledge. Significant amount of such knowledge has appeared in the literature; only a representative sample is included here. Further details on each of the terms and full references can be found in (Nof, 2004.)

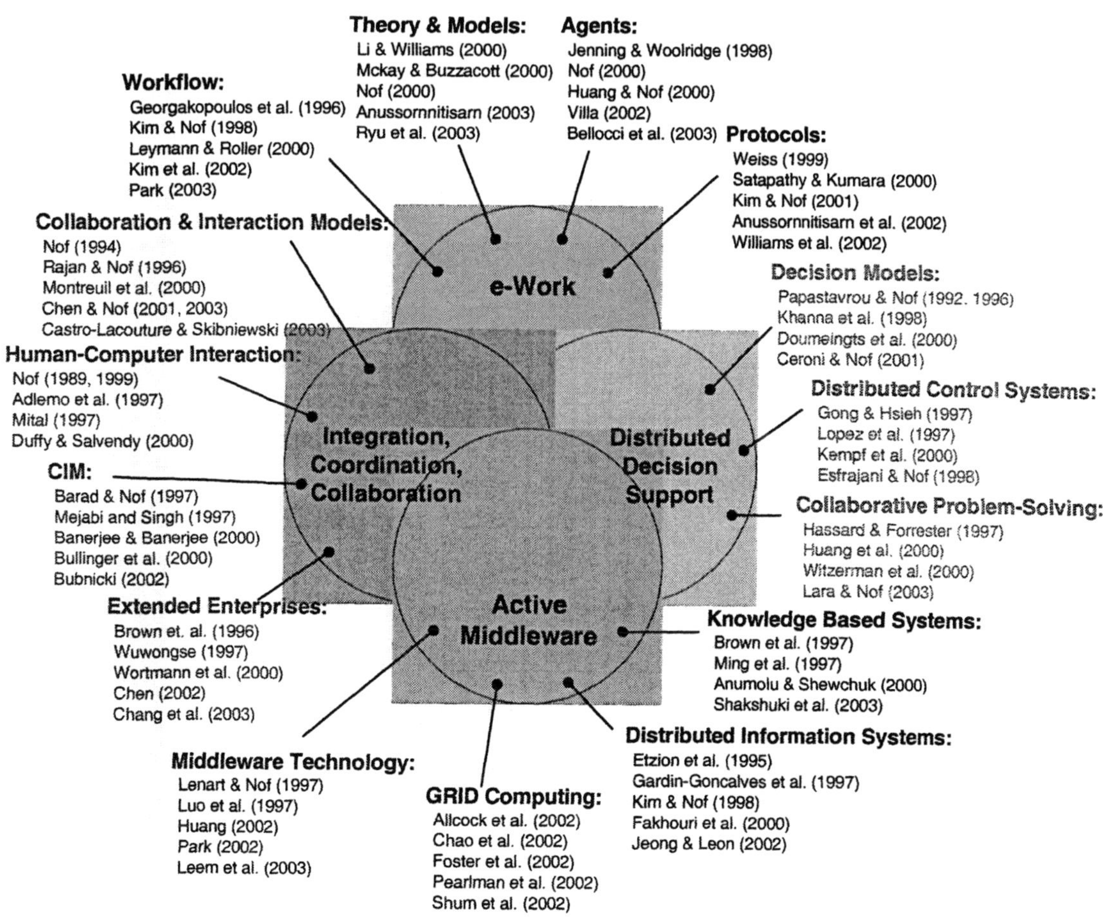

Figure 3. e-Work: Foundations, Tools, and Emerging Discoveries

Smart team collaboration is enabled by 15 e-dimensions from the four clusters (Table 2). Two important ones are agents and protocols. **Intelligent agents** (knowbots, softbots, and robots) perform tasks on behalf of users with a degree of autonomy. They apply in a variety of domains, including: Internet-based information systems, adaptive (customizable) software, autonomous mobile or stationary robots, data mining and knowledge discovery, decision support systems, and intelligent design and manufacturing systems. In future research, development of agent systems and human-agent integration will assist interaction processes, e.g., business workflow, automated negotiation and error recovery, leading to agent based solutions.

Protocols at the work application level (as opposed to computer communication protocols), such as Task Administration Protocols, are defined as the logical rules for workflow control. Protocols enable effective collaboration by communication and resource allocation among production tasks. In **active protocols**, "active" implies that the protocol can, under coordination logic, trigger and initiate necessary, timely interaction tasks and decision-support functions to further improve performance.

In smart distributed systems, e.g., multi-agent systems, coordination protocols are used to guide automated interaction of each agent or an autonomous entity to achieve a common system goal. Efficient coordination will lead to effective collaboration, for example, time-out protocols have improved task allocation in client-server work-cells, and adaptive time-out protocols have rationalized resource-sharing in assembly-and-test facilities with TestLAN (Fig. 4). Auction-based protocols were analyzed for distributed task and resource allocation in multi-agent systems.

In the future, multi-participant negotiation processes, group decision mediation, and conflict management processes will be developed. In addition, the dynamic characteristics of the protocol which acts or reacts to the stochastic nature of distributed environments will be of interest.

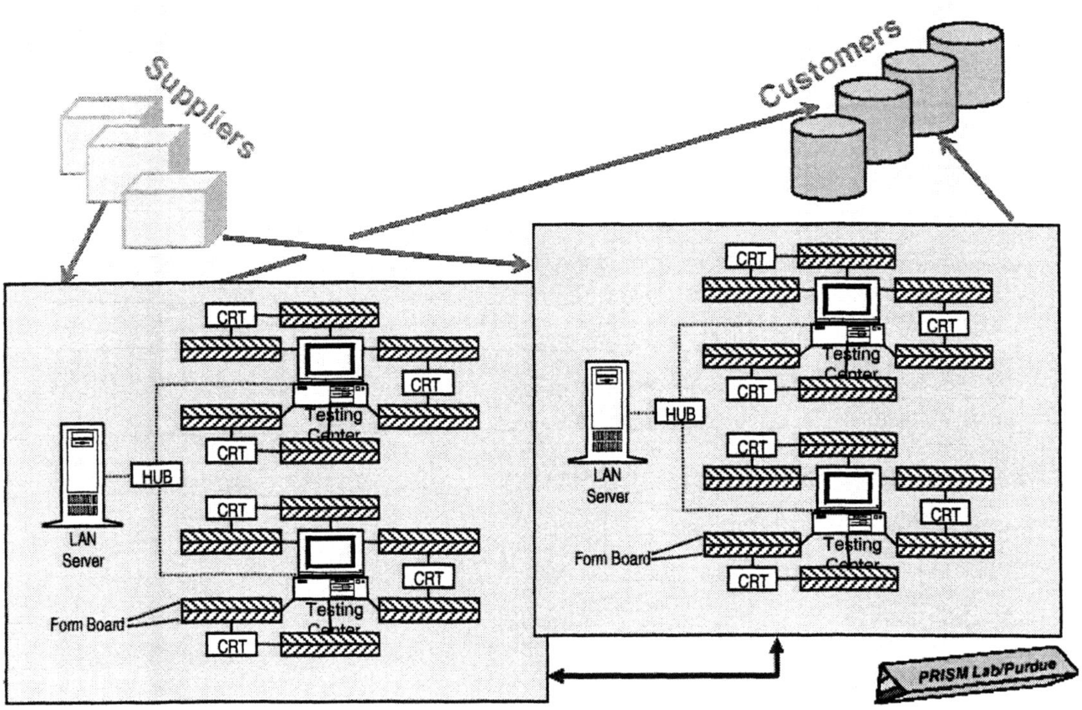

Fig. 4. From e-Work inside assembly-and-test facilities with TestLAN, to e-Work across supply networks and emerging supply grids (Nof, 2003)

Table 2 The Fifteen e-Dimensions of "e" in e-Work and e-Mfg. (Nof, 2003)

Area	Emerging Thrust	Role	Future Challenges
A. e-Work			
A1. Theory and Models	Collaborative, computer-supported and communication-enabled operation in highly distributed organizations of humans/ robots/ autonomous systems.	Augment human abilities to work.	Development and realization of the theory and models for specific applications.
A2. Agents	Software and hardware entities that Perform a set of tasks on behalf of a user with some degree of autonomy.	Useful for systems that are flexible, adaptive and able to integrate heterogeneous components and tasks.	Development of agent system and human-agent integration to assists interaction processes.
A3. Coordination Protocols	Guide and support the automated Interaction of each agent or an autonomous entity to achieve the common system goal.	Efficient coordination processes will lead to effective collaboration, especially in large distributed systems.	Coordination of multi-participant negotiations, group decision mediation, conflict management.
A4. Workflow	Helps guiding and processing activities in a dynamic environment, especially when processes extend beyond organization's boundaries.	Enabling process scalability, availability and performance reliability.	Better collaboration effectiveness, including detecting and solving errors when they occur.
B. Integration, Coordination, and Collaboration			
B1. Theory and Models	Computer-supported collaborative work tools and Models of information exchange among distributed team members.	Helping developers identify link between information resources and processing entities, improve performance of collaborative activities.	Decision making by distributed and autonomous team members, and formulation of resolutions among distributed collaborators.
B2. Human-Computer Interaction (HCI)	Advanced interfaces to help users efficiently and effectively apply computer systems.	Develop systems that support people in their role as learners, explorers, and workers.	Beyond exploring installed bodies of information, discover new knowledge via intensive information exchange, similar to human brainstorming.
B3. Computer Integrated Manufacturing (CIM)	Automated manufacturing and materials handling, computer application to design, machining, and manufacturing of products and services, including quality management and process control.	Integration and management of the entire product life-cycle.	1) Distributed support tools governing flow under time, budget, manufacturing, and environmental constraints; 2) Embedding MEMS and nano sensors for real-time control and inter-active decision making.
B4. Extended Enterprises	Models and methods for better delivery of real-time services in collaboration with customers, employees and business partners in a connected economy.	Integration of business processes beyond a single enterprise, including end-to-end functions, internal operations, and entire supply chains.	Real-time decision making, execution, and conflict management among collaborators over the supply network.
C. Distributed Decision Support			
C1. Decision Models	General-purpose knowledge representation languages with normative status and well-understood computational properties of decision-modeling formalisms and algorithms.	Computer assistance in modeling can significantly reduce the model construction time while increasing model quality, and contribute to a wider	Interaction-based Decision models with distributed knowledge sources enabling adaptive and more

		applicability of decision theory in on-line decision support systems.	responsive decision structures.
C2. Distributed Control Systems	Systems with low level of autonomy among partners, such as job shop scheduling, where global objective functions are typically given.	Allow the parties associated with a system to negotiate for their tasks and their benefits with some degree of autonomy.	Systems integration With higher level of autonomy, such as supply networks.
C3. Collaborative Problem-solving	Methods and quality of solutions to understand, define, and solve problems.	Information exchange to influence other parties' local decisions to act coherently and achieve better results for an entire organization.	Consider collaboration among loosely-coupled entities where less benefits are acceptable or expected by collaborators.
D. Active Middleware			
D1. Technology	Middleware, a layer of services below the application and above the operating system and network layer, operates to foster application portability and distributed interoperability, and provides active services to the clients of a system.	Facilitating development of distributed applications among heterogeneous, legacy and advanced environments.	A very large scale computing platform serving highly heterogeneous, autonomous, and distributed components, beyond typical enterprise networks.
D2. GRID Computing	Computing services that enable sharing, selection, and aggregation of a wide variety of geographically distributed computing resources, for solving la scale and data intensive computing applications.	Computational economy And the development of grid computing architecture and its components, scheduler and resource allocator.	Explore the potential of grid computing paradigm in both scientific and business applications (e.g., supply grid, grid optimization).
D3. Distributed Knowledge Systems	Techniques and standards to design distributed information and knowledge systems, e.g., search robots, semantic webs, coordinate the distributed operation, provide security and access to millions of sources with wide variety of information.	Maintain high reliability, integrity, and quality of service for large scale distributed information and knowledge networks.	Better understanding of e-Trust among distributed information producers and consumers.
D4. Knowledge Based Systems (KBS)	Knowledge-based techniques to support human decision-making, planning, learning, and action.	Construction and Extraction of knowledge for of information needs.	Integration with other technologies to assist a human user or group of users in making better decisions, especially for complex, unstructured problems involving high levels of uncertainty.

4. PRISM CENTER DEVELOPMENTS OF e-WORK DESIGN PRINCIPLES

During research projects for over a decade (see examples in Figure 5), it has become clear that e-Work represents a new, exciting and challenging model of work systems. With an ever increasing scope of distributed and decentralized work activities, increasing distribution and magnitude of world-wide markets, e-Work brings opportunities for better work methods, outcome and yield by augmenting human physical, cognitive, temporal, and locational abilities to work. But it has also been evident that e-Work is associated with new complex needs and new design challenges. Table 4 summarizes a sample of projects attempting to understand the new requirements and challenges, and develop useful theories and solutions for them.

As shown in the Table 3, dimensions of e-Work appearing in Table 2 have been addressed, for various application areas in manufacturing, production, and service. These efforts have been inspired by "role models" from successful, proven models of distributed and parallel computing. The theories, models, and tools that have been developed and implemented are described in detail in the given references. Each of the projects has resulted with recommended design principles for effective e-Work. Several key design principles, based on observations about the new needs and solution approaches, are summarized briefly below.

(1) The Principle of Cooperation Requirement Planning, CRP (by Rajan and Nof)

One of the most powerful augmentations of work abilities by e-Work is in the area of collaboration, varying from minimal information sharing and exchange, to fully collaborative enterprises. Effective collaboration requires advanced planning, as stated in this principle. The *Principle of Cooperation Requirement Planning*, CRP, includes two phases (Rajan and Nof, 1996). In the first phase, CRP- I, a plan of "who does what, how, and when" is generated based on the work objectives and available facility resources. In the second Phase, CRP-II, during execution, the plan is revised in real time, adapted to temporal and spatial changes and constraints. The adaptation responds to changes and new constraints both in the internal facility and its components, and in the external design, logistics and market interactions.

A recent effort in robotic assembly and disassembly has been to integrate CRP with error diagnostics, recovery, and conflict resolution (Chen and Nof, 2003). The principle indicates that effective e-Work requires both advanced planning and adaptive, real-time planning of the necessary cooperation and collaboration.

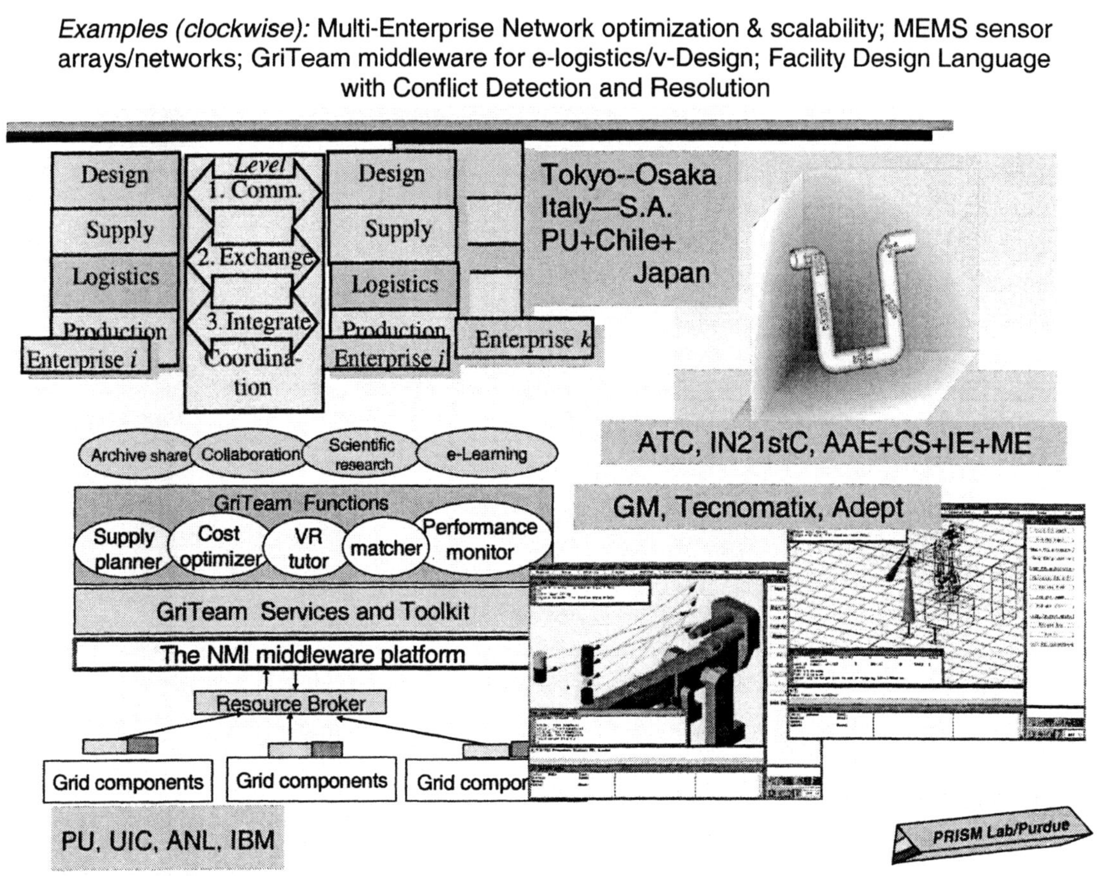

Fig. 5. Four recent examples of e-Work, e-Mfg. and e-Logistics research and the participating universities and sponsors (Nof, 2003).

Table 3 PRISM Center Development of e-Work Theories, Models, and Tools (Examples)

e-Work Design Feature	"Role Model"	Production Application	Tool/Model	Ref.
• Workflow and Coordination	Distributed Database Data Flow Object Orientation	Computer Integrated Mfg. Aerospace Mfg. Laser Machining Cell	• DAFNet & AIMIS • •	Kim (1994) Sauvaire et al. (1998) Lenart (1993)
• Information Sharing and Collaboration	Task Graphs Network Computing Internet/Intranet Internet Internet/Intranet	Distributed Designers Distributed Teams Supply Networks Construction Supply Chain Production and Sales	• IDM • Co-X Tools • Shared Process • Rebar Portal • T-C-M	Khanna et al. (1998) Eberts& Nof (1995) Park (2003) Castro (2003) Matsui et al. (2003)
• Collaboration Planning	Resource Planning	Multi-Robotic Assembly	• CRP	Rajan (1993)
• Agent Design	Agent theory	Manufacturing Operations	• ABMS	Huang (1995)
• Collaboration Protocol Design	Telecom. Protocols Adaptive Control	ERP Applications Electronics testing	• TIE/P • TestLAN/a	Anussornnitisarn (2000) Williams et al. (2002)
• Middleware Protocols	Client-Servers	Automotive Electronics Assembly	• RAP • TOP	Esfarjani (1994) Peralta (1996)
• Parallelism	Parallel Computing	Global Shoe Design&Mfg.	• DPIEM	Ceroni (1999)
• Resource and Task Allocation	Local Area Networks Internet	Electronics Assembly&Test Global Automotive Supply	• TestLAN • M.E.N. Method	Williams (1999) Chen (2002)
• Synchronize / Re-Synchronize	Agent Theory	Robotic Maintenance	• ServSim	Auer (2000)
• Information Assurance	Total Quality	Agent-Based Mfg.	• (MERP)	Bellocci (2001)
• Fault Tolerant Integration	Sensor Fusion	Flow MEMS Sensors	• FTTP	Liu (2001)
• Error recovery	Computer Recovery	Robotic Assembly	• NEFUSER	Avilla (1999)
• Conflict Resolution	Telecom	Facility Design by Team	• FDL/CR	Lara (1999)
• Multi-Enterprise Optimization	Network Flow	Distributed Enterprises	• M.E.N. Opt.	Chen (2002)
• e-Learning	Learning Theory	ERP Applications	• (MERP)	Chen et al. (2001)
• Organizational Learning	Enterprise Computing	Mfg./Assembly Corp.	• CMS	Prytz et al. (1995)
• Viability Measures	Artificial Life	Human-Robot Facilities	• TIE/A	Huang (1999)
• e-Work Scalability	Distributed Computer	Supply Networks	• M.E.N. Model	Chen (2002)

ABMS: Agent Based Management System; **AIMIS**: Agent Interaction Management System; **CMS**: Corporate Memory System; **Co-X**: Collaborative Tool for Function X; **CRP**: Collaboration Requirements Planning; **DAFNet**: Data Activity Flow Network; **DPIEM**: Distributed Parallel Integration Evaluation Method; **FDL/CR**: Facility Design Language / Conflict Resolution; **FTTP**: Fault-Tolerance Time-Out Protocol; **IDM**: Iterative Design Model; **M.E.N.**: Multi-Enterprise Network; **MERP**: ERP e-Learning by MBE Simulations; **NEFUSER**: Neuro-Fuzzy Systems for Error Recovery; **RAP**: Resource Allocation Protocol; **ServSim**: Maintenance Service simulator; **TestLAN**: Testers Local Area Network; **TIE/A**: Teamwork Integration Evaluator/Agent; T-C-M: Two-Center Model; **TIE/P**: Teamwork Integration Evaluator/Protocol; **TOP**: Time-Out Protocol.

(2) The Principle of e-Work Parallelism (by Ceroni and Nof)

Every model of e-Work implies interactions among software work-spaces and human work-spaces. The *principle of e-Work Parallelism* (Ceroni and Nof, 1997- 2003) is concerned with how to optimally exploit the fact that work in these respective spaces can, and must be allowed to advance in parallel. In other words, to be effective, e-Work systems cannot be constrained by linear (sequential) precedence of tasks. A simple example: tasks can be performed in parallel by software agents preparing information for human decision makers, while the latter are busy with other tasks, away, or asleep. In e-Work, workflow parallelism has a deeper meaning, since work activities can be widely distributed, locationally and over human- and software-spaces; there can be widely (and in effective e-Work also wisely) distributed human-human interactions; human-machine and human-computer interactions; and machine-machine,computer-computer interactions.

"KISS": Keep It Simple, System!

The essence of the "KISS" is that the computer and communication support system can be designed as complex as necessary, as long as it can work autonomously, in parallel to and supportive of humans, subject to their inputs and instructions. "KISS" is associated with the principle of e-Work parallelism by the notion that new versions and advanced functions of software systems must minimize the need for repeated human retraining, by using internal, parallel e-functions to simplify the transition to the evolving system's interfaces and logic.

Distributed Planning of Integrated Execution Method, DPIEM

DPIEM is a method for planning how to satisfy a CRP-based plan (Ceroni, 1996; 1999). A key issue related to such planning under the e-Work parallelism principle is the ability to determine the optimal Degree Of Parallelism (DOP). DOP is defined as the maximum level of resource and task parallelism, which will balance the trade-off between increased communication, transportation, and equipment costs incurred with a higher DOP, and the increased yield (productivity) gained by that level of parallelism.

In addition to the issue of DOP, there are several basic issues in the design and implementation of any coordinated problem-solving and decision support system for e-Work, which can also be considered as key issues in active middleware development. The principle of e-Work parallelism includes five guidelines, seeking to let the e-Work Support System (EWSS): (1) Formulate, decompose, and allocate problems; (2) Enable applications to communicate and interact under task administration protocols; (3) Trigger and re-synchronize independent entities to act coherently in making decisions and taking action; (4)Enable entities to reason about actions, plans, and knowledge of other agents and coordinate with them; (5) Develop conflict resolution, error recovery, diagnostics and prevention.

For instance, **PIEM** (centralized optimization algorithms) and **DPIEM** (optimization with distributed protocols) were developed for evaluating and planning the communication and coordination trade-offs in e-Work with parallelism in design, manufacturing, supply, and sales functions by Japanese, North American and South American companies; for outsourcing strategies; information systems middleware development; and for robotic cells design.

Another example of the principle of parallelism are the **Co-X tools** (developed by Eberts and Nof, 1995), where "Co-" implies coordinated, cooperative, and/or collaborative support tools, e.g., Co-design, Co-plan, Co-storm (brain-storm), etc. "Co-" can be designed at three levels of interaction and collaboration: Level 1 – Interfaced sharing of information, ideas, and plans; 2 – Parallelism of information processing tasks and of physical tasks; 3 –Addition of active protocols and agents to support (optimize) better effectiveness through parallelism. The Co-X tools have been developed and implemented for teaching and learning the benefits and limitations of different team designs, and for lab experiments to demonstrate and investigate the interface design and usability of collaboration support (EWSS) in e-Work.

(3) The Principle of Conflict Resolution (by Huang and Nof)

This principle addresses the cost of resolving conflicts among collaborating e-Workers. It has been observed and recognized that with a greater rate of interactions, which increases with the number of collaborating parties, there is also a greater occurrence of conflicts and errors. This problem is critical in terms of scalability. It means that beyond the information and task overloads, e-Work must be designed to overcome quickly and inexpensively as many errors and conflicts as required to be effective. Huang and Nof (1999) showed that for practical values of the probability to resolve a conflict in one of several iterations of negotiations, the total resolution cost would depend on the relative portion of human intervention in the resolution process, as follows:

a. With more the cost of resolution grows exponentially and it is unbounded. Implication: Under this design conditions, the e-Work system will collapse because of ineffectiveness.

b. With less human interaction and involvement, approaching zero, meaning IT services are designed and applied for resolution-support by automating functions of negotiations for corrections or compromises, then the total cost of resolution reaches an upper bound. Implication: In this case, e-Work systems can be effective.

The design objective, following this principle, is to develop better and more powerful IT support for detection, prevention, and resolution or recovery of errors and conflicts, leading to lower overall costs, better quality results, and more effective e-Work.

Conflict resolution for collaborative facility design by FDL/CR, and knowledge-based diagnostics and recovery by NEFUSER were also developed (see Table 4). Control Protocols for such diagnostics, prevention, resolution, and recovery are important design issues. Future collaborative e-Work will depend on cheaper, redundant groups, arrays, networks, and grids of interacting peers, e.g., business partners, human designer teams, DNA-sensor arrays. Enabling effective e-Work in each of these examples requires effective handling of errors and conflicts.

(4) New Measures and Evaluation

New work models, as those illustrated in Table 3 for the design of effective e-Work, require new measures in addition to traditional measures of effectiveness. Some of the measures have already emerged with the advent of CIM and networked enterprises: Flexibility, agility, connectivity, integration-ability, scalability, reachability, and so on. Two recent measures of particular interest to e-Work design are level of autonomy, and viability (e.g., Huang, 1999).

The level of **autonomy** relates to the delegation of authority, decentralization, and to how *active* the agents, protocols, and other e-Work system participants can be. **Viability** in the context of e-Work has been defined as a relative measure of the ratio between the cost of operating and sustaining agents, and the rewards from their services. Theories of agent design and interaction have been developed using measures of autonomy and viability. Anussornnitisarn (2003) discovered that e-Work rationalized by viability-based protocols performs significantly better than by protocols ignoring viability. Examples of other emerging measures for e-Work design are learning-ability, collaboration-ability; related to errors and conflicts, interesting measures are prevention-ability, detect-ability, diagnosis-ability, and recovery-ability. A critical area is information assurance, including measures of security, integrity, and information significance (Bellocci, 2001).

5. CONCLUSIONS

The scope of production, logistics, and other business and service operations is changing, and the way people and systems can operate and collaborate is also being transformed. The challenge to managers of production and logistics is how to design, implement and exploit effective e-Work. Models of e-Work have been developed to address design issues. Design and decision tools have also been developed, with the objective of understanding the emerging requirements, limitations, and potential capabilities of e-Work and e-Mfg.

Modeling challenges for collaborative e-Work design include enhancements of models, to represent e-Work functions along the fifteen e-dimensions in Table 2. Other challenges include the evaluation and assessment of new measures, mentioned in the previous section, such as collaboration-ability, conflict prevention-ability, and error detection-ability.

Design principles for effective, collaborative e-Work systems have been discussed in four main areas:

- cooperation requirement planning
- e-Work parallelism
- Error and conflict handling, and
- e-Work effectiveness measures.

Based on these principles, it is envisioned that smart collaborative, e-Mfg. and e-Logistics will:

1. Optimize coordination of interactions as a key to competitiveness;
2. Manage errors, conflicts, and interactions' complexity among team machines/robots;
3. Apply CSS (Collaboration Support Systems) for machine and human teams;
4. Depend on protocols for fault-tolerant, time-out integration, e.g., FTTP;
5. Employ smart machines and robot teams that can interact better with human teams.

In addition, next generation collaborative e-Work and e-Mfg. will have to answer the following questions:

- How can we implement and benefit from networked, decentralized decision support?
- Will workers and managers trust the results delivered by autonomous agents and support systems?
- Can we develop a collaborative control theory to help solve these problems?
- What are the relations between e-Mfg. mechanisms and collaborative group behavior?
- What are the impacts on re-sequencing workflow cycles (e.g., with parallelism)?
- What are the impacts on education and training?

ACKNOWLEDGEMENT

Research reported in this article has been developed at the PRISM Center with NSF, Indiana 21st Century fund for Science & Technology, and industry support. Special thanks to my colleagues and students at the PRISM Lab, who have worked with me to discover the e-Work knowledge. More details about the PRISM Center: http://gilbreth.ecn.purdue.edu/~prism.

REFERENCES

Nof, S.Y. (1999). Modeling of e-Work with TIE, a Teamwork Integration Evaluator, *Research Memo 99-20,* School of Industrial Engineering, Purdue University.

Nohria, N., et al., What really works, *Harvard Business Review*, Online Version, July 2003.

Complete details of all the other references in this article are available in

Nof, S.Y., (2003). "Design of Effective e-Work: Review of Models, Tools, and Emerging Challenges", *Production Planning & Control*, Vol. **14**, No. 8, pp. 681-704.

Copyright © IFAC Information Control Problems in
Manufacturing, Salvador, Brazil, 2004

ELSEVIER
IFAC
PUBLICATIONS
www.elsevier.com/locate/ifac

Project-driven management support for web-enabled manufacturing

Zbigniew A. Banaszak
Laboratory of Knowledge and Artificial Intelligence Systems, Systems Research Institute of Polish Academy of Sciences, Podwale 75 Wrocław, Poland, e-mail: z.banaszak@iizp.uz.zgora.pl

Marek B. Zaremba
Département d'informatique et d'ingénierie, Université du Québec, Gatineau, Québec J8X 3X7, Canada
e-mail: marek.zaremba@uqo.ca

Abstract

The paper addresses the issue of decision-making support for small and medium size enterprises operating within a virtual project-driven enterprise environment. The problem considered here can be defined in terms of finding a feasible schedule that satisfies the constraints imposed by the work order duration, the price, and the time constrained resource availability. In other words, we are looking for an answer whether a given work order can be accepted for processing in a manufacturing system. The problem belongs to a class of multi-mode case problems of project scheduling, where finding a feasible solution is NP-complete. The proposed method, implemented in an interactive software package, matches the project tasks and the manufacturing system capacity. It is illustrated here on an example of a virtual construction enterprise. *Copyright © 2004 IFAC*

Keywords: project management, virtual enterprise, scheduling, production planning, project-driven manufacturing, branch and bound search.

1. Introduction

Internet-based communication technologies now make it possible to effectively integrate the entire manufacturing chain, from design through manufacturing and supply management to maintenance and service over the entire product life-cycle, from the initial concept of a product to its final end use and disposal. Virtual enterprise can be treated as an example of such a web-enabled organization.

In this general context, the paper addresses the issue of decision-making support for small and medium size enterprises operating within a virtual enterprise. The objective is to find a computationally effective method of planning the production flows in a virtual enterprise. The problem considered here can be defined in terms of finding a feasible schedule that satisfies the constraints imposed by the work order duration, the price, and the time constrained resource availability. In other words, we are looking for an answer whether a given work order can be accepted for processing in a virtual enterprise. The problem belongs to a class of multi-mode case problems of project scheduling, where finding of a feasible solution is NP-complete (Kerzner, 1984; Lova, et al., 2000).

Analysis of the research in the area indicates that 84% small and medium size enterprises work with multiple projects. Another result points to the predominance of projects with less than 50 activities (84%) and about 95% of projects have less than 100 activities (Lova, et al., 2000; Payne,1995).

In order to decide whether a new production order (i.e., a new project) can be executed in a given production system, the producer capabilities and the customer needs have to be taken into account. This issue belongs to the resource-constrained project management problem.

The resource-constrained project management problem has been addressed using two approaches: the scheduling of a single project and the scheduling of multiple simultaneous projects (Anavi-Isakow and Golany, 2003; Brucker, et al., 1999). Most of the publications on project management have been dedicated to a single project. In recent years there has been a growing interest in problems related to project scheduling in multi-project environments. In the single project problem, satisfying time constraints is one of the dominant criteria. In contrast, scheduling of several projects with common and constrained resources takes into account other criteria such as: idle resources, resource levelling, in-process inventory, and project splitting (Lova, et al., 2000).

The rest of the paper is organized as follows. In the next section the main problem is formulated. In Section 3 the solution approach is proposed. Section 4 presents the software package as an implementation of the method. Results and some concluding remarks are presented in Section 5.

2. Statement of the problem

The main objective of a production flow management systems is the coordination of processes and activities related to work order processing, transportation, inventory management, warehousing and production. In other words the goal is to achieve a well-synchronized behaviour of dynamically interacting components, where the right quantity of the right material is provided in the right place, and at the right time (Banaszak, Saniuk, and Zaremba, 2001).

The following problem can be now formulated. Consider a manufacturing system that provides a given production capability while processing some other work orders. That means that only a part of the production capability, specified by the time-restricted resource availability, is available in the system. A given work order is represented by an activity-on-node (AON) network, where the nodes and the arcs represent the activities and precedence relations. The work order is specified by project duration deadline T_h, which is equivalent to a presumed completion time (the work order cycle), as well as a total project cost constraint K_h (selling price). Each activity may be executed in one out of the set of $M_{i,j}$ modes (system resources). Also, each activity may not be pre-empted and the mode once selected may not be changed. Considering a time horizon, which an upper bound on the project's makespan, there is an amount of p units of available renewable resources in the considered period. The cost of using a resource unit is specified in the map of accessible renewable resources.

The problem considered regards of finding of a makespan - feasible schedule that satisfies the constraints imposed by the precedence relations and by the time-constrained resources availability. The objective is to find an answer for the following question: whether there exist a potential virtual enterprise which is able to do the work order (i.e. the project) in agreement with enterprise technical capacities and the client's requirements? The considered approach can be applied to the production flows in virtual enterprises as well as to project-driven small and medium size enterprises.

3. The heuristic method

The problem belongs to a class of multi-mode case problems of project scheduling, where finding a feasible solution is NP-complete (Anavi-Isakow and Golany, 2003; Dvir, et al., 2003). In order to cope with this issue, let us consider a heuristic method based on the branch and bound scheme. The searching procedure is driven by an upper bound evaluation policy.

3. 1. The searching strategies

In order to avoid costly exhaustive enumeration of possible schedules, the explored cases are limited to those possessing the lowest margins of cost and time. In other words, first all the cases that could lead to an unfeasible schedule are considered. The search is thus aimed at proving that a feasible schedule does not exist.

A difference between the assumed project duration deadline and a project makespan obtained in the case of absence of resource time constraints is applied as the upper bound. It means at the beginning a difference between the assumed project duration deadline and the project makespan (i.e., the critical path) obtained in the case of the absence of resource time-constraints is calculated. The same refers to the cost evaluation (i.e., the cost of resources in the critical path). In the case a feasible schedule exists, the makespan taking into account the availability of time-restricted resources is calculated. For such a newly obtained critical path the cost and time margins are calculated again. In the case one of the margins is less than zero, the feasible schedule does not exist.

In order to continue the searching process, a modified project network and a modified resource availability constraints have to be considered as a new data. By removing from the project network the activities assigned to the critical path, we obtain a set of sub-networks. Each sub-network has its own time deadline derived from the former makespan.

Consequently, an ordered set of sub-networks $\{AN_1, AN_2, AN_3, \ldots, AN_n\}$ can be observed with the following sufficient conditions for a feasible project schedule:

i) $AN_i \prec AN_j \Leftrightarrow UB_i \leq UB_j$, $UB_i = TL_i - CP_i$ and $UB \geq 0$

ii) $\forall i \in \{2,\ldots,n-1\}$ $CP_i(RC_{i-1}) \leq M_j(CP_{i-1})$, and
$CP_1(RC_0) \leq M_1(CP_0) = T_h$ and $M_j(CP_{i-1}) \geq 0$

iii) $\forall i \in \{1,2,\ldots,n-1\}$ $M_j = M_{j-1} - CPL_j \leq 0$, and
$M_0 = AC$ and $M_j \geq 0$

where:
AN_i – the i-th activity network,
UB_i – the upper bound of AN_i,
AC – the assumed project cost limit
TL_i – the time limit for the duration of the AN_i sub-network,
CP_i – the makespan of the AN_i sub-network (the time constraints imposed on the resources that are not taken into account),
$RC_i = f(RC_{i-1}, CP_{i-1})$ –the actualisation of the i-1-th resource constraints following exclusion of resources associated to the (i-1)-th critical path,
$CP_i(RC_{i-1})$ – the makespan of the AN_i sub-network taking into account the time constraints RC_{i-1} imposed on the resources,
$M_i(CP_{i-1}) = TL_i - CP_i(RC_{i-1})$ – the time margin of AN_i taking into account the actualisation of the (i-1)-th resource constraints following the exclusion of resources associated with the (i-1)-th critical path,
M_i – the cost margin after taking into account the cost CPL_j, i.e., the cost of the utilization of the resources associated with the critical path of AN_i,
T_h – the assumed project duration and/or joint type nodes of the critical path in the project network.

For each sub-network the corresponding upper bound can be calculated. Finding the sub-network with the lowest value of the upper bound allows us to repeat the main procedure, i.e., to calculate the cost and time margins, and then to consider a new sub-network. It means finding from the extended set of sub-networks the element distinguished by a smallest value of the upper bound, then calculating the margins, and so on. The procedure ends either when one of margins is less than zero or when the set of sub-networks is exhausted.

The applied heuristics focuses on searching on the upper bound value, and results in an order to which the resource constraints are then applied. In other words, the heuristic rule can be treated as a set of sufficient conditions. However, a feasible solution may also exist in the cases the sufficient conditions are not satisfied. The negative result simply means that a feasible schedule cannot be considered due to the conditions build-in the searching procedure. In other words it means there is no guarantee the feasible schedule does not exist. Such a evaluation could be enhanced in the case the set of sufficient conditions will extend.

In order to illustrate the way a new heuristics could be considered (i.e., new conditions could be added), let us assume some modification of the previously introduced conditions i) – iii). Let us replace condition i) by the following one:

$AN_i \prec AN_j \Leftrightarrow UB_i \geq UB_j$; $UB_i = TL_i - CP_i$ and $UB \geq 0$

The conditions encompassing the cost and time constraints together with the resource cost and availability constraints provide a natural framework for the implementation of the constraint logic programming methods.

3. 2. The resource allocation heuristics

The proposed method includes four resources allocation heuristics. It is based on the estimation of the value of resource capability and the average of the resource usage cost in a given period. The proposed heuristics are following:

- the smallest resource capability,
- the biggest resource capability,
- the smallest average resource cost,
- the biggest average resource cost.

The two first heuristics are based on estimating the value of the resource capability in a given time horizon (sub-network duration time). From the set of resource alternatives the one is selected that is characterized by the smallest/biggest capability and then is assigned to the given activity. The value of the resource capability is computed as follows:

$$d_i = \sum_{i=g}^{DD} y(r_i) \quad , d_i \in \{0,1\} \tag{1}$$

where:

d_i – the i-th the resource capability,
g – the beginning of planning horizon,
DD – the end of planning horizon,
$y(r_i)$ – binary variable which determine the availability the r-th resource in the i-th unit time. $y(r_i) = 0$ means that the cost of resource utilization equals to zero (the given resources is busy). $y(r_i) = 1$ means that the resource cost is bigger than zero (the resource is eligible).

In the third and the fourth heuristics, the priority is given to the resource characterized by the smallest/biggest cost in a given planning horizon. The cost value is estimated as following:

$$\overline{k_i} = \frac{1}{l} \sum_{i=g}^{DD} k_i \tag{2}$$

where:

$\overline{k_i}$ – the average cost of the i-th resource utilization,
k_i – the resource utilization cost,
d – the time of the resource availability in i-th unit time in a given time period.

The computed parameters, i.e., the resource capability and rte resource cost usage depend on time. They are determined in a given period from the duration of the executed activities of the sub-network.

4. Computer experiments

The presented method has been implemented in a software package "Simulator of a Virtual Project" available at ***http://www.republika.pl/iwonapisz***. The package permits to respond to a question whether a given work order can be completed for processing in a virtual enterprise? A positive response for the question provides a feasible schedule as Gantt's chart and report of the project cost. Let us consider the following example. The aim of the experiment is to find answers to the following questions: Does there exist a potential structure of a virtual enterprise which is able to complete the work order according to the client's requirements and guaranteeing the due time of already executed work orders remains unchanged? How is the completion time of the project considered? What is the total cost of the work order? Whether there exist alternative variants of the workflow?

The experiment presented refers to an example of a construction developer enterprise, which is considering a new order. The project consists in the construction of a detached house. The client wants to spend no more than 70 000 cost units. The client is to pay for the materials too, but their costs are not included. He requires that the house be built in seven months. The work order may start by April 1 and finish November 30, 2003 at the latest. The summarized time horizon amounts to 244 days. In order to respond to the client's bid, the firm should take into account the production capabilities and the customer needs. The Tab. 1 shows the specification of the project activities.

The developer has to determine what sets of resources (sets of partner firms) are necessary to complete the work order. Then the operations of the manufacturing processes have to be established. Based on the given operation time and the cost of resource usage the relevant resources are selected.

Let us assume that the structure of the virtual enterprise consists of a developer and 12 firms. The firms are specialized in their business process. They have a core competence to do the work best. It is assumed that in the planning horizon (following 7 months) a part of production capability is available to use for the new work order.

In the program the work order parameters have been defined. The client's requirements have been entered into the software package. The network diagram of the work order is shown in the Fig. 1.

Tab. 1. Specification of the project activities

No.	The name of activity	Symbol	Oper. time [day]
1	Transfer and developing of the field	O1	2
2	Execution of measuring job	O2	1
3	Assembly of land	O3	2
4	Execution of hole under foundation	O4	3
5	Connection of energy	O5	4
6	Connection of sewerage	O6	4
7	Execution of foundation	O7	5
8	Pouring of foundation plate	O8	10
9	Construction of design wall	O9	7
10	Execution of ceiling	O10	6
11	Execution of nodal wall	O11	3
12	Execution of construction of roof	O12	6
13	Cover of roof	O13	11
14	Execution of partition	O14	13
15	Execution of electric installation	O15	16
16	Execution of water installation	O16	12
17	Execution of heating installation	O17	18
18	Plastering of internal walls	O18	7
19	Inserting of windows and external doors	O19	5
20	Site of glazing	O20	9
21	Execution of stairs	O21	4
22	Warming of external walls	O22	17
23	Painting of walls	O23	13
24	Plastering of external walls	O24	8
25	Execution of fencing	O25	11
26	Site of floor panel	O26	7
27	Inserting of internal door	O27	5
28	Developing of field	O28	15
29	Investment receipt	O29	1

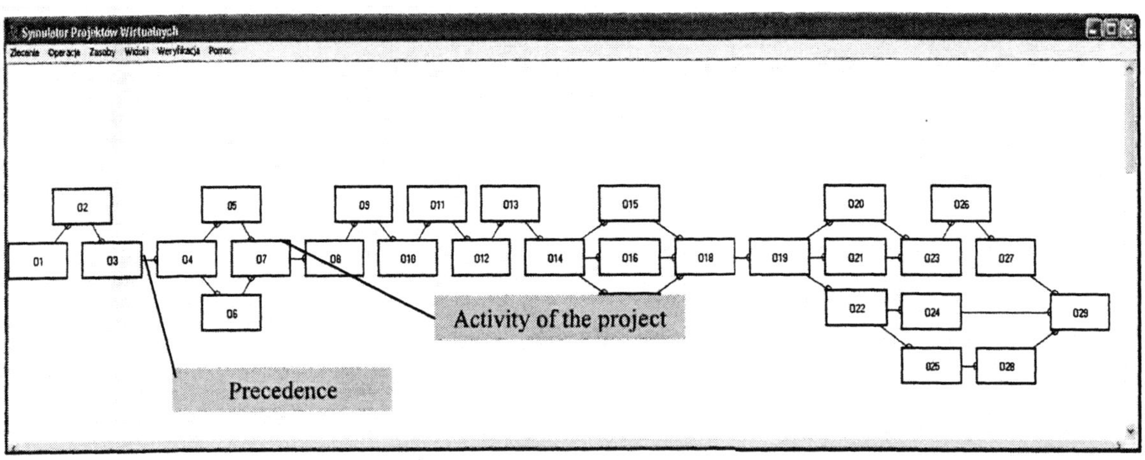

Fig. 1. Network diagram

The specification of the potential co-operators and their production resources is presented in Table 2.

Each of the production resources is characterized by the name, the type, the capability and by the cost of usage. It is assumed that the each of project members defines the costs of resources. The value of the resource cost depends on the cost calculation done by each cooperator. The resource capability results from the schedules planned by the firms. The parameters of all necessary resources have been entered into the software package.

The resources availability constraints are shown in Fig2. The figure specifies the resource availability, more precisely the cost of usage. The zero cost means that a resource is not available (is busy). A feasible variant of the resource allocation is shown in the Fig. 3. It is assumed that in the business process renewable resources can be used and there are alternative resources. It means that it is possible to execute a given activity for different resources (in different enterprises – co operators).

Table 2. Set of production resources

The firm name	The resource name	The symbol
Firm 1	Worker	Z1
Firm2	Geodesist	Z2
	Heavy equipment	Z3
Firm 3	Concrete mixer	Z4
	Worker	Z5
Firm 4	Electrician	Z6
Firm 5	Bricklayer	Z7
	Plumber	Z8
Firm 6	Worker for heating plant installation	Z9
Firm 7	Roofer	Z10
Firm 8	Worker for structuring tile	Z11
Firm 9	Joiner	Z12
Firm 10	Painter	Z13
	Bricklayer	Z14
Firm 11	Worker for warming building	Z15
Firm 12	Worker for inserting windows	Z16
Developer	Main executive	Z17

After entering the input data the process verification starts. In the first step the smallest upper bound strategy is used. It is assumed that there are alternative resources. Therefore the one of the resource allocation heuristics should be chosen. From four implemented heuristics only two lead to obtaining a feasible schedule of the work order. In case of using the biggest resource capability heuristic and the smallest average resource cost heuristic the software package gave a response. The response means that there is enough resource availability to execute the project accurately according to customer's requirements and the technical ability of a production system. The margin time of the critical path is less than zero (-8 time units).

The program checked the existence of the feasible schedule with the two new heuristics, i.e., the smallest production capability heuristics and the biggest resource cost heuristics. The results of the resource allocation variations are two different Gantt's charts. In the first case the completion time equals to 215 days (the time margin is 29 days). In the second case the completion time is longer and equals to 217 days. The obtained schedules are different. The difference is a result of the selected resource allocation variant. A fragment of the feasible Gantt's chart is show in Fig. 4. The project costs are equal to 66.000 cost units and 67.770 cost units respectively.

Mapa dostępnosci zasobów

Dni

Zasoby	1	2	3	4	5	6	7	8	9	10	11	12	13	14	15	16	17	18	19	20
Z1	150	150	150	150	150	150	150	150	150	150	150	150	150	150	150	150	150	150	150	150
Z2	0	0	0	350	350	350	350	350	350	350	350	350	350	350	350	350	350	350	0	0
Z3	0	0	0	0	0	0	0	0	0	0	0	0	0	0	320	320	320	320	320	320
Z4	0	0	0	0	400	400	400	400	400	400	400	400	400	400	400	400	400	400	400	400
Z5	0	0	0	170	170	170	170	170	170	170	170	170	0	0	0	0	0	0	0	0
Z6	130	130	130	130	130	130	130	130	130	130	130	130	130	130	130	130	130	130	130	130
Z7	0	0	0	0	0	0	0	0	0	440	440	440	440	440	440	440	440	440	440	440
Z8	0	0	0	0	0	160	160	160	160	160	160	0	0	160	160	160	160	160	160	0
Z9	0	180	180	180	180	180	180	180	180	180	180	180	0	0	0	0	0	0	0	0
Z10	0	0	0	0	0	0	0	0	0	0	0	580	580	580	580	580	580	580	580	580
Z11	300	300	300	300	300	300	300	300	300	300	300	300	300	300	300	300	300	300	300	300
Z12	0	0	0	0	0	0	0	0	0	0	230	230	230	230	230	230	230	230	230	230
Z13	0	0	0	0	0	0	0	0	0	0	0	0	0	0	0	0	0	0	0	0
Z14	420	420	420	420	420	420	420	420	420	420	0	0	0	0	0	0	420	420	420	420
Z15	0	0	0	0	0	0	0	0	0	0	0	0	0	0	0	0	0	0	0	0
Z16	0	0	0	390	390	390	390	390	390	390	390	390	390	390	0	0	0	0	0	0
Z17	200	200	200	200	200	200	200	200	200	200	200	200	200	200	200	200	200	200	200	200

Fig. 2. The map of accessible of renewable resources

The gain is 2 days. The second advantage is the reduction of the investment. The spared resources are equal to 1.170 cost units. The first heuristics is more efficient. It permits to obtain better results with respect to two project parameters, i.e., time and cost. Adoption of the two heuristics leads to obtaining a feasible variant of the realization of the house-building project.

In the second step, a new strategy has been used, i.e., the biggest upper bound strategy. The four implemented resource allocation heuristics do not lead to a feasible solution. The negative result means that a feasible schedule cannot be obtained due to the conditions built in the searching procedure. Table 3 presents the statistics of the experiments: I – the smallest upper bound strategy, II – the biggest upper bound strategy, 1 – the smallest resource capability heuristic, 2 – the biggest resource capability heuristic, 3 – the smallest average resource cost heuristic, 4 – the biggest average resource cost heuristic.

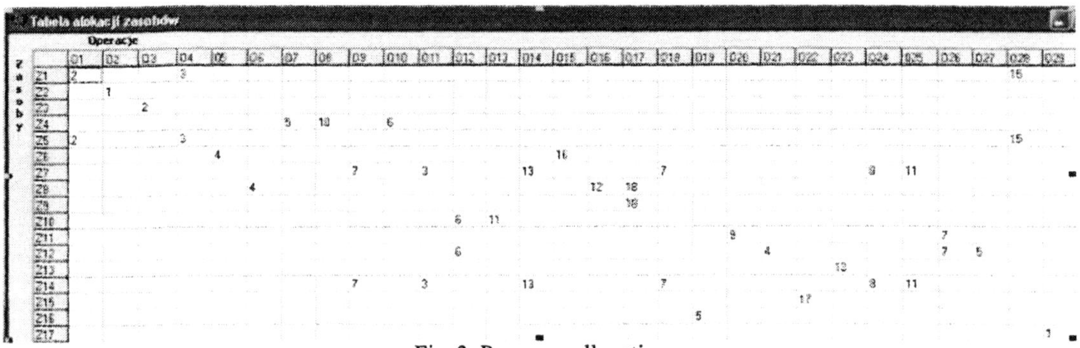

Fig. 3. Resource allocation.

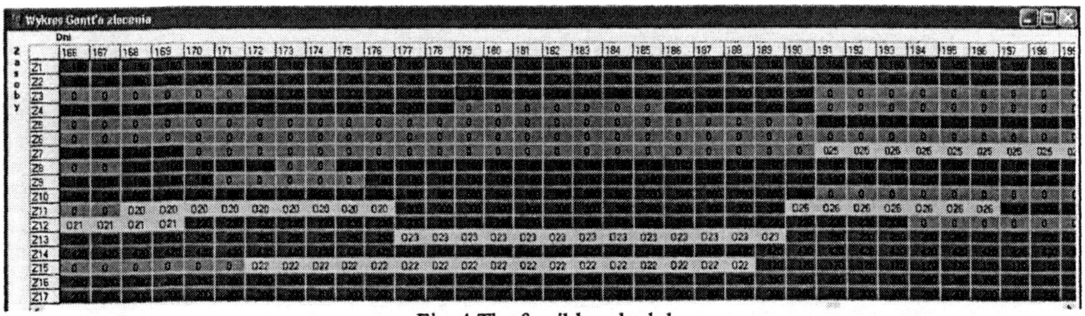

Fig. 4 The feasible schedule

Tab. 3. Results of work order verification

The way of the verification		Schedule	Exit data	
Number of strategy	Number of heuristic	yes/no	Prise [cost units]	Completion time [days]
I	1	yes	66.600	215
I	2, 3	no	–	–
I	4	yes	67.770	217
II	1, 2, 3, 4	no	–	–

5. Concluding remarks

Based on the critical path method and the branch and bound scheme, a new heuristic method has been proposed. The theoretical results obtained have been implemented in a software package. The verification of the method has been presented using the developed software package. The package permits the user to investigate the effect of a new work order in a given manufacturing system. The user can find the answer whether a given work order can be accepted for processing in a given structure of the virtual enterprise. The method proposed can be useful for project-driven production flow management in small and medium size enterprises and for prototyping of the virtual enterprise structure. It provides a good platform for consistency checking between the work order requirements (client requirements) and enterprise capabilities offered (single enterprise or virtual enterprise).

References

Anavi-Isakow S., and B. Golany (2003). Managing multi-project environments through constant work-in-process. *International Journal of Project Management*, 21, pp. 9–18.

Banaszak Z., Saniuk S., and Zaremba M. (2001). Rapid prototyping of constraint-based production flows in a virtual manufacturing enterprise, *Manufacturing Integrated Management*, Vol.1, (1), pp. 3-24

Brucker O., et al. (1999). Resource-constrained project scheduling: Notation, classification, models, and methods. *European Journal of Operational Research*, 112, pp.3–41.

Dvir D., T. Raz, and A. Shehar (2003). An empirical analysis of the relationship between project planning and project success. *International Journal of Project Management*, 21 pp.89–95.

Kerzner H. (1984). Project Management – A system approach to planning, scheduling and controlling. *Van Nostrand Reinhold Company Inc.*, New York.

Lova A., C. Maroto, and P. Tormos (2000). A multicriteria heuristic method to improve resource allocation in multiproject scheduling. *European Journal of Operational Research*, 127, pp. 408–424.

Payne J.H. (1995). Management of multiple simultaneous projects: State-of-art review. *International Journal of Project Management*, 13, pp.163-168.

Copyright © IFAC Information Control Problems in Manufacturing, Salvador, Brazil, 2004

ELSEVIER
IFAC
PUBLICATIONS
www.elsevier.com/locate/ifac

OPTIMIZING THE ORDERING POLICY IN A SUPPLY CHAIN

Yasemin Arda and Jean-Claude Hennet

LAAS-CNRS, 7 Avenue du Colonel Roche, 31077 Toulouse Cedex 4, FRANCE
Tel :+33 561336313, Fax :+33 561636936
e-mail : yarda (hennet)@laas.fr)

Abstract: This study analyzes an inventory management problem within a supply chain. An enterprise whose stock is replenished by outside suppliers faces random demands for one of its products. The objective is to minimize the expected total cost for the enterprise, which includes the holding and shortage costs. The enterprise must choose the suppliers to whom the orders are sent while minimizing the costs. If the enterprise sends all the orders to the same supplier during the planning period, the supplier who has the maximum service rate must be chosen. In the case of choosing possibly different suppliers for different orders, the supplying system is modelled as an open queuing network and explicit expressions are obtained for the average total cost as a function of the decision variables. Numerical results measure the advantage for the producer to use different suppliers instead of a single one. *Copyright © 2004 IFAC*

Keywords: Inventory Control, Queuing Network Models, Manufacturing Systems, Supply Chain Management, Stochastic Modelling.

1. INTRODUCTION

A supply chain is classically defined as a set of enterprises interacting in the processes of manufacturing, delivery and sale of products, from primary suppliers to final customers. Supply Chain Management (SCM) is in charge of organizing the network, planning the tasks and monitoring their execution. As shown in the survey (Aksoy and Derbez, 2003), many commercial software products provide integrated solutions to SCM. However, only few studies have attempted to demonstrate the economic efficiency of SCM.

In supply chain organization, the main difficulty relates to system complexity. Many different viewpoints have to be considered, from legal agreements to technical constraints. To deal with such a complexity, most decision support systems are based on simulation tools. But such tools are difficult to use for systems involving different time scales. They are also limited to evaluation tasks and have no generative ability in architecture and policy design.

In the class of semi-formal models for enterprise modelling, object-oriented models such as the ones based on UML tools (Vernadat, 2001), can provide a good understanding of the structure and components of a supply chain. These models can also help generating formal models of activities and processes.

Contrarily to models based on simulation, analytical models are better adapted to design issues, in particular through optimization. Unfortunately, they also involve oversimplification of reality, but this limitation can often be compensated through over-parameterization, tuning and validation.

A supply chain generally involves several sources of disturbances, and the coordination of product flows is fragile since the randomness in external supply and demand may be amplified through interconnections between partners. Many real supply chains, in particular in the automotive industry, have met serious economical problems because of shortages or excessive inventories. One of the greatest challenges in SMC is to control inventory and production costs along the supply chain in a robust and integrated way, to maximize the service provided to customers. The difficulty is to organize integration in a non-compulsory manner, so as to maintain the autonomy of partners. A possible approach for combining integration and autonomy is through partly automated negotiation processes (Jennings et al., 2001, Besembel et al., 2002).

Most inventory management problems within a supply chain are characterized by the following properties:

- an incomplete control of input and output product flows,

- heterogeneous lot-sizes,
- constraints due to legal agreements between partners,
- use of different policies and decision support systems by the supply chain partners.

In order to take these properties into account, new inventory control models have to be constructed. In particular, it is essential to combine the effects of random fluctuations on demand flows and delays of deliveries from suppliers. Random demands have often been considered in the existing models of inventory control, specially the ones based on the Newsvendor paradigm (Arrow et al., 1951, Porteus, 1990). On the contrary, random delays in part deliveries have not often been integrated in models explicitly. Exceptions are the works of Gurnani et al. (1996), Ould-Louly and Dolgui (2002). In their works, different suppliers provide different parts to be assembled by the producers. Then, there is interdependency between the inventory positions of the end product and of its components. However, the inventory positions of the different components can be independently controlled through the information system. On the contrary, the case of a centralized inventory of a single component does not offer the same possibility of decomposition (Ramasesh et al., 1991).

In this paper, an enterprise is facing a random external demand for a product which is manufactured from identical parts supplied by one or several external suppliers with random delays. The basic case of unitary customer orders with identical probability distributions is considered. Random delivery delays from suppliers are supposed exponentially distributed with different mean values. Manufacturing times in the producer plant are considered negligible with respect to delivery times from suppliers. In the model, the producer then acts as a simple retailer. Under these assumptions, a centralized inventory control model is constructed to combine supply and demand randomness. The objective of the producer is to minimize its average cost by constructing an ordering policy, defined by an optimal reference inventory position and selection of a supplier for each inventory replenishment order.

Section 2 presents the basic ordering problem for one producer and one supplier, then section 3 formulates the single producer - multi supplier optimization problem. In section 4, the case of 2 suppliers is developed and an example illustrates the application of the proposed ordering policy. Finally, in section 5, some conclusions are presented.

2. THE ONE PRODUCER - ONE SUPPLIER CASE

A basic one supplier-one producer model will be used to construct a reference inventory policy that will be easily extended to the N supplier case.

2.1 The reference inventory policy.

The current inventory level of the product considered at time t is denoted I(t). It is defined as the difference between the on-hand inventory and the amount of backorders. In general, an ordering decision should not be based only on the inventory level. One should also consider the number of replenishment orders which have been placed earlier and not yet been delivered, denoted u(t). The global state of the system can then be characterized by the inventory position, denoted P(t) defined by:

$$P(t) = I(t) + u(t) \qquad (1)$$

Depending on the information system available, an inventory position may be controlled at any time through a *continuous review* policy, or at periodic times through a *periodic review* policy. Then, the control policy determines when and how much to order. Different control policies may be applied, within the limits of the legal agreements between producer and supplier.

One of the most popular continuous review policy is the (s,S) policy, in which s stands for the inventory position reorder point and S for the inventory position replenishment level. The base-stock policy can be seen as a variant of the (s,S) policy, for which an order is placed whenever a demand comes, so as to permanently maintain the inventory position S. The (s,S) policy has been shown to be optimal for the problems having independent items and stationary stochastic demand, whenever the cost criterion only depends on the inventory position and has a single local minimum (Axsater, 2000). Moreover, under unitary demands with (i.i.d.) arrival times, the optimal base stock policy reduces to the policy (s,S) with s = S-1. This policy is denoted the *reference inventory policy*. It will be studied in the sequel, first in the single supplier case, then, in section 3, in the multi-supplier case.

When an order comes to the producer, it is immediately satisfied if its amount is available in the stock. If not, it has to wait until the inventory has been sufficiently replenished by the arrival of products from suppliers. In both cases, an order is placed from the producer to the supplier whenever a demand comes and has the same amount (1 in the unitary case). Such a base stock control policy can also be interpreted as a Kanban mechanism (Bollon, 2001). After an initial inventory replenishment stage, the reference inventory policy maintains constant the inventory position of the producer:

$$P(t) = S \quad \forall t \geq t_0 \qquad (2)$$

In the single supplier case, policy optimization then reduces to optimizing the reference inventory position, S.

Two classical techniques are currently used to determine the optimal value of S. The first one is the famous Economic Order Quantity (EOQ) technique due to Harris (1913) and popularized by Wilson (1934). This approach achieves a trade-off between ordering costs and holding costs in the purely deterministic framework of a constant demand rate and a constant and quantity independent delivery time. The second method is based on the "Newsvendor" approach (Arrow et al., 1951). It applies to the case of a random demand for a product subject to obsolescence (such as newspapers). The reference inventory level optimally balances the average stock-out cost with the average holding cost.

2.2 The optimal inventory level under random demand and supply

Many extensions of the newsvendor approach can be found in the literature. In (Khouja, 1999), different criteria have been considered. Lippman and McCardle (1997) have combined newsvendor models and game theory to represent competition between enterprises. A similar combination has been constructed by Lariviere and Porteus (2001) to study the negotiation process between a producer and a supplier.

In spite of the fact that most stochastic inventory models have focused on randomness of demand, several studies have also analyzed the effects of random delivery times. In particular, Song (1994) has represented the demand as a compound Poisson process and analyzed the effects of the probability law of the delivery time on the optimal base stock level and on the total cost. Several authors (Wang, 1996; Dolgui and Louly, 2001; Zheng and Zipkin, 1990; Altiok, 1989) have represented the demand as a simple or as a compound Poisson process, studied the system as a Markov chain, and optimized the system in steady state using the decision variables available.

If the random processes of demand arrivals and supply deliveries are stationary, then, under a stationary *(S-1,S)* base stock policy, the system (producer + suppliers) reaches stationary conditions characterized by stationary probabilities of the number of orders placed by the producer and not yet delivered by the suppliers. In the sequel, these probabilities will be computed in the case of exponential distributions of demand arrivals and delivery times of the suppliers.

From the producer viewpoint, the cost function to be minimized is the sum of the average holding cost and the average stock-out cost. Consider the following notations:

I is the random variable representing the producer inventory level in stationary conditions

u is the random variable representing the number of uncompleted orders from the producer to the supplier in stationary conditions

h is the unit holding cost,

b is the unit stock-out cost.

Under the *(S-1,S)* base stock policy, random variables I and u are related by the following equality, derived from (1) and (2):

$$I=S-u \qquad (3)$$

Using notation $(x)^+$ for $max(x,0)$ and $(x)^-$ for $max(-x,0)$, the average cost criterion takes the form:

$$C(S) = E[\,h\,(I)^+ + b\,(I)^-\,]\,. \qquad (4)$$

Using equation (3) and considering the probability distribution of u, the criterion is re-written:

$$C(S) = h\,E[\,(S-u)^+\,] + b\,E[\,(S-u)^-\,] \qquad (5)$$

$$= h\sum_{w=0}^{S}(S-w)\,\text{Prob}(u=w) \;+ b\sum_{w=S+1}^{\infty}(w-S)\,\text{Prob}(u=w).$$

With $P_w = \text{Prob}(u=w)$ the expected value of u is:

$$Z = E[u] = \sum_{w=0}^{\infty} wP_w \qquad (6)$$

and the criterion can be re-written:

$$C(S) = (h+b)\sum_{w=0}^{S}(S-w)P_w + b\,(Z-S)\,. \qquad (7)$$

Then, considering the incremental function $G(S) = C(S+1) - C(S)$, one obtains:

$$G(S) = (h+b)\,\text{Prob}\,(u \le S) - b. \qquad (8)$$

The PDF $F(S) = \text{Prob}\{\,u \le S\,\}$ being a monotonous increasing function, so is G(S). Then, the value S^* for which $C(S^*)$ is optimal satisfies:

$$\begin{cases} C(S^*) \le C(S^*-1) \Leftrightarrow G(S^*-1) \le 0 \\ C(S^*) < C(S^*+1) \Leftrightarrow G(S^*) > 0\,. \end{cases} \qquad (9)$$

Therefore, a necessary and sufficient condition for optimality is given by the condition:

$$\text{Prob}\,(u \le S^* - 1) \;\le\; \frac{b}{h+b} < \;\text{Prob}\,(u \le S^*) \quad (10)$$

2.3 The case of exponential distributions

Consider now the case of a unitary demand occurring according to a Poisson process with rate λ. The supplier delivery time is exponentially distributed with mean service time $1/\mu$, satisfying the stability condition $\rho = \lambda/\mu < 1$. Under the *(S-1,S)* base stock policy, the inventory position is a constant with value S and the number of uncompleted orders, $u(t)$, represents the queue length of orders for the supplier. It is a simple M/M/1 system with birth-death coefficients (λ, μ). Accordingly, due to equations (1) and (2), the inventory level $I(t)$ is a Markov chain described by the transition graph of Fig. 1. Note in particular that when the inventory level is 0, the number of uncompleted orders is S.

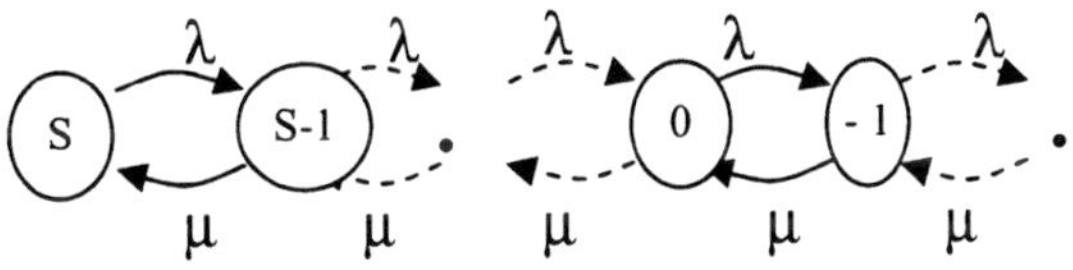

Figure 1. Transition diagram of the Markov chain of inventory level

In stationary conditions,

$P_w = \text{Prob}(\,u = w) = \text{Prob}(\,I = S - w) = \rho^w(1-\rho)$.

The expected value of u, defined by (6), corresponds to the expected number of customers in an M/M/1 queuing system, $Z = \dfrac{\rho}{1-\rho}$.

Moreover, $\text{Prob}\,(u \le S) = \sum_{w=0}^{S}\rho^w(1-\rho) = 1-\rho^{S+1}$.

The cost function then takes the form:

$$C(S) = hS + \frac{\rho}{1-\rho}[(h+b)\rho^S - h] \qquad (11)$$

and the optimality condition for the case of exponential distributions is:

$$1-\rho^{S^*} \;\le\; \frac{b}{h+b} < \; 1-\rho^{S^*+1} \qquad (12)$$

or equivalently,

$$\rho^{S^*+1} < \frac{h}{h+b} \le \rho^{S^*}. \qquad (13)$$

It can be noted that since $\frac{h}{h+b} < 1$, the left part of condition (12) implies $\rho^{S^*+1} < 1$ and thus $S^* \ge 0$.

2.4 *Application to selection of one among N suppliers*

In some economical contexts, contracts between producers and suppliers may involve exclusive rights or high penalty costs to limit the variations of the quantities to be delivered. In these cases, one supplier among N possible ones has to be chosen, and this supplier should receive all the supply orders. Then, the analysis of the preceding paragraphs remains valid but the choice of the supplier should be made beforehand.

The following result, established by (Song, 1994), can be proved very easily in the case of exponential distributions.

Property 1

When the supplier should be chosen once and for all the orders, then the supplier k who should be selected is the one with the highest service rate, satisfying: $\mu_k \ge \mu_n$ *for* $n = 1,...,N$.

Proof

Condition $\mu_k \ge \mu_n$ for $n = 1,...,N$ implies $\rho_k \le \rho_n$ for n = 1,...,N. Then, with $S_k^* \ge 0$, $\frac{h}{h+b} \le \rho_k^{S_k^*} \le \rho_n^{S_k^*}$.

And since S_n^* is the maximal value of S satisfying condition $\rho_n{}^S \ge \frac{h}{h+b}$, then $S_n^* \ge S_k^*$. From expression (12), we obtain:

$$C_n(S_n^*)\text{-}C_k(S_n^*) =$$

$$(h+b)\left(\frac{\rho_n}{1\text{-}\rho_n}\,(\rho_n^{S_n^*} - \frac{h}{h+b}) - \frac{\rho_k}{1\text{-}\rho_k}\,(\rho_k^{S_n^*} - \frac{h}{h+b})\right).$$

Then, using $\frac{\rho_k}{1\text{-}\rho_k} \le \frac{\rho_n}{1\text{-}\rho_n}$, non-negativity of the quantity above can be derived as follows:

$$C_n(S_n^*)\text{-}C_k(S_n^*) \ge (h+b)\frac{\rho_k}{1\text{-}\rho_k}\,(\rho_n^{S_n^*} - \rho_k^{S_n^*}) \ge 0.$$

This condition, together with optimality of S_k^* with respect to criterion $C_k(S)$ implies:

$C_k(S_k^*) \le C_n(S_n^*)$. Clearly, the supplier to be selected is the one with the highest service rate, μ_k, since he (or she) satisfies condition: $C_k(S_k^*) \le C_n(S_n^*)\,\forall n = 1,..., N$. ■

3. THE SINGLE PRODUCER-MULTI-SUPPLIER OPTIMIZATION PROBLEM

The model of the preceding section is now extended to the multi-supplier case. Like in section 2.4, it is assumed that N different suppliers with different service times can deliver the same product, but now the enterprise may choose a different supplier at each inventory replenishment order.

3.1 *Problem description*

As in the preceding section, it is assumed that demands are unitary. They enter the system as a Poisson process with rate $\lambda > 0$.

When a unitary demand arrives at time *t*, the producer serves it immediately if $I(t)>0$. It waits if $I(t)\le 0$. In both cases, he (or she) applies the *(S-1,S)* ordering policy by sending a corresponding order to supplier *i* with probability α_i satisfying $0 \le \alpha_i \le 1$ and $\sum_{i=1}^{N} \alpha_i = 1$. Each supplier has an exponential service rate and treats the requests in the FIFO (First In First Out) order. Let μ_i denote the mean service rate of supplier $i = 1,...,N$, with $\mu_i \ne \mu_j$ for $j \in (1,...,i\text{-}1,i+1,...,N)$.

Such an order dispatching policy from the producer to the supplier is known as a Bernoulli splitting process. A well known property of this process is stated below.

Property 2 :

Consider the Poisson process $\{N(t),\ t \ge 0\}$ *with rate* λ *where* $N(t)$ *represents the number of events in the time interval* $[0,t]$. *Suppose that each event is assigned to class* $i = 1,...,N$ *with probability* p_i *satisfying* $\sum_{i=1}^{N} p_i = 1$. *Let* $N_i(t)$ *be the number of type i events having occurred in* $[0,t]$. *Then,* $N(t) = \sum_{i=1}^{N} N_i(t)$ *and* $\{N_i(t),\ t \ge 0\}$ $i = 1,...,N$ *are Poisson processes with rates* $p_i\lambda$. *Moreover, these processes are mutually independent.*

The proof of this property can be found in (Ross, 2000).

As a consequence of this property, each supplier can be modelled as an M/M/1 queue with arrival rate $\alpha_i\lambda$ and service rate μ_i. The set of suppliers is represented as a network of N independent M/M/1 queues in parallel.

The probability of having k_i orders in queue *i* is given by:

$$P(k_i) = \Pr\{k_i \text{ orders in } i\} = \left(\frac{\alpha_i\lambda}{\mu_i}\right)^{k_i}\left(1 - \frac{\alpha_i\lambda}{\mu_i}\right) \qquad (14)$$

The necessary and sufficient condition for stability of queue i is $\alpha_i\,\rho_i < 1$, with $\rho_i = \lambda / \mu_i$. The probability for the network of queues to be in state $K = \{k_1,k_2,...,k_i,...,k_N\}$ is given by the product-form expression (Baskett *et al*, 1975):

$$P(k_1,...,k_N) = \prod_{i=1}^{N} (\alpha_i\rho_i)^k (1 - \alpha_i\rho_i) \qquad (15)$$

In the order dispatching problem, Bernoulli parameters $\alpha_1, \alpha_2 \ldots \alpha_N$ are decision variables. Their optimal values express the optimal assignment probabilities in steady state. The considered optimization criterion is the sum of the mean holding cost and the mean stock-out cost per time unit. The objective of the study is to compute the optimal Bernoulli parameters and the optimal base stock value, S^* minimizing the average cost criterion.

In the case of N suppliers, the number of orders not yet delivered to the producer is equal to the total number of orders in the open queuing network composed by the queues of orders coming from the producer. Let K_i denote the number of orders sitting in the ith supplier queue. $P(k_i) = Pr\{K_i = k_i\}$ is defined by equation (14). Then, the probability of having w orders waiting in the suppliers queue is given by (16):

$$P_w = \Pr\{K_1 + K_2 + ... + K_N = w\} \quad (16)$$

The average cost function to be minimized has the same forms (5) and (7) as in the one supplier case but the difference is in the expression of the probability P_w. In the N supplier case, the sum of the mean holding cost and the mean stock-out cost per time unit is a function of the base stock level, S and of Bernoulli parameters α_1, α_2... α_N. And since the queues are independent, the mean number of orders in the open network, Z, can be computed as follows:

$$Z = \sum_{i=1}^{N} \frac{\alpha_i \rho_i}{1-\alpha_i \rho_i} = \sum_{i=1}^{N} \frac{\alpha_i \lambda}{\mu_i - \alpha_i \lambda} \quad (17)$$

3.2 Formulation of probability Pw

Probability P_w can be obtained by composition of the probabilities related to the N queues. Such a composition can be computed from the probability generating function defined as follows (see e.g. Stirzaker, 1999):

$$G_X(z) = \sum_x p(x) z^x = E[z^X] \quad (18)$$

where $p(x) = Pr\{X=x\}$. Note that the sum (18) converges for $|z| \le 1$, with $\sum_x p(x) = 1$.

Independence of variables K_i for $i=1,...,N$ implies that the probability generating function of the sum $D=K_1+...+K_N$ is the product of probability generating functions for variables $K_1,..., K_N$.

$$G_{K_1+K_2+...+K_N}(z) = G_{K_1}(z) \times ... \times G_{K_N}(z) \quad (19)$$

with, for $i = 1,..., N$,

$$G_{K_i}(z) = (1-\alpha_i \rho_i) \sum_{k_i=0}^{\infty} (\alpha_i \rho_i z)^{k_i} = \frac{1-\alpha_i \rho_i}{1-\alpha_i \rho_i z} \quad (20)$$

Therefore,

$$G_D(z) = \prod_{i=1}^{N} \frac{1-\alpha_i \rho_i}{1-\alpha_i \rho_i z} \quad (21)$$

Assuming $\alpha_i \rho_i \neq \alpha_j \rho_j \ \forall i \neq j$, the product form (21) can be transformed into a sum as in (Kleinrock, 1975) :

$$G_D(z) = \sum_{i=1}^{N} \frac{A_k}{1-\alpha_k \rho_k z} \quad (22)$$

with coefficient A_k obtained by :

$$A_k = H(N) \times (\alpha_k \rho_k)^{N-1} \times \prod_{\substack{j=1 \\ j \neq i}}^{N} b_{kj} \quad (23)$$

with $H(N) = \prod_{i=1}^{N} (1-\alpha_i \rho_i)$, $b_{kj} = \dfrac{1}{\alpha_k \rho_k - \alpha_j \rho_j}$.

The case $\alpha_i \rho_i = \alpha_j \rho_j$ for some $i \neq j$ cannot be excluded a priori. However, in numerical evaluations, it can be approximated by continuity, replacing the preceding equality, when it is tested, by $\alpha_i \rho_i = \alpha_j \rho_j + \varepsilon$ with $\varepsilon << \max_{i=1,...,N} \rho_i$.

From expressions (22) and (23), we derive :

$$G_D(z) = H(N) \times \sum_{i=1}^{N} \left(\left(\prod_{\substack{j=1 \\ j \neq i}}^{N} b_{ij} \right) \times \sum_{w=0}^{\infty} \alpha_i^{N+w-1} \rho_i^{N+w-1} z^w \right)$$

and finally

$$P_w = H(N) \times \sum_{i=1}^{N} \left(\left(\prod_{\substack{j=1 \\ j \neq i}}^{N} b_{ij} \right) \times (\alpha_i \rho_i)^{N+w-1} \right). \quad (24)$$

3.3 Expression of the cost function

The cost function can then be re-written as follows.

$$C(S, \alpha_1, \alpha_2, ..., \alpha_N) = (h+b) \times H(N) \times$$

$$\sum_{i=1}^{N} \left(\left(\prod_{\substack{j=1 \\ j \neq i}}^{N} b_{ij} \right) \times \left(\frac{S \alpha_i^{N-1} \rho_i^{N-1}}{1-\alpha_i \rho_i} - \frac{\alpha_i^N \rho_i^N (1-\alpha_i^S \rho_i^S)}{(1-\alpha_i \rho_i)^2} \right) \right) +$$

$$b \left(\sum_{i=1}^{N} \frac{\alpha_i \rho_i}{1-\alpha_i \rho_i} - S \right). \quad (26)$$

Convexity of criterion (26) with respect to variables $\alpha_1, ..., \alpha_N$, S is not guaranteed in general. Therefore, minimization of criterion (26) subject to constraints $0 \le \alpha_i \le 1$ for $i=1,...,N$ and $\sum_{i=1}^{N} \alpha_i = 1$ appears to be a hard optimization problem. A decomposed approximation scheme is currently under study. Evaluation of this scheme with respect to the optimal solution will be performed on simple examples such as the two supplier case, analyzed in the following section.

4. EXAMPLE

Consider the case of a producer receiving a sequence of unitary orders according to a Poisson law with mean rate one order per time unit (λ=1). He (or she) has only one supplier with mean service rate μ=1.25. The unit storage cost is h=1 and the unit stock-out cost b=1000. Then, by the optimality condition (13), the optimal base stock level is S*=30 and the average total cost is: $C(S^*) \approx 30.96$.

Suppose now that a new supplier has appeared on the market of the same component product, with mean service rate $\mu_2=0.8$. Considered alone, the new supplier would not be able to satisfy the demand. However, if the producer has the contractual possibility to use both suppliers, he (or she) can split the demand between the 2 suppliers. In the particular case of 2 suppliers, the cost function (26) takes the following form:

$$C(S,\alpha_1,\alpha_2) = hS + (h+b)\times$$
$$\left(\frac{\alpha_2^2\rho_2^2(1-\alpha_1\rho_1)^2(1-\alpha_2^S\rho_2^S)-\alpha_1^2\rho_1^2(1-\alpha_2\rho_2)^2(1-\alpha_1^S\rho_1^S)}{(1-\alpha_1\rho_1)(1-\alpha_2\rho_2)(\alpha_1\rho_1-\alpha_2\rho_2)} \right)$$
$$+b\left(\frac{\alpha_1\rho_1}{1-\alpha_1\rho_1}+\frac{\alpha_2\rho_2}{1-\alpha_2\rho_2}\right).$$

A numerical evaluation of this function with respect to variables α_1, α_2, S is represented on Fig. .2. The optimal solution is: α_1*=0.626, α_2*=0.374 S*=12, C(S*, α_1*, α_2*)=11.449.

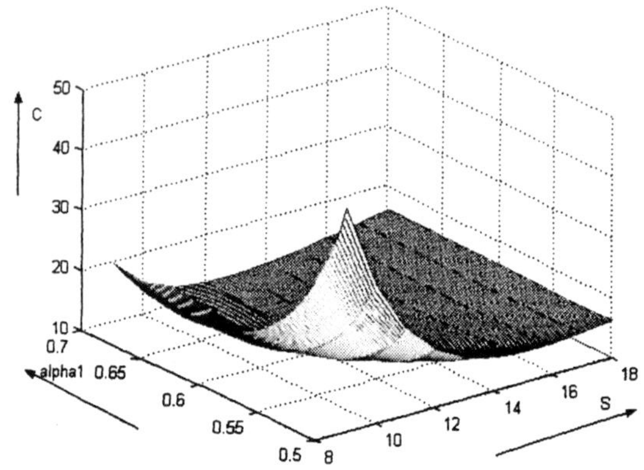

Fig.2 The cost curve around the minimum point

The important decrease of the average total cost with two suppliers instead of one, is mainly due to a lower risk of stock-out for the producer, leading to a decrease of the base-stock level from 30 to 12.

5. CONCLUSION

Supply chain management can be analyzed from several viewpoints, corresponding to different actors (producers, suppliers, customers, etc.) and different criteria (costs, delays, quality, reliability). In the context of a supply chain network made of enterprises with a fair degree of autonomy, global optimality would be purely theoretical, each actor having his economical constraints and his strategic preferences. This paper has considered the cost viewpoint of one actor, a producer, in a case of random demand for a product and random delivery times for a basic component of this product. It has been shown on an example that important savings in stock-out and inventory costs can be obtained by the producer by splitting its supply orders between several suppliers rather than limiting his choice to the most efficient one.

REFERENCES

Altiok, T. (1989). (R,r) Production/inventory systems. *Operations Research*, **37**, 266 – 276.

Aksoy, Y., A. Derbez (2003). Supply chain management. ORMS Today, **June 2003**, 34–41.

Arrow, K., T. Harris, J. Marschak (1951). Optimal Inventory Policy. *Econometrica*, **19**, 250-272.

Axsater, S. (2000). *Inventory control.* Ed. Kluwer.

Baskett, F., K. M. Chandy, R. R. Muntz, F. G. Palacios (1975). Open, closed and mixed networks of queues with different classes of customers. *J. ACM*, **22**, No. 2, 248 – 260.

Besembel, I., J.C. Hennet, E. Chacon (2002). Coordination by hierarchical negotiation within an enterprise network. *Proc. ICE 2002*, Roma (Italia), 507-513.

Chiang, C. (2001). Order splitting under periodic review inventory systems. *International Journal of Production Economics*, **70**, 67 – 76.

Gurnani, H.B., R. Akella, J. Lehoczky (1996). Optimal order policies in assembly systems with random demand and random supplier delivery. *IIE Transactions*, **28**, 865-878.

Jennings, N.R., P. Faratin, A. R. Lomuscio, S. Parsons, C. Sierra, M. Wooldridge (2001). Automated negotiation: prospects, methods and challenges. *Int. J. of Group Decision and Negotiation*, **10 (2)**, 199-215.

Khouja, M. (1999). The single period (news-vendor) problem : literature review and suggestions for future research. *Omega International Journal of Management Science*, **27**, 537 – 553.

Kleinrock, L. (1975). *Queueing systems. Volume 1: Theory*. Ed. John Wiley & Sons.

Lariviere, M.A., E. L. Porteus (2001). Selling to the newsvendor: an analysis of price-only contracts. *MSO&M*, **3**, 293–305.

Lippman, S.A., F. McCardle (1997). The competitive newsboy. *Operations Research*, **45**, 54 – 65.

Ould-Louly, M.A., A. Dolgui (2002). Newsboy model for supply planning of assembly systems. *Int. Journal of Prod. Research*, **40**, 4401-4414.

Porteus, E.L. (1990). Stochastic Inventory Theory, in *Handbooks in operations research and management science. Volume 2: Stochastic Models*, Ed. North-Holland.

Ramasesh, R.V., J.K. Ord, J.C. Hayya, A. Pan (1991). Sole vs dual sourcing in stochastic lead-time (s,Q) inventory models. *Management Science*, **37**, 428-443.

Ross, S.M. (2000). *Introduction to probability models.* Ed. A Harcourt Science and Technology Company, Academic Press.

Song, J.S. (1994). The effect of leadtime uncertainty in a simple stochastic inventory model. *Management Science*, **40**, 603–613.

Stirzaker, D. (1999). *Probability and random variables: a beginner's guide.* Cambridge University Press.

Vernadat F. (2001). UEML: Towards a Unified Enterprise Modelling Language, *Proc. MOSIM'01*, Troyes, 3-9.

Wang, P. (1996). Markovian queueing models with periodic-review. *Computers Operations Research*, **23**, 741–754.

Zheng, Y.S. and P. Zipkin (1990). A queueing model to analyze the value of centralized inventory information. *Operations Research*, **38**, 296–307.

Copyright © IFAC Information Control Problems in
Manufacturing, Salvador, Brazil, 2004

ELSEVIER
IFAC
PUBLICATIONS
www.elsevier.com/locate/ifac

MODELING DEMAND FOR INVENTORY MANAGEMENT OF SLOW-MOVING ITEMS IN CASE OF REPORTING ERRORS

Alexandre Dolgui[1], Anatoly Pashkevich[2], Maxim Pashkevich[3]

[1]*Scientific Methods for Industrial Management Dept. (MSGI)*
Industrial Engineering and Computer Sciences Center (G2I)
Ecole de Mines de Saint Etienne, 158, Cours Fauriel, 42023 Saint Etienne, France
e-mail: dolgui@emse.fr

[2]*Robotic Laboratory, Department of Automatic Control*
Belarusian State University of Informatics and Radioelectronics
6 P.Brovka St., Minsk 220027, Belarus
e-mail: pap@bsuir.unibel.by

[3]*Department of Mathematical Modeling and Data Analysis*
Faculty of Applied Mathematics and Computer Science
Belarusian State University, Skarina Av. 4, Minsk 220050, Belarus
e-mail: pma@omegasoftware.com

Abstract: The paper is devoted to the problem of modeling demand for inventory management of slow-moving items in the case of reporting errors. It is proposed a generalization of the beta-binomial demand model that takes into account possible reporting errors in the learning sample. For the new model, there are developed identification and forecasting algorithms that provide consistent estimators of the model parameters and mean square optimal forecasts. The efficiency of the proposed approach is illustrated by an application example for slow-moving car parts. *Copyright © 2004 IFAC*

Keywords: inventory control, error correction, statistical inference, parameter estimation, prediction methods.

1. INTRODUCTION

The goal of the inventory management is balancing service levels against investment over a very large assortment of stock-keeping units (SKU) (Hopp and Spearman, 1995). In contrast to the traditional "one-at-a-time" approach, the inventory management determines all SKU stock levels to fulfil total service and investment constraints or objectives (Azoury and Miller, 1984). It is intensively used in industry and has proved to be very efficient (Harris, 1997).

In contrast to the standard marginal stock level settings, the inventory management requires a probability distribution of SKU demand that is used to generate forecasts when planning reorder points for SKUs (Hopp, *et al.*, 1997). Recent research done by Grange (1998) has indicated that selecting inappropriate demand stochastic model may result in poor inventory performance. For this reason, it is important to choose an adequate demand probability distribution that will ensure accurate reorder points forecast.

However, even when the selected stochastic model of the demand is adequate, it still can be misleading in cases of unobserved lost sales and sale reporting errors (Jacobs and Wagner, 1989). A number of researchers investigated the problem of unobserved lost sales and proposed efficient identification algorithms for various stochastic demand models (Nahmias,

1994; Agrawal and Smith, 1996). However, no research has been done for the case of sales reporting errors. In particular, there have not been investigated the influence of such errors on the performance of the beta-binomial demand model (BBM).

In this paper, it is proposed a generalization of the beta-binomial model that can be used to model the demand of slow-moving items in the case of reporting errors in the request records. For the new model, there are developed identification and forecasting algorithms that provide consistent estimators of the model parameters and mean square optimal forecasts. The efficiency of the proposed approach is illustrated via an application example.

2. INVENTORY OPTIMIZATION PROBLEM

A basic inventory optimization problem can be stated as follows (Grange, 1998). Denote by m the number of SKU and let D_l and Q_l be the demand rate and stock guidance ("order-to-up" quantity) of the SKU l over a fixed period of time (one week, for example). Define the total fill rate as

$$f(Q)=\sum_{l=1}^{m}\left(E\{D_l\}\cdot P\{D_l\le Q_l\}\right)\Big/\sum_{l=1}^{m}E\{D_l\}. \quad (1)$$

where $E\{D_l\}$ is the expected demand for the SKU l, $P\{D_l \le Q_l\}$ is the individual service level of the SKU l, and $Q=\{Q_1, \ldots, Q_2, Q_m\}$. Let C_l be the cost of the item l. Then the problem is to minimize the total inventory cost ensuring the target total fill rate F:

$$\sum_{l=1}^{m} C_l Q_l \to \min_Q, \quad f(Q)\ge F. \quad (2)$$

In order to solve the problem (2), one must know the probability distribution of the demand D_l for each SKU.

In real life, up to 90% of the SKUs may be slow-moving items. To achieve the high total fill rate, it is required to rely on adequate SKU demand model, which is often descried by the beta-binomial distribution (BBD) (Grange, 1998). An essential advantage of using the BBM vs. the classical binomial model (BM) is that the BBM takes into account possible correlation of the demand within a time period and between the periods, while the BM assumes that all requests are independent.

However, raw demand data that are used to estimate the probability distribution parameters may be often distorted due to reporting errors (Nahmias 1994). These distortions can be caused by staff mistakes, ineffective management or equipment faults. For this reason, this paper focuses on generalization of the beta-binomial model, which takes into account possible reporting errors.

3. DEMAND MODEL IN CASE OF REPORTING ERRORS

For a slow-moving item, let us assume that over a constant time segment t (one hour, for example), not more then one order can be received. Denote by $n = int(T/t)$ a number of time segments in the total time period T over which the demand is modeled (number of hours in one week, for instance). Let us suppose that the demand data for k time periods of length T (k weeks, for example) are represented as a binary $(k\times n)$-matrix $B=(b_{ij})$, where $b_{ij}=1$ if the item was ordered in the segment j of the period i and $b_{ij}=0$ otherwise. The following assumptions define the beta-binomial demand model of a slow-moving item.

1. The probability p_i that the item was ordered during the time segment j of the period i is constant for all segments $j = 1 : n$.

2. The probabilities p_i, $i = 1 : k$, are the independent identically distributed random variables that have the beta distribution with true unknown parameters α_0, β_0.

Because of the reporting errors, the distorted binary $(k\times n)$-matrix $\widetilde{B}=(\widetilde{b}_{ij})$ is observed, where

$$\widetilde{b}_{ij}=b_{ij}\oplus\eta_{ij},\quad P\{\eta_{ij}=1\}=\begin{cases}\varepsilon_0, & \text{if } b_{ij}=0;\\ \varepsilon_1, & \text{if } b_{ij}=1.\end{cases} \quad (3)$$

Here ε_0 is the probability that a false order was reported, and ε_1 is the probability of not reported order. Let us refer to ε_0, ε_1 as the error levels. Besides, let us denote

$$X=\{x_1,x_2,\ldots,x_k\},\ x_i=\sum_{j=1}^{n}\widetilde{b}_{ij},\ \gamma_0=1-\varepsilon_0,\ \gamma_1=1-\varepsilon_1. \quad (4)$$

Here x_i is the total number of orders for the period i.

The problem is to estimate the parameters α, β of the original BBM and to forecast the demand probability p for the future period of time based on the sample X with reporting errors (3).

4. DISTORTED DEMAND DISTRIBUTION

It can be proved (Kharin and Pashkevich, 2003) that under reporting errors (3), the demand distribution (distribution of the sample X) takes the following form ($r = 0 : n$):

$$p_r(\alpha,\beta,\varepsilon_0,\varepsilon_1)=\sum_{i=0}^{n} w_{ri}(\varepsilon_0,\varepsilon_1)\cdot C_n^i\frac{B(\alpha_0+i,\beta_0+n-i)}{B(\alpha_0,\beta_0)}, \quad (5)$$

$$w_{ri}(\varepsilon_0,\varepsilon_1)=\sum_{l=\max(i,r)}^{\min(n,i+r)} C_i^{l-r}C_{n-i}^{l-i}\varepsilon_0^{l-i}\gamma_0^{n-l}\varepsilon_1^{l-r}\gamma_1^{i+r-l}. \quad (6)$$

Let us refer to this distribution as the distorted beta-binomial distribution (DBBD) with the parameters n, α, β, ε_0, ε_1. Using expressions (5), (6), one can show that the first four moments of the DBBD are expressed as:

$$m(\varepsilon_0,\varepsilon_1)=m^0+\left(1-m^0\right)\varepsilon_0-m^0\varepsilon_1,\quad m_0=\frac{n\alpha}{\alpha+\beta}, \quad (7)$$

$$m_2(\varepsilon_0,\varepsilon_1)=m(\varepsilon_0,\varepsilon_1)+\frac{n^{[2-]}}{(\alpha+\beta)^{[2+]}}\times \quad (8)$$
$$\times\left(\alpha^{[2+]}+\beta^{[2+]}\varepsilon_0^2+\alpha^{[2+]}\varepsilon_1^2+2\alpha\beta\varepsilon_0-2\alpha^{[2+]}\varepsilon_1-2\alpha\beta\varepsilon_0\varepsilon_1\right),$$

$$m_3(\varepsilon_0,\varepsilon_1) = m_3^0 + \left(\frac{n\beta}{\alpha+\beta} + 6n^{[2-]}\frac{\alpha\beta}{(\alpha+\beta)^{[2+]}} + 3n^{[3-]}\frac{\alpha^{[2+]}\beta}{(\alpha+\beta)^{[3+]}}\right)\varepsilon_0 + \left(\frac{n\alpha}{\alpha+\beta} + 6n^{[2-]}\frac{\alpha^{[2+]}}{(\alpha+\beta)^{[2+]}} + 3n^{[3-]}\frac{\alpha^{[3+]}}{(\alpha+\beta)^{[3+]}}\right)\varepsilon_1 + o(\varepsilon_0) + o(\varepsilon_1), \quad (9)$$

$$\begin{aligned} m_4(\varepsilon_0,\varepsilon_1) &= m_4^0 + +\frac{n\beta\varepsilon_0}{\alpha+\beta} + 14\, n^{[2-]}\frac{\alpha\beta\varepsilon_0}{(\alpha+\beta)^{[2+]}} + \\ &+ 18\, n^{[3-]}\frac{\alpha^{[2+]}\beta\varepsilon_0}{(\alpha+\beta)^{[3+]}} + 4\, n^{[4-]}\frac{\alpha^{[3+]}\beta\varepsilon_0}{(\alpha+\beta)^{[4+]}} + \frac{n\alpha\varepsilon_1}{\alpha+\beta} + \\ &+ 14\, n^{[2-]}\frac{\alpha^{[2+]}\varepsilon_1}{(\alpha+\beta)^{[2+]}} + 18\, n^{[3-]}\frac{\alpha^{[3+]}\varepsilon_1}{(\alpha+\beta)^{[3+]}} + \\ &4\, n^{[4-]}\frac{\alpha^{[4+]}\varepsilon_1}{(\alpha+\beta)^{[4+]}} + o(\varepsilon_0) + o(\varepsilon_1), \end{aligned} \quad (10)$$

where

$$y^{[z-]} = y(y-1)\ldots(y-z+1), \quad (11)$$

$$y^{[z+]} = y(y+1)\ldots(y+z-1), \quad (12)$$

for any $z \in N, y \in R$.

5. ESTIMATION OF MODEL PARAMETERS

To estimate the parameters of the BBM, method of moments (MM) and method of maximum likelihood (ML) are traditionally used (Tripathi, *et al.*, 1994). In our previous work (Pashkevich and Dolgui, 2003), it was proved that these estimators become inconsistent in the case of data distortions of type (3). For this reason, there are proposed new consistent estimators for the cases of both known and unknown error levels ε_0, ε_1.

5.1 Estimation in the case of known error levels

In practice, it is a common situation when the estimates of the error levels ε_0, ε_1 are known in advance. The source of the prior information can be the past work experience or expert knowledge. In this case, it is possible to develop new consistent methods of the beta-binomial model parameter estimation.

It was proved that the following modified MM-estimators are consistent:

$$\hat{\alpha}(\varepsilon_0,\varepsilon_1) = \frac{a(\varepsilon_0,\varepsilon_1)}{d(\varepsilon_0,\varepsilon_1)}, \quad \hat{\beta}(\varepsilon_0,\varepsilon_1) = \frac{b(\varepsilon_0,\varepsilon_1)}{d(\varepsilon_0,\varepsilon_1)}, \quad (13)$$

where

$$a(\varepsilon_0,\varepsilon_1) = (m^* - \varepsilon_0 n)\times \times\left(m^* n - m_2^* + \varepsilon_0(n-1)(m^* - \gamma_1 n) - m^*\varepsilon_1(n-1)\right), \quad (14)$$

$$b(\varepsilon_0,\varepsilon_1) = (m^* - \gamma_1 n)\times \times\left(m_2^* + m^*\varepsilon_1(n-1) - m^* n - \varepsilon_0(n-1)(m^* - m\gamma_1)\right), \quad (15)$$

$$d(\varepsilon_0,\varepsilon_1) = (1 - \varepsilon_0 - \varepsilon_1)\left(m_2^* n - m^* n - m^{*2}(n-1)\right), \quad (16)$$

$$m^* = k^{-1}\sum_{i=1}^k x_i, \quad m_2^* = k^{-1}\sum_{i=1}^k x_i^2. \quad (17)$$

Another way of using the prior information is filtering the sample X with the reporting errors. Let $\tilde{p}(\varepsilon_0,\varepsilon_1)$ be the vector of the empirical relative frequencies calculated using the sample X. Let $W = (w_{ij})$ be a $(n+1)\times(n+1)$-matrix defined by the expression (6) and let the vector $\tilde{p}_0$ be defined as

$$\tilde{p}_0 = W^{-1}(\varepsilon_0,\varepsilon_1)\cdot\tilde{p}(\varepsilon_0,\varepsilon_1).$$

Then, it can be proved that the vector $\tilde{p}_0$ is the vector of the empirical relative frequencies for the BBM with the parameters n, α_0, β_0. As a result, the model parameter estimation problem with reporting errors is converted to the classical BBM model estimation using the filtered data.

5.2 Estimation in the case of unknown error levels

Let us consider a general case when the error levels ε_0, ε_1 are unknown. Two iterative algorithms of simultaneous estimation of the demand model parameters and error levels are proposed. Both methods suppose that X is the random sample from the DBBD with the parameters n, α, β, ε_0, ε_1.

The first algorithm is based on the method of moments. The idea is to iteratively solve the system of two non-linear equations using the modified method of Newton:

$$\begin{aligned} m_3^* &= m_3\left(\hat{\alpha}(\varepsilon_0,\varepsilon_1),\hat{\beta}(\varepsilon_0,\varepsilon_1),\varepsilon_0,\varepsilon_1\right) \\ m_4^* &= m_4\left(\hat{\alpha}(\varepsilon_0,\varepsilon_1),\hat{\beta}(\varepsilon_0,\varepsilon_1),\varepsilon_0,\varepsilon_1\right) \end{aligned} \quad (18)$$

$$m_3^* = k^{-1}\sum_{i=1}^k x_i^3, \quad m_4^* = k^{-1}\sum_{i=1}^k x_i^4. \quad (19)$$

Here $m_3(.)$, $m_4(.)$ are the third and the fourth theoretical moments of the DBBD with the parameters n, α, β, ε_0, ε_1 (see expressions (7), (8)), and $\hat{\alpha}(\varepsilon_0,\varepsilon_1)$, $\hat{\beta}(\varepsilon_0,\varepsilon_1)$ are the estimates of the parameters α, β when the distortions levels ε_0, ε_1 are fixed. These estimates can be calculated using equation (13).

Let J_0^{cond} be the Jacobi matrix of the system (18) under the assumption that the first two moments are fixed: $m = const$, $m_2 = const$. Then using the modified Newton method results in the following iterative procedure:

$$\begin{pmatrix}\varepsilon_0^{r+1}\\ \varepsilon_1^{r+1}\end{pmatrix} = \begin{pmatrix}\varepsilon_0^{r}\\ \varepsilon_1^{r}\end{pmatrix} + \lambda\cdot\left(J_0^{cond}\right)^{-1}\cdot\begin{pmatrix}m_3^* - m_3^r\\ m_4^* - m_4^r\end{pmatrix} \quad (20)$$

$$\begin{aligned} m_3^r &= m_3\left(\hat{\alpha}(\varepsilon_0^r,\varepsilon_1^r),\hat{\beta}(\varepsilon_0^r,\varepsilon_1^r),\varepsilon_0^r,\varepsilon_1^r\right) \\ m_4^r &= m_4\left(\hat{\alpha}(\varepsilon_0^r,\varepsilon_1^r),\hat{\beta}(\varepsilon_0^r,\varepsilon_1^r),\varepsilon_0^r,\varepsilon_1^r\right) \end{aligned} \quad (21)$$

where $\lambda \in (0, 1]$ is the algorithm parameter that ensures convergence in the case of large error levels ε_0, ε_1.

Let us introduce the following notation for the partial derivatives under the assumption $m = const$,

$m_2 = const$, $\varepsilon_0 = 0$, $\varepsilon_1 = 0$: $(\partial f/\partial x)_0^c$. Then the elements of the matrix J_0^{cond} are calculated using the following expressions:

$$\left(\frac{\partial m_3}{\partial \varepsilon_0}\right)_0^c = 3n^{[3-]}\frac{\alpha^{[2+]}\beta}{(\alpha+\beta)^{[3+]}},\ \left(\frac{\partial m_3}{\partial \varepsilon_1}\right)_0^c = 3n^{[3-]}\frac{\alpha^{[3+]}}{(\alpha+\beta)^{[3+]}}, \quad (22)$$

$$\left(\frac{\partial m_4}{\partial \varepsilon_0}\right)_0^c = 14n^{[4-]}\frac{\alpha^{[3+]}\beta}{(\alpha+\beta)^{[4+]}} + 6\left(\frac{\partial m_3}{\partial \varepsilon_0}\right)_0^c, \quad (23)$$

$$\left(\frac{\partial m_4}{\partial \varepsilon_1}\right)_0^c = 4n^{[4-]}\frac{\alpha^{[4+]}}{(\alpha+\beta)^{[4+]}} + 6\left(\frac{\partial m_3}{\partial \varepsilon_1}\right)_0^c, \quad (24)$$

$$\left(\frac{\partial \tilde{\alpha}}{\partial \varepsilon_0}\right)_0^c = -(\alpha_0 + 2\beta_0 + 1),\ \left(\frac{\partial \tilde{\alpha}}{\partial \varepsilon_1}\right)_0^c = -\alpha_0(\alpha_0+1)/\beta_0, \quad (25)$$

$$\left(\frac{\partial \tilde{\beta}}{\partial \varepsilon_0}\right)_0^c = -\beta_0(\beta_0+1)/\alpha_0,\ \left(\frac{\partial \tilde{\beta}}{\partial \varepsilon_1}\right)_0^c = -(2\alpha_0 + \beta_0 + 1), \quad (26)$$

$$\left(\frac{\partial m_3}{\partial \alpha}\right)_0^c = n^{[3-]}\frac{\alpha^{[3+]}}{(\alpha+\beta)^{[3+]}}\sum_{i=0}^{2}\left(\frac{1}{\alpha+i} - \frac{1}{\alpha+\beta+i}\right), \quad (27)$$

$$\left(\frac{\partial m_3}{\partial \beta}\right)_0^c = n^{[3-]}\frac{\alpha^{[3+]}}{(\alpha+\beta)^{[3+]}}\sum_{i=0}^{2}\left(-\frac{1}{\alpha+\beta+i}\right), \quad (28)$$

$$\left(\frac{\partial m_4}{\partial \alpha}\right)_0^c = n^{[4-]}\frac{\alpha^{[4+]}}{(\alpha+\beta)^{[4+]}}\sum_{i=0}^{3}\left(\frac{1}{\alpha+i} - \frac{1}{\alpha+\beta+i}\right) + 6\left(\frac{\partial m_3}{\partial \alpha}\right)_0^c, \quad (29)$$

$$\left(\frac{\partial m_4}{\partial \alpha}\right)_0^c = n^{[4-]}\frac{\alpha^{[4+]}}{(\alpha+\beta)^{[4+]}}\sum_{i=0}^{3}\left(-\frac{1}{\alpha+\beta+i}\right) + 6\left(\frac{\partial m_3}{\partial \alpha}\right)_0^c. \quad (30)$$

The second algorithm is based on the method of maximum likelihood, which leads to the following problem of the constrained maximization:

$$\begin{gathered} l(\alpha,\beta,\varepsilon_0,\varepsilon_1) \to \max,\ \alpha > 0,\ \beta > 0,\ 0 \le \varepsilon_0, \varepsilon_1 \le 1, \\ l(\alpha,\beta,\varepsilon_0,\varepsilon_1) = \textstyle\sum_{r=0}^{n} f_r \ln(p_r(\alpha,\beta,\varepsilon_0,\varepsilon_1)), \end{gathered} \quad (31)$$

where $\{f_r\}$ are the frequencies for the distorted sample X. The problem of the constrained maximization is solved using the modification of the steepest descent method that is based on the following expressions for the function $l(.)$ derivatives:

$$\frac{\partial l}{\partial \alpha} = \sum_{r=0}^{n}\left(f_r \sum_{i=0}^{n} w_{ri}(\varepsilon_0,\varepsilon_1)\frac{\partial p_i^0(\alpha,\beta)/\partial\alpha}{p_r(\alpha,\beta,\varepsilon_0,\varepsilon_1)}\right), \quad (32)$$

$$\frac{\partial l}{\partial \beta} = \sum_{r=0}^{n}\left(f_r \sum_{i=0}^{n} w_{ri}(\varepsilon_0,\varepsilon_1)\frac{\partial p_i^0(\alpha,\beta)/\partial\beta}{p_r(\alpha,\beta,\varepsilon_0,\varepsilon_1)}\right), \quad (33)$$

$$\frac{\partial l}{\partial \varepsilon_0} = \sum_{r=0}^{n}\left(f_r \sum_{i=0}^{n} \frac{\partial w_{ri}(\varepsilon_0,\varepsilon_1)}{\partial \varepsilon_0}\frac{p_i^0(\alpha,\beta)}{p_r(\alpha,\beta,\varepsilon_0,\varepsilon_1)}\right), \quad (34)$$

$$\frac{\partial l}{\partial \varepsilon_1} = \sum_{r=0}^{n}\left(f_r \sum_{i=0}^{n} \frac{\partial w_{ri}(\varepsilon_0,\varepsilon_1)}{\partial \varepsilon_1}\frac{p_i^0(\alpha,\beta)}{p_r(\alpha,\beta,\varepsilon_0,\varepsilon_1)}\right), \quad (35)$$

$$\frac{\partial p_i^0}{\partial \alpha} = p_i^0(\alpha,\beta)\left(\sum_{j=0}^{i-1}\frac{1}{\alpha+j} - \sum_{j=0}^{n-1}\frac{1}{\alpha+\beta+j}\right), \quad (36)$$

$$\frac{\partial p_i^0}{\partial \beta} = p_i^0(\alpha,\beta)\left(\sum_{j=0}^{n-i-1}\frac{1}{\beta+j} - \sum_{j=0}^{n-1}\frac{1}{\alpha+\beta+j}\right), \quad (37)$$

$$\begin{aligned} \frac{\partial w_{ri}}{\partial \varepsilon_0} = &\sum_{l=\max(i,r)}^{\min(n,i+r)} C_i^{l-r} C_{n-i}^{l-i}(l-i)\varepsilon_0^{l-i-1}\gamma_0^{n-l}\varepsilon_1^{l-r}\gamma_1^{i+r-l} - \\ &- \sum_{l=\max(i,r)}^{\min(n,i+r)} {}_i^{l-r} C_{n-i}^{l-i}(n-l)\varepsilon_0^{l-i}\gamma_0^{n-l-1}\varepsilon_1^{l-r}\gamma_1^{i+r-l}, \end{aligned} \quad (38)$$

$$\begin{aligned} \frac{\partial w_{ri}}{\partial \varepsilon_1} = &\sum_{l=\max(i,r)}^{\min(n,i+r)} C_i^{l-r} C_{n-i}^{l-i}\varepsilon_0^{l-i}\gamma_0^{n-l}(l-r)\varepsilon_1^{l-r-1}\gamma_0^{i+r-l} - \\ &- \sum_{l=\max(i,r)}^{\min(n,i+r)} C_i^{l-r} C_{n-i}^{l-i}\varepsilon_0^{l-i}\gamma_0^{n-l}(i+r-l)\varepsilon_1^{l-r}\gamma_1^{i+r-l-1}. \end{aligned} \quad (39)$$

As follows from our research, both algorithms are rather sensitive to initial estimates. To ensure good convergence, the developed numerical routines include the extremum localization step (using grid-based search) and adaptive tuning of the parameter λ, which gradually increases up to one while approaching to the solution point.

6. FORECASTING OF DEMAND PROBABILITIES

Traditionally, the demand probabilities are forecasted using the BBM Bayes predictor, which due to the conjugate property is reduced to (Collet, 2002)

$$\hat{p}(s) = (\hat{\alpha} + s)/(\hat{\alpha} + \hat{\beta} + n), \quad (40)$$

where $\hat{\alpha}$, $\hat{\beta}$ are the estimates of the BBM parameters, and s is the last observed demand value. This predictor minimizes the mean square forecast error in the case of the consistent estimators for α, β. However, as it was proved in our previous work (Pashkevich and Dolgui, 2003), in case of data distortions (3), the predictor (40) looses its optimality. For this reason, a new predictor that takes into account possible reporting errors was developed.

It was proved that for the beta-binomial demand model with the reporting errors, the mean square optimal forecast has the following pdf

$$\begin{gathered} f_{p_\varepsilon}(x|s,\varepsilon_0,\varepsilon_1) = \textstyle\sum_{i=0}^{n}\omega_{si}\cdot f_{\xi_i}(x), \\ L\{\xi_i\} = B(\hat{\alpha}+i, \hat{\beta}+n-i) \end{gathered} \quad (41)$$

and the predictor function is the weighted sum

$$\hat{p}_\varepsilon(s) = E\{p|s,\varepsilon_0,\varepsilon_1\} = \sum_{i=0}^{n}\omega_{si}\cdot(\hat{\alpha}+i)/(\hat{\alpha}+\hat{\beta}+n), \quad (42)$$

where the weights ω_{si} are defined as follows

$$\omega_{si} = \frac{C_n^i w_{si} B(\hat{\alpha}+i, \hat{\beta}+n-i)}{\sum_{j=0}^{n}\left(C_n^j w_{sj} B(\hat{\alpha}+j, \hat{\beta}+n-j)\right)}. \quad (43)$$

The mean square forecast error of the predictor (42) is expressed as

$$\begin{aligned} r^2(\hat{p}_\varepsilon) = &\frac{\alpha^{[2+]}}{(\alpha+\beta)^{[2+]}} - \\ &- \sum_{i=0}^{n}\left(\left(\sum_{j=0}^{n}\omega_{ij}\frac{\alpha+j}{\alpha+\beta+n}\right)^2 \cdot \sum_{j=0}^{n} w_{ij} C_n^j \frac{\alpha^{[j+]}\beta^{[(n-j)+]}}{(\alpha+\beta)^{[n+]}}\right). \end{aligned} \quad (44)$$

It should be noted that in the case $\varepsilon_0 = \varepsilon_1 = 0$ (i.e. no reporting errors), the weight matrix $\Omega = (\omega_{si})$ becomes the identity, and the proposed predictor (42) is reduced to (40), which confirms validity of the developed technique.

7. APPLICATION EXAMPLE

To validate the efficiency of the developed estimation and forecasting methods for the proposed distorted beta-binomial model (DBBM), there were used the demand data for two Japanese car parts sold in Australia (Snyder, 2002). The review period was assumed to be one month, and values for 36 months were available. The performance of the classical BBM was compared to the performance of the proposed DBBM as follows.

To estimate the model parameters for each car part, the demand data for the first 24 months were used (see Table 1). For the BBM, the MM-estimator was applied, while for the DBBM, the developed in section 5.2 MM-based simultaneous estimator was employed (reporting error levels were limited by 10%). Since the car part demand shows stable seasonality, for both car parts, the Bayes forecast for the third year (months 25-36) was generated using the corresponding month demand of the second year (months 13-24). For the BBM, the predictor (40) was used, and for the DBBM, the proposed predictor (42) was applied. Finally, the performance of each forecast was evaluated by comparing the square root of the mean square forecast error (r).

Tables 2, 3 present the actual demand for the months 25-36 and the forecast results based on the BBM and DBBM for different values of the parameter n. It should be noted that the first n-value is calculated as the maximum demand in the past samples (months 1-24), and then it is slightly increased to consider the case of possible larger demand in the future. As follows from the tables, the proposed DBBM together with the developed estimation and forecasting algorithms ensures essentially lower forecast error for the part 2, and a little lower error for the part 1 when compared to the BBM. For example, for the part 2, $n = 8$, the classical model leads to the error 1.72, while the proposed model ensures value 1.30 that results in 33% performance increase. For comparison, the classical binomial and beta forecasts lead to much worse mean square error 1.86, that also illustrates advantage of the DBBM-based Bayes forecasting technique.

For the considered demand data, there was also investigated the sensitivity of the BBM and DBBM forecast with respect to the parameter n (Fig. 1). As follows from the figure, the classical BBM is rather non-sensitive to selecting the value of n, while the proposed DBBM should be proper tuned to achieve the low value of the forecast error. In particular, it is prudent to select n close to the maximum demand in the learning sample, whereas unjustified increasing of n may reduce the DBBM efficiency.

8. CONCLUSION

The developed estimation and forecasting methods for the distorted beta-binomial model ensure consistent parameter estimates and accurate demand forecasts. The application example has illustrated the advantages of the new model versus the classical one. As a result, the proposed generalization of the beta-binomial model for the case of the reporting errors can be effectively used for modeling demand of the slow-moving items in the inventory management systems.

REFERENCES

Agrawal, N., and S.A. Smith, (1996). Estimating Negative Binomial Demand for Retail Inventory Management with Unobservable Lost Sales, *Naval Research Logistics,* **43**, 839-861.

Azoury, K.S., and B.L. Miller (1984). A Comparison of the Optimal Ordering Levels of Bayesian and Non-Bayesian Inventory Models, *Management Science,* **30(8)**, 993-1003.

Collet, D. (2002). *Modeling binary data.* Champton & Hall/CRC, London.

Grange, F (1998). Challenges in modeling demand for inventory optimization of slow-moving items. *Proceedings of the 1998 Winter Simulation Conference*, pp. 1211-1217.

Harris, T. (1997). Optimized Inventory Management, *Production and Inventory Management Journal,* **38(1)**, 22-25.

Hopp, W.J., and M.L. Spearman (1995) *Factory Physics: Foundations of Manufacturing Management.* Richard D. Irwin, Chicago.

Hopp, W.J., M.L. Spearman and R.Q. Zhang (1997). Easily Implementable Inventory Control Policies, *Operations Research,* **45(3)**, 27-340.

Jacobs, R.A., and H.M. Wagner (1989). Reducing Inventory Systems Costs by Using Robust Demand Estimators, *Management Science,* **35(7)**, 771-787.

Kharin, Yu.S., and M.A. Pashkevich (2003). Statistical Estimation of Parameters for the Beta-Binomial Distribution under Distortions of Binary Observations, *Journal of the National Academy of Sciences of Belarus (Series in Physics and Mathematics)*, No. 1, 11-17.

Nahmias, S. (1994). Demand Estimation in Lost Sales Inventory Systems, *Naval Research Logistics,* **41**, 739-757.

Pashkevich, M., and A. Dolgui (2003). Robust Modeling of Consumer Behavior, *Applied Optimization, special issue on Supply chain design and management.* Kluwer Academic Publishers, 16 pages (in Press).

Snyder, S. (2002). Forecasting Sales of Slow and Fast Moving Inventories, *European Journal of Operational Research*, **140**, 684-699.

Tripathi, R.C., R.C. Gupta and J. Gurland (1994). Estimation of parameters in the beta binomial model, *Ann. Inst. Statist. Math.*, **46**, 317-331.

Table 1. Monthly demand data for two slow-moving car parts (24 months)

Part	1	2	3	4	5	6	7	8	9	10	11	12	13	14	15	16	17	18	19	20	21	22	23	24
#1	3	0	2	0	0	0	0	1	0	0	1	2	0	1	0	0	1	1	2	1	0	2	0	0
#2	8	5	1	2	3	4	4	1	1	0	1	5	4	1	5	2	0	1	1	3	1	1	1	1

Table 2. Car Part #1: Comparison of the forecasts generated using the classical beta-binomial model (BBM) and the proposed distorted beta-binomial model (DBBM)

Month	Demand	$n = 3$		$n = 4$		$n = 5$		$n = 6$	
		BBM	DBBM	BBM	DBBM	BBM	DBBM	BBM	DBBM
25	0	0.3433	0.4082	0.4318	0.5095	0.4760	0.5591	0.5025	0.5887
26	1	0.8586	0.9571	0.8222	0.9055	0.8040	0.8829	0.7931	0.8699
27	1	0.3433	0.4082	0.4318	0.5095	0.4760	0.5591	0.5025	0.5887
28	2	0.3433	0.4082	0.4318	0.5095	0.4760	0.5591	0.5025	0.5887
29	2	0.8586	0.9571	0.8222	0.9055	0.8040	0.8829	0.7931	0.8699
30	2	0.8586	0.9571	0.8222	0.9055	0.8040	0.8829	0.7931	0.8699
31	1	1.3740	1.4776	1.2127	1.2923	1.1320	1.2020	1.0836	1.1483
32	0	0.8586	0.9571	0.8222	0.9055	0.8040	0.8829	0.7931	0.8699
33	0	0.3433	0.4082	0.4318	0.5095	0.4760	0.5591	0.5025	0.5887
34	2	1.3740	1.4776	1.2127	1.2923	1.1320	1.2020	1.0836	1.1483
35	0	0.3433	0.4082	0.4318	0.5095	0.4760	0.5591	0.5025	0.5887
36	0	0.3433	0.4082	0.4318	0.5095	0.4760	0.5591	0.5025	0.5887
Forecast Error		0.7928	0.7716	0.7997	0.7806	0.8073	0.7899	0.8130	0.7970
Parameter α		0.6662	0.7292	1.1058	1.2632	1.4512	1.7014	1.7297	2.0675
Parameter β		2.1552	2.0503	5.1389	5.1569	8.7927	9.1077	12.9221	13.6943
Parameter ε_0		–	0.0000	–	0.0000	–	0.0000	–	0.0000
Parameter ε_1		–	0.1000	–	0.1000	–	0.1000	–	0.1000

Table 3. Car Part #2: Comparison of the forecasts generated using the classical beta-binomial model (BBM) and the proposed distorted beta-binomial model (DBBM)

Month	Demand	$n = 8$		$n = 9$		$n = 10$		$n = 11$	
		BBM	DBBM	BBM	DBBM	BBM	DBBM	BBM	DBBM
25	0	3.4623	2.8508	3.4097	2.6442	3.3688	2.7945	3.3361	2.7751
26	0	1.4302	0.6740	1.4722	0.6421	1.5049	0.6544	1.5311	0.8573
27	1	4.1397	3.7541	4.0556	3.5343	3.9901	3.8030	3.9378	3.5586
28	2	2.1075	1.1864	2.1181	1.0927	2.1262	1.1092	2.1328	1.3502
29	1	0.7528	0.3932	0.8264	0.3913	0.8836	0.4052	0.9294	0.5444
30	0	1.4302	0.6740	1.4722	0.6421	1.5049	0.6544	1.5311	0.8573
31	0	1.4302	0.6740	1.4722	0.6421	1.5049	0.6544	1.5311	0.8573
32	1	2.7849	1.9538	2.7639	1.7877	2.7475	1.8428	2.7344	2.0150
33	0	1.4302	0.6740	1.4722	0.6421	1.5049	0.6544	1.5311	0.8573
34	0	1.4302	0.6740	1.4722	0.6421	1.5049	0.6544	1.5311	0.8573
35	1	1.4302	0.6740	1.4722	0.6421	1.5049	0.6544	1.5311	0.8573
36	1	1.4302	0.6740	1.4722	0.6421	1.5049	0.6544	1.5311	0.8573
Forecast Error		1.7248	1.2954	1.7176	1.2092	1.7129	1.2879	1.7097	1.2788
Parameter α		1.1113	0.5111	1.2796	0.5187	1.4223	0.4683	1.5448	0.7508
Parameter β		2.6989	1.8889	3.6559	2.4127	4.6731	2.3413	5.7378	3.9511
Parameter ε_0		–	0.1000	–	0.1000	–	0.1000	–	0.0700
Parameter ε_1		–	0.0000	–	0.0000	–	0.1000	–	0.0400

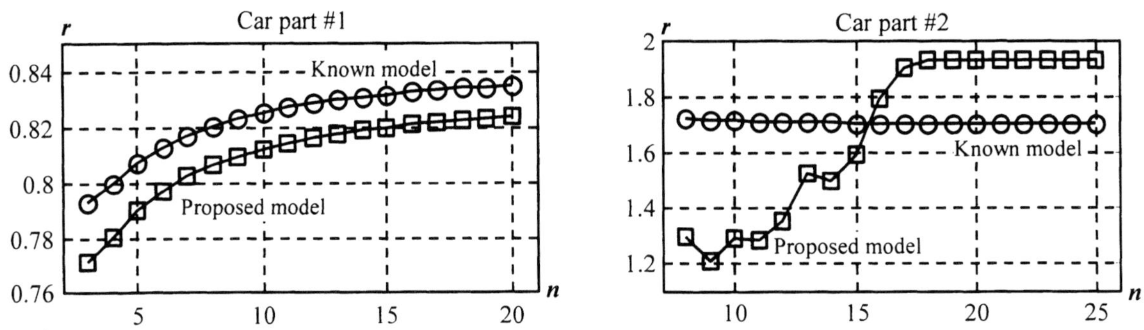

Fig. 1. Comparison of the forecast errors (r) for the classical beta-binomial model (known model, O) and the distorted beta-binomial model (proposed model, □) for different values of the parameter n for two slow-moving car parts.

Copyright © IFAC Information Control Problems in
Manufacturing, Salvador, Brazil, 2004

ELSEVIER
IFAC
PUBLICATIONS
www.elsevier.com/locate/ifac

CENTRALIZED VERSUS DISTRIBUTED DECISIONS, AN INDUSTRIAL CASE

Rémi Pannequin, André Thomas

Research Center for Automatic Control, Nancy, France
{Remi.Pannequin, Andre.Thomas}@cran.uhp-nancy.fr

Abstract: Two tendencies have appeared in manufacturing plants control. On the one hand, there are new management tools, such as APS, which consider the supply chain as a whole, and schedule operations globally and in real time. On the other hand, new control paradigms, such as holonic manufacturing and auto ID seems promising, as enablers of mass-customization manufacturing or agile enterprise. So as to clarify relation between these two systems, we develop two models of an assembly factory, each implementing a different control paradigm. The comparison between them leads us to our future path of research. *Copyright © 2004 IFAC*

Keywords: distributed control, centralized control, simulation

1. INTRODUCTION

Nowadays we can witness the coming out of new tools, which have, or will have, great influence on the way enterprises are controlled. On the one hand, they are optimization and decision-making support systems (Advanced Planning Systems - APS), and on the other hand, "emerging" conceptual systems (distributed artificial intelligence, holonic manufacturing systems, Auto ID systems) or technological advances (Multi agents Systems, RFID, smartCards, etc).

The two categories differ in many aspects. The first ones retain, structure, and process information in a quite centralized way, but the others recommend distributing intelligence toward the material flow. In the first case, it is aimed at optimizing supply chain operations, globally and predictively, whereas in the other, flexibility and reactivity of control is highlighted, with a consideration to a sort of stability in decision-making (Genin, 2003).

With the development of Auto ID, which involves automatic retrieval of products identity, accurate and timely information can be obtained throughout the supply chain. McFarlane (2003) demonstrate how such information enhance both conventional control methods, and also distributed control methods.

Otherwise, in the distributed systems, the role of hierarchy had been studied (Bongaerts *et. al.*, 2000). Hierarchical, heterarchical and holonic control systems are being compared. The holonic paradigm tries to merge best properties of both systems: high and predictable performances, and high robustness against disturbances. A modeling framework was developed so as to describe the behavior of the system, but also to impose performance guarantees.

Two of these architectures are compared by Brennan and Norrie (2003), who developed metrics for evaluating distributed manufacturing control systems. They aim at providing an objective comparison between heterarchical and holonic control systems. Two classes of metric are proposed, those dealing with the manufacturing system, and those dealing with control systems. Results show that both are worthwhile.

So, centralized and distributed decision-making models might coexist. This sets the problem of the interactions between them. In order to validate, and refine this research problematics, we propose a comparison approach.

Consequently this paper is concerned with the comparison between two control modes. The method we have followed consists in building reduced simulation models representing each control mode, for an industrial case. The simulation runs enable us to compare the behavior of the models, using indicators like bottleneck utilization, production rate, etc, for different perturbation levels.

After justifying the use of reduced simulation models, we will present our industrial case, and propose a modeling. Then, we will show and discuss experiment results.

2. REDUCED SIMULATION MODELS

2.1 Different tools the evaluation of performances

Evaluating the performances of a supply chain, or a production system can be achieved in many ways. Amongst these, we distinguish two tendencies. One is an analytical approach, based on the queuing theory, while an alternative approach is the use of discrete events simulation.

The analytical approach begins with the modeling of the system considered in the form of a queuing network. Then the analytical equations corresponding to this network must be solved. But this approach is difficult. It is indeed quite hard to take into account all the properties of the system to study. For instance, blockages, perturbations, complex statistical laws cannot be simply modeled. Then the exact resolution of such a model is often impossible (Papadopoulous, 1996). In spite of some approximate resolution techniques (Baynat, 1996) (Van Djik, 1998) the analytical approach remains difficult, and hardly attainable by practitioner or non-specialists.

In comparison with these difficulties, discrete events simulation can be an alternative. As interesting parameters are measured during simulation run, instead of being analytically calculated, the simulation model can take many aspects of the real system into account, without setting any resolution problems.

So, we have chosen to use discrete events simulation to evaluate the performances of the considered systems.

2.2 Reduced simulation models

Yet, the study with discrete events simulation sets some problems of time, as configuring and running simulation models can take a while. Furthermore, trying to represent a real industrial-sized problem can lead to modeling difficulties. In order to solve this problem we have used reduced simulation models. These reduced models are called that way because they include, for a given level of detail, only a few elements of the system represented (Thomas, 2003).

Therefore, in this reduced model, we represent only the most significant elements. These can be detected by running a complete model that had been done previously (Hung, 1999). But the reduced model can also be directly built. The theory of constraints (Goldratt, 1986), shows us the importance of bottlenecks. According to the type of enterprise to model, some methods of modeling have been presented, which represent bottlenecks, and simplify other elements.

2.3 Validity of these reduced models

The application of this modeling approach to an industrial case has shown that a reduced simulation model could represent the real thing with appropriate accuracy (less than 5% error, in global week productivity, during an 11 weeks test) (Thomas, 2003) (Pannequin, 2003). This result, beyond predictive evaluation of production planning, shows that we properly represented the problem. Moreover, the difficulties met by the control of this production system were preserved in the reduced model.

Therefore it seems to us that we could legitimately use reduced simulation models, so as to compare the two types of control.

3. THE INDUSTRIAL CASE

3.1 Presentation of the case

The firm involved is a subcontractor of automotive industry, which manufactures turbo-compressors.

We only model an assembly site, which consists in modules. A module is a small assembly cell; each producing finished goods independently of the others. The complete assembling factory counts 200 workstations, shared between 10 modules.

Assembly rates are high, since production reaching 13000 units a day.

Assembly of turbo-compressors is done in two main stages:

- first, a sub-assembly is put together, with several operations
- then the final assembly of turbo-compressors, which also needs several operations, is done by three independent parallel production lines.

Both stages are independent. Indeed they are separated by a stock (decoupling point), and therefore constitute each a production loop.

3.2 Decision process

The firm is a subsidiary company of a global-sized group, and the assembly site integrates himself into a European-sized supply chain. Planning and management of this supply chain is done thanks to an integrated management system (ERP - Enterprise Resource Planning).

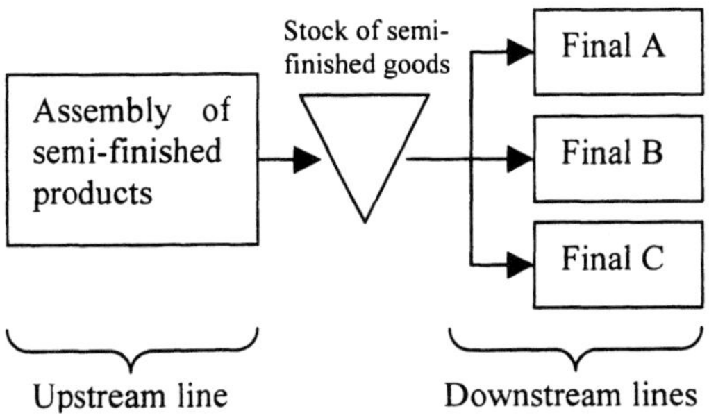

FIG. 1. The assembly module. Reduced model.

This software package is used, in a centralized manner, to establish strategic plans, sales and operations planning, and master production schedules, but some operation management decisions are distributed as near the products flow as possible. According to management policy, the operators of a module are indeed granted to adjust the work within the framework of the week planning. Moreover, customers demands, like modification or constrain on orders, are directly taken into account by the operators.

This is the distributed mode of control. We can highlight the fact that it was possible to let the ERP the responsible for these decisions (through a scheduling module, for instance).

3.3 Centralized versus distributed decisions

In order to precisely delimit the scope of our study, we consider only the control of the assembly line of semi-finished products. At this place it is decided to stop and to release production orders, according to stock levels, and customers' demands.

Indeed, to split production order is a vital issue to control the module. While upstream production loop produces just one type of product at a time, the downstream lines may consume three different types. Assumed that stock levels between both production lines are low, and that production order might be long (table 1), splitting is the only way to avoid shortages in the feeding of downstream lines.

But this splitting is itself costly, in term of production time, while a setup blocks production during about half an hour. So too many changes of reference reduce the productivity of the module, whereas too few changes lead to stock shortages.

Therefore, these setups must be decided, either in a reactive and autonomous way (distributed mode), or in a predictive and hierarchical way (centralized mode).

4. MODELING

4.1 Common elements shared between both models

Two models have been developed in order to simulate the behavior of an assembly module: one includes distributed production control, while the other is configured by a centralized optimization system.

Nevertheless, both models share many common elements. Amongst these, many are concerned with representing the physical system. According to the modeling approach developed in section 2, each assembly line is implemented in the model by its bottleneck (Thomas, 2003).

Moreover, as we only consider the control of the upstream production loop, that sequence of order to process downstream is a given constant for all simulations (see table 1). This sequence is in fact given by the ERP, though the Master Production Schedule (MPS).

Table 1. The simulated Master Production schedule

Production orders	Initial stock	Quantity to be done	Assembling rate (minutes) semi-finished	final products
1	175	1755	1.34	3.60
2	20	192	1.06	3.49
3	30	15	1.14	3.70
4	80	832	1.14	3.57
5	50	448	1.17	3.54
6	30	18	1.17	3.75
7	60	600	1.13	4.09
8	20	192	1.22	3.69
9	25	240	1.11	4.03
10	65	640	1.23	3.62
11	10	128	1.23	3.75
12	25	245	1.11	3.96

4.2 Centralized control mode

In this control mode, the upstream production line is configured before the simulation begins, by an external planning system. ERP data are sequenced with respect to downstream needs, taking into account predetermined splitting rules. The sequencing method is the Manufacturing Resources Planning (Vollmann *et al.*, 1997) (MRP2). We finally obtain the exact sequence of order to be processing (table 2).

4.3 Distributed control mode

In the distributed control mode, the model is able to decide which order must be stopped or released on the upstream production loop. The choice of the new order to launch is done amongst a subset of production order from MPS. In our model, for more simplicity, this subset is constituted by the orders being processed downstream.

Table 2 production sequence for the assembly of semi-finished product

Order	Quantity	Beginning (minute)
2	192	0
4	208	204
1	220	442
5	224	737
4	208	1000
1	220	1238
5	224	1533
4	208	1796
1	220	2034
7	200	2357
4	208	2583
8	192	2821
1	220	3056
7	200	3378
9	240	3604
1	220	3871
7	200	4167
1	220	4393
11	128	4689
10	213	4847
12	245	5110
1	220	5382
10	213	5677
1	220	5940
10	213	6235

Its behavior is founded by the kanban method:
Let J be the subset of productions order within we choose. Let "<" be an order relation, that can be the priority between production orders. So $O_i < O_j$ (where $(O_j, O_i) \in J^2$) if O_i has to be done before O_j.
For each production order O_j of J, we define three Boolean variables $thri_j$, $thrs_j$, end_j respectively true when stock level is lower than a low threshold, greater than a high threshold, or when the order is finished. Indeed, in the kanban method a process is controlled by the levels of the downstream stock, and also obviously by order status (finished or not).

Now, we define for each order O_j of J two higher-level indicators:

$$trigger_j = \overline{end_j} \cdot thri_j$$

$$stop_j = end_j + thrs_j$$

These define the ideal behavior of a *single* order. But we have to coordinate many different orders. So we introduce a Boolean function, which control the change of reference. This function constitutes the core of the decision process:

$$launch_j = (trigger_j \cdot \prod_{\substack{k>j \\ k\in J}} \overline{trigger_k}) + (\overline{stop_j} \cdot \prod_{k\in J} \overline{trigger_k} \cdot \prod_{\substack{k>j \\ k\in J}} stop_k) \quad (1)$$

This means that the production orders O_j is launched when

- O_j is triggered, and no order of greater priority are,
- or, no order is triggered, and O_j is not stopped, whereas all orders of greater priority are.

It can be easily demonstrated that all $launch_j$ are mutually exclusive. So at a time, only one order is launched.

Moreover, so as to stabilize control, we add a rule that limits the spitting. So the minimal size of a production lot is fixed. A reference change is only possible when this minimal quantity has been produced. This minimum lot size is calculated with:

$$Q_{lot} = \frac{Q_{total}}{\left\lfloor \frac{Q_{total}}{200} \right\rfloor + 1} \quad (2)$$

We are aware that these rules don't specify an optimal control, and that they could lead to many problems in certain situations. Nevertheless, it sufficiently fit the present case to be a good comparison base.

5 RESULTS

5.1 Experimental parameters

Many simulations have been conducted, each for a different failure level, others parameters staying constant (table 3). Indeed, we defined three levels of perturbation:

- level 1: there is not failures
- level 2: failure appears following statistical laws deduced from the observation of the real system
- level 3: failure appearance rate is twice as high as in level 2

The failures of level 2 were deduced from real data, and were found to fit Weibull statistical distribution.

The probability density function of the Weibull distribution is

$$f(x) = \frac{\beta \cdot x^{\beta-1}}{\alpha^{\beta}} \cdot \exp\left(-\left(\frac{x}{\alpha}\right)^{\beta}\right) \qquad (3)$$

5.2 Influence of perturbations

We see in the results showed table 3 and figure 3 that the two control modes behave differently in accordance to perturbation levels. So with no perturbation, centralized control behaves better (in term of global production) than distributed one. But as failures rate grows, distributed mode become the best (figure 3). This robustness tally with a relative increase of reference changes, which enable adaptation to perturbations.

Table 3 Results.

Experience model	failure level	Number of setup	Quantity produced
centralized	1	21	4666
	2	20	4324
	3	17	3876
distributed	1	21	4563
	2	20	4393
	3	19	4194

Experience model	failure level	Min	Stock average	max
centralized	1	640	539	449
	2	658	523	400
	3	646	473	335
distributed	1	646	550	473
	2	666	552	439
	3	681	537	407

5.3 Variation of stock levels

With no perturbation, average stock levels are lower in centralized control mode than in distributed one. As perturbation rate increase the two models behave differently, especially:

- in centralized mode, stock levels decrease, and variability remain the same,
- in distributed mode, stock levels stay identical, but variability increase.

This shows that the behavior of distributed control might be unpredictable. Even if the problem presented here were very simple, it would have been not quite easy to predict this result.

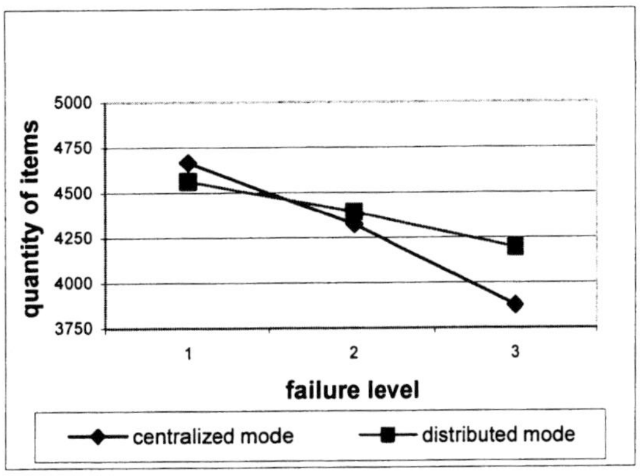

FIG. 3. Produced quantity in function to failure level.

5.4 Discussion

As we experimented on just one industrial case, without any formal verification, these results may be taken with appropriate doubt. With different physical architectures, results might be brought into questions.

Another source of doubt is the control parameters, arbitrarily fixed, like in formula (2). To be rigorous, we ought to try different values, and study their effects on results, with an experiment design.

Nevertheless, this industrial case is interesting, because it enable us to do real experiments on the way a control system behave.

6. PERSPECTIVES

By seeing the results, we note that centralized control mode is optimal, when there are a few perturbations. This way, we can get globally optimized operations, whereas a set of distributed decision can often only be local optimum. Otherwise, when there are more perturbations, distributed control may be the best, because it can adapt itself quickly to variations in the environment.

So, we can see two distinct areas appear:

- first, near product flow, where there are many perturbation, and strong constrain on time, distributed control may be the best,
- secondly, nearer to strategic decisions, where the effects of operating perturbation fade out, and constrains on time are slacker, centralized control may be optimal.

Coexistence of these two different control areas set a number of problems.

First, how can these two areas be delimited? We need some indicators, some measurements that will show that a control element has to be distributed, or centralized. Further way, in the case of a decentralized decision, we have to delimit the scope of this decision, so as to achieve a kind of global optimization (Muhl *et. al.*, 2003).

Then, what kind of relationship can be expected between the two systems? Distributed control is hard to predict, nevertheless, centralized planning systems need to take into account real capacities of the production system. So, on the one hand, centralized systems must have an evaluation of distributed ones. On the other hand, it may be interesting to take into account some characteristics demanded by the centralized system in distributed ones. For instance, some research had been done so as to generate robust sales and operations planning (Genin, 2003). Is it possible to integrate it into distributed control?

REFERENCES

Baynat, B., Y. Dallery (1996), 'A product-form approximation method for general closed queuing network with several classes of customers'. *Performance evaluation*, vol. **24**, pp 165-188.

Bongaerts, L., L. Monostori, D. McFarlane, B. Kádár (2000). 'Hierarchy in distributed shop floor control'. *Computers in Industry*. Vol. **43**. pp 123-137.

Brennan, R. W., D. H. Norrie (2003). 'Metrics for evaluating distributed manufacturing control systems'. *Computer in industry*. Vol. **51**. pp 225-235.

Genin, P. (2003). 'Plannification tactique robuste avec usage d'un APS. proposition d'un mode de gestion par plan de référence'. PhD Thesis (in french).

Goldratt, E. (1986), The Goal The North River Press, second revised edition.

Hung, Y. F., R. C. Leachman, (1999). 'Reduced simulation models of wafer fabrication facilities'. *International Journal of Production Research*. **37**. pp 2685-2701.

McFarlane, D., S. Sarma, J. L. Chirn, C. Y. Wong, K. Ashton (2003). 'Auto ID systems and intelligent manufacturing control'. *Engineering Applications of Artificial Intelligence*. Vol. **16**, pp 365-375.

Muhl, E., P. Charpentier, F. Chaxel (2003). 'Optimization of physical flows in an automotive manufacturing plant: some experiments and results'. *Engineering Applications of Artificial Intelligence*. Vol. **16**, pp 293-305.

Pannequin, R. (2003). 'Obtention de modèles de simulation simples ; une démarche basée sur le Théorie des Contraintes, applicable aux ateliers de type flow shop'. *Masters dissertation*, Henri Poincaré University. (in french)

Papadopoulos, H. (1996), 'Queuing theory in manufacturing systems analysis and design: A classification of models for production and transfert lines', *European Journal of Operational Research*, vol. **96**, pp 1-27.

Thomas, A., P. Charpentier (to appear in 2003). 'Reducing simulation models for scheduling manufacturing facilities', *European Journal of Operational Research.*

Thomas, A., C. Patrick (2002). 'Reduction of simulation model, A sawmill flow shop case'. Communication in international congress ACS2002 Miedzyzdroje, Poland. pp 473-480

Saporta G. (1990). *Probabilité, analyse des données et statistique*, édition technip. (in French)

Van Djik, N.M. (1998), Analytic comparison results for communication network, computer communication. vol **21**. pp 1495-1508.

Vollman, T. E., Berry, W. L., & Whybark, D. C. (1997), *Manufacturing Planning and Control Systems*, Fourth edn, Irwin.

Copyright © IFAC Information Control Problems in Manufacturing, Salvador, Brazil, 2004

ELSEVIER
IFAC
PUBLICATIONS
www.elsevier.com/locate/ifac

MULTIHEAD AND WEB TECHNOLOGY FOR THE IMPLEMENTATION OF A MANUFACTURING EXECUTION SYSTEM

Armando Ramírez-Santaella[1], Arturo Molina [2]

[1] *Solution for Manufacturing Enterprise Systems (SMES)*
[2] *Center of Integrated Manufacturing Systems (CSIM)*
Avenida Eugenio Garza Sada 2501 Sur
Monterrey, Nuevo León, 64849 MEXICO
aramirez@solutionmes.com , armolina@itesm.mx

Abstract: MES (Manufacturing Execution Systems) and Digital Operational Method Sheets (OMS) are vital to control shop floor operations where there is a need to balance manual and automated operations. Specially, Mexican companies face large diversity of products and multiple production lines, with high turnaround rotation of the workers.
A low cost solution that integrates MES with Digital OMS is described, the system has been developed using LINUX servers, Multi-head technology (multiple monitors to a single servant) and a WEB environment. This technological solution offers an affordable, stable, and high performance solution to transform the production into a more balanced system with shorter and faster cycles. *Copyright © 2004 IFAC*

Keywords: Manufacturing Execution Systems, Web Technologies, System Integration

1. INTRODUCTION

The concept of MES (Manufacturing Execution System) has acquired a huge importance in today's modern manufacturing, specially in the IC and SC (Integrated Circuits and Semiconductors) industries. As a matter of fact, MES has a direct relationship with MTO (Make to Order) industrial environment (nevertheless, this fact does not discard MES usage in any other industrial environment).

The characteristics of MTO environments, such as the need for product customization, the necessity to apply fast changes to production lines based on order to conform last-minute customer's order changes make MES systems a requirement for MTO operations [1]

MES is a relatively new concept, introduced in the 1990's. According to MESA (acronym for Manufacturing Execution Systems Association; 1997), Manufacturing Execution Systems deliver information that enables the optimization of production activities from order launch to finished goods. Using current and accurate data, MES guides, initiates, responds to, and reports on plant activities as they occur [2]. MES is a software-based application that grew up in the realm of planners to help schedule optimum production runs, manage inventory, handle regulatory databases and information, and other manufacturing planning chores [1]. MES has the following two characteristics:

- It provides real time information about what is happening in the shop floor, for managers (under a strategic approach), and for direct operation workers (under a purely operative approach).
- It is an information bridge between Planning Systems used in Strategic Production Management (such as ERP) and Manufacturing Floor Control SCADA (Supervisory Control and Data Acquisition). It links the Manufacturing Information System's layers (Strategic Planning and Direct Execution) through the adequate on – line managing and control of updated information related with the basic enterprise resources: people, inventory and equipment [3].

The enormous importance acquired by MES resides, in a significant percentage, on its functionalities (implying people, inventory and equipment) and their interaction with the compounding elements of the industrial plant environment. Authors and organizations, such as McClellan and MESA have given full and detailed descriptions of these functionalities, independently of the industrial application branch of MES. Some other authors, like Kaufmann [4] and Schenker [5] have emphasized some special functionalities and technologies that suits specifically to Semiconductor Industry and general SMEs (Small and Medium Enterprises). Fukuda [6] and Cheng et al [7] have developed important MES-next-generation architecture models. MESA has proposed a hierarchies distinction for MES functionalities. This hierarchy is divided in two stacks: Core Functions and Support Functions. Core Functions include Planning System Interface, Data Collection, Exception Management, Work Orders, Work Stations, Inventory / Materials and Material Movement. On the other hand, MES Support Functions are listed as Genealogy, Maintenance, Time and Attendance, Statistical Process Control, Quality Assurance, Process Data and Documentation Management.

Some important trends have been forecasted for MES immediate, mediate and long term future [8]. According to the authors of this paper the most important are:

- MES will become the engine behind inter and intra manufacturing enterprise collaboration as it is the enabling link for high velocity supply chain performance.
- Due to FDA regulations for cGMP and electronic signatures (Part 11) MES will become standard in Med-Service and Biotech industries.
- MES will raise the average acceptable plant efficiency benchmark from 60-70% to over 90% within 3 years.

This paper describes the implementation of a MES based on multihead and web technology to have digital Operational Method Sheets for controlling manual and semi-automated shop floor operations.

2. THE COMPANY CONTEXT AND PROBLEM

The company started in April 1998 with the vision of being the most advance manufacture site for building lighting fixtures and HID Ballast to supply American market. The first production run was on September of 1999 with a very aggressive ramp-up. The products to be manufacture in the facility are described in figure 1:

Figure 1 – Product Mix to be manufacture at the Mexican Facility

The plan was designed using DFT (Demand Flow Technology), to achieve flexibility in terms of volume changes, product range and production cycles. The company already had implemented with success DFT environment in other facilities, and the use of this production method was encouraged.

OMS (Operational Methods Sheets) were established in order to face the flexibility challenge, due to the number of production lines, the number of operations on each line, the BOM's and the complexity of the product structure. The OMS required to be designed to be used by workers with limited training and multiple assembly skills.

The factory started with OMS printed on paper. However the use of printed OMSs became a problem due to the fact that the number of SKU finish products was very large and product models changed daily. Productivity was not accomplished because the product model change rate was 1 every 10 minutes, even though the setup in the production line, was fast, the problem was to deliver OMS to the workers. It was obvious that an Information Technology solution was needed. But there was a restriction of cost, because the company startup was already with low cash flow.

3. DESIGNING A DEMAND FLOW TECHNOLOGY BASED ON IT

The figure 2 describes the design of the plant and the flow of material.

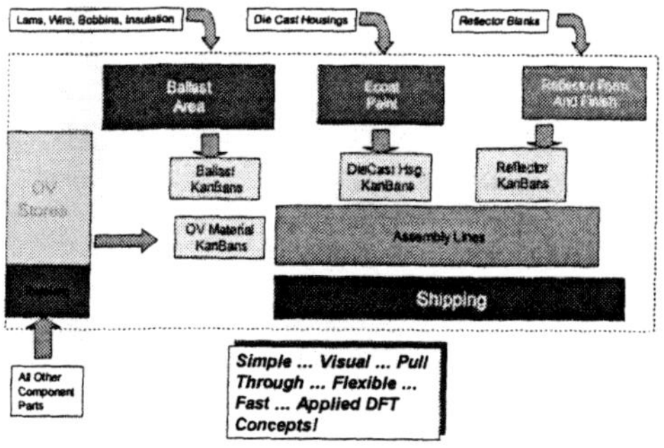

Figure 2 – Plant Layout and Flow of Material

The DFT (Demand Flow Technology) methodology was used to implement the shop floor system. But it was required to be supported but IT (Information Technology) in order to build up a system that could implement an integrated solution which included ERP, MES and Digital OMS. To implement this production system the IT department worked together with an engineering team to design the MES system and to develop the customization and plug-ins to the ERP system. The MES development included all the basic functionality defined MESA [2]. The biggest challenge was the development of the Digital OMS system.

The Digital OMS system was designed and implemented by a group integrated with 2 manufacturing engineers, 2 software engineers and different IT suppliers trying to find the best solution.

The objectives were to have an:

- OMS with high resolution (computer monitors).
- Linked to the BOM's and have online changes and updates (ECN).
- Easy to read, and to follow.
- Synchronized with the flow of material and production.
- Easy to manage and maintain.
- Low cost solution.
- User friendly interface

4. TECHNOLOGICAL SOLUTION TO IMPLEMENT DEMAND FLOW TECHNOLOGY

4.1 Material flow

The material flow is very important in the implementation of DFT; it is a change in the culture of how material is managed. One of the biggest problems was the warehouse, witch was open, no walls, no doors, and all the signals to supply raw material were the empty Kanbans, or the Kanban cards. The material department had to develop certain systems in order to follow the methodology as specified in the DFT method. IT was the most time consuming in the implementation because of the change in the organizational method to be employed.

4.2 Digital OMS (Operational Methods Sheets)

The Digital OMS were the most difficult to design because a suitable technological solution was difficult to find. The major issues to tackle were costs, flexibility and connectivity to the existing ERP. In order to find a suitable low cost solution, the team started to search tools with open source solutions, and the final resolution was to use LINUX as a server driving or handling the monitors. The team also was satisfied with the decision because LINUX code was available for any changed that the development needed. So the first configuration started when the team decided to have OMS in a web browser, having the flexibility of interconnections, the possibility of use XML, and the potential of use any client, cheap and thin client. Based on that assumption, the team design a solution described in figure 3.

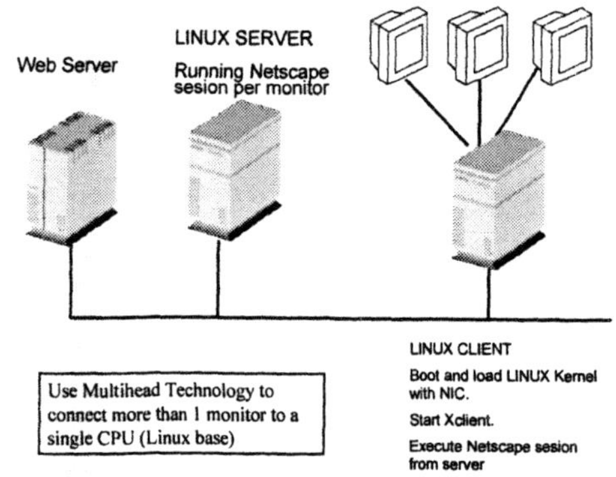

Figure 3 – First design of the system implementation

This first design was implemented with the most important flavors of Linux, but still the solution was unstable. After 5 months of research, the design changed. The previous design was unstable cause it was very heavy solution or needed a big LINUX server to manage all the Netscape sessions, so the team decided to have 1 Linux server with 5 video cards running 5 Netscape's session with in that every Linux Server, as described in figure 4.

5. RESULTS AND BENEFITS

5.1 Solution Cost

The final configuration was tested and stable, Red Hat was the Linux that the team decided as the most stable and reliability. The cost of the solution was very reasonable, considering that the hardware was very low cost. One of the reasons to use Linux is that it work very well even do it is installed in low profile computers, a PC with 2 Mbytes of RAM can load the Linux Kernel. So the servers of the final configuration were the lowest new computers on the market. At that point Celeron 400Mhz were the lowest speed, and 32 Mbytes of RAM, and HDD at least of 2 Gigabytes.

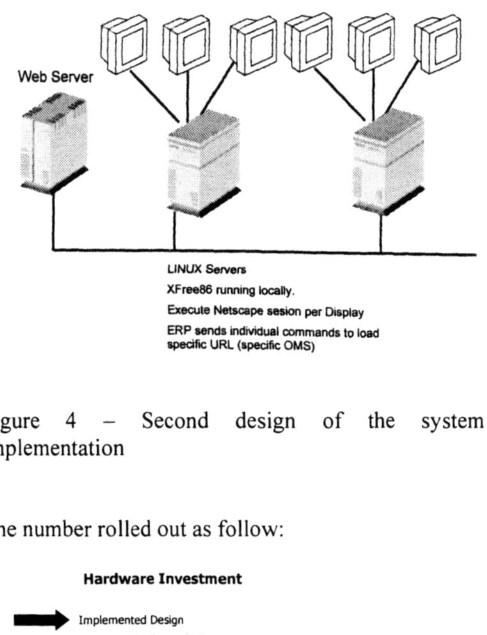

Figure 4 – Second design of the system implementation

The number rolled out as follow:

Hardware Investment

Implemented Design

Hardware Cost
- 400 Dls each CPU
- 120 Dls each Monitor

Cost by individual operation
- (400 / 5) + 120 = 200 Dls (by operation)

TOTAL 7 Prod. Lines X 18 Operations X 200Dls = **25,200Dls**

Software Cost

Licensing	$0.0 (Lunix es Open Source)
Develpment	6 months, 1 developer ($37,000 Dls)

TOTAL Investment	$62,200 Dls

5.2 Benefits

The real benefit of the solution was the Digitization of the process, making a real system linked to the plant schedule, connecting all the department that have production as a client, speeding the model change, to almost seamless, and savings in paper, and time avoiding unnecessary managements. Quality got better, all the company was synchronized.
All these benefits helped to achieve the next results within the first year:

- ✓ Productivity 100K / 60 K = + 67%
- ✓ Increase Throughput to clients
- ✓ Inventory Accuracy 40% -> 95%
- ✓ Backlog Past Due 88K Units -> 5K Units
- ✓ Receive Customer Centricity Award from GE
- ✓ Reduction of the SPAN metric 68 -> 36
- ✓ Reduction Suppliers Lead-time
- ✓ Reduction Scrap 8% -> 2%

All the departments were working as a synchronized clock, because the schedule was the kernel of the solution. The schedule was updated from the supervisors on the assembly line, and the ERP system was in charge of updating the Digital OMS for every single monitor.

The design of the Digital OMS, included the incorporation of the instructions, pictures, and TQC (see figure 5). The management of the BOM was a user friendly system, and was managed by part time students, whom were training to operate the system in 10 minutes.

This concept was implemented in the ballast area, to give instructions to the operator of bobbin machines. This implementation was done in 2 weeks, and the benefits were outstanding. After 2 years of operation, the average production was 1,400 Ballast per day. After implementing the monitors for the bobbin machines, the average production increased to1, 800 ballast per day.

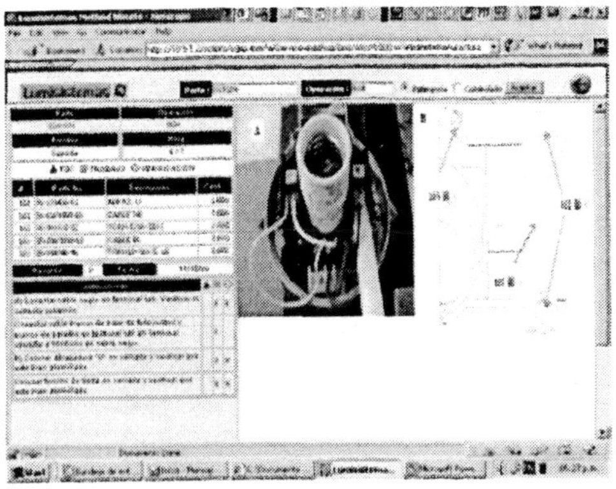

Figure 5 – Example of a Digital OMS

6. CONCLUSION

The purpose of having Digital OMS on line and synchronized with the plant schedule, with a very low cost solution was accomplished. The concept of using Linux as a web monitors server is a solution very stable and with unlimited application on the shop floor, providing information in HTML format (Text, Graphic, Audio, Video) transmitting to the operators information about their work they have on performance, or quality related information, new, access to the intranet, indicators, etc. It is very powerful to have that capability of dispatching information to the shop.

Figure 6 – Example of the production line with Digital OMS

It is very important the control on the shop floor, in order to have all the process synchronized, that is implementations need a lot of discipline, workers need to follow procedures, punching info into the system, to have all the system working as a clock.
The concept is replicable, and right now it is implemented in some other production plants with a diversity of process, like the production line of glass manufacturers and automotive Harnesses. In all these cases, Digitized information is being distributed to the workers in the shop floor to make sure the process is carried out accordingly with the production plan and quality conditions.

Figure 7 – Use of Digital OMS in the production line

7. ACKNOWLEDGEMENTS

The research reported in this paper is part of a Research Chair in Mechatronics of ITESM titled "Design, Manufacturing and Integration of Reconfigurable and Intelligent Machines". The authors wish to acknowledge the support of this grant in the preparation of the manuscript.

REFERENCES

[1] B. Choi, B. Kim: MES (manufacturing execution system) architecture for FMS compatible to ERP (enterprise planning system). International Journal of Computer Integrated Manufacturing Vol. 15, No. 3, (2002) 274 – 284

[2] MESA (Manufacturing Execution Systems Association): The Benefits of MES: A Report from the Field. USA (1997) 1 - 5

[3] M. McClellan: Introduction to Manufacturing Execution Systems. MES Conference & Exposition, June 12-14; Phoenix, AZ, USA (2000) 3 – 11

[4] T. Kauffmann: The Paradigm Shift for Manufacturing Execution Systems in European Projects and SEMI Activities. Semiconductor FABTECH, 8th Edition. Fraunhofer Institut Manufacturing Engineering and Automation, Stuttgart, Germany (1997) 17 – 25

[5] F. Schenker, D. Cilia: Dynamic Scheduling and Simulation in a Job Shop Environment. Project Notes from Carnegie Mellon Software Engineering Institute and U.S. Department of Defense. Pittsburgh, PA. USA (2001)

[6] Y. Fukuda: Production System Modeling and Manufacturing Execution System. Project Notes from JOP / Production System Modeling Technical Committee, Department of Industrial and System Engineering, Hosei University, Japan (1999)

[7] F. Cheng, S. Wu, C. Chang: Systematic Approach for Developing Holonic Manufacturing Execution Systems. IECON'01: The 27th Annual Conference of the IEEE Industrial Electronics Society. Institute of Manufacturing Engineering, National Cheng Kung University. Taiwan R.O.C. (2001) 261 – 266

[8] MESA International: White Paper Number 7. Justifying MES: A Business Case Methodology. USA (2000) 1-4.

Copyright © IFAC Information Control Problems in
Manufacturing, Salvador, Brazil, 2004

ELSEVIER
IFAC
PUBLICATIONS
www.elsevier.com/locate/ifac

BALANCING OF TRANSFER LINES WITH SIMULTANEOUSLY ACTIVATED SPINDLES

Alexandre Dolgui[1], Nikolai Guschinsky[2], Genrikh Levin[2], Mohamed-Aly Louly[3] and Sana Belmokhtar[1]

[1] *Scientific Methods for Industrial Management Dept. (MSGI)*
Industrial Engineering and Computer Sciences Center (G2I)
Ecole des Mines de Saint Etienne
158, cours Fauriel, 42023 Saint Etienne Cedex 2, France
E-mail: dolgui@emse.fr

[2] *United Institute of Informatics Problems*
National Academy of Sciences of Belarus
Surganov Street 6, 220012 Minsk, Belarus
E-mail: {gyshin, levin}@newman.bas-net.by

[3] *Laboratory for Industrial Systems Optimisation (LOSI)*
University of Technology of Troyes
12, rue Marie Curie, 10010 Troyes, France
E-mail: louly@utt.fr

Abstract: A balancing problem for transfer lines with parallel spindles is investigated. All spindle heads of the same workstation are activated simultaneously. The relations of the necessity of executing some operations at the same workstation, the possibility of combining the spindles at the same workstation as well as precedence constraints are given. The problem is to choose spindle heads from a given set and allocate them to workstations in such a way that i) all the operations are performed, ii) all constraints are satisfied, iii) the cycle time is provided, and iv) the line cost is minimal. The proposed method for solving the problem is based on its formulation as a mixed integer problem (MIP). *Copyright©2004IFAC*

Keywords: Machining, line balancing, optimization, MIP.

1. INTRODUCTION

The considered transfer lines consists of a few linearly ordered workstations (station) without intermediate buffers. All manufacturing operations executed at the line are grouped in several blocks so that every block can include one or a few operations, which are executed simultaneously by one spindle head. One or some blocks can be assigned at every station and these blocks also are executed simultaneously. An example of such line layout is given in Fig. 1.

The execution time of a block is usually equal to maximal value among operation times of its operations. The cycle time of a station is the maximal value among operation times of its blocks. The line cycle time is the maximal value of station times.

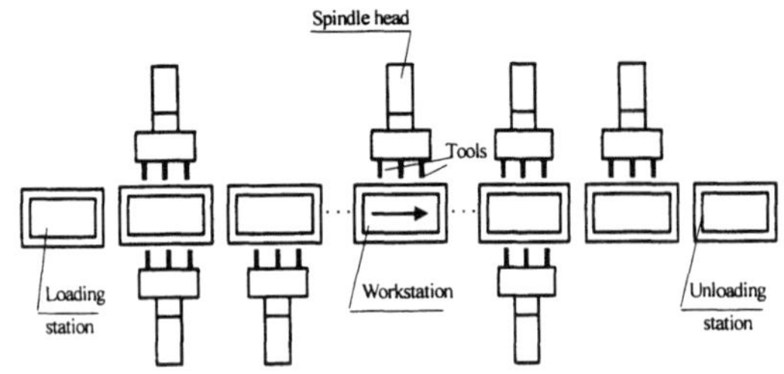

Fig.1. Transfer line layout

Transfer lines are widely used in mechanical industry to manufacture homogeneous products in mass production (Groover, 1987; Hitomi, 1996). They are designed for a long exploitation time and need high investments, but provide high productivity with relatively small exploitation cost. For such a line, concentration of operations in the blocks and at the stations allows to considerably decrease the total number of spindle heads. This results in decreasing the total line cost and the occupied area. The determination of such line structure (the numbers of stations and blocks and an assignment of operations to them) is therefore a very complex combinatorial problem.

2. PROBLEM STATEMENT

In this paper the set of manufacturing operations (tasks) executed at the line is supposed known and it is necessary to define a line structure for given output. Let's suppose the following:

- a set of all possible blocks is known and each block must be used only as a whole;
- each operation may be in several blocks, but any problem decision can contain only one of them;
- station cost, block execution time and cost for each block are given.

The problem is to choose blocks from the given set and to assign them to stations in such a way that all constraints (order and compatibility) are satisfied and the *total cost* of all stations and blocks is as small as possible. This kind of problems is known in literature as *Line Balancing* problems (Scholl and Klein, 1998). For the basic problem, *Simple Assembly Line Balancing* (SALB), each operation is executed separately (each block is composed of only one operation and all blocks at the same station are executed sequentially). The objective function is to minimize the unbalance (cost) of the line under a given line cycle time. The unbalance is minimal if and only if the number of the stations is minimal.

Generally, SALB integer linear programming models are formulated and solved by exact or heuristic methods (Baybars, 1986; Talbot *et al.,* 1986; Johnson, 1988). The exact methods are mainly based on branch and bound algorithms. Since the problem is known to be NP-hard, exact solution of large problems is time-consuming and some heuristics are used to decrease the computational effort. A recent survey of main publications on this subject is given in (Rekiek *et al.*, 2002).

In works (Graves and Lamar, 1983; Pinto *et al.*, 1983; Graves and Holmes Redfield, 1988) various extended problems of *lines balancing* with *equipment selection* and approaches to their solution are investigated. These problems have a number of common features with the studied problem. But, the problem studied in this paper cannot be directly solved by the known methods for the number of essential distinctions between this problem and the balancing and equipment selection problems presented in the literature (see subsections 3.1 and 3.2). The principal difference is the necessity to execute the operations (tasks) for the same part at the same station by block of simultaneously executed (parallel) operations. Therefore, the new balancing and equipment selection problem is more complex.

The closed problems of production line design are studied in (Dolgui *et al.,* 1999, 2000, 2001 a,b,c) where all blocks of parallel operations at the same station are executed sequentially (block by block) and any alternative variants of blocks are not given beforehand (any subset of the give operation set can be considered as a block). Three approaches have been proposed: a constrained shortest path; mixed integer programming (MIP); heuristics. This paper develop the second approach in conformity to the new considered problem where all the blocks at the same station are executed simultaneously and a set of possible blocks is fixed.

3. MIP MODEL

3.1. Input data and notations

The input data are the following:

- the given set **N** of operations that have to be executed at the line ;
- a given set $\mathbf{B}=\{B_r|r\in\mathbf{R}\}$ of the possible blocks of operations (B_r), where **R** is the set of block indices;
- block cost q_r and block time t_r for each block $r\in\mathbf{R}$ and cost q_0 of one station;
- T_0, m_0 and n_0 are the constraints for maximal line cycle time, maximal number of stations and maximal number of blocks at one station, respectively;
- the precedence, inclusion and exclusion constraints.

It is assumed that these constraints are given in the

following way:

1) The order relation over the set **N** is represented by an acyclic digraph $\mathbf{G}^{or}=(\mathbf{N},\mathbf{D}^{or})$. An arc $(i,j)\in\mathbf{N}\times\mathbf{N}$ belongs to the set $\mathbf{D}^{or}$ if the operation j must be executed after the operation i.

2) The exclusion conditions for allocation of blocks at the same station can be represented by a family $\underline{\mathbf{D}}^{bs}$ of subsets $D\subseteq\mathbf{R}$ such that blocks from the set $\{B_r|r\in D\}$ cannot be allocated at the same station all together. These subsets take into account both feasible station structure and manufacturing incompatibility of operations. It's supposed that there is no a subset $D'\in\underline{\mathbf{D}}^{bs}$ such that $D'\subseteq D$ and $\{D\backslash\{r\}|\ r\in D\in\underline{\mathbf{D}}^{bs}\}\subseteq\{D\backslash\{s\}|s\in D\in\underline{\mathbf{D}}^{bs}\}$ for any $r,s\in\mathbf{R}$ if $B_r\subset B_s$.

3) Similarly, the inclusion conditions for allocation of operations at the same station can be represented by a family $\mathbf{D}^{os}$ of subsets of operations from **N** such that all operations of the same subset from $\mathbf{D}^{os}$ must be performed at the same station.

Notes. The inclusion and exclusion constraints on allocation of operations at the same block are taken into account while forming the block set **B**. All blocks dissatisfying these conditions must be rejected from the set **B**. In particular, it is assumed $t_r\leq T_0$ for all $B_r\in\mathbf{B}$.

In Fig. 2. a general schema of the design approach using a MIP model is presented.

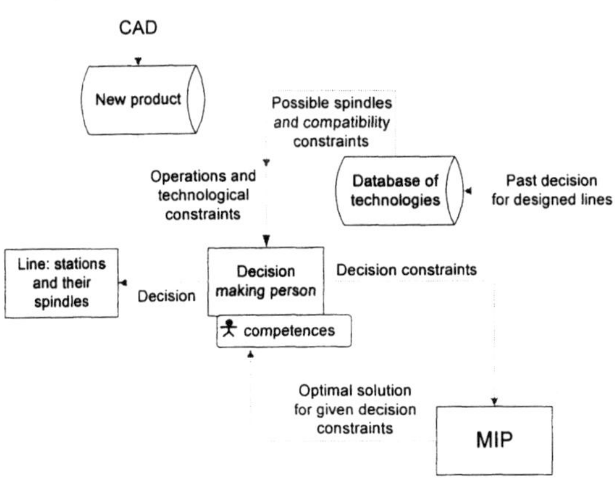

Fig. 2. Iterative approach for decision making

To give a formal statement of the MIP problem, the following variables and additional parameters are introduced:

- binary variables x_{rk}, where $x_{rk}=1$ if the block $B_r\in\mathbf{B}$ is assigned to the station k and $x_{rk}=0$ otherwise, $k=1,...,m_0$;
- variable $y\geq0$ that is equal to the number of stations;
- a lower bound m^* of the number of the stations;
- a segment $Q(r)=[a(r),b(r)]$ of the station's indices where the block $B_r\in\mathbf{B}$ can be assigned;
- family $\mathbf{F}=\{F_1,...F_v\}$ of subsets of the set **R** such that i) $B_r\cap B_t\neq\varnothing$ for any $r,t\in F_s$ and $s\in V=\{1,...,v\}$; and ii) for any $r,t\in\mathbf{R}$ such that $B_r\cap B_t\neq\varnothing$ there is $s\in V$ such that $r,t\in F_s$; i.e. F_s is a subset of alternative blocks (only one block from F_s can be used in a decision) ;
- $F_0=\mathbf{R}\backslash\bigcup_{s\in V}F_s$;
- $w_{rt}=|B_r\cap D_t|$ and $W_t=\{r\in\mathbf{R}|w_{rt}>0\}$ for $r\in\mathbf{R}$ and $D_t\in\mathbf{D}^{os}$;
- $U_t=\{r\in\mathbf{R}|i_t\in B_r\}$, where i_t is certain fixed operation from the set $D_t\in\mathbf{D}^{os}$;
- a set $M(r)$ of operations $i\in\mathbf{N}\backslash B_r$ such that $(i,j)\in\mathbf{D}^{or}$ for $j\in B_r$;
- $\mathbf{H}=\{r\in\mathbf{R}|\ M(r)\neq\varnothing\}$ and $H(r)=\{t\in\mathbf{R}|\ B_t\cap(r)\neq\varnothing\}$ for $r\in\mathbf{R}$;
- $h_{tr}=|B_t\cap M(r)|$ for $r\in\mathbf{H}$ and $t\in H(r)$;
- a set $\mathbf{R}^*$ of the numbers of blocks from **B** that can be assigned at the final station of the line.

3.2. Goal function and constraints

The objective function provides the line cost minimization:

$$q_0y+\sum_{r=1}^{R}\sum_{k=1}^{m_0}(q_r+\varepsilon_rk)x_{rk}\rightarrow min, \qquad (1)$$

where ε_r is a sufficiently small nonnegative value.

The following constraint ensures the execution of every operation from **N** in exactly one station

$$\sum_{k\in Q(r)}x_{rk}\leq1, \quad r\in F_0 \qquad (2)$$

$$\sum_{r\in F_s}\sum_{k\in Q(r)}x_{rk}\leq1, \quad s\in V \qquad (3)$$

Execution of all operations from **N** is given by:

$$\sum_{r\in R}\sum_{k\in Q(r)}|B_r|x_{rk}=|\mathbf{N}| \qquad (4)$$

The order relation on the set **N**:

$$\sum_{t\in H(r)}\sum_{s\in Q(t),s<k}h_{tr}x_{ts}\geq|M(r)|x_{rk},\ r\in\mathbf{H},\ k\in Q(r) \qquad (5)$$

Inclusion constraints for the stations:

$$\sum_{r\in W_t}w_{rt}x_{rk}=|D_t|\sum_{s\in U_t}x_{sk}\ ,\ D_t\in\mathbf{D}^{os},\mathrm{k}\in\bigcup_{s\in U_t}Q(s) \qquad (6)$$

Exclusion constraints for blocs at the same station:

$$\sum_{r\in D_t} x_{rk} \le |D_t|-1,\ D_t\in\underline{\mathbf{D}}^{bs},\ k\in \bigcap_{s\in D_t} Q(s) \qquad (7)$$

Constraint of the maximal number of blocks at station:

$$\sum_{r\in\{t\in R|k\in Q(t)\}} x_{rk}\le n_0,\ \ k=1,\dots,m_0 \qquad (8)$$

The constraints determining the number of stations:

$$y\ge kx_{rk},\ \ r\in\mathbf{R}^*,\ \ k\in Q(r),\ k\ge m^* \qquad (9)$$

In this model, the number of stations y depend of variables x_{rk}. The model does not demand that the variable y takes only integer values. It is provided by the constraint (9) and by the objective function (1).

In next section an approach is proposed to minimize the number of decision variables.

4. MODEL REDUCTION

The following procedures allow to transform the set $\mathbf{B}$, families $\underline{\mathbf{D}}^{bs}$ and $\mathbf{D}^{os}$ in order to decrease the model size.

1) Block B_r is deleted if it satisfies one of the following conditions:

- there are operations $i,j\in B_r$ such that the arc $(i,j)\in\mathbf{D}^{or}$;
- there are a set $D\in\mathbf{D}^{os}$, operations $i\in B_r$ and $j\in D$ such that $B_r\cap D\neq\varnothing$ and the arc (i,j) or (j,i) belong to $\mathbf{D}^{or}$;
- there are a set $D\in\mathbf{D}^{os}$, operations $i,j\in B_r$ and operations $k,l\in D$ such that arcs (i,k) and (l,j) belong to $\mathbf{D}^{or}$;
- there is not a set $\mathbf{B}'$ of mutually disjoint blocks in $\mathbf{B}$ such that $B_r\in\mathbf{B}'$, the union of the blocks from $\mathbf{B}'$ coincides with $\mathbf{N}$, and $|\mathbf{B}'|\le m_0n_0$. In particular, there is an operation $i\in\mathbf{N}\backslash B_r$ such that $\{B\in\mathbf{B}|\ i\in B\}=\{B\in\mathbf{B}\backslash B_r\ |B_r\cap B\neq\varnothing\}$.

In the firsts three cases, precedence constraints are violated for the block B_r. In the fourth case, the operation i cannot be executed at the line if the block B_r is used in a decision.

2) A subset D is deleted from $\underline{\mathbf{D}}^{bs}$ if $|D|>n_0$ or there are $r,s\in D$ such that $B_r\cap B_s\neq\varnothing$. Such subsets are taken into account by constraints (8) and (3) respectively.

3) Intersecting subsets from $\mathbf{D}^{os}$ are merged.

It is easy to see that the problem (1) – (9) has a decision if the following conditions hold:

- $\bigcup_{t\in R} B_t=\mathbf{N}$;
- a digraph $(\mathbf{D}^{os},\mathbf{E}^s)$ is acyclic, where a pair (D',D'') of subsets from $\mathbf{D}^{os}$ belongs to the arc set $\mathbf{E}^s$ if there are operations $i\in D'$ and $j\in D''$ such that $(i,j)\in\mathbf{D}^{or}$.
- for any set $D\in\mathbf{D}^{os}$ there is a set $\mathbf{B}'$ of mutually disjoint blocks from $\mathbf{B}$ such that $|\mathbf{B}'|\le n_0$, $D\subseteq N(\mathbf{B}')$, $D'\not\subset\mathbf{B}'$ for any $D'\in\underline{\mathbf{D}}^{bs}$, and there are no operations $i,j\in N(\mathbf{B}')$ such that $(i,j)\in\mathbf{D}^{or}$, where $N(\mathbf{B}')$ is the union of the sets from $\mathbf{B}'$.

The model size can be reduced by determination of appropriate bounds $a(r)$ and $b(r)$ for all blocks $B_r\in\mathbf{B}$ as well as the value of m^* and set $\mathbf{R}^*$.

For estimation $a(r)$ and $b(r)$, the following approximate algorithm is proposed. It takes into account constraints (2)-(9) only partially. The calculation technique for $a(r)$ and $b(r)$ are similar.

***Algorithm* 1**

$g=1$, $R_1=\mathbf{R}$, $R_2=\varnothing$, $\mathbf{D}_0=\mathbf{D}^{os}$, $M_1(r)=M(r)$ and $M_2(r)=\varnothing$ for all $r\in\mathbf{R}$

Repeat while $R_1\neq\varnothing$ and $g\le m_0$

Select a maximal subset $R^+\subseteq R_1$ such that the following conditions hold for any $r\in R^+$:

a) $M_1(r)=\varnothing$;

b) For any $D_t\in\mathbf{D}_0$ $w_{rt}=0$ or there is $R'\subseteq R^+$ such that $r\in R'$, $D_t\subseteq\bigcup_{s\in R'} B_s$, $|R'|\le n_0$, and $R'\notin\underline{\mathbf{D}}^{bs}$;

c) $M_2(r)=\varnothing$ or there is $R'\subseteq R_2$ such that $M_2(r)\subseteq\bigcup_{t\in R'} B_t$, $|R'|\le n_0$, and $R'\notin\underline{\mathbf{D}}^{bs}$;

if $R^+=\varnothing$ **then** break repetition **endif**

$a(r)=g$ for all $r\in R^+$; $R_1=R_1\backslash R^+$; $R_2=R^+$; $g=g+1$

$\mathbf{D}_0=\mathbf{D}_0\backslash\{D_t\in\mathbf{D}_0|D_t\subseteq\bigcup_{r\in R^+} B_r\}$

$M_2(r)=M_1(r)\cap\bigcup_{t\in R^+} B_t$ and $M_1(r)=M_1(r)\backslash M_2(r)$ for all $r\in R_1$

End repeat

$R_1=R_1\cup\{r\in\mathbf{R}\backslash R_1|a(r)>b(r)\}$

(in the algorithm for $b(r)$ only)

if $R_1\neq\varnothing$ **then** $\mathbf{R}=\mathbf{R}\backslash R_1$

if $\bigcup_{r\in\mathbf{R}} B_r\neq\mathbf{N}$ **then return** FALSE (the initial problem cannot be solved)

else Delete all $D\in\underline{\mathbf{D}}^{bs}$ such that $D\cap R_1\neq\varnothing$

endif

endif

return TRUE

For calculating $b(r)$, the following substitutions are necessary:

a) $g=m_0$, $g\ge1$ and $g=g-1$ instead of $g=1$, $g\le m_0$, and $g=g+1$, respectively;

b) $M(r)$ is replaced by the set of operations $i\in\mathbf{N}\backslash B_r$ such that $(j,i)\in\mathbf{D}^{or}$ for $j\in B_r$;

c) $b(r)$ instead of $a(r)$.

The lower bound m^* of the number of stations can be obtained as the minimal value among all m such that there is $R' \subseteq R(m) = \{r \in \mathbf{R} | a(r) \leq m\}$ satisfying the following conditions:

a) $\bigcup_{r \in R'} B_r = \mathbf{N}$;

b) all blocks from R' are mutually disjoint;

c) $|R'(m)| \leq n_0$, where $R'(m) = \{r \in R' | a(r) = m\}$;

d) $R'(m) \cap D = \varnothing$ for any $D \in \underline{\mathbf{D}}^{bs}$;

e) For any $D \in \mathbf{D}^{os}$ either $N(R'(m)) \cap D = \varnothing$ or $D \subseteq N(R'(m))$, where $N(R) = \bigcup_{r \in R} B_r$.

Value m^* can simply be calculated as $\max\{\min\{a(r) | r \in R, i \in B_r\} | i \in \mathbf{N}\}$.

In turn, a set $\mathbf{R}^*$ of blocks without successors satisfying the following conditions can be accepted as a "upper bound" of the set of blocks that can be located at the last station at the line:

for any $D \in \mathbf{D}^{os}$ either $B_r \cap D = \varnothing$ or there is $R' \subseteq \mathbf{R}^*$ such that:

a) $r \in R'$, $|R'| \leq n_0$, and $D \subseteq \bigcup_{r \in R'} B_r$;

b) all blocks from $\{B_r | r \in R'\}$ are mutually disjoint.

5. A NUMERICAL EXAMPLE

Let us consider an example that illustrates calculation of the model parameters for the following input data:

- the operation set $\mathbf{N} = \{1,2,\ldots,12\}$;
- the block set $\mathbf{B}$ is represented in Table 1;
- $m_0 = 5$ and $n_0 = 2$;
- the digraph $\mathbf{G}^{or}$ is given in Fig. 3;
- the family $\underline{\mathbf{D}}^{bs} = \{D_1 = \{5,6\}, D_2 = \{2,9\}, D_3 = \{2, 10\}, D_4 = \{7,9\}, D_5 = \{8,9\}\}$;
- the family $\mathbf{D}^{os} = \{D_1 = \{3,4\}, D_2 = \{6,10\}\}$.

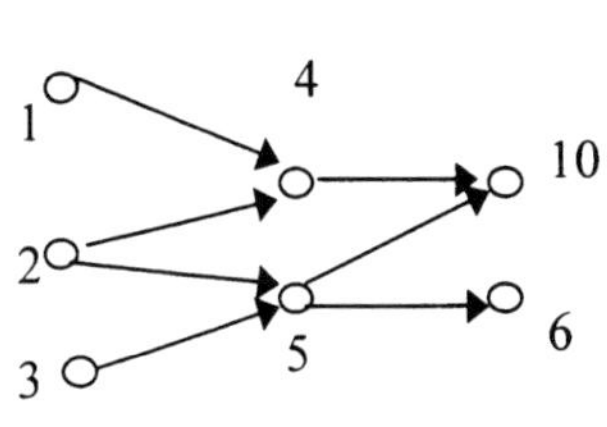

Fig. 3. The precedence digraph $\mathbf{G}^{or}$

The input data analysis proposed in Section 4 allows to away block B_{12} because arcs (1,4) and (3,5) belong to the digraph $\mathbf{G}^{or}$ and the operation subset $\{3,4\} \in \mathbf{D}^{os}$. The remaining blocks satisfy all conditions of Section 4.

Let us calculate above mentioned model parameters:

- the set $F_0 = \{3,4,5\}$;
- the family $F = \{F_1 = \{1,2\}, F_2 = \{6,7\}, F_3 = \{7,8\}, F_4 = \{9,10\}, F_5 = \{10,11\}\}$;
- the parameters w_{rt} are represented in Table 2;
- the sets $W_1 = \{4,5\}$ and $W_2 = \{7,8,10,11\}$;
- $U_1 = \{4\}$ and $U_2 = \{7,8\}$ for $i_1 = 3$ and $i_2 = 6$;
- the sets $M(r)$ and $H(r)$ are given in Table 3;
- the set $H = \{5,6,7,8,10,11\}$;
- the parameters h_{rt} are given in Table 4;
- the bounds $a(r)$ and $b(r)$ obtained by Algorithm 1 are represented in Table 5 ;
- the bound $m^* = 4$ and the set $\mathbf{R}^* = \{7,8,9,10,11\}$.

Table 1: Blocks and their compositions

r	1	2	3	4	5	6	7	8	9	10	11	12
B_r	1	1, 12	2	3	4, 11	5, 8	6, 8	6, 9	7	7, 10	10	1, 5

Table 2 Parameters w_{rt}

r	1	2	3	4	5	6	7	8	9	10	11
w_{r1}	0	0	0	1	1	0	0	0	0	0	0
w_{r2}	0	0	0	0	0	0	1	1	0	1	1

Table 3: Sets $M(r)$ $H(r)$

r	1	2	3	4	5	6	7	8	9	10	11
$M(r)$	∅	∅	∅	∅	1,2	2,3	5	5	∅	4,5	4,5
$H(r)$	∅	∅	∅	∅	1,2,3	3,4	6	6	∅	5,6	5,6

Table 4: Parameters h_{rt}

	t / r	1	2	3	4	5	6	7	8	9	10	11
	5	1	1	1	0	0	0	0	0	0	0	0
	6	0	0	1	0	0	0	0	0	0	0	0
h_{tr}	7	0	0	0	0	0	1	0	0	0	0	0
	8	0	0	0	0	0	1	0	0	0	0	0
	10	0	0	0	0	1	1	0	0	0	0	0
	11	0	0	0	0	1	1	0	0	0	0	0

Table 5: Bounds $a(r)$ and $b(r)$

r	1	2	3	4	5	6	7	8	9	10	11
$a(r)$	1	1	1	2	2	3	4	4	1	4	4
$b(r)$	2	2	2	3	3	4	5	5	5	5	5

The problem solution is presented in Fig. 4. This solution has 4 stations and 7 blocks. For each block its number and composition are given.

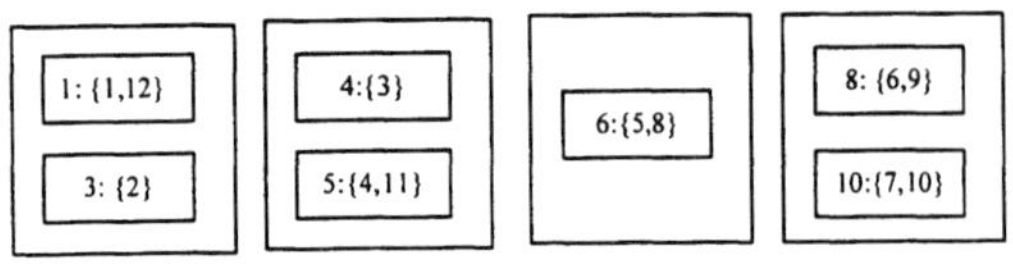

Fig. 4. The problem solution

6. CONCLUSION

A new problem of line balancing and resource planning (equipment selection) is investigated. The economic benefit, which can be achieved justifies the search of an optimal design decision for each specific line. The search of such a solution is time-consuming, but it is not so important since the design of such lines is made once in the life of the line.

In this study, a mixed integer model is proposed to search an exact solution. Some techniques for calculation time decreasing and reducing the model's size are also introduced. At the next step, the model should be tested with a larger set of benchmarks.

One feature of this problem's structure is the large number of constraints, this lead us to think that a promising extension could be the coupling of our MIP model with constraint programming techniques. Another perspective is to test heuristics like genetic algorithms which can provide a good bound to an exact method.

REFERENCES

Baybars, I. (1986). A survey of exact algorithms for the simple line balancing problem. *Management Science*, **32**, 909-932.

Dolgui, A., N. Guschinsky and G. Levin (1999). On problem of optimal design of transfer lines with parallel and sequential operations. *Proceedings of the 7th International Conference on Emerging Technologies and Factory Automation (ETFA'99)*, (J.M. Fuertes Ed.), IEEE, Spain, **1**, 329-334.

Dolgui, A., N. Guschinsky and G. Levin (2000). *Approaches for transfer lines balancing*. Preprint N 8, Institute of Engineering Cybernetics/University of technology of Troyes, 43 pages

Dolgui, A., N. Guschinsky and G. Levin (2001a). Decomposition methods to optimize transfer lines with parallel and sequential machining. *Management and Control of Production and Logistics: A Proceedings Volume of the Conference* (Grenoble, France, MCPL'2000), 983-988.

Dolgui, A., N. Guschinsky, G. Levin and Y. Harrath (2001b). Optimal design of class transfer lines with blocks of parallel operations. *Manufacturing, Modeling, Management and Control: A Proceedings Volume of the Symposium* (Patras, Greece, MIM'2000), Elsevier Sciences, 36-41.

Dolgui, A., N. Guschinsky, G. Levin and Y. Harrath (2001c). Un modèle de programmation linéaire pour l'équilibrage des charges des lignes de transfert. *European Journal of Automated Systems*, **36**, 4-31.

Graves, S.C. and C. Holmes Redfield (1988). Equipment selection and task assignment for multiproduct assembly system design. *International Journal of Flexible Manufacturing Systems*, **1**, 31-50.

Graves, S.C. and B.W. Lamar (1983). An integer programming procedure for assembly design problems. *Operations Research*, **31**(3), 522-545.

Groover, M.P. (1987). Automation, production systems and computer integrated manufacturing, Prentice Hall, Englewood Cliffs, New Jersey.

Helgenson, W.B. and D.P. Birnie (1961). Assembly line balancing using Ranked positional weight technique. *Journal of Industrial Engineering*, **12**, 394-398.

Hitomi, K. (1996). Manufacturing system engineering, Taylor &Francis.

Johnson, J.R. (1988). Optimally balancing large assembly lines with FABLE. *Management Science*, **34**, 240.

Pinto, P.A., Dannenbrin, D.G. and B.M. Khumawalw (1983). Assembly line balancing with processing alternatives: an application. *Management Science*, **29**, 817-830.

Rekiek, B., Dolgui A., Dechambre, A. and A. Bratcu (2002). State of art of assembly lines design optimisation. *Annual Reviews in control*, **26/2**, p.163-174.

Scholl, A. and R. Klein (1998). Balancing assembly lines effectively: a computational comparison, *European Journal of Operational Research*, **114**, 51-60.

Talbot, F.B., J.H. Paterson and W.V. Gehrlein (1986). A comparative evaluation of heuristic line balancing techniques, *Management Sciences*, **32**, 430-454.

Copyright © IFAC Information Control Problems in Manufacturing, Salvador, Brazil, 2004

ELSEVIER
IFAC
PUBLICATIONS
www.elsevier.com/locate/ifac

TOWARDS AN APPROACH GATHERING PRODUCT AND PROCESS CONFIGURATION AS A CONSTRAINT SATISFACTION PROBLEM

Michel Aldanondo, Khaled Hadj-Hamou, Elise Vareilles

Ecole des Mines d'Albi-Carmaux – Centre Génie Industriel
Campus Jarlard Albi 81A81013 Albi CT Cedex 09 - France

Abstract: The goal of this communication is to propose a global approach for product and process configuration taking into account structural aspects in order to be able to develop and manufacture customized products. Configuration basics are recalled, then an approach to take into account process configuration with respect to product configuration is proposed. An example illustrates all the proposition all along the paper.
Copyright © 2004 IFAC
Keywords: Configuration, Constraint Satisfaction Problem, Product Modelling, Process Modelling, Manufacturing process,

1. INTRODUCTION

Dealing with product diversity is nowadays a strong requirement for companies fighting on a world market. This increasing diversity comes from the need to be customer specific or trying to match the exact requirement of each customer without providing too much unnecessary elements. As the decomposition of the total demand in various demand segments is not always enough, many companies try to customize or configure their products. Most of theses companies use configuration software, mainly based on artificial intelligence techniques, to define the customized product that will fit the customer requirements.

If many works have been achieved around product configuration and configuration software (or configurator), as reported in the special issues on configuration (IEEE 1998, AIEDAM 1998 and 2003); as far as we know, the configuration of the relevant manufacturing process has not been studied.
Our first goal is to propose a global configuration approach that deals with both product and process configuration.
Many authors (Mittal and Frayman 1989), (Sabin and Freuder 1996), (Aldanondo *et al* 2003) have shown that product configuration could be efficiently modeled or aided when considered as a Constraints Satisfaction Problem (CSP, Montanari 1974). Most of them do not consider the hierarchical or structural aspects of product description.
Our second goal is to take into account structural aspects for both product and process configuration.

The goal of this communication is therefore to propose a global approach for product and process configuration taking into account structural aspects that permit to develop and to manufacture customized products.
The first section introduces product configuration and proposes a CSP based approach that can handle structural aspects. The second section, thanks to the elements described in the first section, proposes CSP based elements that allow process configuration with respect to product configuration. An example runs all along the paper in order to illustrate the proposed approach.

2. PRODUCT CONFIGURATION

The first sub-section recalls and discusses product configuration definition and introduces the example. Then, well-known CSP based elements are proposed in order to show how they can support configuration. The last sub-section deals with structural aspects.

2.1 Product configuration and configurator.

From all the previous works achieved concerning configuration, it seems that some common features defining configuration could be:

- hypothesis: a product is a set of components,
- given: (i) a generic model of a configurable product able to represent a family of products with all possible variants and options, that gathers (1) a set of component groups (2) a set of product properties and (3) a set of various constraints that restrict possible combinations of components and property values ; and (ii) a set of customer requirements, in which each requirement can be expressed by a selection of a component or a property value,
- configuring can be defined as "finding at least one component set that satisfies all the constraints and the customer requirements".

These elements can be found and discussed in the definitions proposed in (Mittal and Frayman 1989), (Soininen, *et al* 1998), (Aldanondo *et al* 2003). It is important to note that, according to these definitions, the configuration result is always a set of components (or a bill-of-materials).

A configurator is software that assists the person in charge of the configuration task. It is composed of a knowledge base that stores the generic model of the product and a set of assistance tools that help the user finding a solution or selecting components or property values. In any case, the basic common requirement, in terms of assistance, is to guarantee that the configured product is consistent (all constraints are respected) with the generic model and the requirements, during and at the end of the configuration task.

The proposed example is a simplified Custom Storage System (CST) shown in figure 1. Four component groups exist in our example: the Book Case (BC), the High Cabinet (HC), the Low Cabinet (LC) with up to four Roll-out-shelves (RO). The Book Case is available in two heights: 72 cm or 216 cm. All components exist in two finishes: Painted (P) or Wood (W). Therefore the groups of components are composed as follows :

- Book Case (BC): {BC72P, BC72W, BC216P, BC216W}.
- High Cabinet (HC): {HC144P, HC144W}.
- Low Cabinet (LC): {LC72P, LC72W}.
- Roll-out-shelves (RO): {ROP, ROW}.

In addition, some constraints exist and reduce configuration possibilities :

- The assembly must be with the same finish,
- The Book Case must be in all configured Storage Systems (CST),
- The High Cabinet exists if and only if a Low Cabinet exists and if the Book Case is 216 cm high,
- The quantity of Roll-out-shelves can be {1,2} or {1,2,3, 4} wrt Book Case height = 72 or 216.

Product properties can be defined as follow :

- Book Case height (BCH): {72, 216}.
- Product Finish (PFH): {Wood, Painted}.
- Low Cabinet Existence (LCX): {Yes, No}.
- High Cabinet Existence (HCX): {Yes, No}.
- Roll-out-Shelves quantity (QRO): {1,2,3,4}

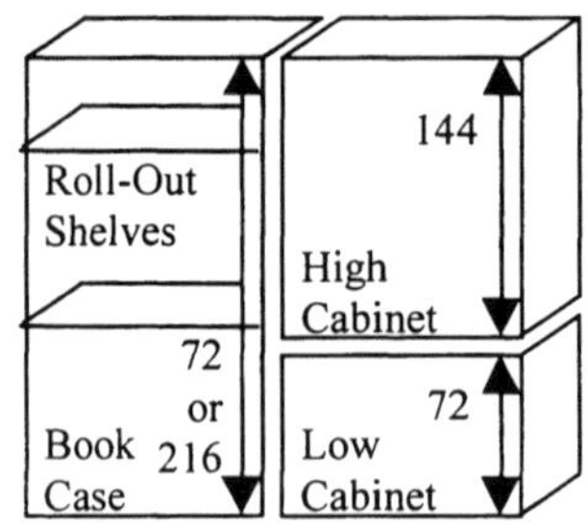

Figure 1 – Custom Storage System example

2.2 Product Configuration as a CSP Problem

CSP, defined by (Montanari 1974) as a triplet {X,D,C} where X is a set of variables, D a set of finite domains (one for each variable) and C a set of compatibility constraints (defining the possible or excluded combinations of variable values), matches partially this problem. Each group of components and each product property is associated with a variable. Each component and each property value corresponds with one value of the variable. The constraint represents the allowed (solid lines of figure 2) or excluded (dot lines of figure 2) combinations of components and property values.

The Dynamic extension of the CSP, DCSP proposed by (Mittal and Falkenhainer 1990) and frequently called Conditional-CSP, introduces : (i) Initial variables: variables that exist in any configured product, (ii) Compatibility constraints: equivalent to the CSP constraints defined by Montanari and (iii) Activity constraints (arrows on figure 2): allowing to control the variable existence in the following ways: (1) Require: a specified value of a variable "X" implies the existence of variable "Y", (2) Not Require : a specified value of a variable "X" implies the non existence of variable "Y". DCSP allows to modulate the existence of any variable corresponding with a group of components or a product property.

With these elements, the example of the previous section can be modeled, as described in figure 2, and shows :

- Four variables corresponding with the four component groups: BC, HC, LC, RO.
- Five variables corresponding with the five product properties: BCH, PFH, LCX, HCX, QRO.
- Eight compatibility constraints between the followings variables :
  - four constraints reducing component selection wrt finish : (PFH,BC) , (PFH,HC) , (PFH,LC) , (PFH,RO),
  - one excluding the RO quantity possibilities wrt the height of the book case : (BCH,QRO),
  - one selecting the correct BC wrt the height of the book case : (BCH,BC),

- one excluding HC when the height of the book case = 72 :(BCH, HCX)
- one excluding HC if LC does not exist : (LCX, HCX)

- Two activity constraints implying the existence of the component group :
  - Low Cabinet (LC) if low cabinet existence (LCX) = yes
  - High Cabinet (HC) if high cabinet existence (HCX) = yes

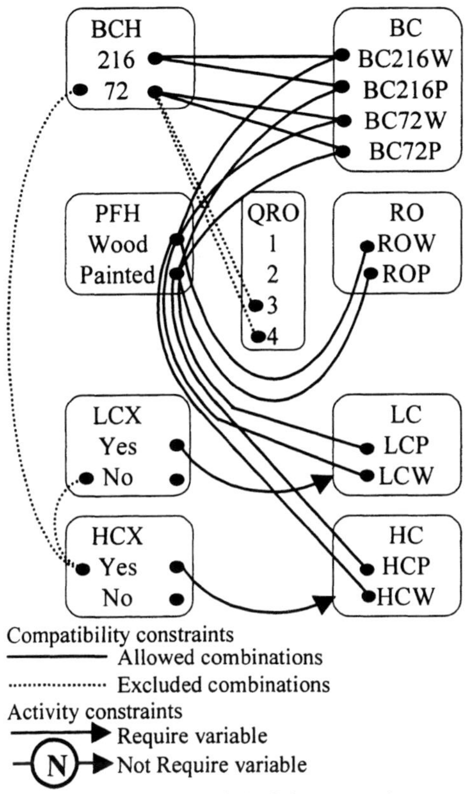

Figure 2 – Model of the example

The left and center parts of figure 2 gather the variables where customer requirements can be expressed : BCH, PFH, LCX, HCX and QRO.
The right and center parts of figure 2 show the resulting component list with quantity if necessary : BC, RO and QRO, LC and RO.
Configuration assistance is provided by constraint propagation mechanisms as those explained by (Mittal and Falkenhainer 1990) and compared in (Gelle and Sabin 2003).
This model represents a solution space of 32 configured CST. An example of customer requirements could be : BCH="216", PFH="Wood", LCX="Yes", HCX="Yes" and QRO="2" and would provide the component list : {BC216W, LCW, HCW, and 2 ROW}.

2.3. Hierarchical aspects

It is clear that the previous problem is not structured or hierarchical, meaning that the product is just a set of components or a single level bill-of-materials. When products gather different sub-assemblies and when each sub-assembly can be present in different upper level products, it is necessary to have some kinds of modular approach for the generic model description of the product.
In our example, we now assume that the Custom Storage System is in fact composed of two sub-assemblies : the first one that gathers the Book Case and the Roll-out-shelves, called BC-RO, and the second one that gathers the Low Cabinet and the High Cabinet, called LC-HC.
Each sub-assembly should be "self-configurable" meaning that it can be configured as a spare part and could be re-used in other upper level products.

As far as we know, this problem was addressed by very few authors. Composite CSP (Sabin and Freuder 1996) considers that a value of a variable can represent a CSP ; when this value is selected, it is replaced by the sub-model. State CSP (Véron and Aldanondo 2000) proposes to associate a state variable to any problem variable; the authors have shown that this approach permits to take into account simultaneously Dynamic CSP and Composite CSP. Recently (Junker and Mailharro 2003) propose some elements mixing description logic and CSP in a constraint toolbox.

As the mechanism proposed in the Composite CSP is very close to our structural need, we propose to adapt it with a DCSP approach. This adaptation requires, as shown in figure 3 :

- to define for each sub-assembly, a configuration sub-model (set of variables plus constraints that describes all its configuration possibilities),
- to define in the model of the upper level product that can use the sub-assembly, at least one variable allowing to select the lower level sub-model,
- to define a new kind of activity constraint that modulates the existence of the sub model gathering the variables and constraints corresponding with the sub-assembly.

More formally, given the model of sub-assembly, SA, gathering :

- a set of variable $v1 \in V1$ initially existing,
- a set of variable $v2 \in V2$ initially not existing,
- a set of compatibility constraints,
- a set of activity constraints conditioning the existence of $v2 \in V2$ written as follow :
  $\forall v2 \in V2 : (vk = Cste\text{-}k) \rightarrow v2$, with $vk \in V1 \vee V2$.

When the existence of the sub-assembly SA is conditioned in the model of an upper level product by the constraint V-SA-X = "Yes", all variables of SA should be constrained as follow :

- variables $v1 \in V1$ becomes initially not existing and should be conditioned by :
  $\forall v1 \in V1 : (\text{V-SA-X} = \text{"Yes"}) \rightarrow v1$,
- activity constraints acting on variables $v2 \in V2$ are cumulated with the selection of the sub-assembly as follow :
  $\forall v2 \in V2 : (\text{V-SA-X}=\text{"Yes"}) \wedge (vk=Cste\text{-}k) \rightarrow v2$, with $vk \in V1 \vee V2$.

We call this constraint : "sub-assembly activation constraint" : (V-SA-X="Yes") → SA. It is important to note that : (i) the proposed Boolean variable triggering the existence of a unique sub-assembly can be replaced by a symbolic variable which permits to select a sub-assembly in a list (ii) all constraint propagation techniques valid for DCSP can still be used without any adaptation.

The previous example can be modeled in figure 3. This model shows the two sub-assemblies "BC-RO" and "HC-LC", the two variables, BC-RO-X and HC-LC-X, that permit to trigger the existence of each sub assembly and the two sub-assembly activation constraints : (HC-LC-X= "Yes") → HC-LC and (BC-RO-X = "Yes") → BC-RO. As each sub-assembly can be self-configured, it must be noticed that the finish variable, PFH, has been duplicated in each sub-assembly but all remaining properties and groups are exactly the same.
The configuration solution space of this model is equivalent to the one shown in figure 2. The customer requirements presented in figure 2 would become : (i) at CST level : HC-LC-X="Yes" and "BC-RC-X=Yes", (ii) at BCRO level : BCH="216", PFH="Wood" and QRO="2" (iii) at HC-LC level : HCX="Yes" and PFH="Wood". The resulting hierarchical bill-of-materials would be : (i) CST composed of 1 BC-RO and 1 HC-LC, (ii) BC-RO composed of 1 BC216W and 2 ROW, (iii) HC-LC composed of 1 LCW and 1 HCW.

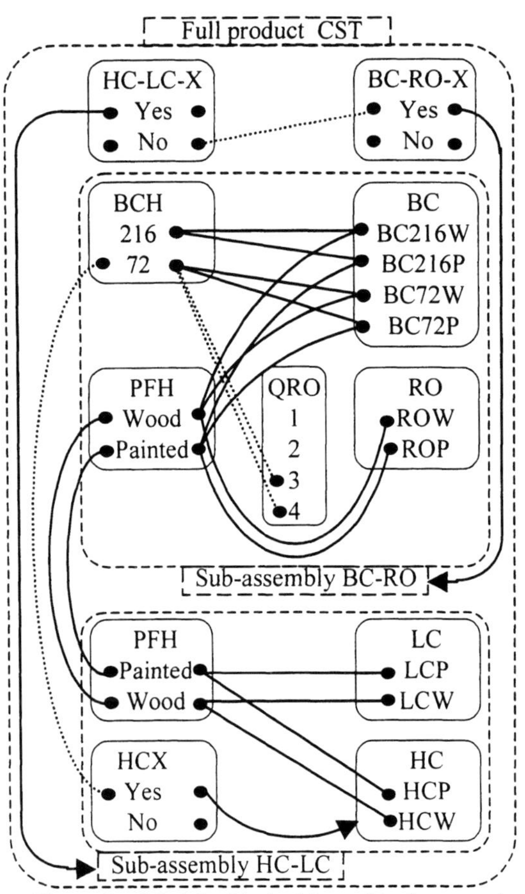

Figure 3 – Model with the hierarchical approach.

3. PROCESS CONFIGURATION

This section is concerned with product and process configuration thanks to the previous elements. The first sub-section describes what we mean by process configuration and introduces the notions of routing and operation. Then it is shown how routing configuration (section 3.2) and operation configuration (section 3.3) can be considered as a constraint satisfaction problem. In each section the links with product configuration model will be explained.

3.1 *Process configuration definition*

The process is associated to what we call a "Process Routing". A process routing is a set of "Operations" linked with "Anteriority constraints". An operation is defined by a set of "Required Resources" that permit to process the operation, each required resource is characterized by a "Resource quantity".
This single level approach, Routing / Operations, has been chosen for clarifying the message, but of course, it is possible to consider multi-level decompositions as for example, Process / Macro-operation / Operation / Instructions, but resource would only appear at the lowest level, instructions in this example. Therefore we will consider a generic model of the Process Routing and generic models of Operations.

3.2 Routing Configuration as a Constraints Problem

Each variable of the relevant constraint based model corresponds with a set of generic operations (Op), where a set can contain a single generic operation.
The representation of generic sequencing, or generic anteriority constraints, is achieved with the well-known "port" notion (Po) early proposed by (Mittal and Frayman 1989). This notion allows to represent links between components, "ports variables" are associated to components and constraints reduce connections possibilities between components.
Therefore, we propose to associate some "Operation-Port" (OpPo) variables to each set of generic operations. These Operation-Port variables are either "Previous" (OpPo-) or "Next" (OpPo+), meaning that the values of each of theses variables are the ports of next (OpPo+) or previous (OpPo-) operation-port that can be connected.
In order to link the routing model with the previous product model :

- Any variable, generic operation (Op) or operation port (Op-Po), can be linked thanks to constraints with the product model.
- Each set of generic operation and relevant ports are gathered in a variable set, the "sub-assembly activation constraint" principle is re-used as an "operation-ports activation constraint" and permits to modulate the existence of the variables relevant to the operation.

Routing model is linked with generic operation models (object of the next section) thanks to "operation activation constraints".

All these elements are used for CST example as follow :

- Four sets of generic operation : Op-BC-RO for cutting and assembly BC-RO (Ph1), Op-HC-LC for cutting and assembly HC-LC (Ph2), OP-CST for assembling the CST (Ph3) and Op-Pack for packaging the whole product (Ph4-1 and Ph4-2). The two sets, Op-HC-LC and Op-CST, exist only if the sub-assembly HC-LC exists in the product model (two operation-ports activation constraints between product and routing models, upper part of figure 4).
- Each set of generic operations contains a single generic operation except the packaging where two operations : Ph4-1 and Ph4-2 are different according to the finish of the product (compatibility constraint between product and routing models, upper part of figure 4). Each generic operation is linked to an operation generic model described in the next section (lower part of figure 4)..
- Each set of generic operations is characterized by some ports :
  - one "next" port (Ph1-Op-Po+) for Op-BC-RO, with two possible values that show that it can be connected to Op-CST or directly to Op-Pack (if HC-LC does not exist). This shows an "OR" possibility for "next" operation anteriority constraints,
  - one "next" port (Ph2-Op-Po+) for Op-HC-LC, with a single possible value that shows that it must be connected to Op-CST,
  - two "previous" ports for Op-CST with a single possible value for each : (Ph3-Op-Po1-) that must be connected to Op-BCRO and (Ph3-Op-Po2-) that must be connected to Op-HCLC. This show an "AND" possibility for "previous" operation anteriority constraints,
  - one "next" port (Ph3-Op-Po+) for Op-CST, with a single possible value that shows that it must be connected to Op-Pack
  - one "previous" port (Ph4-Op-Po-) for Op-Pack, with two possible values that show that it can be connected to Op-CST or directly to Op-BCRO (if HC-LC does not exist). This show an "OR" possibility for "previous" operation anteriority constraints.

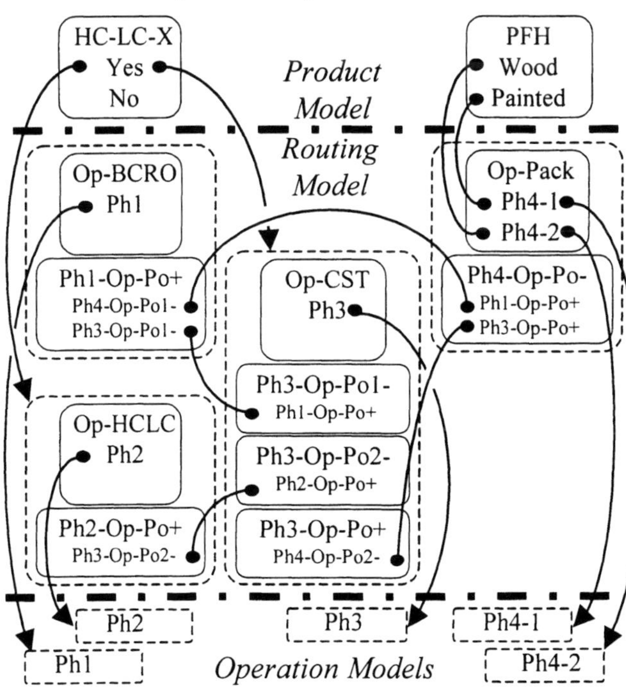

Figure 4 – Process routing model

3.3 *Operation Configuration as a Constraints Problem*

Each variable of the relevant constraint model corresponds with a set of resources (Re), where a set can contain a single resource. Each set of resources is associated with a numerical quantity (Qt) of required resource. A generic operation is therefore a set of couples of variables (resource set, quantity). When there is no "quantity" variable, it is assumed that a single unit of resource is required.

In order to link the operation model with the product model :

- Any variable : resource set (Re) or resource quantity (Qt) can be linked thanks to constraints with the product model.

The existence of each generic operation is triggered by an "operation-ports activation constraint" that permits to configure the necessary operation.

All these elements can be used for the generic operations of our CST example, but only the models of (i) the cutting and assembly operation of BCRO or Ph1, and (ii) the packaging of CST, or Ph4-1 and Ph4-2, are shown in figure 5. Constraints between product model, routing model and operation models are shown.

- The generic operation ph1 requires two sets of resources :
  - Assembly Table (Ass-Table) that gathers a large one (L.As.Tble) and a small one (S.As.Tble), respectively used when the book-case height is 216cms or 72cms. The required quantity of resource is always one.
  - Assembly operator (Ass-Op) that contain a single competency (Op-1). The possible required quantities of resource are 1 and 2 according with the quantity of roll-out-shelves (QRO) 1 or 2 and 3 or 4.
- The generic operation ph4-1 requires one set of resources (Pack-Tbl.P) containing two resources tables dedicated to paint finish, a large one (L.P.Tble) and a small table (S.P.Tble) respectively used if the sub-assembly HC-LC belongs to the product or not.
- The generic operation ph4-2 is modeled in a same way with one set of resources (Pack-Tbl.W) gathering tables dedicated to wood finish, a large one (L.W.Tble) and a small one (S.W.Tble) respectively used if the sub-assembly HC-LC belongs to the product or not.

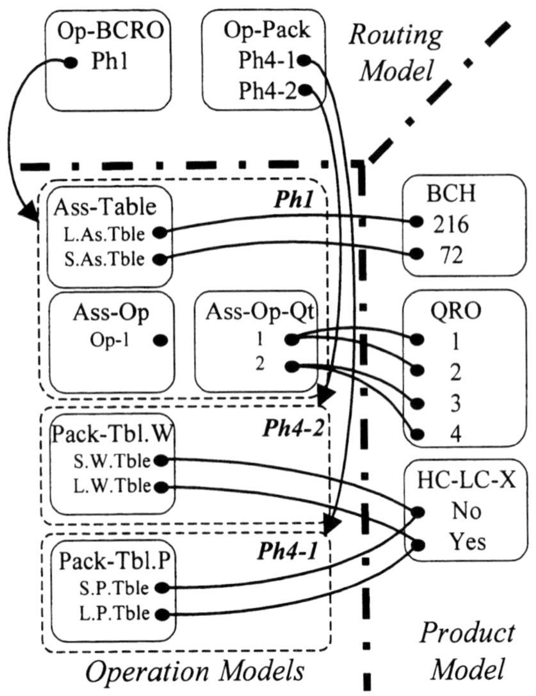

Figure 5 – Operation models

4 CONCLUSIONS

The goal of this communication was to propose a global approach for product and process configuration taking into account structural aspects that permit to develop and to manufacture customized products.

After a quick introduction, the first section has recalled configuration basics and introduced constraint based modeling elements for product configuration. Thanks to the introduction of a new kind of activity constraint, the "sub-assembly activation constraint", it has been shown that the proposed approach could deal with some structural aspects that permit multi-level bill-of-materials and spare part configuration.

The proposed elements have been re-used to model the process. A two levels modeling approach gathering a routing level and an operation level has been proposed. It has been noticed that this approach is valid for multi-level process decompositions. The Port notion frequently used for component location in configuration has been used with success for generic modeling of the anteriority constraints between operations. Manufacturing operations have been modeled as a simple set or resources, more detailed model are of course possible. The links between the process model (gathering routing and operation models) and the product model have been identified and permit process configuration with respect to the configured product.
The proposed elements can be considered as a first step that targets to fill an empty scientific space between product configuration and the real configured product : the process configuration. Many commercial configuration software gather these two approaches, but it is very hard to get detailed informations about reasoning approaches and relevant models used by these systems. As scientific works dealing with this aspect are very rare, we therefore hope to stimulate research in this area in order to have a global workflow concerning the customizable product area. Following process configuration, an other domain where configuration can become an interesting issue is packaging and/or logistic configuration.

REFERENCES

Aldanondo, M., K. Hadj-Hamou, G. Moynard, and J. Lamothe (2003). Mass customization and configuration: Requirement analysis and constraint based modeling propositions. *Integrated Computer-Aided Engineering* IOS Press. **Vol. 10**, n° 2, pp 177-189.

Gelle E. and M. Sabin (2003). Solving Methods for Conditional Constraint Satisfaction. *IJCAI 03 Workshop on Configuration*, Acapulco, Mexico.

Junker U. and D. Mailharro (2003). The Logic of ILOG (J)Configurator : Combining Constraint Programming with a Description Logic. *IJCAI 03 Workshop on Configuration,* Acapulco, Mexico.

Mittal S. and B. Falkenhainer (1990). Dynamic Constraint Satisfaction Problems. *9th National Conference on Artificial Intelligence AAAI*, Boston, USA, pp. 25-32.

Mittal S. and F. Frayman (1989). Towards a generic model of configuration tasks. *IJCAI 89,* Detroit, USA, **Vol.** 2, pp 1395-1401.

Montanari H., (1974). Networks of constraints: Fundamental properties and application to picture processing. *Information Sciences.* **Vol. 7**, pp 95-132.

Sabin D., and E. Freuder (1996). Configuration as Composite Constraint Satisfaction. *Technical Report FS-96-03, Workshop on configuration,* AAAI Press, pp. 28-36.

Soininen T., T. Tiihonen, T. Männistö, R. Sulonen (1998). Towards a General Ontology of Configuration. *AIEDAM,* **Vol. 12,** n°4, pp. 357-372.

Special issue on configuration (1998a). *IEEE Intelligent Systems*, **Vol. 13**, n°4.

Special issue on configuration (1998b*). Artificial Intelligence in Engineering Design and Automation* (AIEDAM), **Vol. 12** n°4.

Special issue on configuration *(2003). Artificial Intelligence in Engineering Design and Automation* (AIEDAM), **Vol. 17** n°1.

Véron M. and M. Aldanondo (2000). Yet another approach to CCSP for configuration problem. *ECAI2000 Workshop on Configuration,* Berlin, Germany, pp. 59-62.

Copyright © IFAC Information Control Problems in Manufacturing, Salvador, Brazil, 2004

ELSEVIER
IFAC
PUBLICATIONS
www.elsevier.com/locate/ifac

TRANSIENT SCHEDULING BASED ON MULTI-AGENT APPROACH

Olfa BELKAHLA DRISS[1], Ouajdi KORBAA[2], Khaled GHEDIRA[1], Pascal YIM[2]

[1]*Stratégie d'Optimisation de l'Ingénierie des Informations et de la connaissancE, ISG de Tunis, Cité Bouchoucha, Bardo, 2000 Tunisia, {Olfa.belkahla, khaled.ghedira}@isg.rnu.tn*
[2] *Laboratoire d'Automatique et Informatique industrielle de Lille (LAIL), Ecole Centrale de Lille, BP48,59650 Villeneuve d'Ascq France. {Ouajdi.Korbaa, Pascal.Yim}@ec-lille.fr*

Abstract: This paper deals with the cyclic scheduling problem for Flexible Manufacturing Systems (FMS) and presents a new Multi-Agent Model for computing the transient states between successive cyclic productions (called transient inter-productions). It is composed of cooperating agents aiming to minimize the global makespan while reducing temporal complexity. The use of Artificial Intelligence techniques, Multi-Agent Systems and Production management makes our model very original. Indeed, the planning phase allows, in the cyclic context, determining the cyclic productions to respect the initial demand. These cyclic productions have to be sequenced each one in relation to the others. Once this operation done, the transient state allowing going from a cyclic state to the following one has to be determined and optimized.

Keywords: flexible manufacturing systems, scheduling algorithms, steady states, transient states, distributed artificial intelligence, agents. *Copyright © 2004 IFAC*

1. INTRODUCTION

In this paper, our interest is about Flexible Manufacturing Systems (FMS) which are able to produce several types of parts simultaneously. The term flexibility means the opportunity to use different procedures to produce a part type (such as non linear operating sequences and identical machines). Furthermore, FMS scheduling involves several assumptions and constraints: tasks are non-preemptive and one machine which begins an operation on a part must finish the operation before freeing the part, the operations are supposed to be known (duration, number, precedence constraints ...). In addition, we assume that parts are transported by pallets with a given transportation order and that these pallets remain stuck to the part until the end of its process. The fact that the operating times are deterministic allows determining a complete deterministic command for further development of the control, such as the on-line control which has to meet the off-line computed schedule as close as possible. The determination of a cyclic deterministic command of FMS interests us because cyclic behavior reduces the complexity of the general scheduling problem.

1.1. Cyclic Scheduling

The cyclic scheduling problem has been quite well studied in reason of the temporal complexity reduction of the problem compared to the general scheduling problem complexity. In fact, it avoids the schedule of all the operations by choosing a set of parts to constitute a cycle and optimize it at best. This cycle will be repeated until the initial demand is respected. Hence, the Cyclic Steady state (CS) is obtained. Few methods take into account different FMS flexibilities such as non linear operating sequences, identical machines ... These flexibilities make problem modeling and resolution very hard. Cyclic scheduling is a NP-hard problem (Serafini and Ukovich, 1989). Hence, only heuristics can be used to solve it.

The deterministic cyclic scheduling was studied by several authors who developed cyclic scheduling heuristics such as the bottleneck machine method (Erschler, et al., 1988), the minimization of the work in process (parts in the system) while respecting the maximal throughput (Hillion and Proth, 1989; Korbaa, et al., 2002; Ohl, et al., 1995; Valentin, 1994; ...), ... Hillion and Proth (1989) gave a resolution method based on progressive placement. Due to the shared resources, the obtained schedule can be not admissible (bigger than the optimal cycle time). In this case, they keep the obtained operations sequencing on the machines and launch a second heuristic which goal is to minimize the work in process in the obtained Marked Graph. Valentin (1994) improved this algorithm by introducing the machines available intervals which allows the algorithm to obtain better results. Ohl et al. (1995) gave a method, based on a progressive placing of the parts, which guarantees obtaining an admissible schedule in one run of the algorithm. Korbaa et al. (1997) improved the last method by introducing the machines available intervals, new placement policies and allowing overlapping cycles: this means that an operation can be launched in one cycle and finished in the following one. More recently, Lee et al. (2001) have proposed an approach based on Petri nets analysis of the shared resources. The algorithm allows obtaining a schedule respecting the optimal cycle time. The result is obtained quite quickly. As for the work in process, its minimization is not guaranteed. Also, Hsu et al. (2002) have proposed a method based on genetic algorithms where, via an indirect coding of the operating sequences, the optimal cycle time is guaranteed. The main advantage of such a method is that it allows the interruption of the algorithm at any time while providing a feasible solution. Of course, to obtain the best possible solution, the resolution must end. Our study focuses on the cyclic scheduling respecting the maximum speed (optimal cycle time) because of its throughput performance and also on minimization of the makespan. Hence, in the following, a cyclic scheduling provided by one of the last methods: Hillion, Valentin, Ohl, Camus/Korbaa, Lee, Hsu … will be considered. In the next subsection, the transient states and especially the transient states between successive cyclic productions will be introduced. Then, in section 2 our multi-agent model for scheduling these transient inter-productions is presented. The third section is devoted to the presentation of an example. The performance of the proposed model is exhibited through an experimental comparison in section 4. Finally, a conclusion is given.

1.2. Transient states

Right now, most of cyclic scheduling approaches considered that the demand is large enough (sometimes infinite) to neglect transient states and to assimilate the optimization of the makespan to the minimization of the cycle time. Nevertheless, to load such a cyclic schedule, a launching phase is needed to lead the system from the initial state to functioning point of the steady state (cyclic schedule). The system is led from the steady state to a neutral position (coming out of all the parts). In addition, FMS are known to be well adapted to small or medium demand. Hence, it is no way to neglect transient states which have to be computed and optimized.

The philosophy of cyclic production is so simple to present and can be summarized as follows:

- The number of parts is large enough not to use acyclic production (huge complexity and combinatory).
- In a first phase, the cyclic behavior is optimized (minimal cycle time = maximal throughput).
- In a second phase, the transient states are computed in order to minimize the makespan.

The transient states scheduling is still weakly studied. In fact, the steady state of job-shops is studied by Erschler et al. (1988) and the transient state is limited for flow-shops. No extension was proposed for job-shop transient states computing. As job-shop interests us, this method can't be applied. In addition, it is based on the scheduling of the bottleneck machine or in FMS several machines can be bottleneck. Munier (1991) has studied this problem for the information processing context but the transient state was reduced to duplicating the non useful operations without makespan optimization. In this context, if an operation op1 precedes op2 in the same operating sequence, then the first occurrence of op2 is done after the first occurrence of op1. This must be verified for all operation's occurrences. So, the transient states schedule is simply determined without search of optimality. (Korbaa, et al., 2002) is principally the first approach to consider the transient states as entire problem to solve and optimize. It's demonstrated that cyclic production can be very competitive if the transient states are optimized even if the initial demand is small (Hsu, et al., 2002). Two different transient states have been distinguished: the **pre-production**, to launch the production and reach the steady Cyclic State (CS) and the **post-production**, to finish the production as given in Gantt diagram of Figure 1 where BDPL corresponds to Beginning Date of the Periodic Loading, EDPL to End Date of the **Periodic Loading** and CT to Cycle Time.

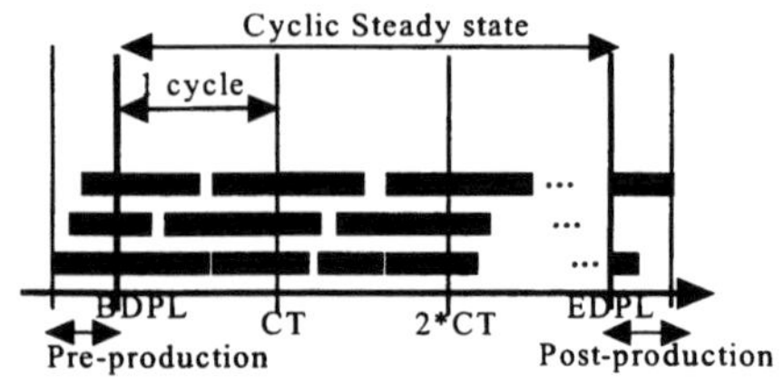

Figure 1. Gantt diagram of a finite cyclic schedule

For the study of transient states, the chosen performance criteria were the minimization of the makespan (because for the production management it doesn't matter the way to do but the duration it needs in order to respect the customers time delays) and the transient duration (because it's better to reach the periodic loading and to end the production as fast as possible). In order to reduce the complexity of the problem, Korbaa imposes that during the pre-production there is no entire Operating Sequence (O.S), i.e. no part type is finished. In addition, if necessary, the O.S may be delayed to the end of the steady cyclic state and the post-production one. This assumption avoids coming back to the general NP-hard problem of FMS scheduling by limiting the transient length. He also assumes that all the operating times are unitary. The Korbaa's approach starts from the given Petri net model and then the transient scheduling problem consists in the reachability of a final marking from an initial marking using the constraint programming, see (Korbaa, et al., 2000). The weakness of the Korbaa's approach is that it was limited to the case of only one cyclic production which starts with an empty system and end with emptying the workshop. Meanwhile, productions cannot be isolated. In the case of many cyclic productions, a new transient state merges: the *Transient inter-production State* which is an interlacing of two transient states corresponding to two successive productions.

This paper proposes to extend this study to the transient inter-productions, because our goal is to reach the periodic loading of each production and to end the corresponding production as fast as possible. This problem (study of the transient inter-productions) was only tackled by Varnier (1996) in the context of the Hoist Scheduling Problem (HSP). In fact, he has solved the transient inter-productions between two **cyclic mono-product** productions using constraint programming tools. So the HSP differs from our problem which considers simultaneously the production of different types of parts. This paper studies the **transient inter-production state** and presents a **multi-agent model**, called **MATIS**: **M**ulti-**A**gent model for **T**ransient **I**nter-production **S**cheduling, for its schedule involving cooperating agents (an agent created to each transient inter-production state) in order to minimize the global makespan while reducing the temporal complexity. Multi-Agent Systems (MAS) are used because of their advantages in many different domains due to the cooperation between agents. In fact, MAS allow solving problems in terms of cooperation and concurrence within a society of agents. Each agent, concurrently and asynchronously acquires information from its environment and from other agents to reason on the basis of these information and to act consequently. See the studies of Ghédira (1994) distributing the simulated annealing for Partial Constraint Satisfaction Problems (P-CSP) and the Extended Forward Checking for Maximal-CSP (Ghédira and Belkahla, 2000). For more details on MAS, see the studies of Ferber and Wooldridge (Ferber, 1999; Wooldridge, 2002).

2. MATIS

2.1. Global view

The *Transient Inter-Production State* is considered as the end of the current production and the beginning of the next one. So, it's an interlacing of two transient states: post-production state (end) and pre-production state (launch). The goal is to schedule many cyclic productions and to minimize the whole makespan. Face to **n** productions (Cyclic Steady states) CS_1, CS_2 ... CS_n, the sequence of producing these steady states have at first to be chosen. Face to **n!** possible sequences which are all the possible permutations of the n productions. Each sequence has:

- **one** pre-production transient (launching of the first cyclic production),
- **one** post-production state (finishing of the last cyclic production),
- **n-1** transient inter-production states (to link up each two successive cyclic productions).

For example, for n=4 productions, the sequence CS_1-CS_2-CS_3-CS_4 has one pre-production to launch CS_1, one post-production to finish CS_4 and 3 (=4-1) transient inter-production states to schedule which are CS_1-CS_2, CS_2-CS_3 and CS_3-CS_4. So, for the n! (=24) sequences, there are **n!(n-1)** transient inter-productions (=72). The computation of all the n!(n-1) scheduling may cause a combinatory explosion. That's why a solution is proposed avoiding computation of such a huge complexity: we remark that many successions are the same in different sequences (each succession exists in n!/n = (n-1)! sequences). For the example of 4 productions, CS_1-CS_2 exists in 6 sequences:

- **CS_1-CS_2**-CS_3-CS_4
- CS_3-**CS_1-CS_2**-CS_4
- CS_3-CS_4-**CS_1-CS_2**
- **CS_1-CS_2**-CS_4-CS_3
- CS_4-**CS_1-CS_2**-CS_3
- CS_4-CS_3-**CS_1-CS_2**

Then, why re-compute them at each time? In fact, due to this existence of a same transient inter-production state in (n-1)!, there are only n(n-1) different transient inter-productions in n! sequences. For the example of n=4, there are only 12 different transient inter-production states. So, instead of scheduling 72 transient inter-production states, only 12 must be scheduled. Hence, The problem becomes:

- Optimize **n*(n-1)** schedules (all possible inter-production)
- Compute all the **n!** sequences: this phase, even if it has big combinatory, is very simple because it just consists on making the sum of the n-1 durations of the already computed inter-production.

Having optimized the n cyclic steady states (defined above), our solution consists in computing all possible successions between these productions: CS_i–CS_j / $i \neq j$ which are of n*(n-1) transient inter-production states. Each time a transient inter-production duration is found, the sequence duration is as and when computed until obtaining all sequence duration's and the best one corresponds to which having the minimal duration.

The problem to solve here is *distributed*, so, there is no reason to compute sequentially the different transient inter-production states and then the sequences. Also, an implementation using a distributed system can give a solution but cannot benefit from information exchange between entities. That's why; *multi-agent systems* are used rather than a distributed system. So, our multi-agent model allows not only minimizing the transient duration's and consequently the total production duration but also accelerating the resolution and the research of the optimal solution. Indeed, MAS will avoid computing successively all the inter-production transient states before beginning computing the possible sequences by allowing the parallel resolution and information exchange between the agents. MATIS does benefit not only from the cyclic scheduling of each production but also from multi-agent systems that allow solving problems in terms of cooperation and concurrence within a society of agents. Each agent, concurrently and asynchronously acquires information from its environment and from other agents to reason on the basis of these information and to act consequently.

In the following subsection, our multi-agent model is detailed. Its goal remains in the minimization of the total makespan by optimizing the executing-production order and starting every production as early as possible without waiting the end of its previous one.

2.2. Global dynamics

The user introduces to the Interface agent the n cyclic steady scheduling corresponding to the n productions to schedule. Normally, as the Interface agent has only the Cyclic Steady (CS) scheduling, it creates as many PrePost agents as CS. Having n PrePost agents, each one, having the CSi, computes its pre-production scheduling and its post-production scheduling using the Korbaa's method. Possible sequences which are of n! relative to the n productions need an enormous temporal computation. As seen above, because of the existence of transient inter-productions in many sequences, we propose to reduce computation by scheduling only one time each transient inter-productions. So, the Interface searches all possible transient inter-production scheduling, of all the productions "taken by couples", which are of n(n-1). It creates as many agent called ***Agent for Transient of***

Inter-productions as transient inter-productions and denoted ATI, i.e. n(n-1) ATI agents. Each ATI_{ij} ($i \neq j$ and $\in \{1\ldots n\}$) is responsible of the determination of the transient state linking CS_i to CS_j. For this, each ATI_{ij} receives the post-production transient from the $PrePost_i$ agent and the pre-production transient from the $PrePost_j$ agent. Each ATI agent will execute its algorithm on a transient of inter-productions which is different from the others. Two ATI can have the same pre-production or the same post-production but not the two at the same time. Each ATI minimizes the duration of its transient inter-productions by trying to draw at best the second production up to the first one while satisfying pallet and machine constraints. After scheduling its transient inter-productions, each ATI_{ab} sends the transient duration's d_{ab} to its acquaintances ATI_{ij} (having production i followed by j) such as $i \in \{1,..n\}\backslash\{a\}$ and $j \in \{1,..n\}\backslash\{a,b\}$ which are of $(n-2)^2$. Hence, the work of duration's computation between ATI agents is equitably shared. Each ATI_{ij} is in charge of computing the duration of all the sequences CS_i- CS_j-... , i.e. $(n-2)!$ sequences, by making the sum of the transient and steady durations. So, the $n(n-1)$ ATI agents allow computing all the n! sequences ($(n-2)!*n*(n-1)$). This combinatory of (n-2)! corresponds to making "mathematical additions" which is obviously less complex and complicated than optimizing a scheduling problem (NP-hard in general). In fact, each time a transient inter-production duration d_{ab} sent by an ATI_{ab} and received by an ATI_{ij}, the ATI_{ij} adds it to the current duration of each sequence containing CS_a - CS_b. For this, each ATI must receive from the PrePost agents the last post-production's duration in each sequence. So, each ATI communicates with all PrePost agents.

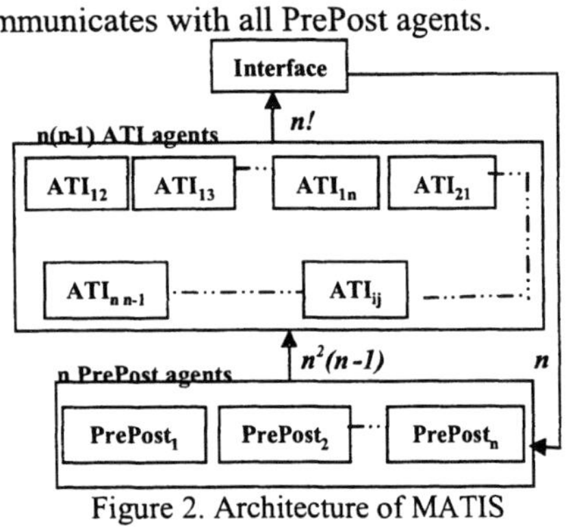

Figure 2. Architecture of MATIS

When an ATI_{ij} finds a current sequence's duration greater than the optimal one found by another ATI, it ignores this sequence. At that way, the n! scheduling sequences are obtained. Each ATI agent, computing the duration of a sequence (the global makespan corresponding to the total demand of the n productions) sends it to the Interface. Each time it receives a makespan from an ATI agent, the Interface maintains the minimal one as well as the production sequence and its schedule. Finally, it sends them to the user. For a clarity purpose, our multi-agent model is schematized in Figure 2.

2.3. ATI behavior

The behavior of an ATI_{ij} is composed of the following steps:

<u>*Assigning dates t_0 and t_0'*</u>. The ATI_{ij} has to schedule the transient inter-productions CS_i-CS_j. In figure 3, the transient post-production of CS_i and transient pre-production of CS_j are schematized where the End Date of the Periodic Loading of CS_i ($EDPL_i$) is named t_0' and the Beginning Date of the Periodic Loading of CS_j ($BDPL_j$) is initialized to a variable t_0 which must be minimized. $EDTpost_i$ corresponds to the End Date of the Transient post-production of CS_i and $BDTpre_j$ corresponds to the Beginning Date of the Transient pre-production of CS_j. Drawing the jth production up to the ith reverts to minimize t_0 ($BDPL_j$), i.e. the end date of pre-production transient of production j.

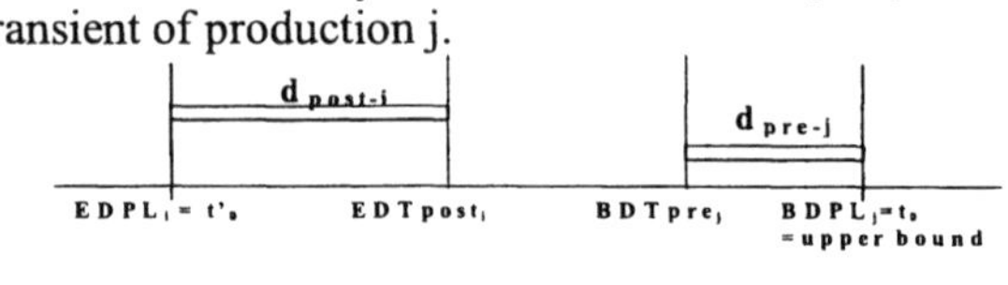

Figure 3. The transient post-production of CS_i and pre-production of CS_j

That's why it's necessary to affect to t_0 a great value which can be an upper bound of t_0 and try to minimize this value in order to find the best transient inter-production scheduling while it is greater than a lower bound. So, each ATI_{ij} computes two bounds: a lower bound (lb) and an upper bound (ub). If it reaches the lower bound then it's the optimum. Thus, the lower bound constitutes a performance criterion. In figure 3, we see that d_{post-i} corresponds to the duration of the transient post-production of CS_i and d_{pre-j} corresponds to the duration of the transient pre-production of CS_j. Then, the upper bound of t_0 is computed so: *upper bound* $(t_0) = t_0' + d_{post-i} + d_{pre-j}$ and denoted ub_{ij}. The ATI_{ij} sends this upper bound to its acquaintances. Each ATI receiving this bound constitutes an upper bound for the total makespan corresponding to each sequence.

t_0 is initialized to the upper bound, then, we reason on machines to draw operations executing on every machine M in pre-production transient state up to the first one. The lower bound, denoted lb_{ij}, is computed so: *lower bound* (t_0) = max_M *{Workload of M during transient state}*. Thus, we obtain $t_0' \leq lb_{ij} \leq t_0 \leq ub_{ij}$. At the same manner, the ATI_{ij} sends the found lower bound to its acquaintances in order to compute a lower bound for total makespan corresponding to each sequence. These bounds serve to stop ATI agents in case of reaching the lower bound or getting a higher value than the upper bound. These values are modified each time a new makespan is found better than the ones already found.

<u>*Scheduling*</u>.

- The ATI draws the 2nd schedule up to the first one until intersection of the two-production operations: by affecting to t_0 the value of the upper bound.
- It schedules operations of the two transient states while satisfying machine and pallet constraints.
- When t_0 reaches the lower bound, thus it's the optimum.
- It obtains so the minimal transient inter-production duration.

<u>*Result communicating*</u>. After scheduling its transient inter-productions, each ATI_{ij} agent sends its duration to

its $(n-2)^2$ acquaintance ATI_{kl} agents (having production k followed by l) such as $k \in \{1,..n\}\backslash\{i\}$ and $l \in \{1,..n\}\backslash\{i,j\}$.

Message processing. Each time an ATI_{ab} receives a transient inter-production duration from its acquaintance ATI_{ij}, it adds it to the current duration corresponding to sequences containing CS_i followed by CS_j until optimizing duration of each sequence. When it finds a current sequence's duration greater than the optimal one found earlier by an ATI, it ignores this sequence. In fact, the optimal duration is transmitted to all ATI agents from the Interface each time the later receives a better duration from an ATI. So this avoids additional computation of duration of some sequences. Finally, the agent ATI_{ab} sends each sequence with its schedule and its duration to the Interface. When all ATI finish their process and stop, the Interface will communicate the schedule of the best sequence (having the minimal makespan) to the user.

2.4. Agent knowledge

Interface agent. The Interface agent has as acquaintances all ATI and PrePost agents. Its static knowledge is the n steady states. Its dynamic knowledge consists of the best solution found: the optimal order and its duration.

ATI agents. Each ATI_{ij} has as acquaintances all the ATI agents, the Interface agent and all PrePost agents. Its static knowledge is the post-production transient of the all the productions (except of CS_j) and the pre-production transient of the two productions. Its dynamic knowledge is the transient inter-production state, the upper bound, the lower bound, transient's duration and optimal duration's of its sequences which are of $(n-2)!$

PrePost agents. Each $PrePost_i$ has as acquaintances the Interface agent and the ATI agents. Its static knowledge is the scheduling of CS_i. Its dynamic knowledge is the final scheduling (pre and post-production transient and steady state) and its optimal duration. Each $PrePost_i$ agent, having the cyclic steady state CS_i, computes its pre-production scheduling and its post-production scheduling to get the final scheduling of CS_i using the Korbaa's method and then sends them to the concerned ATI agents.

3. EXAMPLE

For the clarity's sake, a simple illustrative example is developed (see Figure 4) of respectively two cyclic productions CS_1 and CS_2.

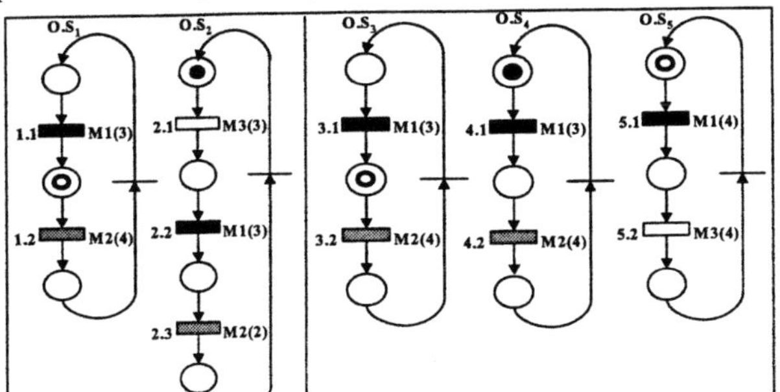

Figure 4. Operating sequences of productions CS_1 and CS_2

For the first production CS_1, 2 operating sequences and 3 machines are considered. The pallets P_1 and P_2 are dedicated to $O.S_1$ whereas P_3 and P_4 are dedicated to $O.S_2$. This production is composed of 3 parts of each type. For the second production CS_2, 3 operating sequences and 3 machines are considered. The pallet P_5 is dedicated to $O.S_3$, P_6 to $O.S_4$ and P_7 to $O.S_5$. This production is composed of 4 parts of each part type.

The figures 5 and 6 give the Dual Gantt Diagram of the whole schedule of each production.

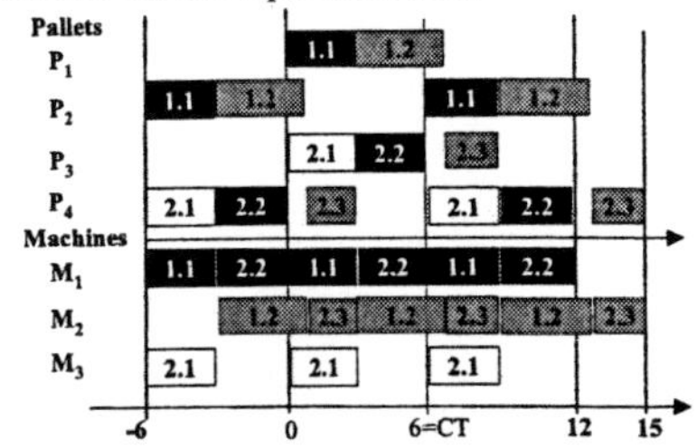

Figure 5. Dual Gantt Diagram of production CS_1

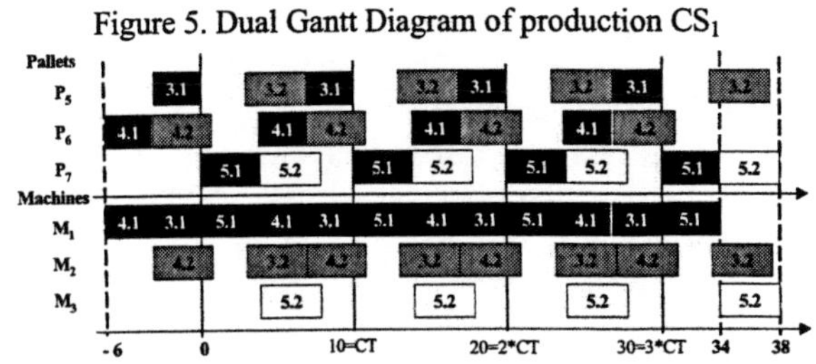

Figure 6. Dual Gantt Diagram of production CS_2

The run of our program gives the transient inter-production scheduling of each sequence: CS_1-CS_2 at Figure 7 and CS_2-CS_1 at Figure 8.

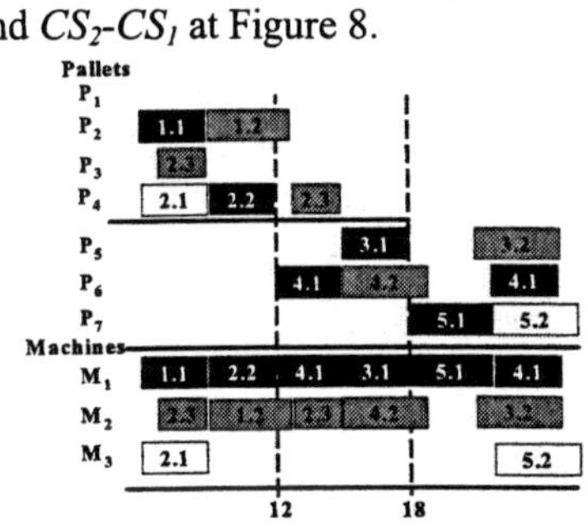

Figure 7. Transient inter-productions of CS_1-CS_2

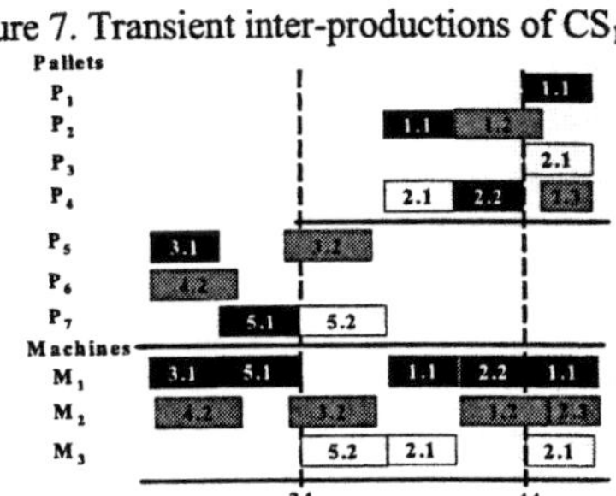

Figure 8. Transient inter-productions of CS_2-CS_1

4. EXPERIMENTATION

In order to evaluate our model, the solution found by our model is compared to the one found by starting every production exactly at the end date of its previous one (Solution of Korbaa 2000) and this in terms of transient inter-production duration and consequently of global makespan and in term of CPU-time. The implementation has been developed with Actalk (Briot, 1989) an object-based concurrent programming language in the Smalltalk-80 environment.

Our solution is better than the one found by starting every production exactly at the end date of its previous one in term of makespan. In fact, for the first sequence CS_1-CS_2, the transient inter-production duration found by our model is 6 and then the makespan is of 62. For the second sequence CS_2-CS_1, the transient inter-production duration found by our model is 10 and then the makespan is of 65. The first sequence is better than 65 which is the one found by Korbaa's method (the transient inter-production duration for the first sequence is 9 and for the second is 10).

The performance of our model is so proved, it improves results and minimizes the loosed time. Here, the difference between the 2 solutions is low because of simplicity of the problem (to better understand the model). In fact, more pallets are used by productions, more the difference becomes great. For this example, we don't mention the CPU-time because of its little size. It's sure that a distributed method in our case is far better than a sequential one due to the hard work that has each ATI agent. But in distributed systems, there is no exchange of information and consequently no cooperation between entities. That's why, in our multi-agent model, all agents solve their problems at the same time and then benefit from each other by acquiring pertinent information which allow them reducing their load of work and avoiding inutile computation.

5. CONCLUSION

In this paper, a new multi-agent model is presented for computing the transient states between successive cyclic productions (called transient inter-productions) by creating an Agent for Transient of inter-productions for each transient inter-production state. It aims to minimize the global makespan while reducing temporal complexity by scheduling at the same time all transient inter-production states and by communicating, on the one hand, upper bounds and lower bounds for the transient inter-production's duration in order to constitute respectively upper bound and lower bound for the total makespan (corresponding to each sequence of productions) and on the other hand, the transient inter-production's duration in order to avoid inutile computation. The advantage of our model remains in the minimization of the total makespan (corresponding to the total demand) by optimizing the executing-production order and starting every production as early as possible without waiting the end of its previous one. Our model benefits from information exchange between agents which allows cutting ways do not lead to a better solution.

REFERENCES

Briot, J.P. (1989). Actalk: a tested for Classifying and Designing Actor Languages in Smalltalk-80 Environment, *European Conference on Object-Oriented Programming.*

Erschler, J., D. Leveque and F. Roubellat (1982). Periodic loading of flexible manufacturing systems, *IFIP Congress*, APMS, Bordeaux, France, p. 327-339.

Ferber, J. (1999). Multi-Agent Systems: An Introduction to Distributed Artificial Intelligence, Harlow, UK, Addison Wesley Longman.

Ghédira, K. (1994). Partial Constraint Satisfaction by Multi-Agent-Simulated Annealing approach, *International Conference of AI*, KBS, ES and NL, Paris.

Ghédira, K. and O. Belkahla (2000). Distributed Extended Forward Checking for Max-CSP, *Intelligent Systems and Applications*, Wollongong, Australie.

Hillion, H.P. and J.-M. Proth (1989). Performance Evaluation of Job-Shop Systems using Timed Event-Graphs, *IEEE Trans. Automatic Control*, **34(1)** : 3-9.

Hsu., T., R. Dupas and G. Goncalves (2002). A genetic algorithm to solving the problem of flexible manufacturing system cyclic scheduling, *IEEE SMC'02*, **Vol 3**, TA1K2, Hammamet, Tunisia.

Korbaa, O., H. Camus and J. C. Gentina (1997). Transient states study for cyclic schedules: bounds and optimization, *IEEE International Symposium on Assembly and Task Planning, ISATP'97*, pp.188-193, Marina del Rey, California, USA.

Korbaa, O., P. Yim and J. C. Gentina (2000). Solving Transient Scheduling problems with constraint programming, *European Journal of Control (EJC)*, **vol.6, n°6**, pp.511-524.

Korbaa, O., H. Camus and J-C. Gentina (2002). A New Cyclic Scheduling Algorithm for Flexible Manufacturing Systems, *International Journal of Flexible Manufacturing Systems (IJFMS)*, **Vol. 14, N. 2**, pp. 173-187.

Lee, J.K., O. Korbaa and J.C. Gentina (2001). Scheduling Analysis in FMS Using the Unfolding Time Petri nets, *INCOM, 10th IFAC Symopsium*, Vienne.

Munier, A. (1991). Résolution d'un problème d'ordonnancement cyclique à itérations indépendantes et contraintes de ressources, *RO-RAIRO*, **vol. 25, n°2**, p. 161-182. *In French.*

Ohl, H., H. Camus, E. Castelain and J. C. Gentina (1995). Petri net modeling of ratio-driven flexible manufacturing systems and applications on the WIP for cyclic schedules, *IEEE SMC'95*, Vancouver, **vol. 4, n°5**, p. 3081-3086.

Valentin, C. (1994). Modeling and analysis methods for a class of hybrid dynamic systems, ADPM '94. Brussels, Belgium, p. 221-226.

Varnier, C. (1996). Extensions de 'Hoist Scheduling Problem' cyclique - Résolution basée sur un traitement des contraintes disjonctives en programmation logique avec contraintes, Thèse de Doctorat, Université de Franche-Comté, n°502.

Serafini, P. and W. Ukovich (1989). A mathematical model for periodic scheduling problems, *SIAM J. Disc. Math*, **Vol. 2, n°4**, p. 550-581.

Wooldridge, M. (2002). An Introduction to Multiagent Systems, John Wiley and Sons Ltd.

Copyright © IFAC Information Control Problems in Manufacturing, Salvador, Brazil, 2004

ELSEVIER
IFAC
PUBLICATIONS
www.elsevier.com/locate/ifac

COMMUNICATION IN INDUSTRIAL AUTOMATION –WHAT IS GOING ON?

Peter Neumann

Institut für Automation und Kommunikation Magdeburg
Steinfeldstraße 3, 39179 Barleben, Germany
peter.neumann@ifak-md.de

Abstract: The fieldbus systems have been successfully introduced in the industrial automation. Nowadays, a large community is inventing the usage of Ethernet-based local communication systems in this domain ensuring the real-time, safe and secure behaviour of these systems. Future scenarios of geographically distributed production plants or services require the use of heterogeneous networks consisting of local and wide area, and wired and wireless communication systems operated by different authorities. Thus, behaviour has to be supplemented by context awareness realised by location-based communication services and context-sensitive applications. This paper will mainly address the ongoing activities in the field of using heterogeneous networks within the automation domain.. *Copyright © 2004 IFAC*

Keywords: Ethernet, Communication Systems, Decentralised Control Systems, Industry Automation, Real-time, Performance, Safety, Security.

1 INTRODUCTION

For the last 20 years a lot of effort has led to the decisive usage of digital communications in distributed computer control systems within the factory as well as the process domain. The proprietary communication systems within SCADA systems were supplemented and partially displaced by the fieldbus systems and sensorbus systems etc. At the same time, Ethernet won the battle as the most used communication technology within the office domain resulting in low component prices caused by the mass production of these components.

Nowadays, there is a large community inventing the usage of Ethernet based communication systems to be used in the industrial automation domain, e. g. in the real-time and safety-critical world. However, opposite to that fieldbus systems are the most important communication systems used in commercial control installations.

Since there is an economic interest to widely introduce the Ethernet-based and Radio-based technologies, the inventors have to guarantee that the domain-specific requirements can be fulfilled. These requirements are:

- Guaranty for real-time behaviour. There are different classes (non real-time applications: diagnosis, maintenance, commissioning, slow mobile applications; soft real-time applications: processes in manufacturing and process automation, data acquisition; hard real-time applications: control applications, fast mobile applications, machine tools, motion control).
- Guaranty for functional safety. This means protection against hazards caused by incorrect functioning including communication via heterogeneous networks.
- Guaranty for security. This means a common security concept for distributed automation using heterogeneous networks.

Future scenarios of distributed automation leads to desired mechanisms for geographically distributed automation functions due to various reasons:

- centralised supervisory and control of (many) decentralised (small) technological plants
- remote control, commissioning, parameterisation and maintenance of distributed automation systems
- to include remote experts or external machine-readable knowledge for the plant operation and maintenance

This means that heterogeneous networks consisting of local and wide area and wired and wireless communication systems will play a raising role. However, there is not only a need for real-time, safe and secure communication. The desired context awareness lead to the usage of location-based communication services and context-sensitive applications. Thus, the functionalities offered by the entertainment electronics will influence more and more the complexity of the communications approach within the automation domain.

Since the fieldbus technology has reached a stable phase in using it in the industrial automation, this paper will mainly deal with the ongoing activities in the field of using heterogeneous networks.

2. STATE-OF-THE-ART

2.1 Local Industrial Communication

Digital communication has been an important driving force of the computer control systems for the last 25 years. To realise the access to data in various layers of an enterprise information system by different users, there is a need to merge the different digital communication systems within the plant level, control level, and device level of an enterprise network. On these different levels, there are different requirements dictated by the nature and type of information being exchanged (Jasperneite *et al.*,2001). Network physical size, the number of devices supported, network speed, response time, frequency of exchange and payload size are a few of the performance characteristics used to classify and group specific network technologies. The real-time requirements depend on the type of messages to be transmitted: meeting deadlines for data transmission, restricting jitter for audio and video stream transmission. Otherwise, the offered resources at the various network levels are different. At the device level, there are extremely limited resources (hardware, communications), but at the plant level there are powerful computers allowing comfortable software and memory consumption.

Fieldbus Systems

In particularly over the last 15 years, fieldbus systems have required a lot of work in the area of definition, specification, implementation and market dissemination. Nowadays, the fieldbus systems are standardised and widely used within the industrial automation (IEC 2002, IEC 2003). Recently, the fieldbus transmission technology has been completed by a radio front end, realised by the European project RadioFieldbus (Rauchhaupt, 2002; Rauchhaupt, 2003), see also (Decotignie, 2002). The real -time behaviour depends on:

- Efficiency of the Protocol Stack,
- Implementation,
- Physical Layer Transmission Speed (bit rate, handover).

The RFieldbus technology requires additional protocol mechanisms and contains the scheduling of real-time and non real-time traffic (Figure1).

The RFieldbus Radio Physical Layer Main Properties are:

- ISM-Band 2,4 GHz, Synchronous Octet Transmission
- Direct Sequence Spread Spectrum (DSSS) Coding, code lengths 11...63 chips
- Radio Data Rate 350kbit/s -1Mbit/s -2Mbit/s
- Two PhL Mode Options:
  Mode I compatible with IEEE802.11 MAC-less
  - IEEE802.11 Header
  - Extensions for Identification, Mobility, Power Management

  Mode II optimised for Real-Time transmission
  - Shortened Preamble
  - Optimised Synchronisation
- Signal Quality- and Power-Management for
  - Real-Time Handover
  - Transmission Quality Supervision.

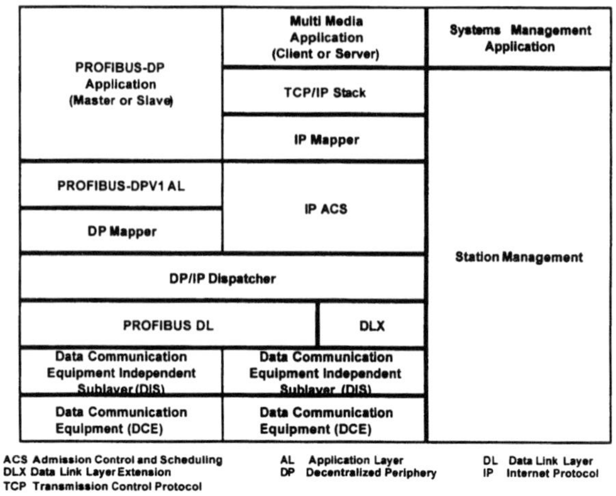

Figure 1. Protocol Suite of a Radio-based Field Device

There are other Radio add-ons for wired systems and other wireless Radio communication systems

- WLANs (IEEE 802.11): IEEE 802.11b, IEEE 802.11a, IEEE 802.11g
- Bluetooth 1
- Bluetooth 2, ZigBee (IEEE 802.15)
- Ultra Wide Band Systems.

Most of these Wireless Radio Networks can be used in non real-time applications, some of them in soft real-time applications (but industrial environments and ISM band limit their applications).

The WLAN technology is more and more used in the higher architecture levels of the automation hierarchy, but also in the shop floor. ZigBee should be introduced to connect the automation devices at the field level, especially in the process automation, because it will operate on a lower baudrate. Ultra Wideband Systems are becoming more and more important for sensors and indoor location-based services.

Ethernet-based Approaches in Industrial Automation

The Local Area Networks LAN based on Ethernet-TCP(UDP)/IP has been standardised and widely introduced in the office domain and also in the automation domain, using shared Ethernet as well as Switched Ethernet (star and tree topology). There are system-immanent limits regarding the real-time behaviour, because most of the solutions use the TCP(UDP)/IP functionality and a middleware above TCP(UDP)/IP to schedule the (soft)real-time traffic and the non real-time traffic. Examples of this category are PROFInet V1 (PNO, Siemens) (PROFIBUS, 2003; Feld, 2003), IDA (Interface for distributed Automation; Schneider, Phoenix Contact etc.) (IDA, 2002), Ethernet/ IP (ODVA, Rockwell), High Speed Ethernet HSE (Fieldbus Foundation,

Emerson). (Poeschmann *et al*., 2003) compares these approaches. There are many investigations regarding temporal behaviour related to Ethernet-TCP/IP based local networks. They include mainly the response aspect of data packet transmission, which is very important within the industrial automation domain. The synchronous video or audio stream transmission is of secondary interest. But the infrastructure of LAN, e. g. switches, offers a high priority of stream transmission. In contrast to that in the industrial automation application, the data packet transmission has to have the highest priority.

A lot of research activities deal with a middleware on top of the MAC layer of Ethernet scheduling the hard real-time and soft real-time/ non real-time traffic, see e.g. (Seok-Kyu Kweon et al., 1999; Agarwal *et al*., 2000; Caponetto *et al*., 2003; Carpenzano *et.al*., 2003; Alves *et al*., 2000; Bonaccorsi *et al*., 2003). (Baek-Young Choi *et al*., 2000 ; Jasperneite , 2002; Xinggang Fan et al., 2002; Georges *et al*., 2002) deal with the usage of Switched Ethernet in the automation domain. The investigated real-time mechanisms are very important for synchronous data transmission in the Motion Control area. Examples of this category are Power Link (B&R, Lenze etc.), EtherCAT (Beckhoff), PEAC (ifak), and in the future PROFInet-IRT (Feld, 2003).

Investigations have shown that at present the switched Ethernet itself is not the bottle neck of data transmission within local automation networks (for star topology as well as for line topology – this means each control device assesses its own switch, and all traffic has to pass through many or all switches). The present bottle neck is the communication stack within the end devices (Jasperneite *et al*., 2001a; Jasperneite *et al*., 2001b; Jasperneite, 2002).
Summarising: Mostly object-oriented approaches are used; mostly middleware concepts are used to realise the separation of SRT and NRT traffic; migration concepts are available to introduce the data of fieldbus-oriented sub-systems.

2.2 Local Soft Real-time Approaches

As mentioned above there are industrial approaches used in real-time automation applications. They are using shared and/ or switched Ethernet and TCP (UDP)/ IP mechanisms. They can be distinguished by different functionalities on top of TCP (UDP)/ IP as well as by their object models and application process mechanisms. The systems which are using Ethernet-TCP/IP offer response time in the lower Millisecond range. The systems are performant but not deterministic. The data transmission is based on the best effort principle. To use these systems within the automation domain, mechanisms are needed to monitor time limits, to use substitution values, to optimise the transmission (using records of many values within one MAC-PDU) as well as time- and event-triggered data transmission.

Ethernet/ IP (Rockwell, ControlNet International, Open DeviceNet Association) uses a *Control and Information Protocol CIP* (Ethernet/IP, 2003). CIP represents a common application layer for all physical networks of Ethernet/ IP, ControlNet and DeviceNet. Data packets are transmitted via CIP router between the networks. The application process is based on a Producer/ Consumer model.
High Speed Ethernet HSE (Fieldbus Foundation). A *Field Device Agent* represents a specific Foundation Fieldbus application layer function (including *Fieldbus Message Specification*, originally specified by the PROFIBUS user organisation). Additionally, there are HSE communication profiles to support the different device categories: host device, linking device, I/O gateway, field device. These devices share the tasks of the system using distributed Function Block applications.
Interface for Distributed Automation IDA (IDA Groupe e. V., Schneider, Phoenix Contact).
The layer functions allow three types of communication channels: Client/ Server Messaging for Engineering data exchange for real-time traffic using *Real-time Publish Subscribe* middleware RTPS from RTI (Real-Time Innovations, 2002), Modbus/ TCP as a widely spread protocol.
PROFInet V1 (PROFIBUS user organisation, Siemens) uses the *DCOM Wire Protocol* with the Remote Procedure Call mechanisms (DCE RPC) to transmit the soft real-time data. An open source code is available on the PNO Website and various exemplary implementations/portations for different operating systems.

All the mentioned approaches are able to support the protocols of the office domain, e. g., SMTP, SNMP, HTTP, some of them BOOTP, DHCP, for Web access and/ or for Engineering data exchange.

The object models of the approaches differ. Ethernet/ IP defines in the CIP specification the user layer including an *Application Object Library* (many of standard objects of industrial devices) and *Device Profiles* (defining the device features).

HSE (HSE , 2002) is based on the Function Block technology (IEC 61499) to design the distributed application as a Function Block network. IDA and PROFInet V1 define a comfortable object background.

IDA (IDA, 2002) defines an object-oriented common Runtime and Engineering model consisting of an *Application Model* (design of modular applications), an *Engineering Model* (description of the specific automation applications and their connections), a *Process Model* (mapping the application model elements to the physical topology), a *Presentation Model* (description of the external behaviour of the application model elements) as well as a *HMI Model* (browser-based supervisory & control). The Engineering Model allows two views: the *Functional view* (Representation of the hierarchical structure of the IDA system's functionality), the *Topological*

View (physical network structure including all segments and connected devices, routers, switches). PROFInet V1 (PROFIBUS , 2002) also defines an object oriented common *Runtime and Engineering Model* (figure 2).

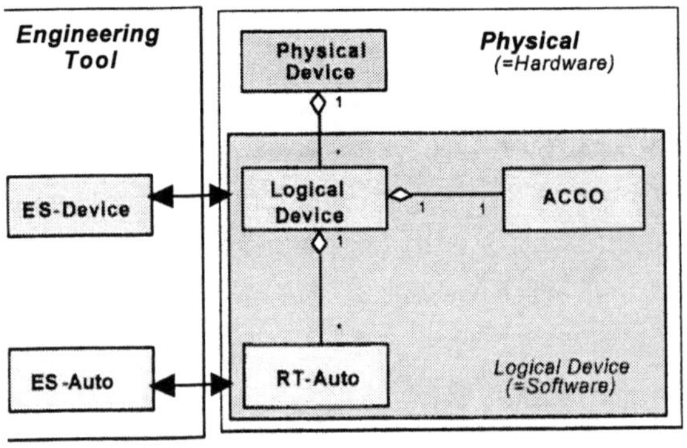

Figure 2. PROFInet V1 Object Model

It allows the pre-configuration of the automation application within the Engineering Tool and mapping it to the physical architecture easily using a powerful Tool (iMap). Based on this infrastructure, the Engineering workflow as shown in Figure 3 could be realised.

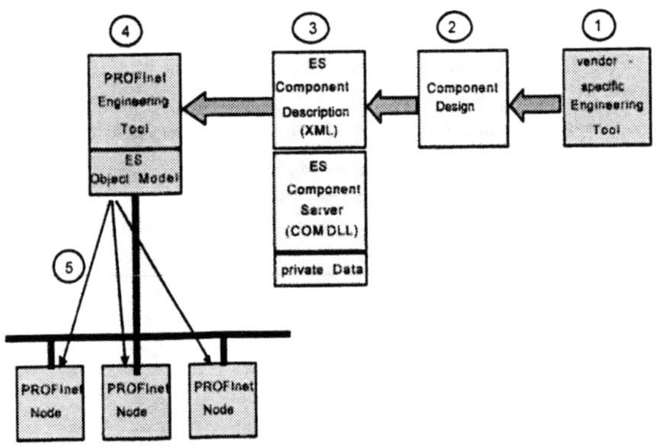

Figure 3. PROFInet Workflo w

2.3 Local Hard Real-time Approaches

As mentioned above the use of middleware concepts on top of TCP/IP leads to limits regarding real-time behaviour caused by the best effort features of TCP. That is why many investigations are directed to use a middleware concept on top of the MAC Layer. In academic and industrial research, different scheduling strategies and smoothing concepts are investigated (Caponetto *et al.*, 2003; Carpenzano *et al*, 2002; Bonaccorsi *et al.*, 2003). Following the results we can distinguish

- deterministic real-time behaviour enabling cycle times in the range of 5 to 10 milliseconds. This is comfortable for most industrial communication in a factory.
- isochronous real-time behaviour enabling cycle times in the range of 250 microseconds to 1 millisecond and with jitters of less than 1 microsecond.

Deterministic real-time approaches

The first concept uses a middleware on top of the MAC layer to realise the scheduling and smoothing functions. The middleware is represented by a software implementation. Examples are:

- *Powerlink* (Ethernet PowerLink Standardisation Group EPSG, Bernecker & Rainer), developed for Motion Control
- *EtherCAT* (Beckhoff) developed as a fast backplane communication system
- *PROFInet V2* (PROFIBUS user organisation, Siemens) developed for any industrial applications.

Powerlink (Meindl , 2002) uses a proprietary real-time protocol on top of the *shared* Ethernet. The scheduling mechanism is a time-division scheme. Every node uses its own time slot (*Slot Communication Network Management SCNM*) to send its data. This mechanism avoids the collisions on the Ethernet. Using 100 Mbps Ethernet Powerlink allows real cycle times of 400 microseconds or less in applications. The network jitter has been proven to be below 1 microsecond. The drives (less than 50 with cycle times in the range of 2 ms) can communicate synchronously with each other using broadcast services. Remarkable is that only hubs can be used for such real-time requirements. Switches do not meet these demands.

EtherCAT uses Bus Terminal Controllers to support active (event-driven) sending and receiving of data via Ethernet. This enables a lateral communication independent of a higher-level master, supported by Function Blocks. The routing functionality ADS (*Automation Device Specification*) enables local and remote communication via any connection route. The *Bus Terminal Controllers* can exchange data between themselves, with PLCs or with bus couplers of Fieldbus systems. The IP sockets can be opened and closed at runtime in order to communicate with additional devices.

PROFInet V2 also uses the basic functionality of PROFInet V1. This means, that on top of TCP/IP the non real-time traffic could be handled by the well-known IT applications (e.g. FTTP, SNMP, DHCP ...), and the soft real-time traffic could be handled by the specific DCE-RPC middleware mentioned above. The soft real-time traffic can be used for device parameterisation, reading of diagnostic data, loading of interconnections, and negotiation of the communication channel for user data. The additional functionality is a *SRT middleware* on top of Ethernet MAC layer to enable high-performance transfer, cyclic data exchange and event-controlled signal transmission. The layer 7 functionality is directly linked to that middleware. The middleware itself contains the scheduling and smoothing functions. This means: the PDU structure is not influenced by TCP/IP. A special ethertype is used to identify real-time PDUs (only one PDU type for real-time communication). That enables an easy *hardware support* for the real-time PDUs (see PROFInet V3) (Feld, 2003). The real-time behaviour of PROFInet V2 is comparable with that of the Fieldbus

PROFIBUS: cycle times in the range of 5 to 10 msec (instead of 100 msec reached by PROFInet V1).
The technical background is a 100 Mbps full duplex Ethernet (switched Ethernet). The middleware software is based on Win 2000 (for PC) or Tricore (for embedded devices).

Isochronous real-time approach

Vicariously for this class of real-time approach the **PROFInet V3** approach should be mentioned (Feld, 2003). This version of PROFInet adds a isochronous real-time channel to the above mentioned V1 and V2 channels. This IRT channel enables a high-performance transfer of cyclic data in an isochronous mode. The time synchronisation and node scheduling mechanism is located within and on top of the Ethernet MAC Layer. The offered bandwidth is separated in bandwidth for cyclic hard real-time and soft/non real-time traffic. This means, within a cycle there are separate time domains for cyclic hard real-time, for soft /non real-time over TCP/IP traffic, and for the synchronisation mechanism. The cycle time should be in the range of 250 μsec (35 nodes) to 1 msec (150 nodes) when simultaneously TCP/IP traffic of about 6 Mbps has been transmitted. The jitter is less than 1 μsec. PROFInet V3 uses switched Ethernet (full duplex). Special 4 Port (later: 2 Port) switch ASICS are under development and will allow the integration of the switches into the devices (nodes) substituting the legacy communication controllers of the Fieldbus systems. Distances of 100 m per segment (electrical) and 3 km per segment (fibre-optical) will be bridged.

2.4 Wide area communications

Allowing remote mechanisms (tele-supervisory, tele-operation, tele-service) using Wide Area Networks, the stock of existing communication technology becomes broader:

- all appearances of the Internet (mostly with best effort quality of services)
- public digital wired telecommunication systems (ISDN, DSL etc.)
- public digital wireless telecommunication systems (GSM-based, GPRS-based, UMTS-based)
- private wireless telecommunication systems, e. g. trunk radio systems.

Using these technologies within the automation domain there are many private protocols over leased lines, tunnelling mechanisms etc. Most of the wireless Radio Networks can be used in non real-time applications, some of them in soft real-time applications (but industrial environments and ISM Band limit the applications). The behaviour of the end-to-end connection via these telecommunication systems depends on the recently offered quality of service and cannot be guaranteed in many cases. It strongly limits the use of these systems within the automation domain.

3 PROBLEMS TO BE SOLVED

As mentioned in section 1, the following problems have to be solved:

- Definition of real-time classes and guaranty of suitable behaviour for Non Real-Time, Soft Real-Time, and Hard Real-Time Applications
- Guaranty for functional safety
- Guaranty for security
- Local awareness in the industrial automation.

3.1 Real-time behaviour of heterogeneous networks

There are different levels of real-time behaviour within a heterogeneous network used in the automation domain depending on the level of the enterprise network. In contrast to a high priority of stream transmission, in the industrial automation application the data packet transmission has to have the highest priority
Within this automation domain, there are other classes of real-time behaviour of the WAN than for other domain (office, e-commerce etc). The most important requirement is to guarantee application-specific deadlines, whatever the required value of the deadline in the various applications is. Using Wide Area Networks, the actual offered mechanisms support mainly the best effort QOS and privilege the video and audio stream transmission. There is a need for offering QOS levels which are required for a suitable response behaviour of data packet transmission. The IPv6 approach offers real-time mechanisms, user priorities and a broader range of addresses. This is very important for the overboarding need of address space using embedded Web servers within the automation devices. In the last few years, a stable IPv6 network supported by different providers was established, e.g. IRCNet. However caused by the misusage there is a very restricted access to these networks.
The usage of wide area networks within the automation domain has to be investigated just as the uninterrupted commercial availability. Since the Internet or other telecommunication systems are general-purpose communication systems, the infrastructure and business model preconditions for the selection of requested QOS within a spectrum of available communication services of various providers have to be developed (Figure 4). This means, in analogy to the "switched" Ethernet in LANs, a WAN switching mechanism has to be developed for this selection, i.e. choosing dynamically the network type and/ or network provider which guarantees the required QOS. There are many investigations to use the IPv6 functionality for real-time services, mostly directed to the usage for audio/video stream transmission (Mahmoodian *et al.*, 2001; Wen-Tsuen Chen *et al.*, 2000; Moon *et al.*, 2001; Chien-Chao Tseng *et al.*, 2001).

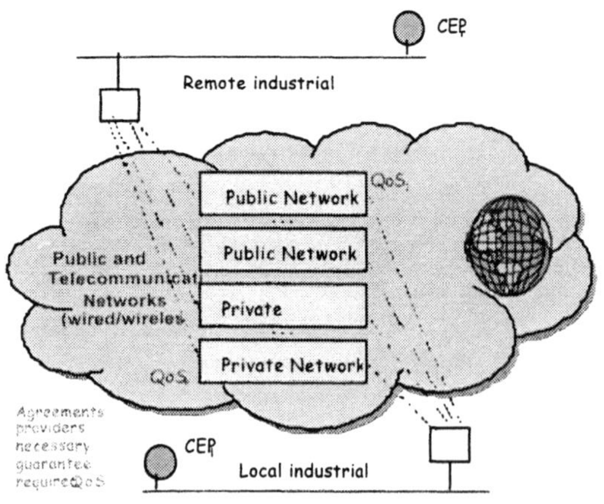

Figure 4. Needed Provider Switch Mechanism

3.2 Functional Safety

Safety means protection against hazards (movement, heat, radiation, electrical shock, etc.) (IEC 61508). "Functional Safety" means protection against hazards caused by incorrect function. Safety includes the communication via heterogeneous network. (EN 50159; Halang et al.; 2002; Mataix et al., 2003 ; Diedrich et al., 2003; Halang et al., 1998) deal with different aspects of safety in communication and computer systems. Caused by the distribution of data via the communication networks, the safety of these networks becomes more and more important regarding the functionality of an automation system.
There is a need to meet defined Safety Integrity Levels (SIL, see IEC 61508), e.g. Residual Error Probability <= 10-7 errors/h for SIL 3 (Neumann, 2002). The communication part requires a residual error probability of <= 10-9 errors/h for SIL 3 (1% of 10-7; the other 99% is required for sensors, PLCs, and actuators etc.).
Measures to avoid the influence of the failures (falsified addresses, loss or damage of data, delay) could be: Numbering of transmitted data,
observation of transmission time (time expectation), authentication using passwords (identification), optimised CRC (redundancy).
One principle to use communication systems for safety applications is to consider the communication channel as a so-called "Black Channel" (IEC 65C/307/NP, 2003). That means all safety-related functions will be realised outside (on top of) the communication layers. The safety measures will be realised in a separated safety layer that is situated between communication protocol and application. With this principle, an existing communication system can be used as it is – and with existing components like ASICs, cables, connectors, repeaters, links etc. as far as they meet the electrical requirements. Such solutions exist for several fieldbus protocols. The background has to be mapped to Internet-based Virtual Automation Networks using heterogeneous networks. Investigations are necessary on how TCP/IP-oriented networks can be extended for the transfer of safety-related data in automation systems.

IEC 61508 defines Safety Integrity Levels (SIL) for electrical, electronic and programmable electronic devices, which contain the permissible residual error probabilities and error detection rates. It is necessary to prove that the used VAN safety codes fulfil these requirements - especially for different types of communication media. The results should be introduced in the proposed new IEC standardisation activities on "Profiles for functional safe and secure communications in industrial networks" (IEC 65C/307/NP, 2003).

3.3 Security

Today's automation islands are relatively secure against attacks. In the case of Internet usage, the reasons of growing attack probability are:

- remote access to these systems is available to many people (advantage),
- but also the remote access enables attacks (disadvantage) by using public telecommunication systems and Internet.

The usage of Internet-based technologies compared with today's automation islands requires a common security concept for distributed automation using Virtual Automation Networks. (Schneider, 2002; Höne *et al.*, 2001; Eugene Schultz , 2002; Schwaiger *et al.*, 2003; Palensky, 2000; Naedele *et al.*, 2001; Fritz *et al.*, 2002; Fuhrberg, 2000) deal with several aspects of security in communication systems.
Security criteria are

- Confidentiality of information (to protect against access by unauthorised third parties)
- Integrity of information (Detection of unauthorised modification or manipulation of message contents with a specified level of confidence)
- Timeliness or sequentiability of message delivery (Detection of any unauthorised message re-timing, re-sequencing or replay of prior messages)
- Authentication and authorisation of communication peers (Verification of identity of a communication peer and that peer's entitlement)
- Availability of information.

The question is how TCP/IP-oriented networks have to be secured under specific automation conditions such as low resources and easy to use configuration in production environments. For that purpose security risk analysis for production networks is necessary. Typical types of assaults are: eavesdropping, hacker attacks, data corruption, unauthorised access, espionage, sabotage, loss, theft, and operator errors.
Many arguments are given why defence-in-depth with multiple, staged, complementary security mechanisms is the suitable approach. There are different abstraction levels and security zones within a Virtual Automation Network. Different types of mechanisms against attacks by remote access are used concurrently around and inside each security zone to defend it. The outer zones contain less valuable targets. The safety-critical automation system is in the innermost zone. Investigations

should lead to the designation of VAN security zones (e.g. Remote Access Zone, Local Plant Operator Zone, Automation Device Zone), and to existent security mechanisms used in all or specific zones (e.g. authorisation, intrusion detection, response, mechanism protection, VAN specific security mechanisms). (Naedele, 2003) describes the approaches in principle.

In addition to defence mechanisms, also detection mechanisms could be helpful which allow the automation system operators to detect attacks, and reactive mechanisms and processes to actively defend against them. Each zone also has to spend time to detect and fend off the attacker.

3.4 Local Awareness in Industrial Automation

The usage of Information Technologies within the life cycle of an automation system is becoming broader. The Engineering in the various phases (design, development, manufacturing and commissioning of automation devices and systems, maintenance of the system) is becoming more efficient using formal methods to develop information models and description languages (describing the features of the components for all the mentioned phases) and computer aided tools to handle these features. An important requirement is a context-sensitive offer of the needed information at these tools . Since the needed database is distributed through the decentralised automation system (connected by the heterogeneous network) it also requires location-based services. The recent approaches use web technologies to organise access to the remote data. Since the remote access raises the probability of attacks to the system, the security becomes more important. Additionally, a unified data structure describing the features of the installed devices has to be standardised. At the moment, there are various profiles in the fieldbus domain. These profiles have to be mapped to the heterogeneous networks.

4 CONCLUSIONS

Future scenarios of distributed automation require more mechanisms for the geographical distribution of automation functions. The resulting automation system has to offer location-based and context-sensitive services to guarantee suitable local and remote functions for different user needs (tele-operation, tele-service) and requires real-time data transmission, safety and security mechanisms.

It requires a special communication network – a Virtual Automation Network. A Virtual Automation Network (VAN) is a heterogeneous network consisting of wired and wireless local Area Network, the Internet, and wired or/and wireless telecommunication systems. This means that geographically distributed application programmes co-operating to fulfil a control application are connected via this VAN accessed by remote connection endpoints. Worldwide distribution of Internet offers the Automation domain a good infrastructure but introduces many additional problems which need to be solved.

REFERENCES

Agarwal, Anjali and Kang Bin Wang (2001): Supporting Quality of Service in IP multicast networks. in Computer Communications, Volume 26, Issue 14, Pages 1533-1540.

Alves, M.; Tovar, E.; Vasques, F. (2000) : Ethernet Goes Real-Time : a Survey on Research and Technological Developments. Techn. Rep. HURRAY-TR-0001, Polytechnic Institute of Porto, Jan. 2000.

Baek-Young Choi, Sejun Song, Birch,N. and Jim Huang (2000). Probabilistic approach to switched Ethernet for real-time control applications, Seventh International Conference on Real-time Computing Systems and Applications, 2000, Proceedings.

Bonaccorsi, A.; Lo Bello, L.; Mirabella, O.; Neumann, P.; Pöschmann, A. (2003): A Distributed Approach to Achieve Predictable Ethernet Access Control in Industrial Environments. 5th IFAC International Conference on Fieldbus Systems and their Applications (FET 2003), Aveiro, Proceedings, pp. 173 - 176.

Caponetto, R.; L. L. Bello; O. Mirabella (2003): Experimental Assessments of Fuzzy Smoothers for Ethernet Networks. 15th Euromicro Conference on Real-Time Systems, Porto 2003. Proceedings, pp. 57-60

Carpenzano, A.; Caponetto, R.; Lo Bello, L.; Mirabella, O. (2002): Fuzzy Traffic Smoothing: an Approach for Real-Time Communication over Ethernet Networks. 4th IEEE International Workshop on Factory Communication Systems. WFCS'02, Västeras.

Chien-Chao Tseng; Gwo-Chuan Lee; Ren-Shiou Liu (2001): HMRSVP: a hierarchical mobile RSVP protocol, International Conference on Distributed Computing Systems, 2001, Page(s) 467 -472.

Decotignie J.-D. (2002): Wireless Fieldbusses - A Survey of Issuses and Solutions. 15th Triennial World Congress of the International Federation of Automatic Control, Barcelona 2002

Diedrich, Ch.; Wollschläger, M.; Hintze, E.(2003): UML specification and developement of safety-relevant industrial communication systems. 4th MATHMOD - 4th IMACS Symposium on Mathematical Modelling, Proceedings.February 5-7, 2003, Vienna University of Technology, Austria.

EN 50159: Railway Applications: Requirements for Safety-Related Communication in Closed Transmission Systems

Ethernet/IP specification (2001), Release 1.0. June 5, 2001, ©2000, 2001 by ControlNet International and open DeviceNet Vendor association

Eugene Schultz (2002): Security Views. In Computers & Security, Volume 22, Issue 5, Pages 368-377.

Feld, J. (2003): Real-time Communication in PROFInet V2 and V3 Designed for Industrial Purposes. 5th IFAC International Conference on Fieldbus Systems and their Applications (FET 2003), Aveiro, Proceedings, pp. 291 - 296.

Fritz, R.; Halang, W. (2002): Sichere Abwehr von Viren. 2002, ISBN 3-89577-266-6

Fuhrberg, K. (2000): Sicherheit im Internet. Published as Online documentation of Bundesamt für Sicherheit in der Informationstechnik. www.bsi.bund.de/literat/doc/sinetdoc/sinetstd.htm. 1997, Rev. 02.03.2000.

Georges, J.-P., E.Rondeau and T. Divoux (2002). Evaluation of switched Ethernet in an industrial context by using the network calculus. IEEE 4th Workshop on Factory Communication Systems WFCS'02, Västeras, 2002,

Halang W.; Konakovsky, R (1998).: Sicherheitsgerichtete Software. Automatisierungstechnik 46, Sonderheft Sicherheitsgerichtete Automatisierungstechnik 2, S. 93 - 103, Oldenbourg Verlag, 1998.

Halang, W. A.; M. Colnaric (2002): Dealing with Exceptions in Safety-Related Embedded Systems. 15th Triennial World Congress of the International Federation of Automatic Control, Barcelona 2002

Höne, Karin and J. H. P. Eloff (2001): Information security policy -- what do international information security standards say? In Computers & Security, Volume 21, Issue 5, Pages 402-409.

HSE: High Speed Ethernet (2001). Specification documents FF-801, 803, 586, 588, 589, 593, 941. Fieldbus Foundation, Austin 2001

IDA (2002): Interface for Distributed Automation, Architecture Description and Specification, Revision 1.1, Nov. 2002, IDA Group e. V., Blomberg

IDA Group (2002): IDA - Interface for Distributed Automation, Architecture Description and Specification. Revision 1.1, November 2002

IEC 2002: IEC 61158 series: Digital data communication for measurement and control - Fieldbus for use in industrial control systems.

IEC 2003: IEC 61784 First edition: Profile sets for continuous and discrete manufacturing relative to fieldbus use in industrial control systems.

IEC 61158-5 Ed 3: Digital data communication for measurement and control. Fieldbus for use in industrial control systems. Part 5: Application layer service definition

IEC 61158-5: Digital data communication for measurement and control. Fieldbus for use in industrial control systems. Part 6: Application layer protocol specification

IEC 61508: Functional Safety of Electrical/Electronic/Programmable El. Safety-Related Systems

IEC 65C/307/NP (2003): Digital data communications for measurement and control - Profiles for functional safe and secure communications in industrial networks, New Work Item Proposal, 2003

Jasperneite, J., P. Neumann (2001). Switched Ethernet for Factory Communication. In: Proceedings of the IEEE conference on Emerging Technologies and Factory Automation (ETFA'01). Antibes Juan-les-pins, France, pp. 205-212

Jasperneite, J.; Neumann, P. (2001a): Switched Ethernet for Factory Communication. Proceedings of EFTA2001, 8th International Conference on Emerging Technologies and Factory Automation, Antibes, Oct. 2001.

Jasperneite, J.; Neumann, P. (2001b): Performance Evaluation of Switched Ethernet in Real-Time Applications. Proceedings of the 4th FET2001, International Conference on Fieldbus Systems and their Applications, Nancy, Nov. 2001, pp. 144-151.

Jasperneite,J. (2002): Leistungsbewertung eines lokalen Netzwerkes mit Class-of-Service Unterstützung für die prozessnahe Echtzeitkommunikation. Shaker Verlag, Aachen 2002. ISBN 3832208321.

Mahmoodian, A.; Haring, G.(2000): Mobile RSVP with dynamic resource sharing, IEEE Wireless Communications and Networking Conference, 2000, Page(s) 896 -901.

Mataix, C.; P. Martin; F.-J. Rodriguez; M. J. Manzano ; J. Pozo ; P. G. Donato (2003): Design and Implementation of a Safety Communication Network in Railways with Intelligent Fault Diagnosis. IEEE Conference on Emerging Technologies and Factory Automation, Lisbon 2003. Proceedings, pp. 109-112

Meindl, A. (2002): Ethernet Powerlink, Application in Practice. PRAXIS ProfiLine - Industrial Ethernet. April 2002, pp.55-57

Moon, B.; Aghvami, H. (2001): RSVP extensions for real-time services in wireless mobile networks, IEEE Communications Magazine, Volume: 39 Issue: 12 , Dec. 2001, Page(s) 52 -59.

Naedele, M.(2003): IT Security for Automation Systems, Motivations and Mechanisms. Automatisierungs-technische Praxis 45 (2003), H. 5, S 84-91

Naedele, M.; D. Dzung; M. Stanimirov (2001): Network Security for Substation Automation Systems. In Computer Safety, Reliability and Security. Proceedings Safecomp 2001, LNCS 2187. Springer, Berlin 2001

Neumann, P. (2202): Merging Fieldbus and Telecommunication Systems in the Industrial Automation Domain. IEEE AFRICON 2002, 6th AFRICON conference in George, ZA, Proceedings Vol.1, pp.197

Palensky, P.; T. Sauter (2000): Security considerations for FAN Internet connections. International Workshop on Factory Communication Systems, 2000

Pöschmann, A.; Werner, Th. (2003): Studie "Ethernet in der Automation", ZVEI, Frankfurt, 2003

PROFIBUS Guideline (2002). PROFInet Architecture Description and Specification. Version V 1.9, PNO Karlsruhe, 2002.

PROFIBUS Guideline (2003), "PROFInet Architecture Description and Specification", Version V 2.0, Januar 2003, PNO Karlsruhe
Rauchhaupt, L. (2002): System and Device Architecture of a Radio Based Fieldbus - The RFieldbus system. 4th IEEE International Workshop on Factory Communication Systems, 2002, Västeras, Proceedings pp. 185-192.
Rauchhaupt, L. (2003): RFieldbus a Survey. 5th IFAC International Conference on Fieldbus Systems and their Applications (FET 2003), Aveiro, Portugal, 2003, Proceedings, pp. 105-114.
Real-Time Innovations Inc (2002). Real-time Publish-Subscribe Wire Protocol Specification. http: //www.rti.com/products/literature.html
Schneider, B. (2002); Secrets and Lies - Digital Security in a Networked World. Wiley (2000)
Schwaiger, C.; A. Treytl (2003): Smart Card Based Security for Fieldbus Systems. IEEE Conference on Emerging Technologies and Factory Automation, Lisbon 2003. Proceedings, pp. 398-406
Seok-Kyu Kweon; Shin,K.G. and Zin Zheng (1999). Statistical real-time communication over Ethernet for manufacturing automation systems, 1999 fifth real-time technology ans applications symposium, Proceedings, 2-4 June 1999.
Wen-Tsuen Chen; Li-Chi Huang (2000): RSVP mobility support: a signaling protocol for integrated services Internet with mobile hosts, IEEE INFOCOM2000, Page(s) 1283 -1292.
Xinggang Fan, Zhi Wang and Youxian Sun (2002). How to guarantee factory communication with switched Ethernet: survey of its emerging technology, Industrial Electronics Society, IEEE 2002 28th Annual Conference, 3, Nov. 2002

Copyright © IFAC Information Control Problems in
Manufacturing, Salvador, Brazil, 2004

ELSEVIER
IFAC
PUBLICATIONS
www.elsevier.com/locate/ifac

THE MOBILE CONTROLLED RFID SYSTEM

Mikael Soini, Teppo Eeva, Lauri Sydanheimo, Markku Kivikoski

Tampere University of Technology, Rauma Research Unit
Kalliokatu 2, FIN-26100 Rauma, FINLAND
E-mail: mikael.soini@tut.fi

Abstract: This paper introduces the concept of mobility when controlling and monitoring RFID (Radio Frequency Identification) systems. Mobile controlled RFID systems enable the remote tracking of RFID tags. The RFID system automatically sends information about the tag movements to the mobile phone. The user can also find out which tags are within the operating range of known RFID readers. Mobile connection between the user and the RFID system is made through the GSM (Global System Mobile communications) network's packet-switched GPRS (General Packet Radio System) service. IP-based (Internet Protocol) communication format allows the system to be applied in different cases and enables the use of other wireless or wired connections. *Copyright © 2004 IFAC*

Keywords: Automation, Communication interface, Control, Mobile, Monitoring, Telecommunication, Tracking

1. INTRODUCTION

When developing supply chain management in order to improve efficiency and cost-effectiveness, automation is a dominant factor. An essential part of automation is identification. Radio Frequency Identification (RFID) is a potential solution for the near future and will in time at least partially replace manual solutions, e.g. bar codes. RFID tag features such as non-line of sight operation, reprogram ability, simultaneous reading of multiple tags, fast read speeds, increased data storage capacity and operating distances of several metres are the parameters that makes RFID superior to the bar code (Raza 1999). On the other hand the RFID tag price is still much higher than the price of the bar code which justifies continued use of bar codes in basic situations such as identification in consumer goods sales. RFID is not a direct replacement of the bar code because it can be used in other, more intelligent applications. These include solutions associated with material flow, warehouse or supply management, wireless commerce, logistics, user identification, access control, burglar alarms, door and gate openings, toll payments and tracking equipment or materials (Stanford, 2003).

A traditional RFID system (figure 1) contains normally three parts: RFID tags, a RFID reader and a computer that controls the RFID reader. The implementation is usual local with a serial port connection between the reader and the host.

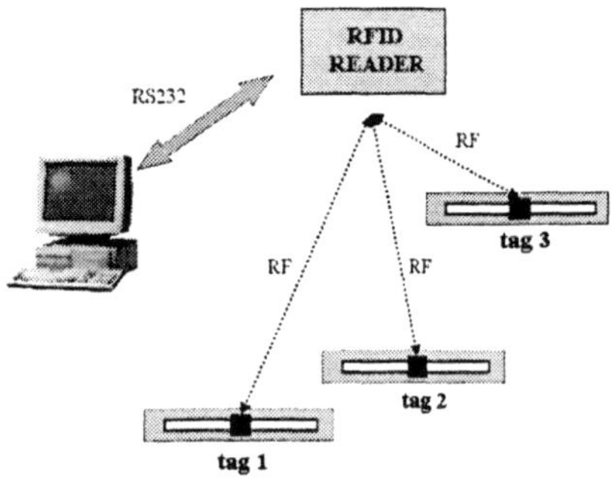

Fig. 1. In a typical RFID system the control and monitoring by a PC connected to the RFID reader by serial port (RS-232).

This paper presents the properties of the mobile controlled RFID system that differs from the above-mentioned traditional system in that it can be controlled and monitored remotely with a mobile station such as a mobile phone (figure 2). In this system the control commands and responses are transmitted through the GSM (Global System Mobile communications) network and using the GSM network's GPRS (General Packet Radio System) service. The packet-switched GPRS increases the data transfer capacity of the GSM network and so improves the system's usability. The developed universal set of control commands and responses between the RFID system and mobile control device flow through the Internet. The host server that is the mobile gateway server handles the user's commands and controls the RFID reader. The RFID reader communicates with multiple passive backscattering RFID tags.

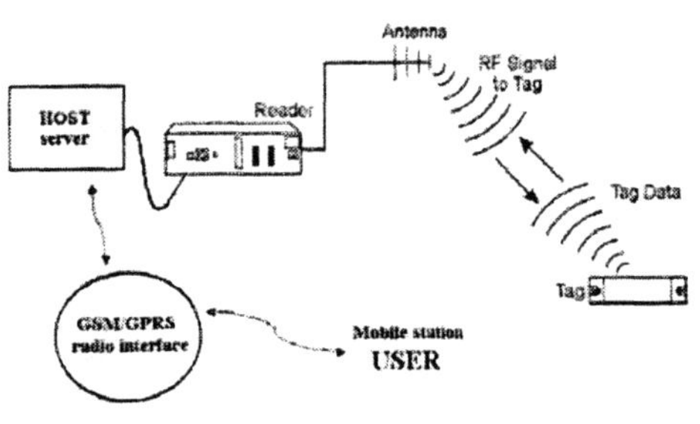

Fig. 2. The basic structure of the mobile controlled RFID system. The RFID reader communicates with passive backscattering tags. The host server or mobile gateway server controls the reader. This server receives commands remotely from the user.

This paper introduces the mobility in this system, system structure and its operation together with the newly developed communication interface. Applications based on these standard and global approaches allows the developed systems to be used in variety of ways. It is also the only way to achieve common acceptance of the RFID technology. In the final section the conclusions and future prospects are presented.

2. TELECOMMUNICATIONS AND MOBILITY IN THE CONTROL AND MONITORING APPLICATIONS

GSM (Global System Mobile communications) is the world's most widespread digital mobile phone standard. GSM is a European standard for cellular circuit-switched communications developed by the European Telecommunications Standards Institute (ETSI). GPRS is a technology which uses the existing GSM networks radio interface offering packet-switched networks such as the Internet. GPRS is the first step towards wireless all-IP (Internet Protocol) or 3G systems and is a developing environment with improving quality and reliability in wireless mobile applications. (Ruohonen, 2004)

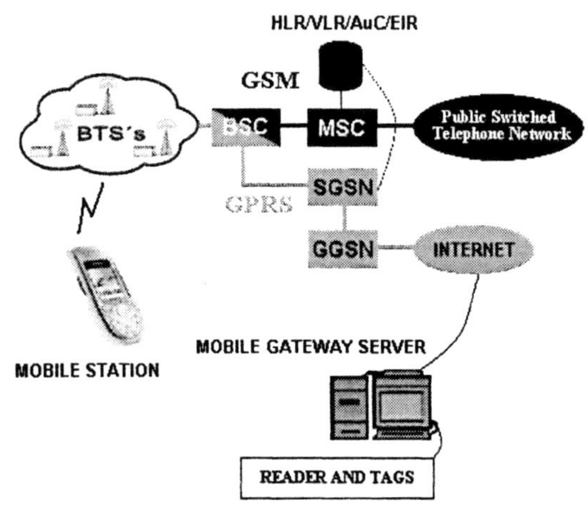

Fig. 3. Mobile control through the GSM network's GPRS service. The GPRS connection between the parties uses the basic building blocks of the GSM network; base transceiver stations (BTS), base station controllers (BSC) and different registers. Service GPRS support nodes (SGSN) and gateway GPRS support nodes (GGSN) bring IP-based Internet features to the GSM mobile phone system. (Samjani, 2002)

Packet-switched GPRS service improves the usability of mobile solutions. GPRS devices are always connected to the network, which saves the time consumed with dial-up-connection formalities in GSM applications. GPRS increases data transmission speed compared to GSM data which also enhances usability. Recently developed, more user-friendly mobile phones with big and colourful displays, fast-running applications and devices permanently connected to the network are the qualities which make the remotely controlled solutions important in logistics and manufacturing environments.

Figure 4 below shows the user interface (UI) developed for this system. The user can choose to follow-up particular tags, remove some tags from the follow-up or read the tag list from the mobile gateway server. Adding to or removing from the follow-up list is achieved by entering the tag's serial number in the appropriate place. This system contains an additional feature that enables the user to control and monitor digital outputs and inputs. This feature gives more opportunities to the user for all-inclusive control and monitoring of automation systems. Different qualities of systems mentioned above are discussed in detail in section 3.

Fig. 4. Displays of the mobile station used in this mobile controlled RFID system. The photo on the left shows the main display of the UI and that on the right shows how the user can add a tag (serial number) to the follow-up list.

GPRS is a time division multiple access-based cellular system that enables the packets to be transferred in end-to-end packet mode communications. A user who wants to transmit data from a mobile station to server must compete with the other to gain access to the medium. The GPRS service then allocates channel resources consistent with the amount of data to be transferred (Chitre 2003). In this way the GPRS does not need a "fixed" connection or constant time slot from the GSM radio interface for the communication because data is sent in small addressed packets or bursts that can use different time slots. This is why invoicing is not based on connection time as in GSM but on the amount of transferred data (volume based charging). This leads to savings in phone costs.

A GPRS device receives a random IP-address from GPRS-backbone in every data connection phase. The GPRS device works as a normal IP-network device, which enables the unconstrained use of mobile solutions that are connected to the Internet or intranet. This quality enables control and monitoring different systems to be connected to a company's LAN (Local Area Network) from a distance through the IP network. GPRS is transparent from the user point of view so only the IP-network connected is apparent to the user.

GPRS is a step towards wholly packet-switched mobile systems that offer more transmission capacity than circuit-switched systems such as GSM. This leads to improved and more versatile mobile solutions, which can be used also in remote control and monitoring applications. The next steps in mobile technology are EDGE (Enhanced Data rates for GSM Evolution) and UMTS (Universal Mobile Telecommunications System). UMTS offers transmission speeds that enable mobile video transfer to be used in, for example, monitoring, identification and access control applications.

The use of mobile networks and in this case GPRS is one way to control and monitor this advanced RFID system remotely. Bluetooth and WLAN (Wireless Local Area Network) devices can also be exploited in this system. With these techniques the data does not flow through the mobile phone operator networks so the data transmission is free of charge. Naturally the transmission distances with Bluetooth and WLAN are only tens or hundreds of metres in length which limits the operating range to perhaps inside a building. In some cases it is practical to use a desktop or laptop connected to the Internet in the company's own local area network (LAN) or from another location. When using LAN or WLAN for data transmission the use of a graphic web-based user interface can be exploited for control and monitoring of the RFID system. The use of PDA (Personal Digital Assistance) devices are possible when using WLAN. In the case of Bluetooth and GPRS it is possible to use a mobile phone or a computer for control and monitoring. The use of these short-range wireless techniques or wired Internet (figure 5) can be put into effect easily because the communication in this system is independent of used transmission techniques. This communication format will be discussed in more detail in section 4.

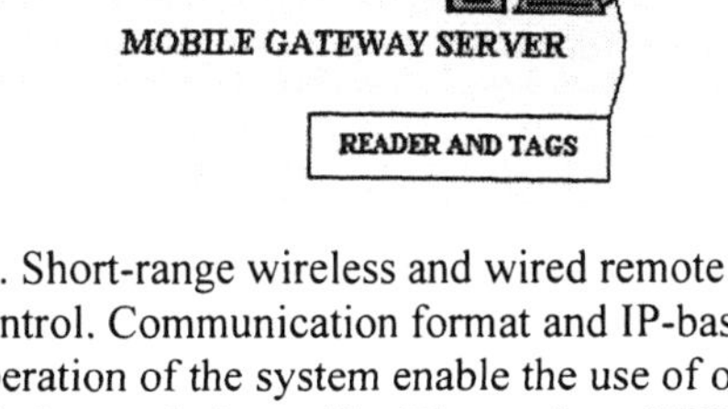

Fig. 5. Short-range wireless and wired remote control. Communication format and IP-based operation of the system enable the use of other wireless techniques like Bluetooth and WLAN or wired Internet in the control procedure.

3. THE MAJOR PARTS OF THE SYSTEM AND SYSTEM OPERATION

This mobile controlled RFID system consists of a mobile station, mobile gateway server, RFID reader and passive RFID tags. RFID tags used in this system operate on 915 MHz frequency band. These tags communicate with RFID readers according to the Auto-ID Center's UHF (Ultra High Frequency) specification. This open specification enables the development of features and performance required in track and trace, logistics, and supplying chain applications. RFID tag operating distances are a couple of metres and tags obtain operating energy from the reader's signal (passive backscattering technique) (Alien, 2003). The use of these UHF frequencies gives larger operating ranges and smaller antenna sizes compared to lower operating frequencies (135kHz or 13,56 MHz). Also electrical interference sources like the electro motors in industrial environments can be avoided in UHF frequencies (Nilsson 2002). On the other hand careful reader antenna placement and orientation with these high frequency radio waves is important because metals and liquids strongly disturb the communication between the tags and the reader.

The reader-tag network architecture supports point-to-multipoint communication. The reader integrates seamlessly into existing IT infrastructure using standard network protocols. An RFID reader is connected to the intranet and it can be controlled with a mobile gateway server through a TCP/IP (Transmission Control Protocol/Internet Protocol) network or serial port (RS-232) connection. In simple systems where the distance between the server and the reader is not too great it is possible to use a RS-232 connection. In situations where there are multiple readers connected to the server and/or the distances between these readers and the server are more than some tens of metres the use of a TCP/IP connection is inevitable.

The RFID reader keeps a list of every RFID tag detected in its reading range. This list is updated continuously when a tag arrives or leaves the reading range. All changes are sent immediately to the mobile gateway server. The user can operate the RFID reader through the mobile gateway server.

Consequently the mobile gateway server works as a connector between the user or mobile station and the RFID reader. The server transforms the commands from the mobile station and transfers them to the RFID reader. As mentioned above the RFID reader sends tag list changes to the server. In addition to this the server reads a tag list from the RFID reader every 30 seconds to ensure that the system is working properly. Using the tag list data, the server looks for any changes in state-of follow-up tags. If there are changes a message is sent automatically to the user.

When the mobile station user has a connection to the mobile gateway server the following operations are possible: Reading the reader's up to date tag list from the server, setting follow-ups to certain tags or removing follow-up tags, reading digital inputs, and controlling or reading digital outputs. This user operation with mobile gateway server is discussed further in section 4.

The sequence chart (figure 6) below shows an example of the operation of this system. Five parties are shown participating in the system operation. The user can define a set of tags to follow-up. Tags are entered on the follow-up list by writing the serial number of the tag to be followed (figure 4) and sending it to the server. After this the user receives information from the mobile station automatically when a tag or tags leave from or arrive in the area. If a tag is not in the user's follow-up list then the update is only visible at the server end. Information about this event can be obtained from the server by requesting a tag list.

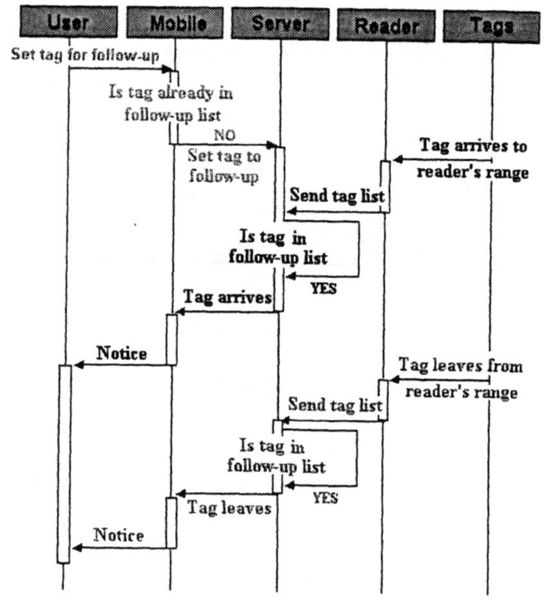

Fig. 6. Sequence chart when setting a RFID tag to a follow-up list. The server keeps a list of the follow-up tags set by the user. The server informs the user when a tag or tags arrive in or leave from the readers operating area.

The tag follow-up list is also maintained in the mobile station. This assists in removing the tags from the follow-up list. The user simply picks the serial number of the removed tag from the menu of the mobile station. This is sent to the server that updates its follow-up list. It is also possible to keep the whole tag list in the mobile phone to help enter tags on the follow-up list. Thus the user could choose a tag to be followed-up straight from the mobile phone's menu. This however does not work if a tag is not at the moment of set-up in the reader's operating range. In the following example the tag serial number must be notified in advance.

A person A sends a packet to a person B. B is spending time in the countryside. Packet delivery is critical so A sends an e-mail to B that a packet (tag serial number X) will be delivered to the local postal office some time tomorrow. B regularly reads his e-mails by GPRS-enabled phone and when he gets A's message he labels the packet (the tag) for follow-up. When the packet arrives at the post office's loading platform reader, B is notified automatically by a message to his mobile phone from the mobile gateway server that his packet is ready for collection.

4. THE GLOBAL COMMUNICATION INTERFACE

When planning the interface or communication format between the mobile station and the RFID system the main goal was to achieve compact, simple and comprehensive communication format. Compact format minimizes the amount of transferred GPRS data and thus it reduces costs of the use. This comprehensive communication is flexible that makes the addition of new commands possible in the future development.

The mobile gateway server works as an interface between the mobile station and RFID system. Communication between these parties is based on connection-oriented TCP/IP (Transmission Control Protocol/Internet Protocol). The set of universally used commands over this IP-network enables the use of IP-capable Bluetooth, WLAN and wired LAN in the system control.

The mobile gateway server transforms universal control commands from the mobile station suitable for the RFID reader and vice versa. It is possible to build the server to communicate with different manufacturers' readers. In this case the system forms a universal communication format from top to bottom, that is, from the tags to the user. This is a very important quality of this developed system because it makes the use of the system possible with different manufacturers devices and their various communication formats.

The following table 1 introduces the commands that can be used for control and monitoring of the RFID system. The use of connection-oriented TCP eliminates the need to transmit any acknowledgements to commands.

Table 1. Messages in tag management, tag follow-up, I/O port controlling and connection failure situations in the system operation

Tag management:	**Explanations:**
GETLIST	Get tag list from the mobile gateway server
GETCOUNT	Get list of active tags that are followed-up from the gateway server
Tag follow-ups:	**Explanations:**
SETFUP	Set a tag to follow-up list of the server. Serial number as a parameter.
RMTFUP	Remove a tag from follow-up list of the server. Serial number as a parameter.
FOUNDFU	Notification by the server of a follow-up tag's arrival in the reader's range. Serial number as a parameter.
DEPFU	Notification by the server of a a follow-up tag's departure from the reader's range. Serial number as a parameter.
I/O ports:	**Explanations:**
SETEXTOUT	Set reader's digital output to on/off
GETEXTOUT	Read reader's digital outputs
GETEXTINT	Read reader's digital inputs
Errors:	**Explanations:**
CERROR	Communication failure. Parties cannot identify a command. Error code as a parameter

The following tables 2,3 and 4 introduce tag management and tag follow-up commands between the client (the user) and the server (mobile gateway server) in detail.

Table 2. Tag management command. The user requests a tag list from the mobile gateway server. The server replies with a list of tags in the RFID reader's operating range

Client:
SendCommand (GETLIST, 0)

Server:
SendCommand (GETLIST, "8000 8004 0000 0004, 8000 8004 F103 0010, 8000 8004 0100 023F")

Table 3 Tag follow-up commands. The client sets *(SETFUP)* or removes *(RMTFUP)* a tag *(8000 8003 12FE 1ACE)* to or from a server's follow-up list.

Client: *SendCommand (SETFUP, "8000 8003 12FE 1ACE")* **Client:** *SendCommand (RMTFUP, "8000 8003 12FE 1ACE")*

Table 4 Tag follow-up commands. The server informs the user that the reader has found *(FOUNDFU)* a tag *(8000 8003 12FE 1ACE)*, set for a follow-up or that the reader detected the tag departure.

Server: *SendCommand (FOUNDFU, "8000 8003 12FE 1ACE")* **Server:** *SendCommand (DEPFU, "8000 8003 12FE 1ACE")*

5. CONCLUSION

This paper introduced the mobile controlled RFID system that can be used in many control and monitoring procedures. The user can operate the RFID system remotely with a mobile station through the GSM networks IP-based GPRS service. The connection is TCP/IP-based also from the mobile gateway server to the RFID reader. Communication on top of connection-oriented TCP/IP connection is made with a developed universal set of commands that can be used with different manufacturers' devices. Reader and passive backscattering tags communicate with the Auto-ID Center's open specifications. This standard approach, from top-to-bottom, allows the system to be used in various cases. Standard approach also makes the use of different IP-based transmission techniques (Bluetooth, WLAN and LAN) possible in some cases as an alternative to GPRS.

This tag tracking solution can be exploited in many cases such as manufacturing processes, warehouse management, logistics, and security. The next development will be to enable multiple users in the system. Further improvements may be made by increasing utilization of the growing number of RFID readers and tags in the system. The creation of chains of RFID readers makes more accurate and efficient positioning applications possible.

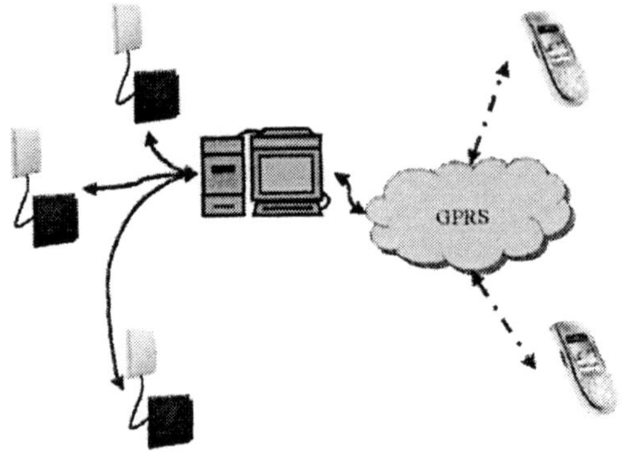

Fig. 7. Multiple users control and monitor multiple tag readers and tags in their operating range.

REFERENCES:

Ruohonen, T., Raumonen, P., Soini, M., Sydanheimo, L. and Kivikoski M. (2004). Quality and reliability of GPRS connections. To Appear: 2004 IEEE Consumer Communications and Networking Conference (CCNC2004), Las Vegas, Nevada, USA, 2004.

Samjani, A. (2002). General Packet Radio Service (GPRS). *IEEE Potentials*, **April/May 2002**, 4 pages.

Alien technologies, RFID implementing (2003). http://www.alientechnology.com/mambo_alien/index.php.

Stanford, V. (2003). Pervasive Computing Goes the Last Hundred Feet with RFID Systems. *IEEE Pervasive Computing*, **April-June 2003**, 9 pages.

Nilsson, M., Hallberg, J. (2002). Positioning with Bluetooth, IrDA and RFID, pp. 16-21. *Master Thesis*, Lulea University of Technology, Department of Computer Science and Electrical Engineering.

Raza, N., Bradshaw, V., Hague, M. (1999). Applications of RFID Technology. *RFID Technology* (Ref. No. 1999/123), IEE Colloquium on, 25 Oct. 1999. Pages: 1/1 -1/5

Chitre, V.A., Daigle, J.N. (2003). Performance of IP-based services over GPRS. *IEEE Transactions on Computers*, June 2003, **Volume 52**, Issue 6, Pages 727 -741

Copyright © IFAC Information Control Problems in
Manufacturing, Salvador, Brazil, 2004

ELSEVIER
IFAC
PUBLICATIONS
www.elsevier.com/locate/ifac

SERVER-BASED REAL-TIME COMMUNICATION ON CAN

Thomas Nolte, Mikael Nolin, and Hans Hansson

Mälardalen Real-Time Research Centre
Deptartment of Computer Science and Engineering
Mälardalen University, Västerås, SWEDEN
Email: {thomas.nolte, mikael.nolin, hans.hansson}@mdh.se

Abstract: This paper presents a share-driven scheduling protocol for networks with real-time properties. The protocol provides fairness and bandwidth isolation among predictable as well as unpredictable streams of messages on the network. The need for this kind of scheduled real-time communication network is high in applications that have requirements on flexibility, both during development for assigning communication bandwidth to different applications, and during run-time to facilitate dynamic addition and removal of system components.
The share-driven scheduling protocol is illustrated by applying it to the popular Controller Area Network (CAN). Two versions of the protocol are presented together with their associated timing analysis. *Copyright © 2004 IFAC*

Keywords: fieldbus networks, real-time, scheduling, CAN, controller area network

1. INTRODUCTION

In optimising the design of a real-time communications system it is important to both guarantee the timeliness of periodic messages and to minimize the interference from periodic traffic on the transmission of aperiodic messages. Therefore, the usage of *server-based scheduling techniques* is proposed to handle streams of both periodic and aperiodic traffic. The servers proposed in this paper are using the Earliest Deadline First (EDF) scheduling strategy, since EDF allows optimal resource utilisation. Examples of such server-based scheduling techniques are the Total Bandwidth Server (TBS) (Spuri and Buttazzo, 1994; Spuri *et al.*, 1995), and the Constant Bandwidth Server (CBS) (Abeni, 1998).

In this paper, server-based scheduling is applied to the Controller Area Network (CAN) (CAN, 1993), which is one of the more common communication networks used today in the automotive industry as well as in several other application domains that have real-time requirements. CAN is implementing fixed-priority scheduling using a bit-wise arbitration mechanism in the Medium Access Control (MAC) layer. This mechanism avoids collisions in a deterministic way and is amenable to timing analysis (Tindell *et al.*, 1995).

Using server-based scheduling techniques for CAN, fairness among users of the network is guaranteed (e.g., "misbehaving" aperiodic processes cannot aversively interfere with well-behaved processes). This is desirable in, e.g., an open architecture where the adding and removing of network users is controlled by an online admission control. In contrast with other proposals for CAN scheduling, aperiodic messages are not sent "in the background" of periodic messages or in separate time-slots (Pedreiras and Almeida, 2000). Instead, aperiodic and periodic messages are jointly scheduled using servers. This substantially facilitates meeting response-time requirements, both for aperiodic and periodic messages.

Share-driven communication on CAN is presented previously (Nolte *et al.*, 2003). The main contributions of this paper are:

- A worst-case response-time analysis for the previously proposed protocol, bounding the maximum delay for a server to send a message.
- An improvement of the previously proposed protocol that, under many network configurations, can give lower worst-case response-times.
- A worst-case response-time analysis for the improved protocol.
- A comparison of the worst-case response-times of the two protocols.

The paper is organized as follows: Section 2 presents related work. In Section 3, server-based CAN is presented together with its corresponding timing analysis, and in Section 4 the modified mechanism is presented together with its corresponding timing analysis. Finally, conclusions are presented in Section 5.

2. BACKGROUND AND RELATED WORK

In this section, CAN and previously proposed CAN scheduling methods are presented.

2.1 The Controller Area Network

The Controller Area Network (CAN) (CAN, 1993) is a broadcast bus designed to operate at speeds of up to 1Mbps. CAN is extensively used in automotive systems, as well as in other applications.

CAN is a collision-avoidance broadcast bus, which uses deterministic collision resolution to control access to the bus (so called CSMA/CA). The frame identifier is required to be unique, in the sense that two simultaneously active frames originating from different sources must have distinct identifiers. Besides identifying the frame, the identifier serves two purposes: (1) assigning a priority to the frame, and (2) enabling receivers to filter frames.

CAN guarantees that the highest priority active frame will be transmitted. Hence, CAN behaves like a priority-based queue.

2.2 Scheduling on CAN

Several different types of scheduling methods exists for CAN, where FPS scheduling is the most natural scheduling method since it is supported by the CAN protocol. Response-time tests for determining the schedulability of CAN frames under FPS have been presented (Tindell *et al.*, 1995). This analysis is based on the standard fixed-priority response-time analysis for CPU scheduling (Audsley *et al.*, 1993). TT-CAN (TT-CAN, 2000) provides *time-driven* scheduling for CAN, and Flexible Time-Triggered CAN (FTT-CAN) (Almeida *et al.*, 1999) supports priority-driven scheduling in combination with time-driven scheduling. FTT-CAN is presented in more detail below. The share-driven approach has for CAN been introduced earlier (Nolte *et al.*, 2003). Later, in section 3 and 4, server-based scheduling of CAN is presented in detail.

2.2.1. Flexible Time-Triggered Scheduling To combine event-triggered traffic with time-triggered traffic in CAN, FTT-CAN (Flexible Time-Triggered communication on CAN) is proposed (Almeida *et al.*, 1999; Pedreiras and Almeida, 2000). In FTT-CAN, time is partitioned into Elementary Cycles (ECs) which are initiated by a special message, the Trigger Message (TM). This message contains the schedule for the synchronous traffic that shall be sent within this EC. The schedule is calculated and sent by a master node. FTT-CAN supports both periodic and aperiodic traffic by dividing the EC in two parts. In the first part, the asynchronous window, a (possibly empty) set of aperiodic messages are sent using CAN's native arbitration mechanism. In the second part, the synchronous window, traffic is sent according to the schedule delivered in the TM.

2.2.2. EDF Scheduling Several suggestions for scheduling CAN using EDF have been presented (Livani *et al.*, 1998; Natale, 2000; Zuberi and Shin, 1995). These solutions are all based on manipulating the identifier of the CAN frame to reflect the deadline of the frame, and thus they reduce the number of possible identifiers to be used by the system designers. Restricting the use of identifiers is often not an attractive alternative, since it interferes with other design activities, and is even sometimes in conflict with adopted standards and recommendations (CiA, 1996; J1938, 1998). FTT-CAN can also schedule messages according to EDF, using the synchronous window.

3. SERVER-BASED SCHEDULING ON CAN

In order to provide guaranteed network bandwidth for both synchronous and asynchronous real-time messages in CAN, server-based scheduling techniques provide a solution.

Server-based scheduling of CAN has been presented earlier (Nolte and Lin, 2002; Nolte *et al.*, 2003). However, in this paper response-time analysis is presented for (Nolte *et al.*, 2003) together with modifications that, for some system configurations, limits the worst-case response-time while retaining fairness. Furthermore, equations that bound the worst-case response-time for each server are presented.

Using servers, the whole network will be scheduled as a single resource, providing bandwidth isolation as well as fairness among the users of the servers. However, in order to make server-scheduling decisions, the scheduler must have information on when messages are arriving at the different nodes in the system, so that it can assign them a deadline based on the server policy in use. This information should not be sent to the scheduler using message passing, since this would further reduce the already low bandwidth offered by

CAN. The method, presented below, will provide a solution to this.

3.1 Server Scheduling (N-Servers)

In real-time scheduling, a *server* is a conceptual entity that controls the access to some shared resource. Sometimes multiple servers are associated with a single resource. For instance, in this paper multiple servers are mediating access to a single CAN bus.

The server used in this paper is a simplified version of the Total Bandwidth Server (TBS) (Spuri and Buttazzo, 1994; Spuri *et al.*, 1995).

3.1.1. Server Characterization Each node on the CAN bus is assigned one or more *network servers* (N-Servers). Each N-Server, s, is characterized by its period T_s, and it is allowed to send one message every server period. The message length is allowed to be of worst-case size. A server is also associated with a relative deadline $D_s = T_s$. At any time, a server may also be associated with an absolute deadline d_s, denoting the next actual deadline for the server. The server deadlines are used for scheduling purposes only, and are not to be confused with any deadline associated with a particular message. (For instance, the scheduling method presented below will under certain circumstances *miss* the server deadline. However, this does not necessarily make the system unschedulable.)

3.1.2. Server State The state of a server s is expressed by its absolute deadline d_s and whether the server is *active* or *idle*. The rules for updating the server state are as follows:

(1) When an *idle* server receives message n at time r_n it becomes *active* and the server deadline is set so that

$$d_s^n = \max(r_n + D_s, d_s^{n-1})$$

where $d_s^0 = 0$.

(2) When an *active* server sends a message and still has more messages to send, the server deadline is updated according to

$$d_s^n = d_s^{n-1} + D_s$$

(3) When an *active* server sends a message and has no more messages to send, the server becomes *idle*.

3.2 Medium Access (M-Server)

The native medium access method in CAN is strictly priority-based. Hence, it is not very useful when scheduling the network with servers. Instead a *master server* (M-Server) is introduced, which is a piece of software executing on one of the network nodes. Scheduling the CAN bus using a dedicated "master" has been previously proposed (Almeida *et al.*, 1999; Pedreiras and Almeida, 2000). In this paper, the M-Server has two main responsibilities:

(1) Keep track of the state of each N-Server.

(2) Allocate bandwidth to N-Servers.

The first responsibility is handled by *guessing* whether or not N-Servers have messages to send. The initial guess is to assume that each N-Server has a message to send (i.e., initially each N-Server s is assigned a deadline $d_s = D_s$). How to handle erroneous guesses is shown later.

The N-Servers' complete state is contained within the M-Server. Hence, the code in the other nodes does not maintain N-Server states. The code in the nodes only has to keep track of when bandwidth is allocated to them (as communicated by the M-Server).

The M-Server divides time into Elementary Cycles (ECs), similar to the FTT-CAN approach presented in Section 2.2.1. T_{EC} denotes the nominal length of an EC. Moreover, T_{EC} is the temporal resolution of the resource scheduled by the servers, in the sense that N-Servers can not have their periods shorter than T_{EC}. When scheduling a new EC, the M-Server will (using the EDF principle based on the N-Servers' deadline) select the N-Servers that are allowed to send messages in the EC. Next, the M-Server sends a Trigger Message (TM). The TM contains information on which N-Servers that are allowed to send one message during the EC. Upon reception of a TM, the N-Servers allowed to send a message will enqueue a message in their CAN controllers. The messages of the EC will then be sent using CAN's native priority access protocol. Due to the arbitration mechanism, it is unknown when inside an EC a specific message is sent. Hence, the bound on the delay of message transmissions will be a multiple of the length of an EC.

Once the TM has been sent, the M-Server has to determine when the bus is idle again, so the start of a new EC can be initiated. The M-Server does this by sending a stop [1] message (STOP) with the lowest possible priority. After sending the STOP message to the CAN controller, the M-Server reads all messages sent on the bus. When it reads the STOP message it knows that all N-Servers have sent their messages [2].

After reading the STOP message the EC is over and the M-Server has two tasks to complete before starting the next EC:

(1) Update the state of the N-Servers scheduled during the EC.

[1] A small delay before sending STOP is required. This message is not allowed to be sent before the other nodes have both processed the TM (in order to find out whether they are allowed to send or not), and (if they are allowed to send) enqueued the corresponding message.

[2] Another way of determining when the EC is finished would be that the CAN controller itself is able to determine when the bus becomes idle. If this is possible, there is no need for the STOP message. However, using a STOP message is suitable with standard CAN controllers.

(2) Decide how to reclaim the unused bandwidth (if any) during the EC.

The following two sections describe how these tasks are solved.

3.2.1. Updating N-Server States By observing the actual transmissions during an EC it is possible for the M-Server to verify whether or not its guess that N-Servers had messages to send was correct, and to update the N-Servers' state accordingly. For each N-Server that was allocated a message in the EC two cases exists:

(1) The N-Server sent a message. In this case the guess was correct and the M-Servers next guess is that the N-Server has more messages to send. Hence, the N-Server's state us updated according to rule 2 in Section 3.1.2.
(2) The N-Server did not send a message. In this case the guess was incorrect since the N-Server was idle. The new guess is that a message now has arrived to the N-Server, and the N-Server state is set according to rule 1 in Section 3.1.2.

3.2.2. Reclaiming Unused Bandwidth It is likely that not all bandwidth allocated to the EC has been completely used. There are three sources for unused bandwidth (slack):

(1) An N-Server that was allowed to send a message during the EC did not have any message to send.
(2) One or more messages that were sent was shorter than the assumed worst-case length of a CAN message.
(3) The bit-stuffing that took place was less than the worst-case scenario.

To not reclaim the unused bandwidth would make the M-Server's guessing approach of always assuming that N-Servers have messages to send extremely inefficient. Hence, a mechanism to reclaim this bandwidth is needed.

In the case that the EC ends early (i.e., due to unused bandwidth) the M-Server reclaims the unused bandwidth by reading the STOP message and immediately initiating the next EC to make sure that no bandwidth is lost.

3.3 Timing Analysis

One property of the above method is that the protocol overhead increases when the network utilisation is low. When the network is not fully utilised, the M-Server will send more TM and STOP message than it would during full network utilisation. This is illustrated in Fig.1, where Fig.1(a) shows a fully utilised network where all servers that were allowed to send a message did so. Fig.1(b) shows what happens if not all servers send messages. Note the increase of TM and STOP messages. In fact, the proposed method will always consume all bandwidth of the network, regardless of the number of server messages actually sent.

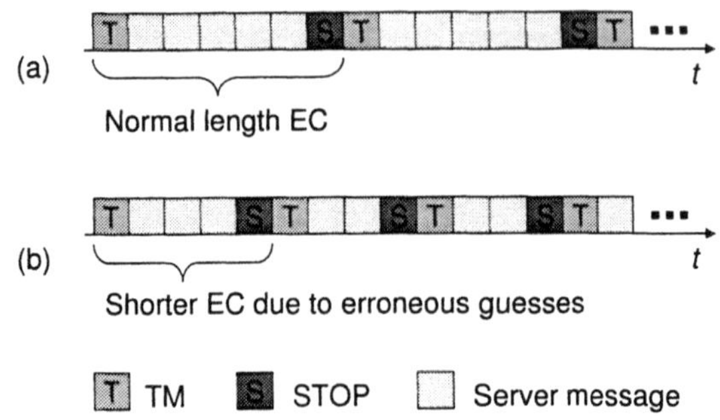

Fig. 1. Erroneous guesses increases the protocol overhead.

This increase in overhead when no N-Server messages are sent makes the analysis more complicated than in the original TBS, and the worst-case analysis of TBS cannot be reused.

As the scheduling mechanism used in this paper is based on EDF, looking at an N-Server x, the server might have to wait for other servers to be scheduled before it will be scheduled. Since the scheduling is completely based on guessed server-deadlines, x might be the only server in the system having messages to send, yet x may have to wait for all other servers to be scheduled due to their earlier deadlines. In order to derive an upper bound on the worst possible delay x can experience (setting time to 0 when a message arrives to x and its deadline is set to D_x) the other N-Servers in the system are put into two sets:

- A – all servers with periods shorter than x's:

$$A = \{s : D_s < D_x\}$$

- B – all servers with periods equal to or longer than x's:

$$B = \{s : s \neq x \wedge D_s \geq D_x\}$$

If the number of servers in A (denoted by $|A|$) is greater than the number of messages fitting in one EC, x may potentially not be scheduled for the duration of its deadline. For instance, if servers in A have no messages to send they will still be scheduled for message transmission for each EC until their deadline becomes greater than D_x.

Hence, N-Server x may not be scheduled until the EC which overlaps D_x. Since, at this point each of the other N-Servers (in sets A and B) may have deadlines arbitrary close to (and before) D_x, x may have to wait for each of the other servers to send one (1) message. No server will after time D_x be scheduled twice, since the first time a server is scheduled (after D_x) its new deadline will be set to a time greater than D_x. Hence, the message from server x may in the worst-case be the $|A| + |B| + 1$-th message to be sent after D_x.

Thus, an upper bound on the response time for N-Server x, R_x is:

$$R_x = D_x + \left\lceil \frac{|A| + |B| + 1}{N_{EC}} \right\rceil * T_{EC}$$

where N_{EC} denotes the number of messages in an EC and T_{EC} the longest time an EC can take (including sending TM and STOP messages).

4. PROTOCOL FOR LOWER RESPONSE-TIMES

The previously presented protocol has a worst-case response-time for a server x that has a large part which is independent of x's deadline (D_x). This can for many network configurations (e.g. configurations with many servers) give unreasonable worst-case response-times. In this section a modification to the previous protocol is presented, based on a dynamic priority version of the Polling Server (PS) (Lehoczky *et al.*, 1997). Using this new approach, the worst-case response-time of server x is mostly dependent on the server deadline D_x (which is equal to the server period T_x).

In the previous protocol, the increased protocol overhead caused by erroneous guesses mostly penalise servers with long periods (since servers with short periods will still receive good service even if they are causing erroneous guesses). The proposed modification will instead penalise the servers which causes erroneous guesses (i.e. the servers that do not send messages when they are allocated bandwidth).

4.1 New Rules

In the modified protocol, the M-Server will always treat an N-Server that has been allocated bandwidth as if the N-Server actually did send a message (regardless whether it did send any message). Hence the rules of Section 3.2.1 will be changed to a single rule:

- Treat the N-Server as if it sent a message. The next guess is that the N-Server has more messages to send. Hence it updates the N-Server's state according to rule 2 in Section 3.1.2.

In order for an N-Server not to cause more than one erroneous guess during its period, the M-Server will apply the following rules when deciding the schedule for the EC starting at time t:

(1) Only servers that are within one period from their deadline will be eligible for scheduling. Formally the following condition must hold for a server x to be eligible for scheduling:

$$d_x - t \leq T_x$$

(2) Among the eligible N-Servers, select N-Servers in EDF order to be scheduled during the EC.

(3) To make the protocol work-conserving, when there are not enough eligible N-Servers to fill an EC, also non-eligible N-Servers are selected in EDF order to be scheduled during the EC.

4.2 Timing Analysis

In order to provide a worst-case response-time analysis, the first step is to establish that the modified protocol will behave no worse than if pure EDF scheduling was deployed. Using that result the worst-case response-time can be derived.

Lemma 1: If all N-Servers that are allocated messages do send messages, no erroneous guesses will be made and the protocol will emulate pure EDF (and all deadline will be met if load is less than 100%).

Proof: When all N-Servers send their allocated messages they will behave like periodic processes and according to the scheduling mechanism they will be scheduled in EDF. The following equation is true if the system load (including protocol overhead when all servers send their allocated message) is less than 100% (Nolte *et al.*, 2003):

$$\sum_{\forall s}\left(\frac{M}{S \times T_s}\right) \leq 1-\left(\frac{TM + STOP + S \times T_{sched}}{S \times T_{EC}}\right)$$

where s is an N-Server in the system, M is the length of a message (typically worst-case which is 135 bits), S is the network speed in bits/second, T_s is the period of the N-Server. TM and STOP are the sizes of the TM and the STOP messages in bits, typically 135 and 55 bits, T_{sched} represents the computational overhead of updating the N-Server deadlines and encoding the next TM after receiving the STOP message, and T_{EC} is the length of the EC. □

Lemma 2: If an N-Server that is allocated a message does not send a message, no other N-Server will miss its deadline because of this.

Proof: When an N-Server does not send a message it has been allocated, the EC will be terminated early (as can also happen if not all messages are of maximum length). However, since the N-Server will not be scheduled again until its next period it cannot cause more overhead (in terms of TM and STOP messages being generated). And, since the current EC is terminated early, subsequent messages from other N-Servers that are to be scheduled in forthcoming ECs will be server earlier than they would have been if the N-Server did send its message (and, hence, they cannot miss their deadlines). □

Theorem: The worst-case response-time for N-Server x is not greater than $2 * T_x + T_{EC}$.

Proof: Lemmas 1 and 2 tells us that N-Server x will be scheduled at least once during its period T_x. There is however no restriction on when, during T_x, the server is scheduled. In the worst-case, the distance between two consecutive occasions of scheduling of x then becomes $2 * T_x$ (if the first message is scheduled immediately in its T_x and the second message is scheduled as late as possible in its T_x).

The second message will be scheduled no later than the EC in which its deadline is. However, the order of transmission within that ECs is unknown. Hence, the pessimistic assumption is that the message is scheduled as the last message in the EC and that the deadline is at the start of the EC. This will incur an extra maximum delay of T_{EC} for the message. □

4.3 Comparison of Mechanisms

The main difference between the new mechanism and the old one is that with the new mechanism a server with a short period will not have to wait for a time longer than twice its server period. Using the old mechanism the maximum time a server could have to wait for is determined by the number of servers in the system together with the size of the EC. In that case a server might have to wait for a period longer than twice its server period.

If the system consists of a large set of servers with short server periods, the new mechanism is better, since the worst-case response-time for these servers are limited by their periods rather than the total number of servers, whereas if the system consists of a few servers with long server period, the old mechanism is giving lower worst-case response times.

5. CONCLUSIONS

In this paper a new mechanism for scheduling of the Controller Area Network (CAN) is presented. The difference between the new mechanism and existing methods is the usage of server-based scheduling techniques. Furthermore, the mechanism is based on Earliest Deadline First (EDF) scheduling to achieve high utilisation. The approach allows a more flexible way to utilize the CAN bus compared to other scheduling approaches such as native CAN and Flexible Time-Triggered communication on CAN (FTT-CAN). Servers provide fairness among the streams of messages as well as timely message delivery.

Two different server-based scheduling mechanisms are presented, suitable for different system configurations. It is also shown how to derive upper bounds for the response-times for both mechanisms.

The main strength of server-based scheduling for CAN, compared to other scheduling approaches, is the good service to streams of aperiodic messages. On native CAN, streams of aperiodic messages can starve other streams of messages. In FTT-CAN the situation is better, since periodic messages can be scheduled according to EDF using the synchronous window of FTT-CAN, thus separating periodic streams from aperiodic ones, guaranteeing real-time demands of the latter. However, no fairness can be guaranteed among the streams of aperiodic messages sharing the asynchronous window of FTT-CAN.

However, each scheduling policy has both good and bad properties. To give the fastest response-times, native CAN is the best choice. To cope with fairness and bandwidth isolation among aperiodic message streams, the server-based approach is the best choice, and, to have support for both periodic messages with low jitter, and aperiodic messages (although no fairness among aperiodic messages), FTT-CAN is the choice.

ACKNOWLEDGEMENTS

The work presented in this paper was supported by the Swedish Foundation for Strategic Research (SSF) via the research programme ARTES, the Swedish Foundation for Knowledge and Competence Development (KK-stiftelsen), and Mälardalen University.

REFERENCES

Abeni, L. (1998). Server Mechanisms for Multimedia Applications. Technical Report RETIS TR98-01. Scuola Superiore S. Anna. Pisa, Italy.

Almeida, L., J. A. Fonseca and P. Fonseca (1999). A Flexible Time-Triggered Communication System Based on the Controller Area Network: Experimental Results. In: *Proceedings of the International Conference on Fieldbus Technology (FeT'99).*

Audsley, N. C., A. Burns, M. F. Richardson, K. Tindell and A. J. Wellings (1993). Applying New Scheduling Theory to Static Priority Pre-emptive Scheduling. *Software Engineering Journal* **8**(5), 284–292.

CAN (1993). Road Vehicles - Interchange of Digital Information - Controller Area Network (CAN) for High-Speed Communication. *International Standards Organisation (ISO).*

CiA (1996). CANopen Communication Profile for Industrial Systems, Based on CAL. CiA Draft Standard 301, rev 3.0.

J1938, SAE (1998). Design/Process Checklist for Vehicle Electronic Systems. *SAE Standards.*

Lehoczky, J.P., L. Sha and J. Strosnider (1997). Enhanced Aperiodic Responsiveness in Hard Real-Time Environments. In: *Proceedings of* 8th *IEEE Real-Time Systems Symposium (RTSS'87).* San Jose, California, USA. pp. 261–270.

Livani, M. A., J. Kaiser and W. J. Jia (1998). Scheduling Hard and Soft Real-Time Communication in the Controller Area Network (CAN). In: *Proceedings of the* 23rd *IFAC/IFIP Workshop on Real-Time Programming.* Shantou, ROC.

Natale, M. Di (2000). Scheduling the CAN Bus with Earliest Deadline Techniques. In: *Proceedings of the* 21st *IEEE Real-Time Systems Symposium (RTSS'00).* Orlando, Florida, USA.

Nolte, T. and K. J. Lin (2002). Distributed Real-Time System Design using CBS-based End-to-end Scheduling. In: *Proceedings of the* 9th *IEEE International Conference on Parallel and Distributed Systems (ICPADS'02).* Taipei, Taiwan, ROC.

Nolte, T., M. Sjödin and H. Hansson (2003). Server-Based Scheduling of the CAN Bus. In: *Proceedings of the* 9th *IEEE International Conference on Emerging Technologies and Factory Automation (ETFA 2003).* Calouste Gulbenkian Foundation, Lisbon, Portugal. pp. 169–176.

Pedreiras, P. and L. Almeida (2000). Combining Event-triggered and Time-triggered Traffic in FTT-CAN: Analysis of the Asynchronous Messaging System. In: *Proceedings of the* 3rd *IEEE International Workshop on Factory Communication Systems (WFCS'00).* Porto, Portugal. pp. 67–75.

Spuri, M. and G. C. Buttazzo (1994). Efficient Aperiodic Service under Earliest Deadline Scheduling. In: *Proceedings of the* 15th *IEEE Real-Time Systems Symposium (RTSS'94).*

Spuri, M., G. C. Buttazzo and F. Sensini (1995). Robust Aperiodic Scheduling under Dynamic Priority Systems. In: *Proceedings of the* 16th *IEEE Real-Time Systems Symposium (RTSS'95).*

Tindell, K. W., A. Burns and A. J. Wellings (1995). Calculating Controller Area Network (CAN) Message Response Times. *Control Engineering Practice* **3**(8), 1163–1169.

TT-CAN (2000). Road Vehicles - Controller Area Network (CAN) - Part 4: Time-Triggered Communication. *International Standards Organisation (ISO).*

Zuberi, K. M. and K. G. Shin (1995). Non-Preemptive Scheduling of Messages on Controller Area Network for Real-Time Control Applications. In: *Proceedings of the* 1st *IEEE Real-Time Technology and Applications Symposium (RTAS'95).* Chicago, IL, USA. pp. 240–249.

Copyright © IFAC Information Control Problems in Manufacturing, Salvador, Brazil, 2004

ELSEVIER
IFAC
PUBLICATIONS
www.elsevier.com/locate/ifac

DESIGNING SUITABLE SWITCHED ETHERNET ARCHITECTURES REGARDING REAL-TIME APPLICATION CONSTRAINTS

Jean-Philippe Georges, Nicolas Krommenacker
Thierry Divoux, Eric Rondeau

Centre de Recherche en Automatique de Nancy (CNRS UMR 7039)
Université Henri Poincaré, Nancy, Faculté des Sciences
BP 239, F-54600 Vandœuvre lès Nancy, Cedex, France
<firstname.name>@cran.uhp-nancy.fr

Abstract:
Since the Ethernet network is not deterministic, its use in factory environments (which are strongly time constraints) can not absolutely guarantee that the applications requirements will be respected. This paper presents a method based on genetic algorithms to minimise end-to-end delays, by providing a good distribution of the devices on the network switches. The objective function is defined by using the network calculus, which is a deterministic theory and enables to bound these delays. A case study is described. The results are compared with those obtained by using a network simulator. *Copyright © 2004 IFAC*

Keywords: Ethernet. Genetic Algorithms. Real-time communication

1. INTRODUCTION

The industrial communications are currently based on specific networks called fieldbuses such as FIP, Profibus (CENELEC, 1996), CAN. They interconnect Programmable Logical Controllers (PLC), CNC, robots, ... to exchange technical data for monitoring, controlling, and synchronising industrial processes. Their main characteristic is to ensure that the end-to-end delays of messages remain limited, compared with the time-cycle of the applications. Thus, these networks are deterministic and some protocols satisfy the integrity constraints of the information. In opposition, the Ethernet networks based on the CSMA/CD protocol are increasingly used to interconnect industrial devices. Some applications confirm this evolution in different industrial areas : cars (Jaguar), pharmaceutics (Boehringer Ingelheim), avionics (Grieu *et al.*, 2003)... Moreover, different organisations such as the Industrial Automation Networking Alliance (IAONA), the Industrial Ethernet Association (IEA) promote Ethernet as "the standard in the industrial environment". Finally research projects such as CIDER (Alves *et al.*, 2000) study the use of Ethernet as an enabling technology for future dependable real-time systems.

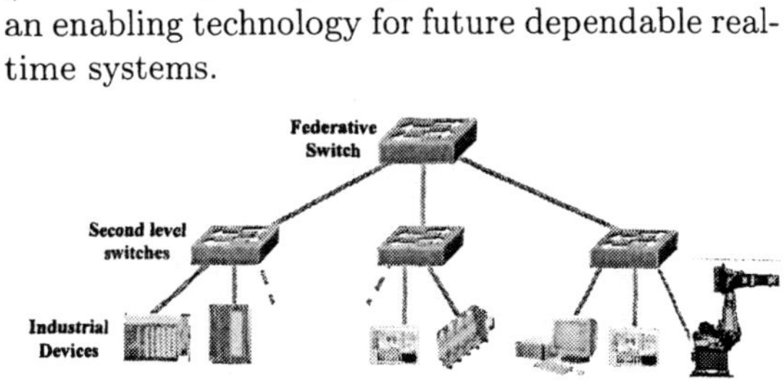

Fig. 1. Hierarchical topology for switched Ethernet networks

To interconnect Ethernet switches, different topologies can be used. (Rüping *et al.*, 1999) shows that the hierarchical topology (figure 1) is the most efficient and therefore mainly used in factory networks. Industrial devices are distributed on second level switches which are directly connected to a federative switch.

In the first part of this paper, it is proposed to design a switched Ethernet architecture which satisfies the time requirements. Our objective is to optimise the network organisation at the physical layer level. Switched networks design consists in distributing industrial devices on the different Ethernet switches in order to reduce the load of backbone and switches as shown in the figure 2. The segmentation of a network is an NP-complete problem for which several heuristics have been developed. The size of the search space for an arbitrary problem instance is defined by the number of industrial and network interconnection devices. Two types of approaches exist. The constructive methods elaborate a solution from an empty solution while the improvement methods try to improve a given solution with the help of a heuristic. In this study, the use of such a heuristic, the Genetic Algorithms, is suggested.

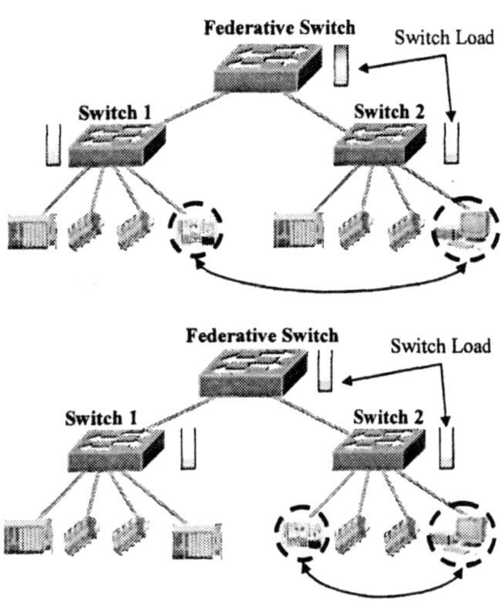

Fig. 2. The load of the federative switch depends on the distribution of the industrial devices on the second level switches

In the second part, we present the network calculus theory introduced by R. Cruz (Cruz, 1991*a*)(Cruz, 1991*b*). It enables to determine upper-bounds of the end-to-end delays and would be used as the objective function in our Genetic Algorithm. Also, the optimisation criteria to minimize will be the maximum end-to-end delay.

Finally, the network design approach described in this paper is illustrated on a factory communication scenario. The obtained solution is evaluated with a network simulator. The results show the segmentation impacts on the temporal performances. Therefore, it is important to have a method to design a switched Ethernet industrial network in order to make it deterministic under some precise hypothesis.

2. GENETIC ALGORITHM APPROACH

2.1 Introduction

The issue is to determine on which second level switch an industrial device has to be connected. The objective is to propose a segmentation scheme for which the maximum of the end-to-end delays are bounded in order to guarantee deterministic communications. To do it, the Genetic Algorithms (GA) will be used.

That is a meta-heuristic inspired by the genetic processes of biological organisms. It uses a direct analogy with the natural behaviour and its mechanisms such as selection, reproduction, crossover or mutation. In opposition to other methods like tabu search or simulated annealing which manipulate a single solution, GA work with a population of individuals. Each individual corresponds to a candidate solution of the search space and is represented by a string, called chromosome. Each chromosome has an objective function value, called fitness. The population evolves into different populations for several iterations, so named generations. For each generation, the evolution is ensured by three main steps which are parents' selection, crossover and mutation. A typical structure of genetic algorithms is defined as follows :

```
Create initial population
Do{
    Choose parent1 and parent2 from population
    Compute offspring = crossover (parent1, parent2)
    Compute mutation (offspring)
    If adapted (offspring) then
            replace (population, offspring)
}
Until (stopping criterion)
```

The characteristics of the GA (representation, initialization, evolutionary strategies, fitness function and genetic operators) vary according to the application kind. In the next parts, we describe the use of the GA for the optimisation of a switched Ethernet topology.

2.2 Coding structure

For the segmentation problem, several encoding methods could be defined. The most often used is the N-string representation. Each chromosome corresponds to a vector in which the ith element of an individual is j if the ith industrial device is connected to the second level switch labelled j. There are as many elements in the vector as factory devices in the network. The k-ary alphabet $[1, 2, \ldots, k]$ is used to code the switch label. For instance, the string [112332231] represents the mapping that assigns devices 1, 2, 9 to switch 1, devices 3, 6, 7 to switch 2 and devices 4, 5, 8 to switch 3. The figure 3 illustrates the encoding of this solution and the decoding of this solution.

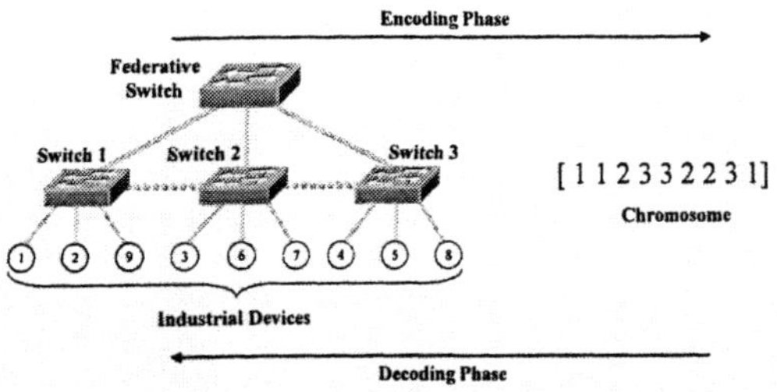

Fig. 3. Encoding network topology

2.3 Crossover and Mutation operators

In GA, these operators are the two most important space exploration operators. A crossover operator creates a new offspring chromosome by combining parts of the parents' chromosomes. A two-point crossover is used in our algorithm. Two cut points are randomly selected. Each chromosome is also divided into three disjoint parts. A new chromosome is formed randomly copying a part from one parent to the corresponding part of the offspring (figure 4). The crossover operator can generate illegal chromosomes. In the figure 5, an offspring violated the number of partitions constraint (3 groups are expected). A repairing procedure is defined : permutations are applied out of the cut-points in order to preserve the scheme of parent chromosomes. The mutation operator enables to introduce unexplored search space to the population. The most often used method is the *bit-flip* mutation. But it can generate illegal chromosomes. So a swap mutation has been defined for preserving the switch number constraint (figure 6).

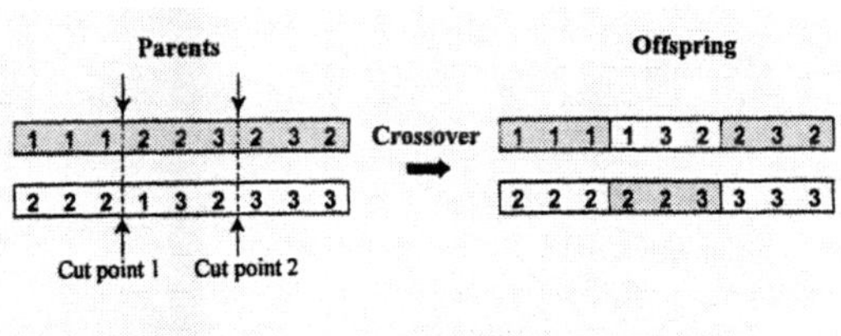

Fig. 4. Two-point crossover operator

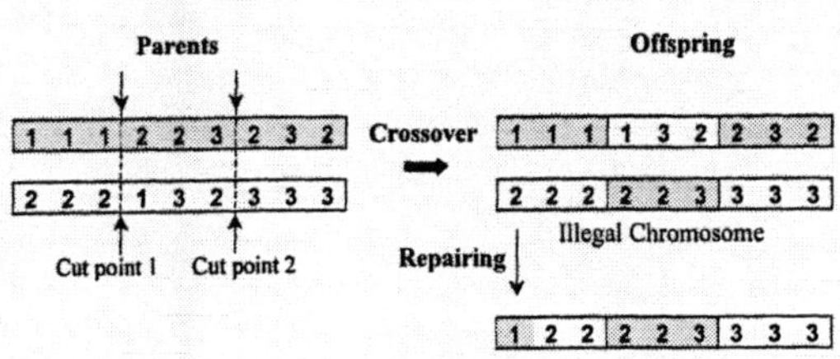

Fig. 5. Repairing step

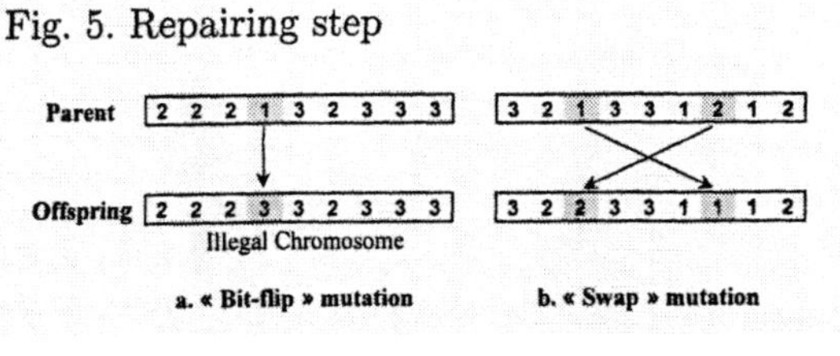

Fig. 6. Mutation operators

2.4 Parameters setting

(Goldberg, 1989) defined conventional GA parameters such as population size (pop_{size}), crossover (p_c) and mutation (p_m) probabilities. They influence the performance of the algorithm. For example, the size of the initial population influences on the quality of the final solution to the detriment of the execution time. For our problem, their values have been computed in (Krommenacker, 2002).

2.5 Objective function

The quality of a solution will be defined according to the respect of the time-constraints. So, to determine the quality of segmentation, the evaluation function has to take into account the end-to-end delay. In previous works (Kamen *et al.*, 1999) (Torab, 2000), the objective consists to reduce the average end-to-end delays which are computed by using stochastic theory. As the question is to know if switched Ethernet guarantees the temporal constraints, a deterministic theory is necessary. The objective function will be formulate as :

$$f = \max_i \{\overline{D_i} - T_i\}$$

with i the message index, $\overline{D_i}$ the message maximum end-to-end delay and T_i the message time constraint). It has to be minimised until it becomes negative (for example, a control message has to be periodically received in a time interval lower than the PLC time-cycle). Now, the network calculus theory is presented. It enables to determine upper-bound delays $\overline{D_i}$ of each message.

3. NETWORK CALCULUS THEORY

The network calculus is a deterministic theory of the queuing systems. The main contributions are (Cruz, 1991*a*),(Le Boudec and Thiran, 2001),(Chang, 2000). The Network Calculus is also more and more used to study Ethernet networks temporal performances (Georges *et al.*, 2002), (Jasperneite *et al.*, 2002).

Since the industrial case gives concise information about communications (communications of a programmable logical controller are defined in its cycle), deterministic hypothesis on the traffic arrival can be supposed. In fact, the arrival of data is assumed to be constrained by a (σ, ρ) envelop called leaky bucket (figure 7). If $R(t)$ represents the instantaneous rate of data, C the capacity of the links on the network., σ the maximum amount of traffic that can arrive in a burst (initially, it would be the maximum frames length) and ρ an upper bound on the long-term average rate of the traffic flow, then $\forall x, y;\ y \geq x,\ x \geq 0$, the burstiness constraint is written $R \sim b$ which means :

$$\int_x^y R(t)\,dt \leq \min\{C(y-x), \sigma + \rho(y-x)\} \quad (1)$$

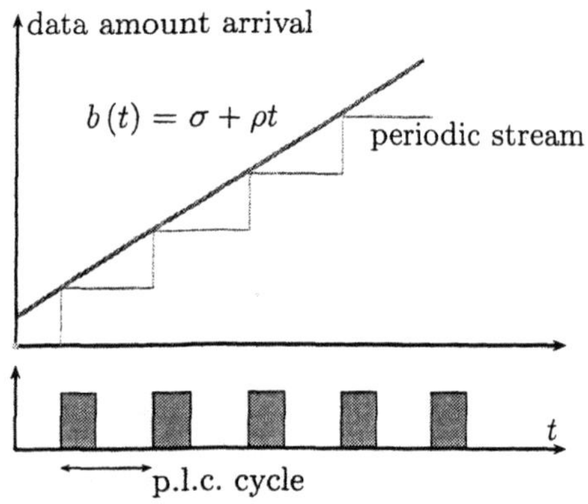

Fig. 7. Traffic upper bounding

Switches are modelled as presented on the figure 8. It is the result of a study described in (Georges *et al.*, 2003) in which several models are built and compared. It is constituted of a sequence of three basic elements : one multiplexer, one queue and one demultiplexer. In order to take into account the internal speed of a switch, the three following capacities will be used : C, the throughput inside the switch, C_{in}, the throughput of arrival of data on input ports and C_{out}, the output speed.

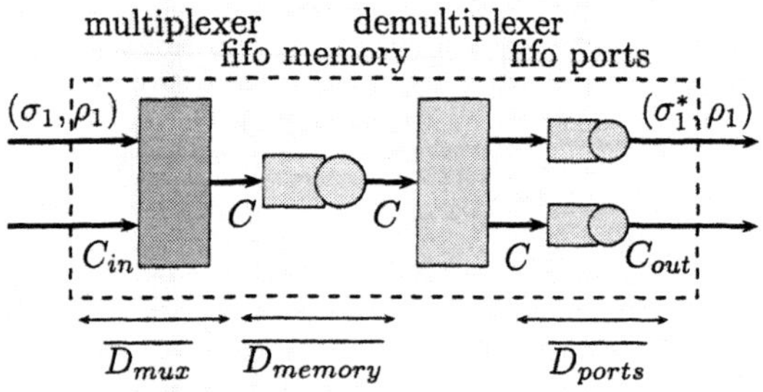

Fig. 8. Switch model

Since the study is about Ethernet networks, the scheduling policy has to be non preemptive. Moreover, it is considered that the components are work-conserving systems, i.e. they cannot have vacations and their forwarding policy is the best-effort. In this paper, the FIFO forwarding policy is adopted in order to respect the fairness of Ethernet : a FIFO memory to model the switching process and FIFO queues to model the transmission function of the output ports.

The quantity of data processed in the element is called the *backlog* (Le Boudec and Thiran, 2001, definition 1.2.1) and could be seen as the congestion value of the element. It is defined at time t by the amount of data which is already arrived at time t, minus by the amount of data processes at the same time. In the network calculus theory, delays correspond to the time needed to process the backlog. So, upper bounded delays depend in the worst case on the maximum backlog expression. For instance, (Cruz, 1991*a*, theorem 4.1) has defined that the transit delay of a 2 input-ports multiplexer FIFO is upper bounded by

$$\overline{D_{mux,1}(t)} = \frac{1}{C_{out}} \max_{t \geq 0} [b_1(t) + b_2(t + L/C_2) - C_{out}t]$$

And for queues, he gives :

$$\overline{D_{fifo}} = \frac{1}{C_{out}} \frac{(C_{in} - C_{out})}{C_{in} - \rho_{in}} \sigma_{in}$$

(Georges *et al.*, 2002) extends this work and calculates the upper-bounded delays for each of the basic elements of the switch model selected in this paper.

In complete architectures (several switches interconnected with links), the problem is that they depend on the inputs parameters σ and ρ. But they are not always known, as for example, after having crossed a switch. To solve that, one of the theorems presented in (Le Boudec and Thiran, 2001) says that a stream constrained by an arrival curve $\alpha(t)$ transiting in a system with a service curve $\beta(t)$ is constrained on output by $\alpha^*(t)$, where

$$\alpha^*(t) = \alpha(t) \oslash \beta(t) = \sup_u \{\alpha(s+u) - \beta(u)\}$$

Cruz completes this study by showing that if $R_i \sim b_i$ et $\overline{D} < +\infty$, then $R_{out} \sim b_{out}$ with $b_{out}(x)$ defined for all x positive by :

$$b_{out}(x) = b_{in}(x + \overline{D}) \quad (2)$$

And if we develop the previous equation by using the relation (1), we have :

$$\sigma_{out} = \sigma_{in} + \rho_{in}\overline{D} \quad (3)$$
$$\rho_{out} = \rho_{in}$$

The peakedness analysis is based on the fact that the arrival rate stays constant and that the delay is translated in a supplementary burst. Indeed, while frames are waiting other jobs to be finished, data still arrive, and in the worst case, the system will forward successively all of these frames. To illustrate how end-to-end delays are computed, it is possible to define that the upper-bound delay for crossing a switch with :

$$\sigma_{switch}^{out} = \sigma_{switch}^{in} + \rho\underbrace{(\overline{D_{mux}} + \overline{D_{memory}} + \overline{D_{ports}})}_{\overline{D_{switch}}}$$

End-to-end delays over the complete network are also obtained by computing $\overline{D} = \frac{\sigma^h - \sigma^0}{\rho}$ where h represents the number of crossed switches minus one. For the evaluation function included in the Genetic Algorithm, all maxima end-to-end delays are computed and the return value corresponds to the highest delay.

4. A CASE STUDY

The first step is to determine the number of switches on which the application has to be distributed. It can be imposed by the designer who can also be helped by using graph partitioning techniques (Adoud *et al.*, 2001).

In order to show the benefits provided by the optimisation of the segmentation of a switched Ethernet network, a factory communication scenario is studied. Fifteen industrial devices have to be interconnected on three switches. The exchanges are represented on the figure 9.

	1	2	3	4	5	6	7	8	9	10	11	12	13	14	15
1						2									
2	1											3			
3					1						4		1		
4							1		1						
5				1								1			
6	1			2	1					2	4				
7															
8			3												4
9															
10															
11					1			1				1			2
12	3			1											
13		4								3					
14		1		2											2
15					1				1		1	3			

Fig. 9. Exchange matrix

The weights defined on the exchange matrix represent the amount of data send by a device to another. For example, device 3 sends to device 11, $4*(basic\ length)$ per $time-cycle$ bytes per second. If it is supposed that the basic length is 64 bytes (minimum size of an Ethernet frame) and that a device sends a packet each 2 ms (time-cycle). So the rate of arrival of data from device 3 to 11 is 5, 12 kbytes/s.

In order to verify that a hierarchical switched Ethernet architecture could be used in this example, with end-to-end delays always lower than the time-cycle, the Genetic Algorithm with the objective function previously presented is launched. An arbitrary distribution and an optimised solution are compared. The first one distributes devices $\{1, 2, ...5\}$ on the first switch, $\{6, ...10\}$ on the second, $\{11, ...15\}$ on the third. The GA provides the following distribution called 'optimised' architecture and places the industrials devices $\{1, 4, 6, 7, 14\}$ on the first switch, $\{8, 9, 11, 12, 15\}$ on the second, $\{2, 3, 5, 10, 13\}$ on the third. The results are confronted with those provided by a network simulation tool : *Comnet III* (Fig 10).

In Comnet III, the modelling of a network device such as a switch is completed by using buffers to represent input and output ports which have

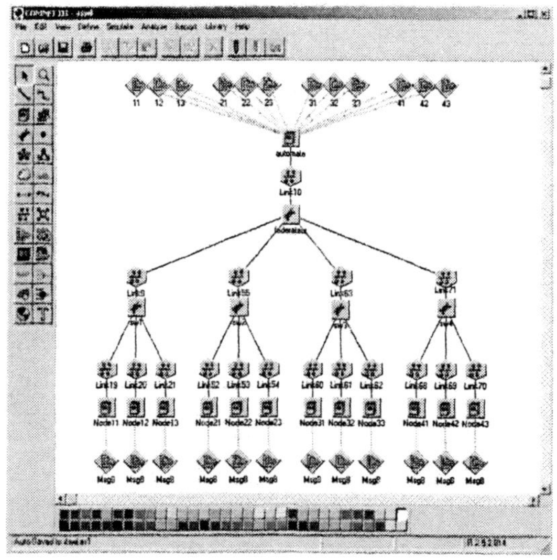

Fig. 10. The network simulator *Comnet III*

associated buffer sizes and buffer processing times (CACI Products Company, 1998). It also uses internal buses for moving packets from ports to ports.

The obtained results are presented on the table 1.

Table 1. Performance of the architectures (values are in millisecond)

architecture	network calculus	comnet
arbitrary	2, 55	1, 68
optimised	1, 86	1, 53

The table 1 shows that delays can be reduce if a methodology for distributing the industrial devices on the switches is applied. Indeed, in the optimised solution, the network calculus theory guarantees that maximum delays will be smaller than the time-cycle, when the non-optimised distribution exceeds it. The simulator results also show that the end-to-end delays are reduced. Nevertheless, the results provided by the simulator are lower than the time-cycle. So the arbitrary solution could be considered as suitable regarding the application constraints. The explanation is the following :

- in the network calculus approach, which always considers the worst case, it is assumed that all frames arrive simultaneously on all switches (especially, on the federative one).
- in the other hand, with Comnet, the frames arrive simultaneously only on the second level switches. The diversity of the incoming traffic logically desynchronises it on the federative switch so that the contention of the frames is reduced.

The network calculus provides pessimistic results, but absolutely guarantees the deterministic behaviour of the communication architecture regarding application requirements. It is not true for the simulation which does not provide such formal validation due to its probabilistic approach.

5. CONCLUSION

In this paper, an algorithm to design hierarchical switched Ethernet networks has been given. The optimisation heuristic used is the Genetic Algorithm for which a new objective function has been developed. This function evaluates the maximum end-to-end delays of the messages over the network. Bounds are computed by using the network calculus theory, which considers the worst case. So, from a well-known incoming traffic, it is possible to guarantee deterministic performances for the communication architecture. This approach enables to valid the network topology and proposes a distribution of the industrial devices which respects the temporal application constraints, even if this heuristic does not provide the best solution.

In future works, the presented method will be analysed in order to improve the convergence of the algorithm. Indeed, its running time is important (2 minutes for the case study described in the paragraph 4). It is not a problem for the design of static applications. However, it has to be reduced for applications which dynamically evolve in terms of number of interconnected devices, traffic flows and default occurrences.

Supplementary constraints could be taken into account such as the cabling cost, the location of the end equipments which limits the number of possible solutions (and the algorithm execution time, since the search space is reduced). The method could be also recursively processed in order to get a n levels hierarchical topology as it is done with the spectral algorithm (Simon, 1991).

REFERENCES

Adoud, H., E. Rondeau and T. Divoux (2001). Configuration of communication networks by analysing co-operation graphs. *Computer Communications* **24**(15–16), 1568–1577. ISSN 0140-3664.

Alves, M., E. Tovar, G. Fohler and G. Buttazo (2000). CIDER - envisaging a cots communication infrastructure for evolutionnary dependable real-time. In: *Proceedings of the 12th Euromicro Conference on Real-Time Systems*. pp. 19–22. Work in progress session.

CACI Products Company (1998). Comnet III reference guide. Release 2.0.

CENELEC (1996). Fieldbus; vol.1 p-net, vol.2 profibus, vol. 3 WorlFip. European Standard EN50170.

Chang, C.S. (2000). *Performance Guarantees in Communication Networks*. Springer Verlag. ISBN 1-85233-226-3.

Cruz, R. (1991*a*). A calculus for network delay, part I : Network elements in isolation. *IEEE Transactions on Information Theory* **37**, 114–131.

Cruz, R. (1991*b*). A calculus for network delay, part II : Network analysis. *IEEE Transactions on Information Theory* **37**, 132–141.

Georges, J.-P., E. Rondeau and T. Divoux (2002). Evaluation of switched Ethernet in an industrial context by using the network calculus. In: *4th IEEE International Workshop on Factory Communication Systems*.

Georges, J.-P., T. Divoux and E. Rondeau (2003). Comparison of switched Ethernet architectures models. In: *IEEE Conference on Emerging Technologies and Factory Automation Proceedings*. pp. 375–382.

Goldberg, D.E. (1989). *Genetic Algorithms in Search Optimization*. Addison Wesley.

Grieu, J., F. Frances and C. Fraboul (2003). Preuve de déterminisme d'un réseau embarqué avionique. In: *Colloque Francophone sur l'Ingénierie Des Protocoles*. Hermes Lavoisier. pp. 287–303.

Jasperneite, J., P. Neumann, M. Theis and K. Watson (2002). Deterministic real-time communication with switched Ethernet. In: *4th IEEE International Workshop on Factory Communication Systems*. pp. 11–18.

Kamen, E. W., P. Torab, K. Cooper and G. Custodi (1999). Design and analysis of packet-switched networks for control systems. *Proceedings of the 38th Conference on Decision and Control* pp. 4460–4465.

Krommenacker, N. (2002). Heuristiques de Conception de Topologies Réseaux : Application Aux Réseaux Industriels. PhD thesis. Université Henri Poincaré Nancy 1.

Le Boudec, J.-Y. and P. Thiran (2001). *Network Calculus, A Theory of Deterministic Queueing Systems for the Internet*. Springer Verlag. ISBN 3-540-42184-X.

Rüping, S., E. Vonnahme and J. Jasperneite (1999). Analysis of switched Ethernet networks with different topologies used in automation systems. In: *IFAC International Conference on Fieldbus Systems and their Applications*. pp. 351–358.

Simon, H.D. (1991). Partitioning of unstructured problems for parallel processing. *Computing Systems in Engineering* **2**, 135–148.

Torab, P. (2000). Performance Analysis of Packet Switched Networks with Tree Topology. PhD thesis. School of Electrical and Computer Engineering, Georgia Institute of Technology.

Copyright © IFAC Information Control Problems in
Manufacturing, Salvador, Brazil, 2004

ELSEVIER
IFAC
PUBLICATIONS
www.elsevier.com/locate/ifac

HOW TO GUARANTEE REALTIME BEHAVIOR USING ETHERNET

Jürgen Jasperneite *, **Peter Neumann** **

* *Phoenix Contact GmbH & Co. KG., R&D Department, Dringenauer Strasse 30, 31812 Bad Pyrmont, Germany, Fax:+49-5281-946212, e-mail:jjasperneite@phoenixcontact.com*

** *IFAK Institute for Automation and Communication Magdeburg, Steinfeldstr. 3, 39179 Barleben, Germany, Fax:+49-39293-81100, e-mail: peter.neumann@ifak-md.de*

Abstract: Ethernet is now conquering the field level in automation technology. Realtime behavior is a knock-out criterion therefore. This paper deals with the key concepts for guaranteed realtime behavior using Ethernet, based on the requirements of factory automation. A classification of the vendor-driven approaches that are currently under discussion for realtime Ethernet is provided.
Copyright © 2004 IFAC
Keywords: Realtime, Ethernet, factory automation

1. INTRODUCTION

In recent years the growing use of Ethernet in field communication has led to many discussions in automation technology. This can be attributed in part to a series of recent extensions to the relevant IEEE 802 standards for Ethernet, which suggest that it may now even be used for time-sensitive data transmission (IEEE, 1998*a*), (IEEE, 1998*b*). Most importantly, these extensions include switching technology and priority support, which make the CSMA/CD method that was previously used in Ethernet applications obsolete.

The gap between office technology and automation technology is narrowing all the time to achieve company-wide information processing (Clatha, 1998). The "connectivity gap" describes the problem, namely that highly complex programming or parameterization is required at present in order to access information from field devices through the communication pyramid (Pöschmann, 2001) . One reason for this is the lack of continuity between different, heterogeneous communication systems in the various levels of an automation solution.

At the same time, thanks to developments in microelectronics (lowering costs while increasing performance), field devices have a growing range of functions and consequently are becoming "more communicative" (Ricot, 2001). This results in an increase in throughput-oriented communication tasks in automation devices (e.g., for parameterization, archiving, and diagnostics), which in turn affects the original realtime data transmission. This is why many modern, intelligent automation devices already have an Ethernet connection, because fieldbuses do not provide sufficient data throughput. Additional interfaces for fieldbuses are required in these devices for the transmission of time-critical process data and integration into the world of programmable logic controllers.

Before a communication technology originating in the office world can make the transition to the industrial automation sector, several specific requirements must be met. The requirements range from standard industrial connectors and simple installation to a widely accepted application layer for interoperable communication. Furthermore, the question of realtime behavior must be answered, which is the aim of this paper.

The paper is organized as follows. Section 2 covers the realtime requirements in factory automation applications. Section 3 goes on to describe the factors, which influence the time response of a bridged Ethernet system. The key concepts for guaranteed realtime Ethernet are described in Section 4. The vendor-driven approaches, which are currently under discussion, can be found in Section 5. Finally, we conclude the paper in Section 6.

2. REALTIME REQUIREMENTS IN FACTORY AUTOMATION

Realtime behavior means that specific system functions are executed in a time interval $[R, D]$, which is determined by the application environment. Here, R (release) [1] denotes the earliest permissible starting time and D (deadline) the latest permissible completion time. Realtime behavior is guaranteed when the requirements for timeliness und simultaneity are met. According to (Jensen *et al.*, 1985), these requirements are best illustrated using time/utility functions (see Fig. 1). A hard realtime condition is when the utility of a data transmission abruptly drops to 0 when the deadline (D) is exceeded. Exceeding the deadline (D) may result in damage to product parts or equipment. This situation is aggravated by current developments in the transmission of safety-critical data over fieldbuses (INTERBUS, 2003), (*AS-Interface*, 2003), (PNO, 2003), (ODVA, 2003). Exceeding time limits in this case can endanger lives.

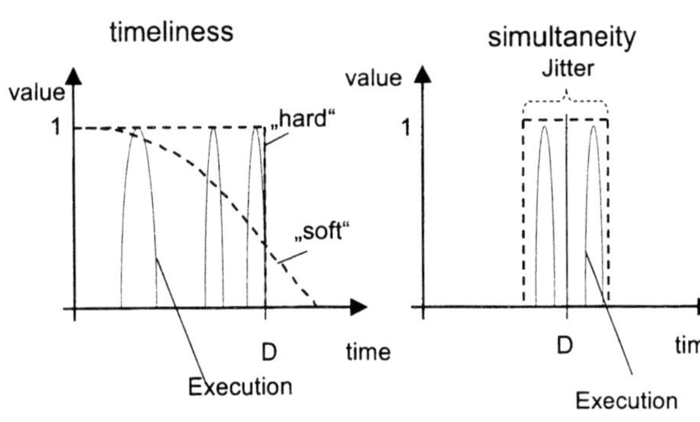

Fig. 1. Realtime criteria

[1] Typically the value of R is determined implicitly by the instant of a function call

A soft realtime condition is when the utility gradually falls to 0 when the deadline (D) is exceeded. In this case a violation of the deadline D is acceptable for some events.

In addition to timeliness, a realtime system must also meet the criterion of simultaneity.

Table 1. Quality of Service (QoS) requirements in field communication

QoS Class	Application	QoS Latency	Jitter
1	Controller-to-controller	100 ms	./.
2	Distributed I/O devices	10 ms	./.
3	Motion control	$\leq 1ms$	$\leq 1\mu s$

Here it is essential for the correct operation of a technical process that different events in a distributed system are executed quasi-simultaneously. Deviation from an ideal simultaneity is referred to as jitter. The requirements in terms of communication quality in factory automation for the area near to the process can be roughly divided into three classes (see Table 1). The definition of these rough classes is based on the experiences with existing classes of applications in the field bus technology.

3. FACTORS, WHICH INFLUENCE THE REALTIME BEHAVIOR OF A BRIDGED ETHERNET SYSTEM

Figure 2 shows a bridged Ethernet system with several connected devices. A bridged Ethernet system can be modeled as a network of queueing systems. A range of factors influence the time response and thus the realtime behavior with regard to latency and jitter. These factors have been examined in the form of a sensitivity analysis in (Jasperneite, 2002).

Starting with the devices, the properties of the packet generation process (packet sizes, burstiness, etc.) and the dwell time, which is the result of runtimes in the protocol stacks, must be considered.

The dwell time of a TCP/IP protocol stack can quickly become a multiple of the MAC runtimes in a bridged Ethernet system (Jasperneite and Neumann, 2001).

Another influencing factor is the scheduling strategy in the devices and the bridges. According to the IEEE standard, FCFS and non-preemptive priority queuing (PQ) can be found in a bridged Ethernet system. Furthermore, when using PQ it is still impossible to prevent high priority packets being delayed by lower priority packets, which are just served in a bridge or device. This blocks the forwarding operation of the high priority packets, which results in higher transaction times and jitter values, especially when using long chains of bridges. This property means that Ethernet cannot be used directly in time-critical synchronous applications, e.g., motion control. This is why intervention is required in the scheduling, which is part of various vendor-driven approaches.

The combination of information flow distribution and network topology also has a great influence on the realtime behavior. The more connections common subsections have along their individual paths, the greater the temporal influence of packet ordering in the relevant bridges. In addition, long chains of switches increase the achievable transaction times (Krommenacker *et al.*, 2001), (Rüping *et al.*, 1999). This is a crucial point when Ethernet has to be used to simulate the line structures of today fieldbuses.

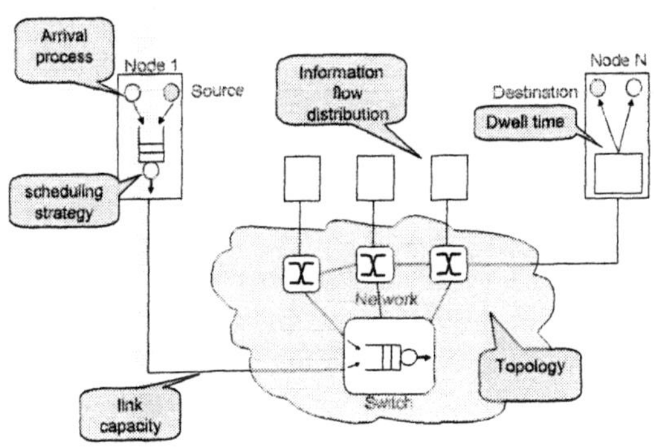

Fig. 2. The various factors, which influence the realtime behavior of a bridged Ethernet system

4. KEY CONCEPTS FOR GUARANTEED REALTIME ETHERNET

In the past, a lot of research work deals with modifications of the MAC-layer of Ethernet because of the statistical nature of the CSMA/CD procedure (Pöschmann, 1999), (Chiueh and Venkatramini, 1994), (Shimokawa and Shiobara, 1985), (Kweon *et al.*, 1999), (Bello and Mirabella, 2001). Today Realtime Ethernet is based on the exclusive use of bridging technology (IEEE, 1998*a*) and the simultaneous use of full duplex transmission (micro segmentation) (IEEE, 1997). This ensures that no more collisions can occur on the medium and therefore makes the CSMA/CD method that was previously used in Ethernet applications obsolete. Using priorities according to IEEE 802.1D/Q create an effective method for isolating time-critical and best-effort data. As mentioned above, the end-to-end delay can be reduced by factor 10 by optimizing the runtimes in the devices. These measures can be used to achieve QoS Class 2, as shown in Table 1.

For additional improvement in terms of latency and jitter, the TDMA (Time Division Multiple Access) method and time synchronization play a very important role.

4.1 TDMA method

Especially in a chain of bridges the latency and the jitter for real-time data transfer will be increased significantly. This leads to a restriction of using standard Ethernet in time-critical synchronous applications, e.g. Motion Control. To use Ethernet also for this class of applications (QoS class 3 regarding to table 1) extensions to the standard are necessary which often based on TDMA mechanism, where pre-allocated time slots are defined for the transmission of time-critical and non realtime data (cf. Fig. 3). The application of such a scheduling scheme has the advantage that the transmission times can be computed simply a priori, however the use of standard components is strongly limited. The replacement of the

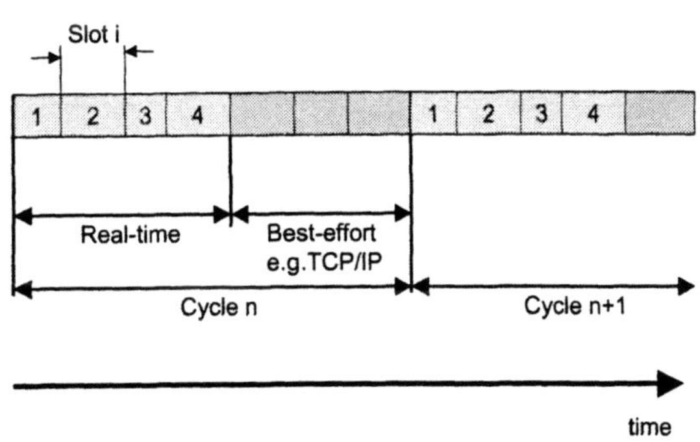

Fig. 3. TDMA method

CSMA/CD method with a token passing procedure was suggested as early as 1985 (Shimokawa and Shiobara, 1985) for the transmission of time-critical data in industrial applications. There are several ways to implement a TDMA method. The easiest option is a poll-select procedure, as is the case for Powerlink (Powerlink, 2003). Profinet IRT (IEC, 2003) uses a completely distributed TDMA method.

4.2 Time Synchronization

In order to provide isochronous conditions in a distributed system, time synchronization between the relevant devices is required. One option is to differentiate between data transfer and the execution of resulting actions. In this way, the actions in the system are only based on temporal criteria. This method is used, for example, by CIP Sync (Ethernet/IP, 2003) .

On the other hand, a distributed TDMA method such as that used in Profinet IRT, also requires appropriate time synchronization for scheduling the realtime data and best-effort data phases, as well as for packet ordering within the realtime data phase.

In both cases, the distributed clocks must be synchronized precisely. An emerging standard for this is the Precision Time Protocol (PTP) (IEEE, 2002). This standard is based on a master/slave procedure. The local clocks are adjusted using special synchronization telegrams. A defined algorithm is used to select the most accurate clock in the network as the master clock. Synchronism of $< 1\mu s$ can be achieved in a switched Ethernet network with the appropriate hardware support. For the use of PTP in chains of Bridges enhancements such as bypass clocks are suggested (Mueller and Weber, 2003). More detailed information on this standard can be found in the proceedings of the last IEEE 1588 workshop (Kang, L. and Eidson, J., 2003).

5. CLASSIFICATION OF VENDOR-DRIVEN APPROACHES UNDER DISCUSSION

The vendor-driven Real-time Ethernet approaches, which are currently under discussion, can be divided into 3 classes with regard to the used architecture (see Figure 4). The development target of these commercial approaches is to fulfill primarily the temporal requirements represented in the Table 1.

Class 1 describes the use of standard Ethernet TCP/IP as it is. In this case the different real time protocols and the best-effort protocols, like HTTP, SNMP, FTP etc., uses the services of the TCP/IP protocol suite. This includes examples such as CIP Sync (Ethernet/IP, 2003), ModBus/TCP (ModBus, 2003), and IDA (IDA, 2001). The class 1 has the largest conformity to the Ethernet TCP/IP standard and can thereby use standard hardware and software components.

Class 2 introduces optimizations, whereby the realtime data bypasses the TCP/IP stack and thus considerably reduces the dwell time in the node, and increases the achievable packet rate. The

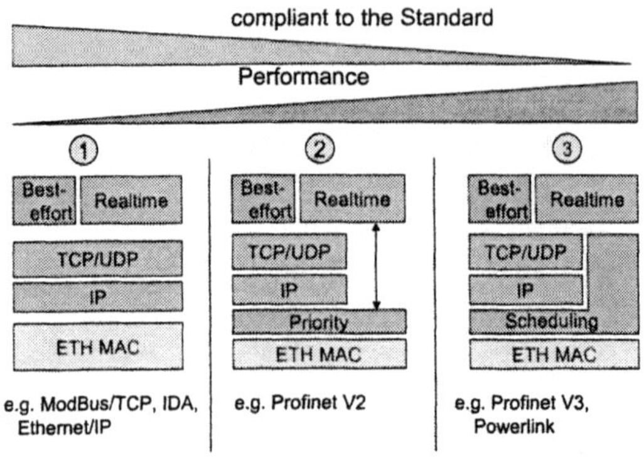

Fig. 4. Structural classification of the different vendor-driven approaches for realtime Ethernet

dwell time of the node is one of the substantial influence factors for the realtime performance and has for embedded devices typical values of $1ms$ (Jasperneite and Neumann, 2001). The soft realtime version of Profinet SRT (PNO, 2003) should be assigned to this class, for example.

In Classes 1 and 2, the priority support described in IEEE 802.1Q can also be used depending on the approach (IEEE, 1998*b*).

In Class 3 the scheduling on the MAC level is again modified through the introduction of a TDMA method. For example, a central poll-select procedure is used in Powerlink (Powerlink, 2003) and a procedure that is distributed in the switches is used in the Profinet IRT version. Class 3 can be used in applications that require maximum latency in the range $1ms$ and a maximum jitter of $< 1\mu s$. In this class there are strong restrictions for the use of standard components or the necessity for special components, like switches.

As shown in Figure 4, conformance with the Ethernet standard decreases from left to right, while the achievable realtime performance increases in the same direction.

6. SUMMARY

Realtime behavior is guaranteed when the requirements for timeliness and simultaneity are met. Realtime Ethernet is based on the exclusive use of bridging technology and the simultaneous use of full duplex transmission (micro segmentation). This ensures that no more collisions can occur on the medium and therefore makes the CSMA/CD method that was previously used in Ethernet applications obsolete. The class of service concept according to IEEE 802.1D/Q create an effective method for isolating time-critical and best-effort applications. The end-to-end delay can be reduced by factor 10 by optimizing the runtimes in the

TCP/IP protocol stacks of the relevant devices. As a result of these standardized concepts, a MAC latency in low double figures (ms) can be achieved.

When using Ethernet in applications with latency in the range less than $1ms$ and a jitter of less than $1\mu s$, such as in motion control, intervention in the scheduling of the MAC level remains unavoidable. Here time slot methods are used in the vendor-driven approaches currently under discussion, whereby one time slot is reserved for the transmission of time-critical data and another is reserved for the transmission of TCP/IP applications, for example.

The price to pay for this improved performance is in the limited use of standard Ethernet components (chips, switches, etc.).

REFERENCES

AS-Interface (2003). http://www.as-interface.com.

Bello, L. Lo and O. Mirabella (2001). Design Issues for Ethernet in Automation. In: *Proceedings of the 8th IEEE international Conference on Emerging Technologies and Factory Automation (ETFA'01)*. Antibes, France. pp. 213–222.

Chiueh, T. and C. Venkatramini (1994). Supporting Real-Time Traffic on ethernet. In: *Proceedings of IEEE Real-time Systems Symposium.* pp. 282–286.

Clatha, A. (1998). Automation Technologies & Strategies for the Millenium. In: *ARC Automation Technologies Forum 98.* Orlando, U.S.

Ethernet/IP (2003). *Ethernet Industrial Protocol (EtherNet/IP).* www.ethernet-ip.org.

IDA (2001). *IDA Architecture Description and Specification V1.0.* IDA-Group. www.ida-group.org.

IEC (2003). *Digital data communications for measurement and control - Profiles for ISO/IEC 8802-3 based communication networks in real time applications MT9 / WG1.* new work item proposal.

IEEE (1997). *Specification for 802.3 Full Duplex Operation.* IEEE. New York. ANSI/IEEE Std. 802.3x.

IEEE (1998*a*). *Media Access Control (MAC) Bridges.* IEEE. New York. ANSI/IEEE Std 802.1D-1998.

IEEE (1998*b*). *Virtual Bridged Local Area Networks.* IEEE. New York. ANSI/IEEE Std 802.1Q-1998.

IEEE (2002). *IEEE Standard for a Precision Clock Synchronization Protocol for Networked Measurement and Control Systems.* IEEE. New York. ANSI/IEEE Std 1588-2002.

INTERBUS (2003). *INTERBUS Club.* http://www.interbusclub.com.

Jasperneite, J. (2002). *Performance Evaluation of a Class-of-service based Local Area Network for using at the Field device level (only in german).* Shaker Verlag. Aachen, Germany. ISBN 3-8322-0832-1, Dissertation.

Jasperneite, J. and P. Neumann (2001). Switched Ethernet for Factory Automation. In: *8th IEEE International Conference on Emerging Technologies and Factory Automation.* Antibes - Juan les Pins, France. pp. 205–212.

Jensen, E.D., C.D. Locke and H. Tokuda (1985). A Time-Driven Scheduling Model for Real-Time Operating Systems. In: *Real-Time Systems Symposium.* IEEE.

Kang, L. and Eidson, J., Ed.) (2003). *Workshop on IEEE-1588, Standard for a Precision Clock Synchronization Protocol for Networked Measurements and Control Systems.* National Insitute of Standards and Technology (NIST).

Krommenacker, N., E. Rondeau and T. Divoux (2001). Study of Algorithms to define the cabling plan of switched ethernet for real-time applications. In: *Proceedings of the 8th IEEE international Conference on Emerging Technologies and Factory Automation (ETFA'01).* Antibes, France. pp. 223–230.

Kweon, S.-K., K. G. Shin and Q. Zheng (1999). Statistical real-time communication over ethernet for manufacturing automation systems. In: *IEEE Real-Time Technology and Applications Symposium.* pp. 192–202.

ModBus (2003). *MODBUS Users Web Site:.* http://www.modbus.org.

Mueller, M. and K. Weber (2003). Impact of Switch Cascading on Time Accuracy. In: *Workshop on IEEE-1588, Standard for a Precision Clock Synchronization Protocol for Networked Measurements and Control Systems.* National Insitute of Standards and Technology (NIST).

ODVA (2003). *Open DeviceNet Vendor Association.* http://www.odva.org.

PNO (2003). *Profibus User Organization.* www.profibus.com.

Pöschmann, A. (1999). Profibus-DP-Protokoll über Ethernet. interner Projekt-Bericht. Institut für Automation und Kommunikation e.V. Magdeburg (IFAK). Barleben.

Pöschmann, A. (2001). Chancen und Nutzen von Ethernet TCP/IP in der Automation. In: *GMA-Kongress 2001: Automatisierungstechnik im Spannungsfeld neuer Technologien, VDI-Berichte Nr. 1608.* Baden-Baden. pp. 617–626.

Powerlink (2003). *Ethernet Powerlink Standardization Group.* www.ethernet-powerlink.org.

Ricot, C. (2001). Challenges and opportunities in industrial automation. In: *Invited Talk at the* 8th *IEEE international Conference on Emerging Technologies and Factory Automation (ETFA'01).* Antibes, France.

Rüping, S., E. Vonnahme and J. Jasperneite (1999). Analysis of Switched Ethernet Networks with different Topologies used in Automation Systems. In: *Field Bus Technology* (D. Diedrich, P. Neumann and H. Schweinzer, Eds.). pp. 351–358. Springer-Verlag.

Shimokawa, Y. and Y. Shiobara (1985). Realtime Ethernet for industrial applications. In: *Proceedings of the international Conference on Industrial Electronics, Control and Instrumentation (IECON'85).* San Francisco. pp. 829–834.

Copyright © IFAC Information Control Problems in
Manufacturing, Salvador, Brazil, 2004

ELSEVIER
IFAC
PUBLICATIONS
www.elsevier.com/locate/ifac

STABILITY OF NETWORKED CONTROL SYSTEMS: AN ANALYSIS OF CONTROL DELAY BASED ON PROBABILITY DENSITY FUNCTIONS

Edmar José do Nascimento *
Antonio Marcus Nogueira Lima *
Francisco Marcos de Assis *

* *Departamento de Engenharia Elétrica*
Universidade Federal de Campina Grande
Caixa Postal 10.105 CEP 58.109-970 - Campina Grande
PB - Brazil
Emails: {jnedmar,amnlima,fmarcos}@dee.ufcg.edu.br

Abstract: In this paper, we analyze the stability of a Networked Control System with random delay. We consider a first order continuous process that is sampled at constant intervals, a discrete controller and a control delay less than one sampling period. Analytical expressions are obtained for the probability density function (pdf) of the eigenvalues of the closed loop system. Our objective is to determine the probability of the system become unstable for given design parameters and for a given pdf model for the control delay. *Copyright © 2004 IFAC*

Keywords: NCS, Stability, Random, Delay analysis, Probability density function

1. INTRODUCTION

Real-time networks are being more and more used to close feedback control loops in a distributed control scenario. This kind of system is often called a Networked Control System, NCS for short, see (Branicky *et al.*, 2000). Some advantages of a NCS are: reduced system wiring, easy of system maintenance and increased system agility. In a NCS, network nodes share a transmission medium in order to exchange information among them. The network nodes of interest in a control application are sensor nodes, controller nodes and actuator nodes. Sensor nodes are responsible for gathering data from a physical process and transmit them through the network. Controller nodes receive data from sensors and calculate control signals that are sent back to the process through the network and finally, actuator nodes are responsible of applying control signals to the process.

One of the main problems of a NCS is the time-delay which is mainly related to network protocols and scheduling techniques. Such time-delays may degrade performance or even destabilize the system. In addition to that, if the time-delay is random, the closed-loop system is no longer time-invariant. Consequently, the standard computer control theory can not be used in the analysis and design of distributed real-time control systems.

Besides the traditional industrial networks like Profibus and CAN, it has been of great interest to use some cost-effective network standards like Ethernet to provide real-time communication, see (Alt and Lages, 2003) and (Kweon and Shin, 2003). Furthermore, recent technological advances in wireless devices have made possible to

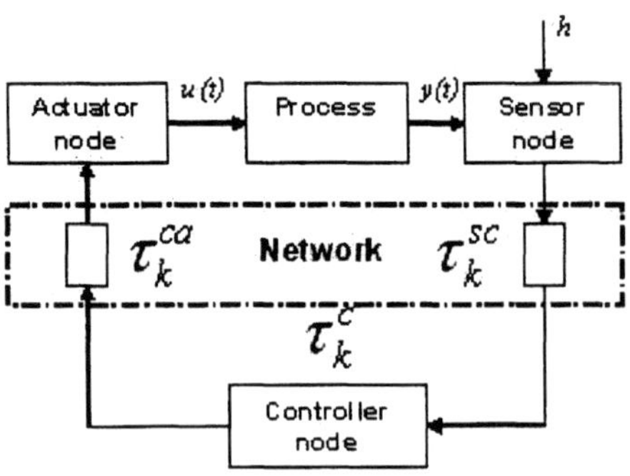

Fig. 1. A NCS with one sensor node, one controller node and one actuator node. Feedback loop is closed through the network.

integrate common control networks with wireless devices. Given the wide range of available network technologies and protocols, it is difficult to provide general models to analyze time-delay effects in the performance of a NCS.

(Branicky *et al.*, 2000) has analyzed the effect of constant delays in the stability of the system; the trade-off between the sampling rate and the maximum tolerable delay has been investigated and the use of estimators for compensating for network delays from sensor to controller has been proposed. In (Xiao *et al.*, 2000), delay was represented as a finite-dimensional, discrete-time jump linear system, with the transition jumps being modelled as finite-state Markov chains. In (Nilsson, 1998), a framework for a NCS with time-delay was proposed. It was considered a system model where there was one sensor, one controller and one actuator, see Figure (1). Three kinds of delay models were analyzed there: constant delay, random memoryless delay and random delay with memory with the distribution selected from an underlying Markov chain. For the random memoryless delay model, design techniques for designing an optimal controller were provided as well as methods for evaluating mean square stability. In order to validate the proposed models, experiments were done with two commercial networks, Ethernet and CAN. This study has revealed that the delay patterns are strongly dependent on parameters like priority level of nodes, network load and inner-characteristics of the network protocol, but in general, the delay varies in a random fashion. A scheme proposed by (Luck and Ray, 1990) allows to get rid of the undesirable effects of randomness by introducing buffers designed for the worst case delay. However the price to be paid is performance degradation because the controller is always using old data to calculate control signals, see (Nilsson, 1998).

In this paper we develop our analysis in a framework that is similar to the one adopted by (Nilsson, 1998). However, we consider that the

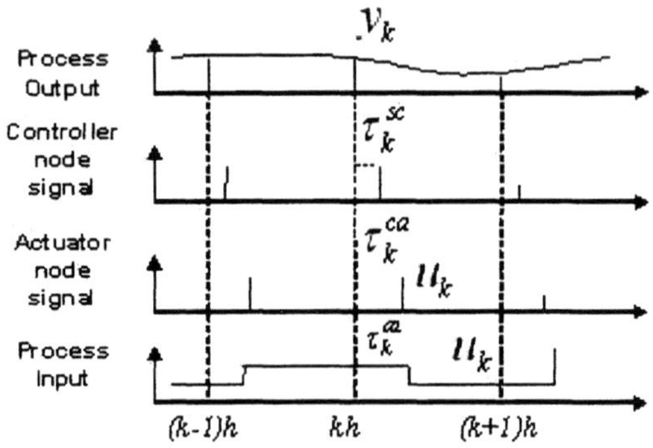

Fig. 2. Timing diagram where is shown the sampling instants in the sensor node, the arrival instants of packet data at the controller node, the arrival instants of data from controller at the actuator and the control signal being applied to the process input respectively.

delay is random and its randomness can be described by a probability density function (pdf). In order to analyze the random behavior of the time-delay induced by the network, we will analyze the probability density function (pdf) of the closed loop system poles. The pdf of the poles allows to calculate the probability for the closed-loop system to become unstable under given design, process and network parameters. This paper is organized as follow. In section 2 we describe the framework we will adopt. In section 3 the pdf of the closed loop eigenvalues is calculated. In section 4 simulation results are presented. Finally, in section 5 we present some conclusions.

2. MODEL DESCRIPTION

We consider the system model proposed in (Nilsson, 1998) where there is only one sensor, one controller, one actuator and one continuous-time process, see Figure 1. The sensor we are considering is time-driven, i.e., its readings are sampled periodically each h time-units while the controller and actuator are event-driven. This means that the control signal is calculated as soon as data from sensors are received and that the actuator applies the control signal to the process just after receiving the data, see Figure 2.

As it can be seen in Figure 1, in order to reach the controller node, sensor data must be sent through the network. The same must be done for the control signal calculated by the controller node in order to reach the actuator node. Thus, there is a delay induced by the network that may be split into:

- Communication delay between the sensor and the controller, τ_k^{sc}
- Computational delay in the controller, τ_k^c

- Communication delay between the controller and the actuator, τ_k^{ca}

The execution time in sensor and actuator nodes can be included in τ_k^{sc} and τ_k^{ca} respectively, without loss of generality. We consider in our model that these delays are random and its respective pdf are known. Furthermore, successive delays are considered to be independent. The control delay τ_k is defined as being the total delay, i.e, $\tau_k = \tau_k^{sc} + \tau_k^{c} + \tau_k^{ca}$. From now on, we consider only the control delay, i.e., we disregard that it is composed by three terms. For our analysis, this assumption is not strong because we are interested only in the closed loop stability and, for such goal it is sufficient the fact that pdfs can be combined in just one and the resulting pdf is known. We also consider a gamma distribution for τ_k with parameters μ and b given by

$$f_{\tau_k}(\tau_k) = \frac{\mu^b}{\Gamma(b)}\tau_k^{b-1}e^{-\mu\tau_k}, \quad \tau_k \in [0,\infty) \qquad (1)$$
$$\Gamma(b) = \int_0^\infty s^{b-1}e^{-s}ds.$$

The gamma distribution was chosen because of its flexibility. In fact, when b is large the gamma distribution closely approximates a normal distribution with the advantage that the former has density only for positive real numbers. The model which includes a continuous process, a discrete controller and the control delay is given by:

$$\dot{x} = Ax(t) + Bu(t), \qquad (2)$$
$$y(t) = Cx(t), t \in [kh + \tau_k, (k+1)h + \tau_k)$$
$$u(t^+) = -Kx(t - \tau_k), \quad t \in \{kh + \tau_k\} \qquad (3)$$

where $u(t^+)$ is piecewise continuous and only changes value at $kh + \tau_k$. As can be seen, we dropped out disturbances of our model in order to simplify the analysis.

Sampling the continuous process with period h and assuming that the control delay in one control cycle is less than h, we obtain the following discrete time model, see (Åström and Wittenmark, 1997).

$$x((k+1)h) = \Phi x(kh) + \Gamma_0(\tau_k)u(kh) \qquad (4)$$
$$+\Gamma_1(\tau_k)u((k-1)h)$$
$$y(kh) = \quad Cx(kh)$$

where

$$\Phi = e^{Ah},$$
$$\Gamma_0(\tau_k) = \int_0^{h-\tau_k} e^{As}Bds,$$
$$\Gamma_1(\tau_k) = \int_{h-\tau_k}^{h} e^{As}Bds$$

If the delay is larger than the sampling period, the state-space description of the sampled system will have a higher dimension, see (Åström and Wittenmark, 1997). Limiting the maximum delay to one sampling period guarantees that successive samples will not arrive in a corrupted order. In order to bound the delay, one could set a timeout in the controller so that the control signal is calculated from a prediction if the sensor data are not available on time, see (Nilsson, 1998). Defining

$$z(kh) = [x^T(kh), u^T((k-1)h)]^T$$

as the augmented state vector, the augmented closed-loop system is given by

$$z((k+1)h) = \tilde{\Phi}(k)z(kh) \qquad (5)$$

where

$$\tilde{\Phi}(k) = \begin{bmatrix} \Phi - \Gamma_0(\tau_k)K & \Gamma_1(\tau_k) \\ -K & 0 \end{bmatrix}$$

In order to obtain analytical expressions for the eigenvalues pdf we consider, from now on, a first order unstable ($a > 0$) open loop process and a linear feedback control law, i.e.,

$$\dot{x} = ax(t) + bu(t), \qquad (6)$$
$$u(t^+) = -Kx(t - \tau_k) \qquad (7)$$

Similarly, if this system is sampled with period h, its closed loop matrix will be given by

$$\tilde{\Phi}(k) = \begin{bmatrix} e^{ah} - \frac{K}{a}(e^{ah-a\tau_k} - 1) & \frac{1}{a}e^{ah}(1 - e^{-a\tau_k}) \\ -K & 0 \end{bmatrix}$$
$$= \begin{bmatrix} (K\varphi + \beta) & \varphi \\ -K & 0 \end{bmatrix} \qquad (8)$$

where

$$\varphi = \frac{1}{a}e^{ah}(1 - e^{-a\tau_k}), \; \beta = e^{ah} - \frac{K}{a}e^{ah} + \frac{K}{a}$$

3. EIGENVALUES ANALYSIS

The eigenvalues of the closed loop system are obtained by solving the characteristic equation

$$\det\left(\lambda I - \tilde{\Phi}\right) = \lambda^2 - (K\varphi + \beta)\lambda + K\varphi = 0$$

whose solutions are

$$\lambda_1 = \frac{1}{2}\left[(K\varphi + \beta) + \sqrt{(K\varphi + \beta)^2 - 4K\varphi}\right] \quad (9)$$
$$\lambda_2 = \frac{1}{2}\left[(K\varphi + \beta) - \sqrt{(K\varphi + \beta)^2 - 4K\varphi}\right] (10)$$

Introducing $\eta = \frac{1}{2}(K\varphi + \beta)$ we get a more compact expression for 9 and 10, as shown below

$$\lambda_1(\eta) = \eta + \sqrt{\eta^2 - 2\eta + \beta} \tag{11}$$

$$\lambda_2(\eta) = \eta - \sqrt{\eta^2 - 2\eta + \beta} \tag{12}$$

Note that expressions (11) and (12) may be complex-valued. However, if we are interested in closed loop stability of a discrete system, we need only to check the absolute value of the eigenvalues. The closed loop system is stable if the eigenvalues are inside the unit circle, see (Åström and Wittenmark, 1997).

Considering the case where $\eta^2 - 2\eta + \beta$ has real roots, i.e., $\eta_1 = 1 - \sqrt{1-\beta}$ and $\eta_2 = 1 + \sqrt{1-\beta}$ are real-valued. In this case, the absolute values of λ_1 and λ_2 are given by

$$|\lambda_1(\eta)| = \begin{cases} \sqrt{2\eta - \beta} & \eta_1 < \eta < \eta_2 \\ |\eta + \sqrt{\eta^2 - 2\eta + \beta}| & \text{otherwise} \end{cases} \tag{13}$$

$$|\lambda_2(\eta)| = \begin{cases} \sqrt{2\eta - \beta} & \eta_1 < \eta < \eta_2 \\ |\eta - \sqrt{\eta^2 - 2\eta + \beta}| & \text{otherwise} \end{cases} \tag{14}$$

If $\eta^2 - 2\eta + \beta$ has not real roots the analysis is simpler than we did just before because λ_1 and λ_2 are real-valued for all η.

3.1 PDF Calculation

As it was mentioned before, τ_k is considered as a random variable with a known pdf. From probability theory it is known that φ, η, $|\lambda_1|$ and $|\lambda_2|$ are also random variables because they are functions of τ_k. Our objective is to find the pdf of $|\lambda_1|$ and $|\lambda_2|$. The procedure used to find a pdf of a function of a random variable is described as follow, see (Viniotis, 1998).

For a given random variable X, with pdf $f_X(x)$ and a given transformation $Y = g(X)$, the pdf of the random variable $Y = g(X)$ is given by

$$f_Y(y) = \sum_{l=1}^{n} f_X(x_l) . \frac{1}{|dg(x_l)/dx|} \tag{15}$$

where the x_l are the real roots of the equation $y - g(x) = 0$ solved for x.

Following the procedure we have just quoted in (15), and assuming that τ_k has a gamma distribution given by (1), we get the following distribution for η.

$$f_\eta(\eta) = C(\ln \frac{g(\eta)}{K})^{b-1} g(\eta)^{\frac{\mu}{a}-1} \tag{16}$$
$$\eta \in \Big[\frac{\beta}{2}, \frac{1}{2}(\frac{K}{a}e^{ah} + \beta)\Big)$$

where

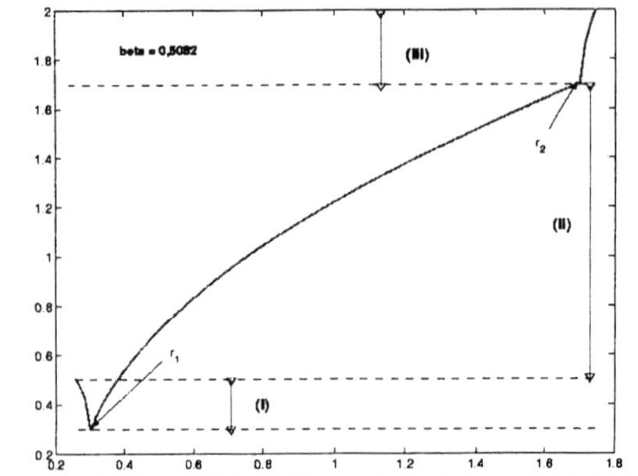

Fig. 3. $|\lambda_1|$ versus η. $|\lambda_1|$ is split into three intervals.

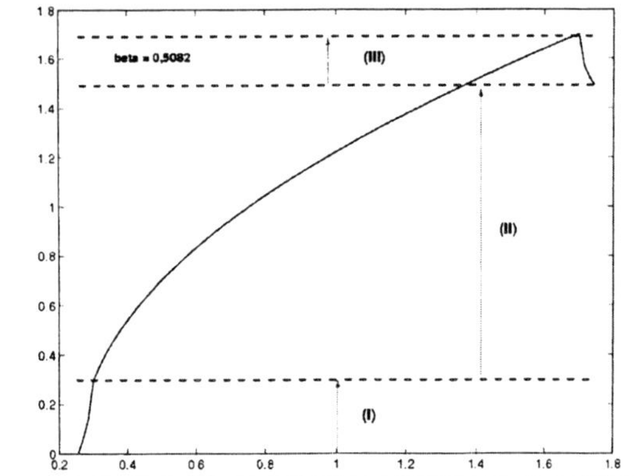

Fig. 4. $|\lambda_2|$ versus η. $|\lambda_2|$ is split into three intervals.

$$g(\eta) = K - ae^{-ah}(2\eta - \beta),$$
$$C = (-1)^{b-1} \frac{2\mu^b e^{-ah}}{K\Gamma(b)a^{b-1}}$$

To determine analytical expressions for the pdf of $|\lambda_1|$ and $|\lambda_2|$ it is important to observe the plots of $|\lambda_1| \times \eta$ and $|\lambda_2| \times \eta$ over the domain where η is defined. As can be seen in Figures 3 and 4, we can calculate the pdf of $|\lambda_1|$ and $|\lambda_2|$ by taking into account three different intervals, each one being defined by a specific expression. Thus, $|\lambda_1|$ domain, $\mathcal{D}_{|\lambda_1|}$, is then partitioned in $\mathcal{D}_{|\lambda_1|} = \mathcal{D}^1_{|\lambda_1|} \cup \mathcal{D}^2_{|\lambda_1|} \cup \mathcal{D}^3_{|\lambda_1|}$ where

$$\mathcal{D}^1_{|\lambda_1|} = \Big[|\eta|\big|_{\eta=\eta_1}, |\eta + \sqrt{\eta^2 - 2\eta + \beta}|\big|_{\eta=\frac{\beta}{2}}\Big),$$
$$\mathcal{D}^2_{|\lambda_1|} = \Big[|\eta + \sqrt{\eta^2 - 2\eta + \beta}|\big|_{\eta=\frac{\beta}{2}}, |\eta + \sqrt{\eta^2 - 2\eta + \beta}|\big|_{\eta=\eta_2}\Big),$$
$$\mathcal{D}^3_{|\lambda_1|} = \Big[|\eta + \sqrt{\eta^2 - 2\eta + \beta}|\big|_{\eta=\eta_2}, |\eta + \sqrt{\eta^2 - 2\eta + \beta}|\big|_{\eta=\frac{K}{a}e^{ah}+\beta}\Big).$$

Similarly for $|\lambda_2|$ we have $\mathcal{D}_{|\lambda_2|} = \mathcal{D}^1_{|\lambda_2|} \cup \mathcal{D}^2_{|\lambda_2|} \cup \mathcal{D}^3_{|\lambda_2|}$ where

$$\mathcal{D}^1_{|\lambda_2|} = \Big[|\eta - \sqrt{\eta^2 - 2\eta + \beta}|\big|_{\eta=\frac{\beta}{2}}, |\eta|\big|_{\eta=\eta_1}\Big),$$
$$\mathcal{D}^2_{|\lambda_2|} = \Big[|\eta|\big|_{\eta=\eta_1}, |\eta - \sqrt{\eta^2 - 2\eta + \beta}|\big|_{\eta=\frac{K}{a}e^{ah}+\beta}\Big),$$
$$\mathcal{D}^3_{|\lambda_2|} = \Big[|\eta - \sqrt{\eta^2 - 2\eta + \beta}|\big|_{\eta=\frac{K}{a}e^{ah}+\beta}, |\eta|\big|_{\eta=\eta_2}\Big).$$

Then, we can obtain the pdf of $|\lambda_1|$ and $|\lambda_2|$ by using equation (15) and by solving equations (13) and (14) for η. The pdf of $|\lambda_1|$ is given by the equations (17), (18) and (19).

$$f_{|\lambda_1|}(|\lambda_1|) = f_\eta(\eta).\sqrt{2\eta - \beta}\Big|_{\eta=\eta^a} + f_\eta(\eta).\Big|\frac{\sqrt{\eta^2 - 2\eta + \beta}}{\eta - 1 + \sqrt{\eta^2 - 2\eta + \beta}}\Big|\Big|_{\eta=\eta^b}, \quad |\lambda_1| \in \mathcal{D}^1_{|\lambda_1|} \tag{17}$$

$$f_{|\lambda_1|}(|\lambda_1|) = f_\eta(\eta).\sqrt{2\eta - \beta}\Big|_{\eta=\eta^a}, \quad |\lambda_1| \in \mathcal{D}^2_{|\lambda_1|} \tag{18}$$

$$f_{|\lambda_1|}(|\lambda_1|) = f_\eta(\eta).\Big|\frac{\sqrt{\eta^2 - 2\eta + \beta}}{\eta - 1 + \sqrt{\eta^2 - 2\eta + \beta}}\Big|\Big|_{\eta=\eta^c} \quad |\lambda_1| \in \mathcal{D}^3_{|\lambda_1|} \tag{19}$$

where

$$\eta^a = \frac{|\lambda_1|^2 + \beta}{2}, \eta^b = \frac{(|\lambda_1|^2 - \beta) - |\lambda_1||\lambda_1|^2 - \beta|}{2(|\lambda_1|^2 - \beta)}$$
$$\eta^c = \frac{(|\lambda_1|^2 - \beta) + |\lambda_1||\lambda_1|^2 - \beta|}{2(|\lambda_1|^2 - \beta)}$$

The pdf of $|\lambda_2|$ is given by the equations (20), (21) and (22).

$$f_{|\lambda_2|}(|\lambda_2|) = f_\eta(\eta).\Big|\frac{\sqrt{\eta^2 - 2\eta + \beta}}{1 - \eta + \sqrt{\eta^2 - 2\eta + \beta}}\Big|\Big|_{\eta=\eta^e} \quad |\lambda_2| \in \mathcal{D}^1_{|\lambda_2|} \tag{20}$$

$$f_{|\lambda_2|}(|\lambda_2|) = f_\eta(\eta).\sqrt{2\eta - \beta}\Big|_{\eta=\eta^d}, \quad |\lambda_2| \in \mathcal{D}^2_{|\lambda_2|} \tag{21}$$

$$f_{|\lambda_2|}(|\lambda_2|) = f_\eta(\eta).\sqrt{2\eta - \beta}\Big|_{\eta=\eta^d} + f_\eta(\eta).\Big|\frac{\sqrt{\eta^2 - 2\eta + \beta}}{1 - \eta + \sqrt{\eta^2 - 2\eta + \beta}}\Big|\Big|_{\eta=\eta^f}, \quad |\lambda_2| \in \mathcal{D}^3_{|\lambda_2|} \tag{22}$$

where

$$\eta^d = \frac{|\lambda_2|^2 + \beta}{2}, \eta^e = \frac{(|\lambda_2|^2 - \beta) - |\lambda_2||\lambda_2|^2 - \beta|}{2(|\lambda_2|^2 - \beta)}$$
$$\eta^f = \frac{(|\lambda_2|^2 - \beta) + |\lambda_2||\lambda_2|^2 - \beta|}{2(|\lambda_2|^2 - \beta)}$$

It must be remarked that the shape of Figure 3 and 4 changes for some values of the β-parameter. In this case the expressions for the pdf of $|\lambda_1|$ and $|\lambda_2|$ will be a little different of the ones we got for this particular shape. However, the procedure to get them still remains the same. It is also not difficult to extend the earlier results to pdfs other than gamma because the expressions relating $|\lambda_1|$ and $|\lambda_2|$ to η remain the same for the model given by (6) and (7).

4. STABILITY OF A NCS

A gamma pdf with parameters μ and b as defined by (1) has expected-value b/μ and variance b/μ^2, so we can interpret b/μ as being the average delay. We want to know, under given process parameters and a given average delay, what is the probability of the system to become unstable if we are free to change the design parameters K and h.

Simulations were carried out for the following scalar system.

$$\dot{x} = x + u(t)$$
$$u(t^+) = -Kx(t - \tau)$$

As a rule of thumb, see (Åström and Wittenmark, 1997), reasonable values for h vary from 0.1 to 0.25 for $a = 1$. In a first analysis we take $h = 0.25$, $\mu = 125$ and $b = 10$ resulting in an average delay of $b/\mu = 0.08 = 32\%h$. To be consistent with our model that assumes a control delay less than one sampling period, we need to calculate the probability of τ_k be larger than h, i.e., the probability our sampled model fails. For the parameters chosen, this probability is calculated by integrating (1), then, in this case $P[\tau_k > h] = 2.90 \times 10^{-6}$. This value is considered to be acceptable in our analysis.

In Figures 5 and 6 it is shown the pdf of $|\lambda_1|$ for $K = 2$ and $K = 5$, respectively. In table 1, we have the probability of having eigenvalues of magnitude larger than one and the expected-value of $|\lambda_1|$ and $|\lambda_2|$ for different values of K. In a second analysis, we fixed the same average delay and a proportional gain of $K = 8$, but we varied the sampling period between $h = 0.1$ and $h = 0.25$. The resulting probabilities are shown in table (2).

As it was expected, if we increase the gain K, the probability of having unstable eigenvalues also

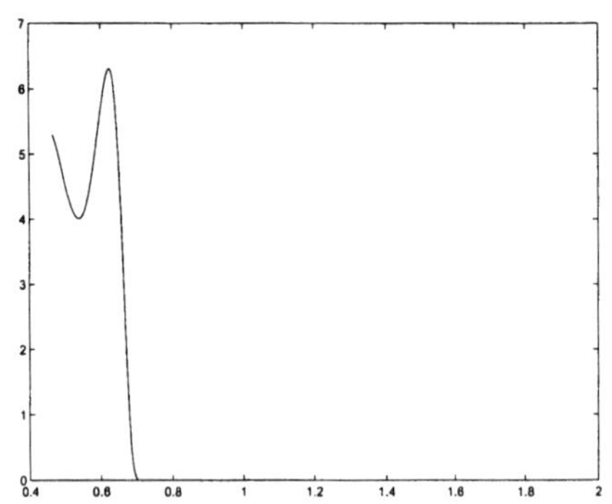

Fig. 5. The pdf of $|\lambda_1|$ for $K = 2$, $b/\mu = 0.08$ and $h = 0.25$

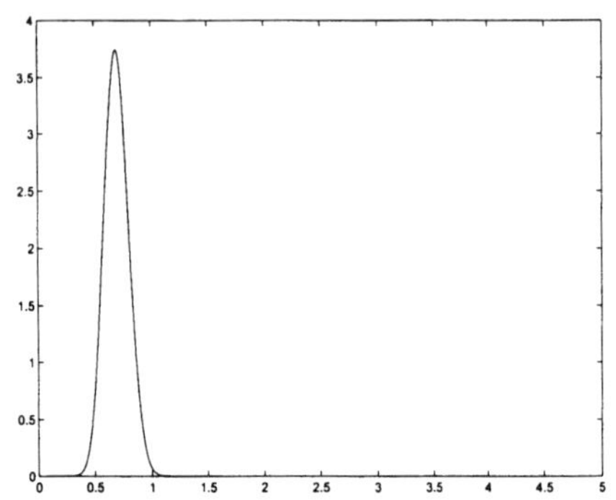

Fig. 6. The pdf of $|\lambda_1|$ for $K = 5$, $b/\mu = 0.08$ and $h = 0.25$

increases. In fact, as shown in Figures 5 and 6, the pdf acquires a regular shape and moves towards right as K increases. We can observe that the probability of having unstable modes can be decreased, even for large controller gains ($K = 8$), if we choose a small sampling period. These observations suggest that one could combine higher gains with faster sampling rate in order to get rid of transients and smaller gains with slower sampling rate in a steady state.

Table 1. For different values of K it is calculated the expected-value and the probability of having unstable modes

K	$E[\|\lambda_1\|]$	$E[\|\lambda_2\|]$	$P[\|\lambda_1\| > 1]$	$P[\|\lambda_2\| > 1]$
2	0.5729	0.3549	6.81e-17	6.82e-017
3	0.5371	0.5371	2.57e-08	2.58e-08
5	0.6933	0.6933	0.0025	0.0025
7	0.8204	0.8204	0.0786	0.0786
8	0.8770	0.8770	0.1792	0.1791
10	0.9799	0.9816	0.4412	0.4418

Table 2. For $K = 8$ and $b/\mu = 0.08$ the probability of having unstable modes is calculated for a sampling period between $h = 0.10$ and $h = 0.25$

h	$P[\|\lambda_1\| > 1]$	$P[\|\lambda_2\| > 1]$
0.10	0.0698	0.0698
0.15	0.0990	0.0990
0.20	0.1355	0.1355
0.25	0.1792	0.1791

5. CONCLUSIONS

We have obtained analytical expressions to analyze the stability of a NCS composed by a first order continuous-time process, a discrete controller and a control delay less than one sampling period. Our expressions allows to compute the probability of having unstable modes in closed-loop. Furthermore, the expected-value and variance of the absolute values of the closed-loop modes can be directly computed from the pdf by integration. It is therefore a practical analysis tool if we know the pdf of the time-delay or even if we can estimate it.

In this paper we did not present examples of our analysis being applied to real physical systems. This was mainly due to our model be rather simple. For higher order continuous-time processes and for time-delays larger than one sampling period the analysis is a little more complex. In this case, analytical expressions for the eigenvalues are difficult to find. However, it is possible by using numerical methods to determine the pdfs as we did in this simple model.

REFERENCES

Alt, Gustavo H. and Walter Fetter Lages (2003). Avaliação de Sistema de Controle em Tempo-Real Através da Rede IP. In: *VI Simpósio Brasileiro de Automação Inteligente.* Bauru, SP, Brazil. pp. 355–360.

Åström, K.J. and B. Wittenmark (1997). *Computer-Controlled Systems.* 3rd ed.. Prentice Hall.

Branicky, M.S., S.M. Phillips and Wei Zhang (2000). Stability of networked control systems: Explicit analysis of delay. In: *Proceedings of the American Control Conference.* Chicago,Illinois. pp. 2352–2357.

Kweon, Seok-Kyu and Kang G. Shin (2003). Statistical real-time communication over ethernet. *IEEE Transactions On Parallel And Distributed Systems* **14**(3), 322–335.

Luck, R. and A. Ray (1990). An observer-based compensator for distributed delays. *Automatica* **26**(5), 903–908.

Nilsson, Johan (1998). Real-Time Control Systems with Delays. PhD thesis. Lund Institute of Technology.

Viniotis, Yannis (1998). *Probability and Random Processes for Electrical Engineers.* McGraw-Hill.

Xiao, Lin, Arash Hassibi and Jonathan P. How (2000). Control with random communication delays via a discrete-time jump system approach. In: *Proceedings of the American Control Conference.* Chicago,Illinois. pp. 2199–2204.

Copyright © IFAC Information Control Problems in Manufacturing, Salvador, Brazil, 2004

ELSEVIER
IFAC
PUBLICATIONS
www.elsevier.com/locate/ifac

CHALLENGES AND POTENTIAL SOLUTIONS FOR COMMUNICATIONS IN THE ELECTRONICS PRODUCTION INDUSTRY

Ivan M. Delamer, Jose L. Martinez Lastra and Reijo Tuokko

Institute of Production Engineering,
Tampere University of Technology, Finland.

Abstract: The CAMX Framework is presented as a set of standards for communications in the electronics assembly industry. The communications layer is studied, and challenges involving scalability, security, and Quality of Service are discussed. Possible solutions that address the presented issues are proposed. The solutions are intended to maintain backwards compatibility. *Copyright © 2004 IFAC*

Keywords: Manufacturing Systems, Communications Systems, Data Transmission, Factory Automation, Information Systems, Information Integration, Protocols.

1. INTRODUCTION

As manufacturing and assembly equipment are improved and get more sophisticated, there is an increasing need to develop and enhance communications interfaces. The present inadequacies are not only limited to production equipment, but also to associated software applications and the supporting enterprise information systems.

The CAMX (Computer Aided Manufacturing XML) series of standards, endorsed by IPC, have been developed to support communications in the electronics assembly industry, leveraging the shortcomings of proprietary and in-house solutions developed by equipment vendors and large manufacturers. CAMX is the result of the work initiated by the NEMI[1] (National Electronics Manufacturing Initiative) Plug & Play Factory Project (Dugenske *et al.*, 2000), and later assumed by a conglomerate of companies and academic institutions. CAMX defines a framework that fosters interoperability among heterogeneous equipment and software applications, thus reducing deployment costs and time.

The CAMX communication protocol has been designed to be open and flexible. However, many aspects have not been addressed within the specification, and shall be discussed throughout this paper. Nevertheless, the flexible approach adopted by the standard enables specific implementations to provide answers to these issues offering enhancements, some of which are also proposed in the latter sections.

[1] http://www.nemi.org/

2. THE CAMX FRAMEWORK

The CAMX framework is defined by the IPC-25xx series of standards[2]. The standards propose a set of messages in XML format (Bray *et al*, 2000) that represent data models for products, assembly processes, assembly equipment, events, etc. It also defines a communications protocol for the exchange of CAMX messages among equipment and software applications, such as Human-Machine Interfaces (HMI) or Supervisory Control and Data Acquisition (SCADA) applications.

The scope of the standardized data models ranges from areas such as the manufacturing shop floor and manufacturing execution systems (MES), to product design (GenCAM[3]), supply-chain management (PDX), and enterprise communications. The XML format of the messages provides for a convenient extensibility mechanism, which will allow future enhancements to be made to the standards without sacrificing backwards compatibility.

Among the IPC-25xx standards, IPC-2501 is the standard that defines the CAMX communication protocol, which is based on a Message-Oriented Middleware (MOM) approach. The standard defines a web-based message exchange protocol based on SOAP (Box *et al*, 2000), itself based on XML and HTTP, the protocol used to access resources on the World Wide Web. All messages go through a central *server*, referred to as the *Message Broker*, that provides all the *clients* with a connectionless messaging service based on publish/subscribe or store-and-forward.

[2] http://webstds.ipc.org/
[3] http://www.gencam.net/

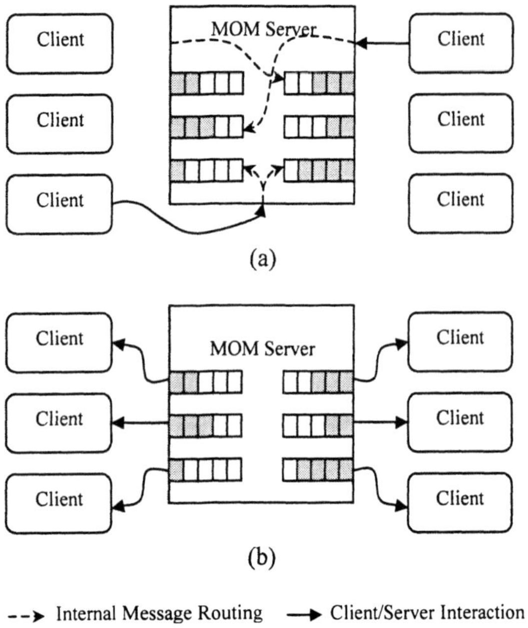

Figure 1. MOM message distribution scheme. Clients send messages to a server, which keeps them in queues for other clients to retrieve (a). Messages are asynchronously retrieved (b).

3. MESSAGE ORIENTED MIDDLEWARE

Message Oriented Middleware first became available in the mid-to-late 1980s as a service layer for distributed applications, and has since increasingly gained popularity in the scope of Enterprise Information Systems (EIS). MOM is defined as a client/server infrastructure that increases the interoperability, portability, and flexibility of an application by allowing the application to be distributed over multiple heterogeneous platforms (Rao, 95). It has become the preferred mechanism to merge heterogeneous enterprise resources, such as ERP software, data warehouses, and others. An attractive feature of MOM is that it allows the integration of legacy enterprise applications.

3.1 MOM Operation

Figure 1 illustrates a typical MOM scheme. Clients send messages to a MOM server through an *acknowledged-datagram* service. The server routes copies to internal queues assigned to each client. The messages can be routed according to a destination field provided by the sender (store-and-forward service), or alternatively messages can be associated with a *topic*, to which interested clients subscribe and each receive a copy (publish/subscribe service). In any case, clients must asynchronously query the server for messages on hold in their associated queue.

MOM infrastructures usually define the format of messages to be platform-independent. This feature enables communication among clients running on heterogeneous operating systems or written in different programming languages.

The asynchronous communication provided by MOM services eliminates the need for linked processes to be synchronized. This feature is known as loosely-coupled communications, where different processes don't necessarily block waiting for other processes to complete a request, as it happens in remote procedure call infrastructures such as CORBA (Emmerich, 1997). The loosely-coupled approach makes algorithms within the clients more robust, as well as making the whole system robust against temporary network failures.

3.2 CAMX MOM in Electronics Production

The CAMX communications protocol is based on a MOM system that uses SOAP to format and exchange messages. SOAP is based on two technologies: HTTP for transport, and XML for syntax. HTTP enables the use of any TCP/IP enabled network for transport, including Intranets, the Internet, Virtual Private Networks (VPNs), and IP enabled industrial buses. XML is a tag-based, platform independent, extensible language designed to represent semi-structured data that describes itself. By using XML, not only transported data is extensible, but the CAMX transport protocol itself too.

CAMX thus defines a framework that, using the Internet, can span from the manufacturing shop-floor to management offices to suppliers across the globe, maintaining a simple, lightweight protocol that allows integration of heterogeneous devices and software running on any platform or OS. The scope of CAMX enabled devices ranges from high complexity assembly equipment with integrated PCs, to small PTCs with embedded RT-Java controllers (Martinez Lastra, 2003).

3.3 CAMX Messaging Scenario

To illustrate the scope of usage of CAMX, an example scenario is represented in the following sequence of events[4]:

1. A piece of equipment reports a part shortage.
2. It also reports the production cycle has been stopped.
3. An execution control application recognizes the information, and halts the production of upstream equipment.
4. Another application issues a command to restore components to the equipment.
5. The equipment detects manual component replenishment, and reports it is operational again.

[4] Other messages may be produced in this scenario, and have been left out for simplicity.

6. The MES system issues commands to resume production to all of the halted equipment.
7. Simultaneously, after replenishing the equipment, a low component stock is detected and a supply-chain order is issued.

As seen from the previous example, CAMX data is exchanged within local environments such as a single assembly line, but also to higher enterprise levels such as MES or MRP (Manufacturing Resource Planning), and even across businesses through the supply-chain. Unsurprisingly, messages of such disparate nature require differentiated treatments, but such treatments are not specified in the standard.

4. CAMX OPEN ISSUES

In achieving a flexible and lightweight protocol, CAMX has left a series of unresolved issues that must be addressed in order to deploy *dependable* systems. These issues are presented in the following subsections, and further discussed throughout the rest of this publication.

Although the hereafter-discussed issues are not resolved within the standard's definition, the flexible CAMX approach enables specific implementations to include features that cover these areas transparently. Because CAMX is built on XML, a language designed to be extensible, backwards compatibility is inherent.

4.1 Scalability

The MOM infrastructure proposed by CAMX defines a central web server, the *Message Broker*, which provides the messaging service to all the clients in the framework. This approach introduces serious scalability issues, since it is clearly unfeasible to have one server handle requests from a whole enterprise as the amount of clients and traffic increases. It also establishes a single point of failure topology.

In addition, significant performance drawbacks are introduced by having a centralized Message Broker when the framework spans across physically dispersed locations. Multi-hop paths containing switches and routers can introduce bottlenecks and delays.

4.2 Security

The HTTP layer of CAMX ensures clients will have no problems in going through firewalls, enabling inter-network communications. However, in opening the manufacturing enterprise to the Internet, the manufacturing enterprise is opened to security breaches and attacks.

Dangers include critical information being stolen, unauthorized publishing and retrieving of messages, and even spamming. CAMX defines no features regarding security issues.

4.3 Quality of Service

Because of the critical importance of product and process data, CAMX defined Guaranteed Message Delivery (GMD) as a compulsory Quality of Service (QoS) policy. This scheme implies that all messages are kept in persistent storage until delivered, meaning that no messages are lost even if there is an unexpected system crash or power failure.

However, there are no QoS parameters regarding the timing of delivery. This means that time-critical or even real-time messages, such as alarms or more importantly control messages, are not necessarily promptly delivered. There are no delay or latency guarantees, nor priority settings.

5. SCALING THE BROKER

The need for scaling the Message Broker system was presented in section 4.1. Since Enterprise Information Systems (EIS) based on web services require scalable solutions, several techniques that have been developed for these areas could be applied to the Message Broker architecture.

5.1 Clustered processing

Having multiple servers working in parallel within a cluster is conceptually the simplest way of scaling a web-centric resource. From a client viewpoint, the server cluster is viewed as a single entity, so no changes need to be done to the communications protocol or to the clients.

However, although well suited for serving static web content such as web pages, parallel processing introduces several difficulties when serving dynamic content, as is the CAMX Message Broker case. The main issue is replicating the data across server instances when messages are constantly being posted and retrieved.

Moreover, increasing the capacity of a server by clustering does not improve the delays introduced by network connections from remote locations. Not only are delays and bottlenecks introduced by switches and routers, but there is also an increased cost in networking hardware that must efficiently the merged traffic from all the serviced areas.

5.2 Distributed Broking

A contemporary approach for making an application scalable, whether its enterprise-wide software or federated control, is to distribute it into multiple, autonomous, collaborative components. The context of CAMX communications provides a natural way to divide message broking among sections of the enterprise, exploiting logical divisions such as assembly lines, factories, departments, etc.

Distributing the broker application to service localized environments can improve performance by eliminating network delays and potential bottlenecks. A distributed service also eradicates single point of failure issues, while maintaining the same protocol from the clients' viewpoint.

In integrating the CAMX sub-domains to form the larger framework, several existing routing techniques can be implemented to transport messages across sectors. Analogies to the tree structured topologies of packet switching as done in IP/Ethernet networks, or message relaying as in IRC conferencing, are among the simplest forms of scalable inter-sectoral integration. Other multiple-path topologies of increased complexity could add robustness to the system, since a link failure in tree topologies can cause segmentation of the larger domain.

5.3 Hybrid approach

Both the clustered and distributed processing approaches bring advantages and disadvantages. A distributed Message Broker seems to provide enhanced benefits. However, some scenarios inevitably concentrate large amounts of localized traffic, compelling the use of clustered processing.

A hybrid framework with a distributed Message Broker topology where nodes can optionally be clusters is the largest scale messaging system envisioned.

6. SECURING DATA

A set of 5 requirements must be followed in order to ensure the safety of both exchanged messages and the supporting system (Nakamur *et al*, 2002). These are:

- *Authentication* of clients.
- *Authorization* to access resources.
- *Confidentiality* of data.
- *Integrity* of data.
- *Non-repudiation* of sent messages.

The integrity requirement is intrinsically fulfilled by the compulsory Guaranteed Message Delivery QoS policy. The other requirements can be met simply by having a closed network populated with well-behaved clients. However, the system becomes vulnerable when opened to the Internet, or in the presence of potentially ill-behaved clients.

6.1 Authentication & Authorization

Authentication is the process by which a client's identity is verified and established to be valid within the framework's domain. Authorization is the mechanism by which an authenticated client is either granted or refused access to a resource.

Authentication requires some form of client-server interaction. Identity data can be included in numerous ways without sacrificing compatibility with the standard. Authorization can easily be implemented through the use of internal tables or databases without a need to extend the communications protocol.

A simple authentication scheme utilizes the *HTTP Basic Authentication* protocol (Franks *et al* 1999). Since this method works at the HTTP layer of communications, on top of which CAMX is founded, it does not affect the CAMX SOAP protocol itself. Since HTTP authentication data is included as extra header fields, a Message Broker that doesn't support this feature would disregard the data, maintaining compatibility.

An alternative approach could feature a custom XML element containing authentication data, included as an extension in the CAMX SOAP message header. The element could provide the same *user-password* data as HTTP Basic Authentication, or include any additional information, and would also be intrinsically ignored or discarded by non-compliant brokers.

Ascertaining both authentication and authorization processes, together with guaranteed message delivery as required by the standard, results in satisfying the non-repudiation requirement for secure communications. By determining identity of a sender, and guaranteeing that the message will be delivered, a client cannot deny having sent it. Non-repudiation is a critical requirement in e-Business, and by extension to eManufacturing, not only for business transactions such as supply-chain orders but also for synchronized process control sequences.

6.2 Confidentiality

Authentication on its own doesn't guarantee safety of communications, since external agents could still intercept data. Both the messages and the authentication methods must be somehow protected, and the only way to achieve this is by encryption.

Using SSL, which is a layer of encryption that sits on top of TCP, messages can be safely transported (Naedele, 2003). As a result of using SSL, HTTP becomes HTTP-S (Secure), and secures both transmitted data and identity information. However, higher layers such as CAMX SOAP are not modified, allowing backwards compatibility and interoperability between with SSL-enabled and non-enabled clients and brokers.

Nonetheless, although the use of SSL effectively protects transported data, as soon as the Message Broker assimilates the message, it is de-encrypted and becomes vulnerable again. This approach, referred to as hop-by-hop encryption, is not satisfactory in some scenarios where data needs to be protected along the whole transmission path (Curbera *et al* 2002).

End-to-end encryption could be provided by directly encoding the XML payload of messages, using the XML Encryption Syntax (Imamura *et al* 2002). In this way, sending and receiving clients are in charge of encoding and decoding data, effectively protecting data across the whole transmission course while maintaining the same communications infrastructure.

6.3 Certificates

The basic authentication methods described in 6.1 may not be enough proof-of-identity for some applications. In that case, it is convenient to use a more robust identity certification method, such as XML Digital Signatures (Hada *et al*, 2002).

Digital signatures are only granted by Digital Certificate Authorities. Recognizing a digital signature issued by a trusted certificate authority greatly reduces the chances of errors in authentication. This technique is of special value in B2B transactions, such as supply-chain orders, but could also be applied to critical messages such as recipe set-up. It also strengthens the non-repudiation confidence quality.

6.4 Implementation and Performance Issues

Most client-side and server-side implementations of HTTP protocol already provide optimized support for both authentication and encryption. However, encryption is a computationally expensive process even when optimized. The encryption of data through SSL, and also encrypted digital signatures, can reduce throughput by factors of over 5 (Kant *et al*, 2000), and the additional overhead can increase latency by several orders of magnitude (Apostolopoulus *et al*, 1999). Encryption should therefore not be used compulsively, but only when data must travel through open or insecure networks. A scenario where encryption is a requirement is that of supply-chain management through the Internet; in a scenario of factory equipment communications through an Intranet, it is not.

7. QUALITY OF SERVICE

In a broad scenario like that of CAMX communications, where the nature of transported data varies greatly, Quality of Service parameters must be provided in order to accommodate specific needs. Higher priorities need to be established for messages carrying critical information, and real-time behavior must be ascertained for messages carrying control data.

7.1 QoS parameters

The CAMX standard specifies that all messages sent to a particular destination be kept in a FIFO queue, from where they are retrieved in the same order as the Message Broker received them. Although orderly delivery is very important to keep consistency of sequential events, some critical messages, such as alarms, must be immediately delivered. Waiting until all previous messages are delivered might not be acceptable in some circumstances.

Having multiple internal queues with different priorities assigned is a possible solution. Priority information could be contained in an extension element of the CAMX SOAP message. Multiple queues would remain transparent to clients, since the Broker internally selects the message with the highest priority waiting to be delivered when queried. Segregating messages into separate queues eliminates the possibility of having a low-priority data blocking high-priority messages.

In addition to priority settings, time-sensitive information such as control and process synchronization data must be delivered within well-defined timing constrains. Parameters defining maximum latency and delay tolerances, as well delivery deadlines might be included once again in an extension element of the CAMX SOAP message.

7.2 Implementing QoS

Because of the extensible nature of XML, QoS parameters can be included in extension elements within the delivered CAMX messages while maintaining full backwards compatibility: a non QoS-aware broker would simply ignore the extended fields. However, delivering messages with real-time constrains is not a straightforward process, since the polling-based architecture is directly in contradiction with the event-driven model of real-time events.

A possible solution could make use of the connection-oriented approach of the TCP layer. According to the specification, clients must constantly be polling for messages. In turn, clients could request that if there are no messages in queue waiting to be delivered, the connection be kept alive, remaining in a wait state, and receive a reply only once a message is ready to be delivered. A timeout should be introduced to prevent deadlock. This approach is also backwards compatible with non-implementing Brokers that don't keep clients in wait mode, returning an empty message and forcing clients to keep on polling as specified by the standard.

A fundamental issue in setting QoS requirements is that, within the publish/subscribe paradigm where it may be possible for a message to have multiple recipients, not all clients will necessarily require the same QoS delivery assurance. For example, monitoring equipment might require that an alarm be delivered in a "as-soon-as-possible" fashion, while a data logging application might only require a "whenever-possible" approach. QoS therefore needs to be demanded by subscribers from publishers that are able to provide the solicited requirements. Since the subscription interests of clients in CAMX are defined in an XML file, subscription descriptions can be extended to provide QoS requirements. Similarly, publishing capabilities can be extended to provide QoS capabilities.

7.3 Determinism

Although there has been much debate in the past about the use of Ethernet for real-time applications, the upgrades in speed up to 100Mbit or 1Gbit, together with the introduction of the switched hub, have made it possible to disregard the lack of determinism introduced by datagram collisions and to deem real-time possible (Kweon *et al*, 1999; Choi *et al*, 2000; Fan *et al*, 2002). However, the conditions for guaranteeing timing determinism using TCP for transport, as well as for the message queue paradigm, need still to be determined.

TCP, over which HTTP and SOAP work, is a connection-oriented protocol that guarantees transferred data integrity, and will retry any failed data transfer until successful. The reason why TCP is not the preferred protocol for real-time data transfer is that packet retries introduce indeterminism, and can have timeouts of up to minutes. However, it provides advantages when reliability is a requirement. Under certain network load conditions, retry-free connections could be statistically guaranteed with determined confidence intervals (Kweon *et al*, 1999).

The MOM architecture also introduces obstacles for real-time, since servers may become congested and develop into bottlenecks. Analysis such as those for packet collisions in a common Ethernet bus might be extended to understand the impact of message collisions in the broker.

7.4 Spanning QoS to the distributed broker

Guaranteeing QoS requirements across a distributed Message Broker introduces new challenges. Although priorities can easily be transferred along routing paths, guaranteed timing parameters must be respected for the entirety of every multi-hop routing path. This might only be possible if a rigid topology is constructed, with defined QoS guarantees for each segment. In a dynamic, multi-path scenario, real-time might only be supplied for a local environment, which should still suit most cases.

8. CONCLUSIONS

CAMX defines an open, extensible communications framework that provides seamless integration not only of multi-vendor equipment within an assembly line, but also of the whole of the manufacturing enterprise. While essentially simple and based on widely used web technologies, it provides openness to supply extended features in implementation specific ways.

The Message Broker, the central piece of the MOM architecture, can be extended to be clustered or distributed, to provide data security and protection from malicious agents, and to guarantee several QoS requirements. Extensions maintain full compatibility to any non-extended, CAMX compliant devices.

REFERENCES

Apostolopoulos, G., Peris, V. and Saha, D. (1999). Transport layer security: how much does it really cost?, *INFOCOM '99, Eighteenth Annual Joint Conference of the IEEE Computer and Communications Societies, Proceedings*, **2**, IEEE, March 1999.

Box, D., Ehnebuske, D., Kakivaya, G., Layman, A., Mendelsohn, N., Nielsen, H.F., Thatte, S. and Winer, D. (2000). *Simple Object Access Protocol (SOAP) 1.1, W3C Note*, online at http://www.w3.org/TR/2000/NOTE-SOAP-20000508, May-00.

Bray, T., Paoli, J., Sperberg-McQueen, C.M. and Maler, E. (2000). *Extensible Markup Language (XML) 1.0 (Second Edition), W3C Recommendation*, online at http://www.w3.org/TR/2000/REC-xml-20001006, 6 October 2000.

Choi, B.Y., Song, S., Birch, N. and Huang, J. (2000). Probabilistic approach to switched Ethernet for real-time control applications, *Seventh International Conference on Real-Time Computing Systems and Applications, 2000, Proceedings*, IEEE.

Curbera, F., Duftler, M., Khalaf, R., Nagy, W., Mukhi, N. and Weerawarana, S. (2002). Unravelling the Web Services Web: An Intoruction to SOAP. WSDL, and UDDI, *Internet Computing*, IEEE, **6(2)**, March-April 2002.

Dugenske, A., Fraser, A., Nguyen, T. and Voitus R. (2000). The National Electronics Manufacturing Initiative's (NEMI) Plug and Play Factory Project, *International Journal of CIM*, **13(3)**.

Emmerich, W. (1997). An overview of OMG/CORBA, Distributed Objects - Technology and Application *(Digest No: 1997/332), IEE Colloquium on*, 22 Oct. 1997.

Fan, X., Wang, Z., and Sun, Y. (2002). How to guarantee factory communication with switched Ethernet: survey of its emerging technology, *Industrial Electronics Society, IEEE 2002 28th Annual Conference of the*, **3**, Nov. 2002.

Franks, J., Hallam-Baker, P., Hostetler, J., Lawrence, S., Leach, P., Luotonen, A. and Stewart, L. (1999). *RFC-2617 HTTP Authentication: Basic and Digest Access Authentication*, The Internet Society, June 1999.

Hada, S. and Maruyama, H. (2002). Session authentication protocol for web services, *Applications and the Internet, Proceedings of the 2002 Symposium on*, IEEE, 2002.

Imamura, T., Dillaway, B. and Simon, E. (2002). *XML Encryption Syntax and Processing, W3C Recommendation*, online at http://www.w3.org/TR/2002/REC-xmlenc-core-20021210/, 10 December 2002.

Kant, K., Iyer, R. and Mohapatra, P. (2000). Architectural impact of secure socket layer on Internet servers, *Computer Design, Proceedings of the 2000 International Conference on*, IEEE.

Kweon, S., Shin, K.G. and Zheng, Z. (1999). Statistical real-time communication over Ethernet for manufacturing automation systems, *1999 Fifth IEEE Real-Time Technology and Applications Symposium, Proceedings of the*, 2-4 June 1999.

Martinez Lastra, J.L. (2003). *Information-Intensive Shop Floor Automation in Electronics Production: Challenges in Current Equipment Control Devices and Potential Solutions*, Tutorial at the 1st IEEE Conference on Industrial Informatics, August 2003, Alberta, Canada

Nakamur, Y., Hada, S. and Neyama, R. (2002). Towards the Integration of Web Services Security on Enterprise Environments, *2002 Symposium on Applications and the Internet, Proceedings*, IEEE.

Naedele, M. (2003). Standards for XML and Web services security, *Computer*, **36(4)**, IEEE, April 2003.

Rao, B.R. (1995). Making the Most of Middleware. *Data Communications International 24*, **12**, pp. 89-96.

Copyright © IFAC Information Control Problems in
Manufacturing, Salvador, Brazil, 2004

ELSEVIER
IFAC
PUBLICATIONS
www.elsevier.com/locate/ifac

A CAN-Based Application Level Error Detection and Fault Containment Protocol

Juan Pimentel, John Kaniarz
Department of Electrical and Computer Engineering
Flint, Michigan USA

Abstract

Fault tolerant unit (FTU) management is the most important process for a dependable system with replicated components. A protocol is necessary to properly manage error detection and fault containment regions in a safety-critical application on a CAN network. A distributed protocol is detailed with an efficient and accurate method to switch between a primary node that fails and a number of replicated nodes. Upon node failure, the protocol contains relevant information to accurately determine the new primary node in the system. This causes failed components to have a fail silent behavior. The result is that when a component fails there is no loss of data on the network. With the protocol engaged, there is a seamless execution of distributed applications in systems with multiple replicated nodes under the presence of failures. The protocol has been implemented, tested, and evaluated as part of a distributed safety-critical architecture.
Copyright © 2004 IFAC

I. INTRODUCTION

Error detection and fault containment for safety-critical systems has been extensively studied at the communication architecture level [4]. Examples of well known safety-critical architectures include TTA (time triggered architecture) [4-6] and FlexRay [7]. Critical failure modes at the communication architecture level is another area that has been extensively studied. In [4], the way the TTA architecture handles various critical modes is summarized.

It is well known that some communication services for safety-critical applications can be offered at layers 1 and 2 (i.e., the communication controller) or above, including the application (i.e., the host computer). For example the TTP/C protocols offers several safety-critical oriented services (e.g., group multicast) at layers 1 and 2 while leaving the responsibility of other services for the host computer. Unfortunately much less is known about offering services for safety-critical applications at the application level. The area of application level safety-critical protocols needs considerable more development and experience particularly solutions that meet the following requirements: scalable, flexible, simple, and cost effective. The contribution of this paper is an application level error detection and management protocol for safety-critical applications that meets these requirements. In addition, the proposed protocol is designed on top of CAN, a highly successful and widely deployed protocol for real-time applications. Application level protocols can be improved by taking into account the nature of actual applications. For our protocol, we have chosen a steer-by-wire system as our target application. A steer-by-wire system is considered a safety-critical application because component, communication, or control failures may lead to injury or loss of equipment and/or life [1,9].

The specific objective of this paper is to present an application level, CAN-based error detection and fault containment protocol for safety critical applications and to implement and evaluate the protocol in the context of a simulated steer-by-wire system (See Fig. 1). The protocol is part of a ultra-dependable distributed architecture developed at Kettering University. Details of the architecture will be provided in other publications.

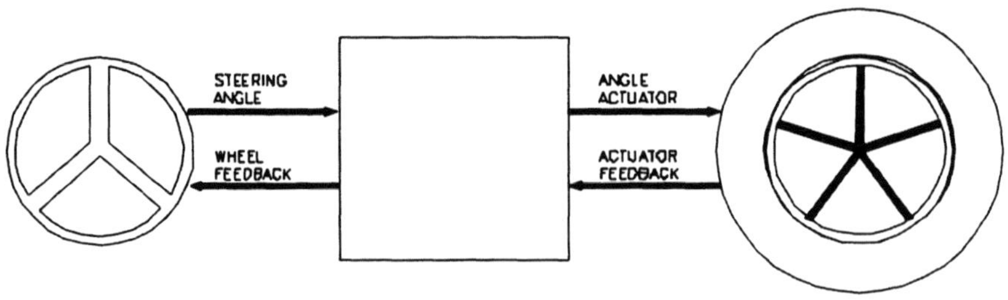

Figure 1. Components of a steer-by-wire system: hand-wheel (left), controller (center) and road wheel (right).

Motivation

There are several trends that provided motivation to seek simple and cost effective solutions to the development of architectures and protocols for safety-critical applications. First, there are already some manufacturers that provide multiple CAN communication controllers for some of their microcontrollers. For example the S12 microcontrollers offers 5 independent CAN channels and it is expected that this trend will continue. Second, there are already some simple and inexpensive sensor and actuators analog interfaces with incorporated CAN interfaces (e.g., Microchip MCP250XX). Again, this trend is expected to continue. Third, recent advances in mechatronics indicate that there will be important developments in distributed electromechanical components (e.g., several replicated steering-wheel actuators performing the function of a single larger actuator). Last but not least, it is expected that the current trend on miniaturization of digital systems and microcontrollers will continue yielding more functionality at a lower cost.

Main Features of the Safety-Critical Protocol

The main features of the of the safety-critical protocol include the following:

1. Massive replication of components
2. CAN2.0B-based (29 bit)
3. Scaleable, transparent (compatible with normal CAN applications)
4. Simple, flexible, cost-effective

Other relevant work

The idea of using a dependable time-triggered paradigm driven by the application is new in the proposed architecture and protocol described in this paper. However the idea of using a dependable time-triggered paradigm driven by the communication controller is well known [2,6,8]. Two examples of time-triggered architectures at the communication level is TTA (time triggered architecture) [4-6] and Flexray [7]. Whereas the former is a pure time-triggered communication protocol, the latter is a combination of time-triggered and event triggered. There are also architectures that have taken an event-triggered communication protocol (e.g., CAN) and made it appear as a time-triggered communication protocol. Two architectures in this category are TTCAN (time triggered CAN) [10] and FTT-CAN (flexible time-triggered CAN) [11]. The major difference in prior approaches from the one considered in this paper is that prior architectures still focus on the communication controller (i.e., layers 1 and 2 in the ISO reference model) whereas our architecture uses an event-triggered communication protocol (i.e., CAN) and a time-triggered synchronization protocol at the application layer.

II. PROCESS MODEL

As noted, the architecture and protocol are specific to certain safety-critical applications that could be event based or time-triggered based. Example of event based applications include body bus automotive functions such as opening a window or door. The actual time of opening or closing a window is completely random and is best modeled as an event based system. Other functions such as engine control, and steer-by-wire systems are continuous in nature and are best modeled as a time-triggered system.

Whether the application is event based or time-triggered based, it can be modeled by the system in Fig. 2 involving the system to be controlled, sensors, actuators, and controllers interconnected by a communication network. Node 1 collect measurements from sensors, send the measured values to a controller (Node 2) which in turn send control values to actuators through Node 3. Fig. 3 depicts the events and parameters of the communication process associated with nodes 1, 2, and 3. In particular we note the following significant communication events:

1. E1: sensor node queues message for transmission (beginning of cycle)
2. E2: sensor message is sent on the bus
3. E3: sensor message is received by controller node
4. E4: controller node finishes control algorithm and queues control message for transmission
5. E5: control message is sent on the bus
6. E6: control message is received by actuator node (end of cycle)

Assumptions

Regarding the process model and its communications shown in Figs. 2 and 3 we make the following assumptions:

1. A source (i.e., sensor) node will send messages with a minimum inter-departure time. If periodic, the source node will send messages with a minimum period. This assumption is needed in order to guarantee a deterministic behavior of the underlying CAN protocol.
2. A controller node (Node 2) will receive one or more messages corresponding to sensor values and generate controlled signals configured as one or more output messages.
3. Source nodes will generate time-triggered cycles (source nodes dictate the beginning of each cycle via events E3 in Fig. 3) that are specific to an application. Time triggered cycles are defined from the viewpoint of an application rather than from the viewpoint of communications. From a timing perspective then, source nodes are master nodes and all remaining nodes are slave nodes.
4. The communication system can support multiple different applications each with their own communication cycles (or equivalently sample periods Tn). That is the system is multi-rate.
5. A multiple application system assumes that each application has a sensor, controller (optional), and actuator.
6. Deterministic behavior can be guaranteed (using CAN schedulability analysis) since the traffic on the network is carefully controlled thanks to assumption 1 above.
7. Messages of one cycle are not inter-mingled with messages from a different cycle. This is possible if schedulability analysis is used to ensure that all possible messages in one cycle will be sent before the next cycle begins.

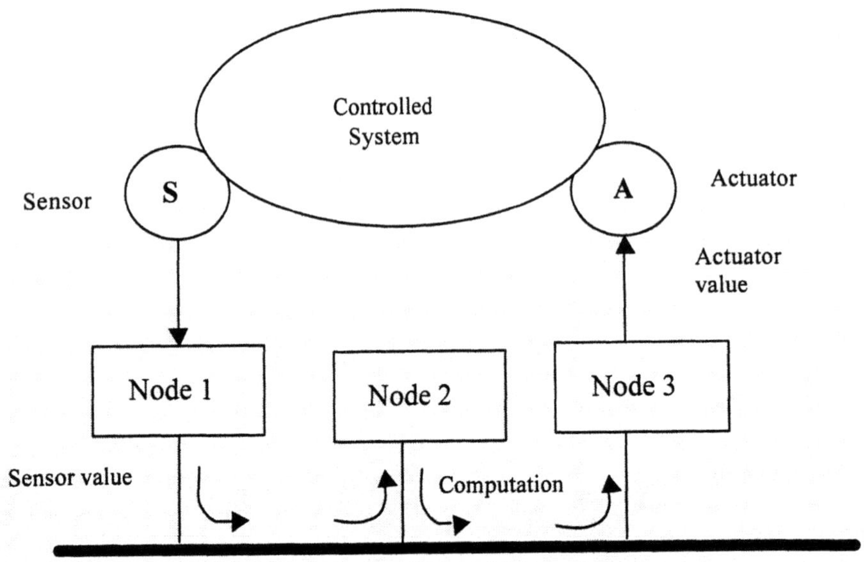

Figure 2. Model for a typical control system

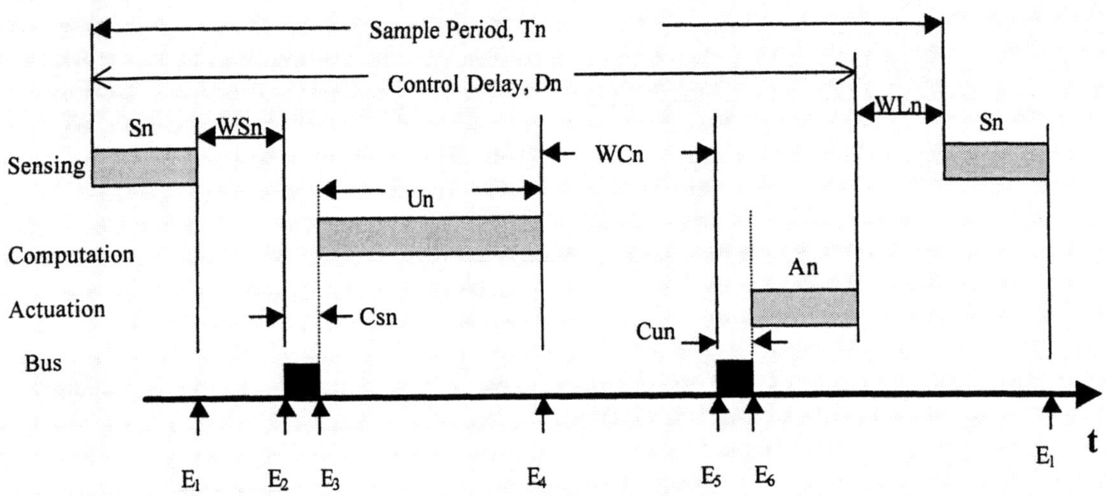

Figure 3. Events and parameters of network communications for a typical control system.

III. FAULT MANAGEMENT PROTOCOL

To deal with component, communication, and control failures the architecture and protocols makes use of massive replication of sensors, controllers, and actuators. Component failures are detected by the absence of messages generated by the component. Communication failures are detected by absence of messages on the communication line. Detecting control failures depends on details of the architecture, other protocols, and application software which are outside the scope of this paper. Each communication node (i.e., sensor, controller, or actuator) has one or more replicated components and configured as an FTU (fault tolerant unit). Fault tolerant units are well known mechanisms to deal with fault containment of safety-critical systems. The regions protected by fault containment mechanisms are referred to as fault containment regions (FCR). Fig. 4 depicts the system of Fig. 2 where each node is replaced by an FTU consisting of two internal nodes.

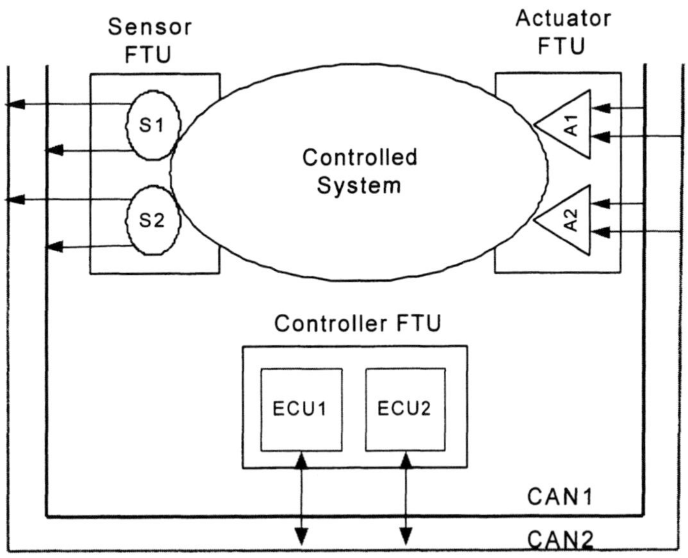

Fig. 4. A safety-critical system with replicated components.

If each replicated component sends an independent message corresponding to sensor signals and each message is treated independently from one another (even though they correspond to the same sensor signal) then there could be up to 32 different messages arriving at the actuator FTU. Obviously, the system would be extremely complex if it were to handle all possible messages. To reduce complexity, the main idea behind the protocol is to manage all components in an FTU in such a way that it would appear as just one node to the other FTU's. Inside each FTU, replicated components would communicate with one another to elect the so-called primary node with the remaining nodes acting as secondary, tertiary, etc. For Fig. 4, the number of independent messages arriving at the actuator FTU is reduced from 32 to 4 thus simplifying the complexity of the system.

Although the main idea behind the protocol is the same, some details are different for sensor FTU's, controller FTU's, and actuator FTU's. This is because sensor FTU's are producers of information, controller FTU's are both consumers and potentially producers and actuator FTU's are consumers. The protocol will be detailed for controller FTU's. The details of the protocols when applied to sensor and actuator FTU's vary slightly.

Fault Management Protocol for Controller FTU's

For some application domains (e.g., automotive) controller nodes are also known as electronic control modules (ECU's) thus we will use these terms interchangeably. As noted, the fault management protocol is only executed only by members (primary and all its replicas) of a certain FTU, that is, it is an intra-group protocol. It is envision that there could be multiple applications on the same system thus the application (App) type is encoded in the protocol header of Fig. 5. For efficiency reasons, this header is encoded using some of the 29 bits of the message ID as specified in the CAN2.0B standard.

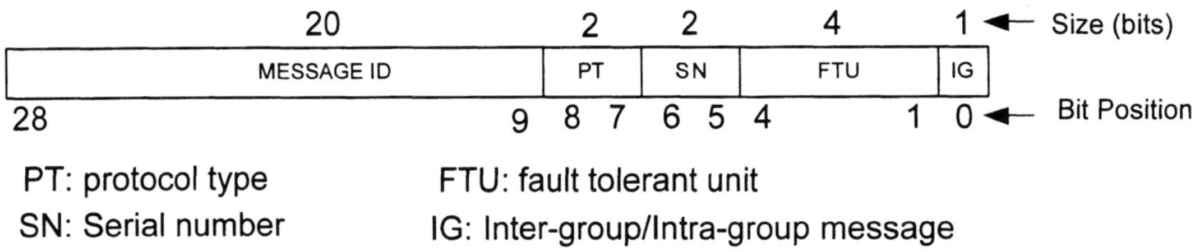

Figure 5. Protocol fields encapsulated in the 29-bit identifier of CAN 2.0B.

The main assumption made by the fault management protocol is that there is always a node acting as the Primary node. If this condition is not true, the entire system is considered to have failed and appropriate messages are generated to indicate a human user of this event. There are several protocol mechanisms that help the system converge to a state involving a Primary node and several active backup nodes. At initialization, the primary node is specified. Thereafter, each node (other than the primary) within an FTU executes the fault management protocol to determine the secondary, and additional backup rankings of all nodes in the FTU. For example, in a system with three controllers, both backup controllers will negotiate which one will act as the secondary node, and which one will act as the tertiary node.

At startup, all controllers belonging to the same FTU request that a ranking of all replicated components be started by sending the *Rank* message. The ranking procedure is controlled by a timer and ends with a ranked set of replicated components. In case more than one Rank message is received by a node before the associated timer expires, the Rank message is ignored. All other nodes in the same FTU respond to Rank messages by sending their serial numbers on the bus via the *Serial* message. As a result of all controllers sending their serial numbers, all components in the particular ECU know of all the other components on the same ECU. The backup nodes are ranked based on their serial numbers into secondary, tertiary, etc. If the primary node component fails the other controllers detect this by noting the absence of the missing message that corresponds to the Primary. Upon failure of the primary component, the secondary component will take over immediately, transmit the message and declare itself primary via a message called *Active*. If a component returns after a failure, a ranking procedure is invoked via the *Rank* message to determine a new ranked set of backup nodes (i.e., secondary, tertiary, etc.) If a new component joins the network, the same ranking procedure is performed. Thus, the primary node is not disturbed when failed nodes are repaired or when new nodes are added to the network.

Main Protocol Rules

The main protocol rules are summarized as follows:

1. A primary node is designated when the system is initialized
2. If the primary node fails, the secondary node becomes primary automatically
3. If the secondary node fails, a ranking procedure is invoked immediately to find a new secondary node.
4. If the primary node goes into a CAN error passive mode (as a result of noise on the bus) it forces a reset thus acting as it has failed to let other backup nodes to take over.
5. If a node acting as primary detects that there is another primary, the first node will issue an Identify message requesting the identity of the Primary node. If a valid response is obtained, the inquiring node ceases to be Primary and forces a ranking immediately.

The following are the main functions of the primary and secondary nodes.

Primary Node

The primary node is responsible for sending inter-group data messages. For the hand wheel node, this involves sending the hand wheel position to the ECU FTU.

Secondary Node

The secondary node is responsible for monitoring the primary node. The secondary node will check that the primary node sends its inter-group messages on time. This is achieved by monitoring event E4. If the secondary node detects that the primary node has missed sending a message (i.e., event E4 did not occur), the secondary node will immediately send the late message, becomes primary and force a ranking of all similar nodes.

Features of the Ranking Mechanism

- Ranking occurs on powerup, and when a failure is detected
- Ranking does not stop or pause a currently functioning system (A primary node does not stop sending inter-group messages during ranking. It will stop only when a new primary node takes over and is ready to send inter-group messages)
- Allows for a node to be deactivated, then return to service once fixed at a later time without disruption
- Allows for nodes to be added or removed without disruption
- When the primary node fails, messages are never completely missed, but are only delayed (for example if primary message failure timer is 100ms, the secondary will wait 110ms, the tertiary will wait 120ms (assuming that there is no secondary), etc, to send the message instead and take over as primary)

Protocol Messages

Communication on a CAN network is based on a set of unique messages each with a specific identification known as message ID. Since the fault management protocol is based on CAN, it involves a number of messages listed below.

Active Message

This message declares that the sender is the Primary, Secondary, or Subsequent node. The type of node is one parameter data of the message. The Active message indicating Primary is automatically sent by the secondary node upon detecting that the primary node has failed. The Active message is also sent by a node as a response to an Identify message indicating that the node is Primary, Secondary, or Subsequent node according to its current ranking. Nodes receiving this message will disable sending inter-group messages, and should have already been ranked by the fault management algorithm. This action also provides additional checks in case any nodes may have missed the ranking algorithm and to prevent two nodes attempting to simultaneously assume the role of Primary node.

Rank Message

This message requests that all nodes in a certain FTU (other than the primary node) start their rank procedure. If a node is already executing its ranking procedure this message is ignored. The *Rank* message is sent whenever the following events happen:

1. On power-up initialization
2. When a failure has been detected on any backup node (i.e., other than the Primary node)
3. When a new backup node enters the system

4. When a failed node re-enters the system after it has been re-initialized or repaired

When the Rank message is received by a node, the receiving node will send its serial number on the bus. During the ranking procedure, all nodes will also listen on the bus a specific amount of time and compare the all serial numbers received with that of its own. When that time period expires, the node will decide whether it is a primary, secondary, tertiary, etc. node based on the relative value of its serial number relative to the serial numbers of all remaining nodes in the FTU (other than the Primary node). Nodes with lowest serial number is declared secondary, the next lowest is tertiary, etc.

Serial Message

The *Serial* message is used during the ranking procedure and is sent immediately after a Rank message is received. As the name implies, the Serial message contains the serial number of the sending node.

Identify Message

The identify message is used whenever there is a perceived inconsistency on what nodes are primary, secondary, tertiary, etc. For example, a primary node receiving a message with the same ID as that just sent in the current cycle will conclude that there is another primary node and will issue the identify message to resolve the inconsistency.

IV. REPLICATED MESSAGES ON MULTIPLE CAN BUSES

To deal with communication failures, every node in the system listens on all replicated CAN buses, and will also send messages over all replicated CAN buses. This creates a problem due to the need to replicate messages on each bus. To illustrate the problem consider the situation in Fig. 4 involving two sensors, two controllers, and two buses. When a controller node receives a message; it will receive two versions of the same message denoted $m_1(k)$ and $m_2(k)$ each on a separate bus corresponding to the kth cycle. The controller will perform its control functions and generates messages $C(m_1(k))$ and $C(m_2(k))$. The controller then will proceed to send these two messages on each CAN bus thus generating a total of four messages $C_1(m_1(k))$, $C_2(m_1(k))$, $C_1(m_2(k))$, and $C_2(m_2(k))$ where $C_i(m_j(k))$ stands for controller message on the ith bus corresponding to message $m_j(k)$. This mechanism creates redundant traffic, and a potential cause for congestion. In order to alleviate the problem, the protocol has a *message discarding* mechanism. Since all messages are sent in duplicate, once on each bus, when a message is received by the controller, the message with the same message ID is ignored on the other CAN bus within the same cycle. For example, if a message is received on CAN bus 1 with message ID 0x00001000, the node will ignore the next message with ID 0x00001000 on CAN bus 2. Effectively, this mechanism will eliminate the problems posed by the replicated messages created by sending messages on two CAN buses.

Sequence Numbers

The message discarding protocol is implemented using sequence numbers. Each message contains a sequence number that is sent as the first data byte. This number is used to encode the order in which the producer node put the same message on the various buses. The consumer will only keep the first message received with the same message ID; the other messages with the same ID on the other buses will be ignored. This guarantees that no old data will be passed along to another node group. Sequence numbers also increase the likelihood of ignoring any errant message that may show up. Sequence numbers add to the robustness of the protocol.

CAN Message ID

CAN Message IDs are used to identify the sender of the message as well as the type of message being sent. The bit assignments for message identification is shown in Fig. 5.

Intra-group Message

Messages for nodes within the same FTU are referred to as intra-group messages. For example, all hand wheels nodes (primary and backups) send messages to each other during arbitration. These messages are considered intra-group messages since the intended targets are also hand wheels.

Inter-group Message:

Messages for nodes in different FTU's are referred to as inter-group messages. For example, the hand wheel FTU will send a wheel position message to the controller FTU. Since the message is being sent from one type of node to another, it is an inter-group message.

V. IMPLEMENTATION, TESTING, AND VERIFICATION

We have implemented the proposed fault management algorithm in CANoe using CAN2.0A (11 bit) for the steer-by-wire application described in section I. We have extensively tested and verified the protocol behavior as well as the steer-by-wire behavior. Since CANoe is also a simulator, we simulated the detailed operation of the hand-wheel as well as road wheels. We tested failures in the simulation by simply turning off the corresponding node (using a simulated switch in CANoe). Because the entire system was simulated using CANoe, we could not extensively test failures of the CAN bus (unless the entire bus was disconnected). The CAN2.0A implementation was our first and we are now in the process of simulating the protocol on CANoe using CAN2.0B (29 bit) and implementing an actual steer-by-wire system on a golf cart using Motorola S12 microcontrollers as target ECU's and Microchip MPC 250XX devices as sensor and actuator nodes.

Failure Modes

Failure modes can be dealt with at the architectural level, at the protocol level, and at the application level. In this paper we only deal with failure modes at the protocol level. In the following, we describe the behavior of the protocol under single-point and multiple-point failures.

Single-Point Failure

- We tested that the system recovered from the following single point failures and verified their fault-silent behavior.
  - One of the Hand Wheel node fails.
  - One of the Road Wheel node fails.
  - One of the ECU controllers fails.
  - One of the CAN buses fails.

Multiple-Point Failure

- Any combination of the above single-point failures (taken two at a time) was tested and verified their fault-silent behavior.
- We identified and detected several multiple point failures and dealt with them in an appropriate manner. For each of the multiple-point failures listed below, we straightened the road wheels and applied the brakes.
  - Both of the Hand Wheel nodes fail.
  - Both of the Road Wheel nodes fail.
  - Both of the ECU controllers fail.
  - Both of the CAN buses fail.

VI. SUMMARY AND CONCLUSIONS

This paper has presented a CAN-based application level error detection and management protocol for safety-critical applications. The time-triggered paradigm is important for safety-critical applications. From the viewpoint of the OSI reference model, there are two main approaches for incorporating the time-triggered paradigm in safety-critical applications: bottom-up (i.e., starting at layers 1 and 2 and moving to higher layers) or top-down (i.e., starting at the application layer and moving to lower layers). Most approaches use the former, this paper describe one example of the latter. In addition, application level protocols add additional degrees of freedom for the design of ultra-dependable systems by taking into account the nature of actual applications. For our protocol, we have chosen a steer-by-wire system as our target application. The protocol has been simulated and implemented in CANoe and initial test and evaluation results are very encouraging.

VII. REFERENCES

[1] Bretz, E.A., "By-Wire Cars Turn the Corner," IEEE Spectrum, Vol. 38, No. 4, pp.68-73, April 2001.

[2] Poledna S., *Fault-Tolerant Real-Time Systems*, Kluwer Academic Publishers, 1996.

[3] L. Lawrenz, *CAN System Engineering: From Theory to Practical Applications*, Springer, 1997.

[4] Kopetz, H. "Fault Containment and Error Detection in the Time-Triggered Architecture," Proc. IEEE Int. Symp. On Autonomous Descentralized Systems, ISADS 2003, pp. 139-148, April 2003.

[5] H. Kopetz, G. Grunsteidl, "TTP – A protocol for Fault-Tolerant Real-Time Systems". IEEE Computer, pages 14-23, January 1994.

[6] Kopetz H., *Real-Time Systems, Design Principles for Distributed Embedded Applications,* Kluwer Academic Publishers, 1997.

[7] http://www.flexray-group.com

[8] Pimentel, J.R, and Sacristan, T. "A Fault Management Protocol for TTP/C," Proc. IECON'01 (Int. Conf. On Industrial Electronics), pp. 1800-1805, Denver, CO., Nov. 2001.

[9] Amberkar, S., D'Ambrosio, J.G., Murray, B.T., Wysocki, J., and Czerny, B.J., A System-Safety Process for By-Wire Automotive Systems, SAE Congress paper, 2002.

[10] Muller, B., Fuhrer, T., Hartwick, F., Hugel, R., and Weiler, H., "Fault Tolerant TTCAN Networks," Proc. 8th Int. CAN Conference, Las Vegas, 2002.

[11] Ferreira, J., Pedreiras, P., Almeida, L., and Fonseca, J., "Achieving Fault Tolerance in FTT-CAN," Proc. 2002 Int. Workshop on Factory Communication Systems (WFCS2002), pp. 125-132, Vasteras, Sweden, Aug. 2002.

APPENDIX

IG and FTU codes

IG (Bit 0): 0: Intra-group Message
1: Inter-group Message

FTU (Bits 4,3,2,1: 0000: Reserved
0001: Hand Wheel
0010: ECU
0011: Road Wheel

Message Definitions

Message	**Message ID**	**PT**	**SN**	**FTU**	**IG**	**Message Data, DLC**
Active	0x00001	00	MM	NNNN	0	No, 0
Rank	0x00002	00	XX	NNNN	0	No, 0
Serial	0x00003	00	MM	NNNN	0	No, 0
Identify (req.)	0x00004	00	XX	NNNN	0	Yes, 1
Identify (resp.)	0x00004	00	MM	NNNN	1	Yes, 1
Application	0x00010 to 0xFFFFF	00	MM	NNNN	1	Yes, P

MM Serial number bits
NNNN FTU bits
P Length of data (number of bytes) in CAN message (DLC in the CAN standard)
XX Don't care. The values of the bits are not important.

First data byte on Identify message (request and response)

0x00 Primary
0x01 Secondary
0x02 Tertiary
0x03 etc.

Model Notation

Tn Sample period of nth control loop
Tc Length of TDMA cycle
Dn Delay of nth control loop
N Number of control loops in the system
s(n) n^{th} cycle in the time triggered system
Csn Transmission time of a sensor message
Cun Transmission time of a control message
Un Execution (computation) time for computing task of nth control loop.
An Execution time for actuation task of nth control loop.
Sn Execution time for sensing task of n^{ih} control loop.
SMn Communication time to send measured values in the nth control loop
CMn Communication time to send computed values in the nth control loop
WSn Wait time to send measured values in the nth control loop
WCn Wait time to send computed values in the nth control loop
WLn Wait time of nth loop added to Dn to make it periodic

Copyright © IFAC Information Control Problems in
Manufacturing, Salvador, Brazil, 2004

ELSEVIER
IFAC
PUBLICATIONS
www.elsevier.com/locate/ifac

MINIMIZING THE END-TO-END LATENCY IN MULTI-SEGMENT TIME-TRIGGERED NETWORKS

Paulo Pedreiras , Luís Almeida
{pedreiras,lda}@det.ua.pt

DET-IEETA
University of Aveiro
3810-193 Aveiro, Portugal

Abstract: Distributed applications complexity is steadily increasing. A well-known technique used to manage such complexity consists in decomposing the whole system in different quasi-independent subsystems, which are also frequently distributed. Inter-subsystem communication, when necessary, is performed via gateway nodes that filter in and outgoing traffic. For real-time systems, this architecture poses additional design challenges, since it becomes necessary to consider both intra and inter-network message exchanges with real-time constraints. This work addresses multi-segment time-triggered networks, and, for this class of systems, presents a methodology, which allows minimizing the end-to-end delay for the inter-network traffic. This goal is achieved by bounding the transmission windows of the messages in each one of the intervening networks, and, then, taking advantage of the properties of time-triggered systems, making such transmission windows starting as soon as possible. In particular, this is based on the determination of the phase of a message that allows maximizing its best-case response time. *Copyright © 2004 IFAC*

Keywords: real-time systems, real-time scheduling, real-time communication, time-triggered communication, phase control, best-case response time.

1 INTRODUCTION AND RELATED WORK

Distributed applications are becoming increasingly complex, due to both an increase in the number of interconnected devices and to the increased amount of data to be shared between them (Decotignie, 2002). A well-known technique used to manage such complexity consists in decomposing the system in distinct subsystems, comprising e.g. sets of sensors, actuators and controllers that cooperate closely to achieve a particular goal. Real-time networks interconnect the components of these subsystems, and communication among subsystems is made possible, when necessary, via gateway nodes.

Having a complex system broken in smaller quasi-independent subsystems may bring several benefits (Saba, *et al.*, 1995). One of such is the possibility to design, build and test separately each one of the subsystems, which not only allows to reduce the development time and effort, as also eases the maintenance. Another potential benefit is having networks with arbitrary physical lengths, independently of the particular constraints imposed by the underline communication infrastructures (e.g. Controller Area Network (Bosch,1991) at 1Mbps imposes a theoretical maximum length of approximately 40m). A set of different networks interconnected via gateway nodes do not suffer such length restrictions since, from the physical point of view, the networks are independent and thus may be freely combined to form longer logical networks. Another important benefit that may arise from performing the system decomposition as above referred is an increase on the overall schedulability level, since this architecture allows restricting the traffic propagation range to the subsystems in which the data is required.

In the scope of real-time applications, multi-segment networks pose additional design challenges, since, besides guaranteeing the schedulability of the real-time messages within each one of the networks (intra-network traffic), it is also necessary to guarantee that the messages that have to cross networks (inter-network traffic) are also delivered to their respective recipients within the specified timeliness constraints, in all anticipated circumstances.

The scientific community, in different contexts, has addressed this category of problems. (Aras, *et a.l*,1994) present a good survey about multi-hop message scheduling, both with hard and soft real-time

This work was partially supported by the European Commission (accompanying measure ARTIST, IST-2001-34820).

requirements. However the framework therein considered is mainly related with Wide Area Networks carrying voice and video real-time traffic. (Kandlur, *et al.*, 1991) have considered a similar challenge, but targeting multi-computer systems with point-to-point network connections (mesh structures with multi-paths). Also (Rexford, *et al.*, 1996) have addressed this issue, but for router-based real-time networks.

In wireless networks the nodes are often required to relay the data messages to extend the potential range of connectivity. Some of these messages may have real-time requirements, and thus adequate scheduling policies must be implemented at the intervening nodes (e.g. velocity monotonic scheduling (Blum, *et al.*, 2002)). A related problem is also found in switch-based Ethernet networks, due to the poor real-time behavior of the standard 8-level fixed priority scheduling. (Hoang, *et al.*, 2002) propose scheduling messages within the switch by the Earliest Deadline First (EDF) policy. In this case each one of the switch ports behaves like an independent network, and the switch scheduler must select, in each instant, which messages should be forwarded to each network.

Frameworks closer to the one addressed in this paper are dealt with in (Saba, *et al.*, 1995), concerning the interconnection of distinct worldFIP networks, and in (Ryu, *et al.*, 1998), concerning generic distributed real-time systems. The former addresses mainly protocol issues, like addressing resolution and the gateway internal structure. The latter presents a methodology that allows deriving component operational parameters from the system-level requirements, which are subsequently used to perform the individual subsystem configuration, respecting the global system constraints.

This paper addresses multi-segment time-triggered networks, frequently used in distributed real-time applications. A methodology is presented that allows minimizing the end-to-end delay for the inter-network traffic. This goal is achieved by bounding the transmission windows of the messages in each one of the traversed networks, and then, taking advantage of the properties of time-triggered systems, making such transmission windows starting as soon as possible.

The remainder of this work is organized as follows: Section 2 presents the system architecture addressed in this paper; Section 3 a technique allowing the computation of the message parameters that guarantee a minimum end-to-end latency for inter-network traffic; Section 4 presents the results of an illustrative example, showing the improvements that may be achieved with the technique presented in Section 3; finally Section 5 presents the conclusions.

2 SYSTEM ARCHITECTURE

Many real-time distributed applications are based on time-triggered (TT) systems (Kopetz, 1997) in which actions occur at pre-defined time instants. Examples of networks that support this type of traffic are FTT (Almeida, *et al.*, 2002), WorldFIP (EN 50170, 1996), PowerLink (Powerlink) and FlexRay (FlexRay, 2001).

In order to follow the compositional approach referred to in Section 1, data interchange between different subsystems using TT networks, when necessary, is achieved via gateway nodes, as depicted in Figure 1. The global interconnection topology can be any but, for simplicity, we will only consider situations in which there exists a single path between any two networks. The different networks that compose the system may operate synchronously, sharing a common time base, or may operate asynchronously.

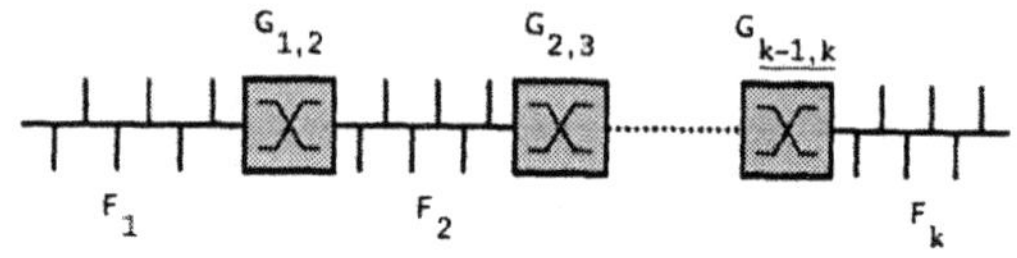

Figure 1: System architecture

In the remainder of this section we briefly present the generic architecture of the gateway nodes, and analyze the impact of using, or not, system-wide synchronization.

2.1 Gateway nodes

The TT networks referred to above impose transmission control mechanisms that must be strictly followed by all the network nodes, gateways included. Therefore, Gateway nodes must act as regular nodes in each one of the interconnected networks, but keep track of the messages that should be relayed from one to the other network (Figure 2). For each one of these messages, the Gateway behaves like a receiver in the source network, and as a transmitter in the target network.

These messages are kept in a shared database. Depending on their nature, the data contained in the messages may be stored in a buffer or may be queued. The first approach is used when the messages carry state data, in which case only the most recent value is important, whilst the queue approach is used when all the message instances are relevant, and thus should be transmitted.

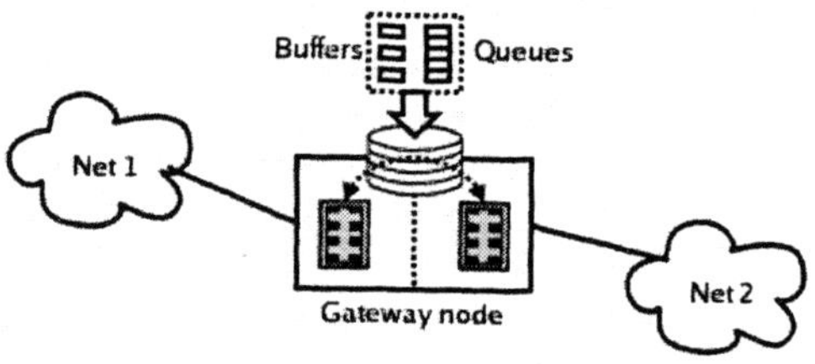

Figure 2: Internal architecture of a gateway node

2.2 Isochronous architecture

In the isochronous architecture, all the networks share a common time-base and the message scheduling and dispatching are synchronized. This framework allows knowing in advance the precise time windows where messages will be scheduled. When a message has to transverse a set of those networks, this knowledge may be used to tune the relevant message attributes (e.g. periods, deadlines and offsets) in order to control the flow of the data between the source and the target nodes.

An example of the application of such technique is depicted in Figure 3, where a message produced in network 1, having a period of 7 time units, must be relayed to network 3, via network 2. If the transmission windows in each one of the intermediate networks are properly bounded, the offset of the message in the involved networks may be set in a way that the successive transmission windows immediately follow each other (Figure 3). This result is particularly relevant since it contributes to minimize the network-induced latency.

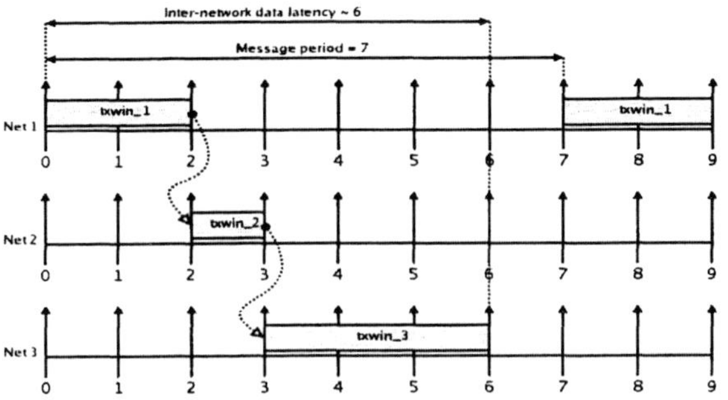

Figure 3: Message routing example: isochronous approach

2.3 *Anisochronous architecture*

In anisochronous systems, not only the scheduling and dispatching activities are not synchronized in the distinct networks, as also the time-base used to trigger the periodic activities may also be different. This architecture renders useless to manage the message phasings in the different networks since, in the general case, the phasings will vary during system run-time, and cannot be predicted.

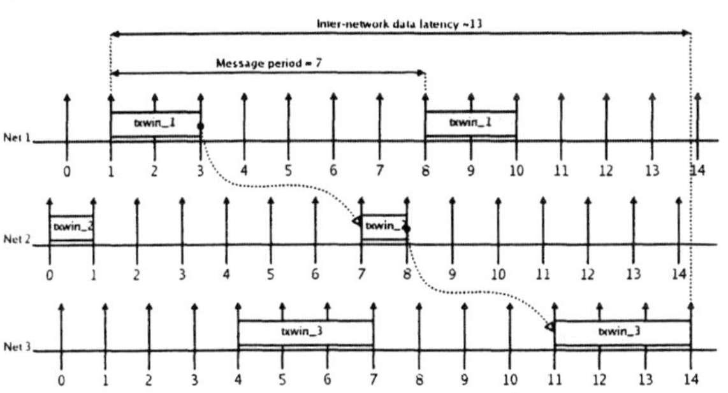

Figure 4: Message routing example: anisochronous approach

Consider the scenario depicted in Figure 4, which is conceptually equivalent to the one depicted in Figure 3, except for the absence of synchronization that results in uncontrolled relative phases of the message transmission cycle in the three networks. As can be observed, due to the lack of synchrony, a given datum may take up to two periods to travel from the producer to the consumer nodes, which in this case is more than twice the latency for the case of the isochronous system.

3 MINIMIZING THE NETWORK LATENCY

For several classes of applications network-induced latency is a parameter of paramount importance. It imposes a lower bound on the corresponding message period, which in turn is an important factor in the Quality of Service delivered to some applications classes, such as control (Buttazzo *et al.*, 2000; Yuan and Yu, 2003) and multimedia (Lee, *et al.*, 1996). The performance of this type of systems also depends strongly on the capability of the network to deliver data with low latency.

Relaying a message from a producer node to a consumer node in a different network implies that the message should be also scheduled and transmitted in each one of the intermediate networks. The following sections present an analysis of the end-to-end latency suffered by a message that is routed through several networks according to the architecture described previously. It allows computing the communication parameters, namely the message offsets at each intermediate network, which minimize the maximum end-to-end latency.

3.1 *System model*

As discussed in Section 2, isochronous systems, in which the different subsystems share a common time base and have synchronized message scheduling and dispatching, can outperform anisochronous ones in what concerns the end-to-end network-induced latency. Therefore, in the remainder of this paper we consider a system architecture composed by isochronous TT networks, with synchronized clocks and where the message scheduling and dispatching are also synchronized to a common cyclic framework. These networks are interconnected by gateway nodes, which behave like common nodes in each of the interconnected networks, except in what concerns the sharing of memory buffers, which are used to transfer the data from one network to the other.

The generic time-triggered message i belonging to a particular network j is characterized by a transmission time ($C_{i,j}$), a period ($P_{i,j}$), a relative deadline ($D_{i,j}$), a priority ($Pr_{i,j}$) and a phase ($\varphi_{i,j}$), as in Definition 1. The period, deadline and phase are expressed as integer multiples of a coarse time unit δ, such as the duration of a cycle in WorldFIP or FTT-CAN. On the other hand, the transmission time is always less than one time unit.

$$M_{i,j}= M_{i,j}(C_{i,j},P_{i,j},D_{i,j},Pr_{i,j},\varphi_{i,j}) \quad (1)$$

The activation instants of message i in network j are given by expression 2, where k is an integer greater than or equal to 0.

$$A_{i,j} = \varphi_{i,j} + k*P_{i,j} \quad (2)$$

The respective response time $RT_{i,j}$ is defined as the difference between the instant in which the message is effectively transmitted and $A_{i,j}$, rounded up in terms of the time unit δ. We further define the minimum and the maximum response times as $BCRT_{i,j}$ and $WCRT_{i,j}$ respectively. The interval of time between $BCRT_{i,j}$-1 and $WCRT_{i,j}$ is called the transmission window $TW_{i,j}$.

Moreover, we assume that the message set for each network is schedulable, i.e. $\forall_{i,j}\ WCRT_{i,j} \leq D_{i,j}$.

At this point, we will consider the specific case of message *i* which is transmitted across a set of networks. Its end-to-end latency is defined as Lee_i and this, as shown in the previous section, is influenced both by the period at every network, which is constant i.e. $\forall_{j,m}\ P_{i,j} = P_{i,m}$, as well as by the phases $\varphi_{i,j}$. The reasoning behind our approach to reduce the Lee_i is to specify a set of phases to apply at each network so that $TW_{i,j}$ follows $TW_{i,j-1}$ as close as possible for the same instance of message *i*, i.e. all consecutive transmission windows are chained (Figure 5). The ideal situation would occur with the phases determined as in expression 3 (networks 1 and 2 in Figure 5). However, depending on the traffic in network *j*, it might happen that the pair ($\varphi_{i,j}$, $BCRT_{i,j}$) given by expression 3 is not possible. In that case, we should look for the earliest $\varphi_{i,j}$ that causes the earliest $BCRT_{i,j}$ that is larger than $\varphi_{i,j-1}+WCRT_{i,j-1}$ (network 3 in Figure 5). This is further explained in section 3.4. The resulting Lee_i can be upper bounded using expression 4, where *n* is the number of traversed networks.

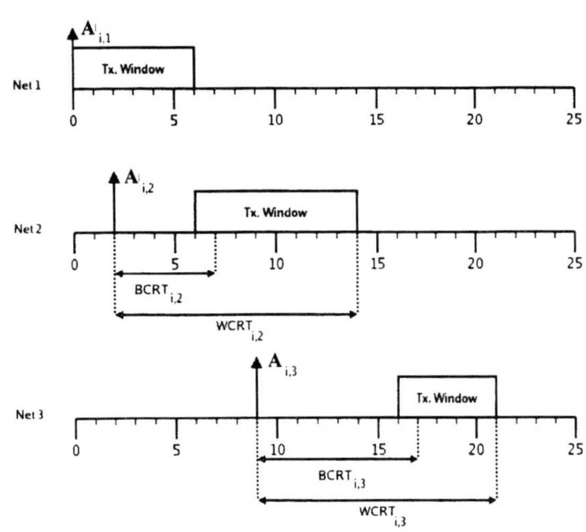

Figure 5: Reducing the inter-network traffic latency

$$A_{i,j} = A_{i,j-1} + WCRT_{i,j-1} - (BCRT_{i,j}-1)$$

$$\Leftrightarrow \varphi_{i,j} = \varphi_{i,j-1} + WCRT_{i,j-1} - (BCRT_{i,j}-1) \quad (3)$$

$$Lee_i \leq \varphi_{i,n}+WCRT_{i,n} - \varphi_{i,1} \quad (4)$$

This arrangement minimizes Lee_i for a given activation instant in the source network $A_{i,1}$ because shorter values imply that at least two consecutive transmission windows would overlap. This would mean that a message instance could be triggered in the successor network before the predecessor one leading to an extra delay of one period or even to losses of instances in case of overwriting.

3.2 Bounding the message transmission window

Minimizing the inter-network end-to-end latency, as discussed in the previous section, requires the computation of the message transmission window, or equivalently, the computation of the best and worst-case message response times (BCRT/WCRT, respectively), for all the traversed networks.

The computation of worst-case response time is a very well studied topic, both for task and message scheduling, and thus readily available solutions may be used to compute its value (e.g. (Almeida, *et al.*, 2002) for FTT; (Almeida, *et al.*, 2002b) for WorldFIP).

With respect to the computation of the best-case response time, however, to our best knowledge existing research results concern pre-emptive task scheduling (e.g. (Redell, *et al.*, 2002)) or a mix of pre-emptive task scheduling with message scheduling (e.g. (Gutierrez, *et al.*, 1998)).

In the work referred to above, the computation of the BCRT is made following a reciprocal reasoning with respect to the well-known WCRT analysis. That is, it searches for the combination of initial phases that minimize the interference caused by higher priority tasks, assuming that all task phases are possible. However, this approach does not suit the system model considered in this paper. In fact, in non-preemptive systems, if arbitrary phases are allowed, the BCRT is always lower bounded by the execution time of the task/message itself, as shown in Theorem 1.

<u>Theorem 1.</u> *In a schedulable time-triggered non-preemptive system, if arbitrary initial phases are allowed, the best-case response time experienced by an arbitrary message i belonging to a message set Γ is given by its transmission time C_i.*

<u>Proof.</u> The non-preemption assumption implies that once started, the message is transmitted until completion, without interruption. This can be stated as in i) where $s_i(k)$ is the starting instant of an instance *k* of message *i* and $f_i(k)$ its finish instant.

i) $\forall_k,\ f_i(k) = s_i(k) + C_i$

Moreover, consider $\Psi_{hep_i}(t)$ a function that returns the number of ready messages in Γ, at instant *t*, having higher or equal priority than message *i* (hep_i). Assuming that Γ is schedulable, then for every activation instant of message *i* ($a_i(k)$) there is necessarily some time instant t_l in the interval $[a_i(k), a_i(k)+D_i]$ in which the bus is not being used by messages with higher or equal priority than its own. Otherwise, message *i* would not be schedulable, as assumed. This is formally stated in ii).

ii) $\forall_k,\ \exists_{t_l \in [a_i(k),\ a_i(k)+D_i]}:\ \Psi_{hep_i}(t_l)=\{\ \}$

Since initial phases may have arbitrary values, it is always possible to set $a_i(k)=t_l$ so that ii) is always verified. In this case the activation time $a_i(k)$ and the start time $s_i(k)$ coincide and thus, by i), the response time equals the message transmission time C_i which is, thus, the best-case response time.

In real applications, however, the message attributes are defined according to specific design constraints beyond temporal ones. For example, if there are precedence constraints, several messages will exhibit the same period. In other cases, there are several messages with harmonic periods, or at least non-prime periods. In such cases, the BCRT of a given message may have a lower bound significantly higher than the message transmission time, contrarily to the situation stated in Theorem 1.

Consider, for example, the message set defined in Table 1, transmitted on a system using elementary-cycles (EC) (e.g. WorldFIP, FTT). The time unit is 1 EC.

This set generates the traffic pattern depicted in Figure 6. Note that the message set has a macro-cycle with duration of 6 time units (minimum common multiple of all message periods), and thus the traffic pattern repeats itself every 6 time units. When a new message is added, depending on its period and phase, distinct situations may occur. Two examples are given below.

ID	*C*	*P*	*φ*
1	1/3	1	1
2	1/3	2	1
3	1/3	3	1
4	1/3	3	1
5	1/3	6	2

Table 1: Message set characterization

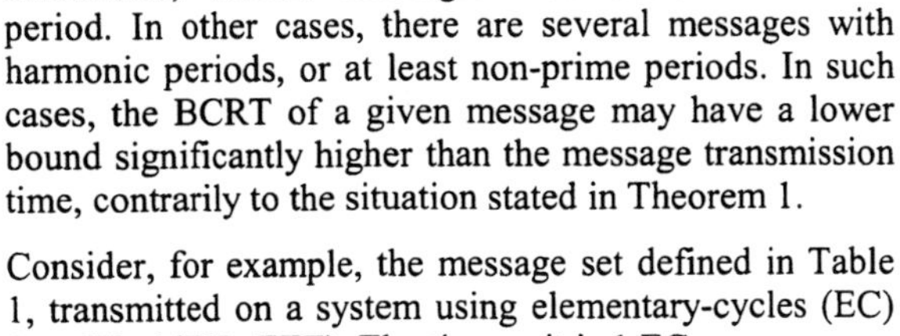

Figure 6: Traffic pattern generated by the message set defined in Table 1 where the vertical axis represents transmissions in the same micro-cycle.

i) The new message period is 7, i.e. prime relative to the macro-cycle. In this case, BCRT=1 since some message instances are transmitted in the EC in which they become ready (e.g. EC 3). WCRT=3 and occurs when the message is activated together with messages 1 to 4 (e.g. first EC). Thus, TW=3.

ii) The message period is 3, a sub-multiple of the macro-cycle. In this case, there are particular phases for which BCRT>1. For example, an initial phase of 1 leads to BCRT=2 and WCRT=3, resulting in a TW of 2. On the other hand, if the initial phase is 3, then it will be transmitted always in the EC where it becomes ready. Thus WCRT=BCRT=TW=1.

3.3 Computing the BCRT

In order to determine the response time of a given transmission request for message *i* within one single network, it is necessary to determine the interference that such request suffers from higher or equal priority messages (hep_i) in that network. The $BCRT_i$ occurs when the subset (S_i) of higher or equal priority messages ready to be transmitted at the time of the request of *i* is minimal. As may be trivially verified, messages belonging to S_i are always ready when message *i* is ready, too. Otherwise, there would be instances of message *i* that would not suffer interference of such messages in hep_i, and thus, by definition, those messages would not belong to S_i.

The computation of the subset S_i is carried out scanning hep_i according to Algorithm 1.

This algorithm allows determining a lower bound to $BCRT_i$ given a set of higher or equal priority messages and with known offsets, including the one of M_i. It is a lower bound and not the exact value because it does not account for possible load that could have been activated before but delayed until the current instant in which M_i is activated. In fact, it accounts for the interference caused by messages that are activated together with M_i, only. Nevertheless, in sake of clarity, we will still use $BCRT_i$ to refer to such lower bound.

```
1-  Si={}
2-  For all j so that Prj >= Pri {
3-    /* For all messages in hepi */
4-    If Pj=1
5-        Add Mj to Si
6-    Else if Pi>1 {
7-        Gij=gcd(Pi,Pj) /*greatest common divisor*/
8-        If Gij>1 /* non-prime periods */
9-            If (φj mod Gij) = (φi mod Gij)
10-             /* messages are in phase */
11-               If Pj<=Pi
12-               /* and Mj always occur with Mi */
13-                  Add Mj to Si
14-    }
15- }
16- BCRTi=response time of Mi given Si
```

Algorithm 1: Computation of the BCRT

To determine the response time of M_i given S_i (line 16), an adequate method must be used, depending on message transmission times and on the specific way the traffic is scheduled inside each elementary (micro) cycle, e.g. whether there is inserted idle-time, or a phase shorter than the cycle. Adequate analysis for such cases are described for example in (Almeida and Fonseca, 2001; Almeida *et al.*, 2002). The example shown in Section 4 was computed with the usual iterative process to derive the response time for fixed priority non-preemptive systems.

3.4 Minimizing the end-to-end latency

The methodology proposed to minimize the end-to-end latency of M_i is to iterate equation 3, starting from the specified initial phase $\varphi_{i,1}$ and $WCRT_{i,1}$ and deriving for j=2..n the phase $\varphi_{i,j}$ of $M_{i,j}$ in the successor network that minimizes $\varphi_{i,j}+BCRT_{i,j}$ subject to the constraint 5.

$$\varphi_{i,j}+BCRT_{i,j} \geq \varphi_{i,j-1}+WCRT_{i,j-1}+1 \qquad (5)$$

The process is described by the Algorithm 2. Basically, for each network $j>1$, it first assigns $\varphi_{i,j}=\varphi_{i,j-1}+WCRT_{i,j-1}$ and computes the respective $BCRT_{i,j}$ using Algorithm 1. Then, the algorithm enters a cycle to decrement $\varphi_{i,j}$ one by one, computing the respective $BCRT_{i,j}$ and testing the constraint 5. The last value of $\varphi_{i,j}$ that fulfils the constraint is the one that minimizes the interval between the transmission windows $TW_{i,j}$ and $TW_{i,j-1}$, as required to minimize the end-to-end latency Lee_i.

```
1- Compute WCRTi,1 given φi,1
2- For all networks j=2 to n {
3-   Compute WCRTi,j
4-   Assign φi,j=φi,j-1 and compute BCRTi,j
5-   While φi,j+BCRTi,j>=φi,j-1+WCRTi,j-1+1
6-       φi,j=φi,j-1 and recompute BCRTi,j
7- }
8- Leei = φi,n + WCRTφi,n - φi,1
```

Algorithm 2: Computing $\{\varphi_{i,j}\ j=2..n\}$ to minimize Lee_i

Notice that the minimization of Lee_i is not optimal, in the sense that the methodology herein proposed does not guarantee to find its absolute minimum. Nevertheless, it is relatively easy to compute and may yield good results, as shown in the next section.

4 ILLUSTRATIVE EXAMPLE

To point out the potential benefits that may be achieved by using the methodology presented in Section 3, this section presents an example where the end-to-end latency is computed both including and ignoring the best case response time estimation.

Consider a multi-segment network, composed by 3 different networks interconnected by two gateways. The networks are of a similar type as the one referred in Section 3.2, i.e., time-triggered traffic based on elementary-cycles (EC). Network 1 is lightly loaded, and each EC has at least 1/3 of its capacity free. Networks 2 and 3 have the traffic load depicted in Table 2. Transmission times, periods and initial phases are specified in a time unit corresponding to the EC duration.

ID	*C*	*P*	*φ*	
			Net 2	Net 3
1	1/3	1	1	3
2	1/3	2	1	3
3	1/3	3	1	3
4	1/3	3	1	3
5	1/3	6	2	4
6	1/3	6	3	5

Table 1: Example's message set characterization

Consider that a new message (k), with initial phase 4, period 12 and transmission time 1/3, originating in a node in network 1 and required by a node in network 3, is added to the system. The WCRT is 1 EC for network 1 and 5 ECs for networks 2 and 3.

If the BCRT computation is ignored, the end-to-end latency corresponds to the sum of the WCRT in each network, and thus totalizes 11 time units. If the BCRT, computed as described in Section 3 is included, the phases of the message in networks 2 and 3 are 1 and 3, respectively. For these activation instants the BCRT is 4 for both networks. The end-to-end latency is 4 time units.

By analyzing Algorithm 1 it is clear that a non-negligible reduction in the end-to-end latency is more likely to occur when a substantial part of the whole message set presents periods that are not relative primes. This, nevertheless, seems to be a common characteristic of many real applications for which many periodic activities require either harmonic periods or periods with a large common divider. We thus believe that this methodology has considerable potential for practical use.

5 CONCLUSIONS

This paper presents a contribution to reduce the communication latency in multi-hop time-triggered networks. The segmentation of networks may be advantageous, or even mandatory, in a variety of applications, because it allows managing the complexity of the design, may increase the schedulability level and also allows the physical extension of the network to limits that would not be possible with a single-segment.

However, in real-time applications the inter-network traffic may also be subject to time constraints that must be met in all anticipated circumstances. In this sense, the isochronous approach, in which the message scheduling and dispatching in all networks is synchronized, presents a high potential to provide inter-network communication with low latency.

This paper presents a technique to reduce the end-to-end latency suffered by inter-network traffic, which consists in computing both the best and worst-case message response times in each one of the intermediate networks. This technique, allows narrowing the transmission windows of a message in all networks and reducing the end-to-end latency by imposing the adequate phases that cause transmission windows to follow each other in tight succession. This method is particularly relevant when a substantial part of the traffic is harmonic or, at least, has non-prime periods.

REFERENCES

Luis Almeida, P. Pedreiras, and J. A. Fonseca (2002). "The FTT-Can protocol: Why and how". In IEEE Transactions on Industrial Electronics, volume 49. IEEE Press.

C. Aras, J. Kurose, D. Reeves, and H. Schulzrinne (1994). "Real-time communication in packet-switched networks". In Proc. of the IEEE, number 82 in 1, pages 122-139.

J-D. Decotignie (2002). "A perspective on Ethernet as a fieldbus". In Proceedings of the 4th IFAC International Conference on Fieldbus Systems and their Applications. IFAC Publications, Elsevier Science, Oxford, England.

H. Hoang, M. Jonsson, U. Hagstrom, and A. Kallerdahl (2002). "Switched real-time Ethernet with earliest deadline first scheduling- protocols and traffic handling". In Proceeding of Intl. Workshop on Parallel and Distributed Real-Time Systems'02, Fort Lauderdale, FL, USA. IEEE.

D. D. Kandlur, K. G. Shin, and D. Ferrari (1991). Real-time communication in multi-hop networks. In Proceedings of the 11th International Conference on Distributed Computing Systems (ICDCS), pages 300-307, Washington, DC, 1991. IEEE Computer Society.

C. Lu, B. Blum, T. Abdelzaher, J. Stankovic, and T. He. Rap (2002). "A real-time communication architecture for large-scale wireless sensor networks". In IEEE RTAS.

Jennifer Rexford, John Hall, and Kang G. Shin (1996). A router architecture for real-time point-to-point networks. In ISCA, pages 237-246.

Robert Bosch GmbH (1991). CAN Specification version 2.0, 1991. , Stuttgart, Germany

Hermann Kopetz (1997). Real-time systems: design principles for distributed embedded applications. Kluwer academic publishers.

European Standard EN 50170 (1996): Fieldbus: Vol. 1:P-Net; Vol.2: PROFIBUS; Vol.3: WorldFIP. CENELEC, European Committee for Electrotechnical Standardization.

(Powerlink) Ethernet Powerlink Standardization Group. Ethernet Powerlink protocol, available at www.ethernet-powerlink.org

FlexRay Requirements Specification Version 1.9.7 (2001). FlexRay Consortium.

G. Buttazzo, L. Abeni (2000). "Adaptive Rate Control through Elastic Scheduling". 3th IEEE Conference on Decision and Control. Sydney, Australia.

C. Lee, R. Rajkumar, C. Mercer (1996). "Experiences with Processor Reservation and Dynamic QOS in Real-Time Mach". Multimedia Japan 96. Japan.

M. Z. Yuan and H. B. Yu (2003). "Challenges and opportunities of implementing parallel computation in networked control systems". Information and Control, pp. 97—102.

O. Redell, M. Sanfridson (2002). "Exact best-case response time analysis of fixed-priority scheduled tasks". Proceedings of the 14th Euromicro Conference on Real Time Systems, pp. 165--172. Vienna, Austria.

J. C. Gutierrez, J. J. Garcia, M. G. Harbour (1998). "Best-case analysis for improving the worst-case schedulability test for distributed real-time systems". Proceedings of the 10th Euromicro Workshop on Real-Time Systems, pp. 35—44. Berlin.

L. Almeida, E. Tovar, J.A.G. Fonseca, F. Vasques (2002). "Schedulability Analysis of Real-Time Traffic in WorldFIP Networks: An Integrated Approach," IEEE Transactions on Industrial Electronics, USA, Vol. 49, No. 5, pp. 1165-1174.

L. Almeida, J.A. Fonseca (2001). "Analysis Of A Simple Model For Non-Preemptive Blocking-Free Scheduling," Proc. ECRTS'01-13th Euromicro Conference on Real-Time Systems, Delft, Holland, 13-15.

G. Saba, Z. Mammeri, J. P. Thomesse (1995). "Some Solutions for FIP Network Interconnection". Proceedings of 1st International Workshop on Factory Communication Systems, Laysin, Switzerland, October 4-6.

M. Ryu, S. Hong (1998). "End-to-end design of distributed real-time systems". Control Engineering Practice 6, pp.93-102. Elsevier Science Ltd.

Copyright © IFAC Information Control Problems in
Manufacturing, Salvador, Brazil, 2004

ELSEVIER
IFAC
PUBLICATIONS
www.elsevier.com/locate/ifac

AN EVENT FILTERING SCHEME FOR DISTRIBUTED REAL-TIME EMBEDDED SYSTEMS

Carlos Mitidieri *,** **Jörg Kaiser** * **Carlos Eduardo Pereira** **

* *University of Ulm, Germany*
** *Federal University of Rio Grande do Sul, Brazil*

Abstract: Event filters provide the information selectivity needed to regulate the functional aspect of the coordination in publish/subscribe systems. This paper tackles the question of filtering events throughout a publish/subscribe middleware, which is aimed to support the interaction among autonomous entities in networked embedded systems. Therefore, the tradeoff between, first, the accuracy whereby subscribers can specify the events that are relevant to them, and second, the efficiency and predictability of the filtering execution, has to be adjusted to cope with resource constraints. The proposed filtering method and the related algorithms are aimed to provide an adequate solution to this tradeoff.
Copyright © 2004 IFAC
Keywords: communication protocols, embedded systems, real-time

1. INTRODUCTION

The development of industrial automation systems shows up a clear trend towards distributed architectures. This reflects the decentralized nature of the control tasks concerned in automation, and the necessity to adapt to dynamic conditions, as well as to evolving processes and ever changing components. For dependability reasons, dependencies have to be avoided whenever possible. Hence, the autonomy of the distributed system components has to be enforced. Moreover, system responsiveness can be improved, if the autonomous components are fine-grained on a hardware level. This can be accomplished by deploying intelligent sensors and actuators at the input/output interface. These are dedicated devices, comprising a computational block (usually presenting low-performance), a network interface and a transducer element. On a higher level of abstraction, the intelligent transducers have to coordinate with other autonomous objects (e.g. controllers) to perform useful tasks. This is accomplished through the underlying communication system.

Fieldbus technology has a central role in systems like these. Their properties allow to support reliable and timely coordination of networked computing nodes. On the other hand, the issues relating to fieldbus communication are on a very low level of abstraction, and should be hidden from application developers handling high level concerns such as the coordination of autonomous entities. A middleware layer providing appropriate abstractions is required, as previously discussed in previous papers (Kaiser and Mock, 1999; Pereira *et al.*, 2001). Briefly reviewing, the coordination model has to support spontaneous dissemination of events, which are generated at the sensor interface or in association to internal state transitions. Moreover, the information transported in association with events usually has to be disseminated efficiently to multiple destinations. Finally, the dissemination of events has to be decoupled from the individual control flows of the distributed objects.

The event-based communication entailed in the publish/subscribe (P/S) model is well known for supporting the requirements mentioned above (Oki *et al.*, 1993; Rajkumar *et al.*, 1995; Eugster *et al.*, 2001; Meier and Cahill, 2002). However, the P/S model does not implicitly provide any level of reliability or timeliness for disseminating events. Thus, the COSMIC middleware (Kaiser *et al.*, 2003*a*) has been introduced, which enriches the P/S model with abstractions and semantics that support the real-time dissemination of events in distributed embedded systems. First, it recognizes that "events" may represent occurrences in the sensed environment, so they must be associated to a context, which is described by attributes such as a location and the occurrence time. Second, it includes real-time event channels (Kaiser *et al.*, 2003*b*), which are abstractions of the communication infrastructure, providing a high level interface for programming the quality level required to disseminate events. Third, it includes a dynamic binding mechanism (Kaiser and Mock, 1999), which is crucial for the applicability of P/S systems in embedded systems.

This paper reports on mechanisms and algorithms for filtering events, which have been designed to integrate COSMIC. The proposed approach addresses the fundamental tradeoff existing between, on one side, the accuracy of the subscription scheme, and on the other side, the computational effort required to match events to subscriptions. This has to be solved considering the resource constraints existing in embedded systems.

The paper is organized as follows. Sec. 2 presents a review of the event filtering approach considered and further elaborated in this paper. The algorithms that have been developed for implementing this approach are analyzed in Sec. 3. Related work is discussed in Sec. 4. Conclusions are given in Sec. 5.

2. EVENT FILTERING

2.1 Overview

Two broad approaches exist for routing events to subscribers. In *subject-based* systems (Oki *et al.*, 1993; Cabrera *et al.*, 2001), a single identifier related to the events' contents has to be verified. In contrast, the entire functional data carried by events has to be looked over in *content-based* systems (Carzaniga *et al.*, 1998; Pietzuch and Bacon, 2002). In this process, some of the parsed items have to be checked against predicates defined by subscriptions. Although the latter approach is more flexible and may select information more accurately, its computational cost is higher.

For efficiency reasons, the subject-based approach has been adopted in COSMIC. However, it turned out from practice that it is usually possible to find groups of subscriptions that, although requiring events characterized by distinct attributes, can be tracked down to a common topic. Of course, distinct subjects could be assigned to each of these groups of subscriptions to distinguish them. This implies that many subjects would become related to the same contents. Hence, the semantic relation between subject and contents would be lost. Moreover, it would be impossible to exploit the natural occurrence of overlapping interests when striving to optimize the filtering performance. As explained in Sec. 2.2, structural properties of events representation can be exploited for distinguishing the events on a subject, and support the detection of overlapping interests with a low computational cost.

2.2 Conformity-based filtering

Events present attributes which complement the contents. They relate, e.g., to the context in which events are generated and the required quality of service. Attributes are included (or not) in events according to applications' semantics. Hence, the structure defined by the presence/absence of attributes partially reveals the meaning of an event, and can be exploited in a method for filtering events. Conformity-based filters (Mitidieri and Kaiser, 2003) consist of such a method for exploiting events' structure in a subscription scheme.

Consider that events are typed by a subject and enclose a collection of $\langle attribute, value \rangle$ pairs. Events pertaining to the same subject are allowed to present different collections of attributes, so distinctive structures. An event that includes no attributes is also valid. When subscribing, an object must specify the subject and the attributes that an event has to include to be of interest. An event matches a subscription if:

(1) The event's subject matches the subject specified in the subscription;
(2) And the event's structure *conforms* to the structure specified in the subscription.

A given element is said to *conform structurally* to another element, if it includes *at least* the same attributes that are included in the latter (Cardelli, 1988). If an event includes every attribute that is specified in a subscription, plus any other, then this event conforms to (matches) this subscription. An event which lacks just a single attribute that is required in a subscription does not conform to (match) this subscription.

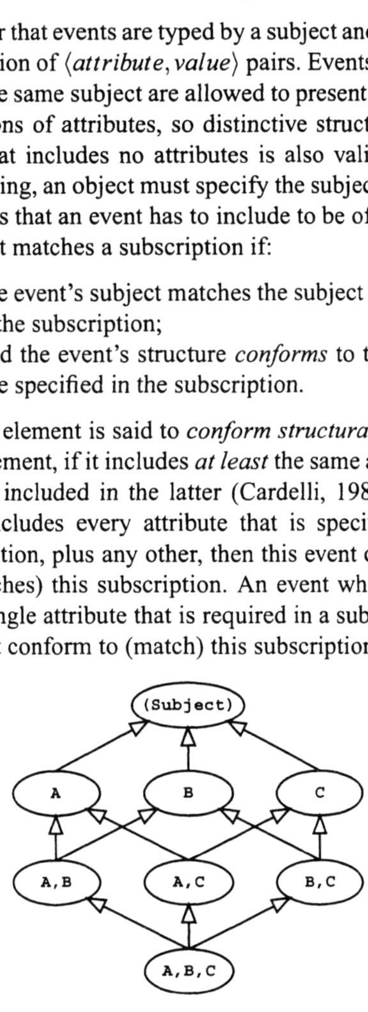

Fig. 1. Structural conformance relationships.

The structural conformance relationship is transitive:

$$(\alpha \Subset \beta) \wedge (\beta \Subset \gamma) \Rightarrow (\alpha \Subset \gamma)$$

where the symbols α, β and γ represent structures and the symbol $\Subset$ stands for the `conforms-to` relation. Hence, an event can match an hierarchy of filters specifications. For instance, the graph in Fig. 1 represents such a hierarchy for three attributes, which are represented by A, B and C. The structures shown in the nodes may belong to an event, or to a subscription. The directed edges represent the conforms-to relation. Note that the subject tagging the root node (which includes no attributes) applies to every node in the graph. Then follows these examples:

- $(Subscriptions: \langle S, \langle B \rangle\rangle)$ is-matched-by $(Events: \langle S, \langle B \rangle\rangle, \langle S, \langle A,B \rangle\rangle, \langle S, \langle B,C \rangle\rangle, \langle S, \langle A,B,C \rangle\rangle)$;
- $(Event: \langle S, \langle A,C \rangle\rangle)$ matches $(Subscriptions: \langle S, \langle A,C \rangle\rangle, \langle S, \langle A \rangle\rangle, \langle S, \langle C \rangle\rangle, \langle S, \langle\, \rangle\rangle)$.

3. FILTERING EXECUTION

3.1 Event frames and attributes vectors

Each subject is associated to a restricted set of specified attributes. E.g., Fig. 2 relates to a particular case whereby a subject is associated to three attributes. Each attribute is mapped to a fixed position in a row. This row has a bit-vector representation, so called *attributes vector* (*AV*). If an event includes a specific attribute, the respective position in the associated vector is turned "on", otherwise it is turned "off". The attributes vector that represents a *conformity filter specification* (CF_{spec}) is formed in the same way.

Events are laid into a frame for transmission. The layout of this frame has been designed to cope with the execution of the developed algorithms. The event frame has fields for the subject, for the attributes vector, and for the values of the attributes, as shown on the top of Fig. 2. The order of the attributes' values in the frame is the same as in the attributes vector. In addition, the data type and the length of every attribute are known in the nodes that publish or subscribe to the related subject. Hence, the location of each attribute value in a frame can be calculated. Event frames are compact in size, because the vectors' mappings and the attributes' data types are known in every node. This is beneficial regarding fieldbuses, which often present very narrow message frames.

3.2 Location of filters

The assumed communication infrastructure is composed of fieldbuses, which are connected via *gateways* to a wide area network (WAN). Fieldbuses are often embedded in an equipment (rovers, manufacturing cells, etc.), whose safe operation requires tight coordination of the local computing nodes. These nodes can be viewed as application *components*, since each of them is dedicated to a specialized activity (smart transducers, controllers, etc.). Supported by gateways, components located in distinct fieldbuses may also coordinate across the WAN. This typically comprises non critical interaction, e.g., cooperative ambient monitoring.

Filters are distributed over components and gateways. At components, filters must be light weight and execute timely, to cope with severe resource constrains and control requirements. Conversely, gateways usually do not have to support stringent temporal requirements, since the WAN can barely support soft real-time communication. Moreover, since the filters in the gateways have to match every event that arrive at their network interfaces [1], they may require processors with relative performance.

3.3 Filtering events at subscribers' nodes

3.3.1. Data structures The data structure that support event filtering at subscribers' nodes is depicted in Fig. 2. It includes a two-dimensional array, which stores the references to *subscriptions lists*. The primary dimension of the array is indexed by subject identifiers. It is labeled as "SID" in Fig. 2. The secondary dimension is indexed by conformity filter specifications (CF_{spec}). Each subscriptions list contains every local subscription that is conformed by the indexing CF_{spec}. Hence, filtering an event requires one simple access to this table. This is a light weight operation that practically presents no jitter.

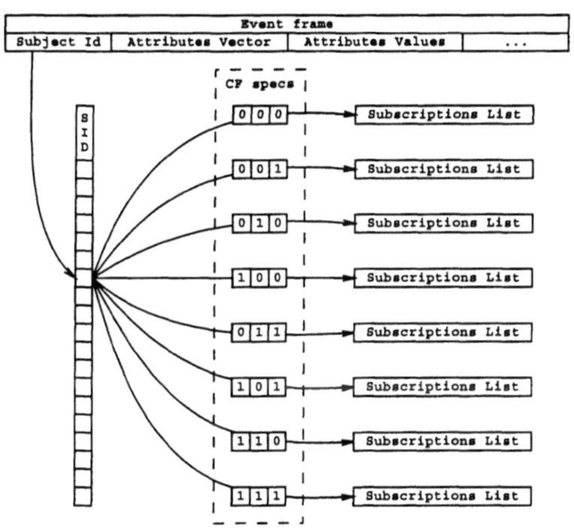

Fig. 2. Data structures for subscribers nodes.

3.3.2. Implementation The data structures are statically allocated and configured, accordingly to the local subscriptions. This static configuration is possible, since the subscriptions that can be issued in a node are known in advance. And it is necessary, if the microcontrollers that are embedded in the smart transducers are not powerful enough to support dynamic memory allocation.

3.3.3. Insertion of subscriptions The subject identifier and the conformity filter specification (CF_{spec}) are read from each submitted subscription. Then, the

[1] The dynamic binding mechanism (Kaiser and Mock, 1999) avoids this need for component nodes.

subscription's CF_{spec} is matched against each of the AV indices of the secondary array, accordingly to the conformity relation. If the CF_{spec} is conformed by an index, a reference to the subscription is inserted in the respective list. An index conforms to a CF_{spec} if the following condition is true:

$$(AV_{index} \ \& \ CF_{spec}) == CF_{spec} \tag{1}$$

where & is the bitwise-AND operator. The algorithm is summarized below.

Algorithm 1

```
For each subscription s
  hs = hash_function(subject_(s))
  cf_spec = attributes_vector(s)
  For(av=0,av<MAX_NUMBER_OF_AVs, av++)
    If((cf_spec & av) == cf_spec)
      *subscriptions_list = filtering_table[hs][av]
      insert_subscription(s,subscriptions_list)
```

3.3.4. Events filtering The subject identifier is obtained for every received event, and applied as the input for a hash function (Vitter, 1982). The attributes' vector is also read from the event frame. Then, both the hash value and the attributes vector are used to index the filtering table, which returns a reference to a *list of subscriptions*. If this list is not empty, it contains all the subscriptions that have been matched. Hence, it is passed to a dispatching module, together with the event. This procedure is summarized in Alg. 2.

Algorithm 2

```
01 For each event e
02   hs = hash_function(subject_id(e))
03   av = attributes_vector(e)
04   *subscriptions_list = filtering_table[hs][av]
05   If(subscriptions_list is not empty)
06      dispatch(e,subscriptions_list)
```

3.3.5. Analysis The proposed algorithms were designed to ensure efficiency as well as predictability when filtering events, at the cost of a higher complexity when inserting and deleting subscriptions. Considering that subscriptions are inserted and deleted outside of critical control paths, while event filtering should usually be performed in real-time control paths, this is a reasonable solution.

Accordingly, the filtering algorithm's execution time has complexity *O(1)*, both in relation to the number of attributes and the number of stored subscriptions. On the other hand, when matching the CF_{spec} of a subscription against the AV indexes (lines 04 and 05 in Alg. 2), the time complexity is $O(l)$, where l is the length of the array. In addition, the insertion and deletion of a subscription in a list is $O(\log_2 s)$, where s is the number of subscriptions, since the subscriptions' lists are implemented as binary search trees. The order in the trees is defined by the subscriptions' identifiers.

The required memory for each secondary array is $O(2^n)$, where n is the number of attributes. Nevertheless, this scheme is addressed for fieldbuses, particularly for the CAN-Bus (CAN, 1991). Hence, the number of attributes that can be included in events is already limited by the reduced frames sizes. For instance, the CAN-Bus frames may carry a maximum of 8 bytes. Thus, both the memory requirements and the execution of CF_{spec} matching loop are expected to be feasible, even for small micro-controlled boards.

3.4 Filtering events at gateways

When filtering events at gateways, events do not have to be passed to a notification dispatcher. Hence, the data structures have to store only flags signaling which events should be forwarded. In addition, the subscriptions' identifiers are registered together with the corresponding flags, to manage deletions.

3.4.1. Data structures Gateways' filters include a filtering table, which is indexed by subject identifier (*SID*) and the number of attributes (*NA*) on events, as indicated in Fig. 3. The stored elements are references to lists of CF_{spec} entries. Each list contains only entries presenting the number of checked attributes corresponding to the indexing *NA*. The presence of a CF_{spec} indicates that the conforming events have to be forwarded. No duplicate of a CF_{spec} has to be stored in a list. Hence, these lists can be structured as height-balanced binary search trees.

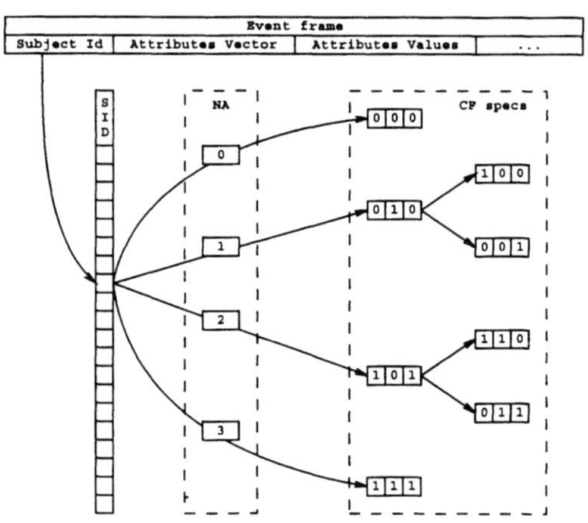

Fig. 3. Data structures for gateways.

3.4.2. Implementation While the *SID* array is static, the *NA* arrays and the addressed lists are dynamically created and destroyed, in response to the insertion and deletion of subscriptions. Dynamic maintenance of this structure is necessary, because it is not possible to know in advance which hosts (i.e., subscribers) will be connected to a network. In fact, it is assumed that they vary, as the system evolves.

3.4.3. Insertion of subscriptions As outlined in Alg. 3, the subject identifier, the CF_{spec} and the respective number of checked attributes are read for each submitted subscription (Lines 02, 03, 04). Then, the applicable tree is retrieved (Line 05) and an attempt to insert the CF_{spec} is carried out (Line 06). If an identical CF_{spec} is already inserted, then only the subscription identifier has to be recorded in the corresponding tree node. This is necessary to manage the deletion of subscriptions. If an insertion effectively occurs, an operation for balancing the height of tree is performed. The algorithm for deleting a subscription includes comparable steps.

Algorithm 3

```
01 For each subscription s
02    hs = hash_function(subject_(s))
03    cf_spec = attributes_vector(s)
04    na = number_of_checked_attributes(s)
05    *bst = filtering_table[hs][na]
06    insert = insert_binary_tree(cf_spec,bst)
```

3.4.4. Events filtering The subject identifier, the *AV* and the number of checked attributes (*NA*) in the *AV* are obtained (Lines 02, 03, 04, Alg. 4) from each received event. Then, a loop is executed (Line 05) to retrieve CF_{spec} lists. This loop ranges from *NA* to zero. The first retrieved list contains only CF_{spec} entries with the same *NA* as the processed event. Hence, a binary search is performed in the first list, since at most one entry may be conformed by the current *AV*. If a match does not occur, the remainder lists are retrieved (one at a time) and traversed, looking for a match. If a match occurs when traversing a tree, the event is forwarded and the search stops. In addition, a CF_{spec} identical to the *AV* of the current event is inserted in the corresponding tree. Therefore, the complexity for matching subsequent events with comparable *AV* will be much lower, as discussed in Sec. 3.4.5.

Algorithm 4

```
01 For each event e
02    hs = hash_function(subject_id(e))
03    na = number_of_checked_attributes(e)
04    av = attributes_vector(e)
05    For(i=na, i>=0, i--)
06      *bst = filtering_table[hs][i]
07      If(i == na)
08        binary_search(av,bst)
09      Else
10        traverse_tree(av,bst)
```

3.4.5. Analysis Gateways concentrate subscriptions. Hence, the algorithms described in this section are designed to support a larger number of subscriptions, as well as more frequent insertions and deletions, in comparison to the previous one. As a matter of fact, the space complexity has been managed to be $O(l)$, where l is the *AV* length. In consequence, it is unavoidable to increase the time required to filter an event. This tradeoff is acceptable, since the interaction that occurs through the gateways is less stringent regarding timeliness, as already observed.

The time complexity for inserting or deleting a subscription is bounded to $O(\log_2 n)$, where n is the number of nodes in the respective binary search tree. The determination of such number of nodes is related to the problem of generating combinations of k elements (i.e. the number of checked attributes) taken from a set of l elements (i.e. the total length of the *AV*), without repetitions. The maximum number of combinations (corresponding to tree nodes) is given by :

$$\frac{l!}{k!(l-k)!}$$

which is a polynomial function on l. Hence, considering the following inequality:

$$O(\log_2 n) \equiv O\left(\log_2 \frac{l!}{k!(l-k)!}\right) < O\left(\log_2 2^l\right) \equiv O(l)$$

it is possible to conclude that the complexity for inserting or deleting a subscription is lower than $O(l)$, where l is the *AV* length.

The time complexity for filtering events presents two cases (Lines 08 and 10 of Alg. 4). When an event with an unequaled *AV* is received for the first time in a gateway, the complexity is bounded to $O(s)$, where s is the number of subscriptions on the related subject. However, the time complexity for filtering the subsequent events of the same flow (which will present the same *AV*) will be lower than $O(l)$, exactly the complexity of insertions and deletions of subscriptions.

4. RELATED WORK

Structural properties of events representation are implicitly considered in the filtering schemes of many P/S systems, although in these cases always tied to the evaluation of predicates (Carzaniga *et al.*, 1998; Starovic *et al.*, 1995; Meier and Cahill, 2002; Mühl *et al.*, 2002; Gelernter, 1985). The omission of such association on the presented scheme allows to break up the filtering process in a number of stages. Such division of labor may improve the overall efficiency of the filtering process, since the searched space (i.e., the number of subscriptions to be matched) can be progressively restricted, while the complexity of the matching mechanism increases. For instance, predicates could be evaluated subsequently to conformity-based filters, but then only for the subscriptions which have been pre-selected so far.

Structural properties reflecting the relations among information categories have been exploited in (Kulik, 2003) for speeding the matching and forwarding of events in P/S systems. That scheme was addressed for disseminating information in the Internet, thus it assumes the existence of a centralized information collector, which would manage to establish the categories of the incoming information flows. Such an infrastructure is not available and is actually undesired in the systems addressed in this work.

5. CONCLUSIONS AND FUTURE WORK

This paper builds up on the exploitation of the publish/subscribe model for coordinating autonomous components in distributed real-time embedded systems. The question tackled relates to the fundamental compromise existing between the flexibility of a subscription scheme and the computational cost associated to the related implementation. This tradeoff has been addressed considering the resource constraints typically found in distributed embedded systems. It has been argued that structural properties of event representation can be exploited to improve subscriptions' selectivity on subject-based systems, while still conserving low the time complexity of filtering execution. The scheme has been implemented and its performance is being evaluated under a variety of conditions. In the future, the research will be focused on the exploitation of event attributes on the high level coordination of autonomous devices.

ACKNOWLEDGMENTS

This work has been supported by the EC, through project IST-2000.26031 (CORTEX: COoperating Real-time senTient objects: architecture and EXperimental evaluation) and by CNPq, and FINEP, the Brazilian research agencies.

REFERENCES

Cabrera, L. F., M. B. Jones and M. Theimer (2001). Herald: Achieving a global event notification system. In: *8th Workshop on Hot Topics in Operating Systems (HotOS-VIII)*.

CAN (1991). *CAN Specification version 2.0*.

Cardelli, Luca (1988). Structural sub-typing and the notion of power type. In: *Conference Record of the Fifteenth Annual ACM Symposium on Principles of Programming Languages*. San Diego, California. pp. 70–79.

Carzaniga, A., D. S. Rosenblum and A. L. Wolf (1998). Design of a scalable event notification service: Interface and architecture. Technical report. Department of Computer Science, University of Colorado.

Eugster, P. Th., P. Felber, R. Guerraoui and A. M. Kermarrec (2001). The many faces of publish/subscribe. Technical Report DSC ID:200104. EPFL. Lausanne, Switzerland.

Gelernter, D. (1985). Generative communication in Linda. *ACM Trans. Prog. Lang. Syst.* **7**(1), 80–112.

Kaiser, J., C. Mitidieri, C. Brudna and C.E. Pereira (2003*a*). COSMIC: A middleware for event-based interaction on CAN. In: *IEEE Conference on Emerging Technologies and Factory Automation*. Lisbon, Portugal.

Kaiser, Joerg and M. Mock (1999). Implementing the real-time publisher/subscriber on the controller area network (CAN). In: *2nd Interantional Symposium on Object-Oriented Real-time distributed Computing*. Saint-Malo, France.

Kaiser, Joerg, Cristiano Brudna and Carlos Mitidieri (2003*b*). A real-time event channel model for the CAN-Bus. In: *International Workshop on Parallel and Distributed Real-Time Systems (WP-DRTS'2003)*. Nice, France.

Kulik, Joanna (2003). Fast and flexible forwarding for internet subscription systems. In: *Second International Workshop on Distributed Event-Based Systems (DEBS'03)*.

Meier, René and Vinny Cahill (2002). Steam: Event-based middleware for wireless ad hoc networks. In: *International Workshop on Distributed Event-Based Systems*.

Mitidieri, C. and J. Kaiser (2003). Attribute-based filtering for embedded systems. In: *Second International Workshop on Distributed Event-Based Systems (DEBS'03)*. San Diego, California.

Mühl, G., L. Fiege and A. P. Buchmann (2002). Filter similarities in content-based publish/subscribe systems. In: *International Conference on Architecture of Computing Systems (ARCS)*. Karlsruhe, Germany. pp. 224–238.

Oki, Brian, Manfred Pfluegl, Alex Siegel and Dale Skeen (1993). The information bus - an architecture for extensible distributed systems. In: *ACM Symposium on Operating System Pronciples*. pp. 58–68.

Pereira, C. E., J. Kaiser, C. Mitidieri, C. Villela and L. B. Becker (2001). On evaluating interaction and communication schemes for automation applications based on real-time distributed objects. In: *4th Int. Symposium on Object-Oriented Real-Time Distributed Computing (ISORC'01)*. Magdeburg, Germany.

Pietzuch, P. and J. Bacon (2002). Hermes: A distributed event-middleware architecture. In: *1st International Workshop on Distributed Event Systems (DEBS'02)*.

Rajkumar, Ragunathan, Mike Gagliard and lui Sha (1995). The real-time publish/subscribe interprocess communication model for distributed real-time systems: Design and implementation. In: *IEEE Real-Time Technology and Applications Symposium*. IEEE Real-Time Technology and Applications Symposium.

Starovic, Gradimir, Vinny Cahill and Brendan Tangney (1995). An event-based object model for distributed programming. In: *OOIS (Object-Oriented Information Systems) '95*. Springer-Verlag. London. pp. 72–86.

Vitter, J. S. (1982). Implementations for coalesced hashing. *Communications of ACM* **25**(12), 911–926.

Copyright © IFAC Information Control Problems in
Manufacturing, Salvador, Brazil, 2004

ELSEVIER
IFAC
PUBLICATIONS
www.elsevier.com/locate/ifac

ON ADDING SUPPORT FOR REAL-TIME AND DISTRIBUTED APPLICATIONS TO THE MATPLC

Mário de Sousa, Adriano Carvalho

Departamento de Engenharia Electrotecnica e de Computadores
Faculdade de Engenharia da Universidade do Porto
Rua Dr. Roberto Frias s/n, 4200-465 Porto, Portugal

Abstract: The MatPLC is an open-source control application, consisting of a core, generic modules, and tools for creating custom modules. Since many control and monitoring systems are distributed in nature and require strict time determinism, features not currently supported, the addition of hard teal-time capabilities as well as support for distributed applications is required. The paper includes an outline of the MatPLC's architecture, and details the design changes required to add hard real-time capabilities to the MatPLC framework. The IEC 61131-3 compiler is presented, and a discussion is made regarding how it may be extended to support the IEC 61499 standard for distributed applications. *Copyright © 2004 IFAC*

Keywords: Programmable logic controllers, Distributed computer control

1. INTRODUCTION

The technology used in PLCs (Programmable Logical Controllers) have evolved with the times, to the point that many modern top of the range PLCs are actually full fledged computers in disguise, executing modern operating systems. More importantly, and in order to take advantage of economies of scale, many PLC vendors have started to adopt hardware similar to PCs (Personal Computers).

The MatPLC project was started with the intention of eliminating the lock the vendors have on the end-users, by taking advantage of open standards as well as open source operating systems running on the *de facto* standard PC platform. It has currently been successfully tested on a DIN rail mounted PC platform marketed by SixnetIO. Most industrial communication networks are supported (Modbus, Devicenet, Profibus, ASI, etc.), either directly or through the use of a network interface card manufactured by Hilscher.

Software standards are also being taken into account. Currently the project includes a compiler for the IL (Instruction List) and ST (Structured Text) programming languages defined in IEC 61131-3 (International Electrotechnical Commission, 2001). This paper discusses future support for the IEC 61499 standard (International Electrotechnical Commission, 2000), which describes a framework for development and support of distributed industrial applications on industrial PLCs. Additionally, the current MatPLC architecture is analyzed, and the changes required to make it time deterministic are discussed.

2. MATPLC OVERVIEW

The MatPLC is an open-source control application, composed of autonomous but cooperating modules that execute in separate user-space processes, and access a common shared state stored in shared memory. Each of these modules is free to decide whether or not to execute in a standard PLC scan loop, which means that the industrial control

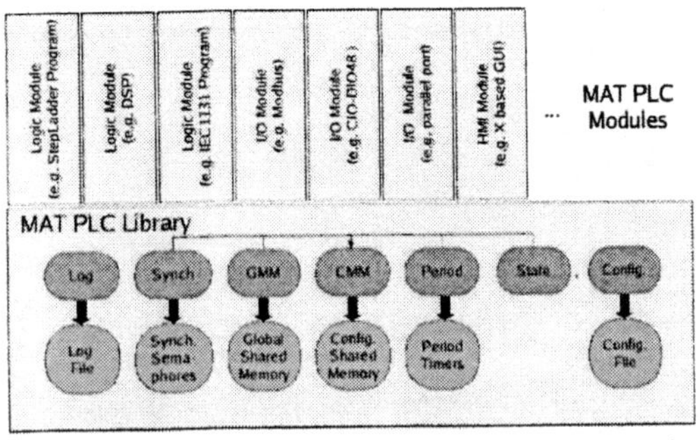

Fig. 1. MatPLC Architecture, showing a selection of modules in the top part and the sections of the library in the bottom.

application may be developed using either asynchronous programming common to most PLCs, or synchronous programming common to most other computer architectures. Both methods may even be used simultaneously, as long as they are not mixed within the same MatPLC module. The application builder may choose to let the modules execute autonomously, or to synchronize their activities.

Access to all shared resources is made through the MatPLC library routines which offer PLC-like semantics for the modules that wish to use them, such as inputs that only change at the beginning of the logic and outputs that are only written at the end of the logic. The library is divided into several sections (Fig. 1):

- configuration memory manager (cmm) - manages the shared memory area that stores core configuration data, guaranteeing that all modules share the same configuration;
- global memory manager (gmm) - manages the shared memory area used to store the state of the plc points;
- synch section - handles the synchronization between modules;
- period section - enforces scan loop timings;
- state section - handles module execution state;
- configuration section - parses the configuration files;
- log section - allows every module to produce log messages in a consistent manner. These are timestamped and written to a logging file.

A more detailed explanation of the internal mechanisms used within the more complex sections follows.

2.1. The Global Memory Map

Since the MatPLC architecture hinges on the simultaneous execution of several modules, every access to the global memory map by a specific module must be made to be atomic with respect to the other modules. Several modes are available for

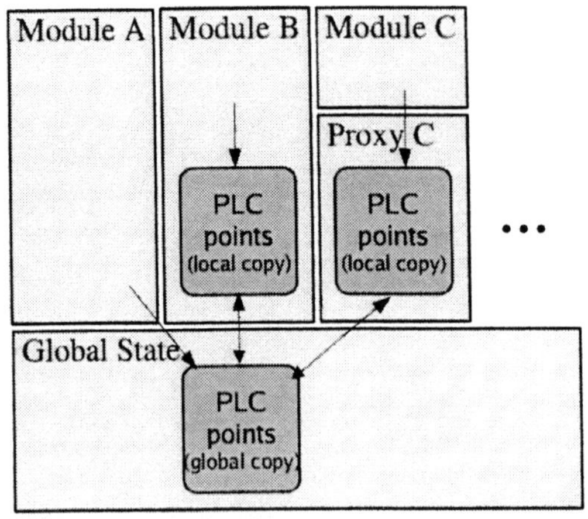

Fig. 2. Synchronization of gmm memory maps (local, isolate and shared)

enforcing this constraint (Fig. 2), none of which is optimal for all possible scenarios in which the MatPLC is expected to be used. The application builder may choose any of the access modes for each executing module.

For the default 'local' mode, a local copy of the memory map is created for the module (Fig. 2 - Module B). When a module accesses a plc point, it is actually accessing its local memory map. Local and global memory maps are synchronized by calling the plc_update() function, which is protected by a semaphore, providing atomic updates with respect to other modules.

The second mode, `isolate' (Fig. 2 - Module C), completely isolates the module from the shared memory used by the PLC, and is mostly used by untrusted modules (e.g. still in a debugging stage). It uses sockets to forward the plc_update() function call from the module to the proxy, introducing significant overhead.

The third and last mode, `shared' (Fig. 2 - Module A), assumes that simultaneous access to the shared map will never occur, and therefore gives the module direct access to the global memory map with no synchronization enforced. This may be guaranteed if all modules are running a scan loop with each loop executing in turn (possible due to the synch section, described below), or if the modules access disjoint portions of the global map.

2.2. The Synch Section

The synch section allows the application builder to specify the sequence of execution of the running modules. This is achieved by specifying a Petri net (Fig. 3), taking into account that particular synchronization points in each module (usually beginning and end of scan, but optionally others) are associated with the firing of a transition. When such a point is reached during the module's

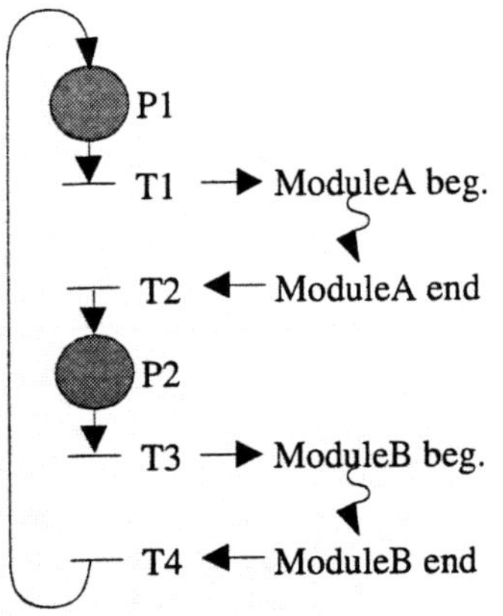

Fig. 3. Synchronization Model

execution, the module blocks until the transition fires (assuming it cannot fire immediately). Note that a transition will not fire unless a module is waiting on it; in this the semantics differ from those of a standard Petri net.

The synch section leverages the standard SysV semaphores to simulate the synchronization Petri net, using a single SysV semaphore set, with a semaphore for each place. Each transition is implemented by simultaneously waiting on all the appropriate semaphores, a functionality of SysV semaphores not supported by POSIX semaphores.

2.3. The Period Section

This section enforces maximum scan rates for each module. It uses POSIX timers to set an alarm that goes off at every multiple of the desired scan period, at which time an alarm counter is incremented. When a module is ready to start a new scan it decrements the alarm counter and continues with the scan. If there are no outstanding alarms, the scan is delayed until the next alarm goes off.

2.4. The State Section

This section of the library handles RUN/STOP modes both for the MatPLC as a whole and for each module. A module will only execute a scan if both the whole MatPLC and the module itself are in RUN mode.

In order for these modes to work correctly with the synch library, they are implemented by adding hidden places to the synch petri net. These places are connected with arcs to the begin of scan synch transitions in such a way that the transition will only fire if both the PLC and the module are in RUN mode, with the tokens being replaced after firing. A module is thus able to atomically verify the conditions required by both the state and the synch sections.

3. REAL-TIME MATPLC

In order to support RT (Real-Time) applications, the MatPLC framework will need to be augmented to become time-deterministic and take advantage of the facilities provided by a RTOS. Considering that a real-time control application is commonly composed of both hard (e.g. control) and non-hard real-time (e.g. graphical interface) components, these components were mapped onto MatPLC modules. The revised framework therefore requires that the modules may be allowed to execute under a real-time priority, chosen by the application builder. This has been achieved by adding a new RT section to the MatPLC core which, during startup, sets the RT priority and locks the process into RAM (so as to eliminate any delays due to swapping memory out to disk).

3.1. CMM and GMM Sections

The cmm and gmm sections each use shared memory during run mode, and a single semaphore each for controlling access to those memory sections. However, unlike the access to the cmm memory area which is only made during initialization, access to the gmm memory area is made during run-time by modules under differing execution priorities. This will lead to the occurrence of priority inversion. To work around this a scheduling protocol that bounds priority inversion (e.g. PCP – Priority Ceiling Protocol) will have to be used.

Another issue stems from the sockets used by the ` isolate' mode of the GMM section, which introduce additional overhead and will probably not be used in a RT setting. For this reason, support for this option under a RT MatPLC has been deferred to a later date.

3.2. Synch Section

The synch section is currently implemented using SysV semaphores, making extensive use of their richer semantics. Since no RTOS currently supports SysV semaphores, this section will have to undergo significant changes.

The choice was made to implement semaphores with SysV semantics, using POSIX semaphores, as a user library function. This may be achieved with a global data structure, protected by a POSIX semaphore or mutex, containing the current state of the emulated SysV semaphore set, including an integer counter for each semaphore in the set and some additional data.

The emulated wait-and-signal operation would involve acquiring the POSIX semaphore and checking whether it needs to wait. If it does, it would record its requirements in the data structure, release it, and block. Once the wait has finished, the signal portion of the function would adjust the data structure and scan it to check whether any of the other modules has become ready to run. It would then wake up the highest priority one of those modules.

3.3. Log Section

The log section currently writes all logs to a file during run time, which is not time deterministic. This will be changed so all logs will be sent to a RT FIFO / message queue, where they may be later removed by a non real-time logging process. This process may then send these logs to a file, to a terminal or even to the UNIX system log.

3.4. Modules

Since the bulk of the code in the MatPLC is in the modules, one would expect that they will require the bulk of the porting effort. However, this does not appear to be entirely the case. The majority of modules fall into two classes: those which are never going to be real-time (file loggers, graphical user interfaces, etc.) and those which require no change at all or very little (e.g. DSP module, modules generated by the IEC 61131 ST/IL compiler). This stems from the fact that the RT modules only execute asynchronous logic, with all synchronization activities residing in the MatPLC library calls that were already discussed above. However, these modules do need to be linked to time deterministic versions of the C, thread and math libraries.

The only exception to the above broad division are the I/O modules, which will require significant work. In order to operate in real-time, such modules must only make use of time deterministic hardware drivers, besides being themselves time deterministic. This means that these modules will need to be considered on a case-by-case basis.

4. IEC STANDARDS OVERVIEW

4.1. IEC 61131-3

The IEC 61131-3 standard [1] defines both a structured framework and several programming languages that PLCs should support. In this framework applications are composed by a single 'configuration' organization unit, and any number of 'function', 'function block' type, or 'program' type organization units. The configuration creates any required program and function block instances, and assigns these to tasks. The tasks themselves may be configured to execute periodically, or upon the occurrence of an event.

Function block types are similar to classes in object oriented languages, however with a single member function, and no inheritance allowed. Function blocks are instantiated as variables from a specific function block type.

Functions have similar semantics to those in traditional procedural languages, with the important distinction that they be idempotent, i.e. all invocations of the same function with the same input values should always yield the exact same result. This implies that a function may not instantiate nor call a function block instance. It may, however, read the current values of the output parameters of a function block.

Program types are very similar to function blocks, differing mainly in where they may be instantiated. Only a configuration may create a program instance, unlike function blocks which may be instantiated from within other function block types, program types or configurations.

The three organization unit types may be programed in one of two textual languages (IL - Instruction Language; ST - Structured Text), or two graphical languages (LD - Ladder; FBD - Function Block Diagram). Programming in LD is similar to designing a relay based electrical circuit, while the IL language is a low level programming language similar to assembly. ST may be considered a high level procedural programming language, somewhat similar to Pascal, including iteration statements as well as flow control statements. On the other hand, a FBD is a graphical network of organization units, with the passing of data between these units being defined by lines connecting output parameters to input parameters.

The standard also defines a graphical language for specifying state machines (SFC – Sequential Function Chart), mostly based on Grafcet, that may also be used in programming function blocks or programs.

4.2. IEC 61499

The IEC 61499 standard [2] defines a framework for configuring co-operating devices connected by a data network, thereby constituting a distributed system. In this framework, each PLC is considered a device, which may in turn contain several resources (i.e. CPUs). The novelty stems from allowing applications to span multiple resources and multiple devices, which really means that an application may be distributed among several PLCs, and the same PLC may be simultaneously executing several applications or sub-applications.

This new framework defines a set of organization units distinct from those used in IEC 61131-3. These are, in descending hierarchical order, the application, the sub-application, the composite function block and the basic function block. Applications and sub-applications may be divided among several resources by allocating their component function blocks to distinct resources. The function block is therefore the atomic unit of distribution, that must be wholly contained within the same resource. It should be emphasized that besides sharing the same name, the IEC 61499 function blocks are very distinct from the IEC 61131-3 function block.

All IEC61499 organization units, besides having inputs and outputs for data, may also receive and send events. Also, the IEC 61499 framework does not seem to imply a cyclic execution scan, as is common in many PLCs. It is the arrival of an event at one of the input ports of an organization unit that will provoke the execution of that organization unit.

Basic function blocks (i.e. those that are not composed of a network of other function blocks) may have several execution states. An execution state chart defines the state transitions, as well as the actions or algorithms to be executed when a each state is active. Only one state may be active at any time, and for states that have more than one associated action, these are executed sequentially. When an action terminates its execution, it may generate an event that will be sent out from the function block. The actions or algorithms may be specified using one of several programming languages, including the languages specified in IEC 61131-3, such as ST and IL. From these characteristics, it may be concluded that a basic function block will have a single thread of execution. Nevertheless, a composite function block, containing several basic function blocks, may have several simultaneous threads of execution, one for each basic function block.

Applications interact with the real world through specialised function blocks known as service interface function blocks. The standard defines service interface function blocks to access the physical inputs and outputs of the PLC, to access the data network connecting the distributed PLCs, and to manage the state of the user' sapplications and the PLC itself. Unfortunately the standard does not completely specify the interface and semantics of the process and communication service interface function blocks. It does, however, include an informative annex detailing how the communication service interface function blocks may provide a publish/subscribe type of service, as well as the client/server paradigm.

The standard includes an extra mechanism, the ' adapterinterface' ,consisting of the ' plugadapter' and the ' socketadapter' ,by which polymorphism may be implemented. A function block implementing a specific plug adapter may be connected to any function block implementing a socket adapter of the same interface adapter type.

5. THE MATPLC' S IEC 61131-3 COMPILER

As previously stated, the MatPLC includes an IEC 61131-3 compiler for the IL and ST languages. This compiler executes in four plus one stages: lexical analyzer, syntax parser, semantics analyzer, code generator, and binary code generator.

The lexical parser analysis the source code and breaks it up into lexical tokens, removing on the way all comments and white-spaces between the tokens. The syntax parser groups the tokens into syntax constructs, and builds an equivalent internal abstract syntax data structure. The semantic analyzer walks through the abstract syntax and determines whether all semantic rules have been obeyed. The code generator, based on the abstract syntax once again, produces C++ code which, when further compiled becomes the final executable.

The lexical analyzer and the syntax parser were implemented using the flex and bison utilities respectively. Many issues had to be overcome, most of them related to inconsistencies in the standard itself (de Sousa *et al*, 2003). Due to a lack of time and resources, the semantic checker has not yet been implemented.

IL and ST code transcription to C++ is rather straightforward as many of the constructs used in ST and IL are also available in C++. Basic data types are either mapped onto C++ basic data types (e.g. integers, reals), or will be mapped onto C++ structures (e.g. date, time) with overloaded operators. IEC 61131-3 extended data types are also neatly mapped onto C++ (enumerations, structures, arrays). Functions are mapped onto C++ functions, with the same input and output parameters. Nevertheless, calling functions with an incomplete parameter list has to be turned into a C++ function call using default values for the unspecified parameters. The function block and program type constructs are both mapped onto a class. However, considering that function block parameters may retain their values between function calls, all input and output parameters are mapped onto public data members of the C++ class. Calling function blocks therefore involves setting the input parameters to their appropriate values, calling the now parameterless function, followed by reading the output parameters.

Complete mapping of configuration constructs onto the MatPLC has not yet been attempted, mainly due to the fact that the expected semantics are not completely defined in the standard.

6. FRAMEWORK FOR IEC 61499

Allowing the MatPLC to support IEC 61499 based distributed industrial applications first requires that a mapping be made between the entities defined in the IEC 61499 framework and the existing MatPLC framework. In our case, and although it is expected that IEC 61499 resources be mapped onto distinct PLC CPUs, it seems more appropriate to map each resource onto a single MatPLC module. The MatPLC module may then be configured to only execute on a specific CPU of an SMP (Symmetrical Multi-Processor) machine if required. IEC 61499 organization units (applications and function blocks) will be implemented as C++ code, which, when further compiled, will generate the MatPLC module. The default function blocks will be available as a set of libraries to which each MatPLC module will link. This will include the implementation of the process service interface function blocks through which the applications will access the MatPLC points, as well as the communication service interface function blocks.

Using the above mapping between the frameworks, communication between IEC 61499 functions blocks residing on separate resources (i.e. on separate MatPLC nodules) will not use the MatPLC (network) I/O modules, nor go through the MatPLC shared data areas, but rather communicate through the service interface functions blocks that will have direct access to the network hardware.

6.1. Communication Protocols

The IEC 61499 standard requires communications over a network to link function blocks allocated to separate devices. This link will be used to transfer the occurrence of events, which requires a one-to-one asynchronous communication, as well as the state/values of all output variables, requiring a one-to-many asynchronous communication.

The occurrence of events may be implemented over a publish/subscribe protocol, or a client/server connection. On the other hand, the transfer of state of output variables may too be implemented over a publish/subscribe protocol, or over a client/server connection oriented protocol with an intermediate event notification service.

The above considerations lead us to conclude that the IEC 61499 framework can be implemented over either a publish/subscribe or client/server protocol. It is therefore our intention to do so using the RTPS (Real-Time Publish Subscribe) protocol over UDP (Real-Time Innovations, 2002), as well as an OMG compliant ORB (Object Management Group, 2002) that includes the event service. By using a common API to access the underlying communication network, the code generated by the future IEC 61499 compiler may use either the RTPS or the CORBA version of linking function blocks over the network.

7. CONCLUSIONS

Code for the MatPLC can be obtained from the project' s cvs server and on the website, http://mat.sf.net. The ' Bespin' version, which includes the IEC 61131-3 IL and ST compiler, was released in March 2004. In the meantime, the project is evolving in many other directions, including support for other hardware and communication protocols. Although the project is still in its infancy, people in the industry are beginning to try it out and making enquiries. Examples of real/world applications using the MatPLC are expected for the near future.

Porting to a Real-Time platform is currently under way, as well as the preparations to test the implementation alternatives for IEC 61499 support.

8. ACKNOWLEDGEMENTS

The authors would like to thank Curt Wuollet for taking the initiative of starting the MatPLC project. The contributions to the project made by Jiri Baum, Juan Carlos Orozco, and Andrey Romanenko are also highly appreciated.

9. REFERENCES

de Sousa, M.J., A.S. Carvalho (2003), An IEC 61131-3 Compiler for the MatPLC, *Proceedings of the 9th International Conference on emerging Technologies and Factory Automation.*

International Electrotechnical Commission (2000), *Function Blocks for Industrial-Process Measurement and Control Systems, Part 1 – Architecture (Voting Draft - 28th March 2000).*

International Electrotechnical Commission (2001), *IEC 61131-3, 2nd Ed. Programmable Controllers – Programming Languages, (Final Draft - 10th December 2001).*

Object Management Group (2002), *CORBA 3.0 Specification.*

Real-Time Innovations (2002), *Real-Time Publish-Subscribe (RTPS) Wire Protocol Specification (Ver. 1.0, Draft Document Version 1.17).*

Copyright © IFAC Information Control Problems in
Manufacturing, Salvador, Brazil, 2004

ELSEVIER
IFAC
PUBLICATIONS
www.elsevier.com/locate/ifac

ON SATISFYING REAL-TIME CONSTRAINTS IN REAL-TIME MANUFACTURING SYSTEMS

Dania A. El-Kebbe

Paderborn University / Heinz Nixdorf Institute
Fürstenallee 11, 33102 Paderborn, Germany
E-mail: elkebbe@uni-paderborn.de

Abstract: This paper presents a predictable real-time scheduling schme upon multiprocessor production stages in manufacturing systems underlying real-time constraints. The preemptive scheduling of systems consisting of periodic and aperiodic tasks on a platform comprised of several uniform multiprocessors is considered. *Copyright © 2004 IFAC*

Keywords: Multiprocessor real-time scheduling, production planning and control.

1 Introduction

Our research interest focuses on proposing a new methodology for developing a Manufacturing System under Hard Real–Time Constraints (MaSHReC). The study was motivated by the conception of a novel manufacturing approach, which emphasizes the temporal restrictions of the system.

The basic approach of a manufacturing System under Hard Real-Time Constraints and its architecture are presented in (3). The Modeling of MaSHReC is presented in (2) and (1). In this paper, a real-time hybrid scheduling scheme upon a multiprocessor platform is developed. The application of the proposed scheduling technique to the control of production systems is presented.

2 State-of-the-Art

Task scheduling in hard real-time systems can be *static* or *dynamic*. A static approach calculates schedules for tasks off-line and it requires the complete prior knowledge of tasks' characteristics. Although static approaches have low run-time cost, they are inflexible and cannot adapt to a changing environment or to an environment whose behavior is not completely predictable. Several uniprocessor on-line algorithms, such as the Earliest Deadline First algorithm (8) and the Least Laxity algorithm (9) are known to be optimal in the sense that if a set of jobs that can be scheduled such that all jobs complete by their deadlines, then these algorithms will also schedule these sets of jobs to meet all deadlines. However, no on-line scheduling algorithm in multiprocessor systems can be optimal: this was shown for the simplest (identical) multiprocessor model by (6) and the result from (6) can be directly extended to the more general (uniform or unrelated) machine models. To cope with various unexpected events in production planning and control, production engineers adopt a rescheduling policy (10). Such rescheduling policies yields to the following drawbacks:

1. Rescheduling policies are feasible for small-sized and simple manufacturing systems. As manufacturing systems grow in size and complexity, a rescheduling policy becomes impracticable.
2. Additionally, no rescheduling policy is made on-line, in the sense that rescheduling policies are unfortunately executed at the end of a production shift.
3. Furthermore, no prediction can be made concerning unexpected arriving requests.

(11) explored the use of *resource augmentation* techniques[1] for the on-line scheduling of real-time tasks. They considered two problems in dynamic scheduling: scheduling to meet deadlines in a preemptive identical multiprocessor setting, and scheduling to provide good response time in a number of scheduling environments. Using the resource augmentation approach, they established that several well-known on-line algorithms, that prove poor performance from an absolute worst-case perspective, are optimal for the problems in question when allowed moderately more resources. (5) extended this method to be applied upon uniform parallel machines. However, results derived from their work are to be applied only to periodic task systems.

The idea of this paper is based on (5). However, they are extended to include changeover time costs. Furthermore, the results of this paper are applied to both hard periodic and *hard* aperiodic task systems. It is meant by hard aperiodic tasks, that once deadlines are assigned to aperiodic tasks, they must not be missed.

3 The On-line Parallel Model

Based on the following assumptions and terminology, the scheduling of hard real-time systems upon a uniform multiprocessor platform comprised of $m \in \mathbb{N}^*$ machines (there is at least one machine) is considered.

[1] A method of analysis introduced by (7) for uniprocessor scheduling, comparing the performance of an on-line algorithm to the performance of an optimal off-line algorithm when the on-line algorithm is given extra resources.

- $\pi = \{s_1, s_2, ..., s_m\} \mid m \in \mathbb{N}^*$ represents the m-machine uniform multiprocessor platform in which the machines have *speeds* or *production capacities* $s_1, s_2, ..., s_m$ respectively; without loss of generality, it is assumed that these speeds have positive values and they are indexed in a decreasing manner: $s_j \geq s_{j+1}$ for all $j, 1 \leq j < m$.

- $\tau = \{\tau_i \mid i \in \mathbb{N}\}$ A set of periodic tasks with hard deadlines.

- $J = \{J_j \mid j \in \mathbb{N}\}$ A set of active aperiodic tasks ordered by increasing deadline, J_1 being the task with the shortest absolute deadline.

- Each Job τ_i or J_j is characterized by an arrival time r_i, a production time c_i and a deadline d_i, respectively r_j, c_j, d_j. Whereas the arrival times, production times and deadlines of the periodic jobs τ_i are known in advance, it is assumed that, for the aperiodic job set J_j, these relevant information about the jobs are known when a job arrives.

- The preemptive multiprocessor scheduling model is considered. In the preemptive scheduling model presented in this paper, a job may be interrupted and subsequently resumed *with* a penalty. $O = \{O_{j,m} \mid j, m \in \mathbb{N}^*\}$ represents the switch time or changeover time caused by the arrival of the part from type j at the machine m. Changeover time is derived by a static analysis on the machine.

- $u_i = c_i/p_i$ The utilization u_i of a task is the ratio of its execution requirement to its period. Without loss of generality, the tasks in τ and J are indexed according to a decreasing utilization: $u_i \geq u_i + 1$ for all $i, 1 \leq i < n$.

- In the context of uniform multiprocessor scheduling, a work-conserving scheduling algorithm is defined to be one that satisfies the following conditions (5): i) no machine is idled while there are active jobs awaiting execution and, ii) if at some instant there are fewer than m (the number of processors in the uniform multiprocessor platform) active jobs awaiting execution then the active jobs are executed upon the fastest machines. More formally, at any instant t and for all $k > j$, if the j'th-slowest processor is idled by the work-conserving scheduling algorithm, then the k'th-slowest processor is also idled at instant t.

- *Job preemption is permitted.* That is, a job executing on a machine may be preempted, prior to completing execution, and its execution may be resumed later. Unfortunately, state-of-the-art real-time multiprocessor scheduling techniques assume that there is no penalty associated with such preemption. It is obvious that disregarding this assumption is inappropriate for manufacturing systems where changeover time overhead may have a considerable time value.

- *Job parallelism is forbidden.* That is each job may execute at most one processor at any given instant in time.

4 Total Bandwidth Server on Uniform Multiprocessors

Recall that the Total Bandwidth Server (TBS) technique is used to schedule jointly hard periodic and hard aperiodic tasks under dynamic priority systems upon uniprocessor platforms. One main benefit of this technique is that it guarantees both periodic and aperiodic task sets. An extension of the TBS technique to include changeover time costs is developed in (4). Each aperiodic request receives a deadline

$$\overline{d}_j = \overline{r}_j + \frac{\overline{C}_j + 2\overline{O}_{j,m}}{U_s}$$

where

$$\overline{r}_j = max\ (r_j, \overline{d}_{j-1}, f_{j-1})$$

A TBS algorithm to be implemented upon uniform multiprocessor systems according to the following rules is defined as follows:

- No machine is idled while there is an active job awaiting execution.

- When fewer than m jobs are active, they are acquired to execute upon the fastest machines while the slowest are idled.

- Higher priority jobs are executed on faster processors. More formally, if the j'th-slowest processor is executing job J_g at time t under the TBS implementation, it must be the case that the deadline of J_g is not greater than the deadlines of jobs (if any) executing on the $(j+1)$'th-, $(j+2)$'th-, $(j+3)$'th-, ..., m'th-slowest machines.

- Whenever the j-th aperiodic task arrives at time $t = r_j$, it receives a deadline

  $$\overline{d}_j = \overline{r}_j + \frac{\overline{C}_j + 2\overline{O}_{j,m}}{U_s}$$

  where

  $$\overline{r}_j = max\ (r_j, \overline{d}_{j-1}, f_{j-1})$$

The utilization of the server U_s will de defined later. Unavoidably, some additional notations are given in the following.

Definition 1 (W(A,π,I,t)). *Let I denote any set of jobs, and π any uniform multiprocessor platform. For* *any algorithm A and time instant $t \geq 0$, let $W(A,\pi,I,t)$ denote the amount of work done by algorithm A on jobs of I over the interval $[0,t)$, while executing on π.*

Definition 2 (S_j). *Let π denote an m-processor uniform multiprocessor platform with processor capacities $s_1, s_2, ..., s_m, s_j \geq s_{j+1}$ for all j, $1 \leq j < m$. S_j is defined as follows:*

$$S_j = \sum_{l=1}^{j} s_l \textit{ for all } j,\ 1 \leq j \leq m$$

Definition 3 (λ_π). *(5) Let π denote an m-processor uniform multiprocessor platform with processor capacities $s_1, s_2, ..., s_m, s_j \geq s_{j+1}$ for all j, $1 \leq j < m$. λ_π is defined as follows:*

$$\lambda_\pi = \max_{j=1}^{n} \left\{ \frac{\sum_{k=j+1}^{m} s_k}{s_j} \right\}$$

The parameter λ_π measures the "degree" by which π differs from an identical multiprocessor platform. Consequently λ_π becomes progressively smaller as the speeds of the processor differ from each other by greater amounts.

Lemma 1. (5) *Let π denote an m-processor uniform multiprocessor platform with processor capacities $s_1, s_2, ..., s_m, s_j \geq s_{j+1}$, for all j, $1 \leq j < m$. Let π' denote an m-processor uniform multiprocessor platform with processor capacities $s'_1, s'_2, ..., s'_m, s'_j \geq s'_{j+1}$, for all j, $1 \leq j < m$. Let A denote any m-processor uniform multiprocessor algorithm, and A' any work-conserving m-processor uniform multiprocessor algorithm. If the following condition is satisfied by π and π':*

$$S'_m \geq \lambda_{\pi'}.s_1 + S_m$$

then for any set of jobs I and at any time-instant $t \geq 0$

$$W(A', \pi', I, t) \geq W(A, \pi, I, t)$$

Lemma 1 specifies a condition under which any work-conserving algorithm A' (such as TBS) executing on π' is guaranteed to complete at least as much work as any other algorithm A (including an optimal algorithm) executing on π, when both algorithms are executing on any set of jobs I. This condition expresses the additional computing capacity needed by π' in terms of the $\lambda_{\pi'}$ parameter, and the speed of the fastest processor in π. The smaller the value of $\lambda_{\pi'}$ (the more π' deviates from being an identical multiprocessor), the smaller the amount of this excess processing capacity needed.

The processing of aperiodic tasks can be integrated into a periodic environment by introducing one or more periodic tasks to execute the aperiodic tasks. Therefore, we may deal with aperiodic tasks in a similar way with periodic tasks. As a result, the following theorem (5) uses Lemma 1 to deduce whether a work-conserving algorithm can feasibly schedule a task set: it states that any collection of jobs I that is feasible on a uniform multiprocessor platform π will be scheduled to meet all deadlines by algorithm TBS on any platform π' satisifying the condition of lemma 1.

Theorem 1 *Let I denote an instance of jobs that is feasible on m-processor uniform multiprocessor platform π. Let π' denote another m-processor uniform multiprocessor platform. Let the parameter $\lambda_{\pi'}$ of π' be as defined in Definition 3:*

$$\lambda_{\pi'} = \max_{j=1}^{n} \left\{ \frac{\sum_{k=j+1}^{m} s'_k}{s'_j} \right\}$$

If the condition of Lemma 1 is satisfied by platforms π and π':

$$S'_m \geq \lambda_{\pi'}.s_1 + S_m$$

then I will meet all deadlines when scheduled using TBS algorithm executing on π'.

Thus, Theorem 1 characterizes a uniform multiprocessor platform π' according to its parameter "$\lambda_{\pi'}$" (as defined in Definition 3), and relates the TBS-feasibility of a system, known to be feasible on some platform π, to the cumulative capacities of π and π', the speed of the fastest processor in π, and this parameter $\lambda_{\pi'}$ of platform π'.

As an immediate result of Theorem 1, the results of (11) concerning TBS-scheduling on identical multiprocessor platforms are obtained:

Corollary 1. *TBS is a preemptive, $(2 - \frac{1}{m})$-speed algorithm for hard real-time scheduling on parallel machines.*

5 Schedulability Analysis of Hybrid Task Systems on Uniform Multiprocessors

In this section, the theory developed in Section 4 is applied to study the deadline-based scheduling of hybrid (hard periodic and hard aperiodic tasks) task systems on uniform multiprocessor platforms. Although scheduling hybrid task systems is partly an "on-line" problem in the sense that the periodic task parameters are assumed known beforehand, the results in section 4 concerning hybrid tasks nevertheless turn out to be useful towards developing a framework for scheduling hybrid task systems on uniform multiprocessors.

The method of analysis developed in this section proceeds as follows:

1. an exact test for determining whether a given hybrid task system is feasible on a particular uniform multiprocessor platform is developed and

2. this exact feasibility test along with the results obtained in section 4 are used, to design a schedulability analysis for determining whether a given

hybrid task system will be successfully scheduled by TBS on a specified uniform multiprocessor platform.

(5) identified a uniform multiprocessor platform upon which a given periodic task system τ is schedulable. They determine a sufficient condition for τ to be successfully scheduled by EDF on any given multiprocessor platform π' (Theorem 3).

Definition 4 ($P(r, m, k)$). *Let $\tau = \{T_1, T_2, ..., T_n\}$ denote a collection of periodic tasks indexed according to decreasing utilization (i.e. $u_i \geq u_{i+1}$ for all $1 \leq i < n$, where $u_i = \frac{c_i}{p_i}$). Let m be any positive integer. For $1 \leq k \leq m$ define P(r,m,k) recursively as follows:*

- *If $u_1 > \frac{U_n}{m}$, let $P(r, m, 1) = u_1$ and $P(r, m, k) = P(r - \{T_1\}, m - 1, k - 1)$ for k = 2, ..., m.*
- *If $u_1 \leq \frac{U_n}{m}$, let $P(r, m, k) = \frac{U_n}{m}$ for k = 1, ..., m, where $U_n = \sum_{i=1}^{n} u_i$.*

Theorem 2. (5) *Consider a set $\tau = \{T_1, T_2, ..., T_n\}$ of periodic tasks indexed according to decreasing utilization. Let π denote the m-processor uniform multiprocessor platform in which the processors have speeds $s_k = P(\tau, m, k), 1 \leq k \leq m$:*

$$\pi = [P(\tau, m, 1), P(\tau, m, 2), ..., P(\tau, m, m)].$$

Periodic task system τ is schedulable on this uniform multiprocessor platform π.

Corollary 2. *The m-processor uniform multiprocessor platform*

$$\pi = [P(\tau, m, 1), P(\tau, m, 2), ..., P(\tau, m, m)]$$

of theorem 3 satisfies the following properties:

1. *The fastest processor has computing capacity*

$$max\left\{u_1, \frac{U_n}{m}\right\}$$

2. *The sum of the processor capacities is U_n.*

Theorem 3 (5) *Let $\pi' = [s'_1, s'_2, ..., s'_m]$ denote any m-processor multiprocessor platform, and let $\lambda_{\pi'}$ be as defined in Definition 3:*

$$\lambda_\pi = \max_{j=1}^{n}\left\{\frac{\sum_{k=j+1}^{m} s_k}{s_j}\right\}$$

Periodic task system τ will meet all deadlines when scheduled on π' using EDF if the following condition holds:

$$S'_m \geq \lambda_{\pi'} * max\left\{u_1, \frac{U_n}{m}\right\} + U_n$$

Below we show how we can transform the problem of scheduling periodic tasks on uniform multiprocessors to the scheduling of periodic and aperiodic tasks. Whereas EDF is a scheduling policy trying to schedule up to the whole capacity of the multiprocessor platform, the TBS algorithm upon multiprocessor platforms aims at using the whole capacity of the system, while assuring that a fraction of this capacity is dedicated to aperiodic requests. This necessitates the computation of the server utilization of the multiprocessor platform given by Theorem 4.

Theorem 4 *Let $\pi' = [s'_1, s'_2, ..., s'_m]$ denote any m-processor multiprocessor platform, and let $\lambda_{\pi'}$ be as defined in Definition 3:*

$$\lambda'_\pi = \max_{j=1}^{n}\left\{\frac{\sum_{k=j+1}^{m} s_k}{s_j}\right\}$$

The aperiodic task system J has a utilization

$$S'_m = \lambda_{\pi'} * u_1 + U_n + U_s$$
$$U_s = S'_m - \lambda_{\pi'} * u_1 - U_n$$

As a result, deadlines of aperiodic jobs may be computed as defined in the following Corollary.

Corollary 3. *The aperiodic jobs $J(r_j, c_j)$ of a m-processor uniform multiprocessor platform are scheduled using TBS and a total bandwidth as defined in Theorem 4 with a deadline*

$$d_j = max(r_j, d_{j-1}) + \frac{c_j}{U_s}$$

Furthermore, the Corollary 4 follows directly from the results of (4) involving the changeover time costs in the TBS algorithm and allowing resource reclaiming.

Corollary 4. *Aperiodic jobs $J(r_j, c_j)$ of a m-processor uniform multiprocessor platform are scheduled using TBS and a total bandwidth as defined in Theorem 4 with a deadline*

$$\bar{d}_j = \bar{r}_j + \frac{\overline{C}_j + 2\overline{O}_{j,m}}{U_s}$$

where

$$\bar{r}_j = max\ (r_j, \bar{d}_{j-1}, f_{j-1})$$

Theorem 4 is now illustrated by an example.

Example. *Consider a task system τ comprised of five periodic tasks (c_i, p_i)*

$$\tau = \{(15, 10), (4, 5), (14, 20), (6, 15), (2, 10)\}$$

and an aperiodic task (r_i, c_i)

$$J = \{(5, 7)\}$$

for this system, $u_1 = 1.5$, $u_2 = 0.8$, $u_3 = 0.7$, $u_4 = 0.4$, $u_5 = 0.2$. *Suppose that* τ *and* J *are to be TBS-scheduled upon the uniform multiprocessor platform* $\pi' = [3, 1, 0.5]$ *- will all deadlines be met?*

By Definition 3, the value of $\lambda_{\pi'}$ *for the uniform multiprocessor platform* π' *is*

$$\lambda_{\pi'} = max\left\{\frac{1+0.5}{3}, \frac{0.5}{1}\right\} = \frac{1}{2}$$

and the total computing capacity is

$$3 + 1 + 0.5 = 4.5$$

The total computing capacity needed for τ *is*

$$1.5 + 0.8 + 0.7 + 0.4 + 0.2 = 3.6$$

and with the fastest processor having a computing capacity

$$s_1 = 1.5$$

By theorem 4, the aperiodic requests J *are feasible on the 3-processor uniform multiprocessor platform with a total bandwidth of*

$$\begin{aligned} 4.5 &= 0.5 * 1.5 + 3.6 + U_s \\ 4.5 &= 4.35 + U_s \\ U_s &= 0.15 \end{aligned}$$

The aperiodic job $J = \{(5, 3)\}$ *is to be scheduled on the 3-processor multiprocessor platform with a deadline equal to:*

$$d_j = max\{5, 0\} + \frac{3}{0.15} = 25$$

and τ *and* J *can consequently be scheduled by TBS to meet all deadlines on* π' *with* $U_s = 0.15$ *and* $d_j = 25$.

6 Performance Evaluation

The purpose of this section is to evaluate the scheduling scheme proposed in this paper. Implementation issues for TBS scheduling upon uniform multiprocessor platforms allow to evaluate the effect of aperiodic task generation over various total bandwidth utilizations and the effect of aperiodic processor utilization over various uniform multiprocessor platforms.

In the first experiment, an evaluation of tasks scheduled with TBS upon a uniform multiprocessor platform with different periodic utilizations is achieved. The model consider a 3-multiprocessor platform with aperiodic task utilization varying from 0.15 to 2.

Figure 2 depicts changeover time considered in the experiment. Changeover time may achieve an overhead approximately equal to production time. This permits a moderately realistic representation of changeover time considerations in production systems.

Figure 3 demonstrates the significance of aperiodic server utilization when scheduling a set of aperiodic

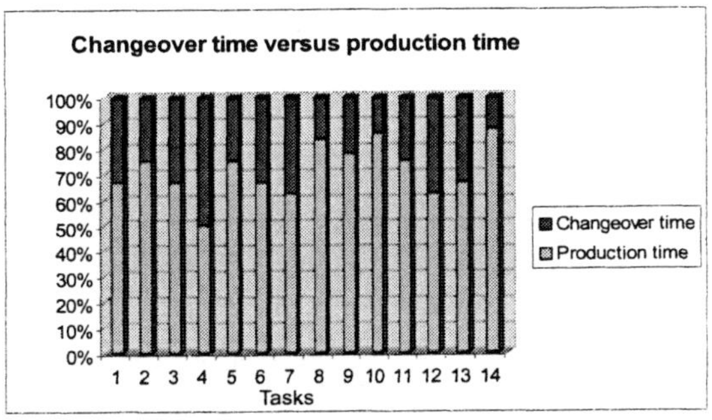

Fig. 1: Changeover time versus production time

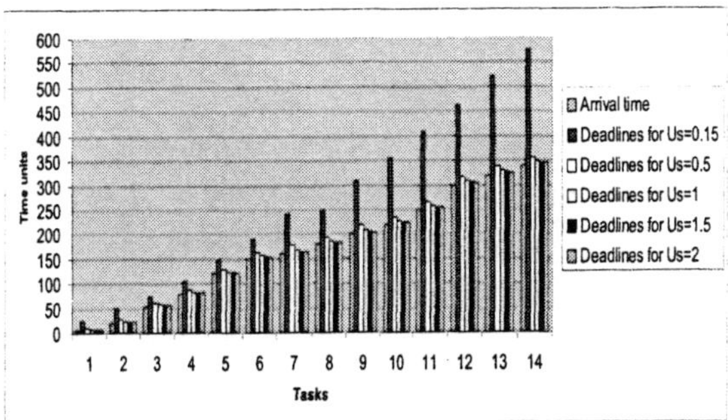

Fig. 2: Performance evaluation of aperiodic tasks scheduled with TBS upon a uniform multiprocessor platform

tasks upon a uniform multiprocessor platform. As seen in this figure, tasks produced with a low server utilization suffer from a broad deadline assignment. This effect may be inacceptable in applications, where aperiodic tasks are expected frequently.

The second experiment aims at studying the effect of periodic and aperiodic task utilizations, when a set of periodic and aperiodic tasks are scheduled upon uniform 3-multiprocessors platforms with different production capacities. Aperiodic server utilizations are computed using Theorem 4 considering a periodic utilization of 3,6.

Figure 4 shows the variation of production capacities of different platforms considered in this experiment. The first multiprocessor platform is an identical platform. The production capacity of the second machine is invariable for all platforms. Production capacities of the first machines are augmented gradually by 0,2 time units. Production capacities of the third machines are reduced gradually by 0.2. The purpose of this representation of production capacities is to achieve progressive representation of results and therefore, to provide clarity for the reader. By comparing Figure 4 and Figure 5, the second experiment has established that as a multiprocessor platform differ from an identical platform, server utilizations become progressively smaller. This is due to the fact that when a multiprocessor platform differ from an identical platform, λ_π becomes smaller and consequently aperiodic server utilization (computed by Theorem 4) becomes

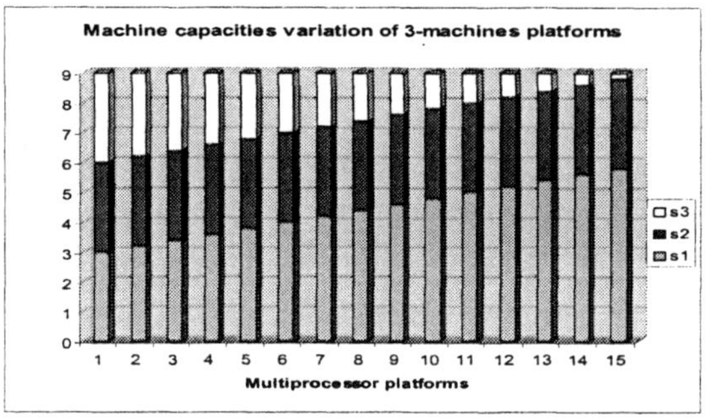

Fig. 3: Machine capacities variation of 3-machines platforms

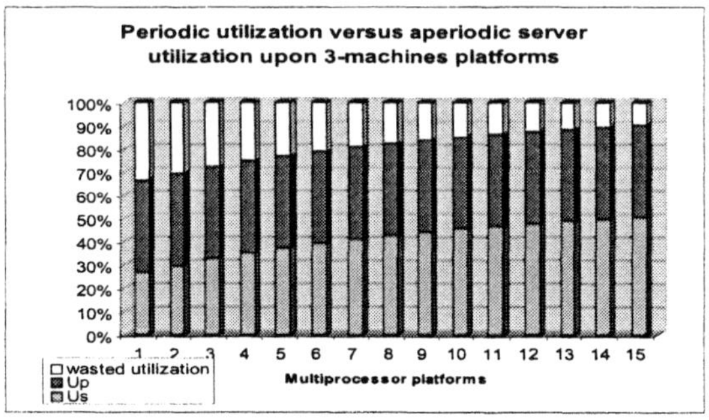

Fig. 4: Periodic utilization versus aperiodic server utilization upon 3-machines platforms

greater. Furthermore, it should be stated that as multiprocessor platforms differ from an identical platform, the wasted utilization of machines becomes smaller. This is due to higher aperiodic utilizations.

7 Conclusions

One of the most important properties, that differentiates real-time systems from other conventional systems, is predictability. The system must be able to predict the consequences of any scheduling decisison. If some task cannot be guaranteed within its timing constraints, the system must notify this fact in advance, so that alternative actions can be planned in time to cope with the event.

Based on the TBS algorithm, a predictable scheduling scheme upon a multiprocessor production shift is presented. It includes the novel features of accounting for changeover time overheads and accounting for *both* periodic and aperiodic tasks. The high performance of the system is proved through experimental simulations.

References

[1] D. A. El-Kebbe. Integration of On-Line and Off-Line Systems in Real-Time Manufacturing. In *Proc. of the Workshop of the Informatics Graduate Colleges*, Schloss Dagstuhl, Germany, May 2000.

[2] D. A. El-Kebbe. Modeling the Manufacturing System under Hard Real-Time Constraints Using Real-Time UML. In *Workshop on Formal Design Techniques Using Real-Time UML*, York, UK, October 2000.

[3] D. A. El-Kebbe. Towards a Manufacturing System under Hard Real-Time Constraints. In *Informatik 2000: 30. Jahrestagung der Gesellschaft für Informatik*, Berlin, September 2000.

[4] D. A. El-Kebbe. Aperiodic Scheduling in a Dynamic Real-Time Manufacturing System. In *Proc. of the IEEE/EE Real-Time Embedded Systems Workshop (Satellite of the IEEE Real-Time Systems Symposium)*, London, December 2001.

[5] S. Funk, J. Goossens, and S. Baruah. On-line Scheduling on Uniform Multiprocessors. In *Proceedings of the Real-Time Systems Symposium*, pages 183–192, London, December 2001.

[6] K. Hong and J. Leung. On-line Scheduling of real-time tasks. In *Proc. of the Real-Time Systems Symposium*, pages 244–250, Huntsville, Alabama, December 1988. IEEE Computer Society Press.

[7] B. Kalyanasundaram and K. Pruhs. Speed is as powerful as clairvoyance. In *36th Annual Symposium on Foundations of Computer Science*, pages 214–223, Los Alamitos, Oktober 1995. IEEE Computer Society Press.

[8] C.L. Liu and J.W. Layland. Scheduling algorithms for multiprogramming in a hard real-time environment. *J. of the ACM*, 20(1):46–61, January 1973.

[9] A.K. Mok. *Fundamental Design Problems of Distributed Systems for the Hard Real-Time Environment.* PhD thesis, Massachusetts Institute of Technology, 1983.

[10] Tatsushi Nishi, Akihiro Sakata, Shinji Hasebe, and Iori Hashimoto. Autonomous Decentralized Scheduling System for Just-in-Time Production. In *Proc. of the 7th International Symposium on Process System Engineering*, pages 345–351, 2000.

[11] C. A. Philips, C. Stein, E. Torng, and J. Wein. Optimal Time-Critical Scheduling via Resource Augmentation. In *Proc. of the 29th Annual ACM Symposium on Theory of Computing*, pages 140–149, El Paso, Texas, May 1997.

Copyright © IFAC Information Control Problems in Manufacturing, Salvador, Brazil, 2004

ELSEVIER
IFAC
PUBLICATIONS
www.elsevier.com/locate/ifac

TIMED AUTOMATA APPROACH TO CAN VERIFICATION

Jan Krakora and Zdenek Hanzalek

Czech Technical University in Prague,
Faculty of Electrical Engineering,
Department of Control Engineering
Karlovo namesti 13, Prague 2, 121 35, Czech Republic
{krakorj,hanzalek}@fel.cvut.cz

Abstract:
This article deals with verification of real time distributed system focusing on CAN model by timed automata and specification of verified properties by temporal logic. Such system, based on several CPUs, consists of an application SW running under real-time operating system (e.g. OSEK) and using standard broadcast communications based on the Controller Area Network (CAN). The crucial problem is to verify both, the time properties (e.g. message response time) and logic properties (e.g. deadlock) of such complex applications. The approach presented in this article is based on the modeling the discrete event system by Timed Automata and on verification by model checking tool (e.g. UPPAAL). In contrast to classical approaches dealing either with shared bus (guarantee message latencies approached by Tindell and Burns) or shared processor (rate monotonic analysis), our approach deals with both kinds of resources. *Copyright © 2004 IFAC*

Keywords: Controller Area Network, Real-time, Timed automata, Medium Access Control

1. INTRODUCTION

Let us assume the distributed real time control system consisting of application processes (designed by application developer) running under Real-Time Operating System (RTOS e.g. OSEK (Geischeder *et al.*, 2000)) while using several processors interconnected via standard broadcast communication based on the Controller Area Network (CAN) (Etschberger *et al.*, 2001). Structure of the application under consideration is depicted in Figure 1. The crucial problem is to verify both, the time properties (e.g. message response time, schedulability of periodic processes, response time) and logic properties (e.g. deadlock, mutual exclusion, priority based access) of the applications incorporating two kinds of shared resources - the processor and the bus. Classical approaches deal separately either with the processor sharing (studied for example by RMS (Klein *et al.*, 1993)) or with the bus sharing (e.g. CAN message latency studied by Tindell (Tindell and Burns, 1994)).

The task schedulability on monoprocessor and multiprocessor systems is widely studied subject ((Buttazzo, 1997),(Liu, 2000)). For example Rate Monotonic Scheduling (RMS) can be used to guarantee schedulability, when the application consisting of periodic processes is running on monoprocessor with priority based preemptive kernel and the processes have their deadlines at the end of period. RMS assigns fixed priorities to the processes according to their request rate (inverse

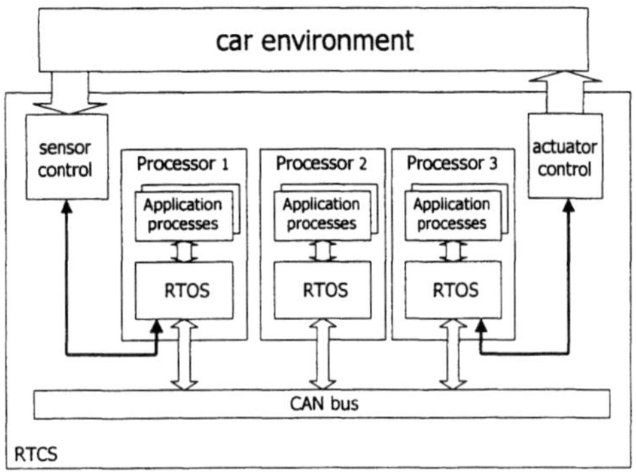

Fig. 1. Real time control system structure

to their period deadline), therefore the highest priority is assigned to the processes with highest frequency. Schedulability of such processes can be verified using Utilization bound theorem or Completion time theorem.

Prediction of the worst-case message latencies for CAN was presented by Tindell and Burns in (Tindell and Burns, 1994). This method is a direct application of the scheduling theory where the common bus is considered as shared resource. In similar way, CAN operates the fixed priority scheduling algorithm and authors assume the rate monotonic priority assignment. The message worst case response time is influenced not only by its length but also by the maximal length of one lower priority message since a high priority message cannot interrupt the message that is already transmitted. Moreover due to the priority based bus arbitration method the message worst-case response time is influenced by all higher priority messages, each of them considering their occurrence ratio.

This article presents an alternative approach based on model checking while using timed automata (Alur and Dill, 1994) and temporal logics (Katoen, 1999). Using this approach we model parts of the distributed system (application SW, operating system and communication bus) by automata. The automata use synchronization primitives enabling their interconnection. Please refer to (Berard *et al.*, 2001) on issues related to implementation and complexity of verification algorithms.

Modeling and verification of concurrent processes sharing one processor have been shown in (Waszniowski and Hanzalek, 2003), (Corbett, 1996). These works incorporate priority based preemptive and non-preemptive scheduling, intertask communication primitives and interrupt handling. These models can be directly combined with CAN model shown in this article while using synchronization primitives. That is why the operating system part is neglected and one application processes per one processor is assumed in this paper. Verification of the CAN model developed here can be directly compared to the results in (Tindell and Burns, 1994) and it can be simply enlarged to deal with internal structure of the application processes, timing parameters of different operating systems, sporadic processes etc. Moreover, while using the model checking approach, one can verify not only the schedulability, but also rather complex properties linked to logic and timing behavior of the distributed system. On the other hand the complexity is a drawback of the model checking approach in contrast to straightforward equations of the scheduling theory.

The paper is organized as follows: in Section 2 basic behavior of CAN is modeled incorporating models of transceiver, bus and application process. Sections 3 provides verification of timing and logic properties and Section 4 shows a case study with periodic and sporadic processes including comparison with Tindell's approach.

2. BASIC CONCEPT OF CAN MODEL

The aim of this section is to model CAN by timed automata. Up to the standard the CAN is a message oriented transmission protocol. Due to the bus topology only one processor can transmit at a given time. Therefore the message response time (i.e. the length of time from the release time of the message to the instant when it is completely received) is given not only by the message length, but also by the access to the shared communication media (so called Medium Access Control - MAC).

The message priority is given by the message ID. The priorities are laid down during the system design in the form of corresponding binary values and cannot be changed dynamically. The identifier with the lowest binary number has the highest priority.

MAC problem is resolved by bit-wise arbitration on the identifiers performed by each station observing the bus level bit by bit. The resolution is in accordance with the "wired and" mechanism, by which the dominant level overwrites the recessive level. The transmission is denied for all processors with recessive transmission and dominant observation. All those processors automatically become receivers of the message with the highest priority and do not re-attempt transmission until the bus is idle again.

In contrast to the results achieved in (Tindell and Burns, 1994) our approach can be simply enlarged to deal with internal structure of the application processes, timing parameters of different operating systems, and other timing and logic properties of the real time control system. Therefore it allows to check the response time, i.e. the actuator to sensor reaction time including not only the message latency but also the latencies introduced by RTOS and application processes. Moreover, while using the verification approach, one can verify not only the schedulability, but also rather complex properties linked to logic and timing behavior of the distributed system. The model is composed of timed automata described in the following subsections.

2.1 Bit-wise arbitration model

The model of CAN arbitration designed in timed automata (Pettersson and Larsen, 2000) is shown in Figure 3. The model describes MAC mechanism for one message accessing the bus. The location *no_trans_needed* represents a situation when the arbitration model is waiting for *trans_request* from the application process. The locations *send_bit_to_bus*, *listen_bus*, *check_next_bit* represent the arbitration process. The locations *request_denied* and *request_success* give result of the arbitration process.

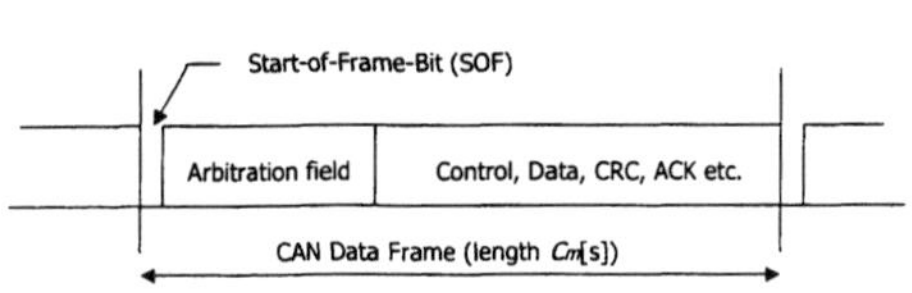

Fig. 2. CAN message frame format

After processing of the Start of Frame Bit (SOF) (see the CAN message frame format in Figure 2) the first bit from the arbitration filed is sent to the bus (transition *send_bit_to_bus* → *listen_bus*). At the same time the transmitting processor senses the bus and both transmitted bit (local variable *id*) and sensed bit (global variable *signal*) are compared. If they are identical and the end of the Arbitration field (*nsigi* states for the length of the Arbitration field) was not reached the next bit is proceeded (*check_next_bit* location) when nominal bit-time elapses (deterministically given as *tbit* constant). If the sensed bit is not identical to the transmitted one, the transmission is denied (*request_denied* location). If they are identical and the end of the Arbitration field was reached the processor wins the arbitration (*request_success* location). The CAN Arbitration model includes the information about the duration of each bit-time given by invariant $t \leq tbit$ in *listen_bus* location and guards $t \geq tbit$, $t \geq 0$ on outgoing transitions. When *tbit* is not deterministic, i.e. *tbit* is bounded on interval $\langle tbit_l, tbit_u \rangle$ then the duration of each bit-time given by invariant $t \leq tbit_u$ and guard $t \geq tbit_l$.

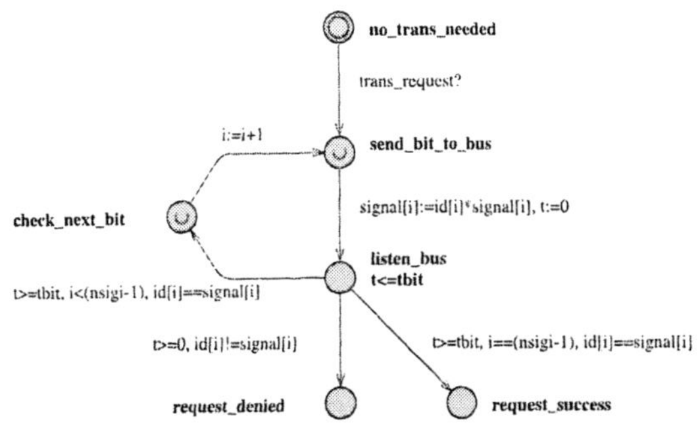

Fig. 3. Arbitration model (in UPPAAL like notation)

2.2 Transceiver model

Above explained bit-wise arbitration is a part of the transceiver model. The implementation of the complete transceiver model is depicted in Figure 4, and its interconnection with other automata is shown in Figure 6. It is composed of the three sections:

(1) the arbitration section described already in Figure 3
(2) synchronisation section (*waiting_for_free_bus* → *send_bit_to_bus* transition) that is used to synchronize all transmitting processors prior to arbitration (this part realises broadcast communication) and
(3) data transmission section given by *trans_section*, *trans_section_finished* and *trans_finished* locations.

The function of transceiver is the following: after receiving the transmission request, the processor is in the waiting state (*waiting_for_free_bus*) until the bus is free. Bus becomes idle, the arbitration processes start (synchronization by urgent *broadcast_synch* channel). If the transmission was denied (*trans_denied* location), the transmission request is immediately repeated and the processor is waiting for free bus again (*waiting_for_free_bus* location). Otherwise the processors message is sent. The duration of message is given by deterministic time C_m. When the transmission is finished (*trans_section_finished*) the bus becomes idle (*bus_trans_finished* channel) and the application process is informed about the end of transmission (*trans_compl_status* channel).

2.3 Bus model

Figure 5 depicts the physical bus model. The model is in idle location when there is no activity on the bus and it is in busy location when any processor transmits. The *trans_vote* global variable is used to detect that at least one processor

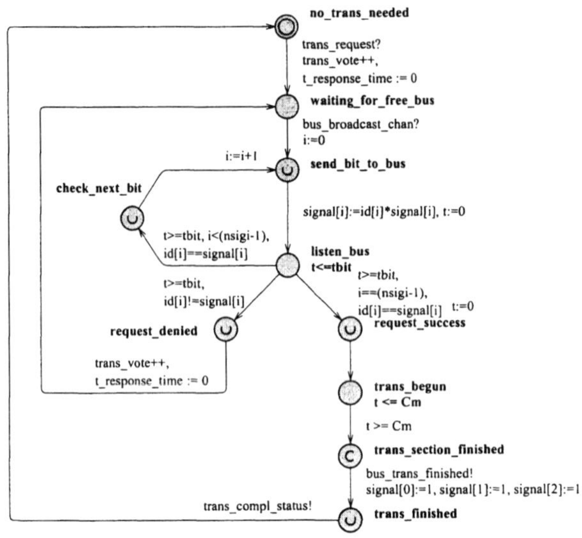

Fig. 4. Transceiver model

is willing to start the transmission. If this is the case than the global synchronization is realized via *bus_broadcast_chan* from the bus model.

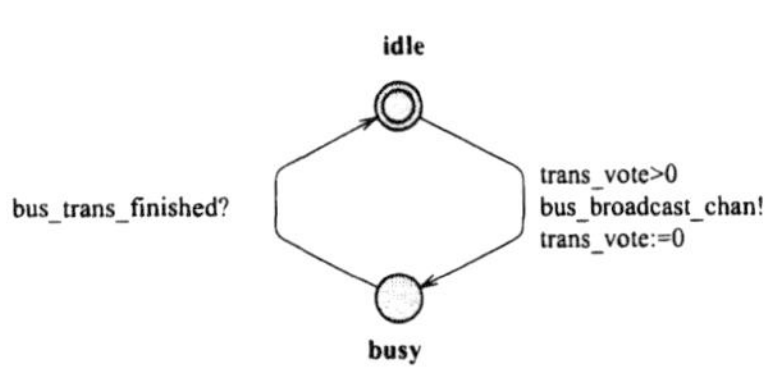

Fig. 5. Bus model

2.4 Application process model

As seen from Figure 6 the case study assumes 4 processors to be connected via CAN. Each processor is running one application process transmitting the messages of the same identifier. The application processes 1, 2, and 3 are periodic processes transmitting messages with identifier 1, 2, and 3 respectively. The application process 4 is a sporadic process transmitting the lowest priority message with identifier 4.

The periodic application process, with period T_m, is depicted in Figure 7. Afterwards each message is delayed by an operating system delay - the time between zero and J_i (called jitter in (Tindell and Burns, 1994)). Then the transmission request is done by *trans_request* channel. When the message is transmitted the process is informed by *trans_compl_status* channel. Location *no_transmission_activity* represents a situation when the process does not perform transmission, i.e. it performs for example computations. Location *init_location* starts the first task period, delayed by time between zero and T_m in order to represent the phase shift of the task.

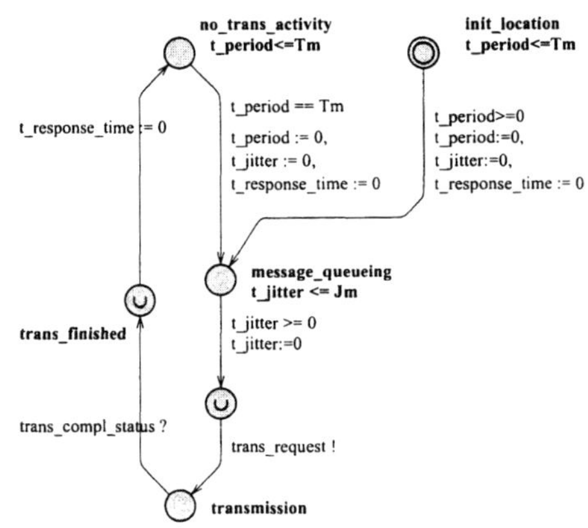

Fig. 7. Periodic application process model

The sporadic process model is depicted in Figure 8. Locations *no_trans_activity_1* and *no_trans_activity_2* represent a situation when the process does not perform any transmission. The process resides an arbitrary time in location *no_trans_activity_1*, then the transmission request is generated and when the message is transmitted the process is informed by *trans_compl_status* channel, and then the process has no influence on the bus. Local variable *t_response_time* in both models is used in properties to be verified.

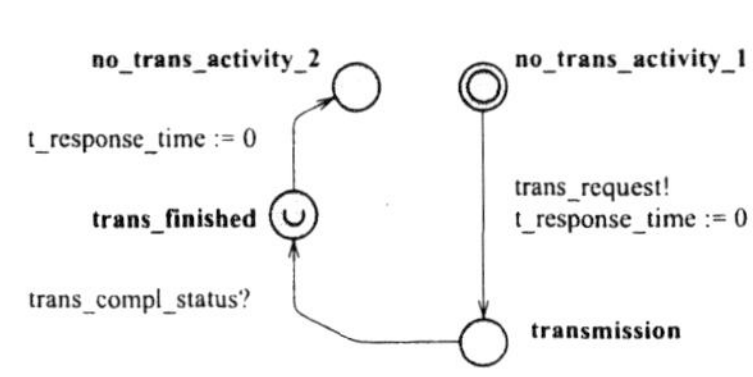

Fig. 8. Sporadic application process model

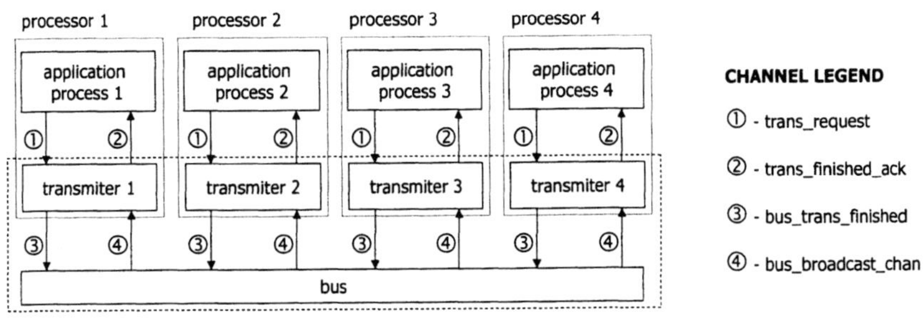

Fig. 6. Case study system configuration

3. VERIFICATION OF THE MODEL

The section presents the case study with periodic and sporadic processes including comparison with Tindell's approach (assuming 125kbps baudrate).

Table 1. Process parameters table

Msg.ID	Type	Period T_m [μsec]	$Deadline$ [μsec]	C_m [μsec]
1	periodic	2000	2000	504
2	periodic	3000	3000	504
3	periodic	5000	4000	504
4	sporadic	-	-	1040

Timing and logical properties to be verified can be for example the following ones:

(1) Is the system deadlock free?
(2) Is there any state in which processor 1 and processor 2 are in the data transmission section?
(3) Is there any situation in which the highest priority message does not win the arbitration?
(4) Are all periodic messages transmitted prior to their deadlines?
(5) What is the worst-case response time R_m of the message with identifier m (for m=1, 2 or 3)?

These properties are formulated in the temporal logic based formalism used in the UPPAAL verification tool UPPAAL (Pettersson and Larsen, 2000) as follows:

(1) A □ (not deadlock)
(2) E ◇ (Transceiver_1.request_success and Transceiver_2.request_success)
(3) E ◇ (Transceiver_1.request_denied)
(4) A □ (Process_m.trans_finished) ⇒ (Process_m.t_response_time < *Deadline*)
(5) A □ (Process_m.trans_finished) ⇒ (Process_m.t_response_time < R_m)

The verification results of timed automata tool are as follows:

(1) Property is satisfied
(2) Property is not satisfied
(3) Property is not satisfied
(4) See the section bellow
(5) R_m found by iteration (using bisection) see the section bellow

4. CASE STUDY

In this section we assume the configuration depicted in Figure 6 where each processor is running one application process transmitting one type of message (the message ID is equal to the application ID is equal to the processor ID). Table 1 shows parameters of three periodic and one sporadic process. The aim of the case study is twofold:

- to verify whether the response time satisfies a given deadline of the message (corresponding to property 4)
- to find the worst-case response time R_m iteratively by repeating the verification for different values of deadline (corresponding to property 5).

Table 2. Results of the experiment related to property 4 and 5

Message ID	J_m[μsec]	formula 4 result	R_m[μsec]
1	0	satisfied	1544
2	0	satisfied	2048
3	0	satisfied	3056
4	-	-	-

Table 2 shows verification results of the experiment related to property 4 and 5 without operating system delay . The response time of each periodic message is shorter then corresponding deadline assuming also relatively long sporadic message.

Table 3. Results of the experiment related to property 4 and 5 with operating system delay

Message ID	J_m[μsec]	formula 4 result	R_m[μsec]
1	456	satisfied	2000
2	0	satisfied	2552
3	0	satisfied	3056
4	-	-	-

Table 3 shows results of the experiment related to property 4 and 5 with operating system delay (J_m).

Values of R_m in tables 2, 3 are identical to those calculated by iterative algorithm (Tindell and Burns, 1994) based on equation

$$R_m = C_m + J_m + w_m$$

where

$$w_m = B_m + \sum_{\forall j \in hp(m)} \left\lceil \frac{w_m + J_j + \tau_{bit}}{T_j} \right\rceil C_j$$

The term B_m presents the longest time that the given message m can be delayed by lower priority messages, the τ_{bit} is the bit time of the bus. The set $hp(m)$ is the set of messages of higher priority then message m.

5. CONCLUSIONS AND FUTURE WORK

This article shows a way of communication protocol modeling by timed automata. Model of entire distributed application can be obtained simply by interconnection with real-time operating system (Waszniowski and Hanzalek, 2003) and application software automata. The resulting model is suitable for verification of desired/undesired states in the time critical distributed applications like automotive IT based on CAN and OSEK. Due to this modular approach, the model is simply extensible to TT CAN.

Verification results of the CAN model can be directly compared to the results achieved by (Tindell and Burns, 1994), when obtaining identical results from both approaches (try to compute R_m for values in Tables 2 and 3). Moreover our approach can be simply extended to deal with internal logic structure of the application processes, processor sharing managed by operating system etc. On the other hand, high complexity is a drawback of the model checking approach in contrast to quite straightforward equations of the scheduling theory. For example verification of system deadlock, which is the most time consuming among tested properties, took 20 minutes for parameters in Table 1. on AMD-Athlon XP 1,8GHz computer with 1,3GB RAM.

Therefore our future work is related to hierarchical modelling and verification of distributed systems, i.e. parts of the system will be modelled and their timing parameters will be verified. Consequently the part will be replaced by a location with upper and lower bound on its timing parameters and it will be used for verification of upper layer. This approach certainly gives a little pessimistic result, but when tailored to the application logic, it can be rewarding.

Acknowledgements: This research work has been supported by the research program No. J04/98:212300013 "Decision Making and Control for Manufacturing" of the Czech Technical University in Prague and by IST project 35102 OCERA.

REFERENCES

Alur, Rajeev and David L. Dill (1994). A theory of timed automata. *Theoretical Computer Science* **126**(2), 183–235.

Berard, B., M. Bidoit, A. Finkel, F. Laroussinie, A. Petit, L. Petrucci and P. Schnoebelen (2001). *Systems and Software Verification: Model-Checking Techniques and Tools.* Vol. 7 of *ISBN: 3-540-41523-8.* Springer.

Buttazzo, G.C. (1997). *Hard Real-Time computing systems.* Kluwer Academic Publishers.

Corbett, James C. (1996). Timing analysis of Ada tasking programs. *IEEE Transactions on Software Engineering* **22**(7), 461–483.

Etschberger, K., Roman Hofmann, Joachim Stolberg, Christian Schlegel and Stefan Weiher (2001). *Controller Area Network: Basics, Protocols, Chips and Applications.* ISBN: 3-00-007376-0. IXXAT Automationpress.

Geischeder, Manfred, Klaus Gresser, Adam Jankowiak, Jochem Spohr, Andree Zahir, Markus Schwab, Erik Svenske, Maxim Tchervinsky, Ken Tindell, Gerhard Gser, Carsten Thierer, Winfried Janz and Volker Barthelmann (2000). Osek/vdx: Specification 2.1.

Katoen, Joost-Pieter (1999). Concepts, algorithms, and tools for model checking.

Klein, M.H., T. Ralya, B. Pollak, R. Obenza, M. Gonza and L. Harbour (1993). Practitioners handbook for real-time analysis: Guide to rate monotonic analysis for real time systems.

Liu, Jane W.S. (2000). *Real-Time Systems.* Prentice Hall.

Pettersson, Paul and Kim Guldstrand Larsen (2000). Uppaal2k.

Tindell, K. and A. Burns (1994). Guaranteed message latencies for distributed safety critical hard real-time networks.

Waszniowski, Libor and Zdenek Hanzalek (2003). Analysis of real-time operating system based applications. *FORMATS 2003.*

Copyright © IFAC Information Control Problems in Manufacturing, Salvador, Brazil, 2004

ELSEVIER
IFAC
PUBLICATIONS
www.elsevier.com/locate/ifac

IMPROVING AUTOMATION SOFTWARE DEPENDABILITY: A ROLE FOR FORMAL METHODS?

Timothy L. Johnson, PhD.

GE Global Research, K-1, 5C30A
P.O. Box 8
Schenectady, NY 12301
(johnsontl@research.ge.com)

Abstract: The growth of manufacturing control software from simple NC and PLC-based systems to concurrent networked systems incorporating PC's, PLC's, CNC's, and enterprise databases has created new challenges to the design, implementation, and maintenance of safe and dependable manufacturing systems. Key milestones in this evolution, and the prospects for the use of formal verification methods in achieving enhanced dependability of future manufacturing software, are examined in this paper and presentation. *Copyright © 2004 IFAC*

Keywords: Automation, manufacturing systems, system engineering, programming theory, reliability theory, safety analysis, computer software, testability.

1. INTRODUCTION

In the US, the Denver International Airport is often used by the Federal Aviation Administration (FAA) as a test site for new technologies. A perennial dread of the air traveler is the baggage handling system: lost bags, delayed bags, and worst of all, bags transferred to the wrong airline, and ending up in remote places like Brazil (well, at least remote from Denver!). So, the FAA and the Denver businesses and politicians decided that the brand new airport would be a wonderful place to showcase new baggage handling technology. The system requirements were duly prepared, the contract awarded, and millions of dollars committed to a network of computer-controlled conveyors that would whisk luggage immediately to its intended destination (deNeufville, 1994). But then came the control system. The initial indication that something was wrong occurred when the rest of the airport, and conveyors, were in place, but the software design had barely begun. The project became a laughing-stock when it was over two years late on delivery (the rest of the airport could not be used without it). Finally, the time for initial testing arrived: It was a disaster! The system could not do even the most basic luggage transport correctly. Patience wore thin. Political and business reputations were ruined. Finally, the system was scrapped and a "semi-automated" (viz., conventional) system was used instead! From the control engineer's perspective, the most serious consequence of this type of failure is that the public was left with the impression that automation itself was at fault, and not that (as was undoubtedly the case) the project was mis-managed. Dozens of airports around the country will now opt for less automated systems in favor of more automated ones, and control engineers will have less to do.

As a whole, only about 30-40% of large software projects that are initiated will run to completion (Brooks, 1995), and this was one that didn't. Even though the record in manufacturing systems – which are highly structured – is probably better than this average, it still could benefit from substantial improvement (Place and Kang, 1993 – selected references from older literature have also been repeated here). Start-ups of new manufacturing and process plants are often notoriously delayed. And increasingly software development is at the heart of most of the problems. With the rapid decrease in cost, and even more rapid increase in the capabilities of computers over the last decade, the computing hardware components of automation have become less costly, more versatile, and more reliable. So the drive to shift hardware functions into software has accelerated over the last decade. Manufacturing software itself has expanded from isolated, carefully designed PLC logic systems that operate for months without interruption, to PC-based platforms, where even in the absence of an application, the operating systems must be rebooted every few days!

Not only have manufacturing control applications become rapidly more complex, but also the expectations of timely response have grown increasingly more demanding. At the same time, other design requirements have also grown more demanding. Availability targets have expanded from 95% to 99.99% or higher in some applications (e.g., network broadcasting). The numbers of measurement and control points and size of control programs have exploded. Networks are a part of almost every system (Perrow, 1984). Enterprise integration, as well as sensor-level integration is expected.

In spite of the increasing level of dependence of manufacturers on automation software that is expected to be safe and reliable, very little rigorous statistical data concerning manufacturing software mishaps is available. The best publicly available data in the US appears to be in the area of Occupational Health and Safety incident investigations, and in documented court cases involving software failures. However, in the case of Occupational Health and Safety, many accident root causes may be traced to process, sensing, or display irregularities – even when software is involved. In court cases, e.g., those involving personal injury in manufacturing operations, the legal profession is frequently challenged to differentiate between error on the part of a software user, and errors in the software itself: until very recently, end responsibility for safety-related functions has often been delegated by the courts to users or operators of software, even in cases of software malfunction. The vast majority of unscheduled outages are "routine", and the appropriate unit or subunit is investigated, and then reset or restarted within a few minutes; nevertheless, part production runs below capacity during this time interval.

The advent of web-based and distributed software, often with multithreaded (concurrent) operation, and the contemplated use of wireless links for factory networks will create system-level fault modes of a complexity that could only be imagined a few years ago. In spite of this, it is not likely that computing progress will be reversed by these considerations. Instead, what may be required is a host of much more powerful verification and validation methods. The study of more powerful verification methods is bound to become more important as software becomes more complex. The purpose of this presentation is to review some of the fundamental factors underlying manufacturing software dependability, to survey the state of the art in current products and research related to verification, validation, and safety of such systems, and to provide a brief preview of some more recent research that shows promise in improving the quality of service (QOS) of manufacturing automation software. With an understanding of feedback processes and manufacturing system dynamics, control engineers and scientists are well qualified to play vital role in the future of dependable manufacturing systems.

2. MANUFACTURING CONTROL BACKGROUND

Many of us are familiar with the development of manufacturing automation equipment – but have we ever thought about the evolution of test and verification for such equipment? The purpose of this brief sprint through history is to trace the growth of test and verification processes and issues that have accompanied the better-known improvements in manufacturing automation and computation. This provides a context in which to assess the context for growth of formal methods in this field.

The earliest machine tool control languages, such as APT, that were invented in the late 1940's (Alford and Sledge, 1976), were deliberately designed for ease of test. The core elements of the language were instructions such as:

START
MOVE (position)
GO TO (line number)
STOP
RESET

In fact, sometimes START and STOP were instantiated in hardware! But, even in these relatively simple cases, program verification could be difficult. The developer would be required to compute any coordinate transformations with a slide rule! Running them through the machine tool, determining if the part was right, and then modifying the program if necessary tested the early programs. The data and logic of the programs were combined in the MOVE statement. The position data for the MOVE command was a set of coordinate values specifying the next set-point for a servo; the entire mechanical structure and coordinate system of the machine tool was presumed to be known to the user. The only transfer of control was via an unconditional GO TO. Instructions were executed according to a fixed, precisely timed clock cycle. No safety checking was performed, so the machining head could collide with any jig or guide, or indeed with a part of the machining table itself. Such machines were commonly found in a state of considerable damage after a few years of use! Nevertheless, numerically controlled machine tools became very popular and were used to make very complex parts with a much higher degree of consistency than human operators could produce. The machine and software could be verified, and even concurrently calibrated, by running a set of simple calibration routines and then gauging the resulting work-pieces to compare them with their intended values. Today's machine tools, of course, support a much higher degree of complexity, and may include built in 3-D simulators with collision detection capabilities.

Figure 1: Machine Tool with CNC (early models used APT)

In the late 1950's and throughout the 1960's and 1970's, a complementary machine was born: the programmable logic controller (PLC). Such machines originally came about to replace racks of relays that had been developed in the 1920's to 1940's to govern the sequential control of complex operations such as telephone line switching, railroad signalling, and some early automation equipment. The development of the PLC was stimulated by several problems with relays racks:

- Relays would fail mechanically after 30,000 or 40,000 operations.
- Reprogramming required re-wiring – and knowledge of the original logic!
- Large systems became limited by reliability of individual components and wiring.

Early PLC's like CNC equipment, used punched paper tapes (or front panel switch settings) to read in programs. Their role in the factory was different, and in some sense complementary to that of the CNC machine:

- Their inputs and outputs were normally logical (binary) rather than numerical (integer)
- They operated asynchronously with respect to the process
- They used a programmed sequence of logic steps, normally executed in a loop repeated at a rate much higher than the rate of occurrence of new events in the controlled manufacturing process
- The essence of the program was in the cycle-to-cycle changes in the results of each ladder logic "rung", not in perfect or unconditional repeatability.

Rather than becoming expert at the wiring of a relay rack, the PLC programmer became expert at "Relay Ladder Logic" (RLL) programming, which was normally interpretive rather than pre-compiled (since programs frequently were modified on the shop floor). Still, the program provided only very implicit reference to the desired properties of the system under control, making it extremely difficult to debug. Often, such programs would be developed first for the desired "correct" sequence of operations, and then conditional sections would be added to the program (often after controller installation) to provide for correct operation during unexpected events such as part jams or power failures. The "main execution loop" structure of such programs was also frequently confounding to the un-initiated, since in its pure form, the PLC would rely on the controlled system to store "state" information. For instance, an external relay would be set on one cycle and then its state could be read on the next cycle as a primitive form of "state transition". This fact, that the developer of an RLL program might scheme to store part of the system state in the plant itself, was often a big impediment to verification, particularly when the verifier was not intimately familiar with the equipment under control. At times, dummy states were even used to delay a transition, based on the known timing of a loop (even though PLC's later contained timers that could be set and tested with a similar effect).

The PLC is also of great interest because it was developed as a "universal" controller, where the program itself *was* the control law and could be changed directly without any intervening "re-design" process. Soon programs having thousands of lines of PLC code were being written. Verification of PLC programs became a cottage industry! As the programs became large and complex, the "side effects" of changing even a single line of code, rather than the likelihood of a bug occurring in the change itself, came to dominate the decision to make program changes; experienced PLC programmers would become more and more wary of changing code as programs became larger. Various methods of simulating plant behavior came into use; in some cases manufacturing equipment itself might contain "test" modes that could be used to assist in verification, since "open loop" testing of the PLC often was not meaningful. Testing in the presence of timing changes in the plant (or the controller) is also quite difficult with PLC's, and usually requires an actual manufacturing equipment installation to verify. For this reason, many "programming" systems were eventually made portable, so that they could be attached to equipment *in situ*. Methods for formal verification of PLC code are now coming into commercial use, and will be discussed a bit later.

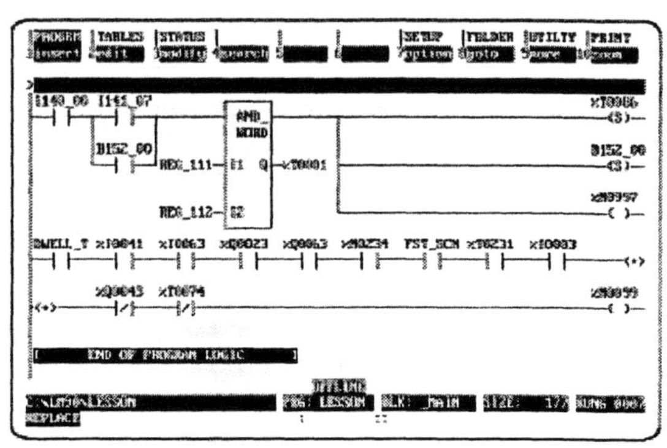

Ladder Diagram (LD)

Figure 2: Relay Ladder Programming (GE Fanuc LogicMaster 90)

Several subsequent developments have now also occurred:

- Factory Local Area Networks (LANs) were developed in the 1970's and 1980's, initially to transmit real-time control data between different PLC's.
- A "supervisory" or "enterprise" layer, initially based on mainframe computers, was added "above" the PLC layer, to support program download and factory-level synchronization and performance data collection.
- A (typically separate, and often non-real-time) monitoring and fault-reporting network was added, and tied in separately to the supervisory layer.

The supervisory and monitoring functions, being essentially "open loop", were less critical to operation, and in many cases would operate for years with serious deficiencies. They were often only tested carefully when someone actually needed to make use of the data that they produced! This was not true, though, for the Factory LAN system, which was typically required to synchronize operations between different work areas in a plant (e.g., two ends of a conveyor system!), so hardware, software, and timing of verification of such LANs became extremely important. Although these systems are being widely replaced with Internet technology today, the Factory LAN evolution during the 1980's and 1990's made important contributions to reliability, data encoding, and protocol development that eventually benefited Ethernet based technology. Since PLC's were by nature asynchronous in concept, the use of Factory LANs was a natural development (it was as if I/O bits were simply set from a more remote location in the factory), and most PLC systems could still be verified on a machine-by-machine basis. In other words, little additional effort was normally required to establish tight concurrency between multiple PLCs ("islands of automation"): where this was required, an input or output line could simply be run directly to both systems. At the same time, factory-level start-up now required more attention when large PLC subsystems became coupled. A common verification approach would be to work backwards from the end of a line forward in order to verify proper interactions of PLC subsystems, e.g., during re-start of a plant following an unscheduled outage.

Today, many individual items of manufacturing equipment have become quite a bit more accurate and sophisticated, often incorporating their own PC-based controllers, rather than relying on a PLC to perform all of the necessary non-mechanical logic. Factory LANS may link PC's as well as PLC's, and there is ample opportunity for inconsistency in these interconnections. In the late 1980's and early 1990's, the "data driven" factory became a popular concept. In the extreme interpretation of this concept, all measurement data is deposited in a database (nominally, the present "state" of the factory), and any local controller can use any subset of the data to determine the outcome of a control action.

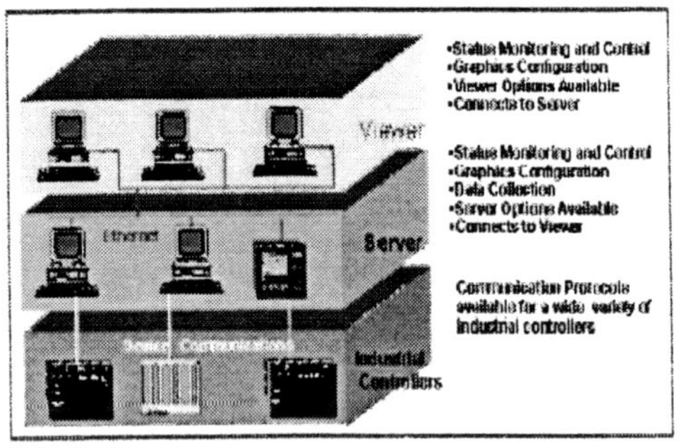

Figure 3: Factory LAN Hierarchy

This paradigm could be successfully applied to factory data collection networks, but was not usually successful when applied to PLC data, since the implementation of sequencing constraints along the production line was difficult to implement; it would be easy to create control loops with non-local feedback that would oscillate at frequencies that were dependent of material transport delays through the system; hence, the "data driven" paradigm was rarely applied to plant wide control systems (with the possible exception of "open loop" controls at the enterprise level). The most common reflections of inconsistent network connections of this type are limit cycles (set/reset of the inconsistent variables), or deadlocks (e.g., two types of equipment each waiting for the other to initiate an action). A common practice in avoiding such conflicts is to force all synchronizing actions to occur through PLCs, while having a separate data, monitoring, and/or "enterprise" network which is merely used for data collection, polling, or remote diagnostics of PC-based equipment controllers, but never for control.

The most significant development within the last decade has been the advent of web-based applications. The ability to remotely view, re-program, re-configure, and sometimes even service equipment from a desktop has enormously improved factory productivity: manufacturing control engineers do not have to personally visit each item of equipment during an outage in order to reprogram, reset or restart it, resulting in much shorter outages, and routine fault monitoring and diagnostics can be done at the desktop. With web-based tools, the remote diagnostics and servicing of equipment by OEMs also eliminates travel and delays associated with unexplained trips of more complex equipment (Locy, 2001). Standards for security and e-business interfaces to support remote monitoring and diagnosis are under development in several industries, as OEM's take on more responsibility to diagnose complex equipment faults. The advent of more complex patterns of data transfer has made system level verification still more difficult, even when dependability at the subsystem level may have improved (sometimes due to very reliable computing hardware replacing much less reliable mechanical relays and interconnects).

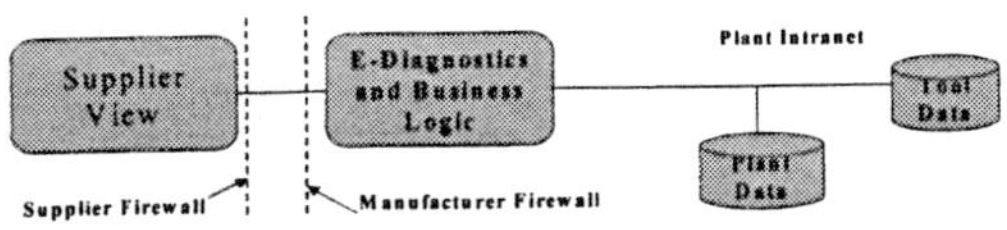

Figure 4: e-Diagnostics Logical Layer

The next generation of factory computing technology can be expected to include wireless links, particularly for monitoring and diagnostics. Generally, control decisions can be made on the basis of much more complex combinations of measured variables. The role of PLC's in the factory can be expected to continue to decline, or at least the re-programming of PLC's can be expected to occur via remote "programmer" interfaces. Model-based control of discrete manufacturing processes can be expected to expand as the ability to visualize equipment state at a remote location is perfected. From a control perspective, two of the most substantial threats to dependability may become external network security threats (particularly on wireless links), and the advent of "applet" style (or downloadable) dynamic software modifications. Rigorous factory software configuration management will be necessary if the impact of these changes is to be confined. Factory-level regression test suites will probably become a reality. Robustness to interruptions of wireless transmission will pose new challenges to controls (and verification) designers!

The replacement of mechanical with electronic controls during this revolution in manufacturing technology has been driven largely by the need for controls to be more reliable than the controlled process, and the fact that typically electronics have (hardware) failure rates about three orders of magnitude lower than mechanical equipment. The threat, however, is that now *software dependability* may limit further automation progress at the enterprise level. Even today, in some industries such as semiconductors and automobiles, system level availability remains moderate, in spite of very high dependability at the unit operation level. Enterprise systems accumulate so much data that they may be forced to "stumble" from one data deficiency to another, only operating as intended for brief periods of time!

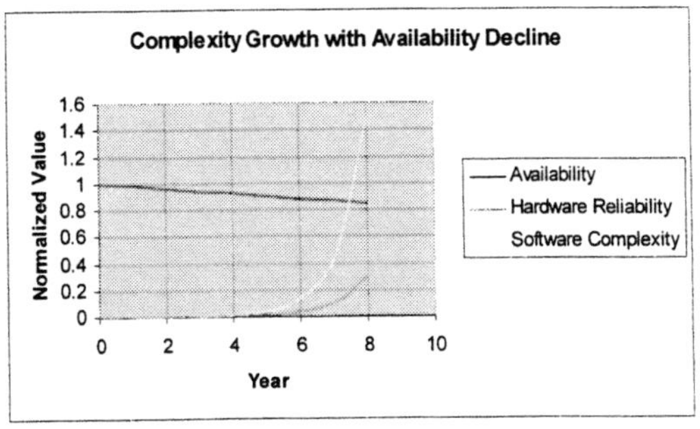

Figure 5: Software Complexity Growth and System Availability Decline (illustration only)

3. SYSTEM ENGINEERING AND DEPENDABILITY

Dependability is a system engineering requirement, and normally involves performance requirements on both the severity and frequency of occurrence of unscheduled outages or production interruptions due to production equipment, and particularly automation equipment, failures. Severe outages usually involve occupational safety, environmental damage, or equipment damage (Crocker, 1987). A recent summary of root causes of industrial accidents based on US Occupational Health and Safety incidents reported that failure to perform adequate Preliminary Hazard Analysis and Change Management studies (particularly for environmental control systems) were among the most common root causes of severe accidents (Belke, 1998). Less severe, but much more frequent outages are caused by loss of primary facility power or by nuisance trips and erroneous fault indications of manufacturing equipment or controls. Commonly, the controller either instigates such trips when production part data or timing is out of specification, or when manufacturing equipment triggers a fault condition. In these cases, primary effort is invested in resolving or re-setting the fault trigger in a minimum time, and then in re-establishing production rates (Farrell, et al., 1993). Recent advances in system level control of supply chains and equipment buffers have suggested that "hedging point" strategies (implemented in the design and control of material handling systems, typically) are an optimum way to accommodate temporary disruptions to production (Bullock and Hendrickson, 1994; Mourani, Hannequin & Xie, 2003).

Standards are playing a major role in the advance of dependability in public systems in general, and manufacturing software in particular. International standards organizations such as the ISO and IEC; professional societies such as IFAC, IEEE, ASME, SAE; and privately sponsored industrial entities such as EPRI, International Sematech, Underwriters

Laboratories, and many others have successfully promoted the development and adherence to standards. Standards have a significant impact on the formulation of performance specifications and are a means by which the manufacturer (by specifying that certain standards shall be met when production equipment is purchased) and the public (through occupational and safety standards) can assure safe operation and high quality products.

Either product or process specifications may affect the dependability of a manufacturing system. A "well toleranced" product design will match part dimensional specification accuracy to manufacturing system capability (Phillips, 1994), so that the dimensional tolerance is not tighter than the capability of cutting equipment, for instance. If this is not done, then standard quality tracking methods such as statistical process control (SPC) will generate gauging alarms more often than necessary. Among numerous process specifications are common variables such as power quality, ambient temperature, vibration levels, humidity, vapor pressures, and particulates in air, that represent potential "common mode" sources for large numbers of out-of-tolerance events. Measurement and control of these system level variables is critical for the reduction of nuisance alarms.

As our focus of interest will be on unit-level control and operations, the primary interest in system level methods concerns the manner in which unit level requirements are developed from system level dependability-related requirements, and in potential areas for improvement of current design processes. Of course, unit level requirements are based on a functional flow-down from system level requirements. Commercial manufacturing practices, at this level, are usually less rigorous than corresponding practices in safety-critical military or transportation products, and a few of the differences are instructive (Neuman, 1995; DoD, 1984b). Practices such as (software) requirements trace-ability and Preliminary Hazard Analysis (PHA) are becoming common in safety-critical systems (Parnas, et al, 1990), but are not yet common in manufacturing systems. Also, perform-ability, statistical reliability/availability and life cycle cost analyses are now becoming common practice in critical systems, but are not yet common in manufacturing control systems - except for part gauging and statistical quality control, both of which are frequently done off-line rather than on-line (Abbott, 1988). The use of a system level simulation during preliminary design, particularly to validate requirements, is also missing from manufacturing design practice, or if present may be done only on parts of the system (e.g., material flow balance at the top level, individual unit operation sequencing at the unit level). Finally, consistent standards are often missing for acceptance testing, with the focus being on "working demonstration" rather than a coherent approach toward extreme-value testing. These deficiencies in system level design practice often result in very long start-up times for new plants, where many equipment items and software must be re-designed.

One benefit enjoyed by many manufacturing systems is the common practice of standards that apply to various classes of equipment: communications, voltage and current levels, terms of reference for PLC equipment (IEEE, 2000; IEC, 1986; CENELEC, 1997; Suyama, 2003). To some extent, standards circumvent the need for "custom" testing of components, with "plug and play compatibility" being a common objective. At the same time, there is a danger that many standards are only specified "down to some level" or "under certain conditions", beyond which, implementation details are left to the supplier's discretion. A frequent consequence of this is that equipment that nominally meets certain standards will in fact fail to interface correctly due to differences in OEM supplier practices in using lower level options. Standards such as the ISO 7-layer protocol model of communications, and data exchange standards are notorious for these problems. A more insidious version of this problem occurs when equipment appears to interface correctly, but fails to communicate when an exception condition is raised – i.e., in precisely the condition where communication is most critical!

Recent developments in the automotive and semiconductor industries (Locy, 2001; Schoop, et al, 2001) have motivated the development of improved methods for equipment monitoring and diagnostics. As noted above, these innovations are motivated by the higher frequency of brief, non-disruptive outages in manufacturing equipment, and by the rapidly increasing costs of diagnosis and repair for very sophisticated precision manufacturing equipment. The Internet provides a significant opportunity for remote monitoring, diagnostics, and even repair of equipment. Expert designers of OEM equipment can remotely access, inspect, and diagnose equipment condition, and in some cases instigate software repairs remotely, or provide remote guidance to on-site repair staff. E-Diagnostics has become an important new technology for reducing false alarms and downtime due to nuisance faults. With higher availability targets for automated factories, brief, repeated outages could significantly impact system availability (Chen & Trivedi, 1999).

At the unit operation level, top-level requirements are often reduced to the sequencing of sub steps, accommodation of faults and alternate modes of operation, on-line part monitoring, and plant level synchronization and reporting. As indicated previously, two forms of hardware are most common for the implementation of control logic: The PLC and the PC. We now proceed to our main task of addressing dependability at this level.

4. DEPENDABILITY AT THE UNIT OPERATION LEVEL

At the unit level, the dependability of control hardware, except in safety-critical applications, has nearly ceased to be an issue (Bryan and Siegel, 1988). Dedicated, embedded code runs in millions of applications daily, with digital hardware failure rates, even for entire CPU boards, commonly in the range of 10^{-6} to 10^{-8}/hr. Or better. This is the result of many semiconductor and circuit design innovations that are too numerous to begin to describe here. A few system level hardware developments during the 1990's are noteworthy:

- The development of highly reliable low-cost non-volatile memories
- The advent of the "safety PLC" (Allen-Bradley, 2001)
- The development of highly reliable multi-layer high-speed asynchronous communication protocols and devices.

While hardware dependability has improved dramatically, software complexity has exploded, and may be on the verge of driving many applications – while very fully functioned – toward lower levels of dependability!

In first examining general purpose embedded applications, frequently implemented on the PC or via PLC co-processors, certain dependability issues become apparent at the "unit operation" control level (Boasson, 1993). The first issue is the *lack of good validation models*. The manufacturing "plant" is usually described either in qualitative terms or sometimes by reference to process equipment supplied by a particular supplier. Even in the rare instances when a nominal operating sequence is pre-specified, the absence of a model of the equipment operation makes the *identification of extreme performance limits or fault conditions difficult* to determine (e.g., Frachet, et al., 1997). Occasionally tools such as the fault tree, FMEA or FMECA are used to provide qualitative information about fault conditions and corrective actions, but even in these cases, the *likelihood of occurrence of various faults is rarely known* in advance (DoD, 1984a; Greenberg, 1986). Similarly, queuing models may have been used to establish operating limits and baseline product flow rates, but this is relatively rare except in cases such as large new plants. Due to the lack of performance limits and system-level models of desired performance, *preliminary hazard analyses (PHA) also cannot be performed completely*, so that extreme test cases are very difficult to define and construct (Gruman, 1989).

Verification (i.e., that a proposed design will meet requirements) is also subject to limitations. In most cases, formal requirements only "cover" a very small subset of anticipated operating conditions, and while this makes verification easier (in the formal sense), it often leads to manufacturing control systems that meet formal specifications but fail to meet informal or unstated requirements. Requirements in commercial applications tend to focus on intended normal operation of a system, and not on how it is expected to behave in the event of specific types of resource limitations. Requirements often fail to address transient performance (e.g., during start-up, restart, or shutdown), and may also omit to mention timing synchronization errors between subsystems, or even within as subsystem (Gorski, 1986). This has become more significant as concurrent but loosely synchronized software subsystems have come to characterize most manufacturing applications.

Improvements can be made in several areas, but with reference to the above discussion some specific topics that require additional research results are as follows: (1) notification of "hardware" designers of operating system performance and certification requirements, (2) use of "ontologies" to capture the extended "meaning" of a formal requirement statement; (3) use of software-integrated fault tolerance, inspectability, and built-in test, (4) specification of opportunities to use standardized, configurable, pre-tested "middleware" libraries. More advanced design methods, based on UML, for instance, lead the designer through event sequence diagrams or similar paradigms to identify the end results of design decisions (Grimson and Kugler, 2000). But as often as not, the PLC programmer will be required to generate PLC code "on the fly" without proper opportunity for test and debugging.

5. OPPORTUNITIES FOR FORMAL VERIFICATION

The IEC61508 standard has provided for the possibility of pre-certification of certain product software (IEC, 2000). For safety critical systems, this may require an Independent Verification and Validation by an external authority, in addition to the submission and review of design and test documentation (RTCA, 1985; CENELEC, 1997; IEEE, 2000; MoD, 1991). The standard also for the first time allows for the use of formal verification methods to be used in the certification process, particularly where the collection of operational test data is not feasible (e.g., on a space flight, or in a system based on new technology (Lions, 1996)). The same is true for corresponding military standards such as the British DO-55, and safety critical system standards such as CENELEC. In general, such methods require a formal requirements analysis, use of certified operating systems, and then formal verification of code (Pilaud, 1990).

The Grafcet standards and subsequent specification methods (Arzen, 2002; Silva, et al., 2001) have attempted to extend logic verification to address certain elements of data flow and timing. The use of IEC 61499 function block representations as a starting point for formal verification methods has been considered by Schnackenbourg, et al. (2003).

Rausch and Krogh (1998) reported some early results in formal verification of PLC programs. A broader overview of prior work is provided by Frey and Litz (2000). De Smet and Rossi (2003) consider formal controller verification of RLL's with and without model checking, and find model checking to be significant in reducing the computational effort required to verify realistic safety and liveness properties for a realistic pick-and-place case study. Encouraging results on robustness of concurrent computational algorithms to inter-process timing variations have been proposed (Ushio and Wonham, 2001). Formal methods for developing and verifying diagnostic codes have been proposed (Paoli and LaFortune, 2003). Results for Petri nets illustrate the possibility of deducing system level properties such as liveness and safeness, from subsystem properties. However, even for networks consisting exclusively of interconnected PLC's, system level verification remains elusive (Rushby, 1986; Sennett, 1989). The SCADE tools (see next section) allow a system level Stateflow™ diagram to be selectively verified between certain points in the diagram.

Certain operating systems such as the OSE™ operating system (www.ose.com) and VxWorks™ operating system have been certified for certain applications, although certification of applications based on these operating systems cannot in most cases be obtained purely based on the operating system certification.

To date, very few opportunities for formal verification have been realized in commercial software. Although formal verification has been used in certain critical military and space applications, the process still involves many manual stages, a good deal of expert knowledge, and an order of magnitude increase in time and/or cost in comparison to commercial applications. Within the commercial domain, perhaps formal verification of correctness of communication protocols is among the most widely used application of formal verification; and this is often performed at the theoretical level, rather than on the software that implements a protocol. Although communication protocols may form a (small) part of manufacturing automation systems, the vast majority of such systems rely on heritage code that (even when the source code still can be found, after years of use) is very difficult to subject to formal verification; from the historical background of Section 2, the reasons for this are evident. Manufacturing applications may be seen to have two properties that are favorable for formal verification – the use of relatively simple programming languages (such as RLL), and the relatively high cost of logical and coding errors. They lack one important pre-requisite for early use of formal verification: both the products and processes of developing manufacturing software are extremely cost-sensitive, and the substitute of low-cost manual programming is readily available.

Initial attempts at PLC program verification, in particular De Smet and Rossi (2003) and related efforts, illustrate some important limitations of presently available verification methods:

- The PLC code itself does not capture enough of the system definition (or "model") or requirements to restrict the size of the verification problem to a practical size. This requires the manual formulation of a number of additional constraints, which at this time requires deep knowledge of both the application and of formal verification methods.

- Presently available formal verification methods do not readily distinguish between errors in logic, errors in coding, and inconsistencies in requirements, although all of these sources may lead to verification failures and counter-examples.

Two factors that may accelerate the resolution of these issues are: (1) improvements in system specification methods, including several cited previously, that ultimately provide a complete set of constraints, requirements, and system models that will permit the complete automation of the verification process in practical applications, and (2) the adoption of more rigorous certification requirements for safety-critical systems by public agencies and standards organizations.

6. PRESENTLY AVAILABLE VERIFICATION TOOLS

A fledgling commercial industry has begun to develop around verification and validation needs, and both commercial and well-tested university software is available. In this section, we provide a brief synopsis of some recent tools that are suitable for improving the dependability of manufacturing and closely related critical systems. This list is representative, and not in any way comprehensive: it contains primarily tools that are suitable for commercial applications.

The following tools are summarized:

Praxis Critical Systems – SPARK
(http://www.praxis-cs.co.uk/flashcontent/our-unique-products-1.htm) Spark is a subset of Ada used for high-integrity program development. It has been used in some commercial applications such as railway interlock safety programming.

Reqtify:
(http://www.tni-world.com/reqtify.asp) The Reqtify toolset allows a user to mark up a requirements document and to construct a requirements traceability matrix from a natural language requirements document.

Reactis:
(http://www.reactive-systems.com/products.msp)
Reactis is a new product that allows automatic model checking of Matlab™Simulink diagrams.

SCADE
(http://www.esterel-technologies.com/v3/?id=13281)
SCADE is a code generator for safety-critical military applications that is a part of a life-cycle code generation and maintenance system.

SlideMDL:
(http://www.ece.cmu.edu/cecs/main/projects.html)
This program-slicing tool developed at CMU operates on Matlab™/Simulink diagrams to trace data dependencies forward or backward from a given point in a data flow diagram.

C-BMC:
(http://www-2.cs.cmu.edu/~modelcheck/cbmc/)
C-Bounded Model Checker, also developed at CMU, is an extension of the SMV (hardware) verification concepts to apply to programs written in ANSI "c" language. See also Kroening, et al., 2003.

Codecheck:
(http://www.abxsoft.com/) Codecheck uses an extended first-pass compiler analysis to flag violations of programming rules at the program development stage.

SPIN:
(http://spinroot.com/spin/whatispin.html) SPIN is one of the earliest and most mature formal verification tools, and can now be applied to distributed computing applications.

Codewizard:
(http://www.parasoft.com/jsp/products/) Codewizard can check coding standards such as IEEE 1483 for railway interlocking, and supports predefined rule bases.

Polyspace Auditor:
(http://www.polyspace.com/) This award-winning product can be applied to formal verification of C source language programs.
As experience is gained with these tools through selective applications, typically by universities or industrial research laboratories, certain leading concepts are expected to emerge. We can certainly look forward to many success stories where formally validated software "saves the day" and avoids safety and environmental hazards, while facilitating the next level of automation – diagnostics and maintenance.

7. A FUTURE VISION FOR DEPENDABLE MANUFACTURING CONTROL

A vision for future advances in dependable manufacturing software is beginning to emerge, but it has many gaps that are not addressed by formal methods alone.

The development of formal methods for the statement and application of system requirements is still a key bottleneck (Grimson and Kugler, 2000; Jaffe, et al, 1991; Svedung, 2002, Machado, et al, 2003). A key issue is that natural language statements of system level requirements are incomplete and may be inconsistent. Not only are the "terminals" (nouns) in the grammars used to express such requirements undefined, but also many of the relationships expressed by the requirements are difficult to translate into quantitative terms. When a requirement is expressed in natural language, the author often infers a host of relationships that are implicit with the language. Not only do the nouns need to be associated with physical entities on the manufacturing floor, but also relationships need to be related to unit operations or modes. A manufacturing control *ontology* is needed so that the inferred relationships among entities in a requirements document can be explored automatically (e.g., by traversing relationship graphs and inferring implied requirements from them).

Often, it is ironic that the controller or controlled object may not be mentioned at all in requirements documents: it is *assumed* to exist! This gap may be filled by the use of standardized manufacturing simulation languages, practices, or graphic representation paradigms (Vain and Kyttner, 2001). For instance, many libraries of process control primitives, and application-specific modeling packages now exist for steam pipe systems, power plants, motors, batch process operations, circuit design, and signal processing. By associating the entities and relationships in these packages with ontologies, one can infer the existence of certain system components (e.g., a boiler control) when only the (controlled) system is cited in the requirements. In this way, a requirement can be associated with a set of preconditions on a high-level plant model, a set of operations (or controlled modes), and a set of outcomes. By using inference (or perhaps fuzzy inference; Holmes and Ray, 2001) to traverse such graphs or ontologies, a much more complete set of inferred requirements (and also test cases) can be automatically generated. Another irony is that in spite of the existence of qualitative requirements, it is often difficult to define precise quantitative behaviors that are expected for specific quantified inputs to a system: not only are the test cases difficult to derive, but the expected performance in any specific test case may be difficult to derive. In fact, tracing through various requirements that apply in a given test case, may allow one to define – by intersecting a number of qualitative performance conditions, each derived from a different path through the requirements network – a much more precise statement of expected behavior. At the same time, this approach may prompt developers to state requirements with greater precision and consistency than can now be accomplished.

A system strategy is needed for certification itself! Verification and validation activities, today, require a combination of process certification, formal verification (which is still optional), traditional case-by-case testing (even when this is known to be incomplete), and system level validation tests (Musa, 1993). This process is very costly and not very strategic. If problems are encountered during system level validation, one must often undertake extensive re-design and re-validation before the system can progress. The case of long start-up delays for new manufacturing plants was mentioned previously. Incomplete requirements, particularly hazard requirements, as indicated above, are the single most serious and most costly sources of loss of dependability (Kletz, 1982; Knutson and Charmichael, 2003). Detailed attention is needed to provide the documentation and information structures during the *early* stages of design that will lead to very low defect rates in *later* stages of design. Concepts such as *late-point identification* and *design of experiments* need to be applied to the determination of system validation test plans. Since system demonstration and validation are generally very expensive, particularly if fault conditions must be verified, great care should be taken to optimize the testing that is done at this stage so that a maximum amount of critical information is extracted during the system validation process, and so that sufficient "tunability" exists in the system level parameters that a complete redesign is only very rarely needed.

A third area worthy of note is design for serviceability and life cycle management of a manufacturing facility. The life cycles of most products today are much shorter than that of the equipment needed to manufacture them. This technological progress has led to enormous waste, and a glut of slightly used but highly specialized manufacturing equipment (with the rapid growth of an attendant world wide market for capital equipment re-use). Before it is placed in service, almost every manufacturing facility is already scheduled for capital equipment upgrades. Dependability is no longer a static concept: it is a dynamic requirement that needs to be updated and re-interpreted throughout the life cycle of a plant. Thus, verification needs to be integrated with the development of diagnostic methods that can be dynamically adapted as equipment is rearranged, tuned, or reconfigured during the life of a plant (e.g., extensions of fault tree analysis, as explored in Henry and Faure, 2003). The continuity of data from the initial concept through operation, and finally disposition of a plant, should be given active consideration. This includes the possibility of recovery, reconfiguration, and dis-assembly or recycling of OEM equipment as it reaches the end of its useful life. Built-in diagnostics and test software, based on logical concepts that are independent of a particular controller architecture are needed. The advanced concepts developed for diagnostics and remote servicing are very promising, but they require too large an investment of engineering effort to be cost-effective when equipment has short life cycles. The preceding design and verification steps, if properly formulated, can provide much data that can be used during later stages of the manufacturing life cycle, so that dependability is not a characteristic that only characterizes "the bottom of the bathtub curve" for a plant. In many cases, most of the profit is made by a business either during the falling (early) or rising (late) stage of the bathtub curve – so this is where attention is needed!

Returning to the example cited in the introduction, there is a sister example that is a success story: The use on a recently implemented, agent-based system at Daimler-Chrysler (Schoop, et al, 2001). This is a well-designed system with demonstrated availability benefits, based on rigorous and consistent application of modern agent-based factory software integrated with PLC-based factory automation software. Although formal verification was not used explicitly in this system, the analysis suggests that appropriate design documentation and practices could be applied to verify important parts of the system, such as the messaging protocol and PLC code. This is an appropriate challenge upon which to close this discussion!

8. REFERENCES

Abbott, H. (1988) Safer by Design: The Management of Product Design Risks Under Strict Liability. *The Design Council*, London.

Alford, C. O., and R. B. Sledge (1976), Microprocessor Architecture for Discrete Manufacturing Control, Part I: History and Problem Definition, IEEE Trans. Manufacturing Technology, Vol. **MFT-5**, No. 2, pp. 43-49 (et seq).

Allen Bradley Co. (2001) Safety PLCs: How They Differ from their Traditional Counterparts. *White Paper* (1755-WP001A-EN-E), Rockwell Automation.

Arzen, K-E., Rasmus Olsson, and Johan Akesson (2002). Grafcet for Procedural Operator Support Tasks. *Proc. 15th IFAC Congress*, Barcelona, July.

Belke, J. C. (1998) Recurring Causes of Recent Chemical Accidents, Proc. *Intl. Conf. and Workhosp on Reliability and Risk Management*, San Antonio, TX, Sept.

Boasson, M. (1993) Control Systems Software. *IEEE Trans. Auto. Control*, Vol. **38**, No. 7, pp. 1094-1106.

Brooks, F. P. (1995) *The Mythical Man Month*, 20th Anniversary Edition. Reading, MA: Addison-Wesley.

Bryan, W. and S. Siegel. (1988) Software Product Assurance—Reducing Software Risk in Critical Systems. In COMPASS '88 *Computer Assurance*, pages 67-74, Gaithersburg, MD, July.

Bullock, D. and C. Hendrickson (1994) Roadway Traffic Control Software, *IEEE Trans. On Control Systems Technology*, Vo. **2**, No. 3, pp 255-264.

CENELEC (1997) Railway Applications: Software for railway, control, and protections systems, *Standard EN 50128* (June)

Chen, D. and K. S. Trivedi (2002), Reliability Engineering and System Safety. *RESS 3012* (in press)

Crocker, S.D. (1987) Techniques for Assuring Safety—Lessons from Computer Security. In COMPASS '87 *Computer Assurance*, pages 67-69, Washington, D.C., July.

De Smet, O. and O. Rossi. (2002). "Verification of a controller for a flexible manufacturing line written in Ladder Diagram via model-checking," *Proc. 21st American Control Conference, p. 4147-4152.*

DeNeufville, R. (1994) The Baggage System at Denver: Prospects and Lessons. *Journal of Air Transport Management*, Vol. **1**, No. 4, Dec., pp. 229-236.

Department of Defense (US). (1984a) Procedures for Performing a Failure Mode, Effect and Criticality Analysis. *Military Standard 1629A.*

Department of Defense (US). (1984b) System Safety Program Requirements. *Military Standard 882B.*

Ehrenberger, W. D. (1987) Fail-Safe Software—Some Principles and a Case Study. In B.K. Daniels, editor, *Achieving Safety and Reliability with Computer Systems*, pages 76-88.

Farrell, J, T. Berger, and B. Appleby. (1993) Using Learning Techniques to Accommodate Unanticipated Faults, *IEEE Control Systems Magazine*, pp. 40-49 (June).

Frachet J.-P., Lampérière S., Faure J.-M., "Modeling discrete event systems behaviour using the hyperfinite signal", *European Journal of Automation*, **Volume 31 n°3**, pp. 453-470, (1997)

Frey, G., and L. Litz, (2000). "Formal methods in PLC Programming", *Proceedings of the IEEE International Conference on Systems, Man and Cybernetics (SMC2000)*, pp. 2431-2436.

Gorski, J. (1986) Design for Safety Using Temporal Logic. In IFAC SAFECOMP '86, pages 149-155, Sarlat, France.

Grimson, J. B., and H-J. Kugler. (2000) Software Needs Engineering – A Position Paper, Proc. *ICSE*, ACM, Limerick, Ireland, pp. 541-544.

Greenberg, R. (1986) Software Safety Using FTA Techniques. In *Safety and Reliability of Programmable Electronic Systems*, pages 86-95, Essex, England, Elsevier.

Gruman, G. (1989) Software Safety Focus of New British Standard, Def Std. 00-55. *IEEE Software*, **6**(3): 95-97.

Henry, S., and J. M. Faure. (2003), Elaboration of invariants safety properties from fault-tree analysis, Proc. IMACS-IEEE Computational Engineering in Systems Applications (CESA'03), Paper S2-1-04-0372.

Holmes, M. and A. Ray. (2001) Fuzzy Damage-Mitigating Control of a Fossil Power Plant, *IEEE Trans. On Control Systems Technology*, Vol. **9**, No. 1, pp. 140-147.

IEEE. (2000) Verification of Vital Functions in Processor-Based Systems Used in Rail Transit Systems. *STD 1483-2000.*

International Electro-technical Commission. (1998-2000) *IEC 61508: Functional safety of electrical/electronic/programmable electronic safety related systems.*

Jaffe, M. S., N.G. Leveson, M. Heimdahl, and B. Melhart. (1991) Software Requirements Analysis for Real-Time Process-Control Systems. *IEEE Transactions on Software Engineering*, March.

Kletz, T.A. (1982) Hazard Analysis—A Review of Criteria. *Reliability Engineering*, **3**(4): 325-338.

Knutson, C., and S. Carmichael. (2003) Safety First: Avoiding Software Mishaps, in *Embedded Systems Programming.*

Kroening, D. R., Clarke, E., and Yorav, K. (2003) Behavioral Consistency of C and Verilog Programs Using Bounded Model Checking, Proc. *DAC 2003*, pp. 368-371, ACM Press.

Krogh, B. (2003) SliceMDL URL: http://www.ece.cmu.edu/cecs/main/projects.html

Lampérière-Couffin S., Rossi O., Roussel J.-M., and Lesage J.-J. (1999), Formal verification of PLC programs: a survey", Proc. *ECC'99*, paper n°741.

Lampérière-Couffin S. and Lesage J.-J. (2002), Formal verification of the Sequential Part of PLC programs. Proc. *IFAC 2002 World Congress*, Barcelona.

Leveson, N. G. (1991) Software Safety in Embedded Computer Systems. *Communications of the ACM*, **34**(2): 34-46.

Leveson, N. G. (1995) *Safeware: System Safety and Computers*. Reading, MA: Addison-Wesley.

Leveson, N. (2002) A New Accident Model for Engineering Safer Systems, *MIT Engineering Systems Division Symposium*, Cambridge, May 2002 (to appear in *Safety Systems*)

Lions, J-L. (1996) Ariane 5, Flight 501 Failure, Report of the Inquiry Board. http://www.esrin.esa.it/htdocs/tidc/Press/Press96/ariane5rep.html

Locy, M. (2001). The impact of e-diagnostics –one year later. Proc. *2001 IEEEE Intl. Semiconductor Manufacturing Symposium*, pp. 435-438, San Jose, CA.

Machado, J.M., B. Denis, J. Lesage, J.M. Faure, and J. F. DeSilva. (2003). "Model of the Mechanism Behavior of PLC Programs, 17th Int'l. Congress of Mechanical Engineering (COBEM), Paper 0831.

Ministry of Defence (UK). (1991) Hazard Analysis and Safety Classification of the Computer and Programmable Electronic System Elements of Defence Equipment. *Defence Standard 00-56*, Ministry of Defence, Great Britain, April.

Mourani, I., Hannequin, S. Xie, X (2003). Optimal discrete-flow control of a single-stage failure-prone manufacturing system. Proc. 42nd *IEEE Conf. On Decision and Control*, pp. 5462-5467. Maui, HA.

Musa, J. (1993) Operational Profiles in Software-Reliability Engineering. *IEEE Software*, March.

Neumann, P. (1995) *Computer related risks*. Addison Wesley.

Paoli, A., and S. Lafortune (2003). Safe diagnosability of discrete event systems. Proc. *42nd IEEE Conference on Decision and Control*, pp. 2658-2664, Maui, HA.

Parnas, D. L., G.J.K. Asmis, and J. Madey. (1990) Assessment of Safety-Critical Software. *Technical Report 90-295*, Queens University, Kingston, Ontario, Canada, Ontario, Canada, December.

Perrow, C. (1984) Normal *Accidents: Living with High Risk Technologies*. Basic Books.

Phillips, R.G. (1994) Use of Redundant Sensory Information for Fault Isolation in Manufacturing Cells, *IEEE Trans. Industry Applications*, Vol. **30**, No. 5, pp. 1413-1425.

Pilaud, E. (1990) Some Experiences of Critical Software Development. In *12th International Conference on Software Engineering*, pages 225-226, Nice, France, March.

Place, P.R.H., and K. C. Kang. (1993) Safety-Critical Software: Status Report and Annotated Bibliography. *CMU/SEI-92-TR-5*.

Radio Technical Commission for Aeronautics. (1985). Software Considerations in Airborne Systems and Equipment Certification, *Standard DO-178a*, Washington, D.C.

Rausch, M., and B.H. Krogh (1998), "Formal Verification of PLC Programs," *Proc. 1998 American Control Conference*.

Rushby, J. M. (1986) Kernels for Safety? In T. Anderson, Ed., *Safe and Secure Computing Systems*, pages 210-220, Glasgow, Scotland, October.

Schnakenbourg, C., J.-M. Faure, and J.-J. Lesage. (2002). "Towards IEC61499 Function Blocks Diagrams Verification", *Proc. IEEE Int'l. Conference on Systems Man & Cybernetics, Paper TA1C2.*

Schoop, R., R. Neubert, and B. Suessmann (2001). Flexible manufacturing control with PLC, CNC and Software Agents, Proc. *5th IEEE Intl. Symp. On Autonomous Decentralized Systems*, pp. 265-371, Dallas, TX.

Sennett, C. T. (1989) *High Integrity Software*. Pitman.

Silva, B. I., O. Stursberg, B. H. Krogh and S. Engell (2001). An assessment of the current status of algorithmic approaches to the verification of hybrid systems. Proc. *40th IEEE Conference on Decision and Control.*

Suyama, K. (2003). Safety integrity analysis framework for a controller according to IEC 61508. Proc. *42nd IEEE Conference on Decision and Control.*

Svedung, I. (2002) Graphic representation of accident scenarios: Mapping system structure and the causation of accidents, *Safety Science*, vol. **40**, Elsevier Science Ltd., pages 397–417.

Ushio, T., Y. Li, and W. M. Wonham. (1992) Concurrency and State Feedback in Discrete-Event Systems, *IEEE Trans. Auto. Control*, Vol. **38**, No. 8, pp. 1180-1184.

Vain, J. and R. Kyttner (2001) Model Checking – A New Challenge for Design of Complex Computer-Controlled Systems, *Proc. 5th Int'l Conf. On Engineering Design and Automation*, Las Vegas, pp. 593-598.

Copyright © IFAC Information Control Problems in Manufacturing, Salvador, Brazil, 2004

ELSEVIER
IFAC
PUBLICATIONS
www.elsevier.com/locate/ifac

PRACTICAL FORMAL SPECIFICATION FOR EMBEDDED CONTROL SYSTEMS

Frantz Iwu, Andy Galloway, Ian Toyn and John McDermid

Department of Computer Science,
University of York, Heslington,
YORK YO10 5DD, United Kingdom
{iwuo, andyg, ian, jam}@ cs.york.ac.uk

Abstract: Control systems are typically specified in two parts: continuous and discrete. The continuous part is described using control laws; the discrete part is described using statecharts. The models are usually analysed for properties such as stability, and executed (or simulated) to validate them. An overview of a programme known as Practical Formal Specification (PFS) is presented. PFS is intended to extend classical approaches to developing control systems using formal techniques. The paper gives an outline of the principles underlying PFS and describes the realisation of those principles in a formal analysis tool integrated with Simulink and Stateflow. *Copyright © 2004 IFAC*

Keywords: Formal Methods, Control Systems, Software Engineering and Software Safety

1. INTRODUCTION

The Practical Formal Specification (PFS) project (Galloway, et al., 1998; McDermid, et al., 1998) arose out of a desire to apply formal techniques to the development of control systems, in a way that added value to the established development practices.

It was also motivated, in part, by the need to meet the spirit of standards (DS 00-55, 1997), which mandate the use of formal methods. As control system development, for example using tools such as Matlab Simulink/Stateflow (MSS), is mature, the challenge faced by PFS was how to add value, and build on current good practices, not to try to replace those practices.

PFS began in 1996. The primary financial support has been from the UK Ministry of Defence (MoD). There have been several industrial inputs to the programme, mainly from Rolls-Royce and BAE SYSTEMS. The most important part of the work with these companies has been the access to practical examples of advanced control system designs, e.g. from demonstrator projects.

The early stages of the programme focused on gaining an understanding of control system specification and identifying areas where formal techniques could add value. In essence the technical aim of PFS is to detect errors early in the design process and avoid costly rework in the development of control systems.

At this stage, several case studies were undertaken, as "pencil and paper" studies. These were inevitably limited in scope but nonetheless found problems in some specifications, which had escaped design reviews and testing. This provided the stimulus for the current phase of the project, which is mainly concerned with providing tool support for the method.

Firstly, the background to the work is presented, and then an outline of the state of the project and the tool support is given.

2. BACKGROUND

A major tenet of computer science research, especially so-called formal methods, is that benefit can be achieved by abstraction – deliberately omitting (implementation) level detail. Typical abstraction mechanisms include:

- data abstraction;
- non-determinism ;
- implicit (non-executable) specification.

In practical control system specifications these are of little use – and certainly the last point conflicts with the need to execute models for validation purposes. PFS has adopted a different approach, more suitable to the domain. In essence the approach adopted by PFS has been to introduce explicit assumptions about the behaviour of the embedding system (system being controlled). This is a rather unusual form of abstraction, but it gives "isolation" between the model and the embedding system.

A second issue is the way that the behaviour of software is specified. The classical formal methods approach is that software is represented by flat (0th-order) pre and post-conditions. These say that the software achieves the post-condition if the pre-condition is satisfied. Conversely, the control specialists' view is that software implements a transfer function.

A third issue is that engineers find the notations typically adopted by formal methods, e.g. Z, B and CSP, difficult to learn and to apply (despite the fact that the mathematics underlying control theory is rather more intellectually demanding). This can be addressed by adapting notations familiar to control engineers to formal specification. Overall, these issues have been fundamental in shaping the PFS method.

Initially, the case studies performed on PFS, although based on actual project requirements, were performed by hand. This, at the time, severely limited the scope that could be achieved. For case studies with greater industrial significance, a larger degree of automation was essential. Hence, the most recent phase of the project has concentrated on providing prototype tool-support for the method. The tools developed are now at a level of maturity where a larger industrial trial could be undertaken. This will provide the basis for quantifying the benefits of the technique in an industrial setting, and should lead to the method being used on real projects in industry.

The paper now describes the development process context for PFS, outlines the method, and illustrates the scope and capability of the current support tool.

3. SOFTWARE DEVELOPMENT PROCESS

Software development process generally starts from the *requirements*, which capture what the software is intended to do. The *specification* expresses the functions of the software that is what it does rather than how it does it, and the *design* is a logical description of how a system will work. In practice, for the embedded control systems' domain, the software development process is usually part of a wider development process for the embedding product. For control systems, the requirements are often the results of decisions made in the broader engineering process. If tools such as MSS are used the software requirements are usually derived from control system models, once they have been validated through simulation. So what can formal methods add?

During the validation process at the requirements and specification level the simulation shows that the system does what is required in a range of scenarios. However, these scenarios cannot be exhaustive, and it is possible for there to be undetected omissions and/or inconsistencies in the specifications. These problems may only be found late in the process, e.g. during rig testing, after full software implementation. This "late discovery" of errors is very costly, as there is iteration around much of the development process to rectify the problems.

There is considerable empirical evidence that control system development projects (in the aerospace industry) suffer from high costs due to late discovery of errors and rework. Based on our experience, the problems arise more in the discrete part of the specification, e.g. in a state machine used to change control modes in response to failures, rather than in the continuous part, e.g. the control laws themselves. PFS has addressed both the continuous and discrete elements of control system specifications, but the focus is on the discrete elements.

The aim of PFS is to eliminate some of the avoidable costs of control system development by finding problems early, through validation of the requirements and specifications before design and implementation. To make this possible, as well as developing the method itself, a tool to support PFS has been developed. Although PFS is not "tied" to MSS a choice was made to implement our tool as an extension to MSS, hence it is known as Simulink/Stateflow Analyser (SSA).

4. CAPTURING REQUIREMENTS IN PFS

As indicated above, one of the main "weapons" in formal methods is the use of abstraction, but classical approaches to abstraction are of limited use. The two main ways that PFS encourages abstraction are:

- By allowing engineers to state assumptions explicitly and to specify required behaviour *only within* such assumptions (later, behaviour can be added to deal with other situations such as degraded performance due to failure);
- By allowing engineers to specify sub-system *contracts* and (perhaps later) providing designs,

which meet these contracts. This mechanism can be used, for example, for abstracting away from the details of continuous, or empirical models, such as control schedules, and PID (proportional, integral, differential) control requirements.

However, in contrast to existing formal approaches, assumptions can be stipulated in a form more compatible with the *transfer function* view of software. Hence, the pre-conditions, which describe assumptions and the pre/post-conditions, which define contracts, may refer to more than one successive instance of the variables of interest. This allows the engineer to capture information about differentials as well as their present value.

When capturing requirements using PFS, it is necessary to decompose behaviour into *state-based* (discrete) and *non-state-based* (continuous) components. The state-based requirements are described using hierarchical state-machines. The continuous components, such as those that abstractly characterise control law requirements, are expressed using PFS *contracts*, implemented by pipelines of standard control blocks (e.g. sum, integrator etc.). To make PFS accessible to control engineers and industrial software engineers, SSA extends MSS in the most "natural" way possible, for example adding further annotations to models. When specifying contracts and other assumptions, PFS advocates the use of a variety of (hierarchical) tabular forms.

In the remainder of this paper, the focus will be on state-based requirements, although other aspects of the PFS programme are briefly considered.

4.1 Essential Requirements

Often development starts by assuming an ideal environment, or context, for the control system. Real-world considerations, such as noise on signals and environmental failures, are deferred until later in the evolution of the model.

Essential requirements refer to behaviour in an ideal environment. Essential requirements are validated with respect to explicitly documented sets of assumptions upon which they rely. These assumptions often concern the embedding system. For example, the value or rate of change of a particular signal is always within a particular range. One of the contributions of PFS is to provide mechanisms for including such assumptions, e.g. by stating the possible change of external values between states in a state machine.

Later, as the ideal requirements are refined toward an actual design, assumptions may be "weakened" to accommodate real-world considerations, e.g. the rate of change assumptions may be violated when sensors fail. When this happens, new functionality must be added to the requirements to handle the additional behaviours admitted by the weakened assumptions.

This is, of course, part of standard development practice, but PFS enables this by means of refinement of assumptions. The refinement process systematically addresses the complexities of the real world, and may be strongly informed by other processes, such as the safety process. At the end of the refinement process, a comprehensive set of requirements exists that rely on a *realistic* set of environmental assumptions. Of course, this realistic set must be validated by independent means.

4.2 State-based Requirements

State-based requirements describe how a component's behaviour changes over the history of its inputs. They are represented in PFS as hierarchical state-machines akin to statecharts. Hierarchical state-machines allow the states of a machine to be organised into super-states and sub-states. In PFS, if a machine is in a super-state, it must be in one and only one of its sub-states. Hence, each transition into a super-state corresponds to a transition into one its sub-states (as characterised by an initial transition on a lower-level machine describing the super-state). An identical transition out of every sub-state can be aggregated into a single transition out of their super-state.

PFS state-machines are syntactically and semantically simpler than statecharts. However, they incorporate additional information that gives them advantages related to abstraction and validation. State-machines document the states of a component, and define, for each state, the response of a component to conditions on its inputs. These responses are characterised by *transitions*. In PFS, each transition carries a trigger condition, which must hold true over the component's input values in order for the transition to be taken. Trigger conditions for each of a state's exiting transitions must be pair-wise disjoint (non-determinism is illegal in PFS state-machines). One of the purposes of SSA is to check these "well-formedness" rules for PFS statecharts. It does this by extracting a formal model (which is intended to be invisible to the user) and carrying out analysis to check whether or not the rules have been observed.

4.3 Assumptions on Requirements

Assumptions are especially useful for abstraction and for validating requirements. For abstraction purposes, the specifier need only specify within stated assumptions. For validation, a model must be shown to respect assumptions wherever they are stated.

PFS state-machines permit several kinds of assumption annotations. Assumptions may be associated with the leaf states and compound states of the machine. In validation, all such assumptions must be shown consistent with the top-level assumptions for the whole state-machine (the assumptions on the root super-state).

An outline of the nature of assumptions is given, but it is important to describe briefly how PFS state assumptions aid abstraction. The root state of the machine carries two forms of assumption on the inputs. Firstly, the root state may carry assumptions about the valid range of inputs. Secondly, it may carry assumptions about the valid rate of change of the inputs. Note that the annotations form a *first-order* pre-condition on the inputs (involving rate of change as well as range). Each sub-state in the state-machine may also carry two forms of assumption. Firstly, it may assume something about the last set of input values (which took the machine into that particular sub-state). Secondly, it may assume something about the *next* set of input values (which will take the machine to a new state).

Ultimately, all sub-state assumptions can be proven contextually valid given the entering transitions to the sub-state, and the rate of change assumptions on the root state. Crucially, the specifier must provide exiting transitions for expected input values, and expected input values only. Whilst this is asking the designer to do a little more than just specify what the control system does, this is all information, which the designer should understand; what PFS is doing is providing a means of documenting this information and using it to check specifications.

4.4 Continuous Specifications

The continuous specifications, i.e. the classical control elements, carry much more conventional specifications of pre- and post-conditions. From a practical point of view the most significant issue is analysing feedback loops, and integrating the analysis of the continuous and discrete parts of the specification.

5. CASE STUDIES

The PFS method has been validated on a number of case studies from the aerospace domain, primarily military aircraft engine control. Case studies to date have been undertaken using "pencil and paper"; nonetheless some potentially significant issues have been found.

Part of the specification for an advanced engine demonstrator was analysed, including staged combustion. The system as specified would have worked correctly with the expected pilot input. However, unexpected pilot input (rapidly toggling a switch) could have put the system in a hazardous state due to inappropriate control over the combustion system. The specification made an implicit assumption, which admitted this unsafe behaviour. In this case making the assumption explicit showed how the software could be made more "robust" against pilot inputs, and thus safe.

A second case study involved an integrated flight and propulsion control system. The specification incorporated complex mode control logic. To ensure safe transition between modes the system initiated a built in test (BIT) on critical actuators before making mode transitions. However, the BIT functions could take a considerable amount of time to execute, and it was possible for the conditions for safe mode transition (air speed etc.) not to hold after BIT, even though they had at the beginning. The specification implicitly assumed that this was not possible, and thus admitted unsafe behaviour. In PFS this can be made explicit in terms of next state assumptions – although it requires knowledge of the physics of the aircraft flight to realise that these assumptions are invalid. Interestingly, in this case, a lab demonstrator had been operated for a considerable time without finding this potential problem, confirming the power of the formal analysis.

These and other case studies confirmed the potential of PFS in clarifying specifications and identifying problems early in the development process. However, it was clear that the techniques could not be applied to real-world problems without tool support – and that is the current focus of the project.

6. TOOL SUPPORT

The early stages of PFS were deliberately "neutral" with respect to notations and tools used for the control system specification. However, for tool development it was necessary to choose a "host" tool as the basis for the PFS analyses. Due to its widespread use, the choice was made to build support for PFS on MSS, producing a tool known as SSA.

SSA is integrated with MSS so that it appears as an extension to Simulink/Stateflow. Where practical, it extends existing MSS dialogues to provide extra annotations; in other cases, new dialogues and facilities have had to be added. It customises Simulink/Stateflow to allow PFS annotations to be recorded on the model. Secondly, it provides automated checks to ensure that the annotations are mutually consistent in all circumstances, not merely those explored by simulation. The tool supports textual and tabular representations of assumption annotations. It is capable of organising constants and common definitions into lexicons (project dictionaries of definitions) and performing well-formedness checks on the annotated model.

6.1 Annotating Stateflow Charts

The annotations on a statechart are conditions on the values of the input and output signals to the chart. These signals correspond to the input and output ports of the encapsulating Simulink block. Annotations associated with a state include: *last* value assumption, *initial* value assumption, *next* value assumption, *state preservation conditions*, and *output definitions*.

The properties of a statechart that can be analysed given the necessary annotations include: establishing a state's *next* value assumption given the combination of its *last* value assumption and its parent state's *next* value assumption; establishing a state's *last* value assumption given the combination of a transition trigger and its source state's *last* and *next* assumptions; establishing a state's initial annotation given the combination of a transition trigger and its source state's *last* and *next* annotations, and establishing that a state's *state preservation conditions* and exiting transitions cover all situations permitted by the combination of the state's *last* and *next* annotations (transition completeness).

6.2 *Annotating Simulink Subsystems*

The behaviour of a simulink subsystem is characterised by a *contract*. A contract is an annotation that describes what a sub-system is intended to achieve. It emphasises what will happen, rather than how it will be achieved. Contracts are usually expressed in terms of pre- and post-conditions on the interface of a component. Assumptions contained in a Simulink contract constrain the values of the inputs to the subsystem. The input/output relation describes how the outputs of the subsystem are related to its inputs. PFS contracts may refer to several successive instances of input and output variables, allowing them to represent *loose* transfer functions (e.g. a difference equation) rather than merely relating single input and output instances, i.e. to fit into transfer functions.

The entire system's *contract* captures the assumptions on which the system depends, and the safety properties it must ensure. Annotations may also be placed on signals within a Simulink subsystem. The annotations that can be associated with a signal include an *invariant* and an *initial value*. SSA can perform analysis of the specification, but the validity of environmental assumptions, e.g. about initial states must be manually checked – for example initial values may be different if a system is initialised in flight instead of on the ground.

The properties of Simulink subsystems that can be analysed, given the necessary annotations, include: contract feasibility (that the input/output relation is well-defined everywhere within the assumptions); contract satisfiability (that the behaviour of the subsystem model conforms to the contract and that the assumptions of the contract guarantee any embedded assumptions) and initial satisfies invariant (that a signal's initial value is consistent with the invariant of that signal).

6.3 *Implementation of PFS and SSA*

In order to support PFS and to realise the SSA functionality some additional blocks have been added to the MSS libraries. These give template blocks for the continuous and discrete elements of the PFS method, and a signal annotation block, which allow initial state and invariant to be specified (the standard MSS signal annotations cannot readily be extended this way). Figure 1 shows an example of a model created using the SSA Toolbox library.

The SSA Contract block is used for annotating and analysing Simulink subsystems. The SSA Well-formedness block checks the subsystem in which it is placed. The Lexicon block allows definitions to be prepared as tabular formulae. The SSA State machine block is for annotating and analysing Stateflow diagrams. The Signal block is for annotating Simulink signals.

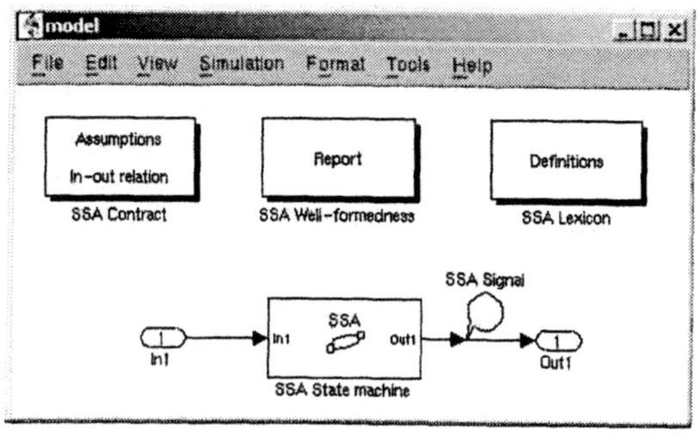

Fig. 1. The SSA Example Model

These "stylised" forms of specification are intended to lead engineers to introduce the necessary contracts and assumptions into the models, as they are constructed.

The analysis is based on existing tools, most notably CADiZ (Toyn and McDermid, 1995) which supports specification and proof in the Z notation. SSA works by translating the control system specification, together with its annotations into Z. It then uses proof techniques to show the healthiness of specifications, etc. (technically these properties are encoded as proof obligations). Generally, doing such proofs is a complex manual task. However, to make the method practical (the P in PFS) it was important to make the analyses automatic. Because of the nature of PFS and the structure of MSS specifications the proof obligations are very regular and it has been possible to develop tactics for automating the majority of the proofs.

7. STATUS OF THE WORK AND FUTURE PLANS

PFS has developed a relatively mature theory for specifying and analysing "synchronous" systems, for those where it can be assumed that all parts of the system operate to a common clock. Some specific technical issues remain, such as arriving at an optimal approach for formal analysis of continuous models with feedback (technically this involves finding the best approach to doing weakest-precondition analysis in the presence of loops).

The emphasis of our theoretical work is now more on issues of asynchronous systems and models, to enable more effective analysis of distributed control systems (Galloway and Blow, 2001). This poses a number of significant challenges, but our work will build on the existing PFS theories.

Several references to safety properties of systems have been made. One of the aims of the associated MATISSE project is to integrate formal methods and safety analysis. In particular, this would mean being able to derive safety requirements, e.g. through fault trees, in a form that they could be turned into PFS post-conditions and proven using SSA. In many cases the derived safety requirements are constraints over sequences of values, not on single outputs. These can be represented using PFS assumptions over previous and next states – although there may be cases where it is preferable to use a state machine to show safe behaviour. At present SSA has limited capability for reasoning about nth-order formulae, so further work is required to support this form of analysis.

Previously, tools have been developed for automatically generating safety analyses from MSS models (Papadopoulos and McDermid, 1999). A longer-term aim is to link these techniques with the PFS notations so that the PFS-enhanced MSS models can be used as a basis for analysis of safety as well as the soundness of the specifications. Also, there is value in using models as a basis for test generation (Burton, 2002), so our long-term aim is to build a coherent method and toolset, around the PFS-MSS models. The above are long-term plans. The next phase of PFS work will involve trials of the tool on industrial scale problems. Hopefully, evidence of the effectiveness of the approach on such problems will be published.

8. CONCLUSIONS

The PFS method has been developed over a number of years, and shown to assist in finding potentially serious errors in control law specifications. A 'proof of concept' support tool for PFS, known as SSA, has been produced. It enables recording and revision of annotations on MSS models, and supports the generation and discharge of proof obligations from the annotated model. SSA is now capable of analysing small specifications, but further work is needed to enable the tool to be used on industrial scale applications.

The automated analysis provided by SSA checks the internal consistency of the specification, and that it conforms to certain assumptions about the behaviour of the embedding system, such as safety properties. PFS and SSA are complementary to other formal analysis methods and tools such as ClawZ from QinetiQ (Arthan, Caseley, et al., 2000) Whilst ClawZ is primarily concerned with ensuring that programs conform to their specifications, PFS is concerned with ensuring that the specifications are correct with respect to the engineers' requirements. Indeed some of SSA's analyses borrow from the ClawZ formalisation, so the two toolsets should be capable of being used together.

In the long run our hope is that a contribution can be made to the production of a comprehensive support environment for the development of safety critical control systems. Developing safety critical embedded control systems and software is a difficult task. It is unlikely that any one-research group can address all the issues necessary to provide effective methods and toolsets. Hence, it is our believe that it is essential to build on best-in-class tools, such as MSS, and other analysis tools such as ClawZ; only in this way will control engineers be given effective tools, which support both control and software engineering.

9. ACKNOWLEDGEMENTS

Acknowledgements for this work are due to the MoD under contract number FBG/01046 and the EPSRC-funded MATISSE project Grant No. GR/R/70590/01 for the support they have provided.

REFERENCES

Arthan, R., P. Caseley and et al. (2000). ClawZ: Control Laws in Z. *In proc. Of ICFEM 2000, IEEE Computer Society Press.*

Burton, S. (2002). Automated Generation of High Integrity Test Suites from Graphical Specifications, PhD Thesis, *University of York.*

DS 00-55 (1997). UK Ministry of Defence, Defence Standard 00-55. *The Procurement of Safety Critical Software in Defence Equipment.*

Galloway, A. J. and J. Blow (2001). Multi-layered Domain Specific Formal Methods. *Joint Workshop on Formal Specification of Computer Based Systems (FSCBS).*

Galloway, A., T. Cockram and J. McDermid (1998). Experiences with the application of discrete formal methods to the development of engine control software. *Proceedings of Distributed Computer Control Systems (DCCS), IFAC.*

McDermid, J. A., A. J Galloway et al (1998). Towards Industrially Relevant Formal Methods: Three small steps, one giant leap. *Proceedings of The International Conference on Formal Engineering Methods (ICFEM), IEEE Press.*

Papadopoulos, Y. and J. A. McDermid (1999). A New Method for Safety Analysis and the Mechanical Synthesis of Fault Trees in Complex Systems. *The 12th International Conference on Software and Systems Engineering and their Applications, Paris.*

Toyn, I. and J. A McDermid (1995). CADiZ: An architecture for Z tools and its implementation. *Software Practice & Experience* **25 (3)** 305-330.

Copyright © IFAC Information Control Problems in
Manufacturing, Salvador, Brazil, 2004

ELSEVIER
IFAC
PUBLICATIONS
www.elsevier.com/locate/ifac

OPEN OBJECT-ORIENTED MODELLING AND VALIDATION FRAMEWORK FOR MODULAR INDUSTRIAL AUTOMATION SYSTEMS

Valeriy Vyatkin, Hans-Michael Hanisch, Gustavo Bouzon

Dept. of Engineering Sciences,
Martin Luther University Halle–Wittenberg,
06099, Halle, Germany

Dept. of Automation and Systems, CTC
CP 476, Federal University of Santa Catarina
88040-900, Florianópolis, SC, Brazil

Valeriy.Vyatkin@iw.uni-halle.de
Hans-Michael.Hanisch@iw.uni-halle.de
gbouzon@das.ufsc.br

Abstract: This paper introduces a framework for formal modelling and validation of automation systems intended to be used by control engineers. The framework is based on a graphical, modular, and typed formalism of Net Condition/Event Systems.
This allows for modelling of realistic hierarchically organized industrial automation systems in a closed loop. The framework consists of methodologies and tools which enable formal analysis of automation systems.
The framework will be used to improve safety, reliability and robustness of automation systems predicting potential faults and deadlocks. *Copyright © 2004 IFAC*

Keywords: Verification, PLC modelling, manufacturing systems, discrete event systems.

1 INTRODUCTION

Modern production systems need to be more flexible and re-configurable. For this reason they are built from standardized processing modules. Their software is also organized in a modular form and is executed on distributed control devices.
When new configurations of production systems are formed from the modular components, the testing becomes a bottleneck for quick commissioning. Formal validation can reduce the time-consuming testing and commissioning phases of system's development and deployment. As the functionality of such systems is determined by cooperation of entities of heterogeneous domains, e.g. mechanical, electric, automation hardware and software, the validation has to take into account the relevant properties from all these domains.
Formal modelling of automation systems proved to be helpful for validation of automation systems by simulation or by formal verification of static and dynamic properties. In automation systems the software represents a variable part, while the models of equipment can be reused through the engineering cycle. Once developed by a machine vendor, the models may follow the equipment, enabling the machine users (e.g. system integrators) to validate new configurations of the machines re-using the models of their components.
This vision, however, requires a more systematic approach to the modelling, than that can be seen now. Models in most of the formalisms such as Petri nets or finite automata lack integrating capabilities: while they may cope well with the modelling of a particular process, building the overall model of a system comprising several processes is difficult.
An opposite example make the modelling techniques based on the Unified Modelling Language (UML). The UML is getting increasingly popular also in automation for its ability to describe systems in object-oriented form. However, the UML lacks a formal background and can hardly be used for deep analysis of the systems.
This paper tries to sketch another approach for systematic modelling of systems by means of a modular modelling formalism.
The paper is organized as follows: Section 2 discusses some extensions to the formalism. Section 3 introduces the approach to modelling and describes the framework of tools supporting it. Section 4 elaborates more on the application scenarios of modelling for better automation systems. Section 5 considers an example of a machine modelling, and Section 6 provides some considerations on the validation. The paper is concluded with discussion of future research.

2 STRUCTURAL EXTENSIONS AND LIMITATIONS

2.1 *Modelling formalism*

In this work we use the formalisms of Net Condition/Event Systems (NCES) (*H.-M. Hanisch and A. Lüder, 1999*). The modular capabilities have encouraged further development of a systematic approach to modelling of industrial systems. The formalism is:

Modular, i.e. it provides encapsulation of place/transition models into modules, connected to each other by condition and event arcs.

Graphical, that simplifies understanding of the model's semantic and facilitates application by engineers.

Distributed state, that helps to cope with the complexity of model-checking, especially when decentralized systems are modelled.

Discrete time, that allows to add new, time dimension to the discrete modelling.

Supported by the model-checking tool SESA (*Starke, Roch et al.*), NCES were applied in a number of studies on formal validation of automation systems (e.g. *H.-M. Hanisch, T. Pannier et al, 2000*).

However, the borders of the NCES formalism have been reached. In particular, questions of creating nested modular models (i.e. encapsulation of a network of modules inside a module) theoretically consideration of which started in (*Thieme, 2002*), needs to be continued and reflected in implementations.

The automation becomes a field of distributed architectures (e.g. IEC61499) where automation systems are represented as networks of function blocks. Evident similarities between the NCES modelling concept and the function block architectures have motivated further development (*Vyatkin V., Hanisch H.-M. et al. 2003*) of the NCES formalism that targets:

- Definition of model types to simplify encapsulation and reuse of the models;
- Less restricted rules for interconnections of modules (multiple input and output links are allowed).

The formalism is supported by the corresponding tools and methodologies.

2.2 *Model type definition*

In the formalism discussed in this paper a basic unit of model is called module. A module is defined by its **interface** and **content**. The interface contains a type name, names of event inputs and outputs, and of condition inputs and outputs. The content can be either a NCES, i.e. places, transitions and arcs as described in the previous section, or a network of model instances, i.e. of other modules, interconnected via event and condition arcs.

A NCES has to be encapsulated in a module. The corresponding model type is called **basic**. A **basic** model type is defined by a graphical representation of a NCES module, i.e. by definition of its interface (event and condition inputs and outputs), and of its place/transition model. The module identifier serves as the type identifier. The values of inputs influence the model's dynamic as they are connected to the model's transitions by condition and event arcs. The model's states and transitions may be reflected at the outputs as they are connected to places/transitions of the model by condition and event arcs.

A **composite** model type includes a description of block's interface, instances of constituent modules that are instances of other model types (basic or composite), and connection arcs.

The definition above makes the NCES "compatible" with other kinds of object-oriented modelling, for example using Unified Modelling Language (UML). Several works appeared recently on application of object-oriented modelling to machines and production systems (e.g. *M. Bonfè, C. Fantuzzi, 2003*). The UML class diagrams are used in these works to represent the structure of production objects as composed from more elementary ones, that also paves the way to hierarchical models. However, application of UML for formal analysis is difficult as it lacks formality. Thus, the approach presented in this paper bridges the gap between the expression power of UML and the formal semantics of NCES.

3 MODELLING APPROACH AND TOOL FRAMEWORK

3.1 *Closed-loop modelling*

The control systems is considered as composed of two independent components: object and controller, connected in a closed loop by control signals and process data. Modelling according to this view requires to model uncontrolled reactive behaviour of objects.

It is worth mentioning that the closed-loop approach to the modelling enables expression of the specifications directly in terms of the machine behaviour (not only I/Os of the controller).

3.2 *Integrated tools for model creation, editing and analysis*

The modelling approach explained in this paper is supported by a number of software tools:

The graphical editor provides full graphical authoring and editing of the models. The editor uses an open XML-based data format for basic and composite NCES models. The data format of composite model blocks intentionally was made identical with that of IEC61499 function blocks, supported by tool (*FBDK*).

The integrated environment for Model Assembly (iMA) inputs the model type files given in XML and is capable of:

1) Assembling a composite, hierarchically organized model from modules contained in different libraries. The component model types are instantiated into NCES modules.
2) Translating the model into a "flat" NCES with the through numbering of places and transitions. The inter-module connections are converted into event and condition arcs between places and transitions. Thus the module boundaries are removed and the model-checking tools can be

applied. In particular, the translator generates files in the input format of SESA model checker.

4 MODELLING OF AUTOMATED PLANTS

Benefits of the typed modelling are well visible in the following example of object modelling. The automated lifter (product of Flexlink Automation Oy., FINLAND) as shown in **Figure 1** is used in production of electronic components. The lifter can be controlled by two different controllers: an OMRON PLC programmed in ladder logic and Nematron SoftPLC programmed in Visual Flow Chart language. Though both controllers achieve similar control goals, the internal logic of control algorithms and even the logic of program execution are completely different (cyclically scanned vs. sequential). However, both controllers eventually deal with the same object.

When the closed-loop plant-controller systems are validated, the model of the lifter can be reused over and over again in connection with models of controllers of different types.

The lifter consists of three transporters, one of which is mounted on a vertically moving platform driven by a step motor as schematically represented in **Figure 2**.

Figure 1. The lifter.

The structure of the model type "Lifter" is defined by means of UML class diagrams as shown in Figure 3.

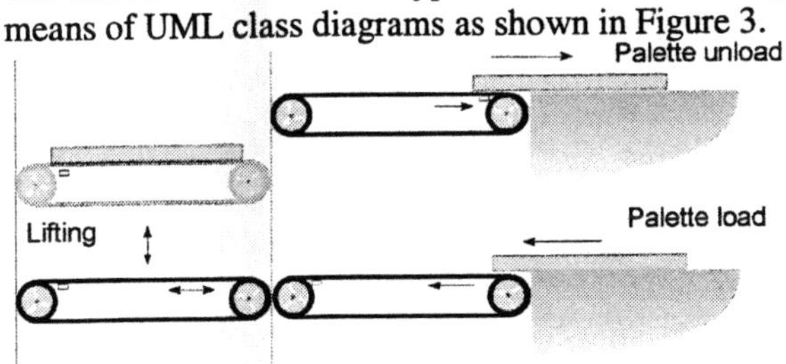

Figure 2. Structure and operation sequence of the lifter.

The definition literally says that the object "Lifter" consists of 4 elements. The loading and unloading one-directional conveyors are identical but turned in opposite directions. The corresponding models are of type **Conveyor**. The vertically moving platform (an object of type **StepMotor**) has a moving belt that moves pallets in both directions (modelled as an object of type **Conveyor2D**).

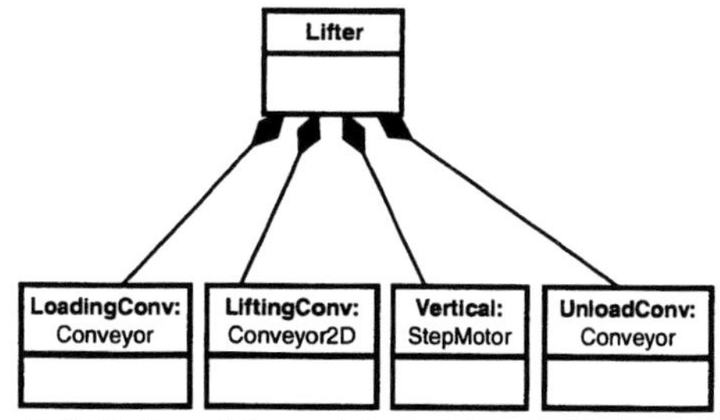

Figure 3. Definition of the model type (class) "Lifter" by means of UML class diagrams.

Note that the model in **Figure 3** does not define an interface of the lifter, nor dependencies between its constituent parts. These dependencies can be reflected in modular models by event and condition connections between the corresponding modules as exemplified in **Figure 4**.

Let us consider the model of a Conveyor. In our example two different types of conveyors are used – capable to move only in one direction, and those moving in both directions. The model of a more complex conveyor can be created based on the simple model using the mechanism of inheritance.

The interface of the model type "**Conveyor**" can be seen in **Figure 4**. The model itself can be conceptually divided onto three elements: Status, Position, and Sensor as shown in the class diagram in **Figure 5**, left. The Status element of type **MovingStatus** models the behaviour of the motor that drives the conveyor and converts the logic control signals into one of the states "Moving" or "Standing still" (that corresponds to the one-directional conveyor). Input "PRESENT" indicates if a pallet is present, and input "FORCED" is used to indicate influence of a neighbour belt on the movement of the pallet.

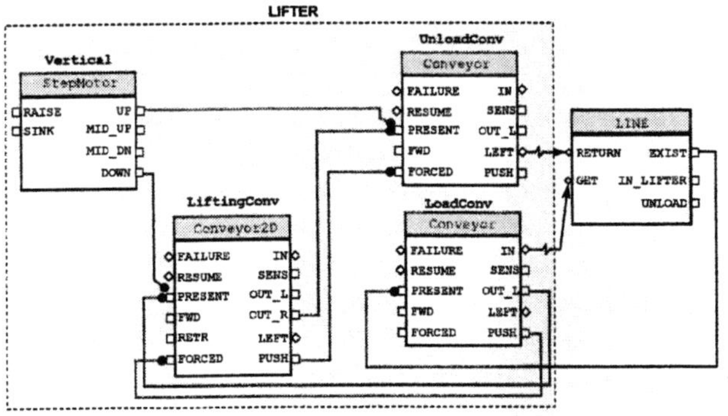

Figure 4. A model of Lifter represented as a network of NCES modules.

The output condition FW_ST is used by the model of belt position.

The structure of the model of the bi-directional conveyor is identical to that of the uni-directional. The difference is in the model **Status** that has type **MovingStatus2D** that inherits the interface properties of the one-directional **MovingStatus** and extends them with one more input and output for the retracted movement. This is shown in **Figure 5** (right). All transporters are equipped with a single position sensor indicating presence of the pallet (fully loaded on the conveyor).

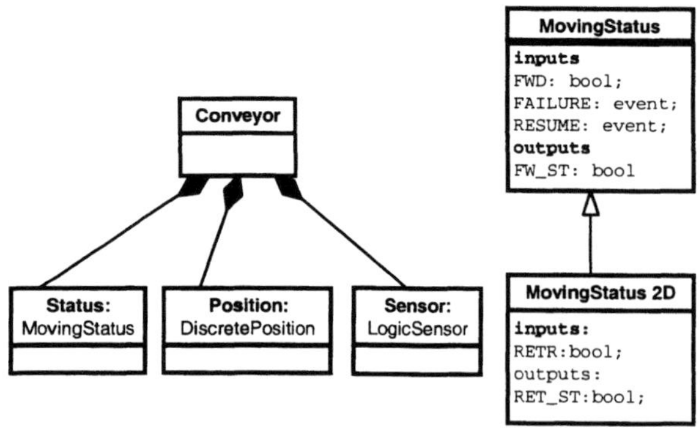

Figure 5. Model type definition of the conveyor and inheritance of the MovingStatus model types.

The data/event flow connections between the sub-models constituting the model of the conveyor are represented in **Figure 6**.

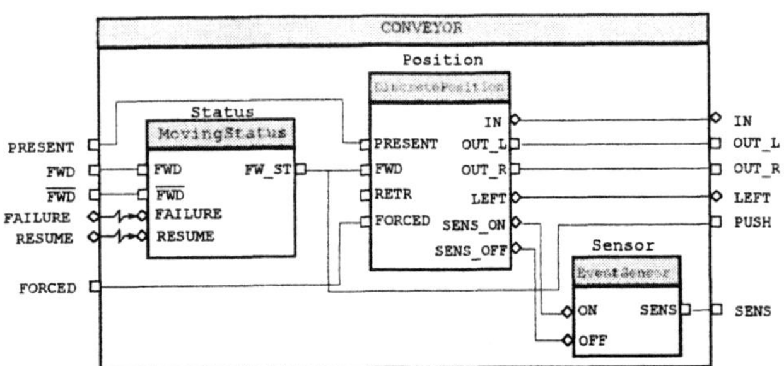

Figure 6. Modular view of the model of conveyor.

The basic models can be described further in form of NCES modules. **Figure 7,a** shows an implementation of the **MovingStatus** in NCES. The model receives the control signal FWD and transforms it into the state of the belt: place p_2 corresponds to the state "belt stands still", place p_1 – belt moves and p_3 to the state indicating a failure. The belt moves when the control signal FWD is ON, and stops when the signal goes OFF (i.e. negation of the signal FWD goes on).

An occurrence of a failure is indicated by an external event that may come from the corresponding model. For example, it can be a stochastic model of failures. Note that the model is sensitive to failures only when the belt moves, i.e. when the place p_1 is marked. It is assumed that the failure can be fixed by an external interaction indicated by event input RESUME.

The model **MovingStatus2D** for the bi-directional moving belt is shown **Figure 7**. It models one additional state for moving backwards, and correspondingly more transitions between all possible states.

The position of a pallet on the belt can be modelled with different precision. A qualitative model in **Figure 8** distinguishes only 3 states of a pallet on the belt: no pallet, pallet on the belt with its front edge between the belt's ends, and pallet's front edge is beyond the right end of the belt.

A more precise modelling of the position can be done using the arc-timed version of NCES. Let us assume that the belt is 3 units long and the pallet is two units long as shown in **Figure 9**. The speed of the belt is 1 unit of length per second. Then it will take 3 seconds for a pallet to reach the right end of the belt and 2 more seconds to leave the belt completely.

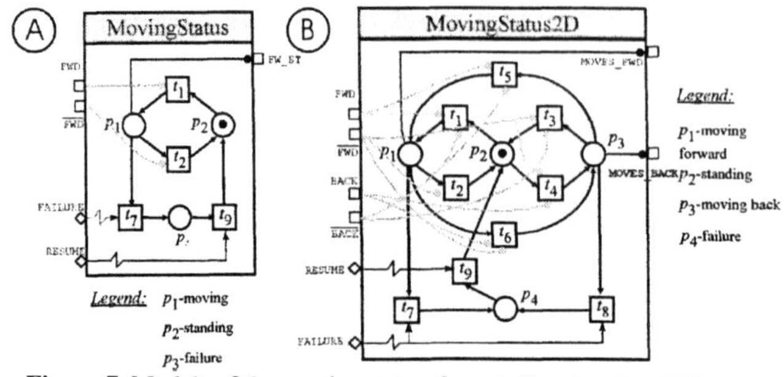

Figure 7. Models of the moving status for uni-directional and bi-directional belts.

Place p_1 corresponds to the state "No pallet". When a pallet appears (input condition "Present") and the state of the moving belt is "Moving forward" (indicated by the input condition FWD) then the transition t_1 occurs and the token goes to place p_2.

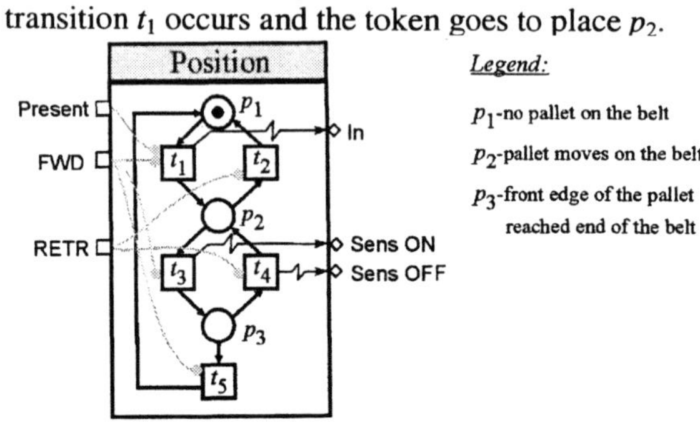

Figure 8. A qualitative non-timed model of the pallet's position.

This place indicates the state "Front edge of the pallet is in the interval 1 of the conveyor". Another reason to transfer to this state is the presence of the input condition "Forced". This condition indicates that the pallet is pushed onto the belt by some external force that maybe another moving belt positioned backwards to this one. This option is modelled by transition t_{12}. In general, moving in this case is slower than if driven by the own motor of the belt. The presented model, however, does not cover with enough precision the case when both forces are present simultaneously. Note that the transition from p_1 to p_2 (either via t_1 or t_{12}) is a qualitative one and does not take time (more precisely has zero delay).

The places p_2-p_4 correspond to the location of the pallet (again front edge!) in the intervals 1-3 respectively. A transition from interval i to interval i+1 occurs in either case "FWD" and "Present" or "Forced" and "Present".

The latter, however, works only till less than the half of the pallet is on the belt – beyond this point friction would not let the pallet move driven only by the external force. The moving to the next interval takes 1000 ms if driven by the own motor of the belt or twice as long under the external force. The backward moving from interval i+1 to interval i occurs if the combination of input conditions "RETR" and "Present" are true. It also takes 1000ms under assumption that the speed of the moving belt in both directions is the same.

Arriving of the pallet to the 3rd interval is indicated by the sensor. This is modelled by two event outputs "Sens ON" and "Sens OFF" associated with firing of transitions t_8 and t_9 or t_{11}, respectively. The sensor goes off when either the front edge of the pallet moves backward to the interval 2, or when the back edge of the pallet leaves the belt in forward direction (and the pallet completely disappears from the belt).

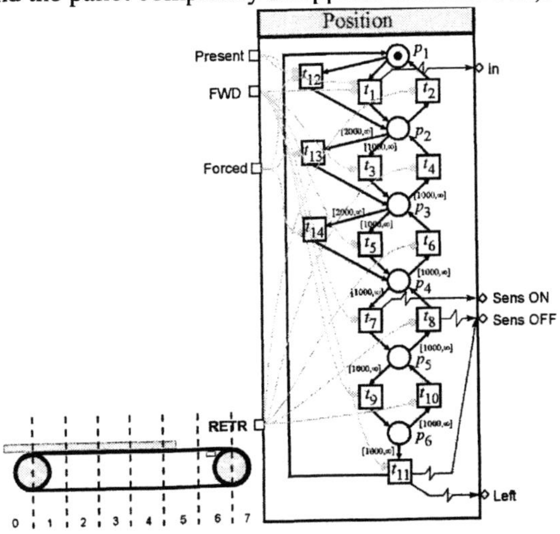

Figure 9. Model of the position of the pallet on the conveyor discretized on 3 intervals.

This model can represent the state of the pallet on the belt with better precision. However, it has its own limitations and drawbacks. In particular, let us consider how the alternative kinds of movement are modelled. A place indicating a position (e.g. p3 indicating interval 2) has several outgoing arcs (p3-t4, p3-t5 and p3-t14) marked with non zero time delays ([1000,∞], [1000,∞], [2000,∞]). Transitions that are targets of these arcs have condition input signals that represent alternative control signals (RETR, FWD, Forced). Either of transitions will fire when it is enabled by marking, conditions and time. It is important that all these conditions are mutually orthogonal (alternative) and they never change within the minimum delay of the place (1000ms in our case). Otherwise the model will not work as intended.

5 VALIDATION

The validation of automation systems modelled by NCES can be performed by simulation and formal verification via model checking.

Simulation usually follows a limited number of scenarios in the system's behaviour. In contrast, the model-checking studies multiple scenarios caused for example by some unpredictable factors, such as variable durations of some operations, communication delays, malfunctions, etc.

In the described framework, the model-checking can be conducted either by means of SESA, or by an embedded model-checker. In the latter case, the results of the model-checking, such as a reachability space (full, or generated until an example/counterexample is found) can be visualized as state/time diagrams of relevant values (e.g. represented as marking of certain places, or firing of certain transitions).

The verification consists in proving specifications with respect to the dynamic behaviour of the model. The specifications can be given either in form of second order predicates, or in form of temporal logic expressions. Terms of the expressions can be formed from inputs, outputs and internal variables of the controller or variables of the model of plant. The latter have to be eventually expressed via marking of places in the model.

More examples of formal verification can be found in (*Vyatkin V., Hanisch H.-M, 2003*). In the example mentioned in the previous section of particular interest were:

1. Robustness of the system in case of malfunctions of some sensors;
2. The control programs in VFL are branching. Formal verification helps to prove that the response time is never exceeded in any feasible IO combination.
3. Quality assurance: the lifter must never allow the situations when the pallet leans or jumps. It may be caused by inexact synchronization of conveyors' levels which may be a result of synchronization of control programs;

The overall model after assembly encountered 571 places and 828 transitions. However, the model-checking of a normal behaviour (without modelling malfunctions in sensors) resulted in the reachability space not exceeding 40000 states. This result reflects the efficiency of distributed state modelling with NCES.

6 GRAPHICAL SPECIFICATION

Although specification of properties for model-checking in temporal logic may be easily created by experts on verification, engineers would benefit of having user-friendly means of specifying the desired behaviour of a module. Inspired by timing diagrams, well-known in the hardware branch, a graphical language for describing the dependency of interface signal changes has been proposed (Vyatkin2001, Bouzon 2002).

A graphical description of a specification is illustrated in **Figure 10**. Signal changes at the beginning or ending of the diagram are implicitly simultaneous. Nevertheless, no further ordering is determined by the horizontal position of signal changes – therefore, a timing diagram usually specifies a partial ordering among signal changes.

The semantic associated to the diagram is as follows: when the set of levels specified at the beginning of the diagram is achieved, it is required that the sequence of changes at the signals does not violate the partial ordering specified at the diagram, until a final state is reached.

Translation and verification approaches differ slightly depending on whether the verified module has inputs. When verifying NCES (autonomous) modules, each signal specification is translated into a NCES supervisor module comprising two basic submodules: an **event generator** creates sequences of transitions, one for each change of level specified for the signal.

Each transition stimulates, through an event arc, the corresponding event input of a **signal generator**, which causes the output of the signal generator to recreate the signal according to the input stimulated. Ordering operators are translated into special places and transitions that create interdependency of event generators.

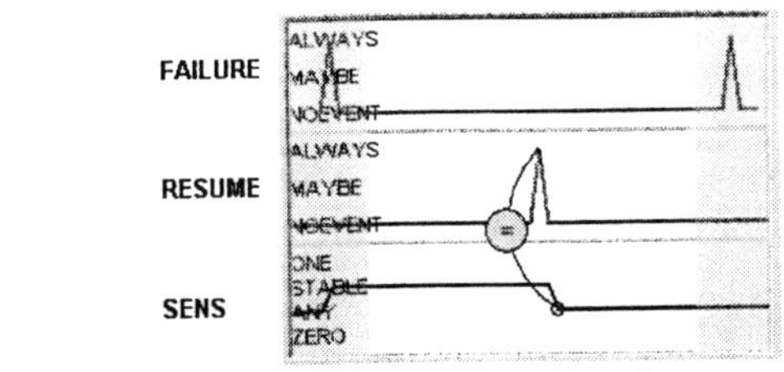

Figure 10. Specification including two event inputs, one condition output and a simultaneity operator.

The verified module is then connected through event arcs to the event generators of the corresponding signals, in such a way that every change of signal in the first is reported to the latter. Along with the translation of the specification into NCES modules, a set of automatically generated temporal-logic statements is created. The composite module is then model-checked against these statements to verify if each transition at the supervisor always fires whenever the corresponding transition at the verified module is fired.

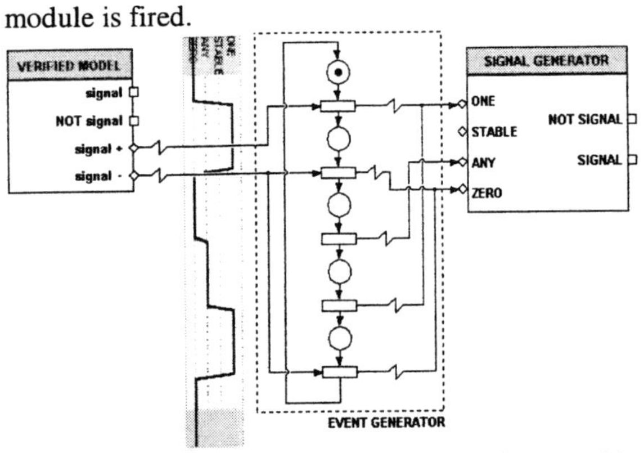

Figure 11. Translation of a single specification for a condition output, and linking to the verified model.

The graphical specification also provides automatic test possibilities for input/output behaviour or non-autonomous NCES modules. In this case, the NCES supervisor modules that describe input signals are used for generating the specified sequences of input signal changes, while the output signals are again verified as described before.

7 CONCLUSION

This paper provided a short overview of an extended modular modelling paradigm which combines the ideas of object-oriented typed modelling of the mainstream UML with benefits of modular place-transition nets. Though the UML support is not integrated to the tool framework supporting NCES so far, however, the similarities with IEC61499 function blocks allow for using the CORFU environment (*Thramboulidis, 2002*) in connection with Rational Rose in order to define the model's structure and interfaces and convert them to the form of interconnected NCES modules. This approach allows for taking advantage of formal verification methodologies and tools with NCES.
Generation of NCES models from UML state charts will be a subject of future research.

8 ACKNOWLEDGEMENTS

The work was supported by the Deutsche Forschungs gemeinschaft under reference Ha 1886/12-2.
The closed-loop model of Lifter was developed following the methodology presented in this paper (Section 4) in MOVIDA-1, a project funded by the National Technology Agency in Finland -TEKES. (A. Lobov, et al.).
The authors thank Reijo Tuokko, Jose L Martinez Lastra and Andrei Lobov from Tampere University of Technology (TUT), as well as all MOVIDA-1 partners, for providing the opportunity to apply the object-oriented version of NCES to real automated equipment and for fruitful discussions on the presented modelling approach.

9 REFERENCES

H.-M. Hanisch and A. Lüder: *Modular Modelling of Closed-Loop Systems*, Colloquium on Petri Net Technologies for Modelling Communication Based Systems, Berlin, Germany, October 21-22, 1999, Proceedings, pp. 103-126

P. Starke, S. Roch, K. Schmidt, H.-M. Hanisch, A. Lüder: Analysing signal-event systems, Technical report, *Humboldt Universitat zu Berlin*, Institut für Informatik, http://www.informatik.hu-berlin.de/lehrstuehle/automaten/tools/, July, 1999

H.-M. Hanisch, T. Pannier, D. Peter, S. Roch, and P. Starke: *Modelling and verification of a modular lever crossing controller design*, Automatisierungstechnik, 48, 2000.

J. Thieme: *Symbolische Erreichbarskeitanalyse und automatische Implementierung struktuirter, zeitbewerter Steuerungsmodelle*, Dissertation zur Erlagung des Grades Dr.-Ing., Berlin: Logos Verl., 2002

IEC 61499 - Function Blocks for Industrial Process Measurement and Control Systems. Publicly Available Specification, International Electrotechnical Commission, Tech. Comm. 65, Working group 6, Geneva, 1998.

Vyatkin V., Hanisch H.-M., Pfeiffer T., "*Modular typed formalism for systematic modelling of automation systems*", 1st IEEE Conference on Industrial Informatics (INDIN'03), Proceedings, Banff, Canada, August 2003

Vyatkin V., Hanisch H.-M. *Verification of Distributed Control Systems in Intelligent Manufacturing,* Journal of Intelligent Manufacturing, special issue on Internet Based Modelling in Intelligent Manufacturing, *vol.14, N.1, 2003, pp.123-136*

Vyatkin V.: *Intelligent Mechatronic Components: Control System Engineering using an Open Distributed Architecture*, IEEE Conference on Emerging Technologies in Factory Automation (ETFA'03), Proceedings, Lisbon, September, 2003

M. Bonfè and C. Fantuzzi: *Design and Verification of Industrial Logic Controllers with UML and Statecharts*, submitted to the IEEE Conference on Control Application 2003, June 23-35, Istanbul, Turkey

FBDK - Function Block Development Kit at www.holobloc.org

K. Thramboulidis: *Development of Distributed Industrial Control Applications: The CORFU Framework*, 4th IEEE International Workshop on Factory Communication Systems, August 2002, Vasteras, Sweden

Vyatkin, Valeriy and Hanisch, H.-M. "Application of Visual Specifications for Verification of Distributed Controllers". Proceedings of the 2001 IEEE Systems, Man, and Cybernetic Conference.

G. Bouzon: *Development of a visual specification language for verification of Distributed Controllers*, Final Paper (Bachelor Degree), Dept. of Automation and Systems, Federal University of Santa Catarina, Florianópolis, May 2002 (in portuguese)

A. Lobov, J. L. Martinez Lastra, R. Tuokko, V. Vyatkin: *Methodology for Modelling Visual Flowchart Control Programs using Net Condition/Event Systems Formalism in Distributed Environments*, IEEE Conference on Emerging Technologies in Factory Automation (ETFA'03), Proceedings, Lisbon, September, 2000

Copyright © IFAC Information Control Problems in
Manufacturing, Salvador, Brazil, 2004

ELSEVIER
IFAC
PUBLICATIONS
www.elsevier.com/locate/ifac

CONTROLLER DEPENDABILITY ANALYSIS BY PROBABILISTIC MODEL CHECKING

Marta Kwiatkowska, Gethin Norman and David Parker

School of Computer Science, University of Birmingham, Birmingham, B15 2TT, United Kingdom

Abstract: We demonstrate how probabilistic model checking, a formal verification method for the analysis of systems which exhibit stochastic behaviour, can be applied to the study of dependability properties of software-based control systems. We provide an overview of these techniques and of the probabilistic model checking tool PRISM, illustrating the usefulness of the approach through a small case study. By using existing formalisms and tool support, we show how it is possible to construct large and complex Markov models from an intuitive high-level description. Furthermore, we are able to take advantage of the efficient implementation techniques which have been developed for these tools.
Copyright © 2004 IFAC
Keywords: Formal verification, performance analysis, Markov models

1. INTRODUCTION

Industrial use of software-based control systems is now increasingly widespread. Programmable controllers can be found in domains as diverse as the manufacturing, transport, energy and power industries. In many of these cases, the safety and reliability of the system is a crucial issue. This fact is evidenced, for example, by the existence, and indeed prominence, of the IEC 61508 and ANSI/ISA S84 standards, which include strict guidelines as to the functional safety of such systems.

A key feature of these standards is the definition of Safety Integrity Levels (SILs). To adhere to a particular SIL, it is necessary to accurately quantify the probability or rate of all safety-related faults which can occur in the system under study. Two common techniques for performing such dependability analysis are reliability block diagrams and fault tree analysis. Both, however, are relatively simplistic. For example, it is often necessary to assume that the probability of a certain fault arising is independent of the occurrence of other failures in the system, whereas in practice, this may be unrealistic and can lead to inaccuracies in the analysis.

A more sophisticated approach which resolves this problem is Markov modelling, where a more accurate, state-based model of the system is derived and analysed. This does, however, have two disadvantages. Firstly, it is more complex for the end-user to implement. Secondly, the size of model required to represent a system accurately has a tendency to increase exponentially. This is a phenomenon often known as the state space explosion problem. Not only does it make the process more expensive in terms of the time, computing power and memory required, it can make the reliability analysis infeasible.

In this paper, we show how formal verification methods can be applied to the problem of analysing the dependability of controller-based systems. In particular, we focus on probabilis-

[1] Partially supported by EPSRC grants GR/S46727 and GR/S11107

tic model checking, an automatic technique for checking whether or not probabilistic models satisfy certain specifications. We show how existing formalisms and tools can be used to specify and analyse these Markov models. In addition, we can then make use of the efficient implementation techniques which have been integrated into these tools in an attempt to curb the effects of state space explosion.

2. PROBABILISTIC MODEL CHECKING

Model checking is a successful and well established technique for formally verifying the correctness of finite-state systems. In recent years, such methods have become increasingly prevalent in industry. Model checking involves the construction of a formal model of the real-life system which is to be verified. This is usually a labelled state transition system, which represents all the possible configurations which the system can be in and all the transitions which can occur between them. The properties of the system to be verified are then also formally specified, usually as formulas of a temporal logic, and passed to a model checker, which automatically determines whether or not each property is satisfied via a systematic exploration of the model.

An extension of this approach which has seen a significant amount of development of late is *probabilistic model checking*, a technique which permits automatic formal verification of systems which exhibit stochastic behaviour. Potential candidates for such analysis include randomised algorithms, which use probabilistic choices or electronic coin flipping, and unreliable or unpredictable processes, such as fault-tolerant systems or communication networks.

Probabilistic model checking is again based on the construction and analysis of a formal model of the system. In this case, the model is enriched with probabilistic information, typically by labelling each transition of the model with information about the likelihood that it will occur. The type of model which we shall focus upon in this paper is *continuous-time Markov chains* (CTMCs). A CTMC comprises a set of states S and a transition rate matrix $\mathbf{R} : S \times S \rightarrow \mathbb{R}_{\geq 0}$. The *rate* $\mathbf{R}(s, s')$ defines the delay before which a transition between states s and s' is enabled. The delay is sampled from a negative exponential distribution with parameter equal to this rate, i.e. the probability of the transition being enabled within t time units is $1 - e^{-\mathbf{R}(s,s') \cdot t}$. When $\mathbf{R}(s, s')>0$ for two states s', a *race* occurs and the transition which becomes enabled first is the one taken. Exponentially distributed delays are often suitable for modelling component lifetimes and inter-arrival times. Furthermore, they can be used to approximate more complex probability distributions.

Other models commonly used for probabilistic model checking are discrete-time Markov chains (DTMCs), which specify the probability of moving between states in discrete time-steps, and Markov decision processes (MDPs), which can model systems which exhibit both probabilistic and nondeterministic behaviour, for example a system which comprises a number of probabilistic processes operating asynchronously in parallel.

Given a probabilistic model, it is then necessary to specify its required properties. Traditionally, in the model checking paradigm, properties are expressed using temporal logic, which provides a concise and unambiguous specification. We will use the temporal logic CSL (Continuous Stochastic Logic) (Aziz *et al.*, 1996; Baier *et al.*, 1999) which is designed for specifying properties of CTMCs. We do not present the full syntax and semantics of the logic here, instead providing a number of illustrative examples with their natural language translation:

- $\mathcal{P}_{\leq 0.01}[\Diamond\ shutdown]$ – "shutdown eventually occurs with probability at most 0.01"
- $\mathcal{P}_{\geq 0.95}[\neg repair\ \mathcal{U}^{\leq 200}\ complete]$ – "with probability 0.95 or greater, the process will successfully complete within 200 hours and without requiring any repairs"
- $\mathcal{S}_{>0.75}[num\_sensors \geq min]$ "in the long-run, the probability that an adequate number of sensors are operational is greater than 0.75"

Note that the use of probability bounds (≤ 0.01, ≥ 0.95, > 0.75) ensures that the properties above constitute questions which can be verified either to be true or false, as is traditionally the case in formal verification. In practice, though, it is often more useful to request the actual values, writing for example:

- $\mathcal{P}_{=?}[\Diamond^{\leq T} shutdown]$ – "what is the probability that the system shuts down by time T?"

Furthermore, the most useful way to analyse the model and to gain insights into its reliability may be to compute and plot such a value as some parameter is varied (e.g. T in the formula above, or a constant in the model itself).

Additional properties can be specified by adding the notion of *rewards*. We can assign each state (or transition) of the model a real-valued reward and then write queries such as:

- $\mathcal{R}_{=?}[\Diamond success]$ – "what is the expected reward accumulated before the system successfully terminates?"

Rewards can be used to specify a wide range of measures of interest. Of course, conversely, we can

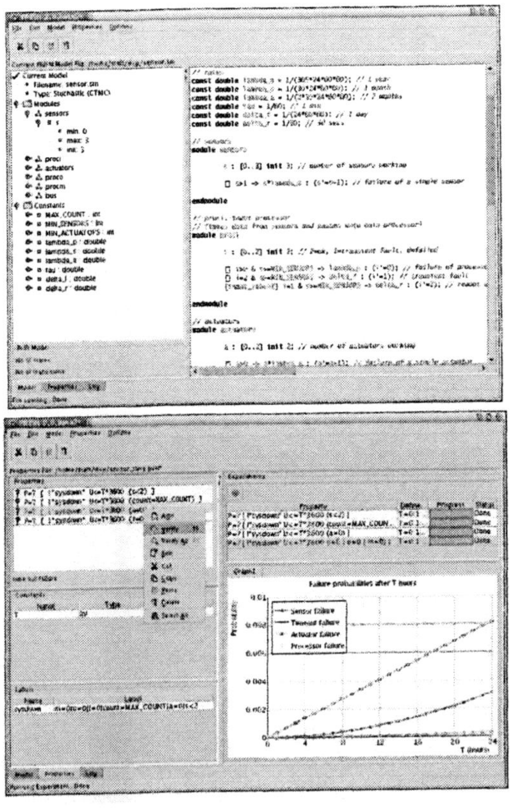

Fig. 1. Screenshots of PRISM running

also consider the rewards to be costs. We can then analyse, for example, expected power consumption, expected number of failures, etc. Suitable temporal logics for expressing cost- and reward-based properties can be found in e.g. (Baier *et al.*, 2000; de Alfaro, 1997).

A *probabilistic model checker* applies algorithmic techniques to analyse the state space of a probabilistic model and determine whether its specifications are satisfied. Typically, this involves computation of one or more probabilities or performance measures. The operations required are graph-based analysis and methods for solving either linear equation systems or linear optimisation problems.

3. PRISM

PRISM (Kwiatkowska *et al.*, 2002) is a probabilistic model checker developed at the University of Birmingham. It provides automatic verification of CSL properties for CTMCs. In addition, it supports analysis of DTMCs and MDPs. The tool has been used to analyse a wide range of case studies, including: randomised algorithms for problems such as leader election, mutual exclusion and consensus; real-time communication protocols; security protocols; communication networks; and dynamic power management schemes.

Probabilistic models to be analysed in PRISM are specified in the PRISM language, which is based on the Reactive Modules formalism of Alur and Henzinger (Alur and Henzinger, 1999). A model is described as a number of ***modules***, each of which corresponds to a component of the real-life system. Each module has a set of finite-ranged ***variables***. These determine the possible states of each module. The whole model is constructed as the parallel composition of these modules.

The behaviour of an individual module is specified by a set of guarded commands. For a CTMC, as is the case here, a command takes the form:

$$[]\ \texttt{<guard>}\ \rightarrow\ \texttt{<rate>}\ :\ \texttt{<action>}\ ;$$

The ***guard*** is a predicate over the variables of all the modules in the model. The ***update*** comprises `<rate>`, an expression which evaluates to a positive real number, and `<action>`, which describes a transition of the module in terms of how its variables should be updated. The interpretation of the command is that if the guard is satisfied, then the module can make the corresponding transition with that rate (see Section 2 for a definition of rate). A simple command for a module with one variable x might be:

$$[]\ (x = 0)\ \rightarrow\ 4.5\ :\ (x' = x + 1)\ ;$$

which states that, if x is 0, it is incremented by one and this action occurs with rate 4.5.

The overall functionality of the PRISM tool is as follows. First, it reads and parses a system description in the PRISM language. It then constructs, from this, the corresponding probabilistic model, in this case a CTMC (although the same description language can be used for both DTMCs and MDPs). PRISM also computes the set of all states which are reachable from the initial state and identifies any deadlock states (i.e. reachable states with no outgoing transitions). If required, the transition matrix of the probabilistic model constructed can be exported for use in another tool. Typically, though, PRISM then parses one or more temporal logic properties (e.g. in CSL) and performs model checking, determining whether the model satisfies each property. The graphical user interface of the tool also allows automatic construction of graphs to visualise results more easily. Figure 1 shows two screenshots of the tool running.

Another important feature of the tool is its implementation. PRISM is a "symbolic" model checker, meaning that it has been developed using data structures base on binary decision diagrams (BDDs). BDDs are reduced, directed acyclic graphs which can be very effective for compact storage of models which exhibit structure and

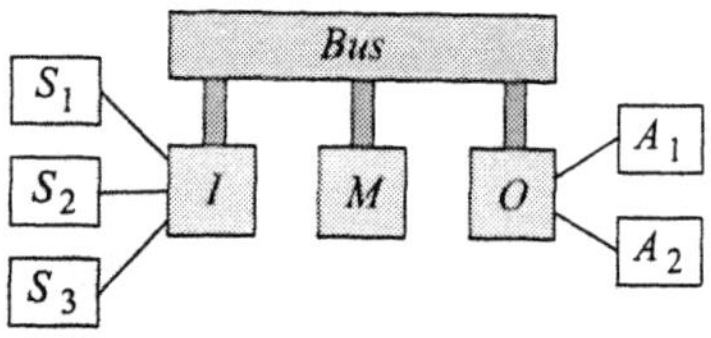

Fig. 2. The embedded system

regularity (e.g. those which have been specified in a high-level description language such as the PRISM language). In practice, this allows construction and analysis of larger models than would otherwise be possible. It also allows efficient BDD-based algorithms developed for non-probabilistic model checking to be easily integrated.

The tool and its source code can be freely downloaded under the GNU General Public License from www.cs.bham.ac.uk/~dxp/prism. Tool documentation, relevant research papers, and detailed information about a wide range of previous case studies can also be found here. In this paper, we have made use of a prototype extension of PRISM which supports model checking of the cost- and reward-based properties mentioned in the previous section. These features will be integrated into the main version in due course.

4. CASE STUDY

To illustrate the applicability of probabilistic model checking and the PRISM tool to the process of dependability analysis, we now present a small case study. We model an embedded system, closely based on the one presented in (Muppala *et al.*, 1994), the structure of which is shown in Figure 2. The system comprises an input processor (I) which reads and processes data from three sensors (S_1, S_2, S_3). It is then polled by a main processor (M) which, in turn, passes instructions to an output processor (O). This process occurs cyclically, with the length of the cycle controlled by a timer in the main processor. The output processor takes the instructions it receives and uses them to control two actuators (A_1, A_2). Finally, a bus connects the three processors together. A concrete example of such a system might be a gas boiler, where the sensors are thermostats and the actuators are valves.

Any of the three sensors can fail, but they are used in *triple modular redundancy*: the input processor can determine sufficient information to proceed provided two of the three are functional. If more than one becomes unavailable, the processor reports this fact to the main processor and the system is shut down. In similar fashion, it is sufficient for only one of the two actuators to be working, but if this is not the case, the output processor tells the main processor to shut the system down.

```
// a sensor fails on average once a month
const double λs = 1/(30 · 24 · 60 · 60);
module sensors
    // number of sensors working
    s : [0..3] init 3;
    // failure of a single sensor
    [] s > 0 → s · λs : (s' = s − 1);
endmodule
```

Fig. 3. PRISM code for the sensors

The I/O processors themselves can also fail. This can be either a permanent fault or a transient fault. In the latter case, the situation can be rectified automatically by the processor rebooting itself. In either case, if the I or O processor is unavailable and this leads to M being unable either to read data from I or send instructions to O, then M is forced to skip the current cycle. If M detects that the number of consecutive cycles skipped exceeds a limit, K, then it assumes that either I or O has failed and shuts the system down. Unless specified otherwise, we assume a value of 2 for K. Lastly, the main processor can also fail, in which case the system is automatically shut down.

The mean times to failure for a sensor, actuator and processor are 1 month, 2 months and 1 year, respectively. A transient fault occurs in a processor once a day on average. The mean times for a timer cycle to expire and for a processor reboot to complete are 1 minute and 30 seconds, respectively. We assume that all these delays are distributed exponentially; hence the system can be modelled as a CTMC.

Our PRISM model of the system comprises 6 modules, one for the sensors, one for the actuators, one for each processor and one for the connecting bus. In Figure 3, we show the section of the PRISM language description which models the sensors. This constitutes a single module *sensors* with an integer variable s representing the number of sensors currently working. The module's behaviour is described by one guarded command which represents the failure of a single sensor. Its guard "$s > 0$" states this can occur at any time, except when all sensors have already failed. The action $(s' = s - 1)$ simply decrements the counter of functioning sensors. The rate of this action is $s \cdot \lambda_s$, where λ_s is the rate for a single sensor and s is the PRISM variable referred to previously which denotes the number of active sensors.

In Figure 4, we show a second module which is the PRISM language description of the input processor. The module has a single variable i with range [0..2] which indicates which of the three possible states the processor is in, i.e. whether it is working, is recovering from a transient fault, or has failed. The three guarded commands in the

```
const double λp = 1/(365 · 24 · 60 · 60);
const double δf = 1/(24 · 60 · 60);
const double δr = 1/30;
module proci
    // state: 2=ok, 1=transient fault, 0=failed
    i : [0..2] init 2;
    // failure of processor
    [] i > 0 & s ≥ 2 → λp : (i' = 0);
    // transient fault
    [] i = 2 & s ≥ 2 → δf : (i' = 1);
    // reboot after transient fault
    [input_reboot] i = 1 → δr : (i' = 2);
endmodule
```

Fig. 4. PRISM code for the input processor

module correspond, respectively, to the processor failing, suffering a transient fault, and rebooting. The commands themselves are fairly self explanatory. Two points of note are as follows. Firstly, the guards of these commands can refer to variables from other modules, as evidenced by the use of $s \geq 2$. This is because the input processor ceases to function once it has detected that less than two sensors are operational. Secondly, the last command contains an additional label *input_reboot*, placed between the square brackets at the start of the command. This is used for synchronising actions between modules, i.e. allowing two or more modules to make transitions simultaneously. Here, this is used to notify the main processor of the reboot as soon as it occurs.

The full version of the PRISM code for this case study is available from the tool website.

5. RESULTS

We have used PRISM to construct the CTMC representing the embedded system described in the previous section and to analyse a number of dependability properties using probabilistic model checking. First, we consider the probability of the system shutting itself down. Note that there are four distinct types of failure which can cause a shutdown: faults in (1) the sensors (2) the actuators (3) the input/output processors (4) the main processor. We can analyse how likely each of these is to be the cause of the shutdown, as time passes. We use the CSL property:

- $\mathcal{P}_{=?}[\neg down\ \mathcal{U}^{\leq T}\ fail_j]$

where $j = 1 \ldots 4$, refers to one of the four failures above and *down* denotes that any of the failures has occurred, i.e. $down = fail_1 \vee fail_2 \vee fail_3 \vee fail_4$. The atomic propositions which make up the property would in practice be predicates over the variables from the PRISM language description. For example, $fail_1$, the failure of more than one sensor, is written "$s < 2 \wedge i = 2$", meaning that the number of working sensors has dropped below 2 and the input processor is functioning (and so can report the failure).

The above property denotes the probability that failure j is the ***first*** to occur. Note that, for example, if an actuator fails, the sensors, unaware of this, will continue to operate and may subsequently fail. Hence, we need to determine the likelihood of each failure occurring, before any of the others do. This is a good illustration of how non-trivial properties can be captured using temporal logic.

In Figure 5(a), we plot the results of this computation over two ranges of values for T: the first 24 hours and the first month of operation. We can see, for example, that while initially the I/O processors are more likely to cause a system shutdown, in the long run it is the actuators which are most unreliable. By omitting the bound $\leq T$ from the CSL formula:

- $\mathcal{P}_{=?}[\neg down\ \mathcal{U}\ fail_j]$

we can ask the model checker to compute the long-run failure probability (i.e. as $T \to \infty$). The results are: (1) 0.6216 (2) 0.0877 (3) 0.0484 (4) 0.2423. Note that these sum to one, i.e. the system will eventually shut down with probability 1.

Second, we consider a number of cost-based properties. We classify the states of our model into three types: "up", where everything is functioning, "shutdown", where the system has shut down, and "danger", where a (possibly transient) failure has occurred but has yet to cause a shutdown (e.g. if the I or O processor has failed but the M processor has yet to detect this). We can reason about the time spent in each class by assigning a cost of 1 to those states, 0 to all others, and then computing the total cumulated reward. We use the properties:

- $\mathcal{R}_{=?}[\mathcal{C}^{\leq T}]$
- $\mathcal{R}_{=?}[\mathcal{F}\ down]$

These give us the expected time spent in each class, up until time T and before the system shuts down, respectively. We plot the results for the first value in Figure 5(b), again for two ranges of T: the first day and the first month of operation. In the table in Figure 5(c), we give the results of the second property, as we vary K, the number of skipped cycles which the main processor waits before deciding that the input/output processors have failed. We see that increasing the value of K increases the expected time until failure, but also has an adverse effect on the expected time spent in "danger" states.

Finally, to illustrate a different type of cost structure, we compute the expected number of processor reboots which occur over time. This is done by assigning a cost of 1 to all transitions which

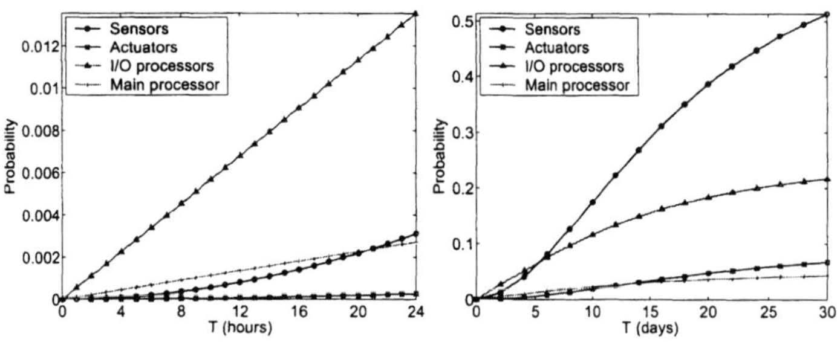

(a) Probability of each cause of shutdown

K	Expected time	
	danger (hrs)	up (days)
1	0.236	14.323
2	0.293	17.660
3	0.318	19.100
4	0.327	19.628
5	0.330	19.809
6	0.331	19.871
7	0.332	19.891

(c) Expected time

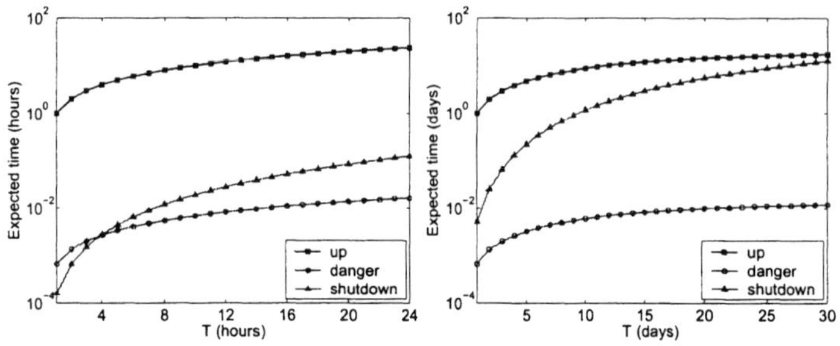

(b) Expected time spent in "up"/"shutdown"/"danger"

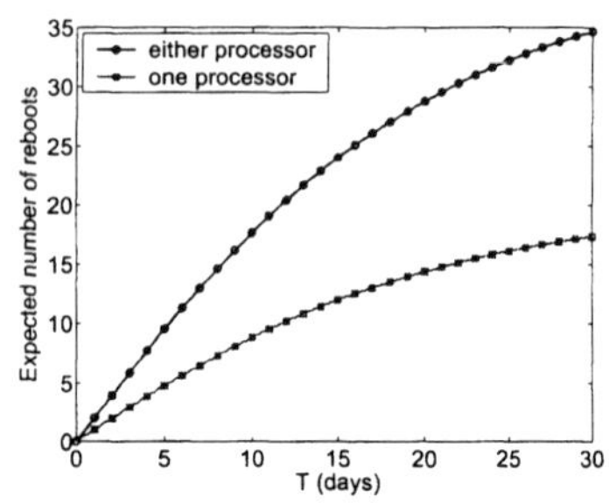

(c) Expected reboots

Fig. 5. Reliability results for the case study obtained using the probabilistic model checker PRISM

constitute a reboot and 0 to all others. The results are plotted in Figure 5(d).

In the experiments we have presented in this section, the number of states in the CTMC, which depends on the parameter K, varied between 2,633 and 7,703. The models were all constructed in less than second. For the model checking itself, the time required to compute each value varied from a few seconds to a few minutes, using a standard workstation.

6. CONCLUSION

We have illustrated how probabilistic model checking, a formal verification technique which has already been applied to a wide range of domains including distributed randomised algorithms, real-time protocols, security protocols and dynamic power management, can be used to analyse dependability properties of controller-based systems. We have used the probabilistic model checker PRISM whose simple system description language provides an intuitive way to construct complex Markov models. For specifying properties to check, we used temporal logic, which allows us to reason about non-trivial behaviour. We were also able to make use of the efficiency improvements which have been developed in this area. A further advantage of this formal verification approach is that it could easily be combined with more traditional non-probabilistic verification processes, which are becoming increasingly common, particularly in safety-critical areas.

For more detailed information about probabilistic model checking, PRISM and its application to the case study described in this paper, see the tool website: www.cs.bham.ac.uk/~dxp/prism

REFERENCES

Alur, R. and T. Henzinger (1999). Reactive modules. *Formal Methods in System Design* **15**(1), 7–48.

Aziz, A., K. Sanwal, V. Singhal and R. Brayton (1996). Verifying continuous time Markov chains. In: *Proc. CAV'96*. Vol. 1102 of *LNCS*. Springer. pp. 269–276.

Baier, C., B. Haverkort, H. Hermanns and J.-P. Katoen (2000). On the logical characterisation of performability properties. In: *Proc. ICALP'00*. Vol. 1853 of *LNCS*. Springer. pp. 780–792.

Baier, C., J.-P. Katoen and H. Hermanns (1999). Approximate symbolic model checking of continuous-time Markov chains. In: *Proc. CONCUR'99*. Vol. 1664 of *LNCS*. Springer. pp. 146–161.

de Alfaro, L. (1997). Formal Verification of Probabilistic Systems. PhD thesis. Stanford University.

Kwiatkowska, M., G. Norman and D. Parker (2002). PRISM: Probabilistic symbolic model checker. In: *Proc. TOOLS'02*. Vol. 2324 of *LNCS*. Springer. pp. 200–204.

Muppala, J., G. Ciardo and K. Trivedi (1994). Stochastic reward nets for reliability prediction. *Communications in Reliability, Maintainability and Serviceability* **1**(2), 9–20.

Copyright © IFAC Information Control Problems in Manufacturing, Salvador, Brazil, 2004

ELSEVIER
IFAC
PUBLICATIONS
www.elsevier.com/locate/ifac

Modelling and Verification of PLC-based Systems Programmed with Ladder Diagrams

Andrei Lobov[◊], **Jose L. Martinez Lastra**[◊], **Reijo Tuokko**[◊], **Valeriy Vyatkin**[‡]

[◊] *Institute of Production Engineering, Tampere University of Technology*
P.O .Box 589, Tampere, Finland

[‡] *Martin Luther University Halle-Wittenberg, Dept. of Engineering Science*
06099 Halle, Germany

Abstract: This paper reports the results on verification of automation software systems that were developed during MOVIDA[1] project. In particular, PLC-based control systems implemented in Ladder Diagrams (LD) were modelled and verified. A formalism of Net Condition/Event Systems, derived from Petri Nets, was applied to model both the behaviour of a PLC (programmed in ladder logic) and the dynamics of the controlled plant. A model-checking was applied to verify system properties. Software tools supporting model generation from the source code automate the process of model generation and verification. The paper describes the details of modelling of ladder logic structures and an example of the model-checking application for verification of programmable control systems. *Copyright © 2004 IFAC*

Keywords: Model-checking, PLC modelling.

1. INTRODUCTION

A Programmable Logic Controller (PLC) is a device that is commonly used for industrial and home automation. The automation substitutes the human labour in order to increase the productivity or substitute human in dangerous working environments. However, programs for the PLCs are subject to errors that could lead to erroneous behaviour of the controlled equipment.

The verification of PLC performance in critical industrial processes, where misbehaviour may lead to the damage of machine equipment, product, environment or cause human injuries, has been active research topic for the last decade. A survey on formalization of PLC programs for most-known industrial programming languages may be found in (Bani Younis and Frey, 2003). The work (Heiner and Menzel, 1998) describes general steps for PLC software verification and validation that is depicted in Figure 1, and applied these in model-based validation of the Instruction List programming language.

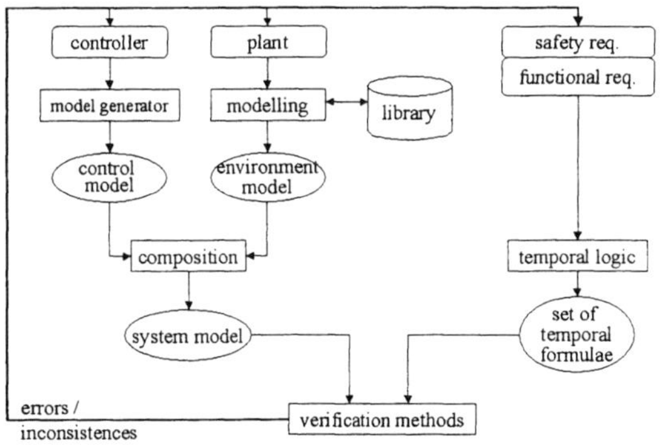

Fig. 1. Model-based validation of PLC software (Heiner and Menzel, 1998)

This paper introduces the application of verification to industrial systems programmed in another

[1] This work was supported by National Technology Agency of Finland (TEKES) and industrial partners (FlexLink Automation Oy, Metso Panelboard, Omron Electronics Oy, PMJ Automec Oy, Swisslog Oy)

programming language – Ladder Logic Diagrams. In addition, this work uses a modular formalism of Net Condition/Event Systems that is suitable for systematic structured modelling of automated objects, machines and processes. The paper is structured as follows: Section 2 describes a formalism that is used to model the control programs and the controlled plant. Section 3 shows application of the formalism to LD programming language – generation of the model of a controller and highlights plant modelling. Section 4 describes model-checking procedure of the system model consisted of the plant model and the model of controller. Section 5 concludes the paper with overview of future work.

2. TIMED NET CONDITION/EVENT SYSTEMS

Timed Net Condition/Event System (TNCES) is Petri nets derived formalism that allows representation of hierarchical models and provides a way to express exchange of events and the state information.

TNCES has been introduced by (Rausch and Hanisch, 1995). It is a tuple (Hanisch, *et al.*, 1997):

$$TNCES = \{P, T, F, m_0, \psi, CN, EN, DC\} \quad (1)$$

where $P = \{p_1, p_2, \ldots, p_n\}$ is finite set of places; $T = \{t_1, t_2, \ldots, t_m\}$; $F \subseteq (P \times T) \cup (T \times P)$, finite set of flow arcs between places and transitions; m_0 is initial marking; ψ is input/output structure of TNCES module; $CN \subseteq (P \times T)$ is a finite set of condition arcs; $EN \subseteq (T \times T)$ is a finite set of event arc. Input/output structure of TNCES module, ψ, represented by the following tuple (Hanisch, *et al.*, 1997)

$$\psi = \{C^{in}, E^{in}, C^{out}, E^{out}, Bc, Be, Cs, Dt\} \quad (2)$$

where C^{in} is a finite set of TNCES module condition input signals; E^{in} is a finite set of TCNES module event input signals; C^{out} is a finite set of TNCES module condition output signals; E^{out} is a finite set of TCNES module event output signals; $Bc \subseteq C^{in} \times T$ is a set of TNCES module input condition arcs; $Be \subseteq E^{in} \times T$ is a set of TNCES module input event arcs; $Cs \subseteq P \times C^{out}$ is TNCES module output condition arcs; $Dt \subseteq T \times E^{out}$ is a set of TNCES module output event arcs.

Time intervals may be assigned to the pre-transition flow arcs $F^- \subseteq P \times T$, which imposes time constrains to the firing of the transition (Hanisch, *et al.*, 1997):

$$DC = \{DR, DL, D_0\} \quad (3)$$

defines a set of delay time DR representing the minimum times that the token should spent at particular place before the transition can fire; DL is the set of limitation time that defines maximum time that the place may hold a token (if all the other conditions for transition firing are met). D_0 is the initial set of the clocks associated with the places (Hanisch, *et al.*, 1997).

An example of TNCES module is depicted in figure 2.

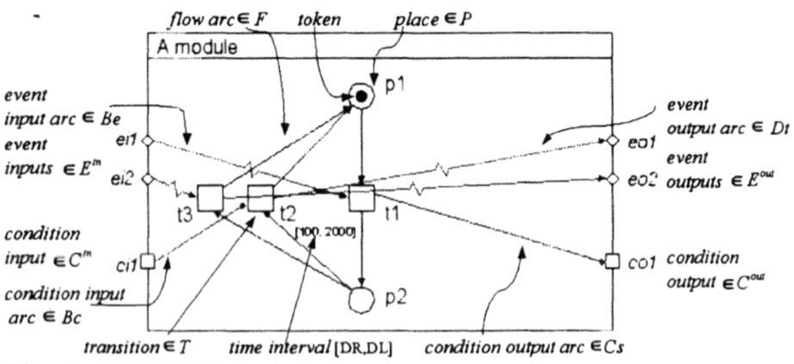

Fig. 2. TNCES Module

Hierarchical models can be built using TCNES, a TCNES, in addition to holding just *TNCES* structure formulated before, can hold a set of interconnected modules. The TNCES composed out of the modules may be formulated as follows:

$$TNCES_{composite} = \{M, Cm^{inp}, Cm^{out}, Em^{inp}, Em^{out}, Ic, Ie\} \quad (4)$$

where M is the set of modules that may be composed of *TNCES* and $TNCES_{composite}$ modules; Cm^{inp} is a set of composite TNCES module condition inputs; Cm^{out} is a set of composite TNCES module condition outputs; Em^{inp} is a set of composite TNCES module condition inputs; Em^{out} is a set of composite TNCES module condition outputs;

$Ic \subseteq C_k^{out} \times C_l^{in} \cup Cm^{inp} \times C_i^{in} \cup C_i^{out} \times Cm^{out}$, where $i, k, l \in [1, n]$, where n is a number of elements (member modules) in M, is a set of condition arcs interconnecting the modules, composite module condition inputs Cm^{inp} with the condition inputs of member modules C_i^{in}, and condition outputs of member modules C_i^{out} with condition outputs of $TNCES_{composite}$-Cm^{out};

$Ie \subseteq E_k^{out} \times E_l^{in} \cup Em^{inp} \times E_i^{in} \cup E_i^{out} \times Em^{out}$, where $i, k, l \in [1, n]$, where n is a number of elements in M, is a set of event arcs interconnecting the modules, composite module event inputs Em^{inp} with the event inputs of member modules E_i^{in}, and event outputs of member modules E_i^{out} with event outputs of $TNCES_{composite}$ - Em^{out}.

3. APPROACH TO PLC-BASED SYSTEMS MODELLING

3.1 Languages for PLC programming

Special industrial programming languages are applied for implementation of the control algorithms. The most of known programming languages in the field were standardized in IEC 61131-3 in 1993. The standard includes four programming languages: Instruction List (IL), Function Block Diagrams (FBD), Ladder Diagrams (LD) and Structured Text

(ST), and a common element Sequential Function Chart (SFC) that serves for program organization into logical steps and expressing the transitions between the steps. Despite of the standard appearance that has aimed to simplify life of programmer, some of PLC vendors have not introduced IEC61131-3 compatible software development tools. The development of the program with one of standardized languages in the rapid application development environment (RAD) for the particular PLC does not create a compatible code for the other vendors PLC. This is understandable due to different PLC hardware properties, e.g. memory allocation and performance. The standard defines the look of the languages and does not specify dedicated hardware platforms as well as the format in which to store the program data. The latter creates a situation when the source code generated in one IEC61131-3 RAD tool cannot be opened in another tools and as a result no parts of the code may be imported to the other tool automatically, the solution is to manually copy the program between RAD tools.

The absence of the common format of the source code representation in the mass storage devices is an additional problem to the automation of modelling and verification of the PLC software projects. The model generators or compilers that create models out out of the source code have to deal with project data represented in the format of particular vendor. The solution may be seen in unified representation of project data. TC6 at PLC Open[2] association has started work on XML format for providing IEC61131-3 logic information in Spring 2003. This work may give a common representation of PLC information and as result a compatibility between different vendors RAD tools and an ease of automated source code/control system verifiers.

3.2 Modelling of Ladder Logic and PLC performance

Ladder Diagrams (LD) is a widely used industrial programming language. It resembles relay diagrams from the times when the control system were built as hardware circuits from relays. Nowadays LD programs may contain complex mathematical functions executed on PLC hardware.

This section discusses modelling of PLC behaviour with TNCES. Thanks to their inherited modularity, the TNCES can be straightforwardly applied to modelling of discrete event control systems. However, the application of TNCES for continuous process modelling requires finding a trade off between adequate representation and discrete nature of TNCES. The possible approach to the problem is discretization of the continuous system behaviour.

In general the problem of PLC-based system modelling can be split into three sub problems.

- PLC execution modelling
- Modelling of basic Boolean functions
- Modelling of non-Boolean functions

The first one is related to the performance of PLC hardware represented by scan-cycle times, instructions execution times, etc. The PLC programs are executed in the cyclic way. One cycle consists of the following phases: first the inputs are read, then the program logic is executed and then the outputs are written. Figure 3 depicts a TNCES skeleton for a PLC model. Place *p1* holds a token representing the initial state of the PLC execution, if the PLC program is enabled (condition input to the *t1* transition) the cycle is started by the update of outputs and acquisition of input values. The firing of t1 transition generates these events. When a token is placed to *p2*, it resides there until a signal notifying about the change in the system enables transition *t2*. The monitoring of the changes in the systems is needed in order to not start a new PLC cycle unless something has changed in the system.

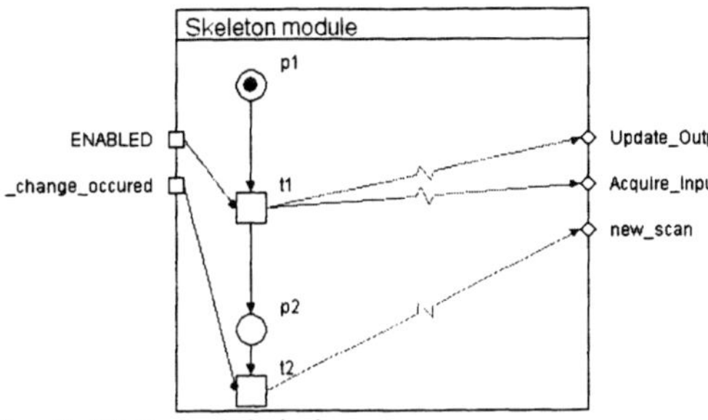

Fig. 3. PLC model skeleton

The state of the TNCES model is distributed and is defined by the marking of all places. Additionally to the marking, the state is characterized by the time stamp, e.g. the time the certain state (marking) is valid. Thus the two states representing the same marking but holding different times are different states.

A special module that monitors the change in the system has to be added to the model (Fig. 4). The module has two event inputs for retrieving information about any change of the PLC program variables during the scan cycle. It does not make sense to run the model over the new scan-cycles if the markings in the model remain unchanged, i.e. when nothing would change during the next scan. The marking may change in the model of the plant or if the time dependent transition fires in a timer TNCES module in the model of controller.

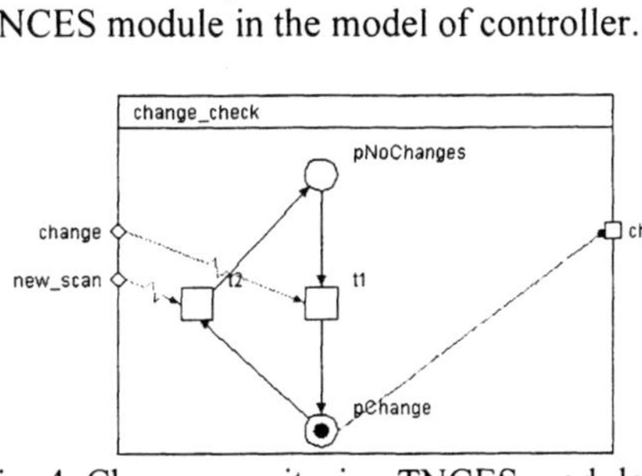

Fig. 4. Change monitoring TNCES module

In paper (Hanisch, *et al.*, 1997) the main artefacts in modelling of PLC behaviour were identified. Additionally to the standard representation of

---
[2] PLC Open association: http://www.plcopen.org

Boolean variable defined in (Hanisch, *et al.*, 1997) an extra *change* event output is added to the TNCES module (Fig. 5). The event is generated when the variable changes its value. This signal is supplied to change monitoring module defined in figure 4.

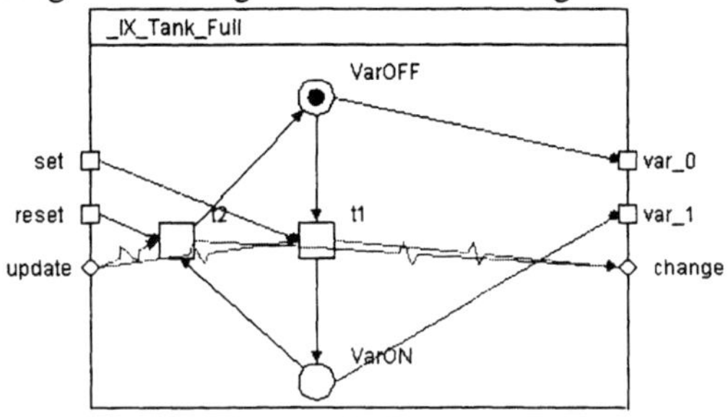

Fig. 5. Input Boolean variable representation

The same is true for the Boolean output module (Fig. 6).

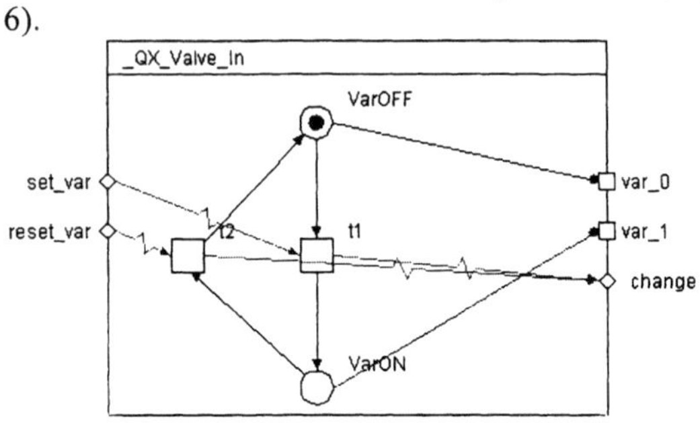

Fig. 6. Output Boolean variable representation

Let us consider a simple process – a filling of the tank with some liquid (Fig. 7). The tank has input valve that supplies the liquid into the tank. Once the tank is filled the valve should be closed. A level sensor (*L*) notifies when the filling process should terminate.

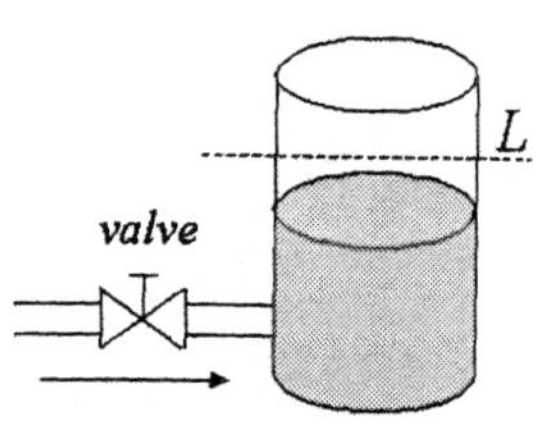

Fig. 7. A tank – controlled object

A single-rack LD program is shown on the left in figure 8. *IX_Tank_Full* variable represents PLC input signal coming from the level sensor of the tank *L*. *QX_Valve_In* is PLC output that controls input valve of the tank. Representation of both input/output variables in TNCES is defined in figures 4 and 5 correspondingly.

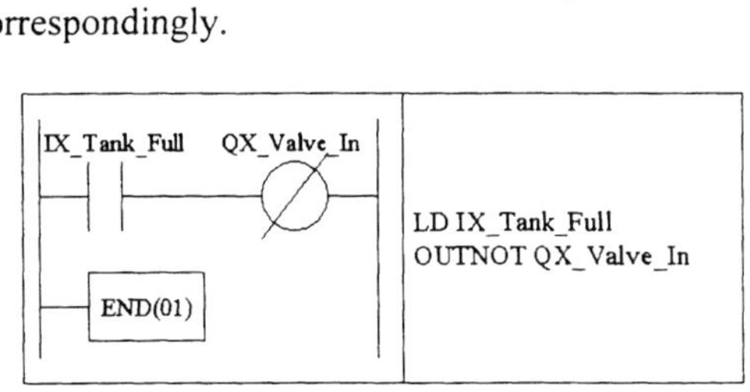

Fig. 8. Tank control program

The program is written for Omron™ CPM1A PLC. Omron control software development tool (CX-Programmer) is made to support a set of dozens of Omron™ PLCs. The same environment and the same programming language is used for each controller, the difference may be in the amount of supported functions. One of the features of CX-Programmer software is an ability to save project's data in a text file. The file would contain all possible data related to the project and controller. This file may serve as an input for a model generator. In particular, in MOVIDA project *MOVIDA NCES Generator* was developed that converts project information into TNCES model. An example of rack 0 textual representation is given on the right in Figure 8. The textual representation of LD may resemble the Instruction List programming language. An LD instruction (that stands for *Load*) loads the value at *IX_Tank_Full* variable into accumulator. At the second row, the negated value of accumulator is stored to output variable *QX_Valve_In* that controls the valve. So, once the tank is filled, *IX_Tank_Full* becomes TRUE, the FALSE value is stored into the *QX_Valve_In*, that will close the valve.

The model generator loads the PLC project information and starts the conversion into the TNCES rack by rack. When a rack has been read, the generator checks for Boolean expressions included into the rack. In the given example it is just an evaluation of a single bit variable *IX_Tank_Full.* If the variable was not met in previous racks its model will be added to the PLC model. The PLC model skeleton is extended by the rack representation. In the current example, which has only one rack, *p3* place along with *t3* and *t4* transitions represent the rack 0. Figure 10 represents the PLC logic model. The whole PLC model consists of interconnected I/O modules (Fig. 5 and 6), a change monitoring module (Fig. 4), and a PLC logic model (Fig. 9).

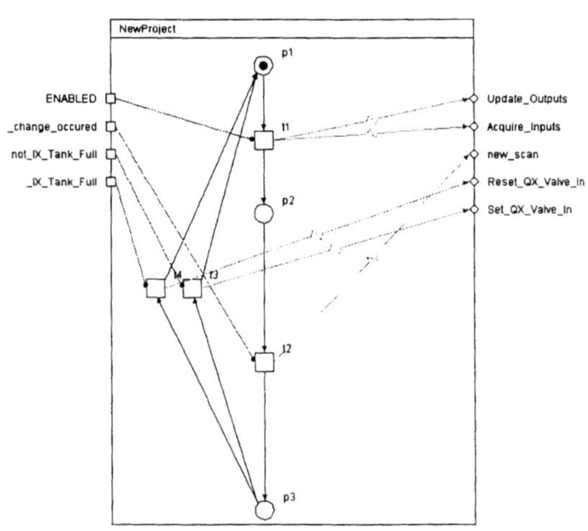

Fig. 9. PLC logic model for tank control program

The model of a controller can be generated by the MOVIDA NCES Generator (Fig. 10). The generator takes as an input Omron™ LD project represented as a textual file, converts it into TNCES, and saves the data in XML based format. The openness and self-explanatory XML representation simplifies the development of the tools that may work with TNCES.

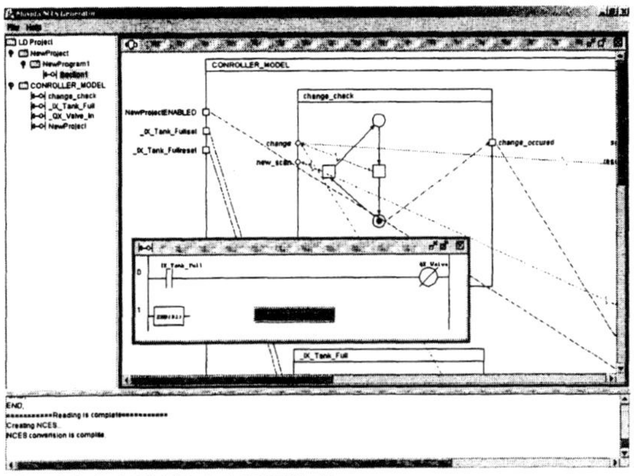

Fig. 10. MOVIDA NCES Generator

3.3 Plant modelling

The approach to validation presented in this paper relies on the closed-loop modelling of plant-controller systems. Additionally to the model of the controller, model of the plant is required as shown in Figure 15. The model of the plant limits the set of possible states of the entire system by limiting a set of possible inputs to the controller.

More details on the plant modelling can be found in (Vyatkin V., *et al.,* 2003). However, a more systematic methodology of modular plant modelling is yet to be formulated. Currently the modelling of plant is supported by MS Visio Template developed at Martin Luther University of Halle-Wittenberg (Figure 11). The template allows manual assembly of TNCES modules.

Both models of the plant and the controller utilize hierarchy and reusability principles. Common modelling components may be reused in the same model and across different models. An evident example of the reusability is modelling of Boolean input or output. Figure 12 represents a plant model of the tank. The notation of the blocks is done according to IEC61499 standard.

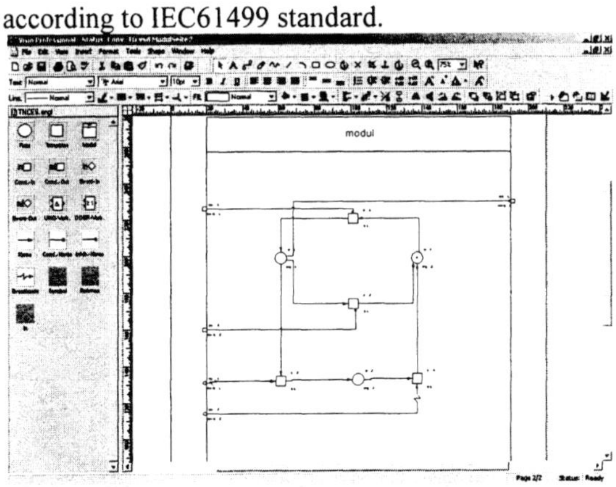

Fig 11. Visio TNCES Template

As the model of the controller, the model of the plant has two Boolean variables corresponding to the valve and the level sensor. The valve is modelled by an input variable while the level sensor by an output variable. This is opposite to the model of the controller, where PLC program has a valve related variable as an output and the level sensor as an input.

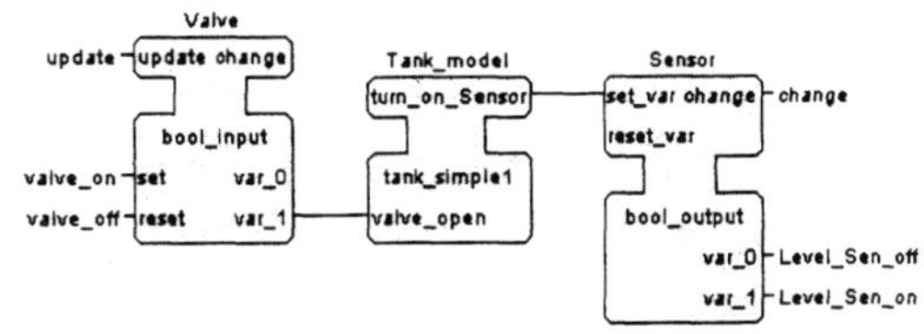

Fig. 12. A model of filling process with valve input and level sensor output

A filling process model of the tank shown in the middle of figure 12 has one condition input *valve_open* and one event input *turn_on_Sensor*. Figure 13 shows the model of the process in detail. This is a trivial abstraction of the real filling process that has only two *states* (*p1* and *p2*) that may be seen as not filled (*p1*) and filled (*p2*). Once the incoming condition signal of the transition *t1* is TRUE for at least of 10000 time units (here 1 unit = 1ms) the *t1* fires generating output event that actually turns sensor on.

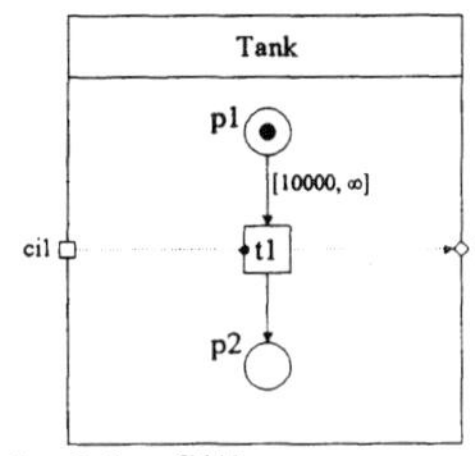

Fig. 13. Model of the filling process

Figure 14 shows the model of the plant at higher level. Model of the plant embodies all the low level modules (Fig. 12) and is ready to be interconnected with the model of the controller.

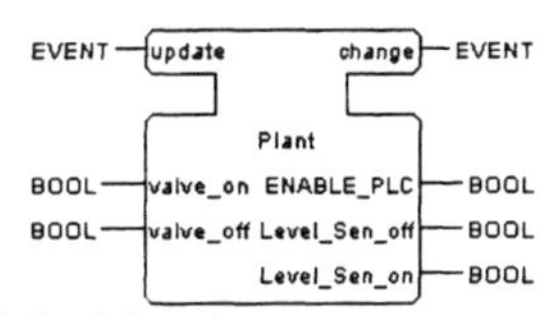

Fig. 14. Model of the plant

3.4 Model composition

Models of the plant and model of the controller are interconnected into the closed-loop providing the representation of entire system that consist of the controlled equipment and control device. For the tank example, figure 15 depicts the model of interconnected controller and plant. The combined model is subject to model-checking for judging about modelled system properties.

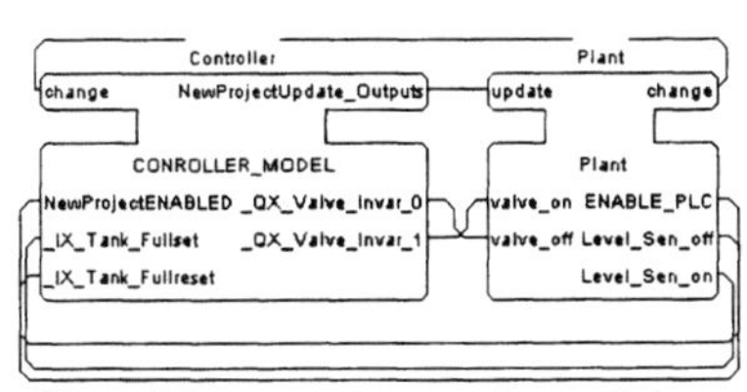

Fig. 15. Closed-loop representation of the system at highest level

4. APPLICATION OF MODEL-CHEKING

The model-checking is an evaluation procedure of finite state model, where the model is searched for particular properties of interest. We utilize the iMATCH tool (Fig. 16) for analysis of the modelled system. iMATCH inputs a hierarchical model of the system (Fig. 15). It converts it into a flat model that is used by an embedded model checker or may be an input to SESA model checker (Starke, *et al.*, 2002). In the reachability graph each state is associated with the certain marking (location of the tokens) of the model. The graph is searched for particular markings that represent allowed/forbidden states of the system. The iMATCH embedded model-checker may check as an input simple predicates (e.g. "*p1* AND *p14*") as well as CTL formulae that are evaluated in SESA. The result may be displayed graphically to the user. The user may observe the marking of each TNCES module at each state and trace along the time the system behaviour.

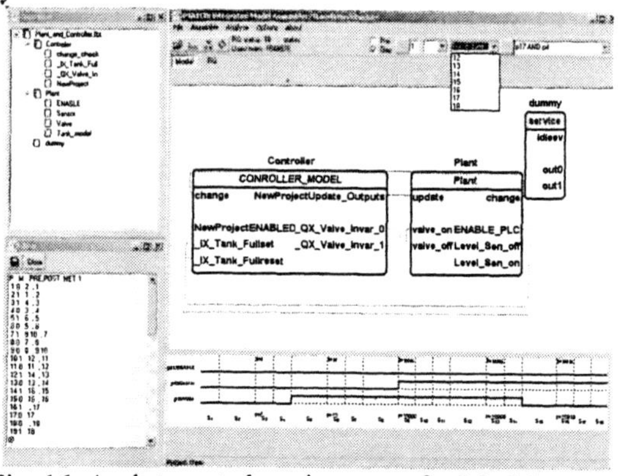

Fig. 16. An integrated environment for Model Assembly, Translation and Checking (iMATCH)

Figure 17 shows the time diagram generated by iMATCH for the discussed example of the liquid tank. The are 18 possible states in total in the system. But we are interested only in plant variables - real behaviour of the physical object, particularly in the relation between sensor detecting that the tank is filled and the valve that inputs the liquid. As may be seen in Figure 17, the valve is opened at state 5 (*s5*) and in 10000 time units (*s9*) the sensor detects the level, and then the valve is closed at *s15*. Taking into account that scan-cycle of the PLC is modelled to be 5 time units, the closing of the valve occurs in less than 2 scan cycles of PLC (10 time units) after the physical sensor has detected the level. This behaviour is considered to be adequate, since between the sampling of input signal and responding with the output is at least one and not more than two PLC scan-cycles.

5. CONCLUSION

The paper has illustrated the application of TNCES formalism that allows reusability and representation of hierarchical models. The modelling and verification of PLC-based systems implemented in Ladder Diagrams were illustrated on a trivial example of a tank. Using the example, we have tried to highlight all steps of the validation process, from the modelling of controller and the plant to the verification of the entire system.

The manual modelling of plant and/or controller for the complex systems may take weeks according to our experience. Industrial engineers in general are not familiar with the formal methods. The difficulty and time consumption of manual modelling is a current barrier to the acceptance of the methods in the industry. However, the development of model generators and verification tools that should automate model creation and verification process has to promote the advantages of the approach.

Currently the formats of data storage for PLC projects are not standardized. Despite the IEC61131 standardization, each PLC vendor supports a number of PLC languages and a number of vendor dependent project data representation formats. In order to automate the model generation, each format has to be investigated before creation of model generators. In general "one vendor – single project data format" relation is valid, however, a single vendor might have different data storage formats for different PLC families, which is an obstacle for development of multi-PLC model generators. The standardization work over the common PLC project data storage and representation format may help to solve the problem.

REFERENCES

Heiner, M., Menzel, T. (1998). *Instruction List Verification Using a Petri Net Semantics*, In Proc. of IEEE International Conference on Systems, Man, and Cybernetics, San Diego, USA

Bani Younis, M., Frey, G. (2003). *Formalization of Existing PLC programs: A Survey*, In Proc. Of Computing Engineering in Systems Applications, Lille, France

Rausch, M., Hanisch, H.-M. (1995). *Net condition/event systems with multiple condition outputs*, In Proc. Of the Symposium on Emerging Technologies and Factory Automation, Paris, France

Hanisch, H.-M., Thieme, J., and other (1997). *Modeling of PLC Behavior by Means of Timed Net Condition/Event Systems*, 6th International Conference on Emerging Technologies and Factory Automation, Los Angeles, USA

Starke, P., Roch, S., and other (2002). *Analysing signal-event systems*, Technical report, Humboldt Universitat zu Berlin, Institut für Informatik

Vyatkin V., Hanisch H.-M., and other (2003). *Modular typed formalism for systematic modelling of automation systems*, 1st IEEE Conference on Industrial Informatics, Proceedings, Banff, Canada

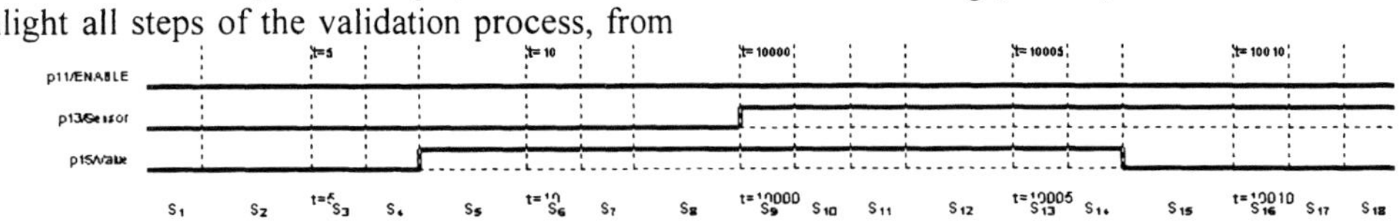

Fig. 17. Time diagram for the tank filling system

Copyright © IFAC Information Control Problems in
Manufacturing, Salvador, Brazil, 2004

ELSEVIER
IFAC
PUBLICATIONS
www.elsevier.com/locate/ifac

AUTOMATIC IMPLEMENTATION AND VERIFICATION OF COORDINATING PLC-CODE FOR ROBOT CELLS

H. Flordal * **M. Fabian** * **K. Åkesson** *

* *Department of Signals and Systems,*
Chalmers University of Technology, SE-412 96
Gothenburg, Sweden
{hugo,fabian,ka}@s2.chalmers.se

Abstract: A method for automatically generating the coordinating PLC-code for a multi-robot cell is presented. Finite state automaton models of the allocation and release of the spatial volumes that the robots share, as well as the sequences of each robot's allocations, are automatically extracted from a 3D simulation environment. This includes explicitly calculating the intersection between the robots' work envelopes, the spatial volumes where collisions may occur. The system model is then analyzed using the Ramadge-Wonham supervisory control theory to verify collision freeness and nonblocking. If necessary, a maximally permissive supervisor that guarantees these properties can be synthesized. The supervisor will include the necessary interlocks but in doing so still permits the robots as much freedom as possible. Thus, if one robot is delayed, the other robots may be free to book and enter a spatial volume that they normally would have to wait for. Finally, from the synthesized supervisor, PLC-code to be run in a controller for coordination of the robots is generated. *Copyright © 2004 IFAC*

Keywords: Robot programming, deadlock avoidance, supervisory control, supervisor synthesis, programmable logic controllers

1. INTRODUCTION

In many industrial applications, particularly in the automotive industry, multiple robots work close to each other, sometimes even inside each others workspaces. In such cases, obviously, there is a risk that the robots will collide and to prevent this, entrance to shared spatial volumes must be coordinated so that only one robot at a time has access to each volume. Typically this coordination is performed by a PLC (Programmable Logic Controller) that interacts with the robots to dynamically administer the sharing.

Implementation and verification of the coordinating PLC software is a tedious and error prone task, most often done manually. Typically, the robots are programmed off-line in a 3D simulation environment, each having its own sequence of paths layed out irrespective of the other robots. Then their interaction is simulated and when a collision is imminent, checks are introduced into the robot programs to signal and wait for the controller to allow one of the robots to proceed. In the automotive industry such checks are called *interlocks*. In this manner the shared workspace is implicitly divided into spatial volumes, *zones*, that the robots need to allocate to perform an operation.

The programming of the PLC controller, on the other hand, is typically performed on-line. The robots are run concurrently, at low speed until two or more robots are halted by an interlock.

Code is then added to the PLC, which checks all prerequisites for one of the robots to continue, and signals this robot to do so. In this way the controller dynamically administers the allocation of the zones.

The problems with manual implementation are many. First, manual work is slow and cost inefficient and it is complicated to manually find the optimal (or even an acceptable) cycle time. Also, essentially the same functionality is programmed twice and by different people, once in the robot programs and once in the PLC, and this adds to the problem. Secondly, the result is a rather rigid and inflexible coordination function that does not allow work to proceed other than in the specified sequence. Thus, if one robot gets delayed, the cycle time of the entire cell will increase, resulting in a production loss. Additionally, simulation can in general only reveal the presence of errors, not guarantee the absence.

The basic problem here is the complexity. The number of combinations grows exponentially with the number of robots and number of alternative paths. The strict sequential coordination allowing minimum, or even no, flexibility, is a means to fight this complexity.

However, if the task of implementing and verifying the coordinating PLC-code could be automated, both development time and flexibility can be significantly improved. The incitement for this increases as the market demands a larger range of customized products produced within the same production system, and that new and old products must be phased in and out much faster than today.

We present a method for automatically generating the coordinating PLC-code for a multi-robot cell. The method makes use of the off-line 3D simulation environment to extract information about the system and based on this generate a supervisor. In the hands of the programmer, we leave the initial programming of the robots and possibly some refining of the final result by adding alternative paths for the robots and rerunning the automatic code generation.

In section 2, the mathemathical preliminaries and modeling abstractions used are described shortly. Section 3 summarizes the software used in the implementation. Section 4 describes the method. Section 5 gives some conclusions and future ideas.

2. PRELIMINARIES

To model the processes and the specifications of their desired behaviour, we use finite state automata (Hopcroft and Ullman, 1979) (finite transition machines) and the Supervisory Control Theory (SCT), introduced by Ramadge and Wonham, see (Ramadge and Wonham, 1989). A finite state automaton A has a set Q^A of *states* of which one, q_0^A, is the *initial state*; a set Σ^A of discrete *events* and a set $\delta^A : Q^A \times \Sigma^A \rightarrow Q^A$ of *transitions* that describes possible states changes on the occurrences of the *events*. See figures 4 and 3 for examples of automata.

There are three types of automata appearing in the SCT and this paper, *plants*, *specifications* and *supervisors*. The plant is a model of the physical behaviour of a system and the specification is a model of the desired behaviour of the controlled system. The closed-loop system is supposed to fulfil the specification, that is, the plant controlled by the supervisor must not violate the desired behaviour defined by the specification. We *verify* if the plant already fulfils the specification and if not, we *synthesise* a supervisor that issues control over the plant so that the specification is fulfilled. In general, there are many ways to fulfil the specification, and the goal is to find the *maximally permissive* supervisor. That is, the supervisor that forces the plant to fulfil the specification but in doing so grants the plant as much freedom as possible.

Furthermore, there are two types of events, *controllable* and *uncontrollable* events. The controllable events are such that when the closed-loop system wants to execute such an event, the execution can be inhibited by a supervisor, thus controlling the system not to evolve in some undesired direction. The uncontrollable events can however not be inhibited and when they occur, the supervisor must be prepared to observe the event or risk losing control over the system. This is a property that must be verified and it must be kept in mind when synthesising the supervisor.

Another important property that we require is *non-blocking*. A non-blocking system is a system such that no matter what state the system is in, there is a sequence of events leading to some *marked* state. A marked state is a state that is more important, in some sense, than other states and therefore must be reachable. Usually, as in all automata in this paper, the initial state is the only marked state. A system that is blocking has deadlocks, non-marked states without outgoing transitions, and/or livelocks, a set of connected non-marked states that has no transitions to any states outside that set, i.e. infinite loops.

3. SOFTWARE

The implementation is based on two pieces of software. A SCT tool is used for the analysis and automatic generation of models, based on the information extracted using the other software, the robot simulator, a 3D environment for simulation of industrial robots.

3.1 SCT tool

Supremica is a software tool for formal verification and synthesis of discrete event systems developed at Chalmers University of Technology. The formal methods in Supremica are based on the SCT.

Among other things, Supremica can be used to verify safety and liveness properties in discrete event systems. If some specified behaviour turns out to be violated, Supremica can synthesize a supervisor that forces the system to fulfil the specification in a maximally permissive manner.

There is also functionality for generating control code that can be executed by standard PLCs. Supremica is presented more thoroughly in (Åkesson *et al.*, 2003) and (Supremica, 2003).

3.2 Robot simulator

The robot simulator is used to supply Supremica with the necessary information about the robot cell. This is used to calculate the spatial volume that the robots occupy and detect collisions between robots and critical zones in the work space.

The software used in this implementation is RobotStudio. It is a software tool from ABB Robotics that supports off-line programming and 3D modelling and simulation of ABB robot systems. See (ABB, 2003).

Supremica communicates with RobotStudio via a COM interface. Supremica thus aquires information by executing the necessary simulations and geometrical manipulations in the simulation environment.

4. METHOD

The input to the method is a simulation environment with a number of robots, each robot supplied with a set of *paths* that the robot can move along and a program specifying the ordering of the paths, in a sequence or supplied with alternatives. The method then consists of four steps.

(1) Zone partitioning
(2) Simulation and automata generation
(3) Supervisor synthesis
(4) PLC code generation

Figure 1 presents a simple cell with two robots that will be used as an example throughout this section. The two robots are programmed to move along a sequence of paths that passes above the two cylinders. The paths are shown as grey lines with small coordinate systems attached. The robots move in opposite directions over the cylinders, the rightmost robot moves I1→A→C→B→I1 and the leftmost robot I2→B→C→A→I2.

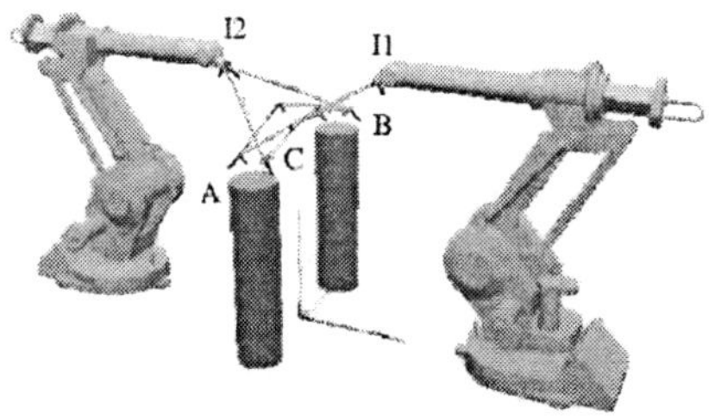

Fig. 1. A simple example of a multi robot cell.

4.1 Zone partitioning

First the zones, that must be accessed under mutual exclusion, must be constructed as 3D objects in the virtual environment. There are a number of conceivable methods for this.

manually The programmer manually specifies zones that several robots occupy during their work cycles. This is not recommended generally, of course, but in certain cases, e.g. where there is a high degree of symmetry, a suitable zone partitioning might be obvious.

space partitioning More or less the entire station is partitioned into (not necessarily) equally sized elements, for example cubes or slabs. The finer the partitioning, the better the flexibility of the final solution.

span intersection The span of the individual robots' motions can be calculated as 3D volumes. These volumes can then be intersected pair-wise, thus resulting in exactly the zones where collisions may occur. These zones can then with benefit be divided e.g. as in the space partitioning strategy, to allow for some additional flexibility.

The span generation is exemplified in figure 2. In the figure, the approximate spans are shown as the shaded region surrounding the robots upper arms. Only the span of the upper arm is shown here for simplicity, the rest of the robot also moves, of course. Note that the spans overlap to some extent. The intersection is the critical zone.

For each zone constructed as above in the station, a finite state automaton is generated automatically in Supremica. Each of these state machines has two states, the initial state, labeled "Free", and one state, labeled "Booked". Between these

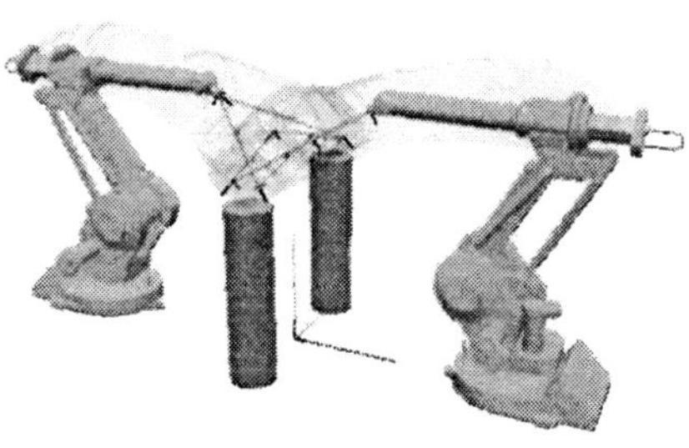

Fig. 2. Two robots and the volumes their upper arms will occupy during the workcycle.

R, M : The robots R, and the set of paths for each robot M_i.
Z : The set of mutual exclusion zones.
foreach $r_i \in R$ **do**
  foreach $m_i^j \in M_i$ **do**
    $z_i^j \leftarrow \texttt{Span}(r_i, m_i^j)$
    $Z_i \overset{\cup}{\leftarrow} z_i^j$

foreach $z_i^j \in Z_i$ *and* $z_k^l \in Z_k$ *where* $k > i$ **do**
  $Z \overset{\cup}{\leftarrow} z_i^j \cap z_k^l$ (* Volume intersection *)

$\texttt{Span}(r_i, m_i^j)$ returns the volume that robot i sweeps through when moving along the path m_i^j.

Alg. 1: Computation of zones.

states, transitions corresponding to the motions of the respective robots will be added. See figure 3.

4.1.1. Main algorithm Let R be a set of n robots, i.e. $R = \{r_1, r_2, \ldots, r_n\}$. Each robot has a program associated with it and each program consists of a set of paths, M_i. Each path describes the motion of a robot between a number of *targets*. A target is for example a welding spot or the home position but may be any location that the robot needs to visit. The paths are assumed to form a loop including anternatives, starting and ending in the home position. Algorithm 1 shows how all zones are computed. Informally, the physical space that the robot is using is computed for all paths that the robot can move along. The zones are computed by intersecting the physical spaces used by different robots.

4.2 Simulation and automata generation

To generate the necessary automata, the robots are simulated individually and each time a robot enters or exits one of the zones, the corresponding automaton gets a new transition representing booking or unbooking of the zone. In addition to this, during the simulation, a finite state automaton is constructed for each of the robots in the simulation. This automaton represents the programmed motion of the robot in terms of discrete events, the starting and finishing of movement along paths. Algorithm 2 describes this formally.

An obvious improvement of the algorithm is to add new targets in the robot program when a zone is entered and exited and split the path in two for more flexibility. This has been implemented but is left out here for simplicity.

Figure 3 shows the automaton representing the zone generated by the span intersection method described above. Note that the event "I1A1_start" also appears in figure 4. The event is disabled if the zone is booked and thus, until the zone has been unbooked, the robot can not enter the zone.

R, M, Z: The robots R, their respective paths M_i and the zones Z from algorithm 1.
P, S : Two sets of automata, P and S representing robots and zones, respectively.
foreach $z_i \in Z$ **do**
  $S_i \leftarrow \texttt{NewAutomaton(Name}(z_i))$
  $q_0^{S_i} \leftarrow free_i \leftarrow \texttt{NewState}(\text{Free})$
  $booked_i \leftarrow \texttt{NewState}(\text{Booked})$
  $Q^{S_i} \overset{\cup}{\leftarrow} free_i \cup booked_i$
foreach $r_j \in R$ **do**
  $P_j \leftarrow \texttt{NewAutomaton(Name}(r_j))$
  $q_0^{P_j} \leftarrow \texttt{NewState(Name(StartTarget}(m_j^0)))$
  foreach $m_j^k \in M_j$ **do**
    $prev \leftarrow \texttt{NewState(Name(StartTarget}(m_j^k)))$
    $move \leftarrow \texttt{NewState(Name}(m_j^k))$
    $next \leftarrow \texttt{NewState(Name(EndTarget}(m_j^k)))$
    $Q^{P_j} \overset{\cup}{\leftarrow} prev \cup move \cup next$
    $start \leftarrow \texttt{NewConEvent(Name}(m_j^k)\text{\_start})$
    $fin \leftarrow \texttt{NewUnconEvent(Name}(m_j^k)\text{\_finish})$
    $\Sigma^{P_j} \overset{\cup}{\leftarrow} start \cup fin$
    $\delta^{P_j} \overset{\cup}{\leftarrow} \langle prev, start, move \rangle$
    $\delta^{P_j} \overset{\cup}{\leftarrow} \langle move, fin, next \rangle$
    foreach $z_i \in \texttt{StartCollisions}(m_j^k)$ **do**
      $\Sigma^{S_i} \overset{\cup}{\leftarrow} start$
      $\delta^{S_i} \overset{\cup}{\leftarrow} \langle free_i, start, booked_i \rangle$
    foreach $z_i \in \texttt{EndCollisions}(m_j^k)$ **do**
      $\Sigma^{S_i} \overset{\cup}{\leftarrow} fin$
      $\delta^{S_i} \overset{\cup}{\leftarrow} \langle booked_i, fin, free_i \rangle$

$\texttt{StartTarget}(m)$ returns the first target of m.
$\texttt{EndTarget}(m)$ returns the last target of m.
$\texttt{StartCollisions}(m_j^k)$ returns the set of zones that the robot r_j entered during the motion m_j^k.
$\texttt{EndCollisions}(m_j^k)$ returns the set of zones that the robot r_j has exited after the motion m_j^k.

Alg. 2: Robot and zone automata generation.

Figure 4 shows the automaton representing the right robot in the figures shown above. It is a straight sequence of starting and finishing of movements representing the different paths in the RobotStudio cell.

4.3 Supervisor synthesis

With all automata generated, the behaviour of the system can be formally examined. The specification of the desired behaviour is simple, the system must never violate the zone automata. Additionally, the system must never block, i.e. the state where all automata are in their respective initial states at the same time must be reachable.

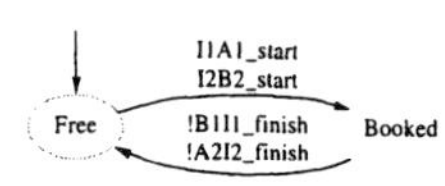

Fig. 3. Automaton model of the zone in figure 2.

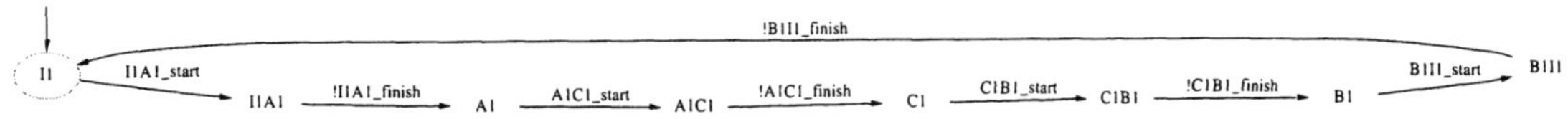

Fig. 4. Automaton representing the right robot in figure 1 as a sequence of states and transitions.

Note that the automatically generated automata are suitable for modular synthesis, i.e. it is not necessarily so that the supervisor must be monolithic, with a potentially huge state space. In the simple example discussed above, verification shows that it is actually enough to use the specification itself as a supervisor.

To show how the synthesis works, a more complicated example including a number of blocking states has been generated. This is similar to the above example but instead of just one zone, this system has three zones, one at the target A, one at target B and one at target C. If both robots move independently and end up booking neighbouring zones, for example zones B and C, a deadlock will arise as the robots will have to switch zones.

Figure 6 shows the closed loop behaviour of this system. Some states are marked by Supremica with a rectangular border and these states must be avoided. Two of these states are deadlocks and the others are states such that there is no way to avoid reaching a deadlocked state. Figure 5 shows the synthesized supervisor that guarantees that the system in figure 6 will never block.

The algorithms used are described in (Åkesson *et al.*, 2003).

4.4 Code Generation

The structure of a modular supervisor is beneficially exploited when generating the PLC-code in order to allow an efficient coding of the supervisor. The idea is to let the PLC do an online synchronization of the modules in the supervised system.

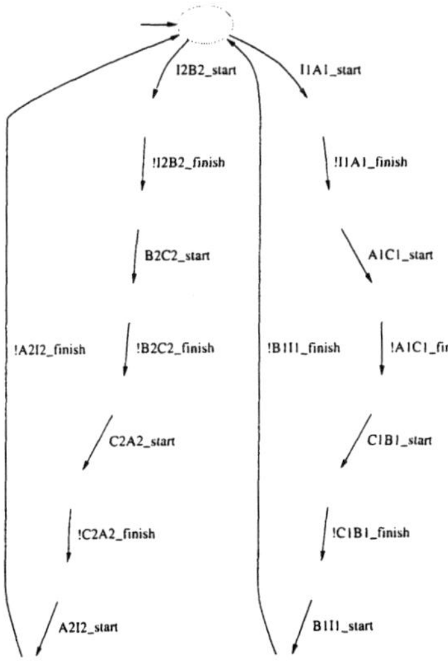

Fig. 5. The synthesized supervisor guarantees fulfilment of the specification and nonblocking.

Events are coded by boolean variables. The variable is true if the event is enabled from the current state and false otherwise. Likewise, states are coded by boolean variables, one for each state in each automaton. These variables are true if the corresponding state is the current state and false otherwise. The connection between the events and the I/O-signals is highly application-specific, and not discussed further here.

Below is an example of a program in Structured Text, a language defined by the IEC 61131-3 PLC programming standard. In the first scan cycle the initial state is set. This is followed by enabling conditions for each event. The variable associated with the event is true if each automaton that has the event in its alphabet is in a state from where it can follow the same event, and false otherwise. Depending on the application and the requirements on the supervisor, zero, one or multiple events may be enabled after evaluating the conditions for all variables. If multiple events are enabled one of them has to be selected for execution. The problem of selecting this event is not straightforward and is discussed in (Dietrich *et al.*, 2001).

Finally, the current states of all the automata are changed. This is done by first checking if the event is enabled; if it is then the current state is updated by executing the event within all automata that can follow the event from the current state. The generated program is executed by one or more PLCs responsible for coordinating the robots.

```
PROGRAM AutomaticallyGeneratedProgram
  VAR (* Internal variables *)
    e_0 : BOOL; (* A1C1_finish uncontrollable *)
    e_1 : BOOL; (* A1C1_start controllable *)
    / code removed /
    enabledEvent : BOOL; (* true if a event is enabled *)
    q_0_0 : BOOL; (* Booked in MutexZone *)
    q_0_1 : BOOL; (* Free in MutexZone *)
    / code removed /
    initialized : BOOL; (* false first scan cycle *)
  END_VAR

  (* Set the initial state *)
  IF (NOT initialized) THEN
    q_0_1 := TRUE;
    q_1_1 := TRUE;
    / code removed /
    initialized := TRUE;
  END_IF;

  enabledEvent = FALSE;

  (* Compute the enabled events *)
  (* Enable condition for event "A1C1_finish" *)
  e_0 := (q_0_0) AND (q_3_1);
  (* Enable condition for event "A1C1_start" *)
  e_1 := (q_2_1) AND (q_3_0);
  / code removed /

    (* Make sure only one event is enabled *)
    (* Priority is given to uncontrollable events *)

  (* Enable condition for event "A1C1_finish" *)
  e_0 = e_0 AND (NOT enabledEvent);
```

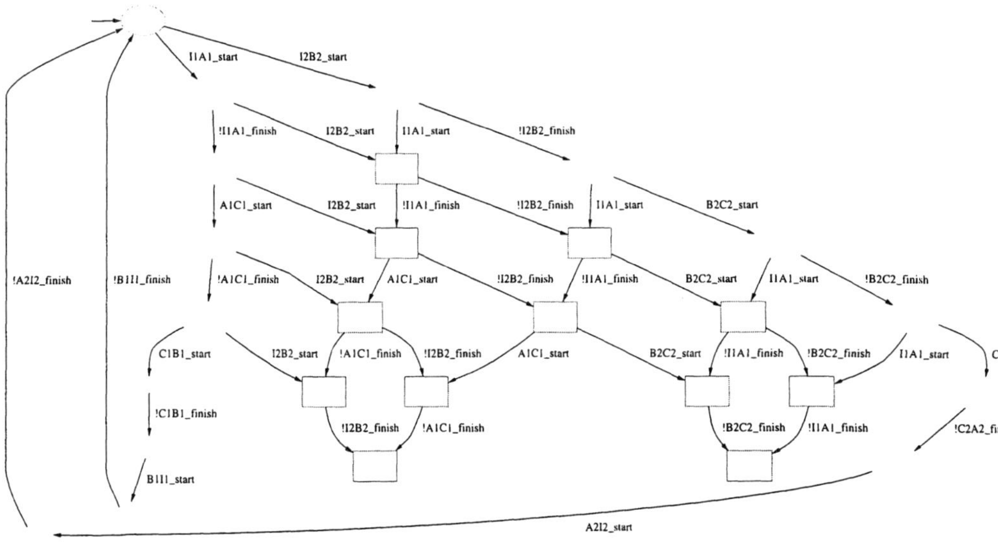

Fig. 6. The synchronous composition of a system with three neighbouring zones and blocking states.

```
  enabledEvent = enabledEvent OR e_0;

  (* Enable condition for event "A2I2_finish" *)
  e_2 = e_2 AND (NOT enabledEvent);
  enabledEvent = enabledEvent OR e_2;
  / code removed /

  (* Change state in the automata *)
  (* Transition for event "A1C1_finish" *)
  IF (e_0) THEN
    (* Transitions in MutexZone *)
    IF (q_0_0) THEN
      q_0_1 := TRUE;
      q_0_0 := FALSE;
    END_IF;
    (* Transitions in RightRobot *)
    IF (q_3_1) THEN
      q_3_4 := TRUE;
      q_3_1 := FALSE;
    END_IF;
  END_IF;
  / code removed /
END_PROGRAM
```

5. CONCLUSIONS

The input to our method is a 3D simulation environment with a number of individually programmed robots. The output is PLC-code that makes sure the robots will never collide or block.

To get a better result, the process above could be iterated with adjustments by the programmer on each iteration. Adjustments could for example be adding of alternative paths where necessary and also adjustment of the mutex zones. Span intersections can give overly restrictive solutions and if so, splitting the zones will help.

Everything described in section 4 is implemented in Supremica using RobotStudio as the simulation environment.

In an extension of this work, we will approach a more general problem setting. Given a 3D simulation environment, a number of robots and a number of targets that should be visited by any robot and in an arbitrary order, find the time optimal sequence of operations and automatically generate robot programs and PLC-code to implement it. This work has already been initiated and some results can be found in the master's theses (Green and Jonsson, 2003; Spensieri, 2003; Stark and Göthberg, 2003).

REFERENCES

ABB (2003). *RobotStudio 3.0 Users Guide.*

Åkesson, K., M. Fabian, H. Flordal and A. Vahidi (2003). Supremica – a tool for verification and synthesis of discrete event supervisors. In: *Proc. of the 11th Mediterranean Conference on Control and Automation.* Rhodos, Greece.

Dietrich, P., R. Malik, W.M. Wonham and B.A. Brandin (2001). Implementation considerations in supervisory control. In: *In B. Caillaud, P. Darondeau, L. Lavagno, X. Xie, editors, Synthesis and Control of Discrete Event Systems.* Kluwer Academic Publishers. pp. 185–201.

Green, E. and A. Jonsson (2003). Optimal coordination of a multi-robot cell. Master's thesis. Chalmers University of Technology, Department of Signals and Systems.

Hopcroft, J.E. and J.D. Ullman (1979). *Introduction to Automata Theory, Languages, and Computation.* Addison-Wesley Series in Computer Science, Addison-Wesley.

Ramadge, P. and W.M. Wonham (1989). The control of discrete event systems. *Proc. of IEEE* **77**(1), 81–98.

Spensieri, D. (2003). Coordination of multi-robot systems using supervisory control and optimization. Master's thesis. Chalmers University of Technology, Department of Signals and Systems.

Stark, H. and T. Göthberg (2003). Definition and implementation of zone interlocking during offline programming – a robcad application. Master's thesis. Chalmers University of Technology, Department of Signals and Systems.

Supremica (2003). `www.supremica.org`. The official website for the Supremica project.

Copyright © IFAC Information Control Problems in Manufacturing, Salvador, Brazil, 2004

ELSEVIER
IFAC
PUBLICATIONS
www.elsevier.com/locate/ifac

AN EVOLUTIONARY PROCESS FOR THE DESIGN OF SAFE COMPUTER-BASED SYSTEMS

Yiannis Papadopoulos[1], Christian Grante[2]

[1]*Department of Computer Science, University of Hull, U.K., y.i.papadopoulos@hull.ac.uk*
[2]*Volvo Cars, Sweden, cgrante@volvocars.com*

Abstract: New design processes in many safety critical industries must address both cost and safety concerns and maximise the potential for automation to address the problem of increasing technological complexity. In this paper, we propose a design process in which techniques for semi-automatic safety and reliability analysis of systems are combined with multi-criteria optimisation techniques to assist the gradual development of designs that can meet reliability and safety requirements within pragmatic cost and profit constraints. The proposed process relies on tools to automate aspects of the design that we believe could be automated, and thus be simplified, without loss of the creative input brought in the process by designers. *Copyright © 2004 IFAC*

Keywords: safety analysis, reliability, multi-objective optimization, fault-tolerance, safety critical systems

1. INTRODUCTION AND BACKGROUND

Safety-critical functions in product and process control systems are often implemented using "stand alone" sub-systems which are isolated and do not share components. In such systems, limited or no interoperation among functions means that failures of non-critical functions are unlikely to cause critical failures. This type of partitioning achieved via modular design has clearly some advantages for safety. However, design via composition of "stand alone" subsystems is clearly not the most effective way of meeting the increasing demands for new advanced functions in product and process control. In vehicles, for instance, by integrating systems and sharing sensory information as well as the basic braking, steering and power-train functionality, it would be possible to deliver functions more economically. In practice though, this would imply distributed architectures in which critical and non-critical data are communicated over the same bus. Integration of functions, though, and its implications, i.e. interaction, interoperation and sharing of data and hardware resources raise serious safety concerns. Common cause failure, and unpredicted dependent failure of critical functions caused by malfunction of non-critical functions are among such safety concerns.

Traditionally, safety assessment in the form of classical safety analysis has provided the means for addressing concerns arising from new technologies. Today, however, increasing complexity in technology seriously questions the applicability of classical risk and safety analysis techniques on new designs. Perhaps the biggest source of difficulties here is the manual nature of the application of those techniques, which makes the classical safety analysis of complex systems increasingly more difficult to complete. Other serious problems include omissions and inconsistencies in the results from the assessment caused by selective and fragmented analysis and a consequent difficulty to relate those results into a coherent view of system failure and recovery.

The automotive but also other transport and manufacturing industries are hence confronted with a situation in which, on one hand economic pressures cause a shift towards function integration, while on the other hand safety concerns push for a re-thinking of the presently established safety assessment processes. Clearly, new development processes will need to emerge in the future which will address both these economic and safety concerns to ensure the viability and success of new integrated technologies. In our view, these processes should address cost and profitability as well as reliability and safety issues from the early stages of the design in order to avoid expensive mistakes and unnecessary costly design iterations. They should also be built upon current achievements in model-based product development, and already established processes in which electronic models of the system are used for system design, prototyping, fault injection and source code generation. A critical factor in dealing with the increasing complexity of systems in the future is automation. Indeed the key in the future for accelerating the development of successful designs, we believe, would be to identify those aspects of the development and assessment process that could be automated and thus simplified without any loss of creativity in the design process. Code generation from abstract models, e.g. using the B method as in (Petit, *et al.*, 2003) and the automated analysis and verification of design specifications are actually two problems in this area where a large body of research is presently focused. Our contribution to this body of research is work towards the development of a model-based safety analysis technique called HiP-HOPS (Hierarchically Performed Hazard Origin and Propagation Studies) that largely automates the development of system fault trees. In this work, we have shown that it is possible to automatically generate safety analyses from design models by augmenting the structure of the model with information in local failure propagation (Papadopoulos, *et al.*, 2001). This has the beneficial effect of simplifying the analysis while keeping the analyses consistent with the design information.

In this paper, we propose a concept in which this work on automated safety analysis is combined with multi-objective optimisation techniques to assist the development of new complex system designs. We describe a tool-supported model-based design process that can address both cost and safety concerns and support the design life-cycle starting from the early stages of design. The proposed process is illustrated in Figure 1. Its three main stages are: a) the optimisation of conceptual functional designs b) the refinement of such designs into models and the semi-automatic safety analysis of such models and c) the optimal allocation of redundancies in models to satisfy given safety and reliability requirements. In the following sections, we present the process focusing on the methods and tools that we develop to automate and simplify aspects of that process. We also outline the current state of this work and draw tentative conclusions based on preliminary results from case studies.

2. OPTIMISATION OF DESIGN CONCEPTS

A new design for a system, for example a vehicle control system, typically starts at conceptual level where the functions to be provided by the system are decided in the context of constraints like the cost of components, the cost of development and the production capabilities of the company. At this stage, there are usually a number of design possibilities that can be translated to different concepts which provide different functions. Issues like potential markets, volumes, costs and ultimately profit have to be addressed for a concept to be translated into a successful design. Typically, the motivation for proceeding further in the design with one or more of those concepts is the potential for profit measured absolutely or more commonly as return on investment.

There is often a wealth of information from past experience or competitor products upon which the decision about the potential of a new design could be evaluated. Such information typically includes knowledge of desirable functions, of the value that customers are prepared to pay for those functions and the components that are likely to be needed to deliver such functions. In practice, though, it is not always obvious how this diverse information can be used to decide which combination of functions and components should be selected for the new design. This is especially true when there is a vast number of viable combinations of functions and components, because a decision on which combination is best would require evaluation of all those design options using, for example, simple or more complicated calculations of profit. Realistically, and for pragmatic designs which may have hundreds of functions and thousands of components, exhaustive evaluation of all options is impossible even if the simplest calculation of profit was adopted.

To address this problem we have developed a program that automates and accelerates this process (Grante and Andersson, 2003). The program takes as input the list of functions considered for the new design, an estimate of the value that customers would be prepared to pay for each function, the components needed to deliver the functions (note that each component can be shared by more than one functions) and the cost of each component. This input is provided in the form of a matrix that relates potential functions to sets of components as this is illustrated in Figure 1. A genetic algorithm is then used to systematically search for those configurations of functions and components that optimise profit within certain budget constraints. The output of the program is a set of candidate concepts for further development that are optimal in the sense that achieve maximum profit for varying (higher or lower) levels of expenditure that fall within the constraints of the budget. To arrive at this set, in the course of an iterative evolutionary optimisation process, the algorithm calculates and then ranks potential solutions according to their degree of dominance (Fonseca and Fleming, 1998) thus forming a *Pareto front* of optimal solutions. In this approach expenditure is calculated as the sum of component costs, while profit is calculated as the sum of function values minus the overall cost.

3. REFINEMENT OF DESIGN AND ANALYSIS

In the context of the proposed design process, the application of the above technique helps to arrive at one or more promising functional configurations for further development. To proceed further in the design, such abstract functional configurations will need to be interpreted into more detailed architectural models (e.g. Simulink models) which can show how

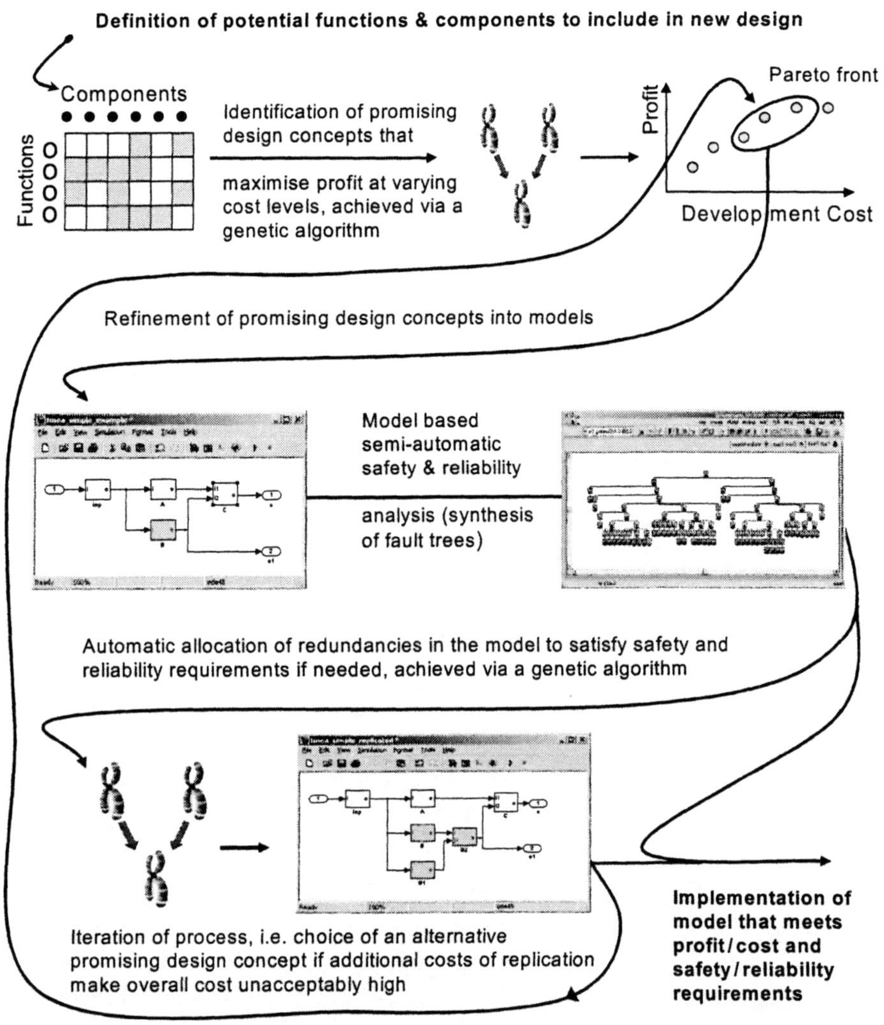

Fig. 1. The proposed design process and supporting tools

components need to interact in order to deliver the desired functions. In the proposed process, the utility of such models is pushed from the domains of simulation and code generation where currently industrial use is focused into a hitherto unexplored area, that of safety and reliability analysis. More specifically, we use our earlier work on automated safety and reliability analysis for evaluating whether the design under development can meet safety and reliability targets.

Application of this work can start once a concept has been interpreted into a Simulink model. Simulink was chosen as a modelling environment because it is both a widely used engineering tool and a tool for which in the past we have developed an automated reliability and safety analysis algorithm. The applicability of the proposed techniques, however, is not restricted to Simulink models. Any model that provides the topology of the system, i.e. components and connections, is suitable for this type of analysis. In (Papadopoulos and Petersen, 2003) for example, we have demonstrated application on models of marine system designs developed in Simulation X.
The proposed analysis can be performed on abstract or more detailed versions of the model as this is progressively refined in the course of the design life-cycle. The first step in the analysis is the establishment of the local failure behaviour of each component in the model as a set of failure expressions which show how output failures of the component can be caused by internal malfunctions and deviations of the component inputs. At this stage, a variant of Hazard and Operability Studies (HAZOP) is used to identify plausible output failures such as the omission or commission of each output and then to determine the local causes of such events. Once this analysis has been completed for all components, the structure of the model is then used to automatically determine how local failures propagate through connections in the model and cause functional failures at the outputs of the system. This global view of failure is captured in a set of fault trees which are automatically constructed by traversing the model and by evaluating the local failure expressions encountered during the traversal. The resultant fault trees may share branches and basic events which arise from dependencies in the model and record common causes of failure, i.e. component failures that propagate through multiple paths simultaneously causing several system failures.

In the context of the proposed process, the application of this technique (i.e. HiP-HOPS) helps to determine whether the design meets given reliability and safety requirements. We should note that credible failure rates are not essential to produce useful results. In the case of software modules or components with insufficient history of use such failure rates would be impossible or very difficult to obtain anyway. In such cases, qualitative application of the technique can still produce useful results. The logical reduction of the synthesised fault trees into minimal cut-sets, for instance, can indicate single points of failure in the system and point out potential design weaknesses.

A HiP-HOPS study will of course in practice indicate that certain requirements cannot be met in which case the design will need to be revisited. Additional safety measures may need to be incorporated and components may need to be substituted by fault tolerant schemes to achieve certain requirements. Design of course is a creative exercise that relies on the technical skills of the design team and draws from previous project experience. However, some further automation could be provided, we believe, to assist some aspects of this process, namely the appropriate decision about the location and level of replication of components in the model.

4. REDUNDANCY ALLOCATION

Critical systems often use replicated components or subsystems to ensure that functions are still provided when components or subsystems fail. In a typical design though, there are many options to replicate at different places in the system and different levels of the design. It may be possible for example to achieve the same reliability by replicating a few sensors in one part of the system or an actuator in another part. Different solutions however will lead to different costs, and the question here is which solution is optimal, i.e. achieves the reliability and safety requirements more cost effectively.

Because the design options for replication in a non-trivial design are typically "innumerable" it is virtually impossible for designers to address the above questions systematically. Thus, people tend to rely on experience and evaluate only a few design options. Some automation in this area would therefore be useful to designers, and could help them achieve successful tradeoffs between cost and reliability in the design of fault tolerant strategies. In our approach, this is achieved by combining the fault tree synthesis concept of HiP-HOPS with recent advances in design optimisation. More specifically, we use synthesised fault trees and genetic algorithms in order to progressively "evolve" an initial design model in which there is no replication of components to a design where replicas have been allocated in a way that minimize the cost of replication while achieving safety or reliability requirements. In the course of the evolutionary process, the genetic algorithm generates populations of candidate designs that incorporate a range of replication strategies based on widely used fault tolerant schemes such as hot or cold standbys and triple-modular redundancy with majority voting. For the algorithm to progress towards an optimal solution, a selection process is necessary in which the fittest designs survive and their genetic make up is passed to the next generation of candidate designs.

The fitness of each design obviously relies on cost and reliability. To calculate fitness, therefore, we need fast ways in which to automatically calculate these two elements. An indication of the cost of a system can be easily calculated as the sum of the costs of its components (although for more accurate calculations life-cycle costs should also be taken into account such as production, assembly and maintenance costs). However, while calculation of cost is relatively easy to automate, the evaluation of safety or reliability is more difficult as conventional methods rely on manual construction of the reliability model. The fault tree synthesis concept that we developed in HiP-HOPS, though, automates the development and calculation of the reliability model, and therefore facilitates the evaluation of fitness as a function of reliability (or safety) thus enabling a selection process through which the genetic algorithm can progress towards an optimal solution. Indeed, in the course of the optimisation process, fragments of fault trees representing modifications in the initial model that are made by the genetic algorithm are mechanically re-constructed, and all fault trees for the current population of models are then re-evaluated in order to determine the fitness of individuals.

One reason for using a genetic algorithm (as opposed to another optimisation technique) is that genetic algorithms carry out an iterative multi-directional search and can therefore be easily designed to generate several good or nearly optimal solutions. In the context of the reliability optimisation problem, this characteristic provides the essential flexibility required in making successful trade-offs between cost, reliability and other factors. Theoretically optimal solutions, for instance, may require unavailable space or imply awkward physical arrangements of components in which case sub-optimal solutions may be preferable. In another scenario, one solution may be optimal in the sense that it achieves a given reliability target with minimal cost. However, it may also be possible to achieve much higher reliability with a rather small increase in cost.

To enable this type of trade-offs among reliability, cost and other parameters, we opted for a *Pareto front* optimization. In this approach, the genetic algorithm is searching not for a single optimal solution but for a set of Pareto optimal solutions, also known as non-dominated solutions. A solution x is said to dominate solution y, if x is as good as y in terms of all objectives, and there exists at least one objective where x is better. Thus, the Pareto set (also called the Pareto front) of optimal solutions in our problem consists of all design concepts that incur a minimal cost which is lower than that of all other concepts that are less or equally reliable.

To calculate and progressively improve the Pareto front, in each iteration, the genetic algorithm determines the relative fitness of candidate design concepts by ranking them according to the scheme described in (Fonseca and Fleming, 1998). In this scheme, ranking is performed using the "degree of dominance" which for each individual equals the number of individuals that is dominated by, plus one. Design concepts with a degree of one, therefore, lie on the Pareto front and represent the fittest designs in each population of candidate concepts. An example of a population ranked according to degree of dominance is illustrated in Figure 2 (each dot in the figure represents a design concept, while the accompanying number gives its degree of dominance). The two parameters of optimisation in this illustrated example are cost and reliability. The figure shows how non-dominated solutions (those with a degree of dominance of one) form an optimal Pareto front.

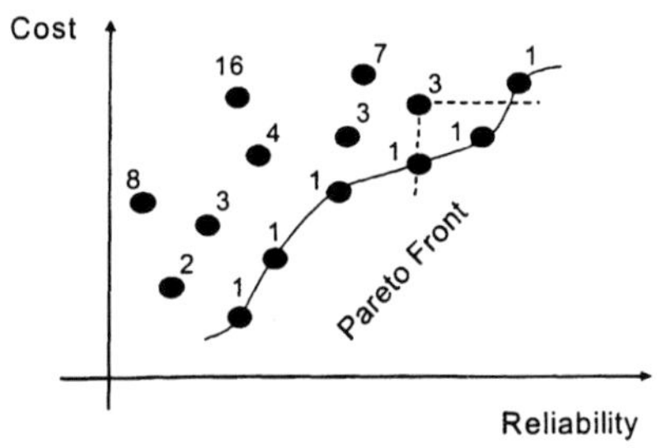

Fig. 2. Example of population ranking

Once more, we should note that quantitative reliability analysis based on failure rates is not necessary to drive this process, which could instead be driven by qualitative criteria, for example the number and criticality of single points of failure in each design. Single points of failure are component failure modes that by themselves cause one or more system malfunctions. Such failures can be determined via automated analysis of the cut-sets of the synthesised fault trees, while the criticality of these failures is established from the severity of their system level effects (i.e. top events in the fault trees).

In the context of the design process, this last stage in the analysis determines whether and where redundancies are needed to achieve the required reliability and safety targets. If the analysis suggests that the cost of replication is unacceptably high then the current design can be rejected, requirements can be reviewed or the process can be iterated for other design concepts. On the other hand, if the analysis shows that safety and reliability requirements can be met with moderate increase in cost then further effort could be committed in the implementation of this design.

5. TOOL SUPPORT AND CASE STUDIES

A set of tools have been developed to support and partly automate the proposed process. The first of those tools, the optimisation tool which assists the search for promising design concepts early in the design has been tested in Volvo Cars on a case study which included 52 active safety functions supported by 48 components (Grante and Andersson, 2003). This configuration could result in more than 70 billion possible design concepts. Manual evaluation of all those concepts with regards to cost and profit would have certainly been impossible. It was only with the aid of the tool that it was possible to arrive at a smaller set of concepts that potentially maximised profit within the given budget constraints. In the area of automated safety analysis, work is developing in technical collaborations with a number of industrial partners which include DaimlerChrysler, Germanisher Lloyd and EADS Airbus. A result of this work is a safety and reliability analysis tool that generates system fault trees from Matlab Simulink or Simulation X models. Case studies using this tool on an advanced brake-by-wire car prototype and a fuel system of a container ship have been reported, for example, in (Papadopoulos, *et al.*, 2001) and (Papadopoulos and Petersen, 2003).

Volvo is currently using the tool to drive the design of an advanced steer-by-wire prototype car. A functional model of the system has been developed in Matlab Simulink and this was deliberately designed without any degraded or fallback modes, in order to test whether the results from the analysis could help in the systematic identification and design of such modes. The model was annotated with information about the local behaviour of functions and then sixteen interconnected fault trees were automatically constructed by the automated safety analysis tool. An indication of the complexity of the failure logic incorporated in this model is that the analysis results to a few thousand cut-sets. It takes a couple of seconds in a standard personal computer to automatically generate and evaluate the fault trees from this model.

Two failure modes were considered during the analysis, the omission and the commission of functions. The synthesised fault trees therefore show how omissions and commissions of input, processing and actuator functions cause system level effects, i.e. omissions or commissions of steering functions. A classification of the severity of those effects into marginal and catastrophic helped to identify the criticality of causes, i.e. failures of input, processing and actuator functions, and this in turn assisted the design of these basic functions. For example, wherever the analysis indicated that the omission of a function had only marginal effects while the commission had catastrophic effects, a design recommendation was made to design the function in a way that it fails silent. This in turn led to the identification of several degraded modes in which non-critical steer-by-wire functions may fail silent with only marginal effects on the system. A hierarchical state-chart was then constructed to show how graceful transition to such modes could be achieved. Driven by these results, a design iteration is currently underway to incorporate new degraded modes in an improved version of the system model.

While the safety analysis tool is relatively mature, the redundancy allocation tool, on the other hand, is still under development. It is therefore currently impossible to comment on performance or make credible comparisons of the proposed approach with earlier approaches to the redundancy allocation problem, at least in terms of speed and scalability. However, we can still identify at this point a few conceptual differences between our approach and earlier work. To the best of our knowledge, for example, earlier approaches calculate reliability using a mathematical function that is derived straight from the system topology on the assumption that the system is formed as a series-parallel configuration of components and that these components and the system itself "work" or "fail" in a single failure mode which typically represents a complete loss of function (Coit and Smith, 1996). It is usually also assumed that a series configuration fails when any of its components fails, while a parallel configuration fails when all constituent components fail. In addition to work on parallel-series systems, some work has also been done to solve the reliability optimisation problem on more complex network topologies in which networks are assumed to function between nodes as long as there is at least one working path connecting these nodes (Denzig, *et al.*, 1997)

The difference between the proposed approach and this earlier work is that, in our approach, the safety and reliability model (i.e. a set of automatically derived system fault trees) is not generated simply from the topology of the system, but from an engineering model which also includes information about the failure behaviour of components. In this model, components do not need to be in series-parallel configurations. The model for example may include bridges between parallel paths, hierarchically nested series and parallel connections as well as complex functional dependencies caused by control loops. A more significant departure from earlier work is that our basic failure assumption goes beyond the classical "success-failure" model. Indeed, in our approach, components can exhibit more than one failure modes which include the *loss* but also the *commission* of functions as well as *value* and *timing* failures. We hope that this type of more detailed and realistic failure modelling, which is possible in the frame of the proposed techniques, will help to improve the accuracy of the automated analysis performed by the tools and ultimately improve the quality of the solutions reached. This, in turn we believe, will help to improve the industrial relevance and uptake of the proposed techniques. Clearly, though, much more work is required in the future to validate this belief.

6. CONCLUSIONS

Development processes in the automotive and other industries need to evolve to address increasing demands for integration of functions over common networked infrastructures. In this paper, we argued that new processes must address both cost and safety concerns and they must maximize the potential for automation to address the problem of increasing technological complexity. We also proposed a process which combines work in automated safety analysis with evolutionary optimisation techniques to achieve some of the attributes that we believe should characterise an effective development process.

The proposed process relies on tools to automate some aspects in the design of critical systems that we believe could be automated, and thus simplified, without loss of the creative input brought in the process by designers. These tools are currently at different stages of development. The automated safety analysis tool is quite mature and allows fault tree analysis of complex design models. On the other hand, the redundancy allocation tool is currently under development in a collaborative project between Volvo and the University of Hull. In this project, a case study is also carried out on a full scale drive-by wire prototype car with individual four wheel steering which has been developed by Volvo cars. The first results from this case study are encouraging and show that the proposed process and tools can be effectively applied from the early stages of the design, integrating safety analysis in the design and potentially elevating it into a tool that can drive the design process. We expect soon to be in a position to report on further results from this study which, we hope, could provide us with further insight into the scalability and limitations of this approach.

REFERENCES

Coit D.W. and A.E. Smith, *Reliability Optimisation of Series-Parallel Systems Using a Genetic Algorithm*, IEEE Transactions on Reliability, R**45** (2):254-260, 1996.

Denzig B., F. Altiparmak and A.E. Smith, *Efficient Optimisation of All-Terminal Reliable Networks Using an Evolutionary Approach*, IEEE Transactions on Reliability, R**46**:18-26, 1997.

Fonseca C. and P. Fleming, *Multi-objective optimization and multiple constraint handling with evolutionary algorithms*, IEEE Transactions on Systems, Man, & Cybernetics, **28**:26-37, 1998.

Grante C. and J. Andersson, *Optimisation of Design Specifications for Mechatronic Systems,* Research in Engineering Design, **14**(4):224-235, 2003

Papadopoulos Y. and U. Petersen, *Combining ship machinery system design and first principle safety analysis*, 8th Int'l Marine Design Conference, pp. **1**:415-426, Athens, May 2003.

Papadopoulos Y., J.A. McDermid, R. Sasse and G. Heiner, *Analysis and Synthesis of the Behaviour of Complex Programmable Electronic Systems in Conditions of Failure*, Reliability Engineering and System Safety, **71**(3):229-247, 2001.

Petit D., G. Mariano, V. Poirriez and J-L. Boulanger, *Automatic Annotated Code Generation from B Formal Specifications,* Symposium on Formal Methods for Railway Operation and Control Systems, pp. 37-44, L'Harmattan, May 2003.

Copyright © IFAC Information Control Problems in
Manufacturing, Salvador, Brazil, 2004

ELSEVIER
IFAC
PUBLICATIONS
www.elsevier.com/locate/ifac

MONITORING AND DIAGNOSIS IN MES OF SEMICONDUCTOR MANUFACTURING WITH "EXTRAKT"

Klaus Kabitzsch[1], Volodymyr Vasyutynskyy[1], Gerolf Kotte[2]

[1] *Faculty of Computer Science*
Dresden University of Technology, D-01062 Dresden, Germany
Fax: ++49 351 463 38460, e-mail: {kk10, vv3}@inf.tu-dresden.de

[2] *SYSTEMA GmbH*
Manfred-von-Ardenne-Ring 6, D-01099 Dresden, Germany
fax: ++49 351 882 4772, e-mail: gerolf.kotte@systemagmbh.de

Abstract: The functionality of Manufacturing Execution Systems (MES) increases tremendously, integrating the modules with sophisticated control logic on different levels. That's why the requirements on 7x24 reliability are much more important at the same time, when the facilities to verify the results decrease. In the presented paper a brief overview of existing problems and solutions in reliability is given. An á posteriori diagnosis based on assumptions about system behavior is described. The architecture and usage of diagnosis tool "Extrakt" is introduced. *Copyright © 2004 IFAC*

Keywords: fault diagnosis, manufacturing systems, event and sequence estimation

1. INTRODUCTION

The semiconductor market experiences an immense turndown in the last years. The pressure on each manufacturer is high, the financial budget is low. To achieve advantages over the competitors, new products and technologies like 300mm wafer fabs were introduced along with quality improving and reducing of the time-to-market. As consequence a variety of hardware and as well as system-to-system interfaces software from different vendors exist for each shop floor automation (SFA) system. In the products now available the software integrates modules of Enterprise Resource Planning (ERP) on the higher abstraction level and of the Equipment Controller (EC) on the tool level. Additionally, the control logic becomes more sophisticated, for instance through Advanced Process Control (APC) and scheduling on wafer level.

The trend of complexity increase goes on. The amount of information is becoming higher, since most data sources are now web-enabled. On the other hand the need for fast reaction on variations at manufacturing level (e.g. APC) and customer demands (e.g. modified orders) forces online rule based business decisions. The next generation of semiconductor manufacturing fabs will likely not be operator-less but with less operators (see fig. 1). All this will lead to systems with:

- much more users and hosts
- much more data with better granularity
- much more software systems (from different vendors) and dependencies between the applications
- much more and more sophisticated, case dependent business rules.

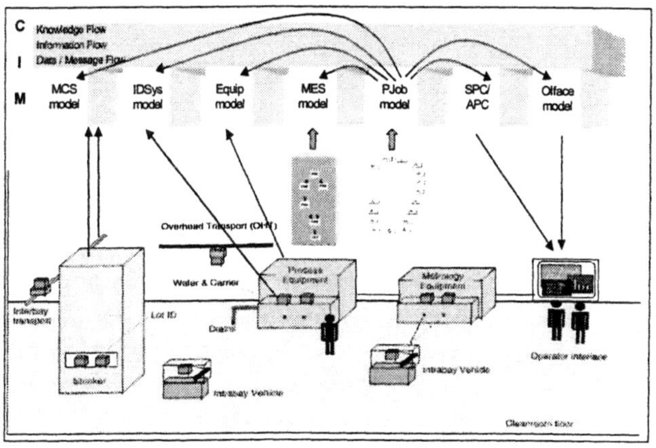

Fig. 1. Layout of a Virtual Fab

The role of the IT system is shifting from a passive "store and retrieve data" to an active controlling system. Without the support by SFA applications (and especially by the MES) the production will stop. Thus, a software downtime is a catastrophic situation with an enormous waste of money. That's why the requirements on 7x24 reliability are much more important at the same time, when the facilities to verify the results decrease.

However, the analysis of the CIM is impeded by its complexity, which increases due to the information flow and control dependencies between the components. Each of such components has its own life cycle, version control and monitoring facility. As a consequence, it is impossible to create a global system specification for each point in time, which would be necessary to detect the faults as deviations from specified behavior. So, the fault detection, identification and removal (or avoidance) are not possible in an automated way, but rely on the qualification and experience of the administration stuff and the response times of the software vendors. The results during partial downtimes, reboot-situations etc. are in most cases unpredictable, so that small causes can result in catastrophic fault situations. The problem becomes even worse, if a core component (like MES) is replaced by a new one, what is planned by many semiconductor manufacturers.

This paper presents the approach to overcoming the difficulties using monitoring (Tsai *et al.*, 1996), which includes storing and analysis of data and messages in CIM. The diagnosis proceeds then using assumptions about system behavior, supporting the administrators with numerous visualization options. The approach is a result of cooperation between Systema GmbH as practitioner in system integration, and Dresden University of Technology, that has a competence in fault diagnosis. The section 2 of this paper contains a brief overview over used methods in reliability achieving. Section 3 pictures the overall proposed architecture for monitoring and diagnosis with diagnosis tool "Extrakt", followed by detailed description of assumption based diagnosis.

2. METHODS TO ACHIEVE RELIABILITY

2.1 Reliability and Fault Management

This article doesn't deal with exact definition of fault, failure, error, and the respective synonyms as stated in (Kopetz, 2001), because it is hard to differ between many reasons that can impact on systems behavior (e.g. limited resources, hardware breakdowns, faulty configuration). That's why the term "undesired technical effect" (Schnieder and Jansen, 2001) is used. It covers not only down-time of components (component is not available), but also incorrect results (not reliable), as well as missing reply-times (not performant), data not fitting the requirements (not usable), and redundant transactions (not optimized). In fact, none of these effects is new. Especially in safety-critical domains, point-solutions cover all facets to keep the system in a stable state. But for semiconductor manufacturing these techniques would mean overkill. Not only the immense efforts to establish safety systems like for a nuclear power plant, also the long time to achieve all certifications for programs would cause the MES to be insufficient for modern production.

Dealing with the results of "undesired technical effects" is not easy. Avoiding the causes, is even more tricky, time-consuming, and in most cases not in the control of the end-user. Depending on system layout and risk assessment, the recovery strategy has to be defined for each used component. Some common methods to achieve reliability are briefly discussed below:

- **Hot stand-by components** avoid the undesired effects by redundancy. However, they lead to hiding of problem's root cause. In most cases, if the application is restarted, the traces of the failing component are erased and so no diagnosis is supported.
- The methods of **quality assurance** (QA) are established at software manufacturers side. But these methods are not applicable for system integrator or administrator, since necessary documentation about products is missing, because it includes intellectual property. Only for a subset of the functionality of each product a use case specification can be done. Techniques for validation and verification are not applicable under these conditions.
- **Bug-tracking systems** are established at each vendor of complex software systems, storing all erased problems. They help the software developer to learn from already detected causes. On the other hand, the customer can view the current status of his problem. For this purpose a Lotus Notes based "Software Problem Report Database" is used at Systema.
- **Excessive testing** is the most extensive (and best usable) method to ensure software quality. Algorithms, typical use cases and typical failure conditions are to be found not only in the foundry book (Haban, 1988) but also in newer case tools. Depending on the resources available (time, hardware, software, persons) there are different approaches may be used. A white-box test of source code checks is used rarely, since source code is not delivered and the techniques (like reviews or coverage tests) are expensive. Black-box tests are based on the specification, from which the test cases are derived. For each use case scenario at least one test case is defined. The expected results have to be defined before test execution. Regarding the control- and the data-flow, error conditions can be forced and the correct failure handling can be tested. As shown above, neither complete specification for the complete system nor for the possible error conditions is available. That's why each test suite will be incomplete, since it can include only known and considered cases.

  Systema has supported its customers doing integration tests. The strategy to establish cyclic, parametric test-scripts at different abstraction levels leads to successful beta-tests for different products. In this context also the tests with negative results are successful. However, this does not necessarily restrict the system from going into productive state. If the circumstances leading to a malfunction are known and understood, they can be avoided in the target environment.

- **Simulation systems** allow to check out the system limits in parallel and concurrent work of different processes. For rare and sporadic faults, long running tests have good chances to hit such situation. Another typical scenario is the creation of a peak load to find out the weakest link, e.g. what is the slowest involved component. These stress tests are used to compare different configurations or tuning options, too.

2.2 Monitoring the system

As shown in previous paragraph, the reliability of the SFA system cannot longer be checked by simple validations of results given to the user. The problems can be detected more easily á posteriori hence it does not need an immediate reaction on a detected fault. The **active monitoring** is used for checking the integrated systems. Depending on systems configurations and parameters, following basic checks are verified:

- Existence (are all service providers running?);
- Availability (are the clients able to request a service?);
- Responsitivity (is the service provider sending a reply?);
- Reliability (is reply or transaction done correctly with respect to time, to contents, to format conformity?).

These checks (so called health checks) are done by an administrative client requests (polling), event driven (based on a known event like an error reply) or on regularly bases (depending on the mean time between failures). If the application is able to support active monitoring, the client does not need to send requests (e.g. ping) but can subscribe for heartbeat messages (watch dog process). Systema has implemented the server classes for this purpose, that are using an abstract message bus interface, including a configurable heartbeat interval.
Unfortunately, active monitoring requires a lot of time and human resources. Neither immediate reaction nor validation of reply (even if checked against expectations) are possible.

If the application is not accessible at a communication media (e.g. batch jobs), **passive monitoring** has to be used. In that case, the trace data are evaluated passively. In many cases, a logging system is already established or can be fairly easy established. If the communication is not hidden, a "sniffer" at fab message bus is an ideal component for it. Systema has developed a monitoring application, that subscribes for many transactions and their replies using an open messaging strategy due to its publish/subscribe paradigm. So any component can subscribe for a name-space-based subset of messages without interfering with the desired destination process. These messages are mirrored then in an Oracle database for ongoing evaluation, which will be described in the next section.

Also the passive monitoring has disadvantages. Only a subset of all occurred faults can be detected, because the information is not specially adapted to monitoring needs. Also detection of partial or temporary down-times is impossible. All these problems should be considered by monitoring implementation.

The next point is evaluation of observation results. State-of-the-art techniques use as a rule the search strategies for known error conditions. In some cases they are passively (the process has to exist) in other cases actively checked (requests the current process state). Even if there are sophisticated methods used to re-detect error conditions like artificial neuronal networks (Cannady, 1998) or expert systems ("solvatio"), it is always a reaction on known faults or anomalies. This work presents a contrary attempt to find out the problematic conditions before they result in critical faults.

3. DESIGN FOR MONITORING AND DIAGNOSIS

The goal of proposed approach is "cognitive diagnosis", that means

- detection of hidden problems before(!) they cause malfunctions,
- detection of redundancies and bottlenecks,
- detection of error-symptoms and not yet classified situations,
- retrieval of context information for certain traces,
- retrieval of an á posteriori model of the observed system (assumptions of normal behavior).

This approach can be used for diagnosing of complex IT systems with incomplete specification. It consists of a collection of observation data in a database and further diagnosis based on different assumptions.

3.1 Instrumentation

"You can not control, what you can not measure."
(Tom de Marco, 1982).

A representative database of log-entries is necessary for diagnosis. Most of the human-readable ASCII logfiles (traces) used up to now do not suit automated evaluation. One uniform **format for each trace record** is needed to be able to include as much different log-histories as possible. So, the biggest common denominator has been determined, that includes compile-time (static) information such as

- module and/or version identifier
- unique record identifier and run-time (dynamic) information
- timestamp (maybe normalized, if a timer jitter between the hosts exists)
- host name
- process-/thread-ID
- optional typed data value for important variables.

The proposed data format was introduced by Systema with full path coverage as logging method in the MES interface components. All data described above are written as binary data with unbuffered write() to hard-disk-drive. The generation is mandatory and not depending on a value of verbose- or debug-level. So it is possible to monitor the components as well, those are not in observation focus.

3.2 *Elements of "Extrakt"*

The methods of cognitive diagnosis are supported by the tool "Extrakt" (Kabitzsch *et al.*, 2001). The diagnosis activities are processed within projects, that manage three basic modules (s. fig. 2).

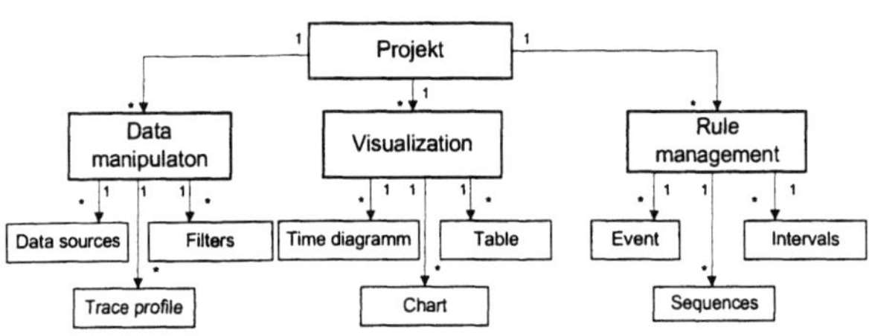

Fig. 2. Elements of "Extrakt".

Data manipulation contains basic mechanisms for data access and acquiring. The data sources describe sources of raw and secondary data, which format is in trace profiles. The filters fulfill necessary data manipulations as well as search requests. The requests are optimized to deal with large data volumes.

Rule management module includes the rules and mechanisms for their control. There are three types of rules. The events define the trace records, that are used in current diagnosis process. Sequences, or complex events, present the sequences of events. The intervals define the time intervals with some properties. For instance, it could be an interval, in which some component configuration is active. The rules can be derived automatically from checkpoint descriptions or be entered by diagnostician manually. The requests for trace records fulfilling rule conditions are executed in data manipulation module. The records that do not fulfill the conditions are marked as outliers and may indicate possible faults.

Visualization module represents the results of trace requests to the human administrator and/or the troubleshooter in form of tables or charts of different types along with further statistics like duration times. It allows:

- Seeing legacy layout (data expansion with context information);
- Seeing overall system state (data condensing);
- Seeing behavior of particular subsystem (data filtering);
- Seeing error conditions (data classification);
- Seeing interaction scenarios (data grouping);
- Seeing timing behavior (data statistics over time);
- Seeing correlations (data statistics over other entries).

The raw trace data as well as found rules and outliers can be viewed. Especially the time-analysis-window (see fig. 3) enables the trouble-shooter to drill down the time point of the problem cause in cases where the timestamp of the effects is already known.

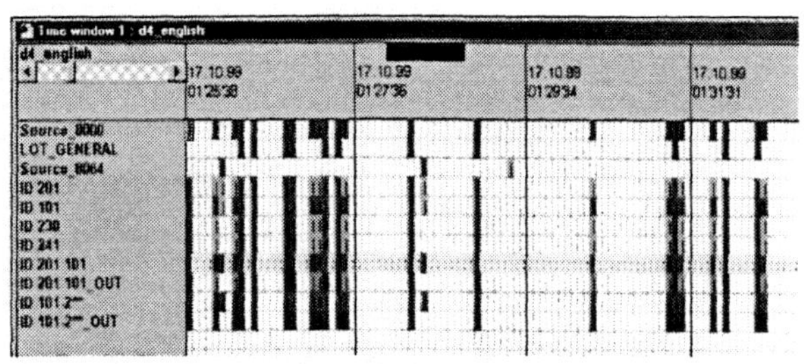

Fig. 3. Time window with events and sequences.

The tool includes some further extensions. For instance, the correlations between different events, sequences or intervals can be proposed automatically.

4. ASSUMPTION BASED DIAGNOSIS

In currently used diagnosis algorithms, the undesired technical effect has taken place and the diagnostician tries to find the cause in the log entries of prior time-period. The knowledge, which component in which environment might be the root cause, is not or only informal determined. The human administrator checks (with the help of a trouble-shooting manual if available) the appropriate files.

The approach of cognitive diagnosis is able to give already some hints in an automated way. The rules for creating such suspicions are described below. The diagnostician stores the found or created rules in a domain based knowledge base. Using this knowledge, a diagnosis is possible without having observed an error pattern. As starting point the given data should be used, like assumptions about production, software or use case scenarios. These assumptions are used for producing the rules in the way described in next sections. The rules are then checked and results are displayed to diagnostician. Depending on these results the diagnostician may reword the requests. In this way the space of possible events is iterative investigated and events are classified as shown in fig. 4.

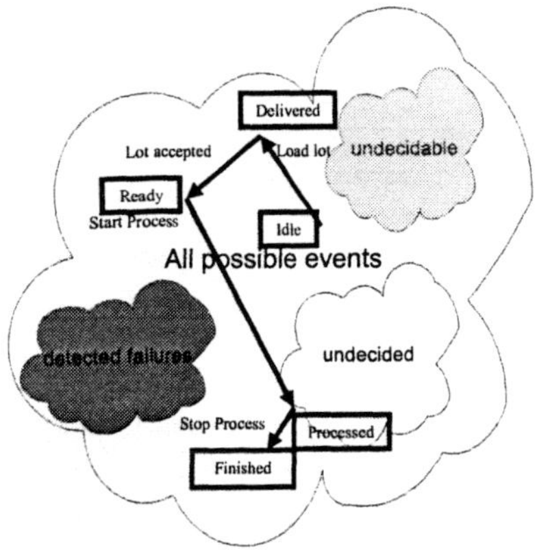

Fig. 4. Event space with evaluated states.

4.1 *Based on assumptions about production*

As mentioned above, no full-covering models for an SFA system exist. They are based on the numbers of process steps to proceed. Starting from the number of waferstarts per week and the route-configurations, there is some **estimation** about the number of "process start" and "process finish" events. An easy test is the **comparison** of these estimations with the real numbers of committed transactions found in the traces. If they match, a normal system state can be assumed. Unfortunately, it is not likely, that the transaction count fits to the first guess. Different deviations can be observed for uncritical reasons:

- More "job-track-in" than "job-track-out" events (cause: rework is not logged as tracking transaction).
- More "job-track-out" than "job-track-in" events (cause: at some inspection step track-in is not required).
- More "job-started" than "job-finished" events (causes: "job-aborted" events, "validation-failed" events, "future-hold-done" events).

The found statements already enlarge the knowledge about normal system behavior. Deviations from this "normal state" can be incorrect functions but as well desired and not yet known. In last case, the estimations have to be adapted to the real system behavior. Based on the global findings, the next step is the **refinement** of time (e.g. count per shift, count per day, count at 17th of each month) and of location (e.g. count per area, count per EC, count per host). The adaptation and refinement **threshold** values for normal behavior will be found in the process of comparison. If the observed numbers are beyond these limits, you can react on that warning sign before a real malfunction would come up.

4.2 *Based on assumptions about software*

Similar estimations can be done for the software. Given the numbers of hosts, users and EC, the raw numbers of transactions, state changes, heartbeats can be calculated. Again it is likely that the estimations do not match exactly with the observed events in the traces. The deviations can be critical or normal, depending on possible cause:

- More heart-beats than expected (cause: more instances than expected, uncritical).
- More administrative messages than expected (cause: human interaction of administrators, uncritical).
- More "start-up" events than "shutdown" events (cause: process crashes with automated or manual restart, critical).
- More "reply-sent" events than "reply-received" events (cause: time-out, has to be evaluated).

Like described above, after matching the estimations with the real data, refinement takes place. The increasing knowledge about system behavior can be used to establish observer processes that create warnings at certain threshold limits. This technique is likely already used for some basic resources like network and disk-space, but can become a much higher granularity (e.g. number of file opens at a host). Researches on that topic are done for instance for intrusion detection, e.g. (Cannady, 1998). A limited forecast can be derived from the data, if system environment is expected to change.

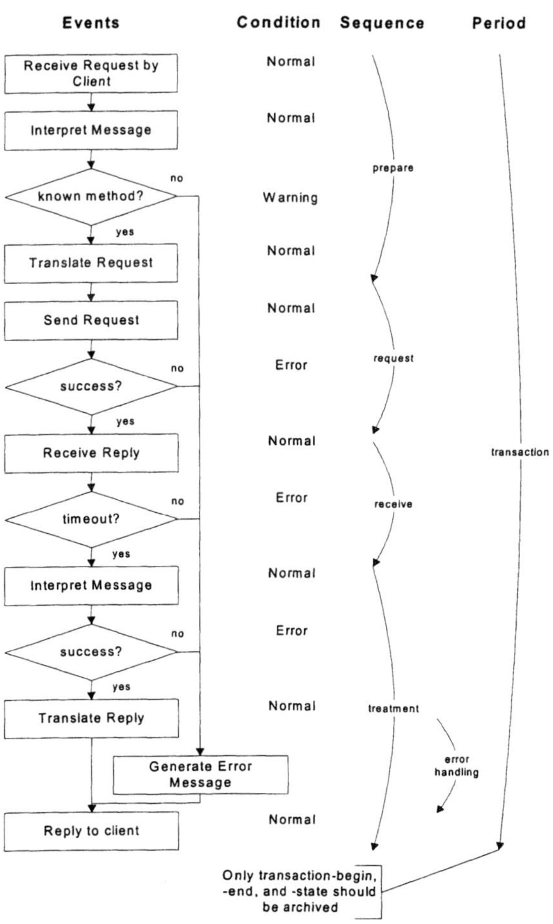

Fig. 5. Typical operations of a Gateway Server

4.3 Based on use case scenarios

The sequencing of the events determines, whether a function was fulfilled successfully. The example in fig. 5 describes the procedures of a gateway process. Some events are tightly coupled (like receive message and interpret message). They can be grouped to sequences that in their turn can be treated as input for higher-level sequences. If the log-entries do not match such sequence, the function was not fulfilled, and a malfunction happened. The sets of observed events can be checked so for semantic correctness.

If disturbances did not allow the completion of the desired function, the method is left through an **exit-point.** This means not necessarily that the complete system is dysfunctional. But especially peaks of such situations should raise warnings to the administrative staff. An identified exit-point means that a requested function was not performed. Based on the nature of the method and the exit-point as well, it is possible to define **recovery barriers**. Between two barriers, all needed data have to be stored (or retrieved again). Depending on the method, there are different options to deal with the exit-point:

- ignore (method is not critical, client will retry),
- retry (method is needed, but client will not force the retry),
- check (method is needed, but retry can cause duplications),
- escalate (method is critical and causes malfunctions).

Sequences have time duration, which can be used for bottle-neck or outlier analysis, benchmarks and configuration experiments. An use-case often used is the verification of user descriptions. The report of a "slow system" is analyzed to find the root cause of the delays. The different response-times can be classified and optimization strategy can be derived.

4.4 Based on combinations and user experience

Sequences do not lie necessarily in small time frames (e.g. customer order to order delivery) nor in same location-frames (e.g. multi-site semiconductor-manufacturers). The next step is to combine the found sequences to higher hierarchy levels. This might cause problems due to different semantics of the data and not always known correlations between software systems. Approach here is to define keys, that are stored as parameter of a found sequence (e.g. lot number and product at MES level). These keys can be combined afterwards with other sequences in other IT systems (e.g. order and customer at ERP level).

Also experience obtained in previous diagnosis activities should be used. Once established, the rules stored in the knowledge base can be

- extended with new use cases(if the business rules in the manufacturing process change),
- adapted to new environment (if the system layout changes),
- reused in other systems (if new systems should be qualified to do same functionality).

5. CONCLUSION

One major factor for the difficulties to establish a reliable software system in the SFA environment is the incomplete knowledge about the complex interaction scenarios. The problem becomes more critical with the automation of production, optimization and quality assurance procedures. That's why the techniques to detect and identify faults are established. The proposed approach is based on a continuously observation of the system. This allows the troubleshooter to find threshold values, anomalies and faults for certain operations and functions, using on assumption about system behavior. The suspicions can be also created in an automated way, giving hints of possible future errors.

REFERENCES

Cannady, J. (1998). Artificial Neural Networks for for Misuse Detection. In: *Proceedings of the 1998 National Information Systems Security Conference (NISSC'98)* October 5-8 1998. Arlington, VA.

Haban, D. (1988). *The Distributed Test Methodology DTM.* Thesis, University of Kaiserslautern.

Kabitzsch, K., G. Kotte and V. Vasyutynskyy (2001). A Monitor Tool for CIM Systems. In: *Proceedings MIM 2001, IFAC Workshop on Manufacturing, Modelling, Management and Control,* Prague, Czech Republic, August 2001, pp. 81 – 85.

Kopetz H. (2001). Real-Time Systems. Design Principles for Distributed Embedded Applications. Kluwer Academic Publishers.

Schnieder E. and L. Jansen (2001). Begriffsmodelle der Automatisierungstechnik – Basis effizienten Engineerings. In: *Proceedings of EKA 2001,* Braunschweig 2001.

"solvatio" http://www.iisy.de/en/

Tsai J.P., Y.-D. Bi, H.-Y. Chen and R. Smith (1996). Distributed Real-Time System: Monitoring, Visualization, Debugging, and Analysis. Wiley-Interscience.

Copyright © IFAC Information Control Problems in Manufacturing, Salvador, Brazil, 2004

ELSEVIER
IFAC
PUBLICATIONS
www.elsevier.com/locate/ifac

DYNAMIC BAYESIAN NETWORKS MODELLING THE DEPENDABILITY OF SYSTEMS WITH DEGRADATIONS AND EXOGENOUS CONSTRAINTS

Weber P.[+], Munteanu P.[*], Jouffe L.[*]

[+] *Centre de Recherche en Automatique de Nancy (CRAN), CNRS UMR 7039*
Université Henri Poincaré, Nancy 1; 2, rue Jean Lamour
54519 VANDOEUVRE-LES-NANCY Cedex – France,
(Philippe.Weber@esstin.uhp-nancy.fr)

* *Centre de Recherche du Groupe ESIEA*
38, rue des Docteurs Calmette et Guérin
53000 LAVAL – France, ({munteanu,jouffe}@esiea-ouest.fr)

Abstract: The work reported here presents an original method to model dependability of systems, taking into account degradations and failure modes governed by exogenous constraints. The component degradation dynamics is considered as a semi-Markov process. Environmental behaviour introduces switching models conditioned by exogenous constraints. Dynamic Bayesian Networks (DBN) are employed to formalise such complex dynamic processes through a compact representation. DBN allow simulating these processes, taking into account events due to the environmental behaviour. A hydraulic system is used to illustrate the reliability estimations obtained by the proposed modelling method. *Copyright © 2004 IFAC*

Keywords: Dynamic Bayesian Network, Markov Switching Model, Hidden Markov Models, Reliability evaluation.

1. INTRODUCTION

One challenge of the enterprise is to dynamically maintain the quality of the services delivered by processes along their life cycle. Thus the goal is to design decision-making aid systems. Nevertheless, most of the current systems do not provide the means for interpretation of information coping with large process disturbances. Moreover, decisions can be taken without a perfect perception of the system state. This partial perception argues in favour of using a probabilistic estimation of the system state. Tools issued from Artificial Intelligence, as described in (Boutillier, *et al.*, 1999), can be used to provide decision-making for manufacturing systems.

Works on system dependability and Bayesian Networks (BN) have recently been developed. Bobbio, *et al.*, (2001), explain how the Fault Tree (FT) can be achieved using BN. Torres-Toledano and Sucar (2003) present the relation between Reliability Block Diagram (RBD) and BN. Weber, *et al.* (2001), proposed a model allowing fault diagnosis using system functioning and dysfunctioning analyses based on Object Oriented Bayesian Network (OOBN). The solutions proposed in these papers are all based on static probabilistic models.

In Aven and Jensen, (1999), a system made of *n* components with positive random lifetimes is called a complex system if components can fail at the same time with positive probability, and the components are allowed to have an arbitrary (finite) number of states (in the following: multistate). The reliability estimation of such systems, taking into account the effects of failure combinations, results in developing more and more scenarios with respect to the system complexity. Then, the classical modelling methods are close to their limits. Moreover, in order to improve the decision-making in maintenance strategies, the goal is to define a dynamic model of the process behaviour. This model allows computing state probability distributions (prognosis) by taking into account the component age, the last executed maintenance operations and the evolution of the environment of the process.

Methods relying on Markov Chains (MC) models lead to a combinatorial explosion of the number of states to specify the system. This modelling method is then unsuitable for complex systems and alternative methods have been proposed. Meshkat *et al.*, (2002) present the relation between MC and Dynamic Fault Trees (DFT). Another method, based on the DFT, is presented by Bouissou and Bon (2003). These authors introduced a modelling formalism that combines concepts from fault trees and MC. Besides, we have shown that probabilistic models such as the Dynamic Bayesian Networks (DBN) allow decreasing effect of combinatorial explosion in reliability estimation by a more synthetic description of the system (Weber and Jouffe 2003).

The aim of this paper is to use DBN as an equivalent model to the MC (Padhraic, 1997). The problems considered here are those whose dynamics can be modelled as stochastic processes. The current system state determine the probability distribution over the next states. The paper is divided into five sections. Section 2 presents the problem statement concerning the MC model and the fast growing of its state cardinality with respect to the modelled system complexity. Section 3 introduces the DBN theory used in the following as an original modelling of complex system reliability. A simulation with a hydraulic system is developed in section 4. Finally, section 5 presents conclusions.

2. PROBLEM STATEMENT

In order to take the uncertainty into account, the process behaviour is represented as a random variable that takes its values from a finite state space corresponding to the possible process states. A MC allows modelling the dynamics of sequences taken by these states (Boutillier, *et al.*, 1999).

2.1. *The Markov Chain notations*

Let X be a discrete random variable that models a process with a finite number of mutually exclusive states $\{s_1,...,s_M\}$. The vector π, then, denotes a probability distribution over these states:

$$\pi = [\pi(s_1) \ \ ... \ \ \pi(s_m) \ \ ... \ \ \pi(s_M)],\ \pi(s_m) \geq 0$$

$$\text{with } \pi(s_m) = p(X = s_m) \text{ and } \sum_{m=1}^{M} \pi(s_m) = 1 \qquad (1)$$

Assuming that the system evolves between states, where the occurrence of an event marks the transition from a state (k) to the next state (k+1), then the process produces the sequence $(\pi_0, \pi_1,...\pi_{k-1}, \pi_k)$ that can be modelled as a discrete MC if:

$$\pi_k(s_m) = p(X_k = s_m | \pi_{k-1}) \qquad (2)$$

The Markov property makes it possible to specify the statistical relationships among states as a transition probability matrix $\mathbf{P}_{MC}$. If the transition probabilities $p_{ij} = p(X_k = s_j | X_{k-1} = s_i)$ are time independent then the MC is said to be homogeneous.

$$\mathbf{P}_{MC}(X_k | X_{k-1}) = \begin{bmatrix} p_{11} & p_{12} & \cdots & p_{1M} \\ p_{21} & & & \cdots \\ \cdots & & & \cdots \\ p_{M1} & p_{11} & \cdots & p_{MM} \end{bmatrix} \qquad (3)$$

The sequence of the probability distribution is estimated by:

$$\pi_k = \pi_0 (\mathbf{P}_{MC})^k \qquad (4)$$

2.2. *Models of system reliability*

The reliability of systems or component is classically modelled with MC. Nevertheless the complexity of the model depends on the assumptions made to approximate the stochastic process. Unfortunately, the modelling complexity increases dramatically as we want to be close to the real behaviour of the system reliability. Indeed, the state space describing the system is made of the Cartesian product of the component states.

Hypothesis of a Markov process

Let's consider the reliability of a component. It is modelled by a discrete random variable X with states $\{o, f\}$, in which (o) indicates an operational state and (f) indicates a failure state.

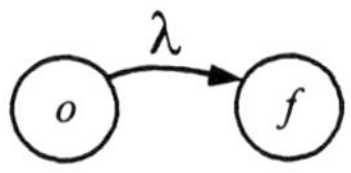

Fig. 1. Markov Chain.

To model the reliability, the matrix $\mathbf{P}_{MC}$ between the states $\{o, f\}$ is defined as follow:

$$\mathbf{P}_{MC} = \begin{bmatrix} 1-\lambda & \lambda \\ 0 & 1 \end{bmatrix} \qquad (5)$$

λ represents the constant failure rate and the density function of the failure time $f(t)$ is:

$$\lambda \equiv \frac{f(t)}{R(t)} = \frac{1}{MTTF},\ f(t) \equiv -\frac{dR(t)}{dt} = \lambda e^{-\lambda t} \qquad (6)$$

where MTTF represents the Mean Time To Failure. The solution of this process is well known and leads to the reliability $R(t)$ of a component with exponential distribution of failure time:

$$R(t) \equiv \exp\left\{ -\int_0^t \lambda(\tau) d\tau \right\} = e^{-\lambda t} \qquad (7)$$

From the MC, the component reliability is given by:

$$R(k) = p(X_k = o | \pi_{k-1}) = \pi_k(o) \qquad (8)$$

The simulation of this behaviour is defined by the recurrence deduced from eq. (4). This equation is justified by systems composed from n multistate components. The complexity of the expression (7) describing the system reliability quickly increases with respect to the number of states. Then eq. (4) estimates the probability distribution over the system states. Assuming that $i \in \{0...l\}$ represents the functioning states, system reliability is defined as:

$$R(k) = \sum_{i \in \{0...l\}} \pi_k(s_i) \quad (9)$$

Hypothesis of a semi-Markov process

When one wants to take into account degradation, the failure rate has to vary over time. This hypothesis is indeed more realistic because component aging implies more and more maintenance actions. To model this behaviour, a generalisation of the Markov Process called Semi-Markov Process (SMP) is introduced (Gertsbakh (2000) pp. 117). The classical way is to model the probability density function of the failure time distribution with a continue distribution as Gamma, Weibull or Normal (Gaussian) distributions. The matrix defined in eq.(5) becomes time dependant. The MC is then not homogeneous anymore. In order to model SMP, an approximation can be made by the transformation of the SMP in an equivalent MP by approaches such as the supplementary variables (Cox, 1955), or Phase Type Distribution (PTD) approximation (also called state space expansion or method of stages). The method consists in increasing the number of system states to model the dynamics or degradations. Then a component is modelled by a discrete random variable X_k with states $\{s_o, s_{d1}, s_{d2}, ...s_f\}$ where s_{di} models the hidden states of degradation. This modelling results in failure rate as a function of the time $\lambda(k)$. For instance, the MP can be defined from constant parameters λ_{ij} of a generalized Erlang equivalent model (Cox 1962) (Fig. 2). To estimate the reliability of the system the following equation is computed from (4):

$$R(t) = \sum_{s_i \neq f} \pi_k(s_i) \quad (10)$$

A general model of this SMP can be formalized from Hidden Markov Model (HMM). Because the observations of the physical process is made in practice through the availability of the function and the states of degradation are usually un observable. Therefore HMM approximates failure time of SMP.

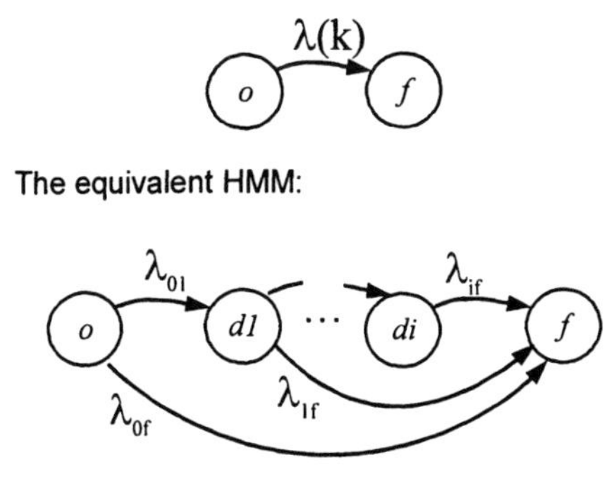

Fig. 2. Hidden Markov Model.

Hypothesis of a semi-Markov process with exogenous constraint

The system environment (exogenous events) impacts on the probability density function of the failure time distribution. Let's consider several system functioning modes. The component degradation follows then several behaviors. To take into account these exogenous events, one has to use several models to represent each situation according to an environmental context, and models become complex when the system is composed by n components with several failure modes.

Markov Switching Models (MSM) are introduced to model this kind of stochastic process with exogenous events. These models are also viewed as conditional HMM where the transition distributions are conditional to an exogenous variable. The MSM models the non-stationary that are due to abrupt changes in the functioning modes of the system (Bengio 1999). Furthermore, the model is represented by Input-Output HMM (IOHMM) if the distribution over the states of the exogenous variable is known. The difference between standard HMM and MSM or IOHMM is that HMM represents the distribution of $p(X_k)$ whereas the MSM or IOHMM represent the conditional distribution $p(X_k|Q^T_k)$ given the input sequence Q^T_k = q_0, q_1,..., q_k. where q_k represents the exogenous constraint with states $\{s_1, s_2, ...s_M\}$.

The computation of π_k with MSM can be carried out only by approximate methods, i.e. simulation. Indeed, it is unfeasible to obtain the analytic solution of this kind of hybrid differential system. In case of IOHMM, it is the distribution over the states of the exogenous variable which condition the solution. The exact resolution of this model is then also unfeasible using classical MC and recurrence equation.

The MC, HMM, MSM or IOHMM are methods well suited to model the reliability of a complex entity/system of low dimension. However, within the framework of general complex systems, the combinatorial explosion of states makes these methods unmanageable. In the following, a method based on DBN is presented. It provides a synthetic representation for modelling complex systems.

3. BAYESIAN NETWORK MODELLING

BN are probabilistic networks. Each node represents a variable and arcs indicate direct probabilistic relations between the connected nodes. DBN allow taking time into account by defining different nodes with respect to different time slices.

3.1. The Bayesian Network notations

BN are directed acyclic graphs (Jensen, 1996) defined as a pair: $\mathcal{G}$=((N, A),$\mathcal{P}$), where "N" is a set of nodes; "A" is a set of arcs; $\mathcal{P}$ represents the set of conditional probability distributions that quantify the probabilistic dependencies.

A discrete random variable X is represented by a node $n \in N$ with a finite number of mutually exclusive states: $\mathcal{S}_n : \{s_1^n, \ldots s_M^n\}$. The vector $x^n = [p_1 \ \ldots \ p_M]$ denotes a probability distribution over $\mathcal{S}_n$ like eq. (1), where p_m is the marginal probability of n being in state s_m^n. In Fig. 3, the nodes n_i and n_j are linked by an arc $(n_i, n_j) \in A$ then n_i is considered as a parent of n_j. The parent set of a node n_j contains all the parents of n_j (*e.g.*, $pa(n_j) = \{n_i\}$ in Fig. 3).

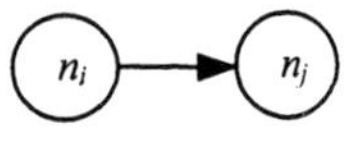

Fig. 3. A basic BN.

In this work, the set $\mathcal{P}$ is represented with Conditional Probability Tables (CPT). Then, each node has an associated CPT. For instance, in Fig. 3, n_i and n_j are defined over the states $\mathcal{S}_{n_i} : \{s_1^{n_i}, \ldots s_M^{n_i}\}$ and $\mathcal{S}_{n_j} : \{s_1^{n_j}, \ldots s_L^{n_j}\}$. Then, the CPT of n_j is defined by the conditional probabilities $p(n_j | pa(n_j))$ over each state of n_j and of its parents. This CPT is represented as a matrix. For instance, the CPT of node n_j of Fig. 3 is:

$$\mathbf{P}(n_j | n_j) =$$

$$\begin{bmatrix} p(n_j = s_1^{n_j} | n_i = s_1^{n_i}) & \cdots & p(n_j = s_L^{n_j} | n_i = s_1^{n_i}) \\ \vdots & & \vdots \\ p(n_j = s_1^{n_j} | n_i = s_M^{n_i}) & \cdots & p(n_j = s_L^{n_j} | n_i = s_M^{n_i}) \end{bmatrix} \quad (11)$$

For the root nodes, *i.e.* the nodes without parent, the CPT contains just a row describing the *a priori* probability of each state.

Various algorithms can be used to do exact or approximate inference, i.e. for the computation of marginal probabilities. The most classical exact inference algorithm relies on the use of a junction tree (more explications can be found in (Jensen, 1996, pp. 76)). Inference in BN allows taking into account any state variable observation (an event) for the updating of the probabilities of each variable. Without observations, the computation is based on *a priori* probabilities. As observations are made, the knowledge is incorporated in the network and the probabilities over the process states are updated.

3.2. Dynamic Bayesian Network

A DBN includes a temporal dimension. This new dimension is managed by time-indexed random variables. X_i is represented at time step k by a node $n_{(i,k)} \in N$ with a finite number of states $\mathcal{S}_{n_i} : \{s_1^{n_i}, \ldots s_M^{n_i}\}$. $x_k^{n_i}$ denotes the probability distribution over these states at time step k. Several time stages are represented by several sets of nodes $N_0, \ldots N_k$. N_k includes all the random variables relative to the time slice k (Hung, *et al.*, 1999; Boutillier, *et al.*, 1999, pp. 38-45).

An arc linking two variables belonging to different time slices represents a temporal probabilistic dependence between these variables. Then DBN allow to model random variables and their impacts on the future distribution of other variables. With this model, the future (k+1) is conditionally independent of the past given the present (k), which means that the CPT eq. (11): $\mathbf{P}(n_{i,k+1} | pa(n_{i,k+1}))$ respects the Markov properties (Kjaerulff, 1995). Moreover, this CPT is equivalent to the MC model of the variable X_i described in the section ***2.1*** if $pa(n_{i,k+1}) = n_{i,k}$ and $\mathcal{S}_{n_{i,k}} = \mathcal{S}_{n_{i,k+1}}$: $\mathbf{P}(n_{i,k+1} | n_{i,k}) = \mathbf{P}_{MC}$. Starting from an observed situation at time step k=0, the probability distribution $x_k^{n_i}$ over X_i states is computed by inference.

To compute $x_{k+T}^{n_i}$, several solutions are proposed in the literature. One of them consists in developing T time slices (Fig. 4), obtaining then a network which size grows proportionally to T (Kjaerulff, 1995). This solution is unsuitable for reliability analysis because the process have to be analysed over a large time horizon. This leads to an explosion of the number of nodes. Nevertheless, this modelling is exactly equivalent to the model described in the above section (Bengio, 1999).

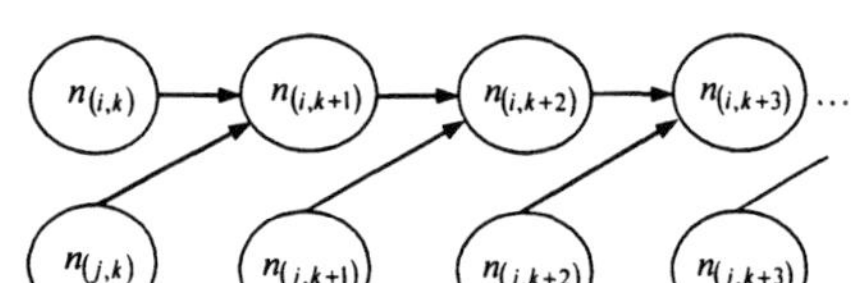

Fig. 4. A un-rollup DBN modelling IOHMM.

Another solution, which keeps a compact network form, is based on iterative inferences. This solution is used in the following. Indeed, it is possible to compute the probability distribution of any variable X_i at time step k+1 based on the probabilities corresponding to time step k. The probability distributions at time step k+2... are computed using successive inferences (Welch and Thelen 2000). Then a network called *2-TBN* (Boyen and Koller 1998) (Fig. 5) with only two time slices is defined. The first slice contains the nodes corresponding to the current time step (k), the second one those of the following time step (k+1). A node is called temporal node when it has a direct link with its temporal clone at time step (k+1), as $n_{(i,k)}$ in Fig. 5.

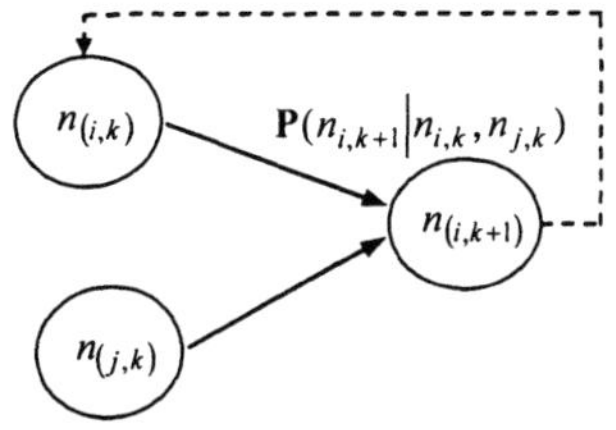

Fig. 5. A 2-TBN modelling IOHMM with temporal node $n_{(i,k)}$ and exogenous observations $n_{(j,k)}$.

Observations, introduced as hard evidence or probability distributions, are only realised in the current time slice k. Then an introduction of the context as exogenous random variable is made by the means of a node $n_{(j,k)}$ which governs the behaviour of X_i thank to the CPT defined in eq. (12). When all the temporal nodes are independent, we can still use an exact inference algorithm. Otherwise, we have to use approximate inference methods, as for example particle filtering (Doucet *et al.*, 2000, Koller and Lerner 2001). When exact inference can be used, the time increment is carried out by setting the computed marginal probabilities of the node at time step k+1 as observations for its corresponding node in the previous time slice.

4. APPLICATION

The proposed method is applied to a example of reliability analysis. This example allows comparing the method based on DBN with the one based on MC. Fig. 6 describes the system. A valve V1 and a pump P2 are used to distribute a fluid.

The valve has two failure modes: remains closed (fRC) or remains opened (fRO). The valve failure rates are the following: $\lambda_{1RC} = 1 \cdot 10^{-4}$; $\lambda_{1RO} = 1.5 \cdot 10^{-4}$ and the failure rate of the pump is time dependent. This dependence follows the Weibull distribution:

$$\lambda_2(t) = \frac{\beta' . t^{\beta'-1}}{\eta'^{\beta'}}, \beta' = 1.5, \eta' = 1000 \ .$$

Parameters of the Weibull distribution are independent of the nature of the fluid. But the valve's parameters are modified by the nature of the fluid. The PH induces oxidation of the valve V1. The parameter λ_{1RO} and λ_{1RC} are modified with factor $\phi = 1$ for the fluid A and $\phi = 4$ the fluid B.

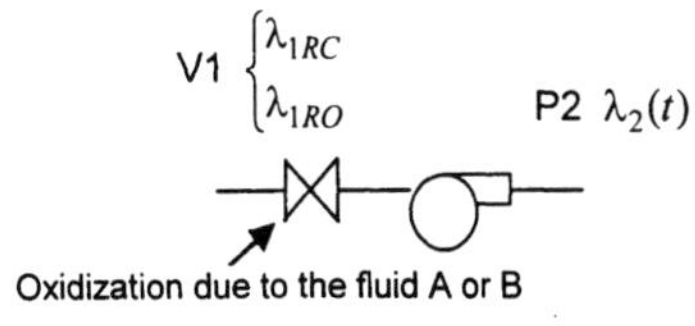

Fig. 6. Hydraulic system.

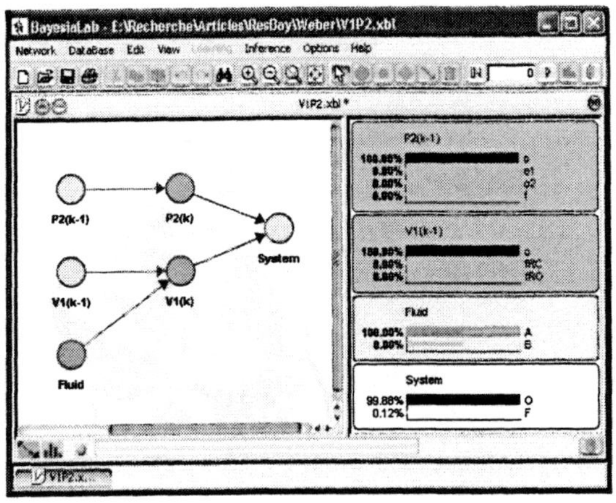

Fig. 7. DBN model of the hydraulic system.

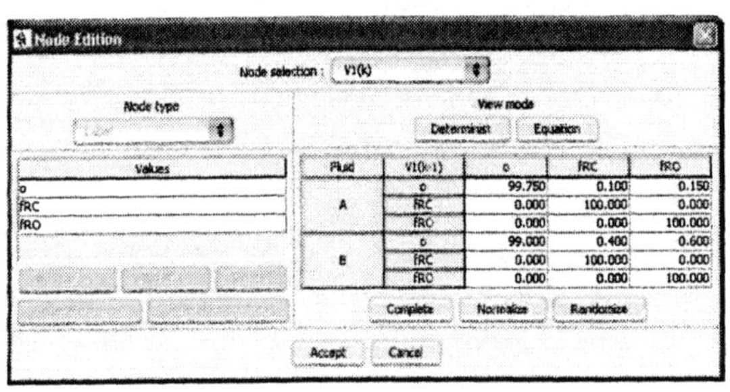

Fig. 8. CPT to model V1(k).

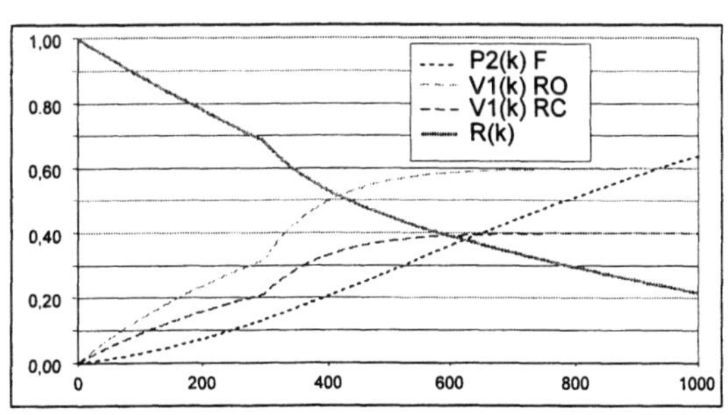

Fig. 9. System reliability estimation.

The CPT is defined as $\mathbf{P}(n_{i,k+1}|n_{i,k}, n_{j,k})$:

$$\mathbf{P}\left(n_{i,k+1}\middle|n_{i,k}, n_{j,k}\right) = \begin{bmatrix} p\left(n_j = s_1^{n_j}\middle|n_i = s_1^{n_i}, n_j = s_1^{n_j}\right) & \cdots & p\left(n_j = s_L^{n_j}\middle|n_i = s_1^{n_i}, n_j = s_1^{n_j}\right) \\ \vdots & & \vdots \\ p\left(n_j = s_1^{n_j}\middle|n_i = s_M^{n_i}, n_j = s_1^{n_j}\right) & \cdots & p\left(n_j = s_L^{n_j}\middle|n_i = s_M^{n_i}, n_j = s_1^{n_j}\right) \\ \vdots & & \vdots \\ p\left(n_j = s_1^{n_j}\middle|n_i = s_M^{n_i}, n_j = s_{M'}^{n_j}\right) & \cdots & p\left(n_j = s_L^{n_j}\middle|n_i = s_M^{n_i}, n_j = s_{M'}^{n_j}\right) \end{bmatrix} \quad (12)$$

A MSM has to be defined to model the impacts of the fluid type (A or B) on the V1 failure rate. A HMM representation of the pump increases the number of states. The approximation is made through the TPD-P2 model:

$$P_{TPD-P2} = \begin{bmatrix} 9.97\cdot10^{-1} & 2.61\cdot10^{-3} & 0 & 1.52\cdot10^{-4} \\ 0 & 9.97\cdot10^{-1} & 2.61\cdot10^{-3} & 7.99\cdot10^{-4} \\ 0 & 0 & 9.97\cdot10^{-1} & 2.61\cdot10^{-3} \\ 0 & 0 & 0 & 1 \end{bmatrix}$$

The combination of component states leads to define the MC model of this system over 11 states of the system. A model of this process is easily realised by means of the DBN depicted in Fig. 7. V1(k) have three states: Available (o), failure Remains Open (fRO) or failure Remains Closed (fRC) and the P2(k) have four states (o,...o2) represent the three degradation states and (f) the failure state. The CPT used to estimate the dynamic behaviour of the components V1(k) is depicted in Fig. 8. The state probabilities of components at current time step can be easily extracted from the nodes V1(k) or P2(k). The reliability of the system is computed from the logical combination: System is unavailable if V1=fRC or P2= f.

The propagation through the DBN model allows taking into account the dependency between the failure modes and the impact of the environment (fluid A or B) for the computation of the system reliability R(k). Inferences are realised thanks to the **BayesiaLab** (http://www.bayesia.com) to simulate the DBN modelling R(k) of the system over 1000 time steps, as depicted in Fig. 9. The fluid changes from A to B at time step 300.

5. CONCLUSION

The proposed method, based on the DBN theory, easily allows designing DBN structures for the modeling of temporal evolution of complex systems. The correspondence between DBN model and MP is presented and applied to the system reliability estimation. The proposed method is a good solution to model the reliability of complex systems. Indeed, the number of states needed to model a complex system with MC increases exponentially. As the DBN representation is based on the modelling of process entities, the obtained model is more compact and readable than MC. This paper shows that DBN represent a very powerful tool for decision-making aid in maintenance.

REFERENCES

Aven T. and U. Jensen (1999) *Stochastic models in reliability*. (Applications of mathematics), Springer-Verlag, ISBN 0-387-98633-2, SPIN 10695247, New York.

Bengio Y. (1999). Markovian models for sequential data. *Neural Computing Surveys*, **2**, pp. 129-162.

Bobbio, A., L. Portinale, M. Minichino and E. Ciancamerla (2001). Improving the analysis of dependable systems by mapping fault trees into Bayesian networks. *Reliability Engineering and System Safety*, **71**, pp. 249-260.

Bouissou M. and J.L. Bon (2003). A new formalism that combines advantages of fault-trees and Markov models: Boolean logic driven Markov processes, *Reliability Engineering and System Safety*, **82** (2), pp 149-163.

Boutillier C., T. Dean, S. Hanks (1999). Decision-theoretic planning: structural assumptions and computational leverage. *Journal of Artificial Intelligence Research*, **11**, pp. 1-94.

Boyen X. and Koller D. (1998) Tractable for Complex Stochastic Processes. *14th Annual Conference on Uncertainty in AI (UAI)*, Madison, Wisconsin, July 1998, pp. 33-42.

Cox, D.R. (1955) A Use of Complex Probabilities in the Theory of Stochastic Processes, *Proceedings of the Cambridge Philosophical Society*, **51**, pp. 313-319.

Cox D.R. (1962) *Renewal theory*. Chapman and Hall, ISBN 041220570X, London.

Doucet A., N. De Freitas, K. Murphy, S. Russel, (2000). Rao-Blackwellised particle filtering for Dynamic Bayesian Networks, *16th Annual Conference on Uncertainty in AI (UAI)*, pp. 176-183.

Gertsbakh I. (2000). *Reliability theory with applications to preventive maintenance*. Springer-Verlag (Ed.), ISBN3-540-67275-3, SPIN 10763074, New York.

Hung K. B., S. Venkatesk, G. West (1999). Layered dynamic probabilistic networks for spatio-temporal modelling. *Intelligent Data Analysis*, **3**, pp. 339-361.

Jensen, F.V. (1996). *An Introduction to Bayesian Networks*. (UCL Press (Ed)). London.

Kjaerulff U. (1995). dHugin: a computational system for dynamic time-sliced Bayesian networks. *International journal of forecasting*, **11**, pp. 89-111.

Koller D. and U. Lerner (2001) Sampling in Factored Dynamic Systems. *Sequential Monte Carlo Methods in Practice*, A. Doucet, J.F.G. de Freitas, and N. Gordon, Eds., Springer-Verlag, pp 445-464.

Meshkat L., J.B. Dugan, F.D. Andrews (2002). Dependability analysis of systems with on-demand and active failure modes, using dynamic fault trees. *IEEE Trans. On reliability*, **51** (2), June.

Padhraic S. (1997). Belief networks, hidden Markov models, and Markov random fields : A unifying view. *Pattern Recognition Letters*, **18**, pp. 1261-1268.

Torres-Toledano J.G., L.E. Sucar (2003). Bayesian Networks for Reliability Analysis of Complex Systems. *Lecture Notes in Computer Science* "Progress in Artificial Intelligence - IBERAMIA'98, Portugal, 1998", Springer-Verlag Heidelberg, ISSN: 0302-9743, **1484,** Jan 1998, pp 195–206.

Weber P., M.C. Suhner and B. Iung (2001). System approach-based Bayesian Network to aid maintenance of manufacturing process. *6th IFAC Symposium on Cost Oriented Automation*, Berlin, Germany, 8-9 October, pp. 33-39.

Weber P., L. Jouffe (2003). Reliability modelling with Dynamic Bayesian Networks. *5th IFAC Symposium SAFEPROCESS'03*, Washington, D.C., USA, June 9-11.

Welch R. and T. Thelen (2000). Dynamic reliability analysis in an operational context: the Bayesian network perspective. *In Dynamic reliability: future directions*. Edited by: C. Smidts, J. Devooght and P.E. Labeau, ISBN 0-9652669-3-1, Maryland, USA.

Copyright © IFAC Information Control Problems in Manufacturing, Salvador, Brazil, 2004

ELSEVIER
IFAC
PUBLICATIONS
www.elsevier.com/locate/ifac

RECONFIGURABILITY ANALYS ER FOR AUTOMATED PRODUCTION SYSTEMS

TOGUYENI Abdoul K. Armand

Laboratoire d'Automatique et d'Informatique Industrielle de Lille (L.A.I.L.-CNRS UMR 8021), Ecole Centrale de Lille, B.P. 48, 59651 Villeneuve d'Ascq Cedex, France

Abstract: Fault Detection and Isolation techniques and alarms' management are generally combined to manage faults in complex Automated Production System such as petrochemical plant. This leads the operators to be surcharged by a lot of data (alarms, corrective action procedures, plant item status ...). This implies that the recovery tasks become very complex since the operators must extract of this amount of data, the most relevant to react to critical faults in a limited time. To address this problem we propose Reconfigurability Analyser, a tool based on the external model of a plant. It performs an interpretation of detected faults to propose recovery actions taking account of available resources. *Copyright © 2004 IFAC*

Keywords: supervision, fault detection, fault isolation, alarms' management, reconfiguration, recovery actions, reasoning.

1. INTRODUCTION

Automated Production Systems (APS) such as chemical or petrochemical processes are complex plants that are difficult to pilot notably in case of fault occurrence. To address this problem, the Intelligent Manufacturing Systems (IMS) consortium has proposed an international project called CHEM [1]. The main objective of CHEM is to develop a set of new Decision Support Systems (DSS) to allow an intelligent supervision of actual plants. The work presented in this paper is a contribution of LAIL in the field of operation support. Indeed, our main objective is to design an automatic tool, which allows real time analysis of plant reconfiguration possibilities in case of fault occurrence. Such analysis must assist the plant operators to select and to apply corrective actions according to the plant status. In CHEM context, the outputs of the Reconfigurability Analyser can also be exploited by other automatic tool that will define a procedure to apply the corrective actions that have been proposed.

This paper is organised as follows. The section two will set the problem of reconfigurability analysis with regard to FDI (Fault Detection and Isolation) techniques and alarms' management. In section three, we will introduce the external model exploited by the Reconfigurability Analyser. Section four will explain how the analyser is implemented and how it operates.

2. THE PROBLEMATIC OF ALARMS' MANAGEMENT

In complex APS, Fault Detection and Isolation (FDI) techniques are generally used to trigger off alarms that are managed by piloting operators. In this case, they must sort the alarms to treat the most important with regard to their critical nature and to the reaction time they have. This task is particularly complex because of the amount of data that the operators must manage. The first reason that explains this amount of data is the number of alarms that generally fill up their control screens. (Kirshen, and Wollenberg, 1992) explains this overload by three main causes: the occurrence of multiple faults in the process, false alarms, and alarm avalanches. FDI algorithms produce a lot of false alarms during plant transient states (when it commutates from one exploitation mode to a new exploitation mode). Indeed, these alarms are often triggered off when signals reach static thresholds. In this case, the problem is due to the fact the thresholds are static. The reason is that the designers of alarms' managers only consider the steady state of the plant (the production exploitation mode) to fix the values of thresholds.

The use of unstructured knowledge explains alarms' avalanches. Indeed, the knowledge that is generally used by Alarms' Processors is based on the knowledge that an expert has on a system. As this expertise increases during the system life, he attempts to complete the alarms' processors knowledge by adding new rules. The result is that after many years

[1] CHEM: "Advanced Decision Support System for chemical and petrochemical processes". Project is funded by the European Community under the Competitive and Sustainable Growth programme of the Fifth RTD Framework Programme (1998-2002) under contract G1RD-CT-2001-00466. See www.cordis.lu or www.chem-dss.org

of new additions their knowledge base can become incoherent. In addition, this knowledge is generally encoded without taking account of physical cause-consequences relations in the plant. So, in complex plant, a lot of alarms can be produced by a common cause. Their occurrence at different instants can lead inexperienced operators to think to a multiple fault situation.

Another reason that explains the complexity of recovery tasks is the necessity for the operators to know well corrective operating procedures. In fact, they have generally several procedures to treat a problem. Depending on the plant status, the operators must select the appropriate procedure. The corollary problem is that this requires to them to know the state of each resource in the plant. For example, they must know the resources of the plant that are currently under maintenance in order to select an available one to replace the breakdown resource. This is difficult in large system because they are made up of many plant items.

A response to this problem is the coupling of Intelligent Alarm Processing System (IAPS) (Naito, and Ohtsuka, 1995; Arzèn, 1996; Ghariani, *et al.*, 2002) and reconfigurability analysis tools. These tools assist the operator by filtering the alarms and by proposing the most relevant corrective actions to apply. They must be based on structured knowledge such as the external model of a plant. This paper focuses on ReconfAnalyser a reconfigurability analysis tool that is developing in the context of CHEM project.

3. THE EXTERNAL MODEL

3.1 An illustrative example

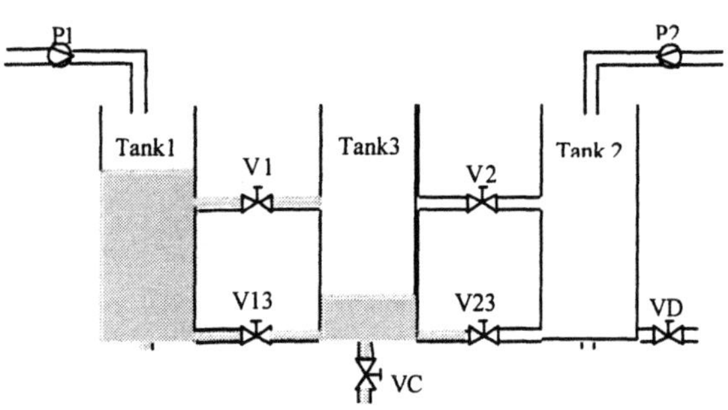

Fig. 1. The plant information diagram of the three-tank system.

To illustrate the different concepts that are introduced in this paper, let us take the example of the three-tank system (Fig. 1) that is a benchmark to study FDI and Fault Tolerance Control (FTC) in APS (Lunze, *et al.*, 2000). The plant is composed of three subsystems. Sub-system1 is composed of the pump P1, the tank T1 and valves V1 and V13. Sub-system2 is composed of the tank T3, valves V1, V13 in one hand, and valves V2, V23 in another hand, and valve VC. Sub-system3 is composed of the pump P2, the tank T2, and valves V2 and V23. The goal of this plant is to supply a constant flow of water to consumers by the valve VC. Sub-system1 supplies the water to the system. In normal functioning, valves V1 and V13 allow transferring the water from tank T1 to tank T3. Level sensors and a regulation controller allow regulating to fix set points, the fluid levels in T1 and T3 by controlling the water flow delivers by pump P1 and the opening or the closure of the valves V1 and V13. Sub-system3 is a passive redundant system with regard to sub-system1.

The plant structural decomposition is summarized by figure 2.

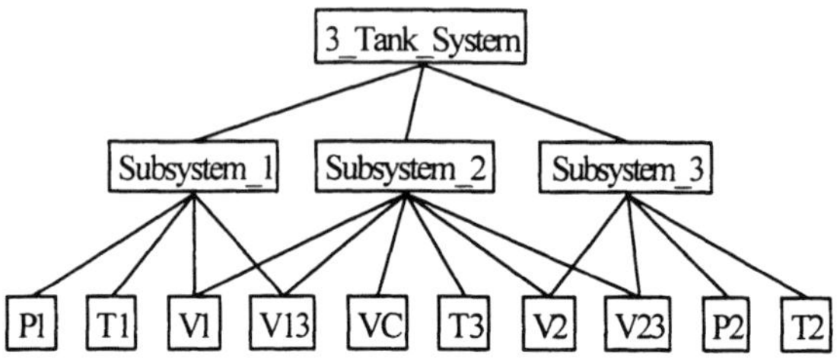

Fig. 2. The architectural model of the three tanks system.

In the rest of the paper the notation "<component>:<service>" is used to specify that <service> is delivered by <component>.

3.2 Basic concepts used by the modelling stage

Our approach distinguishes four basic concepts that are used to define the plant external model: service, version, mission, and exploitation mode.

A service; (Gehin, *et al.*, 1999; Lunze, *et al.*, 2000). It is an external view of a function (Toguyéni, *et al.*, 1993) delivered by a plant item or the plant itself. As an example, let us consider a hydraulic pump. From an external viewpoint, its function is to pump water to feed up a hydraulic circuit. Pump is then considered as a service delivered by the pump component. This concept of service exists at any level of the hierarchy of the architectural model of a plant. Lower level plant items (basic components hereafter) are directly interfaced with the physical process. They provide elementary services. For example, in petrochemical plants, basic components are generally pumps, valves, tanks, reactors, heaters and so on. Pump is then an elementary service. In the same way, the elementary service of a tank is to "stock" a fluid. However some basic components can deliver several elementary services. Since a valve can take two extreme positions, each valve delivers at least two elementary services: "open" or "close" a circuit. Intermediary plant items are composed by elementary components or other intermediary components belonging to lower levels. The services of these components are composed by services of their sub-components.

Let us come back to the example of three-tank plant to illustrate the definition of intermediary services. Sub-system1 delivers two intermediary services: to increase the water level in tank T1 ("subsystem1:increase" hereafter) and to decrease level in tank T1 ("subsystem1:decrease" hereafter).

"subsystem1:increase" is obtained by the composition of four elementary services of the components of subsystem1: "P1:pump", "T1:store", "V1:close" and "V13:close". This example shows us that according to the architectural model of a plant, it is possible to build a graph that explains the inter-dependence between the services of the different levels of the plant components.

A version; It is an abstract concept that enables us to explain that some composed services (intermediary services or missions) are implemented from different sets of lower services. In the rest of the paper, we will use the following notation to distinguish the different versions of a service: <service>_<integer>. The suffix <integer> defines the priority of a version. As an example, let us come back on "subsystem1:to_decrease". This intermediary service belonging to sub-system1 is implemented by three different versions. Let us assume that the flow of pump P1 is inferior to the flow of fluid that goes through V1 or V13. In this case, the first version "subsystem1:decrease_1" is obtained by the opening of V1 and V13. So, this version is summarised by the set {P1:pump, T1:store, V1:open, V13:open}. The second version consists in opening only the valve V1 (subsystem1:decrease_2 ={P1:pump, T1:store, V1:open, V13:close}). The third version consists in opening only valve V13 (subsystem1:decrease_3 ={P1:pump, T1:store, V1:close, V13:open}). The user orders the different versions to reflect his preferences in the way he wants to implement the higher-level service. In the previous example, the versions are ordered with regard the importance of the flow supplied to sub-system2.

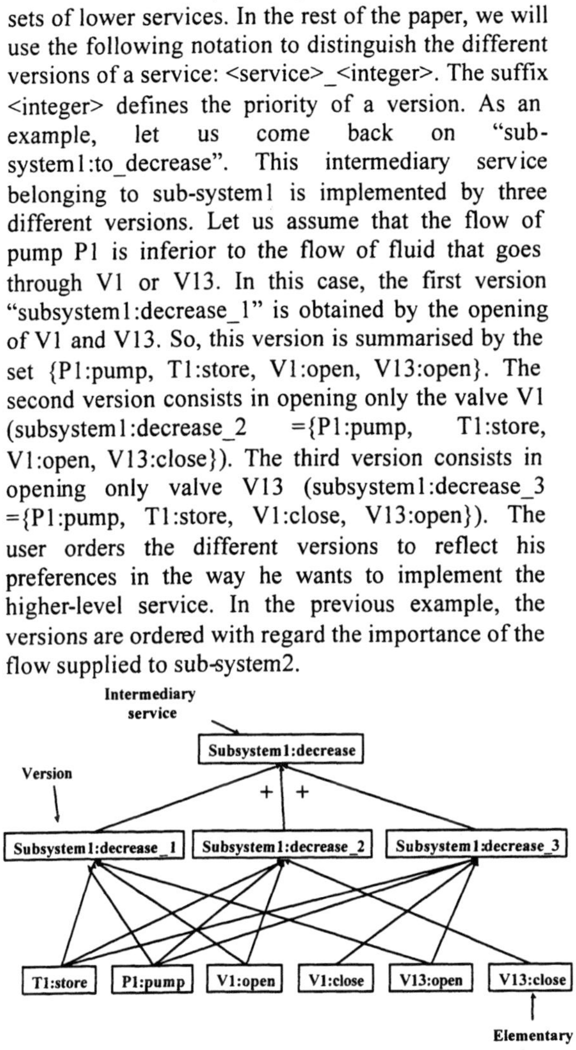

Fig. 3. Graph hierarchy between services and versions.

Figure 3 summarises the concepts' hierarchy between services and versions. When several versions implement a service such as "subsystem1:decrease", the arcs to this service are connected by an "OR" operator. This models a functional redundancy (Toguyéni, and Craye, 2000) to implement a service. Graphically, to simplify the representation of the model, we just represent an "OR" operator near these arcs. From availability viewpoint, this hierarchy means that a composed service is unavailable if all its versions are unavailable. At the opposite, since a version is an "and" composition of lower level services, a version becomes unavailable as soon as one of its lower service is faulty.

A mission (Gehin, *et al.*, 1999); It is a service deliver by the plant itself (top-level service). A mission can be refined in a hierarchy of versions and services, until elementary services. The missions of a plant are established from the principal functions of the plant defined in its functional requirements. For example, the main mission of the three-tank system is to regulate the level in each tank according to the set points defined by the operator ("tanks:regulate_levels").

Exploitation modes and the Exploitation Modes Management Graph; An exploitation mode (EM) is a set of missions that characterized the state of a plant. This set can be empty in order to characterize the idle state of the plant. Let us come back to the three-tank system. Its functional requirements enable us to specify its missions. These missions can be group in five coherent sets that describe the different modes of the plant:

- EM1 (Idle) is composed by a single mission (Mission 1: no operation).
- EM2 (Preparation): In this mode the plant must fulfil two missions: the filling of the system to the level set points (mission 2: "tanks:reach_level") while it preserves the environment (mission5: "tanks:preserve").
- EM3 (Regulation): In this mode the plant must maintain constant the level of each tank by regulating it (mission 3: "tanks:regulate_levels") while it continues to preserve the environment (mission 5).
- EM4 (End of production): The plant reaches this mode when the operator wants to stop its production. In this mode, mission 4 ("tanks:empty") and mission 5 are assumed.
- EM5 (Fall back): This mode is reached when there is a leak. In this case, only mission 4 is assumed.

The Exploitation Mode Management Graph (EMM Graph hereafter) is a state-transitions graph that specifies the different modes of a plant and the transitions between these modes (Fig. 4). The designer defines one of these nodes of the initial mode (The idle mode in our example).

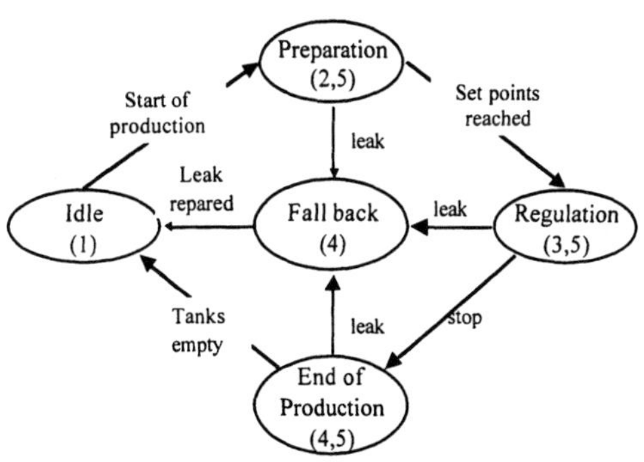

Fig. 4. The EMM Graph of the three tank system.

3.3 The service graph

For the implementation of the analysis tool, the EMM graph and all the mission graphs are compiled in a single graph: the service graph.

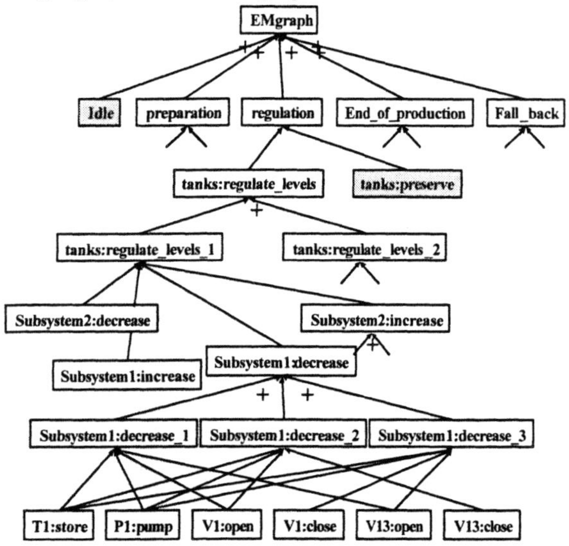

Fig. 5. An extract of the service graph of the three-tank system.

"EMgraph" is a virtual service that defines its root. Its second level is composed of the exploitation modes of the plant. The difference with the versions of a composed service is that the exploitation modes are not pre-ordered. At the beginning, the active mode is the initial mode (The Idle mode in our example). The activation of a new mode depends on the verification by the plant of the condition that constraint the transition between the actual mode and the new selected mode. Each mission of an exploitation mode is linked to the services' graph that describes its implementation (Fig. 5).

4. RECONFIGURABILITY ANALYSER

4.1. The tool architecture

In this section we describe an inference engine for reconfigurability analysis based on the external model. To develop such a tool we use Artificial Intelligent techniques. Our tool is structured in two parts (Fig. 6): a knowledge base and an inference engine (Prolog). The knowledge base contains predicates and rules. The predicates describe the service graph. The rules are used to reason on the service graph. The prolog engine applies the rules to the facts to find a solution to the problem.

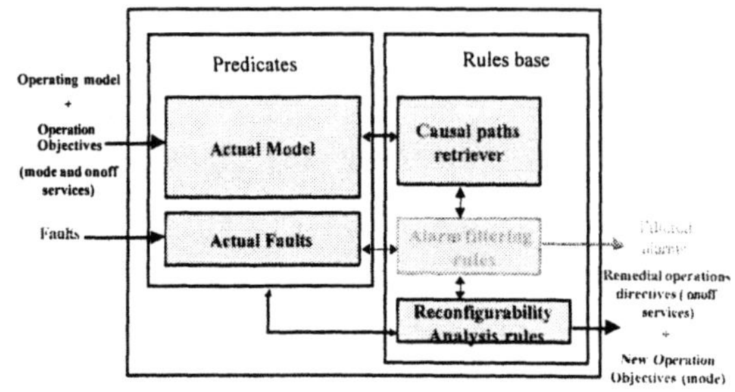

Fig. 6. Architecture of the analyser.

4.2. The knowledge base

Predicates; Structure and properties of the external model are represented by prolog predicates. The predicate "arrow(N_i, N_j, k)" represents the dependency relationship between two adjacent nodes: from N_j to Ni. This predicate enables us to represent both connections from a version to a service and from a service to a version. When Ni is a service and Nj one of its version, in this case "k" is the order of the use of N_j to implement Ni. In the second case, "k" is always equal to 1.

From structure viewpoint, the graph contains two types of nodes: "and" node and "or" node. In figure 5, "subsystem1:decrease" is an "OR" node because it is implemented by several versions. In prolog modelling, by default the nodes are "and nodes" excepted if they are explicitly declared as "or nodes" with the predicate "node_OR(N_i)".

The predicate "transition(e_i, c_{ij}, e_j)" models the transition from the exploitation mode e_i to the exploitation mode e_j. This transition is achievable if the condition c_{ij} is verified.

Two categories of services are distinguished according to operator actions: services that are implicitly active by their connexion to an active version and ON/OFF services. An ON/OFF service must be activated (ON state) to set the plant in a configuration where this service is used. At the opposite, when the service is no more in the current configuration it must be deactivated (OFF state). The predicate "onoff(Ni)" specifies that the service Ni can be activated. In the three-tank system, pump services ("pump") and valve services ("open" or "close") are ON/OFF services. At the opposite, the "store" service of a tank is not an ON/OFF service.

Rules; The analyser functioning is based on five (5) main rules: default_configuration(EM), fault(SE), reconf(X), repairs(SE) and reconf_service(S,SE, V).

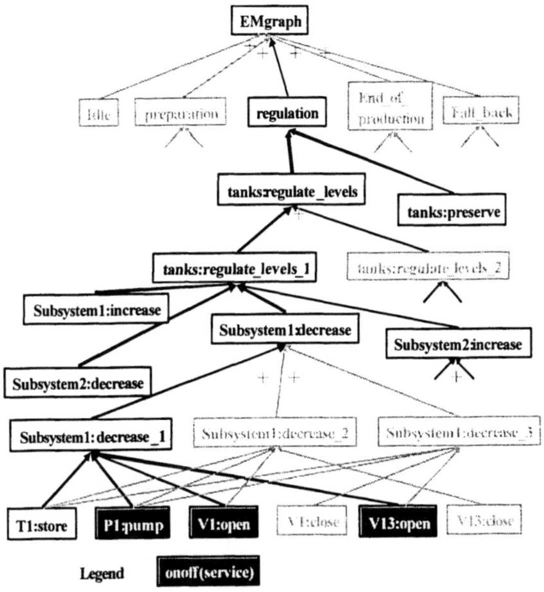

Fig. 7. The three-tank system default configuration for the regulation exploitation mode.

These rules implement essentially two actions: activation (or deactivation) of services and connection (or disconnection) of arrows between versions and services.

Before starting an analysis of faults, the knowledge managed by the analyser must be configured to reflect the current configuration of the plant. This is performed by the prolog rule "default_configuration(EM)" where "EM" is the current exploitation mode of the plant. This rule implements top-down propagation, from EM to the elementary services that implement EM's missions. For each mission or intermediary service, this propagation selects its best available version. This means that only one predicate is created between the service and the selected version. Thus, all ON/OFF services of the current configuration are activated. For example, let us assume that the current state of the three-tank system is the regulation exploitation mode. In the figure 7, the part of the graph of the default configuration is represented in black. The rest of the graph (in grey colour) represents the nodes that are not in the current configuration (they are not connected to the active part of the graph). Since they are ON/OFF services, "P1:pump", "V1:open" and "V13 open" are set in on state.

Let us now suppose that a FDI tool has detected and isolated the fault of valve V1 that remains closed. This means that the elementary service V1:open is faulty. The rule "fault(v1:open,X)" researches all the consequences of the fault. Considering the actual configuration of the system, this rule performs bottom-up propagation in order to propose all the reconfigurations alternatives. The inference engine binds the variable "X" with each "OR" node where it finds an alternative to correct the fault. It performs its analysis by proposing first, the use of a degraded version of the services affected by the fault. The objective is to remain in the current exploitation mode, if possible. In our example, this means that the analyser propose first to substitute the version "subsystem1:decrease_1", by the version "subsystem1:decrease_3" to implement "subsystem1:decrease" service. Let us notice that in this case the second version is not proposed since it needs to use "V1:open" that is faulty (beginning of Fig. 8). In figure 8 we can notice that after the previous proposition, it proposes as second choice to replace the version "tanks:regulate_levels_1" by its alternative "tanks:regulate_levels_2", to implement the service "tanks:regulate_levels". The last alternative is to propose the commutation from the current exploitation mode to another one connected to this mode by a transition. For example, the analyser proposes to commutate from the actual mode "regulation" to "end_of_production" mode. Let us notice that the proposition of commutation to the "fall_back" mode is not appropriate here since it is a safety mode, which may be used only in case of leakage in a tank.

```
?- fault(v1:open, X).
possibility of using subsystem1:decrease_3
to implement subsystem1:decrease
 this implies to activate p1:pump
 this implies to activate v1:close
 this implies to activate v13:open
X = subsystem1:decrease ;

possibility of using tanks: regulate_levels_2
to implement tanks: regulate_levels
X = tanks:regulate_levels;

possibility to commutate from regulation to end_of_production
 if condition stop is verified
X = em_g ;

possibility to commutate from regulation to fall_back
 if condition leak is verified
X = em_g ;

No
?-
```

Fig. 8. Reconfigurability propositions in case of a fault occurrence

The rule "reconf" applies one of the reconfiguration possibilities given by "fault" rule. It can be called to apply the first reconfiguration proposition. In this case, its argument is a variable (example "reconf(Y)", where "Y" is a variable). From the faulty elementary service, this rule performs a bottom-up propagation to find an "OR" service node with an available alternative version. All ON/OFF services of this part of the system configuration are deactivated. When the new version is selected, a top-down propagation activates all its ON/OFF services. The rule "reconf" can be directly apply to a specific "OR" node proposed by the "fault" rule. In this case, two top-down propagations are necessary from this node: one to activate all ON/OFF services of the selected degraded version and the other to deactivate the ON/OFF services of the previous version. In all cases, the previous version is disconnected of the active graph and the new one is connected.

Let us notice that fault occurrences are also taken into account in the parts of the service graph that are not in the active configuration of the plant. This avoids proposing a reconfiguration based on previously faulty service.

When a component (For us its elementary services) is repaired, the information must be taken into account by the analyser to research if this event allows selecting a better configuration. The impact of a reparation is analysed by the rule "repairs(SE)", where "SE" is the name of an elementary service. In this case, the analyser uses a bottom-up propagation to retrieve all versions of services that could be substituted to the current one. For example, in the case where "subsystem1:decrease" service is implemented by the version "subsystem1:decrease_3", the declaration of the repairing of "V1:open" implies the proposition of substitution by the version "subsystem1:decrease_1" or the version "subsystem1:decrease_2". This proposition is applied by the rule "reconf_service(S, SE, V)", where S is the concerned service, SE is the repaired elementary service and V is a version of S. This rule works in a similar manner with when "reconf" is applied to a specific "OR" node. A top-down propagation from S allows disconnecting the faulty "version" and deactivating all ON/OFF services that implement this version. In parallel, version V is connected to service S, and its ON/OFF services are activated.

5. CONCLUSION

In this paper we have presented a prototype of a new software tool to assist operators to pilot Automated Production System. The prototype is implemented in prolog. It analyses in real time, the different alternatives to pilot a process, in spite of the occurrence of faults. It orders the different solutions according to the facility to implement them.

Some of its analysis features must be improved. For example, in figure 8, readers can notice that in case of the selection of the version "subsystem1:decrease_3", the analyser proposes to start the services "p1:pump", "v1:close" and "v13:open". These advices are not suitable in this case because these services are already started. This example shows that it is necessary to refine the behaviour model associate to each node of the service graph. Notably, we must refine the concept of ON/OFF services. Some solutions may be finding in (Dangoumau, *et al.*, 2000). Another improvement concerns the explanations given to the operator. At the moment they are systematically given during the reasoning stage. We think that it is better to give them when they are requested by the operator.

In this paper, we have focused on the analysis process. The results given here prove of our capacity to achieve such analyse. The main problem that we have now to address is how to generate the service graph to be sure that the user obtains a valid model. Our idea is to create a model builder based on the use of a library of basic components. Using syntactic rules, the user would be able to implement the plant instrumentation diagram with respect to physic laws.

REFERENCES

Arzén, E.-E. (1996). High level GraphChart for Alarm Filtering. In: the proceedings of IFAC World Congress. San Francisco (USA).

Dangoumau N, S. El Khattabi and E. Craye (2000). Modes Management for Automated Production Systems. In: the proceedings of WAC'2000, Seventh International Symposium on Manufacturing with Applications (ISOMA), Maui (Hawaii).

Gehin, A.L. and M. Staroswiecki (1999), A Formal Approach to Reconfigurability Analysis. Application to the three Tank Benchmark. In: the proceedings of European Control Conference ECC'99, Karlsruhe, Germany, 31 August-3 September 1999.

Ghariani, A., Toguyéni A.K.A. and E. Craye (2002). A Functional Graph Approach for Alarm Filtering and Fault Recovery for Automated Production Systems. In: *the CDROM proceedings of WODES'2002*, Manuel Silva (Ed.), Saragossa (Spain).

Kirschen, D.S. and B.F. Wollenberg (1992). Intelligent Alarm Processing in Power Systems. In: *the proceedings of the IEEE*, Vol. 80, N°5, pp. 663-672.

Lunze, J., J. Askari-Marnani, A. Cela, P. M. Frank, A.-L. Gehin, B. Heiming, M. Lemos, T. Marcu, L. rato and M. Staroswiecki (2000). Three Tank Control Reconfiguration. In: Control of Complex Systems, K. J. Aström, P. Albertos, M. Blanke, A. Isidori, W. Schaufelberger, T. Sanzed (Ed.). Springer.

Naito N.N. and S. Ohtsuka (1995). Intelligent Alarm-Processing system for Nuclear Power Plants. In: Nuclear Technology, Vol. 109, pp. 255-264.

Toguyéni A.K.A., E. Craye and J.C. Gentina (1993). An approach for the placement of sensors for on-line diagnostic purposes. In: the proceedings of 12th IFAC World Congress. Vol.8, pp83-88. Sydney (Australia).

Toguyéni A.K.A., and E. Craye (2000). Alarm Processing and Faults Recovery Based On Functional Modeling. In: the cdrom of WAC'2000, Seventh International Symposium on Manufacturing with Applications (ISOMA), Maui (Hawaii).

Copyright © IFAC Information Control Problems in Manufacturing, Salvador, Brazil, 2004

ELSEVIER
IFAC
PUBLICATIONS
www.elsevier.com/locate/ifac

UML SPECIFICATION AND DEVELOPMENT OF SAFETY-RELEVANT INDUSTRIAL COMMUNICATION SYSTEMS

Christian Diedrich* (Prof. Dr.-Ing.), Elke Hintze* (Dipl.-Ing.), Arne Neumann* (Dipl.-Ing.)

* *Institut für Automation und Kommunikation e.V.*
Steinfeldstr. 3
D-39179 Barleben
Tel.: +49 (0)39203 810 24
Fax: +49 (0) 39203 811 00
E-Mail: christian.diedrich@ifak-md.de

Abstract: Automation devices fulfilling functional safety is an increasing market area. Therefore methods for the software development are needed which fulfil these requirement. The V model and UML as a semi-formal method can provide support for the implementation and testing of dependable software in this area. This paper gives an overview about a safety related software development process using the example of safety fieldbus communication provided by PROFIBUS. *Copyright © 2004 IFAC*

Keywords: UML specification, Safety fieldbus communication, dependability in networks, Extended finite state machine, Formalised specification and implementation, PROFIsafe, .

1. Introduction

There is a gap between formal languages and methods for the specification and verification of software systems and the product development process in the industry. Many formal methods are very abstract so that engineers are not in a position to use it during the daily work. UML as collection of graphical languages and the possibilty to integrate source code seems to be a means to fill in this gap. Safety software systems are needing a formalised development process. UML language elements /OMG: 2002/ and the V model are suggested for this reason. The V model steps are influeneced by requirements of functional safety which are described in the IEC 61508 standard "Functional safety of electrical/electronic/ programmable electronic safety related systems", mainly in /IEC 61508-1 and IEC 61508-3 : 1998/ (Figure 1). PROFIsafe /PROFIBUS : 2001/ is a safety related communication protocol on top of PROFIBUS DP which is taken as an example because PROFIsafe is used for SIL 3 applications

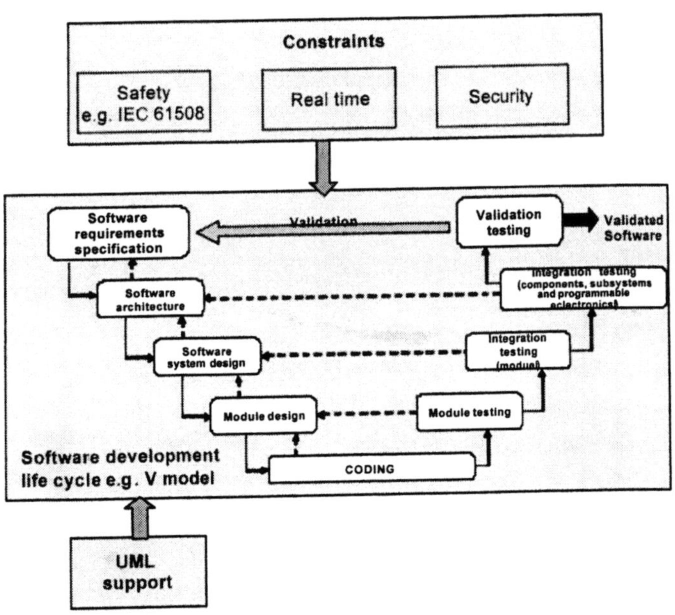

Figure 1 - Aspects of functional safety

(SIL - Safety Integrity Level) according to IEC 61508. A safety integrity level defines the discrete level (one out of a possible four) for specifying the safety integrity requirements of the safety functions of a system.
The process from the definitions of the PROFIsafe profile to certified products has to solve a large scale of hardware and software development steps which have to be transparent for product certifying organisations such as the German TÜV. This paper describes the development process from the requirement definition, design, specification, implementation to the tests along the V model which are using UML language elements. The example of PROFIsafe is used to illustrate the process.

2. Principles of PROFIsafe

Starting point for the development process of the safety functions is the analysis of possible errors and faults of the system. Using the example of digital communication systems in the safety area the analysis results are shown in Table 1. It shows both the possible errors and the means for their recognition mechanism.

The Safety Integrity Level 3 (SIL3) is usually applied for an entire application, this means that every component of the system has to fulfill these requirements. An automation application example is a control loop consisting out of a sensor, controller and actuator. The probability of dangerous errors per hour for SIL 3 has to be between 10^{-7} and 10^{-8} for the entire control loop. Assuming that the sensor, actuator and controller contribute with 98 % to the error rate of the control loop (this is an assumption which is derived from calculations of hardware and software error probabilities of the components), the contribution of the digital communication to the probability of dangerous errors per hour should not be higher then 2 times 1%, i.e. it has to be 10^{-9}. Therefore additional means have to be integrated to the communication protocol.
This analysis results lead to the architecture design, the second step within the V model. In our example the PROFIsafe solutions add a so-called safety layer on top of PROFIBUS layer 7. This gives the opportunity to use the existing industrial communication stacks as well as to operate a mixture of non-safety and safety application on the same segment at the same time (see Figure 2). This

Table 1 Errors and their recognition mechanism

Measure: Failure:	Consecutive Number	Time expectation with acknowledge	Codename for sender and recipient	Data Protection
Repetition	X			
Loss	X	X		
Insertion	X	X	X	
Incorrect Sequence	X			
Corrupted Data				X
Delay		X		
Interconnection of F- and Standard Messages (Masquerade), incl. wrong - and double addressing		X	X	X

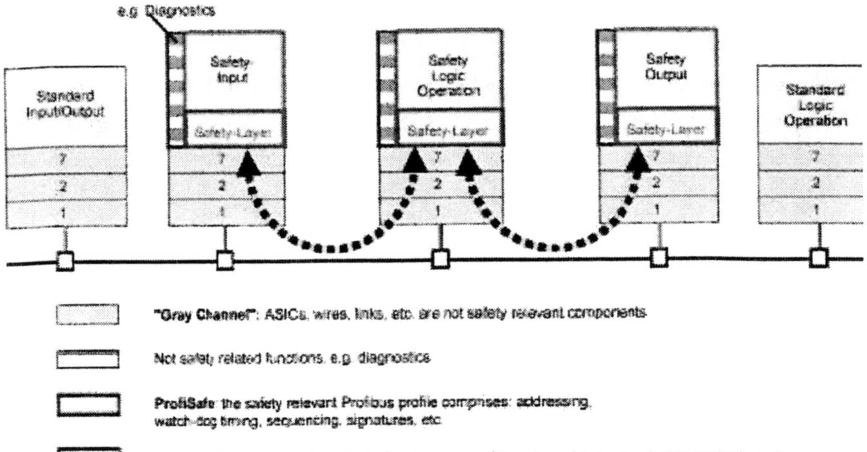

Figure 2 - Location of PROFIsafe in the layer structure of the PROFIBUS communication architecture design can be described using UML component, package and class diagams, which is not shown here because it is quite to large.

The architecture design has to be accompanied by the functional design. The following example shows that this part of the design is an informal text.

PROFIsafe uses the polling mechanism of PROFIBUS between masters and their associated slaves. The safety means are encapsulated in the safety layer. The communication stack and the medium are treated as so-called black channel. The possible errors such as damaged addresses and loss or delay of data are recognised by using the following mechanisms:

- A continuously increasing numbering of the telegrams between master and slave.
- Time supervision both in the master as well in the slave.
- Authorisation by passwords.
- An optimised CRC check sum calculation.

Data for these mechanisms is transferred additional to the cyclic data between masters and slaves.

Both architecture and functional design are the basis for the software system design, the next step of the V model. In our example the system design represent the allocation of the F-drivers in the software sytems of the components of a control loop, i.e. in the safety layer of the devices (Figure 3).

This software system design can be also described using UML package and class diagrams that are not shown here.

The software system is usually composed out of modules. The communication is part of the black channel and do not belongs to the safety related software. But the F driver has to be fulfilling the full SIL 3 requirements. The behaviour of the F driver can be described as automata. Figure 4 shows the state machine of the master F-driver as

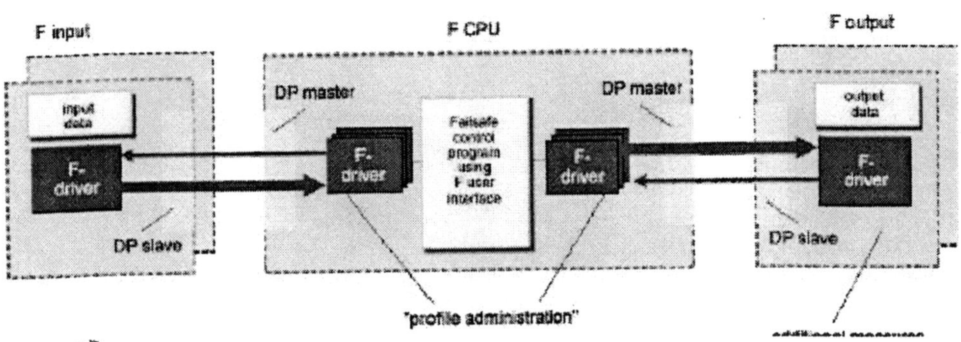

Figure 3 - Communication structure of the Fail Safe functionality

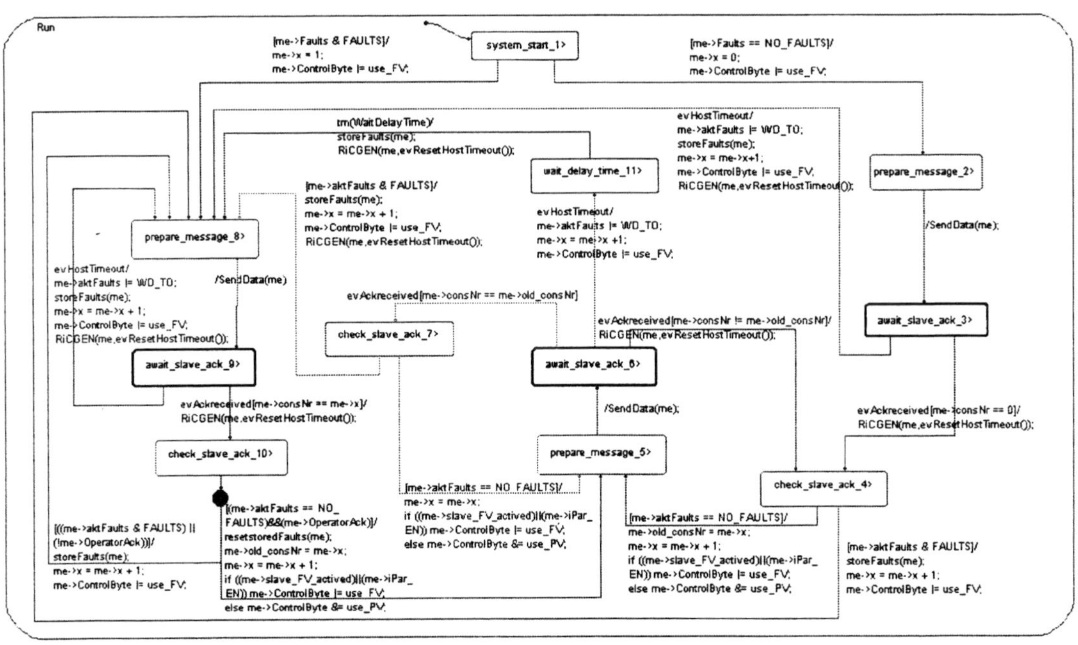

Figure 4 – PROFIsafe master state machine

UML state chart to present the degree of complexity. A detailed description would exceed this paper and are documented in /Reschwamm : 2001/.

The state machine is an extended finite state machine like Harel state machines without hierarchy and orthogonality, i.e. the transition has to calculate conditions so-called guards. Figure 4 shows these guards at the transitions. This type of state machine was chosen because it reduces the number of state implicitly. The PROFIsafe specification received after review by the organisation TÜV a so-called „positiven technischen Bericht" (positive technical report) because among others, the following proofs have had to be done:

- Calculation of the probability of dangerous errors per hour according to the SIL 3 degree (no UML support possible)
- The state machine has had to be specified without hierarchy, the main reason is that execution details of transition actions, pre and post actions crossing different hierarchy levels are not unambiguously specified within UML
- Verification of the state machines, this was done by transferring the extended finite state machine in a flat one and in a textual format. The verification tool VALID Automata /Malik : 2002/, has executed the verification. This manual transformation of extended finite state machines to finite state machines can be replaced using UML in future. According research work is under progress

3. UML specification of the PROFIsafe profile

After module specification coding is the bottom step of the V model. The coding can be derived direct out of the specification if the specification is formalised. This was one of the reasons to specify the PROFIsafe F-driver of our example in UML. We have used UML class diagrams and state charts. Figure 3 shows the package/class diagram. The PROFIsafe F-driver UML specification was implemented for PC and Windows™ 2000. The class diagrams as well as the state charts are transformed direct to C code. The necessary execution environment of the generated code for Windows is a standard library, which is part of the used Rhapsody in C tool from ilogix. The application on top of the F driver and the communication driver are integrated as operations of specific interface classes. The implementation uses the Rhapsody in C run time for PC / Douglass : 1998/.

The software, which is implemented in safety devices, has to cope with special coding and development rules. Therefore there are special specification constraints for the UML languages especially for the state chart language:

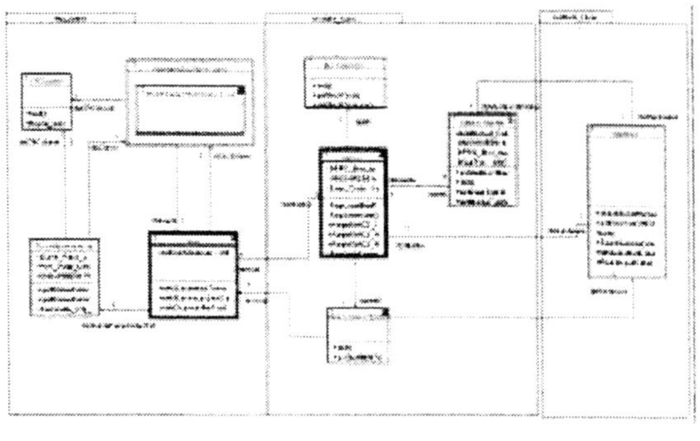

Figure 5 Integration of F driver module in UML in safety application

- Choice of a special state chart subset according to the constraints (e.g. no hierarchy and orthogonality, no entry and exit actions of state)
- Special specification rules e.g. transition conditions (guard) and actions have to be visible at the specification level
- No side effects between transitions, actions and activities, i.e. it is not allowed to manipulate transition variables within actions and activities

These additional rules are necessary for the software development process because the German TÜV and BIA audits and code reviews check these features.

4. Test of the PROFIsafe implementation using the UML specification

Each step of the V model has to be accompanied by verification and tests (see also figure 7). The Test and validation in our PROFIsafe example have the following aspects:

1. The state machines specified in the profile have to be verified. The verification proves the quality of the state machine itself. Examples are correctness, freedom of dead locks and others.
2. Proof of the correctness of the implementation of the code that is generated by Rhapsody in C.
3. Proof if all errors that can occur in practice are covered by the PROFIsafe specification and its implementations.

To 1: There was a verification process in parallel to the specification of the profile using the tool VALID Automata /Malik 2002/. This was one of the pre-requisites that TÜV has granted the positive technical report. This represents the module verification of the V model

To 3: This requirement was covered by the initial error specification and its recognition mechanisms according to table 1. This is an in-formal manual verification made by human beings checking the architecture design and the software system design.

To 2: This step aims to an approval of the compliance between the specified PROFIsafe profile state machines and the implementation. Compliance means that both the implemented PROFIsafe F-driver and its UML specification interact in the same way. This compliance includes the other hardware and software components, because the test is carried out using the entire device. Therefore this test can be seen as integration test (including components, sub-systems, see figure 7). Sequence diagrams were used for the interaction description of the components and modules defined within the architecture design. These sequence diagrams are the intended behaviour of the system. Therefore it is possible to check this behaviour by tracing the sequences through the state machine of the modules in our example the F driver during multiple scenarios. The background is that the sequence diagrams and the state charts can have model concurrency. This is integrated in UML V2.0.
To prove this, the PROFIsafe UML specification is expanded by functions that store actual transition and other state machine information in a log file as well as mechanisms to influence the state machines by test cases. The logging can be started and stopped from an HMI. The test cases are scripts that contain good or error cases. The test scripts as well as the log information are interaction with the internals of the state machine generated from the UML state chart. This approach shows the exact behaviour of the F-driver.
Using this prototype of a test environment simulation of errors, execution of test scenarios and logging can be done for both master and slave devices. One example is shown in Figure 6, which is one requirement specification scenario for the PROFIsafe F-driver module: the first initial cyclic interaction and two successful interactions. A log file registers the transition information (start state, end state) as well as the cyclic data marked if the data is fail-safe or actual data, as well as the Consecutive Number (ConNr), the control and status byte (SteBy, StaBy) and the CRC value. The time difference between two arrivals of new data is also recorded. The comparison of the log file information and the specified scenario sequences is the approval that the implementation behaves according to the requirement specification.
Figure 7 shows the entire software development process for our PROFIsafe example based on the V model and UML. The sequence diagram (Figure 6) is matter of the software architecture design (Figure 2) and the integration test (Figure 7). It is a specialisation of main scenarios derived from the safety functional requirements, which are partly shown for PROFIsafe in Table 1. Figure 3 shows the software system design defining the relations of the F-Driver to the other software components of the devices and the system. The shown PROFIsafe F-driver is one of multiple modules in a safety device. There are the following relations between results of the V model steps and UML language elements:

- Architecture design can be described with the component, package and class diagrams of UML, functional aspects of this design can be described using sequence diagrams
- Software system design can be described with package and class diagrams of UML, functional aspects of this design can be described using sequence diagrams
- Module specification can be described with the package and class diagrams of UML, the behaviour of the modules can be described using UML state charts, the state charts are the parts of the classes and objects

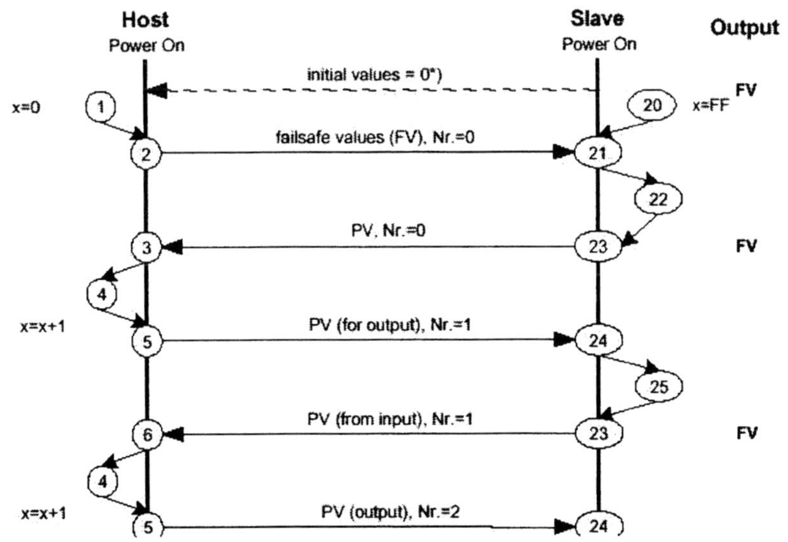

Figure 6 - Scenario Master and Slave at startup phase

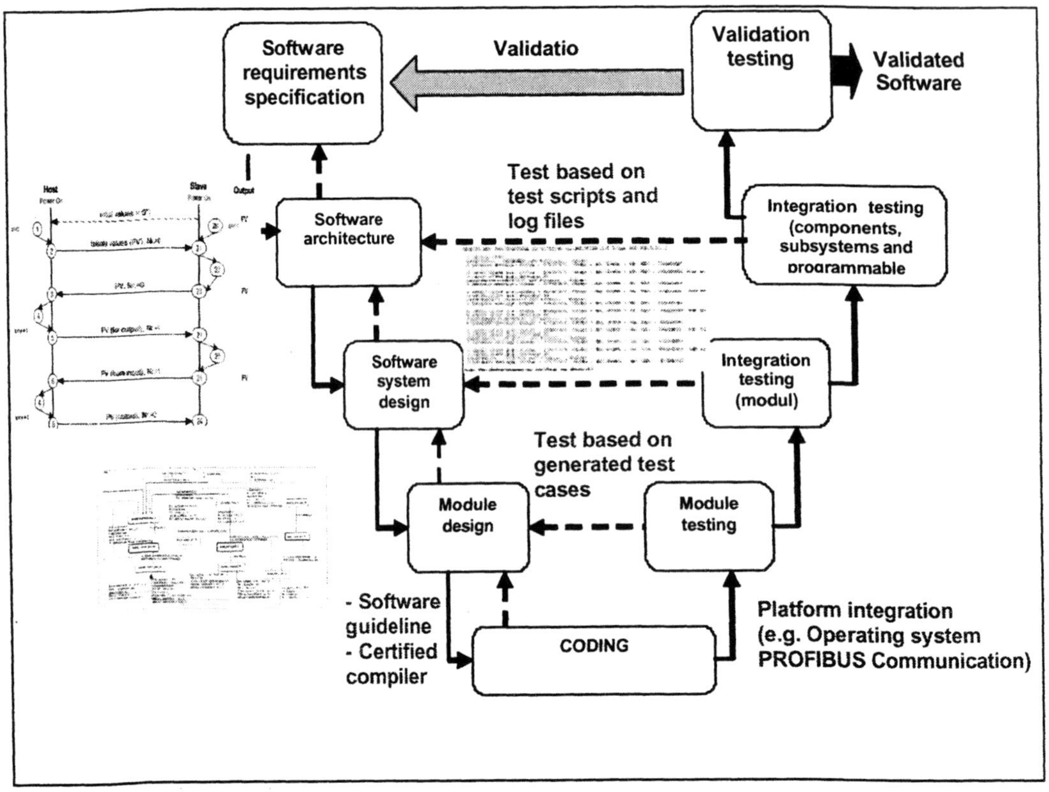

Figure 7 – Safety development steps – example PROFIsafe UML specification and implementation

- Objects are instantiations of classes, state machines are part of objects, and the interactions between the objects are described with sequence diagrams, the interactions influencing the behaviour, i.e. the state machine of the objects. UML V2.0 defines the semantic relations between these model elements.

These UML model elements are the basis for verification and validation activities. It is clear that the UML language elements have to be transformed to formal models in advance, which is scope of multiple activities in the research area.

In future it will be possible to transform UML state charts into finite state machine. Research work for this topic is under progress.

The described device test based on scripts and log files derived from the sequence diagrams supports the integration test of the F driver in the communication part of the device as well as the general device integration.

5. Results and future work

The PROFIsafe profile example shows the steps from the analysis to the implementation and test supported by UML. The use of UML brings the following effects:

- Accompanying the definition work of the PROFIsafe profile with the UML specification from revision V1.1 to V1.11 made some unambiguous elements visible.
- Documentation of the entire design and development process is supported at a high level and provides a transparency to the certification bodies and can be mapped to the requirements of IEC 61508.
- Using a well defined UML language subset avoid dangerous specifications
- It is possible to generate code for embedded systems out of the UML specification
- The opportunity to animate the profile definitions in the tool increase the acceptance of the profile
- Tests can be performed at the level of the UML specification and the generated code.

It is visible that there are manual steps from the specification to the test cases and the test execution. The next steps have to take use of the new standards of UML 2.0 and the related XMI. Then it will be possible to transform UML specifications to existing verification and test case generation tools. Then manual steps are not necessary anymore.

REFERENCES

IEC 61508-1 and -3: Functional Safety of electrical/electronic/programmable electronic safety-related systems. Part 1 General Requirements. Part 3 Software Requirements. Geneva 1998

PROFIBUS Nutzer Organisation: PROFIBUS Profil, PROFIsafe - Profile for Safety Technologie, Version 1.11, July 2001

Reschwamm, St.: Validierung des PROFIsafe-Profils. Diplomarbeit Otto-von-Guericke Universität 2001.

Object Management Group (OMG): Unified Modeling Language Specification, Version 1.5, September 2002

Douglass, Bruce Powel: Real – Time UML Developing Efficient Objects For Embedded Systems, Addison Wesley, 1998, ISBN: 0-201-32579-9

Malik, R., Muehlfeld, R.: "A Case Study in Verification of UML Statecharts: the PROFIsafe Protocol", Proceedings of the 5th Workshop on Tools for System Design and Verification, FM-TOOLS 2002, pp. 89-93.

Copyright © IFAC Information Control Problems in
Manufacturing, Salvador, Brazil, 2004

ELSEVIER
IFAC
PUBLICATIONS
www.elsevier.com/locate/ifac

SDL SPECIFICATION OF A SECURITY ARCHITECTURE FOR THE TS 61158

Miguel León Chávez[1] and Francisco Rodríguez Henriquez[2]

[1]*Benemérita Universidad Autónoma de Puebla*
Facultad de Ciencias de la Computación
14 Sur y Av. San Claudio, CP 72570, Puebla, México
Tel. (52) 222 229 55 00 ext. 7213 Fax (52) 222 229 56 72
email: mleon@cs.buap.mx

[2]*CINVESTAV-IPN*
Sección de Computación
Av. Instituto Politécnico Nacional No. 2508, Col. San Pedro Zacatenco
México, D.F. 07300
Tel: (52) 52 55 5747 3800 ext. 6570 Fax: (52) 555 747-7002
email: francisco@delta.cs.cinvestav.mx

Abstract: This paper discusses the security in Fieldbuses, and presents the specification of a security architecture for the centralized Fieldbuses. The discussion takes into account, on one hand, the security services defined by ISO, and on the other hand, the security services defined by TS 61158. The discussion shows that the TS 61158 provides users with the authentication and control access services for some application objects. Nevertheless, as Fieldbuses become more and more diverse, complex and integrated into other kind of networks, the TS 61158 must provide other security services, such as user and nodes authentication, integrity, and confidentiality because there are three critical points for attacking this network: the Link Active Scheduler, the Delegated Token and the Bus. The paper presents then the specification, using the Specification and Description Language (SDL), of a security architecture for the TS 61158 based on the following cryptographic mechanisms: a reduced version of DES (user-authentication), a node's private key (node-authentication), and the RC4 algorithm (confidentiality). *Copyright © 2004 IFAC*

Keywords: Networks, Fieldbus, System Security, Authentication, Validation.

1. INTRODUCTION

Fieldbuses are special purpose Local Area Networks (LAN) used to connect all kinds of devices into a factory, such as sensors, actuators, transmitters, programmable controllers, (C)NC machines, processors, and so on (Thomesse, 1998). These networks are usually seen as a three OSI layer architecture, which includes the physical layer, the data link layer and the application layer.

Some of the services of the missing layers are still present in some fieldbuses but are merged in the data link or application layers, e.g. support for interoperability, end-to-end control, fragmentation and reassembling, routing, etc.

Typically, fieldbuses are used by the distributed manufacturing systems in order to monitor and control the processes taking place in the factory. Examples of such systems are: factory automation, automotive industry, textile machinery, electronics manufacturing, food and beverage, chemical processing, and so on.

The IEC 61158 Fieldbus Standard (IEC 61158, 1999) can be seen as a collection of Fieldbus Standards because it defines eight types of Data Link Services, numbered from 1 to 8, each corresponding to a Fieldbus Standard as follows: TS 61158, ControlNet, Profibus, P-Net, Foundation Fieldbus, SwiftNet, WorldFIP and Interbus.

Up to now the security in fieldbuses, such as the TS 61158, has been only considered for access protection on some objects. This is not for protection against intentional misuse of the communication facilities of a field device but in order to protect a system of accidental erroneous use of the objects.

However, there exist at least two possible security attacks that centralized fieldbuses can suffer: Non-authorized users gaining access to the communication channel and non-authorized human operators accessing the master node.

Potentially at least, there exists always the possibility for non-authorized users to gain access to the communication channel. If that happens, then the intruders can launch a passive attack by eavesdropping all or part of the information exchanged among the network's entities. Even worse than that, active attacks are also possible as hackers can maliciously modified the data traveling through the communication channel at will.

On the other hand, the master node stores all the network configuration as well as other important system global information, such as presence variables, i.e. variables containing summarized information on the node's global operating state. Hence, mechanisms of user identification should be put in place in order to avoid leakage of valuable data to non-authorized human operators.

As distributed manufacturing systems become more and more diverse, complex and integrated into other kind of systems, potential attacks to the security of the network increase in the same rate.

Hence, it is of the utmost importance to incorporate security mechanisms on fieldbus communication protocols so that such kind of security attacks can be avoided and/or prevented.

This paper discusses the security in fieldbuses, according to the security services defined by ISO. The discussion is focused on the TS 61158 because it is a typical example of the centralized fieldbuses. The paper proposes then a security architecture for those fieldbuses based on node's private key, one-way hash function, and RC4, which is a lightweight cryptographic stream cipher.

The remaining of this paper is organized as follows: Section 2 presents and discusses ISO security services with regard to fieldbuses; section 3 presents TS 61158 and its security services; section 4 proposes a security architecture for the TS 61158; section 5 presents the specification and validation of the TS 61158 security architecture using the Specification and Description Language (SDL). Finally some future work directions and conclusions are drawn in section 6.

2. ISO SECURITY SERVICES

The Security Architecture of the OSI Reference Model (ISO 7498-2, 1989) considers five main classes of security services: authentication, access control, confidentiality, integrity and non-repudiation. These services are defined as follows: The *authentication* service verifies the supposed identity of a user or a system. The *access control* service protects the system resources against non-authorized users. The *confidentiality* service protects the data against non-authorized revelations. Confidentiality is considered an essential role of cryptographic systems. The *integrity* service protects the data against non-authorized modifications, insertions or deletions. The *non-repudiation* service provides certain protection against the sender of a message that refuses to be it, or against the receiver of a message that denies to have received it.

All the security services defined by ISO can be achieved in centralized fieldbuses by using public key cryptography. That can be accomplished by assigning to each active node in the network, a unique private key and a master node's public key. During communication, slave and master nodes may mutually authenticate each other with these keys using well known protocols (such as Diffie-Hellman). To provide confidentiality, nodes may encrypt their contents using a random session key and a symmetric crypto-algorithm specially tailored for constrained environments. Integrity and non-repudiation services can be obtained by signing/verifying all the messages transmitted between a particular active node and the master node. However, strong public key cryptography is in general an expensive fancy solution (Paar, 2002). Moreover, some of the security services defined by ISO are probably not very likely to be useful on the context of centralized Fieldbuses, like the non-repudiation service.

If for a given application, public key cryptography solutions become too expensive, we can still design limited security schemes for fieldbuses at a cheaper price. For instance, we can use a security scheme based on a one-way hash function optimized for heavily constrained environments, as those typically found in fieldbuses (Sarma *et al.*, 2002). Encryption of the hashed message can then be achieved by using lightweight symmetric stream ciphers such as RC4 or A5 GSM (Biryukov *et al.*, 2000), or even a reduced version of traditional symmetric algorithms such as DES or AES, which can be obtained by reducing the size of the encryption key or by limiting the standard number of rounds used during the encryption/decryption processes (16 in the case of DES and 10 for AES).

In this paper we propose the usage of a reduced version of DES, a one-way hash function, and a lightweight stream cipher, as the three main building blocks needed to implant a security architecture for the TS 61158 fieldbus protocol. Next sections elaborate about the details of our proposed security architecture.

3. TS 61158

This section presents the TS 61158 and an analysis of its security mechanisms.

3.1 TS 61158 general architecture.

The Technical Specification (TS) 61158, i.e. the IEC 61158 type 1, defines a Medium Access Control (MAC) protocol based on a centralized access method and uses a timed-token mechanism to organize the access to the common bus. The nodes are classified into the following three functional classes: basic class, link master class and bridge class. The last two classes enable a node to assume the Link Active Scheduler (LAS) role; each other node is called Data Link Entity (DLE), which can be either a Publisher or a Subscriber node.

LAS has an initial schedule, set by the management system, in which periodic activities are scheduled, i.e. LAS transmits a Compel Data message to a Publisher node. Upon reception, the Publisher broadcasts or publishes the data to all the nodes. Any node configured to receive the data is called Subscriber.

On the other hand, LAS uses the notion of Delegated Token (DeT) to transfer the right to transmit to another DLE and to execute either its periodic or aperiodic services for a specified duration time. The DLE which owns the DeT becomes the only master of the link. This right can be delegated to another DLE creating a Reply Token (ReT), with which it can compel the sending of a data or an acknowledgment. When the delegated duration expires, the DeT holding node has to send back the token to LAS.

LAS uses the time between the end of the actual periodic delegation and the next scheduled periodic activity to allow the execution of management services as well as the execution of asynchronous activities by all the active nodes. To do that, LAS first executes the link-related management activities for a predefined duration and then emulates a circulated token mechanism. That is accomplished by sending a DeT to all active nodes, one by one, starting from itself. Each time a node receives the DeT, it has to give it back to LAS after either expiration of the delegation duration, or completion of its asynchronous activities execution.

LAS uses an array of Maximum Token Holding Time (MTHT), with an entry for each node specifying its MTHT during one cycle of token circulation. When it is time for the next scheduled activity, LAS interrupts the circulation of the token, executes the scheduled activity and then resumes once again the token circulation process.

This protocol provides three priority levels, as follows: urgent, normal and time-available. Usually, urgent priority is assigned to time critical data such as alarms. Normal priority is assigned to event based data, diagnosis data, parameter settings, etc. Time-available priority is assigned to trend data, program download data and so on. The permissible priority level is set at the beginning of each token circulation cycle and is used in each delegation.

3.2 TS 61158 security services.

The TS 61158 provides users with the authentication and access control services for some objects. These services are implemented at the application layer, at this level, the application process objects (APOs) are defined, which are components of the application process and are visible across the network. An APO is identified by one or more key attributes, such as access privilege. This attribute specifies the access controls defined by APO, and is composed of the following fields:

- Password, it contains the password for the access rights. Its value is null if it is not used.
- Access Groups, it identifies which of the user-defined access groups are defined for the object.
- Access Rights, it defines the type of access attributes authorized for the object. Legal examples of such attributes are as follows: right to read/write for access groups, for the registered password, and for all communication partners.

The access privilege attribute is defined for the following APOs: variables, events, and load regions. Nevertheless, previous subsection has shown that the management system, invoked by an authorized human operator, configures the initial schedule in LAS. Therefore, the TS 61158 must provide a user-authentication service in order to put in place some mechanisms of user identification in LAS, such as logging and password, to avoid access from non-authorized human operators.

On the other hand, the TS 61158 does not provide the following ISO security services: node-authentication, integrity, confidentiality, and non-repudiation.

Previous subsection has shown that the LAS node uses the notion of Delegated Token (DeT) to transfer the right to transmit to another node for a specified duration time. Usually, the token has two fields for addressing the source and destination nodes. In order to avoid a masquerade attack the TS 61158 must provide a node-authentication service.

The integrity service can be achieved by adding a field into the frame, such as the frame check sequence (FCS), which is calculated when the frame is transmitted and when it is received. If the code received matches the code calculated there is a very high probability that the frame is correct. This service can be also provided by using other cryptographic mechanisms, such as one-way hash functions, which accept a variable-size message as input and produce a fixed-size code, called the hash code. The hash code is a function of all the bits of the message and provides an error detection capability: A change to any bit or bits in the message results in a change to the hash code.

In order to avoid passive and active attacks from intruders that have managed to gain access to the communication channel, the TS 61158 must instrument a data-confidentiality service via feasible encryption/decryption schemes.

Regarding the non-repudiation service, it seems to be not very useful for this kind of networks since the LAS node “gives permission to speak” to each node.

4. SECURITY ARCHITECTURE FOR THE TS 61158

4.1 Security goals.

This section lists the goals that we wish to obtain from the security architecture to be proposed here.

In our model we shall assume that all nodes belonging to the network possess some kind of processing capabilities, and that both of them, LAS and Publisher/Subscriber nodes trust each other. Deceiving from any party should be difficult or impossible in practice. Ideally, fault induction and power disruption should not compromise protocol's security, nor should it open avenues for hijack attempts. LAS and Publisher/Subscriber nodes should be resistant to session hijacking, replay or man-in-the-middle attacks.

As it was mentioned in section 2, all the customary security services for communication protocols can be achieved by using public key cryptography schemes. However, public key cryptography requires a processing power that is typically well beyond the reach of many field devices. Fortunately we can still design limited security schemes for Fieldbuses at a cheaper price by using alternative cryptographic options.

Data confidentiality can be achieved by using some lightweight cryptographic stream cipher such as RC4 or A5 GSM (Biryukov *et al.*, 2000), or even a reduced version of traditional symmetric algorithms such as DES or AES.

Node-authentication can be achieved by using node's private key to sign data to be transferred.

Due to the fact that password-based security mechanisms are needed only for data protection at the LAS node, where typically processing power is not a concern, we can use any of the traditional schemes based on symmetric ciphers.

Summarizing, this paper proposes to achieve at least at a certain degree, the above stated security features by incorporating the following mechanisms into the security architecture proposal:

- A lightweight stream cipher in order to guarantee data confidentiality by encrypting all the relevant data to be transferred by the network's entities.
- Node Authentication via private key.
- A password-based security mechanism to prevent non-authorized users to gain control in the LAS node.

These security mechanisms are discussed in detail in the next subsection.

4.2 TS 61158 Security Architecture.

In this paper we propose to use RC4 as the stream cipher needed to implement data encryption/decryption. RC4 is a symmetric-key stream cipher that was developed in 1987 and kept as a trade secret by RSA Data Security, Inc. In September 1994, the algorithm was anonymously posted on the Internet, and since then it has been studied and used widely in academic circles and is now available for public analysis. RC4 is commonly used as the default cipher for SSL/TLS connections and currently being standardized by the IETF under the name "Arcfour" (Kaukonen and Thayer, 1999).

RC4 uses a 2048-bit key-length needed to initialize a 256-byte state table, called the S-box. If a given application wishes to use a shorter key, then that shorter key is repeated as many times as needed to fill the 2048-bit key. Typically, RC4 is used in a mode where 16-byte (128-bit) keys are repeated sixteen times.

Once the S-box is initialized with the key, the RC4 algorithm enters in a loop that updates the S-box and generates a byte of pseudo-random keystream. That pseudo-random keystream is XOR-ed with the plaintext message to produce the ciphertext.

Thus, the RC4 algorithm can be broken into two phases: initialization and operation. In the first phase the 256-byte state table S is populated using the key K as a seed. Once the state table is setup, it continues to be modified in a regular pattern as data is encrypted. The C-like pseudo-code for the initialization phase of RC4 is as follows:

<u>Initialization Phase</u>:

```
1. j = 0;
2. for i = 0 to 255:
3.   S[i] = i;
4. for i = 0 to 255:
5.   j = (j + S[i] + K[i]) mod 256;
6.   swap (S[i], S[j]);
```

Notice that what the RC4 initialization phase basically does is to swap the location of the numbers 0 to 255 (each of which occurs only once) in the state table.

Let us assume that the initialization phase has been completed and that the input message M consisting of N bytes have been stored in the array $M[0..N-1]$. Then, the following algorithm implements the operation phase of RC4 by encrypting/decrypting the input message M:

<u>Operation Phase</u>:

```
1. i = j = 0;
2. for (k = 0 to N-1) {
3.   i = (i + 1) mod 256;
4.   j = (j + S[i]) mod 256;
5.   swap S[i] and S[j];
6.   D = S[(S[i] + S[j]) mod 256]
7.   output C= M[k] XOR D
}
```

The above algorithm outputs a ciphertext C which is the input message M, XOR-ed byte by byte with the values of the random stream D. As long as the same initial S-box S is used, both of them encryption and decryption processes are completely symmetric. Therefore, if the operation phase algorithm is used to obtain an encrypted ciphertext C from an input plaintext M, then M can be recovered from C by re-executing the same algorithm once again, provided that the same initial state table S is used.

In order to obtain both, integrity and authentication security services, we use the scheme depicted in Fig. 1, which consists on the following steps. First of all, before transmission each delegated token is hashed using the MD5 hash function. The hashed value so produced is then signed by the LAS node using RC4 and the node's assigned unique identifier K_{privAN} as private key. That signed value is then appended to the token for its immediate transmission.

The LAS may obtain the correct private key from a system list of all active nodes' private keys that it has to maintain.

At the other end, the Active Node proceeds to verify both, the authenticity and the integrity of the message just received. To do that, the Active Node uses the following verification procedure. It first decrypts the hashed value attached to the message using its private key. After decryption the hash value so obtained is compared with the hash value of the message and compared to verify or deny authenticity. In this way, only the right node can use the delegated token.

5. SPECIFICATION OF THE TS 61158 SECURITY ARCHITECTURE

This section presents the specification of the TS 61158 Security Architecture, which has been coded using the Specification and Description Language (SDL) (ITU-T Z.100). We selected SDL because it is a language that is intelligible to human beings, but still formal enough to support analysis and comparison of behaviors.

For validation purposes, the term validation model is utilized by SDL. A validation model is a description of the system, which is suitable for validation, i.e. to apply validation techniques such as testing the formal model, exhaustive validation (reachability analysis), non-exhaustive validation (analysis of a random subset of the reachable states), simulation and informal validation techniques (checklist) (Hogrefe, 1996).

A validation model is always executable. In order to construct the validation model of the TS 61158 Security Architecture, shown in Fig. 2, the toolset ObjectGEODE was utilized.

In SDL, behavior is always performed in the context of a *system*, beginning with a top-level description of the system to be validated. Fig. 2 shows the TS 61158 system, which is composed of the following *blocks*: the application and data link layers of the LAS (*LASApplicaLayer* and *LASDataLinkLayer*), the data link layer of the Publisher/Subscriber nodes (*PSDataLinkLayer*), and the physical layer. These blocks are interconnected through *channels*, named ca, cb, c1, c2 and c3. The channel ca connects the LAS with the outside environment. A human operator uses the channel *ca* to invoke the management system.

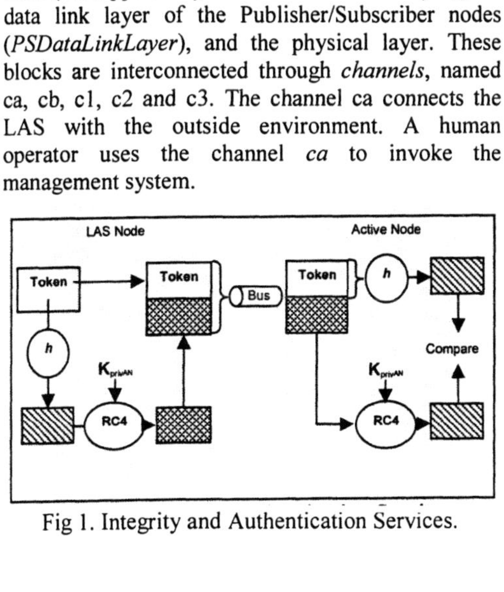

Fig 1. Integrity and Authentication Services.

In addition, Fig. 2 shows the *signals* passed in each direction over the channels, as indicated by the arrows on the channels, and the declaration of the signals as well as the declaration of the data structures.

It can be noted that the operation of the TS 61158 has not changed. At network operation time, LAS continues transmitting a Compel Data message to a publisher node, but now the publisher broadcasts the ciphertext (Data(C)), which is the variable's value encrypted by the operation phase of the RC4 algorithm depicted in subsection 4.2. On the other hand, the subscriber nodes read and decrypt the ciphertext to obtain the variable's value.

In SDL, a *system* occurs only at the top level, while *blocks* only occur inside. The system is decomposed into blocks and channels recursively over as many levels as desired until the basic components, called *processes*, are reached.

Fig. 3 presents the data link block of publisher/subscriber nodes. These blocks are composed of processes. It can be noted that there exist only two new processes, called RC4 and MD5, which executes the node authentication and confidentiality services. The user authentication service was specified into the application layer of the LAS node, but it is not shown for lack of space.

One *SDL process* is a concurrent object with its own control flow, described by an Extended Communicating Finite State Machine (FSM), which is composed of the following four main parts: input port, FSM, timers and variables. This FSM state-transition behavior is expressed in terms of *process diagrams*, which are not presented in this paper for lack of space.

Fig. 4 shows some simulation results of the LAS data link layer block by means of a Message Sequence Chart (MSC). It can be noted two kinds of transactions: LAS transmits a Compel Data message to a Publisher node. Upon reception, the Publisher broadcasts the encrypted data to all the nodes. In the second transaction LAS delegates the token (DeT) for a specified duration time.

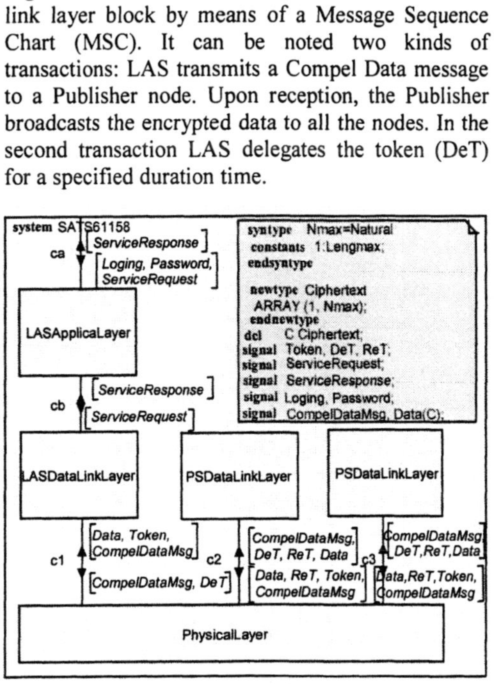

Fig.2. TS 61158 security architecture system.

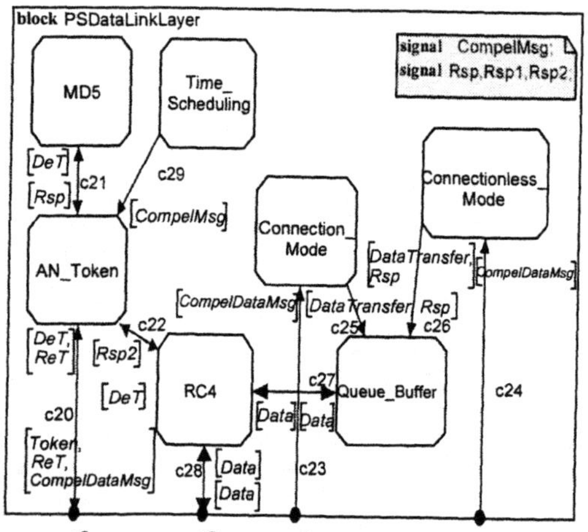

Fig.3. Publisher/Subscriber Data Link Layer block.

6. CONCLUSION

This paper has discussed the security in centralized Fieldbuses, such as TS 61158. The discussion has shown that this Fieldbus provides only limited security, namely the access control security service defined by ISO. However, centralized networks are vulnerable to at least two possible security attacks, as follows: Non-authorized users gaining access to the communication channel and non-authorized human operators accessing the master node.

Both types of security attacks can be avoided using public key cryptography schemes. However, public key cryptography requires a processing power that is typically well beyond the reach of many field devices. To overcome this difficulty, this paper has proposed a Security Architecture that provides the user and node authentication service and the confidentiality service at a cheaper price by using alternative cryptographic options.

We propose to use a reduced version of DES symmetric crypto-algorithms at the LAS node for user authentication. A one-way hash function, such as MD5, and RC4, which is a lightweight cryptographic stream cipher at the LAS and Publisher/Subscriber nodes for the node authentication, and integrity services. Finally, we propose to use RC4 for the confidentiality service.

The proposed TS 61158 Security Architecture has been specified using SDL. For simulation purposes a validation model has been constructed using the toolset ObjectGEODE. From the original TS 61158 Standard, it can be noted that only three processes are required to be added into this Standard: RC4, MD5 processes and password process (not shown in the application layer block of the LAS).

The RC4 process is used by the publisher/subscriber nodes for encrypting/decrypting the variable's value. Therefore, there exists an extra processing into these nodes that must be quantified and measured in order to verify that the timing constraints on the messages are met; and this is our future research work.

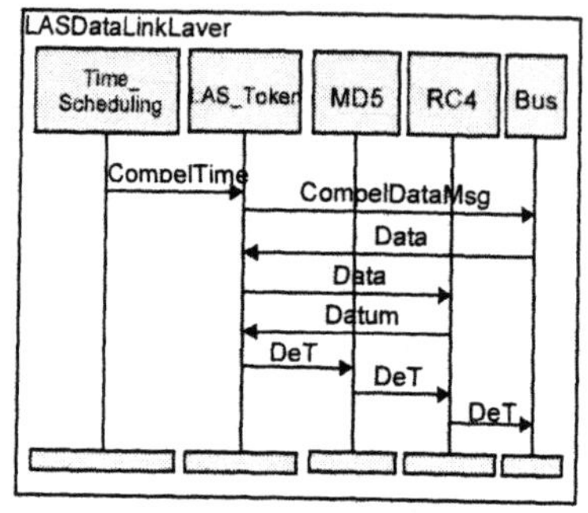

Fig.4. MSC Simulation results.

REFERENCES

Biryukov, A., A. Shamir, and D. Wagner (2000). Real Time Cryptoanalysis of A5/1 on a PC. In *Fast Software Encryption Workshop*, April 10-12, New York City.

Hogrefe, D. (1996). Validation of SDL systems. *Computer Networks and ISDN Systems*, **vol.** 28, no. 12, pp. 1659-1668.

IEC 61158 (1999). International Electrotechnical Commission. Digital Data Communications for Measurement and Control – Fieldbuses for use in Industrial Control Systems. *IEC 61158*.

ISO 7498-2 (1989). International Organization for Standardization. Information processing systems-Open Systems Interconnection-Basic Reference Model - Part 2: Security Architecture.

ITU-T Z.100 (2000). ITU Recommendation Z.100, *the Specification and Description Language (SDL)*.

Kaukonen, K., and R. Thayer (1999). A stream cipher Encryption algorithm Arcfour. IETF, July 1999. Internet draft-kaukonen-cipher-arcfour-03.txt.

Menezes, Oorschot, and Vanstone (2001). *Handbook of Applied Cryptography*. CRC Press, New York, fifth ed.

Paar, C. (2002). Cryptography in Heavily Constraint Environments. In *Proceedings of the first Ad-hoc Workshop*, December 2-3, University of Bochum, Germany.

Sarma, S.E., S.A. Weis, and D.W. Engels (2002). Low Cost RFID and the Electronic Product Code. In *Cryptographic Hardware and Embedded Systems-CHES 2002, Lecture Notes in Computer Sciences*, Springer-Verlag, Heiselberg, Germany.

Thomesse, J.P. (1998). A Review of the Fieldbuses. *Annual Reviews in Control*, **vol. 22**, pp. 35-45.

Trappe, W., and L. Washington (2002). *Introduction to Cryptography with Coding Theory*. Prentice-Hall.

Copyright © IFAC Information Control Problems in Manufacturing, Salvador, Brazil, 2004

ELSEVIER
IFAC
PUBLICATIONS
www.elsevier.com/locate/ifac

ANALYSIS OF HYBRID SYSTEMS: STATE OF THE ART

J. Zaytoon[1], H. Guéguen[2]

1) LAM, Université de Reims Champagne-Ardenne, BP 1039, 51687 Reims Cedex 2, France, Tel: (+33).3.26.91.32.26 ; Fax: (+33).3.26.91.31.06 ; Janan.zaytoon@univ-reims.fr
2) Supélec, B.P. 28 Avenue de la Boulais, 35511 Cesson-Sévigné Cedex, Tel: (+33).2.99.84.45.04 ; Herve.Gueguen@supelec.fr

Abstract: *This paper presents a state-of-the-art of the analysis of hybrid systems. Analysis and verification techniques are usually based on calculation of the reachable state space of a hybrid automaton representing the system under study. Decidability issues related to the verification algorithms require the use of approximation and abstraction techniques. In particular, discrete abstraction of hybrid systems and over-approximation techniques to accelerate the convergence of the reachability-based algorithms are emphasized in this paper. Copyright © 2004 IFAC*

Keywords: Verification, reachability, abstraction, approximations, Hybrid automata, decidability

1. INTRODUCTION

Manufacturing systems are hybrid systems by nature because their modelling and control requires the combination of both continuous and discrete event-based models. The development of systematic methods for efficient and realiable realisation of hybrid systems is a key issue in industrial information and control technology and is therefore currently of high interest to manufacturing engineers. The engineering methods for hybrid systems should deal with issues related to modelling, specification, analysis, verification, control synthesis, and implementation. The aim of this contribution is to present a state-of-the-art related to the analysis of hybrid systems. Analysis and verification techniques are usually based on the calculation of the reachable state space of a hybrid automaton representing the system under study. Decidability issues concerning the verification algorithms require the use of approximation and abstraction techniques to accelerate the convergence of the analysis methods.

1.1. Verification

There are two major approaches to the verification of hybrid systems. In the first, the verification is directly related to the model of the hybrid system; in the second, the hybrid model is first transformed into a discrete-event model to be explored by the verification algorithm. Traditionally, three types of property can be analysed: (i) safety properties that express the non-authorized configurations to be avoided in all the possible evolutions of the system, (ii) liveness properties that describe the possibility or the eventuality of some required system evolutions, and (iii) timeliness properties that express the constraints on the minimum and/or maximum times separating some characteristic events or evolutions.

In the first verification approach, related to the model of the hybrid system, the safety properties are particularly emphasized. These properties are specified in terms of hybrid regions (hybrid state space) that the system should avoid or remain in. Such an expression can be formulated either directly (for example, when a continuous-state variable should never cross a given threshold value) or by an observer—i.e., an automaton (to be composed with the model for verification without modifying the system behaviour) that evolves towards a particular state according to the satisfaction, or non-satisfaction, of the property (Halbwachs, 1993). Timeliness properties can also be expressed for hybrid systems using "forbidden-state observers", as time is represented implicitly in these systems. Conversely, liveness properties cannot easily be expressed in terms of forbidden states if they express the eventuality of occurrence. Only liveness properties related to the possibility of occurrence of a required event at a given time can be expressed in terms of a forbidden state.

The other approach related to verification consists of substituting the hybrid model with a discrete-event model. In this case, the classical verification algorithms of state machines can be used, such as those developed by Clarke *et al.* (2000). The hybrid aspect of these approaches consists of translating the hybrid model into a discrete-event model that is suitable for verification, that is, such that the conclusions of the verifications of the discrete-event system are valid relative to the properties of the original hybrid system. An ideal situation would be one in which each and every evolution of the discrete-event model corresponded to an evolution of the hybrid model, and vice versa (bi-similarity); in this case, any property that was valid relative to the discrete-event system would also be valid with respect to the hybrid system. Unfortunately, in general this goal is elusive, and the commonly used technique consists of determining a discrete-event model that is an abstraction of the hybrid model— i.e., one in which each and every evolution of the hybrid model corresponds to an evolution of the discrete-event model but the inverse is not necessary. The verification of a property characterizing all the evolutions of the discrete-event model guarantees satisfaction of the equivalent property relative to the hybrid model. The properties that can be verified using these approaches are the same as those that can be verified directly with respect to the hybrid model.

1.2. Structure of the paper

The verification approaches presented above require the use of a hybrid model of the system to be verified. Although different modelling formalisms, such as MLD (Bemporad and Morari, 1999), have been proposed in the literature, hybrid automata, which are presented in section 2, are the most commonly used formalism for the specification and verification of hybrid systems. The problem that underlies safety verification for hybrid automata is reachability: can an unsafe state be reached from an initial state by running the system? In practice, many verification problems can be posed naturally as reachability problems. The traditional approach to reachability attempts to compute the set of reachable states iteratively using symbolic model checking. This computation can be automated and is guaranteed to converge in some special cases (Alur *et al.*, 1995) for which the reachability problem is decidable. In general, however, this approach may not be automated, or may not converge. Therefore, one of the main issues in algorithmic analysis of hybrid systems is decidability, because it guarantees that the analysis will terminate in a finite number of steps. Issues related to linear hybrid automata, reachability analysis and decidability results are presented in section 3. Henzinger *et al.* (1995a) showed that checking reachability for a very simple class of two-slop hybrid automata is undecidable. Therefore, other approaches to the reachability problem have been pursued. The most common of these approaches are based on abstracting the hybrid system with a discrete-event- based model for the purpose of analysis (Section 4). To overcome the termination problem, over-approximation analysis techniques, which are discussed in section 5, are also used to enforce convergence of the iterations by computing upper approximations of the state space. The problem related to the continuous expansion of a given hybrid region as a function of the continuous dynamics of the hybrid model is also treated in section 5.

2. HYBRID AUTOMATA

Hybrid automata are finite automata enriched with a finite state of real-valued variables. In each location, the variables evolve continuously according to different flow fields, as long as the location's invariant remains true; then, when a transition *guard* becomes true, the control may proceed to another location and reset some of the variables to new values according to the *Jump* function of the transition. A hybrid automaton (Alur *et al.*, 1995) is given by $H = (Q, X, \Sigma, A, Inv, F, q_0, \mathbf{x}_0)$, where:

- Q is the set of locations including the initial location, q_0;
- $X \subset \Re^n$ is the continuous-state space and $\mathbf{x}_0$ is the initial continuous state;
- Σ is the set of events;
- A is the set of transitions given by the 5-uple (q, *guard*, σ, *Jump*, q'), where q and q' represent the upstream and the downstream locations of the transition respectively, guard is a condition given in terms of the continuous-state vector, σ is an event, and Jump is a function that resets some of the variables to new values when the transition is taken;
- *Inv* is an invariant that assigns a domain of the continuous-state space to each location;
- F assigns to each location a flow field, which is usually given in terms of a differential equation but can also take imperative or less explicit forms, such as differential inclusions.

The state of a hybrid automaton is given by the couple $(q, \mathbf{x})$, which associates a location with the value of the continuous-state vector. This state can advance:

- either by the progression of time in the current location, which results in a continuous evolution of the continuous-state vector according to the corresponding flow field, F;
- or by an instantaneous transition that proceeds to a new location and changes the value of the continuous state according to the *Jump* function of the transition. Such a transition can be taken when the continuous state satisfies the guard and provided that the resulting state vector (through the application of the *Jump* function) satisfies the invariant of the downstream location.

Complex systems can be modelled in a modular fashion by using a number of parallel automata that can be composed to produce the global model of the system. However, modelling with hybrid automata is not always an easy task because the semantics associated to this formalism is behaviour-analysis oriented (Guéguen *et al.*, 2001). A number of approaches have therefore been proposed to generate hybrid automata starting from other modelling formalisms that are more suitable for control engineers (Stursberg *et al.*, 1998), (Guillemaud and Guéguen, 1999).

3. HYBRID REACHABILITY APPROACHES

A region for a hybrid automaton is a pair (q, P), where q is a location and P is a linear predicate on the continuous-state vector. A state $(q, \boldsymbol{x})$ is included in the region (q, P), if $\boldsymbol{x}$ satisfies P—that is, if by replacing each variable in P with its value given by $\boldsymbol{x}$, one obtains a true statement. The reachability problem for hybrid automata is: given hybrid automaton H, and a set P of regions, is there a reachable state of H that is contained in some region in P? The reachability problem can be used to verify safety properties. For example, if the predicate P represents a "good" set of states in which the system should always stay (i.e., a desirable invariant of the hybrid system), then the hybrid automata satisfies this set if and only if all the states of the hybrid automata that can be reached starting from the initial region belong to P. Equivalently, the complementary predicate $B = P^c$ is a "bad" set of states that the system should avoid. Symbolic model checking techniques (Alur *et al.*, 1995), (Henzinger *et al.*, 1995a) can be used to compute automatically the reachable state space of a hybrid system. These techniques are based on verification algorithms that perform reachability analysis to check whether trajectories of the hybrid system can reach certain undesirable regions of the state space. When such computational algorithms are applied to systems with infinite state spaces, they are in danger of never terminating. This makes the issue of decidability, which guarantees termination of the algorithm, very important.

3.1. Hybrid Reachability Calculus

Consider a hybrid automaton for which it is required to determine the set of reachable states, starting from a region characterized by the active location, $q \in Q$, and the state space I that includes the current continuous state. As explained above the possible runs of the hybrid automaton are given as a succession of continuous transitions and discrete transitions. The calculation of the reachable state space, is based on this succession and can be summarized as follows.

Starting from a region (q, I):

- calculate the intersection of the invariant of q, $Inv(q)$, with the expansion of I at q—i.e., the set of states that can be reached from each state in I by applying the flow fields $F(q)$; the state space resulting from this intersection is given by I';
- determine Post(q, I') by iterating the following calculations for each downstream transition of q:
  (i) calculate I_1, by intersecting I' with the set defined by the guard of the considered transition;
  (ii) calculate the image I_2 of I_1 by applying the Jump function of the transition;
  (iii) calculate I_3, by intersecting I_2 with the invariant of location q', $Inv(q')$, where q' is the upstream location of the considered transition;
  (iv) calculate I_4, by intersecting the extension of I_3 at q' with the invariant of location q', $Inv(q')$;
- the resulting set of regions (q', I_4) is then added to the reachable space, and the previous step is reiterated to calculate $Post(q', I_4)$.

These calculations are reiterated until convergence —i.e., until a step is reached where the set of reachable regions does not evolve. To illustrate this calculation, consider the example of the hybrid automaton of figure 1, which models a water-level controller that opens and shuts the outflow of a water tank. The variable x represents a clock of the water-level controller and the variable y represents the water level in the tank. Because the clock x measures time, the first derivative of x is always 1 (i.e., $dx/dt = 1$). In location *shut*, the outflow of the water tank is shut and the water level increases 1 centimetre per second ($dy/dt = 1$); in location *open*, the outflow of the water tank is open and the level decreases 2 centimetres per second ($dy/dt = -2$).

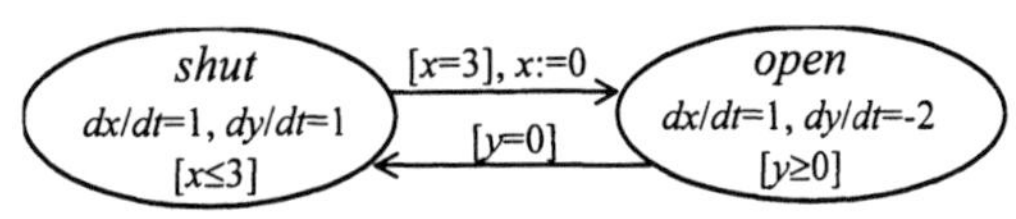

Fig. 1. The water-tank automaton

The transition from open to shut (stop the outflow) is taken as soon as the water tank becomes empty: the guard $y = 0$ on the transition ensures that the transition may be taken only when the water level is 0; the constraint $y \geq 0$ on *open* ensures that the transition to *shut* must be taken before the water level becomes negative. The transition from *shut* to *open* (open the outflow) is taken every 3 seconds: the transition is taken whenever $x = 3$, and the transition restarts the clock x at time zero.

Initially, the tank is empty and the outflow is shut. To tolerate a slight uncertainty in the water level around the zero level, the initial region is chosen to be $S_0 = (shut, I)$ with $I = -0.2 \leq y \leq 0.2 \wedge x = 0$. Now, if it is required to determine whether or not it is possible to attain the target region $(shut, -1 \leq x \leq 3 \wedge x = y + 1)$ starting from the initial region, the reachability calculation will be as follows:

- the initial region is $(shut, I)$;
- the extension of I by the flow field of location *shut* is given by $-0.2 \leq x\text{-}y \leq 0.2$, because $dx/dt = dy/dt$, and its intersection with the invariant of this location is $I' = (-0.2 \leq x-y \leq 0.2 \wedge x \leq 3)$.

The reachable state space is therefore initialized to $R_0 = (shut,\ -0.2 \leq x-y \leq 0.2 \wedge 0 \leq x \leq 3)$. The reachability calculation proceeds as follows:

- the transition leading to location *open* is the only one that can be taken, because the guard of this transition is $[x = 3]$, I_1 is given by $I_1 = I' \cap (x = 3)$; therefore, $I_1 = (x = 3 \wedge 2.8 \leq y \leq 3.2)$;

- by applying the jump function of the transition, $x := 0$, and considering the invariant of location *open*, the sets I_2 and I_3 are given by $I_2 = I_3 = (x = 0 \wedge 2.8 \leq y \leq 3.2)$;
- the flow field of location *open* results in: $I_4 = (2.8 \leq y + 2x \leq 3.2 \wedge x \geq 0 \wedge y \geq 0)$.

The reachable space after this first iteration is therefore given by the set: $R_1 = R_0 \cup (open, 2.8 \leq y + 2x \leq 3.2 \wedge x \geq 0 \wedge y \geq 0)$.

The second iteration consists of applying the *Post* operator to the region (*open*, I_4) to calculate the sets: J_1, J_2, J_3 and J_4, for example $J_4 = (1.4 \leq x - y \leq 1.6 \wedge y \geq 0 \wedge x \leq 3)$. In this case, it is possible to verify the non-convergence of this iterative calculation of the reachable regions. However, a similar backward reachability analysis, starting from $S_{target} = (shut, 1 \leq x \leq 3 \wedge x = y+1)$, allows one to conclude, after two iterations, that the region S_{target} can be reached both from itself and from the region $(open, y + 2x = 2 \wedge y \geq 0)$. Because S_0 has an empty intersection with these two regions, it is possible to conclude that S_{target} cannot be reached from S_0.

This trivial example shows that it is sometimes easier to verify the hybrid system using backward reachability, and that forward and backward analysis are complementary because it is not possible to know in advance which to apply. The example also shows that even for simple examples, the convergence of the reachability calculation is not guaranteed.

3.2. Decidability Considerations

Hybrid systems in which the reachability problem can be solved algorithmically in a finite number of steps are called decidable hybrid systems.

A hybrid automaton is linear if (Alur *et al.*, 1995):

- the invariants of all the locations, the initial region and the guards are given by a conjunction of linear predicates (equalities or inequalities) over the continuous variables with rational coefficients,
- the flow fields (of all locations) are linear differential inclusions given by a conjunction of linear predicates (equalities or inequalities) over the first derivatives of the continuous variables with rational coefficients; and
- the jump functions are given by non-deterministic assignments in intervals the boundaries of which are linear expressions.

The interesting point about these automata is that all the regions of their continuous-state space (invariants, guards) are convex linear regions, and that the images of linear regions by their continuous and discrete transitions are also linear regions (Henzinger and Rusu, 1998). However the reachability problem is not decidable even for this class of hybrid automata (Alur *et al.*, 1995) unless some restrictions are introduced. Indeed the reachabilty problem is decidable for timed automata and for automata that can be transformed into timed automata. The most important criterion for a linear hybrid automaton to be decidable then appears to be that it is initialised, i.e. that the continuous state is always the same when a location is activated. The restrictions on jump functions of the transitions are also very important to be able to extend the class of hybrid automata that are decidable to automata with more complex dynamics (Lafferriere *et al.*, 1999).

A practical consequence of these results is that the calculation algorithms may not converge, and it therefore becomes necessary to use some termination criteria for these algorithms. On the other hand, the user should take the risk of non-convergence into account, and should try to avoid this by adapting a verification strategy that is based on, for example, alternating forward and backward computations.

4. DISCRETE-EVENT ABSTRACTION

To construct a discrete-event model of a hybrid system for the verification of a required property, it is necessary to be able to establish the correspondence between the traces of the hybrid system and those of the discrete-event model. The construction of a discrete-event model is basically concerned with partitioning the continuous part of the state space into a number of domains to each of which is associated a discrete state, then establishing a transition between two discrete states if there exists a continuous trajectory that leads from a point in the domain of the first discrete state to the domain of the second. In practice, partitioning of the entire continuous domain in terms of location invariants is complex, and it is therefore necessary to limit this approach to a certain form of model that can be restricted to some particular domains. For example, CheckMate (Chutinan and Krogh, 2003) treats the hybrid automata whose jump functions are given by the identity function and whose transition guards are given by unions of the boundaries of the locations' invariants. Consequently, these boundaries can be used to construct a discrete-event model.

Starting from a first partition (S) that corresponds, for example, to the faces of the invariants, the partition is refined by applying the following algorithm (Lafferriere et al. 1999), where the Pr function calculates the set of points representing the predecessors of a region according to the vector field:

$$\text{While } \exists P, P' \in S \text{ such that } \varnothing \neq P \cap \Pr(P') \neq P$$
$$P_1 = P \cap \Pr(P')\,;\, P_2 = P - \Pr(P')$$
$$S = (S - \{P\}) \cup \{P_1, P_2\}$$

As shown in Fig. 2, this algorithm is based on the calculation of the set of points representing the predecessors of a region, and then the calculation of the intersection of the predecessors with the other regions of the partition. If this intersection is empty, it is not possible to attain P' from P and, consequently, the discrete-event model does not include a transition between the corresponding states. Conversely, if the intersection is equal to P, all the points of P lead to P', and the firing of the transition between the states associated with P and P' in the discrete-event model corresponds to an evolution of the hybrid system. If neither of the two cases holds, there exists a transition in the discrete model, which relates a subset of P to P'. P is therefore partitioned into P_1 and P_2 that correspond to the two cases above: P_2 has an empty intersection with $\Pr(P')$ whereas the intersection of $\Pr(P')$ with P_1 is equal to P_1.

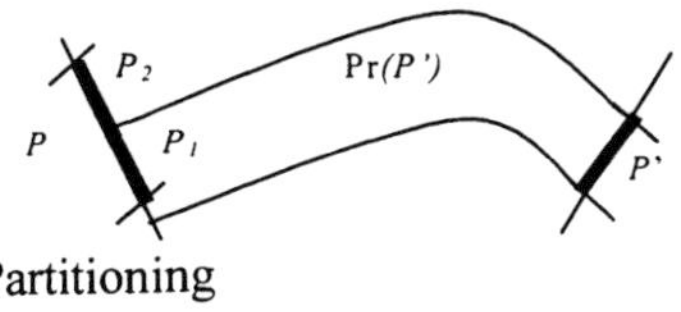

Fig. 2. Partitioning

If this algorithm terminates, the resulting discrete-event model is bi-similar to the hybrid model—i.e., there exists a one-to-one correspondence between the execution traces of both models, and therefore every property that is valid relative to one of these models will also be valid for the other. However, because this algorithm is based on the reachability calculation (Pr), the decidability results and the implementation difficulty are the same as for those in the case of the reachability algorithms presented in section 3.1. Therefore, it is not possible, in general, to construct a bi-similar discrete-event model of a hybrid system, and the alternative solution consists of developing a discrete-event model that is an abstraction of the hybrid model—i.e., to each trace of the hybrid model there exists a trace in the discrete-event model but the inverse does not necessarily hold. To construct such an abstraction, it is sufficient to proceed as follows: (i) choose a partition of pertinent domains (for example, the phases of the invariants), (ii) calculate the space that is reachable by the continuous dynamics starting from each of the elements of the partition, (iii) calculate the intersection of this space with each of the other elements of the partition, and (iv) add a transition relating the corresponding discrete states if this intersection is not empty. Spurious traces that do not exist in the original hybrid model are introduced by such a construction. For example, the partition in Fig. 3 implies that (P1→P3→P5) is a possible trace, which is not the case in the hybrid model.

If it is possible to obtain a partition implying that a given property is verified relative to all the traces of the discrete-event system, then it is possible to conclude that this property is also valid with respect to the hybrid system (Chutinan and Krogh, 2001). On the other hand, owing to the abstraction and the introduction of spurious traces, the invalidity of a property relative to the set of discrete-event traces does not imply that this property is also invalid with respect to the hybrid traces. In this case, it is necessary to refine the abstractions to eliminate some of the spurious traces. Certain strategies can be used in this refinement process to accelerate the iterative abstraction process and to avoid the generation of an excessively complex discrete-event model. One strategy introduced by Strusberg *et al.* (2003) is to identify the traces invalidating a given property and to refine the partition along these traces.

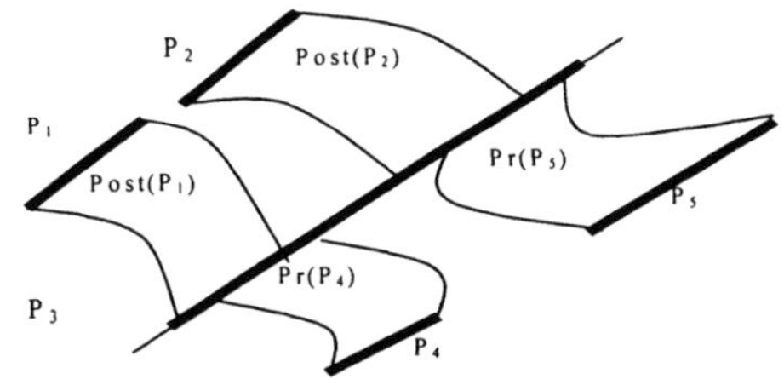

Figure 3. Spurious traces

5. CONTINUOUS EXPANSION COMPUTATION

Property verification, whether carried out through reachability analysis or by discrete-event abstractions, requires the manipulation of regions (sets) in state space (initial region, specification domain, guard conditions, location invariants, ...) and the calculation of the intersections or unions of these regions and their image with the flow fields of the locations and with the jump functions. These manipulations can be performed using, for example, ellipsoidal calculations (Kurzhanski and Varaiya, 2000), but in most of the cases, the sets to be treated are polyhedra involving linear equalities or inequalities over continuous-state variables. However even for these polyhedra it may be useful to use operators such as the convex hull in order to make simpler the sets of inequalities that are manipulated and to get over-approximations of the reachable state to enhance the convergence of the fixed-point computation (Halbwachs *et al.*, 1994).

In the approaches presented in sections 3 and 4, the main issue is the calculation of the reachable space of the system starting from an initial region, on the basis of the continuous dynamics of the system. The region of the state space that is reachable by applying the flow field at location q, starting from region D is given by the following set.

$$R_q(D) = \left\{ x \in X \left| \begin{array}{l} \exists t, \exists y \in D \; s.a. \; x = \Phi(y,t) \\ and \; \forall \, \tau \in [0,t], \\ \Phi(y,\tau) \in Inv(q) \\ \wedge \, \dot{\Phi}(y,\tau) \in F(q, \Phi(y,\tau)) \end{array} \right. \right\}$$

The manipulation of this set—for example, calculating its intersection with a guard condition—requires the elimination of the quantifiers to obtain an expression that is not parameterized as a function of time (Laferriere *et al.* 1999). The set of operations required to characterize the reachable space is rather complex and difficult to implement. Furthermore, even in the case where the initial region is polyhedral, the reachable state is not necessarily polyhedral. The practical implementation of these calculations is only possible in the case where the space reachable from a polyhedron is a polyhedron—i.e. when the automaton is linear and the flow fields are given by differential inclusions. The reachable space is therefore calculated by iterative application of the convex-hull operator, starting from the vertices of the initial polyhedron and the faces of the invariants. If the flow fields of the locations are not given by differential inclusions, the reachable space is not polyhedral. Two approaches can be adopted in this case to over-approximate the reachable space by a union of polyhedra: abstraction of the continuous dynamics by a linear automaton (section 5.1) and over-calculation of the reachable space at different times (section 5.2).

5.1 Hybrid linear abstractions.

Starting from a given hybrid automaton, the aim of this approach is to construct a hybrid linear automaton whose flow fields are differential inclusions and whose reachable state space is an over-approximation of the state space of the original automaton (Henzinger *et al.*, 1998).

Start by considering a hybrid automaton with a single location. The differential inclusion can be calculated by taking the invariant of the location and determining a polyhedron that contains the derivative vector defined by the flow field at each point of the invariant. For a linear system, it is sufficient to consider the convex hull of the derivatives at the points representing the vertices of the polyhedron.
For example, consider the determination of the reachable region that is included in the invariant $I = \{0 \leq x_1 \leq 3 \wedge 0 \leq x_2 \leq 3\}$, starting from region $I = \{1 \leq x_1 \leq 2 \wedge x_2 = 0\}$ for the two-dimensional system whose dynamics are described using the following equation,

$$\dot{\mathbf{x}} = \begin{pmatrix} 0 & -1 \\ 1 & 0 \end{pmatrix} \mathbf{x}$$

The differential inclusion corresponding to its invariant is depicted in figure 4a. Based on this differential inclusion, calculation of the region that can be reached from the region *I*, corresponding to the grey zone in figure 4b, is simple. This region contains the exact reachable region of the original automaton.

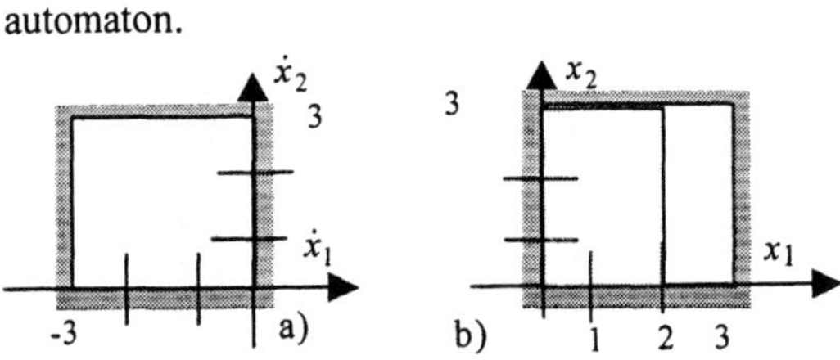

Figure 4. Abstraction by differential inclusion

If the results of this calculation are too coarse for conclusions to be made, it is necessary to refine the abstraction. This can be done by defining a partition for the invariant, associating a location to each element of the partition by defining the associated invariants, and then calculating the differential inclusion for each location by determining a polyhedron that contains the derivative vector for each point of the new invariant. A transition is then included between two locations if there exists a continuous trajectory that crosses the boundary between the corresponding elements of the partition. It is then possible to calculate the reachability on the basis of this abstraction. However some pertinent criteria based on the properties of the continuous dynamics have to be used to refine the partition in a way that improves the efficiency of the reachability calculation (Lefebvre et al. 2002a).

For example Lefebvre *et al.* (2002b) proposed to used half-lines defined by the equilibrium point to determine the characteristic regions defining the partition of the state space for affine planar systems to get a trade-off between the precision of the over-approximation on the one hand, and the simplicity of the automaton and the calculations on the other. For the above example, the simple partition of Fig 5a defines the abstract automaton of Fig. 5b whose reachability-calculation results are given in Fig. 11c.

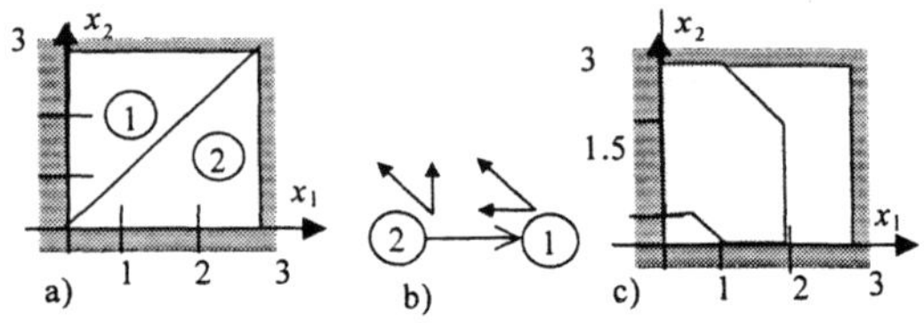

Fig. 5. Improving the partition

This refinement of the partition and, hence, the abstraction can be repeated until a sufficiently precise over-approximation of the reachable space is obtained with the view of reaching a valid conclusion. This approach enables the calculation of the continuous expansion of a given region within a location of the hybrid system. It can be used either to construct the discrete-event abstraction of the hybrid system or as a part of an algorithm for the calculation of the reachable state space of the system. However, it is generally used to construct a new automaton that models the behaviour of the system and can be used as the basis for verification. This construction requires the substitution of each location by a sub-automaton representing the abstraction of its continuous behaviour, the determination of the transitions between these new locations and the original locations. These translations may result in complex automata that are difficult to verify. Again, it is necessary to have a methodology that guides the refinement of the partition, especially for the regions that are considered to be critical for the verification of a given property to reduce the complexity of the resulting automaton. These abstractions enable the use of the existing results and tools such as Hytech or KRONOS (Yovine, 1993), and UPPAAL (Bengtsson *et al.*, 1996) if abstraction by timed automata is possible (Henzinger *et al.*, 1998).

5.2 Reachability in finite time

These approaches proceed by choosing a time step δ and determining a periodic polyhedral over-approximation (R_k) of the reachable space at the instants δ, 2δ, ... kδ, ..., starting from the initial region X_0. The reachable space is then over-approximated by the union of these polyhedra. The aim of this approach is not to calculate an over-approximation of the whole reachable space as is the case of the previous approaches, but instead to calculate an over-approximation of the reachable space in a finite time given by the number of steps calculated. To avoid the accumulation of the approximations through the repeated calculations, the image of the starting region is calculated at the different instants $k\delta$, then the polyhedra R_k are determined on the basis of these images, X_k (Dang, 2000). This principle is illustrated by figure 6. First, the image X_1 of the region X_0 is calculated as a function of the flow field at the instant δ (Fig. 6a). The next step consists of determining the polyhedron R_0, which contains X_0 and X_1 as well as the set of trajectories leading from a point in X_0 to a point de X_1 (Fig. 6b). The calculation is then repeated starting from X_1 to calculate X_2 and then the region R_1 (Fig. 6c).

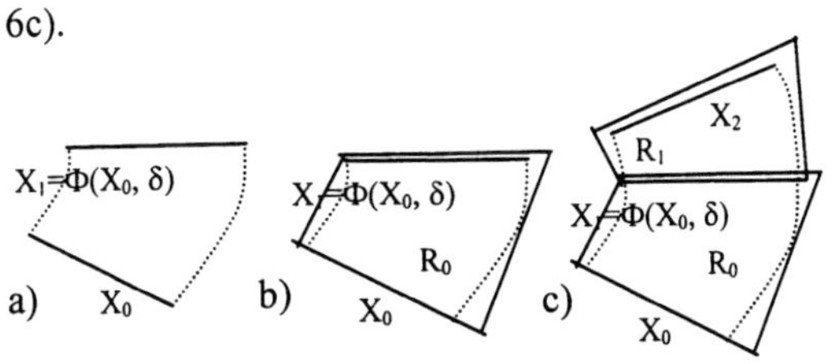

Fig. 6. Sampled approximation

The first difficulty of this approach is related to the calculation of the regions X_k representing the images of the original region at the instants $k\delta$, especially in the case of non-linear dynamics (Chutinan and Krogh, 2003), because it is not sufficient to consider only the vertices of the polyhedron R_{k-1}. The second difficulty relates to the determination of the polyhedron R_k from X_k. The most commonly used approach consists of calculating the first polyhedron as a convex hull of X_k and X_{k+1}, then lifting the faces of the polyhedron until it is guaranteed that all the trajectories are contained (Chutinan and Krogh, 2003). Numerical integration techniques are usually used to calculate X_k. To guarantee that the calculated X_k is an over-approximation of the reachable space at instant $k\delta$, it is possible to use interval analysis techniques (Jaulin *et al.*, 2001) that establish the results in the form of intervals including the exact results if the regions are defined by the Cartesian products of the intervals related to each of the state variables (Henzinger *et al.*, 2000; Flaus, 2001). On the other hand, the precision of the over-approximation is basically related to the time step, δ, used for the calculation; reducing this step results in an improvement in the precision.

One of the limitations of this approach is that the region R_k, calculated from X_k and X_{k+1}, is not an over-approximation of the space that is reachable from R_{k-1}. Therefore, if the repeated calculation determines a region R_i included in a previously obtained region, it does not imply the termination of the reachability calculation. The convergence of the reachability calculation is therefore difficult to determine. This difficulty is more critical in the case where the reachability calculation is used for the verification of a hybrid automaton (section 3) than in the case where it is used to construct a discrete-event abstraction of the hybrid system (section 4). This second case consists of determining the reachable space from a sub-region of the boundary of a given location to calculate its intersection with the other boundaries of the invariant. These intermediate regions are, therefore, not used to determine the termination of the reachability calculation. Consequently, the numerical approaches in finite time are generally used for the calculation of discrete-event abstraction of the hybrid system under study.

6. CONCLUSION

This paper has presented the principles of analysis and verification of hybrid systems. This verification can be performed relative to either the hybrid model by calculation of the reachable state space and its intersection with a forbidden region or a discrete-event model representing an abstraction of the hybrid model. In both cases, the major problem is related to the determination of the continuous expansion of a region, which is generally manipulated in terms of unions of polyhedra. Because this expansion cannot be represented exactly as union of polyhedra, over-approximation techniques may be used for this purpose. These over-approximations can be based on the abstraction of continuous behaviour with a set of differential inclusions, or the calculation of the image of the initial region at certain times.

It is possible to experiment with these verification techniques thanks to the availability of software tools, such as HYTECH (Henzinger *et al.*, 1995c) for linear hybrid automata, KRONOS (Yovine 1993) or UPPAAL (Bengtsson *et al.*, 1996) for timed automata, and Verdict (Stursberg *et al.*, 1998) or CheckMate (Chutinan *et al.*, 2000) for abstraction techniques. Many authors highlight the advantages, the difficulties and the open issues related to the use of these tools (Silva *et al.*, 2001; Stauner *et al.*, 1997). To improve engineering procedures, some of these tools are based on environments that are commonly used by control engineers, such as StateFlow-Simulink for CheckMate. To be efficient in real-sized industrial applications, these tools should be used in a framework of global methodologies that provide high-level modelling facilities to enable the capture of the models of complex systems as well as some practical heuristics related to exploitation of knowledge of the domain and/or some structural system characteristics to guide the underlying algorithms in the choice of the pertinent abstraction and approximation techniques required for the verification of a given property. The development of such an integrated methodology represents one of the main challenges to be faced by the hybrid systems community in the coming years, and requires close collaboration between control engineers and computer scientists.

7. REFERENCES

Alur R., Courcoubetis C., Halbwachs N., Henzinger T.A., Ho P.H., Nicollin X., Olivero A., Sifakis J., Yovine S. "The algorithmic analysis of hybrid systems", Theoritical Computer Science, *n°138, 1995, p. 3-34.*

Bemporad A, Morari M, "Verification of Hybrid Systems via mathematical programming", in (Vaandrager et al., *1999), p. 31 - 45*

Bengtsson J., Larsen K.G., Larsson F., Pettersson P., Yi W., "UPPAL: a tool suite for automatic verification of real-time systems", in Hybrid Systems III : Verification and control, *LNCS vol 1066, 1996, p. 232-243.*

Chutinan A., Krogh B., Milan D., Richeson K., de Silva B.I., "Modeling and verifying hybrid dynamic systems using CheckMate", in *(Engell S., Kowalewski S., Zaytoon J., Eds),* Hybrid Dynamical Systems. Proceedings of 4th Int. Conf. on Automation of Mixed Processes, *Dortmund, Germany, 2000, p 323-328.*

Chutinan A., Krogh B., "Verification of Infinite-State Dynamics Systems using approximate quotient transition systems", IEEE Trans. On Automatic Control *Vol. 46 n°9, sept. 2001 p 1401-1410*

Chutinan A., Krogh B., "Computational techniques for hybrid systems verification", IEEE Trans. On Automatic Control *Vol. 48 n°1 January 2003.p. 64-75*

Clarke E.M., Grumberg O., Long D., Model Checking *Cambridge MA, MIT Press, 2000.*

Dang T., "Vérification et synthèse des systèmes hybrides", Thèse de doctorat de l'INP Grenoble, Verimag, 2000

Flaus J.M., "Stabilité des systèmes dynamiques hybrides", in (Zaytoon 2001), *p. 237-254*

Guéguen H., Valentin-Roubinet C., Pascal J.C., Soriano T., Pingaud H., "Modèles mixtes et structuration des modèles complexes", in (Zaytoon 2001), *p. 157-188.*

Guillemaud L., Guéguen H., "Extending Grafcet for the specification of control of hybrid systems", Proceedings of IEEE SMC 99, Tokyo, Japan, 1999.

Halbwachs N., "Delay analysis in synchronous programs", Proceedings of 5th Int. Conference on Computer-Aided Verification, *LNCS-697, Springer, 1993, p. 333-346.*

Halbwachs N., Raymond P., Proy Y.E., "Verification of linear hybrid systems by means of convex approximations", Proceedings of Static Analysis Symposium, *LNCS-864, 1994, p. 223-237.*

Henzinger T.A., Kopke P.W., Puri A., Varaiya P., "What's decidable about hybrid automata? The algorithmic analysis of hybrid systems", Proceedings of 27th. annual ACM Symposium on Theory of Computing, *1995, p. 373-382.*

Henzinger T.A., Ho P.H., "A note on abstract interpretation strategies for hybrid automata", in Hybrid Systems II, *LNCS, Vol 999, Spinger, 1995, p. 252-263.*

Henzinger T.A., Ho P.H., Wong-Toi H., "A user guide to HYTECH", In: Tools and Algorithms for the Construction and Analysis of Systems, *LNCS-1019, 1995, p. 41-71.*

Henzinger T.A., Ho P.H., Wong-Toi H., "Algorithmic analysis of non-linear hybrid systems", in *(Antsaklis P.J., Nerode A., Eds), Special Issue on Hybrid Systems,* IEEE Transactions on Automatic Control, *Vol. 43, n° 4, April 1998, p. 540-554.*

Henzinger T.A., Rusu V., "Reachability verification for hybrid automata", Proceedings of 1st Int. Workshop on Hybrid Systems: Computation and Control, *LNCS-1386, 1998, p. 190-204.*

Henzinger T.A., Horowitz B., Majumdar R., Wong-Toi H., "Beyond HyTech: hybrid systems analysis using interval numerical methods", in (Lynch et al., *2000), p. 130-144.*

Jaulin L., Kieffer M., Didrit O., Walter E. Applied interval analysis, *Springer, 2001.*

Kurzhanski A. B., Varaiya P., "Elipsoidal techniques for reachability analysis", in (Lynch et al., *2000), p.202-214*

Lafferriere G., Pappas G. J., Yovine S., "A new class of decidable hybrid systems", in (Vaandrager et al., *1999), p. 137-152*

Lefebvre M.A., Guéguen H. Buisson J. "Hybride approximations for continuous systems", Journal Européen des Systèmes Automatisés (JESA), *Vol. 36 n°7 pp959 – 971 2002.*

Lefebvre M.A., Guéguen H. Buisson J., "Structured hybrid abstractions of continuous systems", IFAC world Congress B02, Barcelonne, 2002.

Lynch N. Krogh B.H. (Eds.) Hybrid Systems Computation and Control, HSCC2000, *LNCS Vol 1790, Spinger, 2000.*

Silva B.I., Stursberg O., Krogh B. H., Engell S., "An assessment of the current status of algorithmic approaches to the verification of Hybrid Systems", in Proc. of the IEEE Conf. On Decision and Control, *Orlando, December 2001, pp 2867-2874.*

Stauner T., Müller O., Fuchs M., "Using HyTech to verify an automative control system", Proceedings of Hybrid and Real-Time Systems Workshop, *LNCS-1201, Springer, 1997, p. 139-153.*

Stursberg O. Kowalewski S., Preussg J., Treseler H. "Block diagram based modelling and analysis of hybrid processes under discrete control", In: Zaytoon J. (Editor), Journal Européen des Systèmes Automatisés (JESA), Special Issue on Hybrid Systems, *Vol. 32, n° 9-10, 1998, p. 1097-1118.*

Stursberg O., Fehnker A., Han Z, Krogh B., "Specification-guided analysis of hybrid systems using hierarchy of validation methods", In: Engell S., Guéguen H., Zaytoon J., (Editors), Analysis and Design of Hybrid Systems, Proceeding of the IFAC ADHS'03 Conference, *Saint-Malo, France, June 2003, Elsevier, 2003, p 289-294.*

Vaandrager F.W., van Schuppen J.H. Hybrid Systems, Computation and Control, HSCC99, *LNCS, 1569, Springer, 1999*

Yovine S., Méthode et outils pour la vérification symbolique de systèmes temporisés, Thèse de l'INP Grenoble, 1993.

Zaytoon J. (coordinateur), *Systèmes Dynamiques Hybrides,* éditions Hermès, 2001.

Copyright © IFAC Information Control Problems in Manufacturing, Salvador, Brazil, 2004

ELSEVIER
IFAC
PUBLICATIONS
www.elsevier.com/locate/ifac

PETRI NETS SIMULATION OF COMPONENTS FLOWS IN CAR MANUFACTURING

Fabien Petitjean [1], Alexandre Villeminot [1,2], Jean-Yves Bron [1], Patrick Charpentier [1]

[1] *Centre de Recherche en Automatique de Nancy*
CRAN – Faculté des Sciences BP 239
F-54506 Vandoeuvre-lès-Nancy, France
fabien.petitjean@cran.uhp-nancy.fr

[2] *PSA Peugeot-Citroën*
45, rue Jean-Pierre Timbaud
F-78307 Poissy Cedex, France
alexandre.villeminot@mpsa.com

Abstract: Nowadays, production systems are so complicated that they can not be designed and run without a good understanding of their working. Discrete-events simulation stills the favourite tool to understand and analyse these systems. This paper presents a study carried out by French manufacturer PSA Peugeot Citroën to simulate a representative components flow sample in assembly lines. The aim is to get a global vision of the supply chain and of the impact of derivation from manufacturing forecasts or of one hazard on the different flows. On that big application (5000 supply flows), this paper focuses on the feasibility study of that supply chain simulation by means of Coloured Petri nets. *Copyright © 2004 IFAC*

Keywords: Petri-nets, simulators, delivery systems, industrial production systems, flow control, components.

1. CONTEXT

1.1 The necessity of a components flow control in car production.

Car production systems have considerably evolved from mass production to demand production systems. In order to be competitive car makers have to provide many kinds of vehicles. That results in a new product characterisation: a very strong diversification due to the huge range of optional equipments, a reduced conception to production time and a more and more reliable assembling date. Besides that, the production equipment becomes more distributed between suppliers and manufacturers because of globalization (Barros *et al.*, 2001; Min *et al.*, 2002). In industrial production units, the flow mastery has to be total to avoid considerable profitability losses.
Consequently, car manufacturers have to control their supply chain and need to develop the numerical engineering of the supply chain concept (Charpentier *et al.*, 2003; Muhl *et al*, 2003; Muhl *et al.*, 2001).

1.2 Structure of the supply chain and supplying policies.

The context chosen for this study is French car manufacturer PSA Peugeot-Citroën. The supply chain regarding a car final assembly plant has a traditional "X" structure (Fig. 1).

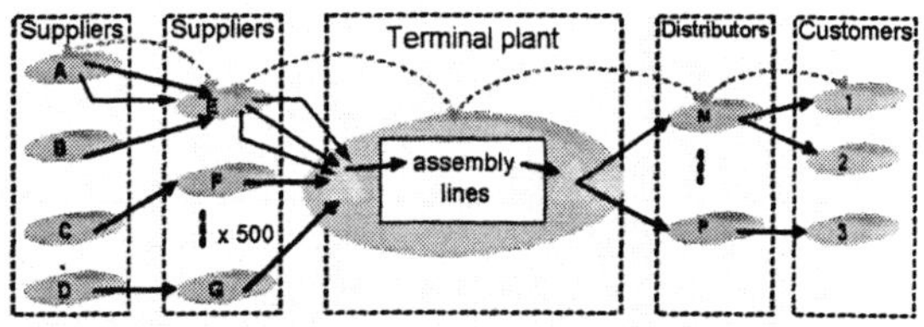

Fig. 1. Supply chain for a car terminal plant structure.

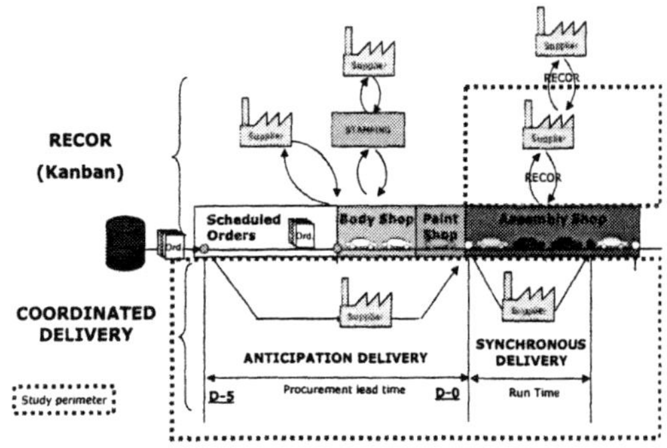

Fig. 2. Supply flow grafted onto the assembly line and cars flow.

Approximately 500 first level suppliers and more than 5000 supply flows are involved when assembling a vehicle. The supply chain is managed by two Just-in-Time policies (Fig. 2):
- The RECOR (REal COnsumptions Renewal), an adaptation of Japanese Kanban by PSA Peugeot-Citroën. It is the customer consumption which activates the sending of one order to the supplier;
- The coordinated delivery. The fixed prevision customer consumption activates the delivery of one order to the supplier.

All this work is part of a more global project in partnership with CRAN (Villeminot, 2003).

1.3 An approach by discrete-events simulation.

Large volume of supplies allows, neither an exhaustive analytic study, nor a mathematical formalization. That is why these studies use discrete-events simulation models to analyse flow between assembly plant and different suppliers.

The main simulation objectives are the confirmation of the supply flow sizing in an uncertain environment and the contribution of a decision-support dedicated to flow pilots in case of random production events. Also, simulation model gives a system description. Lastly, the dynamics are specified in order to consider temporal and stochastic aspects.

As part of a benchmark, it has been decided to develop different simulators from the Meta-Model of the supply flow (Villeminot, 2003). Both of them were programmed with an object-oriented language.

Because of their formalism, Petri nets have also been chosen for a third type of implementation, hoping for a smaller simulation time. A study has been realized to do a benchmark with program implementations. This consisted in the conception and development of a Petri nets simulator for the RECOR policy.

The aim of this task was:
- To determine the feasibility of such a simulator;
- To estimate the possibility of its transposition to an industrial use;
- To evaluate its viability in such a context.

1.4 The objectives of this paper.

This paper shows an experimentation of the setting up and development of a Petri nets model to simulate a huge-sized supply chain. The presented study focuses on Petri nets implementation on such an highly constrained industrial problem (number of flows, simulation time, external data access, results editing). However, the study does not take any interest on obtained nets properties. This approach was envisaged in concurrency with others (using either Arena software or programming).

The approach and the successive steps will be described, emphasizing on the issues encountered on that type of industrial application. The result of this work is an automatic generation of Petri nets for the studied process. The validation phase of the Petri nets simulator will also be shown at the end.

The main interest is to present what can be done today with the current development software environments insofar as the most representative ones have been examined and considered. In that sense, some of the possible implementations of it will be enumerated with their advantages and drawbacks.

2. CONCEIVING THE SIMULATOR, THE CHOICES MADE.

2.1 Presentation of the studied system.

As already stated, this work has been made from the UML Meta-Model of the supply flow released by Villeminot (2003). A systemic approach has been used to provide a modelling frame which allows determining supply flow entities. The studied system has a generic behaviour, a modular character and a large amount of data. Therefore an oriented object approach has been adapted.

In the study case, an order is the modified object and the system is one supply flow (Fig. 3). A systemic representation is made from that analysis. The order has three elementary attributes with values defined by its position in the flow (Le Moigne, 1977).

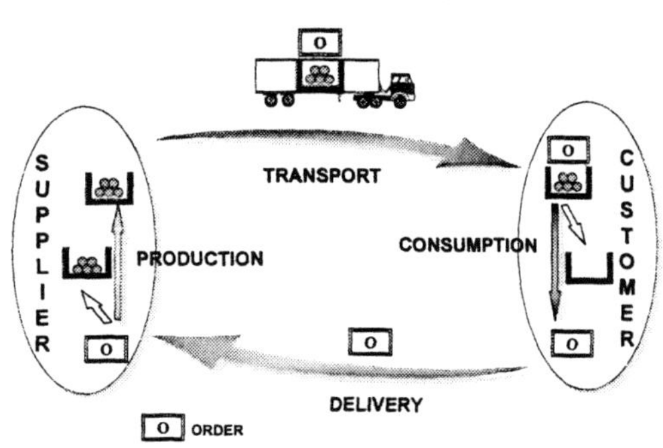

Fig. 3. Considering one supply flow.

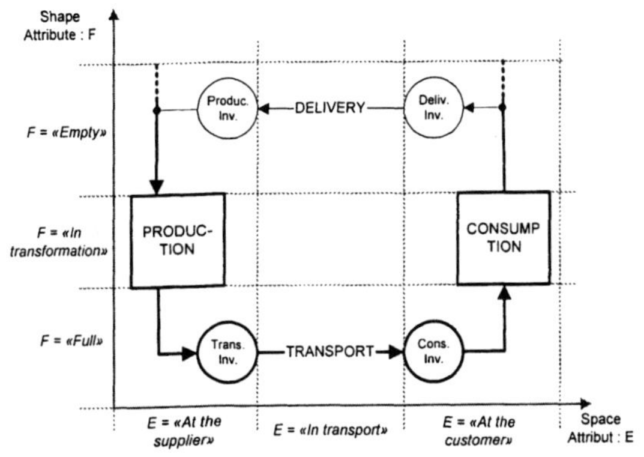

Fig. 4. Systemic representation of a supply flow and order attribute values.

The system operators are (Fig. 4):
- Time and space attribute transformation: "Delivery", sending the "empty" informational order from customer to supplier; "Transport", transport of the full order from supplier to customer;
- Time and shape attributes transformation: "Consumption", transformation from full order to empty order; "Production", transformation from empty order to full order;
- Time attribute transformation: "Inventory before production", "Inventory before Transport", "Inventory before Consumption" and "Inventory before Delivery".

2.2 Algorithmic model / Petri nets model: two different approaches of the same problem.

Whatever the implementation may be, UML Meta-Model has to be translated into the chosen language. In case of an algorithmic model, UML diagrams are translated into codes. Generic classes represent the different processes. For a Petri nets model, Meta-Model is translated into graphs; the interest of such a method is that all the links between processes are visually presented, in other words it is easily understandable.

Another difference concerns the simulator engine. It is necessary to develop it for an algorithmic model, because comportment analysis, calculating and graphical interfaces are dependent on the context. With Petri nets, there is no need to create a dedicated engine because it is integrated to the chosen software tool; that results in a saving of time and a better accuracy for the forthcoming calculus. Petri nets are valued for their properties verification which ensures the model coherency. It is possible and feasible on that study, even if it is not the paper's objective.

2.3 The simulation model: its realisation and technical aspect.

Concerning the simulation model, RECOR code implementation consists in generating model class instances. In an object oriented modelling, a physical object has to be linked to an informational object. Indeed, an object will be attached to each process. Flow is also represented by objects; links between them precise the model functioning.

In Petri nets, many representation equivalences to the Meta-Model are made. Since RECOR is the succession of four processes linked together, it is rather interesting to model them by four sub-nets:
- Consumption sub-net;
- Delivery sub-net;
- Production sub-net;
- And lastly Transport sub-net.

Transport and Delivery have a different comportment, since Transport concerns physical objects and Delivery, informational objects. These sub-nets are hierarchically linked by a main net, called RECOR. The Petri nets used are hierarchical Petri nets; however they are not aggregated during the simulation (He, 2000).

Informational flow is represented by oriented arcs linking one or several places to one or several transitions, vice versa.

Net colouring denotes the relational links between classes; the colour determines in which place will go any particular token in case of embedded token flows. Since tokens will be for example resources, orders or demands, coloured nets will also be used in that context.

One of the main differences is the time representation:
- In a program simulation model, simulation engine controls time; the model comportment drives the temporal evolution of the simulation model;
- Whereas for a Petri nets model, it is essential to implement time. That can be done, either on transitions (these nets are known as T-temporised or T-timed Petri nets), or on places (P-timed Petri nets). In case of a P-timed Petri net [1], two tokens reaching a place at t will stay in until $t + tempo$. In case of a T-timed Petri net, the first token will pass through the transition at t and its time attribute will be increased by *tempo*; the transition will not be passed by the second token before $t + tempo$ and after that, its time attribute will be $t + 2 * tempo$.

Lastly, components types are represented by token types. For instance, every token with a n extension corresponds to the nth supply flow.

The Petri nets simulation model has to authorize an automatic nets parameters setting from supply flows parameters files. It allows the production sequences loading to integrate consumption demands; moreover, results files can be written at the end.

[1] In *DesignCPN*®, it is equivalent to a temporisation on the output arc from a place.

Phase	Approach	Characteristics
Meta-Model	*Both approaches*	UML; Class, collaboration and state diagrams
Model	*Program*	UML into appropriate language Static links and comportments Development of a simulation engine
	Petri nets	Graph translation Integrated simulation engine
Simulation model	*Program*	Class instances
	Petri nets	Process: 1 sub-net Flows: arcs Relational links: coloured Petri nets Time: P-timed Petri nets Supply flow type: 1 token type

Fig. 5. Recapitulating the main differences between algorithmic and Petri nets models.

These characteristics are model specification (Fig. 5).

2.4 The envisaged procedure for the implementation.

The retained approach is:
- Implementation of one supply flow, taking into account at once production constraints (as by example single resources for consumption and production...), packaging units (batch size of the containers) and time. The different sub-nets have to be defined, for a single orders flow;
- Improvements in the first model regarding net colouring (to manage several supply flows) and the reading / writing in a file (to load parameters and production sequences and save the simulation results);
- The third significant phase deals with the automation of nets parameters setting;
- Lastly and following that work, the simulator evaluation will occur, consisting in a comparison with an other simulator. The evaluation will concern the feasibility of that Petri nets simulator, its possible transposition to a production context and its viability in such a context.

3. DEVELOPPING THE SIMULATOR, SOME ISSUES.

3.1 The main Petri nets software tools examined.

Several development environments have been preliminary tested to determine which could possibly be used in the car manufacturing context. Few of them gather the Petri nets extensions useful to the problem, like the time, the colouring, the hierarchy... that were the choice's criterions.

Two important tools were retained: *MISS RdP* ®, a French tool by the "G.R.P." (French Production Research Group) and *DesignCPN* ®, a Danish one, currently one of the bests existing.

The final choice has also taken into consideration the cost of both tools. For that reason and the current impossibility to import files in *MISS RdP* ®, the simulator has been implemented on the software tool *DesignCPN* ®, working on Unix or Linux stations. That tool, developed by University of Aarhus in Denmark (DesignCPN, 1993), has already been used by PSA Peugeot-Citroën for other studies.

It is interesting to notice the lack of tools that manage file import... Perhaps would it be a good basis for future developments by the software suppliers.

3.2 First approach for developing the simulator.

It was originally planned to develop a simulator version for one supply flow, then to colour it by specifying types on the relations and evolving conditions, so that several orders flows could circulate in the same Petri nets. That is of course the principle of the colouring. Basis models are also common and a study can be referred to them even if important modifications are essential and change considerably the original concept (Moore, 1999).

Even if Petri nets and Grafcet are commonly considered as close (David, 1997), the supply flow model has been plainly considered as a Petri net; that results of course in a better representation of the process with the necessary rigour to a strong model. Some of the representations, like mutual exclusion, have been inspired by the literature (Proth, 1997).

Besides that, *DesignCPN* ® sometimes forces one to use non-standard manipulations. Actually, absence of inhibitor arcs, which control place emptiness, prevents the detection of batch consumption (Fig. 6).

However, as *DesignCPN* ® is by definition a Coloured Petri Nets tool, it is compulsory to colour the nets at the beginning of the development. That work, which was forecasted for a second phase, has been advanced to the first conception step. A first version of the supply flow has been non-hierarchically treated, without batch sizes. Validations of each model were preliminary conditions to the improvements or modifications, allowing increasing the model complexity, for instance by taking into account time. Models have also been structured by hierarchical Petri nets.

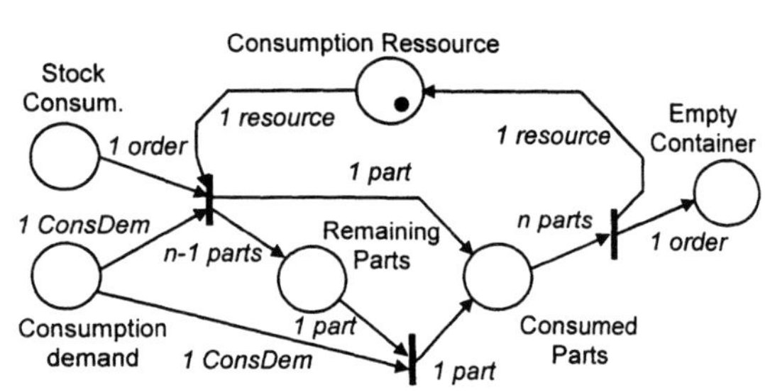

Fig. 6. Representation of an order consumption.

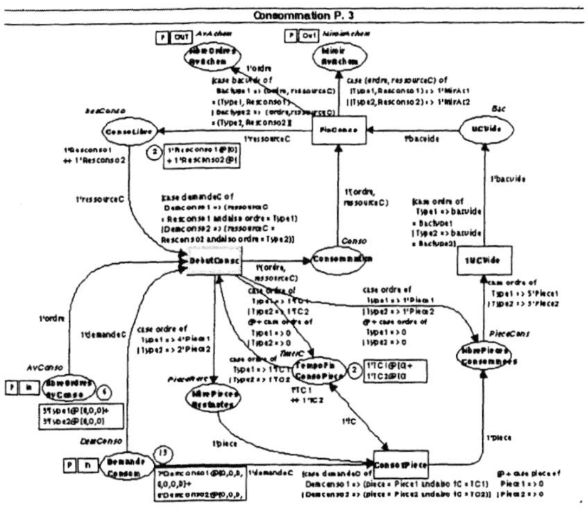

Fig. 7. The first version of the Petri nets simulator; the consumption sub-net.

3.3 *Tests on the first simulator, issues, actions carried out and solutions proposed.*

After the simulator development (Fig. 7), many tests have been made to evaluate its temporal performance. Actually, *DesignCPN*® can simulate in two modes; the results speak for themselves:

- In interactive mode, which shows the evolutions of states along the simulation, it takes about three hours and a half to simulate five hundred steps. In that mode, the simulation is really slow;
- But in the second mode, automatic, it takes less than five seconds to process five thousand steps!

Regarding only the temporal aspect, the simulator is very efficient. However, a problem can be seen and it is linked to the Petri nets formalism: the model is non-deterministic; a random evolution occurs sometimes in processes concerned by batches: Delivery, Production and Transport. Another issue is that it is impossible to initialize the nets marks with different temporised tokens (for instance, consumption demands); in the study context, nets have to be initialized by real data from industrial informational systems.

Different actions have been carried out:

- Firstly, dialogue with the *DesignCPN*® developing staff has permitted the testing of different proposals, unfortunately without success. Some of those solutions were even showing problems of the software tool;
- Secondly, previous researches have brought other ideas, resulting in many extensions and tests on other methods such as code segments which bring program codes in the transitions guards.

These methods were not satisfactory in that context.

Finally, to carry on the development of the simulator, the decision to tolerate indeterminism has been taken, accepting the possible influence on the results. Concerning the import / export, a *DesignCPN* ® characteristic seemed to be interesting: it is possible to export and re-import XML files which are translated by the simulation kernel into Petri nets. Then, one possibility was to do a pre-external parameters setting of the nets, by example in a Visual Basic module dedicated to XML files generation. Lastly, another module could be used for the results exploitation.

4. THE FINAL SIMULATOR.

4.1 *Automatic generation of Petri nets.*

As previously stated, XML files can be exported from existent nets. On the other hand, they can be imported into the software tool. Why could not they be modified between those two operations? That observation brings up to the instances net generation. It is possible to change parts of the nets, like time attributes, but all the nets appearance and structure can also be created from XML. The principle is to generate generic groups of places, transitions and arcs whose parameters and evolution conditions are instanced. That leads to important nets that could not be created manually; for instance it is quite easy to generate a Petri net with fifty components flows, each one with different process times, minimal and maximal batch sizes, initial markings...

All that work is done by a dedicated Visual Basic interface, developed especially to generate the desired nets; it has been called *RecorRdP* (RdP is the French equivalent to PN). It computes in a very short time; the longest occurs during the file reloading into *DesignCPN*®.

4.2 *Simulation and results analysis (Fig. 8).*

The first remark is that automatic mode does not work at all with such important nets as those automatically generated. Also the simulation time will be important as it is interactively done.

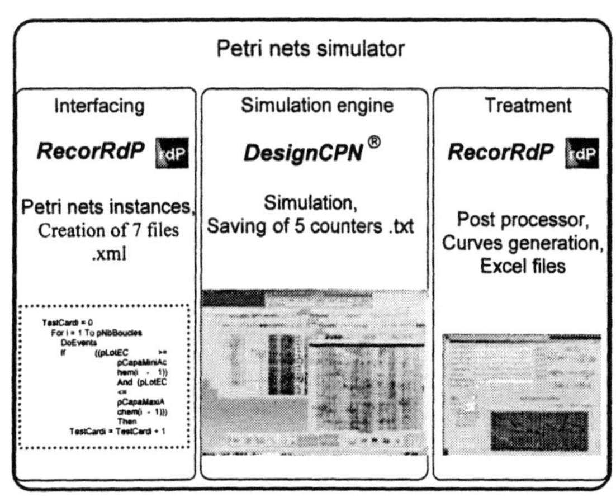

Fig. 8. Architecture of the Petri nets simulator.

To analyse the results, a post-processing module is added to the *RecorRdP* tool. It generates Excel curves and tables which can also be used later.

The simulation is made up of three phases:
- Firstly, the Petri nets generation by XML model files instances;
- Secondly, the nets are loaded into *DesignCPN* ®, then simulation occurs and after that, five counters corresponding to the activity occurrences are saved;
- Thirdly, they are exploited by *RecorRdP* post-processor.

4.3 Validation of the simulator.

The phase of validation is done by comparison with another simulator result (*SimAppro* coded in Visual Basic) and real production results.

There are some significant differences between them, due to indeterminism on transport and delivery demands; in fact, *DesignCPN* ® chose at random which active transition it will shoot. That can sometimes delay supplying from one transport or delivery period (or both of them). Multiplying simulations could reduce indeterminism effects.

The main advantage is the results post-processing; it is possible to zoom on a precise zone of the curve and all the stock curves can be reloaded later. That may be used in another simulator.

5. CONCLUSIONS AND PERSPECTIVES.

After that work on the simulator, it seems that this Petri nets simulator is really feasible, even if *DesignCPN®* is not enough.
Moreover, it can be transposed to a real and big industrial application on condition that interfacing and treatment parts are developed (access to data and parameters, automatic nets generation, editing and shaping of the results for instance). However, it is easier to simulate a lot of flows with few consumption demands; what is more, the higher the number of consumption demand, the longer the time of compilation.

The next step for this research work is to compare, for unbiased basis, the efficiency of the different concurrent approaches implemented and their adaptability to an industrial use.

REFERENCES

Barros L. L., M. Riley and D. Brown (2001). Special millennium issue of the EJOR: A global view of industrial logistics. *European Journal of Operational Research*, **Vol. 129**, 231-234.

Charpentier, P., A. Villeminot, E. Muhl and J.-Y. Bron (2003). Vers l'ingénierie numérique de la chaîne logistique - Exemple d'un constructeur automobile. CPI'2003, Meknès, Maroc.

David, R. and H. Alla (1997). *Du Grafcet aux réseaux de Petri.* Editions Hermès, Paris, France.

DesignCPN ® (1993). DesignCPN programmer's reference and guide, version 4.0., University of Aarhus, Denmark (www.daimy.au.dk).

He, D.W., B. Strege, H. Tolle, and A. Kusiak (2000). Decomposition in automatic generation of Petri nets for manufacturing system control and scheduling. *International Journal of Production Research*, **Vol. 38**, 1437-1457.

Le Moigne, J.-L. (1977). *La théorie du Système Général.* Editions PUF, Paris, France.

Min H. and G. Zhou (2002). Supply chain modelling: past, present and future. *Computers & industrial engineering*, **Vol. 43**, 231-249.

Moore, K.E. and S.M. Gupta (1999). Stochastic coloured Petri net (SCPN) models of traditional and flexible kanban systems. *International Journal of Production Research*, **Vol. 37**, 2135-2158.

Muhl, E., P. Charpentier and F. Chaxel (2003). Optimization of physical flows in an automotive manufacturing plant: some experiments and issues. *Engineering Applications of Artificial Intelligence*, **Vol. 16**, 293-305.

Muhl E., P. Charpentier, P. Raynaud and S. Noiré (2001). Performance robustness for the scheduling of an automotive plant, 10th IFAC Symposium on Information Control Problems in Manufacturing, INCOM 2001, Vienna, Austria.

Proth, J.-M. and X. Xie (1995). *Les réseaux de Pétri pour la conception et la gestion des systèmes de production.* Editions Masson, Paris, France.

Villeminot, A. (2003). Modélisation des relations donneur d'ordres - fournisseurs : application à une boucle en flux tirés. *MOSIM'03*, Toulouse, France.

Copyright © IFAC Information Control Problems in Manufacturing, Salvador, Brazil, 2004

ELSEVIER
IFAC
PUBLICATIONS
www.elsevier.com/locate/ifac

STEADY STATE OF HIGH THROUGHPUT MANUFACTURING SYSTEMS BY CONTINUOUS WEIGHTED MARKED GRAPHS

Isabel Demongodin and Nathalie Sauer

IRCCyN – UMR CNRS 6597
Ecole Nationale Supérieure des Techniques Industrielles et des Mines de Nantes
1, rue la Noe, BP 92101, 44321 Nantes Cedex 3, France
[demongodin, sauer]@irccyn.ec-nantes.fr

Abstract: Continuous Petri nets can be seen as a relaxation of discrete nets, in which the marking of places and firing of transitions are approximated with a fluid model. Several studies have been presented for these models, using linear programming, incidence matrix analysis, or graph theory approaches. In this paper, two approaches for the analysis of time behavior are developed for continuous weighted marked graph structures. Based on linear algebra and graph theory, they are used to compute the steady-state firing speed and the steady-state marking. Such a characterization of the stationary state is applied to a bottling line, i.e. a high throughput manufacturing system. *Copyright © 2004 IFAC*

Keywords: Petri net, graph theory, modeling, formal analysis, manufacturing system, steady state.

1. INTRODUCTION

High throughput manufacturing systems are very widespread in companies of industrial sectors such as the food, pharmaceutical, beauty care, electronics industries and so on. Generally, these systems have quasi - linear structures, composed by elements with high capacities of storing. Treating a great number of products, such systems have their physical components strongly linked by the flow of parts. To represent and evaluate in terms of performances such systems, Continuous Petri nets, defined by David and Alla (1992), are well adapted. They extend the discrete Petri net formalism. They offer powerful concepts and formal techniques for expressing continuous time behavior and for minimizing the representation of reachable states. In the timed continuous model, the dynamic marking is a non-negative real vector, representing part flows of high-speed systems. A maximal firing speed vector is associated to the set of transitions and an instantaneous firing speed vector represents the real quantity of marking passing through transitions. In other terms, maximal speeds represent throughputs of production components while instantaneous speeds represent the real part flows on transfer elements, or in machines. In practice, once the control part of a system has been determined, the global model is without indeterminist situation, i.e. it has a conflict free structure. A system can then be modeled through a marked graph, which constitutes an elementary class of Petri nets where each place has exactly one input and one output transition. Moreover, a system with a global limited capacity but without concurrency can be modeled by strongly connected marked graphs (Proth and Xie, 1997). For a more compact representation of the system, we can use a weighted marked graph, i.e. a marked graph structure where the value of weights can be greater than one. In this context, a high throughput production line can be represented by a neutral continuous marked graph, i.e. a sub-class of weighted marked graphs.

Several procedures for analyzing performances have been proposed for timed continuous Petri nets. For instance, they are based on the evolution graph (David and Alla, 1992) or on linear programming methods (Balduzzi et *al.*, 2000). More specifically, a rich relationship exists between graph theoretic concepts and invariant analysis for neutral continuous marked graphs (Demongodin et *al.*, 2003). The procedure, presented in this study, characterizes the steady state (or stationary state), from the simple knowledge of the structure of the model and its initial state. Two very important characteristics of performances have been formulated: the final firing speeds and the final marking, i.e. the stationary throughput and the final quantities of parts of the production system.

This paper is organized as follows. First, we present a bottling line, in terms of structure, capacity and throughput of machines. After recalling some basic concepts of constant continuous Petri nets and more specifically, some notations used for strongly connected weighted continuous marked graphs, we propose a model for the considered industrial system. Next, we study the timed dynamics and give the main results of our contribution on the steady state of neutral continuous marked graphs. Finally, we derive the analytical expressions of the final speed vector and of the final marking at the steady state and we apply them to the considered industrial model.

2. BOTTLING PROCESS

The considered bottling chain produces spring water bottles. Among high-speed production systems, a bottling line (see Figure 2) is a quasi-linear structure, made of standard elements and storage sections. It consists of machines in series, connected by multi-bands conveyor belts, which transfer the bottles in one-way and are used as buffer zones. The empty bottles arrive by layers on pallets and enter the chain through the "unpalletizer". The bottles are next conveyed by means of the multi-bands conveyors to the washer machine, then towards the rotary filler where they are filled with liquid. Afterwards, the conveyors bring the corked bottles to the packaging machine, where they are gathered by batches.

The physical characteristics, as capacity and throughput of machines and capacity of conveyors are given in Table 1 and Table 2, respectively.

Table 1: Characteristics of machines

Name	Capacity	Maximal throughput (bottles/min.)
Unpalletizer	1 layer of 360 bottles	1800
Washing	80 bottles	1199.88
Rotary filler	133 bottles	800
Packaging	50 packs of 24 bottles	1200

Table 2: Characteristics of conveyors

Number	Capacity (bottles)	Number	Capacity (bottles)
C1	1050	C13	351
C2	75	C14	112
C3	150	C15	40
C4	63	C16	90
C5	10	C17	125
C6	145	C18	24
C7	30	C19	1720
C8	60	C20	400
C9	70	C21	50
C10	1626	C22	338
C11	75	C23	438
C12	275		

3. MODELING

We choose to model the considered system by a constant continuous Petri nets. More particularly, as we assume that the control part has been previously established, the global system can be represented by a neutral continuous marked graph.

3.1. Structure of continuous weighted marked graphs

We assume that the reader is familiar with the basis of Petri nets in terms of structure, firing rules and basic properties (see (David and Alla, 1992) for more details). Let us first consider a constant continuous Petri net, timed on transitions with earliest functioning logics, i.e. a transition is fired as soon as it is enabled.

A *continuous Petri net* is a six-tuple $PN = (P, T, Pre, Post, M_0, Tempo)$ such that:

- P is a finite set of places (dim $P = m$),
- T is a finite set of transitions (dim $T = n$),
- $Pre, Post$: $(P,T) \rightarrow IR^+$ are the incidence functions,
- M_0 is a marking vector that represents the positive real number of initial marks present in each place of the net.
- *Tempo* Φ: $T \rightarrow IR^+ - \{0\}$ is the timed function which associates at each transition a positive real number Φ_j, called *maximal firing speed.*

$°T_j$ (resp. $T_j°$) represents the set of input (resp. output) places of transition T_j and $°P_i$ (resp. $P_i°$) represents the set of input (resp. output) transitions of place P_i. We note C, the *incidence matrix* where each component is given by: $C_{ij} = Post(P_i,T_j) - Pre(P_i,T_j)$. Vector X (resp. Y) which verifies $X.C = 0$ (resp. $C.Y^t = 0$) is called *P-semiflow* (resp. *T-semiflow*).

A *Marked Graph* (MG, also called event graph) is a Petri net in which each place has exactly one input and one output transition and the weight associated to each arc is equal to 1.

A *Weighted Marked Graph* (WMG also called weighted T-graph or weighted event graph) is a marked graph in which a weight (greater than or equal to one) can be associated to each arc.

Let PN be a Continuous Weighted Marked Graph (CWMG) with m places and n transitions. For place P_i we note (see Figure 1):

- $w_i = Post(P_i, T_k)$: the weight of the input arc of place P_i.
- $v_i = Pre(P_i, T_j)$: the weight of the output arc of place P_i.

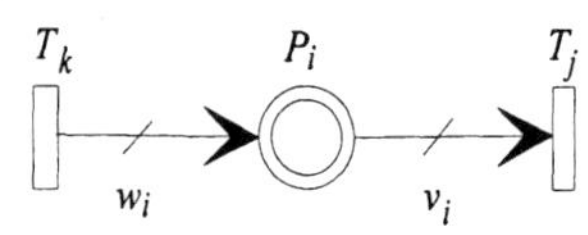

Figure 1: Notations.

Defined for the discrete structure by (Lien, 1976), the *gain of a directed path*, $\rho_{kj} = <T_k, \ldots, P_i, \ldots, T_j>$ from transition T_k to T_j is given by:

$$G(\rho_{kj}) = \prod_{P_i \in \rho_{kj}} \frac{w_i}{v_i}$$

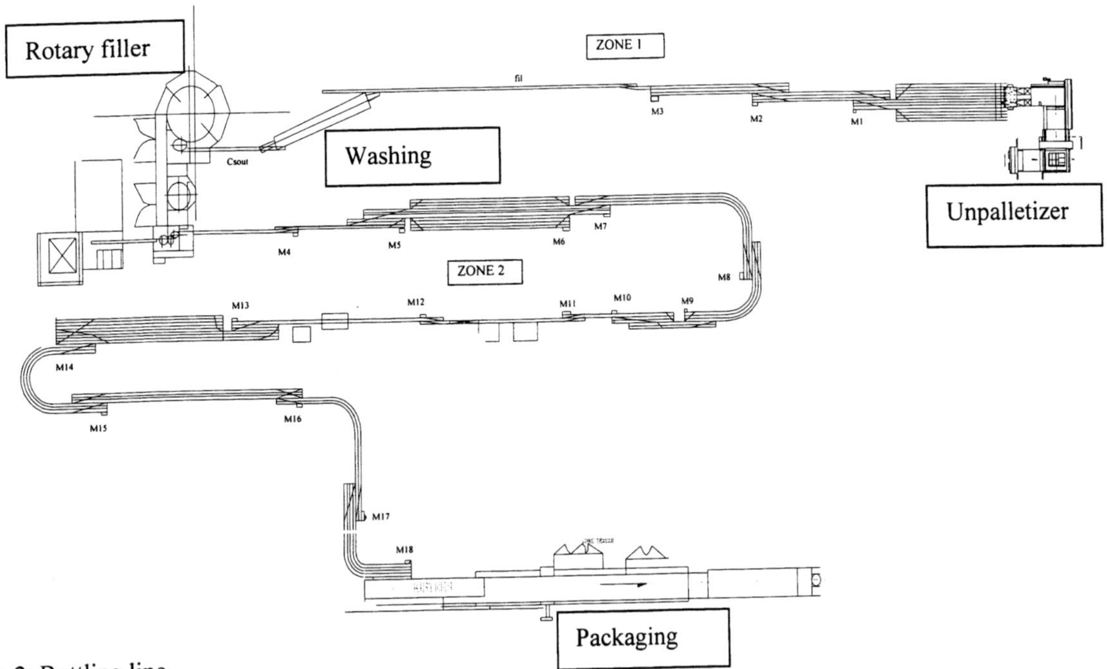

Figure 2: Bottling line.

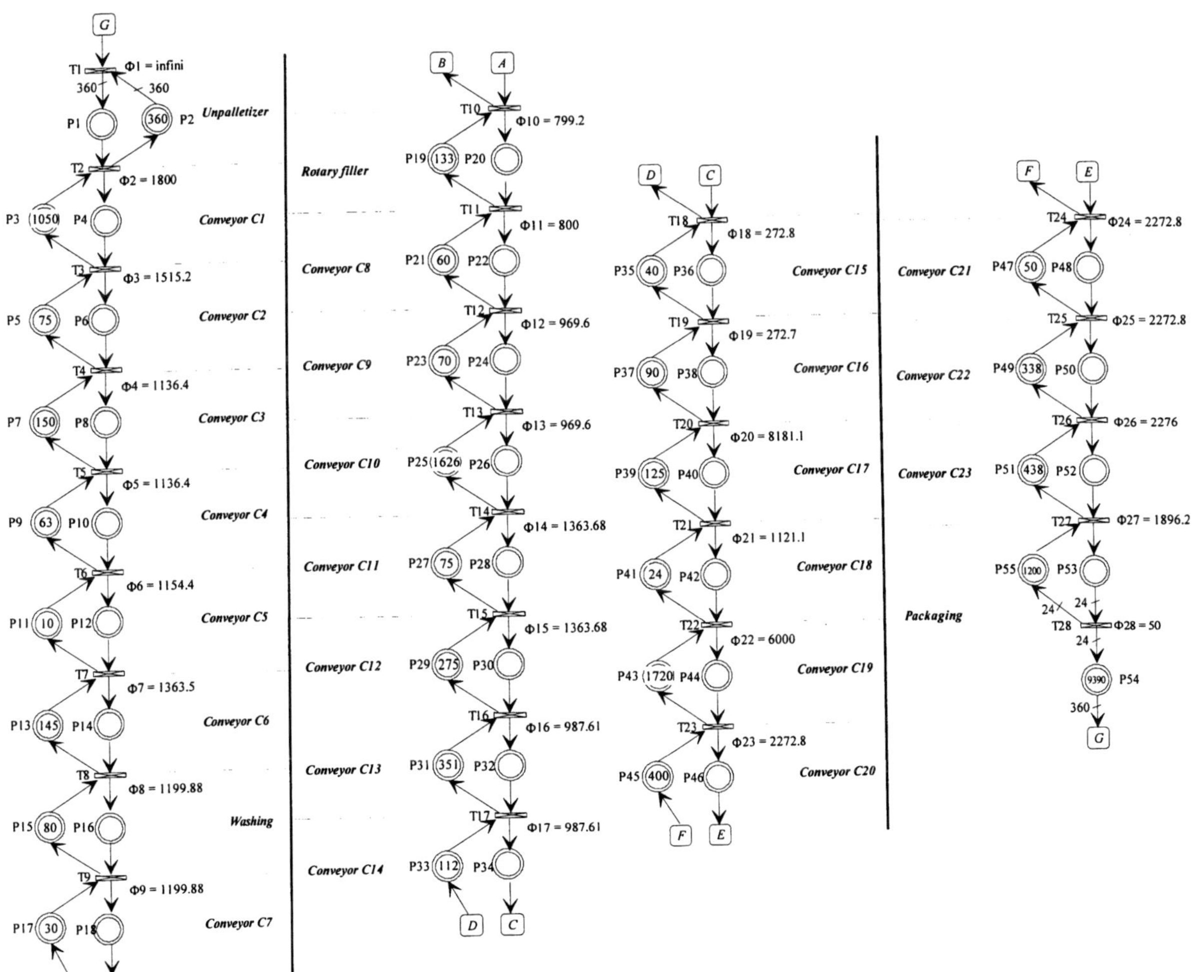

Figure 3: Neutral continuous marked graph of the bottling line.

The *gain of weighted circuit* γ (also called loop gain) is defined by: $G(\gamma) = \prod_{P_i \in \gamma} \frac{w_i}{v_i}$.

Depending on the value of its gain, a circuit can be classified in one of following cases: $G(\gamma) = 1$, the circuit is *neutral*; $G(\gamma) < 1$, the circuit is *absorbing*; $G(\gamma) > 1$, the circuit is *generating*.

In the sequel, we restrict ourselves to strongly connected graphs, i.e. graphs for which there exists an oriented directed path that connects any node (place or transition) to any other node. A non-strongly connected graph can always be decomposed into a finite number of strongly connected sub-graphs. In a more advanced manner, a strongly connected graph can always be decomposed into a finite set of elementary circuits; an elementary circuit being a circuit that contains each node (place or transition) at most once. From these graph theory concepts, Teruel et *al.* (1992) classify a strongly connected weighted marked graph, noted *PN*, as follows.

- *PN* is *neutral* if and only if each circuit in *PN* is neutral.
- *PN* is *absorbing* if and only if it contains at least one absorbing circuit.
- *PN* is *generating* if and only if it contains no absorbing circuit and has at least one generating circuit.

3.2. Modeling of the bottling process

For the dimensioning and performance evaluation of the bottling process, each conveyor is represented as a simple stock with limited capacity. So, we do not consider the transfer delay between machines. Each conveyor and each machine are modeled by two continuous places, connected by two transitions. One place represents the real quantity of bottles on the element while the other one acts as a limited capacity. The maximal firing speed of the output transition corresponds to the throughput of machines or conveyors. The complete model of the bottling line is given in Figure 3, where place P_{54} represents the global limited capacity of the system. According to the previous definition on weighted marked graphs, this continuous model is a neutral graph, composed by 28 neutral elementary circuits.

Each throughput of machines and speeds of transfer elements are represented thanks to continuous transitions with the following maximal speed vector:

Φ = (10000, 1800, 1515.2, 1136.4, 1136.4, 1154.4, 1363.5, 1199.88, 1199.88, 799.2, 800, 969.6, 969.6, 1363.68, 1363.68, 987.61, 987.61, 272.8, 272.7, 8181.1, 1121.1, 6000, 2272.8, 2272.8, 2272.8, 2276, 1896.2, 50).

At the initial time, we consider an empty system. In other words, if we denote the initial marking by $M_0 = M(t_0) = (m_1^0, m_2^0, \ldots, m_m^0)$, i.e. the marking at initial date t_0, due to the capacity of production elements, we have:

M_0 = (0, 360, 1050, 0, 75, 0, 150, 0, 63, 0, 10, 0, 145, 0, 80, 0, 30, 0, 133, 0, 60, 0, 70, 0, 1626, 0, 75, 0, 275, 0, 351, 0, 112, 0, 40, 0, 90, 0, 125, 0, 24, 0, 1720, 0, 400, 0, 50, 0, 338, 0, 438, 0, 0, 9390, 1200).

4. TIMED DYNAMIC

Before studying the timed dynamics, it is necessary to ensure the liveness property of the model, which guarantees the absence of deadlock in the system. We recall that a Petri net is live if and only if every transition can ultimately occur from any reachable marking (David and Alla, 1992). In a more advanced manner, we can say that a strongly connected continuous marked graph is live if and only if each of its elementary circuits is generating or neutral and contains at least one place with a not null marking (Demongodin et *al.*, 1999). Therefor the bottling model in Figure 3 is live since each elementary circuit has a non-empty place.

For the representation of the time dynamics, two vectors are defined on the structure of a continuous Petri net: the marking vector associated to places and the instantaneous firing speed vector associated to transitions.

On one hand, the dynamic *marking vector*, noted $M(t) = (m_1(t), m_2(t), \ldots, m_m(t))$, associates with each place a non negative real value at time t. When a place has a strictly positive marking, it is said to be marked. In the opposite case, it is empty or unmarked.

On the other hand, an *instantaneous firing speed* associated to transition T_j and noted $\varphi_j(t)$, represents the quantity of marks by time unit, which fires through transition T_j. As we consider Constant Continuous Petri Nets, the instantaneous firing speeds between two events have a piecewise constant behavior. Whatever the evolution, the instantaneous firing speed of transition T_j is lower than or equal to the associated maximal firing speed, i.e. $\varphi_j(t) \leq \Phi_j$. Thus, each component of this vector, defined in the IR^+ domain, can be varied inside the interval $[0, \Phi_j]$.

According to notations defined in Figure 1, the *dynamic balance* of place P_i at time t, denoted $b_i(t)$ is given by: $b_i(t) = w_i.\varphi_k(t) - w_i.\varphi_j(t)$, such that $T_k = {}^\circ P_i$ and $T_j = P_i{}^\circ$. More precisely, $b_i(t)$ represents the variation (increasing or decreasing) of marking $m_i(t)$.

- $b_i(t) > 0 \Rightarrow$ the marking of place P_i increases.
- $b_i(t) < 0 \Rightarrow$ the marking of place P_i decreases. $b_i(t)$ can not be negative if $m_i(t) = 0$. This condition guaranties a non-negative marking.
- $b_i(t) = 0 \Rightarrow$ the marking of place P_i is stable.

At a fixed date, a change of the speed vector will occur if there are strictly negative dynamic balances, meaning that some markings decrease and will reach the null value. So, when a place becomes empty, an event occurs, and the new marking vector $M(t)$, dynamic balance vector $B(t)$ and instantaneous firing speed vector $\varphi(t)$ characterize this current state, until the next event.

Under the liveness assumption, we derive an important dynamic property of the instantaneous firing speed. (Proof can be found in (Demongodin et *al.*, 2003)).

Property 1. The dynamics of the instantaneous firing speed of a live strongly connected continuous weighted marked graph follow a non-increasing function. i.e. $\forall\, t_1 \geq 0, \forall\, t \geq t_1, \varphi_j(t) \leq \varphi_j(t_1) \leq \Phi_j, \forall\, T_j \in T$. □

In terms of performances, the throughput of production elements cannot increase.

Finally, in the general case of weighted marked graph, we define the marking of a path, called weighted marking that will be used for characterizing the steady state.

Definition 1. The *weighted marking* of a path, $\rho_{kj} = <T_k, \ldots, P_i, T_s, \ldots P_r, T_j>$, at time t is given by:

$$N(\rho_{kj},t) = v_r \cdot \sum_{P_i \in \rho_{kj}} \frac{m_i(t)}{v_i} . G(\rho_{sj}), \qquad (1)$$

where $T_s = P_i^\circ$, $\rho_{sj} \subset \rho_{kj}$, $P_r = \rho_{kj} \cap {}^\circ T_j$ (i.e. P_r is the last place of ρ_{kj}) and $G(\rho_{jj}) = 1$. □

Remark. If the first transition of the path is not fired after time t, the weighted marking corresponds to the maximal marking that the last place of this path can contain.

5. STEADY STATE

Under the liveness property and according to the dynamic rules, a continuous Petri net where each place has a non-negative dynamic balance, has reached a particular state where the instantaneous firing speed will never change afterwards. This particular state is called the *steady state* (also called *final state, permanent state* or *stationary state*) and is defined as follows.

Definition 2. A live continuous weighted marked graph reaches at a certain instant, noted t^*, a *steady state*. From this final state, all the dynamic balances of place are non-negative, ($B(t) \geq 0$, $\forall\, t \geq t^*$), and the instantaneous firing speed vector remains constant. This last vector is called *final speed vector*, noted $\varphi^* = (\varphi_1^*, \ldots, \varphi_n^*)$. □

Moreover, the following property (Demongodin et *al.*, 2003) can be established.

Property 2. For a live neutral continuous marked graph, in the steady state, the *final marking vector*, noted $M^* = M(t^*) = (m_1^*, \ldots, m_n^*)$ is constant and the dynamic balance vector is null, i.e. $B^* = B(t^*) = 0$.

5.1. Final speeds of neutral continuous marked graphs

For neutral continuous marked graph, the final firing speeds are determined independently of the initial marking and without the use of the evolution graph. The notion of T-semiflows and the maximal firing speed vector are the only characteristics used to determine these final values (Demongodin et *al.*, 2003).

Theorem 1. A neutral strongly connected continuous marked graph (supposed live) reaches a steady state where the final firing speed vector is given by:

$$\varphi_k^* = \Phi_j . \frac{y_k}{y_j}, \forall\, T_k \in T,$$

where transition T_j verifies $\dfrac{\Phi_j}{y_j} = \min_{T_k \in T}\left(\dfrac{\Phi_k}{y_k}\right)$ and y_k is the kth component of any T-semiflow Y. □

Remark. Vector φ^* is independent of the initial marking. Then, in the steady state, the final speed vector of a neutral CWMG (supposed live) is unique whatever the initial marking. Thus, in order to obtain the final firing speed vector of a neutral CWMG, we can choose any initial marking M_0 such that the CWMG is live.

Notation. Let PN be a CWMG with m places and n transitions. In particular, we note T^*, the set of saturated transitions of PN at the final state, i.e. $T^* = \{T_j \,/\, \varphi_j^* = \Phi_j\}$.

Property 3. A saturated transition in the final state is saturated during all the evolution of the graph (also at the initial date). i.e. if $T_i \in T^*$, then $\varphi_i(t) = \Phi_i$, $\forall\, t \geq t_0$. □

Let us return to our example of bottling system (see Figure 3). We recall that the model is live, as each elementary circuit contains at least one place with a not null marking. By solving the equation $C.\, Y^t = 0$, we find the following T-semiflow:

$$Y = (\frac{1}{360},1,1,1,1,1,1,1,1,1,1,1,1,1,1,1,1,1,1,1,1,1,1,1,1,1,1,1,\frac{1}{24}).$$

For finding the final firing speed vector, we apply theorem 1. Transition T_j is given by:

$$\frac{\Phi_j}{y_j} = \min_{T_k \in T}\left(\frac{\Phi_k}{y_k}\right) = \frac{\Phi_{19}}{y_{19}} = 272.7.$$

Therefore, in the final state, transition T_{19}, modeling the junction between conveyor C15 and conveyor C16, is saturated. The final speed of this transition is then equal to its maximal firing speed, i.e. $\varphi_{19}^* = 272.7$.

The other final firing speeds are given by the theorem 1: $\varphi^* = (0.7575, 272.7, 11.3625)$.

5.2. Final marking of neutral continuous marked graphs

As we can observe from property 2, the final marking of a neutral CWMG (supposed live) is constant. So, in (Demongodin et *al.*, 2003) we have derived the analytical expression of the final marking.

Theorem 2. In a neutral continuous weighted marked graph (supposed live), noted PN, the final marking m_i^* of any place P_i is given by:

- For P_i such that $P_i^\circ = T_j \notin (T^* \cup T^{**})$, $m_i^* = 0$.
- For P_i such that $P_i^\circ = T_j \in T^*$,

$$m_i^* = \min_{\rho_{kj} \in \Gamma_j(P_i)} (N(\rho_{kj}, t_0)).$$

- For P_i such that $P_i^\circ = T_j \in T^{**}$,

$$m_i^* = \min_{\rho_{kj} \in \Gamma_j(P_i)} (N(\rho_{kj}, t_0)) - v_i . \min_{\substack{\rho_{kj} \in \Gamma_j \\ / P_s = \rho_{kj} \cap {}^\circ T_j}} \left(\frac{N(\rho_{kj}, t_0)}{v_s}\right)$$

where:

- T^* is the set of saturated transitions of PN in the final state, i.e. $T^* = \{T_j \,/\, \varphi_j^* = \Phi_j\}$.

- T^{**} is the set of transitions of PN which are not saturated in the final state but have more than one input place, i.e. $T^{**} = \{ T_j / T_j \notin T^*$ and $|^\circ T_j| > 1\}$.
- Γ_j is the set of elementary paths ρ_{kj} from a saturated transition $T_k \in T^*$ to given transition T_j, which do not contain another saturated transition.
- For $T_j \in T^*$ and $P_i \in {}^\circ T_j$, $\Gamma_j(P_i)$ is the set of paths $\rho_{kj} \in \Gamma_j$ which contain place P_i. □

Applying the previous theorem to the model in Figure 3, we obtain the following sets: $T^* = \{T_{19}\}$, $T_{28} \notin (T^* \cup T^{**})$, and all the other transitions have more than one input place and are not saturated in the final state, i.e. they compose set T^{**}. Therefore, the final marking of the input place of T_{28} is null, i.e. $m_{53}{}^* = 0$. The input places of transition T_{19} are P_{36} and P_{37}. Let us detail the result for place P_{36}. Set $\Gamma_{19}(P_{36})$ contains two paths: $\rho_1 = <T_{19}, P_{35}, T_{18}, P_{36}, T_{19}>$ and $\rho_2 = < T_{19}, P_{38}, T_{20}, P_{40}, T_{21}, P_{42}, T_{22}, P_{44}, T_{23}, P_{46}, T_{24}, P_{48}, T_{25}, P_{50}, T_{26}, P_{52}, T_{27}, P_{53}, T_{28}, P_{54}, T_1, P_1, T_2, P_4, T_3, P_6, T_4, P_8, T_5, P_{10}, T_6, P_{12}, T_7, P_{14}, T_8, P_{16}, T_9, P_{18}, T_{10}, P_{20}, T_{11}, P_{22}, T_{12}, P_{24}, T_{13}, P_{26}, T_{14}, P_{28}, T_{15}, P_{30}, T_{16}, P_{32}, T_{17}, P_{34}, T_{18}, P_{36}, T_{19} >$. Then, according to definition 1, $N(\rho_1, t_0) = m_0(P_{36}) + m_0(P_{35}) = 40$ and

$$N(\rho_2, t_0) = \sum_{P_i \in \rho_2} \frac{m_i(t_0)}{v_i} . G(\rho_{s,19}) \text{ where } T_s = P_i^\circ,\ \rho_{s,19} \subset \rho_2.$$

Then, $N(\rho_2, t_0) = \dfrac{m_0(P_{54})}{v_{54}} . G(\rho_{1,19}) = \dfrac{9390}{360} . 360 = 9390$

Therefore $m_{36}{}^* = \min\limits_{\rho \in \Gamma_{19}(P_{36})} (N(\rho, t_0)) = 40$.

By the same reasoning, we obtain: $m_{37}{}^* = 90$. The other values are obtained by applying the third case of theorem 2. The final marking vector is then:
$M^* = (360, 0, 0, 1050, 0, 75, 0, 150, 0, 63, 0, 110, 0, 145, 0, 80, 0, 30, 0, 133, 0, 60, 0, 70, 0, 1626, 0, 75, 0, 275, 0, 351, 0, 112, 0, 40, 90, 0, 125, 0, 24, 0, 1720, 0, 400, 0, 50, 0, 338, 0, 438, 0, 0, 4685, 1200)$.

6. CONCLUSION

For manufacturing systems with very great series, the concept of product flow appears. One does not speak any more about operational times associated with the machines but rates or speeds of production. Unfortunately in many cases, production systems are not optimized and are underused. In this paper, we propose a formal analysis for the design and the evaluation of the steady state. In particular, the final firing speeds correspond to the throughput of machines and the knowledge of the final marking permits to determine the inventory level of the system.

Determination of the parameters of the steady state is one of the main properties which characterize the timed dynamic behavior of the system. Using Petri nets, we have proposed analytical methods to compute these parameters without the establishment of the dynamic evolution or the discrete event simulation of the model. In a more detailed manner, we have firstly determined the instantaneous firing speed vector of a neutral continuous weighted marked graph (supposed live) independently of the initial marking, using only the notions of T-semiflows and maximal firing speed vector. Secondly, based on the graph theory, the formulation and the exact value of the marking at the final state have been established.

At present, these results are extended to the analysis of the transient dynamics for continuous weighted marked graphs, in terms of duration and maximal bound of marking. An other perspective is to use these results to extend to continuous case the work done on the marking optimization problem developed for discrete weighted marked graphs (Sauer, 2003). In addition, a more powerful model, called Batch Petri nets (BPN), exists for the study of such high speed systems, including a detailed modeling of conveyor elements (Demongodin, 2001). These models are extensions of continuous and discrete (i.e. hybrid) Petri nets. To analyze performances through BPN models, the most useful approach is the discrete event simulation (see (Demongodin and Prunet, 1997) for a simulation comparison between CPN and BPN models). Nevertheless, some formal results developed for elementary classes of Petri nets can be extended to BPN as it is done, for instance, for the cycle time of batch event graph.

7. REFERENCES

Balduzzi, F., A. Giua and G. Menga (2000). First-order Hybrid Petri nets: a Model for optimization and control. *IEEE Transaction on Robotics and Automation,* **16(4)**, 382-399.

David, R. and H. Alla (1992). *Petri Nets and Grafcet.* Prentice Hall Int., London.

Demongodin, I. (2001). Generalised Batches Petri Nets : hybrid model for high speed systems with variable delays. *Journal of Discrete Event Dynamic Systems : Theory and Applications (DEDS), numéro spécial : "Hybrid Petri nets",* **11**, 137-162.

Demongodin, I., M. Mostefaoui and N. Sauer (1999). Liveness of Continuous Weighted Marked Graphs. *3rd IMACS/IEEE CSCC'99*, Greece, 2640-2649.

Demongodin, I., N. Sauer and M. Mostefaoui (2003). Steady state of neutral Continuous Marked Graphs. *Research report n° 03/14/AUTO* - Ecole des Mines de Nantes.

Demongodin, I. and F. Prunet (1997). Production lines modeling by discrete/continuous aspect in Petri nets. *Conference on Management and Control of Production and Logistics (MCPL'97), IFAC-IFIP,* Campinas, Brésil, 335-343.

Lien, Y.L. (1976). Termination properties of generalized Petri nets. *SIAM J. of Computers,* **5(2)**, 251-265.

Proth, J.M. and X. Xie (1997). *Petri Nets: A tool for design and management of manufacturing systems*, John Wiley & Son (Ed.).

Sauer, N. (2003). Marking optimization of weighted marked graphs. *Journal of Discrete Event Dynamic Systems: Theory and Applications,* **13**, 245-262.

Teruel, E., P. Chrzastowski-Wachtel, J.M. Colom and M. Silva (1992). On Weighted T-System. *Lecture Notes in Computer Science. Proceeding of 3rd international conference on application of Petri nets*, Sheffield, vol. 616, pp. 348-36.

Copyright © IFAC Information Control Problems in
Manufacturing, Salvador, Brazil, 2004

ELSEVIER
IFAC
PUBLICATIONS
www.elsevier.com/locate/ifac

CONTROLLER DESIGN AND VERIFICATION FOR A PARALLEL IMAGE PROCESSOR IN AN FMS USING AN EXTENDED PETRI NET

Akira Nishino* Noboru Kasada* Katsumi Wasaki*
Yasunari Shidama*

* *Shinshu University, Nagano, Japan*

Abstract: We describe a case study of the design and verification of a parallel image processing unit based on an extended Petri net, which is called a Logical Colored Petri net (LCPN). This is suitable for Flexible-Manufacturing System(FMS) modeling and discussion of structural properties. LCPN is another family of colored p/t-net(CPN) with the addition of the following features: integer value assignment of marks, representation of firing conditions as marks' value based formulae, and coupling of output procedures with transition firing. Because the concept of transition firing evaluation orders is omitted from the analysis in conventional CPN, we introduce this concept formally in our proposed net. Therefore, to study the behavior of a system modeled with this net, we provide a means of searching the reachability tree for markings. Finally, we describe a case study of the design and verification for a process controller of an FMS based on the theory of this net.
Copyright © 2004 IFAC
Keywords: Petri nets, Flexible-manufacturing systems, Controllers, Verification

1. INTRODUCTION

There has been a recent rapid expansion in factory automation (FA), making it necessary to organize flexible manufacturing systems (FMSs), comprising robots, transfer machines, automatic storage systems, and image processing systems. Generally, the controllers in FMSs are used to control multiple manufacturing units that work together concurrently or asynchronously while maintaining operation synchronization. Programmable manufacturing controllers (PMCs) have been used as FMS controllers for advanced FA systems and are programmed using sequence ladder languages or diagrams, state charts or decision tables. However, these methods have the following problems (Nagao *et al.*, 1992):

(1) A variety of programming errors are easily introduced.
(2) Verification of the correctness of specifications is difficult.
(3) It is difficult for anyone other than the original programmers to trace control flows or software.

A descriptive model of sequence control by Petri nets has become attractive due to its simplicity (Murata, 1989; Peterson, 1981). A Petri net is a mathematically well defined graphical model that simplifies understanding the control flows. The model of the Petri net can be used to obtain information of systems' behavior when operating in a dynamic environment. In particular, when time or safety factors make actual simulation of a system unfeasible, a study of the relationships between the mathematical objects of a Petri net can reveal conditions such as deadlocks, traps, and reachability of marking states, etc., that aid in the verification of system operation.

In the design of parallel systems, many modeling techniques adopting the concept of Petri net have been described. This is because they are suitable for describing the characteristics of a parallel system and, once a Petri net model has been created, it can be analyzed easily. When designing systems whose characteristics are time independent, an effective approach is the use of place/transition nets (PNs). When the degree of complexity of the target system increases colored Petri nets (CPNs) (Jensen, 1992), or high-level Petri nets, are used in design.

The application of old PNs and CPNs to system design and analysis have been described in such areas as hardware design (1992), deadlock verification of Ada programs et al (1989), and FA control (1992).

The following problem exists when PNs and CPNs are applied to the description of control systems. The operation of these nets (firing conditions of transition and movement of tokens) is uniquely fixed. It is necessary to use many transition and place elements to represent the branching of conditions in processing and, as a result, the net scale increases.

We have already proposed the concept of a logical colored Petri net (LCPN) (Wasaki *et al.*, 2000). This extension is another version of CPNs. LCPN is a colored p/t-net with the following improvements. Tokens have integer value and firing conditions are given by an arbitrary logical formula that is written in terms of the presence of tokens in input places and the data values of the tokens. The token output at firing is decided using a function based on the data values of tokens in the input places. Therefore, transitions and places, which were added as a result of the above condition branching, are unnecessary and the net scale can be reduced (Yamaguchi *et al.*, 2003).

First, the control software for an FMS is modeled by a LCPN. Next, the designed net model is analyzed using a reachability tree. The net model of the system is implemented in hardware and software, the operation is confirmed, and the final version of the LCPN is verified.

We used this technique to develop software for a PMC for an image processing controller in a FMS factory with an LCPN in order to estimate their effectiveness. A desktop personal computer prepared for use as a development tool for LCPNs was used in the laboratory and office settings and a laptop computer was used in the manufacturing rooms.

Using LCPN reduced the cost of design, programming and testing on site to one third of what the development would have cost using conventional procedural language-based methods.

2. LOGICAL COLORED PETRI NET (LCPN)

2.1 Definition of LCPN

Definition 1. LCPN : A logical colored Petri net is a set $N_E = (S_E, T_E, F_E, M_E)$ that fulfills all of the following conditions.

(i) $S_E = \{s_1, s_2, \cdots, s_n\}$ and $T_E = \{t_1, t_2, \cdots, t_m\}$ are sets of place and transition elements, respectively. $F_E = \{f_1, f_2, \cdots, f_l\} \subseteq (S_E \times T_E) \cup (T_E \times S_E)$ is a set of arcs from places to transitions and transitions to places.

(ii) The mark of each place $s_i \in S_E$ $(i = 1, 2, \cdots, n;\ n = |S_E|)$ can take the value of a natural number from 1 to N. This is written $\mu(s_i)$. We assume $\mu(s_i) = 0$ if there is no mark in s_i.

$$\mu : S_E \longrightarrow \{0, 1, 2, \cdots, N\}$$

(iii) The capacity (maximum number of marks) of each place s_i is 1. The set of all possible marks (i) and (ii) is represented as a mapping from S_E to $\{0, 1, 2, \cdots, N\}$.

$$M_E \equiv \{0, 1, 2, \cdots, N\}^{S_E}$$

(iv) ${}^*t_j$ represents the set of all places (input places) that have an arc extending to $t_j \in T_E$ $(j = 1, 2, \cdots, m; m = |T_E|)$. Similarly, $t_j{}^*$ represents the set of all places (output places) with an arc extending from t_j.

$$\begin{aligned} {}^*t_j &= \{s \in S_E :\ (^\exists f)(f \in F_E, f = (s, t_j))\} \\ t_j{}^* &= \{s \in S_E :\ (^\exists f)(f \in F_E, f = (t_j, s))\} \end{aligned}$$

(v) The firing evaluation of a transition t_j for an arbitrary marking $\mu \in M_E$ examines the firing condition Φ^j. $\Phi^j(\mu\ |^*\ t_j)$ is described by a logical expression in terms of the state $\mu\ |^*\ t_j$ of the places which belong to ${}^*t_j$ and is used to determine the next marking $\mu' \in M_E$.

t_j is called firable if Φ^j is evaluated to be true. It t_j is fired, a mark is removed from each place in ${}^*t_j - t_j{}^*$. Places in $t_j{}^*$ depend on the state of ${}^*t_j$ and are modified by the following C^j.

$$C^j :\ \{0, 1, 2, \cdots, N\}^{{}^*t_j} \longrightarrow \{0, 1, 2, \cdots, N\}^{t_j{}^*}$$

$$\text{if } \Phi^j(\mu|^*\ t_j) \text{ is true then}$$
$$\mu' = \begin{cases} 0 & : on\ \ {}^*t_j - t_j{}^* \\ C^j(\mu|^*\ t_j) & : on\ \ t_j{}^* \\ \mu & : otherwise \end{cases}$$

On the other hand, t_j is not firable if Φ^j is false. At this time, the state $\mu' = \mu$ is unchanged.

We show the next marking resulting from a transition evaluation as a mapping f^j from state $\mu \in M_E$ of the places to $\mu' \in M_E$. □

There are two notable characteristics of this firing evaluation. First, the conditions of firing can be given by an arbitrary logical expression and second, the next marking can be decided freely according to the previous state of marks.

2.2 *Introduction of Firing Evaluation Orders*

We introduce the concept of order in firing evaluation in LCPNs and analyze the movement of the modeled system. When an LCPN is implemented in software and hardware, the order of evaluating of each transition is uniquely decided by the clock that calls the associated control procedure or operation. Therefore, when different evaluation orders produce different system behavior and there is a problem, LCPNs become an effective solution. In a previous work, we discussed the issues of Petri net stability and deadlock verification when considering firing evaluation orders using a subclass of PNs called the Boolean Marking Net (BMN) restricted to be a subset of marked p/t-net (Kawamoto *et al.*, 1993*a*; Kawamoto *et al.*, 1993*b*).

Definition 2. The definition of the order of firing evaluation of transitions in LCPNs is:

(i) G is a permutation group of $\{1,\ 2,\ \cdots,\ m\}$ $(m = |\ T_E\ |)$. It is used to decide the order of firing evaluation $\{t_1,\ t_2,\ \cdots,\ t_m\}$ of transitions and each element $g \in G$ corresponds to infinite sequence $t_{g(1)}, t_{g(2)}, \cdots, t_{g(m)}, t_{g(1)}, t_{g(2)}, \cdots, t_{g(m)}, \cdots$ representing a cyclic firing evaluation order.

(ii) The firing evaluation of transition t_j $(j = 1, 2, \cdots, m\ :\ m = |\ T_E\ |)$ is shown by a mapping f^j from state $\mu \in M_E$ to $\mu' \in M_E$.

(iii) The order of firing evaluation concerning m transitions is based on $g \in G$.

$$f^{g(1)g(2)\cdots g(m)} = f^{g(m)} \times \cdots \times f^{g(2)} \times f^{g(1)}$$

(iv) When one cycle g of a firing evaluation for m transitions is given, we write it as:

$$g\ :\ t_{g(1)} \rightarrow t_{g(2)} \rightarrow \cdots \rightarrow t_{g(m)}$$

This is abbreviated as:

$$g\ =\ (g(1),\ g(2),\ \cdots,\ g(m))$$

□

3. MODELING OF THE CONTROL SOFTWARE BY LCPN

3.1 *Outline of the Image Processing System and its Control*

Figure 1 shows the equipment layout of the FMS. This control system is composed of four process

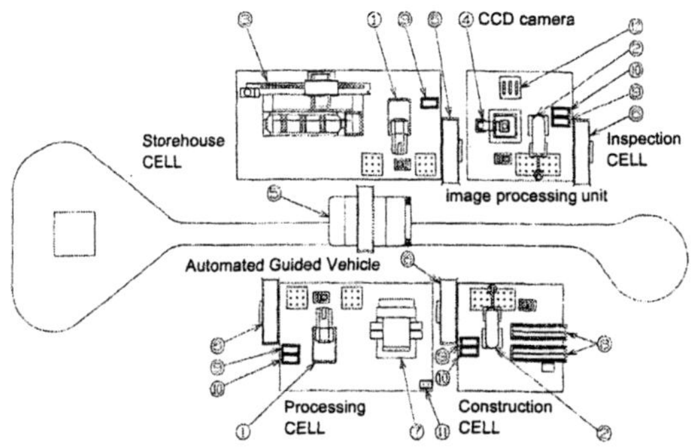

Fig. 1. Equipment layout of the FMS system.

Cells, an Automated Guided Vehicle (AGV), an image processing unit, a CCD-camera, and a system control computer. Figure 2 shows the equipment layout of the image processing system. This system comprised an XY-stage and its controller, CCD-camera, image processing unit and system control computer.

First, the surface of an object placed on the XY-stage is divided into small square areas called cells. Each cell is captured by a CCD-camera one after another as a raw image. The raw image data is compared with several template images prepared beforehand and the quality of the object is inspected. For example, when the object is a printed pattern, the template images consist of the correct patterns to appear in each cell. Sections not corresponding to the template are detected and reported to the control computer as the number of incorrect pixels.

Figure 3 shows the internal hardware composition of the image processor. This device is composed of a template image storage area for holding multiple template data; two template image memory areas (A, B); a matching filter; a pixel counter; and two raw image memory areas (A, B) for storing the image of each cell taken from the CCD-camera. By equipping the system with a dual memory configuration, one pair of raw image data and template data can be matched and the next image can be read in simultaneously. Moreover, because an independent data bus connects the template image storage and the template image memory, transferring from storage to memory can be performed independently. The control of these parallel processes is managed by software in the system control computer.

3.2 *Example of Designing Control Systems with LCPNs*

Figure 4 shows an example of the design of a control system with LCPNs based on the following PMC processing sequence. Each process in this sequence, i.e., reading of raw images, forwarding

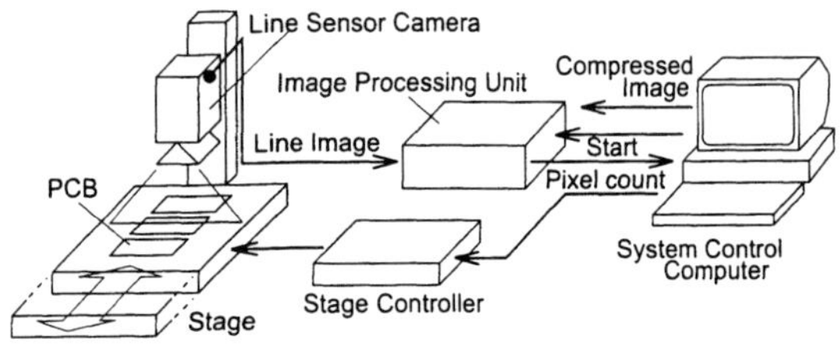

Fig. 2. Equipment layout of the image processing system.

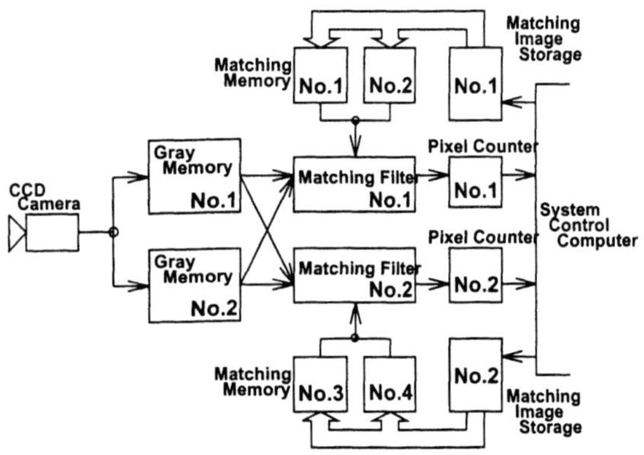

Fig. 3. Internal hardware composition of the image processing unit.

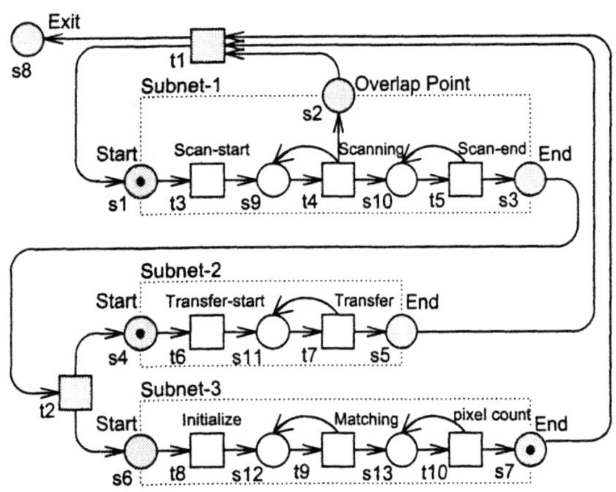

Fig. 4. An example of designing a control system with LCPNs based on the processing sequence.

template images, and matching, is expressed with a sub-net shown in the figure as Subnet-1,2,3, etc. The operation of these sub-nets starts by sending a token to s_1, s_4, and s_6 which are at the entrance of each net. The movement of each net is as follows and the number of the inspected cell is indicated by N in the explanation.

3.2.1. Subnet-1: Reading of raw images Subnet-1 takes the raw images from the camera into the raw image memory. Transition t_3 fires when a token with value $1 \sim N$ enters place s_1 and this places a token in s_9. The value of this token shows the position of the captured image. The actual image capture occurs when t_4 fires and a token updated with the position of the next reading is returned to s_9. In addition, when the updated position reaches the overlap point, tokens are output to s_2 and s_{10}. Readings after the overlap point are performed by the firing of t_5 and when the value of the token reaches the last cell, a token is output to s_3. Finally, once the overlap point has been passed by the firing of t_5, it is possible for another token to enter s_1 again and the next cell image can be read in by the firing of t_4.

3.2.2. Subnet-2: Template image forwarding This section forwards the template images from storage to the template image memory. Transition t_6 fires when a token with value $1 \sim N$ enters place s_4 and a token is placed in s_{11}. The value of this token shows the memory address in template image storage for the forwarding source. The actual forwarding process occurs when t_7 fires and a token is updated with the value of the next memory address in the forwarding source and is returned to s_9. When the forwarding process ends, it outputs a token to s_5.

3.2.3. Subnet-3: Matching and pixel count processing Subnet-3 compares the images and writes the results in the pixel counter. Transition t_8 fires when a token with value $1 \sim N$ enters place s_6 and this places a token in s_{12}. The value of this token shows the position of the image to be processed for matching and pixel counting. When t_9 fires, the actual matching takes place and a token is returned to s_{12}. Once the matching process is complete, a token is output to s_{13} and transition t_{10} fires. At this point, the pixel counter processing is performed and a token is returned to s_{13}. Once the processing is completed, a token is output to s_7.

3.2.4. Synchronization of Subnet-1,2,3 To synchronize the operation of Subnet-1,2,3 the places at the entrance and exit points of each net are connected by synchronization transitions t_1 and t_2. The values $1 \sim N$ of the tokens stored in these places show the number of the cell processed by each net.

When the overlap point is reached during the reading of raw image data, t_1 prepares to read the next cell. It also includes processing for termination when all inspections are complete. The firing condition has two parts, $\Phi^1_{overlap}$ and Φ^1_{exit}, which start the reading process of the next cell and the terminating operation, respectively. t_1 is defined as follows.

$$
\begin{aligned}
{}^{*}t_1 &= \{s_2, s_5, s_7\}\ ,\ t_1{}^{*} = \{s_1, s_8\} \\
\Phi^1_{overlap}(\mu|^{*}t_1) &\equiv (0 < \mu(s_2) < N) \\
&\quad and\ (\mu(s_5) \neq 0) \\
&\quad and\ (\mu(s_7) \neq 0) \\
C^1_{overlap}(\mu|^{*}t_1) &\equiv \begin{cases} \mu(s_2)+1 & : on\ s_1 \\ 0 & : on\ s_8 \end{cases} \\
\Phi^1_{exit}(\mu|^{*}t_1) &\equiv (\mu(s_2) = N)\ and\ (\mu(s_7) \neq 0) \\
C^1_{exit}(\mu|^{*}t_1) &\equiv \begin{cases} 0 & : on\ s_1 \\ \mu(s_2) & : on\ s_8 \end{cases}
\end{aligned}
$$

Transition t_2 is used to perform template image forwarding and matching after reading of the raw images is complete. If the last image has been read, no transfer of the next template is required and only the matching process is started. The firing condition has two parts: 1) Φ^2_{normal}, which is responsible for starting template forwarding and matching operations and 2) Φ^2_{final}, which initiates matching only. t_2 is defined as follows.

$$
\begin{aligned}
{}^{*}t_2 &= \{s_3\}\ ,\ t_2{}^{*} = \{s_4, s_6\} \\
\Phi^2_{normal}(\mu|^{*}t_2) &\equiv (0 < \mu(s_3) < N) \\
C^2_{normal}(\mu|^{*}t_2) &\equiv \begin{cases} \mu(s_3)+1 & : on\ s_4 \\ \mu(s_3) & : on\ s_6 \end{cases} \\
\Phi^2_{final}(\mu|^{*}t_2) &\equiv (\mu(s_3) = N) \\
C^2_{final}(\mu|^{*}t_2) &\equiv \begin{cases} 0 & : on\ s_4 \\ \mu(s_3) & : on\ s_6 \end{cases}
\end{aligned}
$$

The scale of the net for realizing this control system covered three sub-nets, 10 transitions, and 13 places as shown in Figure 4.

4. ANALYSIS AND OPERATIONAL TESTING

4.1 Analysis of LCPN

Whenever a plate arrives, the image processing system should be able to perform processing repeatedly in a stable manner. For this, it is essential that there are no transitions which cannot fire in the system and that the state of markings is maintained when the system performs certain constant repeated operations. For this reason, we generate a reachability tree from a certain initial marking to analyze the net.

Each marking M_i ($i \in \{0, 1, 2, \cdots, 17\}$) in the reachability tree is represented in terms of the number of tokens (0/1) of each place s_i.

$$M_i = (|s_1|, |s_2|, |s_3|, |s_4|, |s_5|, |s_6|, |s_7|, |s_8|)$$

First, when we examine all paths of the marking state we find that all transitions can be firable (Figure 5). Then, we observe that the state at the end of each cycle is always maintained at $M_2 = (1, 0, 0, 0, 1, 0, 1, 0)$. Therefore, the stability of parallel operations in the system is guaranteed

Fig. 5. Part of the reachability tree generated for initial marking M_0.

under the above initial marking for this design example.

Here, we examine states M_1, M_7, and M_{10}. These represent the situation in which the forwarding process and pixel counting process are yet to be completed although the overlap point of the current picture image has been reached. In other words, some process (forwarding of templates, matching, or pixel counting) is not completed in the time allotted for reading in a cell image and this is an abnormal condition in the system. Therefore, it is necessary to correct the net to stop processing when an abnormality occurs and the system falls into state M_1, M_7, or M_{10}.

Therefore, we modified t_1 and added a new place s_{14} to detect a situation where the system fails. The abnormal situation occurs when there is a token in s_2 and no token in s_5 or s_7. The firing condition and next marking of t_1 are changed as follows (see Figure 6).

$$
\begin{aligned}
\Phi^1_{error}(\mu|^{*}t_1) &\equiv (\mu(s_2) \neq 0) \\
&\quad and\ ((\mu(s_5) = 0)\ or\ (\mu(s_7) = 0)) \\
C^1_{error}(\mu|^{*}t_1) &\equiv \begin{cases} 0 & : on\ s_1, s_8 \\ \mu(s_2) & : on\ s_{14} \end{cases}
\end{aligned}
$$

We generated the reachability tree for the corrected net and confirmed that processing stops when the above abnormality occurs.

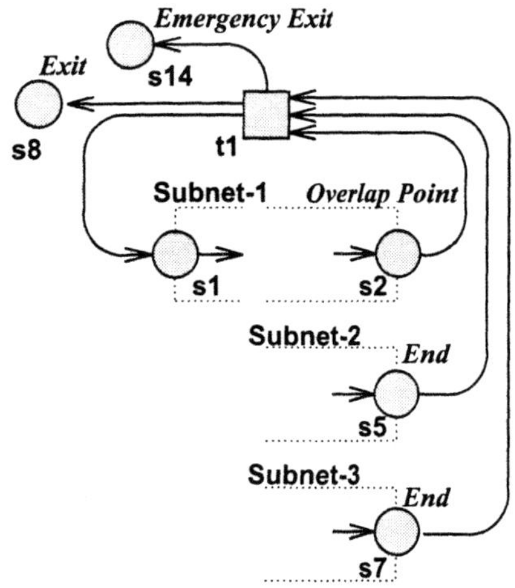

Fig. 6. Correction for the emergency-exit transitions.

4.2 Implementation and Operational Testing of the Control Computer

We implemented the modeled control software on the control computer using LCPNs and confirmed its operation.

The image processing sequence was specified in terms of the LCPN but communication with XY-stage controller, which requires high-speed processing and the basic procedures for numerical computation and comparison as well as screen display, were written in the conventional C programming language. For trial purposes from development to field-testing, we used a personal computer with a Petri net tool that operated on this platform. This tool can consistently perform test operations on (1) LCPN modeling, (2) net verification, (3) basic procedure linking, and (4) on-site simulations.

By using the concept of LCPN, considerable reduction was achieved in the implementation of parallel processing systems compared with older development methods, which lack verification capabilities. This is a result of the following three factors.

(1) The specification can be simplified using LCPN.
(2) The stability of the system can be verified sufficiently by net analysis.
(3) The control program can be united with basic customized procedures.

5. CONCLUSION

This paper presents a case study of the design and verification of a parallel image processing unit based on an extended Petri net (LCPN) proposed previously by our research group. Using this LCPN, we modeled the software for controlling an image processing system and analyzed the net model with a reachability tree. We implemented the control software, generated from a net whose stability was verified, and on the control computer of the system confirmed its operation.

In the future, we plan to study further the operation of the LCPN and improve it as it is applied to CASE tools for developing control systems, then apply the larger control systems in FMS.

REFERENCES

Jensen, K. (1992). *Coloured Petri Nets, Basic Concepts, Analysis Methods and Practical Use : Vol.1 Basic Concepts.* Springer-Verlag. New York, US.

Kawamoto, P.N., Y. Fuwa and Y. Nakamura (1993*a*). Basic concepts for petri nets with boolean markings. *Journal of Formalized Mathmatics.* **4**(1), 1–6.

Kawamoto, P.N., Y. Fuwa, M. Eguchi and Y. Nakamura (1993*b*). The detection of deadlocks in petri nets with ordered evaluation sequences. *Technical Report of IEICE.* **SS92-29, KBSE92-50**, 45–52.

Murata, T. (1989). Petri nets: Properties, analysis and applications. *Proceedings of the IEEE.* **77**(4), 541–580.

Murata, T., B. Shenker and S. M.Shatz (1989). Detection of ada static deadlocks using petri net invariants. *IEEE Trans. of Software Engineering.* **15**(3), 314–326.

Nagao, Y., H. Ohta, H. Urabe and S. Kumagai (1992). Petri net based programming system for fms. *IEICE Trans. Fundamentals.* **E75-A**(10), 1326–1334.

Peterson, J.L. (1981). *Petri Net theory and the Modelling of Systems.* Prentice-Hall, Inc.. New Jersey, US.

Wasaki, K., S. Yamaguchi, Y. Shidama and N. Shimoi (2000). A rapid hdl generation for a pipelined des encipher circuit by using the logical coloured petri net. *Proceedings of the 6th International Conference on Control, Automation, Robotics and Vision (ICARCV2000).* **FM5-2**, 1–6.

Yamaguchi, S., K. Wasaki and Y. Shidama (2003). A high reliability design for nfs server software by using an extended petri net. *Proceedings of the 5th International Conference on Enterprise Information System (ICEIS2003).* **3**, 498–502.

Copyright © IFAC Information Control Problems in
Manufacturing, Salvador, Brazil, 2004

ELSEVIER
IFAC
PUBLICATIONS
www.elsevier.com/locate/ifac

SYNTHESIS AND PLC IMPLEMENTATION OF SUPERVISORY CONTROL VIA PLACE INVARIANTS FOR A MANUFACTURING CELL

Eduardo J. Lima II* Carlos E. T. Dórea**

* *Universidade Federal de Minas Gerais*
Escola de Engenharia, Departamento de Engenharia Mecânica
Av. Antônio Carlos 6627, 31270-901 Belo Horizonte, MG, Brazil
ejlima2@hotmail.com
** *Universidade Federal da Bahia, Escola Politécnica, DEE*
Rua Aristides Novis 2, 40210-630 Salvador, BA, Brazil
cetdorea@ufba.br

Abstract: Based on the concept of place invariants of Petri Nets, supervisors for discrete-event systems can be synthesized so as to have a set of operation constraints respected by the controlled system. The implementation of this technique in a manufacturing cell is presented, with detailed description of modeling, supervisor design and programming, in STL language, of the PLC which controls the cell. Important issues such as: supervisor permissiveness, comparative analysis with techniques based on automata, as well as the systematization of the procedure for application to other systems are also discussed. *Copyright © 2004 IFAC*

Keywords: Supervisory control, manufacturing systems, discrete-event systems, Petri-nets, programmable logic controllers, invariance.

1. INTRODUCTION

Many Discrete-Events Systems (DESs) such as manufacturing cells are composed by sub-systems which have to obey a set of coordination constraints in order to work together. The aim of supervisory control is to coordinate their joint functioning, assuring the respect to such constraints. This control, hence, has a permissive nature, acting just to prevent the occurrence of some events which could lead the system to violate the constraints. When this is performed in a optimal way (inhibiting only the necessary events) the supervisor is said to be maximally permissive (or minimally restrictive).

The Supervisory Control Theory, proposed in (Ramadge and Wonham, 1989) (R&W approach), based on formal languages and finite state automata, establishes necessary and sufficient conditions for the existence of a minimally restrictive supervisor and supplies a formal methodology for its computation. The concept of controllable languages allows to deal with the fact that some events in a DES cannot be disabled by control action. The practical use of this approach is, however, limited by the state-space explosion with the growth of the number of subsystems. Even though this problem can be alleviated with the use of modular control, (Ramadge and Wonham, 1989; Queiroz and Cury, 2000), the controller complexity could nevertheless make it difficult to be implemented in practice.

As an alternative to the complexity of R&W approach, some works have explored the simplicity and the graphic power of Petri Nets. Particularly, in (Yamalidou *et al.*, 1995) the coordination constraints are represented in the form of linear inequalities. The controller which enforces the constraints is also a Petri Net, and is obtained by transforming them into place invariants of the controlled system. If the system has uncontrollable transitions, it may not be possible to directly enforce a given constraint, which is then said to be non-admissible. In (Moody and Antsaklis, 1998) a technique is proposed to transform non-admissible constraints into admissible ones which however does not guarantee the maximal permissiveness of the resulting supervisors. This drawback can be reduced with a change in the firing rules of the Petri Net with the objective of creating controllers defined by constraints disjunction (Moody and Antsaklis, 1998; Lima and Dórea, 2002).

This work presents the results of a practical implementation of supervisory control based on place invariants in a manufacturing cell. The closed-loop system (plant + supervisor), modeled by a Petri Net, is implemented through a program written in STL in a Programmable Logic Controller (PLC) which drives the plant operation. The results are compared to those reported in (Queiroz and Cury, 2002), in which a supervisor for a similar plant has been obtained using a modular approach. The possibility of systematization of the proposed implementation for application in other systems is also discussed.

2. SUPERVISORY CONTROL VIA PLACE INVARIANTS

Discrete processes can be modeled by Petri Nets with n places and m transitions. If the net does not have any self-loop, it may be represented by:

$$\mu_p(k+1) = \mu_p(k) + D_p q(k),$$

with $\mu_p(0) = \mu_{p0}$, $D_p \in Z^{nxm}$ is the incidence matrix, $\mu_p \in Z^n$, with $\mu_p \geq 0$, the marking vector and $q \in Z^m$, with $q \geq 0$, the firing vector.

Supervisory control aims at forcing the plant to obey linear constraints of the following type:

$$l^T \mu_p \leq b, \tag{1}$$

with $l \in Z^n$ and $b \in Z$. This inequality may be transformed into an equality by adding a slack variable $\mu_c \geq 0$, with $\mu_c \in Z$:

$$l^T \mu_p + \mu_c = b. \tag{2}$$

The controller is a place connected to the transitions. Therefore, the controlled system is a Petri Net composed by the original plant and the added controller place, forming a closed-loop system with incidence matrix $D \in Z^{(n+1)\times m}$ and marking vector $\mu \in Z^{n+1}$, which are given by: $D = \begin{bmatrix} D_p \\ D_c \end{bmatrix}$, $\mu = \begin{bmatrix} \mu_p \\ \mu_c \end{bmatrix}$. Because the constraint (2) has to be respected by any marking, it defines a *place invariant*, which is characterized by: $\begin{bmatrix} l^T & 1 \end{bmatrix} \begin{bmatrix} D_p \\ D_c \end{bmatrix} = 0$. The controller and its initial marking are then given by (Yamalidou *et al.*, 1995; Moody and Antsaklis, 1998):

$$D_c = -l^T D_p, \qquad \mu_{c_o} = b - l^T \mu_{p_o} \tag{3}$$

The arcs which connect the controller place to the net transitions are represented in matrix D_c and the initial marking of this place in vector μ_{c_o}.

The same procedure should be applied to each constraint to be obeyed by the net.

In this approach it is also possible to enforce constraints involving firing vectors. Quite often it is necessary to prevent a given transition from firing when a given marking is reached. This type of constraint can be represented in the following general form: $l^T \mu_p + f^T q \leq b, \;\; f \geq 0$.

In (Moody and Antsaklis, 1998) it is shown that the place invariant technique can be easily extended to this type of constraint.

If the controller places points arcs only to controllable transitions ($l^T D_{uc} \leq 0$, where D_{uc} is the matrix formed by the rows of the incidence matrix D_p associated to uncontrollable transitions), then the controller obtained by (3) only acts to disable transitions whose firing would lead the net to a state which violates the constraint (1). Therefore, in this case, the controller is maximally permissive (Moody and Antsaklis, 1998).

Otherwise, the controller would act to disable an uncontrollable transition, which is impossible. In that case, it is still possible to control the system by imposing a transformed admissible constraint such that the original constraint is satisfied for any reachable marking. In (Moody and Antsaklis, 1998) transformed constraints are proposed in the following form:

$$l_n^T \mu_p \leq b_n, \quad \text{with} \begin{cases} l_n^T = R_1 + R_2 L \\ b_n = R_2(b+1) - 1 \end{cases} \tag{4}$$

where: $R_1 \in Z^{1\times n}$, $R_1 \mu_p \geq 0$ for any $\mu_p \geq 0$ and $R_2 \in Z$ is a positive integer.

Any controller which enforces (4) will also enforce (1). However, the enforcement of these new constraints does not guarantee a maximally permissive controller.

In certain cases, an infinite number of transformed constraints can be found using different values for R_1 and R_2. It is then possible to improve the permissiveness of the controller by forcing the disjunction of transformed constraints (Moody and Antsaklis, 1998). In (Lima and Dórea, 2002), an algorithm was proposed which is able to compute transformed constraints, even when the algorithm of (Moody and Antsaklis, 1998) fails, as well as less restrictive disjunction based supervisors.

3. MANUFACTURING CELL DESCRIPTION

The manufacturing cell used here is a module of a didactic plant controlled by a PLC, similar to that described in (Queiroz and Cury, 2002) which has been controlled by a modular supervisor.

The cell is composed by a circular table, a drill and a testing device (figure 1). The circular table can hold parts in four different positions. In position 1, the entering of parts is detected by a capacitive sensor. In the position 2, the drill is mounted on a pneumatic actuator which makes the ascent and descent movements.

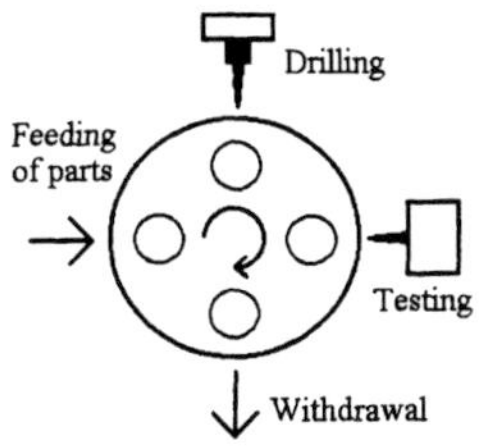

Fig. 1. Manufacturing cell

When the cell manufactures only one part, the sequence is the following: the PLC sends a signal requesting for a part; after the feeding, the table turns 90°, moving the part to position 2; the part is drilled; the table turns moving the part to position 3; the hole is tested; the table turns moving the part to position 4; the PLC sends a signal requesting the other module to pick up the part and waits for a signal indicating that the part has been picked up.

As described in (Queiroz and Cury, 2002), when the cell works on more than one part at the same time some synchronization problems may arise between the various operations: to feed, drill, test or withdraw at the same time as the rotation of the table; to place more than one part at position 1; to turn the table before the parts in positions 2, 3 and 4 have been drilled, tested or picked up; to drill, test or request for picking up parts without parts in position 2, 3 and 4; to drill or test twice the same part; to turn the table when it is empty.

These problems can be solved using a more complete PLC program. It can be done in a traditional way, by an expert programmer, or in a systematic and formal way using supervisory control techniques based on DES models, able to prevent from programming errors and allowing the use of analytical techniques to verify structural and behavioral properties of the controlled system.

4. CELL MODELING

Similarly to (Queiroz and Cury, 2002), the modeling of the cell does not deal with the low level signal sequences of the operations of turning, drilling, testing and communication with other PLCs, since these sequences are always the same. Moreover, the coordination problems to be solved by supervisory control do not arise in such sequences, but in the beginning and the end of operations of each system compound. This approach also simplifies the Petri net model, making the operational procedures transparent to it, so that changes in specific procedures do not necessarily reflect into changes in the Petri net model. It clearly allows for an easier analysis of properties. Here, for the sake of simplicity, these procedures are modeled by state machines.

The transitions corresponding to the operations start (0-turning of the table, 1-feeding, 2-drilling, 3-testing and 4-parts pick up) are controllable for their firing can be disabled by a controller action. The transitions corresponding to the end of operations are uncontrollable for they cannot be disabled by the controller.

The cell is modeled by two subnets. The first one models the operations on the parts and the second one models the turning of the table.

The feeding, drilling and testing are modeled, each one, by three places and two transitions in the first subnet (figure 2(a)): a token in the first place represents a part ready to the operation; a token in the second place represents a part being processed, and a token in the third place means that the operation has finished. The start of operations are represented by transitions α_1, α_2, α_3 e α_4 and the end of operations by transitions β_1, β_2, β_3 e β_4.

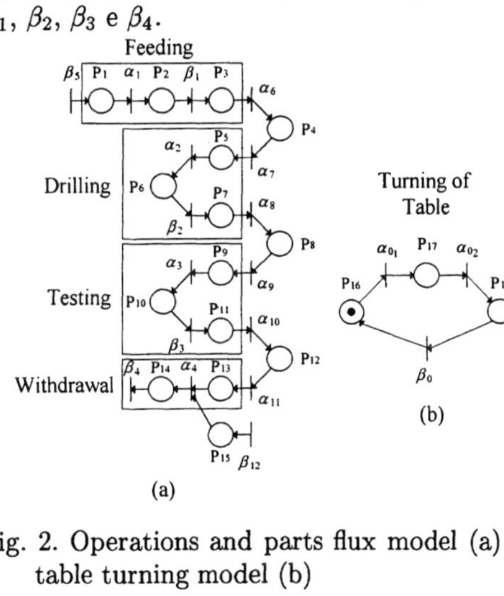

Fig. 2. Operations and parts flux model (a) and table turning model (b)

The table turning is modeled by 3 transitions (α_{0_1}, α_{0_2} e β_0 representing the table turning enabling, the command to turn the table and the end of the turning) and 3 places (p16, p17 e p18 representing the table stopped, enabled for turning and turning). Figure 2(b) shows the subnet which models the table turning.

Transitions α_6, α_8, α_{10} and α_0 correspond to the same event: start of rotation. Transitions α_7, α_9, α_{11} and β_0 correspond to end of rotation. If the second subnet were not used to model the turning, it would be necessary to enforce all transitions in the set $\{\alpha_6, \alpha_8, \alpha_{10}\}$ to fire simultaneously. This problem arises from the impossibility of moving a part from position 1 to position 2 without moving, for example, the part from position 3 to position 4. Since it is not possible to enforce the simultaneous firing of two transitions, the second subnet is necessary to model the turning of the table and synchronize it with the first subnet. It is important to point out that only transitions α_{0_2} and β_0 are linked to the operational procedures, representing the events that are really controlled and observed by the PLC.

The uncontrollable transition β_5 represents the arrival of parts in the feeding buffer. Each firing of this transition adds a token to the place p1. By the model structure, the number of tokens on this place is unbounded, but in practice this number should be limited by the program of the module which feeds the cell. The firing of β_{12} means that the next station is requesting for parts.

The synchronization between these two subnets is done using supervisory control based on place invariants. The uncontrolled net has $n = 18$ places and $m = 19$ transitions (12 controllable and 7 uncontrollable). Therefore, $D_p \in Z^{18\times19}$ is the incidence matrix and $\mu_p \in Z^{18}$, where $\mu_p \geq 0$, is the marking vector. The initial marking has zeros in all places. The tokens are supplied to the net by the transitions β_5 (arrival of parts in the buffers) and β_{12} (requesting for parts by the next station).

5. SYNTHESIS OF THE SUPERVISOR

The synchronization between the first and the second subnet is achieved by forcing the following firing sequence:

$$\alpha_{0_1} \rightarrow \{\alpha_6, \alpha_8, \alpha_{10}\} \rightarrow \alpha_{0_2} \rightarrow \beta_0 \rightarrow \{\alpha_7, \alpha_9, \alpha_{11}\}$$

This means the transitions α_6, α_8 and α_{10} can only be fired after the firing of α_{0_1}. α_{0_2} can only be fired after the firing of all enabled transitions of the set $\{\alpha_6, \alpha_8, \alpha_{10}\}$. Finally, α_7, α_9 and α_{11} can only be fired after the firing of β_0. The strategy to obtain this behavior is: to disable the firing of α_6, α_8 and α_{10} until the firing of α_{0_1}; to disable the firing of α_{0_2} while α_6, α_8 or α_{10} are enabled and to disable the firing of α_7, α_9 and α_{11} while α_{0_2} and β_0 have not fired.

First, the firing of α_{0_1} is disabled while there are parts being processed or to be processed:

$$\begin{aligned} 10q_{\alpha_{0_1}} + \mu_2 + \mu_4 + \mu_5 + \mu_6 + \mu_8 + \\ + \mu_9 + \mu_{10} + \mu_{12} + \mu_{13} + \mu_{14} \leq 10 \end{aligned} \quad (5)$$

After, the firing of α_6, α_8 and α_{10} should be disabled up to the firing of α_{0_1} (this means that place p17 has a token): $q_{\alpha_6} - \mu_{17} \leq 0$, $q_{\alpha_8} - \mu_{17} \leq 0$ and $q_{\alpha_{10}} - \mu_{17} \leq 0$. These three constraints are equivalent to: $q_{\alpha_6} + q_{\alpha_8} + q_{\alpha_{10}} - 3\mu_{17} \leq 0$.

Because by the implementation proposed here two controllable transitions cannot fire simultaneously (see section 6), this constraint is equivalent to:

$$q_{\alpha_6} + q_{\alpha_8} + q_{\alpha_{10}} - \mu_{17} \leq 0. \quad (6)$$

α_{0_2} can only be fired after the firing of the enabled transitions in the set $\{\alpha_6, \alpha_8, \alpha_{10}\}$ (this means p3, p7 or p11 have tokens). This is obtained by the constraint: $3q_{\alpha_{0_2}} + \mu_3 + \mu_7 + \mu_{11} \leq 3$.

But this constraint is not admissible because $\mu_3 + \mu_7 + \mu_{11} \leq 3$ is not admissible. So, it is necessary to enforce a transformed constraint:

$$3q_{\alpha_{0_2}} + \mu_2 + \mu_3 + \mu_6 + \mu_7 + \mu_{10} + \mu_{11} \leq 3. \quad (7)$$

Finally, the firing of α_7, α_9 and α_{11} should be disabled while α_{0_2} and β_0 have not fired (places p17 or p18 have tokens): $q_{\alpha_7} + \mu_{17} + \mu_{18} \leq 1$, $q_{\alpha_9} + \mu_{17} + \mu_{18} \leq 1$ and $q_{\alpha_{11}} + \mu_{17} + \mu_{18} \leq 1$. This is equivalent to:

$$q_{\alpha_7} + q_{\alpha_9} + q_{\alpha_{11}} + 3\mu_{17} + 3\mu_{18} \leq 3 \quad (8)$$

It is still necessary to disable the rotation of the table if it is empty:

$$q_{\alpha_{0_1}} - \mu_3 - \mu_7 - \mu_{11} \leq 0. \quad (9)$$

This constraint allows the rotation of the table only if a part has been fed (a token in p3), a part has been drilled (a token in p7) or a part has been tested (a token in p11).

To prevent more than one part in position p1 the following constraint must be enforced: $\mu_3 + \mu_4 \leq 1$.

Since it is not admissible, the following transformed constraint should be enforced:

$$\mu_2 + \mu_3 + \mu_4 \leq 1. \quad (10)$$

Indeed, to prevent more than one part in p1 it is necessary to avoid that it be requested.

Also, parts cannot be requested while the table turns:

$$q_{\alpha_1} + \mu_{17} + \mu_{18} \leq 1. \quad (11)$$

The control places corresponding to the constraints (5) to (11) where computed using (3). The transformed constraints were computed by the algorithm described in (Lima and Dórea, 2002)

The control place corresponding to constraint (6) is illustrated in figure 3. It can be seen that it only points arcs to controllable transitions.

The controller obtained by (3) only acts to disable transitions when its firing leads the net to a state which violates the constraint. On the other hand, the maximal permissiveness of the controllers cannot be guaranteed when constraints are transformed by (4). However, due to the sequential structure of the cell it is possible to verify that the computed controller is maximally permissive.

Supposing constraints in the form $l^T\mu_p \leq b$, the controller should prevent all the markings for which $l^T\mu_p > b$ and all the markings which enable

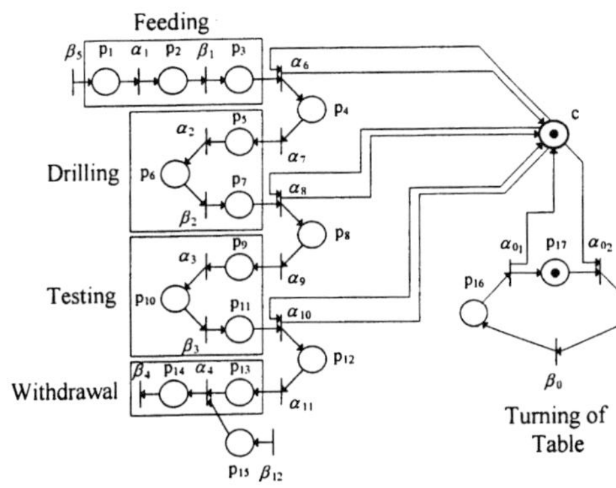

Fig. 3. Controller for constraint (6).

uncontrollable transitions whose firings lead the net to $l^T\mu_p > b$. The constraint $\mu_3 + \mu_4 \leq 1$ only bounds the markings of the places p3 and p4, but any token of the place p2 could be transferred to p3 by the firing of the uncontrollable transition β_1. Hence, the controller which enforces this constraint has to prevent all and only the markings for which $\mu_2 + \mu_3 + \mu_4 > 1$ from being reached, by enforcing the constraint $\mu_2 + \mu_3 + \mu_4 \leq 1$ (constraint 10). This constraint can be enforced with maximal permissiveness by (3). Therefore, the controller implemented here prevents only the markings which violate or could violate $\mu_3 + \mu_4 \leq 1$ from being reached by the firing of β_1.

By the same reasoning it is possible to verify the maximal permissiveness of the other transformed constraints, by operating the following changes in the original constraints: $\mu_3 \rightarrow \mu_2 + \mu_3$; $\mu_7 \rightarrow \mu_6 + \mu_7$ and $\mu_{11} \rightarrow \mu_{10} + \mu_{11}$.

6. PLC IMPLEMENTATION

The program implemented in PLC has been written in the STL (Statement List) language in a form which enables the direct translation to LDR (Ladder Relay Logic). STL has been chosen because it is a text language, which makes easier its automatic generation by a computational tool.

The methodology proposed here differs from most of Petri Nets implementations reported in the literature, which integrate low-level PLC signals into the Petri Net model (see e.g. (Uzam *et al.*, 1999; Frey, 2000) and (Peng and Zhou, 2001) for a survey). Here, the Petri Net deals only with the transitions (in a high level), whereas the control of PLC signals (low-level) is treated by the operational procedures.

The program is composed by three main routines: the *token player* (TKPL), the *executor* (EXE) and the *operational procedures* (OPROC). Table 1 shows in summary the function of each routine. d_c is the vector that determines the disabled controllable transitions, q_c is the firing vector of the controllable transitions and q_{uc} the firing vector of the uncontrollable transitions.

Table 1. PLC program routines.

Routine	Functions
OPROC	Controls the operational procedures Determines $q_{uc}(k)$ Determines $q(k) = q_c(k) + q_{uc}(k)$
TKPL	Calculates $\mu(k+1) = \mu(k) + Dq(k)$ Determines $d_c(k+1)$
EXE	Determines $q_c(k+1)$

Routine OPROC generates the PLC output signals in order to execute the operation cycles of the devices. It implements the "open-loop" (without supervision) functioning of the plant, providing also its interface with the supervisor, which only generates and observes firings, without dealing with the signals that must be generated and read by the PLC. Each device is modeled by a state machine. The firing of controllable transitions (q_c) of the Petri Net starts the cycles, which evolve according to PLC input signals.

Each state machine is implemented as suggested in (Fabian and Hellgren, 1998), with each state being represented by a logical variableand characterized by different values written in the PLC outputs. This routine also generates the firing vector of the uncontrollable transitions (vector q_{uc}) associated to each procedure, when the states resulting from the occurrence of the physical events related to the Petri net transitions are reached.

Routine TKPL verifies the firing of transitions (vector q) and updates the marking of its places. After that, it scans the controllable transitions and computes the vector d_c (*true* for disabled transitions and *false* for enabled transitions). In figure 4 (left) the part of the program corresponding to the firing of α_2 is shown. First it verifies if $Q[2] = 1$. If true, $Q[2]$ is set to zero, a mark is subtracted from p5 and a mark is added to p6. In figure 4 (right) it is shown the part of the program corresponding to the disabling of α_2. This transition is disabled ($D[2] = 1$) if the marking of P5 is less than 1.

```
        A Q[2]
        R Q[2]
        JNB _012
        L M[5]
        L 1
        -I
        T M[5]
        L M[6]
        L 1
        +I
        T M[6]
_012:   NOP 0
```

```
A(
L M[5]
L 1
>=I
)
NOT
= D[2]
```

Fig. 4. STL programs of the firing (left) and disabling (right) of transition α_2

Routine EXE scans the enabled controllable transitions ($d_c(T) = false$) and sets $q_c = true$ for one of them. Only one controllable transition is fired at a time. This avoids problems which could arise when two transitions in conflict were enabled. The routine then establishes a priority list. Figure 5 shows how the firing of α_1 is enforced.

```
          AN D[1]
          S QC[1]
          JC FSP
          ...
FSP:      NOP 0
```

Fig. 5. STL Program commanding the firing of α_1

The size of the PLC program following the presented approach has a direct relation with the number of places and transitions of the Petri Net model of the controlled plant.

In the automata approach of (Queiroz and Cury, 2002), the number of steps of the LDR program is given by the number of states and transitions of the supervisor and the uncontrolled subsystems. In this case, the state-space explosion, even using modular supervisors, can lead to a huge LDR program that could even become impossible to be implemented. The direct application of modular control in this cell would result in a set of supervisors with 116 states and 292 transitions. After reduction these numbers can be reduced up to resulting in a LDR program with 92 steps and 76 logical variables.

The application of the proposed approach resulted, after conversion to LDR, in a program with 76 steps, 25 integer and 30 logical variables.

When the net is safe ($\mu_i(k) \leq 1$), the variables of TKPL can be defined as logical ones. Hence, the used memory can be considerably reduced.

7. CONCLUSION

The synthesis of supervisory control of systems modeled by Petri nets, based on place invariants has shown to be a simple and elegant technique which renders easier the analysis and implementation of controllers which enforce discrete event systems to obey a set of operation constraints. In this work the results of the practical implementation of this technique in a manufacturing cell have been presented. The operation of the plant following the imposed specifications is a proof of its practical viability.

The implementation using automata-based techniques can become impossible depending on the complexity (number of states and transitions) of the overall system, although modular supervision can alleviate this drawback. On the other hand, the size of the program of the PLC implementation of Petri nets depends on the number of places and transitions, which allows for its use in systems with a much larger number of states.

Even though the place invariant approach does not guarantee in the general case the computation of a maximally permissive supervisor, it has been verified that this property holds in the considered cell. The proposed technique has also been successfully implemented to control the joint operation of this cell with four other modules, and the computed supervisor has shown to be maximally permissive as well. From the results presented here, it could be conjectured that maximal permissiveness of place invariant supervisors would hold in general for plants having the same sequential structure.

The structure proposed for the implementation has allowed the development of a software tool for automatic synthesis and generation of the PLC program, which we believe can be applied to quite general manufacturing cells commanded by PLCs.

ACKNOWLEDGEMENTS

The authors are thankful to SENAI-CIMATEC, (http://www.cimatec.fieb.org.br), for the permission to work in its cell. This work was also partially supported by CAPES, Brazil (first author) and CNPq, Brazil (second author, grants # 301439/1997-4 and 471448/01-0).

8. REFERENCES

Fabian, M. and A. Hellgren (1998). PLC-based implementation of supervisory control for discrete-event systems. In: *Proc. 37th IEEE CDC*. Tampa, USA. pp. 3305–3310.

Frey, G. (2000). Automatic implementation of Petri net based control algorithms on PLC. In: *Proc. 2000 American Control Conf.*. Chicago, USA. pp. 2819–2823.

Lima, E. A. and C. E. T. Dórea (2002). An algorithm for supervisory control of discret event systems via place invariants. In: *Proc. 15th IFAC World Congress*. Barcelona, Spain.

Moody, J. O. and P. J. Antsaklis (1998). *Supervisory Control of Discrete Event Systems Using Petri Nets*. Kluwer Academic Publ., Boston.

Peng, S. and M. Zhou (2001). Conversion between ladder diagrams and PNs in discrete-event control design - a survey. In: *Proc. IEEE Int. Conf. System, Man and Cybernetics, 4*. Tucson, USA. pp. 2682–2687.

Queiroz, M. H. and J. E. R. Cury (2000). Modular control of composed systems. In: *Proc. American Control Conf.*. Chicago, USA.

Queiroz, M. H. and J. E. R. Cury (2002). Synthesis and implementation of local modular supervisory control for a manufacturing cell. In: *Proc. 6th WODES*. Zaragoza, Spain.

Ramadge, P. J. G. and W. M. Wonham (1989). The control of discrete event systems. In: *Proceedings. of the IEEE, 77(1)*. pp. 13–30.

Uzam, M., A.H. Jones and I Yucel (1999). A rule-based methodology for supervisory control of discrete event sistems modeled as automation Petri nets. pp. 297–325.

Yamalidou, K., M. Lemmon and P. J. Antsaklis (1995). Feedback control of Petri nets based on place invariants. pp. 15–28.

Copyright © IFAC Information Control Problems in
Manufacturing, Salvador, Brazil, 2004

ELSEVIER
IFAC
PUBLICATIONS
www.elsevier.com/locate/ifac

EFFICIENT SYNTHESIS OF A CLASS OF DISCRETE-EVENT CONTROLLERS FOR FORCED ACTIONS

A. Sanchez [*,1] **J. Douriet** [*]

[*] *Depto. de Ingenieria Electrica y Computacion*
Centro de Investigacion y Estudios Avanzados (Cinvestav)
Apdo. Postal 31-438, Guadalajara 45091, Jalisco, Mexico

Abstract: This paper deals with the synthesis of discrete-event controllers forcing at most one control action in each stage of the control pattern. Supervisory Control Theory (SCT) is employed as the basis modifying the controllability and completeness concepts. BDD-encoded algorithms are presented to carry out all the calculations for synthesizing a controller of this nature. Two examples show the use of the synthesis framework, highlighting its advantages, disadvantages and differences with SCT as well as the performance of the BDD-encoded algorithms. The examples also illustrate how the proposed framework can help to maintain the modeling simple and to synthesize controllers that otherwise could not be obtained. *Copyright © 2004 IFAC*

Keywords: Discrete event systems, controller synthesis, forced actions

1. INTRODUCTION

Supervisory Control Theory (SCT) (Ramadge and Wonham, 1987) is recognized as one of the most solid synthesis theories for discrete-event controllers. The process is modeled using finite state machines (FSMs) and their associated set of finite trajectories (i.e. *-languages). The control mechanism is known as supervisor. Control action is effected by synchronizing in each process state a set of shared transitions. This set must include at least all the uncontrollable transitions enabled by the process. Any other transition is interpreted as disabled. Thus, supervisors are frequently used as part of control schemes with extra decision mechanisms for issuing control commands.

Since SCT is a synthesis theory, there are still many issues to be resolved in the design, implementation and application of controllers of this nature to real problems. For instance, process modeling frameworks compatible with design tools, translation of industrial-standard specifications to automata-based equivalents, construction of correct and complete specifications, computing large supremal controllable languages, realization of controllers in industrial control devices.

In this work we address two of these aspects, namely, the synthesis of a control mechanism capable of issuing (i.e. forcing) control actions and the calculation of these controllers for large systems. In particular we deal with controllers forcing at most one control action in each state of the control pattern. This control technology is frequently found in the manufacturing and process industries. We compare SCT with a particularization of this theory proposed in (Sanchez *et al.*, 1999) aimed specifically at this type of control technology. The proposal capitalized on the solid framework of SCT and introduced the notion of procedural controller, a mechanism capable of

[1] Corresponding author. e-mail address: arturo@gdl.cinvestav.mx

synchronizing with the controlled system either all uncontrollable transitions or at most one control action in each state of the control pattern. Their applicability and benefits in the process industries have already been assessed from both technical and business point of views (Sanchez *et al.*, 2002). In section 2 the definition of procedural control is introduced and the adaptation of SCT is discussed. The concepts of controllability, completeness and control pattern were modified accordingly. In order to attack the synthesis of procedural controllers for large systems, section 3 presents a predicate-based algorithm encoded using Binary Decision Diagrams (Andersen, 1997). The benefits and disadvantages of the modified framework are illustrated in section 4 with two examples. The first example shows in a graphic manner how the proposed approach addresses the synthesis of the proposed controllers whilst SCT can fail. Results of numeric experiments show that the size of systems that can be effectively handled with the current implementations in one shot-calculations is approximately 10^6 reachable states. The second example is taken from the open literature in which SCT is used to synthesize a controller (for an assembly line) of the type discussed in this work. In the original work (Chandra *et al.*, 2001), a controller was obtained using SCT but rendering an uncontrollable closed–loop behavior (in the SCT sense). Using the modified framework, the desired result is obtained in a straightforward manner.

2. FUNDAMENTALS

According to SCT, the plant P is modeled using Finite State Machines (FSMs) and their associated sets of finite trajectories (non-marked and marked *-languages) $L(P)$ and $L_m(P)$. The closed loop behavior is defined as $L(C/P) = L(P) \cap L(C)$ and $L_m(C/P) = L_m(P) \cap L(C)$. In other words, the FSM plant synchronizes with the controller C (that may be realized as another FSM). In this work we use a class of controller, termed procedural controller (Sanchez *et al.*, 1999).

Definition 1. A procedural controller is an FSM $C = \{X, \Sigma, \gamma, x_0, X_m\}$. In the standard notation X is the set of controller states and $\gamma : X \times \Sigma \rightarrow X$ is the transition function. For each $x \in X$ and $\sigma \in \Sigma$ such that $\gamma(x, \sigma)$, one and only one of the following conditions is true

1. $\sigma \in \Sigma_u \wedge (\forall \sigma_c \in \Sigma_c, \gamma(x, \sigma_c)$ is undefined).

2. $\sigma \in \Sigma_c \wedge (\forall \sigma' \mid (\sigma' \neq \sigma) \in \Sigma, \gamma(x, \sigma')$ is undefined).

Thus, C can be in either

(1) a state in which any of the plant-enabled uncontrollable transitions can occur or
(2) a state in which one and only one controllable transition occurs.

Thus, the controller executing synchronously with the plant preempts the execution of uncontrollable transitions from some of the plant states by executing controllable transitions. In a continuos time-driven framework, this means that the execution of controllable transitions are much faster than their uncontrollable counterparts. This is the case of industrial process systems (e.g. pharmaceutical, petrochemicals, food) in which, generally, the dynamics of uncontrollable events is considerably slower than controllable actions (e.g. a microprocessor issuing control commands as a function of a reactor temperature threshold changes)

In order to consider the possible preemption of uncontrollable transitions by controllable transitions, the SCT definitions of completeness and controllability are modified.

Definition 2. **Completeness.** An FSM $M = \{X, \Sigma, \gamma, x_0, X_m\}$ is complete with respect to $P = \{Q, \Sigma, \delta, q_0, Q_m\}$ if for each $s \in \Sigma^*$ and $\sigma_u \in \Sigma_u$, conditions $\gamma(x_0, s)!$ and $\delta(q_0, s\sigma_u)!$ imply that one an only one of the following assertions is true: (1) $\gamma(x_0, s\sigma_u)!$; (2) $\exists \sigma_c \in \Sigma_c$ such that $\gamma(x_0, s\sigma_c)! \wedge \gamma(x_0, s\Sigma_u)$ is undefined.

Definition 3. **Controllability.** A language $K \subseteq L$ is controllable with respect to P if $\forall s \in K$ one and only one of the following assertions is true:

(1) $s\Sigma_u \in L \wedge s\Sigma_u \in \overline{K}$, or

(2) $\exists \sigma_c \in \Sigma_c$ such that $(s\sigma_c \in \overline{K}) \wedge (s\Sigma_u \cap \overline{K} = \emptyset)$.

That is, for every state in an FSM generating a controllable language K either

(1) All the uncontrollable transitions end in one of the states of the FSM, regardless of the existence of controllable transitions (and this is the controllability concept as proposed in SCT (Ramadge and Wonham, 1987) as an invariant with respect to uncontrollable transitions) or,
(2) There is at least one controllable transition leaving the state and ending in another state of the FSM.

This means that even if there is uncontrollable transitions leading outside the FSM, it is possible to synthesize a controller that can preempt them by proper control actions and keep the plant states in the desired set. The aim is to synthesize procedural controllers that are complete, and when synchronized with the plant to be controlled, generate controllable languages. Controllability, as in SCT, is closed under language union. This leads to the notion of supremal controllable language. However, in general, the union of procedural control languages is not closed.

Definition 4. **Supremal controllable language controlable** The supremal controllable language $K^\uparrow$ of a language $K \subseteq L$ is defined as $K^\uparrow = \cup\{K' : K' \subseteq K$ and K' is controllable with respect to $L\}$.

Thus an FSM R generating $K^\uparrow$ can be understood as a superstructure containing all possible procedural controller candidates for a given P. The existence of a procedural controller is based in the following theorem whose proof is given in (Sanchez *et al.*, 1999).

Theorem 5. For an specification language $K(M) \subseteq L(P)$ and $K_m = L_m(P) \cap K$ ($K_m \neq \emptyset$), if M is trim and K_m is closed and controllable, then $\exists C = (X, \Sigma, \gamma, q_0, X_m)$ non blocking and complete, with $L(C/P) \subseteq \overline{K_m}$.

3. SYMBOLIC SYNTHESIS OF A PROCEDURAL CONTROLLER

Predicates can be used to express structural properties of FSMs. They also facilitate the development of synthesis algorithms that can be encoded using efficient tools, such as Binary Decision Diagrams. Following (Zhang and Wonham, 2002), an FSM M can be represented by $M' = \{Q, \Sigma, \|T\|, \|I\|, \|Mr\|\}$ with

- $T : Q \times \Sigma \times Q \longrightarrow \{0,1\}$ as the transition function predicate. The triplet (q, σ, q') satisfies T if and only if $q' = \delta(q, \sigma)$.
- $I : Q \longrightarrow \{0,1\}$ as the initial state predicate. A state q satisfies I if and only if $q = q_0$.
- $Mr : Q \longrightarrow \{0,1\}$ as the marked states predicate such that q satisfies Mr if and only if $q \in Q_m$.

$\|T\|, \|I\|, \|Mr\|$ are the sets associated to the corresponding predicates.

3.1 A Fix-point operator for supremal controllable sublanguages

Let P and E be the plant and specification FSMs in a predicate representation. The starting point for calculating the supremal controllable language can be a composition of these two FSMs

$PE = \{Q \times X, \Sigma, \|T_{PE}\|, \|I_{PE}\|, \|Mr_{PE}\|\}$ with
$\|I_{PE}\| := \{(q, x) \in Q \times X | q \in \|I_P\|$ and $x \in \|I_E\|\}$
$\|Mr_{PE}\| := \{(q, x) \in Q \times X | q \in \|Mr_P\|$ and $x \in \|Mr_E\|\}$
$\|T_{PE}\| := \{((q, x), \sigma, (q', x'))$ defined according to the composition operation $\}$

Thus, the transition relation $\|T_{PE}\|$ is a function with domain $W := Q \times X \times \Sigma \times Q \times X$. Let $\|T_Z\|$ be a transition function of a plant and specification states defined on the domain W. A fix-point operator for the superstructure of procedural controllers $\Omega : 2^W \longrightarrow 2^W$ is given by:

$\Omega(\|T_Z\|) := \|T_{PE}\| \cap \|T_Z\| - \{((q, x), \sigma_u, (q_1, x_1)) \in \|T_Z\| : (q, x) \in K_1, \sigma_u \in \Sigma_u\} -$
$\{((q, x), \sigma, (q_1, x_1)), ((q_1, x_1), \sigma, (q, x)) \in \|T_Z\| : (q, x) \in K_2\} -$
$\{((q, x), \sigma, (q_1, x_1)), ((q_1, x_1), \sigma, (q, x)) \in \|T_Z\| : (q, x) \notin \|Co\|\}$

where:

$K_1 := \{(q, x) : \exists (q, \sigma_u, q_2) \in \|T_P\|, \sigma_u \in \Sigma_u$ and $((q, x), \sigma_u, (q_2, x_2)) \notin \|T_Z\|, \exists ((q, x), \sigma_c, (q_3, x_3)) \in \|T_Z\|$, for some $\sigma_c \in \Sigma_c\}$

$K_2 := \{(q, x) : \exists (q, \sigma_u, q_2) \in \|T_P\|, \sigma_u \in \Sigma_u$ and $((q, x), \sigma_u, (q_2, x_2)) \notin \|T_Z\|, \nexists ((q, x), \sigma_c, (q_3, x_3)) \in \|T_Z\|$, for some $\sigma_c \in \Sigma_c\}$

K_1 is the set of state pairs (q, x) from which at least one uncontrollable transition is enabled from q in the process FSM and is not enabled in the equivalent state of $\|T_Z\|$ but there is at least one controllable transition capable of preemting the ocurrence of any process-enabled uncontrollable transition. K_2 represents the set of state pairs (q, x) from which at least one uncontrollable transition is enabled from q in the process FSM and is not enabled in the equivalent state of $\|T_Z\|$, such taht there is no controllable transition capable of preemting the ocurrence of an uncontrollable transition. $\|Co\|$ guarantees that the surviving states are coreachable. This fix–point operator produces the desired supremal controllable language (Reza, 2002). The FSM obtained is $Z := \{Q_Z, \Sigma, \|T_Z\|, \|I_Z\|, \|Mr_Z\|\}$.

3.2 Calculation of Procedural Controllers

Based in the maximal superstructure, the following algorithm gives as a result a non-blocking complete procedural controller.

Algorithm 1. **Symbolic synthesis of non-blocking complete procedural controller**

Procedural Controller$(R : fsm) : fsm$
  $Z : fsm$
  $X_A, X_B : sets$
  $temp : relation$

  / Maximal superstructure Z as starting point */*
  $Z = \{Q_C, \Sigma, \|T_C\|, \|I_C\|, \|Mr_C\|\} := R$
  / States with ctrble and unctrble transitions */*

$X_A := \{(q,x) \in Q_C : (\exists((q,x),\sigma_u,(q_1,x_1)) \in \|T_C\|,$
$\sigma_u \in \Sigma_u) \wedge$
$(\exists((q,x),\sigma_c,(q_2,x_2)) \in \|T_C\|,\ \sigma_c \in \Sigma_c)\}.$
For each $(q,x) \in X_A$ **do**
temp $:= \|T_C\| - \{((q,x),\sigma,(q',x')) : \sigma \in \Sigma_c\}$
/* *Finding coreachable states* */
If $(Img((q,x)) \subseteq \|Co\|_{temp})$ **then**
$\|T_C\| := \|T_C\| - \{((q,x),\sigma,(q',x')) : \sigma \in \Sigma_c\}$
else
$\|T_C\| := \|T_C\| - \{((q,x),\sigma,(q',x')) : \sigma \in \Sigma_u\}$
Next (q,x)
/* *States with more than one ctrble trans* */
$X_B := \{(q,x) \in Q_C : \exists\sigma_c,\sigma_c' \in \Sigma_c$ s.t.
$((q,x),\sigma_c,(q_1,x_1)),((q,x),\sigma_c',(q_2,x_2)) \in \|T_C\|\}.$
For each $(q,x) \in X_B$ **do**
$\Sigma_X := \{\sigma_c : \exists((q,x),\sigma_c,(q',x')) \in \|T_C\|\}$
repeat
Choose $\sigma_c \in \Sigma_X$
$\Sigma_X := \Sigma_X - \sigma_C$
temp $:= \|T_C\| - \{((q,x),\sigma,(q',x')) : \sigma \neq \sigma_c\}$
/* *Coreachable st with current ctrbl trans?* */
If $(Img((q,x)) \subseteq \|Co\|_{temp})$ **then**
$\|T_C\| := \|T_C\| - \{((q,x),\sigma,(q',x')) : \sigma \neq \sigma_c\}$
Until $((Img((q,x)) \subseteq \|Co\|_{temp}))$
Next (q,x)
/* *Erase unreachable states* */
$\|T_C\| := \|T_C\| \cap \{(q,x,\sigma,q_1,x_1) : (q,x) \in \|Al\|\}$
/* *Create new FSM C from* T_C */
$C := \{Q_C,\Sigma,\|T_C\|,\|I_C\|,\|Mr_C\|\}$
return C

The algorithm terminates obtaining the desired procedural controller. Its complexity is $O(|Q_C|^3)$ (Reza, 2002). The encoding of these algorithms was carried out using BDDs in a standard fashion (Sanchez *et al.*, 2003).

4. EXAMPLES

Two examples taken from the literature are presented to illustrate the synthesis of procedural controllers. Both examples show graphically that applying SCT without modifications, the resulting languages are of no interest from an operational point of view. In the first example the synthesized controllable language using SCT is empty. The modified approach gives rise to a controller satisfying the specification. This example also includes performance results of the synthesis algorithms derived from the modified approach. The second example has been used previously to illustrate the application of SCT as an analysis tool for synthesizing controllers with forced actions. However if SCT is employed in a straightforward manner, the resulting controller halts the process before starting the actual operation. Using the proposed approach, the synthesis of the desired controllers becomes simpler and clearer.

4.1 Pressurization cycle of reactors

A typical unit reactor unit is shown in figure 1. The operation objectives are: i) to pressurize-depressurize the reactor in a cyclic manner and ii) to avoid high-pressure operating states. A description of the desired behavior and its modeling as FSMs can be found in (Sanchez *et al.*, 1999). The process and specification FSM models for one reactor are shown in figures 2 and 3. In the initial process state (state 0 in figure 2), the operator must press the on/off button. The controller responds opening the valve and then waiting for the normal pressure switch to be activated (states 2-5-9). Once this signal is received, the controller shuts the valve and waits for the button to be depressed by the operator in order to reach the desired (marked) state of normal operation (states 9-6-3). Once the reaction is over, the reactor is depressurized and the process returns to the initial state (states 3-0). The synchronization of these two FSM is shown in figure 4. As can be observed, the SCT-supremal controllable language is empty. This is because it is not possible even to start the operation according to the specification FSM. After pressing the push button in state 0 (in figure 2), the specification forces the opening of the valve from state 2. Thus, the push button should not be depressed by the operator (uncontrollable transition) as described by the model. Therefore, the only way of avoiding this uncontrollable conditions is never starting the operation.

Assuming that $L(C/P) = L(C) \cap L(P) = L(C)$, the FSM realizing the procedural controller obtained with the modified framework is shown in figure 5. It can be observed that the closed -loop behavior meets the specification.

Performance experiments were carried out synthesizing the superstructure and a procedural controller for a set of reactor units operating in parallel. Tables 1 and 2 show the results of the synthesis experiments carried out in a PC Pentium 4, 1.4 Ghz with 640 Mb RAM. The largest reachable state set that could be handled in reasonable times was of $2.9(10^6)$, equivalent to 6 reactors operating in parallel.

♯	SSPC				
Bl	$\|Q_P\|$	$\|Q_E\|$	$\|Q_{SE}\|$	T(s)	Nodes
1	12	5	9	< 0.01	133
2	144	25	80	0.02	1,206
3	1,728	125	704	0.19	7,700
4	20,736	625	6,144	1.55	44,683
5	248,832	3,125	53,248	23.56	249,668
6	$2.9(10^6)$	15,625	458,752	245.61	$1.3(10^6)$

Table 1. Time and nodes for calculation of maximal superstructure

4.2 An Assembly Line

This example was originally presented in (Chandra *et al.*, 2001). Its objective was the synthesis of a controller device to carry out the mounting

♯ Bl	$\lvert Q_{SE} \rvert$	$\lvert Q_C \rvert$	SSPC T(s)	SSPC M(Mb)	SSPC Nodes
1	9	7	< 0.01	< 13	105
2	80	35	0.06	< 13	608
3	704	152	1.33	14.5	2,289
4	6,144	605	51.72	17.5	6,445
5	53,248	2,374	338,831	37	16,180
6	458,752	N/D	N/D	128	N/D

Table 2. Memory, time and nodes for calculation of procedural controller

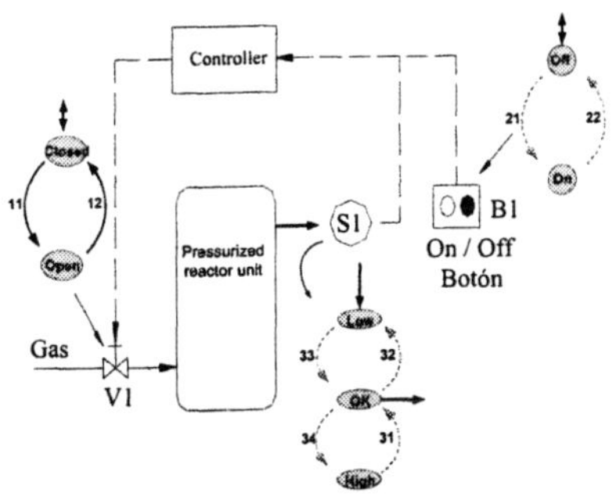

Fig. 1. Pressurized reactor unit

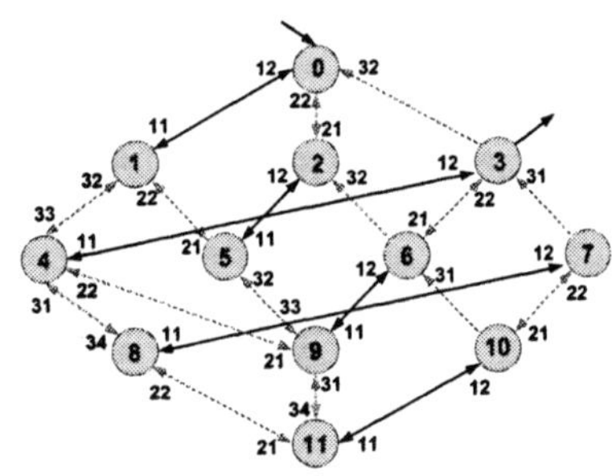

Fig. 2. Process model FSM

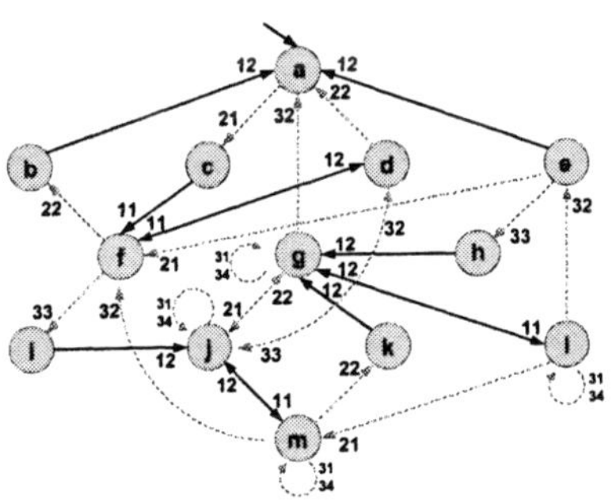

Fig. 3. Specification model FSM

of a roof in a car chassis using the assembly line shown in figure 6. The line is composed of five modules: (1) chassis and (2) roof feeders, (3) press, (4) unload conveyor and a transporter. Ad-hoc models with all-marked states for the plant and specifications were presented and a modular architecture was proposed with the transporter controller as the coordinator for the other four modules. The strategy was to synthesize a supervisor for each of the first four assembly module whilst the fifth was built as a concatenation of the other four. SCT-controllability was verified for

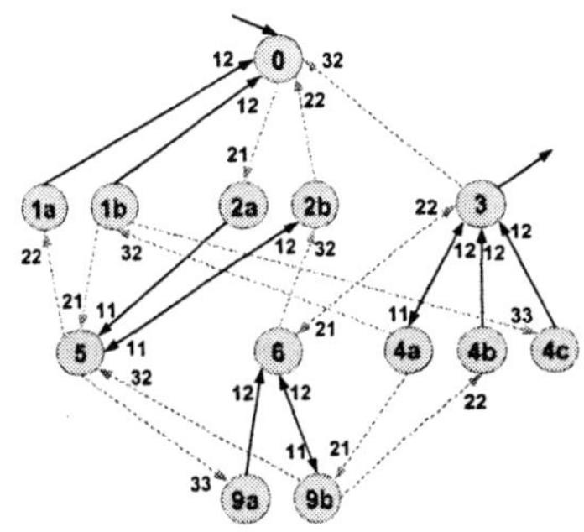

Fig. 4. Process-specification synchronization FSM

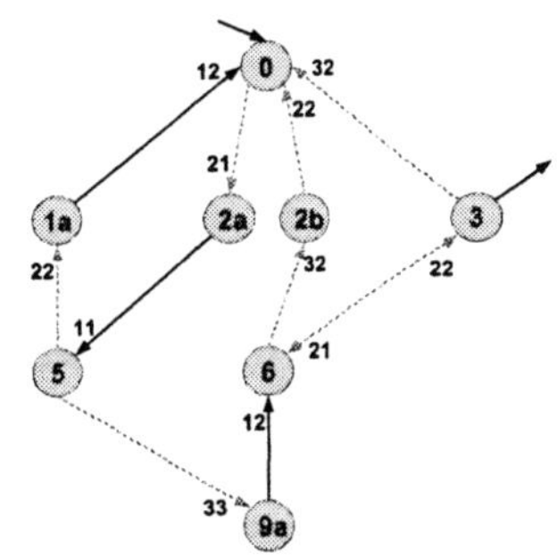

Fig. 5. Controller model FSM

each of the first four modules and it was assumed as a consequence for the transporter supervisor. From the FSM realizing this supervisor, a controller was extracted. No details of this procedure were given. The resultant FSM modeling the closed loop behavior shows a sequential execution of actions in each module. Problems with cyclical execution of process signals were not addressed. However, SCT-controllability is not meet by the transporter closed-loop language. Figures 7 and 8 shows figures of the transporter and its specification whilst figure 9 presents the synchronization of the two previous FSMs. Symbol STi* with i=1..4 models all possible sequences of actions that can be executed in module i with alphabet STi. It can be observed in figure 7 that once the transporter arrives to the roof section (uncontrollable transition h), the execution of uncontrollable transition i (i.e. leaving the roof section) can occur. However, specification 7 drives the system to execute first the roof-tasks grouped in ST2* and uncontrollable transition i is erased from state 11a in figure 9. In order to avoid this uncontrollable condition using SCT, the behavior must be restricted to the arrival of the transporter to the chasis position where it can be safely stopped at state 7 by not executing controllable transition FOff. The same situation occurs with the press section. The reason of not satisfying SCT-controllability is the assumption that when the transporter arrives to a module, it will not depart until the tasks associated to this module are fulfilled. Thus, this situation must be considered either by including this behavior in the corresponding models or using the modification proposed for SCT. We claim that the latter can be a good option since introducing this behavior can be cumbersome and in cases with more complex behavior can make the synthesis process even harder and difficult to automate. Thus, with the proposed framework, plant model building is maintained as a simple task including only equipment behavior and the synthesis of the controller is carried out in a transparent manner. In the case of the assembly line, using the same model of the plant and specifications originally proposed, the synthesis of the controller for the complete line was carried out in one-shot calculation. The result obtained is the FSM shown in figure 9.

5. CONCLUSIONS

The proposed modified framework handles the synthesis of control devices with forced transitions in a simple and effective manner. Since the control pattern is not restricted to always contain at least all the uncontrollable transitions enabled by the plant, simpler closed-loop behaviors can be obtained. The price is paid in the synthesis of the controller from the superstructure, which is a step extra with respect to the required calculations in SCT. However, the BDD-based encoding of the proposed algorithms permit to handle systems of considerable size. The tools used in this paper can be obtained from www.gdl.cinvestav.mx/sspc.

Acknowledgements.
AS kindly acknowledges partial financial support from Conacyt, Mexico project 365935U.

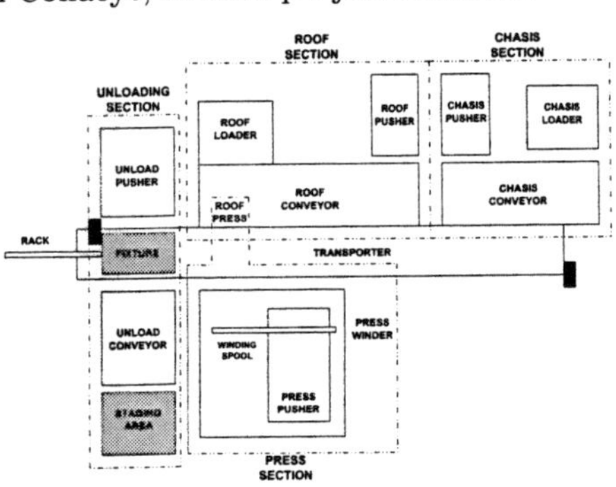

Fig. 6. Assembly line

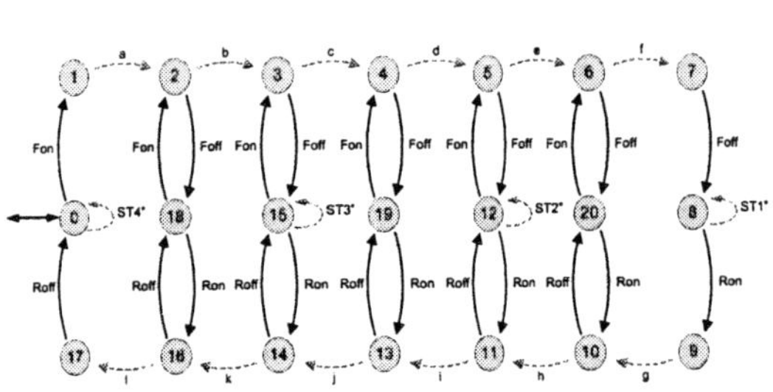

Fig. 7. Transporter FSM model

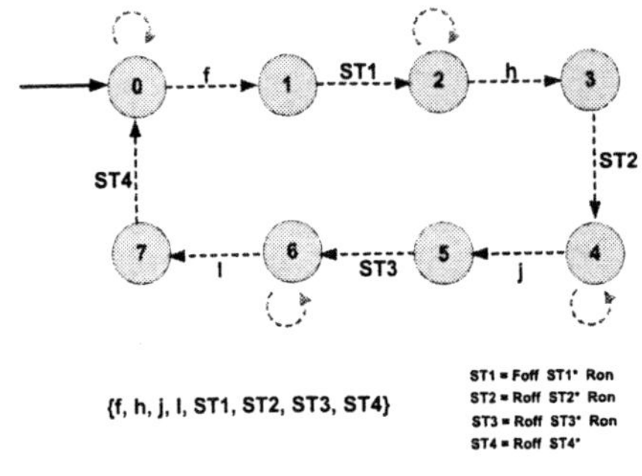

Fig. 8. Behavior specification for transporter

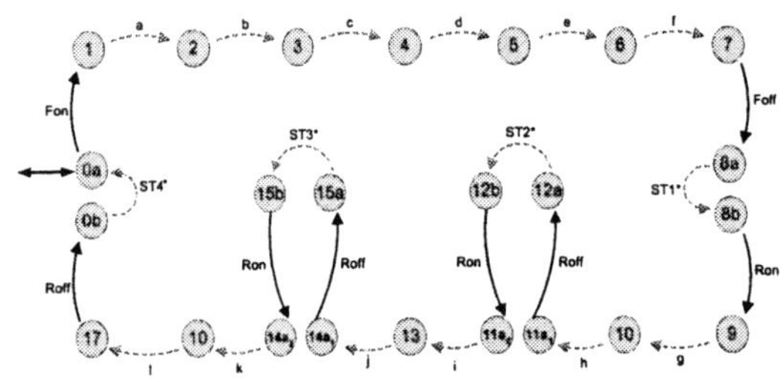

Fig. 9. Synchronization of transporter and specification

REFERENCES

Andersen, H. R. (1997). An introduction to binary decision diagrams. Technical report. Dept. of Inf. Tech. U. of D.

Chandra, V., S. R. Mohanty and R. Kumar (2001). Automated control synthesis for an assembly line using discrete event system control theory. In: *Proc. of American Control Conference*. pp. 4956–4961.

Ramadge, P. J. and W. M. Wonham (1987). Supervisory control of a class of discrete-event processes. *SIAM Journal of Control and Optimization* **25**(1), 206–230.

Reza, J. (2002). Sintesis de controladores de procedimientos utilizando tecnicas de calculo simbolico (in spanish). Master's thesis. Cinvestav.

Sanchez, A., G. E. Rotstein, N. Alsop, J. P. Bromberg, C. Gollain, S. Sorensen, S. Macchietto and C. Jakeman (2002). Improving the development of event-driven control systems in the batch processing industry. A case study. *ISA Trans.* **41**, 343–363.

Sanchez, A., G. Rotstein, N. Alsop and S. Macchietto (1999). Synthesis and implementation of procedural controllers for event-driven operations. *AIChE Journal* **45**(8), 1753–1775.

Sanchez, A., J. Reza, J. Douriet and R. E. Gonzalez (2003). A comparison of synthesis tools for supervisory controllers. In: *Proc. of European Control Conference.*

Zhang, Z. and W. M. Wonham (2002). *Synthesis and control of Discrete-event systems.* Chap. STCT: An efficient algorithm for supervisory control design. Kluwer Academic.

Copyright © IFAC Information Control Problems in
Manufacturing, Salvador, Brazil, 2004

ELSEVIER
IFAC
PUBLICATIONS
www.elsevier.com/locate/ifac

GENERALIZATION OF DES MULTI-MODELING

Oulaid KAMACH, Samir CHAFIK, Laurent PIETRAC and Eric NIEL

Laboratoire d'Automatique Industrielle
Institut National des Sciences Appliquées
Bat St Exupéry 25 av Jean Cappelle-69621 Villeurbanne CEDEX-France
Tel : (33) 4 72 43 62 14
Fax : (33) 4 72 43 85 35
Email: {name}@lai.insa-lyon.fr

Abstract: Abstract: DES multi-modeling appears to be well adapted to management of production system operating modes. Associating a specific model of the process to be controlled and its specifications is in fact natural. However, conceptual problems involving the control aspect may arise, when an admissible distinctive behavior set is specified without considering the ensuing complexity. The aim of this paper is to specify and validate formally operating mode management under generalized conditions. Basically, the paper extends the model commutation problem (process-limited) from one-to-one to one-to-all. Its main results concern the generalized tracking mechanism for a different process behavior combination. *Copyright © 2004 IFAC*

Keywords: multi-models, reactive systems, operating modes, discrete event systems..

1. INTRODUCTION

Today, industrial system efficiency results in not only high productivity rates but also high reactivity performance. This means that, whilst a process is well controlled for a given requirement, an unwanted event must cause it to operate differently (products must continue depending on the system reactivity). When more that one unwanted event is considered, process multiple behavior can be accepted. We assume that the process remains unchanged in an operating mode, but process potentiality (structure and performance characteristics) changes drastically when an exceptional event occurs and this means that the original process has changed. Well structured reactivity will depend firstly on the organization of data emitted from the enterprise level to the execution plant and secondly on control adaptability. Operating mode management offers both an industrial and a scientific challenge in relation to this last point. The main problems encountered in this area are correct specification definition, exhaustive validation and modified process behavior management.

Generally, specification description needs to include nominal and exceptional admissible behaviors. If nominal behaviors are not unique and their definition is laborious (full power operation, downgraded operation, etc.), taking exceptional behaviors into account increases complexity. Reasons for an improperly defined specification set are probably lack of well adapted methodology (even when sectarian methods exist or depend on standards adaptation ISA 88[1]) or insufficient knowledge of legal commutation procedures.

Validation ensures correctness of all predefined requirements at design stage and will establish whether required operating modes are possible, well connected and sufficiently accessible. Thus, validation considers not only internal mode behavior, but also mode commutations, which must establish a set of conditions governing commutation, starting state and recovery state from one mode to another.

Partial contributions to solving this problem have been provided using empirical approaches (GEMMA[2]) but these are limited for small systems. Other contributions offer a more appropriate modeling aspect (Statechart, (HAREL, 1996))

[1] International standard for flexibility in production: www.s88.info
[2] Guide d'Etudes des Modes de Marche et d'Arrêt

involving conciseness, but they remain insufficient for proving properties essential to validation. Automata theory provides a more formal context and its extension will form the basis of our proposal described in the following sections.
Formal commutation has been solved and submitted in terms of model tracking from one mode to another. This paper attempts to generalize the process model commutation problem, when a distinctive operating mode combination is considered. Process model tracking mechanisms are studied especially with respect to starting state recognition, on the first hand, and recovery state recognition on the other hand. Information channels are used to ensure commutation enabling as described in (Lin and Wonham, 1988, Wong et al, 2000). To maintain presentation clarity, two distinct theorems will be introduced; the first associated with starting state search channel, the second with recovery state.
It should be recalled that only distinctive process models are considered.
The paper is structured as follows:
Section 2 presents preliminaries required for understanding the power of supervisory control theory. Section 3 defines the multi-model concept, adapted here solely to distinctive process behavior, and illustrates it using an example. Section 4 presents the information channel for solving the starting state search process and this is formally expressed by theorem 1. Recovery state will be presented in section 5 and is associated with theorem 2.
Finally, this paper provides conclusions and details research prospects.

2. PRELIMINARIES

This section introduces supervisory control theory (SCT) and the problem of considering operating modes.
The original SCT framework is based on distinguishing process and specification models. The process is seen as an uncontrolled Discrete Event System (DES) and is designed by an automaton G. This automaton is an event generator so that $G = (Q, \Sigma, \delta, q_0, Q_m)$ with Q the set of states, Σ the set of event labels, and $\delta : Q\times \Sigma \rightarrow Q$ the partial transition function which is defined at each $q \in Q$ for a subset of events $\sigma \in \Sigma$. q_0 is the initial state while $Q_m \subseteq Q$ is the set of marker states which represent the end of tasks or final states of process. Σ^* contains all possible finite strings over Σ plus the empty string ε. The definition for δ can be extended to a partial function $\delta : Q\times \Sigma^* \rightarrow Q$ such that $\delta(q,\varepsilon) = q$ $(\forall q \in Q)$ and $\delta(q,s\sigma) = \delta(\delta(q,s),\sigma)$ with $\sigma \in \Sigma$ and $s\in \Sigma^*$.
The language generated by G is $(L(G) = \{s\in\Sigma^* \mid \delta(q_0,s)!\})$[3] and its marked language is $L_m(G) = \{s\in\Sigma^* \mid \delta(q_0,s)\in Q_m\}$. $L_m(G)$ can be calculated by Arden's lemma (Wonham, 2002), and this will be used in section 4.
Arden's lemma:

Let A and B, two regular[4] languages.

1- A*B is always a solution of the equation X = AX+B

2- If $\varepsilon \notin A$, then A*B is the unique solution of the equation X = AX+B.

The specification model E is also an automaton, and the controlled DES S/G is obtained by composition of G and E. S/G represents the evolution of the process G restricted by a supervisor S. For further explanation of theory principles, the reader is referred to (Wonham, 2002) or (Cassandras and Lafortune, 1999) (Rudie et al, 1999).
In most cases, the system can be broken down into numerous subsystems. Similarly, process and specification models are the combination of several simple models. Therefore, a current SCT application problem is the explosion in the number of states as the number of components increases. This explosion is often handled by performing horizontal (modular or decentralized) or vertical (hierarchical) break-down of the underlying control problem (Lin and Wonham, 1988, Yoo and Lafortune, 2002, Wong et al, 2000, Chafik and Niel, 2001).
In the other words, production systems must manufacture various productions and react rapidly to failures, if they are to be competitive. Different system use corresponds to different operating modes. Adjustment and maintenance modes are examples of other operating modes that are absolutely necessary for system use. However, a system does not require all components in each operating mode. Furthermore, specifications differ for every operating mode because the objectives of each one are different. Previous approaches are difficult to put into practice on a multi-operating mode system because they consider only one process and because specifications must be in mutual conflict. Based on an example of two operating modes, (Kamach et al, 2002) present a 2-model approach, in which each process model uses different components of the global system and each operating mode corresponds to one model. In next section, we will extend this proposal to the general case of any number of operating modes. In this paper, we restrict ourselves to process models only.

3. DES MULTI-MODELING DESIGN

This section focuses on modeling operating modes by applying a multi-model concept, which involves designing a model process for each operating mode. The problem of commutation between all designed models is formalized by a proposed framework. In this case, commutation is investigated as a channel transmitting information defining the starting state (return state respectively) for each model operating in one specific mode. Commutation will be ensured by

[3] we write $\delta(q,s)!$ as an abbreviation of $\delta(q,s)$ is defined.

[4] A regular expression over Σ is a formal expression obtained by a finite number of applications of operations +, ., *

an information channel formally defined using the notion of projection.

3.1 Example of multi-model process

The aim is to generalize the formalism introduced by (Kamach et al, 2002) to n models (with n > 2). To introduce the proposed approach, we consider a simple manufacturing system, in which four different overall system models are considered: nominal mode is represented by model G_n and there are three downgraded modes G_{d1}, G_{d2} and G_{d3} (figure 3). This system features four machines as shown in figure 1. Initially, buffer B is empty and machines M_3 and M_4 are performing other tasks outside the unit, but which intervene when M_1 (respectively M_2) breaks down (event f_1, respectively f_2 represented by f_i in figure 2). With event b_1 (respectively b_3), M_1 (respectively M_3) takes a workpiece from an infinite bin and enters $q_{1,1}$ or $q_{3,1}$ state of G_n (respectively $q_{2,1}$ or $q_{4,1}$ states). It then deposits it in buffer B after completing its work. M_2 (respectively M_4) operates similarly, but takes its workpiece from B and enters $q_{1,2}$ or $q_{2,2}$ state (respectively $q_{3,2}$ or $q_{4,2}$). It then deposits it in an infinite output bin, when it has finished its task.

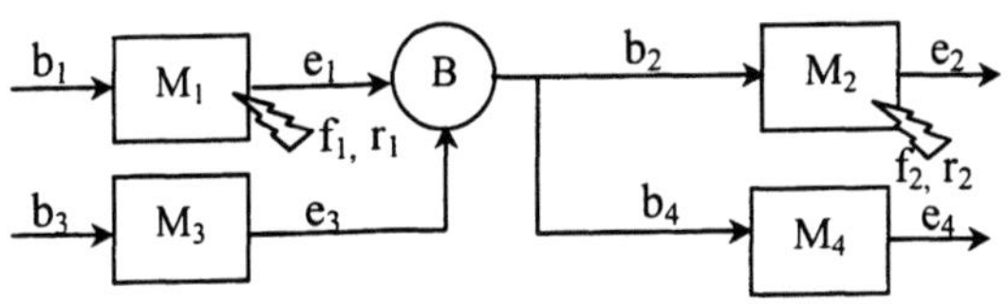

bi : beginning of a task on M_i i ={1,.., 4}
e_i : end of task on M_i : i ={1,.., 4}
f_j : failure of M_j : j ={1, 2}
r_j : repair of M_j : j ={1, 2}

Figure 1.a : schematic description of the production unit example

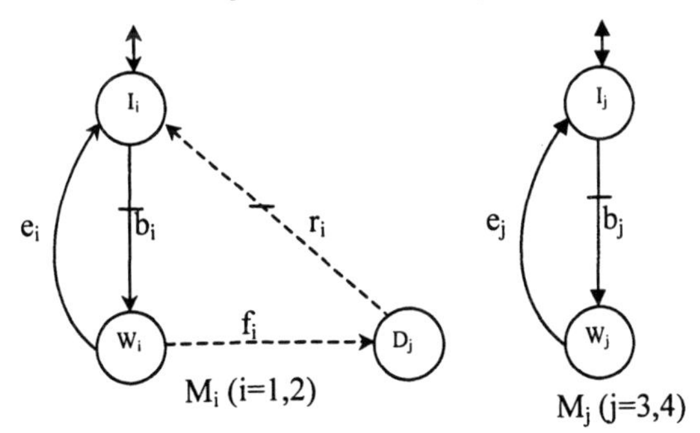

Figure 2 : automata models of machines M_i, M_j

We assume that only M_1 and M_2 can break down and that M_1 (respectively M_2) can not be repaired if M_3 (respectively M_4) is working.
Possible operating modes are represented in figure 3.

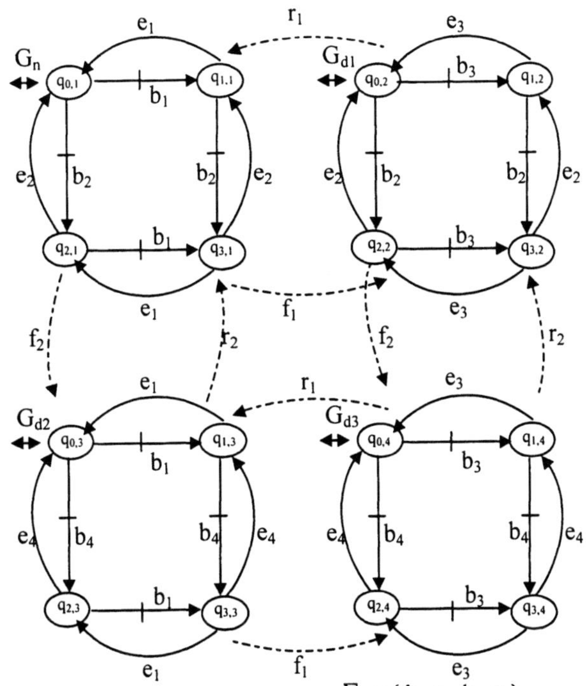

$\Sigma = \Sigma_n \cup \Sigma_{d1} \cup \Sigma_{d2} \cup \Sigma_{d3} \cup \Sigma'$
with $\Sigma' = \{f_1, r_1, f_2, r_2\}$

$\Sigma_n = \{ b_1, e_1, b_2, e_2\}$
$\Sigma_{d1} = \{ b_2, e_2, b_3, e_3\}$
$\Sigma_{d2} = \{ b_1, e_1, b_4, e_4\}$
$\Sigma_{d3} = \{ b_3, e_3, b_4, e_4\}$

Figure.3 : four possible production unit models

3.2 Formal description of multi-model commutation management

The aim is to determine formally each operating mode and the commutation conditions. To do this, we define Λ as a set containing indices of all models composing the global system with card(Λ) = n < ∞. card(Λ) represents the number of models to be designed. In our case, Λ = {n, d1, d2, d3}, so card(Λ) = 4.

Let $\lambda_i \in \Lambda$ with $i \in \{1,..,4\}$. We define $G_{\lambda i}$ as an uncontrollable DES, taken to be an automaton of model λ_i. Formally $G_{\lambda i} = (Q_{\lambda i}, \Sigma_{\lambda i}, \delta_{\lambda i}, q_{0,\lambda i}, Q_{m,\lambda i})$.
We assume that $\Sigma_{\lambda i} \cap \Sigma_{\lambda j} \neq \varnothing$ and initially the system is described by $G_{\lambda 1}$.
Let us define $\Sigma'_{\lambda i,\lambda j} = \cup\{\alpha_{\lambda i,\lambda j}\}$ as the set representing the commutation event from $G_{\lambda i}$ (respectively $G_{\lambda j}$) to $G_{\lambda j}$ (respectively $G_{\lambda i}$). When commutation event $\alpha_{\lambda i,\lambda j}$, occurs, the process model becomes $G_{\lambda j}$. In this case, we must determine the arrival state of $G_{\lambda j}$ after commutation and must direct $G_{\lambda i}$ to an inactive state to disable its action. Intuitively, newly enabled $G_{\lambda j}$ must leave its inactive state and be directed to a state which compatible with the overall system evolution. To do this, we introduce theorem 1 which ensures commutation from $G_{\lambda i}$ to $G_{\lambda j}$ by using the trace or memory of all strings that can occur from $G_{\lambda 1}$ to $G_{\lambda i}$. This memory mechanism is important to ensure overall system tracking. Let us suppose the system is represented by G_n (figure 3). Commutation event f_1 (failure event) is possible from $q_{1,1}$ or $q_{3,1}$. If f_1 is

generated from $q_{1,1}$, then the memorized string occurring in G_n is $(b_1e_1)^*b_3$ and, intuitively, G_{d1} must be directed to $q_{0,2}$. But if f_1 is occurred from $q_{3,1}$ then the memorized string is $(b_1e_1)^*b_1(b_2e_2)^*b_2$ or $(b_2e_2)^*b_2(b_1e_1)^*b_1$. In this case, G_{d1} must be directed to $q_{2,2}$. Theorem 1 formalizes these intuitive results. But before introducing this theorem, two steps are required. Firstly, we introduce an inactive state to any $G_{\lambda i}$ to disable its action. Secondly, we introduce channel information (represented by the projection map) to ensure process tracking. The projection (noted $\pi_{\lambda_i \rightarrow \lambda_j}$ in the remainder of the paper) tells us whether any component belonging simultaneously to $G_{\lambda i}$ and $G_{\lambda j}$ (machine M_2 for G_n and G_{d1}) is working or not, to decide whether to direct $G_{\lambda j}$ to a state in which M_i is working, thereby ensuring system tracking.

a) extension of $G_{\lambda i}$ and $G_{\lambda j}$

Let us extend $G_{\lambda i}$ and $G_{\lambda j}$ by adding an inactive state $q_{in,\lambda i}$ to the state set of the model $G_{\lambda i}$ and an inactive state $q_{in,\lambda j}$ to the $G_{\lambda j}$ state set respectively. Occurrence of commutation event $\alpha_{\lambda i,\lambda j}$ will direct model $G_{\lambda i}$ to its inactive sate $q_{in,\lambda i}$ and activate $G_{\lambda j}$ from $q_{in,\lambda j}$. So, for model $G_{\lambda i}$, the extended model will be defined as follows:

$G_{\lambda i,ext} = (Q_{\lambda i,ext}, \Sigma_{\lambda i,ext}, \delta_{\lambda i,ext}, q_{0,\lambda i,ext}, Q_{m,\lambda i,ext})$ wit h

$Q_{\lambda i,ext} = Q_{\lambda i} \cup \{q_{in,\lambda i}\}$

$\Sigma_{\lambda i,ext} = \Sigma_{\lambda i} \cup \Sigma'_{\lambda i,\lambda j}$

$q_{0,\lambda i,ext} = q_{0,\lambda 1}$ if $\lambda i = \lambda 1$

$q_{0,\lambda i,ext} = q_{in,\lambda_i}$ if $\lambda i \neq \lambda 1$

$Q_{m,\lambda i,ext} = Q_{m,\lambda i}$

$\delta_{\lambda i,ext}$ is defined as follows (Kamach et al., 2002, 2003):

1) ∀∀ θ ∈ Θλι, ανδ ∀ σ Σλι, ιφ δλι(θ, σ)!, τηεν

$\delta_{\lambda i,ext}(q, \sigma) := \delta_{\lambda i}(q, \sigma)$;

2) ∀∀ θ ∈ Θ λι from which $\alpha_{\lambda i,\lambda j}$ can occur (with $i \neq j$ then

$\delta_{\lambda i,ext}(q, \alpha_{\lambda i,\lambda j}) := q_{in,\lambda_i}$.

Similarly, $G_{\lambda j,ext}$ will be defined by the same way that $G_{\lambda i,ext}$.

b) formalization of system tracking

The aim is to define $\delta_{\lambda j,ext}(q_{in,\lambda i}, \alpha_{\lambda j,\lambda i})$. Initially, $G_{\lambda j,ext}$ is in inactive state $q_{in,\lambda j}$. When commutation event $\alpha_{\lambda i,\lambda j}$ occurs, $G_{\lambda j,ext}$ will leave $q_{in,\lambda\varphi}$ to reach a state $q \in Q_{\lambda j}$. As shown in figure 4, when event f_1 occurs, $G_{d1,ext}$, which is in $q_{in,d1}$, can be directed to $q_{0,2}$, $q_{1,2}$, $q_{2,2}$ or $q_{3,2}$. So $G_{d1,ext}$ becomes a nondeterministic automaton. To prevent this nondeterministic situation, we introduce projection map $\pi_{\lambda i \rightarrow \lambda j}$(Kamach et al., 2002) as:

$\pi_{\lambda i \rightarrow \lambda j} : \Sigma_{\lambda i} \rightarrow \Sigma_{\lambda j}$ so that

$$\pi_{\lambda i \rightarrow \lambda j}(\sigma) = \begin{cases} \sigma & \text{if } \sigma \in (\Sigma_{\lambda i} \cap \Sigma_{\lambda j}) \\ \varepsilon & \text{otherwise} \end{cases}$$

We extend $\pi_{\lambda i \rightarrow \lambda j}$ to be defined over a language, so that:

$(\pi_{\lambda i \rightarrow \lambda j})_{ext} : \Sigma^*_{\lambda i} \rightarrow \Sigma^*_{\lambda j}$ such that:

$(\pi_{\lambda i \rightarrow \lambda j})_{ext}(\varepsilon) = \varepsilon$ and

$$(\pi_{\lambda i \rightarrow \lambda j})_{ext}(s\sigma) = \begin{cases} (\pi_{\lambda i \rightarrow \lambda j})_{ext}(s)\sigma & \text{if } \sigma \in (\Sigma_{\lambda i} \cap \Sigma_{\lambda j}) \\ (\pi_{\lambda i \rightarrow \lambda j})_{ext}(s) & \text{otherwise} \end{cases}$$

That is, $(\pi_{\lambda i \rightarrow \lambda j})_{ext}$ is a projection whose effect on a string $s \in \Sigma^*_{\lambda i}$ is to eliminate all events σ of s that do not belong to $(\Sigma_{\lambda i} \cap \Sigma_{\lambda j})$. Projection $(\pi_{\lambda i \rightarrow \lambda j})_{ext}$ allows, from $G_{\lambda j}$, identification of the output states of intersection elements in $G_{\lambda i}$ when $\alpha_{\lambda i,\lambda j}$ occurs. Thus, from $(\pi_{\lambda i \rightarrow \lambda j})_{ext}(s)$, we can determine whether components belonging to $G_{\lambda i}$ and $G_{\lambda j}$ are working or not to direct $G_{\lambda j}$ to a state compatible with a component situation.

Note that in the remainder of this paper we will express $(\pi_{\lambda i \rightarrow \lambda j})_{ext}$ as $\pi_{\lambda i \rightarrow \lambda j}$.

The above demonstration attempts to prove that generalization of commutation resolution assumes a recurrent form. It shows that the information channel, materialized by the projection function, retains only the common components maintained from mode λ_i to λ_j.

4. DETERMINING $G_{\lambda i,ext}$ STARTING STATES

Let us suppose that the set of commutation events produced from $G_{\lambda 1}$ to $G_{\lambda j}$ is $\alpha_{\lambda 1,\lambda l}$, $\alpha_{\lambda l,\lambda k}$,... $\alpha_{\lambda j,\lambda i}$, $\alpha_{\lambda i,\lambda n}$ where $1<l<k<...<j<i<n$, *i.e.* model $G_{\lambda n,ext}$ is now activated. The starting state of this model is determined by transition function $(\delta_{\lambda n,ext}(q_{in,\lambda n}, \alpha_{\lambda i,\lambda n}))$ given by theorem 1.

Theorem 1
∀ n ≥ l, the extended transition function $(\delta_{\lambda n,ext}(q_{in,\lambda n}, \alpha_{\lambda i,\lambda n}))$ *of model* $G_{\lambda n,ext}$ *is given by*

$\delta_{\lambda n,ext}(q_{in,\lambda n}, \alpha_{\lambda i,\lambda n}) =$

$$\delta_{\lambda n}\left[q_{0,\lambda n}, \pi_{\lambda i \rightarrow \lambda n}(\pi_{\lambda j \rightarrow \lambda i}(...(\pi_{\lambda l \rightarrow \lambda k}(\pi_{\lambda 1 \rightarrow \lambda l}(s_1)s_l)..)s_i)\right]$$

□

Theorem 1 permits determination of the next state to be reached in $G_{\lambda n,ext}$ newly enabled by scanning all strings generated from $G_{\lambda 1}$ to $G_{\lambda i}$. To illustrate this theorem, we consider the models shown in figure 4.

The aim is to determine possible starting states of model $G_{d1,ext}$ after generation of failure event f_1 in initial model $G_{n,ext}$. $\alpha_{\lambda 1,\lambda 2} = \alpha_{v,\delta 1} = f_1$ can occur from state $q_{1,1}$ or state $q_{3,1}$ of $G_{n,ext}$ because f_1 failure is possible only when M_1 is working (event b_1).

Case1) First, let us suppose that f_l has occurred from $q_{1,1}$. According to theorem 1, we have

$\delta_{dl,ext}(q_{in,dl}, \alpha_{n,dl}) = \delta_{dl,ext}(q_{in,dl}, f_l) =$

$\delta_{dl}(q_{0,dl}, \pi_{n\rightarrow dl}(s_1))$.

This means we need to determine string s_1. To do this, we introduce language

$$K_{q_{1,1}} = \left\{\omega \in \Sigma_n^* \mid \delta_n(q_{0,1}, \omega) = q_{1,1}\right\}$$

$K_{q_{1,1}}$ is the set of event sequences belonging to G_n, so that f_l is the next event to occur following generation of string ω. $K_{q_{1,1}}$ can be determined from Arden's lemma, as follows:

Let us consider model G_n in figure 3. The aim is to determine ω such that $\delta_n(q_{0,1}, \omega) = q_{1,1}$. To do this, we mark state $q_{1,1}$ of G_n and obtain the following equation system:

$$\begin{cases} q_{0,1} = b_1q_{1,1} + b_2q_{2,1} & (1) \\ q_{1,1} = b_2q_{3,1} + e_1q_{0,1} + \varepsilon & (2) \\ q_{2,1} = b_1q_{3,1} + e_2q_{0,1} & (3) \\ q_{3,1} = e_2q_{1,1} + e_1q_{2,1} & (4) \end{cases}$$

The aim is to determine $q_{0,1}$.

Then (4) $\Rightarrow q_{3,1} = e_1(b_1q_{3,1} + e_2q_{0,1}) + e_2(b_2q_{3,1} + e_1q_{0,1} + \varepsilon)$

$= (e_1b_1 + e_2b_2)q_{3,1} + (e_1e_2 + e_2e_1)q_{0,1} + e_2$ (4')

(4') $\Rightarrow q_{3,1} = A\, q_{3,1} + B$ where

$A = e_1b_1 + e_2b_2$

$B = (e_1e_2 + e_2e_1)q_{0,1} + e_2$.

Since $\varepsilon \notin A$, then (4') allows unique solution A*B, i.e.

$q_{3,1} = (e_1b_1 + e_2b_2)^*[(e_1e_2 + e_2e_1)q_{0,1} + e_2]$

$\Rightarrow q_{3,1} = (e_1b_1 + e_2b_2)^*(e_1e_2 + e_2e_1)q_{0,1} + (e_1b_1 + e_2b_2)^*e_2$ (4'')

Substituting $q_{3,1}$ in (1) and (2) we obtain:

$$\Rightarrow \begin{cases} q_{1,1} = b_2(e_1b_1 + e_2b_2)^*(e_1e_2 + e_2e_1)q_{0,1} + b_2(e_1b_1 + e_2b_2)^*e_2 + e_1\, q_{0,1} + \varepsilon \\ q_{1,2} = b_1(e_1b_1 + e_2b_2)^*(e_1e_2 + e_2e_1)q_{0,1} + b_1(e_1b_1 + e_2b_2)^*e_2 + e_2\, q_{0,1} \end{cases}$$

$$\Rightarrow \begin{cases} q_{1,1} = [b_2(e_1b_1 + e_2b_2)^*(e_1e_2 + e_2e_1) + e_1]q_{0,1} + b_2(e_1b_1 + e_2b_2)^*e_2 + \varepsilon \quad (1') \\ q_{1,2} = [b_1(e_1b_1 + e_2b_2)^*(e_1e_2 + e_2e_1) + e_2]q_{0,1} + b_1(e_1b_1 + e_2b_2)^*e_2 \quad (2') \end{cases}$$

Replacing (1') and (2') in equation (1), we obtain:

$q_{0,1} = [b_1b_2(e_1b_1 + e_2b_2)^*(e_1e_2 + e_2e_1) + b_1e_1]q_{0,1} + b_1b_2(e_1b_1 + e_2b_2)^*e_2 + b_1 + [b_2b_1(e_1b_1 + e_2b_2)^*(e_1e_2 + e_2e_1) + b_2e_2]q_{0,1} + b_2b_1(e_1b_1 + e_2b_2)^*e_2$

$\Rightarrow q_{0,1} = [(b_1b_2 + b_2b_1)(e_1b_1 + e_2b_2)^*(e_1e_2 + e_2e_1) + b_1e_1 + b_2e_2]q_{0,1} + (b_1b_2 + b_2b_1)(e_1b_1 + e_2b_2)^*e_2 + b_1$. (1'')

(1'') $\Leftrightarrow q_{0,1} = Aq_{0,1} + B$.

According to Arden's lemma, $\varepsilon \notin [(b_1b_2 + b_2b_1)(e_1b_1 + e_2b_2)^*(e_1e_2 + e_2e_1) + b_1e_1 + b_2e_2]$, then (1'') allows unique solution $A^*B = K_{q_{1,1}}$ with

$K_{q_{1,1}} = [(b_1b_2 + b_2b_1)(e_1b_1 + e_2b_2)^*(e_1e_2 + e_2e_1) + b_1e_1 + b_2e_2]^*[(b_1b_2 + b_2b_1)(e_1b_1 + e_2b_2)^*e_2 + b_1]$.

We can now determine the starting state of $G_{dl,ext}$ by applying theorem 1. In fact,

$\pi_{n\rightarrow dl}(K_{q_{1,1}}) = [b_2(e_2b_2)^*e_2 + b_2e_2]^*[b_2(e_2b_2)^*e_2] = (b_2e_2)^*$

because $b_2(e_2b_2)^*e_2 = (b_2e_2)^*$, $(b_2e_2)^* + b_2e_2 = (b_2e_2)^*$, then $[b_2(e_2b_2)^*e_2 + b_2e_2]^* = ((b_2e_2)^*)^* = (b_2e_2)^*$, because in regular algebra (Wonham, 2001) $(L^*)^* = L^*$.

i.e. $[b_2(e_2b_2)^*e_2] = (b_2e_2)^*$. Since $L^*L^* = L^*$, then $[b_2(e_2b_2)^*e_2 + b_2e_2]^*[b_2(e_2b_2)^*e_2] = (b_2e_2)^*$

So $\delta_{dl,ext}(q_{in,dl}, f_l) = \delta_{dl}(q_{0,dl}, \pi_{n\rightarrow dl}(K_{q_{1,1}})) =$

$\delta_{dl}(q_{0,dl}, (b_2e_2)^*) = q_{0,2}$ (figure 4).

Theorem 1 therefore confirms the intuitive results referred to in section 3.2.

Case2) Now if f_l occurs from $q_{3,1}$ of $G_{n,ext}$, . Arden's lemma gives us $K_{q_{3,1}}$ so that

$$K_{q_{3,1}} = \left\{\omega \in \Sigma_n^* \mid \delta_n(q_{0,1}, \omega) = q_{3,1}\right\}$$

Marking the state $q_{3,1}$ of $G_{n,ext}$, the above equation system becomes:

$$\begin{cases} q_{0,1} = b_1q_{1,1} + b_2q_{2,1} \\ q_{1,1} = b_2q_{3,1} + e_1q_{0,1} \\ q_{2,1} = b_1q_{3,1} + e_2q_{0,1} \\ q_{3,1} = e_2q_{1,1} + e_1q_{2,1} + \varepsilon \end{cases}$$

As previously, we obtain:

$K_{q_{3,1}} = [(b_1b_2 + b_2b_1)(e_1b_1 + e_2b_2)^*(e_1e_2 + e_2e_1) + b_1e_1 + b_2e_2]^*[(b_1b_2 + b_2b_1)(e_1b_1 + e_2b_2)^*]$.

So $\pi_{n\rightarrow dl}(K_{q_{3,1}}) = (b_2e_2)^*b_2$

Therefore

$\delta_{dl}(q_{0,dl}, \pi_{n\rightarrow dl}(K_{q_{3,1}})) = \delta_{dl}(q_{0,dl}, (b_2e_2)^*b_2)$

$= q_{2,2}$ of $G_{dl,ext}$ (figure 4).

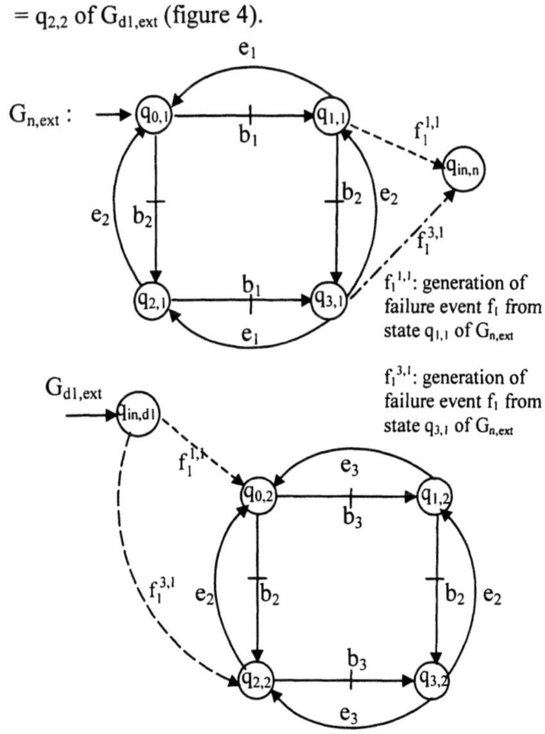

Figure 4 : 2 possible cases of G_{dl} extended model after generating f_l.

With a view to generalizing commutation, suppose the system is represented by $G_{d1,ext}$ and that failure event f_2 has occurred (figure 3). Machine 4 will then replace machine 2. Production will be ensured by machines 3 and 4 and the system will be represented by model $G_{d3,ext}$. In this case, we must determine possible starting states of model $G_{d3,ext}$, knowing that the system is now represented by model $G_{d1,ext}$. From $G_{d1,ext}$, two cases are possible:

Case 3) starting state of $G_{d1,ext}$ is $q_{0,2}$ (figure 4)
In $G_{d1,ext}$, failure event f_2 can occur from states $q_{2,2}$ or $q_{3,2}$. In this case, we compute the language $K_{q_{2,2}}$ ($K_{q_{3,2}}$ respectively). So that

$$K_{q_{2,2}} = \left\{\omega \in \Sigma_{d1}^{*} \mid \delta_{d1}(q_{0,2},\omega) = q_{2,2}\right\} \text{ and}$$

$$K_{q_{3,2}} = \left\{\omega \in \Sigma_{d1}^{*} / \delta_{d1}(q_{0,2},\omega) = q_{3,2}\right\}$$

1) Suppose f_2 is generated from $q_{2,2}$ of $G_{d1,ext}$ ($f_2^{2,2}$ in figure 5.a). In this case, we mark state $q_{2,2}$ and, as above, we obtain

$$\pi_{d1\to d3}(K_{q_{2,2}}) = (b_3e_3)^*.$$

Thus, if the starting state of $G_{d1,ext}$ is $q_{0,2}$ and f_2 has occurred from $q_{2,2}$, then from theorem 1 we obtain:

$$\delta_{d3}(q_{0,4}, \pi_{d1\to d3}(\pi_{n\to d1}(K_{q_{1,1}})K_{q_{2,2}}) = \delta_{d3}(q_{0,4}, (b_3e_3)^*)$$

$=q_{0,4}$ (figure 5.b). The starting state of $G_{d3,ext}$ will then be $q_{0,4}$.

2) f_2 is now generated from $q_{3,2}$ of $G_{d1,ext}$ ($f_2^{3,2}$ of figure 5.a).
We mark state $q_{3,2}$, then

$$\pi_{d1\to d3}(K_{q_{3,2}}) = b_3(e_3b_3)^* = (b_3e_3)^*b_3$$

Therefore $\delta_{d3}(q_{0,4}, \pi_{d1\to d3}(\pi_{n\to d1}(K_{q_{1,1}})K_{q_{3,2}})$

$=\delta_{d3}(q_{0,4}, (b_3e_3)^*b_3) = q_{1,4}$. i.e. if the starting state of $G_{d1,ext}$ is $q_{0,2}$ and if f_2 has occurred from $q_{3,2}$, , then based on theorem 1, $G_{d3,ext}$ will be directed to state $q_{1,4}$ (figure 5.b).

Case 4) Suppose now that the starting state of $G_{d1,ext}$ is $q_{2,2}$ (figure 4), thus

$$K_{q_{2,2}} = \left\{\omega \in \Sigma_{d1}^{*} \mid \delta_{d1}(q_{2,2},\omega) = q_{2,2}\right\} \text{ and}$$

$$K_{q_{3,2}} = \left\{\omega \in \Sigma_{d1}^{*} \mid \delta_{d1}(q_{2,2},\omega) = q_{3,2}\right\}$$

1) Suppose first that f_2 is generated from $q_{2,2}$ of $G_{d1,ext}$ ($f_2^{2,2}$ of figure 5.c).
We mark state $q_{2,2}$, and so $\pi_{d1\to d3}(K_{q_{2,2}}) = (b_3e_3)^*$.
Therefore, based on theorem 1,

$$\delta_{d3,ext}(q_{in,d3}, f_2) = \delta_{d3}(q_{0,4}, (b_3e_3)^*) = q_{0,4}.$$

Thus, from $G_{\lambda1,ext}$ and after $f_1^{1,3}$ and $f_2^{2,2}$, have occurred, $G_{d3,ext}$ will be directed to state $q_{0,4}$ (figure 5.d).

2) If f_2 is now generated from $q_{3,2}$ of $G_{d1,ext}$ ($f_2^{3,2}$ in figure 5.c)
As above, we mark state $q_{3,2}$.
Then $\delta_{d3,ext}(q_{in,d3}, f_2) = \delta_{d3}(q_{0,4}, b_3(b_3e_3)^*) = q_{1,4}$

Thus, after $f_1^{1,3}$ and $f_2^{3,2}$ have occurred, $G_{d3,ext}$ will be directed to state $q_{1,4}$ (figure 5.d).
At this stage we can verify theorem 1. In fact

$$\delta_{\lambda_{4,ext}}(q_{in,d3}, f_2^{3,2}) = \delta_{d3}(q_{0,4}, \pi_{d1\to d3}(\pi_{n\to d1}(K_{q_{3,1}})K_{q_{3,2}})))$$
$$= q_{1,4}.$$

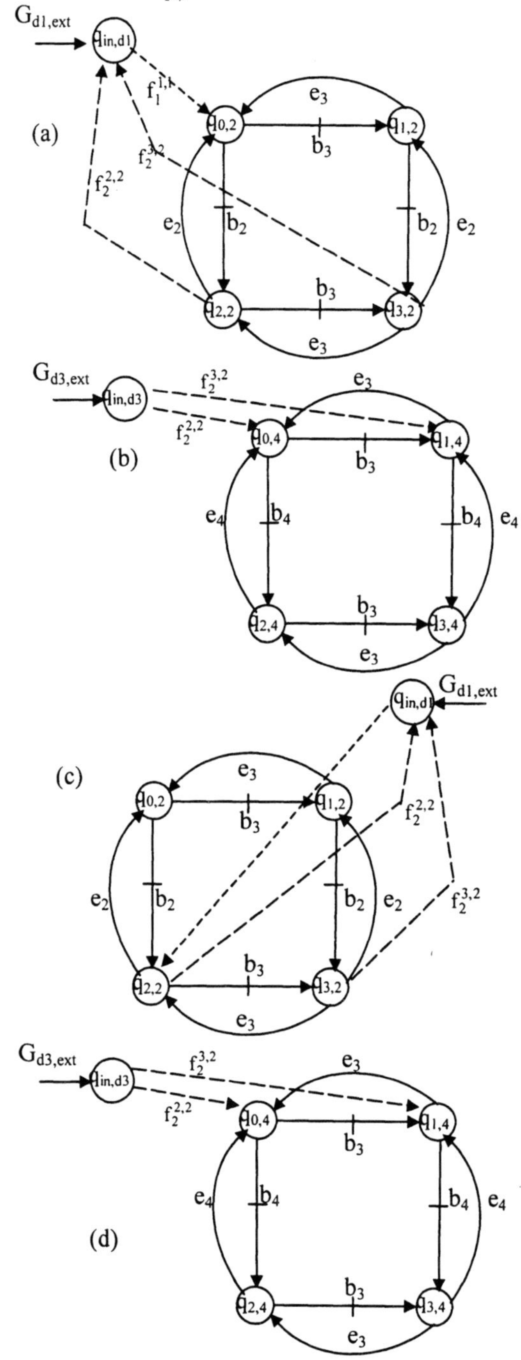

Figure 5 : possible cases of G_{d3} extended model after generating f_1 and f_2.

5. DETERMINING $G_{\lambda i,ext}$ RECOVERY STATES

We assume that after generating commutation events $\alpha_{\lambda 1,\lambda j}$, $\alpha_{\lambda j,\lambda k}$, ..., $\alpha_{\lambda j,\lambda i}$, $\alpha_{\lambda i,\lambda n}$, event $\alpha_{\lambda n,\lambda i}$ (in our case, a repair event) can occur. In this case, $G_{\lambda n,ext}$ will be directed to its inactive state $q_{in,n}$ and $G_{\lambda i,ext}$ will be simultaneously activated, leaving its inactive state $q_{in,i}$ to enter recovery state $q \in G_{\lambda i}$. To do this, the entire system evolution history must be known i.e. all strings generated from $G_{\lambda 1}$ to $G_{\lambda i}$ and all strings that will occur from $G_{\lambda n}$ to $G_{\lambda l}$, which will be reached a second time, must be memorized. Theorem 2 is introduced to ensure this mechanism.

Theorem 2:
$\forall n \geq l$ the extended transition function $\delta_{\lambda\, i,ext}(q_{in,\lambda i}, \alpha_{\lambda n,\lambda i})$ of model $G_{\lambda i,ext}$ is given by:

$$\delta_{\lambda i,ext}(q_{in,\lambda i}, \alpha_{\lambda n,\lambda i}) =$$

$$\delta_{\lambda i,ext}(q_{0,\lambda_i}, \pi_{\lambda n \to \lambda i}(\pi_{\lambda i \to \lambda n}(\pi_{\lambda j \to \lambda i}(\ldots(\pi_{\lambda l \to \lambda k}$$
$$(\pi_{\lambda l \to \lambda 1}(s_1)s_l)\ldots)s_i)s_n)))$$

Let consider the unit production example and suppose that after generation of f_1 and f_2 failure events, unit production is represented by model $G_{d3,ext}$. At this level, repair event r_1 (r_2 respectively) can occur. If this is the case, the system will then be directed to $G_{d2,ext}$ ($G_{d1,ext}$ respectively). On the other hand, when describing our unit production, we have assumed that M_1 can not be repaired if M_3 is working. Consequently in $G_{d3,ext}$, repair event r_1 can occur only from states $q_{0,4}$ or $q_{2,4}$.
However, in cases 3 and 4 described in the last section, we demonstrated that after generating f_1 and f_2, $G_{d3,ext}$ is activated and directed to $q_{0,4}$ or $q_{1,4}$ states (figure 5). Two cases can therefore be distinguished.

Case 5) starting state of $G_{d3,ext}$ is $q_{0,4}$ ($f_2^{2,2}$ in figure 6.a).
r_1 can then occur from $q_{0,4}$ or $q_{2,4}$

1) Suppose that r_1 has occurred from $q_{0,4}$,
$K_{q_{0,4}} = \left\{\omega \in \Sigma_{d3}^* \mid \delta_{d3}(q_{0,4}, \omega) = q_{0,4}\right\}$, we mark $q_{0,4}$ and then
$K_{q_{0,4}} = [(b_3b_4 + b_4b_3)(e_4b_4 + e_3b_3)^*(e_3e_4 + e_4e_3) + b_3e_3 + b_4e_4]^*$
$\Rightarrow \pi_{d3 \to d2}(K_{q_{0,4}}) = (b_4e_4)^*$
So $\delta_{d2,ext}(q_{in,d2}, r_1) =$
So $\delta_{d2,ext}(q_{in,d2}, r_1) =$
$\delta_{d2}(q_{0,3}, \pi_{d3 \to d2}(\pi_{d1 \to d3}(\pi_{n \to d1}(K_{q_{1,1}})K_{q_{2,2}})K_{q_{0,4}}) = q_{0,3}$
(figure 6.b).

2) if r_1 has occurred from $q_{2,4}$, then
$K_{q_{2,4}} = \left\{\omega \in \Sigma_{d3}^* \mid \delta_{d3}(q_{0,4}, \omega) = q_{2,4}\right\}$, by marking $q_{2,4}$ we obtain $\pi_{d3 \to d2}(K_{q_{2,4}}) = (b_4e_4)^*\, b_4$
So $\delta_{d2,ext}(q_{in,d2}, r_1) =$
$\delta_{d2}(q_{0,3}, \pi_{d3 \to d2}(\pi_{d1 \to d3}(\pi_{n \to d1}(K_{q_{1,1}})K_{q_{2,2}})K_{q_{2,4}}) = q_{2,3}$
(figure 6.b)

Case 6) if the starting state of $G_{d3,ext}$ is $q_{1,4}$ ($f_2^{3,2}$ of figure 6.c), applying theorem 2 provides us with the same results as those of case 5.

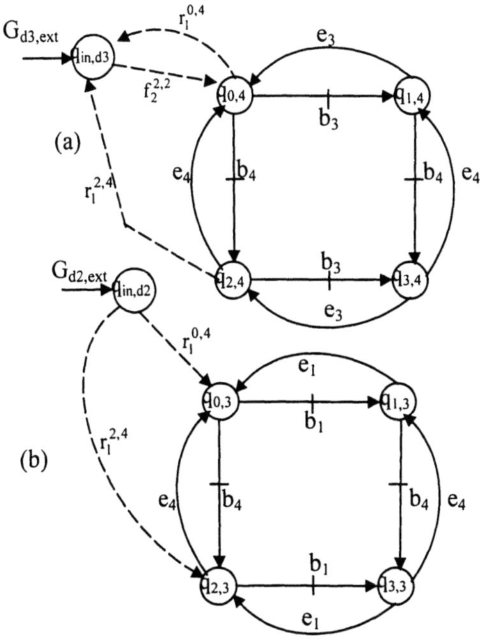

Figure 6 : possible cases of G_{d2} extended models after generating f_1, f_2 and r_1

6. CONCLUSION

We conclude that the proposed method ensures commutation between different models of an overall system reacting to exceptional situations, such as failure event occurrence A major contribution of this paper is that it considers reactive systems with different objectives. Each objective (or operating mode) is represented by a process model. If we assume that different models develop independently, the main problem is then to de-activate model $G_{\lambda i}$ and to commute to model $G_{\lambda j}$, which will be considered as the current process model until an exceptional event occurs. A formal framework based on tracking events is proposed in to ensure commutation. This framework extends for the generalization case the projection definition. Theorems 1 and 2 represent the main contribution of this paper; they allow us to determine with a recurrent form the arrival state of a model after any commutations. Results presented in (Kamach et al, 2003) allow us to obtain controller synthesis for each operating mode.

REFERENCES

Cassandras, B.A. and Lafortune, S (1999). Introduction to Discrete Event Systems. *1st edition: Kluwer Academic Publishers, Boston,* 822 p.

Chafik, S. and E. Niel (2001) Du Contrôle par Supervision Hiérarchique Distribuée. *MSR: modélisation des systèmes réactifs*, pp. 55-70. Toulouse, France,

Harel, D. and A. Naamadl (1996) Statemate semantics of statecharts. *ACM transitions of software engineering and methodology*, vol. 5, no.

4. pp. 293-333.
Kamach, O., S. Chafik., L. Pietrac and E. Niel (2002). representation of a reactive systems with different models. *IEEE SMC*. reference TA2L4 in CDROM, 6-9. Hammamate, Tunisia
Kamach, O., L. Pietrac and E. Niel (2003). multi-model approach for discrete event systems: application to operating modes management. I*EEE CESA*. Reference S2-R-00-0315 in CD ROM. école centrale de Lille, France.
Lin, F. and Wonham, W. M (1988). Decentralised supervisory control of discrete event systems. *Information sciences*, vol. 25, no.5, pp. 1202-1218.
Ramadge, P and Wonham, W. M (1987). Supervisory control of class of discrete event processes. *SIAM Journal of Control and optimisation*, Vol. 25, no.1, pp. 206-230.
Rudie, K. S, Lafortune, and F.Lin (1999). Minimal communication in a distributed Descrete-Event Control System. *In proceedings of the american control conferenc* , pp. 1965-1970.,
Wong, K.C., J.G. Thiste., R.P. Malhame. and H.H. Hoang (2000). supervisory control of distributed systems: conflict resolution. Discrete event dynamic systems, vol. 10, pp. 131-186
Wonham, W.M. (2002). notes on control of Discrete event systems, Dep of electrical and computer engineering, University of Toronto, Canada. URL: http://odin.control.toronto.edu/DES/
Yoo, T. and Lafortune, S (2002). New Results: a new architecture with a dynamic decision fusion rules. *6th international workshop on discrete event systems,* pp. 11-17. Saragossa, Spain,

Copyright © IFAC Information Control Problems in
Manufacturing, Salvador, Brazil, 2004

ELSEVIER
IFAC
PUBLICATIONS
www.elsevier.com/locate/ifac

CONTROLLABILITY REVISITED: A GENERALIZATION FOR THE MODULAR APPROACH

H. Flordal* M. Fabian* K. Åkesson* A. Hellgren*

* *Department of Signals and Systems,*
Chalmers University of Technology, SE-412 96
Gothenburg, Sweden
{hugo,fabian,ka,andh}@s2.chalmers.se

Abstract: In the Supervisory Control Theory, originally presented by Ramadge and Wonham, Ramadge and Wonham (1989), an important property is *controllability*. The definition of controllability as originally presented assumed equal event sets, alphabets, in the specification and the plant. When this assumption does not hold, it has been argued that the differences in the alphabets can easily be removed. However, no matter how we choose to handle the differences, pursuing verification with the original definition will lead to false conjectures in some cases. A few examples on this is presented in this paper. In order to overcome the problem, a generalized controllability definition, allowing non-equal alphabets, is presented. It is of great importance to be able to verify controllability of subsystems with different alphabets when exploiting the modularity of discrete event systems for computationally efficient verification and synthesis. *Copyright © 2004 IFAC*

Keywords: Discrete Event Systems, Modular Supervisory Control Theory, Controllability, Finite Automata, Formal Methods

1. INTRODUCTION

Discrete Event Systems (DESs) are a useful modeling abstraction for certain, mainly man-made, systems, such as manufacturing systems and communication networks. The main characteristic of such systems is that at each time instant they occupy a discrete symbolic-valued state, and perform state-changes on the occurrence of events. Events occur asynchronously and instantaneously at discrete intervals of time. Thus, a DESs behavior is described by the sequences of events that occur, and the sequences of states visited consequently. Typically, this behavior can be described by regular languages and/or finite state automata.

The *Supervisory Control Theory* (SCT) Ramadge and Wonham (1989) is a general approach to the synthesis of control systems for DESs. Given a *plant*, a DES describing the uncontrolled behavior, and a *specification* for the controlled behavior, a *supervisor* can be automatically synthesized to control the plant to stay within the specification. The supervisor controls the plant by dynamically disabling events that the plant otherwise might have generated. However, a subset of the plant events are not subject to supervisor disablement, they are *uncontrollable*, and the supervisor must be such that it never tries to disable an uncontrollable event that is enabled in the plant; the supervisor must be *controllable* with respect to the plant. It is known that a unique controllable and maximally permissive (in terms of event disablement) supervisor does exist for any given plant and specification. The synthesis task concerns finding this supervisor. Given a candidate supervisor, it is also of importance to verify whether this candiate is controllable or not.

Industrially, the SCT has not yet been accepted on a large scale. One reason for this is that industrially interesting cases typically encompass enormous state-spaces. On the other hand, models of industrial systems are typically also *modular* in the sense that they can be described as sets of interacting subsystems. Modular approaches to supervisor verification and synthesis can beneficially exploit the modularity to circumvent the state-space explosion. Approaches showing promising results include among others Queiroz and Cury (2000), Brandin *et al.* (2000), Åkesson *et al.* (2002), Leduc (2002) and Flordal (2001). In these articles, there are a couple of different approaches to the problems of unequal alphabets. In Åkesson *et al.* (2002), the problem is solved by using a generalized definition of controllability. In this paper, this new definition is further examined and its importance stressed.

In its original formulation, the SCT assumes that the plant and the supervisor have the same alphabet. In modular systems, though, this is rarely the case. It has been argued that the modules in such systems can always be treated as alphabetically equal, for example by adding the missing events as self-loops. However, in the modular case this assumption is sometimes contra-productive, as this paper aims to show.

Mainly these problems appear in modular synthesis and verification approaches where different alphabets appear naturally. The problems concern false conjectures that the original definition of controllability Ramadge and Wonham (1989) result in when the assumption of equal alphabets does not hold.

This paper is organized as follows. Section 2 give the notational conventions used. Section 3 then present the original controllability definition together with an alternative definition also found in the literature. In section 4 the revised definition of controllability is presented, a definition that allows the alphabets to be non-equal. In section 5, a number of examples where the false conjectures arise are presented. In section 6, the results are discussed and finally, section 7 presents the conclusions.

2. PRELIMINARIES

This section presents the conventions used in this paper. The universe of discourse is deterministic finite automata and regular languages.

- A deterministic finite automaton is a 4-tuple $A = \langle Q^A, \Sigma^A, \delta^A, q_0^A \rangle$ where Q^A is the set of states, Σ^A the set of events, δ^A the transition function and q_0^A the initial state of A.
- Σ is the universal alphabet of events. Σ_u and Σ_c are the sets of uncontrollable and controllable events, respectively. Also, $\Sigma = \Sigma_u \cup \Sigma_c$. In figures, an exclamation mark (!) before an event's name indicates that the event is uncontrollable.
- Σ^* is the set of all finite sequences of events over the alphabet Σ including the empty sequence, ε.
- For an automaton A, $L(A)$ is the set of all event sequences possible from the initial state of A, that is, $L(A)$ denotes the (prefix closed) language of A. The construct $L(A)\Sigma_u$ represents the concatenation of all traces of $L(A)$ with all events in Σ_u.
- $Proj_{\Sigma^A}$ is the natural projection onto the alphabet Σ^A, $Proj_{\Sigma^A} : \Sigma^* \to (\Sigma^A)^*$; the projection removes all occurences of events in the set $\Sigma \setminus \Sigma^A$. $L_{\Sigma^A}(B)$ will be used as an abbreviation of $Proj_{\Sigma^A}(L(B))$.
- $Proj^{-1}_{\Sigma^A}$ is the inverse natural projection from the alphabet Σ^A onto the alphabet Σ, $Proj^{-1}_{\Sigma^A} : (\Sigma^A)^* \to \Sigma^*$. In the automaton representation of a language, the inverse projection adds selfloops in all states for the events in the set $\Sigma \setminus \Sigma^A$. $L^{-1}_{\Sigma^A}(A)$, or simply $L^{-1}(A)$, will be used as an abbreviation of $Proj^{-1}_{\Sigma^A}(L(A))$. Also, $s^{-1}_{\Sigma} = Proj^{-1}_{\Sigma}(s)$.
- The *synchronous composition* of the automata A and B will be denoted $A||B$, see for example Åkesson *et al.* (2002). This composition operator models that mutual events occur in the composition if and only if they can occur simultaneously in both automata. In general, $L(A||B) = L^{-1}_{\Sigma^A}(A) \cap L^{-1}_{\Sigma^B}(B)$.

3. CONTROLLABILITY

A supervisor controls a plant by dynamically disabling events, as in the Ramadge and Wonham framework, that the plant otherwise might generate. However, the supervisor must be such that it never tries to disable an *uncontrollable* event that is enabled in the plant; the supervisor must be *controllable* with respect to the plant.

Controllability can be used to express safety properties. If safety critical events are modelled as uncontrollable, a controllability test verifies wether they are confined to occur only in states where the control function is prepared, i.e. does not disable the event.

3.1 Controllability definition

Definition 1. If $\Sigma^S = \Sigma^P$, S is *controllable* with respect to P if

$$L(S)\Sigma_u \cap L(P) \subseteq L(S). \qquad \blacksquare$$

Alternatively, using synchronous composition to model the closed-loop system, $S||P$, the following definition can be used.

Definition 2. If $\Sigma^S \subseteq \Sigma^P$, S is *controllable* with respect to P if

$$L(S||P)\Sigma_u \cap L(P) \subseteq L(S||P). \quad \blacksquare$$

Both definitions exist in the literature, and when the alphabets are equal the definitions are equivalent, a simple proof that is left out here.

4. REVISED DEFINITION

We propose a new definition of controllability, suitable for modular approaches. In the next section, motive for using this definition will be given by a couple of examples where intuition suggests that definition 1 is wrong.

Definition 3. S is *controllable* with respect to P if

$$L_{\Sigma^S}^{-1}(S)\Sigma_u^P \cap L_{\Sigma^P}^{-1}(P) \subseteq L_{\Sigma^S}^{-1}(S),$$

where $\Sigma_u^P = \Sigma_u \cap \Sigma^P$. ■

This definition can also be found as definition 5 in Åkesson *et al.* (2002). In Heymann (1990), controllability is defined using the prioritized synchronous composition operator, defined in the same article. This produced a very compact definition - S is controllable with respect to P if $P||S = P_\Sigma||_{\Sigma_c}S$ - this also implied equal languages, but the definition can be modified slightly to match definition 3, by using $P||S = P_{\Sigma^P}||_{\Sigma_c^S}S$ instead.

Note that, trivially, definition 3 is equivalent to definition 1 when $\Sigma^S = \Sigma^P$.

5. DIFFERENT ALPHABETS

Focus will be put on problems associated with the use of definition 1. Using definition 2 solves some, but not all of these problems, as explained in section 6.1.

Definition 1 assumes the alphabets of S and P to be equal. This, however, cannot generally be expected. Discrete event systems are often built up of modules, representing different aspects of a process or a control specification. These modules potentially have very different alphabets. The modules are, of course, intended to interact but verification and synthesis can actually be performed modularly. Requiring that the supervisor and the plant share *all* events would prohibit this optimization. To exploit the modularity efficiently, it is important to have a meaningful definition of controllability for supervisors and plants with different alphabets.

Even if the system is not modular, it is unnecessary to demand a supervisor to contain all events of the plant it is supposed to control. Some events may be irrelevant for control reasons. For example, in a sequence of events without possible branches, it is sometimes enough to consider the initial and final event of the sequence in the supervisor, i.e. the supervisor can be working at a higher level of abstraction.

5.1 Modular verifiction

In modular verification, the general setting is a plant composed of subplants, $P = P_1||...||P_m$ and a specification composed of subspecifications, $S = S_1||...||S_n$, with no general assumptions on the subautomatas alphabets. In general, it can not even be expected that $\Sigma^P = \Sigma^S$. In Åkesson *et al.* (2002), it is shown that if each subspecification S_i is controllable with respect to the plants with which it shares uncontrollable events, then S is controllable with respect to P. Thus, in the best case, there is only need to compose each subspec with a subset of the plant automata and never compose the whole state space. It is also shown that even smaller sets of automata can be composed since it is enough to consider one uncontrollable event at a time and even one plant at a time in a similar manner. This means that it will be necessary to verify controllability of specifications and plants where the alphabets can be very different and not much can be assumed.

In general, it is important to be able to decide whether subautomata S_i and P_j will cause any problems without having to build a model (by synchronous composition) of the entire system. Due to state space explosion, representing such a model may be impossible.

To use definition 1, the alphabets must be adjusted somehow so that the problem fits the definition. When doing this, care must be taken not to introduce or hide uncontrollability in the system. To guarantee that there will be no false conjectures, definition 3 is necessary.

Modular verification much relies on the implicit relations, the modular structure between automata, induced by the way events are shared or not shared. By naively equating the alphabets in one way or another, the modular structure is destroyed. The problem can be said to arise in the relation between languages and automata. When discussing languages, equal alphabets is very important whereas in an automaton approach, it is not. This can be illustrated by the compactness of the "automata based" controllability definition in section 4.

For more on modular verification, see Queiroz and Cury (2000), Brandin *et al.* (2000), Leduc (2002), Flordal (2001) and Åkesson *et al.* (2002). On modular synthesis, see Flordal (2001) and Åkesson *et al.* (2002).

5.2 Adjusting the Alphabets

When the alphabets are not equal, they need to be adjusted in order to fit definition 1. In general, we have two languages, $L^A \subseteq (\Sigma^A)^*$ and $L^B \subseteq (\Sigma^B)^*$ where $\Sigma^A \subset \Sigma^B$.

There are basically three ways to make non-equal alphabets equal; disablement, restriction and expansion. They will be presented in greater detail in sections 5.2.1, 5.2.2 and 5.2.3, respectively.

Note that it might seem that the case of incomparable alphabets has been left out. But without loss of generality, this case needs no special treatment since the problems that arise are the same. To make the alphabets equal, a combination of the above strategies must be used and uncontrollabilities may appear or disappear in the same ways.

5.2.1. Disablement Perhaps the simplest way to get equal alphabets is to simply set $\Sigma^A = \Sigma^B$ without changing the language L^A. This effectively disables all the events in $\Sigma^B \setminus \Sigma^A$.

5.2.2. Restriction Another way to get equal alphabets is to restrict L^B to the alphabet of L^A, making $(\Sigma^B)' = \Sigma^A$. The restriction of L^B is

$$(L^B)' = Proj_{\Sigma^A}(L^B). \tag{1}$$

Note that this operation may cause an automaton to turn non-deterministic but as long as languages are considered, this can be ignored. In practice, it would be necessary to make the automaton deterministic, which is a potentially costly operation.

5.2.3. Expansion The third way to get equal alphabets, is to expand L^A to contain all events in the alphabet of L^B, making $(\Sigma^A)' = \Sigma^B$. The expansion of L^A is

$$(L^A)' = Proj^{-1}_{\Sigma^A}(L^A). \tag{2}$$

Applied to the automaton representing L^A, this operation adds selfloops in all states for the events in the set $\Sigma^B \setminus \Sigma^A$.

5.3 Problem Cases

None of the above strategies are universal solutions. They may work in some cases but not always, as will be shown by the following very minimalistic examples.

Each example starts with a discussion on whether the example reasonably should be considered controllable or uncontrollable, without actually applying any formal controllability definitions. This aims to produce an understanding of what the correct answer should be. After this, definition 1 is applied and the result is discussed.

Note that in all the following examples, in figures 1 to 5, the alphabets of the automata consist of exactly the events that are explicitly represented in each automaton. Uncontrollable events have exclamation marks in front of them.

5.3.1. Disablement of $\Sigma^P \setminus \Sigma^S$ In figure 1, the plant executes a controllable event followed by an arbitrary number of executions of an uncontrollable event. But the supervisor only enables the uncontrollable event once. S might, for example, come from the informal specification "u must happen only once" and in such a specification, the event c is irrelevant. Therefore, S must be considered uncontrollable with respect to P.

If it is simply assumed that $\Sigma^S = \Sigma^P$, S will disable the controllable event c at all times and S is thus falsely considered controllable with respect to P.

There is also the perhaps less obvious case depicted in figure 2, the supervisor never disables u as the event is not represented in S's alphabet. Therefore, S must be considered controllable with respect to P.

If definition 1 is used, by naively treating the alphabets as being equal, it is found that the sequence c in S followed by the uncontrollable event u, which is in $L(P)$, is not included in $L(S)$ and S is falsely considered uncontrollable with respect to P.

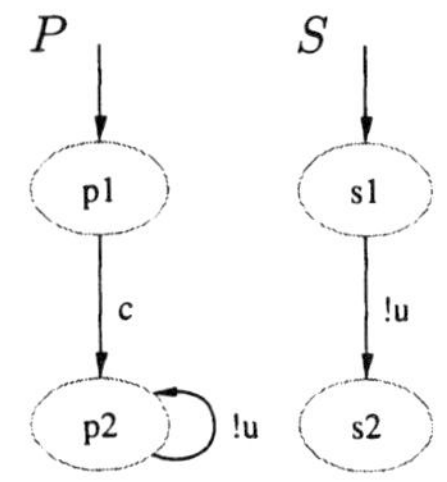

Fig. 1. To make the alphabets of S and P equal, simply set $\Sigma^S = \Sigma^P$.

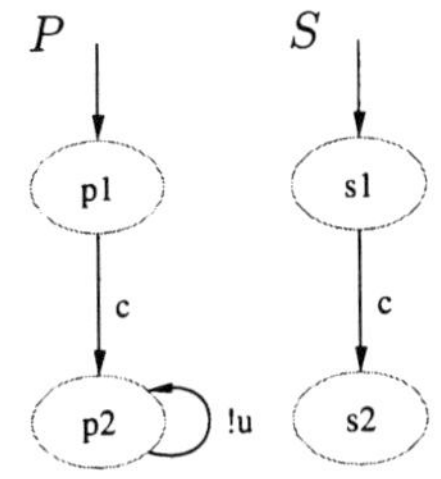

Fig. 2. To make the alphabets of S and P equal, simply set $\Sigma^S = \Sigma^P$.

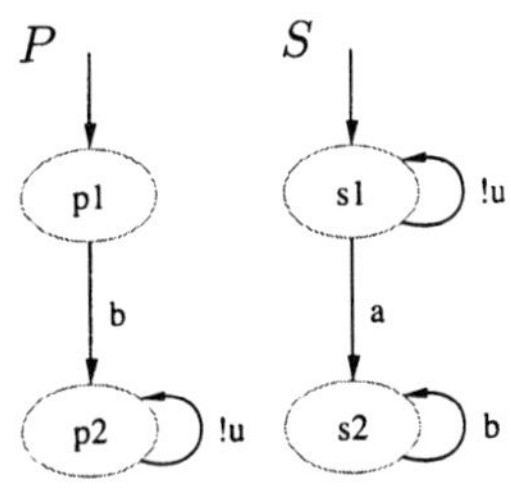

Fig. 3. To make the alphabets of S and P equal, simply set $\Sigma^P = \Sigma^S$.

5.3.2. Disablement of $\Sigma^S \setminus \Sigma^P$ In figure 3, the event a does not exist in P, but S specifies that a must occur before P can be allowed to start its execution. Thus, S disables b until a has occurred and when a has occured, the supervisor disables the uncontrollable event u. Therefore, S must be considered uncontrollable with respect to P.

If definition 1 is used, neither S nor P can leave their initial states since P disables a and S disables b. The automata are blocked in their initial states and nothing bad can happen, thus S is falsely considered controllable with respect to P.

5.3.3. Restriction of Σ^S In figure 4, the supervisor disables the uncontrollable event u in the initial state. Only after the occurence of the event c, is u enabled. Since P can execute the uncontrollable event u in the initial state, S must be considered uncontrollable with respect to P.

When the alphabet of S is restricted, all occurences of events not represented in P are removed from $L(S)$. That is, S only keeps its initial state with the self-loop of u, $L_{\Sigma^P}(S) = \{\varepsilon, u^*\}$. In other words, S now accepts u to occur at any

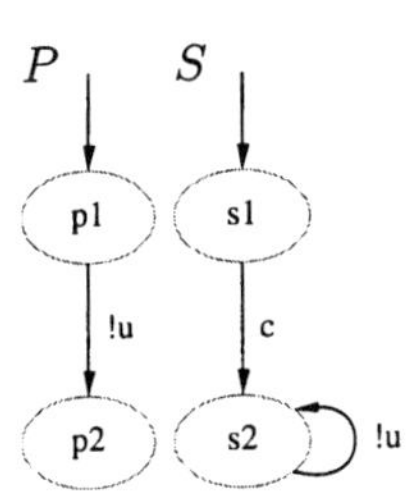

Fig. 4. To make the alphabets of S and P equal, the alphabet of S is restricted by projecting $L(S)$ onto Σ^P.

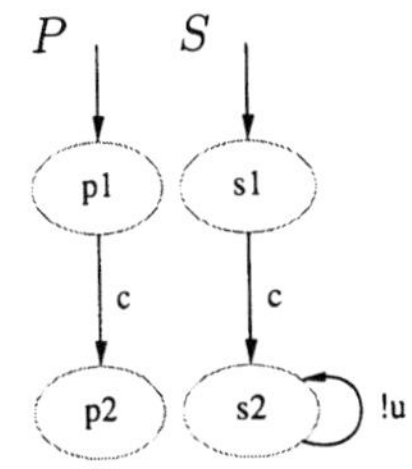

Fig. 5. To make the alphabets of S and P equal, the alphabet of P is expanded by taking the inverse projection of $L(P)$.

time and S is falsely considered controllable with respect to P.

5.3.4. Expansion of Σ^P In figure 5, the supervisor happens to allow an uncontrollable event in the state $s2$, an uncontrollable event that is not present in P. This must be appropriate, since the plant will never attempt to execute the event. Therefore, S must be considered controllable with respect to P.

To make the alphabets equal, we may expand P's alphabet to Σ^S so that $L^{-1}_{\Sigma^P}(P) = \{\varepsilon, u^*, u^*cu^*\}$. Since the uncontrollable event u is now enabled in P's initial state, while it is still disabled in S's initial state, S is falsely considered uncontrollable.

6. DISCUSSION

Most importantly, if the examples in the last section are reexamined using definition 3, this gives the same results as the discussions in all five examples.

6.1 Table of Errors

Table 1 shows that there is no universal solution for using definition 1. It *will* cause problems in some cases. Note that the symbols in the table correspond to the five examples above and actually appear in the same order.

In the case where $\Sigma^P \subset \Sigma^S$, none of the mentioned methods can guarantee a correct result.

If definition 2 is used instead, the two problems in the first row of the table disappears. Simply because the alphabets of $S||P$ and P becomes equal.

Table 1. Table of when incorrect results can be expected from using definition 1. !u stands for risk of falsely uncontrollable result and !c for risk of falsely controllable result.

	disable	restrict	expand
$\Sigma^S \subset \Sigma^P$	!c, !u		
$\Sigma^P \subset \Sigma^S$	!c	!c	!u

But the three problems in the second row remain, as can be shown by the same three examples.

Apparently, as indicated by the two empty cells in the table, when $\Sigma^S \subset \Sigma^P$, both restriction and expansion solves the problem. We will prove this by showing that the projection of interest will yield the same result as definition 3.

Proposition 1. When $\Sigma^S \subset \Sigma^P$, using $L_{\Sigma^S}(P)$ instead of $L(P)$ in definition 1 gives the same result as using definition 3. □

Proof. To prove the proposition, it must be shown that

$$L(S)\Sigma_u \cap L_{\Sigma^S}(P) \subseteq L(S) \quad (3)$$

$$\Leftrightarrow$$

$$L_{\Sigma^S}^{-1}(S)\Sigma_u^P \cap L_{\Sigma^P}^{-1}(P) \subseteq L_{\Sigma^S}^{-1}(S). \quad (4)$$

Since $\Sigma^S \subset \Sigma^P$, we have that $\Sigma_u^P = \Sigma_u$ and $L_{\Sigma^P}^{-1}(P) = L(P)$. Thus, the last expression can be rewritten to $L_{\Sigma^S}^{-1}(S)\Sigma_u \cap L(P) \subseteq L_{\Sigma^S}^{-1}(S)$.

To show the equivalence, it is enough to show that whenever (3) holds or does not hold, the same goes for (4). If (3) holds, it must be so that $\forall s \in L(S)$ and $\forall u \in \Sigma_u$ such that $su \in L_{\Sigma^S}(P)$ it must also hold that $su \in L(S)$. However, since $L(P)$ is prefix closed, it must also be so that $\exists s'u \in L(P)$ such that the projection of $s'u$ onto Σ^S is exactly su. Now, $\Sigma^S \subset \Sigma^P$ implies that $s' \in s_{\Sigma^S}^{-1}$ for all such s', i.e. $s'u \in s_{\Sigma^S}^{-1}u \subseteq L_{\Sigma^S}^{-1}(S)$ which proves that (4) also holds.

Conversely, if (3) does not hold, $\exists s \in L(S)$ and $\exists u \in \Sigma_u$ such that $su \in L_{\Sigma^S}(P)$ but $su \notin L(S)$. Note also that $su \in L_{\Sigma^S}(P) \Rightarrow u \in \Sigma^S$. Similarly to the previous, $\exists s'u \in L(P)$ with $s' \in s_{\Sigma^S}^{-1}$ but now, since $su \notin L(S)$ and $u \in \Sigma^S$, it holds that $s'u \notin L_{\Sigma^S}^{-1}(S)$. This proves that (4) does not hold either and this proves the proposition. ■

Proposition 2. When $\Sigma^S \subset \Sigma^P$, using $L_{\Sigma^S}^{-1}(S)$ instead of $L(S)$ in definition 1 gives the same result as using definition 3. □

Proof. Since $\Sigma^S \subset \Sigma^P$, it is also so that $L(P) = L_{\Sigma^P}^{-1}(P)$ and $\Sigma_u = \Sigma_u^P$. This trivially implies that

$$L_{\Sigma^S}^{-1}(S)\Sigma_u \cap L(P) \subseteq L_{\Sigma^S}^{-1}(S)$$

$$\Leftrightarrow$$

$$L_{\Sigma^S}^{-1}(S)\Sigma_u^P \cap L_{\Sigma^P}^{-1}(P) \subseteq L_{\Sigma^S}^{-1}(S). \quad ■$$

6.2 Explanation

The reason for why definition 3 can be used will now be explained informally. Firstly, by using Σ_u^P instead of Σ_u, as in definition 1, situations where uncontrollable events in S affect P are avoided. This solves problems of the kind represented in figure 5. By using $L_{\Sigma^S}^{-1}(S)$ instead of $L(S)$, figures 1 and 2 type problems are solved and by using $L_{\Sigma^P}^{-1}(P)$ instead of $L(P)$, figures 3 and 4 type problems are solved.

In short, definition 3 expands both alphabets to the union alphabet but avoids situations like the one in figure 5 by using Σ_u^P instead of Σ_u.

7. CONCLUSIONS

In this paper, a generalized controllability definition was presented for the purpose of verification and synthesis of controllability in modular systems. Modular systems appear naturally in many application areas, e.g. manufacturing systems, traffic management systems and telecommunication networks.

The need for a generalized definition was motivated by a few examples of situations where the intuitive controllability status of a system proved to be opposite to what the standard controllability definition would suggest. Conditions were also given for when the original definition *will* work.

Future research will consider approaches to modular nonblocking synthesis and verification.

REFERENCES

Åkesson, K., H. Flordal and M. Fabian (2002). Exploiting modularity for synthesis and verification of supervisors. In: *Proc. of the 15th Triennial World Congress of the International Federation of Automatic Control.* Barcelona, Spain.

Brandin, B., R. Malik and P. Dietrich (2000). Incremental system verification and synthesis of minimally restrictive behaviors. In: *Proc.of the 2000 American Control Conference.* Chicago, USA. pp. 4056–4061.

Flordal, H. (2001). Modular controllability verification and synthesis of discrete event systems. Master's thesis. Chalmers University of Technology, Department of Signals and Systems.

Heymann, M. (1990). Concurrency and discrete event control. *IEEE Control Systems Magazine* **10**(4), 103–112.

Leduc, R. J. (2002). Hierarchical Interface-based Supervisory Control. PhD thesis. University of Toronto, Canada.

Queiroz, M.H. de and J.E.R. Cury (2000). Modular control of composed systems. In: *Proceedings of the American Control Conference.* pp. 4051–4055.

Ramadge, P. and W.M. Wonham (1989). The control of discrete event systems. *Proc. of IEEE* **77**(1), 81–98.

Copyright © IFAC Information Control Problems in Manufacturing, Salvador, Brazil, 2004

ELSEVIER
IFAC
PUBLICATIONS
www.elsevier.com/locate/ifac

DISCRETE EVENT SYSTEMS WITH GUARDS

Claudio de Oliveira * José Eduardo Ribeiro Cury ** Celso Antônio Alves Kaestner *

* *Pontifícia Universidade Católica do Paraná, Curitiba*
** *Universidade Federal de Santa Catarina, Florianópolis*

Abstract: This work presents a model of discrete event systems (DES) consisted of a synchronized deterministic finite automaton with a data structure. The model, denominated DES with guards is realized through automata with guards, where the habilitation of each group of synchronized events may be conditioned to the satisfaction of a predicate related to the discrete state of the automata and to the defined variables by data structure. The most important characteristics of DES with guards are the expression of non-regular behaviors and, depending on the structural properties of the components of the system, the construction of generic models through the parametrization of initial values, final values, etc.
Copyright © 2004 IFAC
Keywords: Discrete Event Systems, Data Structures, Guards.

1. INTRODUCTION

In a similar manner to the classical theory of supervisory control of discrete events systems (Ramadge and Wonham, 1984), the discrete events systems with guards (DESwG) consist of a plant interacting with a supervisor. The plant generates spontaneous events. The supervisor, due to the sequence of events occurred in the past, controls the future evolution of the system by inhibiting the occurrence of some events in the plant. However, differently from the classical theory, the decision of the inhibition or not of events, results from the evaluation of certain logical expressions denominated *guards*. The terms of these guards consist of parameters and variables representing whichever information can be synchronized with the events that occur in the plant.

The major advantages of the DESwG in relation to other models (Ramadge and Wonham, 1984; Li and Wonham, 1993; Sreenivas and Krogh, 1991) that motivated this approach is the possibility of treatment of a greater class of systems, including those whose behaviors are not regular and, mostly, the possibility of the construction of generic models through the parametrization of initial values, final values, etc.

In the same manner as FSMwP model, introduced in (Chen and Lin, 2000), the DESwG can represent efficiently systems that cannot normally be represented by purely finite state machines without huge state space.

In the following section, the model and the concepts related to the synthesis of supervisors are presented formally. Questions related to problems solutions elapsed from the synthesis, algorithms and examples are not explored in this text.

2. DISCRETE EVENTS SYSTEMS WITH GUARDS - DESWG

Basically, the DESwG represent systems like a state transition system equipped with a data structure, plus a set of guards arising from a specification of requisites. The *state transition system* considered, (STS) is the triple $G = \langle Q, \Sigma, \delta \rangle$, where:

(1) Q is a finite set of *states* (discrete).
(2) Σ is a finite set of symbols denominated *events*. The event ξ alone is a event from Σ denominated *initialization event* of the STS. The set of events is divided into controllable Σ_c and noncontrollable events Σ_u. The event ξ is assumed to be controllable.
(3) $\delta : Q \times \Sigma \to Q$ is a partial *function of transition states* established to each $q \in Q$ and a subset of symbols $\rho \subseteq \Sigma$. Suppose there is a state $q_0 \in Q$, denominated *initial state*, such that $q_0 = \delta(q, \xi)$ [1] for every $q \in Q$.

An event is controllable when it can be disabled by an external control agent. The event ξ symbolizes the initialization operation (or re-initialization) of the system. Independently of the discrete state of the system, the initialization (or re-initialization) makes the system change to the (only) state q_0.

The notation $\delta(q, \sigma)!$ indicates that $\delta(q, \sigma)$ is defined. The function δ is extended to the partial function $\hat{\delta} : Q \times \Sigma^* \to Q$ by the rules $\hat{\delta}(q, \varepsilon) = q$, and $\hat{\delta}(q, s\sigma) = \delta(\hat{\delta}(q, s), \sigma)$, since $\hat{\delta}(q, s)!$ and $\delta(\hat{\delta}(q, s), \sigma)!$.

2.1 Data Structures

Informally, a data structure is an arrangement of components to which a meaning can be associated. In the model that follows, a structure component is introduced as a set of operations over a instantiation set I, possibly infinite. Such operations must have some properties seen next.

A *monoid of operations* for the instantiation set I is the monoid $M_I = \langle F_I, \circ, nop\rangle$, where:

(1) the support set F_I is a set of *operations*, or total functions of the form $op : I \to I$, defined on the *instantiation set* I,
(2) $\circ$ is the operation of functions *composition*, and
(3) $nop \in F_I$ is an *identity* (polymorphic) function (corresponding to the operation non-operation) defined to every I as $\forall i \in I,\ nop(i) = i$.

Except for the operation *nop*, the definition of the operations in F_I can have free variables. For example, in the definition $\forall i \in I,\ op(i) = f(i, x)$, x is a free variable. The free variables are called *parameters* and can be used in the construction of generic models. Suppose free variables have compatible dimensions with the instantiation set I.

[1] One of the goals for defining the event ξ and the initial state via a set of transitions ξ is to allow the parametrization of the initial state of the plant.

A *data structure* is an ordered and finite collection of monoids of operations $\mathcal{M}_n = \langle M_{I_1}, M_{I_2}, \ldots, M_{I_n}\rangle$, where $n \geq 0$ [2].

Given the data structure $\mathcal{M}_n = \langle M_{I_1}, M_{I_2}, \ldots, M_{I_n}\rangle$, the set $\mathcal{I}_n = I_1 \times I_2 \times \ldots \times I_n$ is called $\mathcal{M}_n$ *instantiations set* and each n-tuple $\iota \in \mathcal{I}_n$ is called $\mathcal{M}_n$ *instantiation*. The $\mathcal{M}_n$ *data structure support set* is $\mathcal{F}_n = F_{I_1} \times F_{I_2} \times \ldots \times F_{I_n}$. For $\langle f_1, \ldots, f_n\rangle, \langle g_1, \ldots, g_n\rangle \in \mathcal{F}_n$, the composition of $\langle f_1, \ldots, f_n\rangle$ and $\langle g_1, \ldots, g_n\rangle$ is defined as $\langle g_1, \ldots, g_n\rangle \circ \langle f_1, \ldots, f_n\rangle = \langle g_1 \circ f_1, \ldots, g_n \circ f_n\rangle = \langle g_1 \circ f_1, \ldots, g_n \circ f_n\rangle$.

2.2 Extended Discrete Event System

An extended discrete event system, or ExDES, consists basically of a state transition system equipped with a data structure. In formal terms, a ExDES is the quadruple $\mathcal{S} = \langle G, \mathcal{M}_n, \Delta, \mathcal{P}_m\rangle$, where:

(1) $G = \langle Q, \Sigma, \delta\rangle$ is a *state transition system*.
(2) $\mathcal{M}_n = \langle M_{I_1}, M_{I_2}, \ldots, M_{I_n}\rangle$ is a *data structure*.
(3) $\Delta : \Sigma \to \mathcal{F}_n$ is a partial function associating to each event $\sigma \in \Sigma$, a set of *operations* from the data structure support set $\mathcal{M}_n$.
(4) $\mathcal{P}_m : Q \times \mathcal{I}_n \to \mathbb{B}$, where $\mathbb{B} = \{true, false\}$, is a *marker predicate*.

Such as the function δ, the Δ is extended to the partial function $\hat{\Delta} : \Sigma^* \to \mathcal{F}_n$ by the rules $\hat{\Delta}(\varepsilon) = \langle nop_1, \ldots, nop_n\rangle$, and $\hat{\Delta}(s\sigma) = \Delta(\sigma) \circ \hat{\Delta}(s)$.

Let $\mathcal{S} = \langle G, \mathcal{M}_n, \Delta, P_m\rangle$ be an ExDES. A *configuration* in $\mathcal{S}$ is the pair $\langle q, \iota\rangle \in Q \times \mathcal{I}_n$ composed by a discrete state of G and a $\mathcal{M}_n$ instantiation. The *initial configuration* of $\mathcal{S}$ is the configuration $\langle q_0, \iota_0\rangle$, such that, for every state q and instantiation ι, $q_0 = \delta(q, \xi)$ is the initial state of G and $\iota_0 = \Delta(\xi)(\iota)$ is the initial $\mathcal{M}_n$ instantiation. The configuration $\langle q, \iota\rangle$ is *marked* (or *final*) if $\mathcal{P}_m(q, \iota) = true$.

To describe the behavior of an ExDES, we define a *run* of an ExDES as a sequence

$$\xrightarrow{\xi} \langle q_0, \iota_0\rangle \xrightarrow{\sigma_1} \langle q_1, \iota_1\rangle \xrightarrow{\sigma_2} \langle q_2, \iota_2\rangle \ldots,$$

where σ_i is the ith transition and $\langle q_i, \iota_i\rangle$ is the configuration after the ith transition. The sequence $s = \xi\sigma_1\sigma_2\sigma_3\ldots$ of event transitions in the run is a *trace* of a run. For a deterministic ExDES (which we assume from now on), a run is uniquely determined by its trace. The language generated by an ExDES $\mathcal{S}$ is the traces subset $L(\mathcal{S}) = \{\, \xi s \in \Sigma^* \mid \hat{\delta}(q_0, s)! \,\}$ and the *marked language* is the traces subset $L_m(\mathcal{S}) = \{\xi s \in L(\mathcal{S}) \mid \langle\hat{\delta}(q_0, s), \hat{\Delta}(s)(\iota_0)\rangle$ is a marked configuration$\}$.

[2] A data structure can be empty ($\mathcal{M}_0 = \langle\,\rangle$).

The *reachable configuration by a trace* $\xi s \in L(\mathcal{S})$ is $Reach(s) = \langle q, \iota \rangle = \langle \hat{\delta}(q_0, s), \hat{\Delta}(s)(\iota_0) \rangle$. Extending the previous notation to a traces subset $K \subseteq L(\mathcal{S})$, we define $Reach(K) = \{Reach(s) \mid \xi s \in K\}$. For an ExDES $\mathcal{S}$, the *reachable configurations subset* is $Reach(\mathcal{S}) = Reach(L(\mathcal{S})) = \{ \langle q, \iota \rangle \mid (\exists s \in \Sigma^*)\ (q = \hat{\delta}(q_0, s)$ and $\iota = \hat{\Delta}(s)(\iota_0)) \}$. The *coreachable configurations subset* is $CoReach(\mathcal{S}) = \{ \langle q, \iota \rangle \mid (\exists s \in \Sigma^*)$ $\langle \hat{\delta}(q, s), \hat{\Delta}(s)(\iota) \rangle$ is a marked configuration$\}$.

2.3 Specifications

Generally, the configurations of a ExDES must be limited to a subset of safe configurations, non-blocking and/or having specific properties. Let $\mathcal{D} \subseteq Q \times \mathcal{I}_n$ be the subset of desirable configurations. A specification is a description of set $\mathcal{D}$.

In the DESwG, the set $\mathcal{D}$ is described by two types of specifications, both expressed in the predicate form. The first type, called *specification by pre-conditions*, determines the conditions for events enabling. The second type of specification is denominated *specification by invariant* and consists of a predicate that must be satisfied by all configurations of the system.

2.3.1. Specification by Pre-Conditions. Let $\mathcal{S}$ be a ExDES. A *specification by pre-conditions* for $\mathcal{S}$ is the map $\Lambda : \Sigma \to \mathcal{P}$, where $\mathcal{P} : Q \times \mathcal{I}_n \to \mathbb{B}$ is a predicate over the $\mathcal{S}$ configurations set.

Each predicate $\mathcal{P}$ of a specification by pre-conditions is named *guard* and can be parameterized. The set of desirable configurations specified by pre-conditions is denoted $\mathcal{D}_\Lambda$. For a string of events $\xi\omega \in L(\mathcal{S})$, the configuration $\langle q, \iota \rangle = \langle \hat{\delta}(q_0, \omega), \hat{\Delta}(\omega)(\iota_0) \rangle$ is in $\mathcal{D}_\Lambda$ if $\omega = \phi\sigma$, where $\phi \in \Sigma^*$ and $\sigma \in \Sigma$, such that:

(1) the initial configuration $\langle q_0, \iota_0 \rangle$ is in $\mathcal{D}_\Lambda$,
(2) the configuration $\langle \hat{\delta}(q_0, \phi), \hat{\Delta}(\phi)(\iota_0) \rangle$ is in $\mathcal{D}_\Lambda$, and
(3) $\Lambda(\sigma)\langle \hat{\delta}(q_0, \phi), \hat{\Delta}(\phi)(\iota_0) \rangle = true$.

In words, the first condition says that the initial parametrization, parameterized or not, must be desirable. The second condition says that there must be a desirable configuration $\langle q', \iota' \rangle$ preceding the configuration $\langle q, \iota \rangle$. At last, the third condition says that the restriction over the event σ, taking the system from the desirable configuration $\langle q', \iota' \rangle$ to the configuration $\langle q, \iota \rangle$, must be enabled.

2.3.2. Invariant. Let $Q \times \mathcal{I}_n$ be the configurations set of a ExDES $\mathcal{S}$. An *invariant* for $\mathcal{S}$ is the predicate $Inv : Q \times \mathcal{I}_n \to \mathbb{B}$.

An invariant can also be defined with the use of parameters. A specification by invariant describes ordinary properties of all desirable configurations. In other words, the desirable configurations expressed in terms of an invariant are $\mathcal{D}_{Inv} = \{\langle q, \iota \rangle \in Q \times \mathcal{I}_n \mid Inv(q, \iota)\}$.

2.3.3. Equivalence Between Specifications. For every specification by invariant, there is an equivalent specification by pre-conditions. In order to find the pre-condition set equivalent to $Inv(q, \iota)$, the invariant associated to the state q, we begin observing that for each transition σ_i, from the state q_i to state q, there is a predicate of the form $\Lambda(\sigma_i)\langle q_i, \iota_i \rangle$ equivalent to the invariant in q, *i.e.*,

$$\Lambda(\sigma_i)\langle q_i, \iota_i \rangle \Leftrightarrow Inv\langle q, \iota \rangle. \quad (1)$$

Because $q = \delta(q_i, \sigma_i)$ and $\iota = \Delta(\sigma_i)(\iota_i)$, the invariant can be rewritten as

$$Inv(q, \iota) = Inv\langle \delta(q_i, \sigma_i), \Delta(\sigma_i)(\iota_i) \rangle, \quad (2)$$

which is a predicate over q_i, ι_i and σ_i. Replacing 2 in 1, we obtain:

$$\Lambda(\sigma_i)\langle q_i, \iota_i \rangle \Leftrightarrow Inv\langle \delta(q_i, \sigma_i), \Delta(\sigma_i)(\iota_i) \rangle, \quad (3)$$

i.e., $\Lambda(\sigma_i)\langle q_i, \iota_i \rangle = Inv\langle \delta(q_i, \sigma_i), \Delta(\sigma_i)(\iota_i) \rangle$.

Extending this result to all n events labelling transitions for state q, we have:

$$\Lambda(\sigma_1)\langle q_1, \iota_1 \rangle \vee \ldots \vee \Lambda(\sigma_i)\langle q_i, \iota_i \rangle \vee \ldots \vee \Lambda(\sigma_n)\langle q_n, \iota_n \rangle \Leftrightarrow Inv\langle q, \iota \rangle \quad (4)$$

The equivalence 4 can be extended to all configurations of the system, which demonstrates the existence of a specification by pre-conditions equivalent to the specification by invariant Inv.

2.4 Discrete Event System with Guards

A discrete-event systems with guards results from the interaction between a plant and a supervisor. The supervisor is designed to limit the plant behaviors to a subset of safe configurations, non-blocking and/or having specific properties.

Let $\mathcal{D}_\Lambda \neq \emptyset$ be a desirable configurations subset described by a specification (by pre-conditions) Λ such that $\langle q_0, \iota_0 \rangle \in \mathcal{D}_\Lambda$. Then, $\mathcal{D}_\Lambda$ is *controllable* with respect to S if:

$$\forall\, \sigma \in \Sigma_u,\ s \in \Sigma^* \quad Reach(s) \in \mathcal{D}_\Lambda \wedge \xi s\sigma \in L(\mathcal{S}) \Rightarrow Reach(s\sigma) \in \mathcal{D}_\Lambda. \quad (5)$$

Λ is controllable (with respect to S) if $\mathcal{D}_\Lambda$ is controllable.

A *supervisory control* for an ExDES $\mathcal{S}$ is any map $V : \Sigma^* \to \Gamma$, where $\Gamma = \{\gamma \in 2^\Sigma \mid \gamma \supseteq \Sigma_u\}$ is a *control pattern*. Given a specification Λ, we can build a supervisory control V_Λ for Λ as:

$$\forall\ \xi s \in L(\mathcal{S}),\quad V_\Lambda(\xi s) = \{\,\sigma \in \Sigma \mid \delta(\hat{\delta}(q_0, s), \sigma)! \Rightarrow \Lambda(\sigma)\langle\hat{\delta}(q_0, s), \hat{\Delta}(s)(\iota_0)\rangle = true\,\} \quad (6)$$

The role of supervisor is to observe and to enable the occurrence of events in the plant in a way to limit its behavior to a set of desirable configurations. As seen in Section 2.3.3 and without lose of generality, only specifications by pre-conditions are considered. In this way, the commands performed by the supervisor consist of the enabling/disabling of controllable events of the plant according to the satisfaction of the respective guards.

Theorem 1. Let Λ be a specification by pre-conditions. Then, there exists a supervisory control for Λ if and only if Λ is controllable.

A *discrete-event system with guards* DESwG is the system $\mathcal{G} = \langle \mathcal{S}, V_\Lambda\rangle$, where $\mathcal{S}$ is an ExDES, $\Lambda : \Sigma \to \mathcal{P}$ is a specification by pre-conditions and V_Λ is a supervisory control for Λ. We will use the classical notation $V_\Lambda/\mathcal{S}$ to suggest "$\mathcal{S}$ under the supervision of V_Λ".

The closed behavior of $V_\Lambda/\mathcal{S}$ is defined to be the language $L(V_\Lambda/\mathcal{S}) \subseteq L(\mathcal{S})$ described as:

(1) $\xi \in L(V_\Lambda/\mathcal{S})$
(2) if $\xi s \in L(V_\Lambda/\mathcal{S})$, $\sigma \in V_\Lambda(\xi s)$ and $\xi s\sigma \in L(\mathcal{S})$ then $\xi s\sigma \in L(V_\Lambda/\mathcal{S})$
(3) no other strings belong to $L(V_\Lambda/\mathcal{S})$.

Clearly, $\{\xi\} \subseteq L(V_\Lambda/\mathcal{S}) \subseteq L(\mathcal{S})$, $L(V_\Lambda/\mathcal{S})$ is nonempty and closed.

A supervisor can only meet the requisites imposed by a controllable specification. However, in a general way, a specification designed according to a set of desirable configurations is not controllable. In these cases, often, a noncontrollable specification can be transformed in another one more restrictive, but controllable. The following sections explore such transformations and the consequent synthesis of supervisors.

3. SYNTHESIS OF SUPERVISORS

When $\mathcal{D}_{\Lambda_1} \subseteq \mathcal{D}_{\Lambda_2}$, we say that the specification Λ_1 *implies* Λ_2, written as $\Lambda_1 \Rightarrow \Lambda_2$. Besides, if $\mathcal{D}_{\Lambda_1} = \mathcal{D}_{\Lambda_2}$, *i.e.*, if $\Lambda_1 \Rightarrow \Lambda_2$ and $\Lambda_2 \Rightarrow \Lambda_1$, we say that Λ_1 and Λ_2 are *equivalents*, or $\Lambda_1 \Leftrightarrow \Lambda_2$. Clearly, $\Leftrightarrow$ is an equivalence relation on specifications set. The set of all controllable specifications implying Λ is $\mathcal{C}(\Lambda) = \{\Lambda' \Rightarrow \Lambda \mid \Lambda'$ is an equivalence class of equivalents controllable specification $\}$. Following, we will not distinguish a specifications from its equivalence class.

Given two specifications, Λ_1 and Λ_2 in $\mathcal{C}(\Lambda)$, we define the binary relations *meet* ($\sqcap$) and *join* ($\sqcup$) as:

$$(\forall\,\sigma \in \Sigma)\ \ (\Lambda_1 \sqcap \Lambda_2)(\sigma) = \Lambda_1(\sigma) \wedge \Lambda_2(\sigma)$$
$$(\forall\,\sigma \in \Sigma)\ \ (\Lambda_1 \sqcup \Lambda_2)(\sigma) = \Lambda_1(\sigma) \vee \Lambda_2(\sigma)$$

Clearly, $\Lambda_1 \wedge \Lambda_2 \in \mathcal{C}(\Lambda)$ and $\Lambda_1 \vee \Lambda_2 \in \mathcal{C}(\Lambda)$. The meet and join operations on $\mathcal{C}(\Lambda)$ define the partial order $\Rightarrow$ on $\mathcal{C}(\Lambda)$ by the rule $\Lambda_1 \Rightarrow \Lambda_2$ if and only if $\Lambda_1 \vee \Lambda_2 \Rightarrow \Lambda_2$, or, equivalently, $\Lambda_1 \wedge \Lambda_2 \Rightarrow \Lambda_1$. Therefore, the pair $\langle\mathcal{C}(\Lambda), \Rightarrow\rangle$ forms a bounded lattice where the supremal element, *i.e.*, the *supremal specification*, of $\mathcal{C}(\Lambda)$ is $sup\ \mathcal{C}(\Lambda) \Leftrightarrow \bigvee\{\Lambda' \mid \Lambda' \in \mathcal{C}(\Lambda)\}$.

For a noncontrollable specification Λ, the problem of the synthesis of supervisors for DESwG consists in determining $sup\ \mathcal{C}(\Lambda)$. The fundamental idea to be used in the $sup\ \mathcal{C}(\Lambda)$ computation consists of the anticipation of the test of the guards associated to the noncontrollable events to the instants when the respective preceding controllable events are to be enabled. Such anticipations imply an indirect control of noncontrollable events. The concepts and definitions related to this anticipation of guards for a class of non-temporized DESwG satisfying the property "*for every substring s of noncontrollable events, such that $\phi_1 s^* \phi_2 \in L(\mathcal{S})$*, $\hat{\Delta}(s) = nop$" are presented next.

3.1 Guards Anticipation

A string ω is a *path* to σ if $\xi\omega\sigma \in L(\mathcal{S})$. The set of paths to σ is $Path(\sigma) = \{\omega \mid \xi\omega\sigma \in L(\mathcal{S})\}$. The event $\sigma \in \Sigma_u$ is *indirectly controllable regarding to the path* ω if there is $\alpha \in \Sigma_c$, such that, $\omega = \phi_1\alpha\phi_2$. The *indirect control of* σ *regarding the path* ω associates the guard $\Lambda(\sigma)\langle\hat{\delta}(q', \alpha\phi_2), \hat{\Delta}(\alpha\phi_2)(\iota')\rangle$, where $q' = \hat{\delta}(q_0, \phi_1)$ and $\iota' = \hat{\Delta}(\phi_1)(\iota_0)$, to the event α. Informally speaking, the indirect control of σ in relation to the string $\phi_1\alpha\phi_2$ is obtained anticipating the test of $\Lambda(\sigma)$ to the instant when the system is about to enable the event α. Opportunely, it will be shown that if $\phi_2 \in \Sigma_u^*$, the indirect control is *minimally restrictive*.

σ is *indirectly controllable* if it is indirectly controllable regarding each one of its paths. The *minimally restrictive indirect control*, or just, *indirect control*, of σ consists of the set of guards associations necessary to the minimally restrictive indirect control of σ relating to each one of its paths.

After all the anticipations of the guards associated to the noncontrollable events, a new specification by pre-conditions is obtained. This new specification includes the restrictions of the original configuration plus the restrictions added by the anticipations. The main characteristic of this specification is that all restrictions associated to the noncontrollable events are implied by guards associated to the preceding controllable events. Therefore, the noncontrollable restrictions of this new specification can be ignored and so, the resulting specification is controllable.

3.2 Nonblocking Supervisory Control

A system is said nonblocking if, independently of its state, the system can always evolve towards a complete task. In the DESwG, the blocking treatment is performed after the anticipations of the system guards. One of the cases where the blocking can occur is when the system reaches the state q, not marked, where only controllable events can originate. In this case, if none of the guards correlated to the events leaving q is satisfied by current values of the data structure, the system blocks. To eliminate this possibility, an *invariant* must be associated to the state q. The invariant is computed as the logical disjunction of the guards associated to the events leaving q and the marker predicate $\mathcal{P}_m$.

The invariants that avoid blocking can be transformed in a pre-conditions set associated to the events arriving to state q (Section 2.3.3). This new guards set corresponds to a intrinsic specification by pre-conditions that avoids the system blocking. Possibly, this new specification by pre-conditions is not controllable. So, it is necessary to anticipate its noncontrollable guards to the appropriate controllable events. If this new anticipation changes the result of the former anticipations, new invariants must be computed. The procedures of invariants computation, guards computation guaranteeing the satisfaction of the invariants and anticipation of these guards must be repeated until the result obtained in a specific interaction become the same as the former interaction.

4. EXAMPLE

The Figure 1 shows a manufacturing line where parts are processed by two machines, M_1 and M_2, and tested by the tester T_1. The boxes A_1 and A_2 represent buffers capable to store from 0 to N or M items, respectively. The machine and tester states are I-dle and W-orking. The tester can accept an item (event a) or not (event r). If the tester refuse the item, it must be reprocessed by M_2. The operations over the buffers are *initialize* (starts the buffers with 0), *insert* (increments the counter of current stored items) and *remove* (decrements the counter of current stored items).

A regular sequence of activities in the line is:

(1) M_1 process the item and store it in A_1.
(2) When M_2 starts, an item is removed from A_1, processed by M_2 and stored in A_2.
(3) Finally, the tester removes the item from A_2, tests the item, and, if the item is refused, restores it in A_1.

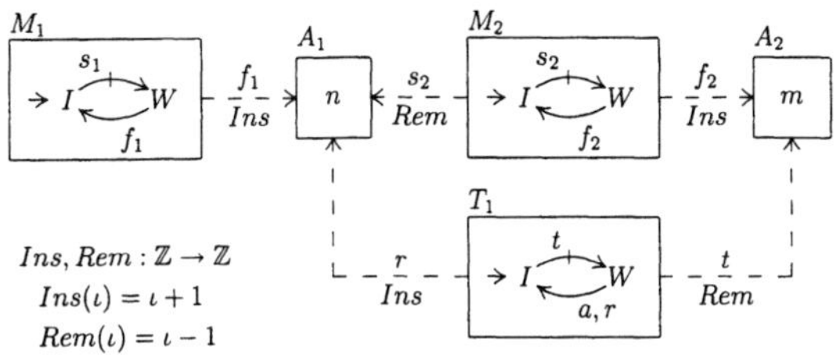

Fig. 1. Manufacturing Line

The goal of this sample is to find a controllable specification to avoid the overflow and underflow in the buffers. The controllable events are s_1, s_2 and t.

The specification by pre-condition to be satisfied is $\Lambda = \{(f_1, n < N), (s_2, n > 0), (f_2, m < M), (t, m > 0), (r, n < N)\}$. Hence, the guards to be anticipated are $n < N$, related to events f_1 and r, and $m < M$, related to f_2. The indirect control of f_1 is made via s_1; the indirect control of f_2 is made via s_2; and the indirect control of a and r is made via t. Therefore, the guard $\Lambda(f_1)\langle\cdot, n\rangle = n < N$ must be anticipated to s_1, the guard $\Lambda(f_2)\langle\cdot, n\rangle = m < M$ must be anticipated to s_2 and the guard $\Lambda(r)\langle\cdot, n\rangle = n < N$ must be anticipated to t ($\cdot$ means don't care).

In (Oliveira *et al.*, 2003) we have showed that these anticipations are:

$$\begin{aligned}\Lambda(s_1)\langle q, \langle n,m\rangle\rangle =& (q = III \wedge n < N) \vee \\ & (q = IWI \wedge n < N) \vee \\ & (q = IIW \wedge n + 1 < N) \vee \\ & (q = IWW \wedge n + 1 < N) \\ \Lambda(s_2)\langle q, \langle n,m\rangle\rangle =& (n > 0) \wedge \\ & ((q = III \wedge m < M) \vee \\ & (q = WII \wedge m < M) \vee \\ & (q = IIW \wedge m < M) \vee \\ & (q = WIW \wedge m < M)) \\ \Lambda(t)\langle q, \langle n,m\rangle\rangle =& (m > 0) \wedge \\ & ((q = III \wedge n < N) \vee \\ & (q = IWI \wedge n < N) \vee \\ & (q = WII \wedge n + 1 < N) \vee \\ & (q = WWI \wedge n + 1 < N))\end{aligned}$$

where, $III \ldots WWW$ are the states of the synchronous product $M_1 \| M_2 \| T_1$.

The new specification is controllable, but it is not nonblocking. The blocking state is III. If III is a marked state and the marker predicate is $m = 0 \wedge n = 0$, the simplified invariant that avoid blocking

Table 1.

	s_1	f_1	s_2	f_2	t	r	a
$\langle III, \langle 0,0\rangle\rangle$	$\langle WII, \langle 0,0\rangle\rangle$	-	-	-	-	-	-
$\langle III, \langle 0,1\rangle\rangle$	-	-	-	-	$\langle IIW, \langle 0,0\rangle\rangle$	-	-
$\langle III, \langle 1,0\rangle\rangle$	-	-	$\langle IWI, \langle 0,0\rangle\rangle$	-	-	-	-
$\langle IIW, \langle 0,0\rangle\rangle$	-	-	-	-	-	$\langle III, \langle 1,0\rangle\rangle$	$\langle III, \langle 0,0\rangle\rangle$
$\langle IWI, \langle 0,0\rangle\rangle$	-	-	-	$\langle III, \langle 0,1\rangle\rangle$	-	-	-
$\langle WII, \langle 0,0\rangle\rangle$	-	$\langle III, \langle 1,0\rangle\rangle$	-	-	-	-	-

in state III is $(m = 0 \wedge n = 0) \vee (n < N) \vee (n > 0 \wedge m < M)$. The equivalent specification by pre-conditions is:

$$\begin{aligned}
&\Lambda^{bf}(f_1)\langle q, \langle n,m\rangle\rangle = (m = 0 \wedge n+1 = 0) \vee \\
&\quad (n+1 < N) \vee (n+1 > 0 \wedge m < M) = \\
&\quad \mathbf{n+1 < N \vee m < M} \\
&\Lambda^{bf}(f_2)\langle q, \langle n,m\rangle\rangle = (m+1 = 0 \wedge n = 0) \vee \\
&\quad (n < N) \vee (n > 0 \wedge m+1 < M) = \\
&\quad \mathbf{n < N \vee (n > 0 \wedge m+1 < M)} \\
&\Lambda^{bf}(r)\langle q, \langle n,m\rangle\rangle = (m = 0 \wedge n+1 = 0) \vee \\
&\quad (n+1 < N) \vee (n+1 > 0 \wedge m < M) = \\
&\quad \mathbf{n+1 < N \vee m < M} \\
&\Lambda^{bf}(a)\langle q, \langle n,m\rangle\rangle = \mathbf{(m = 0 \wedge n = 0)} \vee \\
&\quad \mathbf{(n < N) \vee (n > 0 \wedge m < M)}
\end{aligned}$$

As before, this "intrinsic" noncontrollable set of guards must be anticipated to the relating events. These anticipations produce:

$$\begin{aligned}
&\Lambda^{bf}(s_1)\langle q, \langle n,m\rangle\rangle = \\
&\quad \mathbf{n+1 < N \vee m < M} \\
&\Lambda^{bf}(s_2)\langle q, \langle n,m\rangle\rangle = \\
&\quad n-1 < N \vee (n-1 > 0 \wedge m+1 < M) = \mathbf{true} \\
&\Lambda^{bf}(t)\langle q, \langle n,m\rangle\rangle = \\
&\quad (n+1 < N \vee m-1 < M) \wedge \\
&\qquad ((m-1 = 0 \wedge n = 0) \vee (n < N) \vee \\
&\qquad\quad (n > 0 \wedge m-1 < M)) = \\
&\quad (m-1 = 0 \wedge n = 0) \vee n < N \vee n > 0 = \mathbf{true}
\end{aligned}$$

Finally, by considering the previous guards and the system states, we obtain the following s_1 final guard:

$$\begin{aligned}
&\Lambda(s_1)\langle q, \langle n,m\rangle\rangle = \\
&\quad (q = III \wedge (n+1 < N \vee n < N \wedge m < M)) \vee \\
&\quad (q = IWI \wedge (n+1 < N \vee n < N \wedge m+1 < M)) \vee \\
&\quad (q = IIW \wedge (n+2 < N \vee n+1 < N \wedge m < M)) \vee \\
&\quad (q = IWW \wedge (n+2 < N \vee n+1 < N \wedge m+1 < M))
\end{aligned}$$

Now, the new guards set is nonblocking and satisfy the original requisites. To check this final guards set we can assign values to N and M parameters. For example, if N and M are both initialized with 1 and the initial configuration is $\langle III, \langle 0,0\rangle\rangle$, the configurations mapping is showed in Table 1. From this table, we can observe an equivalence between our solution and that from the classical theory (Ramadge and Wonham, 1984). For others N and M initializations, we can observe the same equivalence. Therefore, in relating to the classical theory for supervisory control, our solution has the advantage of the generality.

5. CONCLUSION

We have proposed a new model for event-driven systems. The described method for the synthesis of supervisors consists of the anticipation of the guards associated originally with noncontrollable events to former controllable events. We conclude our exposition observing that in the expressions we made many simplifications. These simplifications are not so ease in an automatic manner. The algorithmic methods for the automatic computation of the guards anticipations to a class of DESwG will be explored in another opportunity. Regarding to these methods, we can anticipate that, besides of the automatic simplifications for expressions, another great difficulty is the verification of the equivalence of the two specifications (needed to determine the end of the algorithm described before). These practical problems must be better explored in order to build a professional tool for real cases. In spite of this, the already tested cases demonstrate gains compared to the existent methods, certainly contributing with the advance of the state of the art in the treatment of the discrete event driven systems.

REFERENCES

Chen, Y.L. and F. Lin (2000). Modeling of discrete event systems using finite state machines with parameters. In: *Proceedings of the 2000 IEEE - International Conference on Control Applications.* Anchorage, Alaska, USA.

Li, Y. and W.M. Wonham (1993). Control of vector discrete-event systems I – the base model. *IEEE Transactions on Automatic Control* **38**(8), 1214–1227.

Oliveira, C., J.E.R. Cury and C.A.A. Kaestner (2003). Um modelo parametrizado para sistemas a eventos discretos com comportamentos não regulares (2ª parte). In: *VI SBAI'2003.* Bauru, Brazil.

Ramadge, P.J. and W.M. Wonham (1984). Supervisory control of a class of discrete event processes. In: *Lecture Notes in Control and Information Sciences.* Vol. 63. Springer-Verlag. Berlin, Germany. pp. 477–498.

Sreenivas, R.S. and B.H. Krogh (1991). On condition/event systems with discrete state realizations. *Discrete Event Dynamic Systems: Theory and Applications* **1**(2), 209–235.

Copyright © IFAC Information Control Problems in
Manufacturing, Salvador, Brazil, 2004

ELSEVIER
IFAC
PUBLICATIONS
www.elsevier.com/locate/ifac

INFOTRONICS TECHNOLOGIES AND PREDICTIVE TOOLS FOR NEXT-GENERATION MAINTENANCE SYSTEMS

Jay Lee[1], Hai Qiu[1], Jun Ni[2], Dragan Djurdjanovic[2]

NSF Industry/University Cooperative Research Center on
Intelligent Maintenance Systems
(www.imscenter.net)
University of Wisconsin-Milwaukee[1]
Department of Industrial and Manufacturing Engineering
Milwaukee, Wisconsin
University. of Michigan [2]
Department of Mechanical Engineering
Ann Arbor, Michigan

Abstract: In today's competitive market, production costs, lead-time and optimal machine utilization are crucial values for companies. Since machine or process breakdowns severely limit their effectiveness, improved methods are needed to predict products' prediction expectancy. Furthermore, continuous assessment and prediction of product's performance could also enable proactive maintenance to prevent machine from breakdowns. This paper first addresses the paradigm shift in modern maintenance systems. Then, emerging infotronics technologies and predictive tools to transform the maintenance from traditional "*fail and fix*" practices to "*predict and prevent*" methodologies will be presented. Finally, examples will be given to illustrate the lesson learned from the testbeds and implementations. *Copyright © 2004 IFAC*

Keywords: Prognostics, Machine Performance Degradation Assessment, Predictive Maintenance, Remote Monitoring, Infotronics

1. Introduction

Today manufacturing industry are facing unprecedented challenges for global outsourcing to reduce manufacturing costs. This transformation drives companies to continuously increase their competitiveness through relentless lean manufacturing and six-sigma business practices. With the implementation of e-business, it necessitates much shorter time window in product design, manufacturing, delivery, and service in today's supply chain integrated manufacturing environment. This transformation drives near-zero downtime performance for manufacturing equipment and production systems.

Most machine maintenance today is either reactive (fixing or replacing equipment after it fails) or blindly proactive (assuming a certain level of performance degradation, with no input from the machinery itself, and servicing equipment on a routine schedule whether service is actually needed or not). Both scenarios are extremely wasteful. To human beings, it often seems that machines fail suddenly, but in fact machines usually go through a measurable process of degradation before they fail. Today, that degradation is largely invisible to human users, even though a great deal of technology has been developed that could make such information visible. Many sophisticated sensors and computerized components are capable of delivering data about the machine's status and performance. The problem is that little or no practical use is made of most of this data. We have smart devices, but we do not have a continuous and seamless flow of information throughout entire processes. Sometimes this is because the available data are not rendered in useable form. More often, no infrastructure exists for delivering the data over a network, or for managing and analyzing the data even if the devices were networked. When smart machines are networked and remotely monitored, and when their data is modeled and continually analyzed with sophisticated embedded systems, it is possible to go beyond mere "predictive maintenance" to intelligent "prognostics"—the process of pinpointing exactly which components of a machine are likely to fail, and when and autonomously trigger service and order spare parts. This necessitates new development to transform traditional "fail and fix" practices to "predict and prevent" methodology (IMS Center, 2004). Such a paradigm shift towards a continuous assessment and prediction of product's performance will also enable a Collaborative Product Life-Cycle Management (CPLM) (Radjou, 2002) in which products are followed, assessed and improved throughout their life-cycle, starting from product's inception on the design table, and continuing through the entire manufacturing process and product's usage phase. Through such an integrated life-cycle management, degradation information can be used to make improvements.

2 STATE-OF-THE-ART SURVEY

Condition Based Maintenance (CBM) methods and practices have been continuously improved for the last decades. Many diagnostics tools and methods have been developed with good successes. Sensor fusion techniques are also commonly engaged due to the inherent superiority in taking advantage of mutual information from multiple sensors (Roemer, 2001; Hansen, 1994; Reichard, 2000). For example, vibration signature analysis and oil analysis, because of their excellent capability of describing machine performance, have been successfully employed for prognostics for a long time (Kemerait, 1987; Wilson, 1999). In the mean time, alternative approaches in using time/stress, temperature, acoustic emissions (Goodenow, 2000)], and ultrasonic are widely employed as well. However, the development on prognostic have not made major breakthrough since many methods are still based on traditional signal processing (Hardman, 2000).

Typically, prognostic approaches can be classified as following categories: experienced-based prognostics, evolutionary prognostics, feature progression, AI-based prognostics, state estimator prognostics, and physics-based prognostics (Byington, 2003). A large variety of prognostic applications have been reported in the recent technical literatures and many of these were more application specific rather than generic. For example, the development of neural networks had also added new dimensions to solving existing problems of prognostics of a centrifugal pump case (Liang, 1988). Comparison of the results using the signal identification technique shows various merits of neural nets including the ability to handle multivariate wear parameters in a much shorter time. A helicopter transmission prognostic application was presented by Parker (1965), whereby fault detection, isolation, and estimation were conducted by a polynomial neural network. A fuzzy logic-based neural network and decision tree were applied to the prediction of paper web breakage in a paper mill (Bonissone, 1995). Yet another prognostic application presents an integrated prognosis system using a dynamically linked ellipsoidal basis function neural network coupled with an automated rule extractor to develop a tree-structured rule set that closely approximates the classification of the neural network (Brotherton, 2000). That method allows assessment of trending from the nominal class to each of the identified fault classes, which means quantitative prognostics are built into the network functionality. Garga *et al.*, (2001) introduced a hybrid reasoning method for prognostics, which integrated explicit domain knowledge and machinery data. In that approach, a feed-forward neural network is trained using explicit domain knowledge to get a parsimonious representation of the explicit domain knowledge. Vachtsevanos and Wang (2001) present a prognostic approach that is based on the concept of dynamic wavelet neural network and virtual sensors.

Some research activities about infrastructure of prognostics are also reported. From 1998 to 2000, the U.S. army logistics integration agency has funded a project entitled, "Prognostics Framework." (Su, *et al.*, 1999) The project aims to provide an overall architecture to manage the information provided by the individual prognostic techniques. This prognostics framework is generic, open architecture, horizontal technology, and scalable, and it was intended to integrate with embedded diagnostics to provide a "total health management" capability.

Generally speaking, current prognostic approaches can be divided into three basic clusters: model-based prognostics, data driven prognostics, and hybrid prognostics systems. An example of a characteristic model-based prognostics application includes data collected from model-based simulations under normal and degraded conditions. Prognostic models are constructed based on different random load conditions, or modes. However, in the absence of a reliable or accurate system model, another approach to determine the remaining useful life is to monitor the trajectory of a developing fault and predict the amount of time until the developing fault reaches a predetermined level requiring action, which is the so called data driven prognostic method. The Alpha-Beta-Gamma tracking filter and the Kalman filter, two well-known tracking and prediction tools, have been applied to gearbox prognostics (McClintic, 1998; Ferlezm 1998). Both filters have been investigated for their ability to track and smooth features from gearbox vibration data. The literature presents additional discussion on using Kalman filters to track changes in features like vibration levels, mode frequencies, or other waveform signature features and how to estimate the future failure hazard, probability of survival, and remaining useful life (Swanson, 2000; Swanson, 2001). An example of a hybrid method, which fuses the model-based information and sensor-based information and takes advantage of both model-driven and data-driven methods, was proposed by Hansen (1994), by which a more reliable and accurate prognostic result can be generated.

Prognostic information obtained through intelligence embedded into the manufacturing process, or equipment can also be used to improve the manufacturing and maintenance operations in order to increase process reliability and improve product quality. In (Yamada *et al.*, 2002), ability to increase reliability of manufacturing facilities using the awareness of the deterioration levels of manufacturing equipment has been demonstrated through an example of improving robot reliability. In (Seliger *et al.*, 2002) a Life Cycle Unit (LCU) is proposed to collect usage information about key product components, enabling one to assess product reusability and facilitating reuse of products that have significant useful life remaining.

In spite of these progresses, many fundamental issues remain:

1. Most of the developed prognostics approaches are application or equipment specific. A generic and scalable prognostic methodology or toolbox doesn't exist.
2. Currently, methods are generally focused on solving the failure prediction problem. Tools for system performance assessment and degradation prediction has not been well addressed
3. Features used for prognostics need to be further developed.
4. Many developed prediction algorithms have been demonstrated in a laboratory environment without industrial validations.

To address these issues, a toolbox of algorithms for multi-sensor performance assessment and prediction, named Watchdog Agent™, has been developed at the Center for Intelligent Maintenance Systems (IMS) which is a multi-campus NSF Industry/University Cooperative Research Center between the University of Wisconsin-Milwaukee and the University of Michigan. These tools enable one to quantitatively assess and predict performance degradation levels of key product components (Casoetto, 2003; Djurdjanovic, et al., 2000; Lee; 1996), thus offering a possibility to physically realize a closed-loop product life cycle monitoring and management. This great opportunity is currently being pursued by the Center for IMS and the Technical University of Berlin.

3 DEVELOPMENT OF PREDICTIVE TOOLS AND INTELIGENT MAINTENANCE SYSTEMS

Mechatronics systems was first introduced by Yaskawa Electric in Japan in late 1960s. It added *"precision"* and *"intelligence"* to mechanical systems through smart electronics control systems. *Infortronics* technologies transform the paradigm from *precision machine* to *precision information* through the use of intertwined embedded informatics and electronic intelligence in a networked and tether-free environment and enable products and systems to intelligently monitor, predict, and optimize its performance and ultimately perform self-maintenance activities autonomously. For the past three years, the IMS Team has been pursuing frontier research in partnership with over 45 members and sponsors from industry, government, and academe. The Team has pioneered many advances on infortronisc-based product assessment and prediction technologies, including the *Watchdog Agent ™* technologies and *Device-to-Business (D2B)™* platform. Figure 1 shows the system elements of an "Intelligent Maintenance Systems (IMS)."

The key research elements of an Intelligent Maintenance System consists of

1.) transform data to information to knowledge and synchronize decision with remote systems;
2.) intelligent embedded prognostics algorithms for performance degradation assessment and prediction;
3.) infotronics software and hardware platform that enable product to think, link, reconfigure and sustain in a networked and tether-free environment;
4.) embedded product service and life cycle information for closed-loop product design.

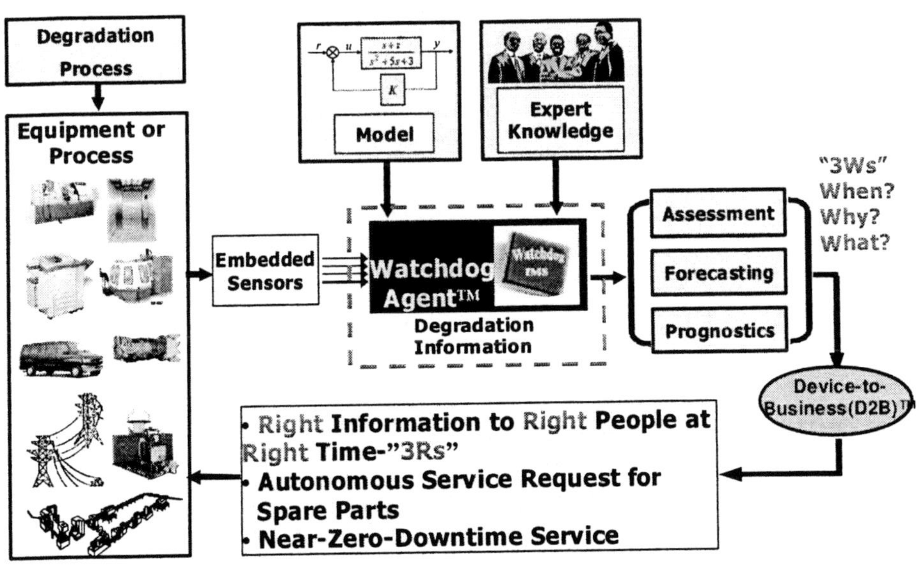

Figure 1. Intelligent Maintenance Systems (IMS)

3.1 Smart Prognostics Algorithms (Watchdog Agent™)

A neural network-based "digital doctor" inspired by biological perceptual systems and machine psychology theory (Lee, 1995), Watchdog Agent™ consists of embedded computational prognostic algorithms and a software toolbox for predicting degradation of devices and systems. The Watchdog Agent™ bases its degradation assessment on the readings from multiple sensors that measure critical properties of the process, or machinery that is being considered. It is expected that the degradation process will alter the sensor readings that are being fed into the Watchdog Agent™, and thus enable it to assess and quantify the degradation through quantitatively describing the corresponding change of sensor signatures. In addition, a model of the process or piece of equipment that is being considered, or available application specific knowledge can be used to aid the degradation process description, provided that such a model and/or such knowledge exist. The prognostic function of the watchdog is realized through trending and statistical modeling of the observed process performance signatures and/or model parameters. This allows one to predict the future behaviour of these patterns and thus forecast the behaviour of the process, or piece of machinery that is being considered.

Furthermore, the Watchdog Agent™ also has the diagnostic capabilities through memorizing the significant signature patterns in order to recognize situations that have been observed in the past, or be aware of the situation that was never observed before. Thus, the Watchdog Agent™ has elements of intelligent behaviour that enable it to answer the questions:

1) When the observed process, or equipment is going to fail, or degrade to the point when its performance becomes unacceptable
2) Why the performance of the observed process, or equipment is degrading, or in other words, what is the cause of the observed process or machinery degradation.
3) What is the most critical object, or process in the system with respect to maintenance, or repair;

In order to facilitate the use of Watchdog Agent™ in a wide variety of applications, with various requirements and limitations regarding the character of signals, available processing power, memory and storage capabilities, limited space, power consumption, personal user's preference etc, the performance assessment module of the Watchdog Agent™ has been realized in the form of a modular, open architecture toolbox. A toolbox that consists of different prognostics tools has been developed for predicting the degradation or performance loss on devices, process, and systems. include neural network based, time-series based, wavelet-based and hybrid joint time-frequency methods, etc. Open architecture of the toolbox allows one to easily add new solutions to the performance assessment modules as well as to easily interchange different tools, depending on the application needs. The tools currently realizing the signal processing, feature extraction, sensor fusion and performance evaluation modules of the performance assessment Watchdog Agent™ functionality are summarized in Figure 2.

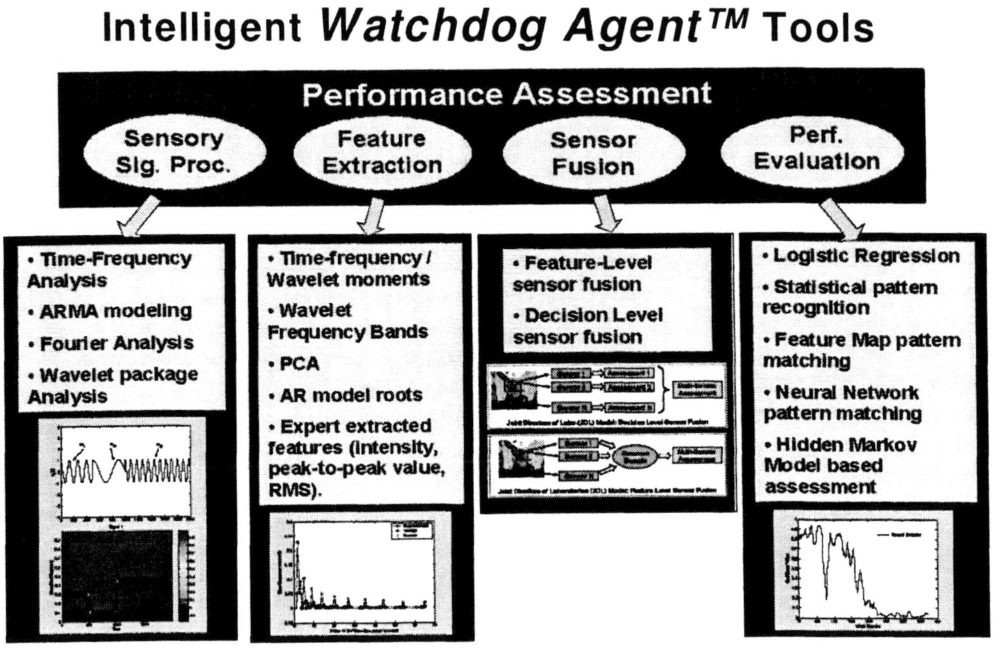

Figure 2 Performance Assessment Modules and Tools

Each of these modules is realized in several different ways to facilitate the use of Watchdog Agent™ in a wide variety of products and applications, with various requirements and limitations with respect to the character of signals, available processing power, memory and storage capabilities, limited space, power consumption etc.

- **Sensory processing module** transforms sensor signals into domains that are most informative of product's performance. Time-series analysis (Pandit *et al.*, 1993) or frequency domain analysis (Marple, 1987) could be used to process stationary signals (signals with time invariant frequency content), while wavelet (Burrus, *et al.*, 1998), or joint time-frequency domains (Cohen, 1995) could be used to describe non-stationary signals (signals with time-varying frequency content). Most real life signals, such as speech, music, machine tool vibration, acoustic emission etc are non-stationary signals, which places strong emphasis on the need for development and utilization of non-stationary signal analysis techniques, such as wavelets, or joint time-frequency analysis.
- **Feature extraction module** extracts features most relevant to describing the product's performance. Those features are extracted from the domain into which sensory processing module transforms sensory signals, using expert knowledge about the application, or automatic feature selection methods such as roots of the autoregressive time-series model, or time-frequency moments and Singular Value Decomposition.
- **Decision-level sensor fusion** is based on separately assessing and predicting process performance from individual sensor readings and then merging these individual sensor inferences into a multi-sensor assessment and prediction through some averaging technique.
- **Performance evaluation module** evaluates the overlap between most recently observed signatures and those observed during normal product operation. This overlap is expressed through the so-called Confidence Value (CV), ranging between zero and one, with higher CVs signifying a high overlap, and hence performance closer to normal. In case data associated with some failure mode exist, most recent performance signatures obtained through the signal processing, feature extraction and sensor fusion modules can be matched also against signatures extracted from faulty behaviour data. This matching allows the Watchdog Agent™ to recognize and forecast a specific faulty behaviour, once a high match with the failure associated signatures is assessed for the current process signatures, or forecasted based on the current and past product's performance. Figure 3 illustrates this signature matching process for performance evaluation.

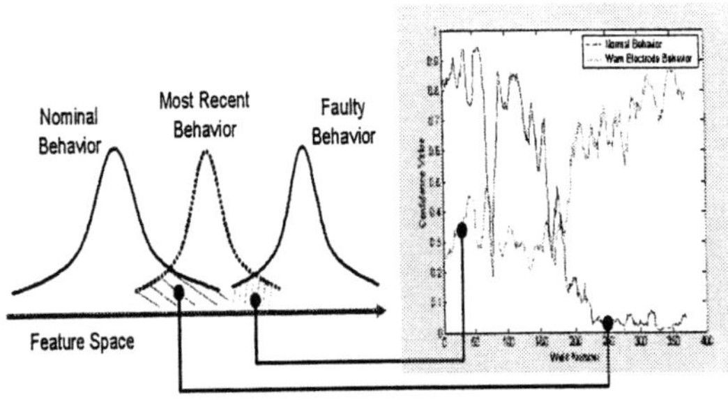

Figure 3. Performance Evaluation using Confidence Value (CV)

Realization of the performance evaluation module depends on the character of the application and extracted performance signatures. If significant application expert knowledge exists, simple but rapid performance assessment based on the feature-level fused multi-sensor information can be made using the relative number of activated cells in the neural network, or using the Logistic Regression approach. For open-control architecture products, the match between the current and nominal control inputs and performance criteria can also be utilized to assess the product's performance. For more sophisticated applications with intricate and complicated signals and performance signatures, statistical pattern recognition methods, or Feature Map based approach are employed. Over time, as new failure modes occur, performance signatures related to each specific failure can be collected and used to teach the Watchdog Agent to recognize and diagnose that failure mode in the future. Thus, the Watchdog Agent(TM) is envisioned as an intelligent device that utilizes its experience and human supervisory inputs over time to builds its own expandable and adjustable world model. Even though the performance CV already bares significant prognostic information about the remaining product's useful life, additional prognostic information can be extracted by capturing the dynamics of the product's behaviour and utilizing it to extrapolate and predict the product's behaviour over time. Currently, Autoregressive Moving-Average (ARMA) modelling and Match Matrix methods are used to forecast the performance behaviour. Figure 4 shows the prognostics and diagnostics methods using the developed tools.

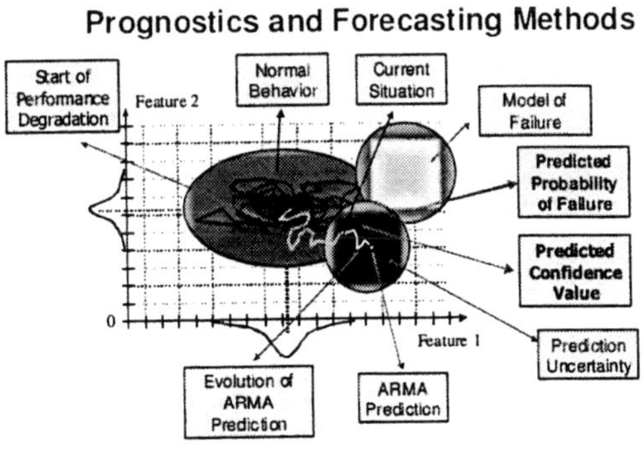

Figure 4. Prognostics and Performance Forecasting

This entire infrastructure realizing multi-sensor performance assessment and prediction could be even further enhanced if Watchdog Agent™ are embedded on identical products operating under similar conditions could exchange information and thus assist each other in building the world model. Furthermore, this communication can be used to benchmark the performance of "brother-products" and thus rapidly and efficiently identify under performing units before they cause any serious damage and losses. This paradigm of communication and benchmarking between identical products operating in similar conditions is referred to as the "peer-to-peer" (P2P) paradigm. Figure 5 illustrates the aforementioned Watchdog Agent™ functionalities supported by the P2P communication and benchmarking paradigm. Figure 6 summarizes a list of developed prognostics tools and their capabilities. Many of these algorithms have been validated on the industry testbeds.

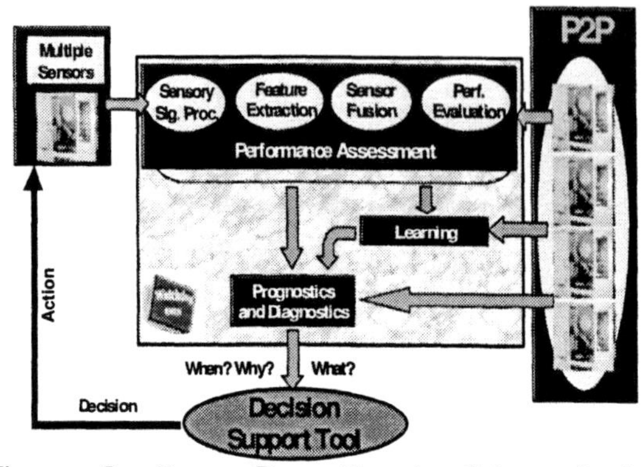

Figure 5: Peer-to-Peer Functionalities of the Intelligent Watchdog Agent™.

	H: Highly Capable / Applicable S: Somewhat Capable / Aplpicable L: Low Capablility / Applicability NA: Not Applicable ? : Still under research	Customer Requirements & Application Conditions										
		Non-stationary Signals	Stationary signals	High Frequency Signal	Low Frequency Signal	Highy Dynamic Systems	Sufficient System Knowledge	Insufficient System Knowledge	Interchangeability	Human Involvement	Low Cost Implementation	Possibility of Cost-Effective Embeddding
Feature Extraction Methods	Analysis	H	L	H	L	H	NA	NA	NA	NA	L	L
	Decomposition	H	L	H	L	H	NA	NA	NA	NA	S	S
	Autoregressive Modeling	L	H	S	H	H	NA	NA	NA	NA	H	H
	Fourier Transform	L	H	H	L	H	NA	NA	NA	NA	H	H
Performance Assessment Methods	Logistic Regression	NA	NA	NA	NA	L	H	S	S	S	H	H
	Feature Map	NA	NA	NA	NA	L	S	S	S	S	S	H
	Statistical Pattern Recognition	NA	NA	NA	NA	L	L	H	H	L	H	H
	Hidden Markov Model (HMM)	NA	NA	NA	NA	H	H	S	S	S	S	S
	Particle Filters Performance Assessment	NA	NA	NA	NA	H	H	S	S	S	S	S
Diagnostics Methods	Support Vector Machine (SVM)	NA	NA	NA	NA	L	H	S	H	S	S	H
	Hidden Markov Model (HMM)	NA	NA	NA	NA	H	H	S	L	S	L	S
	Bayesian Belief Network	NA	NA	NA	NA	S	H	L	L	S	S	S
	ART2	NA	NA	NA	NA	S	H	L	L	H	S	H
ognostics Method	Autoregressive Moving Average (ARMA)	NA	NA	NA	NA	S	S	H	H	L	H	H
	Compound Match Matrix / ARMA Prediction	NA	NA	NA	NA	H	H	S	H	L	H	S
	Fuzzy Logic Prediction	NA	NA	NA	NA	S	H	L	S	H	H	H
	Elman Neural Network Prediction	NA	NA	NA	NA	H	H	L	L	H	S	H

Figure 6. List of Developed Prognostics Tools and Application Conditions

Figure 7 shows the concept and preliminary results of predicting the behavior of performance related process signatures using ARMA modelling techniques. Load sensor readings from a boring machine spindle have been remotely collected and processed into Joint Time-Frequency (TF) distributions. Performance related signatures were extracted from the TF distributions using the TF moments and Principal Component Analysis (Casoetto *et al.*, 2003). ARMA modelling techniques were then utilized to predict the behavior of the extracted principal components, as indicated in Figure 7.

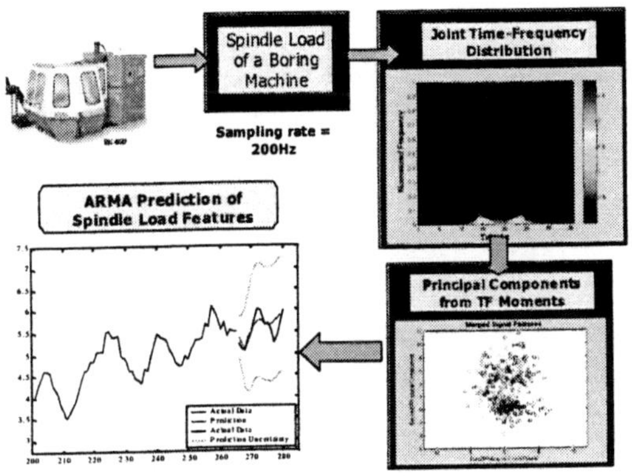

Figure 7: ARMA prediction of features extracted from the spindle load sensor readings.

A system methodology and platform has been developed that enables transformation of product data into more useful formats for remote and telematics applications. An integrated IMS Infotronics platform has been developed to allow scalability and reconfigurablity for many research activities. This platform provides a baseline system for IMS researchers and company members to develop next-generation intelligent product service systems (Figure 8).

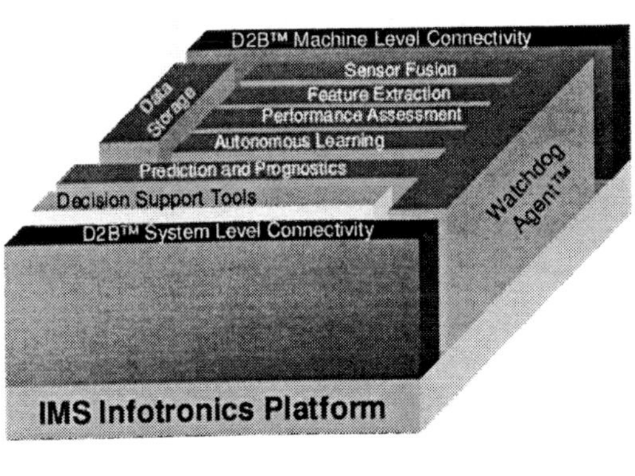

Figure 8 Integrated IMS Infotronics Platform

3.2 Device-to-Business (D2B™) Infotronics Platform

The rapid development of the web-enabled technologies is an important enabler for remote monitoring and prognostics. One critical issue in data transmission is the development of data transmission standards. One of the major barriers is that most manufacturers adopt proprietary communication protocols which led to difficulties to connect diverse machines and products. Currently. The IMS Center is developing a web-enabled remote monitoring Device-to-Business (D2B)(TM) platform for remote monitoring and prognostics of diversified products and systems. According to the standard for Open System Architecture for Condition-Based Maintenance (WWW.OSACBM.ORG), a typical CBM system consists of the following seven layers:

- Sensor Module
- Signal Processing
- Condition Monitoring
- Health Assessment
- Prognostics
- Decision making support
- Presentation

The Watchdog functionality expands this standard topology to a multi-sensor level and realizes sensory processing, condition monitoring, and health assessment and prognostics layers of the CBM scheme. The sensors and decision-making layers within an Intelligent Maintenance System are realized outside the Watchdog Agent.

4 IMPLEMENTATION EXAMPLES

Several Watchdog Agents™ for on-line performance assessment have already been developed and implemented as standalone applications in a number of industrial and service facilities. Below is a list of a few examples to illustrate the developed tools.

Example 1: Remote Monitoring and Predictive Maintenance of An Elevator System

IMS remote monitoring and maintenance platform has been validated and demonstrated in a typical mechatronics system such as an elevator door system. Figure 9 shows the system architecture of the testbed. Data acquisition system collects data from the elevator door controller, switches, encoder and some vibration sensors. The raw data is first processed and transformed into system performance index by Watchdog Agent™. A Watchdog Agent based on the Logistic Regression (Kleinbaum, 1994) and expert extracted features has been implemented on a commercial elevator door shown in Figure 10. Following recommendations from the elevator door manufacturer's maintenance experts, cycle time for door opening and closing as well as the maximum speed of the door are utilized to evaluate the performance Confidence Value (CV) of the elevator door using the Logistic Regression Function of the form

$$CV(x_1, x_2) = \frac{e^{\beta_0 + \beta_1 x_1 + \beta_2 x_2}}{1 + e^{\beta_0 + \beta_1 x_1 + \beta_2 x_2}} \qquad (1)$$

where x_1 and x_2 are the maximal speed of the elevator door motion and the door cycle time, respectively. Firstly, parameters β_0, β_1 and β_2 of the Logistic Regression curve were estimated by finding the nearest (in the least square sense) Logistic Regression function to the function that is equal to 1 in the training set representing the acceptable elevator door behavior, and equal to 0 in the training set representing the unacceptable elevator behavior. Then, performance degradation was assessed by evaluating the Logistic Regression function for the newly arrived elevator door performance features x_1 and x_2. The loser those features are to the normal elevator door behavior, the closer the performance CV calculated using the Logistic Regression would be to 1.

This performance index is then stored in database and transferred to remote monitoring system via the data-transferring layer that consists of the OPC server and the bridge program. When the machine performance index drops below certain threshold or a future possible failure is predicted, the synchronization agent will be triggered to invoke certain business transactions. In this testbed, it is realized by using ORACLE workflow API interface.

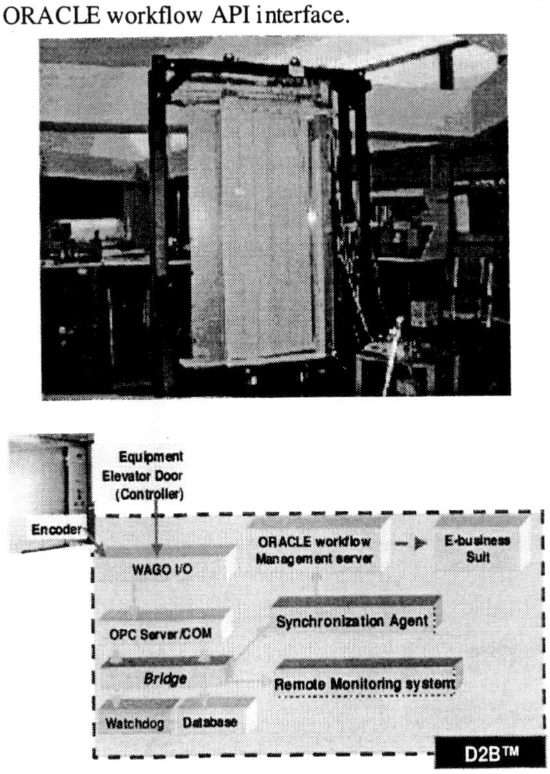

Figure 9 Remote Monitoring and Predictive Maintenance of An Elevator System

Fig.10 shows the daily average open cycle time over 400 days, Fig. 11 shows the change of average maximum angular speed, and then probability of failure at each time point is calculated. At last, performance degradation can be assessed using a confidence value as shown in Fig. 12.

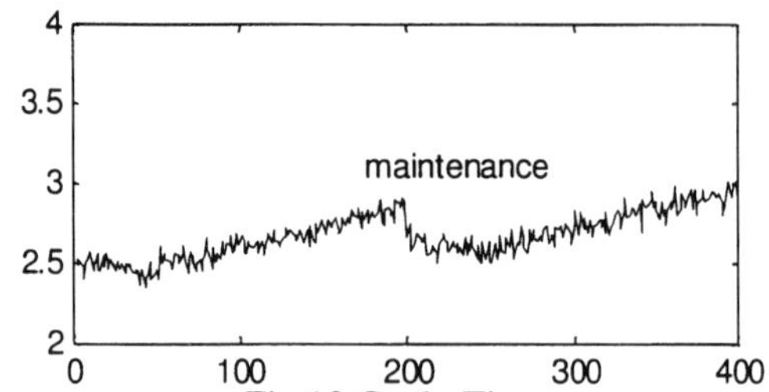

Fig.10 Cycle Time

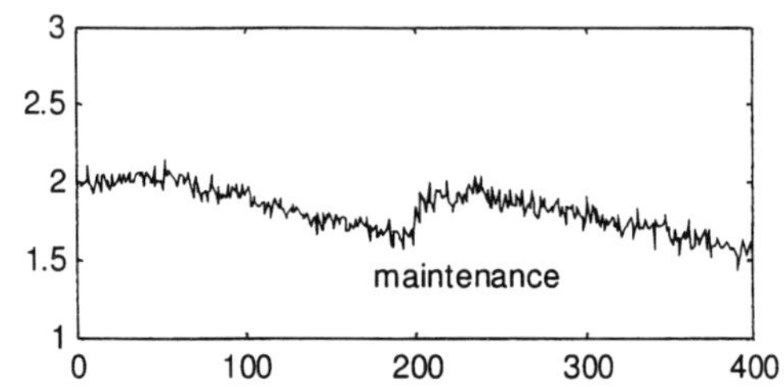

Fig.11 Maximum Angular Speed

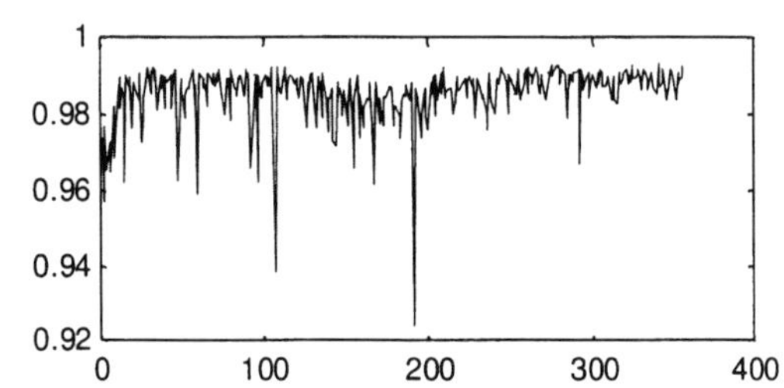

Fig.12 Confidence Value Calculation and Performance Degradation Prediction

Example 2: Prognostics of An AS/RS Materials Handling Systems

A time-frequency based method (Cohen, 1995) has also been implemented for performance assessment of a gearbox in an AS/RS material handling system shown in Figure 13. Four vibration sensor readings have been fused to autonomously and on-line evaluate its performance. The vibration signals were processed into joint time-frequency energy distributions (Cohen, 1995) and a set of time-shift invariant time-frequency moments (Zalubas, *et al.*, 1996; Djurdjanovic *et al.* 2000; Tacer B., 1996) was extracted. Since those moments asymptotically follow a Gaussian distribution (Zalubas, *et al.*, 1996), statistical reasoning was utilized to evaluate the overlap between signatures describing the normal process behavior (used for training) and those describing the most recent process behavior. Figure 14 shows a screenshot of the software application housing this time-frequency based Watchdog Agent used for performance assessment of a material handling system.

Figure 13: Material Handling System for Mail Staging

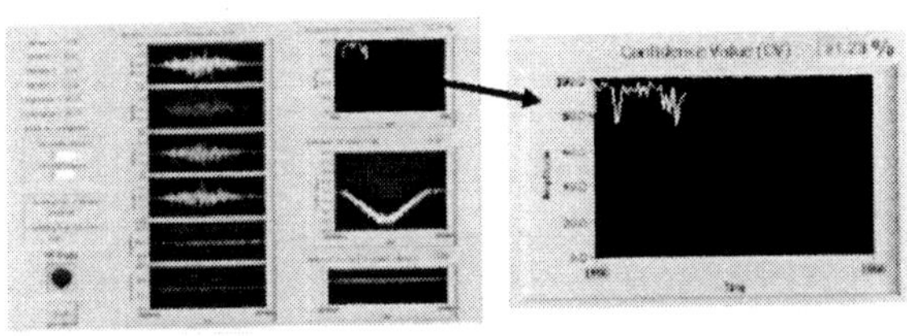

Figure 14: Screenshot of the Time-frequency based Watchdog Agent Assessing Performance of the Materials Handling System Shown in Figure 13.

Example 3: Remote Machine Tool Spindle Performance Assessment and Prediction

Remote monitoring of a vertical machining center is demonstrated in this example. The performance assessment is accomplished using the sensor readings from three vibration sensors mounted near the spindle bearing. Data collection was triggered by the M-code form the CNC machine. Each vibration sensor reading was sampled at 10KHz and stored on the web-server for remote access and subsequent processing Figure 15 schematically shows the experimental setup.

The vibration signals were acquired while spindle was running idle and under normal condition. Each spindle operation cycle consisted of the spindle running without load at 5 different speeds for one second at each speed, with the following sequence of rotational velocities:

1. 2000rpm for one second
2. 4000rpm for one second
3. 8000rpm for one second
4. 4000rpm for one second
5. 2000rpm for one second

In total, 2 hours and 12 minutes (735 continuous running cycles) of spindle operation signatures were obtained. Figure 16 shows typical vibration signals obtained from one CNC machine operating cycle.

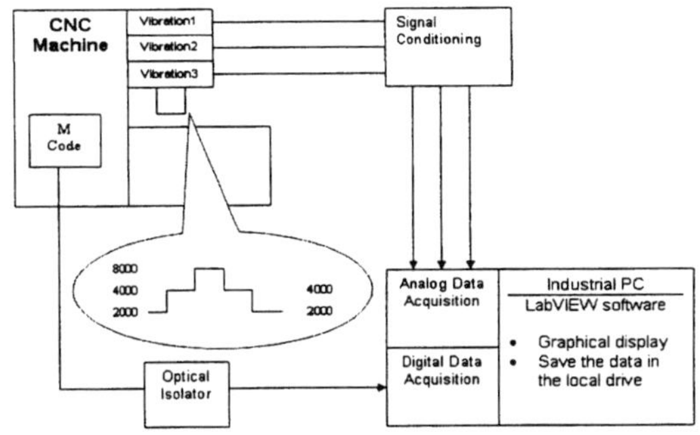

Figure 15: Experimental Set-up for Remote Spindle Monitoring and Performance Assessment

Sensor readings were decomposed into time and frequency localized energy coefficients using the wavelet packet transformation (Yen, 2000). In this example, Daubechies wavelets were used to calculate the wavelet packet transform for each signal of one machine operation cycle. In total, 64 levels of frequency domain resolution were obtained by utilizing a 6-level wavelet packet decomposition. From the three vibration signals decomposed into the wavelet domain, Trial and error technique was used to extract a total of 13 frequency band energies that were indicative of the spindle performance, but were not contaminated by the noise. In essence, if one can acquire failure-related performance signatures, it is possible to compare the energy difference between the normal- and failure-related data sets in order to select an appropriate feature set.

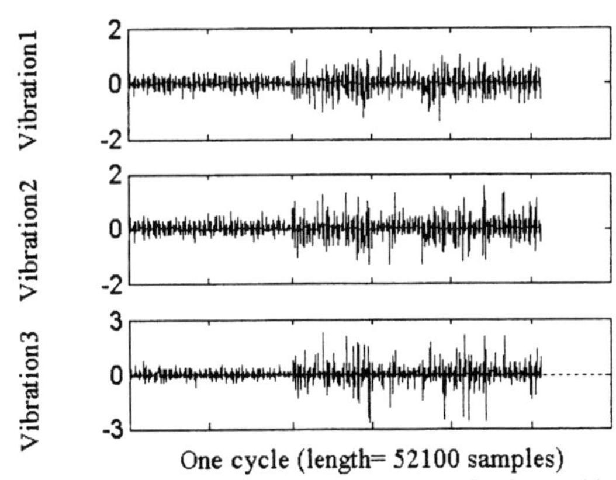

Figure 16. Vibration Signals from one CNC Machine Operating Cycle

Signal energy from the selected wavelet packet frequency bands were used as inputs into multiple Logistic Regression classifier for performance assessment. Logistic regression is a technique for analyzing problems where there are one or more independent variables that determine an outcome that is measured with a binary variable (variable that can take only two values). Logistic linear regression models are of the form

$$y = \frac{\exp(\alpha + \beta_1 x_1 + \beta_2 x_2 + \cdots + \beta_n x_n)}{1 + \exp(\alpha + \beta_1 x_1 + \beta_2 x_2 + \cdots + \beta_n x_n)} \qquad (2)$$

where $\vec{x}(x_1 \cdots x_n)$ denotes signal features and α and $\vec{\beta}(\beta_1 \cdots \beta_n)$ are parameters that are obtained through maximum-likelihood estimation, using each of the known model data sets. where denotes signal features and parameters that are obtained through maximum-likelihood estimation, using each of the known model data sets.

Figure 17 shows the performance Confidence Values (CVs) obtained from the acquired data, using the Logistic Regression classifier.

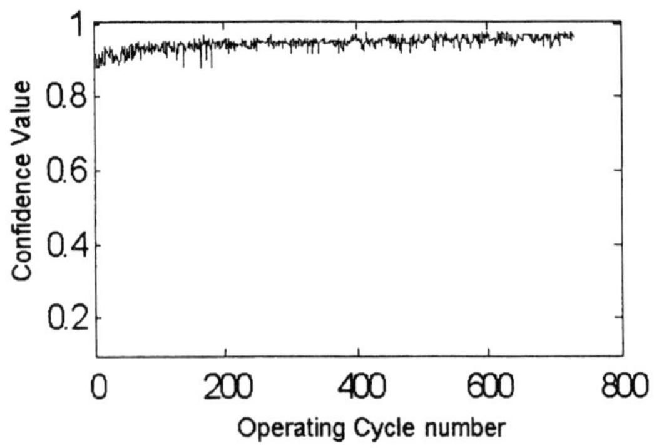

Figure 17: Spindle Performance Confidence Values obtained using the Wavelet Packet Decomposition and Logistic Regression

CONCLUSIONS AND FUTURE WORK

Watchdog Agent™ is a tool for multi-sensor assessment and prediction of machine or process performance. This tool can be utilized to realize predictive Condition-Based Maintenance, Collaborative Product Life-Cycle Management (CPLM) as well as identification of components with significant remaining useful life that could be efficiently and cost-effectively disassembled and reused. A wide variety of Watchdog Agent™ realizations have been devised to address applications of different nature and different levels of complexity and criticality. Innovative sensory processing and autonomous feature extraction methods are developed to facilitate the plug-and-play approach in which Watchdog Agent™ can be set and ran without any need for expert knowledge or intervention. Watchdog Agents™ have been successfully implemented in laboratory and industrial testbeds that include Watchdogs for performance assessment of an elevator door, vertical machining center, gearbox of a material handling device for staging and web-enabled E-Manufacturing testbed of a leading US automotive manufacturer.

Even though preliminary results of Autoregressive Moving Average (ARMA) modelling based behavior prognostics have already been obtained, performance forecasting will be more thoroughly investigated in the future work (knowing "when something is going to happen"). Historical data on the process performance, as well as the performance of similar systems in operation ("brother-systems") will be used to facilitate both the diagnostic and prognostic features through this "peer-to-peer" paradigm. Also, a link should be established with a decision-making module that will enable optimal maintenance action in a timely manner. This link will complete the Watchdog Agent™ functionalities and facilitate the OSA-CBM standard topology in which an Intelligent Maintenance System can operate as a near-zero down-time, self-sustainable and self-aware artificially intelligent system that learns from its own operation and experience (IMS Center, 2004) .

Finally, an embedded infotronics agent is needed to assess and predict product performance, and based on that prediction enable assessment of product's reusability, as well as its proactive maintenance. Embedding is crucial for creating an enabling technology that can facilitate proactive maintenance and life cycle assessment for mobile systems, transportation devices and other products for which cost-effective realization of predictive performance assessment capabilities is not implementable on general purpose personal computers (Lee, 2002). The main research challenge will be to accomplish sophisticated performance evaluation and prediction capabilities under severe power consumption, processing power and data storage limitations imposed by embedding. Successful realization of the embedded infotronics agent will have a profound effect on the society and could fundamentally improve both state-of-the-art technology and quality of life. For example, embedding this technology into manufacturing equipment will facilitate near-zero downtime operation, even when the general-purpose, PC-based Watchdog Agent™ is deemed not cost effective. Similarly, this embedded technology could be well suited to application in the transportation industry where it could reduce the logistical footprint associated with fleet maintenance. Finally, minimal additional research effort would be required to adapt the embedded infotronics agent based technology to perform continuous, advanced tele-care and tele-health functions for the early detection and prevention of diseases and disorders. This advance could thus improve the overall quality of living for the general public, and particularly for those whose abilities are changing due to aging.

ACKNOWLEDGMENTS

This research is supported by the NSF I/UCRC for Intelligent Maintenance Systems at the University of Wisconsin, Milwaukee and University of Michigan.

REFERENCES

Bonissone, Goebel, P.P. K. (1995) "Soft Computing Applications in Equipment Maintenance and Service", ISIE '95., Proceedings of the IEEE International Symposium, vol. 2 , 10-14 July 1995

Brotherton, T.; Jahns, G.; Jacobs, J.; Wroblewski, D. (2000); 'Prognosis of faults in gas turbine engines'Aerospace Conference Proceedings, 2000 IEEE , Volume: 6 , 18-25 March 2000 Page(s): 163 -171 vol.6

Burrus, C. S., Gopinath, R. A. and Haitao, G. (1998). "Introduction to Wavelets and Wavelet Transforms – A Primer", Prentice Hall, Upper Saddle River, NJ.
Byington, C. S., Roemer, M. J. and Galie, T. (2003). Prognostic Enhancements to Gas Turbine Diagnostic Systems, IEEE Aerospace Conference, Big Sky, Montana, March 2003.
Casoetto, N., Djurdjanovic, D., Mayor, R., Lee, J. and Ni, J., (2003). "Multisensor Process Performance Assessment Through the Use of Autoregressive Modeling and Feature Maps," Trans. of SME/NAMRI, Vol. 31, pp. 483-490
Cohen, L., (1995). "Time-Frequency Analysis", Englewood Cliffs, NJ: Prentice Hall.
Djurdjanovic D., et al. (2000). "Computerized Classification of Temporomandibular Joint Sounds", IEEE Transactions on Biomedical Engineering, Vol. 47, No. 8, 2000, pp. 977-984.
Djurdjanovic, D, Ni, J. and Lee, J. (2002). "Time-Frequency Based Sensor Fusion in the Assessment and Monitoring of Machine Performance Degradation", in the Proc. of 2002 ASME Int. Mechanical Eng. Congress and Exposition, paper number IMECE2002-32032.
Ferlez, R.J., Lang, D.C., (1998) Gear-Tooth Fault Detection and Tracking Using the Wavelet Transform, 52nd Meeting of the MFPT, Mar20-Apr 2, 1998
Garga, A.K et al., (2001).; 'Hybrid reasoning for prognostic learning in CBM systems' Aerospace Conference, 2001, IEEE Proceedings. , Volume: 6 , 10-17 March 2001 Page(s): 2957 -2969 vol.6
Goodenow, T., Hardman, W., Karchnak, M.,(2000) Acoustic emissions in broadband vibration as an indicator of bearing stress, IEEE Aerospace Conference Proceedings, v 6, 2000, p 95-122
Hansen, Robert J, . Hall, David L.; Kurtz, Stewart K. (1994) New approach to the challenge of machinery prognostics: Proceedings of the International Gas Turbine and Aeroengine Congress and Exposition, Jun 13-16 1994, p 1-8, American Society of Mechanical Engineers (Paper), 94-GT-3
Hardman, W.; Hess, A.; Sheaffer, (2000). A helicopter powertrain diagnostics and prognostics demonstration, IEEE Aerospace Conference Proceedings, v 6, 2000, p 355-366
IMS (2004), NSF I/UCRC Center for Intelligent Maintenance Systems, www.imscenter.net.
Kemerait, Robert C (1987)., New cepstral approach for prognostic maintenance of cyclic machinery , IEEE SOUTHEASTCON, 1987, p 256-262
Kleinbaum, D. G. (1994), "Logistic Regression", Springer-Verlag, New York.
Lee, J (1995). "Machine Performance Monitoring and Proactive Maintenance in Computer-Integrated Manufacturing: Review and Perspective", International Journal of Computer Integrated Manufacturing, Vol. 8, No. 5, pp. 370-380.
Lee, J. (1996) "Measurement of Machine Performance Degradation using a Neural Network Model", Computers in Industry, Vol. 30, pp. 193-209.
Lee, J. an Ni, J (2002). "Infotronics Agent for Tether-Free Prognostics," Proceeding of AAAI Spring Symposium on Information Refinement and Revision for Decision Making: Modeling for Diagnostics, Prognostics, and Prediction, Stanford Univ., Palo Alto, CA, March 25-27.
Liang, Enju; Rodriguez, Rodrigo J.; Husseiny, Abdo A. (1988), Prognostics/diagnostics of mechanical equipment by neural network, Neural Networks, v 1, no.1., 1988, p 33
Marple, S. L., (1987). "Digital Spectral Analysis," Prentice Hall, Englewood Cliffs, NJ.
McClintic, K., (1998) Feature Prediction and Tracking for Monitoring the Condition of Complex Mechanical Systems, The Pennsylvania State University, MS Thesis, PA, 1998
Pandit, S. M. and Wu, S-M. (1993). "Time Series And System Analysis With Application," Krieger Publishing Co., Malabar, FL.
Parker, B.E. Jr., et al., (1965) "Helicopter gearbox diagnostics and prognostics using vibration signature analysis", Proceedings of the SPIE - The International Society for Optical Engineering, vol. 1965, pp. 531-42, 1993.
Radjou, N. (2002). The Collaborative Product Life_Cycle, Forrester Research, May 2002.
Reichard, Karl M.; Van Dyke, Mike; Maynard, Ken (2000), Application of sensor fusion and signal classification techniques in a distributed machinery condition monitoring system, Proceedings of SPIE - The International Society for Optical Engineering, v 4051, 2000, p 329-336
Roemer, M.J. ; Kacprzynski, G.J.; Orsagh, R.F (2001). Assessment of data and knowledge fusion strategies for prognostics and health management, IEEE Aerospace Conference Proceedings, v 6, 2001, p 62979-62988
Seliger, G., Basdere, B., et al. (2002), "Innovative Processes and Tools for Disassembly", Annals of CIRP, Vol. 51/1, 2002, pp. 37-41.
Su, L.P.; Nolan, M.; DeMare, G.; Carey, D.R. (1999); 'Prognostics framework [for weapon systems health monitoring]' AUTOTESTCON '99. IEEE Systems Readiness Technology Conference, 1999. IEEE , 30 Aug.-2 Sept. 1999 Page(s): 661 –672
Swanson, D.C. (2000), Prognostic modelling of crack growth in a tensioned steel band, Mechanical Systems and Signal Processing, v 14, n 5, Sep, 2000, p 789-803
Swanson, D.C. (2001), A general prognostic tracking algorithm for predictive maintenance, Aerospace Conference, 2001, IEEE Proceedings. , Volume: 6 , 10-17 March 2001 Page(s): 2971 -2977 vol.6
Tacer B. and Loughlin P., (1996). "Time-Frequency Based Classification", SPIE Proc., Vol. 42, 1996, pp. 2697-2705.
Vachtsevanos, G.; Wang, P. (2001); 'Fault prognosis using dynamic wavelet neural networks' AUTOTESTCON Proceedings, 2001. IEEE

Systems Readiness Technology Conference , 20-23 Aug. 2001 Page(s): 857 –870
Wilson, Bary W. etc. (1999), Development of a Modular In-Situ Oil Analysis Prognostic System, International Society of Logistics (SOLE) 1999 Symposium, Las Vegas, Nevada, August 30 – September 2, 1999.
Yamada, A. and Takata, S. (2002) "Reliability Improvement of Industrial Robots by Optimizing Operation Plans Based on Deterioration Evaluation", Annals of the CIRP, Vol. 51/1, 2002, pp. 319-322.
Yen , G. G. and Lin, K.C. (2000) Wavelet Packet Feature Extraction for Vibration Monitoring. IEEE Trans. on Industrial Electronics. 47(3): 650-667, 2000.
Zalubas, E. J. et al (1996). "Shift and Scale Invariant Detection", in Proceedings of IEEE International Conf. on Acoustics, Speech
Casoetto, N., Djurdjanovic, D., Mayor, R., Lee, J. and Ni, J., (2003). "Multisensor Process Performance Assessment Through the Use of Autoregressive Modeling and Feature Maps," Trans. of SME/NAMRI, Vol. 31, pp. 483-490
Cohen, L., (1995). "Time-Frequency Analysis", Englewood Cliffs, NJ: Prentice Hall.
Djurdjanovic D., et al. (2000). "Computerized Classification of Temporomandibular Joint Sounds", IEEE Transactions on Biomedical Engineering, Vol. 47, No. 8, 2000, pp. 977-984.
Djurdjanovic, D, Ni, J. and Lee, J. (2002). "Time-Frequency Based Sensor Fusion in the Assessment and Monitoring of Machine Performance Degradation", in the Proc. of 2002 ASME Int. Mechanical Eng. Congress and Exposition, paper number IMECE2002-32032.
IMS (2004), NSF I/UCRC Center for Intelligent Maintenance Systems, www.imscenter.net.
Kleinbaum, D. G. (1994), "Logistic Regression", Springer-Verlag, New York.
Lee, J (1995). "Machine Performance Monitoring and Proactive Maintenance in Computer-Integrated Manufacturing: Review and Perspective", International Journal of Computer Integrated Manufacturing, Vol. 8, No. 5, pp. 370-380.
Lee, J. (1996). "Measurement of Machine Performance Degradation using a Neural Network Model", Computers in Industry, Vol. 30, pp. 193-209.
Marple, S. L., (1987). "Digital Spectral Analysis," Prentice Hall, Englewood Cliffs, NJ.
Pandit, S. M. and Wu, S-M. (1993). "Time Series And System Analysis With Application," Krieger Publishing Co., Malabar, FL.
Radjou, N. (2002). The Collaborative Product Life_Cycle, Forrester Research, May 2002.
Seliger, G., Basdere, B., et al. (2002), "Innovative Processes and Tools for Disassembly", Annals of CIRP, Vol. 51/1, 2002, pp. 37-41.
Tacer B. and Loughlin P., (1996). "Time-Frequency Based Classification", SPIE Proc., Vol. 42, 1996, pp. 2697-2705.
Yamada, A. and Takata, S. (2002) "Reliability Improvement of Industrial Robots by Optimizing Operation Plans Based on Deterioration Evaluation", Annals of the CIRP, Vol. 51/1, 2002, pp. 319-322.
Yen , G. G. and Lin, K.C. (2000) Wavelet Packet Feature Extraction for Vibration Monitoring. IEEE Trans. on Industrial Electronics. 47(3): 650-667, 2000.
Zalubas, E. J. et al (1996). "Shift and Scale Invariant Detection", in Proceedings of IEEE International Conf. on Acoustics, Speech and Signal Processing, Vol. 5, 1996, pp. 3637-3640.Radjou, N. (2002).

Copyright © IFAC Information Control Problems in Manufacturing, Salvador, Brazil, 2004

ELSEVIER
IFAC
PUBLICATIONS
www.elsevier.com/locate/ifac

KNOWLEDGE BASED SERVICES FOR TRADITIONAL MANUFACTURING SECTORS PROVIDED BY A NETWORK OF HIGH TECH SMES

Paolo Pedrazzoli, Claudio R. Boër

TTS -Technology Transfer System s.r.l.
ITIA-CNR Institute of Industrial Technologies and Automation

Abstract: Manufacturing enterprises, particularly SMEs (Small/Medium Enterprises), are quickly evolving according to market and products fast changes. Manufacturing machines, as a matter of fact, are a core enabling technology in a number of key industrial sectors (metal, wood, leather, stone, plastic, ...), which have common requirements for increased product customization and improved competitiveness in terms of reduced cost, shorter delivery times and improved quality.
In order to pursue these increasing needs of flexibility, manufacturing machines grow more and more complex and the tasks performed by these machines become complex too. ***The greater integration between the machine performances and the related process parameters becomes also a crucial requirement that the user of the machine has difficulties to grasp and control.*** Task planning and all the other operations related to the manufacturing machine (such as maintenance, configuration, training and so on), as a result, grow complex too and new competencies and larger knowledge of the process parameters, machine performances and their interaction are needed.
The approach here described, named KoBaS (after Knowledge Based Services), ***aims to provide a breakthrough in the current practices in the use of manufacturing machines***, by promoting the development of a set of tools of new conception meant to provide, on one hand, an advanced task and process planning and, on the other hand, machine configuration, maintenance, training and management support.
These novel tools, for task and process planning, ***will be provided as services*** (customized for every manufacturing machine) ***for less intensive R&D traditional industrial sectors, by a Network of High-Tech SMEs***, whose constitution is fostered by the actualisation of the approach.
This paper is meant to present the KoBaS approach, its relevance and its foreseen economic impact for the manufacturing sector. *Copyright © 2004 IFAC*

Keywords: Manufacturing Systems, Knowledge Engineering, Simulation

1. INTRODUCTION

The approach core idea is to promote a network of SMEs capable of providing KoBaS services, where the service is a customized knowledge based solution, developed, thanks to the methodologies and tools studied and realized within the approach framework, for the specific machine. (fig. 1)

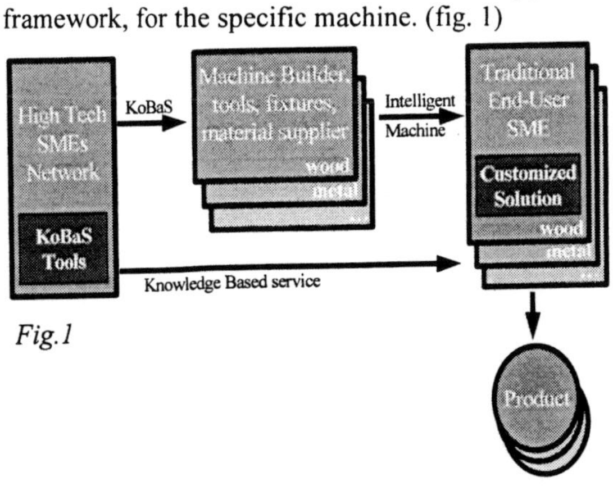

Fig.1

Thanks to this new core of innovative tools, the manufacturing machine itself becomes intelligent, being aware of its state and being able to provide the desired goods, starting from the virtual model of the goods to be produced and naturally interacting with the user.

In order to help the traditional manufacturing SMEs to become more competitive, it is necessary to bring an added value to the machines and systems they design and produce. The proposed approach will bridge the gap: small innovative enterprises organized in a network will bring added value to the traditional manufacturing sectors, providing a more easy to use, customized, cost efficient, open-source, modular but integrated set of software tools specifically designed for traditional manufacturing SMEs.

The basic idea is to promote, on one hand

- ***the evolution of single high tech SMEs towards a network of high-tech SMEs*** (supporting a growth in terms of competencies and resources and therefore creating a critical mass),

and, on the other hand

- ***the evolution of SMEs operating in traditional industrial sectors***, reinforcing and integrating competitiveness, innovation and sustainability and transforming intensive traditional SME sectors to become more RTD conscious. As SMEs actually represent a very large portion of the European manufacturing sector, KoBaS will thus promote European competitiveness.

The approach proposed is based on the realization of simulation and analysis tools, relying on knowledge based system, with high innovative added value, customized and applied to different sectors by the Network of SMEs, meant for PC based platform, thought to be software-license free and to be, above all, easy to use.

Created to be applied to different industrial frameworks, this set of solutions, starting from information related to geometry, materials, process knowledge and environment, will help the user, even not skilled, to produce an optimized task planning, to be aware of maintenance procedures, machine management and training, with a complete visualization of the processes, based on the actual interaction, transparent to the user, of all the above mentioned instruments. The foreseen tools will enhance machine performance under several aspects:

- Man-Machine interface based on low cost Virtual Reality
- Machine task programming based on Knowledge Based Systems
- Machine task Simulation in a three-dimensional customized environment
- Analysis services for the evaluation of the quality of the item produced
- Machine Configuration Support and Mechatronics
- Advanced Machine maintenance with autonomous decision making and leaning capabilities
- Machine training based on Virtual Reality techniques
- Machine management functions

The objective of KoBaS is to make the manufacturing process of a product as efficient as possible through the services proposed. The new technology developed will increase the functionality, flexibility and intelligence of the production process of SMEs, which will presumably lead to greater service value for the customer. The product will most probably be cheaper because of the high functionality of the production process. In this way ***the new product developed with the help of the KoBaS technologies will be improved in quality and improved products always also improve competitiveness, quality of life and working conditions.***

Therefore the KoBaS main breakthrough strategy can be expressed in terms of:

- ***the development of new knowledge based tools for an "intelligent" use and management*** of more and more sophisticated manufacturing machines.
- the above tools will be software ***license-free*** and will offer the possibility to extend and personalize the solutions.
- an innovative ***extended network of high-tech SMEs*** will be developed ***to use, to customize, to support and to make business out of the new developed tools.***
- ***traditional SMEs will evolve***, from simple users, ***to knowledge based users of knowledge base manufacturing machines*** increasing the added value offered to their customers and therefore increasing their market share
- ***the manufacturing machines producers will also improve their offer*** to the market with tools that will make easier the use and management of the sophisticated machines they sell.

2. KoBaS STRUCTURE

Figure 2 reports the paradigm and the future scenario meant for KoBaS, once the approach objectives are actualized. The manufacturing machine builder, confronted with the difficulty of the task and process planning of his equipment and willing to provide a new brand of powerful services to his customer, such as intelligent maintenance, asks to the Network of High Tech SMEs for a set of customized software solutions for his machine. The KoBaS Network thus provides, thanks to the new instruments and approaches developed, a set of KoBaS tools that will be sold, by the machine builder, to the end-user together with the manufacturing machine. The end-user of the manufacturing machine himself can ask to the KoBaS Network a tailored solution for his particular environment and final products.

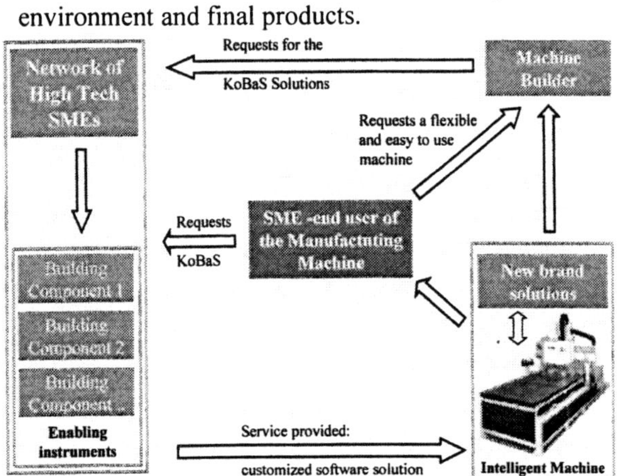

Fig.2

The objective is thus to support high quality, low cost, fault-tolerant, and flexible manufacturing machines with the study and realization of a completely new set of instruments, meant to provide the enabling software core of a new generation of instrumental goods.

The set of tools, or component, are founded on techniques such as Virtual Reality, 3D & Discrete Events Simulation, Knowledge Based Systems and Finite Element Analysis, evolved, combined and optimized in their interaction thanks to **a *new brand***

of innovative technologies that are the main object of KoBaS. The approach is meant to be flexible, adaptable, and scalable, to cope with the needs of a wide range of manufacturing centres for a number of traditional but strategic industrial sectors in Europe.

The basic instruments, developed and used by the Network to provide customized solutions, reflect the main aspects covered by the KoBaS approach in the use of a manufacturing machine:

1- ***Graphical User Interface Construction Component***: with this component the Network will be able to build and promptly configure a customized Man-Machine interface, based on low cost Virtual Reality, in order to make easily and naturally accessible all the functionalities and services to the end-user. The Interface will be capable of learning and adapting to the user environment and characteristic [A].
2- ***Rule-Based Knowledge Core Construction Component:*** this component is meant to provide the Network with the possibility to customize, for every manufacturing sector/machine confronted, a Rule-based Knowledge Base (process and geometry related), in order to support the machine task programming, configuration, maintenance and training through a knowledge based system [B].
3- ***Experience Data Base Construction Component:*** this module is used by the High Tech SMEs to build a Knowledge DB of past experience related with the Process handled. Once the cases are acquired and formalized, this module provides the methods for knowledge access and reuse. This module operates in strict connection with the Rule-Based Knowledge Core Construction Component [B].
4- ***Part Program Creator Component:*** with this module, the High Tech SMEs Network will be able to quickly provide a customized solution for the offline programming of the machine. For the part program creation, this module relies on the information provided by the knowledge base [C].
5- ***3D Simulation Construction Component:*** This module is used to create a customized simulation environment. It is meant to move the machine into the virtual environment as it would do in the real world, providing the framework to display the machine tasks or the training and maintenance procedures [D].
6- ***Analysis Construction Component :*** with this Component, the Network builds a customized FEM analysis. It provides pre and post processing analysis for the process performed by the manufacturing machine under consideration, giving a feedback on the performance of the Production Process [E].
7- ***Maintenance and Diagnosis module Construction Component:*** with this component, the Network will be able to build a customized maintenance service for the studied machine. The solution provided is meant to prevent failures, to suggest intelligent maintenance, to demonstrate maintenance procedures. This module should grow wiser (as the rule related modules) as it's used [F].
8- ***Training module Construction Component:*** the Network will use this component to develop a training module for the machine studied, intended to provide training support for the end-user of the machine. The component will also use the other components capabilities (such as simulation) to provide a training virtual environment for the machine end-user [G].
9- ***Machine Configuration and Mechatronics module Construction Component:*** this component will allow the SMEs Network to develop solutions for the configuration of the machine studied, according to final user needs [H].
10-***Build-in Management Functions Construction Component***: this component is meant to provide the network with the ability to develop customized solutions for the management function of the machine, such as cost planning, scheduling and optimization [B].
11-***Integration Component***: this component is used to build the framework for the solutions proposed, ensuring that all the other modules can communicate, establishing standards for the methods and classes for every customized solution realized. Within this Component also the data structure and data model for the specific solution will be defined.
12-***Web Service Enabler Component:*** this component will provide the means to remotely invoke the KoBaS intelligent-machine functionalities, developed thanks to the other Components, utilizing standard Internet-based protocols

The schema, presented in figure 3, will provide a global view of the KoBaS philosophy.
On the left side of the schema, the basic Components of KoBaS are highlighted. These are tools that must be developed in the framework of the KoBaS approach for the Network of High Tech SMEs. These Components represent the enabling technologies core, used by the Network of SMEs, to provide customized software tools for a large number European key industrial sectors of manufacturing machine builders and users.
On the right side, the schema shows a general architecture of a KoBaS solution implemented for a specific manufacturing machine. The software, realized by the Network with the building Components and customized for the machine, is installed on the machine itself, providing a whole set of solutions: from Part Program creation to Maintenance service and so on.

3. RELEVANCE

The approach will contribute to ***the development of new knowledge based added value products and services in traditional less RTD intensive industries***. The objective is to promote the shift of the industrial

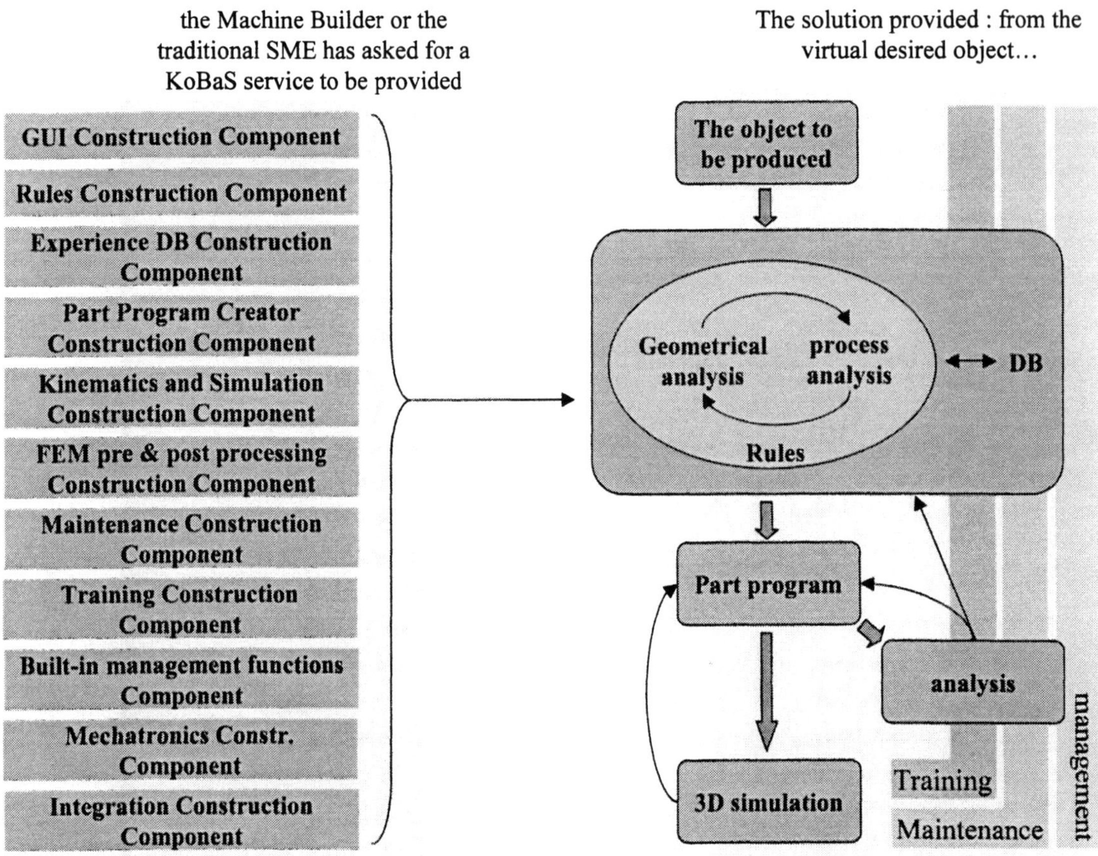

Fig.3 The SMEs Network, thanks to these Components, provide the Solutions ...to the real object, plus Maintenance, Management and Training support

sectors, which are traditionally less RTD intensive, to high-added-value sectors.
To do so, there is a need to reconsider industrial products and processes making them more knowledge based than resources intensive. The actualisation of the approach will increase quality of production and avoid trial-and-error, not depending on intuition, reducing lead time and scrap, designing sustainable product in order to answer to the ever changing consumer and market needs.
The focus of KoBaS is on the future generation of technologies, where high innovative software, supported by new generation of sensors and nanotechnologies, will be integrated into every manufacturing machine, rendering accessible all the functionalities and services through easy-to-use human interfaces. This vision places the user at the centre of future developments for a knowledge based approach to manufacturing.

4. ECONOMIC IMPACT

Knowledge-based services and technology are revolutionizing the economy. Successful companies, whether in manufacturing or services, will derive their competitive edge not from transitorily superior products but from a deep understanding of a highly developed knowledge core competencies. Rarely owning the largest raw materials resource, manufacturing plants, equipment or integrated facilities will provide a maintainable competitive edge for major companies. Such physical properties are too easily cloned or bypassed.
KoBaS will strongly contribute to economic growth in Europe because it will have a ***major impact on European SME and Industry competitiveness and productivity achieved thanks to Knowledge Based Services.***
KoBaS will have an economic macro-impact: ***the tools promoted will be exploited by SMEs belonging to different traditional industrial sectors*** (metal, wood, leather, stone, plastic, ...), establishing a new standard for manufacturing machine productivity and usability. Therefore, a large number of small and medium enterprises will have the chance to use the KoBaS tools with benefits both at economic and technical level. An increase in the competitiveness of the SMEs is thus expected.
On the other hand, the ***IT SMEs will gain a considerable increased know how and experience and will benefit from the knowledge sharing and development synergy promoted by the establishment of the Network of High Tech SMEs, thus settling the bases for a sustainable and lasting growth***.

Traditional SMEs and Industries will greatly benefit from the approach results, ***gaining the possibility, on one hand, to use intelligent manufacturing machines and, on the other hand, to sell them*** .This will result in improved customer relations, increased market share, increase of productivity and optimization.

High Tech SMEs, the core of the constituting Network, will obtain ***increased profits achieved thanks to the new services provided and to market perspective enlargement*** (Statistical data, taken from the "Observatory of European SMEs, 2002 / No. 1", show that the Europe has, in total, about 20.400.000 SMEs and that the 11% are working in the manufacturing area: 2.200.000 SMEs with a average turnover of 0,6 MEuro each. The foreseen potential market for KoBaS is thought to be about the 9-10% of this number) (see Fig.4).

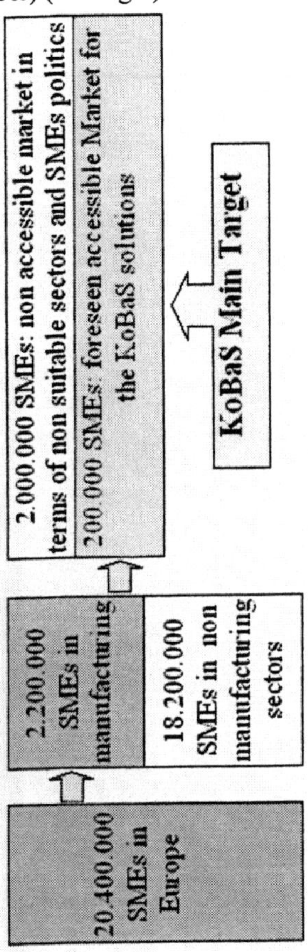

Fig.4

On the other hand, the High Tech SMEs will achieve ***several indirect benefits such as increased know-how thanks to the research work done and to the knowledge sharing within the network.***

5. SOCIAL OBJECTIVES

Strictly connected with the macro-economic impact, the approach will have a social impact. ***The growth of the SMEs competitiveness and, consequently, the capacity of the SMEs to gain new market shares, will create new job opportunities with positive repercussion on the European unemployment rate.***
In addition, due to the introduction of "intelligent machines" supported by the KoBaS tools, a new level of knowledge for workers will be required. As a consequence, each employee will be more involved in the production process. This task enrichment will have positive impact both on production and in the role that each employee plays in the enterprise.
Technological development and social innovation are the base for industrial growth and for the creation of new job opportunities; that is why the results of KoBaS will open new work options and will improve employment prospects. The approach promotes the efficiency, intelligence and flexibility of the manufacturing process and these factors lead to the higher competitiveness of manufacturing companies because they ensure high quality and good market for the product. When a company is more competitive it has the potential to grow and develop and therefore, to open new working places. Moreover, good technology for product development attracts new investors, which again leads to company growth and increase in number of employees.
The KoBaS promotes the high efficiency of product development processes and in this way promotes the conservation of natural resources.

6. STATE OF THE ART

Currently, for the manufacturing industry at large, ***the service and support in the use of a manufacturing machine is strictly dependent on the machine builder*** whatever the nature of the user's needs (programming help, functionality comprehensiveness, quality of results, knowledge acquisition, training, maintenance).
Concerning simulation, analysis, part program creation, training and maintenance activities, ***the current situation is based mainly on equipments without real embedded functionalities.*** The majority of equipment manufacturer presents products with classic support without exploring the association of powerful information technologies coupled to the equipments. ([D] – P.Pedrazzoli)
Nowadays, the continuously growing performance of modern processor enables the implementation of very complex software on PC platform. Then the approach foresees the use of powerful engineering software tools such as 3D process simulation, kinematics and virtual reality as well as finite element analysis, etc. at any stage of the product development and manufacturing process and thus to democratize those high end systems by opening their access and use to SMEs. ([D] – IMTI21, VRML)
Until now, those tools have been reserved only to the heavy applications needed at earlier development stages, such as layout design, collision analysis, assembly checking, design simulation and other time-consuming process flow analysis. Very often they require qualified users, therefore their usage are generally expensive and restricted to skilled engineers.
The KoBaS fosters the realization of very innovative, integrated and very easy to use software solutions for the analysis, diagnosis, simulation, verification, cost calculation and optimization of each stage of the manufacturing process chain. The vision is to shift the use of the tools above mentioned from the design phase to the end-users, integrating the above techniques with emerging

knowledge based systems, in order to obtain a new brand of software tools.
Nowadays there are basically no adequate solutions available for the whole management of a manufacturing machine based on the exploitation of these innovative technologies.
The approach proposes a significant and totally new effort in order to develop new tools meant to make these instruments easy and useful to the end-user. Thus the user becomes aware of the potential of the machine he is using, gaining the possibility to exploit the latent flexibility and capacity.

The KoBaS approach also emphasizes ***the introduction of standardization of product data throughout its entire lifecycle***. Basically the standard is the smooth and seamless exchange of part information from conception to part production. Thanks to the recent worldwide initiatives (in Europe, the USA and Asia) the STEP standard is spreading to the NC manufacturing process in the form of STEP-NC. Therefore within the KoBaS approach, the standard STEP will be analyzed and prototyped in order to encompass the global exchange of product data between the new software tools. ([C] – S.T.Newman)
Although the current Part Program format is based on a standard (ISO 6983), each manufacturing machine builder has extended the G-code to include proprietary functions and adaptations (e.g. cycles). At this time, for the same manufacturing technology, a specific post-processor is needed for each machine type and each Numerical Controller. The end-user is then confronted to the difficulties of porting the program on different machines, changing the tool parameters if needed, editing the vendor specific extensions code, etc. The new international standard ISO 14649 (also known as STEP-NC) defined for multiple manufacturing technologies (milling, turning, EDM, grinding, measuring, etc.) will bring more complete data (features, working steps, technology) to the NC of the machine-tool and give the ability to build more "intelligent" machines.

The great majority of the machines produced by the European manufacturers have not built-in monitoring, diagnosis, simulation and analysis abilities that would help the user to identify the proper course of actions or the cause of a failure. A breakthrough in the state of the art is to offer an intelligent machine capable of assuring its service. Furthermore, the new technologies in sensors and nanotechnologies presently under development, will bring an added dimension to the machine "intelligent" that will need to be deal with.
The main goal of the approach is thus to increase the ability of the machine builders to offer machines with high technology content to match customer purposes and therefore, at the same time, to offer to the same customers the possibility to use more efficiently these high technology machines. Traditional machine builders are not able to provide their customers with the last technologies, because of the related efforts. ***The possibility of having a support from a Network of High-Tech SMEs to offer powerful solutions fully integrated in the CNC to various manufacturing machine type is far beyond the state of the art.***

Finally an aspect that ***needs a breakthrough research is the study, setup and test of methodologies and tools for an efficient management of a Network of High-Tech SMEs*** [I]. Several studies have been conducted and are still in progress on concepts like virtual factory, extended enterprise, fractal factory, etc. The main challenge in KoBaS will be the determination of the rules for the set-up and management of the Network of High-Tech SMEs who have normally a short history, a fast growth and act a in a very dynamic market. These SMEs are generally very concerned of their own knowledge and intellectual properties and they tend to work in a very independent way. Another important aspect is related to the fact that very often such High Tech SMEs are working very close with their parent institution from whom they have spin-off. And, of course, the main challenge will be ***the study of the business set-up of the Network of High-Tech SMEs to service in efficient and fast way the market of traditional SMEs and how they may influence and help the transformation of the traditional SMEs from resource base to knowledge base.*** The business will be mainly oriented towards providing a service rather than selling software and concepts of (mass) customization will need to be studied and applied.

CONCLUSION AND FUTURE WORK

The growing needs of flexibility, performance and user support of the traditional manufacturing machines have been addressed and pointed out.
This paper presents the approach, named KoBaS, conceived and studied in order to answer to these needs: the funding idea is to cast the basis and to promote the constitution of a network of High Tech SMEs capable of providing KoBaS services to the manufacturing sector, where the service is a customized knowledge based solution, developed thanks to the methodologies and tools studied and realized within the KoBaS framework.
Based on this idea, an European Project within the VI Framework Program has been presented and retained.

REFERENCES

[A] VR and graphical user interfaces
B. Koleva, H. Schnädelbach, S. Benford, C. GreenHalgh, *"Traversable Interfaces Between Real and Virtual Worlds"*, 2000, Conference on Human Factors, The Hague, The Netherlands.
B. Ullmer, H. Ishii, *"The metaDESK: Models and Prototypes for Tangible User Interfaces"*, *1997*, 10th Annual Symposium on User Interface Software and Technology, Alberta, Canada.

G. Bishop, H Fuchs, et al., *"Research Directions in Virtual Environments"*, 1992, Computer Graphics vol. 26, p153

G. Viganò, Z. Liao, M. Sacco, S. Mottura, C. R. Boër, *"To Build Object-Oriented Virtual Factory Based on V.R.. and Simulation Tools"*, 2002, Digital Enterprise Technology - 1st CIRP (UK) International Seminar, Durham, UK.

I. Poupyrev, S. Weghorst, M. Billinghurst, T. Ichikawa, *"A Framework and Testbed for Studying Manipulation Techniques for Immersive VR"*, 1997, Symposium on Virtual Reality Software and Technology, Lausanne, CH.

J. Wilson, M. Sacco, H. Bullingher, H. Tamura , et alter, *Proceeding HCI-Conference Heraklion*, 2003, Greece

M. Eisenhauer, B. Hoffman, D. Kretschmer, *"State oj the Art, Human Computer Interaction", 2002*, Final deliverable, Giga Mobile Project Telematica Instituut, The Netherlands.

M. G. Helander, K. Landauer, P. V. Prabhu, *"Handbook of Human-Computer Interaction"*, 1997, Noth Holland

[B] Knowledge Management

A. Berztiss, *"Capability Maturity for Knowledge Management, in proceedings of Database and Expert Systems Applications"*, 2002, DEXA

A. Hopgood, *"Intelligent systems for engineers and scientists"*, 2000, CRC Press

J. Durkin, *"Expert Systems: design and development"*, 1998, MacMillan

J. Giarratano, G. Riley, *"Expert Systems: principles and programming"*, 1998, PWS

J. Kolodner, *"Case-Base Reasoning"*, 1993, Morgan Kaufmann Publishers

K. Parsaye, M. Chignell, S. Khoshafian, H. Wong, *"Intelligent Databases, Object Oriented, Deductive Hypermedia Technologies"*, 1989, John Wiley & Sons

K. W. Darlington, *"The Essence of Expert Systems"*, 2000, Prentice Hall

P. Assogna, *"The Dreaming Knowledge Base, in proceedings of DB and Expert Systems Applications"*, 2002, DEXA

P. Jackson, *"Introduction to Expert Systems"*, 1998, Addison-Wesley

S. J. Russell, P. Norvig, *"Artificial Intelligence, A Modern Approach"*, 1995, Prentice Hall

[C] Part Program and STEP-NC

B.K. Choi, *"Intelligent tool-path generation for high performance NC machining"* 2003, part 1&2, ed.: Computer-aided design Elsevier

C.R. Boër, M. Sacco, L. Dan, *"Distributed CAD/CAPP/CAM integration based on STEP-PDM schema and Corba-compliant PDM Enablers"*, 2001, IceCE2001, Xi'an, P.R. China

D. Kiritsis, *"High precision interpolation algorithm for 3D parametric curve generation"*, 1994, Computer Aided Design; Vol. 26, No. 11

Matthias Müller, *"A new approach for trajectory generation in 5-axis machining"*, EPFL, 2002

N.Karapatis, J.-P.Van Griethuysen, R.Glardon, *"Direct Rapid Tooling: a Review of Current Research"*, 1998, Rapid Prototyping Journal

P. Dejonghe, *"An integrated approach for tool path planning and generation for multi-axis milling"*, 2001, KUL

Peter Smid, *"CNC programming handbook, a comprehensive guide to practical CNC programming"*, 2003, New York : Industrial Press

S.H. Suh, *"STEP-NC technology realizing information-based intelligent manufacturing system"*, 2002, J. of Korean Society of Precision and Engineering, Vol 19

S.T. Newman, R.D. Allen, R.S.U. Russo Jr, *"CAD/CAM solutions for STEP Compliant CNC Manufacture"*, 2002, 1st CIRP (UK) Seminar, Durham UK

[D] 3D Simulation

C. L. Collins, J. M. McCarthy, A. Perez,, H. Su, *"The Structure of an Extensible Java Applet for Spatial Linkage Synthesis"*, 2002, ASME Journal of Computing and Information Science and Engineering

Delmia Igrip approach - http://www.deneb.com.au /simulation-products/IGRIP.asp

Guilin Yang I-Ming Chen, *"A Novel Kinematics Calibration Algorithm for Reconfigurable Robotic Systems"*, 1997, IEEE

http://home.t-online.de/home/Cephalos/home_en.htm

IMTI21, *"Modeling and Simulation for Affordable Manufacturing"*, 2002, http://www.imti21.org/msam/ afworkshop.htm

J. M. McCarthy, *"Mechanisms Synthesis Theory and the Design of Robots"*, 2000,Proc. Int. Conf. Robotics and Automation in San Francisco

M. Rohrmeier *"Web Based Robot Simulation Using VRML", 2000,* WSC

P. Pedrazzoli, R. Castellani, F. Mancini, C. R. Boer, *"A new approach to programming flexibility for CNC machines, An example of application in woodworking machining"*, 2003, EVEN, Gallipoli, IT

S. Anton, *"EASY-ROB, 3D Robot Simulation Tool. Overview & BASIC"*, 2000, http://www.easy-rob.de/main.html

S. Dierssen,, *"Systemkopplung zur komponentenorientierten Simulation digitaler Produkte"(Coupling systems in order to simulate digital products based on component)*, 2002, Zentrum für Produktentwicklung, Diss. Nr. 14762, Zürich, ETH.

Scholz-Reiter, M. Kleiner, G. Spur, *"Integrated simulation of manufacturing processes in CAD systems – Ideas and concepts"*, 1996, CIRP Annals, Vol. 45. "Methods and Tools for Computer Integrated Manufacturing"

T. Hudson, M. Lin, J.Cohen, S Gottschalk, D, Manocha, *"V-COLLIDE: Accellerated Collision Detection for VRML"*,1998, Proc. VRML'97

Tecnomatix approach - http://www.tecnomatix.com/

VRML approach by ListaStudio - http://lista.it/vrml/

Yang Guilin, *"Design and Kinematics analysis oj modular reconfigurable parallel Robots"*, 2001, SIMTech Technical Report

[E] Analysis Service

E.C. De Meter, *"Fast support layout optimization"*, 1998, International Journal of Machine Tools and

Manufacture. v 38 n 10-11 Oct/Nov, p 1221-1239

G. Liao, S. J. Hu, *"Flexible multibody dynamics based fixture-workpiece analysis model for fixturing stability"*, 2000, International Journal of Machine Tools and Manufacture. v 40 n 3, p 343-362

I. M. Deiab, S. C. Veldhuis, M. Dumitrescu, *"Dynamic modeling of face milling process including the effect of fixture dynamics"*, 2002, Technical Paper Society of Manufacturing Engineers.MR. n MR02-150, p 1-8

S. Jayaram, B.S. El Khasawneh, D.E. Beutel, *"Fast analytical method to compute optimum stiffness of fixturing locators"*, 2000, CIRP Annals Manufacturing Technology. v 49 n1, p 317-320

S. Kashyap, W. R. DeVries, *"Finite element analysis and optimization in fixture design"*, 1999, Structural Optimization. v 18 n 2-3 Oct, p 193-201

[F] Maintenance

C. Idhammar, Et Al., *"Preventive Maintenance, Essential Care and Condition Monitoring"*, 1999, IDCON edition

C. Nelms, *"What you can learn from things that go wrong"*, 1994, Failsafe Network, Richmond Virginia

J. Moubray, *"Reliability-Centered Maintenance"*, 2001, Industrial Press, Inc., 2nd Revision edition

P. Silva, A. Soares, *"The Importance of Remote Maintenance for Equipment Manufacturers Competitiveness and their Customer Satisfaction"*, 2000, WWDU 2000 Conference

T. Remus, T. Remus, *"Ultimate Sheet Metal Fabrication"*, 1999, Wolfgang Productions.

[G] Training

B. Robinson, *"Teacher education through open and distance learning"*, 2003, Routledge/Falmer, London

C. Bauer, *"Value oriented use of virtual reality in companies – applications, cost effectiveness, provider"*, 1996, Computerwoche Verlag, Munich

C. Feyten, J. Nutta, *"Virtual instruction : issues and insights from an international perspective"*, 1999, Impressum Englewood, Colo. : Libraries Unlimited

D. Morrision, *"E-learning strategies how to get implementation and delivery right first time"*, 2003, Chichester, Wiley

H. Adelsberger, *"Handbook on information technologies for education and training"*, 2002, Springer, Berlin

M. Meier, M. Bichsel, W. Elspass, U. Leonhardt, M. Wohlgensinger, E. Zwicker, *"Modern tools for an efficient use of product data within the product life-cycle"*, 1999, Konstruktion Heft 9/99, S. 11-18

R. Heger, *"Development of an interactive system for the design and diagnosis of assembling processes"*, 1998, Springer Verlag, Berlin Heidelberg New York

S. Dierssen, *"System coupling for a component-oriented simulation of digital products"*, 2002, Fortschritt-Berichte Reihe 20, Nr. 358; VDI Verlag GmbH, Düsseldorf

[H] Mechatronics

"VDI-guideline 2206 Design methodology for mechatronic systems" (Design methodology for mechatronic systems), 2003, VDI-Verlag, Düsseldorf

A. Günter, C. Kühn, *"Knowledge-Based Configuration - Survey and Future Directions"*, 1999, from "XPS'99: Konwedge-based Systems".F. Puppe. Berlin, Germany, Springer Verlag

B. Faltings, R. Weigel, *"Constraint-based Knowledge representation for configuration systems"*, 1994, Technical Report TR-94/59, Lausanne, Switzerland, Dėpartement d'Informatique, EPF Lausanne.

B. Seybold, *"Techniques for Satisfying Assembly Constraints", 1999,* Diss. Nr. 13227, Zürich, ETH.

C. Puls, *"Die Konfigurations- & Verträglichkeitsmatrix als Beitrag zum Management von Konfigurationswissen in KMU"* (Managing configuration knowledge in SME's using the configuration- & compatibilitymatrix), 2003, Zentrum für Produktentwicklung. Diss. Nr. 14912, Zürich, ETH.

J. Tiihonen, T. Soininen, "Product Configurators Information System Support for Configurable Products", 1997, Helsinki University of Technology, Laboratory of Information Processing Science.

L. Bongulielmi, *"Die Konfigurations- & Verträglichkeitsmatrix als Beitrag zur Darstellung konfigurationsrelevanter Aspekte im Produktentstehungsprozess"* (Presenting configurable relevant aspects using the configuration & compatibilitymatrix), 2003, Zentrum für Produktentwicklung. Diss. Nr. 14904, Zürich, ETH.

The configuration & compatibility matrix http://www.k-v-matrix.info/

U. Leonhardt, *"Digitales Produkt: Beispiel einer Integrationsplattform für Technik- und Verkaufsprozesse mittels Informations- und Visualisierungstechnologien"* (Digital Product: An example of an integrationplatform for sales and development using information and visualization technologies), 2001, Fortschritt-Berichte: Reihe 20 Nr. 340. Düsseldorf, VDI.

[I] Network of HT-SMEs management

C. Boër, F. Jovane, *"Towards a New Model of Sustainable Production: ManuFuturing"*, 1996, CIRP Annals, Vol.45

C. Boer, F. Jovane, *"The organization of the Self-Innovative extended Factory in the ManuFuturing Model"*, 1997, IFIP

C. Boër, R. Fornasiero, A. Avai, *"A performance measurement system in an industrial collaborative environment"*, 2002, 1st CENNET Workshop on Digital manufacturing and business– China World Trade Center, Beijing

E. Filos, E. P. Banahan, *"Will the organization disappear? The challenges of the new economy and future perspectives"*, 2000, E- Business and Virtual Enterprises 3-20

M.T. Martinez, P. Fouletier, K.H. Park, J. Favrel, *"Virtual enterprise - organization, evolution and control"*, 2001, International Journal of Production Economics, pp 225-238 – n°74,

Copyright © IFAC Information Control Problems in Manufacturing, Salvador, Brazil, 2004

ELSEVIER
IFAC
PUBLICATIONS
www.elsevier.com/locate/ifac

IMPROVING E-MANUFACTURING EFFICIENCY THROUGH THE ONTOLOGICAL FILTERING SYSTEM

Raffaello LEPRATTI
Ulrich BERGER

Brandenburg University of Technology at Cottbus
Chair of Automation Technology
P.O. Box 10 13 44, 03013 Cottbus, Germany.
{lepratti, u.berger@tu-cottbus.de}
Tel: +49 - 355 - 69 24 57
Fax: +49 - 355 - 69 23 87

Abstract: In today's global market scenario enterprises unit building collaborative environments, whereby global manufacturing represents a key issue for ensuring market share. An unequivocal communication form among all involved process agents, i. e. human and machines, constitutes a crucial requirement. Referring to the shop-floor level, this isn't easily achievable due to the different professional and cultural backgrounds of the involved machine operators. In this paper an innovative man-machine interaction concept for simplifying human tasks will be presented. A prototype based on the Ontology Filtering System has been developed. It enables formalisation of natural language instructions for multi-purposes in automation technology. Preliminary trial tests have been performed and show suitability towards a future use in the production field. *Copyright © 2004 IFAC*

Keywords: Artificial Intelligence, Knowledge-based Systems, Man-Machine-Interaction, Manufacturing System, Natural Languages, Programming Systems, Robot Programming.

1. STRUCTURE OF THE PAPER

This paper is organized as follows: Communication discrepancies among humans within global collaborative manufacturing environments are in the next section (section 2) underlined, stressing the reasons being connected to. The third section focuses on the human presence in future production scenarios considering trends and changes in their activity competitions as consequence of a growing automation degree, characteristic of the next generation manufacturing systems. Hereby, the need of a natural interaction form is pointed out. In Section 4 the possibility of natural language use as interaction means between humans and machine is further discussed. Section 5 shows how ontologies could contribute to overcome possible differences in natural expressions, whereas in Section 6 a concept for a natural language interface is being presented describing a platform prototype called Ontological Filtering System (OFS), by means of which the concept mentioned above has been tested. In the last section (section 7), first trial tests are briefly discussed and future research purposes are illustrated. The latter substantially consists in the OFS employment for programming purposes within a heterogeneous machine environment in order to prove its suitability towards a future use in a production context.

2. INTER- AND INTRACULTURAL COMMUNICATION ISSUES

In the modern global market scenario markets of different countries become ever more dependent on each other. Every country reacts by opening its economy towards other countries trying so to archive vast economical advantages. However, at the same time, enterprise requirements for market competition are getting more complex as an effect of a strong customising of final products. In order to respond to this situation and ensure market share, enterprises unite and build virtual collaborative environments called *Virtual Enterprises*, which are characterized by a high number of organizations being geographically widespread, that temporary collaborate in a project-oriented manner, in order to achieve common goals (Gebauer and Segev 1998). For external observers, these networks often appear as whole, real enterprises. However, they consist of autonomous parts, which could be added and removed to the network according to changes in customer requirements and collaborative environment dynamics.

A pre-requisite for a smooth co-operation is represented by an unequivocal communication among all involved actors. Unfortunately, there are still a lot of understanding problems, which endanger a successful interoperation. These are mainly due to differences in cultural backgrounds. Indeed, people belonging to different international or even national communities incline to perceive and mentally process same real world situations in different cognitive ways, assigning these to different models (so called mental models). Thus, also two interaction partners speaking the same language and using an identical terminology can misunderstand each other, since vocabulary terms represent merely etiquettes of cognitive categories (Lepratti and Berger, 2003) (Fig. 1). Communication can be therefore successful, not only thanks to knowledge in speaking and understanding partner's foreign languages, but above all because of mutual considering knowledge and cultural standards involved in the communication process (Eckensberger, 1996) Additionally, among same cultural communities the communication processes can still become abrupt. E. g. different professional qualifications are major causes of possible wrong interpretations. One might take into account differences in professional backgrounds between enterprise managers or engineers and personal at the shop-floor level. Therefore, possible deviations could arise at both levels: at the intercultural and at the intracultural one as well.

Fig. 1: Communication barriers in collaborative enterprise environments.

3. HUMAN'S ROLE IN FUTURE EUROPEAN MANUFACTURING SCENARIO

The European Manufacturing Industry is in a transition from being a mass production industry towards a knowledge-based customer- and service-oriented one, providing production on demand, mass customisation, rapid reaction to market changes and quick *time-to-market* of new innovative products.

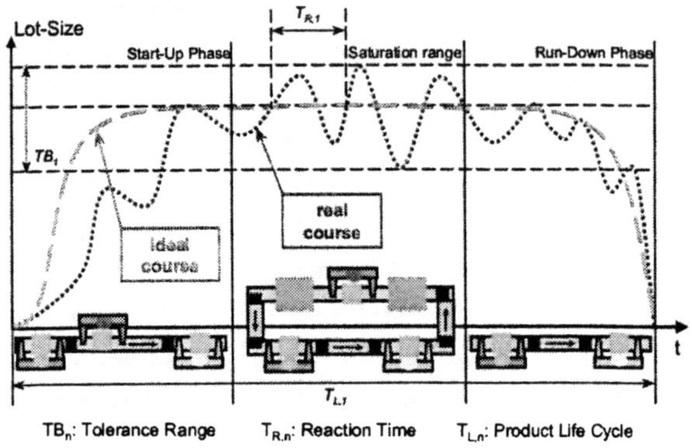

Fig. 2: Customizing effect on the product life cycle.

In this sense, it faces the challenge to produce according to a *lot-size one* paradigm of low cost and high quality (Berger and Lepratti, 2002b). Customising in product development and consequently offering a high variety of product features leads to a strong individualisation in the production domain. Hence, such situation influences the normal course of the product life cycle making it unforeseeable and risky in terms of investment and resource plans (see Fig. 2). In particular two effects should be stressed: On the one hand the *lot-size* value isn't stable enough but rather vacillates between a tolerated range's maximum and minimum ("TB_1" in Fig. 2). On the other hand reaction times to *lot-size* value changes are too slow ("$T_{R,1}$" in Fig. 2). These are consequences of difficulties in recognizing and removing process defaults or adapting parameters, in order to re-stabilising process behaviour according to the actual market circumstances.

In order to overcome these former problems, modern manufacturing industry needs a transition from traditional manufacturing methods into highly flexible knowledge-based global manufacturing. Such a challenging aim involves many difficulties, due to the growing automation degree in production plants. The *United Nations Economic Commission for Europe*[1] announced for instance that the employment of industrial robots (IRs) in the German automotive industry in the year 2002 has reached the 100.000 units leading, thus, to a relation of one IR per every 10 shop-floor operators (Fig. 3). This trend will be confirmed in the last year as well, since 30.000 more IRs have been adopted[2].

3.1 Changes in Shop-Floor Activities.

In connection to this evolution and to new requirements of manufacturing industry, drastical changes in shop-floor activities take place. Personal competences become more important than before and require an higher qualification level (GMA, 2003). If previously worker activities were normally carried out at the same machine -e. g. for monitoring of its

---

[1] UNECE Embargo October 21st, 2003.

[2] Forecast in *Infopack Robotics* of VDMA German Engineering Federation (November 2003).

right operation behaviour-, nowadays an operator has to govern an heterogeneous machine environment absolving various kinds of activities from programming to diagnostic as well as from intervention to maintenance ones (Zühlke, 1999).

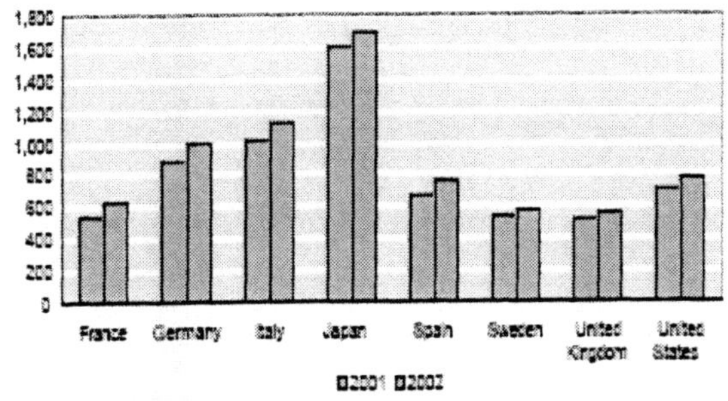

Fig.3: Number of IR per 10.000 production workers in the automotive industry (2001/2002).

Consequently, in order to support this transition process avoiding process breakdowns, researcher and industrial staffs are involved in conceiving and developing new intuitive surface technologies for multiple purposes activities. Major industrial targets related to could be resumed as follows (Berger and Lepratti, 2003) (see Fig.4): (i) reaction times to *lot-size* value changes should be drastically shorted ($T_{R,2} << T_{R,1}$) and (ii) the so called *ramp-on-time* steepness within the *start-up phase* should be raised, in order to faster reach the planned *lot-size* saturation value. Furthermore, as a side effect, the whole product life cycle is remarkable reduced ($T_{L,2} << T_{L,1}$).

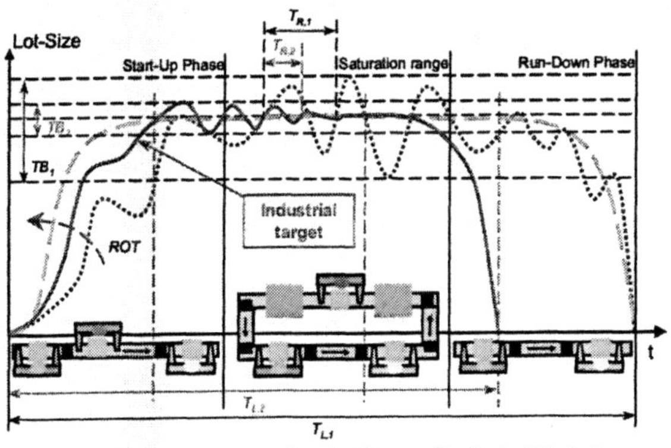

Fig. 4: Aimed targets through employment of an innovative natural man-machine interaction.

3.2 Towards a Natural Feeling Human Interface.

The *European Information Society Technology Programme Advisory Group* (Ahola, 2001) and the *Society for Measure and Automation Technology of the German Engineering Federation* (VDI/VDA-GMA, 2003) underlie the necessity to overcome difficulties concerned with modern man-machine systems characterized by interaction forms that can only vary slightly in a matter according to predetermined functions or predefined syntax and semantic codes. The technology breakthrough will focus on the development of new knowledge for user-friendly and efficient interaction forms between human operators and smart industrial robots, which permit to embed human knowledge and experiences. Thus, the importance of a more *Natural Feeling Human Interface* is hereby empathized.

One of the central challenges consists in the development of front-end interfaces being extremely intuitive in their use by means of e. g. natural language expression as dialogue-based and goal-orientated multi-modal interaction form (multi-user, multilingual, multi-channel and multipurpose). This should be adaptive to both human and machine requirements, embracing the whole spectrum of personnel's cultural, linguistic and professional backgrounds, circumventing communication barriers and providing, at the same time information filtering and its proper presentation in a cross-media environment.

4. NATURAL LANGUAGE AS INTERACTION MEANS

Main goal of the research project described in this paper is the definition of an interaction means for the man-machine communication based on the structure of natural languages. This poses enormous challenges due to their features such as ambiguity, vagueness, possible abbreviations, metaphorical use and its continuous grammatical development and lexical enrichment (Lepratti *et. al*, 2003)..

Still, at the same time, natural language represents the most familiar communication form among humans starting from children with their first expressions up to experts with their special treaties about the universe (Sowa, 2000). So, through its employment relevant and meaningful advantages could be achieved. On the one hand, people aren't compelled to learn special syntax and semantic languages (such as common program languages). On the other hand, enterprises could economize production costs thanks to possible reduction of personnel setting-in period costs.

In order to successfully use natural language as communication, human and machine need a shared information understanding, despite the fact that they communicate at different expression levels (see Fig.3) (Brachman, 1979) (Guarino, 1994).

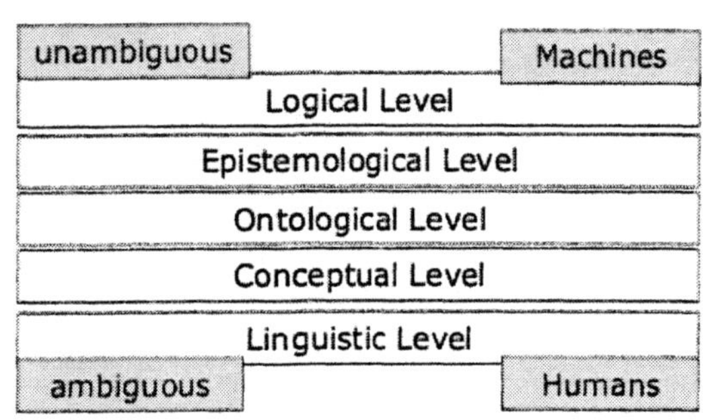

Fig. 5. Communication barriers in collaborative enterprise environments.

While machines communicate at logical level, where information primitives are merely pointers and memory cells being combined in a certain sequence structure by means of logical operators (for instance a bit sequence) and, thus, with a predefined and unambiguous meaning, humans communicate at the linguistic one, where expressions are directly associated to nouns and verbs making meanings ambiguous without knowledge of the context of use.

5. THE ROLE OF ONTOLOGY

An effective mutual commitment, which allows a constrain or at least an approximation of the intended meaning is necessary. Referring to Fig. 5 the *Ontological Level* represents the level where possible interpretation uncertainties of information contents, which arise during a communication process, are removed making information unambiguous and allowing that machines process it. Therefore, the *Ontological Level is* located between the levels of subjectivity and objectivity.
However, *formal Ontologies* have to be distinguished from *linguistic* ones. Formal ontologies derive from Aristotle's metaphysics and, thus, are intended as a study of the nature of being, where all world's things are classificated in categories according to their form assuming this as an abstraction derived from sensory experiences, and so independently of the language being used to describe them. *Linguistic Ontologies* are comparable to large scale lexical resources, that cover most words of a language and describe the various word senses, providing at the same time an ontological structure based on semantic relationships. In (Guarino, 1997) various definitions of *linguistic ontologies* could be found. Noy's and McGuinness's one (Noy and McGuinness, 2001) was identified as the most effective for the purpose of this work. According to them, ontologies *"define a common vocabulary for researchers, who need to share information in a domain of use"*. Perhaps, it is difficult to explain exactly the meaning of a *Linguistic Ontology*, but its definition is less important than the role it plays. In context of this paper project an *Ontology* represents an interface level, on which possible information ambiguities are filtered out and meanings formalized, in order to be correctly understood and processed from machines.

6. THE ONTOLOGICAL FILTERING SYSTEM

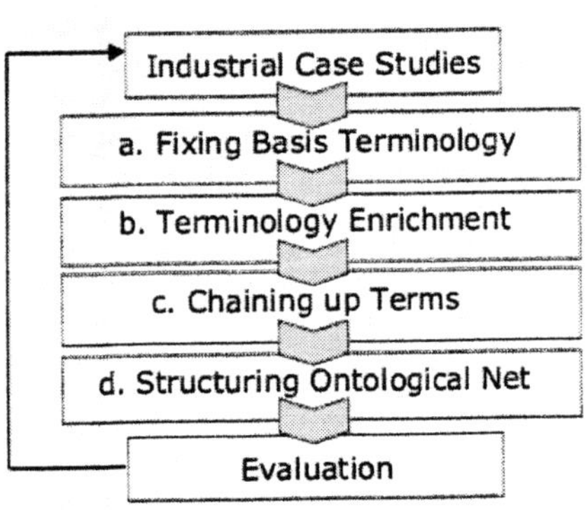

Fig. 6: Workflow structure for the development of ontologies.

A linguistic ontology has been built for a variety of purposes within the automation technology domain. Four steps have been principally performed on the basis of a robot cell for deburring and dressing of aluminium diesel engines (Berger and Lepratti, 2002a) (Fig. 6)

a) Fixing Basis Terminology First of all a basical terminology $\boldsymbol{W}$ consisting of a set of basic terms $\boldsymbol{W_N}$ as well as a set of possible actions $\boldsymbol{W_V}$ among them have been fixed by recognizing principal elements playing a significant role within the considered process domain of context.

$$W_N=\{w_{N1}, w_{N2}, .., w_{Nn}\}=\{robot, gripper, workpiece ..\}$$
$$W_V=\{w_{V1}, w_{V2}, ..., w_{Vn}\}=\{grasp, open, move ..\}$$
$$W= W_N \cup W_V$$

Such terminology is used to forward information to the machines.

b) Terminology Enrichment and Establishment of the Functional Vocabulary Due to the diverse possibilities in natural language expression, a major number of terms, that could be used for the same intended meaning, are found out and included in $\boldsymbol{W_N}$ and $\boldsymbol{W_V}$., defining so the user's vocabulary and consequently the list of primitive procedures that users refer to in their interaction. Since, for instance, the English language bases on a lexicon of more than two hundred thousands different terms, this step was supported using the *WordNet© Electronic lexical Database* (Fellbaum, 1998).

c) Chaining Up Terms As already stressed terms -either Noun or Verbs- could refer to the same meaning but with different degree of granularity in their level of specification. So, e. g. there are terms, which are general in their expression and others that go very deep with their meaning. The first ones could be used for everybody in difference circumstances without special knowledge in the domain of use (the so called *Everyone Terminology*), while the second ones are normally employed only by experts (the so called *Expert Terminology*). Anyway, all terms, which can be adopted within the same domain of knowledge and refer to the same basis terminology, are chained up in so called *Synsets* (*Synonymy-Set*).

Terms belonging to the same *Synset* have, thus, following features:

- They refer to the same basis term within the knowledge domain of application
- They are connected to each other per different semantical relations (synonym, antonym, hyponym an so on)
- They have different granularity in their semantical level of specification
- They have a shared meaning for users.

Figure 7 shows an example of *Synset* for the basic term *Robot.*

d) Structuring Ontology Network The last step -and also the most difficult one- concerns merging all

Synsets defined at point "c" into a whole semantic network call *Ontological Network*. This occurs (1) identifying identical terms belonging to different *Synsets*, (2) aligning identified terms, (3) merging corresponding *Synsets* into each other and (4) building new semantical relations and/or recognizing and removing semantical inconsistencies. (Fig. 8).

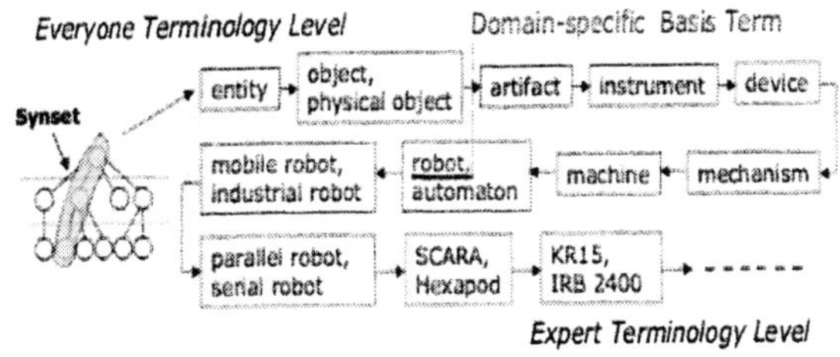

Fig. 7. A sample of *Synset* for the basis term *Robot*. On its top extremity *Entity* represents the correlated term with the most general meaning, while *KR15 (KUKA Robot 15)* is placed on the bottom as it shows the most specific meaning.

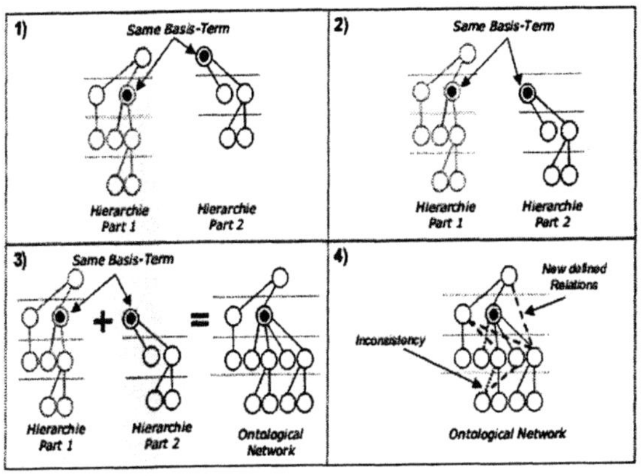

Fig. 8. Finding and merging corresponding *Synsets* into a *Ontological Network*.

Within the syntax and semantics of *Ontolingua* (Gruber, 1995) *Semantic Classes* such as *Humans*, *Machines*, *Processes*, *Failures*, *Work Tools*, *Work Pieces* as well as *Relations* among them have been defined, in order to gather terms together in one group according to their acquired semantic role in the domain of application and constrain their possible semantic combinations. An example of *Classe* and *Relation* between the classes "machine" (?mac) and "workpiece" (?wrk) is given below:

```
(define-class machine (?mac)
   "Any mechanical or electrical device,
   that transmits or modifies energy to
   perform or assist performances of human
   tasks"
   relation_to machine (?mac) = co-operation
   relation_to work piece (?wkp) = handling
   relation_to failure (?err) = restore
)
(define-relation handling (?mac, ?wrp)
   "Manual or mechanical currying, moving,
   delivery or working with something"
   as_result: (?prc)
)
```

In order to further formalize instructions in written English and make machines able to understand them following further subtasks were necessary.

- Linguistic analysis (semiotic and syntax)
- Ambiguity resolution (semantic)
- Dialog Updating

To realize these subtasks at the chair of Automation Technology at the Brandenburg University of Technology a test platform called *Ontological Filtering System* (OFS) was developed (Fig. 9). Its acts as interface between user and machines. The dialog platform converts text inputs into a semantic representation and converts user instructions into machine readable codes.

Within the ontological network information ambiguities can be avoided with respect to the application domain leading used vocabulary terms (step "*b*") to machine language referents (step "*a*"). Therefore, the OFS allows filtering and standardization of natural expression forms and translates entered, syntactically and semantically plausible instructions in machine codes.

7. TESTS AND CONCLUSION

In order to carry out first trial tests, persons have been asked to define (describe) objects, that were illustrated in some pictures. In order to collect different expression forms and to cover the largest terminology, persons working in different professional sectors (e. g. pharmacists, students of different faculties, artists, philosophers, teachers etc.) and specking, in the most of the cases, different languages have been interviewed. An extract of the performed survey is given in Table 1.

Tests within the OFS have been performed. These prove suitability towards a future use in production context motivating the authors to proceed in this research field.

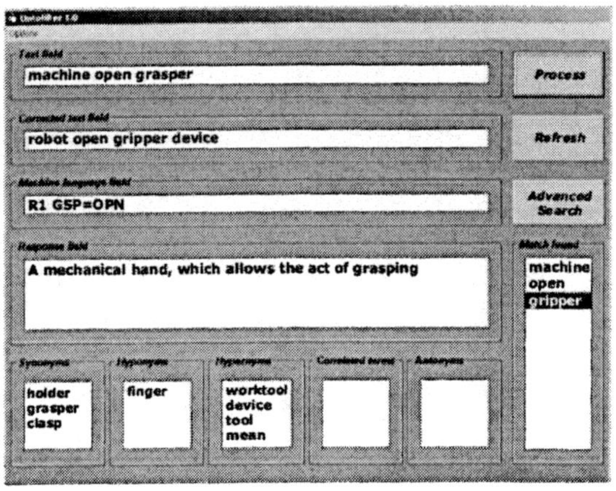

Fig. 9. The *Ontological Filtering System* as test platform for natural man-machine interaction.

Terms	Synset 1	Synset 2	Synset 3
Basis	**conveyor**	**grasper**	**ramp**
Pictures			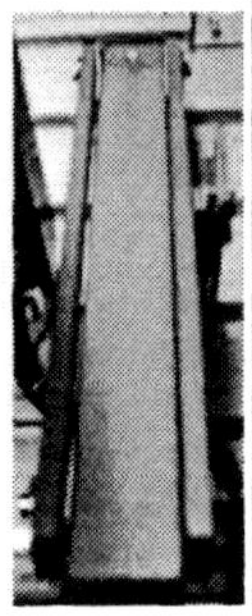
Term 1	conveyer	grip	slide
Term 2	band	gripper	chute
Term 3	***transporter***	holder	***transporter***
Term 4	rail	clasp	panel
Term 5	drive	mounting	descent
Term 6	mechanism	robot	***device***
Term 7	***device***	fixer	rail
Term 8	assembly line	tool	track
Term 9	treadmill	***device***	ramp

Table 1: An extract of performed statistical survey

ACKNOWLEDGMENT

The project described in this paper is performed in the laboratory of the Department of Automation Technology of the Brandenburg University of Technology. Further project trial tests with a *FESTO-Didactic System*, a *KUKA KR15 Robot* and a *SIEMENS Sinumerik 840D* are foreseen.

REFERENCES

Ahola (2001) *Ambient Intelligence.* VTT Information Technology, ERCIM News No. 47, October 2001.

Berger, U.; Lepratti, R. (2002a): *Advanced Factory Automation System for Deburring and Dressing of Automotive Engine Parts.* In Proceeding of the IEEE Int. Conference on Intelligent Engineering Systems, May 26-28, 2002, Opatija, Croatia, pp. 125-130.

Berger, U., Lepratti, R. (2002b) *Intelligent PC-based User Control Interface for On-line Correction of Robot Programs.* In: Proceedings of the 7th International Conference on Control, Automation, Robotics and Vision (ICRACV), Singapore, December 2-5, pp. 276-281.

Berger, U.; Lepratti, R. (2003): *Advanced Men-Machine-Interface for Industrial Robots in the Automotive Manufacturing.* In: Proceedings of the 4th Int. Conference of Industrial Automation (AIAI), Montreal, Canada, June 9-11.

Brachman, R. J. (1979) On the Epistemological Status of Semantic Networks. In: *Associative Networks: Representation and Use of Knowledge by Computers* (N. V. Findler (Ed.)). Accademic Press.

Eckensberger, L. H. (1996). Auf der Suche nach den (verlorenen?) Universalien hinter den Kulturstandards. In: *Psychologische Kulturellen Handelns*, (A. Thomas (Ed.)), pp. 165-197, Hogrefe Verlag, Göttingen.

Gebauer, J., Segev, A. (1998) Assessing Internet-based Procurement to Support the Virtual Enterprise, *Virtual-Organization.net*, *Newsletter (2,3)*.

Gruber, T. R. (1995) *Toward principles for the design of ontologies used for knowledge sharing.* International Journal of Human-Computer Studies, 43, pp. 907-928.

Guarino, N. (1994) The Ontological Level. In: *Philosophy and the Cognitive Science* (R. Casati, B. Smith and G. White (Eds.)), Hölder-Pichler-Tempsky, Vienna.

Guarino, N. (1997) *Understanding, Building and Using Ontologies.* International Journal of Human and Computer Studues (46), pp. 290-310.

Lepratti R., Berger, U. (2003) *Towards Ontology Solutions for Enabling Interoperability in Virtual Enterprises.* In: *Processes and Foundations for Virtual Organisations* (L. M. Camarinha-Matos and H. Afsarmanesh (Ed.)). Proceedings of the 4th IFIP Working Conference in Virtual Enterprises (PRO-VE 2003), pp. 307-314. Kluwert, Boston-London.

Lepratti, R.; Berger, U.; Arnold, K. P. (2003): Die Natürliche Sprache zur Mensch-Maschine-Interaktion im Wissenschaftsmagazin Forum der Forschung der BTU Cottbus, 7. Jahrgang Heft 16, 2003, pp. 104-109.

Noy, N. F. and McGuinness, D. L. (2001) *Ontology Development 101: A Guide to Creating Your First Ontology.* Knowledge Systems Laboratory, Stanford University, CA, USA.

Sowa, J. (1991) Toward the expressive power of natural language. In: *Principles of Semantic Networks* (J. Sowa (Ed.)), pp. 157-189. Morgan Kaufmann, San Mateo, CA, USA.

Sowa, J. F. (2000): *Knowledge Representation: Logica, Philosophical and Computational Foundations.* Brooks and Cole, CA, USA.

VDI/VDE-GMA (2003): *Automation Technology of 2010*, Theory of the Society for Measure and Automation Technology.

Fellbaum, C. (1998) *WordNet: An Elektronik Lexical Database*, MIT Press, USA.

Zühlke, D. (1999): Hardware, Software – Useware. Maschinenbedingungen in vernetzten Produktionssystemen. In: *Elektronik Magazin* Nr. 23, pp. 54-62.

Copyright © IFAC Information Control Problems in Manufacturing, Salvador, Brazil, 2004

ELSEVIER
IFAC
PUBLICATIONS
www.elsevier.com/locate/ifac

CAPACITY PLANNING AND SCHEDULING IN PROJECT ORIENTED PRODUCTION

Bozena Skolud*, Marc Zolghadri**

** Department of Engineering Processes Automation and Integrated Manufacturing Systems, Silesian University of Technology Gliwice, Poland, e-mail: skolud@polsl.gliwice.pl*
*** MCF a l'IUFM d'Aquitaine, LAP/GRAI, Universite Bordeaux 1, e-mail: zolghadr@lap.u-bordeaux.fr*

Abstract: The method presented in the paper guarantees meeting customer demands thanks to the constraint propagation technique applied. The approach is focused on successful problem solution, answering the question if it is possible to accept a new process for realisation in the system where another processes are already realised. In this paper we propose only the decision-making problem solution. The paper is an on-going project. The authors work on the solution that takes advantage of the constraint-based scheduling, which is a framework for solving scheduling problems by stating constraints on the problem variables. *Copyright © 2004 IFAC*

Keywords: production control, distributed control, project driven management

1. INTRODUCTION

Half of everyday activity in business consists in project management or project-driven management; thanks to this clients are treated individually and companies do not fall into routine.
According to Munns and Bjermi (Munns A.K., Bjermi, 1996) a project can be considered as an achievement of a specific objective, which involves a series of activities and tasks, which, in turn, consumes resources. It has to be completed with a definite start and end dates. The function of project management includes defining the requirements of work, allocating the resources required, planning the execution of work, monitoring the progress of the work and adjusting deviations from the plan.
Since the 1950s network planning techniques, such as PERT and CPM enable finding minimum duration schedules for projects, under the assumption that the resources required are not a constraining factor. In practice, especially in the multi-project environment, resources have often limited availability (Demeuleemester, Horroelen, 1994). Recently, the diversification of customer demands has significantly increased industrial competition. The decision- making for a system producing a variety of products with changeable characteristics is one of the most important and difficult tasks of a manager. Such problem is observed mainly in companies characterised by multi-task machinery, production planning based on the receipt of customer orders and the fact that both product lead time and product price which are vital for customer satisfaction are subjects of negotiation with the customer (Hendry, Kingsman 1989). It is also observed that prise is not a subject of negotiation unlike products quality and delivery data. Consequently, the allocation of resources and theirs capacities to activities in order to keep the time regime is one of the most difficult tasks. In other words, the main task is to set up activities schedule in a given system determined by the resources availability in time. Being competitive involves the organisational method of production flow, and, first and foremost, the time at which the method is chosen and applied. Accordingly, in a competitive market SMEs should react to the client or potential client expectations as fast as possible because in any every enterprise, especially the one belonging to the class of SMEs the most important capital consists of customer loyalty. Under such circumstance, enterprises are characterised by single activity connected with very unique product or single – batch production realization of unique

production series, which is never repeated once it is finished. This situation makes it possible for us to suppose that small batch production in SMEs is managed by adopting similar rules as those involved in project management and, thus, will be named as "project driven management."

Most cases of production planning and scheduling in particular belong to the class of NP-hard problems. A combinatorial explosion of possible variants makes it impossible to use an optimal solution in practice. Firstly, a feasible solution does not have to be an optimal one, and, secondly, the set of feasible solutions does not have to contain all possible solutions. Let's stress again that in practice usually the duration of task realisation is not the most important indicator but to assure due date realisation.

The method presented in the paper guarantees meeting customer demands thanks to the constraint propagation technique applied. This approach proposes the creation of sufficient conditions for filtering all solutions and it gives a set of admissible solutions for both the customer and the producer.

The paper is organised as follows. The second section focuses on the case study. The problem formulation is given in section three. The concept of the problem solution based on constraints satisfaction and some conditions of multi-project scheduling are presented in the fourth section. Finally, conclusions and future challenges are discussed in the last section.

2. CASE STUDY

Let us focus on a small firm producing electronic equipment (controllers) and installing them for various applications. The construction of each controller is similar. However, controllers differ in inputs and outputs, depending on the application required. Each order is treated individually in consideration of characteristic parameters of installation conditions. In this case, the following steps are distinguished in the realization of the order:

1. Problem identification (state of the object where the installation is planned),
2. Controller construction and assembling,
3. Programming of the hardware according to expectations and the identified parameters (see point 1),
4. The actual conditions of its application,
5. Setting in motion.

For this case, production organisation depends on the conditions of finished goods distribution and can be achieved as multi-assortment repetitive production in a given, limited planning horizon (see Chapter 4.1). The activities related to this approach give an answer to the question if a new production order can be executed under the conditions of given resource constraints and in the expected time and create the dispatching rules controlling the production flow in the system. Each new production order is treated separately, as well as the activities connected with decision making as to its realisation or its rejection. On the other hand, the production of one product type can be executed separately in batches. In this case, one should notice that the same activities can be executed under the condition of finishing the preceding ones, for example: hardware programming can be executed on condition that the two preceding phases (1st and 2nd) are finished. Other activities may be executed concurrently, i.e. the controller hardware can be assembled simultaneously at the same time at which the object and its state are identified (see points 1 and 2 above). In the small enterprise presented above, not all activities have to be executed sequentially. In turn, the critical value is due time, contracted with the customer or even enforced by him. The simple case study presented above leads to the following observation. Management of the client oriented SMEs that realise make-to-order production is project driven (see Chapter 4.2). This means performance in accordance with the rules of the project management. Each production order, in spite of the product similarity, is different in consideration of product installation conditions.

3. PROBLEM FORMULATION

As already mentioned above, project driven management may refer to the decision on production order acceptance for realisation in the system of simultaneous repetitive production. It may also refer to the acceptance of single or small production order for realisation in the multi-project conditions. In both cases, the production order is one from either multi-assortment or multi-project components. In both approaches, the decision depends on the constraints. In view of this, the Theory of Constraints (TOC) approach is proposed.

One should know that the goal of the system does not relate to the number of calls but to the reply effects, for example: customer satisfaction or new orders. Having realised the goal of the system and the fact of the constraints, the following steps should be executed:

1. Identification the system constraints.
2. Exploit constraints.
3. Subordinate every aspects of the system to this decision.
4. Increase the capacity of the constraints.
5. Go back to step 1.

The following constraints are identified in the system considered in the previous section:

1. *Due time*. Expected realisation time (due date).
2. *Concurrency*. Admissible simultaneous resources occupation by activities belonging to one project and realised concurrently. Successive activity

may be executed before finishing the preceding one.

3. *Potential occupancy.* The resource has limited availability in one time unit that can be more than one. Thus, besides "available" and "not available" „ party available" is also possible.
4. *Structure of resources occupancy.* In order to realise complex operations the constraints of the minimal admissible occupancy of the resource in unity time can appear (for example when the assembly operation has to be carried out by two workers).

One of the producer's constraints taken into account is the productivity of the system. This constraint limits the possibility of the acceptance of the production order for realization in the system.

According to (Lova, Maroto, Tormos, 2000), the most dominant problem is to minimise idle a resource and minimise project duration. Let us stress that in SMEs project driven management the problem does not consist in a concurrent allocation of the resources capacities but in a sequential allocation of the resources to the activities.

Therefore, the problem raises the question: *Is it possible to schedule activities in such a way as to satisfy the constraints that determine the system and client demands under the condition that a new task does not influence the one previously accepted for realisation.*

Additionally, in the case of multi-assortment repetitive production and the case of multi-project realisation case, two approaches are distinguished. The first one is called "all together", where all activities are scheduled at the same time. The second one is called "successive scheduling", where new production tasks (production orders or project) are scheduled under the condition that they do not disturb the processes already accepted and realised.

This paper is focused on the approach of the successive scheduling of both multi-assortment repetitive production and multi-project management.

4. MULTI-ASSORTMENT AND MULTI-PROJECT APPROACHES

4.1. Multi-assortment repetitive production

The problem of production order validation in the system of repetitive production consists in sufficient conditions verification, which guarantees a new production order (Z_{n+1}) realisation in due time determined by given time limit (tz_{n+1}) and the moment of starting the batch realisation (to_{n+1}). This acceptance should not disturb the processes (P_1, P_2,..., P_n) already realised in the system (Skolud, 2000). The decisions is made in 4 steps.

In the *first step* the batch production size of the new production order (B_{n+1}) that does not disturb the already realised processes is determined. The following conditions should be performed:

- resources, shared with new process (P_{n+1}), are not critical,
- at least one single batch production can be executed in the system.

In the situation when the process route passes on non shared resources and the condition $B_{n+1} \cdot t_w > T$ is met (where B_{n+1} is batch size, t_w – operation time on non-shared resource), the production batch delivery period (T'_{n+1}) is determined by:

$$T'_{n+1} = \left\lceil \frac{tpz_{i,n+1} + MAX\ (S_{i,n+1})}{T} \right\rceil T \tag{1}$$

where:
T – the cycle of the system.
T'_{n+1} – production batch B_{n+1} delivery period,
$tpz_{i,n+1}$ – pre-set time of P_{n+1} on the i-th resource,
$S_{i,n+1}$ – realisation time of batch B_{n+1} on the i-th non-shared resource,
MAX ($S_{i,n+1}$) – max element of set $S_{i,n+1}$.

The third step concerns the capacity of buffers allocated to the neighbouring resources M_k and M_p. If condition (2) concerning the buffers capacity allocation holds, then the realisation of the batch size (B_{n+1}), is possible without any disturbance of the production process.

$$Cs(M_k, M_p) = 2 * B_{n+1} - 1, \tag{2}$$

where
$Cs(M_k, M_p)$ – capacity of the buffer allocated to the neighbouring resources
B_{n+1} – the batch size

The fourth step. If condition (4) is achieved, the production order with I_{n+1} number of elements will be finished in due time, in the period between tz_{n+1} and to_{n+1}

$$\frac{I_{n+1}}{B_{n+1}} T'_{n+1} \leq tz_{n+1} - to_{n+1}, \tag{3}$$

where:
I_{n+1} – the production order size,
B_{n+1} – the batch size,
T'_{n+1} – production batch B_{n+1} delivery period,
$to_{,n+1}$ – start time of the production order realisation,
$tz_{,n+1}$ – end time of the production order realisation.

When all the conditions are satisfied, the values describing the system performance, such as the efficiency of resources utilisation can be determined. The decision makes it possible to assign the control procedure in the form of dispatching rules (repetitive sequences of processes) allocated individually to certain resources.

Illustrative example

Let's assume a system of 4 resources (R_1, R_2, R_3, R_4). The following orders Z_1, Z_2, are realised in the system. The routes of processes are presented in

Fig.1 and described by matrixes P_1, P_2. The first row of the matrix contains successive numbers of resources, the second row contains operations times on those resources.

$$P_1 = \begin{bmatrix} 1 & 4 & 2 \\ 2 & 2 & 4 \end{bmatrix}, P_2 = \begin{bmatrix} 2 & 4 & 3 \\ 3 & 1 & 2 \end{bmatrix}$$

Process P_3 is waiting for realisation and is described by the matrix presented bellow. The number of the elements provided for realisation according to process P_3 is equal to 20 elements, due time for realisation is $tz_3 = 100$, available central storage capacity is Cs = 10. Cycle T of the system is equal to 7 and is determined by the occupancy of M_2, which is the critical resource. The batch size for order Z_3 is $B_2 = 2$ according to (3).

The occupancy of the resources is described by a set of vectors. The size of the vector is equal to the cycle time and each vector element corresponds to one time unit. The vector element is equal either to the number of a process that occupies the resource in this time unit, or to 0 when the resource is not occupied.

$$P_3 = \begin{bmatrix} 3 & 4 & 1 \\ 1 & 2 & 2 \end{bmatrix}$$ Matrix describing process P_3

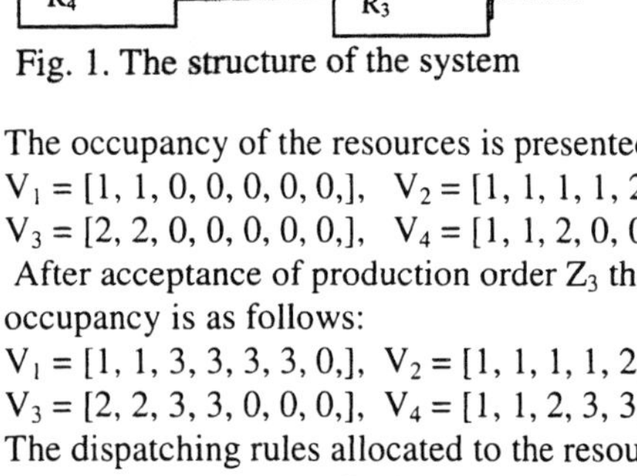

Fig. 1. The structure of the system

The occupancy of the resources is presented below:
$V_1 = [1, 1, 0, 0, 0, 0, 0,]$, $V_2 = [1, 1, 1, 1, 2, 2, 2,]$,
$V_3 = [2, 2, 0, 0, 0, 0, 0,]$, $V_4 = [1, 1, 2, 0, 0, 0, 0,]$.
After acceptance of production order Z_3 the occupancy is as follows:
$V_1 = [1, 1, 3, 3, 3, 3, 0,]$, $V_2 = [1, 1, 1, 1, 2, 2, 2,]$,
$V_3 = [2, 2, 3, 3, 0, 0, 0,]$, $V_4 = [1, 1, 2, 3, 3, 3, 3,]$.
The dispatching rules allocated to the resources are:
$\sigma_1 = (P_1, P_3, P_3)$; $\sigma_2 = (P_1, P_2)$; $\sigma_3 = (P_2, P_3, P_3)$; $\sigma_4 =$

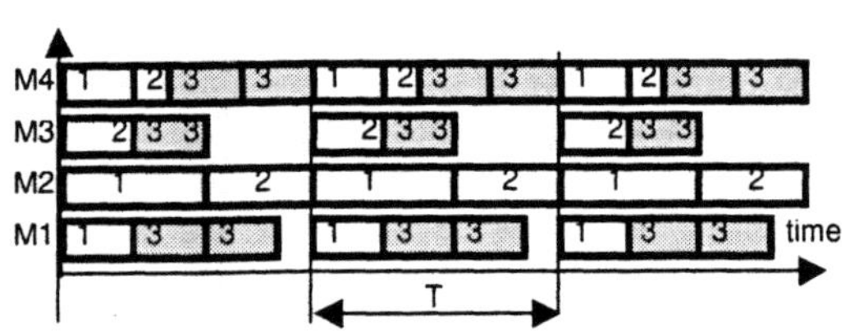

(P_2, P_3, P_3),

Fig. 2. The production flow (Gantt chart)

The dispatching rules describe the sequence of processes that are executed repetitively on the resources. Z_3 realisation time $tr_3=70$ (<100). This means that due time realisation of this production order is possible.

4.2. Project management

The scheduling task is to allocate activities to available resources, respecting all constraints. Two kinds of resources are distinguished (Skolud, Zientek, 2003) unary resources and general discrete resources or cumulative resources.

The first one can process just one activity at a given time; while in the second one the number of activities processed at a given time is limited by the resource capacity. The representative of such resources can be the cell of identical machines, team of workers etc.

The problem of scheduling in both cases basing on the example is presented below.

Unary resources

Illustrative example. Let us assume an organisation of 3 resources (R_1, R_2 and R_3). Two projects P_1 and P_2 were accepted earlier for realization (Fig.3b). These projects cannot be changed as they are already implemented. New project P_3 is waiting for execution. Gantt's chart (Fig.3a) illustrates the resources occupancy for the new project tasks and gives possible delay of each activity realization (according to Forward and Backward Pass) which still ensures due time realization of this project. Figure 3c shows the alternative method of notation, which main purpose of which is to check the possibility of the new project implementation.

The notation applied in this example is the following:

s_i^j – forward scheduling start time for the i-th resource and the j-th project;
e_i^j – forward scheduling end time for the i-th resource and the j-th project;
s_i^k – forward scheduling start time for the i-th resource and the new k-th project;
b_i^k – backward scheduling end time for the i-th resource of the new k-th project;
$i = 1,2,3,...,R_i$ – number of resources;
$j = 1,2,3,...,J$ – number of already implemented projects;
$k = J+1$ number of the new project;
Δl_i – time interval of the idle i-th resource;
$V_i = [l_1, l_2, ..., l_i^t]$ is the vector element of the i-th resource occupancy

$$l_i^t = \begin{cases} 1 & \text{- i-th resource is idle} \\ 0 & \text{- i-th resource is occupied} \end{cases}$$

l_i^t – element of vector V_i and l_i^t; $t = s_i^k, s_i^k+1, s_i^k+2, ..., s_i^k+n$, where $s_i^k+n = b_i^k$

n – dimension of vector element V_i;

In Figure 2 it is easy to notice that idle time units on the resources are the following:

$e_1^k - s_1^k = \Delta_1^k$ (1 time units) on resource R_1
$e_2^k - s_2^k = \Delta_2^k$ (2 time units) on resource R_2

$e_3^k - s_3^k = \Delta_3^k$ (3 time units) on resource R_3

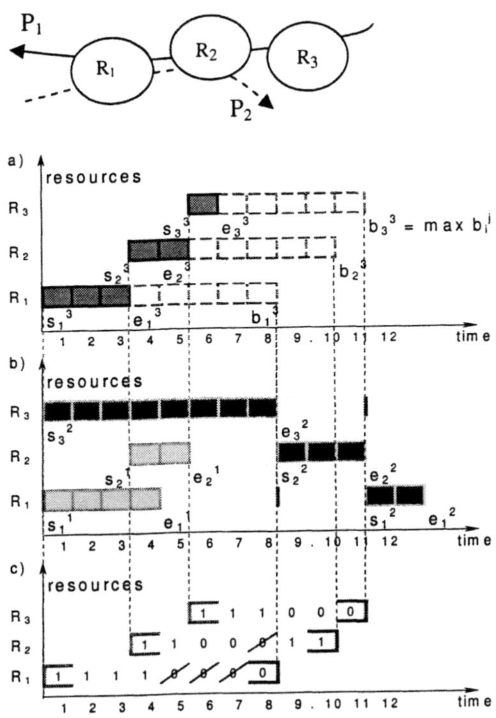

[-forward scheduling start time for the i-th resource and the k-th project

]- backward scheduling end time for the i-th resource of the k-th project

1-resource occupancy

0-idle resource

0-possibility of implementation of the new project

already implemented project

already implemented project

new project

Fig.3. The alternative method of notation

The interdependencies among the activities are described by three conditions of the possibility of a new project acceptance for realisation into the already running system. These three conditions are as follows:

$$e_i^k - s_i^k = \Delta_i^k \quad (4)$$

where Δ_i^k – task duration of the k-th project on R_i:

$$\forall i \in R_i, \exists t_i \in \{s_i^k, \ldots, b_i^k\}, l_i^t = 0 \quad (5)$$

$$\exists r_i \in \langle s_i^k, \ldots, s_i^k + \Delta\, l_i \rangle, \forall t_i \in \langle m_i, \ldots, m_i + \Delta_i^k \rangle, l_i^t = 0 \quad (6)$$

Condition (5) guarantees that there is a period of time on each resource, during which R_i is idle. The satisfaction of condition (6) guarantees that the existing time interval is at least equal to the duration of the new project activity on the idle i-th resource in the considered period of time. When conditions (5) and (6) are satisfied, the acceptance of the new project into the existing system occupancy is possible only if activities on resource R_{i+1} started after the preceding operations on resource R_i are finished. That is why the next condition must be formulated.

General resources

Let us consider resources that are able to work simultaneously. It is useful to think about them as having partial occupancy that enables the realization of more than one task on one resource at the same time.

While talking about resource utilisation planning it is important to consider the state of partial occupancy, which can be represented as a next dimension of resources occupancy.

By checking the previous formulation on each level, which fully shows the possibilities of verifying the partial occupancy state and by adding condition (7), it is possible to state if the resource occupancy guarantees the new project realization.

$$W_i + W_k \leq (b_i^k - s_i^k) \cdot L_i \quad (7)$$

where:

W_i – processing time occupied by the already implemented projects between b_i^k and s_i^k from resource R_i;

W_k – processing time required on resource R_i for new project realization;

L_i – maximum level of R_i resource, assigned to the project; i = 1,2 …, R_i;

Illustrative example. Let us consider the organisation of 3 resources (R_1, R_2 and R_3) depicted in the previous illustrative example. Gantt's chart (Fig.4a) illustrates the time margins for the possibility of the new project tasks realization on each resource and gives possible delay of each activity realization which still ensures that the project is terminated on time. Figure 4c shows an example of resources utilisation planning in view of the state of partial occupancy. It shows how projects may be executed concurrently on the pool of shared resources. We make an assumption that the state of partial occupancy for resources R_1, R_2, R_3 has the following values:

- for already implemented project:
  P_1 – 4 time units (t.u.) on R_1, and 6 t.u. on R_2
  P_2 – 24 t.u. on R_3, 3 t.u. on R_2 and 4 t.u. on R_1
- for the new project:
  P_3 –3 t.u. on R_1, 2 t.u. on R_2, and 1 t.u. on R_3
  L_i – maximum level of resource R_1 equal 3, R_2 equal 2 and for R_3 equal 4; L= max(L_i)=4

In this example condition (5) and (6) are satisfied and also condition (7) is fulfilled, which is why the acceptance of the new project into the existing system already occupied by another task is possible. It is because the operation on R_2 resource can only be started after the preceding operation on R_1 resource is finished, which in this case can be executed earlier.

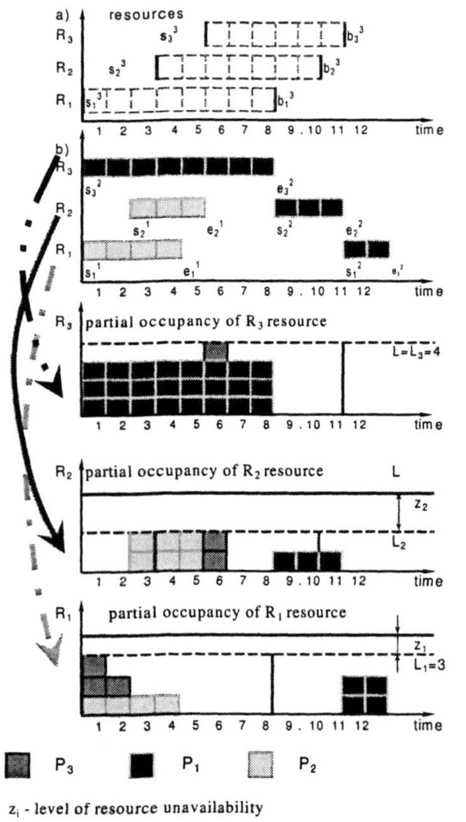

Fig.4. Allocation of activities to partially occupied resources

In such approach to the general resources capacity, scheduling resource levelling is considered as a third dimension (Fig.5).

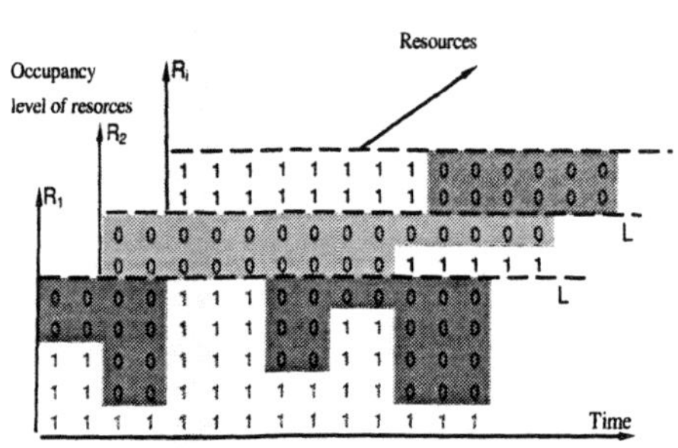

Fig.5. Resource occupancy-consumption profile of a resource

If the free capacity area is larger or equals to the occupancy of the new project, its realisation in the given system is possible. While given project activities are being performed at maximum level L_i of each resource i, on which the project will be executed ($i = 1,2, \ldots, R_i$) are assigned, respectively. If the project that is being executed requires the resource capacity smaller than L_i, the project will use all that it needs and the remaining units will be idle for that resource. If for some processing stage, the considered project requires more resource units, then it will use all L_i units for an extended period of time and thus the total work served in that stage is executed.

5. CONCLUSIONS

The approach presented in the paper is focused on successful problem solution, answering the question if it is possible to accept a new process for realisation in the system where another processes are already realised. In this paper we propose only the decision-making problem solution. The paper is an on-going project. The authors work on the solution that takes advantage of the constraint-based scheduling, which is a framework for solving scheduling problems by stating constraints on the problem variables.

The basic idea is to model the problem as a set of variables with domains (the values for the variables) and a set of constraints restricting possible combinations of the values of these variables. Usually the domains are finite and they are the so-called;" constraint satisfaction problems" (CSP). The task is to find a valuation of the variables satisfying all the constraints i.e., a feasible valuation. In the presence of an objective function defined over the problem variables, the task is to find a feasible valuation minimising or maximising the objective function. Such problems are the so-called "constraint satisfaction optimisation problems" (CSOP). Further research shall be concentrated on the problem of constraints logic programming (CLP) application to the solution of the problem .

REFERENCES:

Demeuleemester E.L., Horroelen W.S., An effective optimal solution procedure for the pre-emptive resource-constrained project scheduling problem. *European Journal of Operational Research*, No. 90, 1996, pp. 334-348

Hendry L.C., Kingsman B.G., Production planning systems and their applicability to make to order companies. *European Journal of Operational Research*, No.40, 1989, pp.1-15

Munns A.K., Bjermi B.F., The role of project management in achieving project success. *International Journal of Project Management*, vol.14, No.2, 1996, pp.81-87

Lowa A., Maroto C, Tormos P., A multicriteria heuristic method to improve rosource allocation in multi-project scheduling. *European Journal of Operational Research,* 127, 2000, pp.408-424

Skolud B., Zientek A., Constraints-based multi-project scheduling in Project Driven Manufacturing WNT, Warszawa 2003

Skolud B., *Planning of multi-assortment production.* Wydawnictwo Politechniki Slaskiej, 2000 (in Polish)

Copyright © IFAC Information Control Problems in Manufacturing, Salvador, Brazil, 2004

ELSEVIER
IFAC
PUBLICATIONS
www.elsevier.com/locate/ifac

ORGANIZING COOPERATIVE WORK FOR THE PRODUCT DEVELOPMENT PROCESS

Michaël David, Zahra Idelmerfaa, Thierry Divoux

Centre de Recherche en Automatique de Nancy – CRAN CNRS UMR 7039
Université Henri Poincaré, Nancy I
BP 239, 54506 Vandoeuvre Lès Nancy cedex, France
Firstname.name@cran.uhp-nancy.fr

Abstract: Research for improving the product development process becomes crucial in new work organizations. Today, design projects involve more and more activities and work groups which must be efficiently coordinated. In this paper, a method for structuring cooperative activities in order to improve its global performance is proposed. The study is based on the recommendations of ISO 9001 standard. Classical methods decompose a project into several work groups which are then scheduled. This first decomposition level is supplemented with a more detailed structuring level in order to reduce the size of the working teams. The steps of the process being defined and planned, the responsibilities inside the working groups are specified. In a last time, the method proposes to find the couples of actors who generate the best interface for the effectiveness of the cooperative work. *Copyright © 2004 IFAC*

Keywords: Concurrent Engineering, Interdisciplinary Design, Project Management, Work Organization, Decomposition Method, Graph Theory.

1. INTRODUCTION

To improve the product development process (PDP), companies lean on cooperative work organisations which enable to collect a great number of skills and thus of actors within the development project. But, these organisations require defining, planning and coordinating the work groups and activities in the most effective possible way. Consequently, it is important to provide methods that enable to structure and to effectively control the PDP.

With the emergence of Concurrent Engineering, numerous works was developed for the organization and the scheduling of cooperative activities (Gebala and Eppinger, 1991; Kusiak, et al., 1994). These approaches are based on tools to describe the design activities and their dependencies (Steward, 1981). They use partitioning methods to decompose a work organization into a set of work groups and to plan them according to their dependencies (Harary, 1962).

A refinement of the groups obtained by the classical partitioning methods is proposed. The objective is to improve the lead time and the workload of these groups. This detailed level allows to determine a division of the activities such as it authorizes a parallel execution of these activities minimizing reviews i.e. repetitions of the work. The method proposes a set of decomposition criteria relative to the balancing of workload and a model for the estimation of development time and cost. The model allows to estimate and to compare various work organizations.

The method comes from the analysis of a complex PDP in order to deduce its optimal organisation. For this, the recommendations of the ISO 9001 standard (ISO, 2000), which defines four points to control and to plan the design and/or development of a product and/or service, are applied:

- the steps of the PDP are defined and planned,
- the required activities for review, verification and validation are well identified,
- the responsibilities and authorities for the PDP are well identified,
- the interfaces between groups involved in the development process are managed.

Through the analysis of the information flow between the actors of a PDP, different techniques based on the graph theory in order to help a quality manager to determine these points are proposed. In section 2, a method for decomposing and planning the PDP in steps is defined. The responsibilities of the obtained groups are specified in section 3. The section 4 describes the interfaces which must be managed to ensure both an efficient communication and to clarify the responsibilities.

2. DEFINING THE STEPS OF THE PRODUCT DEVELOPMENT PROCESS

2.1 General Decomposition of the PDP

The first phase of the proposed method aims at determining an initial organisation of the cooperative activities of a PDP. It enables to identify and schedule the activities that:

- can be executed only after it receives all the information it requires from its predecessors (Series activities),
- do not depend on others tasks (Parallel activities),
- are interdependent and must be executed simultaneously (Coupled activities).

This phase is based on the work relative to the management of complex systems (Steward, 1981). A product design project is modelled by means of a matrix called Design Structure Matrix (DSM). In a DSM, an identically labelled row and column represent a design activity (Fig.1). The elements "1" within each row identify which activities must contribute to information for the proper achievement of the PDP. The order of the rows (or columns) indicates the chronology of activities, i.e. the sequence to start design activities.

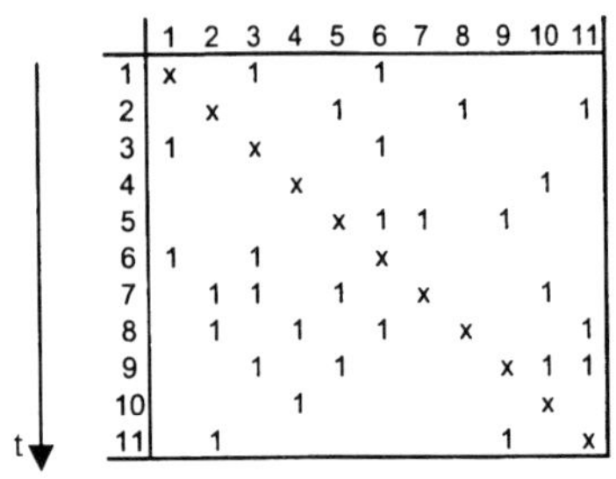

Fig. 1. Initial Design Structure Matrix. Represents the exchanges between eleven actors of a design project. The upper and lower triangle visualise respectively unknown and known information.

The partitioning process consists in rearranging the initial matrix by interchanging rows and swapping the corresponding columns in order to achieve a more organised work sequence i.e. that allows the different actors to work on valid information. The method aims at identifying the strongly related components inside the matrix (Harary, 1962). The actors are then collected within a coupled task, i.e. a work group where the actors must closely cooperate. In the example, three strongly related components constituted of the actors $\{a_1, a_3, a_6\}$, $\{a_4, a_{10}\}$, and $\{a_2, a_5, a_7, a_8, a_9, a_{11}\}$ are identified.

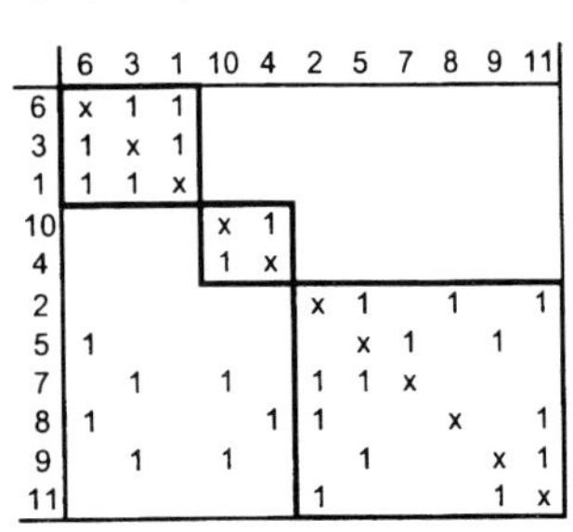

Fig. 2. Partitioned DSM. Three strongly related components called Coupled Tasks.

They are gathered in the matrix in three coupled tasks which are then scheduled in order to remove the "1" on the upper triangle of the matrix and to obtain a partial order (Fig. 2). The two-coupled tasks implying the actors $\{a_1, a_3, a_6\}$ and the actors $\{a_4, a_{10}\}$ can be executed in parallel because they do not have interaction. Then, as soon as these two tasks are performed, the third coupled task implying the actors $\{a_2, a_5, a_7, a_8, a_9, a_{11}\}$ will be able to work since it will have all information it requires, transferred from the first two tasks.

The major interest to define a work organization by partitioning a DSM is to decompose and to schedule the design project. However, the method does not enable to control the size of the groups, i.e. the number of actors within a team. Thus, the application of this method to a complex design project can lead to the constitution of groups with various sizes. In the next level, the interest is related to large groups.

2.2 Detailed Decomposition of the PDP

This phase is the reduction of the number of participants inside groups identified in the first decomposition level. Indeed, to make collaborate in a same work session, a too great number of people can strongly slow down the work performance of this group (Zhao and Liu, 2003).

Performance Estimation of an Organization. The Work Transformation Matrix (WTM) model is used to estimate time and workload of a coupled task and so global duration and cost of a design project (Smith and Eppinger, 1997). It is an extension of the DSM which integrates an assessment of activities in term of estimated time and rework rate (see Fig 3). The diagonal elements represent the time to complete each activity during the first iteration. The off-diagonal elements represent the strengths of dependence between activities, giving rise to the transfer of work, or rework, involved in the iterations. It is allowed that each activity creates a deterministic amount of rework for other activities.

Time matrix *W*: Diagonal elements mean that the activity 10 has an estimated initial time of 4 units and the activity 4 lasts 7 units of time.

Rework matrix *A*: The element ".2" on the first line means that 20 % of the activity 10 must be reworked after the realization of the activity 4. Finally the element ".4" on the second line means that 40 % of the activity 4 must be resumed with the results of the activity 10.

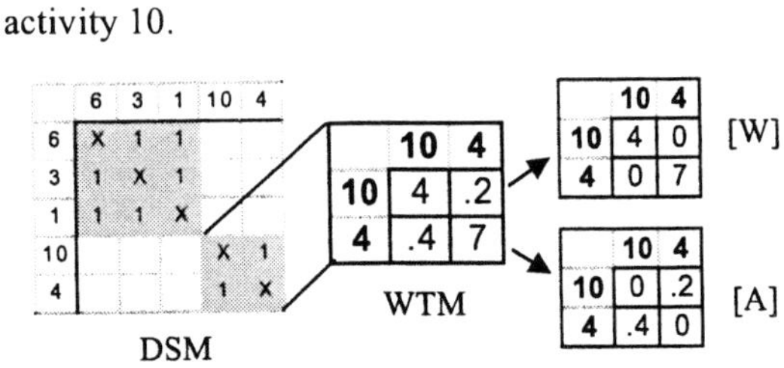

Fig. 3. WTM model (applied on the second Coupled Task of the example)

The working vector u_t represents the remaining work for every activity to the tth iteration. At first, it remains the totality of the work to be executed for each activity, so the initial vector is $u_0 = (1,1,\ldots,1)^T$ with T, number of design activities. After an iteration, the working vector is multiplied by the rework matrix: $u_{t+1} = A.u_t$ or $u_t = A^t.u_0$

A total working vector can also be defined:

$$U = (\sum_{t=0}^{\infty} A^t).u_0 \qquad (1)$$

R is the vector of work times representing the workload of each activity by unit of time (for example men per week) and is expressed by:

$$R = WU = W.(\sum_{t=0}^{\infty} A^t).u_0 \qquad (2)$$

R integrates the time factor. This expression can be reduced. Indeed $\lim_{t\to\infty}(M^t) = (I-M)^{-1}$ if the maximum eigenvalue of a matrix M is lower than 1 (I is the unit matrix). This condition is always true for A, so:

$$R = W(I-A)^{-1}.u_0 \qquad (3)$$

E is the total workload representing the cost of the coupled tasks set. E is the sum of all elements of R:

$$E = \sum_{i=0}^{n} R^{(i)} \qquad (4)$$

n represents the number of activities in the project and the notation $V^{(i)}$ means the ith element of V.

T is duration of a coupled task. T is obtained in adding the duration of the longest activity for each iteration:

$$T = \sum_{t=0}^{\infty} \max[W.u_t]^{(i)} \qquad (5)$$

The WTM model allows estimating the performance of a design project. The optimization criteria only depend on working vector R. Indeed R allows to obtain cost E and duration T for a cooperative process. Then a specific model will be introduced for the coupled task decomposition, i.e. the realization in parallel for a set of workgroups.

The Decomposition of a Coupled Task. The objective is to reduce the complexity of a work task by separating it in sub-tasks. The goal is to find a compromise for structuring and to make collaborating these sub-tasks. For this, the concept of cooperation graphs (CG) is used to study the information flow between the actors and then to deduce the positions of each one (Fig. 4).

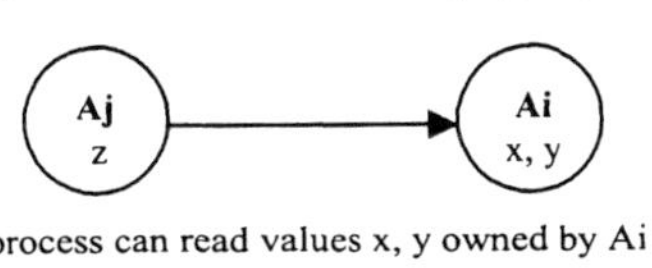

Fig. 4. Cooperation graph. Edge from agent Aj to Ai means that Aj needs information transfer from Ai. Agent Ai cooperates with agent Aj if Ai gives or shares some of its information with Aj.

The concept of CG has been defined by Diaz and al. (1996) who applied it to specify network services and protocols for managing agents in team work. The approach proposes to apply it to cooperative PDP. A cooperative group is composed of agents who are organised according to the relationships between the actors. Vertices represent the agents; edges represent the relations between them. CG provides a formal representation that enables to analyse the kind of cooperation between the actors according to their relationships (Rondeau, et al., 1999):

- if a relation is symmetrical then the cooperation is bilateral,
- if a relation is transitive then the cooperation is hierarchical,
- if a relation is both symmetrical and transitive then the cooperation is total.

Model for the parallelization of a process. A factor K, called cutting matrix, is introduced to calculate the responsibility R (Fig.5). It allows to take into account only activities executed in the first subgroup (R_I). I represents the unit matrix. An example of a coupled task containing 4 strongly connected activities is arbitrarily taken. By identification with (3), a model for the parallel execution of a coupled task after a bi-partitioning is defined, i.e. a dichotomy of the set of activities.

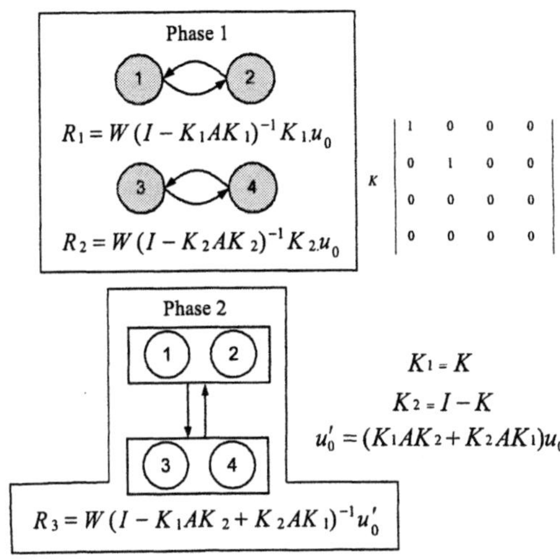

Fig. 5. Model for the decomposition and realization in parallel of a coupled task. K_1 and K_2 specify the composition of each group and are additional. u'_0 defines the information flows to be taken into account in the second phase, i.e. only inter-group exchanges. T= max [t(R_1); t(R_2)] + t(R_3)

The parameters of the WTM determine the values of vectors R corresponding in a parallel decomposition. Phase 2 is called re-connexion phase. The duration of this stage is going to highly influence the total delay. Instinctively, this approach has to bring a decrease of the total duration of design project because all activities are initialized from the first phase. In the second one, it has to make collaborate efficiently the working subgroups together.

This model enables to compare different choices in the project organization. The approach gives the means to estimate in a quantitative and relevant way the quality of the organization of PDP during decomposition as far as the border line is choose. So it remains to define the way of decomposing coupled tasks.

Separation of the Working Groups. In order to minimize reviews and work repetitions in the second phase of the parallel decomposition, the identified subgroups have to be relatively independent (determination of the cutting matrix K). That is the first criterion for a partitioning which will be applied. The second criterion of decomposition is to balance the workload of each group. The analysis of these criteria conforms to the problems defined in the researches relative to the graph partitioning. The general problem of the group decomposition can be expressed as follow.

Given a graph $G = (V, E)$ with V the set of nodes, E the set of edges and $|V| = n$. Decomposition or partitioning of this graph corresponds to the decomposition of the set of nodes V in k sub-groups $V_1, V_2..., V_k$ such as:

- $\sum_{i=1}^{k} V_i = V \quad et \quad V_i \cap V_j = \phi \quad avec \ i \neq j$
- The number of edges connecting its groups is minimal (minimizing inter-group streams)
- The weight of the nodes of every group is appreciably equal (balancing workload)

To solve this problem, several types of algorithms were developed such as spectral (Pothen, 1990), multi-levels (Karypis and Kumar, 1998) or genetic (Todd and Sen, 1999), algorithms. A spectral algorithm is applied in this context. The interest of this algorithm is to be able to arrange nodes on a scale characterizing the force of dependencies of nodes two for two. So the first decomposition criterion is satisfied. Then it remains to choose which place the cut will be applied on. If all nodes have identical weights, the median shall be taken as caesura to respect the criterion of balancing workload. The spectral algorithm allows to obtain the cutting matrix K, which gives the optimal decomposition of a coupled task (David, et al., 2002).

For example, the third coupled task obtained after the preliminary partitioning phase (Fig. 2) is considered: {a2, a5, a7, a8, a9, a11}. The application of the spectral algorithm enables to decompose this group into two sub-groups: {a2, a8, a11} and {a5, a7, a9}.

The analysis of this decomposition shows that the obtained organization leads to an optimal solution:

- the number of actors in the two groups is balanced (3 in each group),
- the interactions between the two groups are minimized (4 exchanges)
- the interactions inside the groups are homogeneous (5 exchanges in the first group and 4 in the second)

The proposed approach to decompose a cooperative design project is based on two phases. First, the analysis of information flows with the DSM allows a general decomposition of the process in coupled tasks. In a second phase, these tasks are separated in relatively independent groups to be able to work in parallel. These two levels define the steps of the PDP and a basic structure for the organization.

3. RESPONSABILITIES FOR THE DEVELOPMENT PROCESS

3.1 Identification of the Actor Responsibilities.

To determine the main responsibilities of the development process, two kinds of actors can be identified by means of the CG analysis:

- the validating members,
- the task managers.

Validating members are actors who are authorised to validate or to reject propositions on models, components, values ... These actors are able to react on a proposition expressed by another one.

That means the kind of interactions is either:

- bilateral cooperation,
- or total cooperation,
- or actors implied in a circuit and not in a hierarchical cooperation.

In a general manner, the following definition (definition 1) is proposed to identify a validating member inside a work group:

Definition 1: The actors implied in strongly related components of a cooperation graph constitute the set of validating members.

Task managers are actors who are able to observe the whole work process and to immediately react by sending new orders or updating information to the other actors. For this, a task manager must have a central position in a work group. The survey to specify this actor is to analyze the different distances between each actor. Concepts of eccentricity and anti eccentricity defined in the graph theory are used.

Definition of the eccentricity: let G be a graph and v be a vertex of G, the eccentricity of the vertex v is the maximum distance from v to any vertex.
That is $e^+(v)=max\{d(v,w) : w \text{ in } V(G)\}$;

Definition of the anti eccentricity: let G be a graph and v a vertex of G, the anti eccentricity of the vertex v is the maximum distance from any vertex to v.
That is $e^-(v)=max\{d(w,v) : w \text{ in } V(G)\}$.

In the context of a cooperative work organization, the eccentricity represents the ability of an actor to capture information inside the group with a minimum of mediators and the anti eccentricity represents the ability of an actor to propagate information inside the group with a minimum of mediators.

Thus, to identify the actor who has a central position in a PDP, i.e. a task manager, the sum of eccentricity and anti eccentricity for each actor is calculated. The actor with the minimum value is the task manager. When two or more actors have the same result, the actor who has the least element corresponding to the max $d(a_j, a_i)$ and the max $d(a_i, a_j)$ is the task manager. Thus, the definition that is applied to identify the task manager is:

Definition 2: The task manager in a CG is identified by the actor a_i with: $\min\{e^+(a_i)+e^-(a_i) : a_i \text{ in } V(CG)\}$. When two or more actors have the same value, the task manager is the actor a_i with: $\min|\{\max\{d(a_i, a_j)\}\cup\{\max\{d(a_j, a_i)\}\} : a_i, a_j \text{ in } V(CG\}|$

The application of these two definitions to analyze co operations in a work group enable to control if the group can efficiently work in a cooperative way, i.e. it collects the right skills, otherwise to restructure it.

The application of the two definitions for the third coupled task depicted in the Fig. 2 allows to identify that this group collects:

- a set of validating members: the whole actors involved in the coupled task;
- a task manager: the actor 2.

4. INTERFACES BETWEEN WORK GROUPS

One of the major difficulties to manage a development project is to ensure the information consistency between the different work groups. The objective required in the PDP as defined in the ISO 9001 is to both identify:

- a communication way which ensures an efficient transfer of the information between groups,
- the responsibilities for the information consistency within each work group.

Firstly, optimizing the information transfer needs to globally minimize the distance between the producer and the consumer of this information. That means it is important to avoid mediators for transfer of information from an actor to another one. The objective is to reduce semantic losses on the information. Secondly, the information consistency must be the responsibility of specific actors named key-members. The key-members are couples of actors who have to control the interface between groups.

The key-members between two groups (for example group α and groupβ) involve two actors:

- one actor of the group α is responsible of the information consistency inside his group;
- one actor of the group β is responsible of the information exchange with the group α.

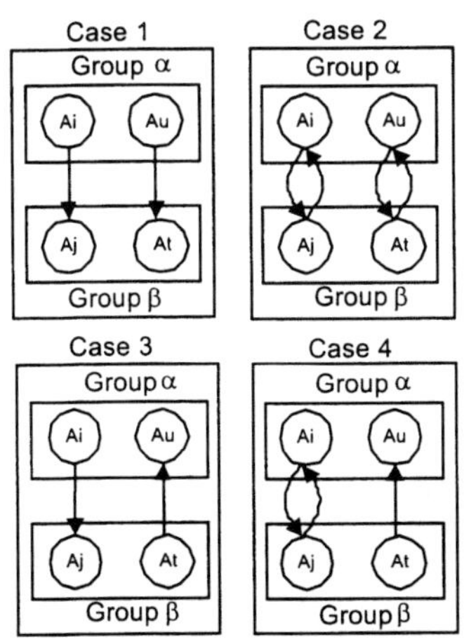

Fig. 6. Cases for two group's interfaces

To identify these actors, the definition 2 for a set of two groups which must be interfaced is applied. Four cases can be analysed (Fig. 6):

Case 1: There is only a hierarchical cooperation between the two groups. In this case, the objective is to determine only one couple of actors to interface the groups. Let's consider two groups α and β. For each couple of actors (a_i, a_j), with a_i belongs to the group α and a_j belongs to the group β, expressions $e^+(a_i)$, $e^-(a_j)$ and, $e^+(a_i) + e^-(a_j)$ are calculated. $e^+(a_i)$ represents the ability of the actor a_i to capture an information in the group α with a minimum of mediators and $e^-(a_j)$ represents the ability of the actor a_j to propagate an information in the group β with a minimum of mediators. The key-members of this cooperation are the couple of actors (a_i, a_j) with: $\text{Min }\{(e^+(a_i)+e^-(a_j))\}$

Case 2: The couples of actors competent to interface two groups are involved in a bilateral cooperation. That means that the abilities of the actors to transfer information in two directions must be taken into consideration. In this case, one couple of actors responsible to manage the communication involving two groups is identified. The same properties defined in the case 1 is used but applied in the two directions. Thus, the key-members of this cooperation are the couple of actors (a_i, a_j) with a_i belongs to the group α and a_j belongs to the group β and:
$\text{Min }\{ (e^+(a_i) + e^-(a_i) + e^+(a_j) + e^-(a_j)) \}$

Case 3: The communication between the two groups is managed by couples of actors in hierarchical cooperation. But, at least one couple of actors transfers information only in one direction and at least one couple transfers information in the other direction. That means that two couples of actors responsible to manage the group interface must be selected. To identify the key-members, the properties defined in the case 1 are applied, firstly to identify the couple of actors responsible to transfer information from the group α to the groupβ, and secondly to identify the couple of actors responsible to transfer information from the group β to group α.

Case 4: This case is a hybrid solution between the case 2 and the case 3. Some couples of actors are involved in a bilateral cooperation and others in a hierarchical cooperation. In this case, the couples involved in a hierarchical cooperation are not taken into account and the properties defined in the case 2 are consequently applied. The reason of this choice is that it is better to have only one couple of actors in a bilateral cooperation, who manages all the communications between two groups than two couples of actors in a hierarchical cooperation.

Remark: When actor couples have the same position in cooperation, the definition 2 described in the previous section is applied. The objective is to decide between the set of all possible couples. Let a_i, a_α belongs to group α and a_j, a_β belongs to the group β, the expression to be evaluated for each couple is:

Case 1&3: $\text{Min}|\{\max\{d(a_i, a_\alpha)\}\}\cup\{\max\{d(a_\beta, a_j)\}\}|$

Case 2&4: $\text{Min}\ |\{\max\ \{d(a_i,a_\alpha)\}\} \cup \{\max\ \{d(a_\alpha,a_i)\}\} \cup \{\max\ \{d(a_j, a_\beta)\}\} \cup \{\max\ \{d(a_\beta, a_j)\}\}|$

To illustrate this study, the example of the interface analysis connecting the second and the third coupled task depicted in the Fig. 2 is used. The analysis is applied to three couples: (a_{10}, a_9); (a_{10}, a_7); (a_4, a_8). The cooperation is hierarchical (case 1), thus only one couple have to be choose. The results are:
(a_{10}, a_9): 1+2 =3 ; (a_{10}, a_7): 1+3=4 ; (a_4, a_8) = 1+4=5

Thus, the actors 9 and 10 have the smallest value and constitute the couple which will be responsible for the interface between these two groups.

5. CONCLUSION

In this article, an organizing method for complex cooperative development process is proposed. For this, the recommendations of ISO 9001 standard are applied. The first point to study is the specification of the process steps. After a first decomposition based on the information flows analysis, several coupled tasks are obtained. If these tasks consist of a too important number of actors, a second decomposition level, which allows to reduce the size of the groups, is defined. This decomposition is based on a spectral algorithm, which takes into account the workload balancing and the minimization of the number of exchanges between groups.

The method needs to evaluate the duration of each step in order to determine the optimal organization. For that purpose, on the basis of the works of Smith and Eppinger (1997) relative to concurrent design systems, a model for parallel approach is used. This model enables to compare different organizations according to criteria like duration and workload. The proposed decomposition allows to find a compromise which optimizes both delays and costs on the basis of the volume of exchange information.

The definition of the responsibilities among the work groups is required in the PDP of ISO 9001 standard. In a complex process involving a lot of exchanges between the different groups (and activities), it is important to clearly define the actors who must ensure the consistency of information produced and consumed by the different groups. The approach proposes the identification of these actors. This technique takes into account criteria that minimize the distance of the information linking the producer and the consumer. The objective is to minimize the semantic losses during the information routing.

REFERENCES

David M., Z. Idelmerfaa and J. Richard (2002). Évaluation et organisation de systèmes coopératifs de conception. *Conference on Integrated Design and Manufacturing in Mechanical Engineering*, ISBN 2-9518169-5-2.

Diaz M., F. Vernadat, Villemur (1996). Spécification et Réalisation Formelles de Systèmes Coopératifs. *Ingénierie des protocoles, CFIP 96*, pp. 357-376.

Gebala D.A. and S.D. Eppinger (1991). Methods for Analyzing Design Procedures. *ASME International Conference on Design Theory and Methodology*, pp. 227-233.

Harary F. (1962). A Graph Theoric Approach to Matrix Inversion by Partitioning. *Journal of numerical mathematics*.

ISO/CD2 9001 (2000). Quality management systems. Requirements. ISO Standard.

Karypis G. and V. Kumar (1998). A Parallel Algorithm for Multilevel Graph Partitioning and Sparse Matrix Ordering. *Journal of Parallel and Distributed Computing*, **Vol. 48**, pp. 71-95.

Kusiak A., Larson and J. Wang (1994). Reengineering of Design and Manufacturing Processes. *Computers and Industrial Engineering*, **Vol. 26**, No.3, pp. 521-536.

Pothen A., H.D. Simon and K.P. Liou (1990). Partitioning Sparse Matrices with Eigenvectors of Graphs. *SIAM Journal of Matrix Analysis and Applications*, **Vol. 11, n°3**, pp. 430-452.

Rondeau E., Z. Idelmerfaa and J. Richard (1999). Identification of Group Organization during a Design Process by Using Cooperation Graphs. *Concurrent Engineering Research Application*, ISSN 1063-293 X, **Vol. 7**, pp. 191-199.

Smith R.P. and S.D. Eppinger (1997). A Predictive Model of Sequential Iteration in Engineering Design. *Management Science*, **Vol. 43, n°8**, pp. 1104-1120.

Steward D.V. (1981). The Design Structure System: A Method for Managing the Design of Complex Systems. *IEEE Transactions on Engineering Management*, **Vol. EM-28**, No. 3, pp. 71-74.

Todd D. and P. Sen (1999). Distributed Task Scheduling and Allocation using Genetic Algorithms. *Computers and Industrial Engineering* **n°37**, pp.47-50.

Zhao J.M. and Z.J. Liu (2003). Team-size Constraint based moderate Decoupling Algorithm for large Coupled Task Sets. *10th International Conference on Concurrent Engineering Research and Applications,* ISBN 90-5809-622-X, pp 123-128.

Copyright © IFAC Information Control Problems in
Manufacturing, Salvador, Brazil, 2004

ELSEVIER
IFAC
PUBLICATIONS
www.elsevier.com/locate/ifac

PROPOSITION OF A SITUATION MODEL IN VIEW TO IMPROVE COLLABORATIVE DESIGN

Farouk Belkadi, Eric Bonjour, Maryvonne Dulmet

Groupe Modélisation et Pilotage des Activités de Conception - Laboratoire d'Automatique de Besançon
Unité Mixte de Recherche CNRS - ENSMM - UFC
25, Rue A. Savary - 25000 Besançon (F)
Tél. (33) 381 402 811 - Fax. 33 381 402 809 - E -mail fbelkadi@ ens2m.fr

The collective character of design activities requires the development of methods and tools in order to control the design process and the collaborative activities. The workgroup is assimilated to a system composed of entities, which are in various interactions managed by a set of coordination and control mechanisms. We present here a model of the design situation in collaborative works using the notion of inter-related entities. We propose also a classification of the different interactions between these entities. The perspectives of this work are to contribute to the improvement of support tools for the collaborative design situation and knowledge capitalization.
Copyright © 2004 IFAC

Keywords: design activity, collaborative work, situation, and knowledge capitalization

1. INTRODUCTION

During the past ten years, the development of new methods and techniques has lead to an improvement of the results of the engineering activities. Design takes an increasingly high place on the level of its processes and activities. Products are designed in a consistently shortening length of time. Almost all the complex artifacts, nowadays, such as cars, are defined with the interaction of many, sometimes thousands of actors, working on different elements of the design. This collaborative design process is challenging because of the strong interdependence between very large various parameters (human, technological, decisional, organizational....). Coordination among many interdependent actors is a key to its success. The challenge is made more difficult in concurrent engineering processes, when activities happen simultaneously and interact with each others.
The purpose of this paper is to present a model of a collaborative design situation using the notion of inter-related entities. The workgroup is assimilated to a system composed of entities, which are involved in various interactions. We propose also a classification of the different interactions between these entities

2. SHORT REVIEW

The design is an activity that requires a collective actor. The problem specifications are progressively defined in the relationships of mutual requirements (De Tersac, 1996) Each individual actor builds his (her) own representation of the problem by taking into account the dimensions that are relevant to his intentions. However, common representations (related to common global intentions) are necessary (Perrin, 1999, 2001) to ensure a coherent integration of individual local results in the collective global actor.
F. Darse (2000) defines the major features of design activities in:

- the problems are not completely defined in the beginning. The context of the design (market, technology, collaborators, managers, process...) creates different sources of uncertainty. Therefore, during the whole project, the solver will require information issued from its environment (tools, partners, customers...) to define its problem and to reduce uncertainty.
- the parameters and their relationships are not split up into independent sub-systems. Sub-systems may correspond to different skills (or competencies). Then, these competencies are interdependent.
- the solution is only satisfactory but not optimal. The designer has to make a choice among acceptable solutions according to performance criteria. There is no completely defined method (or algorithm) to find this solution. The design activity is based on an actions plan that is progressively built during the activity, while at the same time the other actors define mutual requirements according their local solutions and their own actions plan. The solution emerges from different unforeseeable consensus or trade-off.
- these interactions between actors imply numerous iterations and it is difficult to split up the design process into different tasks or responsibility centers.
- the solution evaluations are carried out at different levels: by the designer, by his (her) collaborator separately (from different points of view, heterogeneous criteria), by his collective actor and by external actors who are in charge of representing the customers' voices.

A consequence of this is that the design activity is never an individual activity, but a collective activity. And the role of cooperation is very important in the design process (Boujut 2000). A literature review shows that the collective work is an important subject for many multidisciplinary studies. In Ergonomics, Leplat (1997) considers the collective activity as an activity in which a set of people work for a same goal, and consult themselves, coordinate and cooperate. De La Garza (1998, 2000) distinguishes four types of interactions in a workgroup: cooperation, coordination, collaboration and consultation. However, it clearly exists as a structuring order between the various interaction modes according to the nature of information exchange, individual motivations and participants dependence. In several definitions, cooperation and collaboration are given in the same sense. However, collaboration can not be distinguished, (Dillenbourg & al., 2002).

Later in this paper, we will explain our opinion on these differences

3. THE CONCEPT OF SITUATION

The situation concept is multidisciplinary. In terms of social psychology, the situation represents for an actor, a logical organizational form, according to him, of the perception with which he assembles objects and actions. The situation is defined as a set of interconnected roles that furnish key-links between prospects and behaviors of man. This set of roles allows to an actor to anticipate or to predict his other actions.

William Thomas (1923) asserts that there is not an absolute sense of the situation but that this sense depends upon the culture of the person who gives it as well as his position in this situation. People's actions are not a response to the characteristics of their context but rather the interpretation, the people give to these characteristics.

However, an actor cannot build his situation definition without interference. For example, an actor who will join an organization, which has created its own definition of the situation, cannot produce his definition independently of the existing definition in this organization. The actor's vision is a subjective and a local vision; the organization vision is an objective and a global one. The two visions are important and a minimum of consistency is necessary to obtain a desired performance. Organization management must give to each actor a maximum amount of information about the internal environment.

The difference between a real situation and the individual representation of this situation is the result of two types of factors:

- Sensory factors: they concern the perception capacities of the actor.
- Subjective factors: they concern the interest, which the actor gives to some environment's elements without others and to his interpretation maps relating to his past experiences.

Three representation levels of the situation are distinguished:

Concrete situation S_C
It includes the set of real existing entities able to be observed or not observed by the actors.

Observable situation S_{obi}
It is the set of entities and relations, which can be observed by an actor. Every actor defines his own observable situation; the difference between the concrete situation and the observable situation, can be an indicator of the perception degree. This difference is due to the uncertainty linked to the human sensors.

Useful situation S_u
It is the set of entities and relations observed by an actor and judged as pertinent by this actor in the realization of his task. The difference between the observed situation and the useful situation is that the actor gives indications about the actor behavior.
The degree of consistency given by this actor to some entities is measured by a weight that we will define later.

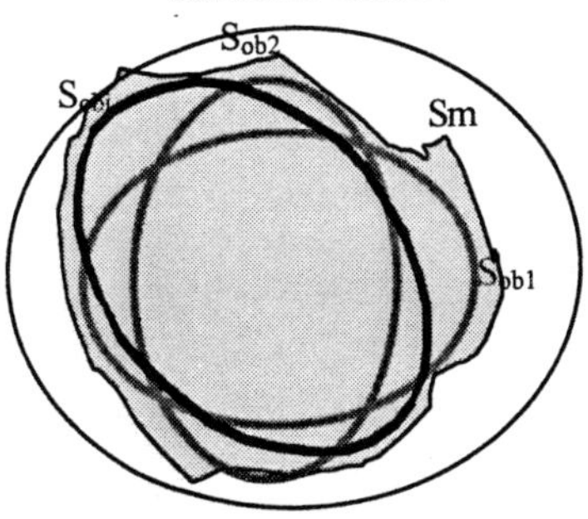

Figure 1: Concrete situation, observable situations and modeling situation

We associate a special form of situation to those levels, we named **modeling situation Sm**. This form includes all the entities and relations observed by an external modeler. The model of this situation includes all entities and relations which can be observed by all the actors. So we assume that an external modeler can perceive the situation in more complete manner.

Note that the useful situation is in reality an external appearance of the cognitive representation of the actor environment, but this latter is not directly observed but deduced of the activity analysis or deduced of the process of distanciation (Piaget, 1973).

4. PROPOSITION OF A SITUATION MODEL

A model is an abstract and simplified representation of a real system, making it easier to analyze. According to the modeler needs and points of view, the situation model gives a description of different relations between its components that take part in the problem description and/or in the problem resolution. Penalva (1999) proposes three basic elements for the model of an activity situation (Cf. Figure 2). These elements are Action, Actor, and Object. In this model, an actor for a given activity can be an object of other activities. The full model is obtained by grouping together all elementary models

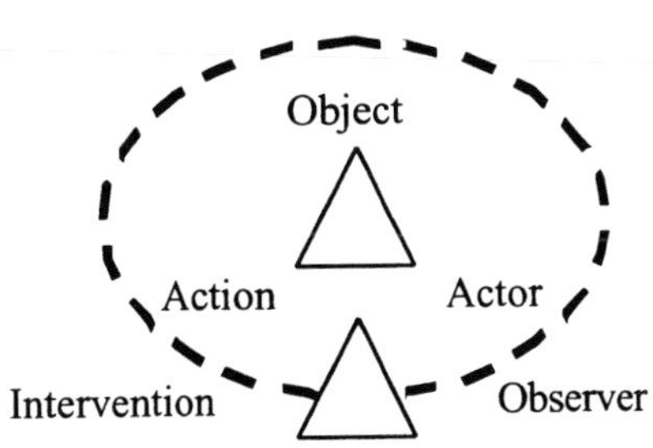

Figure 2: Model of the activity situation according to Penalva

However, the activity is not the only important element in an interaction. The nature of the actor participation in interactions depends on the interpretation, this actor gives to his relations with others, to his relations with his activities and the activities of other actors.

Our model includes three general aspects of the work situation (fig2):

- A structural aspect which gives information about the linking way of environment components
- An operational aspect which informs about what each component do
- A Dynamic aspect which informs about the environment evolution.

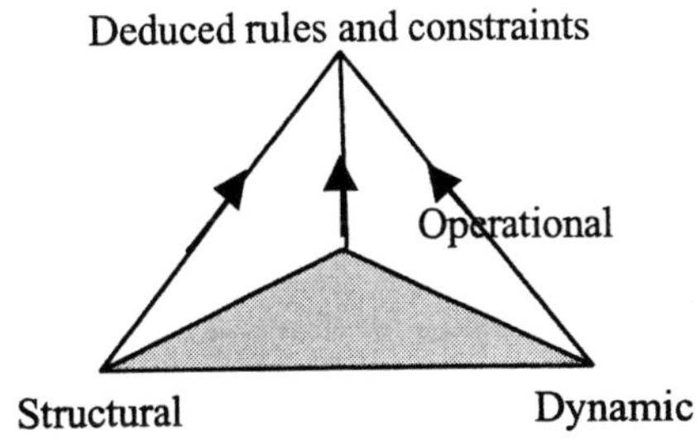

Figure 3: The various facets of the situation model

These various situation's aspects involve a set of rules and constraints which depend on the modeler's point of view. Those rules and constraints are emerging in a new knowledge form. Most knowledge-based systems neglected these types of knowledge, which can give information about the mechanisms of implicit interpretation of an actor

And this induces us to propose our situation definition: **"the situation is a set of various entities and of various interactions describing globally the external environment in which an actor mobilizes his competencies "**

So, the situation model S(t) is defined by a set of entities E linked by a set of relations R:

$$S(t) = \{E(t), R(t)\}$$

4.1. Definition of the entities

We distinguish two kinds of entities

Basic entities (BE) or concrete entities: they gather all the human actors (called individual entities) and the material resources (called physical entities) like product components, work tools, communication tools...

Interactional entities (IE) or abstract entities: they describe links between the entities.

An entity $Ei \in E$ (entities set) is described by the triplet {Type; State; Weight} such as:

Type = {BE ; IE};

State = {0,1} = {Inactive, Active}

Weight : $Wi \in [0,1]$,

measures the importance of the entity "i" for an actor. The weight 0 means that the concerned element is completely neglected by the actor, on the contrary, the weight 1 indicates that this element is very important for the actor

4.2. Typology of interactional entities

4.2.1 Operational interactional entities
These entities describe the various tasks an actor has to perform.

4.2.2 Community interactional entities
These entities establish a membership link between entities and each community entity has its functioning rules. For example, a process is scheduled set of tasks, so it represents a community interactional entity to whom these tasks belong.

4.2.3. Transactional interactional entity
It is a particular form of operational entity. They describe the various mechanisms of information exchange between actors during the realization of their collective tasks However, it is clear that we have various levels of interaction modes according to the nature of information exchange, the individual motivations, and the degree of actors dependence.
So, we propose a structuration between these levels, it is not a standard classification, but its aim is to explain and distinguish the nature of interactions between actors in a given situation.

We propose four transactional interactional entities:

Communication:
It is the simplest form of interaction between actors without the need of a common goal. It is a message transmission process, requiring at the same time, a transmitter, a receiver, a message and a communication support.
The communication goals are either technical goals linked to a task, a sociological need or a professional need.

Cooperation:
The final aim of this interaction is clearly identified. It is the participation of all concerned actors, to a common goal. However, individual goals are not completely dependent. An actor can be concerned by the activity result of an other actor without participating to the same elementary task.
The cooperation is a more complex mechanism than the communication and uses the communication as a mean. So the entity cooperation gives a sense to the entity communication.

Coordination:
The coordination needs the elaboration of rules between actors. It controls the cooperation process and coordination can not exist without a cooperative action. Control mechanisms can be a temporal scheduling of actions, a synchronization between the action or the resources and the common objects performed;
The coordination is more complex and creates a strong dependence between individual goals of actors who have to agree and respect functioning rules during the process. These agreements and these rules can be explicit or implicit, prescribed by an authority or emerged in the action.

Collaboration:
In many definitions, cooperation and collaboration are given in the same sense. However, the collaboration can be used where the different individual tasks can not be distinguished, (Dillenbourg et al., 2002).
The roles in a collaboration action are strongly interwoven. Each actor controls the action of other and there is a strong dependence between actors. No action can by validated completely without the approbation of the other participants.
On one hand, the coordination involves a cooperation process between actors and a coordinator in order to regulate the coordination. On the other hand, the collaboration requires a coordination process, which emerges during action. This mechanism organizes the intervention of every actor during the task's realization.

4.3 Definition of the relations

A relation Rij $\in$ R (set of relations) links a basic or interactional entity to an interactional entity

Rij : Ei $\in$ E (entities set) $\rightarrow$ **Ej** $\in$ {IE};

A relation Rij between two entities Ei; Ej is represented by the quadruplet
Rij = {Ei ; Ej ; Role ; Weigh}

Weight: $\in [0,1]$,
measures the importance of the relation Rij.
Every entity among interactional entities is able to have a role. In fact, the role is a representation of the interpretation that an actor makes of his environment. Roles can have different significance according to the entities, which they refer to.

We propose five kinds of roles (table 1)

Question	Role	Description
Who?	Actor	Entity which directly participates to the interaction
For whom?	Client	Client order is always downstream, and the entity describes the need
About what?	Object	Entity on whom the interaction act...
With what?	Support	Entity which indirectly participates to the interaction.
How?	Manager	Entity which controls the interaction.

Table 1. The roles classification

The below example (figure 4) shows some kinds of relations:

R_{13} = {E1 ; E3 ; client ; Weigh}
The client role signifies that the entity "individual" prescribes the task and waits for its result. At the same time, if a task 3 plays the role of client for an other task 4, it signifies that the result of the task 4 is needed to perform the task 3;

The process has the role of manager for a task because it schedules the daughter entities (tasks) in the mother entity

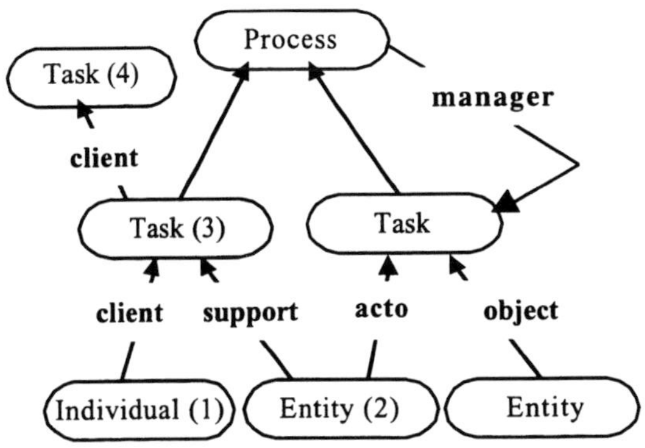

Figure 4: Roles for operational entities

For the community entities we present the following example in figure 5:

In the relation between individual 1 and team 1, the role of client indicates that individual 1 is the prescriptor for the team activities.

Individual 2 has two roles according to the interaction with team 1 or team 2, for the first, he indirectly participates to the activities, for the second he is an actor.

The role of manager allocated to individual 3, indicates that he is the coordinator for team 2.

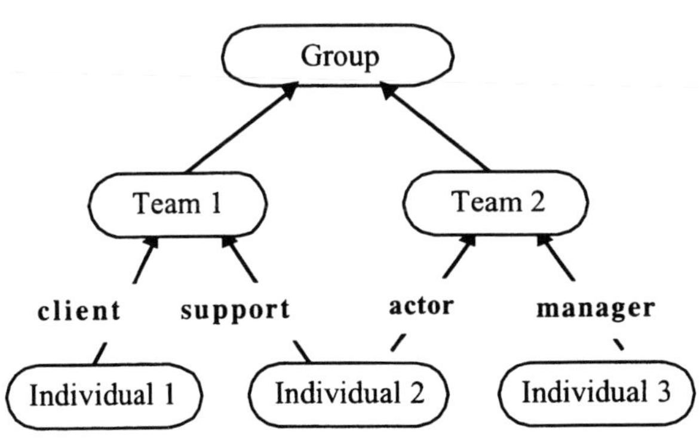

Figure 5: Roles for community entities

In the case of transactional entities, all the active entities participating to the action have the role of actor. The communication channels or the intermediary actors, which participate indirectly to the action have a role of support and the actor who coordinates the action plays the role of manager.

5. MODELING APPROACH

The proposed modeling approach is composed of five steps; A first step of initialization, three steps of knowledge extraction, and a last step of structuration and modeling.

5.1 Step 1: Initialization

It is composed of three stages: needs identification, location of the main sources of information and elaboration of an actions plan.

5.2 Step 2: Identification of the workdomain

During this step, the modeler identifies all the components of the situation: actors, various used objects, organizational structures, and dependencies between actors resulting from these structures.

5.3 Step 3: Prescription analysis

Missions identification:
A mission consists of a task referred to a given action framework. An actor is entrusted with this mission. The prescription does not concern "the how". He has the responsibility to fulfil the mission by determining the relevant organization of finalized actions. A mission is defined by a set of waiting and an action framework.
It is not possible to determine a complete characterization of a considered situation. We have to choose relevant and critical elements according to the task of the mission and to the actor's autonomy (definition of the action framework). Today, due to the situation complexity and the high level of skills needed, the actor has a large autonomy to perform his mission.

Allocations
We identify all the actors who are in charge of performing a mission or a set of missions and who act permanently, temporarily or occasionally. After, we analyze dependencies between actors resulting of organizational structures.

5.4 Step 4: Activity analysis

During this step we direct attention to the various treated objects (intermediary or final). Here the actor becomes the main source of information and knowledge. He has to explain:
- the various plans he elaborates in view to perform his tasks.`
- the various interactions with other actors in which he intervenes;
- the degrees of consistency he gives to every entity in the situation.

This step is realized by the mean of interviews and questionnaire.

5.5 Step 5: Structuration and modeling
During this final step, the various models are built.

6. CONCLUSION

In this paper, we propose a model of a collaborative work situation. The objective of this model is to bring an understanding of various mechanisms, which govern collaborative activities.
In design, mutual dependencies between the various actors generate a need of interactions. The four transactional interactional entities give a first level of characterization of these interactions.

The relations Rij using the notion of roles furnish another kind of information from these interdependencies. The conceptual model based on an entity representation lends itself well to an UML modeling. So we can represent the evolution in situation and furnish a prediction on the action results. The prospects of this work are to contribute to the improvement of the understanding of collective work and to favor the knowledge capitalization.

7. REFERENCES

Bonnevault C., Couffin F., Faure JM., "Méthodes et modèles pour la description des processus de conception dans un contexte de travail coopératif" Journée Recherche Concurrent Engineering, JRCE , Paris, 24 et 25 janvier 2001

Boujut JF., "Enjeux et formes des pratiques coopératives dans la conception", Atelier PROSPER, Ecole des mines de Paris, le 17-02-2000

Darses F., "Assister la conception : perspectives pour la psychologie cognitive ergonomique", Actes de conférences EPIQUE 2001, 129-30 Octobre 2001

De La Garza C. , "Le travail collectif en tant qu'activité de régulation", in Performances humaines et techniques n°96, septembre – octobre 1998

De la Garza C. et Weill-Facina A., "Régulation horizontales et verticales du risque", In Le travail collectif : perspectives actuelles en ergonomie, Chapitre 10, P217-234, OCTARES 2000

De Terssac G., "Le travail de conception : de quoi parle-t-on ?", in coopération et conception, Editions OCTARES 1996

Dillenbourg P., Baker M., Blaye A., O'Malley C., "L'évolution de la recherche sur l'apprentissage collaboratif", Séminaire de recherche Apprentissage collaboratif et co-formation, IUFM Nord Pas de Calais, 2002

Hatchuel A., "Coopération et conception collective. Variété et crises des apports de prescription", in coopération et conception, Edition OCTARES 1996

Leplat J., "Regards sur l'activité en situation de travail : contribution à la psychologie ergonomique", PUF, le travail humain, 1997

Penalva J.M, "Situations et Systèmes Complexes" Actes de l'école d'été Gestion Scientifique du risque, ALBI, 6 - 10 septembre 1999

Perrin J., "Pilotage et évaluation des processus de conception", L'HARMATTAN 1999

Perrin J., "Conception entre science et art", sous la direction de Jacques Perrin, collection Epistémologie, INSA de Lyon, Juin 2001

Piaget J., "Introduction à l'épistémologie génétique", Presses Universitaires de France, Paris 1973

Picard F., "Compétences et processus cognitifs : l'apport de la cartographie cognitive à la formalisation des compétences économiques des agents" in Economie et Société, n° 04/2003.

Simon H.A., "Sciences des systèmes, sciences de l'artificiel", version traduite en français par J-L. Le Moigne, édition Dunod 1991

Thomas W.I., "Standpoint for behavior analysis", Little Brown and Company, Boston, 1923 ref.internet : http://www.umsl.edu/~rkeel/general/010/withomasdefsit.html

Tarpin-Bernard F., "Travail coopératif synchrone assisté par ordinateur :Approche AMF-C", thèse de doctorat, Ecole Centrale de Lyon, 1997

Copyright © IFAC Information Control Problems in Manufacturing, Salvador, Brazil, 2004

ELSEVIER
IFAC
PUBLICATIONS
www.elsevier.com/locate/ifac

PERFORMANCE EVALUATION OF THE COLLABORATIVE DESIGN PROCESS

Vincent ROBIN[1], Philippe GIRARD[1] and David BARANDIARAN[2]

[1] *Laboratoire d'Automatique et de Productique - Groupe GRAI, UMR 5131 CNRS, Université Bordeaux 1, 351 cours de la Libération, 33405 Talence Cedex, France. Tel.: +33 5 40 00 24 07; fax: +33 5 40 00 66 44*
E-Mail: robin@lap-u.bordeaux1.fr , girard@lap-u.bordeaux1.fr

[2] *EADS CCR, Engineering and Information Technologies Department - Enterprise Engineering, Centreda 1- 4 avenue Didier Daurat 31700 Blagnac, France. Tel : +33 5 61 18 87 47*
E-Mail : david.barandiaran@airbus.com

Abstract: Nowadays the design process of a product is more and more collaborative and involves the creation of working groups and the control of collaborative work. This paper presents a model to evaluate design performances. The proposed concept of design environment permits to follow and to manage collaborative design process. Then we present a frame to analyse the as-is design situation, to develop a design environment with an adapted performance measurement system and to follow the progress of the design process. *Copyright © 2004 IFAC*

Keywords: design performance, collaborative design, project management, design environment.

1. INTRODUCTION

Design process is more and more complex and has to satisfy objectives of cost, delay and quality more and more restrictive. Design processes have to integrate more and more expertises and are based now on the collaboration between actors (Poveda, 2001). In this context, one objective of management is to define and to organise the system where the design transformation will take place, according to the needs for collaboration. Therefore, the collaborative process management requires understanding of design processes context (Chiu, 2003) in order to modify the actors' work. This context is defined considering the product and process aspects but also with a view to the human, social and organisational aspects. Ostergaard and Summers (2003) have proposed taxonomy of collaborative design, which is developed around six attributes:

- Team composition
- Communication
- Distribution
- Design approach
- Information
- Nature of the problem

This taxonomy is relevant and brings a new vision of the description of the design situation, since it considers human, organisational and procedural aspects. Human aspects are defined thanks to team composition, communication and distribution. Team composition focuses on the impact of each actor on effective team performance. Communication describes the communication system that is necessary to develop in order to share expertise, ideas, resources or responsibilities. Distribution represents how team members and information are distributed across some variety of boundaries. Design approach, nature of the problem and information permit to take into account process and organisational aspects. Team's design approach affects collaborative design process and concerns the design tools (how progress is evaluated), the structure of the design approach and the process type. Type of design, coupling of sub-tracks, the level of abstraction, the scope and the complexity defines the nature of the design problem. It permits to understand the design problem and consequently to create an adequate multi-disciplinary team. In this taxonomy, the information flow throughout the design process is considered. Ostergaard and Summers judge this flow as a crucial measure of collaborative design effectiveness.

Despite, this taxonomy has to be completed to permit the management of the collaborative design process. Management and evaluation of the design process have to integrate product, process and organisational aspects but have also to place actor on a central place. Actor has to be considered as the principal vector of performance in the design process.

This article focuses on the evaluation of design process performance with a specific view on the actor. This evaluation is proposed through the GRAI model, completed with the design environment concept. The first part describes the GRAI model and defines the concept of design environment. Then the dynamic aspects of management and evaluation of design process is proposed thanks to design environments. The second part underlines the actor's predominance in the evaluation of the design process and describes characteristics of a performance measurement system centred on actors. Finally, an example presents the use of the design environment and the influence of a performance measurement system centred on actor for design process management.

2. GRAI MODEL TO MANAGE DESIGN PROCESS

During design process management, collaborative design context and each element described in the Ostergaard's taxonomy have to be managed. So, it's necessary to use a model that considers all these aspects. Our literature review indicates that no model really takes into account all these aspects. The GRAI model integrates some of these and it could be used (Girard and Eynard, 1999; Girard and Doumeingts, 2003). Despite, it must be completed with the model of design environment, to permit the integration of the collaborative aspects. The GRAI model decomposes the design system in a perfect adequacy with the co-ordination decisions (Figure 1). Consequently, the follow-up of the design situation evolution is easier and really permits to decision-makers to co-ordinate projects. To perform this follow-up, it's necessary to analyze continuously design centre evolution. Objective is to influence the design context to favour collaborative situations between design actors. To achieve this objective, it's necessary to develop a new model that will be able to help decision-makers to create a design environment, adapted to the design actors (Robin *et al.*, 2004). The GRAI model is used because it offers a frame to manage the creation, the implementation, the follow-up and the evolution of the design environments.

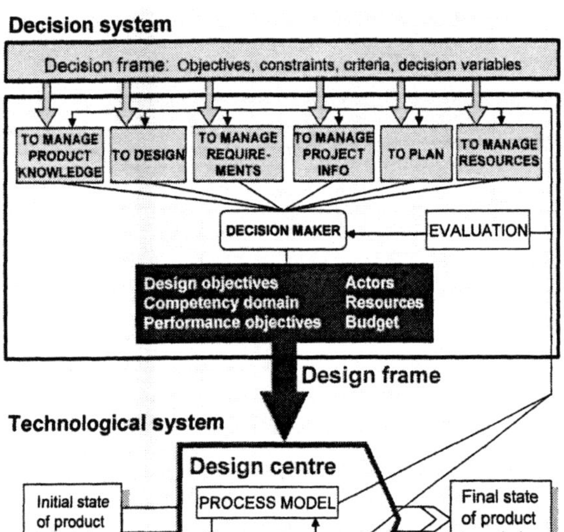

Figure 1: Reference model, structure of a design centre

This model gives a general frame to analyse and to manage design situation. As management of the design environments has to bring about some changes in the design context, according to the current state of the system, it obliges to develop a performance measurement system that considers design objectives, action levers and performance indicators. Performance indicators are necessary to evaluate pertinence of each intermediate situation, according to the global objectives. Identification of the action levers, associated to each performance indicator, permits to decision-maker to adapt the design context, in order to design actors could achieve design objectives. When these interactions are defined, it's possible to create and to implement the performance indicators system, to manage the design process. It permits to follow and to evolve design environment.

The dynamic of the management of the design process, thanks to the design environments, is based on four principal phases (Figure 2):

1) Identification of the need for collaboration,
2) Description of the as-is design situation,
3) Analysis and comparison of this as-is situation with the objectives of the design system, in order to make decision using action levers,
4) Implementation of a design environment, adapted to the design situation, in order to change efficiently the design context.

At each period, decision-maker assesses the necessity to make evolve or not the design frame and the design environment. This need for evolution is based on the description of the as-is situation, on the performance measures and possibly on the identification of a need for collaboration. The decision-maker compares these elements with the objectives, which assigned to them in the decision frame. He decides to create or not a new design environment, to make evolve efficiently the design context. Consequently, according to the action levers on which he could act, he assigns a new design frame to the design centre, completed with:

- Activities that take place in it.
- Organisations that have to be developed.
- Collaboration forms, recommended to make collaborative work easier for actors of the design centre.

The same work will be performing all along the project progress and a priori at each period. Nevertheless, to evolve properly a design environment, the description of the design frame that is defined in the GRAI model has to be completed. Therefore, elements of the context in which actors' collaboration will take place will enrich the design frame. The description of these elements is presented in the following paragraph.

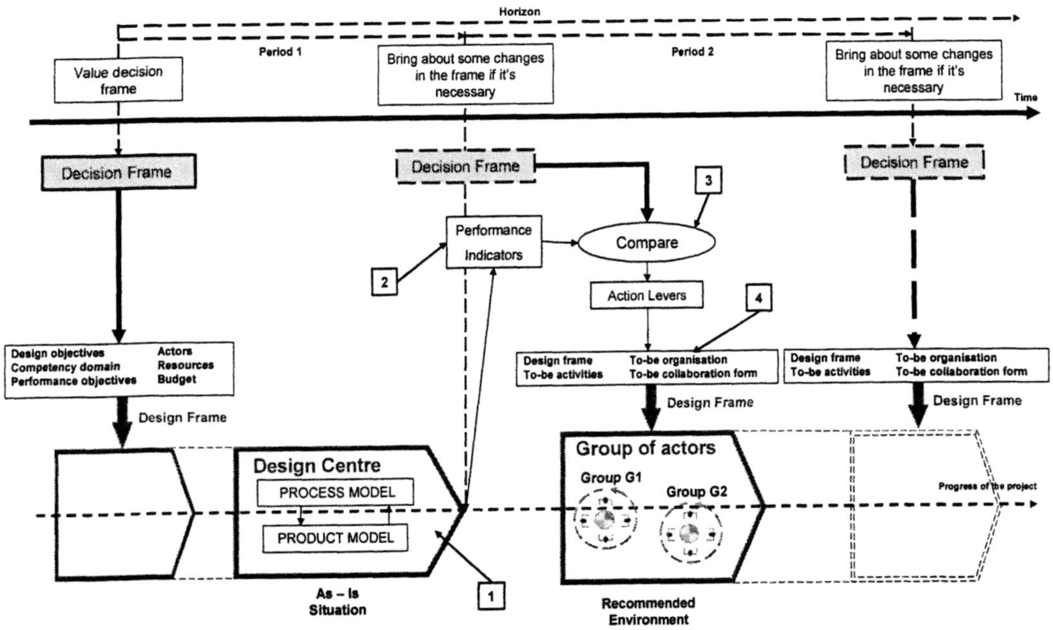

Figure 2: Reference model of the evolution of a design environment

3. PERFORMANCE EVALUATION CENTRED ON ACTORS

To increase design performances and consequently to satisfy customers' requirements and expectations of enterprises, the decision-maker (generally the project manager) could develop a design environment. So, the actors' context of work will be improved and since the project manager will be able to create effective working group according to assigned design objectives, the phase of human resources allocation will be more efficient. Mechanisms that could help project manager to adapt design environment to the observed situation must be based on sociologic aspects (Jeantet, 1996), on the management of the evolution of the production systems (Malhéné, 2000) and on information about the product, the process and the organisation. Creation and deployment of a design environment oblige to analyse the existing design situation. A design situation is defined as the state of the technologic system at a specific point of time (Eder, 2003). It is described according to the six aspects of the Ostergaard's collaborative design taxonomy and it's also characterised by these criteria:

- Actors, particularly their roles and their uses in the design process. Their experience concerning similar projects, their knowledge and their socialisation have to be take account too. Objective is to adapt properly their work environment according to their needs.
- Product, its nature (Suh, 1990), its complexity, its status in the process and its interfaces (Eynard, 1999) have to be take into account.
- Process, particularly the design approach (Perrin, 2001), the design type (routine, innovative or creative) and the collaboration type (Joly and Mangematin, 1995; Dameron, 2000; Girard *et al*, 2003).
- Material and financial resources (business premises, computers, budget,...).
- Constraints of the enterprise and in particular constraints of its environment.

Thanks to this description, it's possible to define parameters of a design environment:

- The design frame (design objectives, competency domains, performance objectives, actors, resources, budget...) (Figure 1).
- The description of the expected design situation:
  - The product model and the process model of the to-be situation.
  - The to-be organisation of the design centre (with a social view).
  - The to-be activities and the performance levels achieved.
  - The to-be collaboration form.

These parameters define a design environment and could evolve during the project progress. Management of the design environments consist in a continuous phase of adjustment and evolution of the environment, according to the design situation evolution. To permit this management, a performance measurement system has to be designed.

Nowadays performance evaluation of enterprises obliges to manage discretionary activities, which are more and more collaborative (Sénéchal *et al*, 2003). Therefore, it is not enough to measure and to manage product data or the progress of the design process. Evaluation should focus on interactions, which are generating the design process. Performance measurement system has to consider interactions between objectives, action levers and performance indicators, to supply pertinent information during the phase of design environment creation. These interactions must be described with a view to the product, the process, the organisation and the management style of the design process.

They influence actors' deeds and decisions and are action inductors during the progress of the design process. It's possible to manage design process thanks to these action inductors. Girard *et al.* (2002) have identified interactions between actor and the elements of the context of his evolution during the design process progress.

These interactions take four forms that depend on the point of view, (Figure 3):

1) The actor and the design object,
2) The actor with an other one,
3) The actor and the group,
4) Evolution of group of actors.

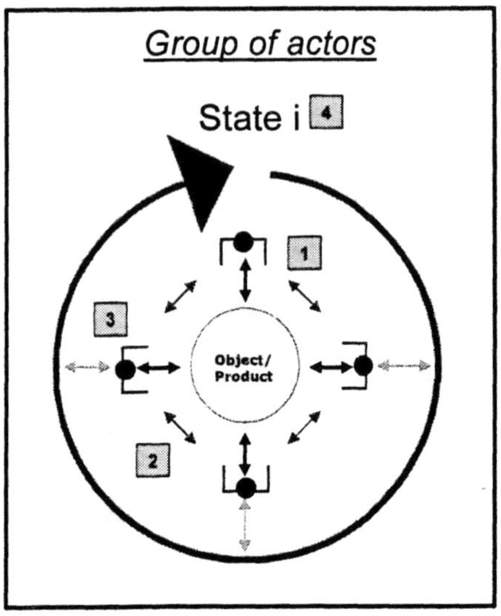

Figure 3: Design actor's interactions with elements of his context

First form concerns interactions between actor and the design object (the design artefact). In this case, it's necessary to analyse actor's impacts on the evolution of the design artefact definition.
An other point of view considers actor's interactions with the group and two forms of evaluation appear. In one hand, evaluation concerns the actor and it is based on his individual relationships with others members of the group. It will be relevant here to identify collaboration vectors and how collaboration between actors appears. Objective is to define collaboration form according to design needs and to favour the actor's awareness to his environment (Girard *et al*, 2003). In the other hand, we describe the general dynamic of the group. It's important here to identify factors that create good state of mind and motivation in the group.
Lastly, last form permits to identify interactions between the group (or the actor) and its environment, the design context. Intern and extern factors, which have an influence on the group and on each actor, and each step of the evolution of the group (or of the actor), have to be defined.
Then the objectives, action levers and performance indicators have to be identified according to these four types of interactions. Following paragraph suggests an example of the design process management in thanks to the design environments.

4. EXAMPLE

This section illustrates the proposed concepts to evaluate design process. The example focuses on the product data process of an enterprise, which is specialised in the manufacturing of material for petrol industry. (Figure 4) shows the modelling of the product data process. It's necessary to define clearly the process to fill properly the GRAI-R&D structure.
At the beginning of the project, strategic level sends design objectives to design project manager to initialise the project. He defines the different phases of the project and creates a design centre adapted to the design situation. After that, he sends the design frame to designers, at the operational level (phase 1, Figure 2). These activities have to supply drawings, technical documentation to permit the selection of 2 or 3 solutions. In the same time, when they receive intermediary results of the design, project committee and design project manager estimate the cost each of different solution. When 2 or 3 solutions are selected, project committee, quality department and design project co-ordinator manage project group to achieve the definition of the final solution. During the phases of selection and definition of the solution, we are in a collaborative design process. To manage these process, design project manager has to evaluate design situation at each period (phase 2, Figure 2), to make converge designers' results and estimation of the project committee. He compares results to design objectives (phase 3, Figure 2). If objectives are not achieved or to improve efficiency of this collaboration between entities, design project manager could create a design environment (phase 4, Figure 2).

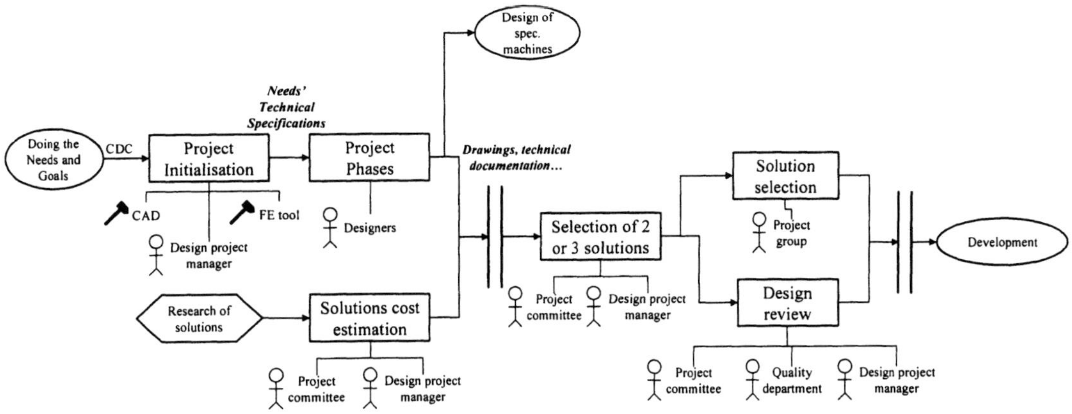

Figure 4: Modelling of the product data process

To illustrate the four influence forms on the actor and then to identify the associated performance indicators system, we use the GRAI-R&D structure describing of the decision system (Figure 5) of the studied enterprise. In this modelling, intersections between lines and columns represent decision centre and biggest arrows (vertical or horizontal) represent decision frames, which are described in paragraph 2.

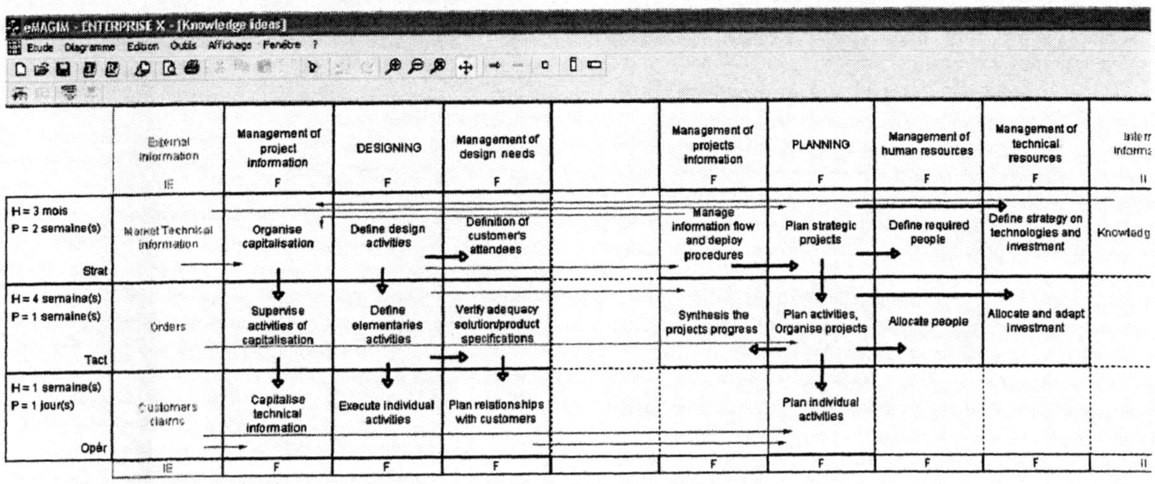

Figure 5: Modelling of the decision -making structure by the GRAI-R&D structure

This modelling permits to define the comportment of each decision centre and relationships between each of them. It corresponds to the first step in the management of the design process. When the GRAI-R&D structure is complete, it's possible to develop a performance measurement system in order to manage design process thanks to the reference model of a design environment in paragraph 2. The performance measurement system has to take into account specificity of each decision centre and of each link between them (Figure 5). Here, we describe the performance measurement system for the three functions: "management of project information", "designing" and "management of design needs". This system evaluates relationships between designers, project group and the context of the design process, i.e. links with design project manager and the entity which doing the needs and goals (Table 1).

Table 1 Interactions between objectives, action levers and performance indicators if considering evaluation focused on actors.

	Objectives	Action levers	Performance indicators
Designer / Design object	Satisfy requirements, define product, reduce complexity, reduce cost, respect quality,...	Product knowledge, product structure, information system, project handbook, ...	Data maturity, product data sharing, work product structure, design methods, ...
Designer / Designer	Exchange data, facilitate work, collaborate, meetings, ...	Procedure to collaborate, task definition, interface of collaboration, IT tools, ...	Number of loops, actor's competence, number of mails, errors analysis,...
Designer / Project group	Co-ordinate work, define project, respect delay, increase innovation, ...	Work breakdown structure, milestones, mechanisms of co-ordination, ...	Number of non planned meetings, number of patents, ...
Evolution of the project group	Increase actor's competencies, use internal or external resources, ...	Standardisation, lecture, enterprise modelling, business plan, ...	Reuse of external solutions, days of training, exchange with others functions, ...

Description of the decision-making structure of the enterprise by the GRAI-R&D structure, the definition of adapted performance measurement system and the reference model of a design environment offer the possibility to project manager to evaluate the design performance according to the objectives. Like this, he takes the good decisions and will have an efficient management.

5. CONCLUSION

Management of the design processes is complex and to improve processes performances, it's necessary to focus not only on the artefact but also on the actor's relations. This paper shows that management has to be develop around three principal phases. In a first time, the decision-making structure of enterprise has to be defined precisely to put in evidence all links between each design centre. Objective of this phase is to underline the rule of each actor and to judge his influence on the design process.
Secondly, a performance measurement system has to be developed according to the decision-making structure. It has to integrate the actors' influences. Lastly, the reference model of a design environment gives the frame to implement the result of the two previous phases.

ACKNOWLEDGEMENTS

This paper is based on research relevant of the working group IS3C: "*Ingénierie des systèmes de conception et conduite du cycle de vie produit*" (Engineering of design system and product life cycle management). The French research pooling supports this working group: GDR MACS of the French academic research committee: CNRS http://www.univ-valenciennes.fr/GDR-MACS/

REFERENCES

Chiu M.L. (2003). Design moves in situated design with case-based reasoning. In: *Design Studies*. **Vol. 24,** p 1-25.
Dameron S. (2000). Processus de coopération dans l'organisation : construction d'une grille de lecture appliquée au cas d'une équipe projet. IXième conférence internationale de management stratégique « perspectives en management stratégique », AIMS 2000. Montpellier.
Eder W.E. (1990). A typology of designs and designing. International Conference on Engineering Design, ICED 03, Stockholm.
Eynard B. (1999). Modélisation du produit et des activités de conception - Contribution à la conduite et à la traçabilité du processus d'ingénierie. Thèse de Doctorat, Université Bordeaux I.
Girard P., Eynard B. (1999). Proposal to control the systems design process: application to manufactured products. In: *Integrated Design and Manufacturing in Mechanical Engineering*. pp. 537-544. JL.Batoz et al. Kluwer Academic Publishers, ISBN 0-7923-6024-9.
Girard P., Doumeingts G. (2003). Modelling of the engineering design system to improve performance. In: *Computers & Industrial Engineering*.
Girard Ph., Robin V., Barandiaran D. (2003). Analysis of collaboration for design coordination. 10th ISPE International Conference on Concurrent Engineering: Research and Aplications. CE'03, Madeira, Portugal.
Jeantet A. (1996). La coordination par les objets dans les équipes intégrées de conception de produit. In : *Coopération et conception*. Terssac G. et Friedberg H.. Octares, Toulouse.
Joly P.B., Mangematin V. (1995). Les acteurs sont-ils solubles dans les réseaux ?. In : *Economies et Sociétés*. N°2, **vol. 9**, p. 17-50.
Malhene N. (2000). Gestion du processus d'évolution des systèmes industriels – Conduite et méthode. Thèse de doctorat. Université Bordeaux I.
Ostergaard K., Summers J. D.. A taxonomic classification of collaborative design process. International Conference on Engineering Design, ICED 03, Stockholm.
Perrin J. (2001). *Concevoir l'innovation industrielle, méthodologie de conception de l'innovation*. CNRS éditions, Paris.
Poveda O. (2001). Pilotage technique des projets d'ingénierie simultanée, modélisation des processus, analyse et instrumentation. Thèse INPG.
Robin V., Girard Ph., Barandiaran D. (2004). A model of design environments to support collaborative design management. 5th International Conference on Integrated Design and Manufacturing in Mechanical Engineering, IDMME 2004, April 5-7 2004, Bath, UK.
Sénéchal O. (2003). Le cycle de vie du système de production. In : *Evaluation de la performance des systèmes de production*. p. 81-104, Hermès.
Suh N.P (1990). *The principles of design*. Oxford University Press, New York.

Copyright © IFAC Information Control Problems in
Manufacturing, Salvador, Brazil, 2004

ELSEVIER
IFAC
PUBLICATIONS
www.elsevier.com/locate/ifac

REQUIREMENTS, STRUCTURES AND CONTENT MODEL FOR WEB-BASED MAINTENANCE PORTALS IN AUTOMATION

Martin Wollschlaeger*, Thomas Bangemann**

* *Industrial Communications, Department of Computer Science,*
TU Dresden, D-01062 Dresden, Germany
Martin.Wollschlaeger@inf.tu-dresden.de

** *ifak Institut f. Automation u. Kommunikation e.V.*
Steinfeldstrasse 3, D-39179 Barleben, Germany
Thomas.Bangemann@ifak-md.de

Abstract: Web technologies are gaining increased importance in automation and control systems, especially for information access in maintenance. However, it is important not to focus on the technology aspects, but to define requirements from the application perspectives. The paper discusses requirements within the focus of maintenance, and presents appropriate structures for a Web portal for maintenance purposes. It is based on a content model, describing the relations between topological and functional entities of the application system and the web technologies used. This hierarchical description model is defined in XML. The structure of this model is described and the use of description information for the realization of different functionality is shown. Special attention is paid to the security features of the solution. This requires specific description elements, ensuring access to description elements is limited to functional roles or user personalization. Different implementation aspects, and experiences derived from prototype implementations will be discussed. Finally, conclusions and an outlook for future developments are given. *Copyright © 2004 IFAC*

Keywords: Maintenance, Fieldbus, Information, Integration, Web

1. INTRODUCTION

Web technology has been adopted in nearly every area, also in automation and communication systems. Besides initiatives like PROFInet (PNO, 2001) or IDA (IDA Group, 2001), dealing with integration of systems from the shop floor up to the business management layer, the adoption of web technologies introduces new concepts and offers new chances for tasks in different stages of an automation system's life cycle. Especially for maintenance purposes, the use of Web technologies suggests advantages for the different users within a complex automation scenario (Wollschlaeger, 2001a). These technologies are state of the art and have developed into de-facto standards. Browsers and servers exist for nearly every platform, they are widely accepted by different classes of users. The browser as a platform-independent application framework should allow to re-use software components. Finally, location independence, multiple client access, and easy installation of software components are in the users' focus. On the other hand, this promised land has some drawbacks, mostly dealing with the heterogeneity of the different technologies.

2. REQUIREMENTS AND STRUCTURES FOR A WEB PORTAL

Web based solutions in automation systems can be described according to the three tier architecture (Fig. 1). Different layers with specific designations can be distinguished. The lower layer provides information from process to controller level of an automation system, usually implemented as a fieldbus (Wollschlaeger, *et al.*, 2002a).

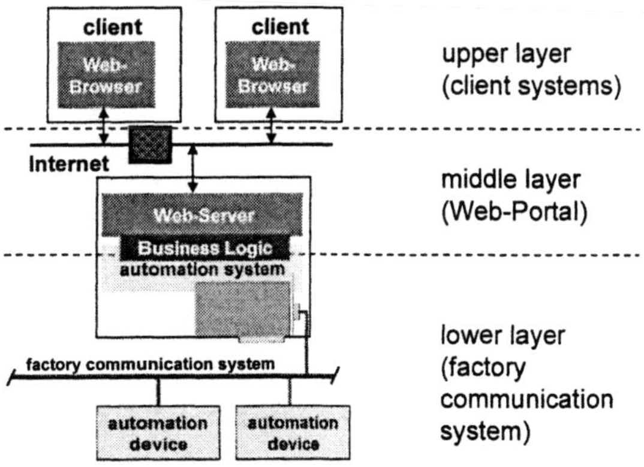

Fig. 1. General architecture of a web portal in automation systems

The upper layer is based on standard IT technology, using web server as data source and web browsers as clients. The functionality of the web portal is implemented as business logic. It provides an application gateway between the different layers. This functionality might be complex, since it is the combination of application-depending and well-chosen, but different technologies. Solutions are requested to describe the interactions of these different technologies and make the interoperable.

Analyzing the business logic, different components can be distinguished (Fig. 2) (Wollschlaeger, *et al.*, 2003b). Besides the data and the interfaces to exchange data between the underlying automation system, different servers can be recognized. Security and administration are integral parts of the portal. The applications implement specific parts of the business logic, dealing with conversions or rearrangements of the data between the layers.

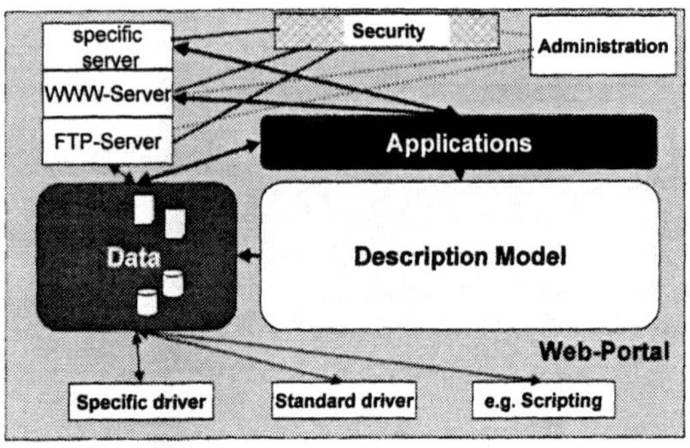

Fig. 2. Internal structure of the web portal.

A Web Portal's architecture should be able to reflect all stages of the life cycle of automation systems. So it is required to have access to the internal structure of a complex automation system, as well as to operations and tools required for planning, configuration, parameterization, operation, diagnosis and maintenance. Each of the components generates a certain amount of data, depending on its function.

- Configuration data are generated by configuration tools and are partly stored in the device
- Resources of a device are described by a set of data (e.g. device description)
- A field device cerates and/or processes data relevant for the technical process
- Functions of devices and processes can be described by data

In addition, a large number of data exists, that are not explicitly of technical nature:

- Documentations extending the facilities, like general descriptions, repairing instructions, lists of spare parts etc.
- General information like order numbers, inventory data, customer relations etc.

The data described above are heterogeneous by nature and have to be characterized into management relevant data and process data. Both together describe the complete data set of the system. The most important part of the portal is a description model, where the relations between the different classes of information are described (Fig. 3).

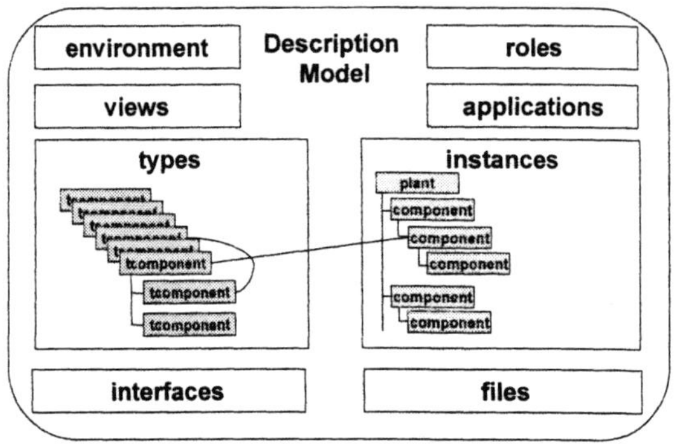

Fig. 3. Logical parts of the description model

Information can be stored at different levels of the system:

- In field devices
  The business logic application requests a managed object from a fieldbus server. This server reads the object from the field device via the fieldbus protocol and returns it to the application. This guarantees consistency, but has less performance.
- In the fieldbus server
  The fieldbus server refreshes a local image of the managed objects. Thus, the business logic application accesses a copy of the real managed objects. With respect to the first method, performance is increased. However, consistency between fieldbus server and field devices has to be guaranteed.
- In a local data base of the web portal
  The web portal acquires managed objects from the underlying system and updates a local image of the managed objects. Every access from an application is a local access with high performance. In addition, local management information can be stored in the same data base. Consistency aspects become more important, since they have to be guaranteed system-wide.
- At referenced locations
  The web portal stores references to static management information. The managed object is accessed via the reference upon request of an application.

The description model has to be able to support all these messages. It has to provide access information to address them via interface constructs. For the model, the data are handled as parameters with different constraints.

Furthermore, the description model has to integrate relations to external information sources. This external information contains details on logical combination of data (views), to file-based resources (pictures, documentation, etc.), and to the environment (clients' capabilities, personalization etc.). Finally, the relations to applications, and appropriate user roles are defined. These user roles represent details of the use cases. So these roles carry important information on access rights to the various sources of information. The individual users defined in the environment may take over one or more roles at several times of usage.

It's necessary to define the details of the description contents by the different use cases. Therefore an object model diagram has been defined. A fragment of this diagram is shown in Fig. 4. Different logical classes can be recognized. These classes are modeled concerning presentation in a web client, and do not necessarily correspond to a device-related or functional view. They represent different logical groups of information (devices, parameters, menus etc.), containing the relevant properties to be presented to a user. A view object is assigned to each class, which will be displayed on the web client when accessing the object instance. The different stereotypes define the relations between several parts of the object model.

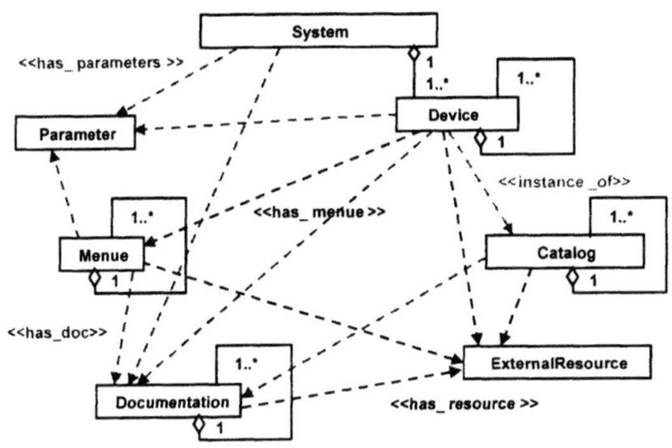

Fig. 4. Object diagram of information in the description model (fragment)

3. WEB PORTAL AND MAINTENANCE APPLICATIONS

The content of the data - and the relations among them - a web portal has to provide differs for specific applications scenarios. Although this does not influence the architectural aspects of a web portal, it does affect the applications (the business logic) running on the portal (or at least accessible through the portal).

In the focus of Maintenance and Online Asset Management applications are such informational elements and devices, which current statuses can be accessed using the plant communication network, e.g. field devices, combined with heterogeneous support information. The tasks are in general the synthesizing of plant data for an economical optimization of technical plant support, for example optimizations of planned maintenance and thus minimization of occurring disruptions, improvement of process diagnosis and control, and the increase of production throughput by using existent reserves (NAMUR, 2001). Several projects can be found dealing with maintenance aspects, for example (Szymanski, *et al.*, 2003). Furthermore, solutions and tools exist, in most cases provided by manufacturers of automation systems or components. However, it is difficult to integrate customer-specific, additional information.

Different maintenance strategies can be found in industrial solutions. The most promising among them is predictive maintenance. However, this is also a complex solution. Several conditions have to be fulfilled for efficient predictive maintenance:

- Devices and systems have to be continuously investigated by extracting an application-specific set of measurement data and events, and by watching their changes and trends.
- A maintenance management has to be defined, providing the required logistic support. It enables remote access to technical documentations or to database systems, and to on-line modeling and expert knowledge.
- A use-case depending data presentation, considering personalization for different user groups, and for different capabilities of user interfaces and access methods.

So a maintenance system has to be able to automatically extract actual values, measurement data, and set-points, but also structural information, from a given automation and control system. These raw data have to be filtered and recombined and then presented in a suitable way for the users. Using this kind of information, predictive inspection and maintenance can be scheduled. As a consequence, this will lead to a minimization of the system's down times. The immanent heterogeneity requires effective, but low-cost solutions. The extraction of the data has to be performed without a significant engineering effort, and – much more important – without consuming additional system resources. An extremely important task is the generation of maintenance information based on the extracted data. Several approaches exist, Different

4. XML-BASED CONTENT MODEL

It should be clear, that the structure described above is quite complex and requires some effort to be managed and to kept consistent. For the implementation, XML (Bray, 1998) was chosen to describe the structure. Besides the general aspects like close relation to other web technologies (in fact, XML is a web technology itself), XML was developed and established as mark-up language for generic purpose.

An XML-based description itself consists of a distributed set of XML-files, which contain specific parts of the description. Typically, different XML files are created, each of them related to specific description contents. In order to prevent multiple definitions, the files are linked together, creating a consistent set of descriptions (Fig. 5). The users, as well as the applications using the description, usually don't recognize the description's distributed character. Transformation rules define mappings between different syntactical information. The transformation rules can be implemented in style sheets using extensible style language for transformation (XSLT) (Boumphrey, 1998).

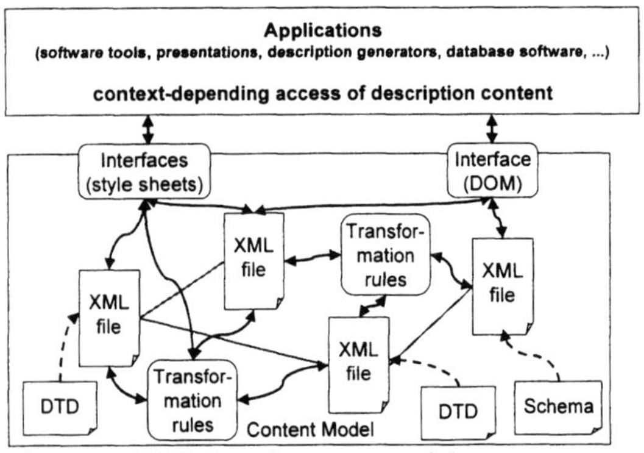

Fig. 5 An XML-based content model

XML can be easily used to create type-depending descriptions useful for role-based access. Since the term "role" in the context of this paper describes a specific use case in an automation context, role-specific information sets with special semantics and syntax have to be generated to access specific parts of the description model (Wollschlaeger, *et al.,* 2001b). Besides XSLT, XML extended links (W3C, 2000) can be used to model context-depending relations (Wollschlaeger, 2000).

The complexity of modern machines and automation systems can be a problem for a homogeneous description. Implementation examples have shown that a unique file containing the whole description is not a promising solution. In contrast, the distributed character of the information fragments is well-suited for implementing a distributed set of small, context-depending description elements, as shown in Fig. 5.

Besides the use of standard security technologies like user authentication, access restrictions and encryption, the use cases and related security policies have to be defined exactly. User roles specify the tasks someone representing a role has to perform during the lifecycle. This role defines which data is allowed to be accessed, and which actions are allowed to be performed. Of course, the user roles depend on the application. It has to be pointed out, that the adequate definition of user roles is a task of general relevance for every automation and control system. Besides the roles, the context a role is fulfilled in is important. It contains information on the stage in the life cycle, the type of user-interface, its position in the architecture of the system, temporal aspects etc. This information is heterogeneous by nature. In order to make it accessible in a generic way, it has to be described by a common methodology. XML can be used here, in combination with unique semantic definitions defined in a schema.

The middle layer of the Web integration concept shown in Fig. 1 can be treated as a specific type of application proxy. It maps the functionality of the underlying automation and control system to web pages or web services. This mapping has to consider the roles described above. Since the business layer also knows context information, the assignment of security policies can be controlled. Furthermore, safety aspects have to be considered here, for example based on a session concept.

5. REALIZATION EXAMPLE

A scalable multi-vendor configuration has been implemented at ifak to proof the concepts described above (Wollschlaeger, *et al.,* 2003a). The following aspects have been in the focus of the installation:

- Integration of a heterogeneous system based on different network structures and with network gateways
- Realization of typical application environments
- Implementation of as products on the market available devices, which are typical in a given application context
- Existence of classical solutions and approaches for the management (including configuration, parameterization, operation, administration), that can be used as a reference
- Access methods to internal details of the used devices, tools and technologies.

The installation contains two separate parts, a continuous process with the adequate equipment used in process control, and a demonstration of a manufacturing process. The network topology of the process control part is shown in Fig. 6.

The PROFIBUS technology used in the installation enables new chances for an online parameterization and for administration and maintenance of field devices. For these tasks supporting solutions are developed at ifak, partly in cooperation with other partners.

An example is the development of an Asset Management Box (AMBox), together with RWTH Aachen (Wollschlaeger, *et al.,* 2002b). This system is used for an extraction of information relevant for service, maintenance and asset management from the underlying process. The complete parameter set is read from the devices, using the block structure definitions in the PROFIBUS PA profile (PNO, 1999). The data are assigned to specific objects of a runtime object system. The semantic meaning of the data is also defined in the profile. The objects represent agents for the managed objects. The manager roles are performed by application modules, which are specific for different management scenarios, e.g. parameterization, administration, service etc. A web access to the objects is implemented as shown above. So the AMBox acts as a management system as well as a maintenance data server.

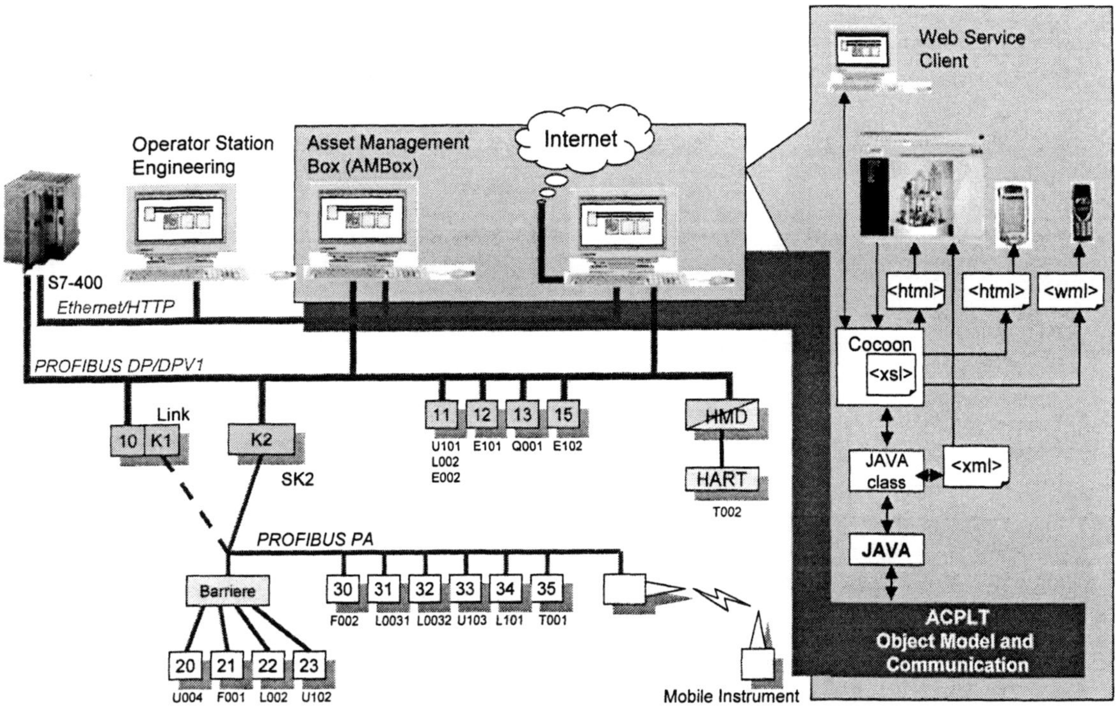

Fig. 6. Network topology and Web Access of the process control system in the realization example.

In the installation described above, Web Services are used for tool-based access from outside. Only HTTP is used as a transport protocol. This overcomes the limitations that still exist in DCOM, CORBA, or even the built-in ACPLT/KS communication services, since these solutions require extra communication ports. The integrative effects of the solution are the providing of Web Services based on the ACPLT object model, which contains meta information on object classes derived from appropriate XML description files.

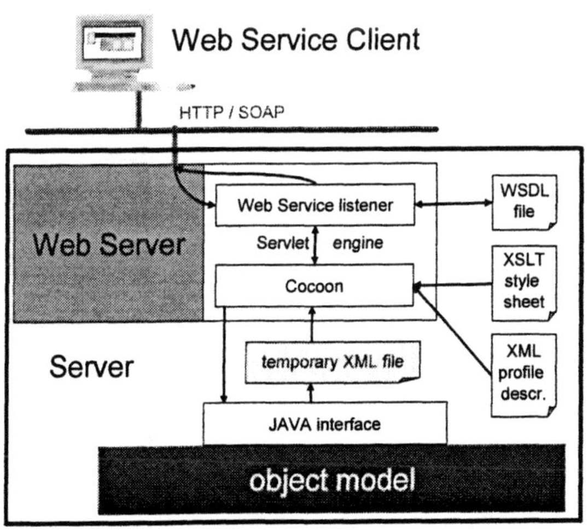

Fig. 7. Exemplary Web-Service implementation

Besides HTML access to the object model, it is possible to implement XML based Web Services. The integration is done as shown in Fig. 7. An appropriate proxy class is implemented to provide access to the object model. It is invoked by function calls issued by a web server. The proxy class returns the information gathered from the object model as an XML stream. This stream can then be transformed into several output formats, ranging from HTML and WML up to SOAP messages required for Web Services. The transformation is performed by the Apache Cocoon engine. It also includes context depending information from the description model. Thus, the appropriate message formats can easily be generated, according to the XML-based service descriptions, which are defined in WSDL format (W3C, 2001). So the web server provides listener functionality for the Web Service implementation, while the description model determines the business logic as described above.

6. CONCLUSIONS

The description model discussed above was derived from an analysis of relations between the actors and the use cases of the web portal solution. The resulting object model can be generalized; the definition of correlations between the components of the description by using typed relations appears to be very flexible. Finally, the solution described above shows the flexibility and the potential of web technologies. The most challenging task is to choose the right technology for the right task, and the definition of an integration core. For this core, XML is the most promising technology. The integrative character of XML allows using the very same infrastructure for both, user interaction in a browser, and web service interface to applications.

ACKNOWLEDGEMENT

The concepts and results described above are derived from a project "Web Portal for Maintenance of machines and systems" funded by "Stiftung Industrieforschung", Germany, as project "S 536", and from the project "Simplification and automation of the engineering of online asset management applications", funded by AIF (Arbeitsgemeinschaft industrieller Forschungsvereinigungen), Germany, under contract 13321 BG/1.

REFERENCES

Boumphrey, F. (1998). Professional Style Sheets for HTML and XML. Wrox Press.

Bray, T., J. Paoli, and C.M. Sperberg-McQueen (1998). Extensible Markup Language (XML) 1.0, http://www.w3.org/TR/REC-xml

IDA Group (2001). IDA –the Internet of Automation Technology. White paper, V 1.0.

NAMUR (2001), NAMUR-Recommendation NE91: Requirements for Online Plant Asset Management Systems, First Issue 11/2001.

PNO Profibus Nutzerorganisation (1999). PROFIBUS-PA: Profile for Process Control Devices, Version 3.0. Karlsruhe, 1999.

PNO Profibus Nutzerorganisation (2001). PROFInet - Architecture Description and Specification. V 0.91. Karlsruhe, 2001.

Szymanski, J.; T. Bangemann, M. Thron, J.P. Thomesse, X. Reboeuf, C. Lang, and E. Garcia (2003). PROTEUS - a European Initiative for E-Maintenance Platform Development. In: *Proc. IEEE ETFA 2003.*

W3C (2000). XML Linking Language (XLink) Version 1.0, W3C Proposed Recommendation. http://www.w3.org/TR/2000/PR-xlink-20001220/

W3C, 2001. Web Services Description Language (WSDL) 1.1, W3C Note, http://www.w3.org/TR/2001/NOTE-wsdl-2001031

Wollschlaeger, M. (2000). A Framework for Fieldbus Management Using XML Device Descriptions. In: *Proc. IEEE WFCS'2000*, pp. 3 10.

Wollschlaeger, M. (2001a). Framework for Web Integration of Factory Communication Systems. In: *Proc. IEEE ETFA 2001*, **1**, pp. 261 266.

Wollschlaeger, M., Ch. Diedrich and M. Thron (2001b). Generation of Role-specific Information for an Enterprise Integration Framework Using XML Descriptions. In: *Proc. IEEE ETFA 2001*, **2**, pp. 237 244.

Wollschlaeger, M., Ch. Diedrich, and R. Simon (2002a). Web Integration of Factory Communication Systems using an XML Framework. In: *Proc. IEEE ISIE'2002.*

Wollschlaeger, M., J. Müller, Ch. Diedrich and U. Epple (2002b). Integration of Fieldbus Systems into On-line Asset Management Solutions based on Fieldbus Profile Descriptions. In: *Proc. IEEE WFCS'2002*, pp. 89-96.

Wollschlaeger, M., and Th. Bangemann (2003a). IT-Based Management of Maintenance Process in Fieldbus-Based Automation Systems. In: *Proc. IFAC FeT 2003.*

Wollschlaeger, M., F. Geyer, D. Krumsiek, and R. Wilzeck (2003b). XML-based Description Model of a Web Portal for Maintenance of Machines and Systems. In: *Proc. IEEE ETFA 2003.*

INFORMATION CONTROL PROBLEMS IN MANUFACTURING 2004

(INCOM 2004)

A Proceedings volume from the 11th IFAC Symposium,
Salvador, Brazil, 5 – 7 April 2004

Edited by

P. KOPACEK
Institute for Handling Devices and Robotics,
Vienna University of Technology, Austria

C.E. PEREIRA
Department of Electrical Engineering,
Federal University of Rio Grande do Sul,
Porto Alegre, Brazil

and

G. MOREL
Research Centre for Automatic Control,
CRAN, Faculté des Sciences,
Vandoeuvre-les-Nancy, France

(In two volumes)

Volume 2

Published for the

INTERNATIONAL FEDERATION OF AUTOMATIC CONTROL

by

ELSEVIER LIMITED

ELSEVIER Ltd
The Boulevard, Langford Lane
Kidlington, Oxford OX5 1GB, UK

Elsevier Internet Homepage
http://www.elsevier.com

Consult the Elsevier Homepage for full catalogue information on all books, journals and electronic products and services.

IFAC Publications Internet Homepage
http://www.elsevier.com/locate/ifac

Consult the IFAC Publications Homepage for full details on the preparation of IFAC meeting papers, published/forthcoming IFAC books, and information about the IFAC Journals and affiliated journals.

Copyright © 2005 IFAC

All Rights Reserved. No part of this publication may be reproduced, stored in a retrieval system or transmitted in any form or by any means: electronic, electrostatic, magnetic tape, mechanical, photocopying, recording or otherwise, without permission in writing from the copyright holders.

First edition 2005

Library of Congress Cataloging in Publication Data

A catalogue record for this book is available from the Library of Congress

British Library Cataloguing in Publication Data

A catalogue record for this book is available from the British Library

ISBN 0-08-044249 8
ISSN 1474-6670

These proceedings were reproduced from manuscripts supplied by the authors, therefore the reproduction is not completely uniform but neither the format nor the language have been changed in the interests of rapid publication. Whilst every effort is made by the publishers to see that no inaccurate or misleading data, opinion or statement appears in this publication, they wish to make it clear that the data and opinions appearing in the articles herein are the sole responsibility of the contributor concerned. Accordingly, the publisher, editors and their respective employers, officers and agents accept no responsibility or liability whatsoever for the onsequences of any such inaccurate or misleading data, opinion or statement.

Printed in Great Britain

To Contact the Publisher

Elsevier welcomes enquiries concerning publishing proposals: books, journal special issues, conference proceedings, etc. All formats and media can be considered. Should you have a publishing proposal you wish to discuss, please contact, without obligation, the publisher responsible for Elsevier's industrial and control engineering publishing programme:

Christopher Greenwell
Publishing Editor
Elsevier Ltd
The Boulevard, Langford Lane
Kidlington, Oxford
OX5 1GB, UK

Phone: +44 1865 843230
Fax: +44 1865 843920
E.mail: c.greenwell@elsevier.com

General enquiries, including placing orders, should be directed to Elsevier's Regional Sales Offices – please access the Elsevier homepage for full contact details (homepage details at the top of this page).

11th IFAC SYMPOSIUM ON INFORMATION CONTROL PROBLEMS IN MANUFACTURING 2004

Sponsored by
International Federation of Automatic Control (IFAC)
IFAC Technical Committee on Manufacturing Plant Control

Co-sponsored by
IFAC Technical Committees on:
- Discrete Event Dynamic Systems
- Computers for Control
- Components and Instruments
- Cost-Oriented Automation
- Human Machine Systems
- Manufacturing Modeling for Management and Control
- Enterprise Integration and Networking
- Large Scale Complex Systems
- Social Impact of Automation

IFIP Working Group 5.7 - Integration on Product Management
IMS - NOE - Network of Excellence - Europe
SBA - Brazilian Automation Society (Brazilian IFAC NMO)
SBA - TC on Industrial Automation
MANET - Brazilian Research Network on Manufacturing Automation

Conference Chairs

Pereira, C.E. (BR)	(General Chair)
Morel, G. (F)	(IPC Chair)
Silva, J.R. (BR)	(Local Organization Chair)
Bernhardt, R. (DE)	(Industrial Chair)

International Programme Committee (IPC)

Albertos, P. (ES)
Banaszak, Z. (PL)
Bititci, U. (UK)
Boer, C.R. (I)
Boverie, S. (F)
Diedrich, C. (DE)
Dolgui, A. (F)
Erbe, H.-H. (DE)
Faure, J.-M. (F)
Florin, F. (RO)
Fonseca, J.A. (P)
Frey, G. (DE)
Garetti, M. (I)
Grabot, B. (F)
Johnson, T. (USA)
Kopacek, P. (A)
Lee, J. (USA)
McFarlane, D. (UK)
Molina, A. (MX)
Monostori, L. (HU)
Morel, G. (F)
Neagu, G. (RO)
Nof, S. (USA)
Pereira, C.E. (BR)
Sanz, R. (ES)
Silva, J.R. (BR)
Stahre, J. (S)
Valckenaers, P. (B)
Vingerhoeds, R. (F)
Zaremba, M. (CAN)
Zaytoon, J. (F)
Zuehlke, D. (DE)

Local Organization
SBA : Brazilian IFAC National Member Organization
MANET: Manufacturing Automation Network – Brazilian Network of Excellence
GCAR/UFRGS: Control, Automation, and Robotics Group - Federal University of Rio Grande do Sul
SENAI/CIMATEC
UFBA
UNIFACS
FTC

Financial Support
CNPq
FINEP
CAPES
MANET
IMS - NOE
KUKA Roboter
FANUC Robotics
FAPESB
USP/EPUSP
UFRGS/EE/DELET/GCAR

PREFACE

INCOM'04 is a tri-annual event of symposia series organized by IFAC under the sponsoring of the Technical Committee 5.1 on ***Manufacturing Plant Control***. Previous editions were held in Vienna (2001), Nancy (1998) and Beijing (1995).

This 11th edition was co-sponsored by 9 others IFAC Technical Committees covering in large theoretical, technological and applications issues of manufacturing plant automation for Shop-Floor Process-Control as well as for Enterprise-Control System Integration.

INCOM'04 main purpose was to point up international researches and developments dealing with all the applications of automation, information and communication technologies in order to control and to manage the manufacturing plant within the e-enterprise, involving all methodological and technological aspects to embed a technical "intelligence" within the components. The symposium addressed automation scientific challenges and issues raised by the IMS paradigm in order to apply MEMS, MECHATRONICS, MES, MAS, HMS and e-technologies to digitally control with more agility the entire manufacturing chain, from supply and design through manufacturing, to maintenance and service, over the whole product and processes life cycle. Information issues at the interface of the technical and business processes were discussed as well as societal and human impact (safety, dependability, usability, knowledge, and experience) of these emerging e-manufacturing technologies..

INCOM'04 Technical Program was organized around 6 technical tracks under track chairs ensuring the scientific relevance of 6 Manufacturing Plant Control Areas:

- **Production & Logistics over Manufacturing Networking**, with an invited plenary lecture on *Collaborative E-Work And E-Mfg.: The State Of The Art And Challenges For Production And Logistics Managers*, and 2 regular sessions on ***Business to Manufacturing Issues***,
- **Manufacturing Automation over Networks**, with an invited plenary lecture on ***Communication In Industrial Automation - What Is Going On?***, an invited session on ***Real-Time Distributed Embedded Systems,*** and a regular session on ***Networked Controlled Systems,***
- **Dependable Manufacturing Systems Control**, with an invited plenary lecture on *Improving Automation Software Dependability: A Role for Formal Methods?*, an invited session on ***Industrial controllers software dependability : formal methods applications***, and a regular session on ***Design and implementation of dependable systems***,
- **Discrete Event Systems in Manufacturing**, with an invited plenary lecture on *Analysis of Hybrid Systems- State of the Art*, and 2 regular sessions on ***Petri nets based approaches*** and ***Design and control synthesis of DES***,
- **E-Manufacturing Technologies and Facilities**, with an invited plenary lecture on *Infotronics Technologies and Predictive Tools for Next-Generation Maintenance*

Systems, and 2 regular sessions on ***Cooperative Manufacturing and Engineering*** and ***Methods and Technologies towards E_Maintenance***,

- **IMS Modelling and Experiments**, with 2 invited plenary lectures on *Industrial Applications of the Agent-Based Technology* and *How Information Technology Enables the Potential Flexibility and Re-configurability of Highly Automated Manufacturing Plants*, and 2 regular sessions on ***IMS Engineering trends and issues*** and ***B2M Performance Issue.***

This core technical programme is completed by 2 other areas:

- **Research Networks**, with 2 invited sessions on ***Remote Experiments for e-learning (Virtual Laboratories)***, and ***B2M Systems Interoperability: Modeling and Benchmarking Issues,***
- Posters, with 3 regular sessions on *Manufacturing Modelling Issues*, *Manufacturing Automation Issues*, and *Manufacturing Assembly and Robotics Issues,* as well as a special session on *Work-In-Progress.*

All the **163 papers submitted** to **INCOM'04,** which came from 25 different countries, have been reviewed by at least 3 reviewers among a panel of **171** relevant scientists and industrialists, leading to the presentation of **7 keynote talks** and **76 papers** in the 6 technical tracks of **INCOM'04** core technical programme and the presentation of *20 papers* in the complementary area of the complete technical programme.

We hope that all readers find the contents of this conference proceedings intellectually stimulating and offering an opportunity to address the above important challenge of defining a coherent framework of concepts, theories, models, methods, methodologies, languages and tools for *Manufacturing Plant Automation and Integration.*

Last but not least, we would like to express our gratitude to all people acting in this promising event, coming from the National Organizing Committee for their enthusiastic and unflagging contribution, and coming from the IPC, IFAC TCs, IFIP TCs, and the Manufacturing community in large for their high scientific contribution.

Salvador da Bahia, April 2004.

Gérard Morel
Carlos Eduardo Pereira
José Reinaldo Silva
Alceu Heinke Frigeri

CONTENTS

VOLUME 1

PRODUCTION AND LOGISTICS OVER MANUFACTURING NETWORKING

Keynote Paper I

Business to Manufacturing Issues I

Business to Manufacturing Issues II

MANUFACTURING AUTOMATION OVER NETWORKS

Keynote Paper II

Networked Controlled Systems

VOLUME 2

B2M Performance Issues

POSTER PAPERS

Manufacturing Modeling Issues

Manufacturing Automation Issues

Manufacturing Assembly and Robotics Issues

RESEARCH NETWORKS

Remote Experiments for E-learning (Virtual Laboratories)

B2M Systems Interoperability: Modeling and Benchmarking Issues

WIP – WORK IN PROGRESS PAPERS

Work in Progress I

Work in Progress II

Work in Progress III

Copyright © IFAC Information Control Problems in Manufacturing, Salvador, Brazil, 2004

ELSEVIER
IFAC
PUBLICATIONS
www.elsevier.com/locate/ifac

A NEW APPROACH TO MAINTENANCE IN THE MANUFACTURING SECTOR

Mario Lopez, Roberto Rinaldi, Claudio Roberto Boer, Roberto Castellani, Paolo Pedrazzoli

ITIA-CNR Institute of Industrial Technologies and Automation
TTS -Technology Transfer System s.r.l.

Abstract: nowadays, the optimization of maintenance activities is becoming more and more important for the manufacturing sector. The increasing requirements to reduce time-to-market, to improve product quality and to keep costs low, point out the research of useful and effective maintenance solutions as topic of common interest.
In order to answer to this growing need, we propose a new approach based on an hybrid solution founded both on scheduling of ordinary maintenance operations as well as on management of extraordinary ones, through advance failure detections thanks to a smart control and management of sensors signals.
Currently, indeed, there are basically no efficient solutions, based on sensors smart signal analysis and on codified behaviour examination, meant to foresee incoming failures.
This paper is meant to present this approach, named NEMO after "New Maintenance hOrizon for 2007", its relevance and its foreseen impact for the manufacturing sector.
Copyright © 2004 IFAC

1. INTRODUCTION

The NEMO Environment, taking into consideration the manufacturing machine in analysis, creates and customizes a software and hardware solution able to recommend a maintenance strategy intended to join ordinary and extraordinary activities. Figure 1 presents a high level view of the suggested approach.

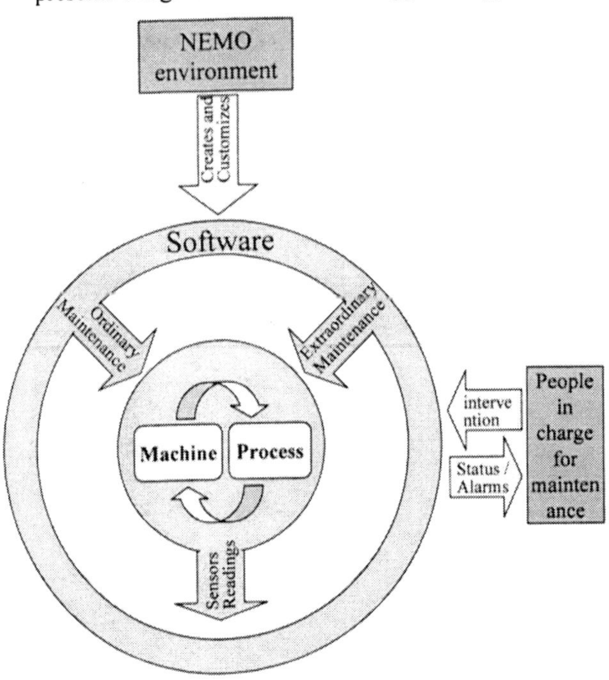

The NEMO environment not only proposes an optimized sensors solution, but also deploys a software core that, thanks to a smart interaction with the machine, provides preemptive maintenance.
This approach fosters the development of tools meant to allow the formalization of a new design paradigm based on a non-invasive integrated solution meant to optimize a machine maintenance strategy. Main advantages for the end users are strictly related to production costs reduction. This advantage is connected to:

- Non foreseen fault number reduction
- Reduction of breakage time losses (optimized management of maintenance procedures, reduction of time loss during diagnostic phase)
- Reduction of warehouse costs (increase of production plants availability)
- Optimization of maintenance strategy (dynamic scheduling of ordinary and extraordinary maintenance activities)
- Optimization of maintenance procedure strategy (effective, timely and automatic interrogation of maintenance dedicated staff, optimized diagnostic phase, optimized information flow)

The design tool proposed is dedicated to machine builders. In this way their products will be innovative if compared to current market standards and consequently strongly competitive.
This environment can be used to sensorize production units already operating, thanks to the low intrusion level of the solutions, thus it foreseen the creation of sensorized solution on demand.

2. NEMO STRUCTURE

The solution proposed is based on cutting-edge technologies like knowledge management, work flow management, real time signal analysis and object oriented approach, joined, developed and optimized in their interaction, in order to achieve a set of new methodologies and instruments.

The accomplishment and set up of the suggested solution requires both: 1- a software environment in order to study the optimal solution; 2- several modules to customize the management software core for the hardware studied.

Therefore, the NEMO approach foresees the following modules (see figure 2):

· FMECA Environment (reliability analysis environment). This data elaboration environment is composed of both a mathematical engine, dedicated to reliability analysis of a generic production unit, and procedures for automatic creation of FMT (Fault Model Tree) and OMT (Ordinary Maintenance Timetable)

· MLCE (Maintenance Lifecycle Cost Estimator). This software environment is meant to evaluate maintenance costs related to the whole manufacturing machine lifecycle.

· Sensorization Environment. This software environment is meant to aid the designer to choose the right set of sensors to monitor health status of the production unit.

Furthermore, the following components have to be created in order to develop and customize a sensorized solution and information management:

· Historical Analysis Component. This software component is composed of analytic tools meant to work out the logical engine to detect in advance incoming failures through signal analysis.

· MSM (Maintenance Strategist Module) This software component is composed of a logical engine, intended to optimize maintenance strategy, and a customizable software solution to manage information maintenance procedures.

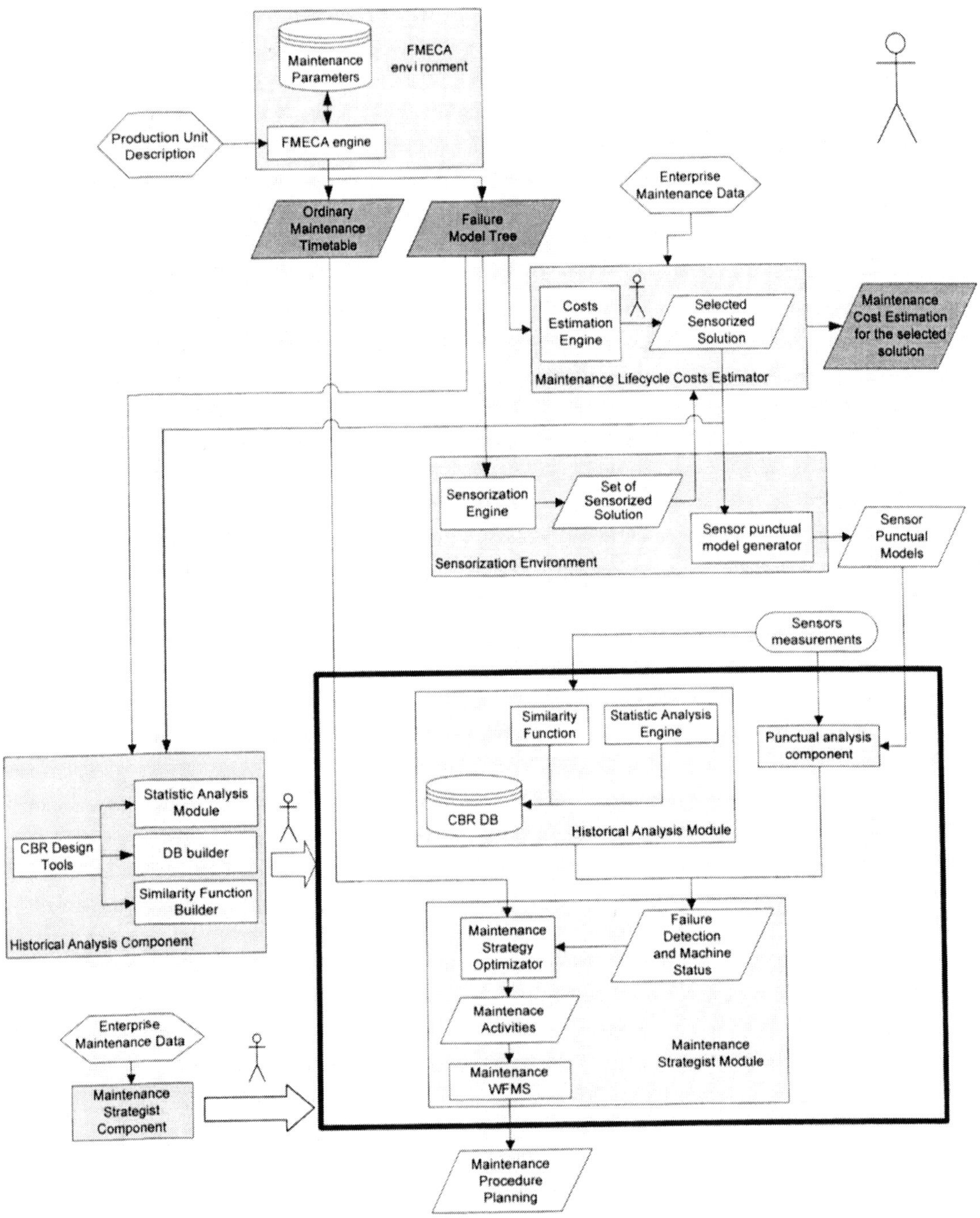

Furthermore, a data collection of different reliability parameters (MTBF, MTTR, etc.), relevant to mechanical components, will be put in the DB "Maintenance Parameters".
These modules have to be entirely embedded in the same graphical interface. Their integration will allow to get a complete software instrument able to work out and maintain the best sensorized solutions to the particular maintenance issue.

3. NEMO SOLUTION DEPLOYMENT STRATEGY

The schema in figure 2 is intended to be a paradigm for the future use of the environments and components foreseen in the proposed approach.
A generic manufacturing machine will be taken into consideration and a project cycle will be executed, in order to exemplify the study and development of a sensorized solution. It will be worked out through NEMO approach to the monitoring and maintenance optimization issues.
The decomposition of the manufacturing machine into its functional units is the first operation required. These units will be analyzed and reduced to a tree structure made up of basic elements reciprocally in parallel or in series.
The components structure have to be reproduced, through "Drag and Drop" technique, within FMECA Environment. Every graphical element, that belongs to the system library, corresponds to a particular record of the Database ("Maintenance Parameters"). This is not a simple data depository containing reliability parameters, actually every database object is provided of information relevant to its behaviour (different fault recorded and possible reparations). This research activity implementation will allow to achieve a really attractive database. It will become a reference for reliability data acquisition concerning mechanical parts. Besides, this data structure could be used in order to create component libraries also for other different industrial sectors.
Once the machine structure has been reproduced within software environment, the mathematical engine to process reliability parameters ("FMECA Engine") will elaborate information associates to the basic elements and will work out two different outputs: the "Ordinary Maintenance Timetable" (OMT) and the "Fault Model Tree" (FMT) [S. Amari]. These documents must be validated; hence, they can be modified by a technical designer of the mechanical plant. If FMT and OMT are modified than they will be reprocessed in order to get the new data structure.

The second project phase deals with the choice of sensors to be set on the plant. They will be chosen in order to select the most proper ones, to monitor the different functional units of the manufacturing machine. The sensorization engine for the manufacturing machine ("Sensorization Environment") requires, as an input, the FMT worked out during the previous phase, and provides, as an output, a group of possible solutions based on the individuation of the most proper set of sensors for the case in analysis; first of all it takes into account the most critical functional units. This kind of operation doesn't modify the manufacturing machine FMT but concerns reliability parameters, reducing the number of unforeseen failures related to the different components and, once operational, it allows to improve diagnostic phase. All these aspects considerably modify costs evaluation relevant to the production unit maintenance. [Z. Vintr]
The assessment of economic benefits deriving from using every different sensorized solution is the main goal of the next design phase. Inserting the required data is the a non-automatic procedure. Data consists of company costs relevant to people in charge for recovering, warehouse costs, time breakage costs etc. This is needed in order to calculate the economic advantage deriving from using a maintenance solution based on signal analysis sampled by sensors, comparing it to a traditional maintenance strategy. Costs will be evaluated referring to the whole manufacturing machine operational lifecycle. Hence, the choice of the best solution will be achieved taking into account economic advantage and selecting the less expensive one between the suggested.
At this point of the project cycle the whole hardware structure (sensors and data acquisition tools) has been made out and the proposed solution has to be validated. Therefore, in the range of possible results, only one is pointed out. This solution has to be provided, as an input, to the "Sensor Punctual Model Generator", that automatically supply the models to analyze signal in real time. In fact, these are simple numerical models related to the typology of selected sensors and oriented to the detection of sudden failures. The verification process is made through mensuration of the differences between sampled data and expected one. These models are part of the software modules to be installed in order to monitor production plant.
A further operational step deals with the achievement of the "CBR" (through "Historycal Analysis Component"). The aim of this software module is to individuate, by historical data analysis, improper functioning descending from a regular degenerative process of machine performances. This software unit accomplishment is achieved through a group of wizard procedures. They will lead to the implementation of the DB regrouping the already solved cases and to the creation of "similarity functions". Hence, at the beginning, the DB could be empty: a learning system, guided by machine user, will allow to enlarge the number of already solved cases contained in the DB, in order to individuate a larger typology of faults. So a large cases solution, able to manage the most presumable critical situations, can be achieved quickly through data collection and sharing.
Last project phase deals with the customization of a software module meant to coordinate maintenance procedure ("Maintenance Strategist Component"). This module allows to optimize maintenance strategy through analysis of condition monitoring and real

time comparison of incoming maintenance activities (ordinary and extraordinary).
Hence, using "Maintenance Strategist", it is possible to join different maintenance activities reducing, in this way, the manufacturing machine breakage times. Last, maintenance procedures will be synchronized by a dedicated software module (such as WFMS). It allows to automatically contact, through internal network (LAN), the people in charge for recovering the best efficiency conditions and the whole staff involved in machine fault. Furthermore, this system allows the management of the sequence of operations composing the maintenance procedure.

The "Sensors Punctual Models", the "CBR", and the "Maintenance Strategist Module" are the software tools to be installed, together with selected hardware instrumentations, in order to manage the sensorized solution worked out for the manufacturing machine.

4. BENEFITS

Straightaway economic benefits are related to the machine end-user. Indeed, the users of sensorized solutions, meant to prevent malfunctioning, will be able to optimize their maintenance strategy, achieving relevant reduction of time breakage, that is directly connected with economical looses due to production stops and maintenance intervention costs. Moreover also scrap percentage is expected to decrease. In fact, malfunctioning prevention will not only reduce machine fault, but will also avoid to produce scrap pieces thanks to a prompt forecast of the machine malfunctioning.
Furthermore, the high availability level of machines will lead to time-to-market reduction, thus promoting an effective answer to market's needs.
The manufacturer of a plant or a machine, designed according to the new paradigms proposed by the NEMO approach, will take advantage of the augmented competitiveness of the machines sold: this is strictly connected with 1) the increased quality of the machine produced 2) reduced costs for after-sale technical assistance 3) reduced cost for intervention.
Among the results expected due to the NEMO implementation, it is foreseen the realization of a Data-Structure meant to catalogue, from a reliability and maintainability point of view, a large number of mechanical components and functional units. This reference tool has not been implemented yet and the research of this kind of data proves to be hard and difficult to achieve.

5. RELEVANCE AND STATE OF THE ART

Among the various documentation attached to the generic production machine, we find the user-manual and maintenance-manual. Within this document, we find detailed all the technical procedures and ordinary maintenance timing related with the machine sold.

The day by day maintenance activities are generally either ordinary or generated by a breakdown; rarely we face preemptive procedures, and even more rarely the operating condition reestablishment processes are driven by preemptive analysis.
Nowadays, indeed, there are basically no efficient solutions, based on sensors smart signal analysis, meant to foresee incoming failures [Z.D. Zhou, L. Monostori].
A preemptive maintenance approach, in fact, has been adopted just in peculiar and specific framework, and, due to the difficultness of such an approach, has not been already adopted as a market standard, while it is still subject to intensive research [O. Ryabov, Y. Hatamura, D. Chen].
Simultaneously, support tools, for the design of a sensorized solution and the implementation of the software modules to manage data acquisition and analysis, have not been developed yet. NEMO objective is to foster a real technological breakthrough in both these fields.

The state of the art analysis conducted is meant to highlight the innovative aspects related to the different modules fostered by the NEMO approach.
In relation to the software tool developed in order to carry out the reliability analysis, the market offers a wide range of choices [HTTP references]. This class of products presents to the user several statistical calculation tools that often are redundant. Moreover, these products are just mathematical engines meant to analyze the reliability of a given machine, while the specification of the NEMO SW module must go further: indeed, the FMECA Environment will not only provide the reliability analysis, but will also enable the creation of two data structures that will be the reference model during all the design phase of a customized sensorized solution. In fact the FMECA environment outcome is the automatic development of the FMT (Fault Model Tree) and of the Ordinary Maintenance Timetable.

Commercial products, developed for reliability analysis, enable the economical evaluation of management costs for the manufacturing machine, during a full life cycle of the plant. The MLCE (Maintenance Lifecycle Cost Estimator) module enables the same function, and moreover assists the definition of a customized sensorized solutions, starting from the relevant data related with the breakage time of the machine and with the cost of the company under investigation. This module will be fully integrated in the system architecture: it will manage automatically the data flows and will be able to evaluate the reduction of breakage time according to the sensorized solution selected.

The effective sensorization of a manufacturing machine is an open issue for applied research. The final goal is to achieve the machine operative condition real-time and to be able to forecast malfunctioning and upcoming failures. An approach meant to create design paradigms aimed to the re-use of preexisting know-how has not been explored yet.

The Sensorization Environment will present an engine intended to define the possible sensorization solution for a given machine, that will be able, thanks to several techniques and approach such as "Sensor Fusion", to fully monitor the manufacturing machine.
The implementation of this software module will be a fundamental step for the set-up of the approach proposed and will represent a significant breakthrough not only thanks to the introduction of a maintenance point of view in the design phase, but also enabling an automatic maintenance strategy for the machine studied.

The automatic sensorization module enables the automatic deployment of a sensors structure to be installed on the machine. Similarly, it will have to automatically produce the software modules for the management of the signal sampled [R.G. Wilhelm, J.H. Suh, K. Jemielniak]. This module is called Historical Analysis Component. The analysis of the state of the art showed that no adequate similar solution exists on the market and that not even a complete research effort has been carried out.

The Maintenance Strategist Module will have to carry out the functions of a Work Flow Management System (WFMS), dedicated exclusively to the maintenance procedures. The market offers a wide range of solution. The NEMO specifications suggest an integrated software meant to customize specific solution in a short delivery time; moreover an engine intended to automatically optimize the maintenance strategy for the manufacturing machine will be developed. The approach proposed foresees the development of a dedicated module for the exclusive management of maintenance procedure, in order to assure both low costs for the software and an easy integration in the approach framework.

6. CONCLUSION AND FUTURE WORK

The growing needs of flexibility, performance and support of the manufacturing machines points out the need of new maintenance strategy and tools.
This paper presents the approach, named NEMO, studied in order to answer to these needs: the approach is based on an hybrid solution founded both on scheduling of ordinary maintenance operations and on management of extraordinary ones, through advance failure detections thanks to a smart control and management of sensors signals.

REFERENCES

http://blocksim.reliasoft.com/

http://www.novell.com/news/press/archive/1998/01/pr98004.html

http://www.reflex.com.au/products_why_availability.aspx

D. Chen, K.S. Trivedi, "Analysis of Periodic Preventive Maintenance with General System Failure Distribution", IEEE 2001

J.H. Suh, S.R.T. Kumara, S.P.Mysore, "Machinery Fault Diagnosis and Prognosis: Application of Advanced Signal Processing Techniques", CIRP 1999

K. Jemielniak, O. Otman, "Catastrophic Tool Failure Detection Based on Acoustic Emission Signal Analysis", CIRP 1998

L. Monostori, "A Step Towards Intelligent Manutacturing: Modelling and Monitoring of Manutacturing Processes through Artificial Neural Network", CIRP 1993

O. Ryabov, K.Mori, N. Kasashima, "An In-Process Direct Monitoring Method for Milling Tool Failures Using a Laser Sensor", CIRP,1996

R.G. Wilhelm, R. Chatndrashekar, R. Sun, M. Hegedus, B. Chu, W.J. Tolone, J. Long, "Finite Schedule Monitoring and Filtering in a Computer Integrated Environment", CIRP 2000

S. Amari, G. Dill, E. Howald, "A new approach to solve dynamic fault trees", IEEE 2003. Annual

Y. Hatamura, K.Iino, K. Tsuchiya, T. Hamaguchi, "Structure of Failure Knowledge Database and Case Expression" CIRP, 2003

Z.D. Zhou, Y.P. Chenn, J.Y.H. Fuh, A.Y.C. Nee, "Integrates Condition Monitoring and Fault Diagnosis for Modern Manufacturing System", CIRP 2000

Z. Vintr, R. Holub, "Preventive maintenance optimization on the basis of operating data analysis", IEEE 2003. Annual

Copyright © IFAC Information Control Problems in Manufacturing, Salvador, Brazil, 2004

ELSEVIER
IFAC
PUBLICATIONS
www.elsevier.com/locate/ifac

WEB BASED SERVICES TO MAINTAIN DCS'S APPLICATIONS IN THE TORERO PROJECT

Christian Schwab[1], Marcus Tangermann[1], Arndt Lüder[1], Luca Ferrarini[2], Carlo Veber[2], Giuseppe Fogliazza[3]

[1] *University of Magdeburg, Center Distributed Systems @ IAF, Universitaetsplatz 2, 39106 Magdeburg, Germany, {christian.schwab, marcus.tangermann, arndt.lueder}@mb.uni-magdeburg.de*
[2] *Politecnico di Milano, Dip.di Elettronica e Informazione, P.zza L. da Vinci 32, 20133 Milano, Italy, {ferrarin, veber}@elet.polimi.it*
[3] *Machining Centers Manufacturing S.p.A., Via Celaschi 19, 29020 Vigolzone (PC), Italy, g.fogliazza@mcmspa.it*

Abstract: Beside the design and implementation of the control application of a Distributed Control System (DCS) and the start-up phase, an important aspect within the life cycle of this system is the maintenance process. This paper will present an Integration Chain for engineering and maintenance purposes for a DCS which is based on web services and the Internet. Based on this Integration Chain the total life cycle of the DCS will be supported. This work is part of the research project TORERO, which is funded by the European Commission under the IST program. *Copyright © 2004 IFAC*

Keywords: Industrial Automation, Distributed Control System, Life Cycle, Maintenance, Web based Services, Internet, Information Technology

1. INTRODUCTION

The current situation in automation is characterised by a strong trend towards distributed systems based on intelligent devices (ARC, 2001; PNO, 2001; Schwab, et al., 2003; Tangermann, 2003). Related to this, several proposals created by different organisations and companies exist, specifying a future way of building industrial automation applications targeting distributed control applications (IEE, 2003; IDA, 2002, PNO, 2003).

In addition, an upcoming trend in automation industry is the increasing amount of Information Technology (IT) used within control systems for several purposes as remote management or engineering support for example (ARC, 2003; Schwab, et al., 2003).

In the near future, the Distributed Control Systems (DCSs) – consisting of networked intelligent field devices – will be connected via an industrial Ethernet network, providing a seamless integration of all levels of the automation pyramid, ranging from the ERP (Enterprise Resource Planning) and MES (Manufacturing Execution System) level down to the field device level (ARC, 2003). In the course of this the adoption of Internet related technologies in the field device level will be supported. A DCS will have direct access to the company Intranet and the Internet, assuming that appropriate security mechanisms are applied. This will open up new ways for managing the DCS via the company Intranet / Internet.

On the one hand there is the use of the Internet for remote access to field devices such as PLCs (Programmable Logic Controllers), I/O (Input/Output) modules, switches, etc., based on standardised or semi-standardised Internet technologies. This is called web based management and is state of the art in industrial automation.

On the other hand there is the use of the Internet for engineering and maintenance purposes, e.g. for semi-automated software download of updated versions and bug fixes. Such functionalities, adopted to DCS, are still under development although this is already state of the art in the office world.

In this respect, the research project TORERO (Total life cycle web-integrated control) (TORERO, 2003a) - funded within the IST (Information Society Technologies) initiative of the European Commission - aims at specifying a DCS, which is called TORERO system, consisting of a new type of Internet connected and web based engineered and maintained field device (TORERO device). Based on this infrastructure and in connection with Software Servers and Tools, an Integration Chain of a TORERO system based on web based services and the Internet, covering the engineering and maintenance phase of the DCS and thus support the total life cycle of the TORERO system, will be specified.

Within this paper first results of the TORERO project will be described. A practically applicable system structure for web based software engineering and self-maintenance of a DCS based on IEC 61499 like software building blocks (IEC61499, 2000; Christensen, 2000; Neumann, 2000) provided by device suppliers and OEMs (Original Equipment Manufacturer) and a special dedicated tool system will be given.
Therefore, this paper is organised as follows. To provide a basement for the further understanding, in the second chapter the life cycle of a control application will be described. Based on this, in the third chapter the web based Integration chain of a TORERO system and the actors involved in this process will be presented. In the following chapters the main parts of the Integration Chain, namely the TORERO Software Server and the TORERO Integrated Development Environment will be described. This will be followed by a chapter containing considerations about safety and security issues of the design and maintenance processes.

2. TORERO APPLICATION LIFE CYCLE

The life cycle of a TORERO application will not differ from the life cycle of other control applications. It consists of the processes (1) engineering, (2) maintenance, and (3) removal/reuse. But in contrast to conventional systems within the TORERO application life cycle a designer will be assisted by the Internet with respect to the provision of pre-designed Function Blocks (FBs) and bug fixes, etc. corresponding to the used devices.
Within the engineering process the control application designer will make the following steps:

- Design of the control application, independent of the underlying hardware,
- Choice of the applied TORERO devices,
- Automatic generation of communication related code after the semi-automated allocation of the control application to the TORERO Devices (TaS, et al., 2003, TORERO, 2003b, TORERO, 2003c),
- Deployment of the whole control application including the communication related code to the TORERO Devices.

Within these steps the designer will be assisted by the Internet. The TORERO Device (TD) supplier will ship with its devices basic pre-designed FBs useable for efficient control of the TD (so called Device Functions e.g. pre-processing of measured data from a Sensor input) together with an information on further Internet based FB sources – the so called TORERO Software Server (TSS). Out of the shipped FBs and the FBs provided via the TSS the control application designer can select the applied FBs. Therefore, the designer will use the TORERO Integration Chain.
Regarding TORERO the term "maintenance" denominates the process of updating existing device functions and/or the operating system of a Torero Device. As a first step the TIDE (TORERO Integrated Development Environment) contacts all available TDs in the network and requests their device description which contains all data about installed software (device functions, operating system etc.). This is done at the start of the TIDE automatically or on request by using the Universal Plug and Play (UPnP) protocol stack (UPnP, 2000) (see step (1) and (2) of Fig. 3).
Within the device description, an URL to the according Torero Software Server (TSS) is given. The TIDE contacts each of the TSS and sends the `GetUpdateListCommand` as described in chapter 3. The servers reply with the `GetUpdateList-Response` message containing information about available updates after processing the version information send by the TIDE. The TIDE user now has to decide whether the update fits to its needs. If the user decides to download the update he has to choose which download method (FTP or SOAP attachment) shall be applied. After the TIDE has downloaded the update, the data is stored within the local database of the TIDE until the deployment to the devices. Also in this case the TORERO Integration Chain will be applied.

3. TORERO INTEGRATION CHAIN

The complete TORERO environment forms an Integration Chain ranging from the TORERO Device (TD) Supplier over the System Integrator to the End User. This chain is based on the Internet as the communication media, the TORERO Software Server (TSS) as a central point for software storage and the TORERO IDE (TIDE) as a client to download this software.
The basic source for data in the TORERO Integration Chain is represented by the TORERO Software Servers of the individual manufacturers/supplier of TDs. It stores updates of device functions including updated device descriptions, bug fixes and additional device functions.
The next element in the chain is the System Integrator who designs complex automation systems containing many different TORERO Devices. During the design phase the System Integrator updates all necessary software elements on the TD to have an up-to-date software basis for the development process. Therefore the TIDE of the System Integrator contacts the TSS of every manufacturer via Internet (the URL is derived from the device description) and downloads the necessary updates on choice of the TIDE user. System Integrators can be present at multiple levels in the system, like a machine builder who integrates components (TORERO Devices) into a machine and a System Integrator building up an entire plant.
The last element in the chain represents the End User. During the maintenance process the TIDE contacts all TSS of the supplier of the TDs used in the system and downloads necessary function updates on choice of the user. Another possible use is the download of updated, modified or new application code from the System Integrator to the TORERO devices of the End User via the TIDE and Internet.

Fig. 1 is a sketch of the basic interactions among the different actors in the TORERO Integration Chain. Two TD suppliers, two System Integrators and one End User are depicted.

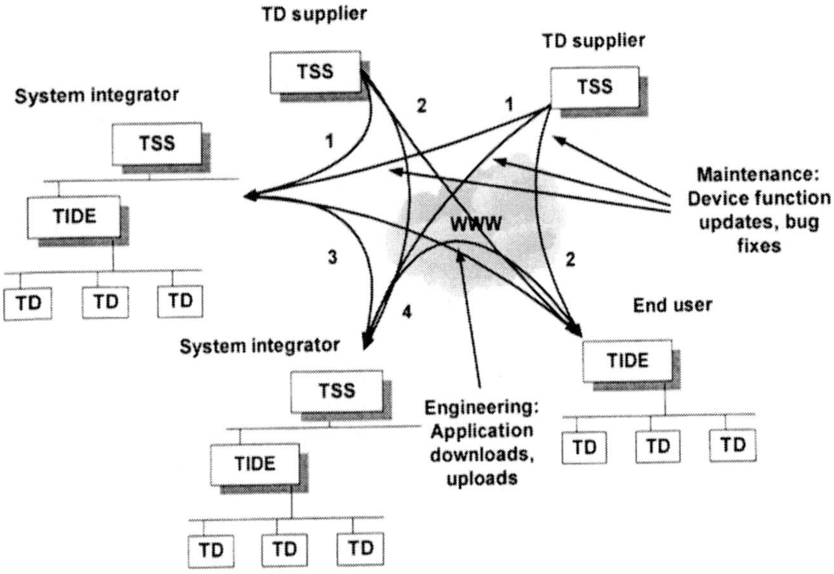

Fig. 1: TORERO Integration Chain

These three main actors, namely the TD Supplier, the System Integrator and End User have different roles within the Integration Chain as presented in the following.

TD Supplier

- Provides TDs to the System Integrator and End User.
- Updates functions and descriptions of his TDs and makes them available via the Internet (updated device functions, bug fixes, etc.).

System Integrator

- Integrates TDs into a machine or an entire system.
- Designs, builds and provides the machine / entire system / control system to the End User exploiting supplier functions.
- May be responsible for the maintenance of the system delivered to the End User (e.g. add some more device functions, add more components, bug fixes, etc.).

End User

- Operates the control system.
- Buys the overall system from the System Integrator and operates it.
- Reports bugs and failures to the System Integrator and the TD Supplier.
- Receives TD code updates from the TD Supplier.
- Receives updates of specific control application from the System Integrator.

To better describe the different interactions among the three actors, Table 1 is provided. It has to be noted that all the interactions between the actors go beyond the content of the table as it only describes interactions assisted by the TORERO environment. The No.s refer to the arrow labels in Fig. 1. The subject of the verbs in the cells refers to the actors in the first column.

Table 1: Interactions among TORERO actors

From: \ To:	TD Supplier	System Integrator	End User
TD Supplier	–	Arrow #1: • Provides TD Data • Provides Software and Description updates	Arrow #2: • Provides Software and Description updates
System Integrator	–	Arrow #3: • Provides aggregated TD Data • Provides Software and Description updates	Arrow #4: • Provides Control Application • Provides Software and Description updates
End User	–	–	–

Based on the functional description of the Integration Chain within TORERO given above, in the following more technical aspects regarding the TORERO Software Server and the TORERO IDE will be given as central points within the Integration Chain.

4. TORERO SOFTWARE SERVER

The TORERO Software Server (TSS) acts as a download server that provides both updates for device functions/operating systems of TORERO Devices (TD) during the maintenance process and generic function blocks during the engineering/reengineering process.

The TSS is strongly based on the application of web services using the Simple Object Access Protocol (SOAP) as the application layer within the ISO-OSI model (Newcomer, 2002). This protocol exchanges well formatted XML messages containing information about remote procedure calls similar to RPC (Remote Procedure Call) or RMI (Remote Methode Invocation). This data can be transported over a wide variety of protocols such as HTTP or SMTP.

A TSS as depicted in Fig. 2 consists of two main parts. The web front-end provides the web service itself receiving and sending the SOAP messages as well as decoding and encoding them. The backend consists of a database storing all kinds of updates and a version management administering the according version information.

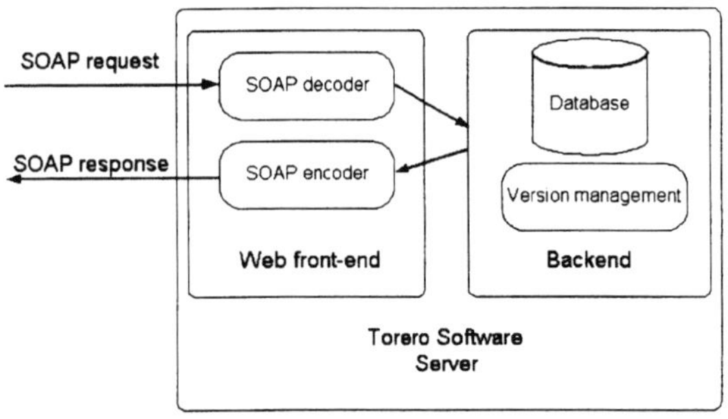

Fig. 2: Schematic structure of the TSS

An incoming SOAP request is decoded by the SOAP decoder and forwarded to the backend. The version management queries the database depending on the incoming command and generates an answer. This answer is forwarded to the web front-end where the encoder builds the according SOAP response which is sent back to the originator of the request.

During the maintenance the TSS offers two different commands. The command `GetUpdateList` is used during the maintenance process to retrieve information about the available updates/bug fixes for device functions and the operating system of TDs (see step (3) of Fig. 3). As parameters, all necessary information that identify the TD itself and the device function or operating system the update is requested for, will be transferred. These are among other:

- product family,
- serial number,
- a vendor specific type key,
- name of the device function,
- actual version of the device function.

The TIDE may request updates for more than one device function and more than one device.
The TSS must respond with a `GetUpdateListResponse` (see step (4) of Fig. 3) providing information about the available updates such as:

- new version number,
- hardware requirements e.g. RAM or processor,
- software requirements e.g. OS version,
- description of changes between the new version and the previous one.

Additionally the TSS provides a Unified File Name (UFN) that identifies the requested update within the server.

Generally, a TSS can provide two access methods for downloading an update. A file may be downloaded using the File Transfer Protocol (FTP) whereas the login data (username, password etc.) is provided by the `GetUpdateListResponse`. Another possibility for transferring a file from the TSS to the TIDE represents MIME attachments. The SOAP specification allows for attaching binary data such as pictures or files to a SOAP message similar to MIME attachments used for E-Mails (Tidwell, et. al., 2001). If the TIDE user prefers the usage of SOAP attachments for downloading the `GetFB` command is issued to the TSS containing the UFN provided by the `GetUpdateListResponse` (see step (5) of Fig. 3). The `GetFBResponse` contains the requested update as a SOAP attachment (see step (6) of Fig. 3).
In both cases the `GetUpdateListResponse` contains information which download methods are supported and provides all necessary information such as the login data for the FTP access.
Finally the accessed FBs will be downloaded by the TIDE to the TD using FTP mechanisms (see step (7) of Fig. 3).

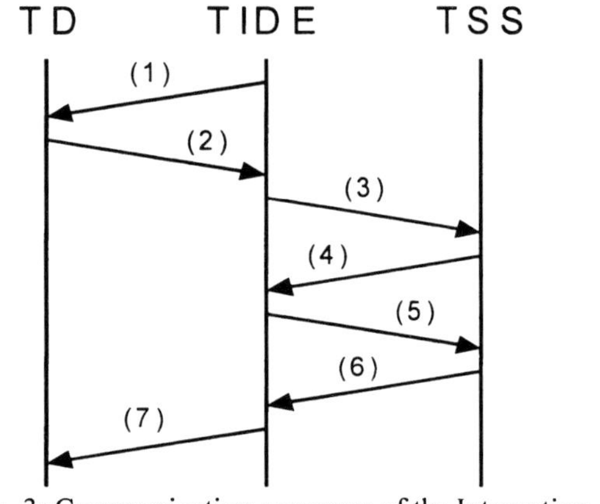

Fig. 3: Communication sequence of the Integration Chain

5. TORERO IDE

The client side for the download process forms the TORERO Integrated Development Environment (TIDE).
The TIDE chaperons the user of a TORERO environment through all engineering and maintenance tasks in the life cycles of a DCS.
Therefore the TIDE is based on a modular structure based on Eclipse (www.eclipse.org) and can be easily extended using plug-ins. Within this chapter we will focus on the elements important for the maintenance process.

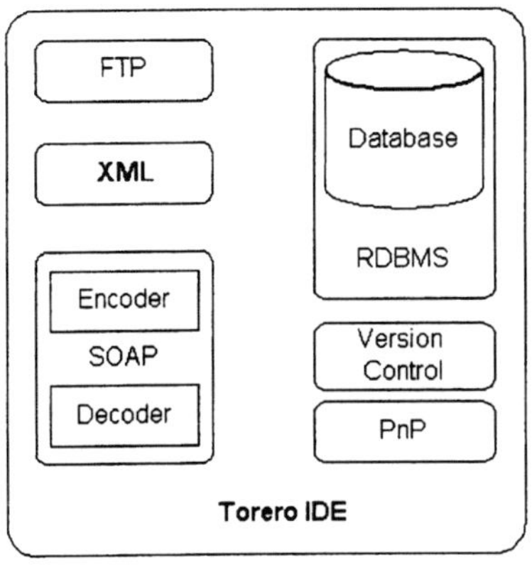

Fig. 4: Schematic structure of the TIDE

As depicted in Fig. 4 the TIDE possesses six maintenance relevant elements.

The first element called Plug and Play (PnP) is used for finding all TORERO Devices that are present in the TORERO network and for retrieving their device description containing all relevant information about the device hardware and the installed software. Within TORERO the Universal Plug and Play protocol stack is used to fulfil these tasks.

Since the device description uses XML for structuring the context the TIDE needs an XML parser. This parser uses the DOM model for easy handling of the documents and can validate these documents against the XML Schema language.

The SOAP encoder and decoder contained in the SOAP plug-in take over the communication with the different TSS. This plug-in encodes or decodes the different commands send to or received from a TSS and, where applicable, forwards them to the target plug-in.

The version control management software has two different tasks. First, it has to keep track of installed versions and compare them with available updates. Second, the version control has to compare the requirements of updates to the device hardware and installed software and give the TIDE user hints whether an update can be installed on a TD or not.

As stated in the chapter 3, the transfer of a file can be done using the FTP protocol which is implemented by the FTP plug-in.

Downloaded updates are stored in a relational database management system (RDBMS) and can be reinstalled on device e.g. in case of the replacement of a device.

6. SAFETY AND SECURITY ISSUES

The application of the Internet within the above described engineering and maintenance scenarios is causing several dangers for the complete system. Well known are the issues concerned with the so called white colour criminality or Hacker attacks. But this are only the most public issues. The upcoming problems are safety and security related.

Security addresses the prevention of illegal access - in the widest sense - to the automation system. Security thus has implications for safety as well. Here, security is used in the sense of IT security and is concerned mainly with securing the hosts and network of the automation system. Within TORERO we follow the security guidelines of IAONA given in (IAONA, 2003).

In detail, security is defined there in terms of these security objectives:

- Data Integrity,
- Authentication,
- Authorization (access control),
- Confidentiality,
- Availability,
- Auditability,
- Non-repudiation,
- Third-party protection.

With respect to TORERO we have to ensure the availability and integrity of necessary data as well as the protection of the included partners TSS, TIDE, and TD with respect to unexpected and unauthorised access.

The solution of the security problem can be achieved by the application of a security system with different stages of expansion. The stages are the following:

1. Application of user/password combinations on the TSS. Thereby only authorised persons can access the server.
2. Application of firewalls outlining TSS on the one side and TIDE and TD on the other side from the Internet were both firewalls can only be actively foraminated by the TIDE. Thereby, unexpected access can be avoided.
3. Application of Virtual Private Networks (VPN). Thereby, a protected connection between TIDE and TSS will be established.
4. Application of encryption mechanisms under the SOAP communication by applying HTTPS or a Public Key Infrastructure (PKI). Thereby data integrity can be ensured.

Beneath the safety issues covered by the security system an additional safety problem can emerge. This problem is given by the necessary guarantee of unchanged FB behaviour during the maintenance process with a guarantee of an unchanged applied interface as well as an unchanged expected behaviour. The version management system of the TIDE currently ensuring the necessary bordering conditions of the application of FBs can also take over the necessary control of the applicability of FBs with respect to interfaces and behaviour. Therefore, a standardises XML based interface and behaviour description has to be defined which will be provided over the TSS together to the FB during the `getFB` request. By a tag based tree comparison algorithm as given in (Deter, 2002) the version manager can decide the applicability of the new version by comparing the provided description of the new FB with a description of the applied features of the old FB.

7. SUMMARY AND OUTLOOK

This paper presents an Integration Chain of a Distributed Control System based on a new web enabled device type and an architecture which allows not only the web based engineering but also the web based maintenance process of the DCS and thus supports the total life cycle of the DCS. The main parts of the Integration Chain, namely the Software Server and the IDE also have been presented. A test implementation containing the Software Server and a module of the IDE which is called "maintenance plug in" are under development using Java according to the Real-Time Specification for Java (RTSJ, 2001; Dibble, 2002) and AspectJ (AspectJ, 2003; TaS, et al, 2003). First prototypical tests, which are currently going on, have shown the applicability of the described scenario.

Future work will include the whole implementation of all aspects described within the Integration Chain in order to support the total life cycle of the TORERO application, covering both the engineering and the maintenance process. Upcoming results will be published on the TORERO homepage (TORERO, 2003a).

REFERENCES

(ARC, 2001): ARC Advisory Group: PLCs: Is there a future? ARC Strategies, Dedham: ARC, 2001.

(ARC, 2003): ARC News, ARC Advisory Group, Dedham, Ma. USA, April 2003.

(AspectJ, 2003): AspectJ, http://www.eclipse.org/aspectj/, Version 1.1.0, June 2003.

(Christensen, 2000): CHRISTENSEN, James H.: Basic Concepts of IEC 61499, Fachtagung 2000 Verteilte Automatisierung – Modelle und Methoden für Entwurf, Verifikation, Engineering und Instrumentierung (Tagungs-band), Univ. of Magdeburg, Germany, 2000.

(Deter, 2002): S. Deter, A. Lüder, J. Peschke: XML und Plug-and-Participate zur Flexibilitätserhöhung in der Fabrik-automatisierung, SPS/IPC/Drives 2002, November 2002, Proceedings, pp. 275-283

(Dibble, 2002): DIBBLE, Peter, C.: Real-Time Java Platform Programming, Prentice Hall Computer Books, 2002.

(IAONA, 2003): IAONA e.V.: The IAONA Handbook for Network Security, Draft/RFC, Magdeburg, 2003, www.iaona.org.

(IDA, 2002): IDA-Group: Interface for Distributed Automation – Architecture Description and Specification, Revision 1.1, Blomberg, November 2002.

(IEC61499, 2000): IEC-PAS 61499-1, Function Blocks for Industrial Process Measure-ment and Control Systems, Edition 1.0, 2000-09, VDI Verlag, Düsseldorf, 2000.

(IEE, 2003): N.N.:Offene Plattformen für den Mittelstand (Transl. Open Platforms for medium-sized and small enterprises) in IEE Automatisierung + Datentechnik, Hüthig GmbH & Co. KG, Heidelberg, March 2003.

(Neumann, 2000): NEUMANN, P: Aktuelle Triebkräfte der Entwicklung von Automatisierungssystemen, Fachtagung 2000 Verteilte Automatisierung – Modelle und Methoden für Entwurf, Verifikation, Engineering und Instrumentierung (Tagungsband), Univ. of Magdeburg, Germany, 2000.

(Newcomer, 2002): NEWCOMER, ERIC: Understanding Web Services, Addison-Wesley Professional, 2002.

(PNO, 2001): Profibus Nutzer Organisation: ProfiNet – More than just Ethernet, Karlsruhe, PNO, 2001.

(PNO, 2003): PNO: PROFInet – Architecture Description and Specification, Version V2.0, Karlsruhe, January 2003.

(RTSJ, 2001): Real-Time Specification for Java (First Public Release), Addison-Wesley, 2001.

(Schwab, et al., 2003): SCHWAB, C., A. P. KALOGERAS, K. LORENTZ, et.al.: Next Generation Integrated Development of Automation Control Code in TORERO, IEEE International Symposium on Industrial Electronics – ISIE, Rio de Janeiro, Brazil, June 2003.

(Tangermann, 2003): TANGERMANN, M: TORERO – Total Life Cycle web-integrated Control, Presentation at OMG Conference, Working Group Real-Time and Embedded Specialised Systems, Paris, France, June 2003.

(TaS, et al., 2003): TANGERMANN, M., C. SCHWAB, K. LORENTZ, et.al.: Aspect-Orientation of Control Application Code for Distributed Automation Systems: The TORERO Approach. In: Proceedings OTM Confederated International Workshops, Java Technologies for Real-Time and Embedded Systems, Catania, Sicily, Italy, November 2003; Lecture Notes in Computer Science – LNCS 2889, Springer-Verlag Berlin Heidelberg, 2003.

(Tidwell, et. al., 2001) TIDWELL, DOUG; SNELL, JAMES; KULCHENK, PAVEL: Programming Web Services with SOAP, O'Reilly, 2001.

(TORERO, 2003a): The TORERO consortium, www.uni-magdeburg.de/iaf/cvs/torero.

(TORERO, 2003b): The TORERO Consortium: Deliverable 2.1 Integrative design and development of web-enabled control system design methodology (Draft), 2003. (www.uni-magdeburg.de/iaf/cvs/torero)

(TORERO, 2003c): The TORERO Consortium: Deliverable 2.2 Design and Development of encapsulation mechanisms, web-service based code download, automatic code distribution mechanisms, and testing mechanisms (Draft), 2003. (www.uni-magdeburg.de/iaf/cvs/torero)

(UPnP, 2000): COHEN, AGGARWAL., GOLAND: Universal Plug and Play Architecture, 2000, www.upnp.org.

Copyright © IFAC Information Control Problems in
Manufacturing, Salvador, Brazil, 2004

ELSEVIER
IFAC
PUBLICATIONS
www.elsevier.com/locate/ifac

PROBABILISTIC VS. DYNAMICAL PROGNOSIS PROCESS_BASED E-MAINTENANCE SYSTEM

Muller A., Suhner M.-C. and Iung B.

Centre de Recherche en Automatique de Nancy (CRAN), CNRS/UHP UMR 7039
ESSTIN - 2, rue Jean Lamour
F-54519 VANDOEUVRE-LES-NANCY Cedex – FRANCE
(alexandre.muller, marie-christine.suhner@esstin.uhp-nancy.fr, benoit.iung@cran.uhp-nancy.fr)

Abstract: Today, the degradation checking of critical components is one of the efficient means to minimise the non expected stops of production system and to reduce its costs. It consists in monitoring and forecasting the degradation evolution and in implementing preventive actions to anticipate the failures and keep production system in normal operation. In that way, this paper proposes the deployment and the experimentation of a prognosis process within the e-maintenance architecture of the CRAN IMS platform. The deployment follows a methodology based on the combination both of probabilistic approach for degradation mechanism modelling and of event one for degradation dynamical monitoring. *Copyright © 2004 IFAC*

Keywords: E-maintenance, degradation, forecasting, probabilistic models, condition monitoring, decision making.

1. INTRODUCTION

In today's market, companies have to face fierce competition and satisfy customers who are more demanding in terms of service, high quality, better performance (Zhang, *et al.*, 2003). These companies must then restructure their production systems, shift their management from a reactive to an anticipative role and provide new product/production concepts which are more flexible, integrated, safe and clean. So, for most of manufacturers and with the emerging applications of web-based technologies, they tried to shift their traditional integration philosophy to an e-manufacturing philosophy which transforms companies from a local factory automation to a global enterprise and business automation. Koç, *et al.* (2003) define e-manufacturing as a transformation system that enables the manufacturing operations to achieve near-zero-downtime performance as well as to synchronise with the business systems through the use of web-enabled and tether-free infotronics technologies. The particularisation of this global manufacturing philosophy to the maintenance specific area, leads to define e-maintenance as a component of the e-manufacturing concept which profits from the emerging information and communication technologies to implement cooperative and distributed multi-user environment. The distance between the actors and the shop floor level is now reduced enabling proactive decision process execution. E-maintenance can be seen as an efficient support of "just-in-time" or predictive strategies opposed to traditional preventive ones (too static...). These new strategies are more able to monitor and master component or function degradation in order to anticipate the system failure and then to minimize the non expected stops, the production costs... To ensure these objectives, the implementation of a complete e-maintenance system must support (condition) monitoring, diagnosis, prognosis and decision making modules (Lee, 1998). Some new works aim at contributing to such implementation. Indeed Nieva, *et al.* (2002) proposed a conceptual model for remote acquisition systems. In the context of collaborative maintenance, Ali, *et al.* (2002) or Ba, *et al.* (2001) developed a software platform allowing remote monitoring and diagnosis. Hou, *et al.* (2003) formalised a methodology for remote diagnosis with an approach of neural networks and rough sets. Furthermore, Adjallah, *et al.* (2002) defined an approach to solve the problem of maintenance resources allocation for distributed systems. Today, most advanced researches in e-maintenance are carried off by IMS Center[1] directed by Jay Lee. For example, his team developed an internet-based predictive maintenance technology called IMS (Intelligent Maintenance System) which consists of intelligent machine degradation assessment, e-prognostics and e-diagnostics (Koç and Lee, 2001). This e-maintenance system is integrated into e-manufacturing and e-business ones. Among the different processes making up e-maintenance, the prognosis process is the support of the anticipative role. It is often considered as the Achilles heel of the e-maintenance architecture (Wang and Vachtsevanos, 1999). According to the OSA-CBM[2] architecture, the prognostic functional scope consists of the prediction of system, sub-system or component future health considering the present health, the employment schedule and failure progression models/reasoners (Thurston and Lebold, 2001).

In many cases, prognosis processes are component-oriented. It is now not sufficient to face the complexity of industrial systems (multi-components) where the degradation modelling is really difficult due to the economic or stochastic degradation dependencies among components (Berk and Moinzadeh, 2000). One challenge consists therefore in developing new methodologies based on realistic assumptions for prognostic implementation which is applied not only at the component level but also at the function or system ones (higher abstraction[3]).

In that way, this paper presents and experiments the deployment of (e-)prognosis process based on a methodology which combines both a probabilistic approach for degradation mechanism modelling and an event one for dynamical degradation monitoring. This process is integrated into a global strategy for remote maintenance decision making able to help the operator, when the system is degraded, to take the right decision keeping the highest production performances while keeping the system and the operators in a safety context.

In relation to this methodology, the e-prognosis modelling concepts (section 2) and its implementation mechanisms within e-maintenance architecture (section 3) are defined. Then an application of this process for the prognosis of a sub-set of our CRAN experimental platform is developed (section 4) ending, in section 5, at proposing conclusion and prospective for this work.

[1] http://www.imscenter.net/

[2] http://www.osacbm.org

[3] The abstraction levels of a system allow to study the "Why" (goal) then the "What" (the function performing the goal) until the "How" (the component supporting the function).

2. E-PROGNOSIS PROCESS MODELLING

At the conceptual phase, the main methodology goal is to formalise, from generic modelling concepts, the e-prognosis process both with a probabilistic model and an event one. It leads, in operational phase, to combine these two models to better predict the future state of the system.

2.1 E-prognosis process concepts.

The formalisation and the genericity of the prognosis process modelling are based on the following concepts:

- Process approach.
- Flow-based performance evaluation.
- Causality.
- Degradation.
- Indicator.
- Inference.

• *Process approach.* The industrial system to be considered is modelled through functional analysis (Chittaro and Kumar, 1998) such as a process approach (ISO, 2000). This step allows defining all the "transformation" processes of the systems at all the abstraction levels[3]. One process is thus broken-up into several sub-processes, and the same mechanism is applied from the sub-processes until the expected level of abstraction is reached (e.g. component).

The proposed definition of a process relies on the four concepts (goal, function, behaviour and structure) introduced by Lind (1994) and interpreted as follow:

- *Support* (or *Structure*) represents an arrangement of physical components.

- *Goal* represents the purpose of the process (what the process is intended to do in relation to the product transformed).

- *Function* is a desired action without consider why or by what means an action is performed.

- *Behaviour* explains how a system does what it is intended to do. The behaviour can be identified as: nominal (function fulfilled), degraded (partial loss of the function) and failing (total loss of the function).

• *Flow-based performance evaluation.* A process is linked to another one by the flows[4] it produces or consumes. By definition, a flow (typed as Mass, Energy or Information) is composed of *objects* characterized by *attributes*[5]. The concept of *Flow-based performance evaluation* means that the performance (goal's achievement) of a process can be directly measured on the produced flows and more precisely, on the value of its attributes.

• *Causality.* The process behaviour is described by a rule-based approach which takes into account the causal relationships between its inputs flows, the support, and outputs flows. The description of the causal relationships is based on the functioning-malfunctioning dual approach proposed by Léger and Iung (1998).

[4] A flow is defined by a quantity of objects in a given time.

[5] An attribute is a property characteristic of object or flow.

• *Degradation.* The deterioration of process behaviour (performance loss) is due to the support degradation according to at least one functioning mode. In the same way, the deterioration of a complex system is due to the aggregation, propagation... of each degradation process.
For example at shop floor level, the components can follow deteriorating processes like wear, crack, corrosion, shrinkage (Grall, *et al.*, 2002). These physical mechanisms can be modelled by stochastic processes and often by Markovian or semi-Markovian processes.
• *Indicator.* Each state of the support or flow should be observed (perfectly or not). It means to identify an indicator for support or flow, as those introduced by Knezevic (1987) called RCP (Relevant Condition Predictor). This type of indicator can measure directly the degradation level or indirectly (by means of a parameter correlated with the degradation).
• *Inference.* The proactive dimension of the prognosis process is introduced by the ability of the model to propagate the effect of a component's degradation on the whole system and to simulate the deterioration process through time. The propagation rules are defined by means of two mechanisms of inference:
- *Causal* inference which is required to determine the impact of single process degradation to the global performance of the system. Actually, the causal inference propagates the entity degradation to the rest of the system thanks to the flows and the mode-cause-consequence relationships.
- *Temporal* inference which is required to predict the future level of deterioration of each process. The temporal inference allows the following of degradation evolution according to its physical deterioration mechanism and considering its future operational conditions of use.

2.2 E-prognosis process development.

Each previous modelling concept is considered as a generic rule for e-prognosis process development. This development is supported by the global methodology, more detailed in Muller, *et al.* (2004), and which is structured in three steps:

• *Process & flow-based approach.* The first step aims at breaking up the complex system to be studied into processes and sub-processes according to the flow based approach already introduced in 2.1.
• *Elaboration of the probabilistic network.* The second step consists in transforming the previous process model into a probabilistic one which represents the causal relationships and temporal degradations in an unified way. The probabilistic model is developed by means of Bayesian Networks (BNs) which offer interesting perspectives on knowledge representation and decision support system (Jensen, 1996). A BN is a directed acyclic graph defined by a set of nodes (representing the flows and the supports), and of edges which characterize condition dependencies between them (e.g. the impact of a degraded support on its produced flows). In this scientific contribution, only discrete variables are considered with a finite number of mutually exclusive states (e.g. the nominal, degraded and failing states for a support). The relationships quantification is defined by a CPT (Conditional Probability Table) which contains probabilities of a child node (e.g. an output flow) being in a specific state given the states of its parents (e.g. support and input flows). A variable without parent is called a root variable (e.g. external variable such as a electrical power source).
In addition, the temporal degradation mechanisms are integrated into the probabilistic model by means of dynamic nodes (Dean and Kanazawa, 1989). A dynamic variable is constituted of two discrete random variables represented in a compact form (Weber and Jouffe, 2003) by two nodes linked with an arc which explains the temporal probabilistic dependence between these variables. Hence, the resulting network is a Dynamic Bayesian Network (DBN) which represents the joint probability distribution over all variables (e.g. the state of the global system).
• *Design of the monitoring architecture.* The development of the monitoring architecture materialising the event approach is directly deducted from probabilistic model: each observable variable is associated to an indicator The values of these indicators are refreshed with the current status of the shop floor data and controlled in real time by means of algorithms (Iung, *et al.*, 2003). These events are displayed and stored in an operating knowledge database adapted to the refinement of the probabilistic model's parameters.

2.3 E-prognosis process execution.

The methodology proposed provides a probabilistic model and a dynamical architecture allowing by combination to deploy the e-prognosis process through two phases:

• *Identification of the current state of the system*
The real time monitoring of indicators supplies evidences[6] with BN (dynamical monitoring) which are propagated in the network in order to update the conditional probabilities (causal inference).
• *Prediction of the future state*
From the updating, the execution of the DBN model forecasts the future evolution of the system's degradation (temporal inference).

So, the result coming from the e-prognostic deployment can be an input of the decision-making procedure: if the future estimated situation is a state considered as "safe" or "successful" to fulfil the goal of the system, no action should be planned. The system will evolve towards an acceptable degraded state (those forecasted) and it can derive under control. In contradictory way, if the degradation

[6] An evidence is an assignment of values to a set of variables

degree can be the cause of risk or insufficient performance, an action must be started immediately (in safety case) or at scheduled time (preventive maintenance).

3. E-PROGNOSIS PROCESS IMPLEMENTATION WITHIN E-MAINTENANCE ARCHITECTURE

In order to demonstrate the feasibility and the added value of the methodology proposed, the implementation and the validation of prognosis processes are performing on the CRAN IMS platform (Intelligent Manufacturing versus Maintenance System). The platform purpose is to regulate a water level process in a closed control loop by means of "intelligent" components (valves, sensors, pump) and an automation structure (Fig. 1).

This platform is characterized by the distribution of a technical form of intelligence directly into the components (in order to be as close to the process as possible). The intelligent components (Pétin, *et al.*, 1998) communicate together by means of heterogeneous networks (WorldFip, Profibus and Ethernet). Some processes are embedded within the physical components which should ensured self-control and self-executing. Therefore, the "intelligence" distribution (e.g. condition monitoring) reduces the degradation detection time in order to implement just in time maintenance (e.g. "the right action, at the right time").

In addition to the physical and the software technical intelligence parts, the platform is composed of an e-maintenance architecture. This last one supports the structure required for e-prognosis process deployment by means of the CASIP[7] platform for the event approach and by the BayesiaLab[8] software for the probabilistic one (Fig. 2). The server CASIP-SAM is connected with the shop-floor data acquisition system to the communication networks (via OPC) in order to acquire, transform, display in real time and store the data coming from the field components. The monitoring process extracts information required to determine the system behaviour:

- The support and flow states are elaborated dynamically thanks to the indicators defined in SAM server.
- The values of these indicators are refreshed with the current status of the shop floor data and controlled in real time by means of condition monitoring algorithm.

The detection of an event (e.g. a state change of a component towards a more degraded state) is considered as a symptom or an alert which can start a request for prognostic. These events (but even so periodic values of shop-floor data or indicator) are stored in an Oracle database.

[7] CASIP: Computer Aided Safety and Industrial Productivity - http://www.predict.fr

[8] http://www.bayesia.com

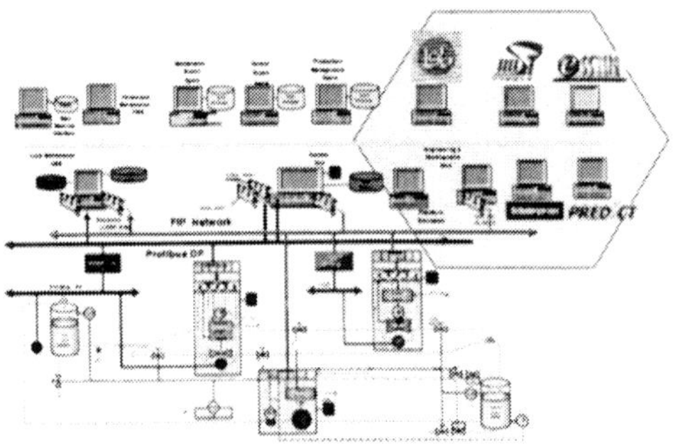

Fig. 1. CRAN IMS Platform

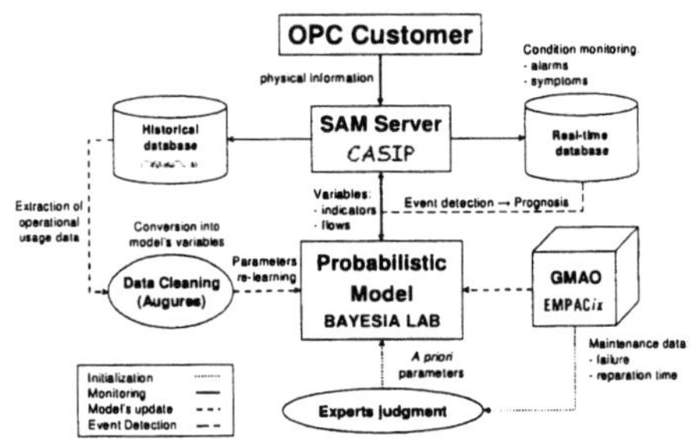

Fig. 2. E-prognosis architecture

The probabilistic model is supported by BayesiaLab, a tool dedicated to BN which implements the causal and temporal inferences required by the prognosis process. The BN parameters can be also refined through learning or updating without model reconfiguration. To integrate dynamic values directly into BayesiaLab a java application called "Augures" has been developed to automatically extract from Oracle database a XML file converted in adequate format.

In addition, each remote partner of the platform (expert, vendor, academic...) is able by means of a client SAM to observe in real-time all available information and to administrate or configure the e-maintenance processes. This type of architecture allows the collaborative maintenance development such as those proposed by Yu, *et al.* (2003) introducing a Problem-Oriented Multi-Agent-Based E-Service System (POMAESS). The purpose of POMAESS is to aid the maintenance decision making by starting from degraded situation, and giving the possibility to remote experts to collaborate by negotiation, coordination... with each other in open and time constraint environment.

Furthermore in the e-maintenance architecture, the (multi-site) enterprise asset management EMPACix[9] collects the data related to maintenance actions (cost, lost time, maintenance time...).

4. E-PROGNOSTIC APPLICATION

In relation to the global IMS platform, the application developed focuses on one sub-process which is the modulation of the pressure of the water flow by means of a motorized modulating valve (Fig. 3).

[9] http://www.indus.com

4.1 Process & flow-based approach.

The function of the process "To modulate the pressure of the water flow" is fulfilled by controlling the angular position of the valve body and consequently modifying the valve port area (Fig. 3a).

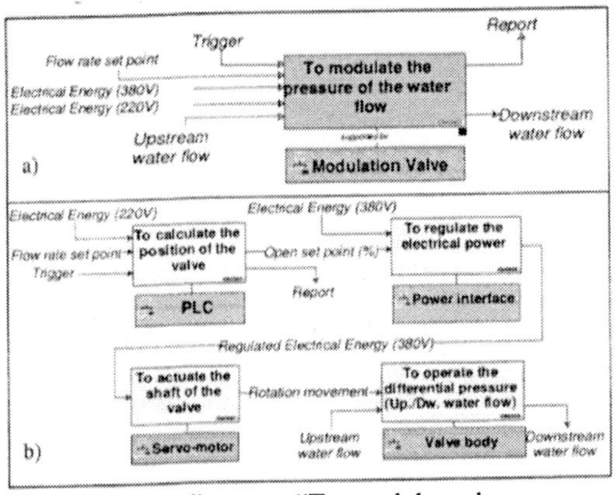

Fig. 3. Process-diagrams "To modulate the pressure of the water flow"

The sub-process is itself broken up into four sub-processes implemented respectively by a PLC, a power interface, a servo-motor and the valve body (Fig. 3b). Only the deployment of the e-prognosis module related to the last sub-process "To operate the differential pressure" is developed in this section (as explained in section 2, the methodology can be applied as the same way on the others sub-processes). This process is defined by its:

- *Support:* valve body.
- *Goal:* to operate the differential pressure (ΔP) between the upstream and downstream water flow across the valve. The goal's achievement of the process is directly measured by attributes of the downstream flow.
- *Function:* to modify the valve port area.
- *Behaviour*: described by all relationships existing between the input flow (rotation movement, upstream flow), the support (valve body) and the output flow (downstream flow).

4.2 Elaboration of the probabilistic network.

This second step aims at transforming the process & flow-based diagram (Fig. 3) into a Bayesian network.

- *Nodes & States.* Valve body is subjected to fouling mechanism modelled as a Markovian deterioration process. The effect of fouling is a modification of valve port area which is related to a variation of the valve flow coefficient Kv_s (defined as the water flow rate (m^3/h) which will flow through the valve with a pressure drop of 1 bar). Thus, the support "Valve body" is represented by one dynamic variable VB with four states (Fig. 4-5): *OK* is the nominal state (no fouling: $Kv_s = Kv_{s\ new}$), D_1 ($Kv_s = \alpha_1\ Kv_{s\ new}$) and D_2 ($Kv_s = \alpha_2\ Kv_{s\ new}$) are the degraded states which do not entail the loss of the function and *HS* is the failing state representative of function loss ($Kv_s = \alpha_3\ Kv_{s\ new}$).

The input energy flow "Rotation movement" is characterised by two static variables: the angular velocity Ω and the angular position X of the valve shaft. The Ω state (either *OK* or *HS*) is determined by the comparison between the current velocity ω_k and the angular velocity set point $\omega_{k\ ref}$:

if $\frac{|\omega_k - \omega_{k\_ref}|}{\omega_{k\_ref}} \leq 10\%$, then $\Omega = OK$ else $\Omega = HS$.

The energy water flows can be represented by two static variables Q_1, Q_2 associated to their volume flow rate and P_1, P_2 associated to their pressure. In this case, Q is constant because there is no leakage in the valve. In addition, the pressures P_1 and P_2 are not represented but only the differential ΔP ($P_2 - P_1$) which is the goal's achievement indicator of the process.

- *Edges.* The causal relationships lead to the definition of oriented arcs which link the "potential cause" variables (Ω, X, Q and VB) to "effect" variable (ΔP). Therefore, the structure of the model contains 4 arcs (Fig. 4).

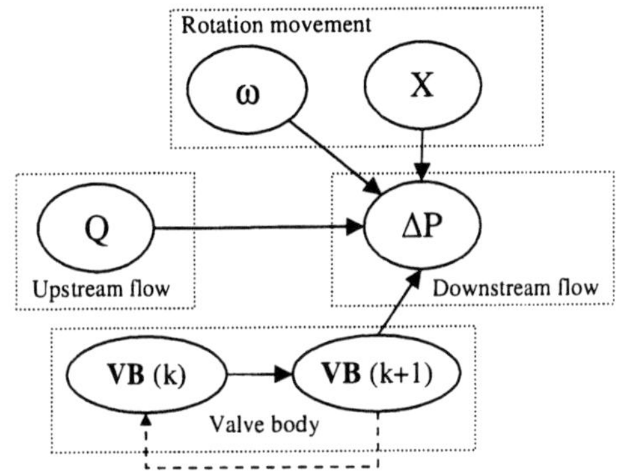

Fig. 4. BN model of the process "To operate the differential pressure"

- *Parameters of the model.* The prior probabilities attached to the input flow "Rotation movement" are fixed by default in this case (actually, they would be generated by the process "To actuate the shaft of the valve"). It is the same procedure for the external input "Upstream water flow" which is generated outside the considered system (environment link).

The definition of the CPT of ΔP relies on the quantification of the causal relationships. It consists in the elicitation of a conditional probability *p* to each instance of the output flow attributes ΔP in each configuration of (Ω, X, Q and VB). This step is based on two major rules: if Ω = *HS*, then it is considered that the process goal is not realised and the value "-1" is affected to ΔP; if Ω = *OK*, the value of ΔP is determined thanks to the physical relation which links the volume flow rate and the differential pressure for a regulation linear valve (Equation 1).

$$\Delta p = \left(\frac{Q}{Kv_s X} \right)^2 \frac{\rho}{1000} \qquad (1)$$

- Q: Volume Flow rate (m^3/h)
- ρ: Water Density (kg/m^3) → supposed constant
- $\Delta p = p_2 - p_1$: Differential Pressure (bar)
- X: angle position of the valve (%)

The CPT of *Valve body* is established thanks to the Markovian process of fouling. The conditional probabilities are equivalent to the transition rates $\lambda(VB_{k+1}/VB_k)$, fixed *a priori* according to the operating knowledge available for similar valve (it is assumed that these parameters are time-invariant).

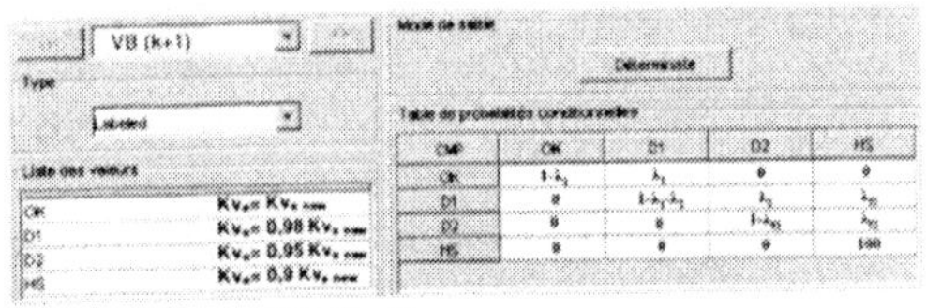

Fig. 5. CPT of VB (k+1)

4.3 Design of the monitoring architecture.

The development of the monitoring architecture is deducted from probabilistic model: as far as possible, each variable should be associated to a real-time indicator. Actually, it is the case only if the variable is physically observable i.e. the instrumentation system is able to determine its current value.

- The monitoring system controls the rotation movement by means of an integrated positionner connected to the PLC. This component provides the current angular position X of the valve. CASIP-SAM acquires continuously this information and determines at the same time the angular velocity ω of the valve's shaft.

$$\omega_k = \frac{|X_{k+1} - X_k|}{t_{k+1} - t_k} \qquad (2)$$

- The physical parameters Q and Δp are acquired in real-time by SAM server from the flow meter and the differential pressure measure transducer associated to the regulation valve.
- The indicator of the physical support "Valve body" is Kv_s which is monitored indirectly thanks to the values of ΔP, X and Q on the basis of equation 3:

$$Kv_s = \frac{Q\sqrt{\rho / 1000\Delta p}}{X} \qquad (3)$$

The monitoring process supported by CASIP environment executes continuous algorithms of processing of these observations and starts the prognosis process automatically if it detects an alarm i.e. a state change for the valve body (e.g. $OK \rightarrow D_I$).

4.4 Prognostic issued from prognosis process

The prognosis process is started in order to evaluate the differential pressure Δp in the future taking into account the initial conditions of Ω, X, Q and VB.

Numerical application

λ_1	10^{-3}	$Kv_{s\ new}$	18.1 m^3/h
λ_2	8.10^{-3}	α_1	0.98
λ_{f1}	2.10^{-3}	α_2	0.95
λ_{f2}	10^{-2}	α_3	0.90

Step 1: Start. If the monitoring process detects a state change, it triggers an alarm which indicates the event to the user (Fig. 6).

Therefore, the user can inject this observation of valve body state to the BN i.e. "VB = D_I" at time k = t_1. In the same time, it is necessary to forecast the states of the parent variables (Ω, X, Q) considering the future operation of the system.

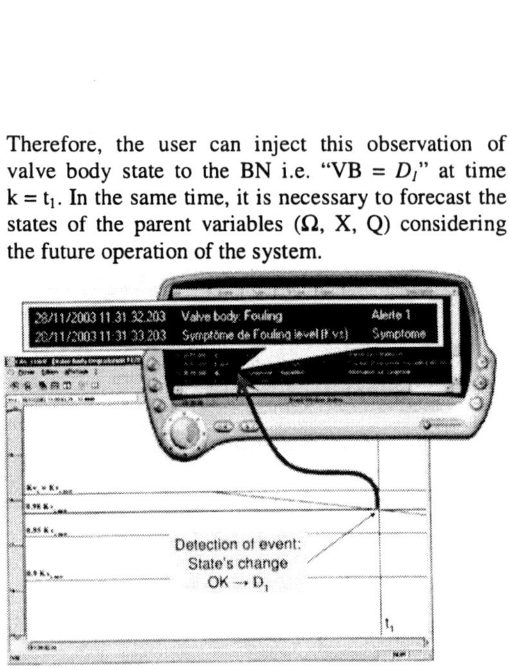

Fig. 6. Degradation monitoring of "Valve body"

In this example, the angular velocity is "*OK*", the angular position is equal to 2/10 of the full-open position and the flow rate is about 2000 dm^3/h. This information is translated as the evidence *e* {*OK, 2, 2000, 1770*} to the variables {Ω, X, Q, VB} as proposed in the figure 7.

Step 2: Inferences. Software BayesiaLab is used for implementing the causal and temporal inferences. Firstly, a causal inference is made to determine the probability of the output node ΔP given the starting condition *e*. Then, from the observed situation at time t_1, the temporal inference computes the posterior probability P [VB (t_1 + T) / VB (t_1) = *1770*] of valve body (Fig. 7) and consequently allows to prognosticate the probability P [ΔP (t_1 + T)] by a causal inference at time t_1 + T.

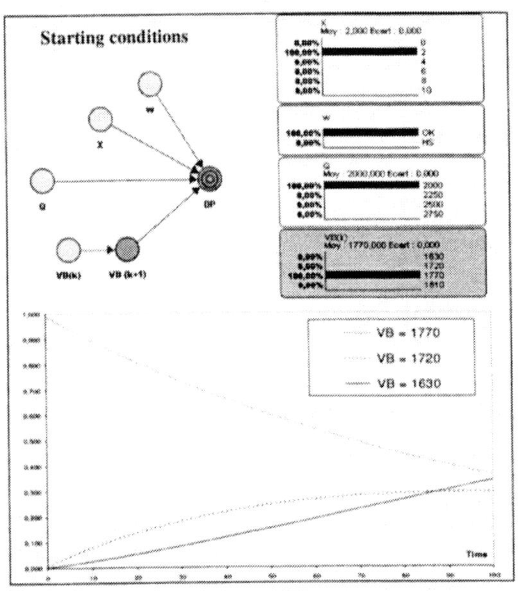

Fig. 7. Temporal evolution of VB

With the same conditions (Q = 2000, X = 2), the performance of the process at time t_1 is evaluated by a differential pressure $\Delta p_{t1} = 319$ mbar while the prognosis at time t_1+100 indicates an expectation of $\Delta p_{t1+100} = 344$ mbar. This information can be propagated to another process (as an input data) or directly used by a decision process. It represents the concrete result of the e-prognosis process.

5. CONCLUSION

The main contribution overviewed in this paper is a methodology structured by generic modelling concepts for e-prognosis process formalisation and deployment based on a combination both of probabilistic model and dynamic monitoring architecture. This e-prognosis process is implemented on a real e-maintenance architecture mainly supported by the tool CASIP. A first experimentation and validation of the feasibility methodology have been made by its application to one regulation subset of the IMS platform. However, the context of this validation has to be extended to a more complex abstraction level in order to really demonstrate the feasibility.

In addition, the model has to incorporate as soon as possible the possible maintenance alternatives (no intervention, systematic actions, conditional ones...) which can be performed on the components (Wang, 2002) to restore the system in a satisfying state. Indeed, the e-prognostic would provide the simulation of scenarios combining degradation mechanisms and maintenance actions. The future work is based on theses perspectives.

REFERENCES

Adjallah, K., K. Adzakpa, A. Ivanov, C. Varnier C., and N. Zerhouni (2002). Approach to Remote Maintenance Management of Distributed industrial Complexes. In: *MIM'2002 Managing Innovative Manufacturing*, 101-108, Milwaukee (USA).

Ali, A., Z. Chen, J. Lee and M. Koç (2002). Web-enabled Device-to-Business Platform for Distributed and Dynamic Decision Making Systems. In: *MIM'2002 Managing Innovative Manufacturing*, 157-169, Milwaukee (USA).

Ba, T., E. Garcia, H. Guyennet, J.-C. Lapayre, N. Zerhouni, R. Zemouri (2001). TEMIC: Industrial Cooperative Tele-Maintenance. In: *29th ICC&IE Int. Conf. on Computers and Industrial Engineering*, 465-470, Montréal (Canada).

Berk, E. and K. Moinzadeh (2000). Analysis of Maintenance Policies for *M* machines with Deteriorating Performance. *IIE Transactions*, **32**, 433-444.

Chittaro, L. and A. Kumar (1998). Reasoning about function and its applications to engineering. *Artificial Intelligence in Engineering*, **12**, 331-336.

Dean, T. and K. Kanazawa (1989). A model for reasoning about persistence and causation. *Computational Intelligence*, **5**, 142-150.

Grall, A., C. Bérenguer and L. Dieulle (2002). A condition-based maintenance policy for stochastically deteriorating systems. *Reliability Engineering & System Safety*, **76**, 167-180.

Hou, T., W.-L. Liu and L. Lin (2003). Intelligent remote monitoring and diagnosis of manufacturing processes using an integrated approach of neural networks and rough sets. *Journal of Intelligent Manufacturing*, **14**, 239-253.

ISO International Organisation for Standardisation (2000). *ISO 9000:2000 Quality management systems-Requirements*. Geneva (Switzerland).

Iung, B., G. Morel and J.B. Léger (2003). Proactive maintenance strategy for harbor crane operation improvement. *Robotica*, **21**, 313-324.

Jensen, F. (1996). *An Introduction to Bayesian Networks*. UCL Press, London (UK).

Knezevic, J. (1987). Condition parameter based approach to calculation of reliability characteristics. *Reliability Engineering*, **19**, 29-39.

Koç, M. and J. Lee (2001). A system framework for next-generation E-maintenance system. In: *2nd International Symposium on Environmentally Conscious Design and Inverse Manufacturing*, Tokyo (Japan).

Koç, M., J. Ni, J. Lee and P. Bandyopadhyay (2003). Introduction of e-Manufacturing. In: *31st North American Manufacturing Research Conference (NAMRC)*, Hamilton (Canada).

Lee, J. (1998). Teleservice engineering in manufacturing: challenges and opportunities. *International Journal of Machine Tools and Manufacture*, **38**, 901-910.

Léger, J.-B. and B. Iung (1998). Methodological approach to modelling of degradation detection and failure diagnosis in complex production systems. In : *9th International Workshop on Principles of Diagnosis*, 209-216, Cape Cod (USA).

Lind, M. (1994). Modeling goals and functions of complex industrial plant. *Applied Artificial Intelligence*, **8**, 259-283.

Muller, A., M.-C. Suhner and B. Iung (2004). Bayesian Network-based Proactive Maintenance. Accepted for *PSAM 7 / ESREL'04 International conference on probabilistic safety assessment and management, 14-18 June 2004*, Berlin (Germany).

Nieva, T. and A. Wegmann (2002). A conceptual model for remote data acquisition systems. *Computers in Industry*, **47**, 215-237.

Pétin J.-F., B. Iung and G. Morel (1998). Distributed intelligent actuation and measurement system within an integrated shop-floor organisation. *Computer in Industry*, **37**, 197-211.

Thurston, M. and M. Lebold (2001). Open System Architecture for Condition Based Maintenance OSA/CBM. In: *IEEE AUTOTESTCON Systems Readiness Technology Conference*, Valley Forge (USA).

Wang, H. (2002). A survey of maintenance policies of deteriorating systems. *European Journal of Operational Research*, **139**, 469-489.

Wang, P. and G. Vachtsevanos (1999). Fault Prognosis Using Dynamic Wavelet Neural Networks, In: *AAAI Spring Symposium in Equipment Maintenance Service and Support*, Polo Alto (USA).

Weber, P. and L. Jouffe (2003). Reliability modelling with dynamic bayesian networks. In: *5th IFAC Symposium on Fault Detection, Supervision and Safety of Technical Processes*, 57-62, Washington (USA).

Yu, R., B. Iung and H. Panetto (2003). A multi-agents based E-maintenance system with case-based reasoning decision support. *Engineering Applications of Artificial Intelligence*, **16**, 321-333.

Zhang, Q., M. Vonderembse and J.-S. Lim (2003). Manufacturing flexibility: defining and analyzing relationships among competence, capability, and customer satisfaction. *Journal of Operations Management*, **21**, 173-191.

Copyright © IFAC Information Control Problems in
Manufacturing, Salvador, Brazil, 2004

ELSEVIER
IFAC
PUBLICATIONS
www.elsevier.com/locate/ifac

MAINTENANCE IMPROVEMENT BY UNIQUE PRODUCT INFORMATION ENABLED BY UBIQUITOUS COMPUTING

Sandra Gross, Elgar Fleisch

Auto-ID Center University of St. Gallen,
Institute of Technology Management,
Unterstrasse 22, CH 9000 St. Gallen, Switzerland

Abstract: This paper investigates how maintenance of maintenance-intensive products can be improved from a service-center's perspective. It analyzes requirements and current approaches. Research results include: The lack of unique product information often causes economic losses for companies. Ubiquitous computing technologies such as automatic identification (Auto-ID) can be used to obtain lifecycle information associated with a product and product parts. They enable benefits in the maintenance process itself such as efficiency gains due to a higher data accuracy and less manual mistakes. The data can provide benefits for future product lifecycles as well: quality improvements and simplified variant management become possible.
Copyright © 2004 IFAC
Keywords: Maintenance, Identification, Information, Integration, Automation

1. INTRODUCTION

Many companies are coming under increasing pressure today due to trends in competition and markets, sustainability demands and regulations. This pressure is directly related to the products on offer by the companies. In many industries services for products are one of the main differentiation feature of competing companies. For instance, in the industrial goods industry 80 percent of the companies state that services and proximity to customers bear future chances for their businesses. Today the relevance of services compared to products is 33 percent, in ten years the relevance shall be 51 percent. There is still a gap between the service offer of service centers and the perceived relevance of services for customers, especially in spare part service, as well as maintenance, inspections and repairs (Backhaus and Kleikamp, 2001).

For each industry and partner of the value chain, there are different perceived problems, which also depend on the product associated services such as reclamations, spare part management, hotline/enquiry line, training, inspections, preventive maintenance, or repairs (PROMISE, 2003; Zollikofer-Schwarz, 1999). The focus of this paper lies on maintenance-intensive products. Maintenance and repair can be carried out by the customer respectively the user of the product, the producer, or a service center (which can be run by the producer or a third party). This paper is written from the perspective of the service-center.

The purpose of this paper is to derive optimization potential for the maintenance process by evaluating the role of unique product information collected in the product lifecycle. For instance, to minimize unexpected breakdowns, the producer ensures that the product is composed of the right product parts and components according functional and customer requirements. Automatic monitoring of products can lead to improvements in maintenance planning, scheduling and execution.

In order to achieve this goal, the authors systematically analyzed the state of the art of common features in maintenance processes in order to identify problems, solutions, barriers and enablers. This research procedure originates from the applied social sciences (Ulrich, 1981). A central characteristic of the research framework is the ability to solve problems occurring in operational and business management practice.

The structure of the paper is as follows: Current problems in maintenance underlie the relevance of the topic. Then the paper examines the maintenance process and service center requirements in order to derive barriers of existing solutions. To overcome them, the authors introduce enabling technologies of ubiquitous computing such as automatic identification (Auto-ID) and present examples of maintenance improvements enabled by this technology. Finally, the paper concludes with a summary of business benefits of the suggested solution, some managerial implications and an outlook on further research.

2. PROBLEMS IN MAINTENANCE

On one hand markets are changing from supply-dominated markets to demand-dominated ones. This forces companies to shorten their innovation and production cycles in order to fulfill their customers' demands. This leads to an increased number of product variants in the market, for which the producer has to provide after-sales services. On the other hand customers demand a longer product lifespan for certain products, especially for durable goods. This results in a number of problems related with maintenance:

- Storage and logistic cost: The producer has to keep all imaginable spare parts in stock. (Schuh and Schwenk, 2001), (Strassner and Fleisch, 2002), (Alexander, et al., 2002).
- Decreasing customer's loyalty: The challenge today is not "just" producing the right product at the right time for the right customer, but also to fulfill the users' requirements during the usage of each individual product. The customer has to be content with the product and the related services. Producers especially fear third party producers which also provide spare parts or after-sales-services. Into the same problem category fall customers who don't know anymore which products from which company they use. This can be the case e.g. with products with a life span of over 30 years which require seldom maintenance.
- Lack of manufacturing solutions: Today producers have to provide manufacturing solutions, not just products and machines. The knowledge and experience in maintenance and associated services is often a main differentiation feature amongst competitors (Hohwieler, et al., 2001). The products and machines have to guarantee a certain level of performance to avoid problems such as compensation demanded by customers for production losses caused by failure of manufacturing equipment. Also, the products and machines have to be re-configurable and recyclable in order to meet the challenges of mass customization and tighter environmental regulations (Lee, 1998).

3. MAINTENANCE PROCESSES

There is a wide variety of maintenance processes depending on industry and product category. However, there are some common features which are discussed in this section. They affect the process, process goals and requirements.

3.1 Process Description

A sample maintenance process is shown in Fig. 1. Usually, companies prefer preventive maintenance to breakdown maintenance. Breakdown could result in serious tangible and intangibles such as operating expenses, penalty, decline of brand image, and future business opportunity losses. Unexpected breakdowns also result in a sharp rise of maintenance cost and reduce customer satisfaction. Preventive maintenance can be time or condition based. They usually involve rules of thumb concerning inspection and replacement cycles. Other approaches include the Taguchi methods, which take the average lifetime of a part into account (Yamashina and Otani, 2001a). Maintenance planning also determines the cost of maintenance for a manufacturer or the service center. It includes activities as work order planning, scheduling, dispatching and execution.

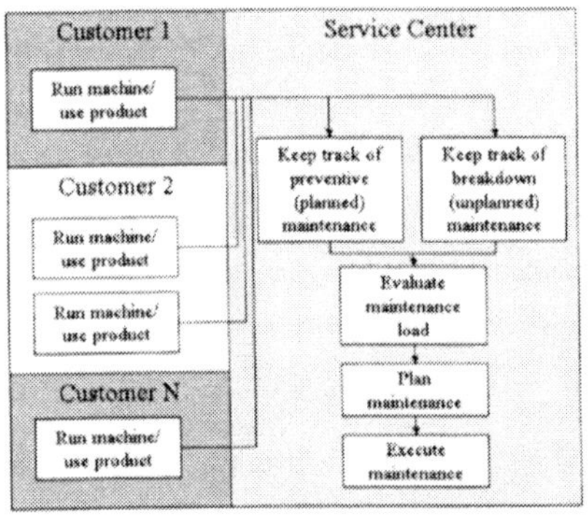

Fig. 1. Sample maintenance process (adapted from Duffuaa, et al., 2001)

3.2 Process Goal and Requirements

For the customer, products and machines cause re-occurring costs during maintenance and support. The cost is composed of spare parts, personnel, supply and service providers. Hidden maintenance cost are production losses due to downtime, cost for maintenance equipment, storage room for maintenance equipment and an increased maintenance workforce (Lewis and Steinberg, 2001). One requirement is the maximization of the capacity of products, as in plant engineering, construction, or car maintenance and repairs. Fig. 2 shows the factors that contribute to the capacity of products (Strub, 2003; Lewis and Steinberg, 2001).

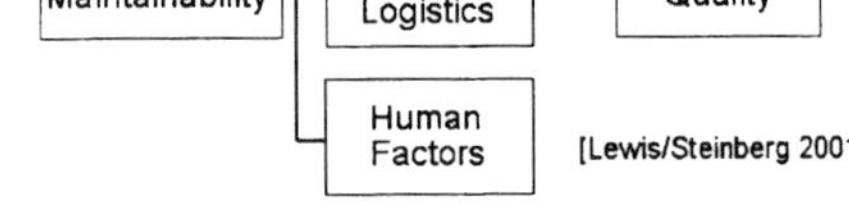

Fig. 2. Maintenance and repair criteria (Lewis and Steinberg, 2001)

To reach this goal economically, the customer and/or the service center have to ensure that the maintenance resources are used efficiently. The resources are: manpower, materials and spare parts, tools and equipment, and standards and procedures. The requirement is thus (Duffuaa, et al., 2001):

- Optimal manpower needed for each trade,
- Effect priority levels on scheduling, and
- Effect of spare parts policies on down time.

4. BARRIERS OF EXISTING SOLUTIONS

The goal of this section is to examine where optimization potential exists in the maintenance process. So the authors first examine existing solutions and second state the gap between them and the process requirements.

4.1 Existing Solutions

Current solutions include e-maintenance tools, which try to fulfill the requirements. The tools try to evaluate the current capability of machines and products without human intervention and some of them also influence the machines automatically (Hohwieler, et al., 2001).

There are several remote diagnostics techniques that collect data from customers' products and plants remotely. This information is used to perform remote machine monitoring and troubleshooting, and to provide service instructions for the service personnel. The e-maintenance systems assess the behavior and degrading level in order to prevent unexpected breakdown. In the case of a breakdown, the history of a component helps in a quick recovery of the machine. All collected information is shared with other user sites in order to transform the collected information into knowledge about the machine. Also, some tools provide multimedia based features for collaborative maintenance assistance, such as smart helmets or gloves (Lee, 1998).

E-maintenance systems usually possess interfaces to other applications such as spare parts management or ERP systems (Hohwieler, et al., 2001). The interfaces connect the organizational unit, which is responsible for maintenance, with different functional areas, require specific information, and use different business information systems (BIS). Sample interfaces with a maintenance system appear in the following functional areas (adapted from Strub, 2003):

- Purchase: e.g. exchange of receipt of parts information and purchase information
- Human Resource: e.g. hourly rate for maintenance personnel and hours worked
- Inventory: e.g. availability of repair parts and confirmation of parts consumed for repair
- Warehouse: e.g. data about warehouse transfer

The management of maintenance activities is often integrated into Product Lifecycle Management (PLM) tools, although they are not necessarily called like that. PLM tools consist of many modules and it is common practice to implement the modules incrementally. The two main reasons for using PLM tools are process integration and data integration. Therefore, they usually consist of a workflow and a data integration component. The data integration module - product data management (PDM), is highly industry specific: In the heavy equipment industry, the core customer values are reliability and longevity of products and therefore provide mainly functions such as configuration management and spare time management. In the pharmaceutical industry, the core customer value is trust so that PDM tools deal with patents, trials and compliance (Burkett, et al., 2002). Moreover, the tools are also product specific: Computer Aided Design (CAD) applications and simulation tools and their contents highly depend on the kind of developed product (Lindemann, et al., 2000).

4.2 The Gap

Maintenance problems and difficulties in sticking to the maintenance process requirements are often based on

- Missing information transparency,
- Lack of integration,
- Lack of standardization.

First, information transparency often does not exist on unique product or product parts level due to insufficient product data, or master data management. Second, real-time information on real world objects are missing about the current state of a product or product part, maintenance equipment and spare parts.

There is a lack of process and data integration in product lifecycle information systems: companies start implementing PLM modules incrementally. This leads to the implementation of parts of the product lifecycle information, but not to the complete picture (Parlikad, et al., 2003).

The lack of standardization concerns all trading partners in the value chain: to derive the most potential benefits from unique product information the whole history of the product has to be known and accessible. One new standardization approach which is based on Auto-ID technology is the electronic product code (EPC) from EPCglobal (EPCglobal, 2003).

5. ENABLING UNIQUE PRODUCT INFORMATION WITH UBIQUITOUS COMPUTING

Unexpected breakdowns have to be avoided in order to enhance the capability of products. At the same time, service centers are interested in increasing maintenance intervals and avoiding unplanned assignment of maintenance personnel due to failure. Preventive maintenance scheduling based on failure characteristics of individual parts, maintenance records, and real-time data about the depreciation state of individual parts are expected to result in the best efficiency gains (Yamashina and Otani, 2001a; Yamashina and Otani, 2001b). Today, often diagnostics is done by analyzing the error codes instead of analyzing the history of each product component.

Table 1 shows how ubiquitous computing technologies contribute to derive unique product specific information. The technology can collect information during the whole lifecycle so that product history and current state of the product are addressed. If products and parts are referenced by a unique number, the data associated with the product can also include instructions for the future state of the product respectively the product part.

Table 1 Functionalities and Technologies of Ubiquitous Computing (adapted from Strassner and Schoch 2002)

Functionality	Description	Sample Technology
Automatic identification	Automatic identification per product identification number	Radio Frequency Identification (RFID), barcode, biometric systems
Localization	Localization of products and their parts and visualization of former and possible future paths, mostly coupled with identification	Global Positioning System (GPS), Global System for Mobile Communication (GSM), ultrasonic
Automatic monitoring	Determination of current state of the object and/or its environment	Sensors / sensor networks e.g. pressure or temperature

The following sections examine how unique product information enabled by ubiquitous computing can lead to improvements in the maintenance process (Gross, et al., 2003).

5.1 Automatic Identification and Localization

Product-ID Products and product parts carry an identification number e.g. by being tagged with a RFID transponder. Either the service personnel has a reader which reads this product-ID while conducting maintenance, or the remote diagnostic system transfers the product-ID together with the report. Business benefits are accurate product/parts identification, less mistakes than in manually written maintenance reports, and easy identification in dark and harsh environments (such as dusty warehouses).

Product Specification Data Product specification data such as configurations, material, and catenation as well as product related data such as planning data, logistics enables an automatic lookup and interface to spare parts and material availability and spare part management. A potential business benefit is a more efficient process due to process and data integration.

Product History Maintenance scheduling can be done automatically by checking maintenance equipment availability and tools availability automatically. This enables a sounder decision making about priorities, working schedules, and planning.

In quality management the quality of individual parts can be checked more reliably. This can lead to shorter reaction times.

The information gap to customer relationship management can be closed e.g. by updating CRM systems when the customer has complaints and/or positive feedback. Potential business benefits are a better understanding of customer requirements and reclamations and thus a better ability to bind the customer.

The information gap to product design can be closed e.g. by informing design teams about frequent failures or special combination of product features in a product's history which leads to failures. This enables better products and product variants for the next generation.

Product Instructions and Decision Support The maintenance execution can become guided and documented electronically. This leads to less errors, and better documentation.

5.2 Automatic Monitoring

Sensors and sensor networks collect product state information such as operational data, the degree of depreciation or damage per part in real-time. This technology can minimize unexpected breakdowns. According to inspection regulations the intervals can be maximized for "good" parts.

6. EXAMPLES OF MAINTENANCE IMPROVEMENTS WITH UNIQUE PRODUCT INFORMATION

There are already examples of maintenance improvements with unique product information in practice. The first two sections describe companies that have already started to use unique product information enabled by ubiquitous computing. The third example describes a maintenance scenario of a white appliances producer, who is in the role of a service center as well. The scenario is based on unique product information.

6.1 Example: ExxonMobile

ExxonMobil uses RFID tags as a key to an asset tracking initiative for Equipment Health Monitoring and daily mechanical integrity inspections. With the RFID tags installed, operators make scheduled plant inspections to collect process and mechanical data about equipment using handheld computers. Additionally, low frequency multi-page RFID tags are used to ensure that operators properly perform procedures and provide verification that operators are at the correct piece of equipment while collecting data. Also, they have to identify the correct part for changing as spare parts such as valves look very similar.

Once inspections are completed, the data is uploaded and synchronized with the plant's Computerized Maintenance Management system over a wireless network. Approximately 30 hand-held computers and thousands of RFID tags (134khz) are used (Construction Industry Institute, 2000).

Benefits are mainly seen in accuracy and control of the maintenance process: ExxonMobile could achieve 40 percent more accuracy than on manual collection methods. The current state of the product helped to identify valves, and the maintenance time and down time could be reduced.

6.2 *Example: Ventilation Flaps Maintenance at Frankfurt Airport*

Regulations in Germany force airport-operating companies to conduct regular ventilation flaps maintenance. Fraport AG, the operator of Frankfurt Airport, set up a pilot with SAP from December 2002 to November 2003 in order to support the control and maintenance of ventilation flaps at the airport in Frankfurt. The maintenance personnel have to maintain about 220,000 ventilation flaps. Currently, technicians identify the vents by barcode, fill in a three-page paper based report manually. Then, they transfer the report information into the companies computer network at the end of the day.

To enhance the efficiency and quality of the maintenance activities the technicians were equipped with handhelds, and RFID transponders were attached to sample ventilation flaps. The RFID chips contain identity numbers and information related to the last inspection. The handheld displays the inspection route for the technicians, reads the RFID information at the ventilation flap, acts as reporting tool and transfers the collected maintenance information at the end of the day to the central computer systems.

The benefits of increased efficiency and data accuracy was already seen in the pilot implementation. The pilot currently investigates further process improvements that could be provided by this technology (Quack, 2003).

6.3 *Example: Miele*

The appliances company Miele and IBM Global Services developed the idea of the smart washing machine. The pilot demonstrated the advantages of a virtual washing machine pool control in an apartment house. The accommodation units often have to share common washing machines. The challenge is to optimize the machine's operating time. This includes scheduling of washing orders and immediate action on breakdown.

The solution is as follows: Every washing machine is connected via the internet with the Miele service center. They possess an integrated firewall and are equipped with a GSM transmitter. The renters can book and change their washing times via internet, via their mobile phone with a WAP browser, or via telephone. If required, the system sends a SMS with the actual status of the washing machine to the renter. Also, the machines perform a self-diagnosis. If maintenance is due or breakdowns occur, they are reported to the warden via SMS and the Miele Service Center by the system immediately. If the warden can perform the repair by himself he gets further instructions. Otherwise the closest Miele service technicians come and repair the washing machine, enter the time and material involved in a program on a laptop and send the reports to the Miele Service Center.

This model provides several benefits that are just possible because unique product information is available: Abnormal behavior can often be fixed in time so that breakdowns can be avoided. Also, service times can be reduced as the washing machine informs the Miele Service Center in advance about the error-state of parts. The efficiency is increased as parts replenishment and repair equipment is ordered automatically via internet. Thus, the supplier can send the needed parts directly to the location of the washing machine (IBM, 2001).

7. CONCLUSION

This paper shows how Auto-ID enabled unique product information contributes to improvements in maintenance due to a higher information transparency. This enables more profound decisions and thus increases benefits for both the customer and the maintenance service center compared to solutions without unique product information:

- The ability to identify products and product components uniquely and automatically increases the efficiency of the maintenance process and increases data accuracy due to less manual mistakes,
- Maintenance information is updated in the producer's systems automatically and can be used for an enhanced maintenance planning and for further product generations,
- Standardized product data with Auto-ID technologies increase process efficiencies.

Managerial implications are: Before integrating unique product data into existing IT-systems, companies should first analyze the level at which they already support individual information. The levels are product description, current state, past state of the product, and future state of the product. Existing solutions such as the three examined case studies already implement part of the product information model. Companies should start with the implementation of applications that provide quick wins such as Exxon and SAP did. In most cases initial wins result from process efficiency gains: less process steps and automated information retrieval from real world objects. For further expansion, the solutions should comply to necessary standards.

Further research should concentrate on a special industry in order to prove the case for unique product information. The product information should be integrated automatically into existing IT-systems enhancing the decision making process. However, research projects of the authors showed, that the integration of RFID data into existing business information systems poses one big challenge for companies. As there is no common reference model for UbiComp infrastructures yet, the authors want to develop an RFID systems application architecture. The aim of the architecture is to support a structured approach of planning, introducing and managing RFID-systems. The research results could also be applied for maintenance applications.

Once unique product information is available, the authors expect that the concept also imposes potential benefits in other after-sales services such as reclamations, spare parts, user inquiries. Potential benefits are a shorter time to market, product and service quality improvement, a simplified variant management, and a reduction of product cost.

REFERENCES

Alexander, K. et al. (2002). *Focus on the supply chain: Applying Auto-ID within the Distribution Center.* Auto-ID Center White Paper, Cambridge.

Backhaus, K. and C. Kleikamp. (2001). Marketing von investiven Dienstleistungen. In: *Handbuch Dienstleistungsmanagement* (Bruhn, M. and Meffert, H. (Ed.)), pp. 73-101. Gabler Verlag, Wiesbaden.

Burkett, M., J. Kemmeter and K. O'Marah (2002). Product Lifecycle Management: What's Real Now. AMR Research.

Construction Industry Institute (2000). *Proceedings ot the Construction Project Improvement Conference*, Austin, Texas. http://construction-institute.org/ cpi2001slides/cpi2001_proc.pdf

Duffuaa, S.O., M. Ben-Daya, K.S. Al-Sultan and A. A. Andijani (2001). A generic conceptional simulation model for maintenance systems. *Journal of Quality in Maintenance Engineering*. **Vol. 7, No. 3**, pp. 207-219.

EPCglobal (2003). http://www.epcglobalinc.org

Gross, S., D. McFarlane, A. K. Parlikad and E. Fleisch (2003). *The Role of the Auto-ID enabled Product Information in a Product 's Usage: A Maintenance Example.* Auto-ID Center White Paper, Cambridge.

Hohwieler, E., G. Schreck and R. Berger (2001). *Bereitstellung elektronischer Dienstleistungen für Produktionssysteme.* Fraunhofer Institut für Produktionsanlagen und Konstruktionstechnik, PTK, Berlin.

IBM (2001). *Miele's intelligente Waschmaschine: mobiles e-business für's Alltägliche.* http://www-5.ibm.com/de/pressroom/presseinfos/2001/010208_1.html

Lee, J. (1998). Teleservice engineering in manufacturing: challenges and opportunities. *International Journal of Machine Tools & Manufacture*. **Vol. 38**, pp. 901-910.

Lewis, M.W. and L. Steinberg (2001). Maintenance of mobile mine equipment in the information age. *Journal of Quality in Maintenance Engineering*. **Vol. 7, No. 4**, pp. 264-274.

Lindemann, U. et al. (2000). Flexible Integration von Produktentwicklung und Montageplanung. *Industrie Management*. **Vol. 16, No. 1**, pp. 23-27.

Parlikad, A.K., D. McFarlane, E. Fleisch and S. Gross (2003). The Role of Product Identity in End-of-Life Decision Making. Auto-ID Center White Paper, Cambridge.

PROMISE (2003). *Product Lifecycle Management and Information Tracking using Smart Embedded Systems* (PROMISE). FP6 Proposal No. 507100.

Quack, K. (2003). RFID-Chip meldet: "Wartung ausgeführt". *Computerwoche online.* http://www.computerwoche.de/index.cfm?pageid=256&artid=52033&type=detail&kw=Wartung%20ausgeführt

Schuh, G. and U. Schwenk (2001). *Produktkomplexität managen.* Carl Hanser Verlag, München Wien.

Strassner, M. and E. Fleisch (2002). *Segment Evaluation Automotive - Applications of Auto-ID technology in the Auto Industry.* M-Lab Report No. 15, St. Gallen.

Strassner, M. and T. Schoch. (2002). Today's Impact of Ubiquitous Computing on Business Processes. In: *Short Paper Proc. International Conference on Pervasive Computing.* (Mattern, F. and M. Naghshineh (Ed.)), pp. 62-74.

Strub, J.J. (2003). *Computerized Maintenance Management Systems.* http://www.technologyevaluation.com/Research/ResearchHighlights/Erp/2003/07/research_notes/

Yamashina, H. and S. Otani (2001a). Cost-optimized maintenance of the elevator - single unit. *Journal of Quality in Maintenance Engineering.* **Vol. 7, No. 1**, pp. 49-70.

Yamashina, H. and S. Otani (2001b) Optimal preventive maintenance planning for multiple elevators. *Journal of Quality in Maintenance Engineering.* **Vol. 7, No. 2**, pp. 128-150.

Ulrich, H. (1981). Die Betriebswirtschaftslehre als anwendungsorientierte Sozialwissenschaft. In: *Die Führung des Betriebs.* (Geist M. N. and Köhler R. (Ed.)). Schäffer-Poeschel, Stuttgart.

Zollikofer-Schwarz, G. (1999). *Die Entwicklung des After-Sales-Management.* Dissertation, University of St. Gallen, St.Gallen.

Copyright © IFAC Information Control Problems in Manufacturing, Salvador, Brazil, 2004

ELSEVIER
IFAC
PUBLICATIONS
www.elsevier.com/locate/ifac

CONDITION PREDICTION FOR MANUFACTURING ELECTRICAL EQUIPMENT

Timothy L. Johnson[1], Shantala A. Roysam, and Kevin E. Weiler

GE Global Research (K-1, 5C30A)
Advanced Computing Technologies
P.O. Box 8
Schenectady, NY 12301

Abstract: This paper addresses the complex issues involved extending prediction intervals for inspection and maintenance of electrical protection equipment on manufacturing lines. Illustrations are based on representative data. *Copyright © 2004 IFAC*

Keywords: Device degradation, diagnostic inference, estimation algorithms, forecasts, industrial automation, maintenance engineering, non-parametric identification, optimal estimation, probabilistic risk assessment, statistical inference.

1. INTRODUCTION & BACKGROUND

In order to maintain product quality, complex manufacturing equipment must be periodically inspected, calibrated, adjusted, maintained, repaired or replaced. Statistical process control (SPC), in this context, is the procedure by which product tolerances are maintained through a periodic product inspection and manufacturing equipment adjustment cycle – and this is usually regarded as a part of normal manufacturing system operation. But there is also a "long cycle" or "life cycle" process that is concerned with the gradual degradation, wear, or aging of equipment over time – and this asset management process (AMP) is concerned with the prevention of manufacturing equipment failures, wear, and verification of protective functions that might not be exercised during extended periods of normal operation. Asset management processes are an "outer loop control" on equipment condition that enables SPC to operate as an "inner loop" process. Condition based maintenance (CBM) is a method of regulating asset inspection and maintenance intervals based on the measured or estimated condition of assets. This is in contrast to earlier "break-fix" approaches, in which assets are simply allowed to break or go beyond calibration limits before they are maintained (intervals too long), or the "periodic maintenance" approach, which is associated with excessive costs and downtime (intervals too short). Normally, this type of inspection and maintenance requires a "scheduled outage" that reduces long term availability and manufacturing throughput, and hence optimum asset management normally seeks to maximize the interval between inspection-maintenance cycles, a characteristic of "impulse control" or "costly control" problems (Bensoussan, 1971).

In practice, the implementation of CBM is often limited by the ability to accurately predict when equipment of a given type will next be in need of inspection and maintenance, based on all prior inspection and operating data. The consequences of over-estimating the maintenance interval are a greatly increased risk of failure, leading to an unscheduled outage, equipment damage, safety risk, and extended as well as unexpected loss of production capacity. If a plant is operating near capacity, this may result in severe additional costs and even penalties to the manufacturer for late deliveries or lost production. The consequences of under-estimating the maintenance interval are excessive time spent in scheduled outages. Chen and Trivedi (2002) have recently examined the impact of scheduled maintenance intervals on over all manufacturing system availability and showed that there is an optimum interval (under certain simplifying assumptions) for CBM based simply on the trade-off between minimizing unscheduled outage time and maximizing expected availability. Although this analysis is idealized, it does indicate that a theoretical basis exists for regulating the CBM interval.

In practice, however, an important issue is the inability to forecast or predict the rate of equipment degradation. If estimates of degradation rates are subject to high variability, then CBM intervals will be driven toward smaller values than are really necessary – simply due to the large uncertainty in the predicted degradation rates. This paper addresses the problem

[1] Corresponding Author.

of improving the accuracy of predicted failure intervals, as experienced in representative GE work with durable capital equipment such as aircraft engines, locomotives, and medical instrumentation, and industrial power distribution and protective equipment.

In considering the problem of predicting future equipment state or condition with the greatest accuracy, many factors must be considered. Figure 1 classifies some of the primary sources of uncertainty in predicting future equipment performance.

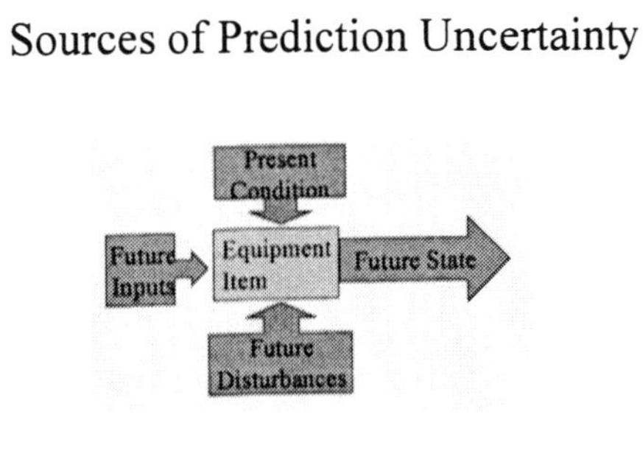

Figure 1: Sources of prediction uncertainty.

The present condition ("initial state") of an equipment item is always unknown to some degree. Part-to-part manufacturing variations, which may include variations in material and its entire history of formation, part dimensions, or even less obvious factors such as impurities, residual stresses, and the effects of past use (aging) are always present. Future inputs refer to variables that exhibit a causal relationship to the behavior of the equipment, and may include future control variables, settings, and values of continuous inputs such as temperature or flow rate. Normally, inputs are assumed to be measurable and to be at least partially under the control of the user. The term "disturbance" is often used for a variable over which the user has no control and which may or may not be measurable. Some disturbance sources may be known in advance, while others are less well known.

Two limiting cases should be noted. To simplify the discussion, assume that future inputs are known perfectly. The one extreme case is where most of the uncertainty resides in the initial condition of the equipment, and where future disturbances are assumed to be very small or negligible. This case may be considered the "*pure prediction*" case, since there are no significant future sources of uncertainty, but only the effects of propagating the uncertainty in the initial state of the equipment item. Although it is not discussed here in detail, the pure prediction case also allows the future prediction to be tuned to the condition of a specific (e.g., specific serial number) item of equipment. At the other extreme is the case where the initial condition is known perfectly, but where un-measurable future disturbances are expected to dominate the future condition of the equipment; this is sometimes called the "*pure forecasting*" case, since the future performance is dominated by un-measurable (and perhaps intangible) variables. Most cases of practical interest lie between these two extremes. Although it is an oversimplification, it is sometimes useful to view the accuracy of a prediction as depending primarily on the inherent dynamics of the equipment condition degradation process in the case of pure prediction, and primarily on the ability to predict the future disturbance process dynamics in the case of pure forecasting.

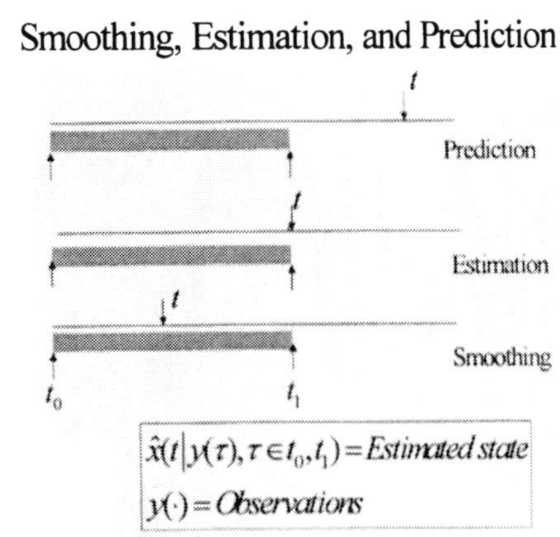

Figure 2: Observation intervals used in smoothing, estimation, and prediction.

The general estimation problem has many variations, depending on the "information set" upon which an estimate is conditioned. An important distinction between smoothing, estimation, and prediction has been made by Kalman (1960), and many others. In "smoothing" the estimate of equipment state at a given time may depend on observations both before and after the time of interest, while in "filtering", only measurements prior to the time of interest can be used. In prediction, the time of the estimated condition is in the future, and only measurements up to the present time can be used. In certain problems, it is also important to distinguish whether or not there is a *plan* of future measurements (measurement "policy"), since (even though the future measurement values may not be known at present), there will be an expected difference in performance under a plan of using future values in contrast to the case where no more future data is expected). This is of particular significance when a system is considered to be repairable, in the sense that future measurements will be used to determine a repair policy that will in turn affect the future equipment condition, such as in CBM.

In the most favorable case where an accurate "model" of equipment degradation or failure processes is available, then there is still a question of what the exact nature of a "prediction" is. In a Bayesian problem formulation, the future state of the system is a random variable characterized by a (conditional) probability density function. Usually, though, some finite sets of variables are used to characterize the nature of the prediction more precisely: The conditional mean is commonly used (along with the conditional variance), but the mode, the maximum a posteriori (viz., after the current measurement), and other a posteriori variables are commonly used to capture the specific nature of the prediction that is needed.

The notion of using a long-term dynamic model for equipment condition prediction was put forth in (Johnson, 2002). In practice, such models are almost always derived on the basis of prior design and measurement data, and a priori information (e.g., product specifications) that may include not only the specific equipment unit of interest, but also a class of units of the same or similar types, in similar applications. Most types of equipment exhibit multiple dominant failure modes, which may or may not interact, and the leading failure modes must be included in a prediction model. Ray and Tangirala (1996) have examined the stochastic differential equations governing mechanical fatigue crack propagation in aluminum sheet metal, and compared theory with experimental results for the Newman model of crack growth (Newman, 1983). They illustrate model validation based on very carefully controlled laboratory data collected by Virkler, et al (1979), that show the inherent growth of part-to-part variability of fatigue growth rates as fatigue life is reached. Not only does this data illustrate the significance of the "initial state" uncertainty in predicting time to failure, but also the fact that for the times (or fatigue cycle counts) of most interest -- namely, the times immediately prior to failure, the variance of the underlying process is growing rapidly, even when inputs are perfectly known and disturbances are kept to an absolute minimum.

When the same data are used to develop models that are used for verifying and testing them – a situation often encountered in practice -- the prediction process may be regarded as less accurate than when these two sets are independent. The problem of model validation and verification is of great significance, because the outcome of this process feeds the prediction process. In practice, one is usually involved in a process wherein *dynamic model identification and condition prediction are done concurrently*. The statistical framework of such a process should be rigorously described, since the problem formulation and subsequent solution may determine whether the prediction is tuned to the sample statistics, to the sample life usage path executed by an individual sample, or to the sample or path statistics of the input or disturbance processes. In practice, the relative accuracy of these observation model formulations can be very difficult to judge.

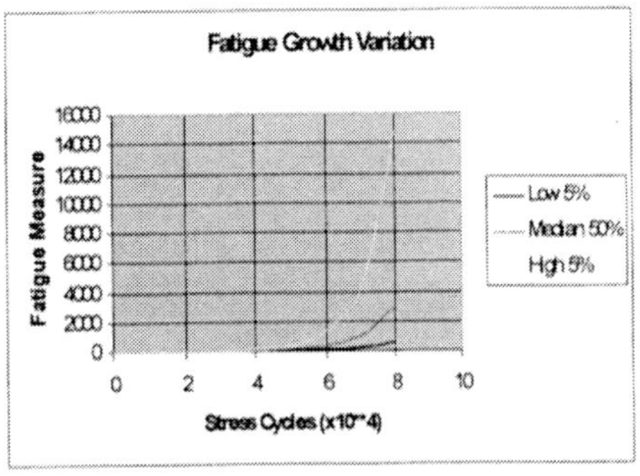

Fig. 3: Fatigue Growth (simplified from Virkler, et al., 1979)

2. METHODOLOGY ISSUES AND APPROACHES FOR PERIODIC INSPECTION DATA

Many types of prediction and forecasting methods have been applied to the problem of estimating failure or degradation times of equipment. In the most ideal circumstances, tribologists have studied the development of fatigue and wear in the laboratory and developed statistical models for a limited number of very simplified geometries and materials. The reader is referred to (Johnson, 2002), (Suresh, 1991), and (Ray and Tangirala, 1996) for examples of such studies. In most equipment of practical interest, one is confronted with alloyed or joined materials, complex part shapes, and multiple and concurrent environmental stress factors that give rise to competing failure modes. So, even though high quality statistical models such as those of Sobezyk, et al (1992) are available for ideal cases where independently verified stochastic models can be developed, the development of practical models has remained an art, and several approaches are in use. In some (e.g., "data-driven" or empirical) approaches, no effort is made to distinguish a physically based model of fatigue from other sources of variation such as correlated measurement error, and only the accuracy of "final results" is sought. Some (but not all) fuzzy logic and neural net models fall in this class. In more structured approaches, such as a Bayesian model-based approach, an explicit identification-filtering-prediction methodology is used, e.g., prediction based on the conditional probability density or conditional mean. In cases where the discrimination among multiple failure modes is as important or more important than the determination of the exact extent of a failure, then approaches such as finite-state Markov models or Bayesian Belief Networks have been employed. Finally, purely or essentially statistical models such as principal component analysis, Weibull models, or ARMA models, based on maximizing the statistical forecast accuracy, have been employed.

Condition prediction should be viewed in the context of the methodology employed to collect data and generate predictions, since this methodology may dominate the achievable results. Figure 4 illustrates the key steps of the CBM process. Practical efforts to apply prediction methods to actual fielded equipment encounter a number of practical limitations, which are discussed briefly in this section. The limitations of data accuracy and quality that are encountered in the steps of this process are discussed briefly. And then examples of applying some of the dominant methods on actual data are illustrated.

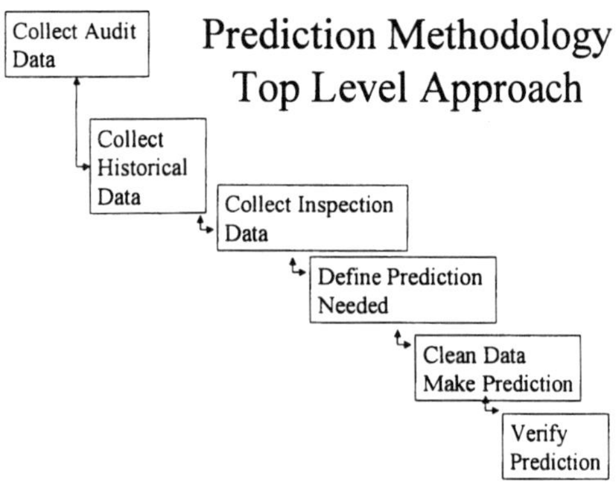

Figure 4: Prediction Methodology

2.1 Classification of equipment

In actual manufacturing plants, equipment of varying age, manufacturer, and model will be found, even for equipment performing similar manufacturing functions (e.g., conveyor systems or segments, bowl feeders, etc.). Since faults or condition degradation may be relatively rare, or occur over a long time period, it is important to structure data to allow for statistical comparisons over sets that grow from the most similar equipment (e.g., same manufacturer, make, model, age, and location), whenever this is available, to larger sets that include successively less similar equipment (e.g., different manufacture, same function, or different model, but the same manufacturer). A hierarchical arrangement of equipment classes facilitates this.

2.2 Classification of data

A number of types of data may be available. It is important to place this on as comparable a basis as possible. The following data set types might be available in a CBM scenario:

- Manufacturer specifications
- Acceptance test data for equipment (when new)
- Warranty inspection or repair data
- Factory Periodic inspection data (walk-around inspections during operation)
- Factory historical outage data
- Scheduled CBM outage data (planned inspections)
- Continuous monitoring data
- Monitoring or logged data from upstream or downstream gages or inspection stations.
- Unscheduled outage incident data
- Maintenance and repair records
- Equipment disposal and retirement records
- Incident root cause analysis (rare)

In practice, it is useful to divide this data into four groups:

- Nameplate data – Asset description
- Historical data
- Periodic inspection data (augmented with unscheduled outage or "incident" data)
- Continuously monitored (time series) data

The choice of equipment classification will limit the achievable resolution of a forecast; if relationships among equipment types are not properly represented, then common factors that may affect concurrent wear (e.g., due to a common load shared by several different items of equipment) will not be properly represented, resulting in a confounding of stress factors and potential limitations on observability of the system. For instance, maintenance and repair records often constitute "input measurements" for such systems, and are very important, as are any variables that significantly change the (mechanical, electrical, chemical, thermal) energy stored in the equipment structure itself. The selection of measurement/inspection data, and the timing of the collection of data impacts the achievable accuracy of prediction; the frequency of measurement must be high enough to track the most rapid condition changes of interest, relative to the most frequent scheduled maintenance interval of interest. *The nature of CBM is to balance these factors, i.e., the inspection interval will be decreased to the point where either condition degradation becomes unacceptable, or to the point where the uncertainty in the prediction will necessitate more frequent measurements.* Either factor may dominate this decision in practice.

Examples of missing data include the potential inability to precisely identify the time at which an equipment item became degraded beyond a pre-specified limit, due to the fact that inspections are carried out too infrequently. Methods such as interval censoring allow this uncertainty to be represented properly. Another common example of missing data is the early history of an equipment item, prior to the advent of digital archiving of inspection data: The installation date of new equipment may be known, but no inspection data may be available until digital record keeping began when the equipment was 15 years old. This introduces a large variation in knowledge of the initial equipment state.

2.3 Determination of Disturbance Sources and Dimension of State

Principal component analysis can be applied to both time series data, and to cross-correlation matrices between measurement records on similar equipment types. When applied to time series data, principal component analysis (factoring the temporal autocorrelation function or matrix) will yield the dominant time constants in the system. When applied to the cross-correlation matrix across a set of similar components, the method will identify sets of equipment having closely correlated behavior (e.g., multiple loads connected in parallel across a common transmission line may show correlated electrical stresses). These methods may also reveal subsets of variables that have little relevance to prediction or forecasting. Care must be taken to avoid artifacts due to over or under-sampling. Both of these factors allow the proper size of the system "state" vector to be estimated.

Classification and regression trees (CART) are attractive since they achieve optimal classification accuracy for a pre-specified number of factors while also ordering disturbance sources according to decreasing importance. In the figure below, the target or response variable is the actual equipment condition recorded during inspection, the factors are age and several inspection variables. The most accurate classification is the one that minimizes the cost function representing the impact of misclassification and also the decision that lead to a simple, compact tree with few nodes as illustrated below in Figure 5 for an electrical equipment example. The decision boundary or query selection (if-then conditions) for equipment failure prediction using Age, Contact condition, and closing delay of the equipment are represented in the CART tree in Figure 5a. Figure 5b shows a Pareto of variable importance. Control and contact condition are the main factors in prediction of the equipment condition.

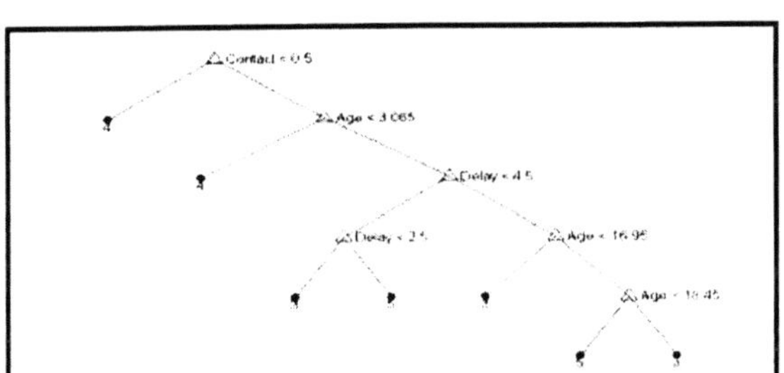

Figure 5a: CART Tree for Fault Factors

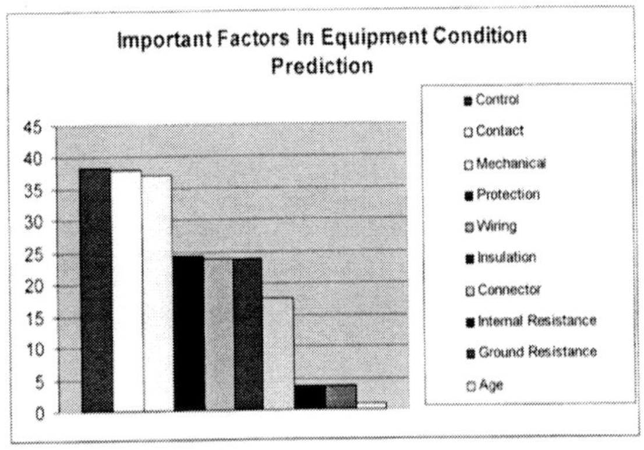

Figure 5b: Factors in Equipment Condition Prediction

Figure 5: **CART Results for electrical equipment: (a) Classification and Regression Tree; (b) Relative importance of various factors in CART classification.**

This data set will grow with time over the life cycle of an individual equipment item. As part of the asset description, it is important to provide links that relate a given equipment item to other mechanically linked equipment (upstream or downstream equipment), or equipment that processes the same part sequences. As indicated in the previous section, a predictive analysis needs to use past data to both estimate the present equipment condition, and also to estimate future input statistics. A common issue with this data is that it will be incomplete (e.g., missing the early portion of a life cycle, prior to when digital data collection was initiated), or contain errors and inconsistencies that can no longer be resolved.

2.4 Definition of equipment condition

Raw inspection data does not provide directly useful information about equipment condition. Most types of production equipment are periodically inspected and re-calibrated based on statistical quality control analysis of the parts that they produce, e.g., using the data from a gauging station that follows the completion of a manufacturing step to periodically readjust cutter offsets. The equipment is considered to exhibit a wear or use-related condition degradation when it can no longer be adjusted to remain within the specified part tolerance of the manufacturing process. These wear-related conditions develop gradually over a long period of time. A second type of fault is the sudden failure, which may have its origin internally to the equipment, or may be related to extreme use (e.g., sudden jamming). In this case, the equipment may suddenly go out of specification and/or produce bad parts. Usually, several different root causes of degradation or failure are operating concurrently, and the expressed failure is simply the one that causes the machine to exceed specification limits first. If this fault is identified and repaired, a different root cause is likely to become the more likely cause of the next failure.

The notion of "equipment condition", which is widely used in prioritizing repair actions, is to define a single (scalar) index of equipment degradation or failure likelihood that in some sense reflects the "envelope" or worst case of all of the failure sources. One advantage of using a single "condition" summary is that it allows equipment to be ranked for repair/replacement priority; its value may be more robust to missing or approximated data then inspection histories for individual tests. The disadvantage of this approach is that the equipment "state" has been aggregated to a single dimension (i.e., "condition") that may conceal clear trends in individual root cause (disturbance) measures. "Condition" may also be viewed as a scalar "response" output or "performance" variable that is a monitor of system effectiveness over time.

2.5 Time Series Construction

Most time series results have been developed for uniform sampling intervals, but this is rarely the case in practice. A well-known approach can be used to represent non-uniformly sampled data into either time-varying system format, or non-uniformly measured ("continuous-discrete") sampled data format (Jazwinski, 1970). Time series may be derived for both continuous and discrete variables, and various models of such processes (e.g., continuous-time systems with random failures, finite-state systems augmented by failure processes, etc.) have been used (Chizeck, et al, 1986). More recently, hybrid models having concurrent continuous and discrete states (bi-states) have been used. Commonly in CBM applications, different equipment will be inspected or maintained at different (but known) intervals, possibly out of synchrony with one another, and so provision must be made for multi-rate and asynchronous sample intervals. Common practices are to project inspection data down onto a much smaller subspace (even a single condition dimension, in some cases) via a linear or non-linear transformation, and then to examine. Inputs, disturbances, states, and output measurements must then be defined for the class of equipment under evaluation. Figure 6 shows the evolution of an integer-valued equipment condition variable vs. time. More generally, multivariate time series methods may be employed.

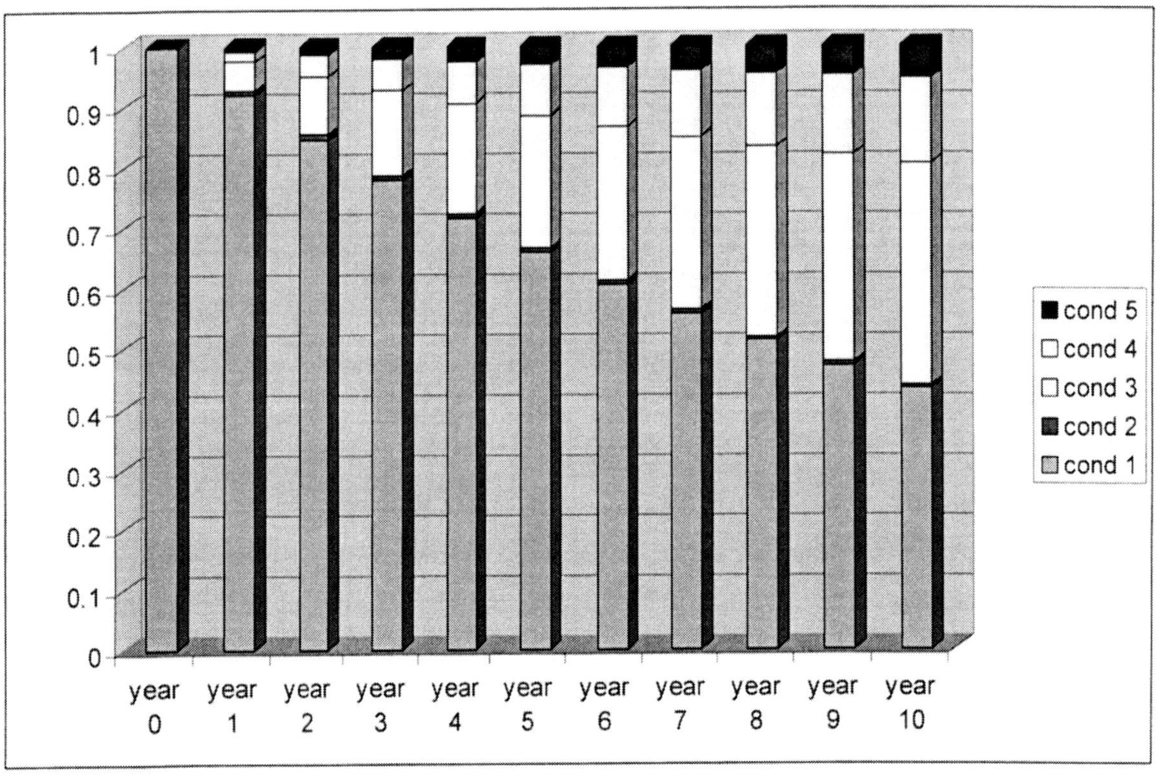

Figure 6: Condition evolution vs. time.

2.6 Modelling and System Identification

Models of system failure or degradation may have both continuous and discrete states. Finite state Markov models are commonly used in practice to represent condition states that change discontinuously among a finite set of values. When condition data is mapped into a finite set of integers, as is often the case, then a history of condition values can be used to estimate a Markov transition matrix.

A representative condition transition matrix, based on analysis of time series data, is shown in Figure 7.

Here, Condition 5 is the failed state, and no repair actions are represented. Hence, the probability of this equipment being ultimately in the trap state is 1.0. This analysis also reveals that State 4 is very rarely used. The function of refurbishment or replacement, in this case, might be to "reset" the system state to Condition 1.

0	0	0	0	0
0.91941	0.00701	0.04905	0.01682	0.00771
0	0.33333	0.33333	0.33333	0
0	0	1	0	0
0	0	0	1	0
0	0	0	0	1

Figure 7: State transition matrix

A crude system identification method that is commonly used for finite state Markov processes is to "count state transitions" and to compute the transition matrix based on sample statistics. This counting may be performed using condition data from all members of an entire category of equipment, even though it may be applied to assess the likelihood of a sample sequence for an individual equipment item. Care should be taken to match the model order to the true statistical order of the system, or the results may lose resolution due to over-aggregation (e.g., if the system has 10 states of actual fault status, but only 5 conditions are used to represent them, then the estimated states in a 5-state model may lose resolution and accuracy relative to a 10-state model).

Returning to continuous system dynamics, which may be more appropriate for modelling of wear-related variables, Bohlin (1976) suggests a useful low-order time series model that, in effect, tracks the three components of a local second order model (offset, rate, and acceleration states). This "generic" model class is very useful in capturing dominant "drift" and "acceleration" of fatigue effects based on individual item data.

$$y(t) = \sigma_0 e_0(t) + \sigma_1 \Delta^{-1} e_1(t) + \sigma_2 \Delta^{-2} e_2(t) + \sigma_3 \Delta^{-3} e_3(t)$$

where t is an integer representing sampling time, $e_0(t),\ldots,e_3(t)$ are orthonormal sequences of Gaussian random variables, Δ is the backward-difference operator, and $\sigma_0,\ldots,\sigma_3$ are unknown constants. The four terms of the equation represent, respectively, a high frequency noise, drift, trend, and acceleration term (of the wear process).

Bohlin also provides very efficient identification and prediction algorithms for this model, by converting it to the "Integrated Moving Average Model" form

$$y(t) = \Delta^{-m} C(q^{-1}) e(t)$$

where m is the highest negative power in the original equation having a nonzero coefficient, and $e(t)$ are Gaussian and orthonormal. The error sequence is representable in the form

$$e(t) + \sum_{j=1}^{n} a_j(t) e(t-j) = \lambda w_0(t)$$

where

$$a_k(t+1) = a_k(t) + \mu w_k(t)$$

are modelled as the parameters of independent random walk processes. He shows that the predictor corresponding to these equations can be written in the form (Bohlin, 1976, eq. 65, 66)

$$\Delta^m (q^{-1}) \hat{y}(t+\tau|t) = C(q^{-1}) \hat{e}(t+\tau|t)$$

$$A(q^{-1}) \hat{e}(t+\tau|t) = \lambda \hat{w}(t+\tau|t) = 0$$

for $\tau > 0$, with start values provided by past observations and estimator errors $(e(t+\tau), \tau < 0)$. Bohlin also derives an expression for the prediction error variance, and associated outlier tests based on likelihood ratio calculations.

This model, as indicated below, is particularly well suited to prediction for many types of time series data. More general models can include both finite and continuous state variables, as in the case of the finite state condition "with reset" cited above.

3. PREDICTION APPROACHES

3. 1 Weibull Analysis

The Weibull distribution has been used to model statistics of "time to next failure" or "time to next repair" (AMSAA, 1981). In the following, the term "failure" is used to designate both types of models. Unlike the more common Poisson distribution (which has a single parameter), it allows independent variability of the expected time-to-failure and the variability in failure time, and is a more accurate model to reflect the effects of several concurrent failure processes. In associating this statistical model with physical processes, it is important to accurately represent the repair assumptions in the processing of data. The "normal equipment baseline" that serves as the reference point for application of the Weibull is important: using a reference baseline of "new" equipment will not provide the same results as using a reference baseline of "20-year old equipment with average use and wear". In particular, the Weibull model is not appropriate for many types of "wear" phenomena, where the system may remain for extended periods of time in a degraded mode of operation. The model is most easily applied to conditions where repairs or restoration are performed near to the time of failure, as is the case with periodic inspections. Another issue, however, with periodic inspection data, is that the time to failure is not accurately known; it is only accurately known to within a certain interval of time. Another issue arises due to ergodicity assumptions, when time-series data on one individual equipment item is mixed with sample data from a population of similar equipment items.

Classification of the equipment into categories can provide significant insight into the manner in which the estimated Weibull statistics are influenced by the equipment type variable.

Figure 8(a) illustrates a hierarchical representation of equipment types for a sample installation. The level to which the classification is considered has direct impact on the analysis. Figure 8(b) shows a Weibull regression plot with representative condition data. For a fixed data set, the use of fewer levels implies that more data points are available for a given equipment type. However, the specificity of such analysis is necessarily low. Of particular importance is handling of the outliers. With fewer data points as in the second case, each outlier that is rejected carries with it significant risk of throwing away a real but rare failure mode or hide manufacturer dependent equipment failure or deterioration.

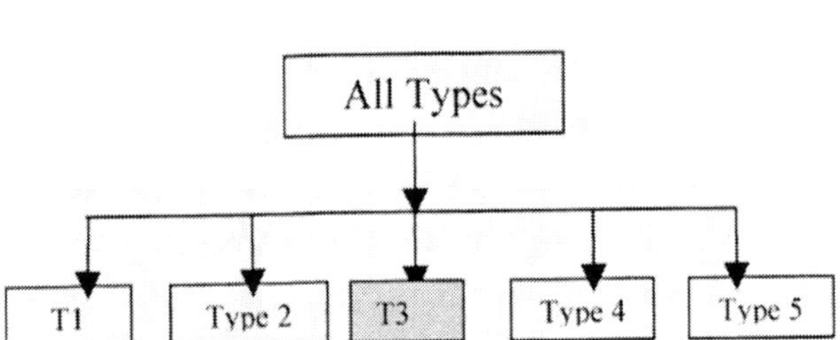

Figure 8(a)

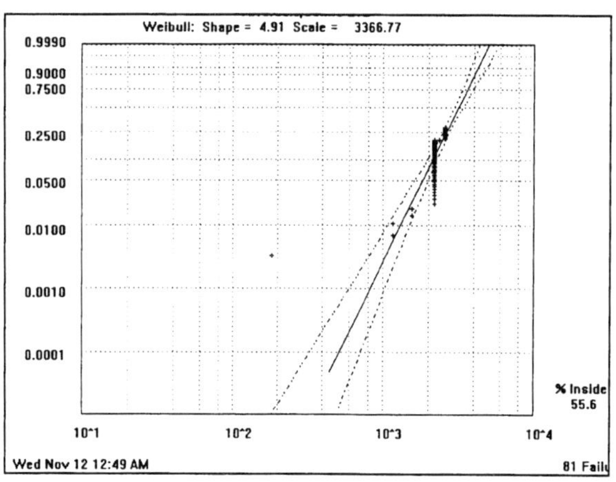

Figure 8(b)

Figure 8: Hierarchical Weibull Analysis

3.2 Time Series Prediction

Time series prediction is best suited to continuously varying conditions such as wear, and is capable of representing periods of time when the system operates in degraded mode without repair. For industrial equipment, the time scales of interest are many orders of magnitude longer than for traditional dynamic models of the type used in control design, and the physical processes to be modelled are also quite different from those of interest in control analysis; however, the same basic physical modelling principles can be applied. In the field of tribology, the independent variable is commonly taken as "stress cycles", "rotations" or "events", rather than time. Conversion of tribology models into dynamic models will require that a proper model be developed for the "stress cycle" variable. For instance, the Rice Level Crossing (Rice, 1945) formula can be used to use a random process (with time as the independent variable) into a counting process (with level-crossings as the dependent variable), and these can be used as inputs to the "stress cycle" variable, an (uncommon) application of the commonly used method of state space augmentation. A common practice is to run a system identification algorithm (see previous section) in parallel with (or one step behind) a prediction algorithm, so that the system model parameters are updated concurrently with the state estimate of the system. Although this ad hoc method cannot be used under all conditions, Bohlin (1976) points out that under certain conditions this is a valid approach. In the following analysis, however, it is assumed that the system aging model is known in advance from physical principles.

The result of this is a random process that can be described as an Ito equation (see below). It is well known that (under appropriate statistical and regularity assumptions) Kolmogorov's forward equation governs the propagation of the density of the state of this system.

The case of interest for prediction in the presence of noisy aperiodic observations is the "continuous-discrete" case (Jazwinski, 1970, Section 6.3), where between observations the conditional density is governed by the Chapman-Kolmogorov equation:

$$\partial p(x,t;Y_t)/\partial t = L(p)dt$$

where

$$L(\cdot) = -\sum_{i=1}^{n} \partial(\cdot f_i)/\partial x_i + (1/2)\sum_{i,j=1}^{n} \partial^2(\cdot(GQG^T)_{ij})/\partial x_i \partial x_j$$

for an n-th order vector stochastic dynamic system described by the Ito differential equation:

$$dx_t = f(x_t,t)dt + g(x_t,t)d\beta_t$$

where G is the gradient of g and Q is the variance parameter of the vector random walk process β_t. At the time of additional a priori information an update of the density based on Bayes rule is performed:

$$p(x,t_k|Y_{t_k}) = p(y_k|x)p(x,t_k|Y_{t_k^-}) / \int p(y_k|\xi)p(\xi,t_k|Y_{t_k^-})d\xi$$

and for prediction of the density for $t > \tau$, given the observation sequence Y_τ,

$$dp(x,t|Y_\tau) = L(p)dt, t > \tau$$

The extended Kalman Filter has been proposed as a means to approximate the mean and variance of the a posteriori density, although with current processor power and memory, and in view of the fact that inspections of industrial systems are relatively seldom (intervals of weeks to years, typically), a more accurate integration of the functional differential equation for the a posteriori density itself can be performed. Ray and Tangirala (1996), however, have illustrated the application of the extended Kalman filter to one representative example.

The preceding questions concerning the relative roles of ensemble and individual equipment measurements in defining limits to the predictability of industrial equipment condition are much easier to resolve in the this context. In the simplest case, the ensemble average can be used to initialize the prior statistics each time the predictor is run (i.e., it plays the role of an "initial condition"), while each individual inspection affects the "update" equation, based on Bayes rule. Since these systems are normally non-linear, it is not easy (as it is in the linear case) to use superposition to distinguish and compare the relative benefits of these effects in the prediction.

CONCLUSIONS

A variety of practical conditions that fundamentally limit the accuracy of prediction achievable in condition based maintenance applications have been discussed. In addition, the fact that the inspections to date are used to affect future inspection dates in the CBM scenario could and should be discussed. Markov decision processes can exhibit similar phenomena. It is in fact possible that certain CBM algorithms, in principle, could exhibit steady state oscillation, or even instability of the inspection interval.

A closer examination of experimental results such as those of Virkler, et al (1979) suggest an interesting phenomenon that may be characteristic of prediction methods applied to wear or fatigue data: In the early phase of damage, the fatigue rates are relatively low, and the sample-to-sample variance is low: therefore, ensemble averages contain useful predictive information for any member of the sample set. However, in the later phases of fatigue, the sample-to-sample variability (ensemble variance) grows rapidly, whereas the fatigue growth pattern of a single sample remains relatively predictable, implying that individual sample inspections should play a larger role in making the prediction. This might be termed a "prediction paradox", and could indeed be a factor in producing oscillations in CBM maintenance/repair intervals as described above.

The fundamental limitations on predictability of wear and breakage processes over the long term are suggested by the previous discussions. The inherent tendency of fatigue processes to accelerate, and also to become more variable with repeated stress events is the most fundamental limitation to prediction interval for such equipment, and further work on improving equipment models is needed. Perhaps the next most important limitation, in many practical situations, is the uncertainty of the future system "input" and "disturbance" statistics. This can be addressed in part by measuring and also predicting input and disturbance signals independently of the equipment condition prediction, but often these variables cannot be regulated and are subject to sudden and unforeseeable changes. A priori bounds on these variables (due to system design, or independent analysis) can be of use in limiting the degree of uncertainty that arises from them.

Other factors influencing predictability are more readily controlled: The accuracy and choice of the inspection and measurement systems, the distribution of measurement resource between periodic inspection and continuous monitoring, and the development of accurate physical models of aging, fatigue, and wear processes are all solvable if sufficient time and resource are available.

REFERENCES

D. Chen & K. S. Trivedi (2002), Reliability Engineering and System Safety, RESS 3012 (in press).

T. L. Johnson, 2002, Simplified Life Cycle Models for Industrial Equipment, Proc. IEEE Mediterranean Control Conference.

T. Bohlin (1976), "Four Cases of Identification of Changing Systems", in System Identification: Advances and Case Studies, by R. K. Mehra and D. G. Lainiotis, Academic Press, New York, pp. 441-518.

A.H. Jazwinski, 1970, Stochastic Processes and Filtering Theory, Academic Press, New York.

D. A. Virkler, B. M. Hiulberry, and P.K. Goel, (1979) "The statistical nature of fatigue crack propagation," ASME J. Eng. Materials Tech., Vol. 101, no. 2, pp. 148-153.

R. E. Kalman, 1960, "A New Approach to Linear Filtering and Prediction Problems", Trans. ASME, Ser. D.; J. Basic Engg., Vol 82, pp. 35-45.

A.Bensoussan (1971), Filtrage optimal des systemes Lineares, Dunod, Paris.

A. Ray & S. Tangirala (1996), "Stochastic Modelling of Fatigue Crack Dynamics for On-Line Failure Prognostics," IEEE Trans. Control Systems Technology, Vol. 4, No. 44, pp. 445-451.

J. C.Newman, Jr. (1983), "A Crack Closure Model for predicting fatigue crack growth under aircraft loading," Methods and Models for Predicting Fatigue Crack Growth under Random Loading, ASTM STP748, pp. 53-84.

S. Suresh (1991), Fatigue of Materials, Cambridge Univ. Press, Cambridge, UK.

K. Sobczyk, K. Spencer, and B. F. Spencer, Jr., Random Fatigue: Data to Theory (1992), Boston, MA; Academic Press.

U.S. Army Materiel Systems Analysis Activity (AMSAA), MIL-HDBK-189, Reliability Growth Management, 13 February 1981 (updated), Appendix A.

S.O. Rice (1945) "Mathematical analysis of random noise," Bell Syst. Tech. Journal, vol. 24, pp. 46--156, Jan.

H. J. Chizeck, A. S. Willsky and D. Castanon, "Discrete-Time Markovian-Jump Linear Quadratic Optimal Control," (1986) Intl. J. Contr., Vol. 43, No. 1, pp. 213-231.

Copyright © IFAC Information Control Problems in Manufacturing, Salvador, Brazil, 2004

ELSEVIER
IFAC
PUBLICATIONS
www.elsevier.com/locate/ifac

INDUSTRIAL APPLICATIONS OF THE AGENT-BASED TECHNOLOGY

Vladimír Mařík

Gerstner Laboratory, Czech Technical University, Prague, Czech Republic
& Rockwell Automation Research Center, Prague, Czech Republic
e-mail: marik@labe.felk.cvut.cz

Abstract: The paper is aimed at summarizing the current state-of-the-art in development of the agent-based solutions in three areas connected with industrial production, namely in holonic control, production management and virtual enterprises. It is shown how the results in these three sub-areas converge and that the agent-oriented technology could provide a good background for developing a unifying approach. Two industrial applications of agent-based systems are described in more details. Both the positive and negative aspects of applying the agent-based solutions in industry are discussed. *Copyright © 2004 IFAC*

Keywords: Agents, distributed artificial intelligence, distributed computer control systems, intelligent manufacturing systems, simulation, industrial control, industrial production systems, system integration, and system architecture

1. INTRODUCTION

Both the complexity of the manufacturing and busisness environments as well as the complexity of tasks to be solved is growing continuously. In many industrial scenarios, traditional centralized and hierarchical approaches applied to manufacturing control, production management, planning and scheduling, supply chain management and business solutions are not adequate and can fail because of the insufficient means to cope with high degree of complexity and practical requirements for **robustness, generality** and **reconfigurability.**

These issues naturally lead to a development of new manufacturing and supply chain architectures and solutions based on the consideration of highly distributed, autonomous and efficiently cooperating units integrated by the *plug-and-operate* approach. This trend of application of multi-agent systems (MAS) techniques is clearly visible at all levels of manufacturing and businesses:

On the lowest, real-time level, we can find autonomous control units tightly linked or even fully physically integrated with the physical manufacturing hardware (e.g. with the conveyor, drilling machine, assembly cell, etc.). Such autonomous units are referred to as **holons** or **holonic agents** (Deen, 2003).

When solving the production management tasks (production planning and scheduling, strategic decision making, data processing etc.) on both the workshop and factory levels the **intelligent agents** exploring the results of the AI research can be used with advantage (Weiss et al, 1999). Each manufacturing unit can be represented by an intelligent agent equipped by a relevant amount of knowledge; the agents communicate, coordinate their activities and cooperate to achieve an optimal solution in a reasonable time (usually not in a hard real-time).

More generic visions of intensive cooperation among enterprises connected via communication network has lead to the ideas of **virtual enterprises** (VE): "A virtual enterprise is a temporary alliance of enterprises that come together to share skills or core competencies and resources in order to better respond to business opportunities, and whose cooperation is supported by computer networks" (Camarinha-Matos and Afsarmanesh, 2004).

The philosophical background of all these highly distributed solutions of control and management problems on different levels is the same: the community of autonomous, intelligent and goal-oriented units efficiently cooperating and coordinating their behavior in order to reach the global level goals (and, maybe, sometimes prioritizing also their private goals, especially in the case of VEs). Agents can be - in general - defined as autonomous, problem solving computational entities capable of effective behaviour in dynamic and open environments (Luck, *et al.*, 2002). Let's talk about the units mentioned above in general as about *the agents*.

There are several typical agent architectures ranging from purely *reactive* agents (agents operate in a simple "stimulus-response" fashion) to more deliberative or *goal-oriented* agents, such as the "belief-desire-intention (BDI)" agents (Rao and Georgeff, 1992), which are proactively reasoning about their goals and actions. *Holons* can be considered as a specific reactive agents which are strongly connected with the physical level devices, operate in hard real-time and can be organized into a holarchical structures (see Paragraph 2 for details).

The terminology used in the field is in the process of development, sometimes confusing. Let's distinguish throughout this paper, for better understanding, between

- **the agents and agent-based systems** in the sense of units/systems vaguely declaring *properties of agency* (autonomy, capability to communicate among themselves etc.) on one side and
- the **multi-agent system (MAS) technology** on the other side. A group of agents organized precisely according to specific, precisely defined principles of community organization and operation (like principles of the agent architecture, style of messaging and negotiations protocols, etc.), being supported by an adequate agent platform (besides others with the registration and deregistration processes enabled) is usually called a multi-agent system.

Thus, an *agent-based solution* mentioned in the paper should be understood as a solution which considers the existence of agents as autonomous units. Multi-agent system and any *MAS solution* in our notation satisfies the strict principles of agent community organization and represents *a kind of technology*. **Intelligent agents** are always considered as part of a MAS solution.

In principle, there are three nearly isolated communities developing highly distributed, agent-based solutions in industrial environment:

- the community of *holonic manufacturing systems*,
- the *community* designing *MAS solutions for knowledge intensive decision-making on both the workshop and factory levels*, that means on the level of the *production management* and, finally,
- the *community* aiming to develop solutions *for virtual enterprises* aiming to coordinate and integrate manufacturing facility with external entities (sales networks, suppliers, customers, cooperators, etc.).

These communities use different terminology and are focused on exploring different agent-based techniques. But, because of the same nature of the background philosophy and the same principle ideas, there is a great chance (and hope) to link all the communities and their solutions to integrate all the processes on different levels, namely on the real-time manufacturing level, production control level, and the VE/supply chain level as well using the same or compatible principles and techniques **based on MAS principles.** Let's illustrate that the MAS philosophy and technology can be used as a joint general denominator and can help to integrate the solutions in the three areas mentioned above.

This integration vision leads to the vision of a **cross-integrated enterprise (CIE),** where the collaborative units are all the members of the control, manufacturing and supply chain considered horizontally (ranging from the control to business processes) across the entire enterprise (manufacturing machines and cells, workshops, departments, enterprise divisions, suppliers and customers) (Mařík and Pěchouček, 2004). In a CIE, the planning department of a supplier can directly "talk" to the controllers of a partner's plant asking for actual data, or the workshop agents can directly negotiate with the workshops or even controllers of the outsourced company etc.

Let's briefly summarize the results achieved by the three different communities solving industrial problems on different levels.

2. DISTRIBUTED REAL-TIME CONTROL AND HOLONS

The problems of decentralized, highly distributed control can be solved by holons as autonomous, cooperative units. (van Leeuwen and Norrie, 1997), (Deen, 2003). Holons can be organized in hierarchical or heterarchical structures.

Quite specifically, any holon can contain other holons that can again – in a recursive manner – contain other holons Such a structure, called *holarchy*, provides a mechanism of encapsulating capabilities of different kinds of holons at a specific level of manufacturing process into a higher-level entities being considered as holons as well. From this viewpoint, the whole factory can be considered as a holon being composed of different kinds of holons such as those representing physical objects like machines, AGVs, conveyor belts, pumps, valves, products etc. as well as non-physical entities like customer orders, production plans and so. Focusing at the lower-level, a holon representing for instance a specific CNC machine is also composed of other holons, one being responsible e.g. for product fixturing, another one for changing a particular tool according to a type of machinery required for a product being currently processed (cutting, drilling, etc.), another holon being e.g. in charge of negotiating with the production planning holon etc.

Holons for real-time control (e.g. machine holons mentioned above) are expected to provide reactive behavior rather than being capable of deliberative behavior based on complex "mental states" and strongly proactive strategies. Their duty is mainly to react to changes in the manufacturing environment (e.g. when a device failure or a change in the global plan occurs). Under "stable circumstances", during routine operation, they are not required to change the environment proactively. The reason for the prevalence of reactive behavior of real-time control holons is that each of them is linked to a physical manufacturing facility/environment (e.g. to a transportation belt or a storage manipulator), changes of which are not very simple, cheap or desirable in a comparatively "stable" manufacturing facility. The physical linkage to physical equipment seems to be a strong limiting factor of the holons' freedom in decision-making.

The standard IEC-61499 known as *function blocks* has been developed by the *Holonic Manufacturing Systems (HMS)* consortium for the holonic real-time control (Christensen, 1994). It is based on the function blocks part of the well-known IEC-1131-3 standard for languages in programmable logical controllers (PLCs). The major advantage is the separation between the data flow and the event flow amongst various function blocks. Multiple function blocks would be logically grouped together, across multiple devices into an application, to perform the desired process control.

While both these standards – IEC-61499 as well as IEC-1131-3 (usually in form of ladder-logic) fit well the real-time control purposes, none of them addresses higher level aspects of communication, negotiation, cooperation and deliberative decision making.

3. PRODUCTION MANAGEMENT LEVEL AND INTELLIGENT AGENTS

On the productions management level (which covers the distributed order pre-processing, production planning and scheduling processes, decision-making on both the shop-floor and workshop levels, financial management, billing etc.), the technology of intelligent agents developed in the area of Distributed AI can be applied. **Each unit** (workshop cell, workshop, storage unit, production planning department, financial department, etc.) is **represented by an intelligent agent**. The agents communicate asynchronously in a peer-to-peer way. They solve mainly planning and scheduling problems, decision-making, non-real time higher-level control tasks etc., that means they are engaged in activities requiring deliberative behavior.

An agent's architecture usually consists of the **agent's body** and the **agent's wrapper**. The wrapper accounts for the inter-agent communication and just-in-time reactivity. The body is an agent's component, responsible for carrying out the main own functionality of the agent. An agent communicates in a standard way with the others and understands the *Agent Communication Language (ACL)*. The *agentification process* provides an elegant mechanism for **system integration**, a tool supporting the *technology migration* from the centralized systems towards the agent-based architectures.

The communication among the agents is usually not just a random exchange of messages, but the message flow is managed by a set of standard **negotiation protocols**. These protocols range from a simple "question-reply", via "subscribe-inform" to more complex negotiation protocols like *"contract-net-protocol" (CNP)* or different versions of *auctions* (Dutch, sealed, Vickery etc.). **Communication among the agents is an important enabler of their social behavior** (Mařík, *et al.*, 2001). The term social behavior means that the agents are able to communicate, to understand the goals, states, capabilities etc. of the others and to respect the general rules and constraints of behavior valid for each of the community member. These rules are e.g. called policies in (Bradshaw, *et al.*, 2003) or norms in (Filipe and Cordeiro, 2004).

One of the crucial questions is how to store and use the knowledge by individual agents locally. For this purpose, different **acquaintance models** are located in the wrappers of the agents (Cao, *et al.*, 1997). These are used to organize, maintain and explore social knowledge about the other agents (about their addresses, capabilities, load, reliability etc.). To keep the temporary and semi-permanent knowledge fresh, several **knowledge maintenance techniques** have been developed to significantly reduce the communication in the period of an urgent decision-making. The other natural requirement is to use as short messages as possible, while keeping the information content rich. For this purpose, **knowledge ontologies** helping the agents to understand and correctly interpret semantic meaning of the messages received (Handler, 2002), (Vrba and Hrdonka, 2002).

The other important technology is **processing of meta-knowledge** by specialized, higher-level agents called **meta-agents** (Pěchouček, *et al.*, 2003), (Tožička, *et al.*, 2003). The meta-agent technology can be explored to monitor the activities of individual agents, to detect patterns of emergent behavior in the community and to propose specific changes in the general policies or particular items of knowledge. Machine learning principles, (e.g. data-mining and inductive inference) are usually used by the meta-agents. The meta-level knowledge potentially enables to introduce self-reflective and self-regulatory principles into the life of the community of agents.

A lot of efforts in the area of intelligent agents have been dedicated to **standardization and interoperability principles.** The interoperability is usually achieved through standardization. There have been several standardization initiatives recognized in the multi-agent area, but the FIPA (*Foundation of Intelligent Physical Agents*) consortium seems to be the most efficient and influential. The FIPA consortium provides a variety of standards for organizing communication and negotiations, for creating and maintaining ontologies etc.

FIPA produces mainly two kinds of specifications: (i) *normative specifications* that mandate the external behavior of an agent and ensure interoperability with other FIPA-specified subsystems, and (ii) *informative specifications* of applications for guidance to industry on the use of FIPA technologies. FIPA standards attempt to be high-level, *neutral abstractions*. The core FIPA specifications are neutral with respect to both (a) the application area and (b) the hardware and software platforms to be used. Of course, the existing software infrastructures should be considered when designing the standards and leveraging the advantages of the existing, non-agent software technology.

The FIPA Abstract Architecture defines a high-level organizational model for agent communication and core support for it. It is neutral with respect to any particular network protocol for message transport or any service implementation. This abstract architecture cannot be directly implemented; it should be viewed as a basis or specification framework for the development of particular architectural specifications.

The FIPA Abstract Architecture contains agent system specifications in the form of both descriptive and formal models. It covers three important areas, namely

a) agent communication
b) agent management
c) agent message transport.

The messages exchanged among the agents must comply with a FIPA-ACL (Agent Communication Language) specification. FIPA-ACL is based on speech act theory (Searle, 1969) and resembles KQML (Knowledge Query Manipulation Language) (Finin et al., 1994).

According to the FIPA philosophy, the agents are grouped into high-level organizational structures called **agent platforms (APs).** Members of each platform are usually geographically "close" – for example, they may be run on one computer or be located in a local area network. Each platform must provide its agents with several mandatory services that are made available by three mandatory components: the Agent Management System (AMS), the Directory Facilitator (DF) and the Agent Communication Channel (ACC).

The AMS administers a list of agents registered with the platform. This component implements the creation and deletion of the running agents and provides the agents with a *"white-page-list"* type of service, e.g. a list of all agents accommodated on the given agent platform and their addresses. Unlike the AMS, the DF supplies the community with a *"yellow-page-list"* type of service, where each agent can register its services.

The ACC provides *message transport services* (MTS) that implement the infrastructure for message transport within one agent platform, and a *message transport protocol* (MTP) that is responsible for interoperability between agents from different agent platforms.

While there is an extensive range of either proprietary or open-source agent development platforms (see http://www.agentlink.org) there are just a few that are fully FIPA-compliant: April Agent Platform (Fujitsu Labs of America), FIPA-OS (Emorphia), Grasshopper (IKV++), JADE (CSELT), Zeus (British Telecom). One of the first FIPA-compliant platform is **FIPA-OS** that has been provided as an open-source licence by Emorphia (http://www.emorphia.com, formerly Nortel Networks). FIPA-OS is a component-based toolkit written in Java enabling rapid development of FIPA compliant agents. The main open-source competitor to FIPA-OS is **JADE** – Java Development Toolkit that has been implemented by Telecom Italia Labs (http://sharon.cselt.it/projects/jade/). JADE is also an open source system implemented in Java. As in the case of FIPA-OS the agent platform can be distributed across machines of different OS (if required). An interesting comparative study of the available agent platforms can be found in (Vrba, 2003).

4. VIRTUAL ENTERPRISE LEVEL

Let's consider each company in a VE is operating autonomously, carrying out its relevant duties locally, communicating and exchanging information with the others just when needed. Such a company, as a rule, wants to have the right to join and leave the VE freely, but it understands that it should satisfy generally valid rules of behavior. In an ideal case, the VE is able to operate in a turbulent environment, to react on unpredictable situations by e.g. automatic re-configuration or extension/reduction of its capabilities and resources.

The researchers working in the field of VEs are looking intensively for technologies which would help them to simulate, understand and control processes connected with creation, life-time and dissolution of virtual enterprises. Multi-agent systems and relevant technologies seem to represent good tools for solutions needed in the VE area (Camarinha-Matos and Afsarmanesh, 2001).

Let's consider each company as an agent able to carry out specific (usually very complex) functions. The agents are registered with a certain platform and communicate in a standard ACL. The processes of the VE formation as well as the joint planning and scheduling activities are based on jointly known *negotiation rules and scenarios*. These are very similar (or identical) to those called *protocols* (like e.g. CNP) or *auctions* (e.g. Dutch or sealed) in the MAS domain. The highly specialized members of a virtual enterprise, like **brokers** or **professional network organizers**, can find easily their counterparts in the MAS community, e.g. various middle agents, brokers etc. The negotiation and brokering algorithms proved to be useful in the MAS area can be used to formalize (and later to automate) the corresponding processes in the field of virtual enterprises. The **meta-agents** could be also introduced as tools helping to detect the less efficient parts or bottlenecks of the network as well as providing advice supporting the self-evolution of the VE in the desired direction.

The processes of the VE creation are analogous to **coalition formation processes** (Shehory and Kraus, 1998) in the MAS domain: Usually there is a group of companies willing to cooperate sharing necessarily a certain part of the company confidential knowledge (this group might be labelled as an alliance and the shared knowledge as a semi-private in the MAS terminology). The alliance members negotiate the opportunity to create a VE for a given specific task (in MAS terminology: *to create a coalition*), there is – as a rule – one company overtaking the role of a VE leader (**the coalition leader** in the MAS terminology) (Pěchouček, *et al.*, 2002). Once the VE is set, it is necessary to develop a joint production plan and, later, schedule (*team action planning*).

The concept of **knowledge sharing**, classification of knowledge into **public, private and semi-private** knowledge in the way as developed in MAS has got a high potential in the field of VEs. The requirements for keeping confidentiality of agents' knowledge and preventing knowledge disclosure as much as applying specific **security principles** used in the MAS area can be re-used in the VE domain as well.

5. LINKS AMONG THE THREE LEVELS

The experts in the three areas mentioned above do feel that their techniques and solutions cover always just a part of the global problem. They tend to extend them towards the "neighboring" areas, developing specific extensions. Let's show that the MAS technologies are general enough to provide adequate unifying theoretical and technological solutions for all the three subareas.

5.1 Extension of holons towards intelligent agents

It is obvious that the holons used for the hard real-time control can be used just for a limited scope of tasks, e.g. for very fast reconfiguration of the control equipment in a pre-determined, "hard-wired" way. To add more intelligence to the holonic decision-making requires application of more knowledge-intensive methods and techniques. Thus, a general architecture combining a low-level RT-control subsystem (function-block or ladder-logic application) with an intelligent agent was presented (Mařík, *et al.*, 2003). A software agent and function block/ladder-logic control application (connected to the physical layer) are encapsulated into a single structure. In such a holon equipped with a higher-level software component, three communication channels should be considered: (i) *intra-holon* communication between the low-level control part and the software agent component, (ii) *inter-holon* communication that is aimed at communication among the agent-based parts of the holons – FIPA standards are used more and more often for this purpose, (iii) a *direct* communication channel between RT-control parts of neighboring holons.

The discussions in the HMS consortium finally lead to the generic architecture of the holonic control devices (HCD Architecture) (Brennan, *et al.*, 2003), in which the higher-level software component is called a *Deliberative Layer*. This layer communicates with *Physical Layer* (actuators and sensors) via the function-block (IEC 61499) oriented *Controls Function Layer*.

The holons defined as mentioned above behave – on the level of inter-holon communication – like standard intelligent agents. They are called **holonic agents** (or agentified holons), as they consist of both the holonic part connected with the physical layer of the manufacturing system (operating in hard real-time) and the software agent for higher-level, soft-real-time intelligent decision-making. It has already been mentioned that the inter-holon communication is usually standardized by the FIPA approach, and direct communication could be achieved by IEC standards (not necessarily IEC 61499). The FIPA standards are not applicable for the low-level real-time control purposes as they don't take account of the real-time control aspects.

The attention of system developers is currently directed mainly at the intra-holon communication, which is usually both application- and company-specific and is usually connected with the solution of the "*migration problem*" (the problem of exploring the classical real-time control hardware for holonic control). McFarlane et al. (McFarlane, *et al.*, 2001) introduced a blackboard system for accomplishing the intra-holon communication, while others (Fletcher and Brennan, 2001) proposed using a special management service interface function block.

The intelligent agent components of holons help to solve the migration problem. The encapsulation approach can be viewed as a bottom-up attempt to extend the capabilities of the low-level, hard real-time control elements by adding some higher-level deliberative software components. This might help in solving certain categories of tasks where the groups of cooperating holons (interacting via the "*cooperation domains*" in the terminology introduced by the HMS consortium) are not too complex and are clearly separated. For more complicated and ambitious tasks to be solved by the holonic approach, a more general, top-down analysis of the control and decision-making specifications is required (Gruver, *et al.*, 2003). In this context, different holonic reference architectures, like PROSA (introducing product holons, resource holons, order and staff holons) (Van Brussel, *et al.*, 1998), HCBA (Holonic Component-Based Architecture introducing resource and product holons) (Chirn and McFarlane, 2000) or MetaMorph (introducing mediator holons using brokering mechanisms to find appropriate holons for establishing collaborative subsystems) (Maturana, *et al.*,, 1999) have been developed to aid the design and development of holonic manufacturing systems.

The PROSA architecture is a typical holonic architecture. The object oriented programming technology (not the MAS technology) is expected to be used for its implementation. The same PROSA philosophy can be used also to design the agent-based solutions on higher levels as it clearly identifies typical agents (holons), like resource agents, staff agents, product agents etc. Considering the product agent as a "standard player" in the communication and negotiation processes opens new possibilities for holonic/agent-based solutions which are still not deployed fully. The same is true for representing a staff by a specific agent type.

The other trends are aimed at stressing the top-down hierarchy of holons in the factory, where each holon can be composed of several holons of the same or different structure (Fischer, *et al.*, 2001). E.g. the factory holon consists of several workshop holons which consist of many machine holons each, etc. Each holon contains several agents which are loosing part of their autonomy entering the holon. But the agents have the right to leave/join holons freely to achieve the optimality of the design in the changing manufacturing environment. This approach transfers the holonic ideas to the *holonic agent-based systems* design.

The more generic holonic ideas and considerations have led to the vision of a **holonic factory** (Shen and Norrie, 1996), (Chirn and McFarlane, 2001), (Shen, *et al.*, 2001). Here all the operations (starting from product ordering, planning, scheduling and manufacturing, to invoicing the customers) are based entirely on holonic principles. A holonic factory contains a group of principal system components (holons) that represent physical manufacturing entities such as machines or products as well as virtual entities like orders. The holons work autonomously and cooperate together in order to achieve the global goals of the factory. Thus, the factory can be managed towards global goals by the activities of individual autonomous holons operating locally.

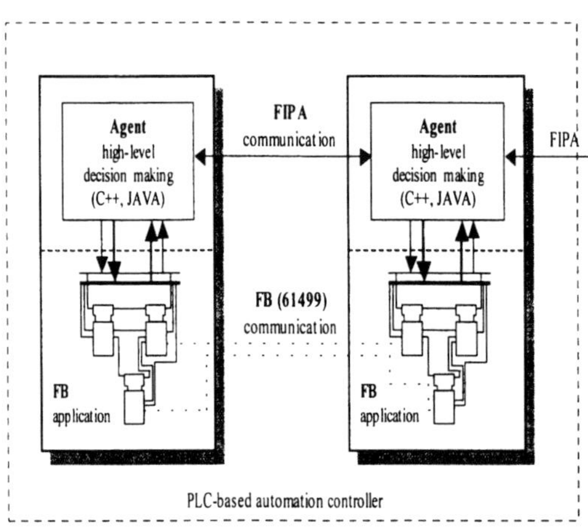

Fig. 1. Holonic agent: combination of function block application and software agent

Only the real-time low-level holonic control systems represent a technology, but with a very limited applicability. The holonic visions above the level of the real-time control have never been fully implemented. Let's remark that "*holonics (understood in the extended way) including the vision of the holonic factory is not a new technology, but it is a system-wide philosophy for developing, configuring, running and managing a manufacturing business. Multi-agent system area is a promising technology provider to accomplish the ambitious holonic visions*" (Mařík, *et al.*, 2002), but maybe in a little bit different, more general and flexible way of CIE.

5.2. Extensions of intelligent agents towards low-level control and virtual enterprises

It seems to be quite desirable to extend the MAS-based decision making on the production management level towards the real-time control. The main problem to be captured is the hard real-time nature of the low-level control. One of the solutions is to locate both the real-time control and the higher-level decision making – deliberative in its nature – on the same controller. The first steps towards such a solution has been proposed e.g. by the Rockwell Automation company (Hall, 2001), where the ladder-logic code encapsulated with the higher agent code (written in C++ or Java) can be run on the same PLC of the Logix family. The exchange of information between the agent and its ladder-logic application can be accomplished via a jointly used data-table (data storage area of the controller). Usually, one controller hosts one or more agents, but not all of them – they have to be distributed in reasonable groups over several controllers and allowed to communicate with each other either within a single controller or among different controllers.

The communication language (JDL - *Job Description Language* encapsulated in FIPA messages in the case of Rockwell solution) is usually simple, without any unnecessary overheads, but efficient enough. From this perspective, compliancy with the FIPA standards is being recognized as a crucial property ensuring the interoperability of holonic agents not only at this lowest real-time control level (allowing e.g. communication of different kinds of holonic agents hosted by PLC controllers from different vendors), but also the interoperability between holonic agents and other agents at higher levels of information processing within the company, e.g. data-mining agents, ERP agents, supply chain management agents and so on.

On the other hand, a direct linkage between the production management level and the supply chain management level seems to be very attractive. Such solutions could e.g. enable to influence the local production planning and scheduling by actual requirements of the customers or by the current capabilities/capacities of the suppliers. The market is usually very turbulent. Some changes in the requirements can be changed on daily/hourly basis.

The direct interactions among the agents responsible for production management (namely for production planning and scheduling) and those representing external bodies (suppliers, customers, cooperating companies etc.) was introduced e.g. within the frame of the ExtraPlant system (Říha, *et al.,* 2002). In this system, the production planning and scheduling (P&S) systems of one manufacturer can be directly linked with the P&S systems of the partners, they communicate among themslves, coordinate their plans and schedules according the current situation. This solution (called "extended enterprise") is a clear step towards the full integration of several bodies into a virtual enterprise using the MAS technology.

5.3. Virtual enterprises seeking for adequate agent-based support

The whole area of the virtual enterprises has been motivated namely by the economic needs to integrate small and medium enterprises into more sigificant production units. The most important enabler for this is the internet communication. Unfortunately, the majority of proposed solutions for technical infrastructures needed for VEs have been usually developed on ad hoc basis. In the latest period, the specialists working in the field are looking for more standardized solutions (Rzevski, *et al.,* 2004). They found the results achieved in the field of MAS applicable in the VE area (Camarinha-Matos and Afsarmanesh, 2001), (Mařík and Pěchouček, 2004). The first trials to used the FIPA compliant platforms, like JADE or FIPA-OS for the cooperation and coordination tasks have been presented. It was shown, that such tasks, like automatic VE formation and dissolution can be solved by using the coalition formation technology developed in the MAS area. Simlarly, simulation of emergent behavior of VEs can be easily simulated by the MAS approach. But, there are still many weak-points of using the MAS technology directly in the VE area.

6. ROLE OF AGENT-BASED SIMULATION IN THE LIFE-CYCLE OF AN AGENT-BASED SYSTEM

The process of developing and implementing an agent-based system relies on several phases and widely explores the simulation prinicples. This is especially true in the case of systems without any central element where unexpected emergent beahvior can appear. The following stages of the design process based on simulation can be gathered like this:

(1) Identification of agents: The design of each agent-based system starts from a thorough analysis of:

- The system to be controlled or manufacturing facility to be deployed.
- The control/manufacturing requirements, constraints, and hardware/software available.

The result of this analysis is the first specification of *agent classes* (types) to be introduced. This specification is based on the application and its ontology knowledge. The obvious design principle is that each device, or each segment of the transportation path or each workcell is represented by an agent. In very up-to-date systems, also the product itself or semi-product can be considered as agent able to negotiate its own processing/assembling with the agents of the manufacturing environment.

(2) Implementation/Instantiation of agent classes from the agent type library. This library is either developed (step 1), or re-used (if already available). Particular agents are created as instances of the definitions in the agent type library. Furthermore, the implementation of communication links among these agent instances is established within the framework of initialization from these generic agent classes (for instance agents are given the names of their partners for cooperation). In such a way, the first prototype of an agent-based control/manufacturing system (or similarly, a supply-chain management system) is designed.

(3) Simulation: Behavior of an agent-based system is not deterministic, but rather emergent: The decision-making knowledge stored locally in the agents invokes the global behavior of the system in a way which cannot be precisely predicted. Yet the direct experimental testing of the global behavior with the physical manufacturing/control environment being involved is not only extremely expensive, but non-realistic as well. Simulation is the only way out. For this purpose, it is necessary to have:

- A good model of the controlled process or the manufacturing facility or the virtual enterprise. This model must depict all the entities within the factory/enterprise and their interfaces to the external world.

- A good simulation tool for running the 'controlled process or manufacturing facility or virtual enterprise' model. Standard discrete event simulation tools like e.g. Arena, Grasp or Silk are not adequate and their usage for this purpose is rather limited.

- A suitable agent runtime environment for modeling the interaction of agents. On the basis of the results of agent platforms comparison, the JADE and the FIPA-OS platforms as open-sources or JACK (Howden, *et al.,* 2001) as a commercial tool can be recommended.

- A suitable simulation environment to model the real-time parts of holons. There are function block emulation tools from Rockwell Automation (Holobloc – see www.holobloc.com) and the modified 4-control platform from the Softing company (www.softing.com). Yet these tools do not adequately handle real-time control problems. A more sophisticated real-time solution would be to use embedded firmware systems like JBED, with its time-based scheduler, to run JAVA objects (that simulate function blocks) and manage events in realistic manner (Zhang et al., 2000). Another approach to such low-level simulation is to encode the function blocks as UML capsules and execute them exploring the Rational RealTime (http://www.rational.com/products/rosert/index.jsp). This solution allows the capsules to be downloaded on specific hardware and run using real-time operating systems like VxWorks.

- System integration strategies developed and implemented in the form of sub-system interfaces. Of particular concern are the interfaces that "glue" together diverse simulation tools and enable the designer to switch from simulation tools to physical manufacturing/control systems or virtual enterprises as appropriate.

- HMI (human-machine interfaces) for all the phases of the system design and simulation.

4. **Implementation of the target control /manufacturing system or virtual enteprise**: In this stage, the target control, manufacturing, production management or supply-chain system is re-implemented into the (real-time) running code. This implementation usually relies on ladder logic, structured text or function blocks at the lowest level of control. However, some parts of the targeted manufacturing systems (such as resource or operation planning subsystems) are often reused as in the phase (3). E.g., in the ExPlanTech production planning MAS system (Říha, *et al.*, 2001), there was 70% of the real code reused from the simulation prototype. Therefore, the choice of the multi-agent platform in the phases (3) and (4) is critical (it is advised to operate with one platform only).

The simulation phase is much more crucial for the development process of agent-based systems than it has been for the development of "classical" centralized systems. It enables besides others:

- to **predict the behavior of the system as a whole**. The fact there is no central unit in the agent-based system represents a critical barrier in a wider applicability of the agent-oriented ideas. The simulation runs help to understand the system behavior and to detect the patterns of emergent behavior. Considering that the behavior of a MAS is emergent, to ensure that all types of possible behavior were explored/covered by simulation still remains a painful problem (the situation is similar to that of system testing).

- to **predict and test the optimal scenario for the agent-based system** development

- to select **the most optimal negotiation framework and strategy** for individual units in the system

- to **link directly the simulation with real-life processes**. That means that whereas a part of the agent-based system is engaged fully in the real-life activities, the remaining part can be just simulated. The shift of the borderline between the simulated (in more precise terms "emulated") and the real part of the system can be carried out in a quite smooth way. This would help to speed-up the initial "commissioning" process significantly.

The requirements on simulation tools or platforms for MAS-oriented solutions require new types of simulation systems (simulation platforms) with embedded MAS principles. One of pioneering systems of this kind is the MAST simulation tool being developed by Rockwell Automation (Vrba and Hrdonka, 2002).

7. AREAS SUITABLE FOR APPLICATION OF THE AGENT-BASED TECHNIQUES

Not all the areas connected with manufacturing and supply chain management are suitable for application of the agent-based approaches. 'Classical" centralized approaches might be more efficient in many situations. The agents can be applied with advantage in such tasks of highly distributed decision-making and control, requiring very high robustness of solutions in the turbulent environment and considering changes/extensions/reconfigurations of the equipment on the fly.

Let's mention the most important industrial areas suitable for application of the agent-based techniques and the relevant agent-oriented techniques they preferably explore:

1. **Flexible real-time control in discrete manufacturing,** especially in the automotive industry. For this or similar types of industries (aimed at mass-production of individually customized products), very variable customization requirements, changes in the plans and schedules, changes in technology as well as equipment failures seem to be quite obvious features of every-day operation. All these requirements and emergency situations can be conveniently handled by the agent technology. The techniques of planning and scheduling and rather reactive system reconfiguration are the most important agent-based technologies used here. The negotiation scenarios are not very complex, meta-reasoning is used in a modest extend.
2. **Intelligent control of highly distributed systems,** namely in the chemical industry and in the area of utility distribution control (distribution of electrical energy, gas, waste water treatment etc.). Many decisions can be done locally, in a very fast way, the communication among the autonomous units is carried out only if really needed. In this area, negotiation scenarios might be more complex, the application of meta-reasoning is highly desirable. The system reconfiguration based on deliberative behavior can be explored with advantage.
3. **Transportation of material / material handling**. The transportation paths (conveyors, pipelines, AGVs etc.) and their sensing and switching elements (diverters, crossings, storages, valves, tag readers, pressure sensors etc.) can be

easily represented by agents, their mutual communication can be defined and organized in a quite natural way. The pioneering testbeds built have documented the viability and efficiency of this approach (Fletcher et al., 2003), (Maturana et al., 2003). Quite a new aspect, which will influence this field significantly, is the RFID tagging technology which enables not only individual identification of the transported pieces, but – in principle - negotiation with them. The other interesting approach allows considering also the individual batches of liquids as agents – in this case, an intelligent exploration of complex pipeline systems can be achieved. The most important MAS techniques applicable in this field are: Planning and scheduling, agent-based simulation, and system reconfiguration.

4. **Production management** in the case of any kind of production. Production planning and scheduling combined with knowledge-intensive negotiation scenarios and agent-based simulations are usually used for this purpose. Both the meta-reasoning in the MAS systems and the (self)reflectivity techniques can be explored (Pěchouček and Norrie, 2001).

5. **Virtual enterprises** represent an extension of the production management tasks towards cooperation with external entities, like supply chain elements, customers etc. They can be seen as a broad potential field for application of the MAS technology. All kinds of MAS techniques (P&S, reconfiguration, negotiation scenarios, meta-reasoning etc.) are expected to be deployed here, but the technologies of the agent-based simulation, coalition formation, handling of private knowledge and secure communication are the most important for this domain (Camarinha-Matos and Afsarmanesh, 2004).

6. **Diagnostics of complex systems** can be considered as an important task. The findings and partial diagnostic conclusions obtained locally can be integrated into global diagnostic decisions. The direct physical linkage of the control and diagnostic hardware and software is a quite natural trend in building agent-based systems: In this way, certain local diagnostic conclusions can be explored locally, by the control algorithm of the same agent. The agent-based diagnostics of complex systems exlpores mainly efficient agentification and system integration techniques in combination with complex negotiation and meta-reasoning procedures.

7. **System integration of legacy systems** is another important application field. The appropriate agentification of any hardware or software module can strongly support the system integration of any kind. The techniques of agentification and system integration are widely used in this domain and do represent a key approach for migration from centralized systems to highly distributed (agent-based) solutions in industry.

8. INDUSTRIAL APPLICATIONS OF THE AGENT-BASED SOLUTIONS

Let's mention several **successful applications of agent-based control and diagnostics:**

Semi-real operation of a small **production line at Daimler Chrysler**, Stuttgart, Germany, demonstrated very high flexibility, increased throughput, robustness and reliability of the agent-based manufacturing facility (Bussmann and Sieverding, 2001). Increased investment costs declared as well.

Agent-based real-time control of **steel production line at BHP, Melbourne, Australia, developed by Rockwell Automation.**

Reconfigurable Shipboard Automation project carried out jointly by the Rockwell Automation and Rockwell Scientific Company resulted in a successful demo of the cooling system of a ship (Maturana et al., 2003). Full-size implementation expected.

Joint **testbed demo developed by CDAC, Univ. of Cambridge and Unilever comp.,** aimed at material handling, storage and package of discrete units. Very successful and instructive (Fletcher et al., 2003).

Agent-based production planning and scheduling systems for manufacturing of components and assembling in the automotive industry (LIAZ Pattern Shop and Gedas for VW-Skoda) as well as for the supply chain management (Chatzapoulos, Greece) developed by the Gerstner Lab CTU and CertiCon (Říha et al., 2001), (Říha et al., 2002) .

Two of them will be described in more detail.

8.1. Reconfigurable Shipboard Automation

The fundamental requirements for the shipboard automation have been aimed at flexible distributed control of a ship equipment to enable reduced manning and reliable and survivable operations of a ship using the commercial-off-the-shelf products. The shipboard automation architecture is divided into three levels (Tichý, *et al.*, 2002): *The Ship-level* is concerned with overall goals of the ship and communicates directly with the ship crew. *The Process-level* is aimed at optimizing the performance of the automation components and ensuring available services. *The Machine-level*, the lowest one in the hierarchy, is responsible for the real-time control, diagnostics, and system reconfiguration.

The first deployment of the shipboard automation architecture has been focused upon the control of a chilled water system.

To solve the technology migration problem, the firmware of the "classical" automation controllers (Logix family controllers) has been extended to enable creation of intelligent agents and constitutes the Distributed Control Agent (DCA) infrastructure. The DCA infrastructure glues the automation components via synchronized multithreads. Several agents (implemented in C++ code) can run on a single Logix controller.

The agent architecture consists of four main components: planner, equipment model, execution control module, and diagnostic module (see Fig.2). The planner reasons about the plans and events emerging in the physical domain and represents the core of the agent. The equipment model provides the decision-making support by describing and evaluating the physical system configurations. The execution control module acts as a control proxy and translates committed plans into execution control actions. The diagnostic part is responsible for detection of local events in the physical systems and evaluates them according to a model. The diagnostic component of an agent may include a suite of data acquisition, signal processing, diagnostic, and prognostic algorithms. The diagnostic information gathered and evaluated locally can be accessed by any agent in the community. In such a way, the global diagnostic of the system based on distributed diagnostic approach can be achieved.

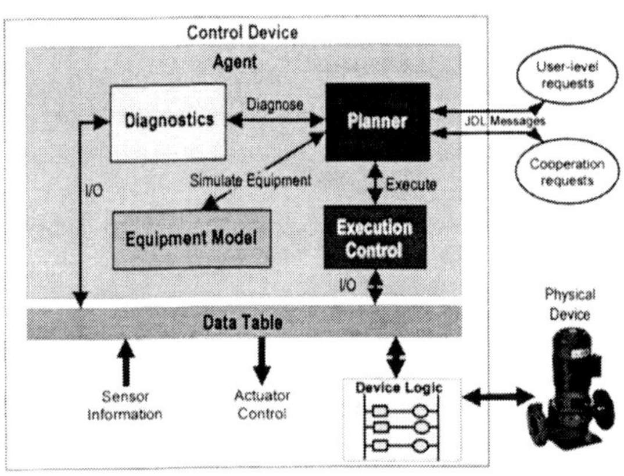

Fig. 2. Shipboard Automation Agent Architecture

Any of the agents can proactively start to plan an action. The planning process is carried out in three phases: Creation, Commitment, and Execution. During the Creation phase, the agent recognizes the current situation, creates a plan template, and possibly negotiates (using the Contract-Net-Protocol) with other agents to cover the whole plan template. In the Commitment phase, the agents commit their resources to fulfil the task. In the last, Execution phase, the plan is executed by all participating agents.

The agents communicate in all the three phases of their activities using a Job Description Language (JDL) as a content language of FIPA-ACL. Any intelligent agent can contact any device in the architecture using a JDL message. The FIPA-compliant part of the architecture explores the FIPA standard techniques to locate the agents. It includes a full implementation of Directory Facilitator (DF) that provides matchmaking (to locate agents by capabilities) and the Agent Management Services (AMS) functionality (to locate agents by addresses). Moreover, the architecture supports multiple DF agents – this enabled to develop the Dynamic Hierarchical Teams architecture to increase the robustness of the system (Maturana, *et al.*, 2003). This architecture ensures availability of matchmaking service despite predefined number of failures since the structure has vertex and edge connectivity equal to size of teams.

A special software tool, called the Development Environment (DE) helping the user to develop the distributed application has been developed. A very important part of this environment is the library of components called the Template Library (TL). The library is editable by the user, and each template of an agent can contain both the low-level control behavior (in the form of a ladder diagram) and the higher-level intelligent behavior (written in JDL). Because of the "object" view of the components in TL, the inheritance is strongly supported.

The capabilities and efficiency of the shipboard automation solution was demonstrated on two vast testbeds. The implementation of the architecture as a commercially available product is in progress.

8.2. ExPlanTech Production Planning System

ExPlanTech (Říha, *et al.*, 2002) is one of the systems of the ProPlanT family of planning and scheduling systems developed by the Gerstner Lab, Czech Technical University. It is focused upon the production planning and scheduling tasks in a factory with a possibility to influence the developed schedules "from outside" (that means by customers or suppliers). It gives advises and views on the production how it could look in the future if some order is going to be accepted, prioritized or refused. It provides the information about workshops' loads, creates plans with respect to deadlines, priorities and precedence dependencies of orders. The ExPlanTech system was built upon the already existing, legacy ERP system of the LIAZ Pattern Shop Ltd., one of the leading European producers of patterns and forms mainly for automotive industry.

There are five types of agents creating the community of agents in ExPlanTech, namely: a configurator agent (CA), a scheduler agent (SA), a database agent (DBA), a resource agent (RA) and a monitor agent (MA). All of them are equipped with a simplified 3bA acquaintance model for storing and maintaining the social knowledge.

The orders are stored in the ERP and are presented to the community via the DBA. The schedules developed by the agent community are being transferred back via the DBA to the ERP later. Simply, the database agent DBA represents a bridge between the ERP system and the agent community: It translates the information from the ERP recorded in the XML format and received in the form of the TCP/IP messages into the FIPA compliant messages and vice versa. This solution has been chosen as a simpler one than those based on developing the appropriate agent wrapper for the ERP.

Each workshop is represented by an individual scheduler agent SA. This is responsible for creating schedules for the workshop and for coordinating them – via negotiations - with the other agents in the community. The CA agent collects the information on general production schedules in relative time as well as on specific plans in absolute time. The acquaintance model of this agent "owns" rich information about the workshops (their load, gaps in plans) and tasks (deadlines of orders etc.). It uses these pieces of information to optimally decompose the tasks and to make preliminarily optimal task allocation (addressing specific requests on carrying out particular parts of the plan to the most suitable SAs). It uses the social knowledge to facilitate and speed-up the process of negotiation and to reduce the communication traffic.

Both the RA and MA types of agents create a group of extra-enterprise agents as they enable to communicate and negotiate on behalf of information sources outside the plant: The resource agents serves as an interface between an enterprise supplier or cooperator (in the case of outsourcing). A partner enterprise can announce the status of services and resources currently available. This shared knowledge should be expressed in a standard form with respect to the agreed ontology. The RA can, e.g. read the data from the partner's ERP to provide the ExPlanTech community with the latest data for computation of different parameters (delivery dates, prices, amounts). The ExPlanTech can contract a cooperator using the RA in the cases when outsourcing is the optimal way how to proceed.

The MA agent is going to serve customers of the factory who can observe the results of scheduling using an Internet browser. According to his/her access right the customer is able to trace his/her orders, watch their statuses and, eventually to invoke a new re-scheduling when changing some parameters of his/her own orders (changing the priority, deadlines etc.). The same agent can be used by the managers of the plant while they are off-site: Unlike the customers they can be provided with the information on loads of the workshops and their future plans.

The ExPlanTech system has been written in JADE and is used in the company since 2002.

9. ADVANTAGES AND OBSTACLES OF AGENT-BASED SOLUTIONS

Why the agent approach and the MAS technology seem to be so important for the area of manufacturing and supply chain management? What are the reasons and advantages of applying them? What they really do bring? Let's try to summarize the current experience. The main advantages can be summarized like this:

a) **Robustness and flexibility of the control/diagnostic systems**:

- Robustness is achieved mainly due to the fact that there is **no central element**, no centralized decision making. Any loss of any subsystem cannot cause any fatal failure of any other subsystem (from the control and decision making points of view).
- The agent technology enables to handle the problems of **production technology failures** in a very efficient fashion. Optimal reconfiguration of the available equipment (which remains in operation) can be carried out in a very fast way. Thus sustainable continuation of the production task or operation or safety stopping of the manufacturing process can be achieved (if the physical constraints given by the real plant facility would allow this, and this might be not always the case).
- **Changes in the production facility** (adding a machine, deleting a transportation path etc.) can be handled on the fly, without any need in re-programming of the software system as a whole. Just couple of messages are exchanged, and the agents are aware of the change and behave accordingly.
- **Changes in production plan or schedule** can be handled easily, without the need in stopping the process or bringing it back to some of the initial states. The changes in the production plan or schedule can be handled in parallel to solving the tasks connected with changes in the the facility equipment and/or failures.

b) **Plug-and-operate approach** is strongly supported. This enables to change/add/delete the hardware equipment as well as software modules on the fly. The migration process from the old to the new technology can be carried out smoothly, without any need in stopping the operation. This makes also the system maintenance costs significantly cheaper.

c) Control and diagnostics are carried out as **near to the physical processes** as possible, control and diagnostic subsystem can cooperate on the lowest level (and in a much faster way). **Control and diagnostics** can be really fully **integrated**.

d) The **same agent-based philosophy can be used on different levels,** in different subsystems of the manufacturing facility and company. The same MAS principles and techniques can be e.g. applied on the hard-real-time level (holonic control), soft-real-time control, strategic decision making for control tasks, for integrated diagnostics or diagnostics running as a separate process aside of control, for production planning and scheduling, for higher-level decision making on the company level, for supply-chain management as well as for purposes of virtual enterprises (viewed as coalitions of cooperating companies). Being even the same communication standards and negotiation scenarios used across all the tasks mentioned above, a very high efficiency resulting from automatic communication and negotiation between units on different levels and located in different subsystems (like e.g. direct negotiation between a controller or a storage manager agent and the sales department of a partner in the virtual enterprise) can be obtained. This fact can lead to the vision of a **Cross-integrated enterprise (CIE).** The idea of CIE can be viewed as an extension of **a holonic factory** (Shen and Norrie, 1996), as a shift from a vision towards implementation, towards exploration of the recent results of the MAS technologies as a unifying background.

e) When the same philosophy, stemming from the MAS approach is used across the manufacturing and business process from the individual machines up to the virtual enterprise as a whole (in the horizontal direction) (Glanzer *et al.*, 2003), **new interesting potential interactions do appear**: E.g. the holonic agents from different companies can talk to each other asking for changes of some manufacturing parameters, the business department of one company can send data directly to the scheduling mechanism of a specific workshop belonging to a partner's company without the need to inform the managerial level, the manufacturing holons can directly warn the purchase/sales department avoiding classical hierarchical communication channels in the plant etc.

There have been many agent-based experimental solutions proposed, developed and in some cases implemented. **But, the real-life industrial applications/prototypes in use are very, very rare and usually with a very restricted functionalities of agents. Why it is so?**

There can be several crucial disadvantages of the current industrial agent-based solutions easily identified:

a) **The investments** needed to implement the agents-based manufacturing system **are much higher** than those for "classical", centralized solutions. Unfortunately, the available flexibility, which is the pay-off for these expenses, is usually so enormous, that the manufacturing process can leverage just very small portion of it (Bussmann, 2003).

b) As there is no central control element present (in an ideal agent-based factory), the unpredictable, **emergent behavior** can be expected. This causes a lot of obstacles for the agent-based solutions to be easily accepted by the company management. The only way out seems to be a very thorough **simulation of the agent-community behavior**. From the authors' own experience, the simulation detects number of patterns of emergent behavior. The "dangerous" patterns can be avoided by introducing appropriate policies and meta-agents.

c) All the current prototypes or applications work with tens or hundreds of agents as a maximum number. The current platforms are not able to handle the **scalability problem** – the problem of running/communication of thousands agents in parallel with the robustness required (and needed) by industry. This is simply limitation given by the current capabilities of the "industrial IT" technology.

d) The current control systems offered by all the important vendors support the centralized control solutions, only. The **migration towards autonomous, independent controllers,** communicating asynchronously (when needed) among themselves in the peer-to-peer way seems to be the necessary technology enabler for wider application of the agent-based solutions. The migration problem is not being solved yet.

e) Nearly all the community **of control and system engineers has been educated to design,** run and maintain strictly **centralized solutions.** This is quite serious obstacle, as the engineers with the "classical", centralized approach (presented in the last three decades under the CIM label) are really not ready and able to support the agent-based solutions. Much more educational efforts will be needed to overcome this serious hurdle.

f) **Standards and platforms** needed to achieve the required and needed interoperability are **in the early stage of their development.** FIPA **standardization** efforts demonstrate one possible way how to achieve the full interoperability. **The FIPA standards developed and approved yet are not ready to support the interoperability in the area of real-time control, distributed control, diagnostics, production management and virtual enterprises**. But, a very intensive cooperation with the FIPA organization might speed-up the process of introducing interoperability principles for industrial systems.

g) More intensive exploration of knowledge seems to be the key technology driver behind the development of industrial MAS-based systems. The real-life industrial applications would require **adequate knowledge capture and management methods** - the extent of the required knowledge handling exceeds the capabilities of knowledge handling by the current MAS systems by several orders. The current knowledge management approaches are just toy examples from the point of view of industrial needs and they one of the most serious obstacles to system scalability.

h) The current experience concerning the **knowledge ontology** representation, maintenance and exploration in MAS just helps to understand how important the common shared ontologies will be for the MAS solutions. The ontologies used in industrial applications are usually very simple, just hierarchical structures of concepts and entities in an XML format. More sophisticated ontology structures, exploring e.g. the DAML+OIL languages are still missing. Simply, **the ontologies in the MAS area are not developed enough to provide a direct support to the industrial solutions.**

i) The **algorithms and metrics evaluating efficiency of the agent-based solution** are still missing. An overview of available metrics (Goranson, 1999) documents that significant development in this direction is needed.

j) Capabilities to **reconfigure and optimize the agent-based system performance** in the case of a failure or partial damage and capability to set-up the new optimization criteria autonomously should become quite natural for industrial systems. The first practical experience in solving similar tasks of system reconfiguration has been gathered in the field of holonic manufacturing (Maturana et al., 2003). For the purpose of self-adaptation and performance improvements in the everchanging environment, the principles of self-understanding, self-consciousness and self-reflection should be intensively researched.

k) Not all the tasks can be solved by the agent-based or MAS approach (the estimates talk about 30% of the control tasks and 60% of the diagnostic tasks to be suitable for application of the agent-based techniques). But, certain areas with a higher degree of applicability of the MAS technology have been identified already (see above). In general, **applying the agent-based technology in inappropriate tasks can lead to frustration.** Identifying clear cut-criteria for suitability of the MAS approach for specific, particular industrial tasks would be highly appreciated.

10. CONCLUSIONS

Both the agent-based visions and the MAS technologies for manufacturing, production management, supply chain management and virtual enterprises is developing in a really fast way. This development trend tightly follows the current trend in multi-agent systems research in the field of Artificial Intelligence as well as all the recommendations of the FIPA standardization consortium. But, a long way is still in front of us: It is necessary e.g. (i) to change the way of thinking of industrial designers and engineers of control systems, (ii) to document the reliability and manageability of the emergent behavior of the MAS systems (for this purpose much more robust simulation tools and clear evaluation criteria should be developed), (iii) to support the migration processes from the centralized to agent-based control which concern both the hardware and software, (iv) to solve the technical problems of interoperability, communication and negotiation among the agents and (v) to solve the scalability problem and (vi) to work towards a widely acceptable ontology structures and languages.

REFERENCES

Bradshaw, J.M. et al. (2003). Making Agents Acceptable to People. In: *Handbook of Intelligent Information Technology*. IOS Press, Amsterdam.

Brennan, R.W., Hall, K., Mařík, V., Maturana, F.P., and Norrie, D.H. (2003). A Real-Time Interface for Holonic Control Devices. In: *Holonic and Multi-Agent Systems for Manufacturing,* LNAI No.2744, 25-34. Springer Verlag, Heidelberg.

Bussmann, S. (2003). Success and Failures of Agent/Holonic Control Technology at Daimler Chrysler. *Invited talk at the HoloMAS'03 Int. Conference*. Prague.

Bussmann, S., and Sieverding, J. (2001): Holonic Control of an Engine Assembly Plant – An Industrial Evaluation. In: *Proc. of IEEE Conf. on SMC*. Tucson, AZ.

Cao, W., Bian, C.G., and Hartvigsen, G. (1997). Achieving Efficient Cooperation in a Multi-Agent System: The Twin-base Modelling. In: *Cooperative Information Agents,* LNAI No 1202, 210-221. Springer Verlag, Heidelberg.

Camarinha-Matos, L.M., and Afsarmanseh, H. (2001). Virtual Enterprise Modeling and Support Infrastructures: Applying Multi-agent System Approach. In: *Multi-Agent Systems and Applications,* 335-364, LNAI No. 2086. Springer Verlag, Heidelberg.

Camarinha-Matos, L.M., and Afsarmanesh, H. (eds.) (2004): *Collaborative Networked Organizations.* Kluwer Academic Publishers, Boston, MA.

Deen, S.M. (ed.) (2003). *Agent Based Manufacturing: Advances in the Holonic Approach*. Springer Verlag, Heidelberg.

Filipe, J., and Cordeiro, J. (2004). Organizational Semiotics: A Normative Agent-Based Approach to VE Modelling. In: *Processes and Foundations for Virtual Organizations,* 271-278. Kluwer Academic Publishers, Boston, MA.

Finin, T., Fritzson, R., McKay, D., and McEntire, R.: KQML as an Agent Communication Language. In: *Proc. of 3rd Int. Conf. on Information and Knowledge Management CIKM'94*. ACM Press.

Fischer, K., Funk, P., and Russ, Ch. (2001). Specialised Agent Applications. In: *Multi-Agent Systems and Applications,* 365-382, LNAI No. 2086. Springer Verlag, Heidelberg.

Fletcher, M., and Brennan, R.W. (2001). Designing a Holonic Control System with IEC 61499 Function Blocks. In: *Proceedings of the International Conference on Intelligent Modeling and Control*. Las Vegas.

Fletcher, M., McFarlane, D., Thorne, A., Jarvis, D., and Lucas, A. (2003): Evaluating a Holonic Packing Cell. In: In: *Holonic and Multi-Agent Systems for Manufacturing,* LNAI No.2744, 246-257. Springer Verlag, Heidelberg.

Glanzer, K., Schmidt, T., Wippel, G., and Dutzler, Ch. (2003). Integration of Shop Floor Holons with Automated Business Processes. In: *Holonic and Multi-Agent Systems for Manufacturing,* LNAI No.2744, 145-155. Springer Verlag, Heidelberg.

Goranson, H.T.(1999). *The Agile Virtual Enteprise: Cases, Metrics, Tools.* Greenwood Publ. Group. Greewood.

Gruver, W.A., Kotak, D.B., van Leeuwen, E.H., and Norrie, D.H. (2003). Holonic Manufacturing Systems: Phase II. In: *Holonic and Multi-Agent Systems for Manufacturing,* LNAI No.2744, 1-14. Springer Verlag, Heidelberg.

Hall, K. (2001). Agent Infrastructure for Use by HMS Consortium. *Technical Report of Rockwell Automation,* www. rockwell.com.

Hendler, J. (2002). Ontologies on the Semantic Web. *IEEE Intelligent Systems.* March/April 2002.

Howden, N. et al. (2001). JACK Intelligent Agents – Summary of an Agent Infrastructure. In: *Proc. of IEEE Int. Conf. on Autonomous Agents.* Montreal.

Chirn, J.L., and McFarlane, D.C. (2000). A Holonic Component-Based Approach to Reconfigurable Manufacturing Control Architecture. In:*Proc. of 11th Int. Workshop on the Database and Expert Systems Applications.* IEEE Computer Society Press, Los Alamitos.

Chirn, J.L., and McFarlane, D.C. (2001). Building Holonic Systems in Today's Factories: a Migration Strategy. *Journal of Applied Systems Studies,* **2**(1).

Christensen, J.H. (1994). Holonic Manufacturing Systems: Initial Architecture and Standards Directions. In: *Proc. of 1st European Conference on Holonic Manufacturing Systems,* 1-20. Hannover.

Luck, M., McBurney, P., and Preist, C. (2001). Agent Technology: Enabling Next Generation Computing. A Roadmap for Agent-Based Computing. http://www.agentlink.org/roadmap.

Mařík, V., Fletcher, M., and Pěchouček, M. (2002). Holons & Agents: Recent Developments and Mutual Impacts. In: *Multi-Agent Systems and Applications II,* LNAI No. 2322, 233-267. Springer Verlag, Heidelberg.

Mařík, V., Pěchouček, M., and Štěpánková, O. (2001). Social Knowledge in Multi-Agent Systems, In: *Multi-Agent Systems and Applications,* LNAI 2086, 211-245. Springer Verlag, Heidelberg.

Mařík, V., Pěchouček, M., Vrba, P., and Hrdonka. V. (2003). FIPA Standards and Holonic Manufacturing. In: *Agent Based Manufacturing: Advances in the Holonic Approach,* Deen, S.M. (ed), 89-121. Springer Verlag, Heidelberg.

Mařík, V., and Pěchouček, M. (2004). Agent Technology for Virtual Organizations. In: *Processes and Foundations for Virtual Organizations,* 243-252. Kluwer Academic Publishers, Boston.

Maturana, F., Shen, W., and Norrie, D. (1999). MetaMorph: An Adaptive Agent-Based Architecture for Intelligent Manufacturing. *International Journal of Production Research,* **37** (10), 2159-2174.

Maturana, F., Tichý, P., Šlechta, P., Staron, R.J., Discenzo, F.M., Hall, K., and Mařík, V. (2003). Cost-based Dynamic Reconfiguration System for Evolving Hierarchies. In: *Holonic and Multi-Agent Systems for Manufacturing,* LNAI No. 2744, 310-320. Springer Verlag, Heidelberg.

McFarlane, D.C., Kollingbaum, M., Matson, J., and Valckenaers, P. (2001). Development of Algorithms for Agent-based Control of Manufacturing Flow Shops. In: *Proceedings of the IEEE International Conference on Systems, Man and Cybernetics.* Tucson, AZ.

Pěchouček, M., and Norrie, D. (2001). *Knowledge Structures for Reflective Multi-Agent Systems: On Reasoning about Other Agents.* Research Report 538, Dept. of Mechanical and Manufacturing Engineering, University of Calgary. Calgary.

Pěchouček, M., Mařík, V., and Bárta, J. (2002). A Knowledge-based Approach to Coalition Formation. *IEEE Intelligent Systems,* **vol. 17, no.3,** 17-25.

Pěchouček, M., Štěpánková, O., Mařík, V., and Bárta, J. (2003). Abstract Architecture for Meta-reasoning in Multi-agent Systems. In: *Multi-Agent Systems and Applications III.,* LNAI No. 2691, 85-99. Springer Verlag, Heidelberg.

Rao, A.S., and Georgeff, M.P. (1992). An Abstract Architecture for Rational Agents. In: *Proceedings of Knowledge Representation and Reasoning KR&R-9,* 439–449.

Rzevski, G., Skobelev, P., Batischchev, S., and Orlov, A. (2004). A Framework for Multi-agent Modelling of Virtual Organisations. In: *Processes and Foundations for Virtual Organizations,* 253-260. Kluwer Academic Publishers, Boston.

Říha, A., Pěchouček, M., Krautwurmová, H., Charvát, P., Koumpis, A. (2001). Adoption of an Agent-Based Production Planning Technology in the Manufacturing Industry. In: *Proceedings of the 12th International Workshop on Database and Expert Systems Applications,* 640-646. IEEE Computer Society, Los Alamitos.

Říha, A., Pěchouček, M., Vokřínek, J., and Mařík, V. (2002). ExPlanTech: Exploitation of Agent-based Technology in Production Planning. In: *Multi-Agent Systems and Applications II,* LNAI No. 2322, 308-322. Springer Verlag, Heidelberg.

Searle, J.R. (1969). *Speech Acts.* Cambridge University Press, Cambridge, MA.

Shehory, O., and Kraus, S. (1998). Methods for Task Allocation via Agent Coalition Formation. *Artificial Intelligence,* **101 (1-2),** 165-200.

Shen, W. and Norrie, D. (1996). Agent-based Systems for Intelligent Manufacturing: A State-of-the-art Survey. *AI Journal,* **101 (1-2),** 165-200.

Shen, W., Norrie, D.H., and Barthes, J.A.(2001). *Multi-Agent Systems for Concurrent Intelligent*

Design and Manufacturing. Taylor and Francis, London.

Tichý, P., Šlechta, P., Maturana, F., and Balasubramanian, S. (2002). Industrial MAS for Planning and Control. In: *Multi-Agent Systems and Applications II,* LNAI No. 2322, 280-295. Springer Verlag, Heidelberg.

Tožička J., Bárta J., and Pěchouček, M. (2003). Meta-Reasoning for Agents' Private Knowledge Detection. In: *Cooperative Information Agents VII*, LNAI 2782, 316-323. Springer Verlag, Heidelberg.

Van Brussel, H., Wyns, J., Valckenaers, P., Bongaerts, L., and Peeters, P. (1998). Reference Architecture for Holonic Manufacturing Systems: PROSA. *Computer and Industry*, **37** (3), 255-274.

van Leeuwen, E.H., and Norrie, D. (1997). Intelligent Manufacturing: Holons and Holarchies. *Manufacturing Engineer*, **76**(2), 86–88.

Vrba, P., and Hrdonka, V. (2002). Material Handling Problem: FIPA Compliant Agent Implementation. In: *Multi-Agent Systems and Applications II*, LNAI 2322, 268-279. Springer Verlag, Heidelberg.

Vrba, P. (2003). Agent Platforms Evaluation. In: *Holonic and Multi-Agent Systems for Manufacturing,* LNAI No.2744, 47-58. Springer Verlag, Heidelberg.

Weiss, G., ed. (1999). *Multi-Agent Systems: Modern Approach to Distributed Artificial Intelligence.* MIT Press, Cambridge, MA.

Zhang, X., Balasubramanian, S., Brennan, R.W., and Norrie, D.H. (2000): Design and Implementation of a Real-time Holonic Control System for Manufacturing. *Journal of Information Sciences*, **127** (1-2), 23-24.

Copyright © IFAC Information Control Problems in
Manufacturing, Salvador, Brazil, 2004

ELSEVIER
IFAC
PUBLICATIONS
www.elsevier.com/locate/ifac

HOW INFORMATION TECHNOLOGY ENABLES THE POTENTIAL FLEXIBILITY AND RE-CONFIGURABILITY OF HIGHLY AUTOMATED MANUFACTURING PLANTS

Giuseppe Fogliazza[1]

[1] *Machining Center Manufacturing S.p.A, Via Celaschi 19, 29020 Vigolzone (PC), Italy,*
g.fogliazza@mcmspa.it

Abstract: This paper will present an industrial approach to the realization of flexible manufacturing systems, based on the strong integration of mechanical design flexible automation and software engineering. Information technology provided a powerful analogy to shape the behavior of such manufacturing systems; from the fields of operating systems we get a strong base to design and implement their control system, and finally from software engineering methodology we derive a new approach to develop manufacturing systems, capable to face new challenges in the competitive global market. *Copyright © 2004 IFAC*

Keywords: Manufacturing Systems, Industrial Automation, Supervisory control, Object-oriented modeling, Control System design

1 INTRODUCTION

Despite the generality of the title we will focus our attention on a very specific area of manufacturing. We will present manufacturing systems built around horizontal spindle machining centers. The hope is to present considerations about their architecture, expected behaviour, and design methods, that could be easily applied to manufacturing systems in general.

Machining centers are flexible production means in the sense that they can switch from a production task to another with zero or very limited setup. So they can be successfully applied in production environments characterized by high mix variability and low volumes.

Figure 1: Horizontal spindle machining center.

From the mechanical point of view, a machining center is composed of a work area with a spindle, three or four working axes and a rotating work table. Flexible automation provides devices for the movement and storage of pallets and tools

Main components of their control system are typically PLC, distributed I/O, numerical control, terminals, dedicated microprocessors.

Using flexible automation we can arrange machining centers in various system configurations from single machine multi-pallet cell (Figure 1), to parallel machines sharing pallet and tool storage like in the following figure.

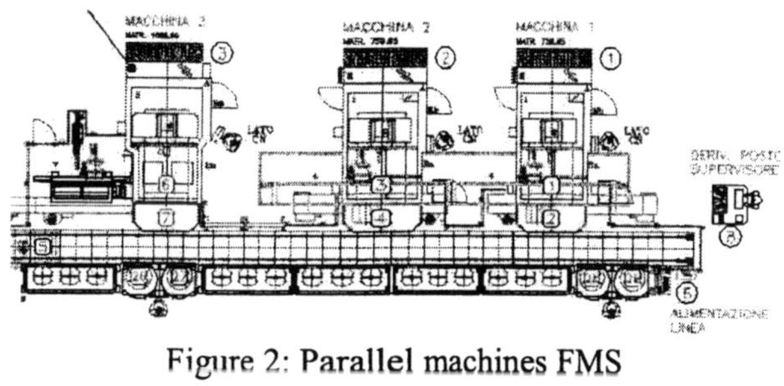

Figure 2: Parallel machines FMS

In the 80's flexible manufacturing systems and their proposed vision of a complete work shop automation, promised a revolution of machining departments. Very rapidly this vision has been disrupted, fundamentally by the incapacity of existing FMSs to manage their own complexity. This led to a very slow return of huge investments and to a reduced acceptance by workshop's operators kept outside the flexible automation vision, and considered only as versatile robots.

We learned that automation alone is not the correct answer. End users do not need automated machinery but production systems. Automation potential need to be supported by management services to reduce the perceived complexity of the manufacturing system and to cope with frequent unexpected events that take place inside workshops.

Workshops, and specially machining workshops, represent a challenging environment. Tool breaks, delayed delivery of rough material, quality check failures, frequently result in a production stop for one or more specific components. This turbulence influences heavily the productive performances of traditional machine tools, requiring costly setups to move production from different part types.

Unfortunately the workshop's turbulence easily disrupts also the carefully planned behavior of poorly managed FMSs.

Software developed to support FMS operations was frequently derived from traditional industrial information technology, strong in accounting, production tracing and performance analysis, and weak to deal with real-time control and specific automation management issues.

2 PRODUCING IN A CONCURRENT ENVIRONMENT

One potential major advantage of flexible automated machine tools is their capacity to move from one part type to another without setups. This flexibility could keep stable the productivity ratio in a turbulent environment. Traditional tradeoff between flexibility and efficiency could be changed.

In order to fulfill the promises of flexible automation there was the need to develop a production environment providing new specific powerful information management services strongly integrated with real-time control.

The original idea that started the approach presented in this paper was to develop a concurrent production environment.

Concurrency is the potential parallel execution of different actions. In a concurrent environment, different production tasks are kept alive in an intermediate execution status. The capability of flexible automation and machining centers to eliminate setup costs is the fundamental prerequisite to realize a concurrent environment managing a huge number of different production tasks, active at the same time.

From the field of operating systems we know that it is possible to improve the overall performance of a processor by executing several processes concurrently instead of sequentially. When a process stops waiting for external resources the processor is kept busy by another process waiting to enter in execution.

With a simple analogy between processors and machine tools, it is easy to see that we can improve overall throughput of machining centers executing more different production tasks in parallel instead of sequentializing them in a fixed schedule. When a task needs to be suspended for problems related to tools, material or quality, the machine is kept busy producing a different part type.

In a concurrent production environment the higher is the production flexibility in term of number of different part types contemporary active and the higher is the execution efficiency.

In order to realize a concurrent production environment we devoted our attention to a frequently neglected aspect of a manufacturing system: supervisory control. Normally, this level of control is added later in the project, keeping its distance from real-time control issues, and providing just accounting, logging and evaluation of performance ratios. The goal to build a concurrent production environment and to deliver manufacturing systems, not just machinery, required a newly designed supervisory software, strongly integrated with lower control levels and shaping the behavior of mechanical devices.

Concurrency has been exploited for a long time in processing system thanks to time sharing operating systems.

A parallel between processing systems and manufacturing systems shows many strong analogies:

Processing Systems	***Manufacturing Systems***
Hardware	Physical devices
Operating System	Supervisory Control
Application Program	Production plan
Processor scheduling	Dispatching rules
Virtual Memory Mgmt.	DNC

Table 1: Analogies between processing and manufacturing systems

Based on this analogy, we pursued the vision of a supervisory control developed as a time-sharing operating system for flexible manufacturing plants.

3 OPERATING SYSTEMS FOR PRODUCTION PLANTS

Software services to support operations and to control flexible automation need to be built like operating systems. In fact, continuing the analogy started in the previous section, they control the execution of production programs, while regulating access to physical resources (clamping fixtures, tools, machines, transport devices) and information entities (part programs, tool descriptions, production orders).

Even if it is attractive the idea is not particularly innovative. (Nof, 1980) introduced the term of Production Operating System and Garetti developed a complete programming language to describe production jobs (Garetti, 1985).

Beyond analogies operating systems provide a rich methodological framework and an integrated approach to develop complex hardware/software systems (Vanneschi, 1984).

Structured approach to the realization of an operating system for a production plant is achieved through different level of abstraction from the physical layer to high level services.

This approach is obviously indipendent from system dimension. Even a single machining center equipped with pallet and tool storage as shown in fig.1, could be configured as a flexible manufacturing system.

Applying structured design, the funcionality of a complex system can be organized in a hierarchy of levels, characterized by a set of visible resources and mechanisms operating on them. At the lowest level there is the physical plant, whose functionality and resources are abstracted (or virtualized) proceeding toward higher levels.

Levels	*Resources*
Production	Manufacturing plant
Supervisor	Machines, transporter
Connected machine	Virtual devices
Real-Time control	Sensors and Actuators
Hardware	Physical Components

Table 2: Structured design hierarchical levels

Resource management policies defined at a certain level are implemented using mechanisms provided by inferior levels. We can see each level as a virtual machine providing a language to write programs, and the run-time support of this language. Structures of the language are implemented as policies written in a lower level language and executed by the lower level virtual machine. In structured design two principles define the vertical organization:

- Entities at a certain level are directly accessed only through the language's mechanisms available at the same level.
- Each level does not know policies used by inferior levels to implement its own run-time support.

Horizontal organization of a level use the concept of abstract data types, that is entities implementing data structures accessible only through well-defined interfaces. This concept can be used to model both active and passive entities. Active entities will be denoted as modules.

A module is an instance of an active abstract data type autonomous and sequential :

- Autonomous in taking decisions on its internal processing flow and on interactions with other modules.
- Sequential is the control algorithm it executes.

Taking this concept in place we can re-define an abstraction level as a set of cooperating modules encapsulating resources directly visible programming the system at that level and an interaction structure to support modules cooperation.

One of the key property a manufacturing systems and its operating system have to exhibit is robustness, that is operability under malfunctions

Manufacturing systems need to operate in a predictable way with proper tools to handle any disturbance from rush orders to human errors to machine failures. Despite their size and complexity they have to be continuously available. This requires overall reliability and maintainability, but also some degree of fault tolerance.

If reliability is a measure of compliance with specification, fault tolerance is the capacity of a system to avoid failures in presence of faults. Fault tolerance requires a certain degree of redundancy in the systems. For example there is the need to add specific funcionalities:

- to detect errors, that is transitions not in compliance with specification
- to localize faults
- to re-configure the system to insulate and to substitute the incorrect components
- to recover the system to a correct status.

Design methodologies for robust systems are all related to the application of 4 general principles, known as Denning's principles from the name of the researcher who formalized them in a famous ACM's paper on fault tolerant operating systems (Denning, 1976)

1. Closed environment
2. Minimal privileges
3. Resource access control with no side-effects
4. Error independent verification of decision

Systematic application of structured design principles enhances the potential robustness of designed systems through the definition of explicit well-defined interfaces between modules and between abstraction levels, that are perfect sites where errors could be located and their propagation can be avoided.

4 EXECUTION OF NON LINEAR PROCESS PLANS

Each single step in the execution of a production task is the run of a complete part program on the CNC of a specific machine. Until now in our discussion we have assumed this level of interaction between supervisor and CNC's.

Normally inside each part program various machining tools are requested following a rigid sequence, fixed once and forever when the program has been written.

Only few of the precedence constraints are related to technological reasons. The majority is there only for the limitation of the language used to represent machining part programs. Tool operations to machine a part are more naturally arranged in a network connecting various threads made by technological sequences.

The concept of Network Part Program, and preliminary development of management modules has been introduced during the research project Mod-Flex-Prod (MFP, 1999).

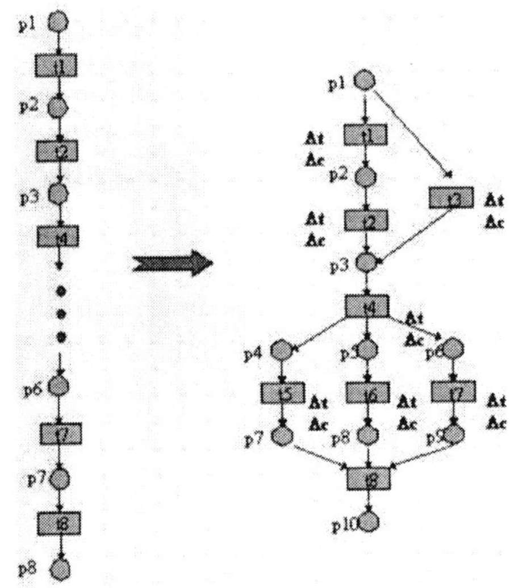

Figure 3 From sequential to Network Part Program

The execution of a part program as a network of tool operations, supported by software tools for the management of such a network, represents a major improvement in the control of a manufacturing system, and let to achieve a new degree of integration between management policy and control.

In fact moving the decision making process on "what tool will be the next", from the Numerical Control of a single machining center, to a management module, with potential visibility of all the resources of a plant, can afford the implementation of new optimization policy.

Flexibility in part process planning is one of the major sources of improvement in the control of integrated manufacturing systems, due to the fact that it represents a bridge between process planning and production planning and control. Traditional process plan are static, linear (strictly sequential) and they assume unlimited factory resources. Indeed process planning is often carried out without considering manufacturing system information, such as resource availability, breakdown of equipment or disruption caused by stochastic bottlenecks. Since static schedules quickly become obsolete as the state of the system moves away from the predicted state, there is a need for schedules to be developed in real time or for re-planning of the old process plan. Re-planning is often done by improvisation and can result in long throughput times.

If alternative ways of producing a single part can be incorporated into the set of options available in determining the product mix, a strong improvement can be obtained for throughput and flow time, due to a better balancing of workloads and to a highly adaptive reaction to unplanned situations. Thus the need for alternatives can be satisfied defining a network (non linear) part program, composed by:

- *alternative operations*: the execution of one operation excludes the execution of its alternative one;
- *parallel operations*: operations which can be executed in any sequence.

Research has pointed out that the consistent improvement of system performance using network part process plans is around 10% when a better use of resources is considered; this improvement can be increased if also reactions to disruptions and breakdowns are considered. Due to its relevance in potential system performance improvement and its connection with production planning, non linear process plan generation has been recently considered one of the key problem within CAPP related research (Colosimo, 1999).

This evolution fits perfectly with our approach of supervisory control as an operating system for manufacturing plant. Representing each single tool operation as a resource managed by the operating system, it becomes possible to achieve:

- easier implementation of tool sharing policies
- optimization of part program execution on very flexible multi-face clamping devices
- possibility to dynamically split the execution of a part program on more spindle for resource allocation optimization
- management of alternative operation to react in real time when a tool is not available.

5 A NEW CHALLENGE: FOCUSED FLEXIBILITY

Flexibility in manufacturing and automation is a very broad concept, covering various dimensions and putting pressure on system development. Generally speaking, flexibility is the capacity of a system to adapt to variation in production requirements.

Moreover productivity is always a ratio between the production obtained and the production costs. So in order to increase productivity we have to improve performances but also to reduce overall costs.

In order to reduce production costs a current trend is to avoid to include in a manufacturing system all the flexibility it will need in its life cycle but to progressively extend and to re-configure the system (Matta, 1999).

A new generation of manufacturing system is emerging, highly customized in order to provide the exact amount of flexibility and productivity required by production requirements.

We use the term focused flexibility manufacturing plant to denote such systems, designed to solve specific production tasks and incorporating new type of machines and new automation devices like robotized arms, and conveyors to transport unclamped part to working bays.

Figure 4 Focused Flexibility plant

To survive in the market of highly customized manufacturing systems it is necessary to perfectly match user requirements. Nevertheless such requirements are often incomplete and moreover they rapidly change during the life cycle of the plant. This trend toward customization and requirements changing ask for systems that are capable of adaptation to an evolving environment (Koren, 1999).

The more focused the systems are the more frequent and deep the requests of re-configuration.

6 MODULARITY AS A PROCESS

In order to cope with the enhanced level of customization required by the new generation of focused flexibility plants, we have to enhance our system development methods, and to push forward the integration between management, control and mechanical design.

Once again information technology comes to an aid, bringing interesting ideas and methods. From software engineering we have learned that in the light of more stringent development requirements the priority should be shifted towards building modular and reusable components, which can be used in many applications. The components must then be integrated to form an application.

Large and complex systems can benefit significantly from a component based development approach where new systems are constructed by composing reusable, documented and previously tested concurrent objects. The re-use of components is a very powerful tool to reduce development time and related costs.

The keyword is reusability, but unfortunately this is a slippery feature of components.

It seems obvious that it is not sufficient to declare to follow a component-based approach to make components arise with all the expected features that will make them reusable. First of all we do not know the context of reuse. If we could look forward to all plants we will realize in the next 5 years that probably we could develop a good set of reusable modules. Instead we can only look back to the last 5 years trying to guess how to adapt existing modules to new requirements.

In order to code the knowledge collected during the development of various installations, there is the need to build a specific application framework.

According to Gamma's definition (Gamma, 1995), a framework is:

"A set of co-operating classes that makes up a reusable design for a specific class of software. A framework provides architectural guidance by partitioning the design into abstract classes and defining their responsibilities and collaborations. A developer customizes the framework to a particular application by sub-classing and composing instances of framework classes".

In 1999 we joined the Italian National Research Program on Innovative Production System, to develop new methodologies to build mechanical machining stations.(project assigned by MURST to Consorzio Produzione 2000 dec. n° 1075/Ric del 29.09.1998).

The project offered the opportunity to develop a framework for the development of focused flexibility manufacturing systems, integrating management, control and mechanical design.

The key points of the approach are: object-orientation, hierarchy and formal models for specification and design (Ferrarini, 2001).

The object-oriented approach has been essentially proposed for software development, even if it is based on quite general concepts such as modularity, abstraction, encapsulation and inheritance. In the late 1980s (Andersson, 1989), the object-oriented approach also started to become popular in the field of modeling, especially with reference to physical system modeling.

One of its most valuable feature is that a modeling module has to correspond to a concrete physical object or component (Maffezzoni, 1999), so that the interfaces of the module with the rest of the system correspond strictly to the physical ports by which the considered component exchanges power or signals with the external world.

An abstraction hierarchy of the system being considered is realized by aggregating more sub-models and their connections within a composite (larger) model, which may in turn be connected to other models.

Modularity, along with the standardization of interfaces, the aggregation of simpler models with abstraction, and the strict correspondence of model objects to real objects are the basic mechanisms by which object-oriented modeling is superior in dealing with complexity. To enhance the reuse of model-ware, object-oriented modeling environments can use inheritance to a certain extent, possibly by defining a specific model type (or class) as a subclass of a more general one.

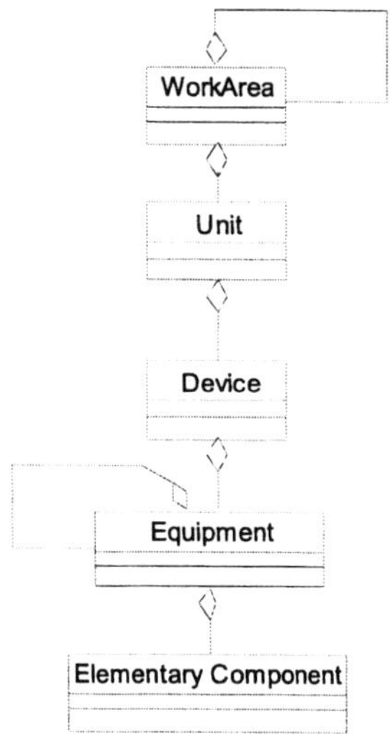

Figure 5Application framework. UML representation of main abstract classes.

The application framework for manufacturing systems, has the central role to define compositional rules for modules.

In principle each new system is a simple instance of the framework. In reality the high market competitiveness, force to develop new solutions, to improve performance, to reduce costs or both. The framework must shape a design environment fluid enough to accept evolution and new ideas, but not too much to avoid anarchy. As always it is a complex tradeoff.

Refactoring, a process and a set of techniques to re-organize code while preserving the external behavior, is gaining popularity in software engineering due to its practical value in creating agile code (Fowler, 1999) (Beck, 2000).

In our context refactoring has the primary role to keep the application framework updated to represent at least all plants developed so far. Existing plants could potentially be re-developed as instances of the current version of the framework.

This approach has been successfully applied in the development of the supervisory control of focused flexibility manufacturing plants.

Now the real challenge is to bring the potential advantages of the framework-based approach to mechanical and control design, influencing procedures and defining requirements for supporting tools: CAD for mechanical designer, IDE for control and software engineer and Product Life-cycle Management for the whole organization.

7 SUMMARY AND OUTLOOK

In this paper we presented the increasing role of information technology in the development of flexible manufacturing system based on horizontal spindle machining centers.

Starting from the development of supervisory control as operating systems for manufacturing cells, down to the deep integration between management modules with real time control tasks, to the proposal of a new integrated methodology to design hardware, control and management services, information technology and software engineering were fruitful source of inspiration and provided promising idea, to cope with the complexity of flexible manufacturing systems.

In order to fully exploit the potential of the methodological approach sketched in this paper, it is necessary to drop barriers between supervisory control, developed with high level languages, and real-time control systems.

Real-time system were historically built using low level programming languages and executed on dedicated hardware and specialized operating system: efficiency, high resource utilization and integration with hardware, were the prime concerns, whereas modularity and reuse came in second.

The upcoming trend in automation industry is the increasing amount of Information Technology used within control systems, and a major attention to reusable component-based approach.

From the software point of view a major milestone is the international standard IEC 61499 (IEC61499, 2000), enabling modular and hierarchical control application design with direct links to devices, resource and communication systems.

In the near future Distributed Control Systems (DCS), consisting of networked intelligent field devices will be connected via an industrial Ethernet network, providing seamless integration of all levels of the automation pyramid, ranging from ERP (Enterprise Resource Planning) to MES (Manufacturing Execution Systems) to supervisory control, down to the field device level.

In the course of this, the adoption of Internet related technologies in the field device level will be supported, opening new ways for managing the DCS via the company Intranet/Internet.

An analysis of possible integration of the proposed modeling and design approach with new trends in distributed control systems are investigated in the on-going research project Torero (Total life cycle web-integrated control) (Torero, 2003) funded within the IST initiative of the European Commision.

8 REFERENCES

(Andersson, 1989) ANDERSSON M.: An object-oriented modeling environment. In Iazeolla et al., Eds., Simulation Methodologies, Languages and Architectures and AI and Graphics for Simulation. European Simulation Multiconference, Rome, 1989.

(Beck, 2000) BECK K., Extreme programming explained:embrace change, Addison Wesley, 2000, IDBN 0201616416

(Colosimo, 1999) COLOSIMO B.M., M. PELLIZZARO, Q. SEMERARO, T. TOLIO, 1999, APPG1: A New approach for alternative process plans generation. 32nd CIRP International Seminar on Manufacturing Systems, Leuven, Belgium, May 24-26 1999.

(Denning, 1976) DENNING P.J., Fault Tolerant Operating Systems, ACM Computing Surveys, vol.8, n.4, Dec. 1976, pp.359-389.

(Ferrarini, 2001) FERRARINI L., G. FOGLIAZZA, Advanced Control System Design for Machining Centres, IEEE/ASME International Conference on Advanced Intelligent Mechatronics (AIM '01), 8–11 July 2001, Como, Italia, pp. 671-676, vol. I.

(Fowler, 1999) FOWLER M., ed., Refactoring: Improving the design of existing code, Addison Wesley Longman, 1999, ISBN 0201485672.

(Gamma, 1995)GAMMA E., R. JOHNSON, J. VLISSIDES, Design Patterns: Elements of reusable Object-Oriented Software. Addison Wesley, 1995, ISBN 0201633612.

(Garetti, 1985) GARETTI M., A. POZZETTI, A. SIANESI, La Struttura di controllo degli FMS: esperienze su di un sistema a programma per la descrizione dei cicli di lavorazione, Proceedings of the ANIPLA Conference on Integrated System for Industrial Automation, Genova, 1985

(Maffezzoni, 1999) MAFFEZZONI C., L. FERRARINI, E. CARPANZANO, "Object-Oriented Models For Advanced Automation Engineering", IFAC Journal Control Engineering Practice, vol 7, 1999, pag. 957-968.

(Matta, 1999) MATTA A., T. TOLIO, High flexible and high productive manufacturing system architecture, Proceeding of the 15th International Conference on Production Research, Limerick, Ireland, Augus t 1999, 1547-1550

(MFP, 1999) BRITE EURAM Mod-Flex-Prod Project, Deliverable 7.2: Part Program Manager Software Architecture, BRPR-CT97-0440, 1999

(Nof, 1980) NOF, S.Y., A.B. WHINSTON, W.I. BULLER: Control and Decision Support in Automatic Manufacturing, AIIE Transactions, vol.12, n.2, June 1980, pp.156-169.

(Koren, 1999) KOREN Y. U. HEISEL, F. JOVANE, T. MORIWAKI, G. PRITCHOW, H. VAN BRUSSEL, A.G. ULSOY, Reconfigurable Manufacturing Systems, CIRP Annals, Vol. 48, No. 2, (Keynote paper)

(Torero, 2003) The TORERO Consortium: Deliverable 2.1 Integrative design and development of web-enabled control system design methodology , 2003. (www.uni-magdeburg.de/iaf/cvs/torero)

(Vanneschi, 1984) VANNESCHI M., F. BAIARDI, L. RICCI, A. TOMASI, Structuring processes for a cooperative approach to fault-tolerant distributed software, 4th IEEE Symp. On Reliability in Distributed Software and Database Systems, 1984.

Copyright © IFAC Information Control Problems in Manufacturing, Salvador, Brazil, 2004

ELSEVIER
IFAC
PUBLICATIONS
www.elsevier.com/locate/ifac

MODULAR DEVELOPMENT OF AGILE MANUFACTURING CONTROL SYSTEMS

Emanuele Carpanzano, Claudio R. Boer, Andrea Cataldo, Paolo Pedrazzoli

Institute of Industrial Technologies and Automation, National Research Council, Milan, Italy
{e.carpanzano,c.boer,a.cataldo,p.pedrazzoli@itia.cnr.it}

Abstract: In the near future, agile control systems will be a key element for the realization of intelligent manufacturing systems (IMS). Therefore, innovative development methodologies for the structured design and testing of complex control systems are mandatory. In the present work a development methodology is introduced, which is based on reference models that are compliant with emerging international standards, and exploit object-oriented concepts. By means of the proposed framework the definition of complex control systems is simplified, and the maintenance and reuse of designed control solutions is enhanced. In particular, the agility of the developed control systems is significantly improved. Furthermore, an application is presented, where the structured design of the control of an intelligent manufacturing system is discussed, so highlighting the benefits of the proposed framework. In particular, an innovative shoe plant is considered as an application example. *Copyright © 2004 IFAC*

Keywords: Manufacturing, Logical Control, Design Methodology, Standards, Object Oriented

1. INTRODUCTION

Reliable and flexible automation systems are a crucial point for competitiveness of modern manufacturing systems (Carpanzano, et all., 2002). Reusability and reconfigurability of developed control solutions are also fundamental elements to reduce the costs and times needed to design and realize a new production systems, or to modify existent ones. Furthermore, intelligent control systems, which are able to self-reconfigure themselves according to varying operating conditions and to deal with non-nominal situations by means of suitable fault diagnosis and recovery methods, are another key point for modern industrial plants (Sanderson, 2000, Ünsal et al., 2000). In such a context, the definition of development methodologies that support the structured design and testing of the whole automation system of a manufacturing plant is mandatory (Ferrarini, et all., 2002). In the present work a framework for the design of control and supervision systems of modern manufacturing systems is discussed. One of the key elements of the proposed approach is the adoption of a *structured design methodology* (Ferrarini, et all., 2002). This is based on formal *reference models*, based on the *object-oriented paradigm* (Maffezzoni, et all., 1999), and compliant with *international standards*, in order to guarantee interoperability and interchangeability. In particular, in the present work an architectural control system model is proposed which has been derived after analyzing different existing standards (OSACA (OSACA Association, 1997), OMAC (Birla, et all., 2001), OSEC (Fujita, et all., 1998)), and the IEC 61499 standard is adopted as a functional control system model (Standard IEC 61499, 2001). In particular, the benefits of the proposed methodology are illustrated with reference to an application example.

The paper is organised as follows. The design cycle model of the adopted development methodology is introduced in section 2. Then, the used reference models for the design of the control system architecture and of the control system functions are outlined in section 3. In section 4 the design of the control system of an area of a manufacturing plant is described as an illustrative example, highlighting the advantages of the presented approach. Finally, concluding remarks are drawn (section 5).

2. THE DEVELOPMENT METHODOLOGY

A structured development methodology for the design of industrial automation systems is based on a design cycle model (DCM) that establishes the order of the phases involved in the development of the control system. In the present work the DCM described in (Carpanzano et al., 2002) is used, such a DCM is based on the following phases.

A. *System definition*: the process to be automated is described, and the activities to be performed and the objectives of the automation system are defined.

B. *Control system specification*: the tasks and the essential functions of the supervision and control system are defined. In order to meet the objectives of unification, reusability, and traceability, the automation system shall observe a common structuring scheme consistent through the subsequent phases of the DCM. Thus, the system can be structured into a hierarchy of different layers (plant, area, cell, unit, control system devices), based upon a successive decomposition of top-level control objectives into elementary control actions.

C. *Control system design*: the supervision and control system are developed through suitable formal reference models for both the architectural and the functional design.

D. *Control system integration and testing*: it is checked if the designed control functions meet the requirements. This is achieved by means of suitable formal analysis and/or simulation methods.

E. *System integration and testing*: the control code is generated, and it is checked if the control system works on its devices, and if all the system's requirements are met.

3. THE REFERENCE MODELS

In the present section the adopted reference models for the design of the control system architecture and of the control system functions are introduced in sections 3.1 and 3.2 respectively.

3.1 Model for the control system architecture

In order to define the reference model for the control system architecture, different international standards for the representation of open control systems architectures have been taken into account. Specifically, the main features of the OSACA (OSACA Association, 1997), OMAC (Birla, et all., 2001) and OSEC (Fujita, et all., 1998) standards have been considered. A comparison among the main features of such standards is presented in (Pritschow, et all., 2001). Even if such standards are not compliant with each other, here a new reference model for the control system architecture is proposed which exploits their common aspects. Such a model is sketched in Fig. 1, where the basic modules are illustrated, i.e.: *HMI and Configuration Tools, Logic Control, Motion Planning and Axis Control.* Such modules are further detailed in Figs. 2, 3 and 4.

The logic control module is structured on different hierarchic levels according to the plants functional decomposition in areas, cells and units. In particular, each hierarchical level is constituted by three parallel functional branches, based on the concept of feedback control, as is shown in Fig. 1.b. First, the forward control functions, responsible for activity decomposition, execution planning and control. Second, the nominal feedback functions for refinement and update of a priori knowledge, based on the actual, but essentially expected, evolution of the process, and consequent formulation of controlled adjustments of the forward control. Third, the non-nominal feedback functions for the monitoring of discrepancies between actual and allowable states, in both the forward control and the nominal feedback functions, diagnosis of their origins, and generation of directives and updates including recovery strategies and constraints for forward control. Suitable self-reconfiguring functions can be defined as non-nominal feedback functions. A single level, area, cell or unit, can be recursively structured on more hierarchic levels, e.g. the logic control module of a robot unit can be structured on the hierarchic levels mission, task and action, and each such sublevel of the level unit is divided in the three above mentioned functional branches according to Fig. 2.

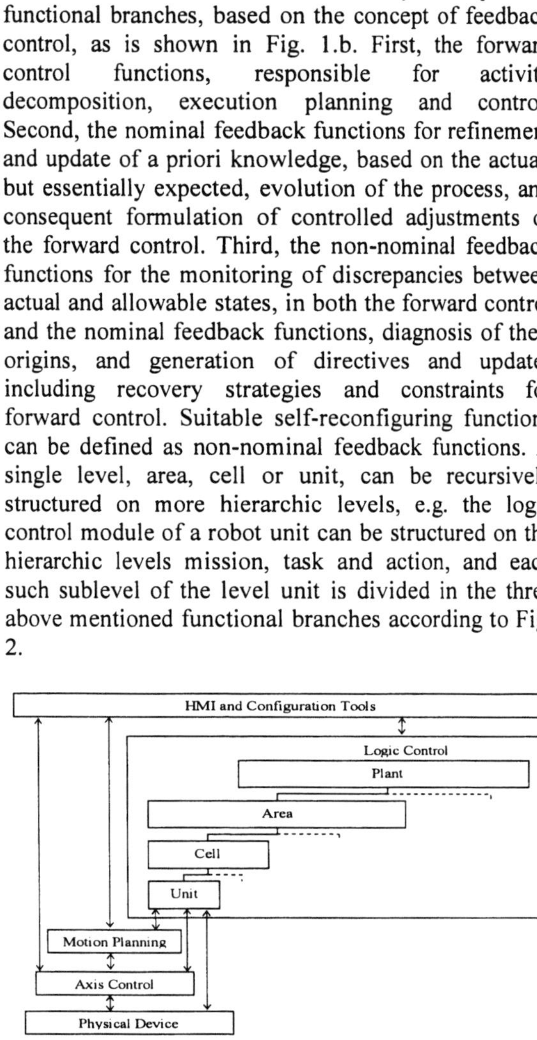

Fig. 1. Reference model for the control system architecture.

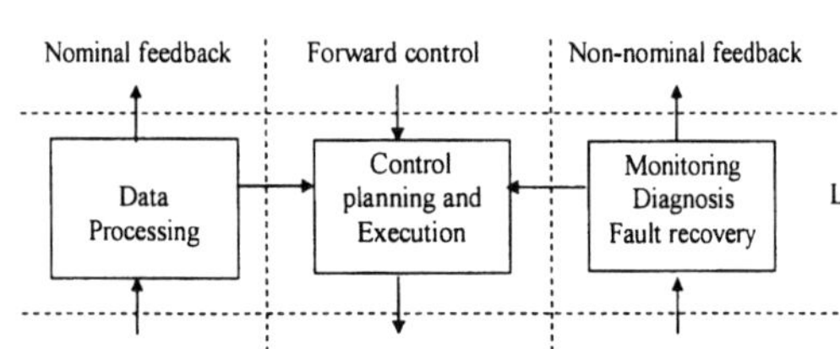

Fig. 2. Generic hierarchic level of the logic control module.

As for the motion planning module, the following submodules can be distinguished (Fig. 3):
- the *Motion planning scheduler* coordinates all other submodules within the considered module and exchanges signals with the upper logic control modules;
- the *Interpreter* is responsible for the translation of the part program into an inner data format;
- the *Trajectory control* submodule is dedicated to the contouring error control;
the *Motion law optimizer* performs a forward analysis of the tool trajectory in order to calculate an optimum end-effector velocity profile;
- the *Interpolator* is responsible for the trajectory sampling, keeping the end-effector as close as possible to the planned path and velocity;
- the *Kinematics* submodule applies the inverse kinematics of the controlled machine tool or robot to the interpolator output.

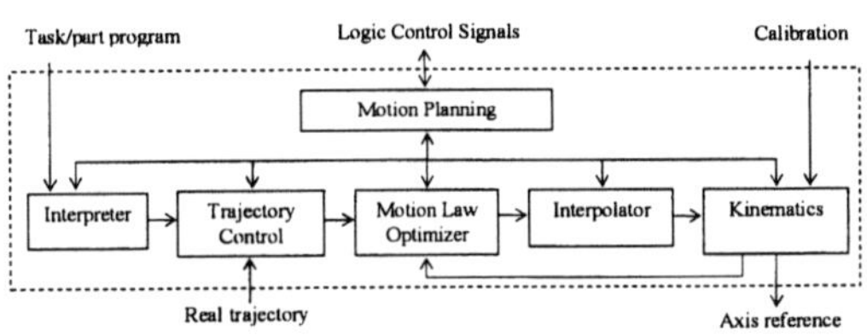

Fig. 3. Motion planning module.

The axis control module is detailed in Fig. 4, with reference to such a figure the following submodules can be distinguished:
- the *Plant dynamic estimation* aims at predicting the dynamic behaviour of the physical process;
- the *Motion Control* submodule takes care of keeping the real trajectories "as close as possible" to the reference ones by means of proper feed-back laws;
- the *Device Driver* provides the actual plant state through sensor input readings, and supplies command references to the controlled plant;
- the *Sensor Processing and Fusion* collects raw sensor data from the device driver and works out a more accurate and noise free estimation of the real trajectory.

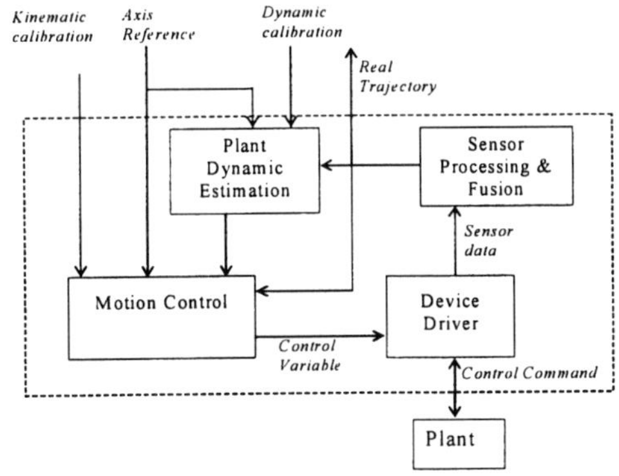

Fig. 4. Axes control module.

The proposed model is based on basic object-oriented modelling concepts (Maffezzoni, et all., 1999) like:
- *Modularity*: a complex control system is divided into smaller and more manageable modules;
- *Hierarchy*: the modules are structured on different hierarchic levels, according to a top-down functional decomposition approach;
- *Encapsulation*: every module hides its internal algorithms and variables to the other components of the control system;
- *Aggregation*: the main modules of the control system architecture are obtained through the aggregation of simpler submodules;
- *Inheritance*: specialized components can be derived from more general ones.
The benefits deriving from the adoption of the proposed models, as will be shown in section 4, are that the definition of complex control systems is extremely simplified, reusability is improved, and reconfigurability is strongly enhanced. In particular, whenever one or more functional modules or submodules have to be modified or replaced, this can be done with reference only to the directly involved components of the control system architecture, so reducing the required engineering times and costs significantly.

3.2 *Model for the functional design*

The IEC 61499 standard, which defines function blocks for industrial process measurement and control systems, is here introduced to develop detailed functional models of the automation system software, that define also the distribution of the supervision and control functions on the physical devices (Scoop et al., 1996, Standard IEC 61499, 2001). The IEC 61499 standard is based on a fundamental module, the function block (FB), which represents a functional unit of software, associated to a hardware resource of the control system. As shown in Fig. 5, a FB instance is characterised by: its type name and instance name, sets of event inputs/outputs and data inputs/outputs, internal data, an execution control chart (ECC), consisting of states, transitions and actions, which invokes the execution of algorithms in response to event inputs; and a set of algorithms, associated with the ECC states.

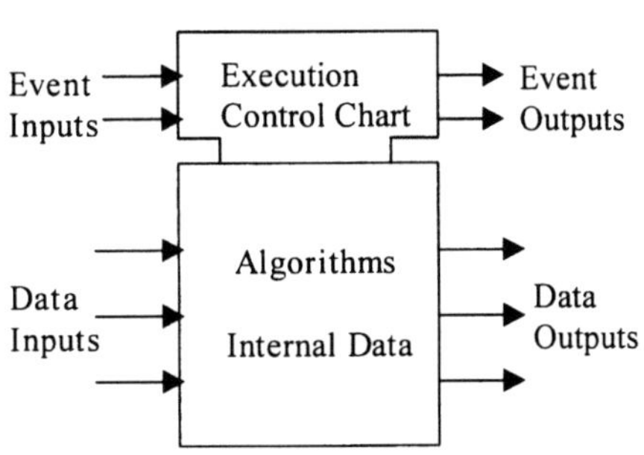

Fig. 5. Functionl block model.

The execution of algorithms is invoked by the ECC (which is basically a Moore automaton) of a FB instance, in response to event inputs. When the execution of an algorithm is scheduled, the needed input and internal data values are read and new values for output and internal data may be computed. Furthermore, upon completion of execution of an algorithm, the execution control part generates zero or more event outputs as appropriate. By properly connecting more FBs an application is defined.
As regards the configuration, an application can be distributed among several control system devices. A device uses the causal relationship specified by the application to determine the appropriate responses to events. Furthermore, in the IEC 61499 standard a resource is considered to be a logical subdivision within the software (and possibly hardware) structure of a device, which has independent control of its operations. Each FB instance is associated to one single resource.
With the given definitions the architecture of a manufacturing control system can be modelled as a collection of devices, divided in resources, interconnected and communicating with one another by means of one or more communication networks, while the functions performed by such a system are modelled as applications.
The adoption of the IEC 61499 standard as formal reference model enhances the definition of reusable and reconfigurable models, since the principles of modularity, encapsulation and standardization of interfaces are strongly exploited (Ferrarini et al., 2002, Mintchell 2000).

4. DESIGN OF THE CONTROL SYSTEM OF AN INNOVATIVE SHOE MANUFACTURING SYSTEM

In this section the presented framework is applied to the design of the control system of an innovative shoe manufacturing plant owned by ITIA-CNR and shown in figures 6 (picture) and 7 (layout). For the sake of brevity a simplified version of the overall IMS is considered (see figure 8). In particular, the design dealt with is illustrated by going through the different phases of the DCM introduced in section 2.

A. *System definition*
With reference to Fig. 8 the considered cell is constituted by the following units:
- 2 conveyor belts: an infeeder and an outfeeder;
- a 3-arms positioner (P1) capable of grasping and releasing workpieces from/to the infeeding/outfeeding belt;
- 2 rotating tables (T1, T2) with 4 slots each (S1,...,S8);
- a 2-arms positioner (P2) responsible for moving workpieces from slot S3 of table T1 to slot S4 of table T2, and viceversa;
- 5 machining/assembly stations (M1,...,M5), that are robots or machine tools performing 4 different operations (A, B, C, D) on slots S2, S5, S7 and S8, respectively. M3 does not operate in standard conditions, while it may perform either operation B or operation C, in case of failures of M2 or M4.

Fig. 6. Shoe IMS: picture.

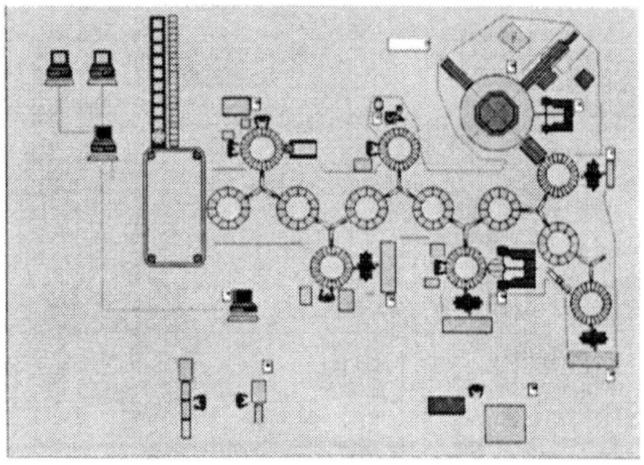

Fig. 7. Shoe IMS: layout.

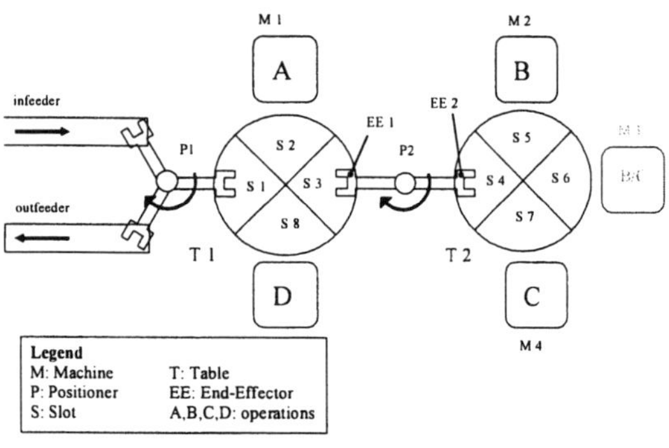

Fig. 8. Cell layout.

B. *Control system specification*
The control system is structured by describing each of the above mentioned components of the cell as a unit. In particular, the desired behaviour of the cell is first given through a natural language specification, and then it is translated into formal rules. As an example, the 2-arms positioner specification rules are reported hereafter.

```
//    2-arms    positioner    formal    control
specification
Table T1,T2;
if(OK){ if (isS3Ready() && isS4Ready()){
      T1.block(); T2.block();
      grasp(); rotate(); release();
      T1.unblock(); T2.unblock();
      } }
else {// errCode handling ...}
```

C. *Control system design*

The automation system is designed using the reference models previously introduced. Specifically, the control system architecture is reported in Fig. 9, where all cell components (machines, positioners, rotating tables and conveyors) are managed by appropriate unit controllers. The control signals are fed through the different hierarchical control layers from logic control down to physical devices and viceversa. The machine tools need a complex control system structure, containing proper motion planning modules (MP) and axis control modules (AC), while the conveyor has a reduced set of control modules, since the MP module is not (needed), and the positioners and rotating tables are directly managed by their unit logic controllers.

Then, the control system functional model is designed through the IEC 61499 standard formalism. Specifically, the cell can be represented as a composite FB and each unit inside the cell can be represented as a basic FB.

Assume that different type of products are produced by the manufacturing cell dealt with. For simplicity 2 different products can be considered. In such a case it is possible to design self-reconfiguring control functions for each unit of the cell, so that the different type of pieces (Piece1 and Piece2) are locally recognised, and the proper corresponding working operation is executed by the unit. As an example consider the FB for machine M1 (Fig. 10), whose ECC is reported in Fig. 11. It may be noticed that a different operation is performed according to the piece to work (drilling for Piece1, milling for Piece2). Notice that whenever a different operation has to be performed by a unit on a different piece, not only the logic control functions are self-reconfigured. In fact also the motion planning control functions and the axis control functions may be properly self-reconfigured; e.g. the task/part program changes for the motion planning module, and the piece data change the plant dynamic estimation module within the axis control module.

The machine M3 is represented by the FB shown in Fig. 12, whose ECC is reported in Fig. 13. This machine starts to work when a failure event occurs on machines M2 or M4 and a piece to work is present in slot S6. This FB triggers also the two events B and C, to point out that the corresponding operation is completed.

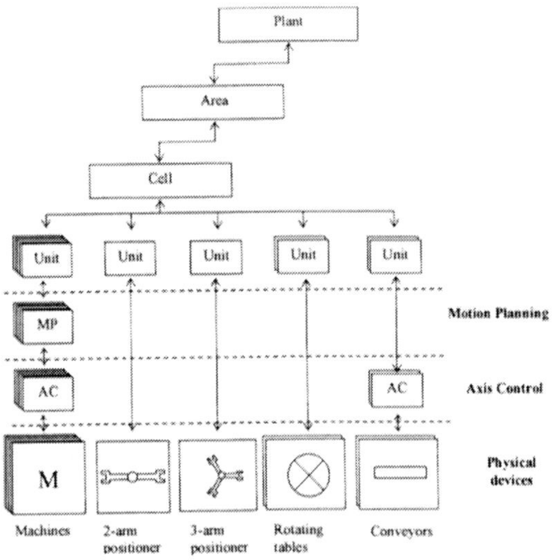

Fig. 9. Control system architecture.

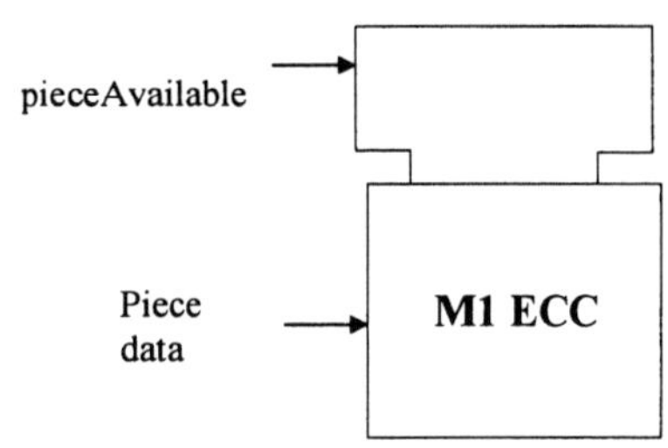

Fig. 10. FB of Machine M1.

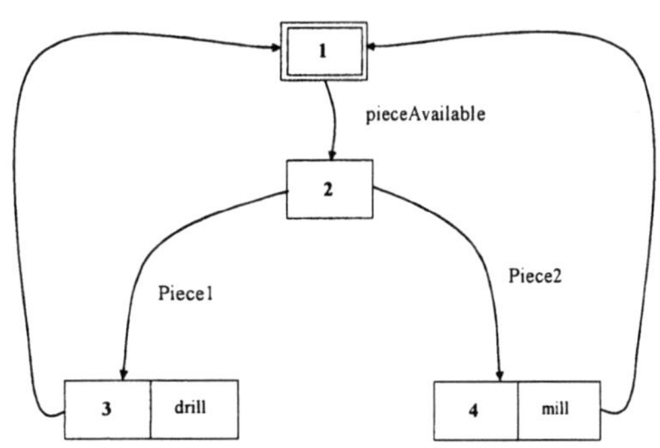

Fig. 11. ECC of Machine M1.

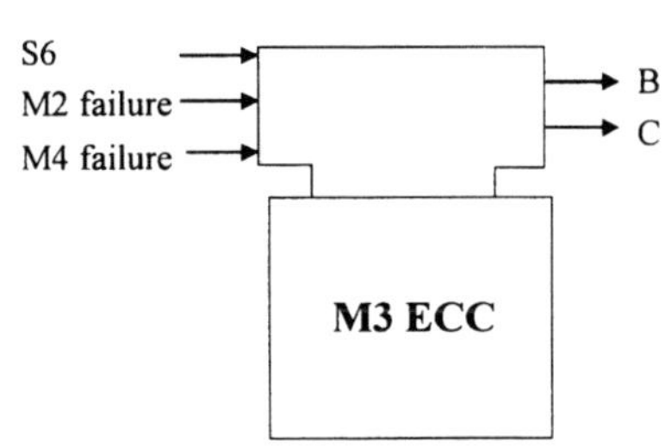

Fig. 12. FB of Machine M3.

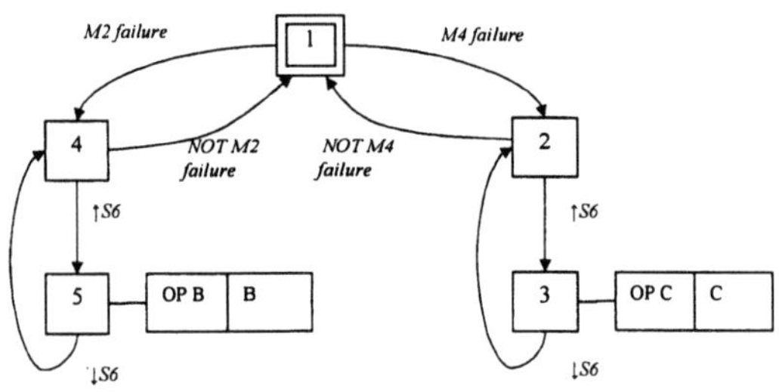

Fig. 13. ECC of Machine M3.

D. *Control system integration and testing*

Once the functional model is defined by means of the IEC 61499 standard, this can be tested through the bottom-up approach presented in (Carpanzano, et al., 2002).

The described testing operations can be performed before final implementation on the target system, so improving the reliability of the control software implemented on the real system, and reducing the overall development times and efforts significantly (Mintchell, 2000).

E. *System integration and testing*

Finally, the control software can be generated automatically starting from the IEC 61499 model, and can be downloaded on the target system. So, it can be verified that the desired requirements are fully met by the considered manufacturing cell.

The use of the considered methodology facilitates clear documentation and easy maintenance of the designed control solutions, it also improves reusability and enhances fast integration of new features as well as easy reconfiguration of the designed control system.

5. CONCLUDING REMARKS

In the present work a development methodology for the design of intelligent manufacturing control systems has been discussed. The adopted architectural and functional reference models have been presented, and the main benefits of the proposed approach have been outlined. Such benefits consist of the reduction of the development efforts and times, and of the reliability of the obtained control systems. Furthermore, the adoption of formal models that exploit object-oriented concepts, simplifies the description of complex systems and enhances control solutions reuse and reconfiguration.

The study of formal methods for the design of self-reconfiguring control functions, and the definition of formal analysis and simulation techniques for the adopted reference models, will be subject of future work. As well as the application of the proposed framework to different manufacturing plants, and the development of CACSD tools that support the control system engineer throughout the whole design cycle.

REFERENCES

Birla, S., Faulkner, D., Michaloski, J., Sorenson, S., Weinert, G., Yen, J. (2001). Reconfigurable Machine Controllers Using OMAC API. *CIRP 1st International Conference on Agile, Reconfigurable Manufacturing*, May 21-21, University of Michigan, Ann Arbor, USA.

Carpanzano, E., and Ballarino, A. (2002). A structured approach to the design and simulation-based testing of factory automation systems . *Proceedings of the IEEE International Symposium on Industrial Electronics*, ISIE2002, July 8-11, L' Aquila, Italy.

Ferrarini, L. and Carpanzano, E. (2002). A Structured Methodology for the Design and Implementation of Control and Supervision Systems for robotic applications. *IEEE Transactions on Control Systems Technology*, **vol. 10, no. 2**, pp. 272-279.

Fujita, S., Kanemoto, M., Asano, H., Inoue, T., Okano, A., Uno, H., Kakishita, N., Noda, A. (1998). OSEC Position Paper, *ITIA Series*, OSE Consortium,

Maffezzoni, C., Ferrarini, L., and Carpanzano, E. (1999). Object-Oriented Models for Advanced Automation Engineering. *Control Engineering Practice*, **vol. 7**, pp. 957-968.

Mintchell, G.A. (2000). Power Up Programming with Graphical Modeling. *Control Engineering Magazine*, December 2000.

OSACA Association. (1997), *OSACA handbook*.

Pritschow, G., Altintas, Y., Jovane, F., Koren, Y., Mitsuishi, M., Takata, S., Van Brussel, H., Weck, M., Yamazaki, K. (2001). Open Controller Architecture – Past, Present and Future. *Annals of the CIRP*, **vol. 50/2/2001**.

Sanderson, A.C. (2000). Reconfigurable and Modular Robotics. *International Mechatronics Forum at the International Conference on Intelligent Robots and Systems*, IROS2000, Takamatsu, Japan.

Schoop, R., and Strelzoff, A. (1996). Asynchronous and Synchronous Approaches for Programming Distributed Control Systems Based on Standards. *Control Engineering Practice*, **vol. 4, no. 6**, pp. 855-861.

Standard IEC 61499 (2001). Function Blocks for Industrial-Process Measurements and Control Systems. IEC TC65/WG6.

Ünsal, C., and Khosla P.K. (2000). Mechatronic Design of a Modular Self-Reconfiguring Robotic System. *Proceedings of the IEEE International Conference on Robotics and Automation*, San Francisco, CA, USA, April 2000, pp. 1742-1747.

Copyright © IFAC Information Control Problems in Manufacturing, Salvador, Brazil, 2004

ELSEVIER
IFAC
PUBLICATIONS
www.elsevier.com/locate/ifac

INVESTIGATING THE ROLE OF PRODUCT INFORMATION IN END-OF-LIFE DECISION MAKING

Ajith Kumar Parlikad, Duncan McFarlane

Centre for Distributed Automation & Control
Institute for Manufacturing
Cambridge University Engineering Department
Mill Lane, Cambridge, UK, CB2 1RX.

Abstract: With the increasing focus of manufacturing companies on providing environmentally sound products, the availability of information to improve product recovery decisions is becoming critical. Infact, a fundamental obstacle in making efficient product recovery decisions is the loss of information associated with the product after the point-of-sale. We show qualitatively that the availability of product information has a positive impact on product recovery decisions, and discuss how automated product identification technologies like RFID can be employed to provide the necessary information. The paper also investigates how recovery decisions can be modelled to represent the impact of product information on those decisions. *Copyright © 2004 IFAC*

Keywords: Data acquisition, Environment, Evaluation, Information analysis, Probabilistic models, Decision making, Uncertainty.

1. INTRODUCTION

In recent years, there has been a tremendous growth in the demand for consumer goods. Greater consumer awareness, constant demand for better products, and rapid development and improvement in technology has resulted in the shortening of product innovation cycles. New products and product variants appear in the market with increasing regularity. An upshot of this phenomenon is the tendency of consumers to discard/return products in lieu of better products with increased functionality long before the old products reach the end of their useful functional life. Recovering value from these returned products has been gaining increasing importance in the past few years due to environmental concerns as well as the widespread recognition of the financial benefits that come along with it.

These issues, coupled with the increasing focus of manufacturing companies on providing products that are environmentally sound, make the need for information enabling the best recovery decisions increasingly essential. Infact, loss of information associated with a product after the point-of-sale is identified in the literature as one of the major obstacles for efficient recovery of value from returned products (Thomas *et al.*, 1999).

This paper investigates the role of information associated with a product in making efficient product recovery decisions. The advent of new automated identification approaches such as the one developed by the Auto-ID Centre (McFarlane *et al.*, 2003) makes it increasingly possible for a product to carry complete information associated with it throughout its lifecycle and ensure flow of this information between the various actors in the supply chain. In this paper, we will provide an insight into how such technologies could provide the information necessary to improve the quality of decisions made during product recovery operations.

Obtaining the necessary information for end-of-life (EOL) decision making has a cost attached to it. Hence, we need to make a quantitative assessment of the impact of product information (un)availability on the performance of product recovery decisions. This paper provides a brief look at various modeling approaches that could be used to represent the impact of product information on product recovery decisions.

In the next section, we discuss the processes involved in recovering value from returned products.

2. END-OF-LIFE MANAGEMENT

End-of-life (EOL) management concerns the processing of products (with the objective of recovering maximum value out of them) after the initial user discards the product. First we take a look at the various motives for improving EOL management of products.

2.1. Motivations for EOL management

Environmental degradation has become a big concern, and governments all around the world are formulating "producer responsibility" laws to put pressure on businesses to manufacture products that minimise Eco-burden. For example, In June 2000, DG XI of the

European Commission issued a "Proposal for a Directive on Waste from Electrical and Electronic Equipment (WEEE)" that makes the producers of electronic and electrical equipment responsible for take-back and recovery management of their products. The directive also specifies collection targets for local authorities as well as recovery and recycling targets for the producers to be met before certain deadlines.

In addition to legal compliance, efficient product recovery management has a financial motive as well. Guide *et al.*, (2000) claim that remanufacturing operations account for sales in excess of $53 billion per year in the US. In addition, the shift from selling *products* to selling sets of *services* makes the reuse of recovered materials, parts, and products desirable. Besides ecological and economic factors, customer awareness is creating opportunities for "green marketing" and new markets for returned goods.

Recently, consumers have become aware of the environment and the potential problems that could arise by neglecting it. Realising this, manufacturers have started to manufacture products that are environmentally friendly (or 'green products') to gain advantage in the marketing platform against their competitors, and to analyse the product lifecycle in order to insert the environmental component into the product design to produce a product that has a low production cost and is environmentally friendly (Gungor and Gupta, 1999).

When a product reaches the end-of-life phase of its lifecycle, there are various options available to recover value from them, which will be discussed in the next section.

2.2. EOL options

Thierry *et al.* (1998) illustrate five options for recapturing value from discarded products as well as components: (1) *Repair and reuse*, the purpose of which is to return used products in working order; (2) *Refurbishing*, the purpose of which is to bring the quality of returned products up to a specified level by disassembly to the module level, inspection and replacement of broken modules; (3) *Remanufacturing*, the purpose of which is to bring used products up to quality standards that are as rigorous as those for new products by complete disassembly down to the component level and extensive inspection and replacement of broken/outdated parts; (4) *Cannibalisation*, the purpose of which is to recover a relatively small number of reusable parts and modules from the used products, to be used in any of the three operations mentioned above; (5) *Recycling*, the purpose of which is to reuse materials from used products and parts by various separation processes and reusing them in the production of the original or other products.

In addition, parts and materials that could not be recovered by any of the above five operations will be *disposed* in accordance with safety and environmental regulations. An alternative for disposal is *energy recovery* by means of incineration, which has its own environmental impact. It is to be noted that more than one of these options may be chosen for any particular product, i.e., a certain proportion of components maybe reused, and the rest may be recycled, incinerated, or disposed.

The hierarchy of product recovery options in order of preference for material sustainability is shown in Fig. 1. The figure, which is adapted from Goggin (1998) also shows the current and expected future scenarios of product recovery in terms of volume of product recovered. If the situation is thus reversed, a source for cost (from disassembly and disposal) can be converted into a potential source for revenue (from recovered products/components).

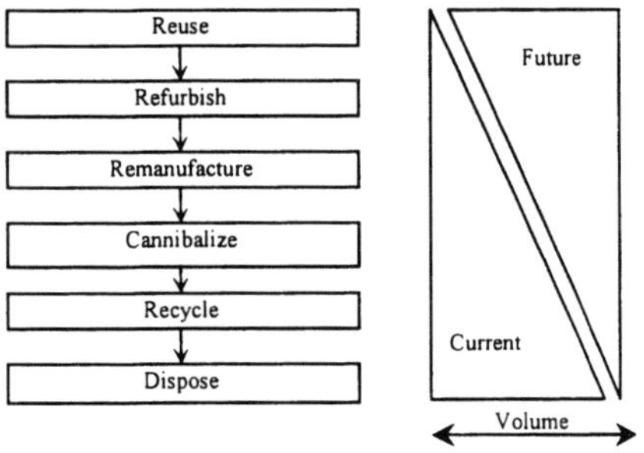

Fig. 1: Product Recovery Scenario: Current and Future

In the next section we describe the various operations involved in recovering value from returned products.

2.3. EOL operations & decisions involved

The EOL phase of a product begins when the user returns (or disposes) products due to a variety of reasons ranging from malfunction to the arrival of new (and better) products in the market. Fig. 2 provides a simplistic view of operations involved in recovering value from returned products.

Collection is performed either by municipal collection facilities or through the manufacturers' take-back programs. After collection, the products are sent either to the manufacturer's facilities or to third-party remanufacturers/recyclers. Here, the products are identified and *sorted* according to product types or other categories. Products are identified using various methods such as the manufacturer's barcode, manuals and specifications attached to the product, or in many cases, expert knowledge of the people handling the products. It is seen that the information required is not normally readily available, and in many cases, the available information is seen to be inaccurate or incomplete. In such cases, extra efforts have to be put in to gather the required information. This could be by way of contacting the manufacturer, or by detailed inspection or testing of the product.

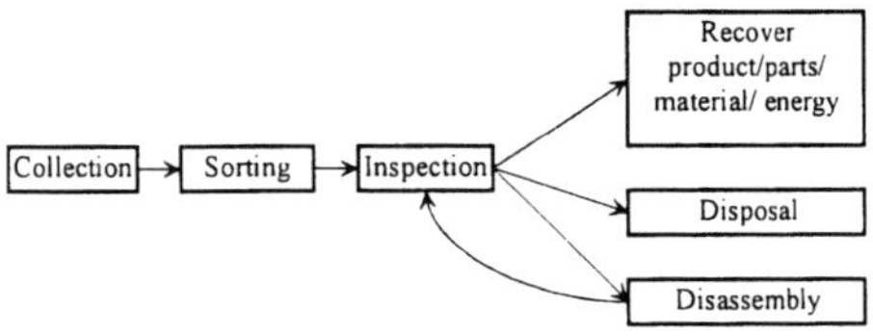

Fig. 2: EOL operations

The next step in the recovery process is to decide the appropriate *recovery option* to be chosen (refer to section 2.2 for the different recovery options). The product recovery decision is actually an iterative process. While selecting the best option, one has to take into account its technical, commercial and ecological feasibility (Krikke *et al.*, 1998). Technical feasibility reflects the technical possibility to realise a particular recovery option. Commercial feasibility points to business potential. Ecological requirements follow from environmental legislation.

In addition to the recovery options, the product could be *disassembled* into constituent modules, or could be disposed. The term disassembly used in this context needs to be clarified. The aim of disassembly is to make separate recovery or disposal possible for every single subassembly. Removal of parts for recovery operations is not considered as disassembly, because it does not have the aim of separate treatment of the subassemblies. These operations are thus processing operations related to the recovery option itself.

Disassembly may be selective (product not fully disassembled) or complete (product fully disassembled). It is necessary to determine the optimum *disassembly level* for each product, as the highest possible value recovery may already have been achieved when only a part of the materials, parts or components are separated. At each level of disassembly, the best recovery option is selected until all the components are either recovered or disposed. The recovered products and components are routed back into the supply chain through secondary markets (as refurbished products) or as low-cost spare parts for maintenance.

Hence, the recovery option decision is intertwined with the problem of deciding the optimal level of disassembly to be performed. In the simplest case, this decision depends on the cost of performing the recovery and disassembly operations and the residual value of the product and its constituent components. The residual value of the product depends on various quality parameters such as the age, functional condition (working, or not working), physical condition, functional age (as a measure of obsoleteness), remaining useful life, etc. The values of these parameters, which define the current state of the product, often have to be determined through inspection or estimated from knowledge gathered from experience. The accuracy of this estimation depends on the variability or uncertainty associated with it. In cases of mass customised products or products that have a moderate to long usage life, the uncertainty factor is seen to be very high. This is due to the fact that for mass customised products, each item is potentially different from other items of the same product type or brand. In most cases, detailed information regarding the constituents of a product or its functionalities would have to be determined through detailed inspection. Products with moderate to long usage life tend to change or evolve over the usage phase. Parts or components might be added or removed during maintenance and upgrade and the manufacturer's specification would not match the actual configuration of the product when it reaches the EOL stage. For example, the product manual might show that the product is ten years old, leading to a low residual value for the product. Nevertheless, certain components might have been replaced or upgraded recently that would increase the product's residual value. Without this information, the product might end up being recycled for material recovery or even disposed, where as it could have been potentially reused. Hence, maintenance and replacement history will lead to greater chance of part or module reuse and reduce the cost of lost opportunity due to the disposal of potentially reusable parts and modules. Such information can be obtained from the maintenance history logbooks that accompany the product. Although, in most cases it is found that they are unavailable or not updated.

It is to be noted that products can be disassembled via various routes or *disassembly sequences*. For reasons such as irreversible joints, degradation during use, changes to the structure brought about by maintenance or upgrade etc, the disassembly process is often not the exact reverse of assembly process. Hence, in addition to design and reliability information, disassembly information - possibly in the form of instructions - should be made available at the point of product recovery.

Summarising, it is clear that product recovery operations are characterised by the high level of uncertainty with regard to the identity and current state of the products, which in turn often prevents the best recovery option being chosen. It was also seen how these decisions depend on information associated with the product at different stages of its lifecycle. The fact that each product is subjected to a different set of conditions throughout their lifecycle makes it evident that product information will be unique to every single product.

By increasing the amount of information available, the uncertainty associated with the state of the system is expected to reduce, facilitating more effective product recovery decisions. In the next section we will show how advanced product identification systems such as Auto ID technologies can make item-level product information readily available to the decision-maker.

2.4. Role of Auto-ID technologies in managing lifecycle information

Two of the most common automatic identification technologies are bar-coding and RFID. Because of the superior automatic identification potential, we focus here on RFID, and in particular the approach developed by the Auto-ID Centre. The reader is referred to McFarlane *et al.* (2003) for a detailed insight on these technologies. The Auto ID approach, also referred to as the "EPC Network ™ " consist of the following features:

1. *A Unique product identity described by an EPC (Electronic Product Code).* The EPC is a 96-bit numerical code that is unique to each item. The

EPC code is sufficiently large to enumerate all objects, and to accommodate all current and future naming methods. It provides for industry coding standards, such as those from the Uniform Code Council (UCC) and the European Article Numbering (EAN) International.

2. *Radio Frequency Identification (RFID) tags.* The EPC is embedded on RFID tags that are attached to individual products. Auto-ID technology thus makes dynamic data collection possible, where product identification is build into a product, giving it a unique footprint at the item level (see Fig. 3).

   Each RFID tag is scanned by a RF reader, which transmits the product's EPC to the Internet, where the real product information is kept. Depending on product characteristics and associated lifecycle information requirements, the tags itself may not retain any information other than the EPC. Nevertheless, where required, 'active tags' that are able to store information can be used to capture dynamic usage history information.

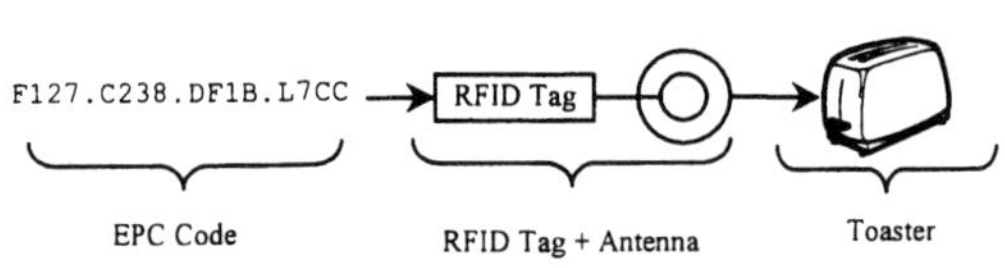

Fig. 3: Unique Identifier for a Toaster

3. *A Name Server that directs queries to information linked to an EPC.* The EPC acts as a pointer to a specific location on the Internet where product data is stored – it is analogous to a web address. An Object Naming Service routes queries to information about an object that carries an EPC code. The ONS is based in part on the Internet's existing Domain Naming Service (DNS), which routes 'addresses' to appropriate web sites.

4. *A Mark-Up Language to describe Product features.* Physical mark-up language (PML) is a new standard language based on XML for describing the form, fit and function of products on the Internet. The advantage of having a standard language to represent product information is that it can be used by all the actors of the value chain irrespective of the information systems and databases in place. This allows product information to be stored by various users in the product's lifecycle in their networks, and allows access to the information to authorized users (see Fig. 4).

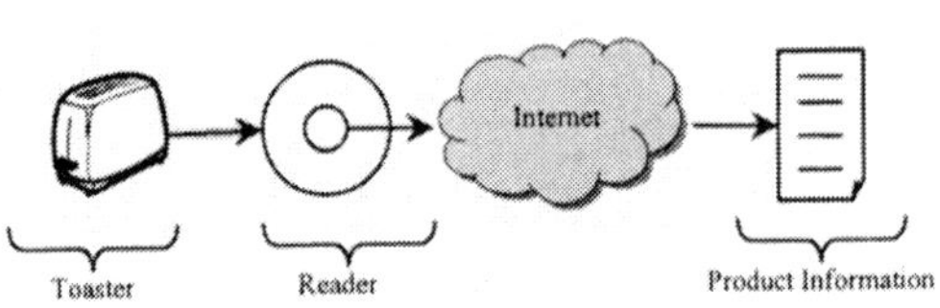

Fig. 4: Accessing product information through EPC code

Parlikad *et al.* (2003) illustrate a product-oriented information model based on the capabilities provided by Auto ID technologies. This information model enables a comprehensive representation of the current and past states of the product, as well as provide capability to also influence the future state of the product. This implies a product capable of being linked to information which influences decisions made about its reuse, rework, disassembly etc, and hence contributing to a highly automated product recovery process. Such an automated process will have the advantages of being cost efficient compared to manual processes and also would eliminate the cost due to lost opportunity and ineffective product recovery arising from wrongful disposal of reusable parts and modules.

2.5. Recap

In previous sections, we described the various operations and decisions involved in recovering value from returned products. It was shown descriptively how availability of product information could possibly result in improved product recovery decisions. We also discussed how the information necessary for making decisions at end-of-life can be potentially delivered using advanced product identification technologies. It is to be noted that Auto ID is only one of the methods that could be used to provide the required information. Parlikad *et al.* (2003) provide a review of other lifecycle information systems available and their shortcomings.

In the next section we look at ways by which recovery decisions can be modelled in order to demonstrate quantitatively how availability of product information changes recovery decisions and subsequently the performance of the recovery process.

3. MODELLING PRODUCT RECOVERY DECISIONS

Obtaining information has a cost associated with it. In order to prove the economic feasibility of deploying information systems for data capturing and retrieval, we need to be able to quantify the cost benefits that the availability of information brings to operations. It is hence required to quantitatively measure the impact of each piece of information associated with a product on performance of product recovery processes.

In this section, we will explore how product recovery decisions can be modelled in order to develop a measure for the impact of product information on product recovery decisions. We will first look at some of the techniques used for modelling product recovery decisions by the research community.

3.1. Modelling techniques

Various techniques for modelling product recovery processes are found in the literature for optimising the different decisions involved in the processes. Two of the most commonly investigated decisions are that of the disassembly level and disassembly sequence. A variety of methodologies such as graph based methods (Lambert, 1997), Petri Nets (Tang *et al.*, 2000), and

Linear Programming (Kanehara, *et al.*, 1993) are used to model the disassembly process in order to determine the optimal disassembly level and sequence. It is seen that all these models assume that the state of the product is completely known prior to disassembly. They do not account for the variability in the condition of returned products and unavailability of complete information to determine the state of the product.

Krikke *et al.* (1998) present a novel approach to determine an optimal recovery strategy of a product. The objective function takes into account, technical, commercial, and ecological criteria as well as uncertainty on these criteria due to lack of information. It is supposed that each assembly is classified according to a particular *quality class* once it is collected or released after disassembly. The uncertainty aspects of quality are modelled by conditional probabilities. The option that maximises the *expected profit* is chosen at each level of disassembly. The drawback of this model is that it fails to incorporate *new information* that often becomes available during the recovery process itself. For instance, disassembly of a product potentially would lead to more information regarding the state of components inside.

This brings our attention to Partially Observable Markov Decision Processes (POMDP) that are used to model situations where decisions have to be made with incomplete information regarding the state of the system (Monahan, 1982). An information or *belief state* represents all the information available at the decision time that is relevant to the selection of the optimal action. In a POMDP, we have a set of observations which provides a hint about what state it is in. The observations are probabilistic and tell us the probability of each observation for each state in the model. After each observation is made, the belief state is updated to reflect the new information obtained. The decision problem in the context of POMDP requires one to find an action or a sequence of actions that maximises the objective reward function. Because of their generality, POMDPs are used in numerous applications such as machine maintenance, quality control, robot navigation, etc.

In the next section, we will look at how this concept can be applied to model product recovery decisions.

3.2. Towards modelling product recovery decisions

When a product is returned after its initial use, it is collected, sent to the remanufacturer and sorted. After sorting, the product is inspected in order to make a decision as to its final destiny. The first step here is to determine the *state* of the system. The state of the system is typically identified by a number of parameters such as age, physical condition, functional condition, residual life etc. of the product. Formally, we can express the state X of the system as a function:

$X = f$(age, condition, residual usage life,...)

It is to be noted that this list of parameters is just a representative list. The actual parameters that influence the state of the system would have to be determined on a case by case basis. These parameters not only specify the state of the system, but also determine the residual value of the product.

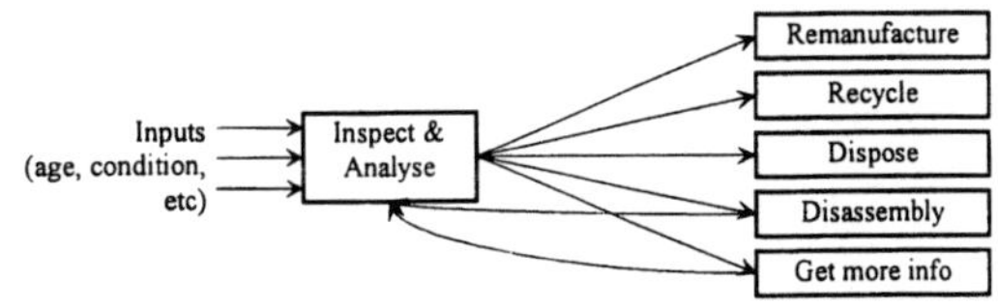

Fig. 5: Product recovery decision

In most cases, the values of these parameters are uncertain and hence have to be estimated by probability distributions. Depending on the values of the parameters, the state of the system could be represented in the simplest form by say, the set $X = \{0,1,2\}$, where '0' represents a bad product, '1' represents a good product, and '2' represents a repairable product. This set of possible states can be represented by a probability distribution π. Thus π is called the *prior knowledge* or *belief* regarding the state of the system.

At this point, various *actions* can be chosen among the available options (such as recover, recycle, incinerate, dispose, or disassemble). If enough information is not available to make the decision, the action taken could be to gather more information (refer to Fig. 5). Each of these actions is associated with a *payoff*, which is a profit or a cost associated with performing that action at that particular state. If, for example, the product is reused, the payoff would be a profit arising from reselling the product in the secondary market. If the product is disposed, the payoff would represent the disposal costs incurred.

If r is the payoff resulting from taking action a when the system is at state x, we can define the payoff function w as $r = w(x, a)$.

The expected payoff U is hence given by

$$U = \sum_{x} \pi(x) w(x,a)$$

The decision policy here is to choose the action that gives the highest expected payoff.

It is seen that the decision taken depends on the probability distribution of the state of the system. At each step of disassembly or inspection/testing, we will obtain *new information* about the state of the system. Marschak and Radner (1972) describe information as the outcome of *information gathering*. A given method of information gathering applied to the true state of the system x results in a particular *information signal*. If Y denotes the set of alternative possible information signals y, to a given information gathering method is associated a function, say η, from X to Y. This function is called an *information structure*. Thus, $y = \eta(X)$.

For a fixed information structure η, each signal y is identified in a natural way with a subset of X, namely the set of states x for which $\eta(x) = y$. It follows that every information structure induces a partition of X into an exhaustive family of mutually exclusive subsets, each subset identified with a particular information signal (Marschak and Radner, 1972).

The following example illustrates the importance of the POMDP concept of considering new information in modelling product recovery decision processes. Suppose for a particular product under consideration, we have no information regarding its current state. As described before, the state of the system can be represented by the set $X = \{0,1,2\}$. Since we have no prior knowledge, we assume equal probabilities across the different possible states. Suppose the product usage information was gathered from a source such as a history log. This potentially would partition the state set into $\{0\}$, $\{1\}$, or $\{2\}$ depending on the information presented in the history log. In other words, the history log would provide information about whether the product is good, bad, or repairable. It can be seen that this 'information signal' changed our 'belief' regarding the state of the system, resulting in a probability of 1 for the true state of the system and probabilities of 0 for other states. Evidently, the new information would change the expected payoff and hence possibly change the decision regarding the recovery option to be chosen.

By comparing the expected payoffs resulting from different information structures, it would be possible to quantitatively evaluate the impact of different information structures on product recovery decisions. It is our contention that POMDPs thus provide a means to model product recovery decisions in an accurate manner.

4. CONCLUSIONS AND FUTURE WORK

Lack of information infrastructure, excessive cost due to manual product recovery operations, and unavailability of product information were cited as the major obstacles for effective end-of-life management of products. This paper showed qualitatively how availability of product information has a positive impact on product recovery decisions, and also illustrated the role of automated identification technologies in providing the required information.

Various modelling techniques were explored in order to address the need to quantitatively evaluate the impact of product information on product recovery decisions. It was shown that POMDPs provide a feasible approach towards modelling product recovery decisions for addressing this issue. Clearly, the model discussed in this paper is in its conceptual stage and needs to be developed and refined in order to evaluate how product information affects product recovery decisions and under which circumstances obtaining complete product information can be cost effective.

REFERENCES

Ferguson, N. and J. Browne (2001). Issues in EOL product recovery and reverse logistics, *Production Planning & Control*, **12**, 534-547.

Goggin, K. (1998). *Modelling end-of-life product recovery*, PhD Thesis, National University of Ireland.

Guide V.D.R., V. Jayaraman, R. Srivastava, and W.C. Benton (2000). Supply chain management for recoverable manufacturing systems, *Interfaces*, **30**, pp 125-142.

Gungor, A. and S.M. Gupta (1999). "Issues in environmentally conscious manufacturing and product recovery: a survey," *Computers and Industrial Engineering*, **36**, pp 811-853.

Kanehara, T., T. Suzuki, I. Inaba, and S. Okuma (1993). On algebraic and graph structural properties of assembly Petri Net: searching by linear programming, *Proc. IEEE/RSJ Intl. Conf. Intelligent Robots and Systems*, pp 2286-2293.

Klausner, M. and W.M. Grimm (1998). Sensor-based data recording of use conditions for product takeback, *Proc. IEEE Intl. Symposium on Electronics and the Environment*, pp 138-143.

Krikke, H.R., A. van Harten, and P.C. Schuur (1998). On a medium term product recovery and disposal strategy for durable assembly products, *International Journal of Production Research*, **36**, pp 111-139.

Lambert, A.J.D. (1997). Optimal disassembly of complex products, *International Journal of Production Research*, **35**, pp 2509-2523.

Marschak, J. and R. Radner (1972). *Economic Theory of Teams*, Yale University Press.

Monahan, G.E. (1982). A survey of partially observable markov decision processes: Theory, models, and algorithms, *Manufacturing Science*, **28**, pp 1-16.

McFarlane, D.C., S. Sarma, J.L. Chirn, K. Ashton, and C.Y. Wong (2003). Auto ID systems and intelligent manufacturing control, *Engineering Applications of Artificial Intelligence*, **16**, pp 365-376.

Parlikad, A. K., D.C. McFarlane, S. Gross, and E. Fleisch (2003). The role of product identity in end-of-life decision making, Auto ID Centre White Paper.

Scheidt, L. and S. Zong (1994). An approach to achieve reusability of electronic modules, *Proc. IEEE Intl. Symposium on Electronic and the Environment*, pp 331-336.

Simon, M., G. Bee, P. Moore, J. Pu and C. Xie (2001). Modelling of the life cycle of products with data acquisition features, *Computers in Industry*, **45**, pp 111-122.

Tang, Y., M. Zhou, E. Zussman, and R. Caudill (2000). Disassembly modelling, planning, and application: a Review, *Proc. IEEE Intl. Conf. Robotics & Automation*, pp 2197-2202.

Thierry, M.C., M. Salomon, J. van Nunen, and L. Van Wassenhove (1998). Strategic issues in product recovery management, *California Management Review*, **37**, pp 114-135.

Thomas, V., W. Neckel, and S. Wagner (1999). Information Technology and product lifecycle management, *Proc. IEEE Intl. Symposium on Electronics and the Environment*, pp 54-57.

Copyright © IFAC Information Control Problems in Manufacturing, Salvador, Brazil, 2004

ELSEVIER
IFAC
PUBLICATIONS
www.elsevier.com/locate/ifac

PRODUCT-DRIVEN AUTOMATION ISSUES FOR B2M-CONTROL SYSTEMS INTEGRATION

David GOUYON *, Jean Marcelo SIMÃO *, **, Khaled ALKASSEM *, Gerard MOREL*

* *Nancy Research Centre for Automatic Control (CRAN)*
Henry Poincaré University (UHP)
{gouyon, simao, kassem, morel}@cran.uhp-nancy.fr
&
** *Post-graduation Program in Electrical Engineer and Industrial Computing (CPGEI)*
Paraná Federal Center of Technologic Education (CEFET-PR)
simao@cpgei.cefetpr.br

Abstract: Enterprise-control system integration between business systems, manufacturing execution systems and shop-floor process-control systems remains a key issue for facilitating the definition, the development and the deployment of plant-wide information-control systems for e-B2M issues. Many rationales, trends and experiments are applying the HMS paradigm and its related agent-based technology in order to meet these B2M integration, interoperability and agility issues in industry. Research works presented in this paper emphasize some formal large-scale discrete-manufacturing product-driven automation issues within a broader modelling framework addressing the HMS paradigm as a pertinent system engineering approach mainly for products traceability purposes. *Copyright © 2004 IFAC*

Keywords: Manufacturing-Enterprise Control-Integration, Manufacturing Execution System, Manufacturing Automation, Holonic Manufacturing Systems, Systems Engineering, Holonic Supervisory Control.

1. INTRODUCTION

Manufacturers have to face the increasing desires of on-demand customized products as well as the legacy rules evolutions for managing and tracking the products life cycle throughout the extended manufacturing chain. Overcoming information-control problems for make-to-order e-manufacturing is challenging the IMS[1] community for facilitating integration, interoperability and agility between business systems, manufacturing execution systems and shop-floor process-control systems. Many works are applying the HMS paradigm and its related agent-based technology for the deployment of plant-wide information-control systems in industry (Deen, 2003)(Valckenaers, 2001).

[1] Intelligent Manufacturing Systems international initiative; http://www.ims.org/

Others works outline the lack of formal methods for controlling large-scale discrete-manufacturing execution-systems (Qiu *et al.* 2003) as well as for controlling products instances life cycle (Fig. 1) within shop-floor manufacturing systems (McFarlane *et al.* in Morel and Grabot 2003).

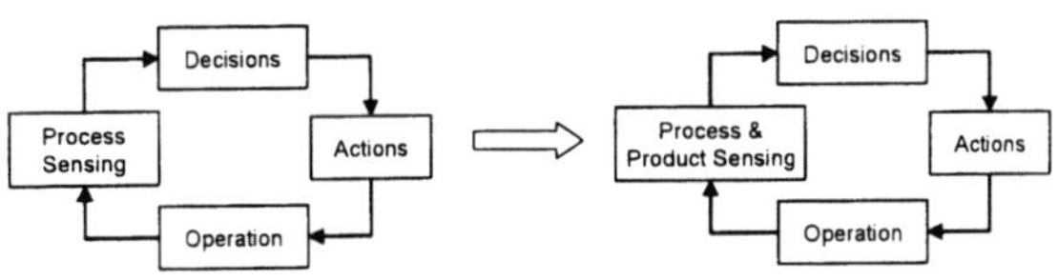

Fig. 1. Moving from process-driven control to product-driven control

Our own practice of object-oriented modelling techniques emphasizes the lack of formal constructs to assemble automation objects over information when considering physical constraints.

2. RESEARCH PROJECT FRAMEWORK

All these rationales, trends and experiments address the HMS paradigm as a pertinent systems engineering approach (Valckenaers *et al.*, in Morel and Grabot 2003)(Christensen in Deen 2003) to cope with product-driven business-to-manufacturing (B2M) management and control integration issues.

2.1 HMESE Project

Our on-going research works deal with the conceptual level of a *Holonic Manufacturing Execution System Engineering* (HMESE) framework (Fig. 2) enabling the *Holonic Modelling* of a manufacturing plant-wide information-control system before its projection on standardized MES and Automation technology.

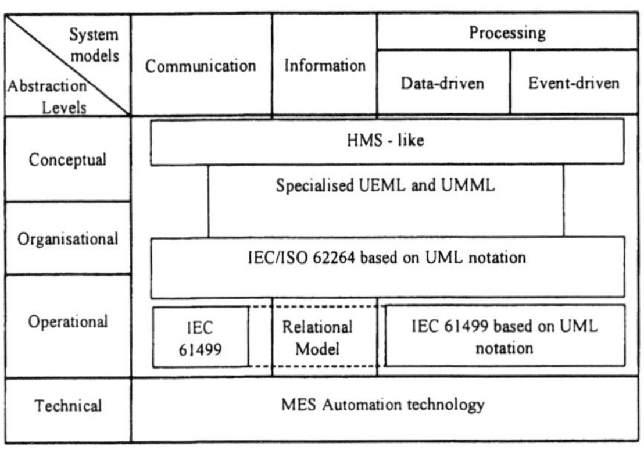

Fig. 2. HMESE framework (Morel *et al.*, 2003)

This HMESE approach is prototyped with an industrial Business Process Modelling case tool[2] which has to be specialized by extension of its meta-model with stereotypes, instances of meta-classes, standard or automation or skill meta-objects ... This specialisation is mainly formalized with ORM[3] for its fact-orientation expressiveness (Halpin 2001) before to be translated into UML constructs for meta-model integration.

HMESE is part of a CRAN research project intituled *Product-driven Controlled Manufacturing Systems* aiming to define, develop and deploy automation methods for the product-driven control and management of industrial processes in order to ensure the traceability of both goods and services all along products life cycle.

Our common project-postulate stating *Holon-driven Modelling* (Fig. 3) as a pertinent abstraction of any product-driven manufacturing-execution engineering issue is going to be validated through industrial experiments related to continuous, discrete and batch manufacturing processes as well as in research works-in-progress.

[2] MEGA case tool: www.mega.com

[3] VisioModeler: www.microsoft.com/visio

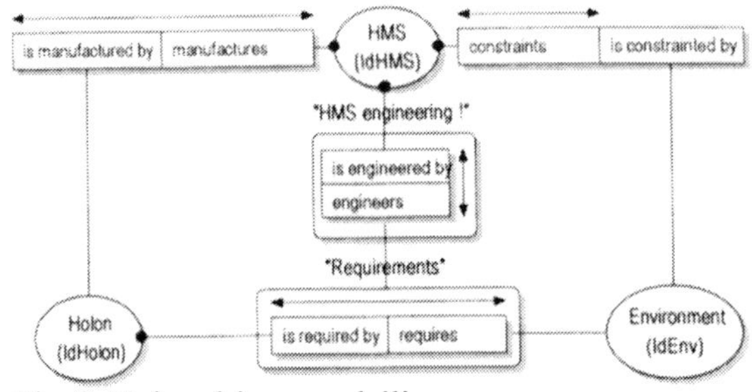

Fig. 3. Holon-driven modelling

PhD-thesis works presented in sections 3, 4, 5 deal respectively with *Holonic Process Modelling*, *Product-driven Control Synthesis* and *Holonic Supervisory Control Simulation* on the same Flexible Assembly Cell case-study.

2.2 Flexible Assembly Cell Case Study

FAC (Fig. 4) is an assembly *Cell* integrated in the manufacturing *Area* of the training AIPL-PRIMECA[4] *Site* enabling to put trainees and trainers in a realistic Business (ERP) to Manufacturing (MES and Shop Floor Automation) e-integration context.

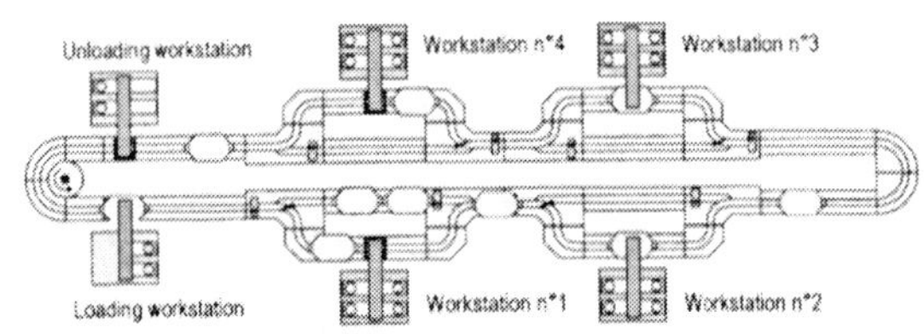

Fig. 4. Flexible Assembly Cell (FAC)

FAC enables flexibility in the routing of products assembly through six Workstations (WSs): one for loading pallets circulating on a conveyor, four similar WSs to make products by parts assembly, and one to unload products on pallets. Six types of products can be assembled by the FAC (Fig. 5) from six part types.

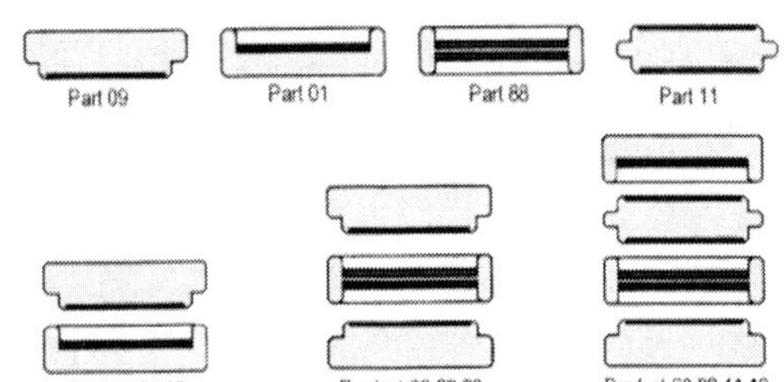

Fig. 5. FAC assembled products

FAC architecture is conceptually considered as a HMS for research works while its technical organization for training works enables only to store products data (i.e. products-composition...) in a digital memory embedded on each pallet carrying four products. Automatic Identification of pallets enables to control the routing of products assembly through the WSs controllers networked by a field-bus.

[4] http://www.aipl.uhp-nancy.fr/lorraine/

3. FAC HOLONIC PROCESS MODELLING

As addressed in previous works in automotive industry (Muhl *et al.* in Morel and Grabot 2003), holonification (reengineering approach) should be applied in order to improve and optimise the overall performance of manufacturing plant because of mass customization (Da Silveira *et al.* 2001) impact.

Among many concerns remains the plant rescheduling dynamics (Sousa & Ramos 1999) and the definition, before production rather than on the fly (Qiu and Russel 2004), of the ordering of products instances, process flows and process operations depending of real-time manufacturing capabilities.

Another concern of current B2M Process Modelling techniques is, on one hand, their lack of semantics to efficiently take into account the physics of the manufacturing area and, on another hand, their lack of abstraction over technology to efficiently check with the complexity of understanding large-scale plant behaviours, especially during the early phases of elicitation and specification.

FAC is conceptually reengineered as an HMS by applying our *Holon-driven Modelling* approach in order to be fully controlled and managed by the products for B2M integration through a MES. *Holonic Manufacturing Process Modelling* enables first to trace within *Holonic Process Diagrams* the Shape / Space / Time transformations of *Product-Holons Flows* and *Resource-Holons Flows* exchanged with external Holonic Business / Manufacturing Processes as well as between internal *Holonic Manufacturing Processes* (Fig. 6).

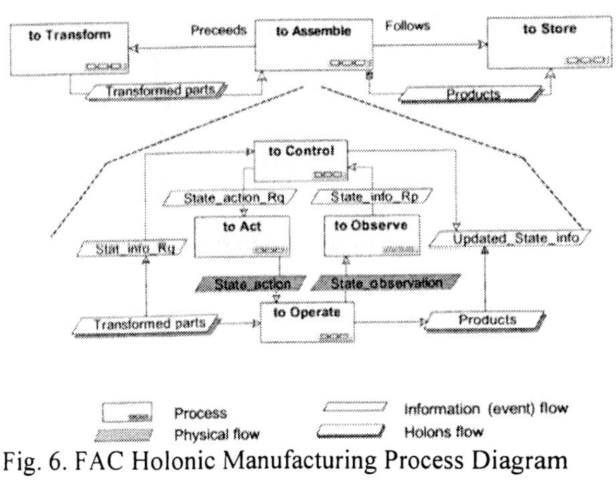

Fig. 6. FAC Holonic Manufacturing Process Diagram

These *Holonic Process Diagrams* can be then clustered in order to describe *Product-Holons Flows* handled by a unique Manufacturing Organizational-Unit. *Product-Holons Flows* and *Resource-Holons Flows* can be split into Physical Flows and Information Flows (data-driven and event-driven) in order to describe within *Manufacturing Flow Charts* the routings of the Physical-Objects and of the Information-Objects handled by different organizational units (Fig. 7).

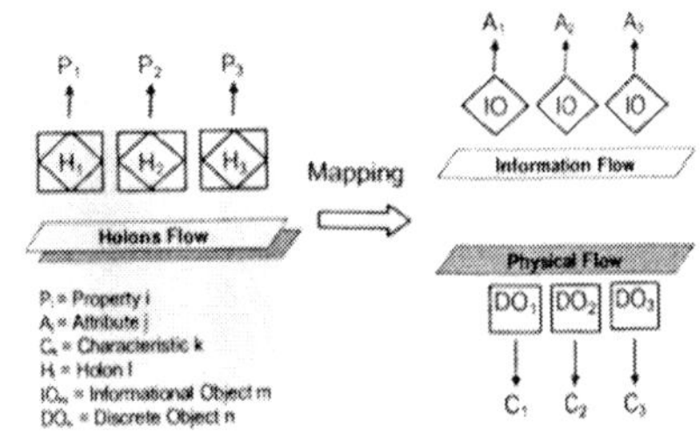

Fig. 7. Mapping of Holons Flows into Information and Physical flows

This systemic approach enables to extract *Product-Holons* and *Resource-Holons* from the *Holons Manufacturing Flows* before their formal interpretation (Fig. 8) to design *Holons Class Diagrams.*

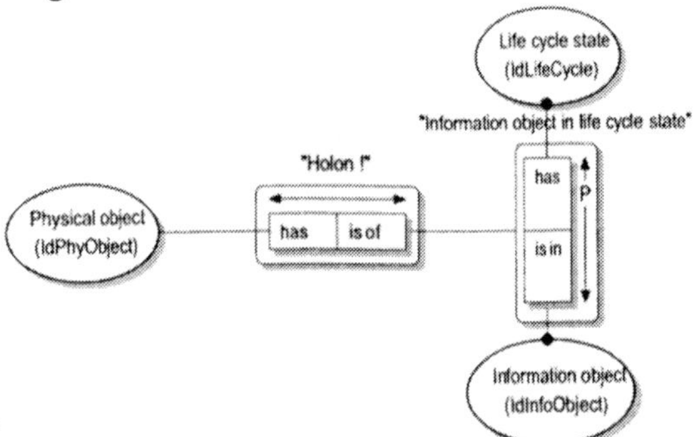

Fig. 8. Holon Meta-Model

Holons Properties which should be observed, controlled and modified for traceability purposes are then specified from *Holons Properties Meta-Model.* Figure 9 illustrates an example of rank and motif properties for product-driven manufacturing flow control and management.

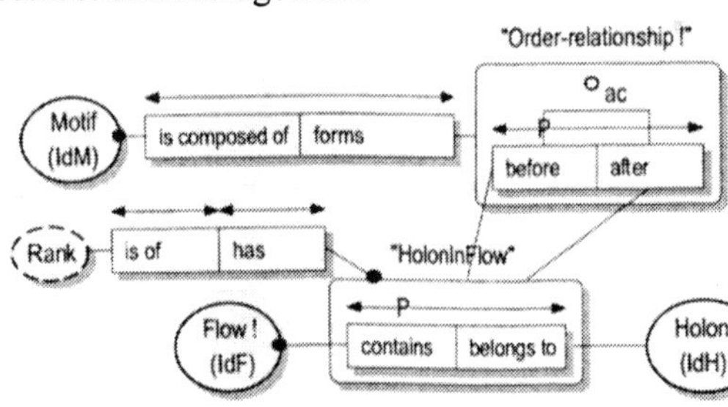

Fig. 9. Holon-rank driven and Holon-motif driven flow meta-model

Holons Operations processed by *Holonic Processes* can be described by Finite State Automata or State Charts in order to define all the flexible routings of *Product-Holons* and *Resources-Holons* through the manufacturing-line work cells.

4. FAC PRODUCT-DRIVEN CONTROL SYNTHESIS

As addressed by (Qiu *et al.* 2003), formal techniques can be applied in order to synthesise product trajectories to execute make-to-order Business-to-Manufacturing. Our going works are compliant with this vision in order to automatically synthesise products-driven routes at the MES level. Our approach should be recursively applied to the FAC case study in order to synthesise both Product-driven routes and Product-driven supervisory controllers. This section is restricted to Product-driven trajectories while results on supervisory control synthesis are described in (Gouyon *et al.*, 2003).

The formulation of the automatic synthesis of a control system proposed by (Fusaoka *et al.* 1983) consists in defining the control rules of the dynamics of a physical system, from the behavioural goals to be met, according to:

$$\textit{Dynamics} \wedge \textit{Unknown control rules} \supset \textit{Goal} \qquad (1)$$

This formulation is compliant with the system theory in automatic control (Cassandras and Lafortune 1999) and with the Supervisory Control Theory (Ramadge & Wonham 1987) of Discrete Event Systems. Nevertheless, as addressed by (McFarlane 2003), both these approaches do not fully meet the requirements for Product-driven control, mainly the product observability through its automatic identification.

Our interpretation aims to synthesise through time the routings of each product instance and to synthesise at time the synchronisation with the Product-driven manufacturing processes according to:

$$\begin{array}{l}\textit{Product-driven manufacturing behaviour} \\ \wedge\ \textit{unknown Product-driven control rules} \\ \supset\ \textit{Product-driven manufacturing planning} \qquad (2)\end{array}$$

Solving this Product-driven Control predicate (2) by applying synthesis algorithms usually used in Supervisory Control Theory (Ramadge & Wonham 1987) consists in ensuring the global consistency between two Product-driven Behavioural steps. Product-driven resources supervisors are synthesised from resources dynamics models and from Product-driven supervised-resources specification models (Fig. 10a). Product-driven routings supervisors are finally synthesised from Product-driven supervised-resources models and from Product-driven manufacturing planning specification models (Fig. 10b).

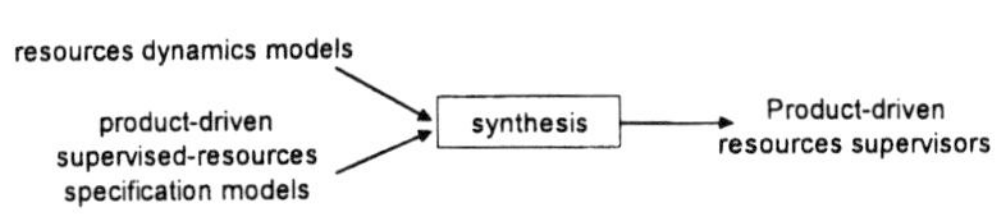

a) Product-driven resources supervisor synthesis

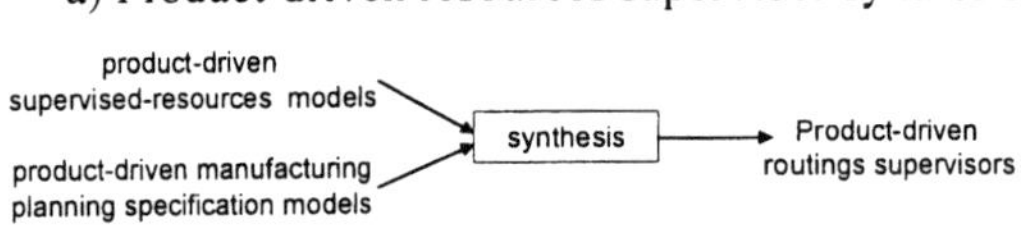

b) Product-driven routings supervisors synthesis

Fig. 10. Product-driven synthesis principle

Product-Holons collaborate with the manufacturing processes composed of various *Resource-Holons*, able to know their state, their position and their capabilities. *Resource-Holons* and *Product-Holons* have to share a common communication protocol composed of functional requests and reports such as "*Request for operation X*" or "*Report of operation X achievement*".

All the specifications and process models are formalised with finite state machines. Fig. 11 illustrates the event-driven model of product manufacturing planning provided by product-process engineering.

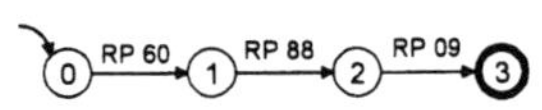

Fig. 11. Example of a product manufacturing planning (events correspond to operations reports)

A projection operation (Fig. 12) enables to automatically derive product-driven manufacturing behaviour models.

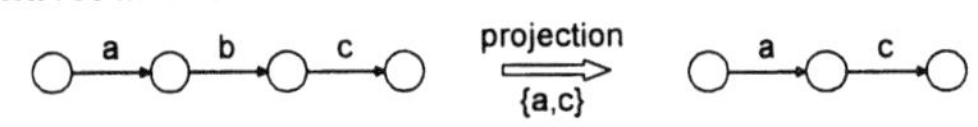

Fig. 12. An example of projection

As FAC case study example of product-driven manufacturing behaviour, let us consider first two workstations, one with pieces 60 and 09 (workst1 at position 1), the other with pieces 88 (workst2 at position 2), and a conveyor for moving one-product pallets between WSs (Fig. 13).

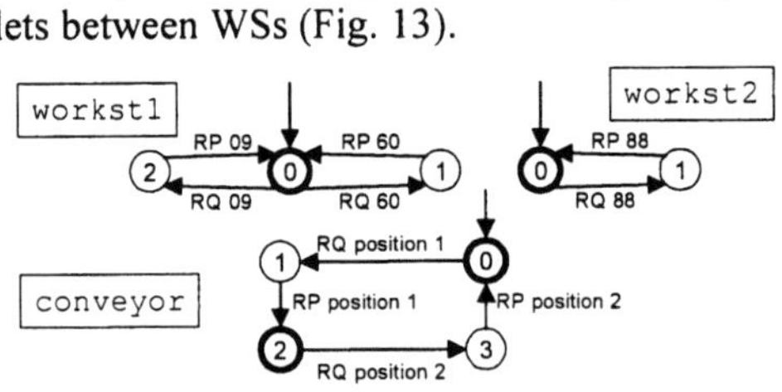

Fig. 13. Event-driven process components models

A Product-driven manufacturing behaviour model corresponding to (2) is illustrated Fig. 14 for the above mentioned example.

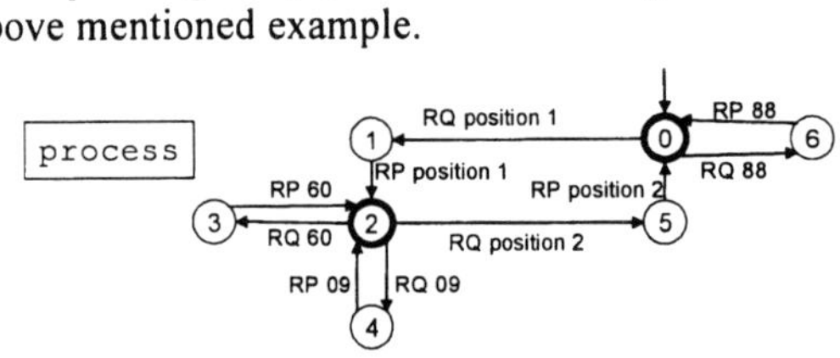

Fig. 14. Product-driven process model

The whole set of Product-driven routings for a product assembled with a sequence 60-88-09 as specified in Fig. 11 is then automatically synthesised (Fig. 15).

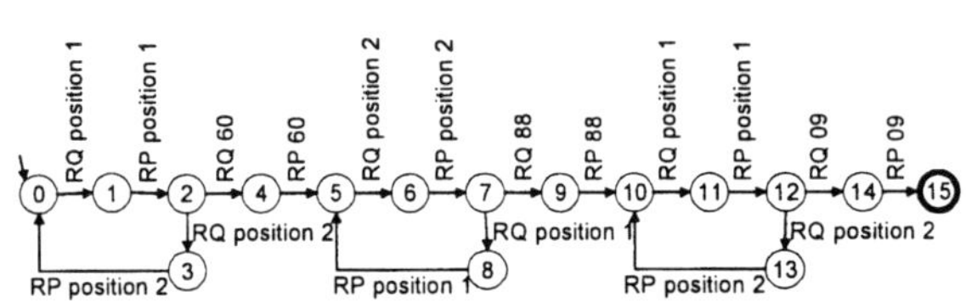

Fig. 15. Synthesised Product-driven routes for product 60,88,09

Over this simple example, another scenario consists in: (a) one-product pallets, (b) WS0 with parts 60 and 01 (first operation for all products), (c) WS1 with parts 10, (d) WS2 with 09, (e) WS3 unused, (f) WS4 with parts 09 and 10, and (g) WS5 unloading all products; one product-cycle through FAC. The supervisor of all product-driven trajectories for the assembly of products 01,09 (Fig. 16) and 60,10 under this scenario has been validated by simulation as described in section 5. Finite state automata are translated into UML state charts.

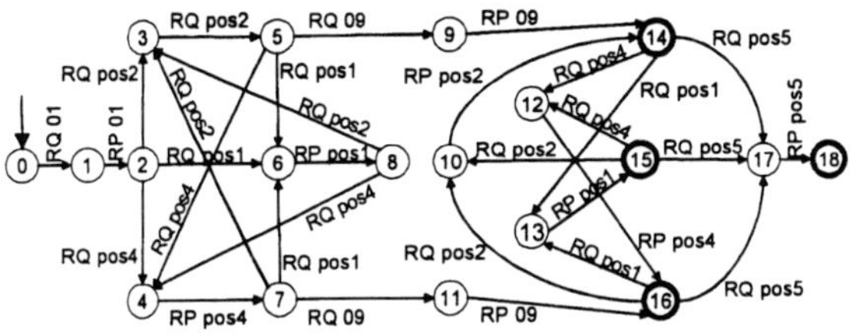

Fig. 16. Product 01,09 driven possible trajectories

Next works towards implementation into IEC 61499[5] function blocks (Christensen in Deen 2003) deals with the integration of the TCT[6] tool for the synthesis phase with Control Build[7] tool for execution phase.

Others works towards product-driven MES supervisory control, should consider these product-driven trajectories as *virtual supervisors* of the real assembly routes.

5. FAC HOLONIC SUPERVISORY CONTROL SIMULATION

Previous researches aimed to define an Agent-oriented Supervisory Control model for Flexible Machining Cell (FMC) simulation purposes (Simão & Stadzisz, 2002).

In order to better fit with the physics of the controlled processes, FMC and FAC are *holonified* according to Fig. 8 meta-model.

These two particular discrete-manufacturing case-studies models enable to generalise a first meta-model of *Holonic Supervisory Control* (HSC) which has been instanced and partially tested in qualitative simulations on ANALYTICE-II (from CPGEI/CEFET-PR).

FAC HSC modelling principles (Fig. 17) are illustrated by a particular simulation scenario considering the synthesised Product 01-09 and Product 60-10 assembly discussed in the above section:

(1) An *Order-Holon* (from B2M management system) comprises information (delivery date, type and client preferences) concerned to make a specific product (e.g. product n°1 of type 60,10 from FAC example).

(2) A *Product.Class-Holon* contains information (e.g. possible process plans and operations list) to produce a type (class) of product (e.g. products type 01,09 with the trajectories obtained by synthesis (Fig. 16).

(3) A *Product-Holon* treats a product instance: (a) it is created when an *Order-Holon* arrives, using information from *Order-Holon* and *Product.Class-Holon*, (b) it encapsulates the trajectory chosen by the *HSC* in function of *Resources-Holons* availability, (c) it records, when needed, traceability information (e.g. *WS1 made assemblage 60-10*).

(4) A *Resource-Holon* encapsulates a production entity (e.g. *WS1* or *Conveyor*) showing its capabilities (e.g. *Assemblages 01-09* or *Transport from Place-0 to Place-2*) and states (e.g. *busy* or *not busy*, and *Product-HLs quantity inside*), as well as receiving (high-level) "commands" (e.g. *Assemble 01-09!* or *Transport from Place-0 to Place-2!*) and up-dating information in the *Product-Holon* that was physically modified.

(5) A *CtrlRule-Holon* is a "*control rule*": (a) it knows the *co-operation condition*, i.e. states of *Resource-HLs* (e.g. *conveyor way-1 is free* and *WS1 is free*) desired by a *Product-Holon* (e.g. *Product-Holons 60,10* and *01,09*) to realise operations (e.g. *to assemble 60-10* or *01-09*), (b) it actives the co-operation *action*, when the *condition* is true, that instigates the cooperation among involved holons to carry out operations, and (c) it has new facts (or states) updated in its *conditions* by notification from concerned *Resource-Holons* and *Product-Holons*, i.e. an *active facts base* notifying only interested *CtrlRule-Holons*, avoiding unnecessary exchanges (a linear processing time).

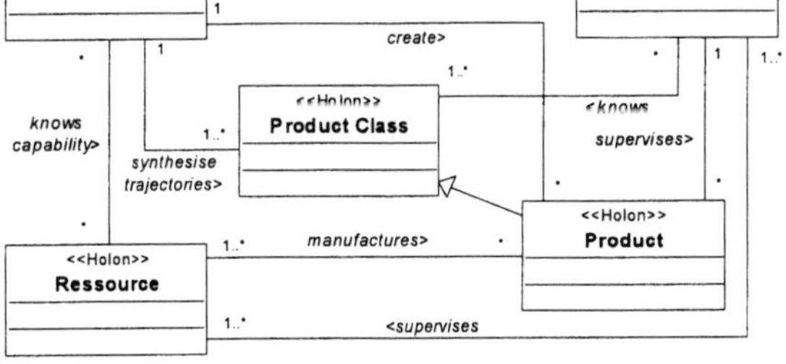

Fig. 17. A partial vision of the HSC in the holarchy.

CtrlRule-Holons "emerge" from interactions between *Resource-Holons* (capabilities) and *Product.Class-Holons* (requirements). In this interaction, *Product.Class-Holons* memorize which *CtrlRule-Holons* are interesting to make their corresponding products. So, when a *Product-Holon* arrives, it chooses a *CtrlRule-Holon* (via related *Product.Class-Holon*) to make its first operation and, after holons cooperation has been made, it chooses another *CtrlRule-Holon* for its second operation, and so on.

[5] IEC 61499 (2000): Functions Blocks, Part1 - Architecture. *IEC PAS 61499-1*, Geneva, Switzerland.
[6] TCT: http://odin.control.toronto.edu/DES/
[7] Control Build http://www.controlbuild.com/

This scenario proposes a HSC model where decisions are in distributed *CtrlRule-Holons* and monitoring, production and traceability come from *Resource-Holons* (as an *intelligent* distributed expert-system). In this context, the product-drive comes from *Product-Holons* that choose good *rules* (among all possible) to make their operations.

The proposed HSC brings agile features, for example some emergence (ctrl-rules creation), flexibility exploitation (rules that looks for all possibilities), some re-configurability (e.g. changing rules priority) and some fault detection (e.g. *Product-Holon* does not find a rule). Additionally, this scenario is being enlarged to contemplate others parts and product, but first results confirm HSC potentiality tough some concepts have not been automated, e.g. *CtrlRule-Holons* emergence.

Nevertheless, many HSC open issues remain and further case-studies should be necessary in order to increase the genericity of the proposed HSC meta-model before to efficiently translate in UML constructs for product-driven Shop Floor Control industrial experiments.

6. CONCLUSION

This paper deals with conceptual issues of an on-progress HMESE approach aiming to facilitate the design, the development and the deployment of Product-driven control and management B2M applications, mainly for products traceability purposes through manufacturing execution systems and shop-floor process-control systems.

Holon-driven Modelling aims to be a core guideline during the early phase of specification of such applications in order to limit the degrees of freedom of the *Software-Holons* from their linkage to *Hardware-Holons*.

This prevalence of manufacturing processes on business processes enables to design *Holonic Process Diagrams* ensuring, by definition as stated in Fig.8, the coherence between material-flows and information-flows as first *Holon-Flows*.

Holon-types are extracted from these Holon-Flows in order to design *Holon Class Diagrams* for products management and tracking purposes.

Holons routings processed by *Holonic Processes* can be described by UML State Charts or synthesized as Finite State Automata, depending of the requested reactivity of decision making before production as usually or on the fly as addressed by (Qiu, 2004).

Finally, *Holonic Meta-Modelling* enables to specialize a Business Process Modelling case tool for industrial experiments or a simulation tool for research experiments.

REFERENCES

Cassandras C.G., Lafortune S., (1999). *Introduction to discrete event systems*. Kluwer Academic Publishers. ISBN 0-7923-8609-4.

Da Silveira G., Borenstein D., Fogliatto F.S., (2001). Mass customization: literature review and research directions. *International Journal of Production Economics*, **vol. 72,** pp 1-13.

Deen, S.M. (Ed.) (2003). *Agent-Based Manufacturing - Advances in the Holonic Approach.* Springer ISBN 3-540-44069-0.

Fusaoka A., Seki H., Takahashi K., (1983). A description and reasoning of plant controllers in temporal logic. *Proceedings of the 8th International Joint Conference on Artificial Intelligence,* Karlsruhe, Germany, pp 405-408.

Gouyon D., Pétin J.-F., Gouin A., (2003). Process and specification models for control synthesis. *Reactive Systems Modelling MSR'2003*, Metz, France (in french). ISBN 2-7462-0778-8.

Halpin T., (2001). Augmenting UML with Fact-orientation. *Proceedings of UML: a critical evaluation and suggested future, HICCS-34 Conference,* Maui, Hawaii.

McFarlane (2003). Product Identity and Its Impact on Discrete Event Observability. *European Control Conference 2003*, Cambridge, UK.

Morel G., Panetto H., Zaremba M., Mayer F., (2003). Manufacturing enterprise control and management system engineering: paradigms and open issues. *Annual reviews in Control*, **vol. 27**, pp 199-209

Morel G., Grabot B., (Eds.), (2003). Intelligent Manufacturing. *Special issue of Engineering Applications of Artificial Intelligence,* **vol. 16 (4).**

Qiu R., Wysk R., Xu Q., (2003). Extended structured adaptive supervisory control model of shop floor controls for an e-manufacturing system. *International Journal of Production Research,* **vol. 41 (8),** pp 1605-1620.

Qiu R., Russel D., (2004). A formal model for incorporating shop floor controls into plant information systems. *Int. J. Adv. Manuf Technology,* **vol. 23 (23),** pp 47-57.

Ramadge P.J., Wonham W.M., (1987). Supervisory control of a class of discrete event processes. *SIAM Journal of Control and Optimization*, **vol. 25, (1),** pp 206-230.

Simão J. M. and P. C. Stadzisz (2002). An Agent-Oriented Inference Engine applied for Supervisory Control of Automated Manufacturing Systems. *Frontiers in Artificial Intelligence and Applications.* **vol. 85**, pp 234-241, IOS Press, Amsterdam - The Netherlands. ISSN: 0922-6389.

Sousa P., and Ramos C., (1999). A distributed architecture and negotiation protocol for scheduling in manufacturing systems, *Computers in Industry*, **vol. 38,** pp 103-113.

Valckenaers P. (Ed.), (2001), Special issue on Holonic Manufacturing Systems, *Computers in Industry*, **vol. 46 (3)**.

Copyright © IFAC Information Control Problems in
Manufacturing, Salvador, Brazil, 2004

ELSEVIER
IFAC
PUBLICATIONS
www.elsevier.com/locate/ifac

SUPPLY NETWORK CONTROL, AN ENGINEERING PERSPECTIVE

**Bart Saint Germain * Paul Valckenaers *
Paul Verstraete * Olaf Bochmann *
Hendrik Van Brussel ***

* *K.U.Leuven - P.M.A.,
Celestijnenlaan 300B, B-3001 Leuven, Belgium,
Bart.Saintgermain@mech.kuleuven.ac.be*

Abstract: This paper presents an engineering perspective on the supply network coordination problem. This manuscript treats the coordination as an ongoing task and takes disturbances and exceptions as business as usual. A manufacturing control and coordination mechanism based on PROSA and ants is used as base for the overall coordination of the supply network. Reputation serves to enforce the social behaviour among the production entities. The interactions from the control mechanism towards the reputation models are given. *Copyright © 2004 IFAC*

Keywords: Supply network, holonic, PROSA, production control, reputation

1. INTRODUCTION

Today, manufacturing industries are aware that designing the right kind of flexibility into their forthcoming production systems is vital to their future competitiveness (Goethals, 2002). For long time, companies have tried to optimize their own internal processes. Nowadays companies do not operate anymore in an isolated world. To achieve a further optimization in their production system, industrial units have to look beyond the "walls" of their company.

This paper discusses a multi-agent coordination and control approach to operate the external and internal logistics in a multi-site/multi-organization topology. Past experience reveals that manufacturing control cannot be implemented through 'one shot scheduling solutions'. On the shop floor, schedules have acquired a reputation for rapidly becoming invalid because of the frequent changes and disturbances. This effect becomes even bigger in supply nets (H.L. Lee and Whang, 1997) and requires until now some "brute-force" solutions (i.e. huge buffers, low product-type variability,). Consequently, the multi-site manufacturing control presented in this paper, treats changes and disturbances as business-as-usual. It considers manufacturing as a non- stop 'going concern'.

The manuscript first defines the vision on the problem, supply network control. An engineering point of view is taken as a start point. Next, the paper describes how a supply network can be controlled starting from a factory control and co-ordination mechanism (Hadeli, 2003). Finally, the interactions to trust and reputation mechanisms are given in order to create a social behaviour into the coordination mechanism.

2. SUPPLY NETWORK CONTROL DEFINITION

This chapter presents a definition of supply network control. Control and especially supply network control can be understood in many ways.

Following an engineering perspective, supply network control is defined by the nature of a supply network and the problems it brings about. The starting point for this paper comes with the insight that a production entity cannot be regarded as a black box. Instead a production entity will be defined as an autonomous and specific unit in the overall supply network. Supply networks on the other hand are multi-level networks of entities.

Supply network control will be defined as a nonstop going concern with as main concern the productivity and responsiveness of the total supply chain. This global productivity has to be understood with respect of the social nature of the problem, win-win situations will be preferred.

2.1 Production entity definition

Production entities are units in a production environment that deliver services. Services as they are carried out by entities have to be understood in the broad sense. Transport services, producing components and repairing machines are only a few examples of possible services. Services are described by means of their capabilities and constraints. Truck transport services are described by the transport capability and by constraints like maximum height/width and weight of the pieces that trucks can carry. Entities have typically two properties that characterize themselves: autonomy and specificity, explained in the following paragraphs.

Autonomy The autonomy stands for the ability to decide how, when, where to deliver services together with which services they provide to the environment. These local decisions are not always without consequences, which have to be taken into account in the decision process. In other words, entities should behave in a socially acceptable manner. This first observation above emerges out of the essential difference between a factory control and the supply network control. Production entities in a supply network do not all belong to a single owner, and therefore the network control cannot impose a single goal and/or strategy. The entities will have varying degrees of autonomy, which the network control has to account for.

Specificity Specificity reflects the uniqueness of an entity. With respect of the technology and strategic decisions enabling to provide services, it is very difficult to model the different entities in general. Finite capacity, disturbances and policy changing in the technological process makes forecasting and planning of production a unfeasible/inaccurate task without having a deeper insight into the entity. In other words, this paper starts from the fact that an entity cannot be adequately modeled by a black box comprising a processing unit with input and output buffers. This novel approach contrasts with more economic and control theoretical approaches where a lot of assumptions are made towards the logic inside the entities: fixed lead-times, no disturbances, etc.

This observation, entities are specific, builds on the research team's experience with manufacturing control - recall that this paper is about scaling manufacturing control technology to supply network coordination and control. When a control system operates in a competitive environment, overconstraining its options by oversimplifying the world model is likely to be detrimental for system performance. And, unlike in business systems, it is not economically feasible to reengineer the production systems to fit the model of the control system. Hence, production entities cannot be reflected in the control system by a single generic model.

2.2 Supply network definition

A supply network can be defined as a multilevel network of production entities collectively responsible for procurement, manufacturing and distribution activities associated with one or more families of related products. Supply networks are composed out of entities in two ways, vertical and horizontal relations. (1) A horizontal connection expresses the ability to use services of the connected entity. (2) Vertical relations model the hierarchy at one point in time from the supply network. Organizations are divided in several plants, plants comprise a couple of divisions, etc. On each of the levels a same pattern of a horizontal network together with aggregations can be found. Both horizontal and vertical connections are subject to frequent changes. New partners are coming into the supply network, organizations restructure, switch to outsourcing, etc.

As mentioned in section 2.1 entities do not always reflect units in the organizational structure. This makes that the hierarchy visible in the supply network is not necessary a mapping of the organizational structure. More powerful structures can be defined to coordinate the supply network without having to change the structure of the organization.

2.3 Supply network control definition

Supply network control copes with the logistics in a supply network as defined in the previous section. Supply network control handles the external factory logistics while interacting with internal factory logistics. Logistics management includes not only decision taking but also and not less important, information handling, monitoring and forecasting.

Controlling supply networks comprises a variety of objectives. The first objective is to cope with the going concern in supply chain control. Treating the control system as a static solution will fail in the dynamic and uncertain environment of the supply network. Disturbances on production level are propagated in both horizontal as vertical direction of the supply network, and the fluctuation on the uncertainty becomes even bigger (ref bullwhip). The second and probably the most important objective of the control system is the productivity over the overall supply network. There should be a tendency to pareto-optimal solutions where win-win situations - out of the pareto-optimal set - over different entities are mostly preferred when taking a local decision.

Internal versus external logistic interactions happen with respect for the autonomy and specificity of each of the entities belonging to the supply network. Specificity makes central coordination and control very hard and not desirable due to the crucial knowledge inside the entity. This knowledge is very hard to transfer in a general way and has to be inevitably handled by the entity itself.

2.4 Discussion

In this paper a definition for supply chain control is given starting from an engineering perspective. Entities from supply networks differ from resources in a factory in two ways. Entities are autonomous and specific. These differences and the dynamics of the problem make central approaches and approaches where assumptions on internal entity logic are made hard to maintain. Supply networks are not defined as a single graph but as a multi level network that represents the structure of the entities at one point in time. Supply network control is defined as a going concern with as main target maximizing the productivity and responsiveness of the total supply network.

3. SUPPLY NETWORK COORDINATION AND CONTROL

The starting point of this chapter is a factory control and coordination mechanism. This factory control is based on a reference architecture called PROSA and uses ant based techniques to coordinate and control the production facility.

PROSA was developed for manufacturing control applications; nonetheless this section shows how this architecture can be adapted towards a supply network architecture. Next a way how the insights on ants can help to coordinate the supply network is described.

3.1 Supply network Architecture

The reference architecture described in this paper is based on the holonic paradigm developed by Arthur Koestler. Koestler introduced the word holon as a composition of the word "holos" (=whole) an the post-fix -on like proton (=part). The two main properties of a holon are autonomy and cooperation. The autonomy provides stability and robustness in the changing world. The cooperation on the other hand gives the holon a place in the bigger system.

PROSA Starting from the holonic concept a reference architecture is build for production systems, called PROSA (Wyns, 1999). Manufacturing control, as a conceptual design, is divided in a number of sub-systems, cooperating with each other, and across the different types of holons, a holarchy is formed. The structure of this architecture is built around tree basic holons: product, order and resource holons.

- Product holons provide the process knowledge of a product. This means that a product holon can identify all possible subsequent production steps for an order in a certain state. The implementation and representation of such a recipe is internal to the product holon.
- The order holon manages the physical product instance being produced, the product state model, and all logistic information processing related to the job. It is responsible for performing the assigned work correctly and on time.
- Each resource holon corresponds to a physical part in the manufacturing system. A resource holon corresponding to a machine provides the prior knowledge of this machine; it knows the supported operations, the production capacity, queuing strategy, etc.

A main advantage of this architecture is that it reflects reality, not functionality. No unstable (reality is stable) constraints are introduced until this point. This fact makes that PROSA will provide a stable basic layer for the total software design.

Supply network architecture extensions As Koestler (Koestler, 1989) stated, systems can consist out of several smaller entities and some systems belong to bigger aggregated systems without having an upfront-defined hierarchy. As described in the previous paragraph, PROSA is a holarchy over the different holon types. The extension towards supply network control will employ the same mechanism and a holarchy will by formed inside each PROSA holon type. The following overview indicates what this extension means for every separate PROSA holon.

Product holons are not longer stand-alone subsystems. Whenever you take a snapshot at a point in time, the structure of different product holons will form a hierarchy. E.g. suppose there is a product holon that corresponds to a specific car model. Such a holon provides the knowledge to manufacture this car type to the other holons in the control and coordination mechanism. As in real life, this product holon is not able to cope with all complexity. In fact, the product holon for that car type consists out of several subholons each of them describing their respective subsystem, the wheels, the engine, the seats, etc.

As the product holon can be composed out of several subsystems, a resource holon can consist out of smaller resource holons each of them responsible for a part of the total, initial problem. In this way the supply network is modeled with as key point that it reflects the reality (see chapter 2). All machines, operators, raw materials, etc are composed into a production facility. By composing production facilities and logistics between the facilities, organizations are modeled.

Order holons can separate their concerns and delegate sub-problems to smaller order holons, responsible for that sub-problem. For instance, an order holon, responsible for a large transportation task, may spawn several new order holons responsible for smaller and more manageable tasks (like a single transport segment covered by a truck), which the original order holon coordinates.

3.2 Supply network Coordination

Ant coordination and control Ant coordination and control is based on the insights from the behaviour of ants searching for food. The main advantage of such an ant-based design is that individual agents/holons are not exposed to the complexity and dynamics of the global situation. None of the agents need a mental map of the environment nor do these agents communicate amongst themselves about the environment. Also the communication is carried out in a very robust way by means of an evaporate-and-refresh mechanism. There are however two requirements for the ant coordination mechanism to be effective: (1) there must be a large number of ants in the system, and (2) it must be cheap and easy to lose some of the ants. In practice, this means that the ants in the network control system must be virtual entities, and cannot be strongly connected to the presence and location of physical entities in the supply network. In this case there are tree types of ants: intention, exploration and feasibility ants. The ants typically originate from a holon that reflects such a physical entity (e.g. an order holon reflecting a car engine being produced) but are virtual such that they can navigate at cyber speed through the models in the network control system.

Feasibility The primary concern of order holons is to get their product instance(s) manufactured. To achieve this goal, the hard constraints in the manufacturing/supply network system have to be accounted for. To spread the hard constraints, resource agents representing production entities create, at a certain frequency, ants and send them upstream. These ants post information (pheromones) on local blackboards, which are attached to resource agents, describing the processing capabilities of the corresponding path discovered so far.(Hadeli, 2003).

The above is the basic feasibility algorithm as it is used in manufacturing control. There are some issues that need to be addressed when this algorithm is used to propagate capabilities throughout a supply network.

- The frequency at which feasibility ants are propagated will be different depending on the level of dynamics.
- Product information could be used in order to limit the ant propagation. It makes no sense to propagate the capabilities of a production path that takes significant more time than the actual production process. It makes no sense to propagate a capability that takes one year in order to bake a cake.

Exploring The exploration services investigate possible routes and processing sequences for the product instances in a factory. The basic exploration service implement the following scheme:

(1) Each order agent creates exploring ants at a regular frequency.
(2) Each exploring ant virtually executes a possible scenario on behalf of its order agent. A scenario is more than only a set of resources to be visited. Possibilities to join, make batches, execution slots for the different operations, etc are part of the total solution space.
(3) At the end, the explorer ant will report the explored solution to the order agent. Based on the results of the exploring ants, the order agent keeps a set of candidate routes and is

refreshed regularly, or evaporates if the information becomes to old. The performance of the scenario that was followed by the explorer ant is not based on static information. Instead it looks at possibilities to join batches, look at queuing time, takes disturbances in account, etc.

Intention propagation An intention reflects to some extend the future behaviour of a holon. The intention emerges out of the set of possible, explored, candidate solutions. One of the possible candidate solutions is regarded as the best and will be executed, with high probability, in the near future. Next the holon triggers an intention ant responsible for informing the different service providers about the intention of its holon. This intention is posted on the resources in the form of pheromones, it will evaporate if the intention is not refreshed regularly. Based on these intention service providers can build a dynamic load forecast, which can be used by explorer ants to make their performance estimate reliable.

Confidentiality/filtering of information The main difference for the ant topology with factory control lies in the holarchical structure of the supply network. Some resource holons represent a set of production entities. These aggregated holons have not only horizontal connections, e.g. suppliers, clients, but also vertical connections to their subholons, e.g. divisions. Ants cannot always access the vertical connections due to two reasons, (1) Some information is confidential for the outside world and (2) it is not always desirable to have all information, it would only trigger information overload. The aggregated holon tackles this problem by hiding the connections and encapsulating the needed information from the underlying production facilities.

3.3 Discussion

The supply network control is build upon an architecture, providing a coherent foundation for the design of the control system. Upon this stable layer a coordination and control system is implemented using ant-based techniques. The advantages of this ant-based coordination system are numerous.

- Disturbances are business as usual. The feasibility that uses the evaporate-and-refresh mechanism, such as ants use, enables to rediscover the capabilities over time.
- Coordination is considered as an ongoing task. Explorers and intentions try to find good solutions and re-evaluate them while the time is going on.

Still there are some questions left. The behaviour of the different holons should be social. In factory control this is a relative easy task, you can hardcode the social behaviour of every holon and force them to use this behaviour. In a supply chain there should be an additional mechanism. The key to that additional mechanism is trust. The trust mechanism is addressed in the following chapter.

4. TRUST AND REPUTATION IN SUPPLY NETWORK CONTROL

Most generally, trust can be considered as the expectation held by each member of a society that the existing natural and moral social orders will persist. This is to say that members of society have an inherent trust or faith that 'the sun will rise tomorrow' and society will exist more or less as it did today. This insight gives the ability to indicate where reputation serves supply network control by enabling social behaviour (Carter *et al.*, 2002).

4.1 Enabling social behaviour in supply network control by means of reputation

In the previous paragraph there was noticed that while scaling the coordination mechanism from factory control to supply network control a social behaviour is desired. Trust and reputation serve as key element to make the network of entities behaving in a social way. Any unsocial behaviour will be punished in terms of a reputation drawback. Reputation as it is, builds upon and affects behaviour by means of all communication channels, interactions. As defined in previous section, interactions should conform to the expectation of social and natural order. The next paragraph explains how and which expectations are held by all entities in the supply network.

In the PROSA with ant coordination and control, interactions and coordination actions happen through indirect communication and more specific by means of explorer, intention and feasibility ants. Intention and explorer ants hold expectations from the social order in contrast to the feasibility that holds expectations from the natural order.

Intentions coming from different entities, will be treated differently according to their reputation profile. In the case where a resource agent receives an intention form order agent 'o1' to book a slot on the respective resource, reputation 'o1' can serve as a priority indicator for the scheduler from the receiver of the intention. E.g.: If there is an excess in demand according to the resource capacity, non-trusted clients will have less priority in the planning and shifted to later time-slots.

Explorer ants rely on reputation information for performance estimating/decision-making. Eg.: if an order agent has to decide where to perform the next production step and there is a choice between tree equal service providers but two of them have a bad reputation, than the third one will be preferred.

Feasibility will spread the capabilities of the underlying resources throughout the network. Resource capabilities, e.g. the services of a service provider, need to conform to the actual capabilities. E.g.: a factory can claim to be capable of hardening with quality 'a' but in fact it is quality 'b'. Reputation profiles can give you an idea of the estimation on how much of the feasibility information conforms to the actual situation.

4.2 Entity roles

Entities like suppliers, clients, etc can act according to different roles. E.g.: entities usually have a client and provider role. In some cases, an entity behaviour can differ depending on its role. In every-day-life, persons can have a working role and a husband role. It is clear that the behaviour in these roles, working or husband role, are completely different. As a consequence, reputation profiles are not exclusive for one entity. Entities can have many different reputation profiles, each one of them attached to a role.

In the PROSA reputation profiles are build according to different interactions. As the reputation profile can differ among the roles, not only an entity id is needed to indicate the players in the interaction but also the respective role plays an crucial role. it.

4.3 Reputation versus contracts

In the PROSA + Ant coordination and control system intentions of different agents in the system are used to build a short term forecast. In a single factory all means are used for the same goal and therefore the intentions will not be misused. In other words, there can be a social behaviour without "forcing", controlling the different entities. In supply chain we encounter a slightly different situation. Entities will try to maximize their profit. Negative feedback on reputation can serve as implication for bad behaviour. Nevertheless there are some situations were reputation is not the only/best mechanism to force good behaviour.

- Contract making out of intentions. In supply networks an intention is not always enough to ensure correct behaviour because of the risks involved. Contract making can evolve out of intentions and gives a juridical assurance of the intention.
- Money compensation for bad behaviour. At some point, risks are compensated not only with reputation drawbacks but also with money. The drawback in the reputation profiles will be less if an entity pays the compensation fees for the harm he did to other entities. This balance between money and damage to ones reputation is not necessarily a stable function.

5. CONCLUSION

This paper presents an engineering point of view on the supply network control problem. Starting from a manufacturing control system based on PROSA and ants, useful insights on several common problems in supply networks are given. Disturbances, the changing environment, uncertainties, etc are treated as business as usual.

Extending the used manufacturing control is done in several steps: (1) extending the PROSA architecture, (2) adapting the ant based coordination mechanism, (3) adding reputation models to enforce social behaviour.

6. ACKNOWLEDGEMENTS

This paper presents work funded by the Research Council of the K.U. Leuven - concerted research action on Agents for Coordination and Control - and the 5th framework programme of the European Commission.

REFERENCES

Carter, Jonathan, Elijah Bitting and Ali A. Ghorbani (2002). Reputation formalization within information sharing multiagent architectures. *Computational Intelligence* pp. 45–64.

Goethals, F. (2002). Extended enterprise infrastructures (www).

Hadeli, P. Valckenaers, C. Zamfirescu H. Van Brussel B. Saint Germain T. Holvoet E. Steegmans (2003). Self-organising in multiagent coordination and control using stigmergy. ESOA.

H.L. Lee, V. Padmanabhan and S. Whang (1997). The bullwhip effect in supply chains. *Sloan Management Review* pp. 93–102.

Koestler, A. (1989). *The Ghost in the Machine.* Arkana Books.

Wyns, J. (1999). Reference architecture for Holonic Manufacturing Systems. PhD thesis. K.U.Leuven.

Copyright © IFAC Information Control Problems in Manufacturing, Salvador, Brazil, 2004

ELSEVIER
IFAC
PUBLICATIONS
www.elsevier.com/locate/ifac

PATTERN BASED ANALYSIS OF FRACTAL MANUFACTURING SYSTEMS

K. Jääger and J. Vain

Department of Computer Science/Institute of Cybernetics
Tallinn Technical University

Abstract: Self-organization, self-optimization, and goal orientation in fractal manufacturing systems requires special treatment when solving planning or decision making tasks in highly dynamic environment. In this paper the parametric model checking technique is proposed as possible candidate for fractal analysis. As a main contribution, a flexible modeling technique is presented that bases on model patterns and pattern transformations. *Copyright © 2004 IFAC*

Keywords: fractal manufacturing system, patterns, model checking, timed automata.

1. INTRODUCTION

Because of the need for operational flexibility and technological adaptability in agile manufacturing systems, new manufacturing paradigms such as bionic/biological, holonic and fractal manufacturing have been proposed (Ueda, 1992; Brussels, *et al.*, 1998; Tirpak, *et al.*, 1992; Tharumarajah, *et al.*, 1996). It is pointed out in (Ryua, *et al.*, 2003) that for distributed and highly dynamic manufacturing environments the fractal architecture is well suited. The paradigm of fractal manufacturing system (FrMS) is based on the concept of autonomously cooperating multi-agents referred to as fractals. Fractals can be described by specific internal features such as self-organization, dynamics, and self-similarity interpreted as similarity of goals among the fractals to conform the objectives in each unit. Fractals are assumed to function as a coherent whole. This is achieved through the process of participation and coordination among fractals, supported by an inheritance mechanism to ensure consistency of goals (Tharumarajah, *et al.*, 1996).

The architectural model of FrMS proposed in (Ryua, *et al.*, 2003) defines the components of the FrMS, *basic fractal units* (BFU). Each BFU consists of five functional modules: Observer, Analyzer, Resolver, Organizer, and Reporter. BFU provides services according to an individual-level goal and acts independently while attempting to achieve the upper level goal. Consequently, fractals may be seen as dynamic with the ability to reconfigure themselves in response to the environment.

Although FrMS has many conceptual advantages (self-organization, self-optimization, goal orientation etc.), these features introduce also fundamental difficulties for analysis and implementation. The inherently non-stationary behavior of FrMS makes concrete functional properties hard to predict and the analysis of behavioral properties (like of recursive programs) by algorithmic methods undecidable in general case. Therefore, for analysis of FrMS two different contexts should be clearly distinguished:

- The global level analysis of FrMS presumes studying properties of the FrMS as a whole through all its evolutionary phases. Global analysis makes sense mostly in the design phase. Because of the complexity issues of global analysis, substantial approximations and abstractions of the FrMS model are necessary.
- The fractal level analysis, that must be performed each time when a fractal is adapting to the changes of its environment, concerns mainly local context (configuration and relations with neighbors) of the fractal and needs less abstract models. In this paper, the fractal level analysis is under focus.

According to the functional model of (Ryua, *et al.*, 2003), the fractal level analysis is performed cooperatively by Analyzer and Resolver modules. The Resolver plays the most important role in a fractal. It is responsible for generating job profiles, goal forming, and decision-making processes. After gathering information on the factory goal and fractal status, the Resolver makes job profiles with the configuration of sub-fractals, and sends them to the Analyzer for evaluation. Resolver provides the Analyzer with goal and the domain of possible structural and parametric variations. The function of the Analyzer is to analyze alternative job profiles with status information, to rate dispatching rules, and to simulate analyzed job profiles in real-time.

In the given architectural model the functional modules of a fractal are implemented by a set of *agents*. The agents for Analyzer are: *Schedule Evaluation Agent* (evaluates job profiles generated by the Resolver); *Dispatching-rule Rating Agent* (chooses the best dispatching rule for achieving its goals among rules such as shortest processing time, earliest due date etc.), and *Real-Time Simulation Agent* (performs simulation in the on-line state with the results of the analyzed job profiles and the best dispatching rules). Each agent has its repository of methods to handle calls from other agents.

As a contribution of this paper a new method is proposed to add in the Analyzer agents, namely, *model checking* (MC). MC combines advantages of deductive and simulation-based analysis methods and has proven to be a promising technique in industrial applications (Hune, *et al.*, 2000).

In its general setting, MC is an algorithmic technique for verifying that a model *M* of the system satisfies a given property specification *R*, formally stated as $M \models R$ problem. The procedure uses an exhaustive search of the state space of the model to determine if given property is true or not. Although originally restricted with finite state models MC can be combined with various abstraction, induction and compositionality principles to handle infinite state systems (Inan and Kurshan, 2000). Parametric model checking (PMC), in addition to checking satisfiability relation, provides the parametric characterization of those behaviors of the model *M* that satisfy or falsify given property *R*. PMC has been used for different manufacturing planning and analysis tasks: for throughput planning, scheduling, assembly line balancing etc. (Hune, *et al.*, 2000; Vain and Kyttner, 2001; Vain, *et al.*, 2002).

Applying PMC in Analyzer agents one faces with following key issues: constructing and (on-line) actualizing models of fractals, specifying and adjusting the analysis tasks to be solved on the model, and choosing methods to ensure efficiency of PMC in the presence of decision time constraints.

The rest of the paper is structured as follows. In Section 2, the pattern-based model construction technique for PMC is introduced. Section 3 addresses the issue of efficient decidability of PMC, the complexity/accuracy-driven model tuning procedure is used with predefined abstraction and refinement transformations. In Section 4, the approach is illustrated using a production cell example. Finally, complexity results are demonstrated and the future work is discussed.

2. PATTERN-BASED MODEL CONSTRUCTION

In addition to analytic and heuristic methods the formal techniques have gained increasingly attention in the modeling and analysis of manufacturing systems. In (Ryua, *et al.*, 2003) the functions, activities, and relationships of agents are modeled using the unified modeling language UML. Algebraic analysis methods and Petri net based formalisms (e.g., Bourrieres, 1998) have been popular over a decade. Still, the usage of formal techniques for manufacturing systems has been focused mainly on modeling and simulation, paying less attention to the cost factors of analysis such as the model size, time complexity of the analysis etc.

When choosing the proper method for fractal level analysis in Analyzer agents one has to consider that the analysis tasks have to be solved regularly using the model and constraints that are changing in real time. Therefore, the algorithmic complexity of the analysis by PMC and accuracy of results are main drivers in the model construction and tuning.

The idea of proposed method is following. The model is constructed using a fixed set of model patterns and composition rules. Each pattern and composition rule contributes to the complexity and accuracy of analysis. Assuming the factor a construction step constitutes is known, the approximate complexity class and precision of analysis can be estimated before running the real analysis procedure. The complexity and accuracy are characterized by approximating estimates O and ϑ respectively. On the other hand, the computational resources of the Analyzer, required accuracy, and available time for analysis are assumed to be known to Analyzer. It transforms the model using abstraction and refinement transformations so that constraints on O and ϑ are satisfied. To minimize complexity, the model with highest level of abstraction among those, allowing the analysis results with required accuracy, is searched for. The first step towards constructive modeling approach is to clarify the set of patterns, their compositions and refinement/abstraction transformations.

In the following, the patterns are defined using timed automata (TA), one of the most popular formalisms used in model checking tools (Uppaal, 2004).

Main behavioral feature of fractals, regardless they represent simple machining unit or a shop floor, is the repetition of their operations. The operations consist of sub-operations the order of which may be deterministic or non-deterministic, performed jointly with many objects or with just one. The simplest model pattern represents a group of operations without any ordering constraint except sequential execution in time (Fig. 1(*a*)).

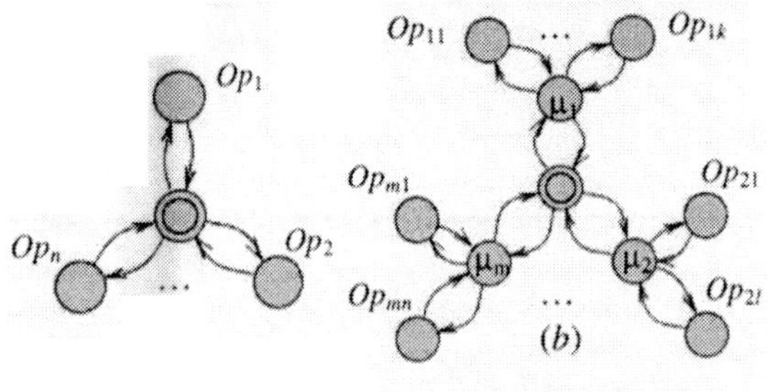

Fig. 1. Structural patterns of fractal behavior: (*a*) Single-mode operations; (*b*) Multi-mode operations

Operations of the production unit may depend on its functioning mode. Switching between the modes μ_1, ... , μ_m is made explicit by the pattern in Fig. 1(*b*). Resource dependencies of operations are modeled by conditional transitions labeled with transition guards (*Grd*(.)) and assignments (*Asg*(.)). In Fig. 2 the resources observable in the model are denoted by the vector of resource variables ***R***. The resources consumed and produced by some operation Op_i are denoted by $\boldsymbol{I}_i$ and $\boldsymbol{O}_i$, respectively. In case the store capacity for resources ***R*** is a critical quantity, the upper bound of store capacity $\boldsymbol{R}\lceil$ is introduced in the operation exit condition (Fig. 2(*b*)).

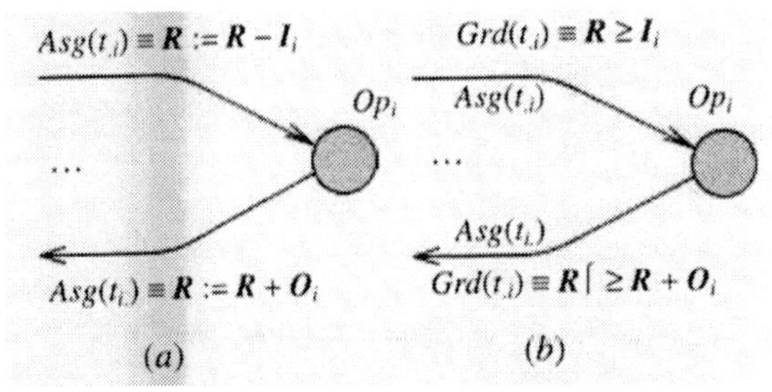

Fig. 2. Resource dependency patterns: (*a*) Resource handling operation (*b*) Resource dependent operation

The strength of TA formalism lies in its capability to represent timing information explicitly. Patterns in Fig. 3 bring out different aspects of timing. In Fig. 3(*a*) the operation duration given by an interval [d_1, d_2] is modeled. Here an auxiliary continuous variable *Clock* is used to measure the duration of operation Op_i. Since the duration of Op_i is non-deterministic the location of Op_i is split into two locations Op_{i1} and Op_{i2} with invariants corresponding to lower bound d_1 and non-deterministic delay within interval [*0*; d_2 - d_1], respectively.

Timing of TA models is feasible only if all locations and transitions have proper timing specification (i.e., no timing gaps exist), because the enabled transitions without timing constraints may be taken at arbitrary time instant. To force guarded transitions to be taken without unbounded delay, the (timeless) guarded transitions are substituted with a fragment consisting of an urgent location (denoted by $\cup$) and a waiting cycle with bounded delay (Fig.3(*b*)). The maximum transition delay, after the guard evaluates true, is parameterized by *d*.

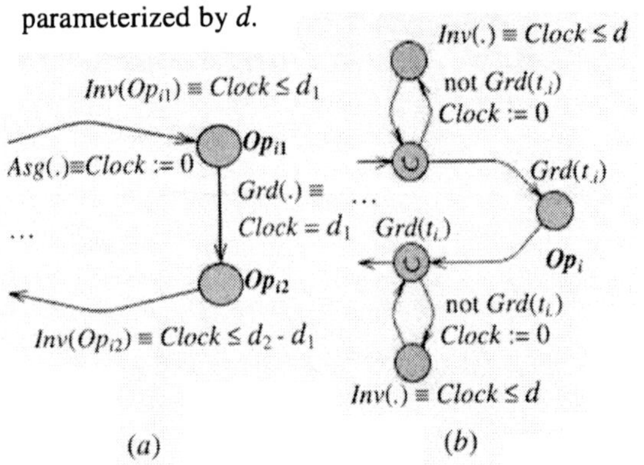

Fig. 3. Timed operation patterns: (*a*) Timed operation with duration interval [d_1,d_2]; (*b*) Conditional time-bounded start (stop) with bounded delay

Often manufacturing systems include conveyors. Since conveyor speed, the positions where pieces are placed/removed on/from the conveyor are of importance, a conveyor pattern parameterized with those characteristics is depicted in Fig. 5(*a*). In the array ***GP*** the relative pickup position is saved by the operation that places the piece on conveyor. The conveyor automaton increases the relative position (variable *pos*) periodically by constant *shift* in each time slot (specified by the parameter *unit*). When the item being closest to its destination reaches its pick-up position (the transition guard ***GP***[0] $\leq$ *pos* evaluates *true*) it is removed and the next pickup position of the closest item (the element of ***GP***[1]) is shifted to ***GP***[0]. The arrival transition in Conveyor is synchronized with pick-up operation of some other automaton by synchronization primitive *arrived*!.

3. MODEL TRANSFORMATION BY REFINEMENT AND ABSTRACTION

Model patterns introduced above allow modeling of fractal behaviors on relatively low level of abstraction. That is appropriate for analysis by simulation. When having large parallel compositions of TA models, the usage of unbounded variables, non-deterministic choice, and cycles in patterns is the main growth factor for global search space of MC. Below, some refinement and abstraction transformations of patterns needed for adjusting the model granularity according to given complexity and accuracy constraints are introduced.

The abstraction-refinement relations can be considered in orthogonal dimensions: operational structure, timing and resource dependencies. Refinement and abstraction of patterns are interpreted in terms of TA refinements, namely, the state space refinement where the existing variable types are substituted by richer types; strengthening of transition guards; reducing non-deterministic choice between transitions; splitting locations and their invariants into sub-locations and sub-invariants.

- Operation refinement $\succ_o$ (Fig. 4) is applicable when the effect of an operation has to be specified in terms of its sub-operations without giving their explicit precedence. An important restriction is that all external relations of the refined operation have to be preserved regardless of which of sub-operations substitutes the original operation in the external relation. Operation refinement is expensive transformation since it introduces an exponent in complexity function *O*.

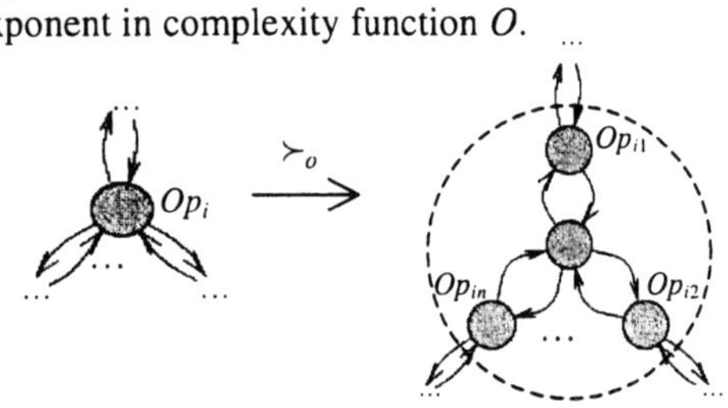

Fig. 4. Operation refinement

- Model state space refinement $\succ_s$. When the effect of an operation has to be specified in terms of resources that change when the operation is performed, the transitions to and from the model fragment representing the operation are supplemented with assignment expressions. An assignment expression *Asg*(.) describing resource consumption and/or creation is defined on the domain of model state variables. The simpler the variable type is, the smaller are model state space and branching factor of the analysis search tree.
- Sequencing refinement $\succ_q$. The precedence relation on a set of fractal operations may be non-deterministic meaning that there is no preference in what order operations are performed. This is common in coarse-grained models where some order determining information is abstracted away. The non-deterministic choice is removed by introducing guard conditions to transitions. The ordering is deterministic if the guards $Grd(t_{ij})$ of all transitions t_{ij} of competing transition groups T_i are mutually exclusive, i.e., for all valuations of state variables occurring in $Grd(t_{ij})$ the condition $\wedge_j\ Grd(t_{ij}) = false \wedge \vee_j\ Grd_n(t_{ij}) = true$ holds. Pattern in Fig 2(*b*) is sequencing refinement of pattern in Fig. 2(*a*). Refinement $\succ_q$ reduces the exponent in complexity function *O*.
- The operation timing refinement $\succ_{to}$ introduces the duration of an operation. The duration can be specified by interval $[d_1; d_2]$ or by some exact value *d*. The smaller the interval is the more refined (deterministic) the behavior of the model is. An application example of the refinement $\succ_{to}$ is transforming the pattern in Fig. 2(*a*) to the one in Fig. 3(*a*).
- The non-determinism of timeless guarded transitions is reduced by transition timing refinement $\succ_{tt}$. It is applied for deriving the pattern 3(*b*) from pattern 2(*b*). In pattern 2(*b*) the transition delay after guard evaluates true is unbounded in time. By adding an urgent location (the location of TA that is left as soon as there is some outgoing transition enabled) and a cycle that includes a time-bounded location, the guarded transition has to be taken within the time interval [0; *d*] after the guard evaluates true. Computationally the pattern 3(*b*) is rather expensive for small *d*-s. On the other hand, increasing the delay parameter *d* increases also behavioral non-determinism in time.
- Pattern in Fig. 5(*a*) has been derived from pattern in Fig. 1(*a*) by successive application of refinement transformations $\succ_o$, $\succ_s$, $\succ_q$, and $\succ_{to}$. Fine grain sate space of Fig. 5(*a*) allows distinguishing single events, related to individual items on conveyor. Since analyzing the model that incorporates the pattern is of high complexity, some abstractions are needed. The abstractions are inverse transformations of refinements used in the pattern construction. In Fig. 5(*a*), when abstracting from individual items and different place and pick-up positions one gets an abstract model depicted in Fig. 5(*b*). The price to be paid is imprecision of the transport time estimate. The margin of error is the duration of one full conveyor cycle.

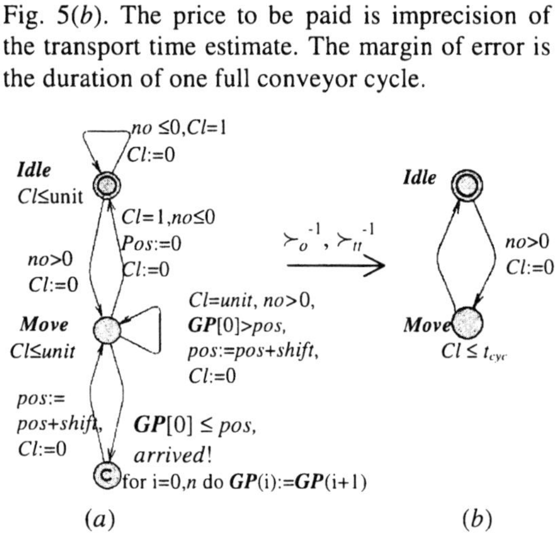

Fig. 5. (*a*) Discrete approximation of a conveyor; (*b*) Abstraction of the conveyor

4. CASE STUDY: A PRODUCTION CELL

Although the model patterns defined above can be applied for modeling fractals of enterprise level as well as of single machine level, the flexibility of the approach is demonstrated on a simple production cell modeling and analysis example.

4.1. System description

The process in Fig. 6. consists of the sequence of tasks prescribed as follows. The loading machine M1 places the blank on the conveyor M2 (task1). The conveyor transports the blank to the robot M3 (task2) that picks it from conveyor and places to conveyor M4 (task3). On the M4 the blank is measured, classified and transported to a certain position (task4) where the robot M6 and mill M5 work it into the finished part (task5). Task5 in detail is: robot M6 takes the good blank and places it to the mill worktable, where the blank is processed by desired program. After completion of milling, the robot M6 removes the item from the mill and places it into the indexing table M7. The process is repeated as long as the cell is supplied with blanks.

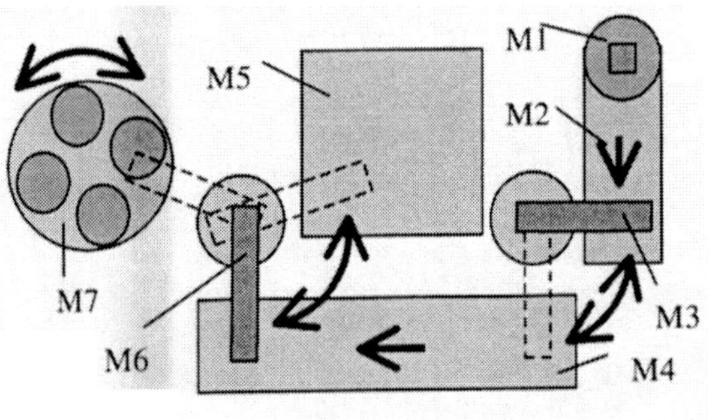

Fig. 6. Production cell

4.2. Modeling of the system

The production cell model is constructed using the model patterns described above. Machines M1 - M7 are considered as active agents in a fractal unit. Each of them is modeled in the root model using pattern 1(*a*) that is refined with pattern 3(*b*) (Fig. 9). That provides fine grain timing specification of the coarse operational view. When solving any PMC task that requires traversal of the full search tree, the model should be made more abstract since 3(*b*) is source of high complexity. Experimental curves in Fig. 7 and Fig. 8 show that an efficient way to reduce time complexity is to bound the time horizon of the model and increase delay parameter *d* of pattern 3(*b*). Results of complexity reduction experiments are summarized below.

4.4. Experimental results

Experiments illustrate how the use of patterns, their refinement and abstraction transformations can help to reduce complexity in the PMC. A number of experiments with different level of pattern abstractions were made. The results showed clearly the need for right balance between refinement and abstraction in order to solve model checking tasks on complicated models.

The diagram in Fig. 7 illustrates an experiment made on the production cell models of different level of abstraction. The point "1" corresponds to the root model presented in Fig. 9, and the third model is the one of the highest level of abstraction. It can clearly be seen, how the complexity of the model is reduced when increasing the delay parameter *d* of pattern 3(*b*) (point "2" in the diagram), and applying then abstraction $\succ_n^{-1}$ (point "3").

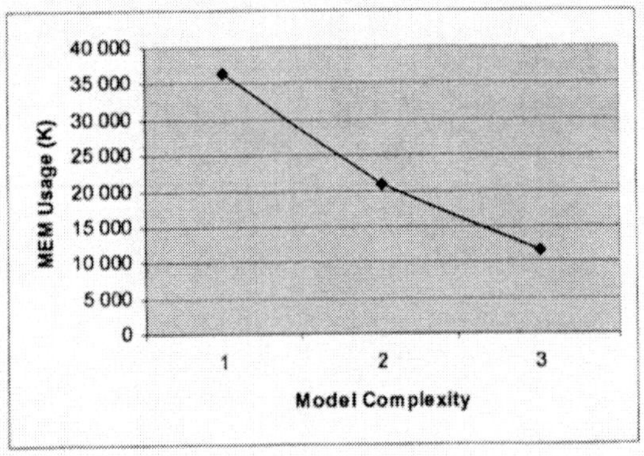

Fig. 7. Complexity of MC depending on the level of model abstraction

The Fig. 8. represents the results of operation abstraction $\succ_o^{-1}$ where the reduction of the number of suboperations in the pattern 1(*a*) provides drastic saving in the memory space.

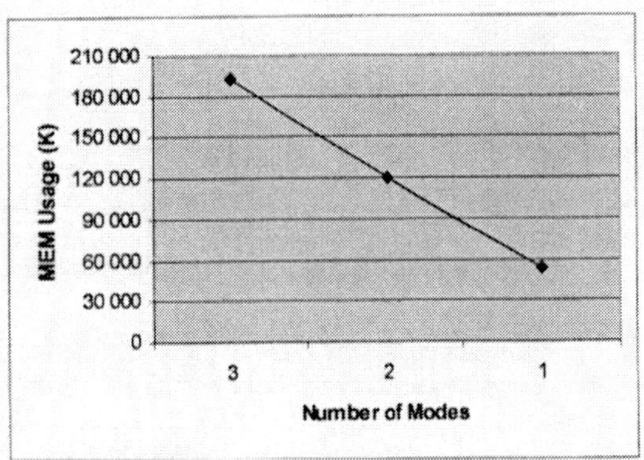

Fig. 8. Complexity reduction gained by abstraction transformation $\succ_o^{-1}$

5. CONCLUSION

The paradigm of fractal manufacturing poses new challenges in the area of modeling and analysis of production systems. Features like self-organization, self-optimization, and goal orientation of fractals requires special treatment when solving any fractal level planning or decision making task. In this paper the parametric model checking technique was proposed as one candidate for analysis in highly dynamic environment. As a main contribution, a flexible modeling technique is presented that bases on model patterns and pattern transformations.

The patterns introduced allow systematic model construction and tuning depending on time and computational resources allocated for analysis. It was shown experimentally on problems that require full state space traversal, that so called complexity and accuracy driven model adjustment allows finding a balance between computational complexity and precision of results. Defining the proper set of adjustment heuristics is the subject of further domain specific research.

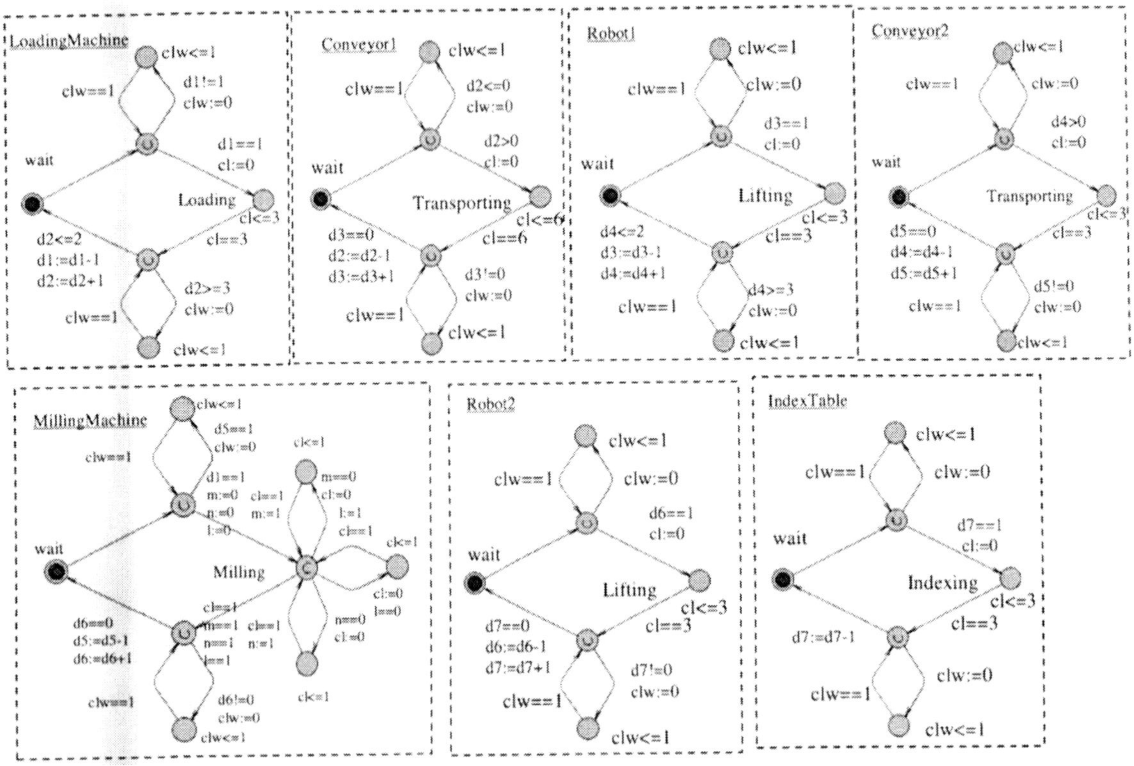

Fig. 9. Root model of the production cell

REFERENCES

Ueda, N. (1992). A concept for bionic manufacturing systems based on DNA-type information, in: Human Aspects in Computer Integrated Manufacturing Proceedings of the IFIP TC5/ WG5.3 8th International PROLAMAT Conference, Tokyo, Japan, pp. 853–863, 1992.

Brussels, H.V., J.Wyns, P. Valckenaers, L. Bongaerts, P. Peeters (1998). Reference architecture for holonic manufacturing systems: PROSA. *Computers in Industry*, **37(3)**, pp. 255–274

Tirpak, T.M., S.M. Daniel, J.D. LaLonde, W.J. Davis (1992). A note on a fractal architecture for modeling and controlling flexible manufacturing systems. *IEEE Transactions on Systems, Man, and Cybernetics*. **22**, pp. 564–567.

Tharumarajah, A., A.J. Wells, L. Nemes (1996). Comparison of Emerging Manufacturing Concepts. *International Journal of Computer Integrated Manufacturing*. **9/96**, pp. 217–226.

Ryua, K., Y. Sonb, M. Jung (2003). Modeling and specification of dynamic agents in fractal manufacturing systems. *Computers in Industry*, **52/03**, pp. 161–182.

Bourrieres, J.P. (1998). Multilevel algebraic modelling of complex manufacturing tasks. *Preprints of the 5th IFAC Workshop on Intelligent Manufacturing Systems–IMS98*. Gramado-RS, Brasil, pp.261-265.

Inan, M.K. and R.P.Kurshan. (Eds.) (2000) *Verification of Digital and Hybrid Systems*, NATO ASI Series F, Vol. 170. 405 p.

Hune, T.S., K.G. Larsen and Paul Pettersson (2000). Guided synthesis of control programs for a batch plant using UPPAAL. *RS-00-37*. BRICS. Dept. of Computer Science, Univ. of Aarhus.

Vain, J., R. Kyttner (2001). Model checking - a new challenge for design of complex computer controlled systems. *5th Conference on Engineering Design and Automation*, LV, pp. 593-598.

Vain, J., I. Randvee, T. Riismaa, and J.Ernits (2002). Solving Line Balancing problems with Model Checking. In. *Proc. Estonian Acad. Sci. Eng.*, **8, 4**, pp. 211-222.

Uppaal homepage: http://www.uppaal.com/.

Copyright © IFAC Information Control Problems in
Manufacturing, Salvador, Brazil, 2004

ELSEVIER
IFAC
PUBLICATIONS
www.elsevier.com/locate/ifac

TOWARDS A PERFORMANCE MEASUREMENT SYSTEM FOR THE EVALUATION OF SCHEDULING SOLUTIONS

P. Brambilla*, S. Terzi, S. Cavalieri***

**University of Bergamo, Department of Industrial Engineering, Italy*
***Politecnico di Milano, Department of Economics, Management and Industrial Engineering, Italy*

Abstract: The scheduling phase is an important production activity. For scheduling problem, many approaches and techniques have been proposed and it is evident the need to evaluate and benchmark them. The paper analyses the requirements for the development of a *Performance Measurement System for the Evaluation of Scheduling Solutions* (PMS-ESS). PMS-ESS is part of a broader project idea that aims to establish a benchmarking approach, instantiated into a *Benchmarking Service*, for supporting industrial practitioners in the evaluation of scheduling systems and solutions. This project idea is currently supported and disseminated by the IMS-NoE SIG 4 community. *Copyright © 2004 IFAC*

Keywords: Scheduling Evaluation, Benchmarking, Performance Measurement

1. INTRODUCTION

In the last years, Production Planning and Control (PP&C) has been increasingly becoming a critical activity, since competition in the markets is leveraging on a multitude of factors ranging from product quality, to delivery times and pre-sales and after-sales services. Among all PP&C activities, scheduling decisions are the final temporal decision-making phase where plant managers have to act for fixing any short noticed variations and maintaining satisfying overall production system performances, "assigning scarce resources to competing activities over a given time horizon to obtain the best possible system performance" (Kempf *et. al.*, 2000).
In literature, several approaches exist for scheduling problem, from the traditional off-line scheduling systems, which simply elaborate a production plan, to the most sophisticated "*Intelligent*" production scheduling systems, which try to continuously elaborate new production plans for following the incoming events (e.g. Multi-Agent systems, Genetic Algorithms...). Within these approaches, more or less "*Intelligent*", industrial practitioners are often in trouble for founding an effective problem-solving scheduling solution for their plant management, also because a comprehensive identification of which performances could be advocated with one control architecture had been neglected in literature during the last decades.
In Europe, a Special Interest Group (SIG 4) on *Benchmarking and Performance Measures* - coordinated by KULeuven and University of Bergamo - has been recently formed within the European IMS Network of Excellence (IMS-NoE, 2003), in order to provide a solution to this lackness. Within this context, the paper aims to present a first proposal of a multi-layered framework specifically addressed to the evaluation of scheduling solutions and systems, developed by the authors taking into account several contributions coming from the SIG 4 community and from a detailed literature analysis. The proposed framework, named PMS-ESS (*Performance Measurement System for the Evaluation of Scheduling Systems*) has been developed specifically for providing support to scheduling researchers and practitioners in assessing the level of applicability of traditional and novel scheduling techniques. The paper is organised as follow:

- par. 2 presents the basic requirements identified for the definition of a performance measurement system for the evaluation of scheduling solutions;
- par. 3 presents a literature review conducted in the area of scheduling evaluation;
- par. 4 reports the considerations gained from the literature survey interesting for the development of the PMS for the evaluation of scheduling systems, expressed in terms of "technical requirements" of the looked PMS-ESS.
- par. 5 shortly illustrates the PLM-ESS itself;
- par. 6 concludes the paper.

2. SCHEDULING PERFORMANCE MEASUREMENT REQUIREMENTS

As well known, scheduling complexity depends on diverse aspects, like the plant structure over which the scheduling solution has to be applied, but also like the process and production rules to be performed. In general, the more sophisticated the system is, the more the scheduling complexity increases. Taking into account that scheduling is highly context-dependent and considering several suggestions gathered from industrial and research members of the SIG 4 community, some basic requirements of PMS for scheduling evaluation have been identified:

I) Open and easily customisable - In the industrial world, a great number of types of production systems exist. A PMS for scheduling evaluation might be a framework open to different typologies of users and production environments, easily applicable and customisable in order to comprehend heterogeneous manufacturing systems and scenarios.

II) Effective comparison of schedule quality – An industrial practitioner could be interested in taking into account various scheduling techniques and policies; consequently, the looked PMS has to be capable to compare alternative planning and scheduling practices.

III) System efficiency measuring – According to the previous requirements, an industrial practitioner could be interested to know not only relevant measures for scheduling, but also all the operational measures that can be gathered from shop floor. For example, in a job-shop the setup-time can be not-dependent on jobs' sequencing, so, during scheduling comparison, setup-time can be ignored. However, for plant manager, the knowledge of total setup-time could be an interesting measure to estimate the necessary labour to support production. Hence, the requested PMS has to include all system performance measures.

IV) Performance evaluation at diverse levels - Every level of an enterprise is interested by performance measures, but moving up to the corporate hierarchy, measures and indicators tend to be expressed more and more in aggregate terms. So, whereas at the lowest level a set of single "atomic" measures can be useful and effective, at upper levels managers need some synthetic indicators.

V) Quantitative data and symbolic judgments - An industrial user could be interested if a certain production schedule can satisfy one or more of his/her needs (e.g. a certain target value of the production makespan could be desired). In this case, he/she is searching for a "boolean" result, like "*the schedule is capable to satisfy the requirement*" or not. On the contrary, a scheduling researcher could need a support in evaluating his/her optimisation algorithm or about a specific rescheduling technique. This way, a symbolic judgement like "*very good in costs, good in plant saturation, bad in responsiveness*" could be useful for a user. Obviously, for the requested PMS, the gained judgement would not be a subjective one but an objective one, since derived from quantitative results.

VI) Focus on organisational goals - Into enterprises, a major problem is that different organisational units involved in a scheduling activity have diverse, often conflicting, goals. Thus, they have substantially different expectations from a schedule. For example, the sales department often looks at a schedule from the point of view of orders being delivered to the customer, while the manufacturing department is under pressure to reduce costs. Hence, the sales department would look for a schedule that has good due-date performance, while the manufacturing department would prefer a schedule with high machine utilisation, few setups, and long production runs. The needed PMS has to highlight these organisational goals.

VII) Real plant measuring – In reality, scheduling is a nervousness and dynamic phenomenon. In fact, at the shop floor level, disruptions like machine breakdowns or new incoming orders are on the daily agenda. For an effective evaluation, the needed PMS might consider dynamic aspects.

VIII) Rescheduling consideration - In the face of unpredictable disruptions, a major approach is the prediction-reaction paradigm (Kempf *at al.*, 2000). When a disruption occurs, a schedule can be revised to limit the degeneration of performances (rescheduling). The looked PMS has to be capable to assess the quality of rescheduling.

Taking into account these requirements, a literature review related to scheduling evaluation has been conducted. A short description of the results of the literature review is reported in the next paragraph.

3. STATE OF THE ART OF SCHEDULING EVALUATION

Performance measurement is a large research topic: performance measures are used to evaluate, control and improve production processes, but are also used to compare the performance of different organisations, plants, departments, teams and individuals. In literature, basic performance scheduling indicators are well accepted and used (e.g. *makespan*, *tardiness*, *lateness*, *flow time*, *setup time*, *working time*...), and usually, even if most of them are defined for a single job, they are used in an aggregate way, in order to calculate mean and total value among all jobs. Nevertheless, although the understanding of what constitutes a "good" production schedule is central to the development and evaluation of scheduling systems, few works have given contributions on this aspect in a more comprehensive way. In particular, only the work of Kempf et al. (2000) provides an exhaustive and theoretical approach to scheduling evaluation. The conducted analysis could be defined along four main categorization classes, which address diverse scheduling evaluation dimensions:

A) Stochastic measurement and disruption facing – As well known, the production floor is not such a static environment, but a larger variety of dynamic events occurs (e.g. machine breakdowns, deliveries delayed, workers), affecting the feasibility of proposed schedules. The inability to accurately respect proposed schedules is referred to as *scheduling nervousness* or *disruption*. To take into account this dynamic dimension, the simulation approach has been applied in literature. Facing with disruptions, a technique can be defined as good if it is capable to guarantee the maintenance of certain desired system characteristics despite fluctuations in the behaviour of its component parts or its environment. In literature, diverse measures have been proposed. In Bongaerts *et al.* (1999), the concept of *predictability* is defined as the degree to which something is known in advance. It may refer to the certainty with which a variable will have a certain value or belong to a certain range. Hence, the stochastic distribution of a value is one way to characterize the *predictability*. Like Bongaerts *et al.* (1999), Mignon *et al.* (1995) proposed measures for investigating *schedule robustness* under uncertainty. A measure of *schedule robustness* has been computed as a function of the variability of the objective function value. The lower this variability, the more robust the schedule is. Ouelhadi *et al.* (2003) investigated the scheduling/rescheduling problem of steel continuous casting. They focus on the modification of predictive schedule to minimise deviation of performance, taking into account the decision to locally repair the schedule or reschedule from scratch (complete reschedule). They introduced diverse measures: *utility*, *stability* and *robustness*. For these authors, *utility* measures the change in schedule objective following schedule revision; the expression is a simple difference between results. *Stability* measures the deviation from the original schedule caused by schedule revision to quantify the undesirability of making large changes to the initial predictive schedule, unless absolutely necessary. *Robustness* combines the *utility* and *stability* measures. Jensen (2001) proposed the concepts of *Robustness* and *Flexibility*. A schedule expected to perform well relatively to other schedules, when facing some set of scenarios and when right-shifting is used for rescheduling is said to be robust. A schedule expected to perform well relative to other schedules, when facing some set of scenarios and when some rescheduling method using search is used is said to be flexible with regard to that rescheduling method. Gören (2002) took into account the terms *Robustness* and *Stability*. A schedule whose performance does not degrade in the face of disruption is called robust. A schedule, whose realized schedule does not deviate from the original one in the face of disruptions, is called stable. Daniels and Kouvelis (1995) refer to schedule *Robustness* focusing on the best worst-case performance compared to the corresponding optimal solution. The schedule's maximum absolute deviation from optimality over all scenarios is used to calculate robustness. Ali et al. (2003) have stated that a mapping is defined to be robust with respect to specified system performance features against perturbations in specified system parameters, if degradation of these features is limited when perturbations occur.

B) Rescheduling effort - Schedules generated in practice cannot be used for a long time period because of unexpected disruptions and random events. Thus, it is necessary to revise the existing schedule at some points in time. A question arises: when-to-schedule? When-to-schedule decision determines the system responsiveness to various kinds of disruptions. There are several alternative ways to decide on timing of schedule decisions: (i) the *periodic scheduling* approach (the period length can be constant or variable), (ii) the *continuous scheduling* (the schedule is revised after a number of events that change system state occur), or (iii) *adaptive scheduling* or *controlled response* (a scheduling decision is triggered after a predetermined amount of deviation from the original schedule is observed), or, finally, (iv) *hybrid approaches*. Any way is chosen, the scheduler has main alternatives to react to the presence of real-time events. It can generate a new schedule from scratch, almost a complete reschedule, or make alterations to the previous schedule, a sort of schedule repair. There is a need of measures capable to evaluate the changes caused by rescheduling, in order to have an idea of rescheduling shop-floor efforts. A plant manager, in fact, during scheduling solutions comparison, could prefer a scheduling policy that has little worse general performance, against very low rescheduling efforts. In Jensen (2001) two types of distance measures for schedules are presented. Firstly, the *Hamming distance*, originally used in computer world to evaluate the difference between two strings of characters; the *absolute Hamming distance* D(S1,S2) between schedules S1 and S2 is equal to the number of precedence relations that differ from S1 to S2. Secondly, the *schedule overlap*, as an attempt to measure the likeness between two schedules as seen on the processing floor. The measure is defined such that two identical schedules will have an overlap measure of one, while two completely different schedules will have an overlap measure of zero.

C) Flexibility – Even if flexibility is one of the most quoted competitive objectives, a search for flexibility in scheduling problem not provide relevant results. It has been found no work detailing the concept of flexibility for scheduling solutions, even if a wider set of definitions occurs. Generally, flexibility could be defined as: (i) the ability of a manufacturing system to cope with changing circumstances or instability caused by the environment, (ii) the quickness and ease with which plants can respond to changes in market conditions, (iii) the ability of the system to quickly adjust to any change in relevant factors like product, process, loads and machine failure, (iv) the ability to change or react with few penalties in time, effort, cost, or performance, (v) the capacity of a manufacturing system to adapt successfully to changing environmental conditions and process requirements.

4. PMS-ESS TECHNICAL REQUIREMENTS

Looking at the literature analysis and merging it with the basic requirements defined in par. 2, some considerations could be elaborated and translated into more PMS-ESS needed features, defined as "Technical Requirements":

A) Taking into account some observations coming from the world of PMS at enterprise level (Terzi *et al.*, 2003), it might be said that the great diffusion of *Balanced Scorecard* (Kaplan and Norton, 1992) is meaningful. According to the simple and concise structure of *Balanced Scorecard*, the needed framework must provide all useful measures but, at the same time, a synthetic view of enterprise performance. The looked PMS-ESS might be a kind of a "cockpit panel", with an indicator for each dimension of relevant performance. Each indicator or measure could be the aggregate result of a number of lower-level measures, appropriate for the production system under examination. It should make it possible to obtain a simple general framework, with a few indicators, but completely customisable and adaptable to every industrial and manufacturing scenario.

B) For scheduling evaluation there are different types of measures: traditional measures of plant, measures for robustness, and measures of difference between schedules. Each type of measures must have a precise arrangement in the framework.

C) In literature on simulation and scheduling, some articles (e.g. Cavalieri *et al.*, 2000) propose the analysis of scheduling techniques under different conditions: in a stable manufacturing system, in a system with external perturbing events and into a system with internal perturbing events. These observations might be taken into account during the PMS implementation.

D) No exhaustive work has been found in literature on topic of scheduling distance or rescheduling efforts. In particular, no authors have presented measures to evaluate one of the most important causes of rescheduling efforts (e.g. the fact of knowing in advance the changes). If a scheduling solution provides good results, but its rescheduling modifications are made at the last moment, just before they must be applied, not only this is an organisational nervousness that causes rescheduling efforts, but also it makes it difficult to respect the schedule (e.g. workers could not have sufficient time to reorganize production).

E) In order to implement a PMS for effective scheduling evaluation into different manufacturing scenarios, measures and variables dependent on scheduling decisions might be highlighted. No works in literature have been found on this topic, which could be named the *dependence on schedule*. It is necessary to determine which measures depend on the production planning. These measures are not completely linked to the physical characteristics of the manufacturing systems, but can vary with the choices made during production programming. Let us explain with an example: White (1996) identified 125 performance measures through his literature survey. In the category *Flexibility*, it is possible to find measures such as (White, 1996: table IV, p.52): (i) *Number of part types processed simultaneously*, (ii) *Production cycle time*, (iii) *Setup time*, (iv) *Time to replace tools, change tools or move fixtures*, (v) *Percentage programmable equipment*, (vi) *WIP*. Into a production system, measures like *Time to replace tools, change tools or move fixtures* or *Percentage programmable equipment* have the same value apart from the scheduling methods that have been chosen. In the evaluation process of scheduling algorithms, they can be ignored. On the contrary, *Number of part types processed simultaneously* or *WIP* have values that depend on the scheduling system. Consequently, their appraisal is a necessary stage. Not so steady statements can be made for measure like *Production cycle time* or *Setup time*. Their dependence on the schedule of operations can be different for each production system and for time horizon to be considered.

Hence, it is possible to translate all the previous observations into more specific technical features. The looked PMS might be developed with:

1) Universal customisable measures - The needed PMS (I and III req.) has to be able to effectively measure every manufacturing system. So a feature is a set of measures, universal since capable to measure every system and customisable to do it effectively.
2) Methodology to customize the framework - Above all from the I, III and V requirements, a user of the PMS might be able to select measures for his/her manufacturing system and to choose the kind of judgement to obtain from the PMS.
3) Diverse dimensions of performance - The PMS is imagined to have some areas where the performance measures can be organised. This feature is deduced from the consideration of V and VI requirements.
4) Aggregation of measures - To satisfy the IV, V and VI requirements, the single atomic measures have to be aggregated to obtain a more synthetic view of the results.
5) Absolute measures and relative measures - To effectively measure the manufacturing system (III req.), the PMS might have a set of absolute measures. Instead, the necessity to have a comparison of schedules (II req.) and a quality judgement (V req.) requires relating the absolute measures of a schedule to the measures of other schedules.
6) Manufacturing system measures - From I and III requirements, it is inferable that the PMS has not only to show scheduling measures, but also manufacturing system measures.
7) Effective schedules' comparison – For providing a benchmarking between diverse scheduling techniques (II req.), it is obvious that the PMS might include specific measures for scheduling evaluation and might highlight these measures in a clear way.
8) Dynamism consideration - To measure a plant (VII and VIII req.) where disruptions can occur and to evaluate the quality of rescheduling, dynamic aspects have to be included in the PMS.
9) Rescheduling consideration - This feature matches the VIII requirement. It includes, though, also the considerations of the II requirement.

5. THE PROPOSED FRAMEWORK (PMS-ESS)

According to the literature review and the defined requirements, a first proposal of a framework for the evaluation of scheduling solutions has been developed. The proposal is a three-layered framework, in order to allow the analysis of a scheduling solution from three different perspectives, named PMS-ESS, which stands for *Performance Measurement System for the Evaluation of Scheduling Solutions*. The three layers are:

- *Effectiveness Domain* - This part of the framework regards measures and indicators for the evaluation of the effectiveness of the production plant according to the control of a scheduling solution. *Effectiveness* it is not merely considered the efficiency of sub-systems (e.g. single machines), but the overall set of measures capable to describe how the manufacturing system works in aggregate.
- *Robustness Domain* - This domain is composed by a set of indicators capable to give an idea of the robustness of scheduling solutions. For *Robustness*, authors stand for the ability of scheduling systems to perform graceful degradation of their performance in face of disruptions.
- *Flexibility Domain* - This domain allows users of the framework to get relevant results of a series of experimental tests conducted with the same scheduling system in different production environments. The set of measures pointed out in this domain tries to provide an answer to *"how the scheduling solution acts in a larger or different type of manufacturing system or according to a different production plan?"*

On developing the framework, the starting idea was to collect the maximum number of measures, in order to build a framework capable to be applied to every production system. As mentioned in par. 3, this is not a feasible idea, since proposed measures in literature are several with no standardised taxonomy, and some measures need to be defined *ad hoc* for manufacturing systems. The reached conclusion is the proposal of a general framework, with the three previously introduced domains, apt to every manufacturing system. In fact, dimensions of *Effectiveness*, *Robustness* and *Flexibility* can be considered universal. Not general and not universally used is the set of measures. For every type of manufacturing system a set of measures might be proposed. This way, the framework can be considered customisable, as requested (1).

According to (2) technical requirement, customisation of PMS-ESS is carried out at diverse levels. Firstly, the user will be able to choose his/her set of measures among those proposed in the library. New measures could also be implemented. Then, the user could choose the type of evaluation (e.g. only comparison with other scheduling techniques or also determination of target values, to verify if the schedule proposed can fit his/her requirements). Then, implementing aggregated indicators, the user could decide which measures have to be included and their weights.

Three areas compose the proposed framework. This is the high level organisation requested by (3) requirement. In *Effectiveness Domain*, single measures are grouped into five categories, the five SCOR attributes (Supply Chain Council, 2002, 2003). In *Robustness Domain*, there are both indicators focused on *Stability* of scheduling technique and indicators determining rescheduling efforts. In *Flexibility Domain*, three aspects are considered: *scalability*, *plan flexibility* and *reconfigurability*.

According to (4) requirement and looking to the results of a literature analysis, the methods more useful for the aggregation need are the *Multiple Attribute Decision Making* (MADM) methods, capable to compare different sets of values. In particular, the *TOPSIS* method has been chosen: it provides an indicator for each set of values, therefore an immediate comparison can be made.

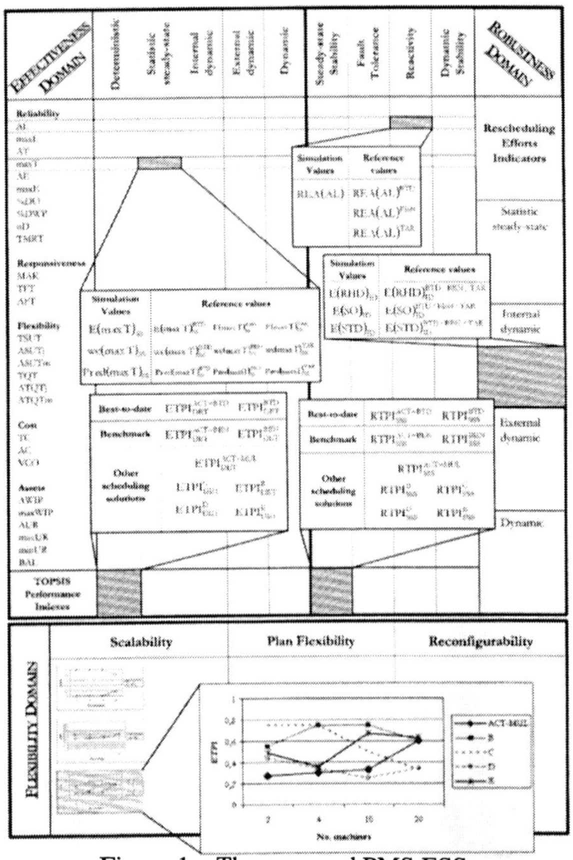

Figure 1 – The proposed PMS-ESS

According to (5) requirement, in the framework, for each measure and for each way to evaluate it, the absolute value is provided. But searching for a PMS apt to a benchmarking service, a comparison with other scheduling solutions is needed. For that, each measure is accompanied by reference values, in order to connect actual results to database's ones. So no relative measures have been included, but each measure is related to reference values.

An effective comparison of schedules (7) is realised through various aspects of the framework: (i) with an appropriate choice of measures, (ii) taking into account different production scenarios, (iii) using indicators of robustness and rescheduling efforts. The effective comparison is carried out in *Flexibility Domain* with graphical tools.

According to the (8) requirement, a real plant might be tested into diverse scenarios (e.g. using simulation or emulation). Consequently, statistical measures have been included into the PMS. The (i) mean, the (ii) worst-case and the (iii) predictability, which is based on standard deviation, are calculated among the values obtained from simulation for each measure. Also robustness and rescheduling efforts indicators take into account dynamic aspects of scheduling techniques.

According to the last requirement (9), through literature analysis, three indicators, named the *Rescheduling Efforts Indicators*, have been included, adapted with some minor modifications or implemented nearly from scratch.

6. CONCLUSIONS

The paper presents the requirements for the development of a PMS-ESS, elaborated taking into account a detailed literature review. In the future, PMS-ESS, after being tested into industrial test cases, will be integrated in a wider project, named *Benchmarking Service*, which is currently under deployment into the IMS-NoE SIG 4 community. Intention of the *Benchmarking Service* is to provide emulations of shop floor systems and their control systems, which will be brought together to execute performance tests, monitoring selected and elaborated performance criteria (Terzi *et al.*, 2003). These tests will address all relevant aspects, including ensuring the quality of the scheduling and control software itself, the deployment effort in a shop floor and the productivity of the shop floor system itself.

ACKNOWLEDGEMENT

The content of the paper derives from the efforts spent by authors within the IMS-NoE community, Special Interest Group 4. In particular, authors wish to thank Paul Valckenaers, coordinator of SIG 4, and Marco Taisch, coordinator of IMS-NoE, for their support.

REFERENCES

Ali, S., A. A. Maciejewski, H.J. Siegel, and J. Kim. (2003). Definition of a Robustness Metric for Resource Allocation. In: *Proceedings of the 17th International Parallel and Distributed Processing Symposium* (IPDPS 2003), Nice, France, 22-26 April.

Bongaerts, L., Y. Indrayadi, H. Van Brussel, and P. Valckenaers, (1999), Predictability of Hierarchical, Heterarchical, and Holonic Control. In: *Proceedings of the 2nd International Workshop on Intelligent Manufacturing Systems* 1999, Leuven, Belgium, pp. 167-176, 22-24 September.

Cavalieri, S., M. Taisch, M. Garetti, and M. Macchi. (2000). An experimental benchmarking of two multi-agent systems for production scheduling and control. In: *Computers in Industry*, Vol. 43, pp. 139-152.

Daniels, R., and p. Kouvelis, (1995). Robust Scheduling to Hedge Against Processing Time Uncertainty in Single Stage Production. In: *Management Science*, Vol. 41, No. 2, pp. 363-376.

Doumeingts, G. (1997). ECOGRAI – A method to design and to implement Performances Indicators Systems for industrial organizations, Internal documentation.

Gören, S. (2002). Robustness and Stability for Scheduling Policies in a Single Machine Environment. M.Sc. thesis, Bilkent University, Ankara, Turkey.

Jensen, M.T. (2001). Robust and Flexible Scheduling with Evolutionary Computation. Ph.D. thesis, Department of Computer Science, University of Aarhus, Denmark.

Kaplan, R.S. and D.P. Norton (1992). The balanced scorecard - measures that drive performance. In: *Harvard Business Review*, January-February, pp. 71-79.

Kempf, K., R. Uzsoy, S. Smith and K. Gary (2000). Evaluation and comparison of production schedules. In: *Computers in Industry*, Vol. 42, pp. 203-220.

IMS-NoE (2003). Network of Excellence on Intelligent Manufacturing Systems. Project reference IST-2001-65001 IMS-02006.

Mignon, D.J., S. J. Honkomp, and G. V. Reklaitis. (1995). Framework for Investigating Schedule Robustness Under Uncertainty. In: *Computers in Chemical Engineering*, Vol. 19, Supplement, pp. S615-S620.

Ouelhadj, D., P. Cowling, and S. Petrovic. (2003). Utility and Stability Measures for Agent-Based Dynamic Scheduling of Steel Continuous Casting. In: *Proceedings of the IEEE International Conference on Robotics and Automation*, Taipei, Taiwan, 14-19 September.

Terzi, S., P. Brambilla, and S. Cavalieri, (2003), Evaluation and validation of scheduling systems: issues and trends, In: *Proceedings of the 3rd International Workshop on Performance Measurement*, Dalmine, Italy, 19-20 June, pp. 155-168.

Supply-Chain Council (2003). Supply Chain Operations Reference Model - SCOR Version 5.0, Supply Chain Council, Pittsburgh, Pennsylvania.

White, G.P. (1996). A Survey and Taxonomy of Strategy-Related Performance Measures for Manufacturing, In: *International Journal of Operations & Production Management*, Vol. 16, No. 3, pp. 42-61.

Copyright © IFAC Information Control Problems in Manufacturing, Salvador, Brazil, 2004

ELSEVIER
IFAC
PUBLICATIONS
www.elsevier.com/locate/ifac

THE USE OF QUALITATIVE INDICATORS FOR PERFORMANCE MEASUREMENT IN MANUFACTURING CONTROL SYSTEMS

Paulo Leitão[1] and Francisco Restivo[2]

[1] *Polytechnic Institute of Bragança, Quinta S^{a} Apolónia, Apartado 134, P-5301-857 Bragança, Portugal, pleitao@ipb.pt*
[2] *Faculty of Engineering, University of Porto, Rua Dr. Roberto Frias, P-4200-465 Porto, Portugal, fjr@fe.up.pt*

Abstract: New manufacturing control paradigms are being developed to improve the ability of enterprises to respond to change more effectively. However, there are no agreed definitions and methodologies for the evaluation and comparison of achievements of different manufacturing control systems approaches, making also difficult the communication and cooperation between manufacturing control systems developers. In this paper some qualitative performance indicators are identified and defined - re-configurability, robustness and agility - and methodologies for their evaluation are suggested.
Copyright © 2004 IFAC

Keywords: Intelligent Manufacturing Systems, Manufacturing Control Systems, Performance Measurement.

1. INTRODUCTION

Enterprises always look for profit, through more productivity, more quality, more flexibility, more agility and better adaptation to unexpected disturbances. In the emerging global economy, where the demand for mass customised products puts new requirements in terms of the ability of enterprises to respond to change, the steady state analysis is becoming less important, while dynamic response is becoming a key issue in the design of new manufacturing systems, with large impact at all levels, both technical and socio-economical.

Several manufacturing control architectures using emergent paradigms and technologies, such as holonic manufacturing systems, have been proposed (see (JASS, 2001) and (Deen, 2003) for more details about some existing holonic work in the area). One of the holonic architectures proposed is the ADACOR (Adaptive Holonic Control Architecture for Distributed Manufacturing Systems) architecture (Leitão and Restivo, 2002), which addresses the need for the agile and fast reaction to disturbances at the distributed shop floor level, increasing the agility and flexibility of an enterprise, when it works in volatile environments, characterised by the frequent occurrence of unexpected disturbances (more details about ADACOR architecture can be found in (Leitão and Restivo, 2003; Leitão et al., 2003).

The analysis and evaluation of the dynamic behaviour of manufacturing systems requires the definition of a methodology to allow the analysis of manufacturing control system properties, the evaluation of its performance, and the comparison between different control approaches. In spite of some research in the area, such as the described in (Cavalieri et. al, 1999) and other within the working group *Benchmarking and Performance Measures* (SIG4) belonging to the Intelligent Manufacturing Systems – Network of Excellence[1], one draw back in the validation of manufacturing control systems is the missing of a standard benchmark framework that allows to compare the dynamic performance of the several control systems.

The traditional performance measurement systems evaluate quantitative indicators directly related to production parameters: throughput, number of delayed orders, WIP, manufacturing lead time, etc.

[1] For more information, please see http://www.ims-noe.org/

The problem is how to evaluate the performance of the systems in the presence of unexpected changes. Here, performance indicators may be of a qualitative nature, since they usually reflect subjective views of the expected behaviour of the systems in those circumstances.

In this paper, the problem is analysed, and some qualitative performance indicators are identified and defined - re-configurability, robustness and agility - and methodologies for their evaluation are suggested.

2. PERFORMANCE MEASUREMENT IN MANUFACTURING CONTROL SYSTEMS

The assessment, evaluation and comparison of achievements of different manufacturing control systems, and the cooperation between manufacturing control systems developers to improve their solutions, require the definition of appropriate performance measurement frameworks.

Performance measurement is the process of using a tool or a procedure to evaluate a concrete efficiency parameter of the system. As an example, a common way of evaluating a car engine performance is to calculate the kilometres per litre ratio, i.e. the ratio of the number of kilometres driven to the number of litres of petrol consumed. The term *measurement* implies that the approach being used is rigorous, systematic and quantifiable.

A performance measurement procedure must be objective, based on scientific evidence and must not affect or distort results. In developing a performance indicator, it must be tested to ensure that it is reliable (i.e. the use of the procedure results in the same reading regardless of who does the measuring or when and where the measurement is taken), valid (i.e. the procedure measures what is intended) and standardised (i.e. definitions of standards, data elements, data collection, and data analysis are sufficiently precise and comprehensible that they can be understood and applied in the same way regardless of who refers to or applies them).

Performance measurement results describe an observed level of performance (such as throughput rate or number of industrial accidents per year) allowing to analyse the performance of a system and to compare the performance of different systems[2], but they don't tell why the performance is as it is. Results may not reveal which factors account for differences in measured levels of performance.

Traditionally, performance measurement approaches used to have a scope focused in the financial aspects. (Neely, 1999) identified that the reasons why these types of measures are criticised is because they:

- encourage short-termism and local optimisation;
- lack strategic focus and fail to provide data on quality, responsiveness and flexibility;
- do not encourage continuous improvement.

In the manufacturing context, the performance parameters can be classified as qualitative or quantitative. The quantitative measures are based on different production performance indicators, such as the manufacturing lead time, the tardiness, the waiting time, the throughput, number of delayed orders and the WIP. The qualitative measures are of a more subjective nature and reflect properties of the manufacturing control solution, such as the agility, flexibility and robustness, which cannot be directly obtained from the production data.

The choice of the performance indicators depends on the business domain and the objective of its use. For some applications the throughput is the major performance indicator, but in other circumstances the robustness or service quality can be the focus of the performance measurement.

A broad number of performance measurement systems (PMSs) has been proposed and in some cases deployed in practice, using indicators more relevant to the manufacturing area and that can be used to drive the production process. Some of the most widely cited PMSs are: the Balanced Scorecard (Kaplan and Norton, 1996), the Performance Measurement Matrix (Keegan et al., 1989), and the Integrated Dynamic Performance Measurement System (Ghalayini and Noble, 1996). These approaches present some disadvantages, some of them identified by (Bititci et al, 2002):

- most of the performance systems are historical and static, and do not reflect the dynamic aspects associated to the manufacturing changes;
- only few performance measurement systems have an IT infrastructure, which leads to cumbersome and time-consuming data collection and reporting;
- these approaches focus in the performance measurement of the manufacturing system, and not in the associated control system.

Since the scope of this paper is the manufacturing control system, our intention is the definition of suitable set of performance indicators addressing the dynamic and qualitative aspects of a manufacturing control system. In these circumstances, the authors propose a methodology to evaluate the performance of manufacturing control applications, which comprises the following main steps:

[2] As an example, nowadays the measure of productivity is in the order of the day, being used this parameter to compare and rank the economic performance level of each country.

- Identification of a set of performance indicators adequate to manufacturing control.
- Definition of each identified performance indicator.
- Definition of a procedure to measure each performance indicator.

In the following sections, the proposed methodology will be exemplified by identifying some performance indicators to measure qualitative aspects of manufacturing control systems, and defining procedures to measure those indicators. Finally, a procedure to calculate the overall performance of a manufacturing control system, allowing to rank different systems, is suggested.

3. MEASURING THE DYNAMIC PERFORMANCE INDICATORS

The parameters associated to the dynamic behaviour of a manufacturing system, such as the capability to support the introduction of a new product in the factory plant or to react properly to the occurrence of a machine breakdown, have considerable importance in the performance of a manufacturing system. However, in performance measurement research, little attention has been devoted to the system dynamics, especially to the quantification of these parameters.

An important aspect when dealing with the occurrence of disturbances is the response effectiveness. In our opinion, the parameters more suitable to assess this effectiveness are: re-configurability, agility and robustness. These major key parameters in the analysis of the dynamic response are difficult to measure presenting a certain degree of subjectivity in their analysis. In the following, these parameters will be defined and methodologies for there measurement will be proposed.

3.1 Re-configurability

Flexibility is related to the capability of the manufacturing system and its control system to support the production change when the consumption or demand changes, and the capability of re-configuration in the case of malfunctions of some resources.

Several flexibility classifications, as discussed by (Gerwin, 1993; Browne et al., 1984), are available in literature, identifying different types of flexibility, such as mix, changeover, volume, product and sequencing. The mix flexibility is concerned with the capability of a system to handle a range of products or variants, supported by the execution of fast set-ups in the process. The changeover flexibility is the capability to change quickly the production system in order to be able to offer new products. The volume flexibility is the capability to deal with production volume variability, facing the demand. The product flexibility is the capability to modify rapidly the product at design level. The sequencing flexibility is the capability to support alternative sequences for the production plan execution by using resources that have capability to execute different operations, organised in a proper way.

In spite of the little attention that has been devoted to the measurement of flexibility of manufacturing systems and their associated control systems, some authors have approached the issue. As an example, (Ibarrondo and Mercader, 2001) presents an approach to measure the volume and mix flexibility (the aggregation of these types of flexibility is designated by operational flexibility) based in measuring the minimal and maximum capacity of the system, and (Tsourveloudis, 1998) presents a fuzzy logic framework to measure several types of flexibility: machine, routing, material handling system, product, operation, process, volume, expansion and labour flexibilities.

None of these types of flexibility is the focus of our work, since we are looking for the flexibility associated to the control system, which we defined as re-configurability.

The re-configurability of the control system is the ability to support different manufacturing system configurations, i.e. different production systems scenarios, with a small customisation effort.

We assume that there are two dimensions on the effort, the quality and the quantity of the effort, and that these two values can be measured/estimated for a certain of previsible scenarios. The measures to determine the re-configurability of the control system are of two different types: degree of complexity of the customisation and development time for the customisation.

The degree of complexity parameter is a subjective parameter that can be evaluated in a relative way and expressed in a scale from 0 to 100 percent, through questionnaires to the software developers and manufacturing system engineers. The development time can be measured counting the time spent when customising the application, or estimated by experts. As these two parameters are highly subjective and not quantifiable for a given set of scenarios, the authors propose a fuzzy approach to this question, reflecting this qualitative nature.

The degree of complexity can be fuzzified into the fuzzy set {Low, Medium, High}. Each one of the previous terms is a fuzzy variable defined on the base variable. Since a fuzzy set A is a collection of ordered pairs $A = \{(x, \mu(x))\}$, where the item x belongs to the universe and $\mu(x)$ is its grade of membership in A, it is necessary to define the membership functions for each fuzzy variable.

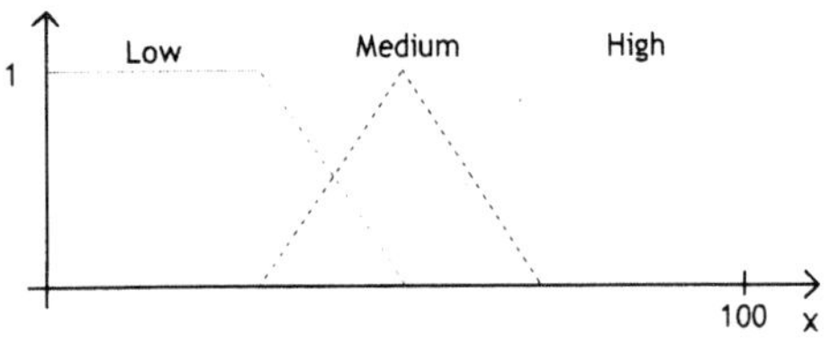

Fig. 1 - Membership Functions for the Degree of Complexity Variable

In case of the degree of complexity fuzzy variable the membership functions are illustrated in Fig. 1. The membership function of {Low} and {High} fuzzy sets use a trapezoidal function type and the membership function of {Medium} fuzzy set uses a triangular function type. The use of triangular and trapezoidal functions type is the simplest way to define membership functions, but they not conduct to a smooth surface in the output parameter. The use of a Gaussian type function allows smoothness in the output parameter.

The development time parameter can also be fuzzyfied applying a similar procedure, and represented in terms of the fuzzy set {Low, Medium, High}.

Table 1 – Fuzzy Rules to Determine the Reconfigurability Parameter

Degree of Complexity	Customisation Time	Reconfigurability
Low	High	Low
-	Medium	Medium
Low	Low	High
Medium	High	Low
Medium	Low	High
High	High	Medium
High	Low	High

The re-configurability parameter is also expressed in terms of the fuzzy set {Low, Medium, High}. The fuzzy inference process is based in a fuzzy rule base. The reconfigurability fuzzy variable is determined according to fuzzy rules, such as those expressed in the Table 1.

3.2 Robustness

The robustness of a control system is the capability to remain working correctly and relatively stable, even in presence of disturbances.

A robustness performance measurement procedure should measure how a system reacts to possible erroneous inputs or environmental factors that could affect the system. Ideally, it is necessary to exercise the system with all possible errors, to guarantee an absolutely robust system. However, in reality, it is not possible to simulate all possible natural errors that can occur in the system (and verifying the system operation and waiting for the occurrence of errors that occur infrequently is too time consuming).

Until now, there has been no effective approach to the quantitative measure of the robustness of a manufacturing control system. Here, a simple and repeatable way to measure the manufacturing control system robustness is introduced.

The robustness performance measurement procedure will give a relative measure of robustness, having not a quantitative value. The procedure to measure the robustness comprises the introduction of possible errors (as many as possible) and the verification if the control system remains working correctly.

What-if Tests to Evaluate the Control System Robustness
T1: During the normal operation the system remains stable?
T2: Introduce a new component in the system (e.g. a new machine). The system remains stable?
T3: Breakdown a centralised component of the system (such as a central scheduler or a supervisor). The system remains stable?
T4: Breakdown a local distributed component of the system (such as a resource, an order or a component that integrates a legacy system). The system remains stable?
T5: Introduce a failure in the network that links the holon responsible for the resource and the physical resource. The system remains stable?
T6: Remove the available resource(s) to execute a job (for example, a transporter device). The system remains stable?
T7: Test the graphical interface by clicking on the buttons and introducing inputs value with different data types (for example, inputing a string in a date field). The system remains stable?
T8: Introduce the occurrence of machine breakdowns. The system react to it?
T9: Introduce configuration files for a holon not respecting the XML format define for the configuration file. The system remains stable?
T10: Include a data type error in the configuration file (for example, introducing a string value in the field related to the estimated processing time). The system remains stable?
T11: Remove all possible resources to execute an operation. The system remains stable?
T12: In this case, the system is able to continue the resource allocation process (for example, waiting a certain time and announcing again the execution of the operation)?
T13: Send messages to the holons with no understandable contents. The system remains stable?
T14: Introduce one rush order. The system respond to it?
T15: Increase the number of production orders. The system (schedulers) remains stable?
T16: Plug in a new scheduling algorithm (or at least change the scheduling heuristic rule). The system remains stable?
T17: Add new rules to the decision component. The system remains stable?
T18: Change the layout configuration or resource skills. The system remains stable?
T19: For the normal number of holons in the system, verify if exists communication overhead (for example, during the resource allocation process, bid proposals are not received).
T20: Duplicate the number of holons in the system. The response remains acceptable?

Fig. 2 – Set of What-if Conditions to Verify the Robustness of the Control System

Fig. 2 illustrates an example of a set of 20 what-if conditions to evaluate the robustness of a manufacturing control system. The degree of robustness is given by $\frac{nP}{nT}.100\%$, where nP is the number of tests passed and nT is the total number of tests.

As an example, a manufacturing control system application that fulfils only 10 what-if tests, has a score of 50%.

The best set of what-if questions is an open issue.

3.3 Agility

The agility of a control system measures its capability to react in a short period of time to the occurrence of unexpected disturbances. Thus, the agility is a time-based property, related to the time needed by the system to recover from the occurrence of the disturbance.

Agility is dependent of the manufacturing system flexibility, especially the routing flexibility, since it is only possible to react with agility to the occurrence of disturbances if exists capacity and alternative resources in the manufacturing system. Thus, a system has low agility in case of low routing flexibility.

The proposed approach introduces the concepts of frequency and duration of the disturbances to quantify the control system agility. Fig. 3 illustrates the evolution of the throughput over the time. This evolution is affected with the occurrence of disturbances.

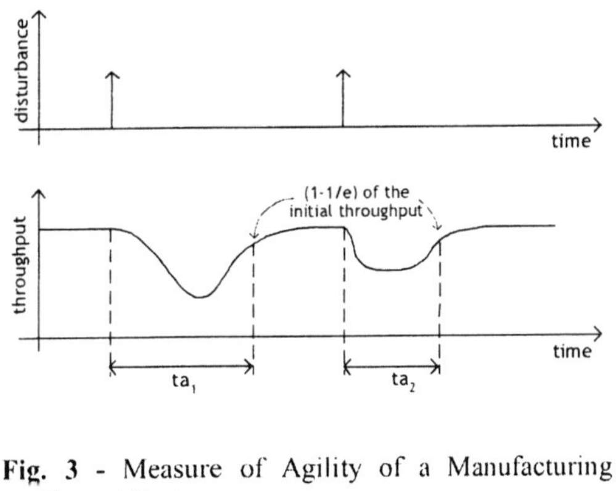

Fig. 3 - Measure of Agility of a Manufacturing Control System

The Dirac's impulse in the disturbance graph represents the occurrence of a disturbance, which degrades the performance criteria, illustrated by the decrease of the throughput value. The parameter *ta* is the required time spent by the system to adapt to the disturbance. This parameter is measured between the occurrence of the disturbance and the time when the system achieves a certain percentage of the initial throughput. A good indicator of this percentage is based in the time constant of RL and RC circuits, which is given by (1-1/e).

The parameter *ta* is a statistical value, calculated by determining the mean value of ta_i, the control system being more agile as smaller the ta parameter is. It is measured in time units per disturbance.

The agility could also be expressed in terms of the maximum number of disturbances that can be handled per time unit, which is given by the inverse of *ta*. It measures the maximum number of disturbances per time unit which do not degrade the system performance below a certain level.

The measurement of agility using the previous approach requires the periodic measurement of an appropriate performance criterium, such as the throughput or the amount of processed material.

4. EVALUATION OF OVERALL SYSTEM PERFORMANCE

Manufacturing control systems can be compared according to the performance criteria that they intend to maximise (or minimise), for example comparing the different manufacturing control systems under the throughput or agility point of view. However, to rank them, it is necessary to use a procedure that takes in consideration several performance parameters. In our approach, a spidergram is used to summarise the main performance vectors defined for the evaluation of manufacturing control applications.

Spidergrams, also known by polar or radar charts, are a form of graph that allows a visual comparison between several quantitative or qualitative aspects of a situation, or when charts are drawn for several situations using the same axes (poles). Normally, between three and eight attributes can be plotted on each chart, and the chart becomes difficult to analyse for more than eight axes. Scales for each attribute are arranged radially and the points plotted on each radius are joined to generate a shape that can be visually compared with the same plot for another situation.

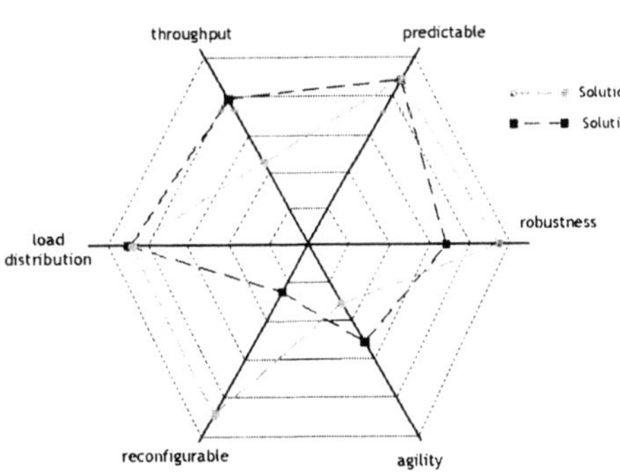

Fig. 4 - Spidergram to Evaluate a Manufacturing Control System under Multiple Criterias

In the example of Fig. 4, the performance evaluation is based in six-labelled poles: throughput, predictability, robustness, agility, re-configurability and load distribution. The performance of different manufacturing control solutions is reflected in the shape of the hexagon drawn to link the plotted points. In this example, the solution A presents better robustness and re-configurability than solution B; on the other hand, the solution B presents better throughput and agility than solution A.

Additionally, one of the merits of using a spidergram is that the area occupied by a manufacturing control system's spidergram prrovides an overall system score (Fletcher et al., 2003).

Finally, the spidergram provides an excellent way of showing the gap in performance between a given manufacturing control system and a world class one (or the best practice in the domain), by representing both data on the same chart.

5. CONCLUSIONS

Nowadays, it is becoming more important to evaluate and to compare manufacturing control applications. This is especially true in the analysis of the system performance associated to the transient aspects of the manufacturing system, usually using some qualitative measurement criteria. These are however more difficult to measure and quantify than the quantitative indicators.

This paper defines a procedure to analyse and evaluate manufacturing control systems, taking in consideration the qualitative indicators. Some qualitative indicators were identified and their definitions were clarified: the re-configurability, robustness and agility. Procedures to measure and quantify these performance indicators were proposed: the re-configurability in terms of effort, the agility in terms of time and the robustness in terms of a percentage of what-if conditions fulfilled.

The proposed procedures can be generalised in the future, for example by introducing more general test verifications in the analysis of the robustness.

ACKNOWLEDGMENTS

This paper has been written within the activities of IMS-NoE (Intelligent Manufacturing Systems Network of Excellence).

REFERENCES

Bititci U., Nuderupati, S. and Turner, T. (2002). Web Enabled Performance Measurement Systems; Management Implications. *International Journal of Operations and Production Management*, 22(1), pp. 1273-1287.

Browne, J., Dubois, D., Rathmill, K., Sethi, S.P. and Stecke, K.E. (1984). Classification of Flexible Manufacturing Systems. The FMS Magazine, pp 114-117.

Brussel, H. Van, J. Wyns, P. Valckenaers, L. Bongaerts and P. Peeters (1998). Reference Architecture for Holonic Manufacturing Systems: PROSA. *Computers In Industry*, **37**, pp. 255-274.

Cavalieri, S., Bongaerts, L., Taisch, M., Macchi, M. and Wyns, J. (1999). A Benchmark Framework for Manufacturing Control. In: *Proceedings of the Second International Workshop on Intelligent Manufacturing Systems*, Leuven, Belgium, 22-24 September.

Deen, S.M. (ed.) (2003). *Agent-Based Manufacturing: Advances in the Holonic Approach*. Springer Verlag Berlin Heidelberg.

Fletcher, M., McFarlane, D., Thorne, A., Jarvis, D. and Lucas, A. (2003). Evaluating a Holonic Packing Cell. In: *Holonic and Multi-Agent Systems for Manufacturing*, V. Marík, D. McFarlane and P. Valckenaers (eds.), Lecture Notes in Artificial Intelligence, vol. 2744, Springer-Verlag, pp. 246-257.

Gerwin, D. (1993). Manufacturing Flexibility: A Strategic Perspective. *Management Science*, **39**, pp 395–410.

Ghalayini, A. M. and Noble, J.S. (1996). The Changing Basis of Performance Measurement. *International Journal of Operations and Production Management*, **16**(8), pp. 63-80.

Ibarrondo, J. and Mercader, J. (2001). Measuring Operational Flexibility. In: *Proceedings of the Fourth Stimulating Manufacturing Excellence in Small and Medium Enterprises International Conference*, Alborg, Denmark, 14-16 May, pp. 292-302.

JASS (2001). Special Issue on Industrial Applications of Multi-Agent and Holonic Systems. Journal of Applied Systems Studies, 2(1).

Kaplan, R.and Norton, D. (1996). *The Balanced Scorecard: Translating Strategy into Action*. Harvard Business School Press.

Keegan, D.P., Eiler, R.G. and Jones, C.R. (1989). Are Your Performance Measures Obsolete?. *Management Accounting (US)*, June, pp. 45-50.

Leitão, P., Colombo, A.W. and Restivo, F. (2003. An Approach for the Formal Specification of Holonic Control Systems. *Holonic and Multi-Agent Systems for Manufacturing*, V. Marík, D. McFarlane and P. Valckenaers (eds), Lecture Notes in Artificial Intelligence, vol. 2744, pp 59-70, Springer-Verlag.

Leitão, P. and Restivo, F. (2003). Towards Autonomy, Self-Organisation and Learning in Holonic Manufacturing. *Multi-Agent Systems and Applications III*, V. Marík, J. Müller and M. Pechoucek (eds), Lecture Notes in Artificial Intelligence, vol. 2691, pp 544-553, Springer-Verlag.

Leitão, P. and Restivo, F. (2002). Holonic Adaptive Production Control Systems. In: *Proceedings of the 28th Annual Conference of the IEEE Industrial Electronics Society*, Sevilla, Spain, 5-8 November, pp. 2968-2973.

Neely, A. (1999). The Performance Measurement Revolution: Why Now and What Next?. *International Journal of Operations and Production Management*, **19**(2), pp. 205-228.

Tsourveloudis, N. and Phillis, Y. (1998). Manufacturing Flexibility Measurement: A Fuzzy Logic Framework. *IEEE Transactions on Robotics and Automation*, **14**(4), pp. 513-524.

Copyright © IFAC Information Control Problems in
Manufacturing, Salvador, Brazil, 2004

ELSEVIER
IFAC
PUBLICATIONS
www.elsevier.com/locate/ifac

TESTING AND VALIDATION OF DETERMINISTIC SCHEDULES IN A SIMULATED STOCHASTIC ENVIRONMENT

Botond Kádár * András Pfeiffer *
László Monostori *

* *Computer and Automation Research Institute, HAS, Kende u. 13-17, Budapest, POB 63, H-1518, Hungary, Phone: (36 1) 279-6195, Fax: (36 1) 4667-503, e-mail: {kadar, pfeiffer, laszlo.monostori}@sztaki.hu*

Abstract: The paper presents a hierarchical architecture of an integrated production planner and job-shop scheduler system with flexible modelling capabilities and powerful, scalable solution methods. The paper tackles the problem of handling uncertainty during the execution of predictive schedules in a dynamic environment. It addresses the simulation module of the architecture highlighting its main functionalities. The simulation serves as an evaluator of the calculated schedules providing support for the planning and scheduling decision makers. The schedules are calculated by a constraint-based deterministic scheduling algorithm. Results of experiments which were achieved by using the model of a real, large job-shop environment are also described. *Copyright © 2004 IFAC*

Keywords: Job-Shop Scheduling, Manufacturing Execution Systems, Simulation

1. INTRODUCTION

The broad goal of manufacturing operation management, such as a resource constrained scheduling problem, is to achieve a co-ordinated efficient behavior of manufacturing in servicing production demands while responding to changes in shop-floors rapidly and in a cost effective manner. Operation scheduling is viewed as a major issue which is a complex task requiring co-ordination. Shop-floor scheduling, like resource constrained scheduling problems in general, is a combinatorially complex, NP hard problem, thus computationally unfeasible to be solved by the sole use of conventional Operations Research (OR) approaches. Artificial Intelligence (AI) based or hybrid techniques using domain specific heuristics are necessary to guide the search and to provide satisfying good solutions in due time. This demand gave a high importance of constraint satisfaction techniques (Kovács and Kis 2003).

The quality of factory scheduling, generally, has a profound effect on the overall factory performance. The advanced generation of factory schedules is necessary to co-ordinate the manufacturing activities in order to meet organizational objectives, and to anticipate potential performance obstacles (e.g., resource contention), thus minimize the disturbing effects on the overall manufacturing system operation. In industrial practice, however, at least two factors confound the use of predictive (advance) schedules as operational guidance (Smith 1994):

- Advance or predictive schedules result from scheduling systems running with static models that ignore important new operating constraints/objectives of live shop operation

(Prosser 1991). It is due to the lack of a close correspondence to the live status of executed processes and the data resulting from their real-time monitoring.

- They cannot cope with the environmental and internal uncertainties such as unexpected production demands raised by changing market conditions, late deliveries, failed operations/break-downs of machines/equipment, unavailability of operators all of which work against efforts to follow predictive schedules (Szelke and Monostori 1999).

Traditionally, the major part of the scheduling research activity deals with off-line scheduling problems. The real world shop-floor scheduling problems, however, are of dynamic nature, which necessitates more complex techniques.

The closer we are to the realization of plans and schedules, the higher the chance of unexpected events is that may render plans and schedules inadequate. That is why practical scheduling is driven by uncertainty, and the methods applied in dynamic job-shops rarely utilize theoretical results (McKay and Wiers 1999).

The performance of a shop-floor control system mostly depends on its ability to rapidly adapt schedules to current circumstances. Scheduling techniques addressing the dynamic scheduling problem are called *dynamic scheduling algorithms.* These algorithms can be further classified as *reactive* and *proactive* scheduling techniques. An additional categorization of scheduling techniques relates to the stochastic or deterministic characteristics of the problem (Bongaerts 1998, Szelke and Monostori 1999, Kádár 2002).

Reactive scheduling (Szelke and Kerr 1994) is, generally, conceived as a real time revision or repair of a complete but execution-time flawed schedule to keep it in line with the live status of shop-floor processes and events and to make it further executable near to the predicted one. In addition, it has been recognized the importance of stable behavior of manufacturing system operation (keep the manufacturing processes smoothly moving). Operational solutions for reactive scheduling are *complete rescheduling, deferred commitment* and *tweaking* (Bongaerts 1998). Some probabilistic representations of scheduling uncertainty have been reported by (Sadeh and Fox 1990, Berry 1992). (Monostori *et al.* 1996) gives a detailed survey on how other possible approaches were applied in scheduling.

The paper illustrates a simulation framework for the testing and evaluation of deterministically calculated advance schedules in a large job-shop environment. The simulation is coupled in a hierarchical planning and scheduling system. The

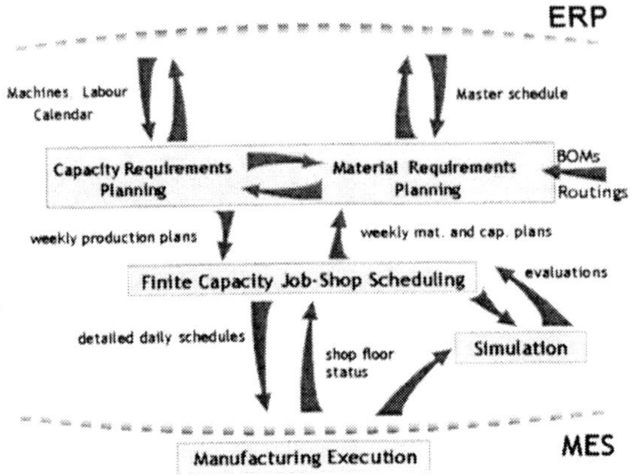

Fig. 1. Main system components: medium-term capacity and production planner, short-term scheduler and the simulator

overall architecture provides the base of a dynamic scheduling system in which the power of the constraint-based scheduler enables complete rescheduling.

2. ARCHITECTURE OF THE SYSTEM

On the base of previous explanatory experiments and basic research, a multi-tiered system structure was defined (Figure 1). The layers of the system are as follows (Kovács *et al.* 2003):

- The solution of medium-term, integrated capacity and production planning problem is provided by an integer-linear programming approach.
- The solution of the short term, detailed finite scheduling problem is calculated by a constraint programming technique.
- The evaluation and analysis of the predictive short term schedules is carried out by a discrete event simulation model.

An important practical requirement is that the system components should be able to work with the data stored in existing - so-called legacy - production information systems. In this respect, the proposed architecture connects the Enterprise Resources Planning (ERP) and the Manufacturing Execution System (MES) of a factory.

The details of the capacity planning module and the finite capacity scheduler are described in (Kis 2002, Kovács and Kis 2003, Kovács *et al.* 2003). In the following sections of the paper the simulation module of the above architecture is described.

3. THE SIMULATOR

Simulation captures those relevant aspects of the Production Planning System (PPS) problem, which cannot be represented in a deterministic, constraint-based optimization model. The most important issues are uncertain resource availability, uncertain processing times, uncertain raw material quality, and insertion of conditional operations into the technological routings.

Here, a multi-purpose application of simulation is depicted. On the one hand, the simulation modelling follows the traditional methodology and results in a *complex decision support tool*, on the other hand, the developed simulator is utilized as a component of a higher level system taking the role of the real production system (see Figure 1) and acting as a *"quasi emulator"*. Quasi emulation means that, contrary to traditional simulation models which integrate the control logic of a real system - generally, utilizing dispatching rules - the model and control of the physical system are separated. In our case the simulator executes the optimal schedules calculated by the constraint-based scheduler.

3.1 Functions of the simulator

The intention to connect the scheduler to a discrete event simulator was twofold. On the one hand, it serves as a *benchmarking system* to evaluate the schedules on a richer model, on the other hand, covers the *non-deterministic character* of the real-life production environment. Additionally, it is expected that the statistical analysis of schedules could help to improve the execution and support the scheduler during the calculation of further schedules. The evaluation of schedules is measured over several runs of the discrete event simulator where the number of replications (independent simulation runs, using different random numbers) depends on the construction of confidence intervals. The main functions of the discrete event simulator are as follows: it

- evaluates the robustness of daily schedules against the uncertainties,
- helps in visualizing and verifying the results of the PPS system,
- supports the systematic test of the pilot PPS system,
- offers a benchmark platform for the calculated schedules,
- supports dynamic rescheduling decisions.

3.2 Uncertainties in the simulation model

The basic types of uncertainties that are modeled in the simulation model are as follows:

- *Start time uncertainty:* Delivery of purchased items (raw materials) is not always reliable. Hence, it is not sure whether some operations can really be started at the planned earliest start dates.
- *Downtimes:* Due to failures and/or the unexpected absence of machines and/or workers may not be available as planned.
- *Processing time:* The actual processing time of some operations may depend on the proficiency and skill of the worker. Processing times may be shorter or longer than planned.
- *Rework and adjustment:* The execution of specific operations depends on the result of quality check operations. Based on the result of the check, they may be repeated or some adjustment operations are performed.
- *Quality uncertainty:* Whether some adjustment operations should be performed or not may depend also on the quality of incoming material. Due to a major quality problem, even the progress of a product may be blocked.

3.3 Architecture of the simulation module

The main requirements for the simulation module were identified as follows:

- Common data-model with the scheduler.
- Short response time, enabling multiple model runs.
- Support for Input/Output inspections.
- On-line and bi-directional connection to the scheduler (Input and Output).
- Playback time horizon 1 week.
- Support for different playback strategies.

To satisfy all the requirements for the flexible simulation system the structure presented in Figure 2 has been developed.

The simulation uses the same meta model as the finite capacity job-shop scheduler. Resources, products, process plans, production information, etc. are transformed exactly to the same form for all system components. Hereby, the complexity of the integration procedure of the simulation module into the system is significantly reduced. None the less, the common data tables ensure data integrity during the creation of the simulation; moreover, the data-model serves as a basis for the more detailed shop-floor model.

Running the simulation by applying the basic data tables results in a waste number of queries during the model run reducing the simulation speed significantly. However, in order to ensure enough number of simulation replications for the evaluation of a short time production schedule, the total response time should be minimized. To resolve the above two contradictory objectives an exhaustive data pre-processing phase is included in the simulation process. The data-preparation

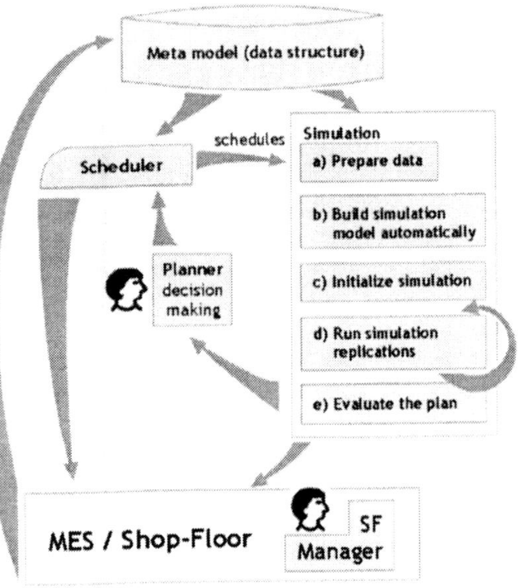

Fig. 2. Architecture and the main process flow in the simulation module

is carried out before the overall simulation study (phase **a** in Figure 2). The advantage of the shorter response time compensates the redundant data storage in the simulation model.

Modelling real production systems frequently brings up the problem of handling hundreds of resources in a simulation model. Having the modelling objects in hand, which were created on the base of the conceptual model, in our architecture the simulation model is created automatically based on the prepared data (phase **b** in Figure 2).

The automatic generation of the model is followed by the initialization phase (phase **c** in Figure 2). In this case, besides classical parameter settings, this procedure involves the generation of input-parameter-specific model components (entities such as products, operators). Contrary to the previous phase, this procedure is carried out for each replication.

The simulation runs are repeated until the required number of replications is reached (phase **d** in Figure 2). Each replication is a terminating, non-transient simulation run, having the same initial parameters and settings, but different parameters for uncertain simulation times and events generated on the base of random numbers.

In the last phase the plan is evaluated using the evaluation criteria and the results of the evaluation process are interpreted by shop-floor managers who are predestined to take necessary actions (the shop-floor manager, phase **e** in Figure 2). The results of the evaluation procedure may also be applied in a rescheduling decision. This scenario requires human decision as well (the planner, phase **e** in Figure 2).

4. CASE-STUDY

We elaborated a case-study at a factory that produces mechanical products by using machining and welding resources, assembly and inspection stations and some highly specialized machines. Production is performed in a make-to-order manner where deadline observance is an absolute must, even regarding unpredicted orders. Since quality assurance is a key issue, tests may result in extra adjustment operations. The planning and scheduling method was validated and tested with the real-life data. First, projects were generated from existing routing tables and Bill of Materials (BOMs), then, using the resource calendars, the planning problem was solved on a 15-week horizon, with a time unit of one week. Then, the production plan was passed to the constraint-based finite job-shop scheduler that worked with a 10 min. time unit.

The object-oriented hierarchical simulation model of the plant is based on the functional decomposition approach. The simulation includes the modeled elements of the real plant and each unit of a production set is identified uniquely and traced during its lifecycle. The simulation model is created following the "flexible" simulation modelling technique described in the previous section.

The deterministic inputs of the simulation are provided in three main data tables. These are tables of *resources, process plans*, and the *short-time-schedule*, passed by the scheduler. The resources of the plant are categorized in two main groups: *machine* and *personnel*. The stochastic inputs are represented by the uncertainty parameters mentioned above. On the base of the resources table, the whole model is generated automatically during the data preparation phase. This is combined with the weekly calendar of the resources.

The simulation model of the case-study implements a dual-frame architecture (*Model* and *SimManager* frames). SimManager is responsible for the data preparation, model creation, initialization and evaluation. The components of the model are created into the Model frame. One of the main advantages of this *server - client* architecture is that it enables the transfer of the application to a distributed simulation environment.

The shop-floor of the case-study includes more than 100 resources, all of which are modeled in the simulation module. The short term schedule table contains approx. 2000 tasks to be executed in one replication. The time frame of one simulation replication is one week. The statistical data are collected both on resource and product side. Figure 3 shows the developed simulation model.

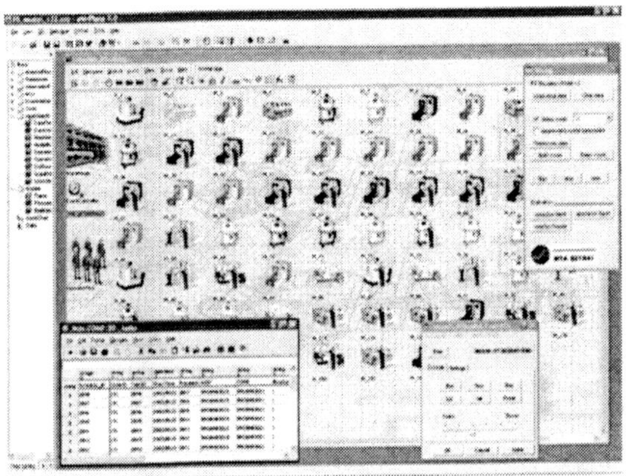

Fig. 3. The visualized simulation model of the case-study

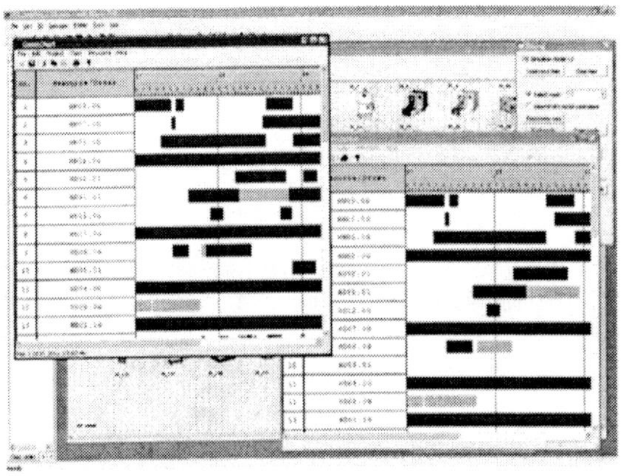

Fig. 4. Visualized Gantt charts of both, the initial and executed schedules in the simulation model

Input	**Size**
No. of tasks in a job	20-500
Working resources/week	80-120
Avg. no. jobs/week	12-15
Avg. no. tasks/week	2000

Table 1. Size of the scheduling problem in the case-study

As a main driver, the simulator should only play back the plan without changing the optimized sequence of the tasks. However, in some cases when - because of the uncertainty in the simulation environment - dead-locks, or delays occur, some predefined dispatching rules might be applied to avoid the "freeze" of the simulation system.

4.1 Schedule execution

Because the time horizon is limited and the orders under process are included in the schedule in question at the initialization, the system does not have a transient phase coming from the warm-up period. This is handled on the level of the short term scheduler that determines the optimal schedule taking the current work in-process (WIP) state of the shop-floor into account.

It is useless to force the execution of a deterministic, rigid schedule in a realistic production environment because the rapidly changing conditions in almost every case will violate the calculated schedule. Dynamic, partly unpredictable environments require reactive scheduling techniques and a mixed-initiative support for schedule execution.

In order to overcome the above dilemma, the predictive short time schedule is relaxed in the sense that we remove the fixed start times of operations and keep only the sequence of the operations on the various resources. We use this control rule in order to follow the optimal solution as far as possible. By default, an operation can be processed if it is in the front of all of its queues. However, since there are not only single, but also alternative resources, we can apply a more liberal execution policy, while keeping all the time the consistency of the overall job-shop schedule. Accordingly, an operation can be processed any time if it does not cause lateness of the operations that are before it in its respective queues.

4.2 Results of the evaluation

The initial schedule and the schedules after the simulation runs can be visualized in the simulation module providing a comfortable user interface for necessary interventions (Figure 4).

Play-back strat.	Avg. tard.	Max. tard.	WIP
Det. proc. times	0	0	47
Stoch. proc. times	9.11	14.89	51
95% machine av.	20.32	55.35	53

Table 2. Illustrative results of the schedule evaluation procedure for one week (average values in time-unit, calculated from 250 simulation replications).

Table 2 shows, how the value of average tardiness changes in deterministic and stochastic cases. As mentioned in the description of the problem, the most important in the *"case-study" factory* is the minimization of tardy jobs and WIP level. Taking this fact into consideration, we selected the mean tardiness, maximum tardiness and the remaining WIP level after the schedule execution (after the executed week) as performance measures in the evaluation of the schedules.

Deterministic execution means that no uncertainty was set in the simulation. As expected, in this case the executed schedule is exactly the same as the planned one. In the *stochastic processing time* scenario (row 2 in Table 2) the processing time of the tasks is set randomly applying the uniform distribution. The lower bound is 90%, while the upper one is 130% of the planned process time. This set-up includes the variation of processing times coming from the difference of the skills of operators, as well. Row 3 in Table 2 refers to the *machine availability* which is 95% in the case-study. The results presented in Table 2 may only give a small sample how the simulation can be applied in the framework. To create a real dynamic scheduler, after the simulation the executed schedules are uploaded in the common database. Those tasks which were not executed are added to the plan of the next week during the rescheduling process. The average time of one simulation run is approximately 10 seconds.

Further performance measures will be included in a further implementation, supporting other decisions as well. Naturally, the system does not want to decrease the role of the human decision making process leaving the selection of the most convenient schedule to the shop-floor managers. Additionally, further developments steps were initiated to reduce the 1 week scheduling-execution cycle and initiate rescheduling in the course of the week.

5. CONCLUSIONS

The paper presented a hierarchical PPS system whose components - a medium-term aggregate capacity and production planner, a short-term jobshop scheduler, and a discrete-event simulator - work on more and more detailed models of a given production environment. While the basic models of the planner and scheduler are deterministic, the simulator can capture non-deterministic events - especially those that may occur on the shop-floor.

The paper gave a detailed description of the architecture and the functionalities of the simulation module. Different experiments can be performed with the overall system supporting the mid- and short-term planning and control decisions on the shop-floor.

6. ACKNOWLEDGEMENTS

The research was supported partially by the project "Digital enterprises, production networks" in the frame of the National Research and Development Programme by the Ministry of Education (Grant No. 2/040/2001). A part of the work was covered by the National Research Fund, Hungary, Grant Nos. T034632 and T043547. Botond Kádár greatly acknowledges the support of Bolyai János Scholarship of the Hungarian Academy of Sciences significantly easing the contribution.

REFERENCES

Berry, P.M. (1992). Scheduling: A problem of decision making under uncertainty. In: *Proc. 10th European Conf. on AI, ECAI'92* (John Wiley, Ed.). Vienna. pp. 638–642.

Bongaerts, L. (1998). Integration of scheduling and control in holonic manufacturing systems. PhD thesis. Katholic University Leuven. Belgium.

Kádár, B. (2002). Intelligent approaches to manage changes and disturbances in manufacturing systems. PhD thesis. Technical University of Budapest. Hungary.

Kis, T. (2002). A branch-and-cut algorithm for scheduling projects with variable-intensity activities. *Mathematical Programming*. Submitted paper.

Kovács, A. and T. Kis (2003). Partitioning of trees with respect to new criteria. *Information Processing Letters*. Submitted paper.

Kovács, A., J. Váncza, B. Kádár, L. Monostori and A. Pfeiffer (2003). Real life scheduling using constraint programming and simulation. In: *Proc. of IMS 2003, 7th IFAC Workshop on Intelligent Manufacturing Systems*. Budapest, Hungary. pp. 233–239.

McKay, K.N. and V.C.S. Wiers (1999). Unifying the theory and practice of production scheduling. *Journal of Manufacturing Systems* **18(4)**, 241–255.

Monostori, L., A. Márkus, H. Van Brussel and E. Westkamper (1996). Machine learning approaches to manufacturing. *CIRP Annals* **45**(2), 675–712.

Prosser, P. (1991). Reactive scheduling. In: *Proceedings of SIGMAN Workshop of AAAI-IJCAI'91*. American Ass. of Artificial Intelligence. Detroit. pp. 57–59.

Sadeh, N. and M.S. Fox (1990). Variable and value ordering heuristics for activity based job-shop. In: *Proceedings of the Fourth Int. Conference on Expert Systems in Production and Operation Management*. Hilton Head Island, USA. pp. 134–144.

Smith, S.F. (1994). *Intelligent Scheduling*. Chap. OPIS: A Methodology and Architecture for Reactive Scheduling, pp. 29–66. Morgan Kaufmann. San Francisco.

Szelke, E. and L. Monostori (1999). *Modeling Manufacturing Systems*. Chap. Reactive scheduling in real-time production control, pp. 65–113. Springer. Berlin, Heidelberg, New York.

Szelke, E. and R.M. Kerr (1994). Knowledge based reactive scheduling state-of-the-art. *Int. Journal of Production Planning and Control* **5, March-April**, 124–145.

Copyright © IFAC Information Control Problems in
Manufacturing, Salvador, Brazil, 2004

ELSEVIER
IFAC
PUBLICATIONS
www.elsevier.com/locate/ifac

IT SOLUTIONS ADVANTAGES AND DRAWBACKS FOR DIFFERENT LEVEL OF PLANT DISTRIBUTED SIMULATION, PLANNING AND CONTROL

R. Bandinelli, M. Tucci

Università degli Studi di Firenze, Dipartimento di Energetica "Sergio Stecco", Via Lombroso 6/17 50134, Firenze, Italy
(r.bandinelli, tucci)@ing.unifi.it

Abstract: Nowadays distributed simulation technologies are often objective or instruments of academic research projects. Many inter process communication (IPC) standards have been proposed and many solutions have been found to supply a distributed simulation framework. Nevertheless these technologies were not adopted by industries. The main aim of this paper is to point out some guidelines and best practices for the implementation of such environments in actual industrial situation in terms of efficiency, robustness and effectiveness. Starting from the objectives of the modelling and simulation of the plants and of their control, the paper presents issues arising from time management approaches, IPC standards and their efficiency due to delay in information propagation, and plant modelling architectures and solutions.
Copyright © 2004 IFAC
Keywords: Remote control, Distributed simulation, Scheduling, Inter Process Communication Standards.

1. INTRODUCTION

The process of globalization and the necessity for industries to manage activities in a global environment force the management to take complex decision in a shorter time then before. Besides, distributed enterprises with several production plants cause the decisional center to manage more information and more unexpected events, like new important orders, delayed material supply, random machine break downs or production failures.
This fact makes standard approach to planning and scheduling no more suitable to achieve high performance in terms of reactivity and efficiency. Nowadays a continuous control on the whole chain is becoming more and more important, in order to minimize delay due to unexpected events.
In order to manage this distributed or extended enterprise, simulation is gaining importance as an instrument either for prediction or for validation. Simulation is used, moreover, at different decisional levels, i.e. strategic designing and planning of the supply network, tactical decision on capacity of the network, operational scheduling and control of production. Stand alone simulation is widely used in the world and therefore it is very well-known. The use of simulation for manufacturing system design and analysis is correctly recognised by scientists and practitioners and the literature is rich in this field.
Nevertheless these technologies was not adopted by industries in a distributed environment.
The main aim of this paper is to point out some guidelines and best practices for the implementation of such distributed simulation in actual industrial situation in terms of efficiency, robustness and effectiveness, with a particular attention at the technologies in respect to the desired control level on the production plants.

2. NEEDS FOR DISTRIBUTED SIMULATION AND REMOTE CONTROL

Distributed simulation demands are closely connected with the performance measurements that we want to assess, thus it's very difficult to highlight some common requirements. Thus it is very important in the preliminary steps to analyze the actual requirements and the performance that have to be assessed.
In order to make this point clear it is necessary to properly classify the plants network.
For our purpose plants network can be classified according to two dimensions. A vertical dimension considers the extension of the supply chain and the number of the different process segments (i.e. manufacturing, assembly, etc..) with reference to the whole production. In this dimension distributed simulation deals with optimization of planner decisional center(s). Referring to a time horizon based classification of the production planning and control, this dimension faces the problem at a strategic level when designing, for example, a supply chain or at a tactical level when defining or validating a supply chain production plan.
The second one is an horizontal dimension which considers, once the focus is on a specific process phase, the possible assignment of batches or jobs to several distributed plants, taking into account the logistics, the capacity and the inventory of them, and controlling in depth the shop floor. In this dimension distributed simulation goal is the optimization at an operational level.
Our work will describe only these two farthest classes, but more complex intermediate cases are possible. Another important point in order to evaluate the needs for distributed simulation is the scheduler/planner requirements.

In order to evaluate these needs we can divide both distributed control and simulation in on-line and off-line solutions. With on-line control we consider a system where the models are linked with the physical plants and where time management is linked with real time. But the controller can be connected to an emulator of the plant obtaining an off-line simulator.
Traditional approach of distributed simulation is an off-line approach and it can be used to validate a production plan over a model representing different physical plants. This is an use of the simulation that just permit a low level of control of the plants. In a more complex view of the network, a virtual environment can be built representing all the actors joining the supply network. In this case only initial conditions (backlog, inventory, machine capacity, etc.) are needed by the software simulator to produce the output. The effort to update information into the scheduler is quite low and depends on the scheduling interval, that is usually fixed.
Recent scheduling software requirements are heavier, as they are usually real-time scheduler, more reactive, rule based, etc.
The Remote Control is a particular way to achieve the control over the plants that came from the idea of the "Remote Factory[1]". In a remote factory environment a huge amount of information have to be exchanged. In particular, if remote scheduling control is desired, all information regarding the state of the machines, the jobs, the queue and the production orders must be available to the scheduler/simulator every time a rescheduling is necessary.
When in such a structure of scheduler/simulators the central scheduler is connected to the actors of the supply chains which expose their actual state in production, stocks and capacity, we can say that we have an on-line distributed scheduler.
Depending on the level of control and the needs of real-time response, a different approach in plant modelling architecture and in the choice of technologies has to be taken.

3. ISSUES IN THE CONTROL OF THE PLANTS NETWORK ARISING FROM THE OWNERSHIP OF THE EXCHANGED INFORMATION

Another important point in choosing the right technologies is the ownership of the information. If we consider a supply chain where different actors works together to achieve a common result (production and delivery of goods) but each one considering personal profit as the primary goal, we have to assume that not all the information about the plants can be shared among the different actors of the network.
Different aspects have to be taken into account, on the other hand, if both the decisional centers and the production plants belong to the same ownership. In this second case, all the information can be shared among the actors of the network, and a prediction closer to reality can be achieved.
A typical example of this situation can be found in distributed environment where production plants are spread all over the world because of lower labour cost. This is a typical situation of the industries of the 21st century. In this kind of environment it is becoming important to have the possibility to reorganize the production and to minimize the period from a communication between the decisional centers and the plants. Traditional off-line scheduling, with scheduling interval of a week or more, are often not able to assure a satisfying level of control on production and are to achieve high level of performance in terms of product customization and respect of the due date.
In order to achieve satisfying performance it would be often necessary to release the constraint of fixed scheduling interval, moving to an almost real-time scheduling which requires the collection of plant state and the dispatching of production order to the plant several times in a day.
This kind of approach is possible only if we consider an infrastructure where all the actors are modelled inside a virtual environment, where each model is physically located in the individual site and it exchange information with all the other models continuously during the day.

4. DIFFERENT APPROACHES IN PLANTS' NETWORK MODELLING

In remote control and distributed simulation of supply chains, the scheduling (planning or control) activity and the amount of information required are heavily influenced by the topology of the plants' network, i.e. the vertical dimension (spanning of the supply chain and number of actors) and the horizontal dimension (number of plants with interchangeable production).
Moreover, the performance measurements that are required as output of the simulation or remote control have a big impact on the emulation/simulation models level of detail, and thus on the exchanged information.
Another important point, as shown above, is the definition of the technology used by the scheduler/planner according to the requirements described in chapter 2.
In literature we can find many schedulers with different logic. For our purposes schedulers/planners can be roughly classified in traditional open loop scheduler and intelligent/reactive scheduler. While for the first type of scheduler it's necessary to have information only about the initial state of the plant, reactive scheduler needs continuous update of information from the field in order to optimize scheduling and maximize performance.
It's easy to understand how this second approach needs a more robust and efficient framework in order to supply good result.

[1] The *Remote Factory* idea deals also with problems different from the remote headquarter control; in particular, at the present the main interest of the *Remote Factory* project and of the SIG4 group of the IMS Network of Excellence (http://www.ims-noe.org) deals with the establishment of such a modelling and simulation infrastructure as a benchmarking service for solving the dichotomy which afflicts the world of scheduling research, where PP&C experts are totally detached from industrial reality.

Another distinction among schedulers can be done according to different aggregation approach. We can consider a centralize approach, where all the decision are taken from a unique decisional center, or a heterarchical approach, where many decisional centers can coexist together and each of them is responsible for a part of the production plan. This second case can be partially found in a spread chain where there are many production plants and also assembly plants. Each production plant can be autonomous in the decision regarding the scheduling of the production plan, but not in deciding the production plan itself, which is imposed by a centralized decisional unit. In the planning activity, even in the real case a lot of negotiation have to be carried on in order to reach a contracted plan. But in this case the information exchange is less frequent (daily or weekly as opposed to real-time).
As it will be shown in the chapter 5, each of this situation needs a different type of IPC standard in order to effectively work.

5. STATE OF THE ART OF IPC STANDARD

5.1 Traditional standards

Distributed simulation projects are very often subject of academic research. Many standards have been proposed, but none of them can be considered the best solution for all the cases presented in the previous chapter.
We can consider formalisms that describe and/or implement simulation models according to set of rules (HLA, DEVS), actual IPC standards (CORBA, RMI, COM+, JavaBeans, SOAP, JINI) and integrated formalism/IPC infrastructure solutions (HLA-RTI, CORBA-DEVS, HLA-DEVS, GRIDS, SIMJAVA).

The most important framework that has been proposed is definitely HLA-RTI. Developed by the American Department of Defence, High Level Architecture (HLA) is a recently approved standard (IEEE 1516) dealing with component-oriented distributed simulation. It defines rules and interfaces allowing for heterogeneous components (called federates) interoperability in distributed simulation. RTI is a software that implement the specification of distributed simulation (i.e. time advance and so on) using a socket based communication infrastructure. It can connect in an synchronous way many agents inside the same physical LAN, using asynchronous connections. It was used in research projects as the framework to allow many plants simulator of a supply chain to exchange information one to each other [Rapaccini et al. 2003].

Another important standard in distributed simulation is CORBA® (Common Object Request Broker Architecture). CORBA is the middleware standard promoted by the OMG (Object Management Group) for connecting distributed component of software and to define a common architecture for the distributed applications. CORBA has become a standard for the communication among objects in a client-server environment. It's completely object oriented and permit to the applications the use of local and remote objects independently from the software or the hardware of the components. It is the most important alternative to RTI environment even if it is a general IPC infrastructure without any provision for simulation time management.

Java Remote Method Invocation® (RMI) [http://java.sun.com/products/jdk/rmi/] enables the programmer to create distributed Java technology-based applications, in which the methods of remote Java objects can be invoked from other Java virtual machines, possibly on different hosts. A Java program can make a call on a remote object once it obtains a reference to the remote object itself, either by looking up the remote object in the bootstrap naming service provided by RMI or by receiving the reference to it as an argument or a return value. A client can call a remote object in a server, and that server can also be a client of other remote objects. RMI uses object serialization to marshal and unmarshal parameters and supports true object-oriented polymorphism.

Enterprise JavaBeans is the server-side component architecture for the J2EE® platform [http://java.sun.com/products/ejb/]. EJB enable rapid and simplified development of distributed, transactional, secure and portable Java applications. The EJB component model simplifies the development of middleware applications by providing automatic support for services such as transactions, security, database connectivity, and more.

DEVS (Discrete Event Systems Specifications) is a Technology independent framework. Previously implemented in sequential and parallel simulations with great result in speed up, it is derived from a System framework [Zeigler, 1997]. The major difference between most Object Oriented frameworks and the System framework is that the latter recognizes that simulation deals with dynamics, the way in which system behave over time. HLA recognizes dynamics (object behaviour) but doesn't provide a formal basis for its description [Zeigler and Doohwan, 1997].

CORBA – DEVS (Discrete Event System Specification) is a formalism Implemented by Zeigler, Kim, Buckley (1999). The main advantage of this formalism is the major execution speed due to the CORBA environment.

HLA – DEVS (Discrete Event System Specification) is a formalism that implements the DEVS formalism over the HLA/RTI. HLA provides two types of communications: attribute updating and interaction among federates. Interactions can have parameters similar to the attributes of objects. An interaction is a non-persistent, time-tagged event generated by one federate and received by others through RTI. The strategy underlying the mapping of DEVS to HLA is to exploit the information contained within DEVS models to automate as much as possible of the programming work required for building HLA compliant simulations. For example, some DEVS specifications are automatically mapped into HLA interactions [Zeigler et al. 1999].

GRIDS (General Runtime Infrastructure for Distributed Simulation) is a formalism used to describe simulation environment on a high level layer. It is built on Java language and it has three distinct stage of execution: Register, Broadcast and Run. This implementation of a formalism with an IPC infrastructure gives an alternative to HLA for the realization on a distributed simulation. Speed is the main feature this solution provides [Sudra et al. 2000].

SIMJAVA® is an application based on RMI architecture that is composed by a client and a server component [Howel, 1998]. These component communicates through the Remote Reference Layer of the RMI system. SIMJAVA is a toolkit for building working models of complex systems. SIMJAVA is a very simple and light application and is designed to be flexible and extendable. The main problem of this formalism is that it doesn't have any specialized method for time management.

JINI® is an application based on the EJB infrastructure. It is based on Java architecture and it was developed as a natural extension of the Enterprise JavaBeans services. JINI is a client application that permit to different clients to publish and subscribe state variables on a Java Virtual Machine and to be informed about their changes. JINI is totally Object Oriented and it is a very fast application. JINI doesn't have any time management support. All the time management is delegated to the simulations software or other software. With JINI it is possible to build a web based simulation system based on XML standard [Huang et al. 2001].

COM+® provides a proprietary framework that allows developers to build transaction-based distributed applications from COM components, managed from a single point of administration. COM+ includes: a) MTS transactions; b) Security Administration and Infrastructure Services; c) Queued Request so that invocations on COM components can be done through Microsoft message-queuing server (MSMQ); d) Publish-and-Subscribe Event Service; e) Load Balancing Service to allow transaction requests to be dynamically spread across multiple servers.

5.2 Web based standard

Nowadays web based technologies has entered into all of the research IT areas. Web based solutions bypasses the firewall problems while providing strong security support, and maps existing content allowing the resource to be made available on the web, without modification, in formats well-suited to integration with diverse business systems. The use of XML to deliver metadata about existing web resources can provide sufficient information to empower non-browser applications to automate interactions with web servers so that they do not need (or at least try) to parse some unstructured HTML documents to find out the information they are looking for.

There are mainly two different ways to implement Web Services, and these are XML-RPC® and SOAP. XML-RPC® is a set of specifications and a set of implementations that allow software running on heterogeneous operating systems, running in different environments to post procedure calls over the Internet. It is a remote procedure calling that use HTTP as the transport and XML as the encoding. XML-RPC is designed to be as simple as possible, and to allow complex data structures to be transmitted, processed and returned.
As stated in the various documents regarding the SOAP protocol, SOAP specifies a Remote Procedure Call (RPC) mechanism using XML as the method for encoding the data to be transported for the call and an arbitrary transport method (most commonly HTTP) for communicating both the request and the response that together constitute the Remote Procedure Call. SOAP in a descendant of the XML-RPC protocol and besides all the XML-RPC features it offers the possibility to communicate among different environment with a very flexible syntax.

Therefore it's useful to consider the possibility of building a SOAP to CORBA or SOAP to RTI bridge that would accept SOAP requests, forwarding them to established or new CORBA or RTI based servers and return the result as a SOAP response. SOAP encapsulates the HLA-RTI simulators in order to manage their synchronized start, the acknowledgement of starting conditions and the objectives of the simulation, overcoming this missing features of HLA-RTI, but when the simulation starts, it is based only on RTI, until some results are available to be communicated.

Referring to the traditional IPC technologies for distributed simulation environment, at the moment the most important project to supply a web based environment is called "The EXMF Project". More details will be shown in the next paragraph.

6. DISTRIBUTED SIMULATION PERFORMANCE EXAMINATION

6.1 Advantages and drawbacks regarding each infrastructure

Many infrastructures for distributed simulation have been reviewed and some of them tested. Most of them are IPC standards and are used in other environments too. As said before, the most important standard in distributed simulation is HLA-RTI and it's the only one that supply a multifaceted and flexible time

Copyright © IFAC Information Control Problems in
Manufacturing, Salvador, Brazil, 2004

ELSEVIER
IFAC
PUBLICATIONS
www.elsevier.com/locate/ifac

SIMULATION OF STEEL INDUSTRY USING SYSTEM DYNAMICS

Ludovic DEGRES, Henri PIERREVAL, Christophe CAUX

Research group in Manufacturing Systems of IFMA
LIMOS, UMR CNRS 6158
Campus de Clermont Ferrand, Les Cézeaux, BP 265
F-63175 Aubière Cedex, France
(ludovic.degres/ henri.pierreval/christophe.caux)@ifma.fr

Abstract: In order to be competitive, Steel Industries (SIs) need to develop their long-term vision to invest in time in the right equipment. The specificity of SI's process renders the long-term management of the production capacities difficult. In this study, Forrester's System Dynamics is proposed as a modelling approach of that type of systems at a strategic level. An overview of Forrester theory and some essential principles to use it to model SIs are introduced. To illustrate the use of this modelling approach, an example of an existing SI plant facing sudden demands variation is given.
Copyright © 2004 IFAC

Keywords: Simulation, System Dynamics, Steel Industry, and Process Industry.

1. INTRODUCTION

In order to better meet more and more complex customer needs and to better respond to increasing challenges from competitors, Steel Industries (SIs) need to develop their long term vision to invest on the right resource at the right time.
As in other fields, it is necessary that the production facilities can be adapted according to the charge in the production demand. This refers to the long-term management of the capacities. This turns out to be a critical issue, because of the high cost of machines (e.g. furnace, forging press, stamping press). Investment strategies on new resources are often based on qualitative and managerial information. One must take in account products diversity, process complexity and the specific features of process industries. Examples of typical questions to answer are as follows:

- Does a given resource risk to be a bottleneck of the system?
- When would the workloads and capacities be unbalanced?
- Would a particular management policy reduce the problem?
- What kind of investment should be done?

Regarding these questions, a simulation approach would be very useful to support decision.

Decision-makers have also to take in account external factors. Indeed, the demand for steel products varies according to such factors as the steel price or the delivery times. As these demand variations depend on such external factors, it is difficult to predict when they can occur and in what volume. In this respect, it is important to understand what would be the behaviour of a SI plant facing sudden order variations. Such an issue seems not to have been addressed in the literature. (Hafeez *et al.*, 1996) determine a minimum reasonable inventory, in the particular case of a two-echelon steel industry supply chain that services the construction industry, using an integrated system dynamics framework. However, they do not focus on the flow of material and the impact of orders on these flows in their modelling approach. In this paper, an approach studying the impact of orders variations on physical flow is proposed.

The article is organised as follows. In the first section, specific features of SI are highlighted. In the second section, some important principles of Forrester's System Dynamics that are useful to model SI, are recalled. In the third section, a way of modelling SI using Forrester's System Dynamics principle is proposed. Then, in section four, the use of this modelling approach is illustrated using the case of a real SI plant facing sudden products orders variations. Finally, a conclusion on the benefits and limitations of Forrester's System Dynamics to model SI is given.

2. SPECIFIC FEATURES OF STEEL INDUSTRIES

According to APICS's definition (APICS, 1984), SI belongs to "process industries", which are defined as "businesses that add value to material by mixing, separating, forming or chemical reactions". It is also indicated that "processes may be either continuous or batch". SI is based on a continuous/discrete/batch processing: parts can be processed continuously (melting), accumulated in queues, processed together in batches (e.g., heating) or processed separately (rolling or machining).
SIs do not manage classical "products" or "articles" as in more conventional industries (e.g. assembling industry). In the SIs considered in this paper, forecasts and production rates are usually estimated in tonnes. For some resources such as arc furnaces, the throughput is estimated in tonnes per unit of time. The production throughput being defined through weight, managers have to deal with flows, rather than other type of physical entities.
Typically, the fabrication of alloy or super alloy products can be described through three main stages, which are depicted in Figure 1, Figure 2 and Figure 3. (Chaussin 1980). This production cycle is quite long (from six weeks to six months). It is usually managed according to a push strategy.

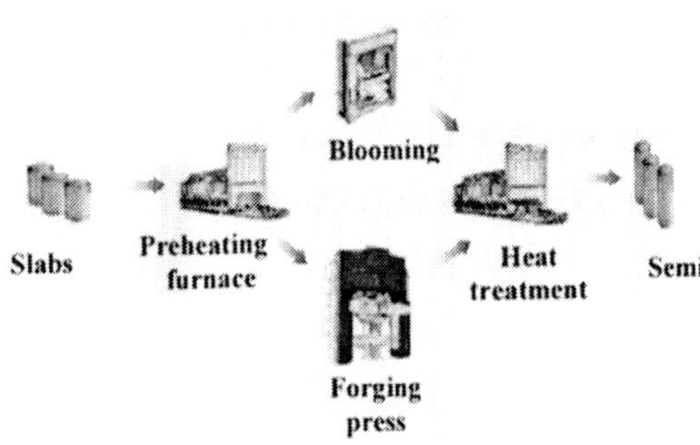

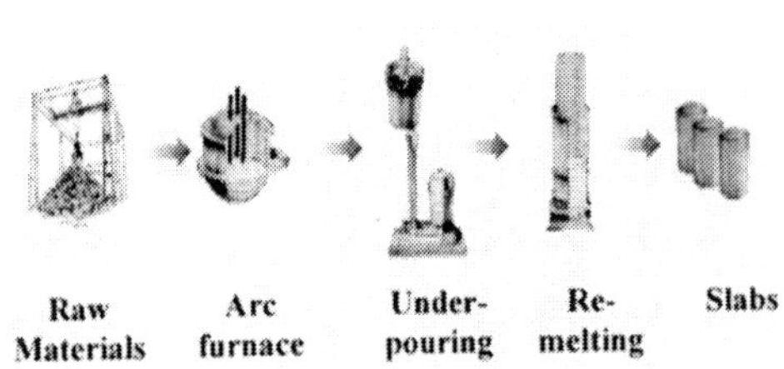

Figure 2: Transformation.

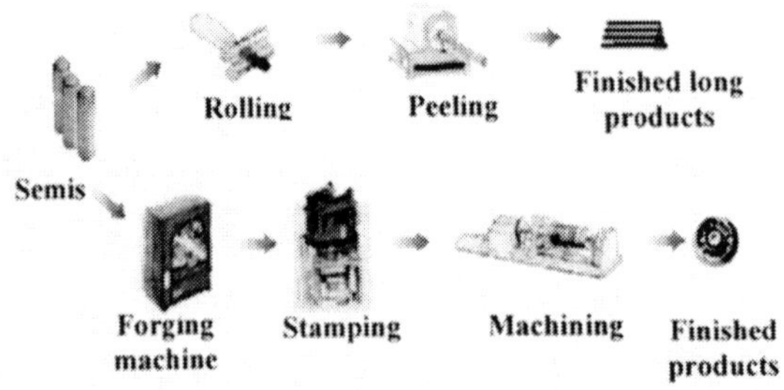

Figure 3: Finishing.

The first step of the process is melting. As steel grade is defined at that stage, it is a quite important step. Indeed steel grade (the classification of an ore according to the desired or worthless material in it) is one of the main characteristics of metallurgical products: the processing time, the product value, the Bill Of Material (BOM) depend on it. Resources used at that stage are not really flexible, it is impossible to change easily the production sequence. For example, it is impossible to melt a low-alloy steel right after a high-alloy steel. Then, a furnace can be a bottleneck because of the product mix (especially the products steel grade mix). The product mix is the proportion of each product that will be produced within the whole list of products sold.
The variations of the product mix make the production planning hard and long term decisions, such as investments, is difficult to make. Furthermore, whatever the size and weight of a final product, it is impossible to produce half a melt in an arc furnace. For example, if a 500-kg slab is needed to make a final product and no stocks are available, then there is no other choice than producing a whole melt (20 tonnes).

Several times during the transformation process, the slabs can be split into different parts. In other words, the material flow appears to be divergent as in other process industries rather than convergent (David *et al.*, 2003). In the long term, a given production management strategy, relying on the concept of components and not taking in account the divergence of the flow, might lead to think that a resource is bottleneck even if it would not have been with another planning strategy.
Heat treatment (or preheating) is a particular step of the transformation. Products at this stage can be batch processed if they have identical quality requirements. As a consequence, the throughput of a resource depends on the product mix (i.e., a pre-heating furnace can be busy only for one product even if it is able to process several product in the same time). On the other resources (hot rolling mill, forging press, stamping press) products are processed separately. Thus, the transformation of slabs is a discrete/batch process. This renders

grouping and sequencing decisions difficult to make, since according to the chosen grouping or sequencing strategy a resource can be bottleneck or not.
According to the quality and the product complexity that are required by customers, a product can have a long or a short transformation and finishing stages: a product can be processed several times on the same resource. As a consequence, the workload of transformation and finishing depends on routings. There are often alternative technical solutions that can avoid critical sectors of the factory to be overloaded.
For most of the transformation and finishing processes, resource-processing times depend on the steel grade: forging one tonne of super-alloy is longer than forging one tonne of alloy. For that reason, resources throughput is not linear and depends on steel grade.

In the considered SI, orders are likely to evolve according to sudden market changes that will create variations of the product mix. As described earlier, product mix is one of the most important factors to consider to manage capacities in SI. For these reasons, one of the major issues to consider for decision making is the behaviour of a system facing changes in the product mix.

3. OVERVIEW OF FORRESTER'S SYSTEM DYNAMICS

One of the key principles of Systems Dynamics is the idea of feedback loop, which is the base of any dynamic organisation (Forrester, 1961). In others words, one of the main idea of this theory is that organisations are better described if more attention is paid to the flows that control it than to the entities that made it. The main goal of Systems Dynamics is to study the behaviour of systems showing cause-effect relationships. Forrester's models are based on a causal analysis of the organisation to study. Causal loop diagrams are built. They allow the multiple feedback mechanisms, which guide the system behaviour, to be understood and formulated.
To develop a model, Forrester uses four main concepts that are as follows:

- Several levels (accumulation within the system) (drawn as tanks). Levels are the present values of the variable they contain that have resulted from the accumulated difference between inflows and outflows.
- Flows, which transport the content of one level to another.
- Decision functions, which control the rates of flows between levels (drawn as valves).
- Information channels, which connect the decision functions to the levels.

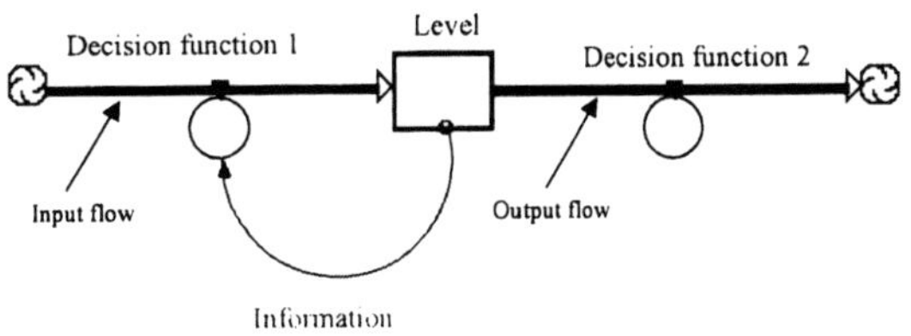

Figure 4: Forrester's System Dynamics basic symbols.

Figure 4 depicts a very simple structure of a reservoir or level, with an inflow and an outflow. To describe this general structure, a system of equations is defined. This system of equations consists of two types of equations, which correspond to levels and rates. Equations control the changing interactions of a set of variables, as time advances. The continuous advance of time is broken into small intervals of equal length *dt*. For example the equations of the level of the figure 4 is:

$$Level(t) = Level(t - dt) + (Decision\_function\_1 - Decision\_Function\_2) * dt$$
$$INIT\ Level = 0 \quad (1)$$

Level at time t depends on its own value at t-dt and the value going in from decision function 1 minus the value going out to decision function 2. It is necessary to give the initial value of it to solve this equation.
There are as many equations as variables. To model the behaviour of the variables, the differential equations system is integrated. Several software tools exists that support this approach.

4. MODELLING SI WITH FORRESTER'S SYSTEM DYNAMICS

Forrester's System Dynamics has been used in many fields such as management, finance, agri-food industries (Thiel, 1998), production systems (Thiel, 1996), automotive industry tactical strategies (Minegishi, 2003) or automotive industrial plant networks (Bruniaux, 2000).

As the model is built for strategic level, the concept of logistic unit described by (Bruniaux, 2000) is used to identify the important components of the model. A logistical unit is a physical area of a plant that is composed of a stock and a group of resources. In the same way, as far as strategic problems are studied, one is not concerned with modelling the effect of batch process.

Using Forrester's comparison with piping networks a logistic unit (e.g., furnace, stamping press) is a network of several parallel elementary flows going through of two reservoirs and a valve between them. The first reservoir is the workload to go through the valve. The valve gives the discharge rate of the flow coming from the valve drains into the last reservoir. From this reservoir, the flow goes through another valve modelling the transport to the following sector. The sequence of valves and reservoirs for a flow is defined by the path of a flow's product.

The final model appears as a network of product flows starting from an order to a final product. It starts from an initial stock of order and finish to a final stock of final products. As an example the equation of a level of products A waiting going through a valve is:

$$\begin{aligned}
&\mathit{Waiting_product_A_Forging_Logistic_Unit(t)} = \\
&\mathit{Waiting_product_A_Forging_Logistic_Unit\ (t\text{-}dt)} + \\
&\mathit{(Flow_in_A_from_Heat_Treatmen_Logistic_Unit\ t\text{-}} \\
&\mathit{Flow_out_A_to_Stamping_Press_Logistic_Unit)} * \\
&\mathit{dt} \\[1em]
&\mathit{Flow_out_A_to_Stamping_Press} = \\
&\mathit{Operating_Time_Of_Forging_Logistic_Unit_for_A} \\
&\mathit{{*}Steel_grade_coefficient_A} \\[1em]
&\mathit{INIT\ Waiting_product_A_Forging_Logistic_Unit} \\
&\qquad = \alpha \qquad\qquad (2)
\end{aligned}$$

For the sake of confidentiality, this equation is simplified. The equation means that the level of product waiting to be forged depends mainly on the difference between the flow coming from the heat treatment furnace and the flow going to stamping press. This outflow is defined as the number of products it is possible to forge in a time period that is why it depends on operating time and steel grade coefficient.

The flow path is determined using the routing of the product through logistic units. Several flows can go through a same logistic unit. Thus, one of the major difficulties is that several flows going through a same logistic unit have to be represented. Thanks to this approach, the flow divergence can be modelled. As a consequence a logistic unit is described as a set of several parallel flows associated to their own decision function. There is one flow for one type of product. Each flow is modelled as either one of the five shown in Figure 5.

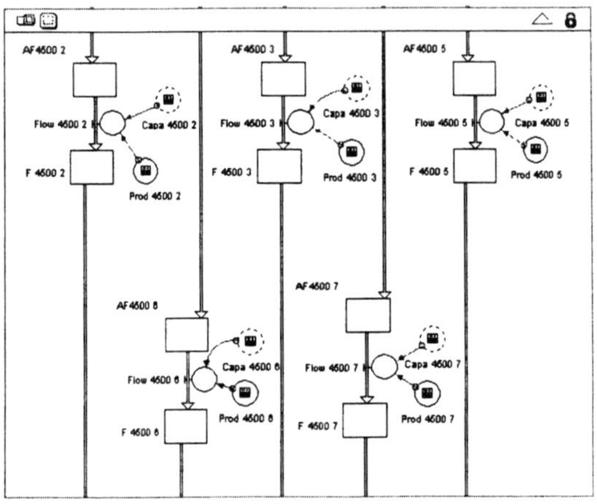

Figure 5: Forging logistic unit with five products flows modelled in Forrester's Dynamics.

The product mix, affected by sudden order variations, is an important factor to take into account. To evaluate the consequence of these variations, the number of products processed per unit of time is measured. This measure is computed for each logistic unit and depends on the steel grade of the products of the flow that goes through this logistic unit. This measure also depends on the duration the logistic unit works in a day (for example number of shifts per day) and the part of this duration dedicated to each flow that go through it.

5. APPLICATION TO AN INDUSTRIAL PLANT FACING SUDDEN DEMAND VARIATIONS

5.1 Plant overview and problems

This approach has been applied to a plant of the firm Aubert & Duval Holding, in France. This firm melt and convert alloy steels, and high-quality superalloys. The studied plant is organised with into main production units: Melting, Forging, and Hot Rolling.

The firm needs to understand the behaviour of the SI facing sudden variations of demand on a long-term horizon (two years). In fact, these sudden order variations modify product mix. The main goal was to identify the effects of product mix on workload/capacity balance. To evaluate it, the stock to process in tonnes is measured. It means that it is tried to define logistic units that let up the product flow and the way they face a peak of product to process (i.e., time to return to an empty upstream stock).

5.2 Modelling approach

Seven representative products families of the whole plant production have been modelled. This choice was made in collaboration with firm experts. For each products family, the method described earlier was applied. Parameters of the model (flow rate, steps of the flows) were defined using experts knowledge and ERP data.
The model developed was build within three stages:

1. Static description of the system with a systemic approach, namely Analyse Modulaire des Systèmes (Mélèze,1982) and causal diagrams (to identify logistic units, stocks and the main variables of the system and their interactions).
2. Implementation using Ithink®.
3. Test and validation with experts.

5.3 Experiments

The behaviour of the plant facing a sudden increase of a certain type of orders was studied. For the sake of confidentiality, detailed results cannot be given here. The model was 600 Ko and developed with I think®. It usually takes 30 seconds to run it.

It is possible to test the model with several input signals (pulse, ramp, and step) as shown on figure 6.

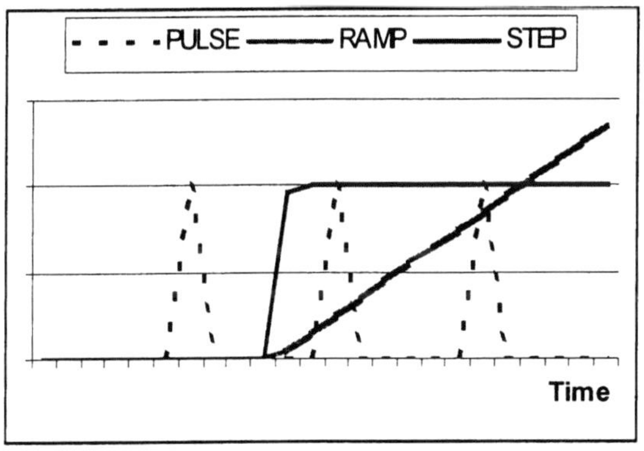

Figure 6: Different possible input signals to test the model.

Pulses represent very disturbing situation that will be used to illustrate the approach. Actually, it represents regular sudden changes of product mix. Furthermore (Forrester, 1961), underlines that pulses seems to be one of the most perturbing input signal for industrial systems.
Several pulses input signals model a sudden increase of a type of orders (figure 7). It is possible to modify the volume, the first date and the interval between two peaks.

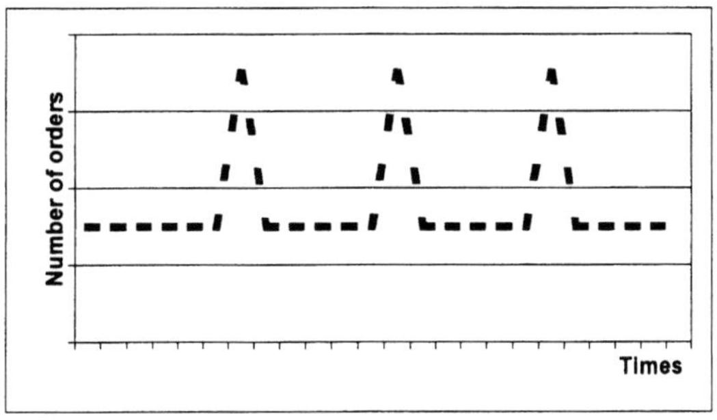

Figure 7: Input signal representing the increase of orders.

The goal was to disturb the system with pulses on different representative products. The initial state of the model was defined so that it is as close as possible to the typical situations of the plant in terms of production rate and capacity. It is assumed that the demand was constant and equal to an average of the recent past demand. There were no seasonality effects observed in the past.

5.4 Results

As explained earlier, steel grade has a strong impact on the throughput of a logistic unit. Since the proportion of steel grades depends on product mix, experiments on variations of product mix have been performed. Then, the behaviour of some logistic units facing pulses of product order was studied. To evaluate the impact of the pulses on the system, the number of delayed product (products waiting to be process) is measured. This allows lacks of capacity in certain logistic units to be detected. It has been observed that, when the interval between two peaks (sudden high variations) is not smaller than a given number of months any logistic unit can return to its initial state in a quite short time (Figure 8). It means that after a time period, there are no more waiting products. Figure 8 illustrates it: after a peak, the portion of waiting products grows up but returns to zero before the next pulse.

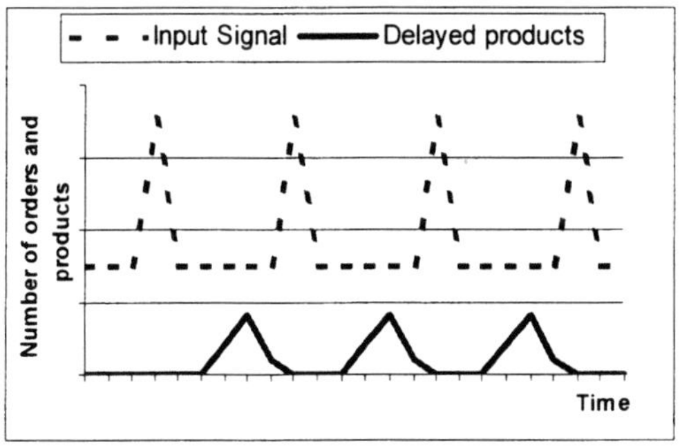

Figure 8: Level of delayed product with a low frequency of pulses.

In the same way, any area of the model can return to a regular state (i.e. no waiting products) even with pulse twice larger (twice the volume) as far as the frequency of the pulse is low. Thus as soon as the limit of x month is exceeded the waiting products grows up. The Figure 9 shows that the number of delayed product does not return to zero. There is an accumulation of products waiting to be processed.

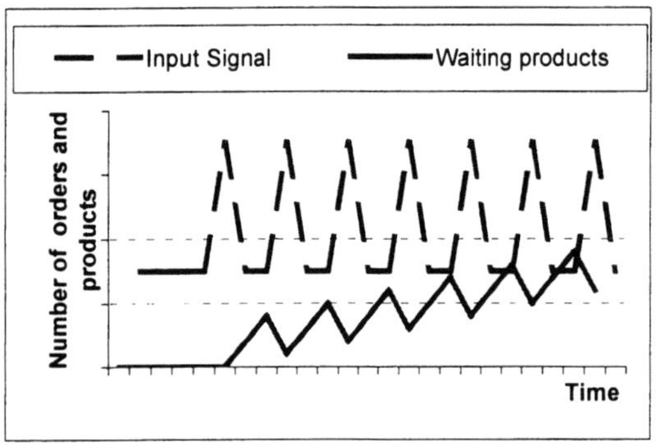

Figure 9: Level of delayed products with a high frequency of pulses.

For the plant studied, surge production is definitely a disturbing factor. The system can cope with high peaks of order but cannot resist to frequent peaks. As those peaks represent sudden changes of product mix, it means that our system can hardly stand frequent sudden variations of product mix.

6. CONCLUSION

In this paper, it has been shown that System Dynamics provide very interesting capabilities to model SI. The approach used has focussed on strategic issues for the firm. In particular, emphasis has been put on how the plant would react in case of sudden variations of the demand. The industrial example presented has demonstrated that, if these changes arose an a regular basis (which reflects excessively bad conditions), then the company would not be much disturbed, except if the volume and frequency of these variations exceeded a given threshold. In such a case, new investments would have to be considered by the decision-makers to avoid the accumulation of late products in the system. System Dynamics turn out to be a convenient approach for such a purpose, both thanks to the systemic methodology associated and thanks to the modelling formalism. Several types of input signals can be used to evaluate the main trends of the system behaviour. Other modelling formalisms such as continuous Petri nets and bound graphs could be also considered. The continuous approach is relevant at this strategic level where macroscopic models are needed, since the facilities are not modelled at a high level of detail. However, at more operational levels where more details have to be included in the models, discrete events simulation or combined simulation worldviews would appear as more appropriated to study the dynamic behaviour of the production facilities that compose the logistic units.

The future works that are contemplated includes three directions. The products families used to simulate the system should be defined using by a suited aggregation method, to improve the model accuracy. The description of the dynamic behaviour of the logistic units will also be improved. Finally, a sensitivity analysis of the different factors likely to affect the dynamic behaviour of the system should be conducted, so as to confirm the trends observed in the simulation experiments.

REFERENCES

APICS. (1984). *Dictionary.* APICS, Falls Church (USA), fifth edition.

Bruniaux R.; Pierreval H.; Caux C. 1999."Simulating supply chains with system dynamics". In *Proceedings of the ESS'99 Conference (Nuremberg, Germany, October 26-28)* 209-221.

Bruniaux R. 2000. "Simulation continue de réseaux de sites industriels, Application aux chaînes logistiques dans le secteur automobile". PhD. Dissertation. Université de Clermont-Ferrand (France).

Chaussin C. and Hilly G. 1980. *Métallurgie.* Tome 1 and 2. Dunod.

David F.; Pierreval H.; Caux C. 2003. "ERP for the management of aluminium conversion industry: what is relevant?" In *Proceedings of IEPM'03 Conference (Porto, Portugal, May 26-28).*

Forrester J.W. 1961. *Industrial dynamics.* Productivity Press.

Hafeez K.; Griffiths M.; Griffiths J. and Main M.M. 1996. "System design of a two-echelon steel industry supply chain." *International Journal of Production Economics*, No.**45**, 121-130.

Mélèze J. 1982. *L'analyse modulaire des systèmes de gestion: AMS.* Edition Hommes et Techniques.

Minegishi S. 2003. "Performance et fiabilité dans les délais de livraison produits industriels assemblés: application à l'industrie automobile. PhD. Dissertation. Ecole Central de Paris (France).

Thiel D. 1996. " Analysis of the behaviour of production systems using continuous simulation." *International Journal of Production Research,* vol. **34**, pp. 3227-3251.

Thiel D. 1998. La dynamique des systèmes. Hermes.

Copyright © IFAC Information Control Problems in Manufacturing, Salvador, Brazil, 2004

ELSEVIER
IFAC
PUBLICATIONS
www.elsevier.com/locate/ifac

VIRTUAL MANUFACTURING AS A WAY FOR THE FACTORY OF THE FUTURE

Mariella Consoni Florenzano Souza[1], Marco Sacco[2], Arthur José Vieira Porto[1]

[1] *Laboratório de Simulação, IFM, Escola de Engenharia de São Carlos-USP, Av. do Trabalhador São-Carlense 400, 13566-590, São Carlos, SP, Brazil.*

[2] *Istituto di Tecnologie Industriali e Automazione, Consiglio Nazionale delle Ricerche, Via Bassini 15, 20133, Milano, Italy.*

Abstract: Virtual Manufacturing represents the emerging approach the enterprises can use to improve their processes, introducing new products more quickly in the market in a cost effective way. The fundamental idea is to create an integrated and synthetic environment, composed of software tools and systems such as Virtual Reality and Simulation to support such processes. This paper presents the Virtual Manufacturing Environment (VME), highlighting the potential application areas and the resulting benefits. Two significant cases are presented as example of the applicability of the concept. *Copyright © 2004 IFAC*

Keywords: Virtual Reality, Data Flow, Simulation, Integration

1. INTRODUCTION

More pressure has been placed on manufacturing companies, as customers are becoming more and more the focus of the market, mainly regarding the delivery of customized and diversified products, quality improvement and time reduction. As products, processes and data are becoming more complex, the decision making process in many disciplines such as product design, manufacturing engineering, and production management involves the consideration and analysis of many variables. These decisions often result in a long-term impact on the determination of the success or failure of the organization. In most cases, making these decisions based just on instinct and experience can be extremely risky for the company.

In order to achieve success in this context, manufacturing companies need to develop capability to quickly react to market changes, making correct decisions through efficient and fast evaluations. In pursuing these capabilities, companies are using ever more frequently simulation technology, software tools and digital model to reduce costs and increase profitability.

Perhaps one of the most recent and interesting solution is Virtual Manufacturing (VM). This new approach provides a way for engineers to develop, evaluate and simulate a complex system entirely on a computer before spending time and money on the construction of physical mock-ups. In this environment, many oriented technologies to support activities of the development process can be used in an integrated way, such as Virtual Reality (VR) and simulation that provide capability to rapidly conduct experiments to predict and evaluate the results of alternative manufacturing decisions. Leading industries in many sectors such as aerospace and automotive companies are making use of VM to support many of the stages of their product development process.

VM represents an important step toward the factory of the future providing ability to make changes in a virtual environment, saving money and time, and resulting in better designs. The purpose of this paper is to discuss the VM environment approach, highlighting its potentiality to many industrial processes, and describe the activities carried out by Simulation Lab and ITIA-CNR in the field.

2. THE APPROACH OF VIRTUAL MANUFACTURING ENVIRONMENT

VM is a new approach that has been emerging in the manufacturing area to improve industrial processes. Because of this, the meaning of the term is still evolving and has an even less agreed upon definition. The fundamental idea behind most of concepts however, is that VM is an integrated and synthetic environment to be used for concurrently simulating all of the activities involved throughout the life cycle of a product. The proposal is to provide a strategy to integrate all the processes, from engineering to manufacturing, to enhance all levels of decisions and controls providing the product development, validation and manufacturing virtually before realizing physically the real product (Lawrence Associates Inc., 1994).

According to Banerjee and Zetu (2001), the Virtual Manufacturing Environment can be used in a wide variety of manufacturing systems contexts, and can be defined as the modelling of manufacturing systems and components with the effective use of computers, audiovisual and sensory displays to simulate or design alternatives for a manufacturing environment. The main objective is to predict potential problems and inefficiencies in product functionality and manufacturability before the real manufacturing is realized.

This virtual environment can be more realistic with the use of Virtual Reality. This technology can provide a better integration among designers and customers requirements, allowing the interaction and immersion of the user in the virtual world based on graphical 3D images created in real time by the computer. Immersed in the virtual world, engineers can create and modify their projects in real time, verifying immediately the effects of these modifications. The main impact of this technology in the Virtual Manufacturing regards the products design and project, processes simulations, operations planning and training, and real time control of shop floor.

Achieving that approach means to develop an integrated solution environment composed of a set of systems and software tools that operate throughout all the phases of the development process aiming at realizing modeling, simulation, analysis and communications. The actual challenge is to develop better and newer technologies and integrate them with current and available ones to create Virtual Manufacturing systems (Committee, 1999). Overcoming the interoperability barriers among systems and engineering data is an important focus of research and development to provide a complete VME. Today, industries as the automotive and aerospace have to deal with the consequences of the lack of data and system interoperability.

Once system and data interoperability problems are solved, VME will overcome engineering inefficiencies and will enable companies to optimize critical factors directly affecting the competitiveness of their manufactured products. The results of developing a product, from concept definition to production steps, in a virtual environment are: the creation of virtual prototypes that can be manipulated in real time, providing the parts interference verification before the physical prototype construction; the execution of manufacturing and assembly processes simulations; and the virtual interaction with the product. It will allow one to rapidly analyze the factory performance, predict results of process changes before the real implementation, reducing time and costs of development and product implementation. Beside VM provides accurate and efficient models to the evaluation and control of the factory. Moreover the same system will be a good communication and training tool once the product will be developed.

The use of advanced techniques to evaluate different scenarios in an analytical way, especially VM, has been observed in some companies as Daimler Chrysler, John Deere, General Motors, Caterpillar and Boeing. Besides that, there are many researches going on in this area and its concepts have been developed in scope, gained a greater importance and international acceptance. Some advantages of VM are (Shridhar and Ravi, 2002): impacts on designers productivity; improves design and manufacturing responsiveness; improves team communication and project understanding; reduces unnecessary re-work costs; improves worker safety and efficiency; reduces the time to market.

The following section highlights the initiatives of two research institutes in the Virtual Manufacturing field, describing two case studies. Some conclusions at the end of the two following paragraph summarize the major outcomes and show the potential steps forward.

3. THE VIRTUAL MANUFACTURING AT SIMULATION LAB AND ITIA

Simulation Lab and ITIA-CNR are two research entities developing studies and projects on VME.

Simulation Lab is a research academic group, part of IFM (Institute Factory of Millennium), in the University of São Paulo - USP, Brazil. The initial purpose of VM research was to achieve a better comprehension of the use of higher interactivity interfaces. This study resulted in the development of a virtual prototype of the interlock of a CNC turning machine. The next step was the creation of a framework to allow the generation of a concurrent and distributed environment.

The Lab's efforts are to organize and systematize all the concepts and experiences developing a method to the application of VME in the product development process. This method will consider the use of many tools and software that can be included in the environment and the analysis of the interoperability question. The interchange with ITIA-CNR started with the aim of exchanging information and experience to conclude the development of a method. The case presented here describes a proposal for the use of neutral format files to provide data interoperability in a VM environment.

ITIA-CNR is the Institute for Industrial Technologies and Automation of the National Research Council of Italy. As a promoter of industrial innovation, performs strategic activities of Scientific Research and Technological Development for the Competitiveness and Sustainability of Italian and European Manufacturing Industries.

The RTD activities of the Institute aim at innovating Products, Processes and Organization: in this area, Virtual Manufacturing plays an important role.

3.1 Simulation Lab Case: Alternatives of utilization of neutral formats files enabling the data interoperability in a virtual manufacturing environment

Many of the activities performed during the product development process such as prototyping and testing, refining, validation and production, can be supported by the use of VM tools and solutions. These applications, most of the times, need to share product and process data and the data input and output formats are frequently different. As a consequence, the development of standards in a neutral format is required. This approach avoids the development and maintenance of many interfaces of specific systems and allows applications to be oriented directly to a shared file.

The purpose of this work is to present alternatives to the use of standard format files aiming at achieving a data interoperability level in a VM environment (Ravelli, 2003). In this research, a VME involves the use of systems in four task groups: elaboration of macro and detailed process plans, shop floor resources management, numerical control programming, and validation (simulation and virtual reality). These task groups need different interoperability levels regarding information access need, system association, information exchange frequency and detail level of data converted among these systems.

The integration of product and process data in this environment will be possible with the presence of a mechanism as a digital standard representation enabling the information sharing between the different VM systems.

In order to analyze the use of neutral standards alternatives regarding information interoperability to these task groups an information data flow was identified (Fig. 1).

Data input requirements, representations and formats are analyzed regarding information such as: **A** (tolerances and specifications), **B** (geometry) and **C** (manufacturing information).

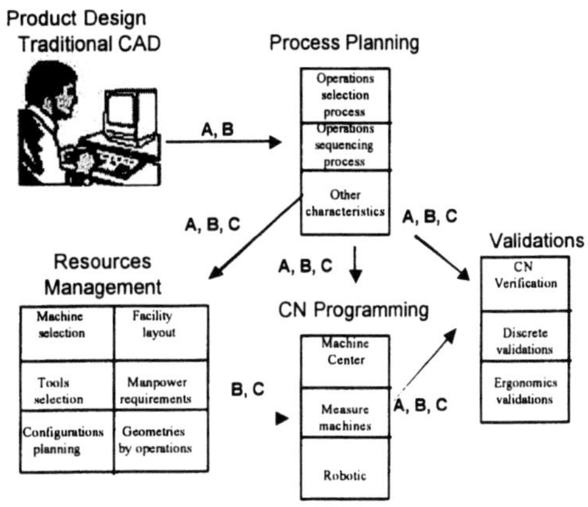

Fig. 1. Research Model: Data Flow in the VME.

Fig. 2 presents the proposed alternatives to allow the interoperability to this data flow between product project and MV tasks. In order to achieve that proposal, studies and researches in this field (Marsan et al., 1998; Greco, 2002a; Greco, 2002b; Schutzer and Henriques, 2002; Mecklenburg, 2001) have been analyzed.

Some of the neutral formats that were considered here are: STEP (Standard for Product data Exchange), neutral formats to 3D imagine visualization, SDX (Simulation Data eXchange), UPR (Universal Product Representation) and VRML (Virtual Reality Modeling Language).

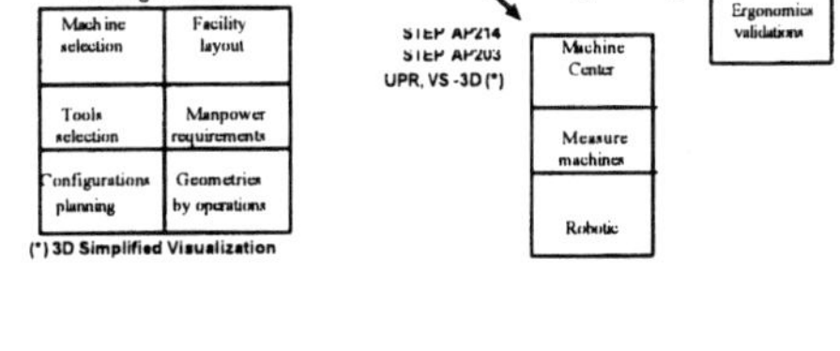

Fig. 2. Neutral formats to interface VM tasks groups

Data created by a CAD system in the product project phase and data created or modified by systems of the VM environment are stored and managed by the database both in their proprietary font format as in neutral formats. Automatic procedures to the generation of these additional files, the control of technical changes and the information access have to be managed by PDM systems pertaining to the product development process.

STEP, also known as ISO 10303, is a standard allowing the interchange and share of product model data between different modules of a development system. One of the most relevant application protocols for traditional CAD systems is AP203, which focus is on the description of shape and data of product configuration. It can support processes as: the development and exchange of design data with other steps of the life cycle of the product, the project development and technical changes. Another important application protocol is AP214. This protocol describes the relevant data structure to represent technical information regarding the mechanical development process. Some of the development activities supported by AP214 are: product definition, project, prototyping, process planning, quality control, tools design and production.

The SDX standard allows data exchange among simulation software, integrating factory project and material handling to simulation and analysis environment. It represents a mean of CAD users send information of facility layout to simulation systems. VRML is an international standard format to represent multimedia and 3D objects with hyperlinks to another object as texts, films and images, and allows the interoperability using the strategy of data integration through database. UPR presents a great potential to data exchange between traditional CAD systems and can be a solution to the exchange or project representation based on features. Neutrals formats of 3D visualization such as STL (Stereo Lithography) and CFL (Cubical Facet List) can be used to store information internally in the VM systems (Ravelli et al., 2003).

The proposed model represents the achievement of a complete integration between tolerances information and project specifications, and manufacturing information, with the correspondent geometry.

3.2 ITIA-CNR Case: The MPA-Virtual Factory

Towards the development of the Digital Enterprise, ITIA-CNR proposed an idea of integrating virtual reality and discrete events simulation in the ManuFuturing project (Boër and Jovane, 1996; Sacco et al., 2000), which focuses on the development of a new production system model for self-innovative and extended factory. A step forward it is applied in MPA project, where a factory is defined from its main modules and virtually reproduced in order to control and test the shop floor performances (Viganò et al., 2002).

The goal of this work is to define and to develop the "Virtual factory": a system that allows the modeling and the simulation of a factory in a virtual reality environment where both the layout design and the production process are taken into account. Moreover the factory can be design in its structural components (i.e. offices, wall, stairs, etc.). The VR system is running on graphical workstation and mainly developed using C++, while the simulation system and the Standard Facility Library (SFL - the DB containing the whole factory resources and building blocks) are based on PC and Java.

In the MPA, the SFL maintains standard modules for virtual 3D design and the designed layout for successive simulation. The VRE (Virtual Reality Environment) provides a friendly 3D immersive interface for the factory building or the shop floor design; thus, the users are able to select the desired resource objects from the SFL, put them on the virtual factory floor, connect outputs to inputs and verify the hypothetical layout against the existing spatial constraints. All the operations can be executed in the virtual room of ITIA-CNR. Due to the high costs and heavy burden of the virtual room, especially for SMEs, a GUI (Graphical User Interface) is envisaged as a complement of 3D interface in case the dedicated hardware platform is not available. The fundamental idea of emulation and control software is to create a simulation tool that can support the analyses and evaluations of factory building and shop floor layout. The simulation data then can be fed-back to the 3D or 2D interface to get emulation effect.

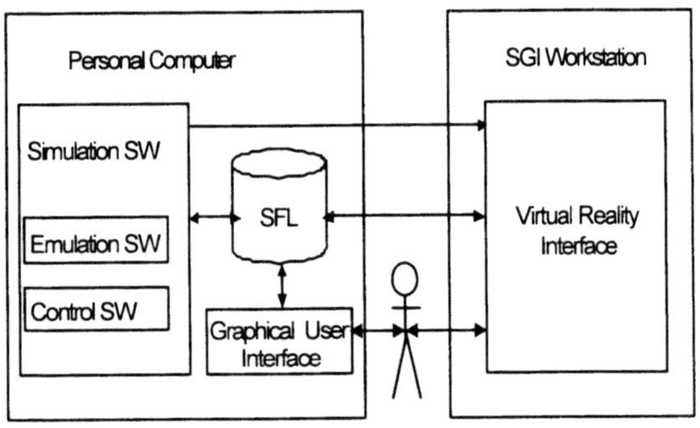

Fig. 3. The MPA logical architecture

The pilot case to test the application and the overall framework is a factory layout plus the factory building modules (both the case come from real companies data); the represented process is for manufacturing and assembling car air-condition system and, since in reality the process is replicated 5 times, in the factory the pilot application deals with just one single line; the building, as well, is reconstructed just in its more meaningful modular component.

For the pilot application almost all the 3D modular components have been modelled from scratch. In future a conversion procedure will be foreseen.

A Brief Introduction to the XML integration of the Simulation and the VRE

The integration between the VRE and the simulation can be fulfilled in different ways: using SOCKET (Haverlock, 1998; Mahmoud, 2001), CORBA (Schmidt and Vinoski, 1995; Vinoski, 1997) or through XML (Mohseni, 1994; Sun Microsystem 2001). In the MPA project the whole possibility have been studied and tested; finally the XML solution has been adopted in order to create a lighter environment that can be used on PC avoiding the workstation and the expensive immersive devices. After the simulation completes the emulation process, it is designed to produce a XML record file to store all the simulation processes, including the necessary parameters for the VRE to replay the simulation effect under 3D environment. This idea is similar to transferring the layout file from the SFL to the simulation. Consequently, the integrative method discussed above could also be used to transfer the record file from the simulation to the VRE. Utilizing XML file to implement data integration has the following advantages:

- The learning curve is very shallow because XML is a natural step up from HTML.
- The programmer should not worry about the complicated interface definition and interaction. What is only required to do are producing and receiving the XML file on the basis of W3C standard.
- XML is good at handling matters that are hierarchical. The SFL database is a real hierarchical object tree that can be represented reliably and conveniently by XML document.
- Some APIs support the development, for example, Java provides powerful tools like JAXP, JAXB.

The disadvantages are:

- XML document is human-readable file whose security is not so good as that of transferring file in binary format, like CORBA.
- The application has to parse or unmarshal the XML document after receiving the XML file. It is less inefficient and needs a more heavyweight client especially when the file is large and works with parser DOM. Anyway, it sacrifices the performance as the return of simplicity.
- The programmer has to know which language (C++, Java) is used in the parsing side. Different languages should have corresponding parsers.

3D factory layout builder

A new step that allows the visualisation of Virtual factory on PC is to create a three-dimensional environment meant to represent a factory layout taking information out of the SFL directly using XML. The environment offers to the user navigation functionalities as an avatar inside the virtual factory.

The tool is characterized by low hardware requirements (PC-based), user-oriented semi-immersive graphical interface and a modular and open architecture.

All the information needed in order to create the virtual factory layout is stored in the SFL. The database contains information related to the resources, their characteristics and spatial information (positions and orientations); the model repository collects all the 3D models in VRML format. In order to create a modular system a "Database Interface" has been created so that the virtual layout builder can work without regards to the internal structure of the Information System. Mainly two modules compose the Layout Builder; a Parser for the layout representation based on the Schema mentioned above and a Loader, based on the VRML specifications, which loads the models pointed out by the XML file. The result of the layout builder is a virtual 3D representation of the factory with a intuitive navigation system and some reports about the resources information in the factory.

The software provides three main functionalities: the representation of the virtual factory, the semi-immersive navigation of the virtual factory and the reports about the factory layout

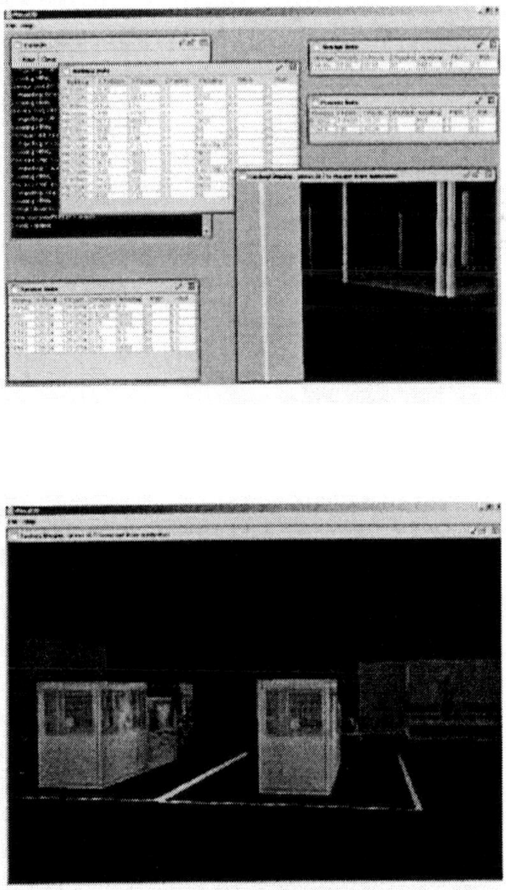

Fig.5. The virtual factory

The user can then navigate this environment as if he is really inside it. Since the mouse and the keyboard, not using stereo-glasses and data gloves, provide the input we can define it a semi-immersive navigation. Moreover the user can view some reports about the layout containing information related to every single resource with a direct link to the virtual resources in the environment.

4. FINAL CONSIDERATIONS

The competitiveness of a company depends more and more on the utilization of new developments on methodologies and environments, supported mainly by computational tools and technologies. The Virtual Manufacturing approach leads to an integrated and synthetic environment to develop a product, from concept to market launching. However, applying Virtual Manufacturing involves an effort to develop a good solution to integrate all the applications and data present in that environment. An analysis has to be performed for each case and alternatives have to be studied. The interoperability barrier exists and a great deal of effort has been placed on this question. The authors have presented just a brief overview of the problem and two case studies where these aspects have been studied and applied to real companies' cases. The work is in progress and some results emerging from the evaluation campaigns will be presented at the conference.

REFERENCES

Banerjee, P., D. Zetu (2001). *Virtual manufacturing.* New York, John Wiley & Sons. 320p.

Boër, C.R., F. Jovane (1996). Towards a new model of sustainable production: ManuFuturing, *CIRP Annals*, STCO, pp. 415.

Committee on advanced engineering environments (1999). *Advanced engineering environments:* Achieving the vision, Phase 1. Washington, D.C.: National Academy Press. 58p.

Greco, J. (2002a). MCAD Up close: interoperability, Part 1. *Desktop Engineering*, v.7, n.7, pp. 14-18.

Greco, J. (2002b). MCAD Up close: interoperability, Part 2. *Desktop Engineering*, v.7, n.12, pp. 22-25.

Haverlock, K. (1998). Object Serialization, Java, and C++ - Exchanging Data Between a Sender and a Receiver. *Dr. Dobb's Journal.*

Lawrence Associates Inc (1994). Virtual manufacturing user workshop. *Technical report*, Ohio. 27p.

Mahmoud, O.H. (2001). Transporting Objects Over Sockets – *Advanced Socket Programming.*

Marsan, A., V. Kumar, D. Dutta and M.J. Pratt (1998). An assessment of data requirements and data transfer formats for layered manufacturing. *Technical Report NISTIR 6216,* NIST, Gaithersburg, MD.

Mecklenburg, K. (2001). Seamless integration of layout and simulation. In Winter Simulation Conference, *Proceedings*, pp. 1487-1491.

Mohseni, P. (1999). An Introduction to XML for Java Programmers - The Extensible Markup Language and Java are a Perfect Duo for Data Exchange and More. JAVAPro *Visual Basic Programmers Journal.*

Ravelli, C.A. (2003). *Análise da interoperabilidade de dados na implementação de um ambiente de manufatura virtual.*Escola de Engenharia de São Carlos, USP (Master Dissertation).

Ravelli, C. A., A.J.V. Porto, M.C.F. Souza (2003). Alternativas de utilização de formatos neutros de arquivos para habilitar a interoperabilidade de dados em um ambiente de Manufatura Virtual. In: Simpósio Brasileiro de Automação Inteligente, Bauru, SP, *Anais.* (CD ROM), pp.138-143.

Sacco, M., S. Mottura, G. Viganò, A. Avai, C.R. Boër (2000). Tools for the innovation: virtual reality and discrete events simulation to build the 2000 factory, *Proc. AMSMA'2000*, Guangzhou, P.R. China, pp. 458-462.

Schmidt, D.C., S. Vinoski (1995). Object Interconnections – Comparing Alternative Client-side Distributed Programming Techniques. *SIGS C++ Report Magazine.*

Shridhar, J.M., S. Ravi (2002). Virtual Manufacturing: an important aspect of collaborative product commerce, *Journal of Advanced Manufacturing Systems*, v.1, n. 1, pp. 113-119.

Shutzer, K., J.R. Henriques (2002). Intercâmbio de dados no desenvolvimento de produto. *Engenharia Automotiva e Aeroespacial*, SAE Brasil, n.3, pp. 26-36.

Sun Microsystems Inc. (2001). The Java Architecture for XML *Binding User's Guide.* White paper.

Viganò, G., Z. Liao, M. Sacco, S. Mottura, C. R. Boër (2002). To Build Object-Oriented Virtual Factory Based on Virtual Reality and Simulation Tools. *1st CIRP Seminar on Digital Enterprise Technology*, University of Durham, p.225-228.

Vinoski, S. (1997). CORBA: Integrating Diverse Applications within Distributed Heterogeneous Environments. *IEEE Communications*, vol. 14, no. 2.

Copyright © IFAC Information Control Problems in
Manufacturing, Salvador, Brazil, 2004

ELSEVIER
IFAC
PUBLICATIONS
www.elsevier.com/locate/ifac

MODELS TO SPLIT UP AND TO INTEGRATE COMPETENCIES SYSTEMS IN DESIGN ACTIVITIES MANAGEMENT

Eric Bonjour, Maryvonne Dulmet, François Lhote

Groupe Pilotage et Modélisation des Activités de Conception- Laboratoire d'Automatique de Besançon
Unité Mixte de Recherche CNRS - ENSMM - UFC
25, Rue A. Savary - 25000 Besançon (F)
Tél. (33) 381 402 811 - Fax. 33 381 402 809 - E -mail : ebonjour@lab.ens2m.fr ; mdulmet@ens2m.fr

Design activities require complex organizations to support the production of technical objects about systems that tend to be more and more complex too. The management of design projects and of their skill-networks requires models to take into account the complexity of competencies systems. Socio-technical systems management has to define an appropriate organization to create synergy between competencies, processes, products and technologies, at different levels within the company or between companies. This paper proposes models to help project managers or skill-networks managers to organize competencies systems, i.e., to split up objectives they have in charge in terms of collective or individual competencies. It contributes to articulate a manager's point of view with an actor's point of view. The validation of this model is achieved in the framework of a research contract with a French car enterprise, in a specific field of car engineering processes. The final aims of our research works are about the design of a management device of the engineering competency system.
Copyright © 2004 IFAC

Key words: *socio-technical systems management, human systems, competency systems modeling, competency management, design engineering*

1. INTRODUCTION

Companies are facing changeable environments, different sources of uncertainty and an increasing importance of human dimensions (cognitive, social, organizational, affective). Considered like one of the main sources of competitive advantages, competency and knowledge become key-notions to increase the enterprise's performance (Argyris, 1995), (Faure & all, 1999), (Hermine, 1996), (Nonaka, 1995), (Spencer, 1993), (Wielinga, 1992).

Today's industrial objectives are to manage a permanent increase in performance: to reduce the "time to market", to reduce the product development costs, to increase innovation, quality, reactivity, organizational learning ... To meet these requirements, some Managers' objectives and actions are then

- to optimize the project organization and the assignment of missions to the actors (project management)
- to manage actors' competencies (human resource management)
- to develop competencies and knowledge in their own domains and skill-networks (knowledge management, skill development)
- to develop new solutions regarding new strategic requirements (innovation management)

The notion of competency has many definitions in the literature (Levy-Boyer, 1996), (LeBoterf, 2001) but several common characteristics have been highlighted. A competency is linked to a finalized action (to reach a goal, to carry out a task, to perform a mission....). It depends on a context. It puts in interactions several kinds of cognitive components (concepts, knowledge, know-how...) and external resources (tools, networks...). Competency may be required and then, referred to a given task (or mission) that needs to be carried out in the future. Competency may be acquired and recognized, embodied by an actor (one or several persons) and then, available to perform a given similar mission.

Concerning the design activities, systems which have to be designed as well as design activity organizations are very complex. Activities are collective (strong relationships between actors) and require co-operation, iterations with discussions to find solutions. Value is added thanks to the contribution of the individual actors to the collective actors (competency integration, results integration). A part of the integration is described thanks to the definition of missions and another part is emerging from the discussions between the actors. The major issue of managers is to design a coherent organization to meet customer's requirements with the highest performance (quality, costs, delay...) in the short term and to manage the development of relevant competencies in the long term. Few works exist about this topic. We call it a bifocal management.

This paper proposes articulated models to help managers to split up and to integrate competencies systems in design activity management. First part points out the industrial context of our research works. Second part presents the conceptual framework and different models to split up and integrate competencies systems. Industrial examples illustrate the different phases of modeling.

2. CONTEXT AND OBJECTIVES

Our works take place in the framework of a research contract that began in 1999, with a French car enterprise. They are aimed at supporting some processes concerning coherent modeling and management of competencies systems. They are based on the analysis of engineering activities. In a middle term perspective, the requirements of this car enterprise concern the characterization of processes which lead to the development of competencies used in the field of concurrent engineering for simultaneous product and process design. We have already defined and validated a framework of competency modeling based on a systemic approach. In the long term, our final aims are (1) to develop a coherent device to manage the design competencies systems as well as (2) to improve the development of knowledge and competencies and to tend to auto-adaptable competencies systems.

This paper presents our works about the identification of the individual or collective competencies (with a deployment approach) and about the definition of the collective competencies systems structure (with an integration approach).

3. SPECIFICITY OF DESIGN ACTIVITIES

Design is often considered as a problem-solving activity that requires a collective actor. The problem specifications are progressively defined in the relationships of mutual requirements (De Tersac, 1996). Each individual actor builds his (her) own representation of the problem by taking into account the dimensions that are relevant with his intentions only. However, common representations (related to common global intentions) are necessary (Perrin 1999) to ensure a coherent integration of individual, local results in the collective, global actor.

The major features of design activities are as follows:
- problems are not completely defined in the beginning. The context of the design (market, technology, collaborators, managers, process...) are different sources of uncertainty. Therefore, during the whole project, the solver will require information issued from its environment (tools, partners, customers, ... to define its problem and to reduce uncertainty.
- the parameters and their relationships are not split up into independent sub-systems. Sub-systems may correspond to different skills (or competencies). Then, these competencies are interdependent. For example, during the design of the architecture of the front axle in a car, the design of the suspension has to meet mechanical requirements to ensure comfortable and stable behavior on the road but its volume depends on the car architecture and on the volume of the other sub-systems of the front axle and its mechanical parameters are correlated with its volume and with elastic parameters of the wheels or of the no-roll bar.
- the solution is only satisfactory but not optimum (Darse 2001). The designer has to make choice among acceptable solutions according to performance criteria. There is no completely defined method (or algorithm) to find this solution. The design activity is based on an action plan that is progressively built during the activity, at the same time other actors define mutual requirements according to their local solutions and their own actions plan. The solution emerges from different unforeseeable consensus or trade-off.
- these interactions between actors imply numerous iterations and it is difficult to split up the design process into different tasks or responsibility centers.
- the solution evaluations are carried out at different levels : by the designer, by his (her) collaborator separately (from different points of view, heterogeneous criteria), by his collective actor and by external actors who are in charge of the represention of the customers' voice.

The design is the progressive determination of the product and its process, which ensure the convergence of several actors' activities towards a satisfactory and coherent solution. It is an organization of individual problem solving integrated thanks to negotiations between actors globally coordinated by the deployment of common objectives split up into several missions.

We will use the term of generic mission to name the missions that the actor has to perform successively with the respect of the same action framework and of the same kind of task. Expectations concerning the product have been modified in an iterative and collective activity due to the progressive process of problem solving. The actor's intermediate results are useful to other actors to decrease their uncertainty level and then, to improve the quality of their current expectations. Thus during a generic design mission, an actor may have to perform several instances of his mission, but with current expectations that are lightly different at each iteration. The allocation and the acknowledgement are not performed for each iteration. The actor is entrusted with a generic mission, he has to perform it several times during his design activity.

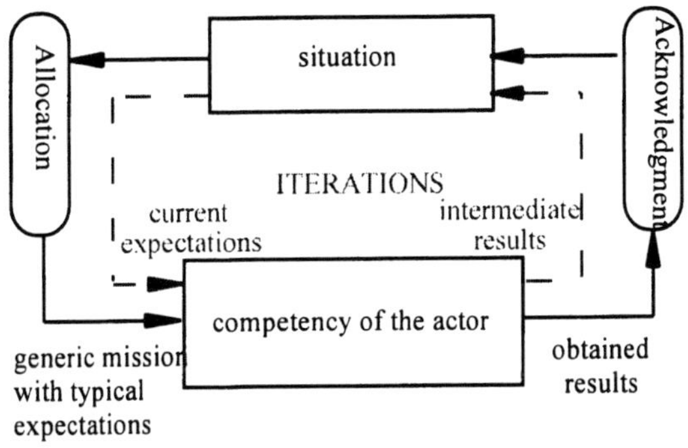

Figure 1: specificity of design activities

We identify three levels in the design activity:
- the actor's relationships with his collective actor. At this level the actor furnishes intermediate results which are the result of negotiations, choices of elements which have to be frozen or redefined, but he has to let some degrees of freedom to the upstream designer. The actors build mutual requirements thanks to their cooperation;
- the actor's action to meet the expectations of his mission inside an iteration. This action is of the same kind for each iteration, i.e. he uses the same cognitive structure to lead his action.
- the acknowledgement of competencies in the field of design assumes that we are able to characterize and quantify the expected contribution of the actor to the collective performance.

4. MODELS FOR DESIGN ACTIVITIES MANAGEMENT

4.1 Conceptual framework

In an enterprise, we distinguish between three levels of competency modeling in relation with three kinds of actors: individual actor, collective actor and the global Organization (enterprise).

Individual actor: This is a person who is in charge of performing a mission or a set of missions and who acts permanently, temporarily or occasionally at the service of a considered enterprise. This actor is characterized by his (her) corpus of competencies (or knowledge and skills) and belongs to a recognized skill-network (design and manufacture of diesel engine, architecture and design of the electronic systems, quality management...). In the skill-network, each actor shares common representations (terminology, knowledge, rules, identity and value).

Collective actor: This is a group (or a team) who is in charge of performing a mission or a set of missions and who acts permanently, temporarily or occasionally at the service of an enterprise. There are collective actors of various levels: binomials, teams and workgroups, project structures, skill-networks, functional departments... A collective actor may be split up into smaller collective actors and/or individual actors. An individual actor may take part in different collective actor. This concise definition is closed to that of Cohen et al. (1997), which is more precise: "a team is a collection of individuals who are interdependent in their tasks, who share responsibility for outcomes, who see themselves and who are seen by others as an intact social entity embedded in one or more larger social systems (for example, business unit or the corporation), and who manage their relationships across organizational boundaries."

Global organization (at the level of the enterprise): it is linked to the strategic management of the enterprise, (Pralahad, 1990]. Strategy is defined by global objectives at different terms. Objectives have to be specified in terms of strategic competencies.

The following concepts concern any type of actor.

Task is " what is to do in the system ". It is a set of processing: physical transformation and/or information processing and/or problem solving. It concerns a set of given planned objects (material or immaterial inputs). The objective of the given task is to produce a set of expected outcomes (resulting objects or outputs).

Process is a scheduled organization of tasks which have similar objectives on the same flow or the same type of flow (material, information, knowledge, decisional objects...)".

Situation: It corresponds to the context of the actions. According to Theureau (1994), situation is what is meaningful for the actor in his environment. It is steadily changing. In a normal, foreseeable situation, it enables the actor to understand, to plan and to perform the mission. But the situation could be characterized by its variability and by the gap between the forecasts and the actual context. For instance, the given actual inputs are different from the given planned inputs (task variability), unpredictable events may occur ... The actor is permanently in interaction with elements of the situation. He has to cope with the actual situation, to identify relevant information flows and to adapt the organization of his action.

Mission: A given **mission** is defined by the Manager and is affected to an actor under his responsibility. It consists of a task referred to a given action framework. An actor is entrusted with this mission. The prescription does not concern "the how". He has the responsibility to fulfil the mission by determining the relevant organization of finalized actions. In the field of design activities, we consider that a mission is an artifact of the Organization. This artifact is complex, evolutive and results from a negotiated building between the actor who is in charge of carrying out this mission and the manager who has received delegation from the Organization. So, during the design activity, the mission definition is likely to be modified, along with the design process, to adapt the collective organization to external and internal events or requirements.

Actions framework: The manager is not able to determine a complete characterization of a considered situation (it is not possible). He has to choose relevant and critical elements according to the task of the mission and to the actor's autonomy. Today, due to the situation complexity and the high level of skills needed, the actor has a large autonomy to perform his mission. Tasks and missions can be allocated to an actor with various degrees of explicitation according to the chosen actor.

Competency (from an external point of view)
An external view considers competency as a black box which provides for a successful finalized action, in answer to two inputs: on the one hand, action aims, i.e. expected results (perhaps with constraints and performance criteria), on the other hand, a determined action framework, i.e. key-elements of the situation. If these inputs are specified by the Organization in terms of a mission, the associated competency is called operational competency. An operational competency is an appreciation or "acknowledgement" made on a given actor, by a determined judge related to a defined mission (or finalized actions). It is related to a successful mission or a class of couples {aims; framework}. (See [BON 01] for more details)

Competency (from an internal point of view)
Competency is supported by a cognitive structure which is called scheme. The scheme activation corresponds to the "mobilization" and dynamic organization of a set of heterogeneous cognitive resources that leads to the production of an acknowledged performance, in the framework of a finalized activity and a particular class of situations. Thanks to its competency, an actor transforms a mission, with which he had been entrusted, into a successful flow of action that will achieve expected results. (See Bonjour, 2002 for more details).

Activity: It corresponds to how the actor performs his mission. Activity has two aspects: mental (not watchable) and sensorimotor (watchable). An observer can watch the concrete action plan that only corresponds to a part of the scheme activation.

4.2 Deployment and integration

Deployment and integration of competencies:
The allocation of a mission to a collective team and the deployment of this mission among the actors of the team needs to take into account two phenomena which are the coordination and the integration;

The coordination allows to guarantee the activity's consistency. A mission of n-level is assigned to an individual or to a team. Through a coordination plan (identified on figure 2 as transfer models), which is architecture of actions frame, n-level mission is split up into n-1 level missions assigned to limited teams and to individual. The coordination carries out the differentiation of the roles among the actors.
During the integration process, local results are integrated in global results (these are also transfer models), and they will put in light operational competencies. During the action we have the equilibration of the schemes of the individual or collective actors, and the activity is translated through actions frames. Then the results of the n-1 level missions are synthesized to provide results of n-level missions.

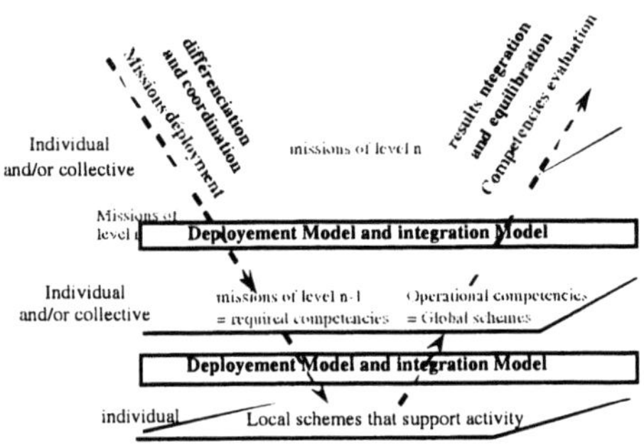

Figure 2: Deployment and integration of competencies

Top-down model: The process
The proposed modeling framework distinguishes beween the point of view of the treated object (task concept) and the point of view of the actor (activity concept). This distinction is similar for a great part with the ergonomics distinction which contrasts prescription (scheduled task) and real action (activity of the actor).
This model represents how a global mission is decomposed into sub missions. The identified actions to be performed by the collective actor will be allocated by the manager. The process is likely to be modified to adapt the organization to external changes or new requirements from the actors; On the figure 3 we present the collective mission "To design a powertrain at the better level with respect to the strategic objectives". At the same time, we show the direct process acting on the expected objects and we show some support processes, which are needed to design in a better manner this object.
- Requirements synthesis process
- Results justification process
- Respect of requirements in cost, quality, delay...
- Capitalization process
- Management process....

Bottom-up Model: The activity graph

Tasks and missions have been allocated to an actor (collective or individual). The activity of this actor can be described by an activity graph. It corresponds to how the actor performs his mission. Activity has two aspects: mental (not watchable) and sensorimotor (watchable). An observer can watch the concrete actions plan that only corresponds to a part of the scheme activation. We have said that design is the progressive determination of the product and its process, which ensure the convergence of several actors' activities towards a satisfactory and consistent solution. On the following figure (figure 4), we can see how a large collective actor is performing his mission with iterations and negotiations. The actors are globally coordinated by the deployment of common objectives split up into several missions.

The last figure (fig 5) focuses on a part of the collective mission described by the figure 4. We can see the activity graph corresponding to the activity of a designer who is in charge of the mission " to develop the architecture of the front axle". This graph was built by analyzing the activity and by talking with the designer. Here, we are nearly at the scheme level and we strongly put in light the negotiation and iteration phenomena.

5. CONCLUSION

In the field of Industrial Engineering, competency management and knowledge management are often based on poor definitions of the used concepts and suffer from a lack of integrated models. The connection between knowledge and the activity that uses it, is not clearly modeled. The internal model of competency which is presented in this paper is very useful to explain on what an actor's competency is based.

In this paper we presented a model of deployment and integration of competencies. We applied this model in the framework of the Car Systems Engineering. Our research work is now in progress to show that this model is useful to understand or build the competencies system structure in regards to the design structure. Section 4.2 shows that competency system has the same organization as the Ve cycle of Concurrent Engineering.

Our further goal is to use this correspondence to define a guideline to organize and to manage competencies system.

6. ACKNOWLEDGEMENT

The authors wish to thank all the actors of the automotive enterprise who help them to validate their conceptual framework and more specially, Mrs. Beretti for her advice, her help and her friendship.

7. REFERENCES

Argyris, C, "savoir pour agir, surmonter les obstacles à l'apprentissage organisationnel", traduction, Paris, Interéditions, 1995

Bonjour, E, Dulmet, M, Lhote, F, "Interests and limits of an external approach of competencies used in the enterprises", CD Proceedings for the 4th International Conference on IEPM, Québec, 2001

Bonjour E, Dulmet M, Lhote F., "An internal modeling of competency based on a systemic approach with socio-technical systems management in view", IEEE SMC System, Man and Cybernetics 02, Hammamet, 6-9 octobre 02, Actes CD, 6p

Cohen, S.G., Bailey, D.E., "What makes teams work : group effectiveness research from the shop floor to the executive suite", Journal of Management, vol 23, N°3, pp 239-290, 1997

Darses F., " Assister la conception : perspectives pour la psychologie cognitive ergonomique ", *Actes de conférences EPIQUE 2001*, le 29-30 Octobre 2001

Faure, A. G. Bisson, "Modeling the experience feedback loop to improve knowledge base reuse in industrial environment", Actes 12th KAW, Banff, CA, 1999

De Terssac G., " Le travail de conception : de quoi parle-t-on ? ", *in coopération et conception*, Edition OCTARES 1996

Hermine, J.L., "Les systèmes de connaissances", Paris, Editions Hermès, 1996

Levy-Boyer, C, "La gestion des compétences", Ed Organisation, Paris, 1996

Le Boterf, G. "Construire les compétences individuelles et collectives", 2ème Ed, Editions d'organisation, 2001

Nonaka, I., Takeuchi, H. "la connaissance créatrice", traduction française 1997, Bruxelles, De Boeck Université, 1995

C.K. Pralahad C.K. and G. Hamel, The Core Competence of the Corporation, Harvard Business Review, 1990

Spencer, L.M., Spencer, S. M., "Competence at Work: Models for Superior Performance", J Whiley & Sons, N.Y, 1993

Theureau J. et al., "Ergonomie des systèmes informatiques - la conception centrée sur le cours d'action", Toulouse, Octarès, 1994

Wielinga, B., J. Breuker, G. Schreiber, "KADS, a modeling approach to knowledge engineering", academic press, London, UK, 1992

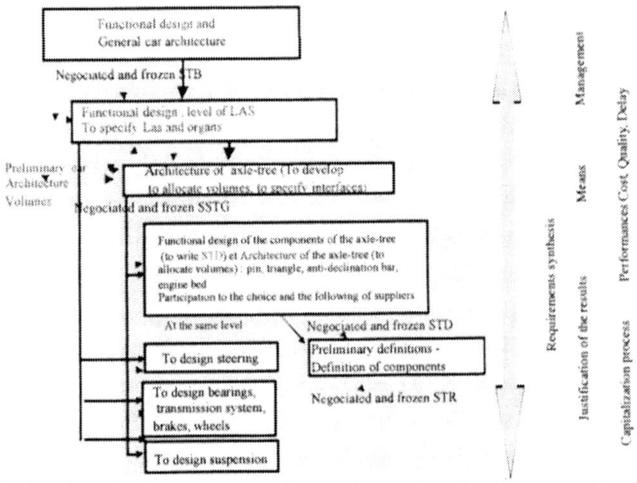

Figure 3: collective mission "To design a powertrain at a better level with respect to the strategic objectives"

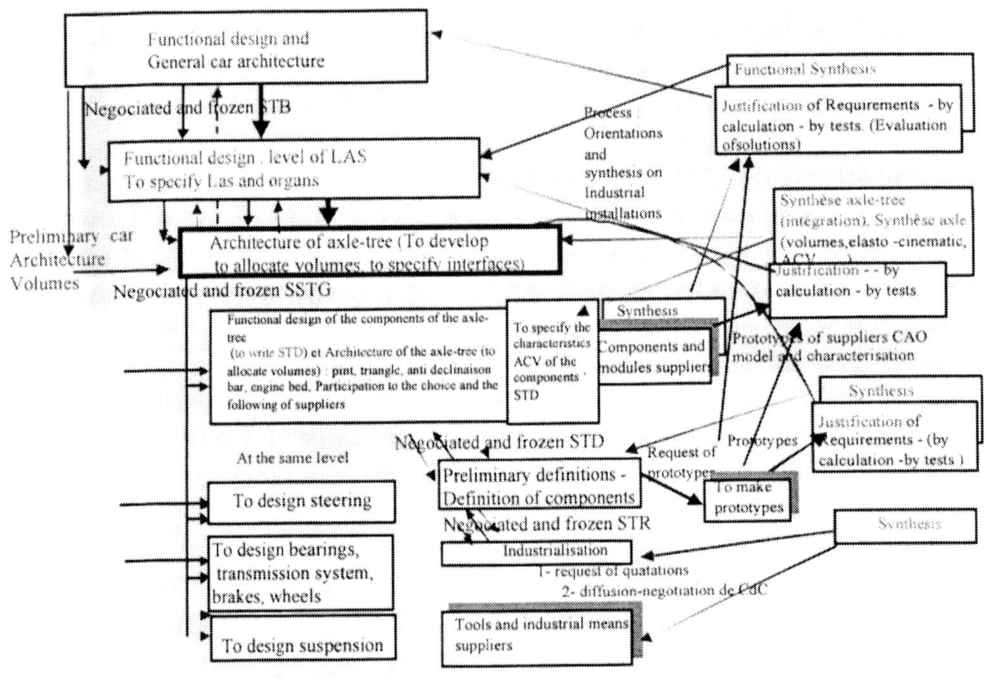

Figure 4: Activity graph of a collective actor

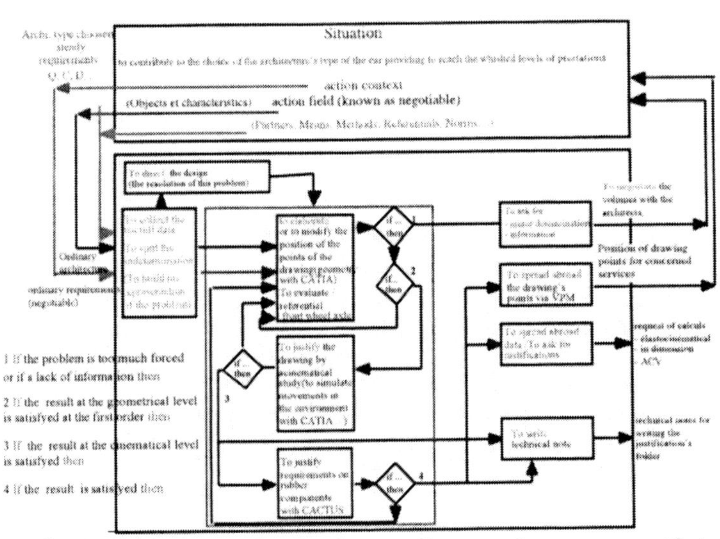

Figure 5: Activity graph of an individual actor: "to develop, to allocate volumes, to specify interfaces"

Copyright © IFAC Information Control Problems in Manufacturing, Salvador, Brazil, 2004

ELSEVIER
IFAC
PUBLICATIONS
www.elsevier.com/locate/ifac

HIERARCHICAL OPTIMIZATION FOR THE FLEXIBLE JOB SHOP SCHEDULING PROBLEM

N.ZRIBI[1] I.KACEM[2] A. EL KAMEL[1]

[1] *École Centrale de Lille*
Cité scientifique - BP 48
59651 Villeneuve d'Ascq Cedex-France

[2] *UTT - Université de Technologie de Troyes*
12, rue Marie Curie - BP 2060
10010 Troyes Cedex - France

Abstract: In this paper we propose a new approach for the flexible job shop scheduling problem. This approach is based on the decomposition of the problem in an assignment problem and a sequencing problem. The first part is ensured by an assignment technique based successively on a heuristic approach and a local search. The second part consists in applying a genetic algorithm to deal with the sequencing problem. Finally a comparison with different existing results in the literature is carried out. *Copyright © 2004 IFAC*

Keywords: Job-shop scheduling problem, assignment problem, local search, lower bounds, genetic algorithms.

1. INTRODUCTION

The flexible job shop scheduling problem is a generalization of the traditional job shop problem. It is defined as a set of M machines and a set of N jobs, where each job consists of an ordered sequence of operations. Each operation is performed on a single machine belonging to an allowable set. Many approaches have been proposed to solve this problem. In (Brandimarte, 1993; Dauzère-Pérès and Paulli, 1997; Mastrolilli and Gamberdella, 2000; Kacem and Borne, 2003), we can find new extensions of this problem where tasks can be performed by several resources at the same time and/or the choice to assign tasks on different machines is possible.

In this paper, we propose a new approach to deal with the flexible job shop problem. The second section of this article presents the specificities of the flexible job shop and gives the mathematical formulation used to deal with such a problem. In section 3, we describe the assignment technique used in order to balance the workloads on the different machines and to pre-optimize the makespan. Then, in section 4, we propose a genetic algorithm to solve the job shop scheduling problem. The last part will be devoted to the presentation of some conclusions concerning this work based on a comparison with existing results in the literature.

		M_1	M_2	M_3	M_4
J_1	$O_{1,1}$	4	3	5	8
	$O_{2,1}$	10	5	4	6
	$O_{3,1}$	8	6	2	3
J_2	$O_{1,2}$	5	9	10	4
	$O_{2,2}$	2	6	9	7
	$O_{3,2}$	4	3	1	11
J_3	$O_{1,3}$	5	9	6	4
	$O_{2,3}$	1	9	10	4

Figure 1: Example of a P_{FJSP}

2. PROBLEM FORMULATION

The problem is to organize the execution of N jobs on M machines. The set of machines is noted U. Each job J_j represents a number of n_j ordered operations (precedence constraint). The preemption is not allowed. The execution of the ith operation of job J_j (noted $O_{i,j}$) requires one resource or machine selected from a set of available machines. The assignment of the operation $O_{i,j}$ to the machine M_k entails the occupation of this machine during a processing time called $d_{i,j,k}$. Thus, to each FJSP (Flexible Job Shop Scheduling Problem), we can associate a table P_{FJSP} of processing times such that: $P_{FJSP} = \{p_{i,j,k} \in IN^* \mid 1 \leq j \leq N; 1 \leq i \leq n_j; 1 \leq k \leq M\}$. An example is presented in Fig.1.

In this problem, we make the following assumptions:

- all the machines are available at $t = 0$
- each job J_j can be started at $t = r_j$
- each machine can only execute one operation at a time

The scheduling problem in a flexible job shop can be decomposed in two subproblems:

- a routing subproblem that consists of assigning operations to machines;
- an operation scheduling subproblem associated with each machine which goal is the minimization of the makespan.This is a job shop scheduling problem.

3. ASSIGNMENT PROBLEM

Definition 1 *To each repartition of the operations on the resources set, we associate an assignment. Each assignment is characterized by a set S such that:* $S = \{S_{i,j,k} \in \{0,1\} \mid 1 \le j \le N,\ 1 \le i \le n_j,\ 1 \le k \le M\}$. $S_{i,j,k} = 1$ *if* $O_{i,j}$ *is assigned to* M_k *else* $S_{i,j,k} = 0$ $(\sum_{k=1}^{k=M} S_{i,j,k} = 1)$.

Our objective is to minimize the workload of the critical machine (the most loaded machine) $Cr_2 = \max_k \{W_k\}$ (W_k is the workload of machine M_k) in order to pre-optimize the makespan.

3.1 *Local search algorithm*

To solve the assignment problem, we have to balance the workloads of the different machines. The literature presents an interesting assignment heuristic (localization approach) proposed by (Kacem, 2003). Such an approach uses a local optimality condition at each step and allows us to construct iteratively a set of good solutions. Such a heuristic is based on the following rules:

1) Let define l as the index of the iterations during the execution of the assignment procedure. Therefore, it represents the index of the lth operation to be assigned, thereafter:

∘ the lth operation $O_{i,j}$ that will be assigned is noted $O_{i,j}(l)$ (for $1 \le l \le NT$),

∘ $W(l,k)$ represents the charge of the machine M_k at the lth iteration of the assignment procedure,

2) At the iteration l, we assign $O_{i,j}(l)$ to the machine $M_{k(l)}$ such that $W(l, k(l)) \le W(l-1, k) + p_{i,j,k} \ \forall \ k \ \ne \ k(l)$. Then: $W(l, k(l)) = W(l-1, k(l)) + p_{i,j,k(l)}$ and $W(l,k) = W(l-1,k) \forall k \ne k(l)$.

3) We exchange randomly two operations and we construct a new assignment based on the new operation list and by following the same procedure.

Such a heuristic has been proved to be effective by comparing its results with a set of lower bounds proposed by Kacem.

In order to ensure a high level of the solution quality, we have chosen to improve the assignments given by Kacem's heuristic. In this end, a local improvement search has been studied. Such search is based on a Tabu algorithm and an adapted *routing move* technique. In this section, we give a description of our Tabu algorithm.

The solution is described as a list of operations with theirs corresponding machines. A *routing move* is defined by the relocation of a critical operation (operation that belongs to the most loaded machine) to a feasible machine position. For a given solution, we consider every such relocation of every reroutable critical operation.

The *routing move* is based on the following steps:

1) Find the critical machine M_{kc},

2) Find an operation $O_{i,j}$ (such that $S_{i,j,kc} = 1$) such that it can be assigned to another machine $M_{k'}$ without increasing the criterion value (Cr_2),

3) Reassign $O_{i,j}$ to $M_{k'}$ if possible.

For a large or highly flexible problem, this neighborhood could potentially grow very large, and a candidate list or elite solution strategy might be considered. We establish a simple tabu attribute for this move type by prohibiting the return of the operation to the machine from which it was removed during the extent of short term memory.

To evaluate the efficiency of such an algorithm, we have implemented it on a set of instances based on practical data. The different simulation results obtained confirm the good quality of the localization heuristic and the effectiveness of the Tabu search since it can enhance the solutions given by kacem's heuristic.

4. SEQUENCING PROBLEM

After the assignment step, the flexible job shop problem is reduced to a classical job shop problem.

∘ to each operation $O_{i,j}$, we define $M_{a(i,j)}$ as the machine on which $O_{i,j}$ will be executed $(S_{i,j,a(i,j)} = 1)$,

∘ to each machine M_k, we associate the set E_k of the operations which will be executed on the above machine $(E_k = \{O_{i,j} \mid a(i,j) = k\})$,

∘ to each JSP (Job Shop Scheduling Problem), we can associate a table D of processing times such that: $P_{JSP} = \{p_{i,j} \in \mathrm{IN}^*, 1 \le j \le N; 1 \le i \le n_j;\}$. An example is presented in Fig.2,

The problem is then to assign the starting time $t_{i,j}$ and the completion time $tf_{i,j}$ to each operation $O_{i,j}$ $(tf_{i,j} = t_{i,j} + p_{i,j})$.

The considered objective is to minimize the makespan or the completion time of the last operation of the last job as defined in the following equation:

$$C_{\max} = \max_j \{tf_{n_j,j}\} \tag{1}$$

		M_1	M_2	M_3	M_4
J_1	$O_{1,1}$	*	3	*	*
	$O_{2,1}$	*	*	4	*
	$O_{3,1}$	*	6	*	*
J_2	$O_{1,2}$	*	*	*	4
	$O_{2,2}$	2	*	*	*
	$O_{3,2}$	*	*	1	*
J_2	$O_{1,3}$	*	*	*	4
	$O_{2,3}$	1	*	*	*

Figure 2: Example of a P_{JSP}

4.1 *Lower Bounds*

In this section, we present two lower bounds for the job shop which make it possible to evaluate the solution quality.

Proposition 2 *If C^*_{max} is the makespan of the optimal solution, then:*

$$C^*_{max} \geq \max_{1 \leq j \leq N} \left(r_j + \sum_{1 \leq i \leq n_j} p_{i,j} \right) \tag{2}$$

Proof. Obvious, according to the precedence constraints on the jobs.

Definition 3 *To each operation $O_{i,j}$, we associate an earliest starting time $r_{i,j}$ calculated by the following formula:*

$$\begin{cases} r_{1,j} = r_j \ \forall\ 1 \leq j \leq N \ , r_{i+1,j} = r_{i,j} + p_{i,j} \\ \quad \forall\ 1 \leq i \leq n_j - 1, \ \forall\ 1 \leq j \leq N. \end{cases} \tag{3}$$

Definition 4 *To each machine M_k, we associate a one-machine problem (Π_k). In such a problem, we have to execute the operations of E_k by respecting the earliest starting time $r_{i,j}$ calculated using 3. The others conjunctive constraints are relaxed. We denote by $C_{FIFO}(k)$ the completion time of the last operation of E_k on M_k in the schedule Ω_{FIFO} with Ω_{FIFO} the schedule obtained by using the FIFO rule.*

Proposition 5 *If C^*_{max} is the makespan of the optimal solution, then:*

$$C^*_{max} \geq \max_{1 \leq k \leq M} \left(C_{FIFO}(k) \right) \tag{4}$$

Proof. To each machine M_k, solving the problem Π_k (using FIFO rule) allows us to obtain a lower bound for the completion time of the last operation of E_k on M_k. Therefore, the maximum value of such a variable represents a lower bound for the makespan.

Proposition 6 *If C^*_{max} is the makespan of the optimal solution, then:*

$$C^*_{max} \geq \max(lb_1, lb_2) \tag{5}$$

such that:
$lb_1 = \max_{1 \leq j \leq N} \left(r_j + \sum_{1 \leq i \leq n_j} p_{i,j} \right)$,
$lb_2 = \max_{1 \leq k \leq M} \left\{ C_{FIFO}(k) \right\}$.

Proof. Obvious according to the two preceding propositions.

4.2 *Genetic Algorithm for JSP*

Real-world optimization problems are generally very hard to solve because of the constraints complexity and the large search space (Sarker, and al., 2001). It is the case for the scheduling problems. In particular, the Job shop Scheduling Problem (JSP) is considered one of the most difficult problems in combinatorial optimization (Carlier and Chretienne, 1988; Garey, and al., 1979). In the most part of cases, the combination of objectives and the constraints (precedence and disjunctive constraints) can drastically increase the complexity. Exact methods such as linear programming, dynamic programming and branch and bound need an important computing time to converge. In this context, meta-heuristic approaches can be applied to construct a good approximate solution (Davis, 1990; Fogel, 1994). In particular, the literature shows that genetic algorithms could be successfully used for scheduling problems (Dasgupta and Michalewicz, 1997; Caux, and al. 1995).

4.2.1 *Initial Solutions*

The problem considered presents many difficulties. The first one is the respect of the different constraints. The second difficulty is the calculation of the starting times $t_{i,j}$ of each operation $O_{i,j}$. To solve such a problem, we apply initially some heuristic rules (SPT, LPT, FIFO,...) to obtain acceptable solutions and to solve the different conflicts (Bel and Cavaillé, 2001). This heuristic makes it possible to calculate the starting times $t_{i,j}$ by taking into account the availabilities of the machines and the precedence constraints. Thus, we obtain a set of schedules according to the used priority rules. Such a set will be completed by some random solutions. To such a set, we apply a genetic approach that will be described in the next section.

4.2.2 *Genetic Algorithm*

The priority rules make it possible to generate a set of feasible solutions. To such a set, we apply an efficient genetic algorithm. Here, we describe such a method: coding, operators and parameters.

Coding: Tasks Sequencing List: (T.S.L) We choose to use a simple linear encoding. It consists in representing the schedule in a list of NT operations ($NT = \sum_{j=1}^{j=N} n_j$). Each chromosome will represent the tasks sequencing in the form of a NT-cells list. Each cell represents a task T_z ($1 \leq z \leq NT$) and it is coded in the following way: $(i, j, a(i,j))$. The computation of the starting times and the completion times $(t_{i,j}, tf_{i,j})$ is obtained by applying the procedure described in Fig.4 according to the order z of each task in the list (see in Fig.3). Such procedure is based on the sequencing of the tasks (i, j) on the machines M_k

T_1	T_2	...	T_z		T_z
$(1,2,3)$	$(2,2,3)$		(i,j,k)		$(3,2,4)$

Figure 3: Coding

$$
\begin{array}{l}
DM[k] = 0 \;\; for \;\; k = 1..M \\
DJ[j] = r_j \;\; for \;\; j = 1..N \\
\\
FOR \;\; z \in 1...NT \\
if(k = a(i,j)) \;\; t_{ij} = max(DM[k], DJ[j]), \\
tf_{ij} = t_{ij} + p_{ij} \\
Update \; DM : \;\; DM[k] = tf_{ij} \\
update \; DJ : \;\; DJ[k] = tf_{ij} \\
END \; FOR
\end{array}
$$

Figure 4: Scheduling procedure

$(k = a(i,j))$ according to the availability date $DM[k]$ of M_k and the availability date $DJ[j]$ of J_j.

Operators In our case, we have to respect the precedence constraints between the operations of each job. Therefore, we can avoid a correction process which will be very expensive in the term of computation time.

In this way, we have proposed to use some operators developed by Lee and Yamakawa to respect precedence constraints for some sequencing problems (Lee and Yamakawa, 1998).

Parameters

- The mutation and crossover probabilities are fixed in a traditional way:
  - $P_{crossover} = 0.95$,
  - $P_{mutation} = 0.05$.
- The population size is fixed to 100 for the two examples presented in the next section.
- The used selection mechanism gives the priority to the reproduction of the best individuals according to their evaluation defined by the equation (1).
- The criteria of stop are the following:
  - the maximum number of the iterations is reached,
  - the threshold of satisfaction is reached.

5. COMPUTATIONAL EXPERIMENTS

To evaluate the efficiency of our approach we have chosen to apply it on two study cases which have been already processed in the literature. The first one concerns a job shop with partial flexibility. The second one, however, concerns a job shop with total flexibility.

		M_1	M_2	M_3	M_4	M_5	M_6	M_7	M_8
J_1	$O_{1,1}$	5	3	5	3	3	*	10	9
	O_{21}	10	*	5	8	3	9	9	6
	$O_{3,1}$	*	10	*	5	6	2	4	5
J_2	$O_{1,2}$	5	7	3	9	8	*	9	*
	O_{22}	*	8	5	2	6	7	10	9
	$O_{3,2}$	*	10	*	5	6	4	1	7
	$O_{4,2}$	10	8	9	6	4	7	*	*
J_3	$O_{1,3}$	10	*	*	7	6	5	2	4
	O_{23}	*	10	6	4	8	9	10	*
	$O_{3,3}$	1	4	5	6	*	10	*	7
J_4	$O_{1,4}$	3	1	6	5	9	7	8	4
	O_{24}	12	11	7	8	10	5	6	9
	$O_{3,4}$	4	6	2	10	3	9	5	7
J_5	$O_{1,5}$	3	6	7	8	9	*	10	*
	$O_{2,5}$	10	*	7	4	9	8	6	*
	$O_{3,5}$	*	9	8	7	4	2	7	*
	$O_{4,5}$	11	9	*	6	7	5	3	6
J_6	$O_{1,6}$	6	7	1	4	6	9	*	10
	$O_{2,6}$	11	*	9	9	9	7	6	4
	$O_{3,6}$	10	5	9	10	11	*	10	*
J_7	$O_{1,7}$	5	4	2	6	7	*	10	*
	$O_{2,7}$	*	9	*	9	11	9	10	5
	$O_{3,7}$	*	8	9	3	8	6	*	10
J_8	$O_{1,8}$	2	8	5	9	*	4	*	10
	$O_{2,8}$	7	4	7	8	9	*	10	*
	$O_{3,8}$	9	9	*	8	5	6	7	1
	$O_{4,8}$	9	*	3	7	1	5	8	*

Figure 5: Case of partial flexibility

5.1 *First Example:Case of job shop with partial flexibility (8 jobs,8 machines, 27 operations)*

This example is introduced in the table of Fig. 5. It has already been studied using temporal decomposition (Chetouane, 1995), classic genetic algorithm (Mesghouni, 1999) and controlled evolutionary approach (Kacem, 2003) which is a multi-objective optimization of FJSP based on evolutionary algorithms. These approaches deal with the routing and scheduling subproblems simultaneously.

5.1.1 *First step: assignment*

We apply Kacem's heuristic to have an initial solution and then the Tabu search procedure, we obtain the following results:

Initial solution (Kacem's heuristic)

Machines workloads (W_k): $\{W_1 = 10, W_2 = 13, W_3 = 12, W_4 = 9, W_5 = 8, W_6 = 9, W_7 = 6, W_8 = 10\}$.

The workload of the most loaded machine= $Max(W_k) = 13$.

If we estimate the makespan for this assignment solution using the lower bound given in section 4 $C_{\max} \geq 13$.

Solution given after Tabu search:

Machines workloads (W_k) : $\{W_1 = 10, W_2 = 10, W_3 = 12, W_4 = 12, W_5 = 8, W_6 = 9, W_7 = 6, W_8 = 10\}$.

The workload of the most loaded machine= $Max(W_k) = 12$.

If we estimate the makespan for this assignment solution using the lower bound given in section 4 $C_{\max} \geq 12$.

In this example, the Tabu search allow us to reduce the workload of the most loaded machine as well as the lower bound of the makespan. Thus solution given by Tabu search assume that the makespan should be enhanced.

		a_{ij}	p_{ijk}	t_{ij},tf_{ij}
J_1	$O_{1,1}$	4	3	0,3
	O_{21}	5	3	3,6
	$O_{3,1}$	6	2	6,8
J_2	$O_{1,2}$	3	3	1,4
	O_{22}	4	2	7,9
	$O_{3,2}$	7	1	9,10
	$O_{4,2}$	5	4	10,14
J_3	$O_{1,3}$	7	2	0,2
	O_{23}	3	6	6,12
	$O_{3,3}$	1	1	12,13
J_4	$O_{1,4}$	2	1	0,1
	O_{24}	6	5	1,6
	$O_{3,4}$	1	4	6,10
J_5	$O_{1,5}$	1	3	0,3
	$O_{2,5}$	4	4	3,7
	$O_{3,5}$	6	2	8,10
	$O_{4,5}$	7	3	10,13
J_6	$O_{1,6}$	3	1	0,1
	$O_{2,6}$	8	4	1,5
	$O_{3,6}$	2	5	9,14
J_7	$O_{1,7}$	3	2	4,6
	$O_{2,7}$	8	5	6,11
	$O_{3,7}$	4	3	11,14
J_8	$O_{1,8}$	1	2	3,5
	$O_{2,8}$	2	4	5,9
	$O_{3,8}$	8	1	11,12
	$O_{4,8}$	5	1	14,15

Figure 6: Solution for the first instance

	Cr_2	C_{max}
Temporal Decomposition	19	19
Classic GA	11	16
Controlled EA	13	14
	11	16
our method	12	15

Figure 7: Comparison for the first instance

5.1.2 *Second step: Sequencing*

If we apply our genetic algorithm, we obtain the scheduling given by the table of Fig. 6.

In comparison with the other approaches, Fig. 7 shows that our approach enables us to construct as interesting solutions as those obtained by the controlled evolutionary approach. The main advantage of this method is a reduction of the computation time (0.2 s for this example).

5.2 *Second Example:Case of job shop with total flexibility (10 jobs,10 machines,30 operations)*

This example is introduced in the table of Fig. 8. It has already been studied using the above approaches (Chetouane, 1995; Mesghouni, 1999; Kacem, 2003).

5.2.1 *First step: assignment*

We apply Kacem's heuristic to have an initial solution and then the Tabu search, we obtain the following results:

Initial solution (Kacem's heuristic)

Machines workloads (W_k) : $\{W_1 = 4, W_2 = 6, W_3 = 5, W_4 = 5, W_5 = 5, W_6 = 5, W_7 = 5, W_8 = 3, W_9 = 4, W_{10} = 4\}$.

The workload of the most loaded machine= $Max(W_k) = 6$.

If we estimate the makespan for this assignment solution using the lower bound given in section 4 $C_{\max} \geq 8$.

Solution given after tabu search

		M_1	M_2	M_3	M_4	M_5	M_6	M_7	M_8	M_9	M_{10}
J_1	$O_{1,1}$	1	4	6	9	3	5	2	8	9	5
	$O_{2,1}$	4	1	1	3	4	8	10	4	11	4
	$O_{3,1}$	3	2	5	1	5	6	9	5	10	3
J_2	$O_{1,2}$	2	10	4	5	9	8	4	15	8	4
	$O_{2,2}$	4	8	7	1	9	6	1	10	7	1
	$O_{3,2}$	6	11	2	7	5	3	5	14	9	2
J_3	$O_{1,3}$	8	5	8	9	4	3	5	3	8	1
	$O_{2,3}$	9	3	6	1	2	6	4	1	7	2
	$O_{3,3}$	7	1	8	5	4	9	1	2	3	4
J_4	$O_{1,4}$	5	10	6	4	9	5	1	7	1	6
	$O_{2,4}$	4	2	3	8	7	4	6	9	8	4
	$O_{3,4}$	4	3	12	1	6	5	8	3	5	2
J_5	$O_{1,5}$	7	10	4	5	6	3	5	15	2	6
	$O_{2,5}$	5	6	3	9	8	2	8	6	1	7
	$O_{3,5}$	6	1	4	1	10	4	3	11	13	9
J_6	$O_{1,6}$	8	9	10	8	4	2	7	8	3	10
	$O_{2,6}$	7	3	12	5	4	3	6	9	2	15
	$O_{3,6}$	4	7	3	6	3	4	1	5	1	11
J_7	$O_{1,7}$	1	7	8	3	4	9	4	13	10	7
	$O_{2,7}$	3	8	1	2	3	6	11	2	13	3
	$O_{3,7}$	5	4	2	1	2	1	8	14	5	7
J_8	$O_{1,8}$	5	7	11	3	2	9	8	5	12	8
	$O_{2,8}$	8	3	10	7	5	13	4	6	8	4
	$O_{3,8}$	6	2	13	5	4	3	5	7	9	5
J_9	$O_{1,9}$	3	9	1	3	8	1	6	7	5	4
	$O_{2,9}$	4	6	2	5	7	3	1	9	6	7
	$O_{3,9}$	8	5	4	8	6	1	2	3	10	12
J_{10}	$O_{1,10}$	4	3	1	6	7	1	2	6	20	6
	$O_{2,10}$	3	1	8	1	9	4	1	4	17	15
	$O_{3,10}$	9	2	4	2	3	5	2	4	10	23

Figure 8: Case of total flexibility

		a_{ij}	p_{ijk}	t_{ij},tf_{ij}
J_1	$O_{1,1}$	1	1	3,4
	O_{21}	3	1	4,5
	$O_{3,1}$	4	1	6,7
J_2	$O_{1,2}$	1	2	0,2
	$O_{2,2}$	4	1	2,3
	$O_{3,2}$	3	2	5,7
J_3	$O_{1,3}$	10	1	0,1
	$O_{2,3}$	8	1	1,2
	$O_{3,3}$	8	2	2,4
J_4	$O_{1,4}$	7	1	0,1
	$O_{2,4}$	10	4	1,5
	$O_{3,4}$	4	1	5,6
J_5	$O_{1,5}$	9	2	0,2
	$O_{2,5}$	9	1	2,3
	$O_{3,5}$	4	1	3,4
J_6	$O_{1,6}$	6	2	0,2
	$O_{2,6}$	9	2	3,5
	$O_{3,6}$	7	1	5,6
J_7	$O_{1,7}$	1	1	2,3
	$O_{2,7}$	3	1	3,4
	$O_{3,7}$	6	1	5,6
J_8	$O_{1,8}$	5	2	0,2
	$O_{2,8}$	2	3	2,5
	$O_{3,8}$	2	2	5,7
J_9	$O_{1,9}$	6	1	2,3
	$O_{2,9}$	7	1	3,4
	$O_{3,9}$	6	1	4,5
J_{10}	$O_{1,10}$	3	1	0,1
	$O_{2,10}$	4	1	1,2
	$O_{3,10}$	5	3	2,5

Figure 9: Solution for the second instance

Machines workloads (W_k) : $\{W_1 = 4, W_2 = 5, W_3 = 5, W_4 = 5, W_5 = 5, W_6 = 5, W_7 = 3, W_8 = 3, W_9 = 5, W_{10} = 5\}$.

The workload of the most loaded machine= $Max(W_k) = 5$.

If we estimate the makespan for this assignment solution using the lower bound given in section IV $C_{\max} \geq 7$.

In this example also , the Tabu search allow us to reduce the workload of the most loaded machine as well as the lower bound of the makespan. Thus solution given by Tabu search assume that the makespan can be enhanced.

5.2.2 *Second step: Sequencing*

If we apply our genetic algorithm, we obtain the scheduling given by the table of Fig 9

Like in the first example, our approach allows us to obtain an interesting solution as obtained by the con-

	Cr_2	C_{max}
Temporal Decomposition	16	16
Classic GA	7	7
Controlled EA	5	7
our method	5	7

Figure 10: Comparison for the second instance

trolled approach with a very short processing time. Fig. 10 summarizes the different results given by each method.

6. CONCLUSION

In this paper, we have proposed a two-phase algorithm to make it easier the optimization of the flexible job shop scheduling problem. Such a method makes it possible to construct a good solution for the assignment subproblem using different heuristic approaches (localization method and Tabu search) and then a genetic algorithm is applied to optimize the sequencing property of the potential solutions. It takes into account all the constraints and the specificities of the problem. Many adapted operators are used. These operators respect the two components of the problem: the precedence constraints and the disjunctive constraints. Preliminary simulations show that small difference between the lower bound and the values of the criterion obtained for the solutions generated by the genetic algorithm is generally satisfactory and promising. This result demonstrates the good quality and the adequation of the proposed method. A future perspective is to study the proposed method on a large set of benchmark data.

REFERENCES

Banzhaf, W., P.Nordin, ,R.E. Keller and F.D. Francone(1998).
Genetic programming An introduction on the Automatic evolution of computer programs and its application Morgan Kaufmann publishers, Inc. San Francisco, California, USA.G.

Bel, G. and J.B. Cavaillé(2001).
Approche simulatoire . In: *Ordonnancement de la production*, chapitre 6 (P. Lopez et de F.Roubellat. (Ed)), Hermès, France.

Brandimarte, P. (1993).
Routing and scheduling in Flexible job shops by tabu search. *Annals of Operations Research*, Vol **41**, 157-183.

Carlier, J., P. Chretienne (1988).
Problèmes d'Ordonnancement: Modélisation / Complexité / Algorithmes. Editions Masson, France.

Carlier, J. (1989).
An algorithm for solving the job shop problem. *Management science*, Vol. **35**, 164-176

Caux, C., H. Pierreval, MC. Portmann (1995).
Les algorithmes génétiques et leur application aux problèmes d'ordonnancement, *Kevie A.P.I.I* (Automatique Productique et Informatique Industrielle), **29**, 4-5, 409-443.

Chetouane, F. (1995).
Ordonnancement d'atelier à tâches généralisées, perturbations, réactivité, DEA Rep., Polytech.Nat.Inst. Grenoble, Grenoble, France.

Dasgupta, D. and Z. Michalewicz (1997).
Evolutionary Algorithms in Engineering Applications. Springer-Verlag, Berlin, GERMANY.

Dauzère-Pérès,S. and J. Paulli (1997).
An integrated approach for modelling and solving the general multiprocessor job-shop scheduling problem using tabu search. *Annals of Oper. Res.*, Vol. **70**, 281-306.

Davis, L. (1990).
Handbook of Genetic Algorithms, Van Nostrand Reinhold, New-York, 1990, USA.

Fogel, D.B. (1994).
An Introduction to Simulated Evolutionary Optimization. *IEEE Transactions on Neural Networks*, Vol.**5**, 1, 3-14.

Garey, M.R. and S. Johnson (1979).
Computers and intractability: A guide to theory of NP-Completeness. W.H. Freeman and Co. , New York.

Kacem, I. and P.Borne (2003).
Hybrid method for evaluating schedule performance in flexible job-shops. In: *Intelligent Sensory Evaluation–Methodologies and Applications*, Springer Verlag.

Kacem, I. (2003).
Ordonnancement multicritère des job shops flexibles : formulation, bornes inférieures et approche évolutionniste coopérative. Ph.D Thesis, Ecole Centrale de Lille, Jan. 6th , 2003.

Lee, KM. ,T. Yamakawa and KM. Lee (1998).
A Genetic algorithm for general machine scheduling problems .*Journal of knowledge-Based Electronics Systems*,Vol **2** ,60-66, Australia.

Mastrolilli,M. and L.M. Gambardella (2000).
Effective neighborhood functions for the flexible job shop problem. *Journal of Scheduling*, Vol **3**, issue 1,3-20.

Mesghouni,K. (1999).
Application des algorithmes évolutionnistes dans les problèmes d'optimisatin en ordonnancement de production, Ph.D. dissertation, USTL 2451,France, 1999.

Sarker R., H.A. Abbas, C. Newton (2001).
Solving multi-objective optimization problems using evolutionary algorithm. Proceedings of International CIMCA Conference, July 9-11, 2001, Las Vegas, Nevada, USA.

Copyright © IFAC Information Control Problems in
Manufacturing, Salvador, Brazil, 2004

ELSEVIER
IFAC
PUBLICATIONS
www.elsevier.com/locate/ifac

CENTRALIZED ARCHITECTURE FOR REAL TIME SCHEDULING OF BATCH SYSTEMS

Michel dos Santos Soares *, **Stéphane Julia** *

* *Faculdade de Ciência da Computação,*
Universidade Federal de Uberlândia,
P.O. Box 593, 38400-902, Uberlândia-M.G-Brazil,
email: michelssoares@yahoo.com.br, stephjl@aol.com

Abstract: The objective of this article is to present a centralized architecture based on a collaboration diagram and on a p-time Petri net model for real time scheduling of production systems. The main idea consists of proposing a software architecture which depends on an object whose purpose is to centralize all the interactions between the different software objects. The internal behaviour of the central object is represented by a p-time Petri net model which shows the global behaviour of the entire system. A specialized inference mechanism called token player is then applied to the p-time Petri net model for the real time scheduling. In particular, for the Petri net fragments involved in conflict situations, symbolic dates assigned to tokens are calculated using a non-conventional (max;+) algebra based on the sequent calculus of Linear Logic. These dates are used to solve conflict situations off-line on an implicit manner in such a way that the resource conflicts can be treated in real time at the central object level. The approach is illustrated through an example of Real Time System used at the global coordination level of a Batch System. *Copyright © 2004 IFAC*

Keywords: Object modelling, Petri-nets, Scheduling, Real Time Systems, Batch Systems

1. INTRODUCTION

Nowadays, the Object Oriented methods seem to be very suitable for proposing approaches to represent Real Time Systems. One of the advantages of using an Object Oriented approach for Real Time Systems is that the real objects of the physical system are naturally transformed into software objects that can be easily implemented and verified. Because it will be very difficult for the destruction of a real object of the physical system to occur, the software objects will persist: once implemented, they will become permanent software components. Among all Object Oriented notations, UML (OMG 1999) is one of the best accepted in the software industry. It offers a good structure for the static part of objects and provides several types of dynamic diagrams for control specification. In particular, with the dynamic diagrams proposed by UML, it is possible to represent the communication mechanisms among several objects for a specific scenario. Therefore, UML notation has its limitations when it is used for specifying Real Time Systems. Actually, it is extremely difficult to show the set of all dynamic interactions of a complex system using a single UML diagram. As a consequence, it will be almost impossible to verify that the simultaneous execution of several scenarios will not lead to a deadlock situation. Another limitation is that when the UML notation is used for real time specification, it supports basic time-outs, but it does not provide formal notations for specifying the time constraints like date intervals or time-limited offerings for example.

Petri Nets (Murata 1989) are very suitable to model Real Time Systems, as they allow for a good representation of conflict situations, shared resources, synchronous and asynchronous communications, precedence constraints and explicit time constraints in the time Petri Net case. Using the tools the Petri Net theory provides, it is possible to verify some good properties of Real Time Systems, like the liveness property for example, and simulate time Petri Net models in order to evaluate the performance of a system. On the other hand, the Petri Net theory is not Object Oriented. Even in the Object Petri Net case (Sibertin-Blanc 1994), the class concept does not exist.

The dynamic behaviour of a system imposes a scheduling of control flow. The scheduling problem (Lee 1994) consists of organizing in time, the sequence of operations considering time constraints (time intervals) and constraints of shared resources utilization necessary for the operation execution. From the traditional point of view of Software Engineering, the scheduling problem is similar to the activity of scenario execution. A scenario execution becomes a kind of simulation which shows the system's behaviour in real time. In the real time system case, several scenarios can be executed simultaneously and conflict situations which have to be solved in real time (without a backtrack mechanism) can occur if a same non-preemptive resource is called at the same time for the execution of operations which belong to different scenarios.

The main idea of this paper is to propose an object oriented software architecture which depends on an object which centralizes all the interactions between the different software objects. The internal behaviour of such an object will be given by a p-time Petri net model and an approach, based on the sequent calculus of Linear Logic (Girard 1987) and on a non-conventional (max;+) algebra (Rivière 2001) will be applied to solve conflict situations during the real time scheduling. In the following sections, the proposed approach will be applied to an example of a Real Time System used at the global coordination level of a Batch System.

2. CENTRALIZED ARCHITECTURE BASED ON A UML/PETRI NET MODEL

Batch Systems (Julia 2000) operate on continuous raw materials (generally fluids) and are composed of continuous flow equipment (a continuous flow of material goes through them, as in thermal exchanges for example) and also discrete equipment

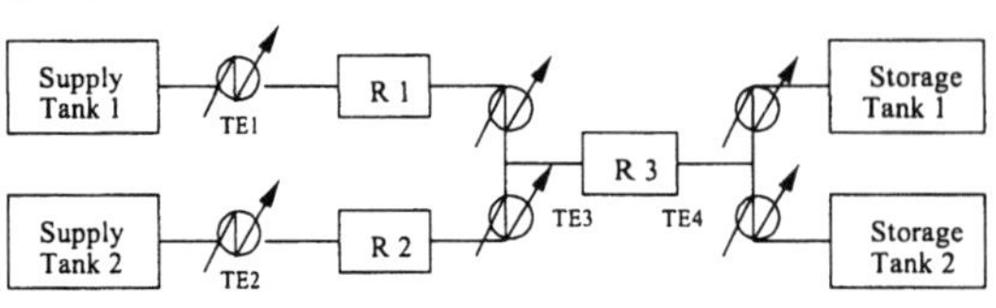

Fig. 1. Batch System

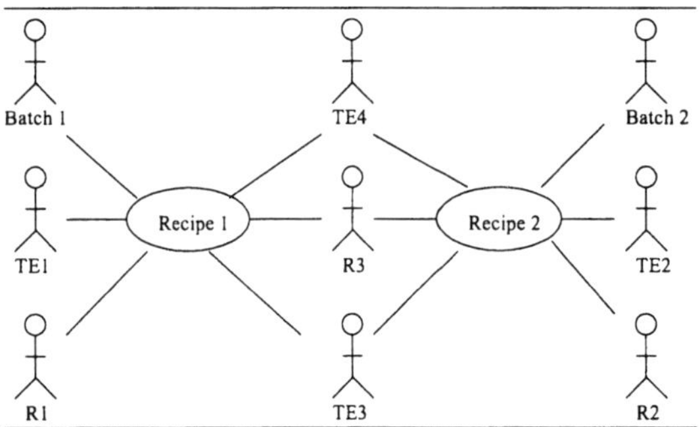

Fig. 2. Use Case Diagram

(reactors, intermediary buffers etc.). A batch is a quantity of material which is transformed passing through different equipment and respecting a specific recipe which defines the sequence of operations. In this kind of system, there exists resource allocation mechanisms. For example, it may be necessary for the same reactor to treat different batches. If the reactor works in a closed mode (a single batch, at the same time, in a single reactor), it can be seen as a discrete resource. This means that the resources are non-preemptive disjunctive ones.

The Batch System of figure 1 executes two different recipes. Initially, two batches (1 and 2) are stored in their respective supply tanks (1 and 2). Recipe 1 consists of transferring batch 1 from supply tank 1 to reactor R1 to a processing stage utilizing thermal exchange TE1. After batch 1 is processed in R1, it is transferred to reactor R3 to another stage in processing, passing through TE3. When finalized, batch 1 is deposited in storage tank 1, passing through TE4 for the final product liberation. Recipe 2 behaves in a similar manner and consists of transferring batch 2 from supply tank 2 to storage tank 2, passing through reactors R2 and R3 and using thermal exchanges TE2, TE3 and TE4. The advantage of considering such a system is that some of the main features which generally appear in Real Time Systems will be considered. The execution of different recipes can be seen as a typical example of parallelism, and the utilization of common equipment (for example, reactor R3 will be used by both recipes) is an example of resource sharing.

A Use Case diagram is used to show the main functions of the Real Time System from a user's point of view and the relationship between the system and the environment. From the Use Case

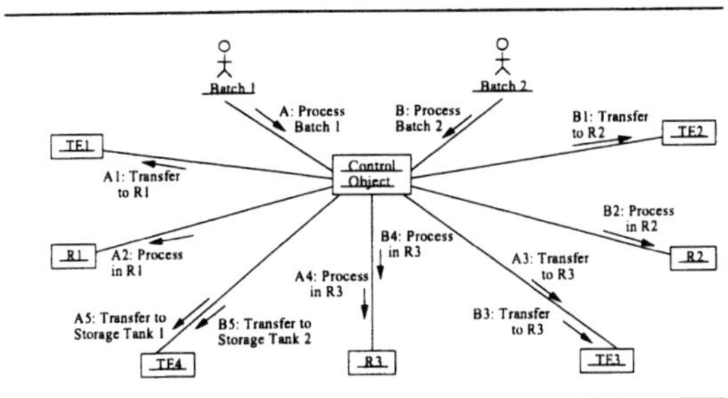

Fig. 3. Collaboration Diagram

diagram of figure 2, it is possible to note that the execution of recipe 1 needs the physical equipment TE1, TE3, TE4, R1 and R3. In the same manner, the execution of recipe 2 needs the physical equipment TE2, TE3, TE4, R2 and R3. As actors are good candidates for objects, it is natural to conclude that each actor of the Use Case diagram will have to communicate with its corresponding software object.

In (Medeiros 1998), different kinds of software architectures were presented in the Real Time System's case. In particular, it was clearly emphasized that centralized architectures were more flexible than hierarchical ones. They are based on an object which contains the global behaviour of the entire system and which centralizes all interactions between the different software objects. Figure 3 represents the collaboration diagram that shows the centralized architecture of the Real Time System used at the global coordination level of the Batch System of figure 1. For example, the scenario corresponding to recipe 1 is iniciated by the actor Batch 1 which sends the synchronous message A. Then, the central object, which can be seen as an active object (the one which makes decisions), sends, as soon as possible, the synchronous messages A1, A2, A3, A4 and A5 in sequence to the objects required to execute the scenario corresponding to recipe 1. Each recipe (scenario) is represented on the collaboration diagram as an independent control flow (A for recipe 1 and B for recipe 2) which will allow maximum flexibility during the real time execution, maintaining the natural parallelism of the system whenever possible. All the callings will be realized through synchronous messages, which will be of great advantage for the implementation. Another advantage of such an architecture is that once the software objects corresponding to physical equipment are implemented, they remain in the software architecture even when new scenarios are introduced (new recipes). It will be only at the central object level that the new scenarios will be implemented.

The internal behaviour of the central object can be given by a p-time Petri net model. Figure 4 shows the p-time Petri net model which corre-

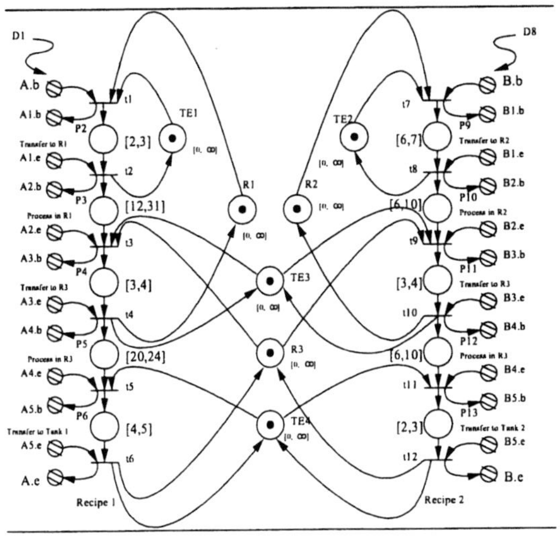

Fig. 4. p-time Petri net

sponds to the internal behaviour of the central object (recipe 1 + recipe 2). Each recipe is represented by a production route (a sequence of transitions and places) and each object utilization is represented by a resource (a shared resource if the object appears in both recipes, like R3 for example). The synchronous communications between the central object and the physical equipment objects are realized through communication places (semaphore type). For example, the place A1.b represents the synchronous calling A1 on the collaboration diagram, and the place A1.e represents the synchronous answer of object TE1 at the end of the transfer operation from supply tank 1 to reactor R1. The static definition of a p-time Petri net (Khansa 1996) is based on static intervals which represent the permanency duration (sojourn time) of tokens in places and the dynamic evolution of a p-time Petri net depends on the time situation of the tokens (visibility intervals associated with the tokens). The static intervals associated to each place of the p-time Petri net in figure 4 represent the duration on each object. For example, the treatment duration on object R1 (processing of batch 1 in reactor R1) will have to be superior to 12 time units and inferior to 31 time units in order to respect the explicit time constraints of the system.

The limitation of centralized architectures has to do with the problem of the overloading of information treatment on the object which centralizes the operations. As a matter of fact, all the resource conflicts will have to be carried out in real time and at the same object level. A solution is then to find a kind of implicit off-line treatment of conflict situations in such a way that real time resource allocations will be realized without additional calculus.

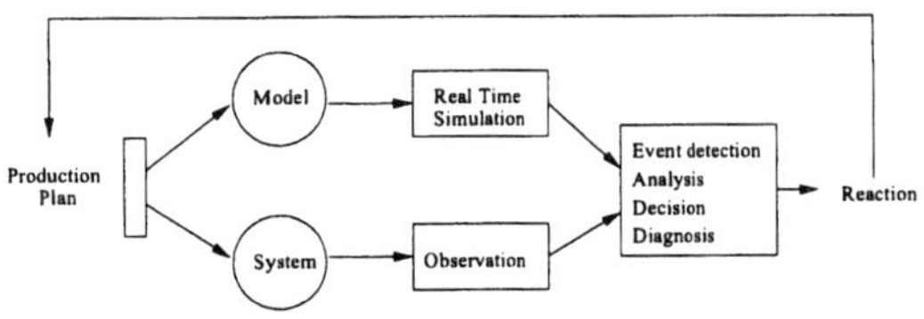

Fig. 5. Global Coordination of production systems founded on a model

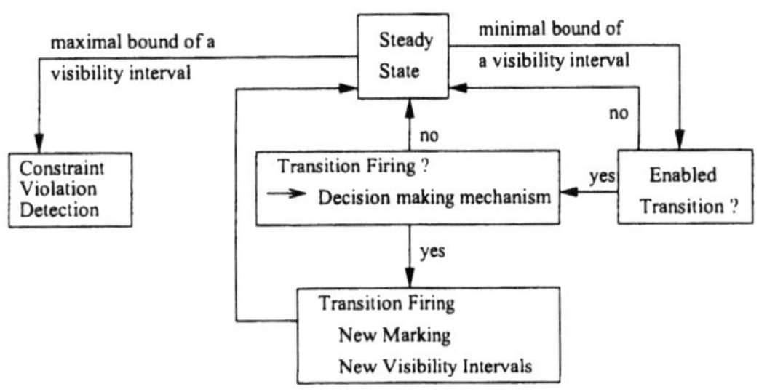

Fig. 6. Token Player for p-time Petri net

3. REAL TIME SCHEDULING PRINCIPLE

3.1 Token Player Algorithm

In the global coordination and real time monitoring of production systems, each time the state of the system changes, a detection module verifies if the state of the simulation model is compatible with the actual state of the system and brings up to date the state of the simulation model. A decision module indicates then which operation has to be initiated as shown in figure 5. When the model of the system is a p-time Petri net, then, the real time scheduling which is carried out in parallel with the real operations of the system, can be executed by a special inference mechanism which allows the firing of certain enabled transitions. Figure 6 shows the working of such an algorithm called token player. Each time a token of the p-time Petri net becomes available (which means the minimal bound of its visibility interval is reached), the token player verifies if a transition is enabled. If it is the case and if the corresponding transition is not in structural conflict with other transitions, then, the transition is fired. If the transition is in structural conflict with others transitions, a conflict resolution mechanism is applied to verify if it can be fired. If the conflict resolution mechanism allows one to fire the corresponding transition as soon as the token becomes available, then, the transition is fired. On the contrary, the enabled transition is not fired. If the maximum bound of a visibility interval is reached (death of a token), then, a constraint violation is detected which could have been caused by bad management of the available resources (wrong decision) or by a physical problem at the system level (the synchronous answer from an equipment object is not received before the maximum bound of the visibility interval associated to the corresponding operation at the simulation model level).

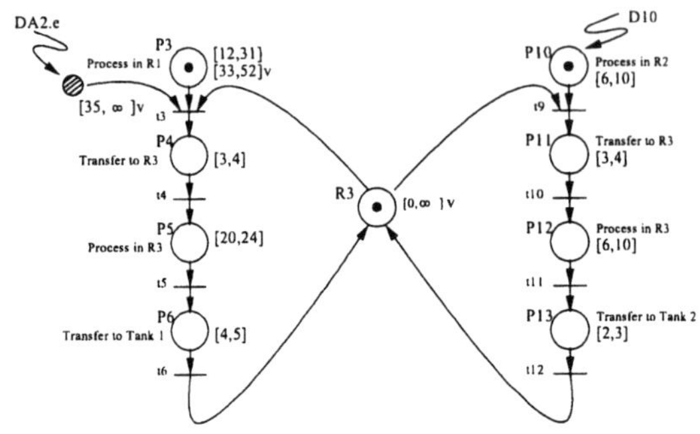

Fig. 7. Useful part of the conflict for R3

The basic difference between the token player proposed in this article and other token player algorithms used in some simulation tools based on Petri nets is that transitions are not fired necessarily as soon as they become available because of the conflict resolution mechanism. In particular, this mechanism must work on a short temporal projection that permits one to emphasize possible future time constraint violations. The main problem when working with a centralized architecture is then to be able to solve the resource conflicts without overloading the central object which contains the real time simulation model.

3.2 Conflict Resolution Mechanism

With a p-time Petri net, conflicts for shared resources are visible during a time interval and not only at a single time point. The definitions (enabling interval of a transition and conflict time interval) necessary to understand the conflict notion of a p-time Petri net were given in (Julia 2000). The p-time Petri net of figure 7 which represents the Petri net fragment involved in the conflict for the shared resource R3 is considered to illustrate the conflict notion of a p-time Petri net. In this figure, transitions t3 and t9 are in structural conflict. The static intervals associated to their input places are [12;31] for P3, [6;10] for P10, $[0;\infty[$ for R3 and $[0;\infty[$ for the communication place $A_{2,e}$. If the resource R3 has not been used yet, its visibility interval is $[0;\infty[$. It is supposed that the token in P3 is produced at date 21 and the token in $A_{2,e}$ at date 35. Consequently, the corresponding visibility intervals are [12+21;31+21]=[33;52] for the token in P3 and $[0+35;\infty+35[=[35;\infty[$ for the token in $A_{2,e}$. If the earliest date for producing a token in P10 when considering the production route of recipe 2 in figure 4 is D10=36, then,

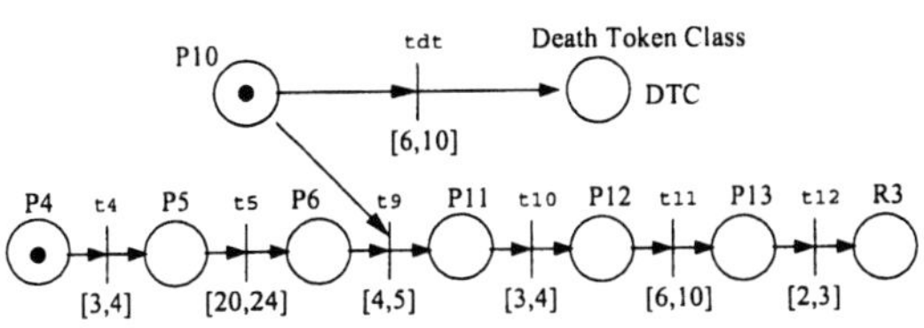

Fig. 8. Watch dog

its visibility interval is [6+36;10+36]=[42;46]. The enabling interval is given by the intersection of the visibility intervals and is $[35;\infty[\cap[0;\infty[\cap[33;52] = [35;52]$ for t3 and $[42;46] \cap [0;\infty[= [42;46]$ for t9. The conflict time interval associated to the pair (t3;t9) is given by the intersection of the enabling intervals : $[35;52] \cap [42;46] = [42;46]$. So, an effective conflict between t3 and t9 is able to occur between date 42 and date 46.

It seems natural to fire t3 as soon as possible at date 35 since there is not a visible token in P10 at that time (the minimum bound of the visibility interval associated to the token in P10 is 42). Using a class graph algorithm (Khansa 1996), it is possible to verify if a special class, called "death token class", can be reached when considering the Petri net fragment in figure 7 and the firing of t3 at date 35. Such an approach was presented in (Julia 2000). A "death token class" in this case would mean that the resource R3 necessary to respect the time interval associated to the place P10, could not be available before the maximum bound of the visibility interval associated to the token of P10. But the class graph has some limitations. The state space can be very large and the duration a token can remain in a same place has to be delimited by real numerical values and not symbolic ones. Each time a new conflict state is reached during the real time execution of a system, new conflict time intervals are calculated and a different class graph has to be generated. As a consequence, the real time conflict resolutions can need too many calculus knowing that all the conflicts are treated at the same object level.

The concept of "death token class" which exists when conflict situations for shared resources are considered in p-time Petri nets, can be represented by a Watch Dog (two transitions in structural conflict) modeled by a t-time Petri net (time intervals associated to transitions) where a specific place represents the "death token class". Watch Dogs are commonly used to analyse behaviours which deviate from their normal evolution. As there exists an equivalence between a p-time Petri Net and a t-time Petri Net in case of the earliest firing strategy (Khansa 1996), it will be possible to transform a conflict state given by a p-time Petri Net model into a t-time Petri Net model corresponding to a typical case of Watch Dog where one of the transitions in structural conflict represents the normal evolution of the system and the other one signals a time constraint violation. For example, considering the conflict situation for the shared resource R3 shown in figure 7, if the decision of firing transition t3 at date 35 (earliest date when considering the enabling interval of t3) before transition t9 is made, then, the problem is to verify that the reactor will be available in place R3 before the maximum bound of the visibility interval associated to the token in P10, which represents the latest date authorized for the firing of transition t9. On the contrary, a "death token" class will be reached. The Watch Dog modelled by the t-time Petri Net in figure 8 represents such a situation. This Watch Dog specifies that, at date D10 + 10 (D10 is the date when the token appears in P10 and 10 is the maximum bound of the time interval associated to transition t_{dt}), if the transition t_9 has not been fired yet (which means that the resource in R3 is not available at this moment for the firing of t_9 in figure 7), then, the transition t_{dt} has to be fired and a death token class is reached. The place R3 does not appear in the Watch Dog after P6 because its static interval is $[0;\infty[$ and, as a consequence, the effect of this place on the temporal reasoning is useless.

In (Rivière 2001), it was shown that applying the sequent calculus of Linear Logic (Girard 1987) to a Watch Dog represented by a t-time Petri Net and using a non conventional (max;+) algebra, production and consumption symbolic dates assigned to each atom (token) involved in the studied scenario can be calculated. In particular, comparing the symbolic dates of the tokens which belong to the places in structural conflict, it is possible to know if the Watch Dog effectively invalidates the normal operations, i.e. if a "death token class" can be reached.

With Linear Logic, a marking M is a monomial in $\otimes$, that is a marking represented by $M = A_1 \otimes A_2 \otimes \cdots \otimes A_k$ where A_i are place names. For instance, the initial marking on the Petri Net in figure 8 is $P4 \otimes P10$. A transition is an expression of the form $M_1 \multimap M_2$ where M_1 and M_2 are markings. For example, transition t_4 on the Petri Net in figure 8 is noted $t_4 = P_4 \multimap P_5$. A sequent $M, t_i \vdash M'$ represents a scenario where M and M' are respectively the initial and final markings, and t_i is a list of non-ordered transitions. A sequent can be proved by applying the rules of the sequent calculus as there exists an equivalence between Petri Nets reachability and the proof of sequents in Linear Logic (Girault 1997). A Linear Logic proof tree is read from the bottom, up and a proof stops when all the leaves of the tree are identity sequents ($P_1 \vdash P_1$, for example). In a proof tree, each transition firing generates a symbolic date associated to each atom (token). In this article, D_i will denote a date and d_i a duration associated to a transition firing. A pair (D_p, D_c) will be associated to each atom of the proof tree; they respectively represent the production and the consumption date of atoms.

From the watch dog of figure 8, two scenarios can be derivated :

$$Scl = P4 \otimes P10, t_4, t_5, t_{dt} \vdash P6 \otimes DTC$$
$$Sc2 = P4 \otimes P10, t_4, t_5, t_9, t_{10}, t_{11}, t_{12} \vdash R3$$

These scenarios are in conflict i.e. one scenario invalidates the other. As a matter of fact, Sc2 represents the normal evolution of the system when Scl represents a time constraint violation (a "death token class").

In a t-time Petri Net model, any enabling duration d_i takes its values within a time interval $[\delta_{i,min}; \delta_{i,max}]$. For example, when considering the scenario Scl, the domain for the production date of atom P5 (the token in P5) is given by the time interval $[D_4 + d_{4,min}; D_4 + d_{4,max}]$. Based on a property shown in (Rivière 2001), it is possible to say that Scl invalidates Sc2 (which means that the death token class will be reached) if the maximum value of the enabling duration of t_{dt} in scenario Scl is smaller than the minimal sojourn time of atom P6 in scenario Sc2. For example, if the transition t3 is fired at date 35 in figure 7 when t3 becomes enabled, then, a token will appear in P4 at date 35 and the production date for atom P4 in figure 8 will be D4=35. From the symbolic production and consumption dates obtained when scenarios Scl and Sc2 are considered, the following results are obtained : the minimal production date of atom P6, which is equal to $D_4 + d_{4,min} + d_{5,min} = 35 + 3 + 20 = 58$, is bigger than the maximal consumption date of atom P10, which is equal to $D_{10} + d_{dt,max} = 36 + 10 = 46$. As a direct consequence of this result, the firing of t_3 in figure 7 as soon as transition t3 becomes firable will not be allowed by the conflict resolution mechanism and the token player will not allow the firing of t3 when it becomes available.

4. CONCLUSION

The principle advantage shown in this article is the possibility of accelerating the conflict resolutions in a p-time Petri net. This is realized by using the sequent calculus of Linear Logic and a non-conventional (max;+) algebra which allows the calculation of symbolic dates, instead of real numerical values. As a direct consequence, conflict situations are implicitely solved off-line and the generation of a class graph, each time a conflict situation for a shared resource is met during the real time scheduling, is not necessary anymore. In particular, such an approach allows to centralize the coordination of all the equipment objects considering a single object whose real time execution is realized through a specialized inference mechanism which can solve the conflict situations in real time.

REFERENCES

[Girard 1987] Girard, Jean-Yves. (1987). Linear Logic. *Theoretical Computer Science.* 50:1-102.

[Girault 1997] Girault, F. (1997). *A logic for Petri nets, JESA Vol. 31, n.3, Eddition Hermes*

[Julia 2000] Julia, S., Valette, R. (2000). Real Time Scheduling of Batch Systems. *Simulation Practice and Theory, Elsevier Science.* pp. 307-319.

[Khansa 1996] Khansa, W., Aygaline, P., Denat, J. P. (1996). Structural analysis of p-time Petri Nets. *Symposium on discrete events and manufacturing systems. CESA'96 IMACS Multiconference.* Lille, France.

[Lee 1994] Lee,D.Y.,DiCesare, F. (1994). Scheduling flexible manufacturing systems using Petri nets and heuristic search. *IEEE Transactions on Robotics and Automation.* 10(2). p. 123-132.

[Medeiros 1998] Medeiros, A.A.D. (1998). A survey of Control Architectures for Autonomous Mobile Robots. *Journal of the Brazilian Computer Society.* Rio de Janeiro-R.J.-Brazil. v.4, p.35-43.

[Murata 1989] Murata, T. (1989). Petri Nets: Properties, analysis and applications. *Proceedings of the IEEE* 77(4). pp. 541-580.

[OMG 1999] OMG Unified Modeling Language Specification version 1.3. *Object Management Group.*

[Rivière 2001] Rivière, N., Pradin-Chezalviel, B., Valette, R. (2001). Reachability and temporal conflicts in t-time Petri Nets. *9TH International Workshop on Petri Nets and Performance Models (PNPM'01).* Aachen-Germany. pp. 229-238.

[Sibertin-Blanc 1994] Sibertin-Blanc, C. (1994). Cooperative nets. *Vol. 815 of Lecture Notes in Computer Science. Springer Verlag.*

Copyright © IFAC Information Control Problems in
Manufacturing, Salvador, Brazil, 2004

ELSEVIER
IFAC
PUBLICATIONS
www.elsevier.com/locate/ifac

HUMAN ADAPTED QUALITATIVE DIAGNOSIS FOR ABRUPT FAULTS

TRIKI Slim, SAYED MOUCHAWEH Moamar, RIERA Bernard, CARRE-MENETRIER Véronique, ZAYTOON Janan

Laboratoire d'Automatique et de Micro-électronique (LAM)
Université de Reims-Champagne-Ardennes
Moulin de la Housse
BP 1039
51687 Reims, FRANCE
{slim.triki;moamar.sayed-mouchaweh;bernard.riera;Veronique.carre;janan.zaytoon}@univ-reims.fr

Abstract: In this paper, an approach to realize a qualitative diagnosis is proposed. This approach is based on the piecewise linear representation, the classification by fuzzy rules, the modelling by bond graph and the fault hypothesis tests. It uses a progressive monitoring to refine the hypothesized fault candidates generated by a temporal causal graph. The functioning of this approach is illustrated in simulating a fault with a three-tanks system. This approach is capable to detect a change, to determine its occurrence instant and to reduce the number of its probable causes if this change is the response of a fault. The goal of this approach is to develop efficient human – supervisory tool cooperation in facilitating the perception of process state and in decreasing the computational complexity performed by human operators to find the causes of a fault.
Copyright © 2004 IFAC

Keywords: Qualitative diagnosis, Segmentation, Fuzzy classification, Human Machine Systems.

1. INTRODUCTION

Instrumentation technology provides a great deal of data about a process functioning for human operators and supervision engineers. Thus they need a hard reasoning task to extract the significant information necessary to identifying unusual conditions and consequently to responding rapidly and effectively by taking corrective actions. Therefore an automated detection and fault diagnosis system can be useful to assist operating personnel in his supervision tasks. (Riera and Debernard, 2003) has shown the importance to develop a prototype of human-machine interaction to monitor a system. The human operator takes the final decision and this prototype has to cooperate with him to perform this task.

Fault diagnosis is based on two major approaches: quantitative and qualitative (Manders, *et al.*, 2000). The quantitative approach is a numerical modelling based on depth knowledge about system structure and its parameters. However the numerical modelling has two drawbacks: the first is that this perfect information is not available in the most of real complex systems and the second is the sensibility of the fault diagnosis to model errors and to noise (Frank, 1990). The qualitative approach is based essentially on the expert knowledge, which is represented by rules (Zadeh, 1992). These rules can be built using fuzzy logic in order to better represent the human experience, which can be both uncertain and imprecise. The major advantage of this approach is its capability to manipulating incomplete data and noisy information. Others advantages are: the use of symbolic source of information and the interface facility between operator and machine. Nevertheless, two problems arise in this approach: its sensibility to membership functions and the no physical meaning to the rules.

In this paper, an approach to realize a qualitative diagnosis is proposed. The goal is to assist monitoring operators in detecting a fault with its possible causes and consequently to help him in determining the necessary actions to remove this fault. In the first part of the paper, the necessity to develop human adapted supervisory tools is revealed. In the second part, an approach to realize a qualitative diagnosis is proposed. This approach uses the piecewise linear representation, the classification by fuzzy rules, the modelling by bond graph and the fault hypothesis tests. This approach is applied on simulated example of three tanks system. Finally, the advantages and the drawbacks of this approach and the perspectives for our future work are discussed.

2. HUMAN ADAPTED SUPERVISORY TOOLS

Supervision of highly automated processes is an activity in which human being, despite the increasing automation of the last years, is still present. Consequently, the design of efficient supervisory systems requires a different approach of automation, being more systemic than analytic (Riera and

Debernard, 2003). That can lead to original solutions in matter of automation based around the concept of human-machine cooperation. Considering only "real time" aspect of process supervision, human supervisory tasks are at a medium level between strategic functions (planning for instance) and tactic functions linked to the command part of the production system. Human operator receives continuous and discrete information coming from the supervisory system and acts on it sometimes. Supervisory tasks necessitate high level of knowledge and can be shared in three classes: command (to act on the process), monitoring (detect abnormal running) and diagnosis (localization, causes and consequences of a failure). An abnormal process state can be detected by human operator either with an alarm (i.e. passive detection) or by his own observation of process running (i.e. active detection) (cf. Figure 1).

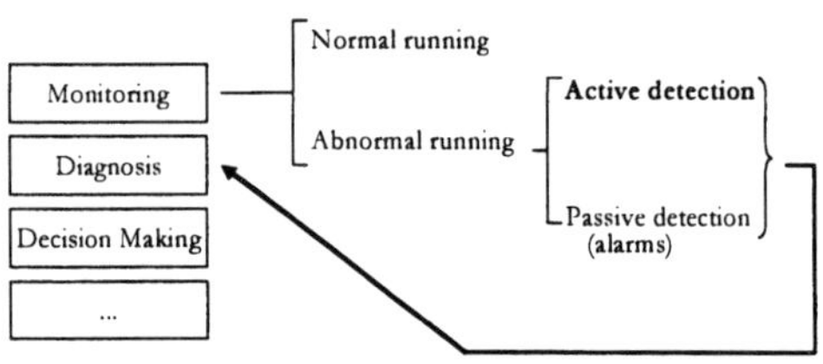

Fig.1 From monitoring to diagnosis

Usually, the designer of supervisory system works only on the improvement of alarms. Another approach is to design supervisory tools capable to facilitate active detection. After detecting an abnormal running, human operator has to diagnose the process state. This consists in determining the primary causes but also the effects on the process. First of all, human operator has to check the validity of the problem (failure of sensor, disturbance compensated or not by automatic controllers ...). Capacity of abstraction and the fact that people may take decisions even in the case of great uncertainties explains partially why it is not possible to withdraw human being from supervisory loop. Supervisory tools for continuous automated processes have not evolved since the development of the first numerical control-command systems. One can find essentially displays (synoptic) copying the physical structure of the process (P&ID diagrams), trends and alarms based on the triggering of a sensor or a simple comparison of a measurement with a fixed threshold. In the approach, proposed in this paper, a supervisory system is seen as a set of supervisory tools cooperating with and designed for human operator. For that, the following principles of design is applied:

- supervisory tools have to facilitate the active perception of the process (situation awareness),
- supervisory tools have to be specific and adapted to the different supervisory tasks (monitoring, diagnosis, ...),
- supervisory tools have to cooperate with human operator (eg : adapted to human reasoning),
- supervisory tools have to propose to human operator information (and not only measurements), obtained by an integration of the Fault Detection and Isolation (FDI) algorithms.

In other words, supervisory tools do not have to cross the boundary between the right level of support, and assistance that removes the human operator from the supervisory loop. In this field, this paper deals with an original human adapted supervisory tool supplying information about the signal trends and performing a qualitative diagnostic for abrupt faults.

3. PIECEWISE LINEAR REPRESENTATION

One of the most important and useful feature that could be extracted from signals is their trends. Trend displays for the supervision by human operator are absolutely necessary. These displays enable him to follow and to learn process dynamics, to anticipate effects of failures and facilitate diagnosis. Hence, a supervisory tool supplying information about signal trends seems to be adapted to human operator.

There are many works in the literature that attempt to symbolically describe the evolution of a dynamic process in using high level representations of time series (Bakshi and Stephanopoulos, 1994; Bakshi and Stephanopoulos, 1996; Cheung and Stephanopoulos, 1990; Janusz and Venkatasubramanian, 1991; Triki, *et al.*, 2003). They extract from time series, or a signal, its trends to go up, down or maintain level. In many cases, these features correspond to signal derivatives. Nevertheless, derivatives are corrupted if the signal is noisy, leading to further imprecision. A possible alternative that may be robust to noise consists in finding a piecewise linear function able to approximately describe the time series data using least squares regression. In this way, the time series data is partitioned in several segments and each segment is represented by a linear function. This alternative is called the Piecewise Linear Representation (PLR) (Keogh, *et al.*, 2001). Its aim is to approximate a time series T, of length n, with k straight lines. Thus it can be called segmentation method. Since k is much smaller than n, this representation makes the storage and the understanding of the data by human operator more efficient. For this reason, PLR is selected to perform the abstraction of the dynamic evolution of a process.

There are three major approaches to time series on line segmentation: Sliding window, Top-Down and Bottom-Up approaches (Heckbert and Garland, 1997). In this paper, the sliding window is chosen to be adaptable with the signal dynamic: fast, medium and slow, as it will be seen later.

The segmentation problem is to produce the best representation of a time series T in using k segments. There are two ways: linear interpolation and linear regression. The linear regression is preferred because it is more robust against noises and consequently more adapted to detect significant changes in time series. The only constraint to the segmentation by regression is to well choose the maximal error that achieves a good trade off between data compression and approximation fidelity. Of course, using regression technique involves possible discontinuities between 2 adjacent segments and the points defining

one segment does not belong to the original time series.
In the next section, a strategy, using the Mamdani fuzzy inference technique (Mamdani, 1977) is presented to determine the segments dynamic (fast, medium or slow) and to detect changes instants. The main idea is to "fuzzyfy" the 2 characteristics of the trend: the magnitude and the duration in order to classify the segment (from slow variation to fast variation).

4. FUZZY CLASSIFICATION

4.1 Fuzzification of episodes features.

The crisp values of magnitude and duration of each linear segment (also called episode) are fuzzified by using two membership functions (see figures 2 a, b). For both, fuzzy partitions, trapezoidal and/or triangular membership functions are used. Orthogonal normalised fuzzy partitions are defined in the universe of each feature (i.e. duration and magnitude).

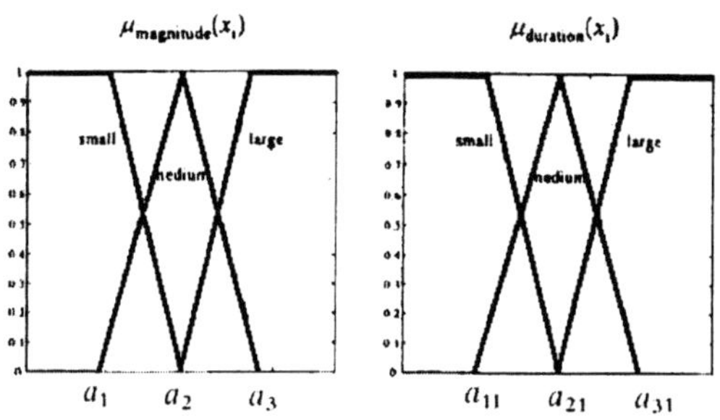

Fig.2 (a) Fuzzy partition of magnitude. (b) Fuzzy partition of duration.

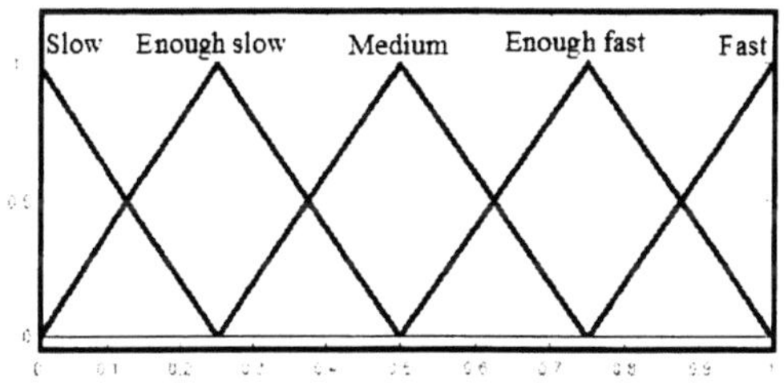

Fig.3 Fuzzy partition of variation.

The fuzzification of segments features leads to the descriptions: $\tilde{x}_1 = [x_1(A_{11})\ x_1(A_{21})\ x_1(A_{31})]$ for magnitude value $x_1 \in X_1$ and $\tilde{x}_2 = [x_2(A_{12})x_2(A_{22})x_2(A_{32})]$ for duration value $x_2 \in X_2$.
The linguistic description of segments in a process variable is represented by a fuzzy proposition written from segments features. Each fuzzy proposition is used as an input to a fuzzy inference algorithm which performs a fuzzy classification of the trend ([slow], [enough slow], [medium], [enough fast], [fast]). Hence, conclusions about the dynamic are expressed from each segment features with fuzzy rules as for instance:

If $\tilde{x}_1$ is "*large*" **AND** $\tilde{x}_2$ is "*small*" **THEN** segment belongs to the class "*fast variation*".

The table 1 represents the symbolic rules bases.

Table 1 Table of inference.

		Magnitude		
		Small	Medium	Large
Duration	Small	Medium Variation	Enough fast Variation	Fast Variation
	Medium	Enough slow Variation	Medium Variation	Enough fast Variation
	Large	Slow Variation	Enough slow Variation	Medium Variation

4.2 Fuzzy Inference Algorithm.

In a theoretical point of view, a linguistic fuzzy model (Mamdani, 1977) has been introduced as a way to capture available qualitative knowledge in the form of if–then rules. A fuzzy rule can be writing as following:

$$\Re_m : If\ \tilde{x}_1\ is\ A_{i1}\ and\ \tilde{x}_2\ is\ A_{j2}\ then\ \tilde{y}\ is\ B_k,\quad m = 1,2,...,9.$$

$\tilde{x}_1$ and $\tilde{x}_2$ are the input linguistic variables (i.e. segment features : magnitude and duration), and $A_{ij} = A_{i1} \times A_{j2}$ are the linguistic terms (small, medium, large) where $i = \{1,2,3\}$ and $j = \{1,2,3\}$. Similarly, $\tilde{y}$ is the output linguistic variable (i.e. variation or trend type) and $B_k : k = \{1,2,...,5\}$, are the resulting linguistic terms (slow, enough slow, medium, enough fast, fast) $B(Y) = \{B_1, B_2, B_3, B_4, B_5\}$ (see figure 3).
The rule evaluation is performed using the fuzzy operator (AND). In our case, the operator AND is the product. The result of the antecedent evaluation is applied to the membership function of the consequent. The consequent membership function is clipped to the level of the truth-value of the rule antecedent.
Next step is aggregation of the rule outputs. It is the process of unification of the outputs of all rules. The idea is to take the membership functions of all rule consequents and combine them into a single fuzzy set (called B'). The last step is the defuzzification.

4.3 Defuzzification.

To obtain a crisp value, the output fuzzy set must be defuzzified. There are several defuzzification methods. One of the most popular is the centroid technique. This method computes the y' coordinate of the Centre Of Gravity of the fuzzy set B':

$$y' = COG(B')$$

Figure 4 shows an example of measured variable in which magnitude gets down around t=2500s. A piecewise linear representation of the variable is shown in the same figure. Fuzzy classification (from slow to fast) of the segments trends is depicted at the bottom of the figure.

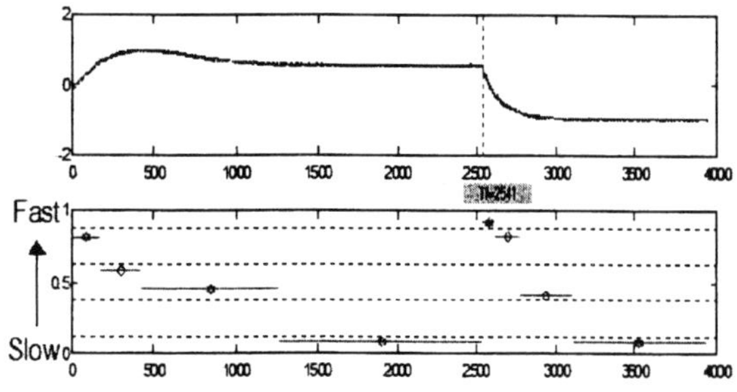

Fig.4 Example of application of fuzzy classification to detect abrupt change.

By applying fuzzy classification based on result of PLR, the instant of significant variation occurring in the measured variable is detected with more accuracy. The example above shows that at the instant where the magnitude gets down, segments change abruptly from "*slow variation*" class to a "*fast variation*" class, this change occurs at T_f=2541s. This abrupt change of class allows detecting an abrupt change in the magnitude of a monitored variable. This method enables detecting changes in signals that must be interpreted by human operator or an algorithm. Only a dynamic model can explain the changes in the signal. In the following of this paper, a qualitative diagnosis tool based on a Temporal Causal Graph (Manders, *et al.*, 2000) is proposed.

5. QUALITATIVE DIAGNOSIS METHOD

The symbolic information issued from the fuzzy classification is devoted to a monitoring operator. He or she needs to know the causes of this abnormal change in order to take the necessary corrective actions. This isolation demands a hard working and long time, which can be expensive or even fatal. A qualitative diagnosis is used to support monitoring operators in isolating the probable parameters, candidates of an abnormal change. This qualitative diagnosis is based on the method proposed by (Manders, *et al.*, 2000). This method is based on a Temporal Causal Graph (TCG) and on a progressive monitoring procedure for the diagnosis. The TCG is derived from a bond graph model in a straightforward manner by applying a causality assignment procedure and tracing the causal paths. Monitoring mechanism means a comparison between actual observations with the predicted behaviours to generate prune hypothesized faults during transient analysis (Manders, *et al.*, 1999). The notion of using higher order derivatives in the analysis of measured values is referred to as "progressive monitoring". Hence, transient behaviour of an observed variable is expressed by (Manders, *et al.*, 1999) as a t-uple (called signature) of qualitative values (+, 0, -) for magnitude, 1st, 2nd and higher order time-derivative.
This qualitative diagnosis can be applied using the results coming from the PLR step. Indeed, using segments seems interesting because the algorithm is faster and uses the benefits of segmentation (robust to noise, easily understandable ...). This method is divided into the following steps:

- construction of a qualitative signature, {mag(j), slop(j), curv(j)}, for each segment. This signature, which indicates respectively the magnitude between 2 following segments, the slope and the curve of each segment, is calculated as following:

$$\forall j = 2,\dots,n \quad mag(j) = \begin{cases} - & if\ x_{init}(j) < x_{fin}(j-1) \\ 0 & if\ x_{init}(j) \approx x_{fin}(j-1) \\ + & if\ x_{init}(j) > x_{fin}(j-1) \end{cases}$$

$$\forall j = 1,\dots,n \quad slop(j) = \begin{cases} - & if\ x_{init}(j) > x_{fin}(j) \\ 0 & if\ x_{init}(j) = x_{fin}(j) \\ + & if\ x_{init}(j) < x_{fin}(j) \end{cases}$$

$$\forall j = 1,\dots,n \quad trend(j) = x_{fin}(j) - x_{init}(j)$$

$$\forall j = 1,\dots,n-1 \quad curv(j) = \begin{cases} - & if\ trend(j+1) < trend(j) \\ 0 & if\ trend(j+1) \approx trend(j) \\ + & if\ trend(j+1) > trend(j) \end{cases}$$

$x_{init}(j)$ and $x_{fin}(j)$ are the crisp values of respectively the initial and the final points of the segment j, n is the number of segments. The sign + indicates an increase case, 0 a steady case and – a decrease case. Thus there is a signature of the form (+, 0, -) for each segment.

- applying a progressive monitoring to take into consideration the dynamic changes in the observed parameters. Indeed when a fault is detected, a set of hypothesized fault candidates are generated in using the TCG. These candidates will be refined in comparing, for each of them, the predictive signature, issued from the segmentation step, and the real measure. The use of segments instead of points accelerates the refinement of the fault candidates (cf. figure 5).

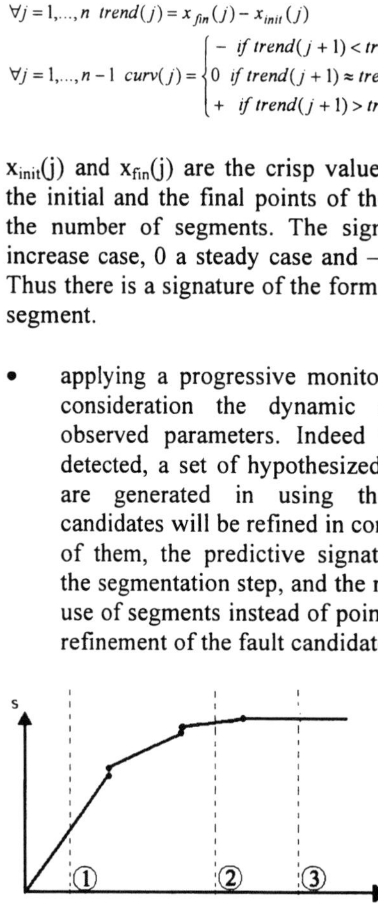

dt	0	1	2
①	0	+	–
②	+	+	–
③	0	0	•

Fig.5 Example of progressive monitoring by crossing by segment.

This method seems very efficient to determine possible candidates after a failure involving abrupt changes in the dynamic. The following simulated example, illustrates the functioning of this method of qualitative diagnosis.

6. ILLUSTRATIVE EXAMPLE

As an example, this method is applied to a three-tank fluid system in a simulation. The tank system and corresponding TCG model are shown in Figure 6. The tank capacities, C_1, C_2, and C_3, and pipe resistances, R_{12} R_{23}, and R_b, constitute the set of parameters of the physical components and thus the possible fault candidates. The e_i and f_i vertices shown in the TCG correspond to tank pressure and pipe flow rate variables, respectively. Circled vertices in the TCG indicate the measured variables. In this example the pressure in the third tank (e_{10}) and the flow rate in the pipe connecting tank 1 and tank 2 (f_3) are measured. Consider the fault situation where the capacity of tank 1 increases abruptly (indicated with C_1^+), as might happen when a leak occur in the tank. The resulting transients in the measurements are shown in Figure 7.

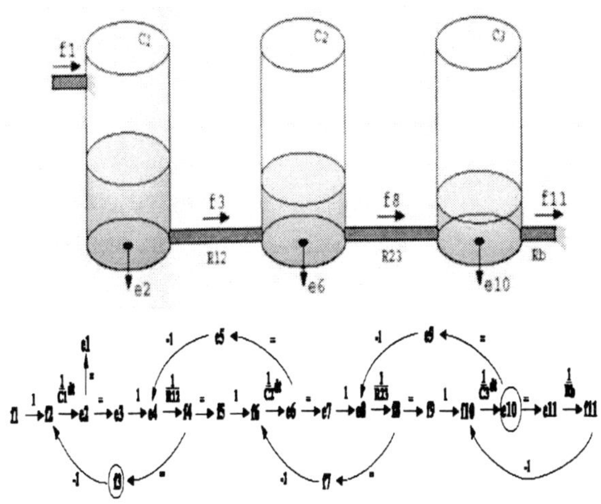

Fig. 6 Three-tank fluid system and its TCG model representation.

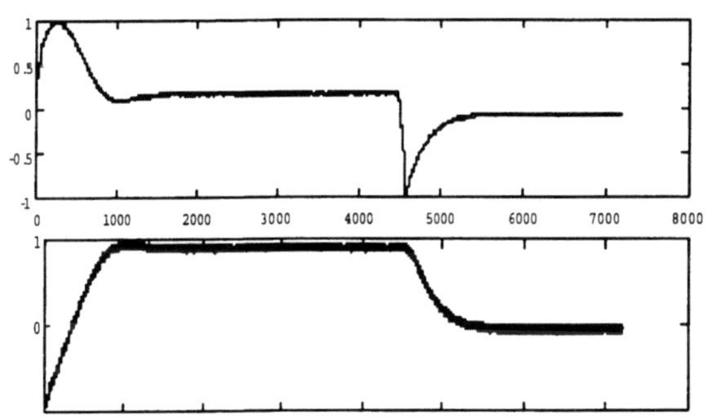

Fig. 7 Evolution data for measurements e_{10} (top) and f_3.

Results of abstraction by PLR and fuzzy classification are shown in Figures 8 and 9. Vertical line in figure indicates the instant of abrupt change. As it is shown in figure 9, human operator may be able to detect, by a visual inspection, abnormal decrease of e_{10} and f_3. But, change occurred almost simultaneously and human operator doesn't know with more accuracy either e_{10} deviate the first or f_3. In (Triki, *et al.*, 2003) it was demonstrated that temporal causality of fault propagation may improve perception of human operator and then support him to build a correct interpretation of potential undesired deviation occurred in measured variables.

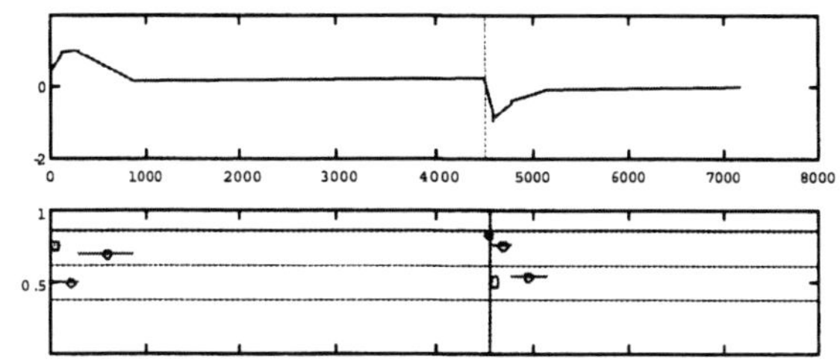

Fig. 8 Results of PLR and Fuzzy classification of measured variable e_{10}.

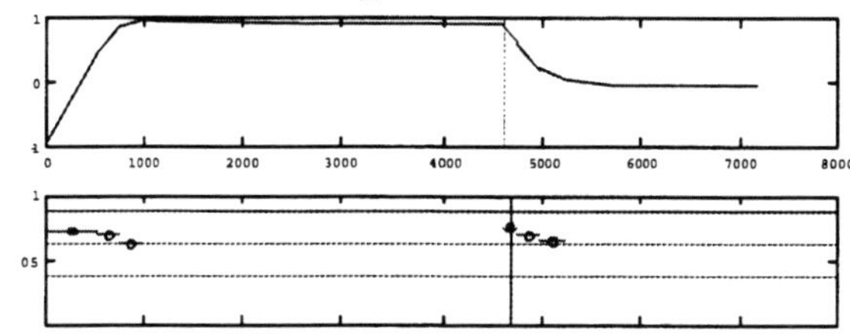

Fig. 9 Results of PLR and Fuzzy classification of measured variable f_3.

The initial deviation detected by applying PLR and fuzzy classification is an abrupt decrease of e_{10}. Once an abrupt change is identified, progressive monitoring is triggered by human operator who has valuable information allow him to distinct abrupt change result from faulty behaviour and change result from response to a set point change of controlled variable. When fault is detected, abnormal decrease of e_{10}, backward propagation generates the fault hypotheses: $\{R_{12}^+, C_2^-, R_{23}^+, C_3^-, R_b^+, C_1^+\}$. Next task consists of deriving the signature of the fault hypotheses which is performed by a forward propagation. Finally, fault hypotheses should be refined by applying progressive monitoring and checking predictive and measured signature (see table 2 and figure 10).

Table 2 Diagnosis results.

Step 0	Step 1	Step 2
e_{10} - . f_3 0 .	e_{10} -+ . f_3 0 . .	e_{10} -+ . f_3 - .
R_{12}^+ e_{10} -+- f_3 00-+	R_{12}^+ e_{10} -+- f_3 00-+	R_{12}^+ e_{10} -+- f_3 00-+
C_2^- e_{10} -+- f_3 0+-+	C_2^- e_{10} -+- f_3 0+-+	C_1^+ e_{10} -+- f_3 00-
R_{23}^+ e_{10} 0-+ f_3 0-+-	C_1^+ e_{10} -+- f_3 00-	
C_3^- e_{10} 0-+ f_3 +-+-		
R_b^+ e_{10} 00- f_3 0+-+		
C_1^+ e_{10} -+- f_3 00-		

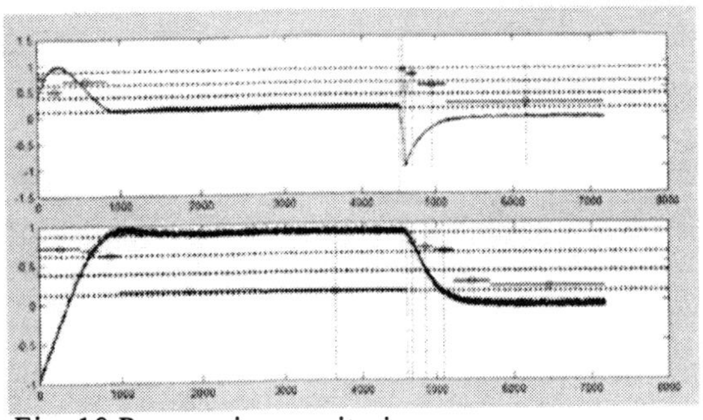

Fig. 10 Progressive monitoring.

It can be concluded from Table 2 that, this method has reduced the probable candidates of the fault from 6 candidates, in the initial step, to 2 candidates in the third step. Thus this method can be used to design Human Machine Interface supporting the operator in his monitoring work to detect a fault, the instant of its occurrence and its causes. The final decision stays at the hand of the operator. The goal of this qualitative diagnosis method is to be integrated in a human adapted monitoring supervisory tool, which does not take the decision instead of operator.

7. CONCLUSION

In this paper, a new strategy to realize a qualitative diagnosis in using the abstraction method Piecewise Linear Representation (PLR) and the progressive monitoring to refine the hypothesized fault candidates generated by the Temporal Causal Graph (TCG) was presented. It has been shown, in simulating a fault with a three-tank system, that this method is capable to detect a fault, to determine its occurrence instant and to reduce the number of its probable causes. The goal of this method is to help monitoring operators in their work and decreases the computational complexity needed to find the causes. Consequently this method may improved the perception of human operators to support them in building a correct interpretation of potential undesired deviation occurred in measured variables. Now, we work on a formal validation of this strategy in order to define its limits knowing process parameters. We hope test our method in using real examples with monitoring operators. The ambition is to extract the best minimum of information necessary to monitor a system by a human operator.

REFERENCES

Bakshi, B. R., & Stephanopoulos, G. (1996). Reasoning in time: modelling, analysis and pattern recognition of temporal process trends. In G. Stephanopoulos, & C. Han, *Intelligent Systems in Process Engineering-Paradigms from Design to Operations* (pp. 487–549).

Bakshi, B. R., & Stephanopoulos, G. (1994). Representation of process trends-3: multiscale extraction of trends from process data. *Computers and chemical Engineering*, 18, (pp. 267–302).

Cheung, J. T. Y., & Stephanopoulos, G. (1990). Representation of process trends-1: a formal representation framework. *Computers and chemical Engineering*, 14, (pp. 495–510).

Frank, P.A., Fault diagnostic in dynamic systems using analytical and knowledge-based redundancy, a survey and some new results, *Automatica* 26 3, 1990, pp. 459–474.

Ghiaus, C., Fault diagnosis of air conditioning systems based on qualitative bond graph, *Energy and Buildings* 30, 1999, pp. 221–232.

Heckbert, P. S. & Garland, M. (1997). Survey of polygonal surface simplification algorithms, Multiresolution Surface Modeling Course. Proceedings of the 24th *International Conference on Computer Graphics and Interactive Techniques*.

Janusz, M. E., & Venkatasubramanian, V. (1991). Automatic generation of qualitative descriptions of process trends for fault detection and diagnosis. *Engineering Applications Artificial Intelligence*., 4, 329–339.

Keogh, E., Chu, S., Hart, D., Pazzani, M. (2001). An Online Algorithm for Segmenting Time Series. *The 2001 IEEE International conference on Data Mining*.

Mamdani, E.H. (1977). Application of fuzzy logic to approximate reasoning using linguistic systems. *Fuzzy Sets and Systems 26*, 1182–1191.

Manders, E.-J., G. Biswas, Narasimah S., P.J. Mosterman, (2000). Combined Qualitative/Quantitative approach for fault isolation in continuous dynamic systems. *4th Symposium on Fault Detection Supervision and Safety of technical processes*, Budapest, Hungary, 14-16 June 2000.

Manders, E.-J., P.J. Mosterman and G. Biswas(1999). Signal to symbol transformation techniques for robust diagnosis in TRANSCEND. In: *Tenth International Workshop on Principles of Diagnosis*. Loch Awe, Scotland. (pp. 155–165).

Riera B. & Debernard, S. (2003), Supervisory Systems for process control, Handbook of Cognitive Task Design, Lawrence Erlbaum Associates, 2003, pp. 255-282.

Triki S., Caulier P., Riera B. (2003). Outil d'aide à la surveillance adapté à l'opérateur humain : Application sur un système de trois cuves. Journées Doctorales en Automatique, Valenciennes, France, 25-27 juin 2003.

Zadeh, L.A., Knowledge representation in fuzzy logic, An Introduction to Fuzzy Logic Applications in Intelligent Systems, Kluwer Academic Publishers, 1992, pp. 1–25.

Copyright © IFAC Information Control Problems in
Manufacturing, Salvador, Brazil, 2004

ELSEVIER
IFAC
PUBLICATIONS
www.elsevier.com/locate/ifac

A CONCEPTUAL MODEL OF MAINTENANCE PROCESS IN UNIFIED MODELING LANGUAGE

Ivana Rasovska, Brigitte Chebel-Morello, Noureddine Zerhouni

Laboratoire d'Automatique de Besançon
UMR CNRS 6596
24, rue Alain Savary 25000 Besançon, France
Tel.: +33 (0) 3 81 40 27 97 Fax: +33 (0) 3 81 40 28 09
Email: {rasovska, bmorello, zerhouni}@ens2m.fr

Abstract: In this paper, a conceptual model of maintenance process is proposed in the aim of releasing different types of help (decision-help systems) to maintenance actors. The maintenance domain is defined and modelled with the aid of knowledge management. The object-oriented representation was chosen in order to propose real domain model for development of information system or prototype. This system should comply with decision support and allow technico-economical knowledge capitalisation.
Copyright © 2004 IFAC

Keywords: maintenance engineering, decision making, decision support systems, knowledge-based systems, knowledge representation.

1. INTRODUCTION

Recently maintenance services in industrial companies have become more and more important because of their large impact on good technical and economic function of production systems. Maintenance has become a strategic issue that brings competitive advantages to the equipment manufacturers. Concentration on the maintenance tasks implicates the outsourcing maintenance and the maintenance process automation.

In this paper, the objective is to study a complete maintenance process, to permit the largest possible automation of this maintenance process and to release different types of help proposed to maintenance specialists in their daily maintenance tasks. We consider that the maintenance specialists could operate on-site or off-site and should be able to control and to manage the manufacturing plant over its whole life cycle. The principal goal of the maintenance services in the manufacturing company is specified in Monchy, (2000) and represents the maximization of the enterprise profit that means especially

- to minimize costs (maintenance costs and breakdown consequential costs),
- to maximize equipment availability (minimize down times, time for preventive maintenance, technical failure times),
- to provide good equipment reliability and security,
- to satisfy general quality, safety and environmental goals.

This paper is organized as follows. In the first section, the analysis of the maintenance process is presented, thus the definition and different types of maintenance. The most important concepts of specific decision techniques are identified in the hierarchical manner, corresponding to the maintenance process phases and concerned actors, and based on the practical experience and domain knowledge. To develop a decision support system for maintenance actors, concrete domain knowledge must be defined and modelled. We use concepts, methods and techniques of knowledge management, that allow to model and/or acquire knowledge for decision-making system. Therefore, in the second section, the domain of knowledge management and principal reasons of the object representation choice are discussed. In the following section, several use case diagrams for the final users are proposed. Finally, the unified modeling language model of the maintenance process is proposed in the form of the class diagram, and conclusions and perspectives of the actual work are discussed.

2. MAINTENANCE PROCESS ANALYSIS

2.1. Maintenance definitions

The definition of maintenance in AFNOR (2001) is the following: combination of all technical, administrative and managerial actions during the life cycle of an item intended to retain it in, or restore it to, a state in which it can perform the required function. The purpose of maintenance function is to minimize production costs (this will be achieved by minimizing maintenance costs), to assure good maintenance service quality and maximize availability and reliability of maintained equipment.

In accordance with European standard AFNOR (2001) and with Kaffel, (2001), three main types of maintenance are defined:

- Corrective maintenance focuses on efficiently repairing or replacing equipment after the failure occurrence. This maintenance cannot be avoided by reason of nonzero failure probability. It often produces high failure costs caused by no planned production breakdowns with unknown downtimes duration. The costs minimization aims to increase the maintainability of equipment by improving the speed of repair, or return to service, after a failure.
- Preventive maintenance focuses on keeping equipment in good condition in order to minimize failures and on repairing components before they fail. This aims to increase the reliability of equipment by reducing the frequency of failures. We have three different approaches of the preventive maintenance. *Systematic (scheduled) maintenance* plans periodical interventions according to time or consumed units, *conditional maintenance* plans interventions according to the wear state monitored by sensors and *predictive maintenance* follows the sensor's indicator evolution by means of experimental measures.
- Ameliorative maintenance represents special interventions aiming to improve equipment functions, as availability, reliability and security one.

2.2. Maintenance process concepts

A physical system of process functions is studied in the view of corresponding decision-making techniques used by concerned actors. Upon this fact, possible decision-support applications are deduced and oriented to their final users and the process operations. One of the contributions of this paper concerns definition of four fundamental technical and business fields, identified in the general maintenance process, as shown in Fig.1. On the basis of this distinction, the conceptual model is proposed below in the paper.

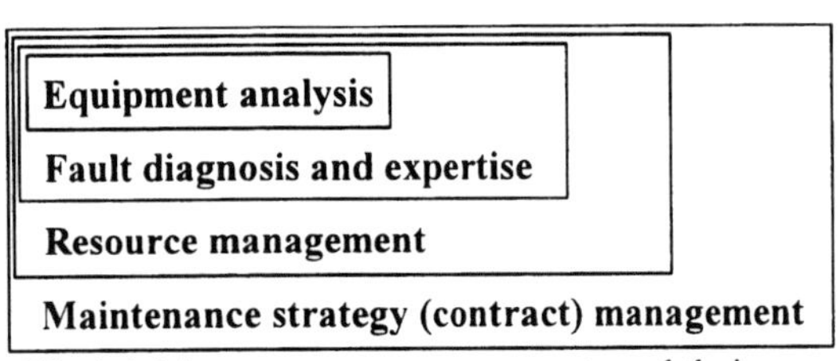

Fig.1. Maintenance process concepts and their interface

- Equipment analysis: This process consists of two main parts – functional analysis (represented by equipment decomposition and failure tree) and failure analysis (represented by FMECA[1]). This is realised by maintenance experts (method engineers). It serves to realize the scheduled maintenance, to propose a suitable sensor installation on equipment for the conditional maintenance and to identify a possible equipment improvement for the ameliorative maintenance. This analysis is necessary for the following technical process.
- Fault diagnosis and expertise: The aim of this process is to help the operator, during his intervention, to diagnose the problem and to solve it without the recourse to an expert. Generally, the maintenance operator does not know where is the problem, which type of failure had occurred, what equipment part is touched etc. The first function – fault diagnosis identifies the failure type, its symptom and origin. Once the failure is known, the second function is called thus the expertise. It gives the answer to questions as how eliminate this failure, which competency, spare parts and tools operator needs etc. This function can propose the equipment documentation and maintenance manuals which contain the repair procedures to the maintenance operator, as described in Auriol, et al., (1999).
- Resource management: This concept deals with the resource planning for all maintenance interventions, especially the preventive and conditional ones. Logistic experts are responsible for logistic tasks as spare parts storage, purchasing and delivery to maintenance operators.
- Maintenance strategy management: This represents decision support concept for maintenance managers. This function should quickly answer or help to answer questions about maintenance strategy choice for one concrete equipment or a set of equipment, about customer requirements in the maintenance contract or about management of maintenance services in general.

To model maintenance process for future decision support system, concepts, methods and techniques of knowledge management are presented in the following section.

3. DOMAIN OF KNOWLEDGE MANAGEMENT

The term "knowledge management" means the set of know and know-how of actors which enables company to reach its objectives and goals. Knowledge management proposes methods and tools (such as cognitive modeling, conceptual modeling, modeling and representation languages, etc.) for different domains as information retrieval, information research on the web, knowledge management and capitalization in the enterprise, management indicators definition and their implementation in information systems, etc. as shown in Charlet, et al., (2000).

[1] Failure Mode Effects and Criticality Analysis

In this paper, a generic model of maintenance process is proposed. For generating an analysis and conception document, an object-oriented approach was chosen. Aamodt, (2001) shows that the object knowledge representation is well adapted to handling composite entities and especially assures two modeling points of view:

- Symbol-level modeling which represents the level of data structures and programming languages and where the behavioral law is sequential interpretation of program procedures. This is the developer's point of view.
- Knowledge-level modeling is proposed as the description of computer systems. The system is described as an intelligent agent with its own goals and with knowledge of how to achieve these goals. This is the domain expert's point of view.

Object-oriented conception method deals better with user needs identification, offers simpler conception architecture and creates more easily maintainable systems. Such methods create a set of models and diagrams in according to domain identification and its functionality specification. These diagrams provide multiple perspectives of the system under analysis or development for users as well as for software developers.

The Unified Modeling Language (UML) was chosen as the modeling language to present our model. The UML specification and notation was adopted by The Object Management Group in the aim of reducing the degree of confusion within the industry surrounding modeling languages. OMG, (2003) defines the UML as a graphical language for visualizing, specifying, constructing, and documenting the artefacts of software systems, as well as for business modeling and other non-software systems, in general distributed object systems. The maintenance process model is based on use case diagrams for each proposed automatic maintenance support system and on class diagrams.

4. UML MODELING OF MAINTENANCE PROCESS

4.1. Use case diagrams

In Larman, (2002), the use case diagram is the illustration of use cases and actors, and the relationships between them. This use case diagram is an excellent picture of the system context that is showing the boundary of a system, what lies outside of it, and how it gets used. It serves as a communication tool that summarizes the behaviour of a system and its actors. The role of actors in the maintenance process and a short characterisation is thus of the primary importance. These participants will be final users of a decision support system, so it's necessary to identify who and how is affected to various process tasks. Since the actor names and their roles could differ in different companies and since there is a lot of specific actors only in our studied case, we had defined three generic classes of process participants and their corresponding roles:

- Maintenance operators – are maintenance specialists (technicians or exceptionally experts) who intervene directly in the repair process. They are responsible for the maintenance intervention realisation and performance.
- Maintenance experts – this class of actors (one speaks about method engineering) could also intervene in the repair as an expertise instance. Otherwise they are responsible for maintenance equipment strategy choice, determination of preventive maintenance interventions, resource planning, equipment functional analysis, equipment improvement study and fulfilment of rules and standards concerning equipments and workshops. In general they are involving to a contract realisation.
- Managers – do not intervene in the repair process. It's the superior level of company management which ensures the tasks concerning project preparation, contract negotiation, engagement to customer, etc. The class covers actors corresponding to site manager, project and contract manager, and commercial.

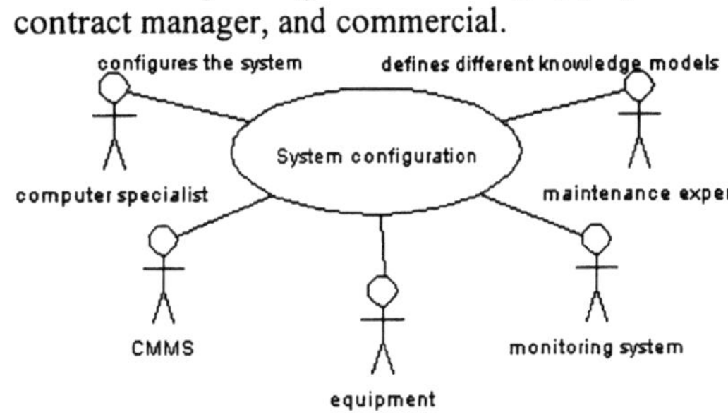

Fig.2. Use case diagram – system configuration

This use case represents, on the one hand, the system configuration, such as developing of the system by a computer specialist. On the other hand, maintenance experts must define maintenance domain knowledge by models used to develop decision-help system for diagnosis and repair and for maintenance strategy management. This system should be connected to the current systems already installed on the site as a CMMS[2], a monitoring system as SCADA[3] and to the equipment.

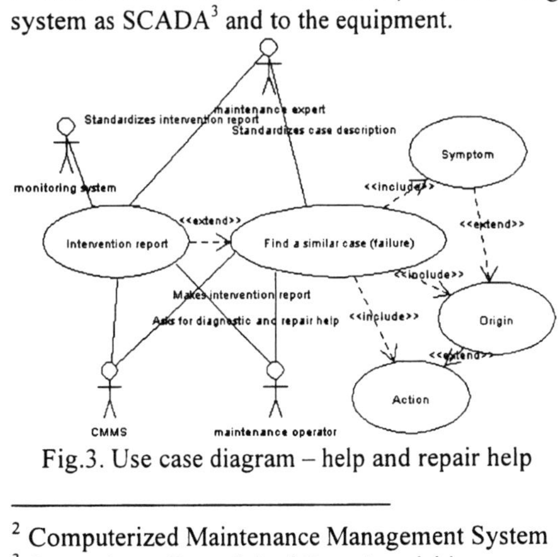

Fig.3. Use case diagram – help and repair help

[2] Computerized Maintenance Management System
[3] Supervisory Control And Data Acquisition, a computer system for gathering and analysing real time data.

Another use case diagram in Fig. 3 represents the phase "find a similar case in the database" in the diagnosis and repair help system. This case represents a failure and is characterized by a triple - symptom, origin and action to do. This failure model is done in the precedent use case and is based on the equipment analysis done by maintenance experts. After the identification of the failure symptom, system shall propose possible origins and choose the right one with the aid of similarity measurements between new problem and the nearest cases found in the database. Afterwards on the basis of proposed diagnosis, it shall propose the repair action and the information retrieval will allow operator to use the necessary equipment documentation and maintenance manual.

The second important function in this use case is the exploitation of the intervention report. Maintenance operator must fill the intervention report after each intervention. This one ensures experience feedback and learning phase in the decision help system for the diagnosis and repair. The information about consumption during the intervention constitutes the basis for the decision support system done for managers and explained further.

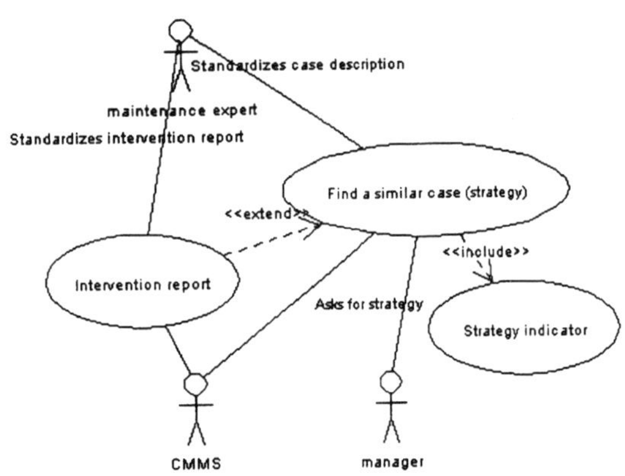

Fig.4. Use case diagram – strategy support system

In Fig. 4 the decision support concept for maintenance managers is shown. This use case represents a situation when a new equipment arrives on the site and the maintenance expert or manager has to make the first propositions concerning possible and effective maintenance strategy and especially maintenance costs for the customer. The choice of maintenance strategy for the equipment is based on strategy indicators such as maintenance costs, maintenance times, production mode etc.

4.2. A conceptual maintenance model

In OMG (2003) class diagrams are defined as follows: they show the static structure of the domain model, in particular, the things that exist (such as classes and types), their internal structure, and their relationships. A class diagram is a collection of static declarative model elements, such as classes, interfaces, and their relationships, connected as a graph to each other and to their contents. Class diagrams may be organized into packages either with their underlying models or as separated packages that build upon the underlying model packages.

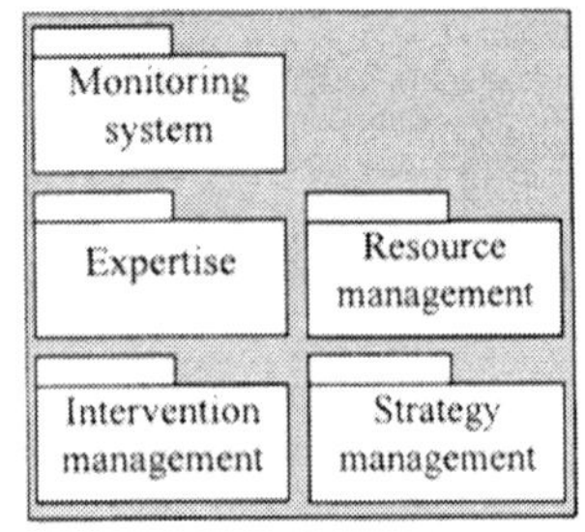

Fig.5. Packages

A conceptual model of the maintenance process involves five packages, as shown in Fig. 5.
- Monitoring system consists of sensors monitoring equipment functions.
- Expertise consists on the one hand, of the equipment functional and failure analysis and on the other hand, of the diagnosis and repair help system.
- Resource management deals with the resource planning for all maintenance interventions, especially scheduled, preventive and conditional ones.
- Intervention management makes part of the CMMS system already installed on the site and which must be connected to the other packages.
- Strategy management represents the decision support concept for maintenance managers.

Consequently, we describe in detail each of these packages. Finally, we conclude with a complete conceptual model diagram representing the relevant concepts that make up the maintenance process, and their relationships. In the Fig. 6 we present a conceptual model of the maintenance process in general.

The monitoring system assures the data acquisition from different sensors installed on the equipment. An example of such a system could be the SCADA system. The SCADA systems are used to monitor and control a plant or equipment in the industries such as telecommunications, water and waste control, energy, oil and gas refining and transportation. Nieva, (2001) deals in his thesis with modeling of remote data acquisition of embedded systems using internet technologies. He specifies role-based generic system specifications and creates a conceptual model of such a system.

The package - expertise phase of the maintenance process is presented as the complex of all equipment analysis. First, it concerns the functional analysis of the equipment based on functional equipment model and using decomposition methods and techniques, failure tree method and other suitable approaches. It consists of the failure analysis corresponding to analysis methods as

FMECA. This analysis will allow creating of the case representation for a decision support system for diagnosis and equipment repair. We refer to Lenz, et al., (1998) where the problem of diagnosis is described and discussed. Using a combination of inductive and case-based reasoning techniques, a decision support system for the technical maintenance of the aircraft engines was developed. Establishing the diagnosis is only the final step of the diagnostic process and it has to be followed by a repair action. The data on which the diagnosis is based are values of certain attributes (e.g. in a feature vector) which are acquired by observations, questions or historic data from the intervention management in CMMS. The case will be associated with a failure which arrives on the equipment and will be represented by a triple - symptom, origin and repair action. This triple is based on the equipment analysis mentioned before.

The resource management represents a support for the scheduled maintenance (especially preventive) tasks as planning of spare parts, tools and operator's availability. The resource planning system could be based on special methods dealing with the optimal planning.

The intervention management is treated in the CMMS system. This system plans preventive interventions, manages intervention reports and capitalizes concerned data. The intervention report constitutes the basis and is comprised of two main parts – *the consumption* which describes intervention duration (time that operator spends on the equipment reparation), costs and competency, changed spare parts and their costs, needed tools etc. Another part of intervention report represents *technical observations* such as intervention type, corrected symptoms, origins and real actions done by operator and eventually his remarks on the system diagnosis. It proposes an effective tool for experience feedback tasks which is based on a standardized intervention report. The consumption influences directly equipment indicators whereas the observations correct the automatic diagnosis proposed by the system.

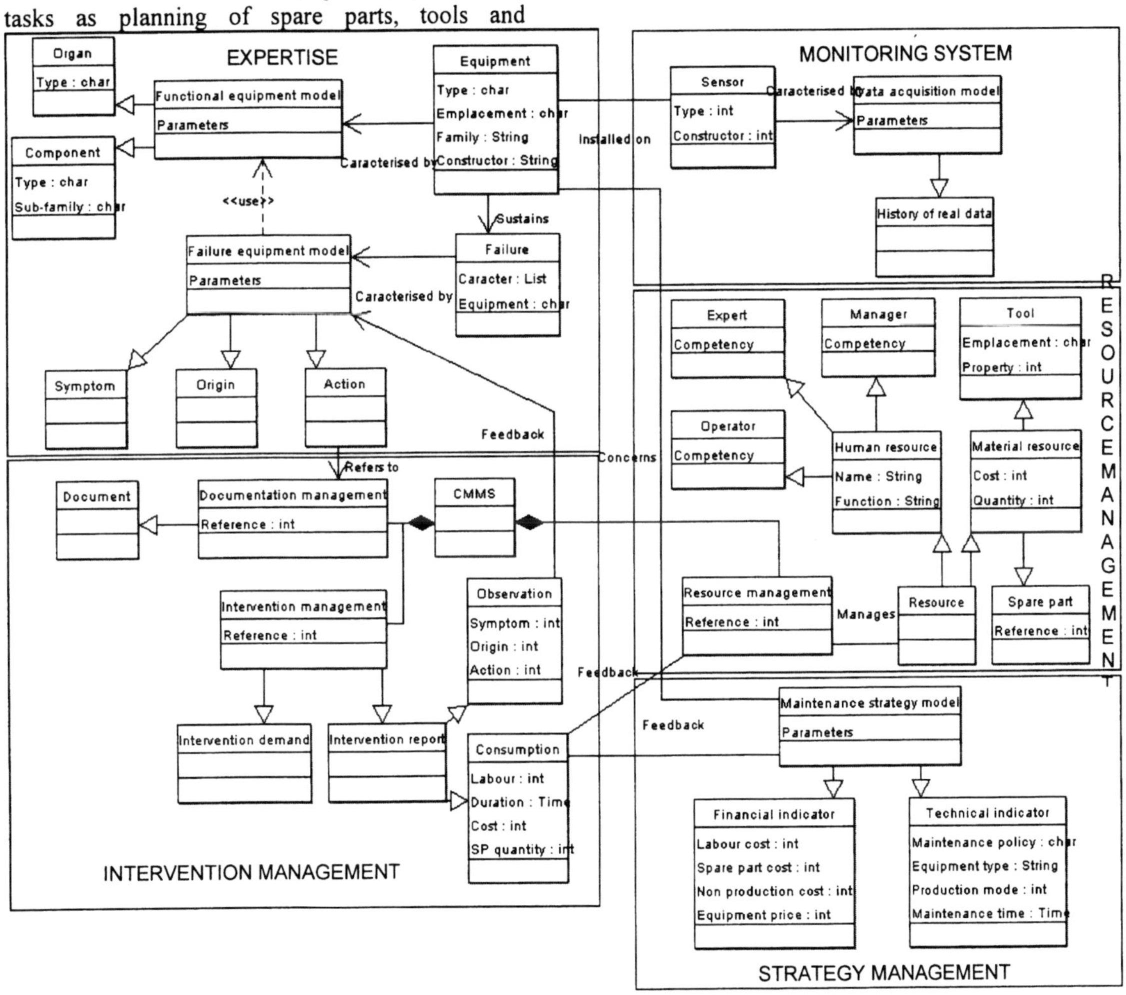

Fig. 6. A conceptual model of maintenance process

The operator effectuates an intervention which is described by its intervention report and saved in the CMMS. In this class diagram we specify also the documentation management that could be managed by the information retrieval. It would be appropriate that it takes part in the diagnosis step as a part of the repair help proposed to a human operator by the system. The problem solving is central and the retrieval of documents is considered as a useful step during the problem solution.

In HELPMATE[4] project described in HELPMATE, (2000), Auriol, et al., (1999) and Althoff, et al., (1995), a computer-based system capable to support the operator in the diagnosis and repair faults within the robot station is developed. This system combines a case-based reasoning approach for diagnosis with context information, as electronic on-line manuals, linked using open hypermedia technology.

The strategy management represents the decision support concept for maintenance managers based on the maintenance strategy model. A decision of possible and effective maintenance strategy and related costs is taken by a maintenance manager and demands a description of equipment indicators and its association with maintenance strategies. This system will be based on the association of the equipment indicators to the maintenance strategies. The equipment indicators are related to the consumption during maintenance interventions which are indicated in the consumption part of the intervention report. We distinguish two types of indicators, i.e. the technical ones that concern equipment type, production mode, chosen maintenance policy etc., and the financial ones that concern different types of maintenance costs.

CONCLUSION

The maintenance process analysis has allowed identifying of the four fundamental technical and business fields in the general maintenance process. On the basis of this analysis the conceptual model of maintenance process in UML was proposed. The object-oriented analysis method was used to propose the real domain model for a future development of decision-help system or prototype. This development is the goal of PROTEUS project programme in which this study was realised. This model will be applied on the concrete equipment of the real site. Our interest will focus on the data capitalisation based on the intervention report in the intervention management. It should be based on the definition and specification of the intervention report form the most appropriated to our tasks.

ACKNOWLEDGEMENTS

This study was funded by PROTEUS project programme sponsored by the French Ministry of the Economy, Finance and Industry and the Federal Ministry of Education and Research of Germany under the label of European Commission Initiative ITEA. It is part of initial study for the second work package of this project.

[4] Hypermedia with Enhanced Linking Providing Maintenance and Analysis Tools for Engineering

REFERENCES

Monchy, F. (2000). *Maintenance - methods and organisation* (Dunod (Ed)), Paris.

AFNOR (2001). Maintenance terminology. European standard. NF EN 13306.

Kaffel, H. (2001). *La maintenance distribuée: concept, évaluation et mise en œuvre.* Thesis, Faculté des Sciences et de Genie. Université Laval, Quebec.

Charlet, J., M. Zacklad, G. Kassel, D. Bourigault (2000). Ingénierie des connaissances: Evolutions récentes et nouveaux défis. Eyrolles.

Aamodt, A. (2001). *Modeling the knowledge contents of CBR systems.* Proc. Of the Workshop Program at the Fourth International Conference on Case-Based Reasoning, Vancouver.

OMG (2003). *Unified Modeling Language Specification.* Version 1.5.

Auriol, E., R.M. Crowder, R. McKendrick, R. Rowe, T. Knudsen (1999). Integrating case-based reasoning and hypermedia documentation: an application for the diagnosis of a welding robot at odense steel shipyard. Engineering Applications of Artificial Intelligence 12, pp. 691-703.

Larman, C. (2002). Applying UML and patterns: An Introduction to Object-Oriented Analysis and Design and the Unified Process. Prentice Hall PTR.

Nieva, T. (2001). *Remote data acquisition of embedded systems using internet technologies: A role-based generic system specification.* Thesis, Ecole polytechnique fédérale de Lausanne, Lausanne.

Lenz, M., B. Bartsch-Spörl, H.D. Burkhard, S. Wess (Eds.) (1998). *Case-based reasoning technology: From foundations to applications.* Springer.

HELPMATE Consortium - Auriol, E., B. Minguy,Ch. Scott, M.J. Tellefsen (2000). *D7.3 Definition of Best Practice.*

Althoff, K., E. Auriol, R. Bergmann, S. Breen, S. Dittrich, R. Johnston, M. Manago, R. Traphöner, S. Wess (1995). *Case-based reasoning for decision support and diagnostic problem solving: The INRECA Approach.* In Proc. Of the 3rd German Workshop on CBR, University of Kaiserslautern.

Copyright © IFAC Information Control Problems in
Manufacturing, Salvador, Brazil, 2004

ELSEVIER
IFAC
PUBLICATIONS
www.elsevier.com/locate/ifac

DISTRIBUTED EMBEDDED SENSORS BASED ON A NETWORKED REPLICATION SYSTEM

Perfecto Mariño, Francisco Poza, Manuel Ubeira
and Fernando Machado
{pmarino, fpoza, mubeira, fmachado}@uvigo.es

Electronic Technology Department, ETSI Industiales
University of Vigo 36200 Spain

Abstract: Power transformers' failures bring great costs to electric companies since they need resources to recover from them and to perform periodical maintenance. To avoid this problem in four working 40 MVA transformers, the authors have implemented the measurement system of a failure prediction tool, that is the basis of a predictive maintenance infrastructure. Applying Data Warehouse techniques, the prediction models have been provided with an abstraction of sensors the authors have called Virtual Acquisition Cards (VAC). By means of these virtual devices, models have access to clean data, both fresh and historic, from the set of sensors they need. Besides, a replication scheme was implemented to allow the distribution of demanding processing tasks and the remote management of the prediction applications. *Copyright © 2004 IFAC*

Keywords: Communication networks, Data replication, Electric power systems, Embedded systems, Preventive maintenance, Sensor systems, Supervisory control

1. INTRODUCTION

To minimize the number of failures in power transformers the authors have implemented the measurement system of a failure prediction tool (Fig. 3), which is the basis of a predictive maintenance project and has been named Predictive Maintenance Measurement System (PMMS) after that functionality.

Several models (Lamela et al., 1999; Rivera *et al.*, 2000) were designed to predict the behavior of transformers under the current and past state of the possible failure sources (humidity, vibrations, thermic energy, gases and aging). To feed models with the data they need to run their algorithms, sensors were installed in every transformer.

The prediction applications were released from data management duties and provided with Virtual Acquisition Cards (Jackson and Zave, 1998), which are an abstraction of the actual sensors that supply models with the data they need.

With the division of the project in three different subsystems: Data Acquisition System (DAS), Data Store (DS) and Prediction Applications (PA), and the creation of the VACs, a scalable and easily reconfigurable system was achieved, and the development time of the whole project was significantly reduced. The system has been complemented with the implementation of a Replication scheme.

2. SYSTEM ARCHITECTURE

Since models need to use past samples captured by any sensor, it must be decided where and how to store them. The obvious approach is letting the application manage its own historic data records,

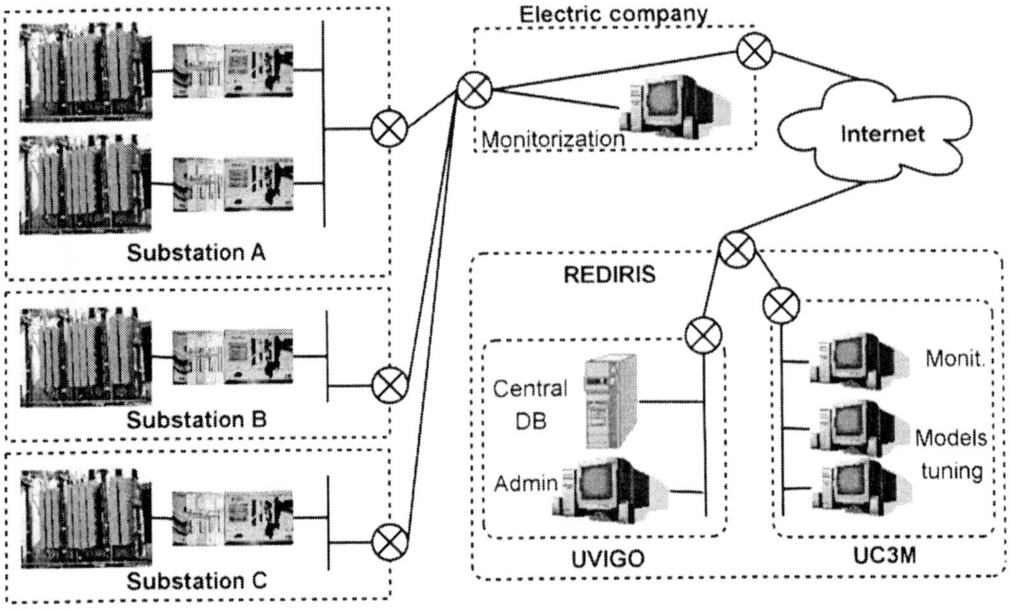

Fig. 1. Communication Networks infrastructure involved in the distributed embedded aplication

which would increase the application complexity unnecessarily, since it has to deal with both real-time and past samples simultaneously. Another solution would be to rely on the sensors themselves or the capture equipment for the handling of past samples. Now, the drawback would be that the capture system is usually subject to hard real-time restrictions, and unpredictable demands for historic samples could degrade performance such as to compromise this requirement (Kopetz, 1997). Besides, if independent memory units were attached to each sensor, the cost per byte would not be competitive with that offered by centralized approaches.

It is the wide availability of standard database interfaces with low cost per byte that makes them such a good choice for data storage. The interposition of a Data Warehouse (DW) between the capture equipment and the consumer applications (Figs. 2 and 3) releases both of data management tasks, hence reducing their complexity (Devlin, 1997; Inmon, 1993). This distribution of functionality (Fig. 3) among system elements eases management, since changes on either side do not affect the others necessarily. As a counterpoint, another subsystem has to be designed and run. It must be evaluated if upgrading from a monolithic system to a reconfigurable and scalable one is worth it, both in economic and development effort terms.

Besides of determining the conditions under which the introduction of a DW pays off, scalability and flexibility (see sections 4.3 and 5) has been kept in mind during the design stages. Therefore, the resulting system anticipates to the eventual need to accommodate new sensors or applications with different needs, making the actualization process fast, easy and much less error-prone.

The networks involved in the real working system are outlined in figure 1. There are three electric power transformation substations, from which four transformers are being monitored and analyzed. Two of them are installed at substation A and the other two at substations B and C. The measures obtained from the different sensors in the transformers are processed and stored in an industrial computer with PXI chassis. All of the subsystems depicted in figure 2 (that is the DAS, the database and the forecasting applications) reside and are executed on every computer.

To ease the adjustment of the models while interferences with the DAS process are avoided, the databases at the substations are replicated (see section 5) to a central database server that resides at the University of Vigo. All the models are adjusted against the central database remotely from the Carlos III University, where the prediction models are executed.

It is possible to perform remote administration and monitoring of the industrial computers at the substations (from the electric company and the Carlos III and Vigo universities).

The protocol stack of all parts involved in the prediction system is depicted in Figure 2. As it shows, the DAS (section 3) inserts data into the DS (section 4), to which it is connected through a switch. To maximize performance, the DAS makes use of a custom made library (C-DLL) that uses the native protocol of the database of choice (see section 4.1). The prediction applications access the DS through a LAN or WAN network using

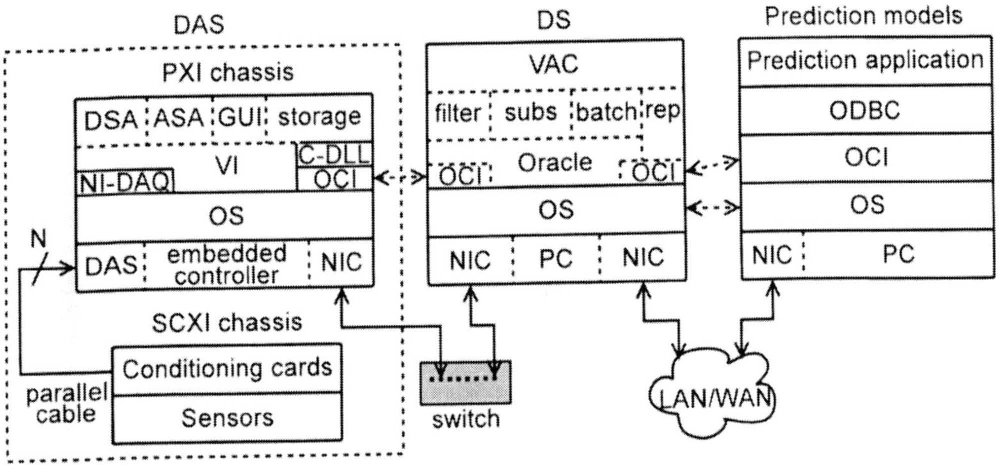

Fig. 2. Protocol stack of the PMMS plus the prediction models

the ODBC protocol. The prediction applications don't have to deal with sensors by themselves, in fact their access points to the measurement system are an abstraction of sensors called Virtual Acquisition Cards (see section 4.3). The rep level that appears in the DS refers to replication mechanism implemented (see section 5), that makes use of database facilities.

3. DAS

The DAS is the subsystem in charge for the measurement and process of variables in the power transformer. It is formed by a set of sensors, pre-conditioning and conditioning equipment and a data acquisition card.

The system uses 16 analog sensors: 3 PT100 for the measurement of temperature, 4 intensity transformers to change the measurement range of currents values in the high voltage (HV) and low voltage (LV) sides of the power transformer, 4 voltage transformers to measure HV and LV, 3 piezoelectric accelerometers for the measurement of vibrations, 1 capacitive hygrometer for measuring the moisture in oil and 1 sensor of gases.

Also 9 digital sensors are used. These sensors are free-voltage relay contacts, and they are used for monitoring the status of High Voltage switch, oil pumps and air fans of the transformer.

Some of the previously commented analog sensors have specific commercial pre-conditioning equipment. Thus the NEXUS equipment (Brüel & Kjaer) is used to convert the charge measurement from the piezoelectric accelerometers in a voltage signal. The DOMINO equipment (Doble) supplies a 4-20mA current signal proportional to the moisture measurement of the capacitive hygrometer. Finally the HYDRAN equipment (GE Syprotec) supplies a 4-20mA current signal proportional to the concertration of gases measured in its built-in sensor (see figure 4).

Several specific cards with Hall Effect sensors have been designed. These cards isolate and condition the voltage and current signals from voltage and current measurement transformers attached in High and Low Voltage sides of the transformer. They supply a ±20mA current signal proportional to the instant value of the variable measures.

After making the signal conditioning, the DAS processes the raw data from sensors before forwarding them to the next processing unit (the Data Store). Depending on models' needs, it performs several operations like FFTs or calculation of the RMS and mean values with some data signals.

4. DATA STORE

The Data Store (DS) is functionally split into: buffers and tables, filters, samplers and batch processor, and Virtual Acquisition Cards (Fig. 3). Tables store all the data coming from sensors, whereas buffers reflect only the last value available from every sensor.

There is a dedicated filter for every model that selects the last value from the required sensors, so that a model-specific view of the buffers is offered.

If multiplexing were the DAS' duty, it should be model aware and would have to be modified after any change on them. But this would not be a good approach, since coupling unrelated tasks, such as data acquisition and prediction, reduces the scalability and reusability of the system (Mariño *et al.*, 1999; Mariño *et al.*, 2000*a*; Mariño *et al.*, 2000*b*). The only potential advantage deriving from this solution is the speed gain that could make its use indispensable (Johnson, 1995).

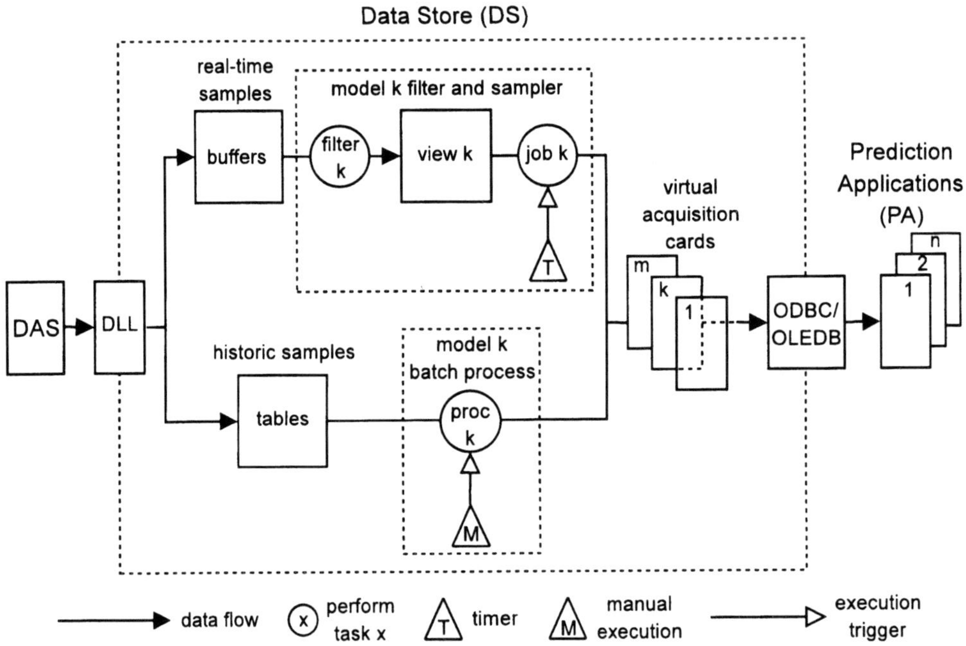

Fig. 3. Data Store (DS) structure into the Predictive Maintenance Measurement System (PMMS)

4.1 DAS/DS interface

The DS provides two access points or interfaces, as figures 3 and 2 show. The leftmost offers storage facilities to the DAS, regarding its real-time restrictions. The current implementation of the interface makes use of a new library (called DLL in figure 3), which uses the OCI (Oracle Call Interface), which is the API provided by Oracle to communicate with its database. This way the DAS will always use this higher level API, without ever noticing what is behind it. With the use of this technique, the same advantages that would come from the use of ODBC (Open DataBase Connectivity) or OLEDB (Object Linking and Embedding Database) are obtained, but with the significant difference that the developer is responsible for the programming of the abstraction layer. The speed gain achieved it is usually worth, if not compulsory, because of the high performance demands of the DAS.

The rightmost interface is used by the prediction applications, offering versatile access to historic and real-time data from the sensors, but suitably adapted to their needs. In this case the choice for communications was ODBC, because DS is application independent and it is wiser to use a widely known data access method, above all if speed is not a critical issue.

4.2 Reconfigurable Filters

The data transformation stage is a critical part of any data warehouse, since it usually determines the overall performance of the system (Vranes and Stanojevic, 1995; Keck and Kuehn, 1998). It is also very important to keep scalability and extensibility in mind during the design stage, for the data warehouse is likely to require additional features and raw-power as the application processing needs grow.

Each one of the filters picks the sensors it needs from the buffers, and arranges their values in the way required by its model. The transformation brings the model the view of the buffers it needs.

The characteristics of the filters can be individually and easily configured by means of several procedures in the DS, that can be called directly from prediction applications via ODBC. The prediction algorithm designer could also create a text file listing the sensors needed by every model and send it to the DS, where a batch process would be executed loading the new configuration into the database.

4.3 Virtual Acquisition Cards

Every view is periodically sampled by its own scheduled job, whose frequency is suitably configured after the prediction algorithm needs, independently of the other models. Afterward, the data is added some identifying information, such as time-stamps and a sequence number, and inserted in the model's tables. These tables along with their SQL (Standard Query Language) interface, reconfigurable filters and samplers form the core of Virtual Acquisition Cards.

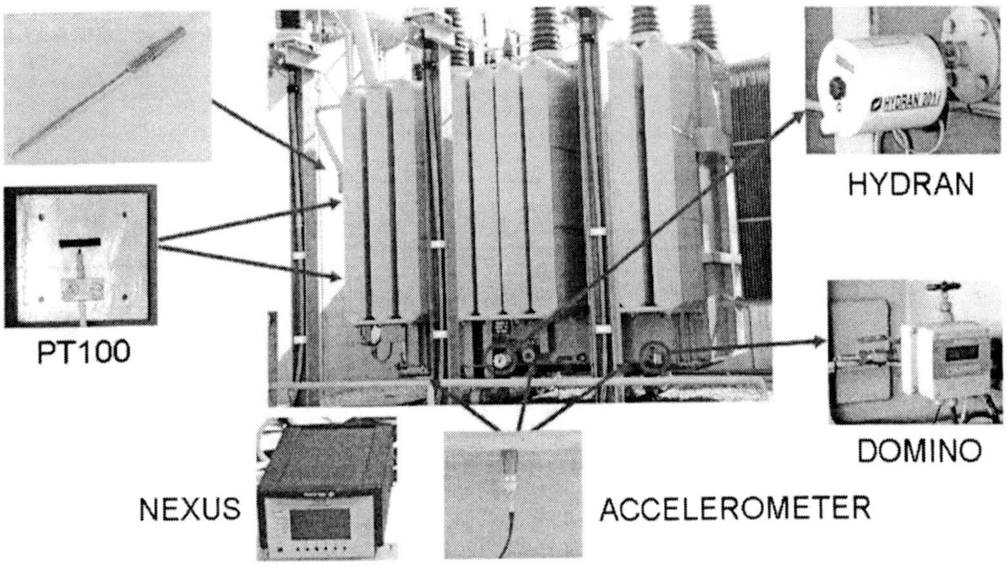

Fig. 4. Placement of sensors in the power transformer

By means of VACs, models have access to clean data, both fresh and historic, from the set of sensors they need. Besides, several characteristics of the data flow coming from the VACs, such as the sample rate or the set of sensors itself, can be dynamically reconfigured (as Fig. 3 shows) with no cost for the prediction application (Cowan and Lucena, 1995).

Models have indirect access to the historic data introduced from the DAS into the sensor tables. For this purpose, the old samples are batch processed (Fig. 3) to fill the model tables, following a configurable frequency and time interval.

The maintenance (installation, removal and modification) facilities for VACs are very powerful and of easy employment, thanks to the use of metadata and stored procedures.

5. REPLICATION AND SCALABILITY

VACs needed by some prediction models consume a lot of computational resources from the DS, which, under heavy loads, could keep out the DAS from inserting new samples into the sensor tables, thus losing data. The obvious solution would have been upgrading the computer, but this would have been too expensive and time consuming, for the system is to be installed in several stations. In addition, it appeared the need to monitor and control the systems at the distribution plants remotely, in a centralized manner.

To overcome these problems it was designed and implemented a network that connected all the distribution plants to a Central Server (CS), so that it could support a centralized and bidirectional replication scheme. Replication of data serves much more than backup needs, being specially useful in an scenario like the one under discussion, in which clone systems are run at different places.

On the one hand, the VAC problem mentioned above is completely avoided when replication is used, for the central server and the DAS are completely decoupled. Thus, running demanding replication models on the replicated environment has no effect in the remote stations (Babin and Hsu, 1996; Kulkarni and Ramirez, 1997). On the other hand, with this approach stations can be easily monitored (upward replication) and controlled (downward replication) remotely. The replication works as follows:

- Every DAS inserts data in its local DS (Fig. 3).
- In the CS there is a scheduled job that periodically connects to the databases at the remote stations to pull the changes occurred in the replicated tables since the last refresh. Thus, incremental actualization is performed, which minimizes bandwidth use (only changes are transmitted) and synchronization times.
- For every distribution plant, the CS performs a one-way replication of the sensor tables, so that changes occurring in the original tables are forwarded to the CS, whereas the others (coefficient and estimation tables mostly) are configured as updatable snapshots, which maintains the original and the snapshot tables synchronized with each other, thus allowing the prediction applications both to manage and to monitor the tables at the remote station. As a consequence, the station itself can be managed remotely, since tables determine the behavior of the maintenance system.

If more computational power was needed, only the central server would have to be upgraded, while systems at the distribution plants would remain unchanged. Hence, all the systems at the distribution plants and the central server form a very scalable system as a whole.

6. CONCLUSIONS

After having installed four clone PMMS the authors have concluded that the use of VACs reduce greatly the complexity and the development time of prediction applications, since they are completely released from preprocessing duties to get clean data, being able to focus on the prediction algorithm exclusively.

Furthermore, VACs can provide models with all the information from sensors they could ever need, since they can be easily reconfigured to offer any extra information. For the same reason, the development stages of the forecasting algorithms, in which the set of sensors and the sampling characteristics hasn't been established yet, speed up.

The division of the system in three decoupled systems (DAS, DS and PA) has simplified their parallel development by three fairly independent teams.

The modularity of the system has allowed the distribution of their parts. In fact, the replication infrastructure created allows the execution of CPU-intensive prediction models without interfering with the data acquisition tasks at the distribution plants. Thus the system is scalable, since it can deal with the increasing complexity of forecasting algorithms.

ACKNOWLEDGMENTS

This work has been sponsored by two R&D projects from the following entities: R&D National Secretaryship and CICYT, Ref. DPI2000-1702, Central Government (Madrid, Spain); and Presidency Department, Ref. PGIDT01TIC30301-PR, Autonomous Government (Galicia, Spain).

REFERENCES

Babin, G. and C. Hsu (1996). Decomposition of knowledge for concurrent processing. *IEEE Transaction on Knowledge and Data Engineering* **8**(5), 786–801.

Cowan, D. and C. Lucena (1995). Abstract data views: An interface specification concept to enhance design for reuse. *IEEE Transactions on Software Engineering* **21**(3), 229–243.

Devlin, B. (1997). *Data warehouse: from architecture to implementation.* Addison-Wesley. MA.

Inmon, W. (1993). *Building the Datawarehouse.* John Wiley.

Jackson, M. and P. Zave (1998). Distributed feature composition: a virtual architecture for telecommunications services. *IEEE Transactions on Software Engineering* **24**(10), 831–847.

Johnson, T. (1995). Approximate analysis of reader/writer queues. *IEEE Transaction on Software Engineering* **21**(3), 209–218.

Keck, D. and P. Kuehn (1998). The feature and service interaction problem in telecommunications systems: a survey. *IEEE Transactions on Software Engineering* **24**(10), 779–796.

Kopetz, H. (1997). *Real-Time Systems: Design Principles for Distributed Embedded Applications.* Kluwer Academic Publishers. Boston.

Kulkarni, U. and R. Ramirez (1997). Independently updated views. *IEEE Transaction on Knowledge and Data Engineering* **9**(5), 798–812.

Lamela et al., H. (1999). Characterization of optoelectronic sensors for temperature and vibration measurement in transformers. In: *14th Design of Circuits and Integrated Systems Conference.* Design of Circuits and Integrated Systems Conference. Palma de Mallorca, Spain. pp. 495–499.

Mariño, P., C. Sigüenza, J. Nogueira, F. Poza and M. Domínguez (1999). A reusable distributed software architecture driven by metadata. In: *Proceedings 1999 Asia-Pacific Software Engineering Conference.* Takamatsu, Japan. pp. 246–249.

Mariño, P., C. Sigüenza, J. Nogueira, F. Poza and M. Domínguez (2000*a*). An event driven software architecture for enterprise-wide data source integration. In: *Proc. Conf. on Information Technology: Coding and Computing.* Information Technology. Las Vegas, Nevada. pp. 140–145.

Mariño, P., C. Sigüenza, J. Nogueira, F. Poza and M. Domínguez (2000*b*). Metadata driven data paths: Improving data warehouse plumbing. In: *Proceedings 2000 IASTED International Conference on Applied Informatics.* Innsbruck, Austria. pp. 640–645.

Rivera, H. Lamela, J. García-souto and J. Sanz (2000). Measurements of mechanical vibrations at magnetic cores of power transformers with fiber-optic interferometric intrinsic sensor. *IEEE Journal on Selected Topics in Quantum Electronics* **6**(5), 788–797.

Vranes, S. and M. Stanojevic (1995). Integrating multiple paradigms within the blackboard framework. *IEEE Transactions on Sofware Engineering* **21**(3), 244–262.

Copyright © IFAC Information Control Problems in Manufacturing, Salvador, Brazil, 2004

ELSEVIER
IFAC
PUBLICATIONS
www.elsevier.com/locate/ifac

INCLUDING ARTIFICIAL INTELLIGENCE TECHNIQUES IN FOUNDATION FIELDBUS DEVICES

Gáudio Vinícius Lopes Besch
Sergio Viana Fialho

Federal University of Rio Grande do Norte (UFRN)
Graduate Program of Electrical Engineering

Abstract: The inclusion of Artificial Intelligence resources in Fieldbus devices increases their accuracy and provides the means for a more efficient operation. It is best done through the development of new function blocks compliant with Foundation Fieldbus standards. This approach just adds new components of software to the highest layer of the FF protocol stack, profiting from the underlying communication system. Also, FF standards provide mechanisms for creating new functional blocks preserving the interoperability, which is an important feature to be considered. So, a new Neural Network functional block is proposed and preliminary issues are discussed. *Copyright © 2004 IFAC*

Keywords: Fieldbus, Artificial intelligence, Neural network, Functional blocks, Process control.

1. INTRODUCTION

The next evolutionary step in control systems seems to be the inclusion of Artificial Intelligence (AI) in field devices. After all improvements, modern Fieldbus devices are now able to perform AI algorithms and cope with new demands from high sophisticated environments of industrial networks. Starting from the pneumatic systems, in the 30's, to digital industrial networks, in the 90's, control systems have evolved significantly (Berge, 2001). Now that Fieldbus equipment integrates computational abilities, such as powerful microprocessors and large memories, it seems natural to go one further step towards really intelligent devices. Besides, AI algorithms are better understood today and have already proven their efficiency in dealing with environments where traditional techniques are not robust enough to match the new challenges (Bartos, 2003).

Nowadays, Fieldbus devices are able to perform sophisticated functions like (Yokogawa, 2001):

- Self diagnosis to provide greater reliability to the plant,
- Processing of traditional control functions, such as PI, PID, Lead Lag Controllers and others,
- Calibration and scaling of inputs and outputs to match engineering units,
- Fully digital two way communication for control and monitoring purposes,
- Device configuration through the network.

Although these functions do not require AI, they imply processors, memories, software and a communication protocol integrated to the field devices (Tian, et al., 1999). These same resources can also permit the implementation of AI techniques within the instruments.

Among many Fieldbus standards, Foundation Fieldbus (FF) is one of the most successful, because it has been designed to be open since the beginning of its development and has also been supported by important international standardization organizations.

FF offers a good platform to allow the integration of AI capabilities into Fieldbus devices. FF compliant instruments provide an excellent processing capability and also native mechanisms to include new high level functions to the standards.

This paper begins with an overview about the main reasons for implementing AI methods in field devices and a brief description of the considered alternatives to achieve this goal with FF instruments. This is followed by a small introduction to FF protocols with special attention to the layer where Function Blocks take place. Finally, a new FF functional block is proposed to implement a Neural Network in field equipment.

2. REASONS FOR AI WITHIN FIELD DEVICES

When dealing with sensors and other input devices it is highly desirable to count on the capacity for self-calibration, validation and compensation.

Usually, common sensor or actuator devices use just a simple calibration curve based on linear interpolation. That may be not enough to provide the levels of precision required nowadays, since nonlinearities may cause significant mismatches not tolerable any more. Particularly in Brazil, regulations over fiscal measurements imposed by the governmental regulatory agency became very strict, after the privatization of oil industries.

A greater accuracy in a larger range of transducer values can be reached through the use of Neural Networks (NN). Calibration curves are also affected by natural shifts with time. So, it is possible to consider time as an additional input for the NN calibration function in order to get more precision. Historical data, supplied by previous calibrations, may allow NN to find the way the shifts depend on time.

Also, the adoption of fixed schedules for calibration and maintenance purposes may not be an intelligent way for asset management, because they do not take into account all the affecting conditions that degrade the device working. A really clever manner of triggering calibration procedures must consider environment conditions like temperature, pressure and humidity and, also and most important of all, the historical measurements done by the sensor or actuator itself (Pagnano, 2003). For instance, a pressuremeter that operates only near central points of its range should be permitted to have longer periods between calibration procedures than another, which works next to the higher limits of its envelope.

Measurement validation is another activity where AI techniques may apply. It requires the comparison of as much redundant data as possible. Again, NN can be used to make comparisons between multivariable input functions based on previous experience. Separating individual signals from mixed redundant sources is also possible in some cases, using, for instance, the method of Independent Component Analysis (ICA). Those methods are generally implemented through NN and may help to validate results.

Besides, some physical processes are too complex to be modeled by traditional analytical equations. AI has demonstrated more robustness to produce better results when an intuitive modeling is in order (Bartos, 2003). Up to now, systems using AI methods such as Neural Networks, Fuzzy Logic, Expert Systems and Genetic Algorithms have been implemented in central processors, due to the great demand of processing capability and other resources that were limited in field devices. Now, as the processing power of field instruments increases, the first cells of AI start to move towards the field.

3. DISCUSSION OF THE ALTERNATIVES

Taking advantage of its personal specific knowledge on computer networks and AI, the Department of Computational Engineering and Automation of UFRN started a new research project. Its long-term objective is to develop Fieldbus devices that support AI resources to improve calibration, validation and compensation procedures for flow measurement instruments.

Two basic approaches were discussed. The first one consists of adding new digital hardware into FF devices with embedded software implementing AI algorithms. This alternative has no links with the protocol stack itself, except, perhaps, for a local interface, which supplies data to be used by the FF stack. The newly developed software would be integrated to the new hardware. Data exchange between the new hardware and the original one would take place through internal physical connections inside the device.

The other alternative intends to profit from the mechanisms already established by the FF protocol. It consists of developing structures called Functional Blocks. These structures are formally modeled and described by the FF specs. They perform functional tasks usually found in control systems, which are mentioned in the first section of this article. The FF standard also describes the way new Functional Blocks can be created and included in the specs. Presumably, there is no need for adding new hardware, since the computational ability installed in actual specific FF devices may be enough to implement algorithms as complex as those used by AI. It is an elegant alternative, because it involves only the use of tools included in the network standards. However, the drawback of creating new Functional Blocks is the need for the development of programs with interfaces to the subjacent protocol layer, demanding a profound knowledge of the FF protocol stack.

4. FF PROTOCOL STACK OVERVIEW

The FF protocol stack is a fully digital industrial network standard. It is based on modern concepts like object oriented paradigm and provides much more than simply replacing the 4-20mA analog standard (Fieldbus Foundation , 1998).

First, it takes its concepts from the OSI network architecture, using some layers of its reference model, and adapting it to the industrial environment, as shown in Fig. 1 (Fieldbus Foundation, 2001).

Actually, there are two kinds of data buses defined by the FF standard: H1 and HSE (High Speed Ethernet). HSE follows Ethernet scheme, while H1 has a particular protocol stack (Sean, 2001). Our subject in this paper will be restricted to H1 issues, since HSE is better explained by Ethernet related material.

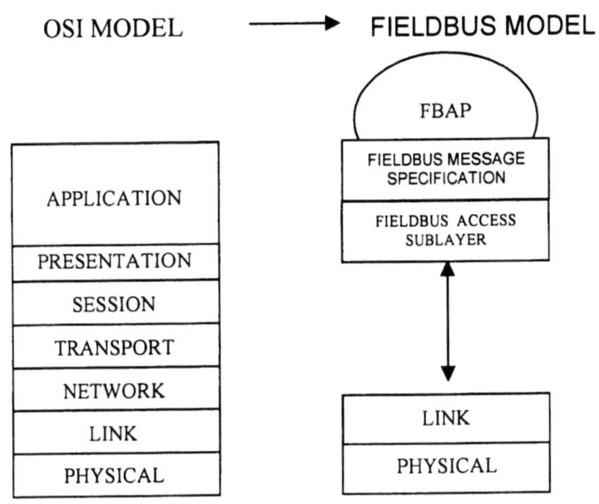

Fig. 1. Comparison between OSI and Foundation Fieldbus models

A User Application Layer was added by the FF standard on top of the OSI stack to provide high level objects, like AI (Analog Input) and PID (Proportional Integral Derivative) blocks, which were designed to perform usual control and monitoring functions in the context of process control.

4.1 Topology

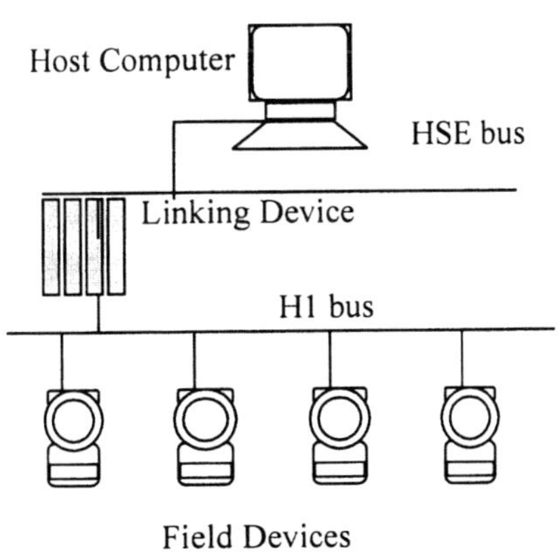

Fig. 2. Typical FF Topology

Fig. 2 shows the typical FF topology: the instruments are connected to a H1 bus and are able to exchange messages with each other, without the presence of a PLC. The communication with the computer hosting the supervisory system is made through the HSE bus, which connects to H1 bus by means of a special device named Linking Device.

4.2 Physical Layer

The physical layer has been developed to cope with the special characteristics of a harsh industrial environment. The main issues are the following:

- Rate – The transmission rate is 31.25 KBps (=1MBps/32). It is fast enough to allow control of typical processes, but sufficiently slow to permit the generation of low power signals of intrinsic safety quality with a reasonable S/N ratio.
- Power supply – Devices can optionally be, and usually are, powered through the same pair of wires that transport data. It saves cabling and space in the congested rooms of a plant.
- Data signal coding – The signal is a Manchester Byphase coded voltage signal of about 1 Vpp, resulting from current drain from field instruments. It permits a bus length of up to 1900 m, with up to 12 devices connected in parallel.
- Standard - The physical layer is defined by IEC 61158-2, which corresponds to FF-816. That guarantees a great acceptance both in America and in Europe.

4.3 Data Link Layer

This layer provides a deterministic strategy to control the access to the physical medium through a master-slave scheme. Every H1 bus must have a device playing the role of a master. It regularly distributes a token to each device, asking for data, so they all have precise instants to send data.

Data Link Layer follows IEC 61158 part 3 and 4.

4.4 Application Layer

This layer is divided into two sublayers: Fieldbus Access Sublayer (FAS) and Fieldbus Message Specification (FMS).

The FAS provides communication channels similar to network connections. They are intended to support three kinds of connection between devices.

The first one is the publisher-subscriber relationship. It is used for publishing control data from one sensor, for example, to one or more other instruments. That is the kind of connection essential to real time control. It permits messages to be sent periodically at exact moments.

The second one is the report-distribution relationship. It is planned to be helpful in alarm conditions. This kind of connection allows alert data to be presented at the bus without the need for polling the sources all the time. That approach increases the bus throughput.

The third one is the client-server relationship. Through this kind of channel, the server only sends messages when asked by the client. It is intended for configuration stage.

The FMS provides a set of standard services designed to supply all the needs of a control system, These services are accomplished through messages like the following ones:

- Read-Request / Read-Response
- Write-Request / Write-Response
- Information-Report
- EventNotification
- InitiateRequest / InitiateResponse

4.5 Function Block Application Process (FBAP).

The FBAP, is composed by many standard objects, which implement common functions in control and monitoring systems. These high level objects, called Function Blocks, provide an easy interface for the user to build control and supervisory systems. The FBAP is the layer where the AI tools may naturally be incorporated to the protocol stack.

In the FBAP model, every field device must have a representation in terms of blocks. So that, each physical device seems to the FBAP user just like a set of blocks, which perform functions and store information. These blocks have configurable parts to provide flexibility to the system. Blocks that exchange data may be located in the same device or in different devices connected by the network. When data is passed through the network, the messages used are those specified by FMS.

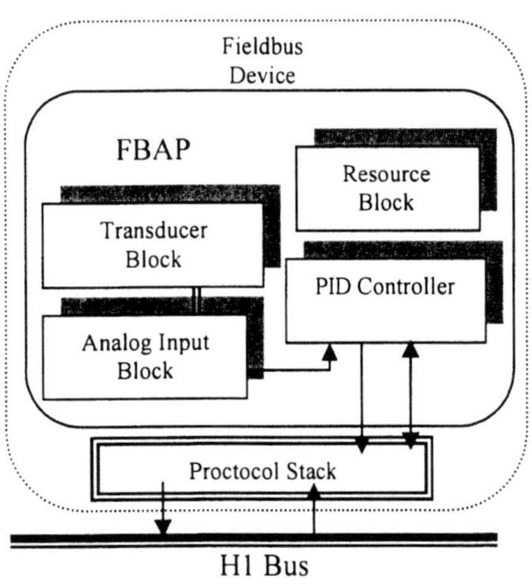

Fig. 3. FBAP example

Fig. 3 shows an example of a fieldbus device as it is represented in the FBAP layer.

Functional Blocks. Functional blocks implement usual control functions. Although new blocks might be included in the standard, the FF 890 Parts 1 and 2 list only ten functional blocks. They are the following:

- Analog Input - AI
- Discrete Input - DI
- Manual Loader
- Bias/Gain Station
- Control Selector
- P, PD Controller
- PID, PI, I Controller
- Ratio Station
- Analog Output - AO
- Discrete Output - DO

The list has already been increased by new parts of the FF specs to include advanced and flexible blocks, but AI blocks are still to be included.

Block Interfaces. Block interfaces are the main feature in the design of a new block. Actually, the FF standard defines only the communication interfaces between the functional blocks. It doesn't matter the way the block is implemented, provided that it responds properly to the FMS messages through standard interfaces. These interfaces are represented by parameters. They contain the data that really matters to the system, like process variables and configuration values. Basically, blocks are modeled by objects, which receive data from input parameters, process them and deliver output parameters. So, standard blocks have well defined parameters, which are addressed by the FMS messages.

These parameters may be implemented by simple variables, like floating point numbers or boolean values, or by structures containing more than one simple variable. There are many standard types that may be used to build parameters. Others types can be constructed from these predefined ones.

As for their usage, the parameters are classified in three groups: input, output and contained.

Block input parameters are able to receive values from output parameters of another block through periodic FMS messages. For instance, during the plant operation, a temperature sensor periodically sends messages, carrying the output parameter from its Analog Input Block to a PID control block possibly placed in another device. This latter receives the transmitted values through its input parameter, processes it and delivers the output control signal to the related actuator. This is done through a FAS sublayer publisher subscriber relationship. The publisher sends its outputs using InformationReport FMS messages and the subscribers interested in those data receive them regularly, without having to ask for them. That is the typical way of building a control loop. This kind of

FMS service cannot be used to convey contained parameters.

The contained parameters cannot be transmitted through a publisher subscriber relationship. They are usually dealt with during the configuration phase. Unlike the previous type, contained parameters are read and written through client server relationships. The messages allowed to convey this type of data are ReadRequest and WriteRequest for the Client and ReadResponse and WriteResponse for the Server.

PID Controllers can be used as an illustration. They receive inputs through IN and SET_POINT parameters, perform the PID algorithm tuned by contained parameters like GAIN, RESET and RATE, and the resulting control signal is output through the OUT parameter.

5. DEVELOPMENT OF INTELLIGENT FUNCTIONAL BLOCKS

5.1 Preliminary Studies

The main objective of the preliminary studies conducted by the project team has been to demonstrate the feasibility of using Neural Networks to model calibration curves and also to separate certain cases of noise from input signals.

The simulated calibration curves were implemented by a NN. The NN was trained in a computer until it reached its final configuration and then loaded into a DSP board, which performed the calculation from then on.

The first simulation tests generated nonlinear functions like logarithmic curves. They have shown that a simple two layer NN easily produces better results than a linear interpolation, considering just five points in a range from 1 to 100. Although the training of the NN demands a lot of data and computational effort, once this phase is completed, the requirements are very loose. The trained NN may be easily loaded into the device, while the process of training would probably have to be made off-line with powerful equipment, during the calibration procedures conducted inside a lab.

The separation of noise from input signals was implemented using the ICA method. This method has been devised to receive two or more combined signals from independent sources, including even noise, and output the individual signals separated from each other. Actually, it assumes that the noise and the signal were submitted to different unknown gains before they were combined. The algorithm tries to find out those gains by means of statistical treatment. Then, reversing the combining process based on the obtained gains, it is possible to separate the signals from each other.

Simulated square and sinusoidal wave signals were successfully separated in tests performed in a PC using ICA techniques, although the requirements on the processing capabilities have not yet been evaluated.

5.2 Issues on the Design of NN blocks

Fundamentally, designing a new FF block consists of determining the central algorithm to be performed, as well as defining the inputs, the outputs and the contained configuration parameters. For the intended purposes, the central process involved in a NN is the calculus based on synaptic gains. The input is the variable to be transformed by the NN and the output is the result obtained from that process.

The designer of the new block should first classify it into one of the standard block classes and determine its particular parameters. Other parameters are inherited, that is, they are mandatory for the class to which the block belongs. So, should a NN block be classified as a functional block, by heritage, it must contain parameters like MODE (mode of operation), primary input, primary output, among others. There are also universal parameters, which every block must include. They provide usual information, like block identification and the ability of being active or inactive.

When defining block parameters, one must also consider that every parameter must belong to a certain class of a defined FF standard type. For instance, output parameters are actually record structures containing a value and a status. The status indicates the quality of the value being output, so that the block that receives this output value can decide whether it should be used or discarded.

Since, in general, a NN is implemented by means of a matrix, the obvious contained parameter is a matrix of floating point numbers whose elements are the values of the synaptic gains. So, a first general approximation for a NN block is presented in the Fig. 4.

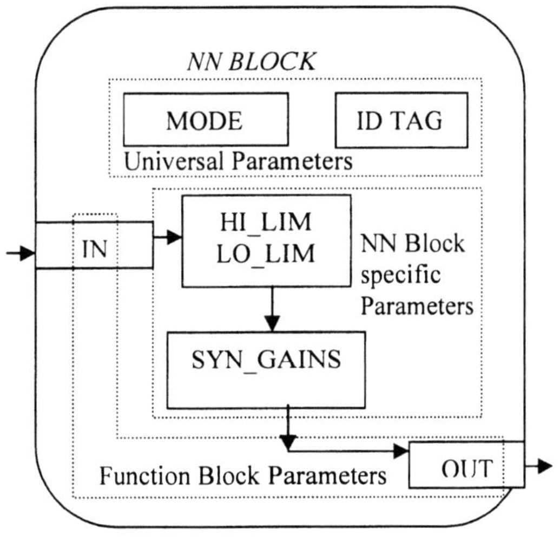

Fig. 4. First general approximation for a NN Block

Apart from the central algorithm, there must be a number of secondary tasks in order to put the block in the desired working mode. It means that the overall algorithm actually exceeds the main function to be implemented. So, in the case of a NN block, besides building a code to implement the network itself, it is also necessary to provide code for many other related operational functions. In the following, there is a list of some of those functions: validating the inputs, deciding what actions to take if the inputs are not reliable, selecting the inputs, simulating values for input and output.

Many other issues, mainly regarding safety aspects, should also be considered. For example, if the inputs are under abnormal conditions, the outputs must go to safe values. That is the reason why, usually, there must be some configuration parameters, like HI_LIM and LO_LIM, defining the limits for the inputs.

6. CONCLUSION

The integration of AI techniques in Fieldbus systems has been investigated, due to the expected benefits of the implementation of new and powerful features in field devices, such as the capabilities for self-calibration, self-validation and self-compensation.

Foundation Fieldbus devices are being specially considered, since its protocol stack is thoroughly specified and the whole documentation is now available. Besides, FF devices are usually complex enough to support the addition of the intended features.

Preliminary studies have shown that one of the best ways to introduce AI into the field devices is to create new standard functional blocks supported by the FF protocol stack. This is an attractive alternative because FF specs are open and offer a high level user interface and flexible tools to allow the creation of new sophisticated block functions. FF compliant instruments may integrate processing and communication capabilities, which support the implementation of complex algorithms and powerful data connections.

A first general scheme for a new functional block was proposed and some of the main related issues were identified. In the sequel, the complete block specification will be produced, following the language and procedures recommended in FF standards.

Further work is envisaged aiming at the validation of the formal block specification, prior to its actual implementation.

REFERENCES

Bartos, F. J. (2003). Artificial Intelligence ...Within. In: *Control Engineering*, Sep. 2003. Reed Elsevier Inc.

Berge, J. (2001). In: *Fieldbuses for Process Control: Engineering, Operation and Maintenance*, Pag 18. ISA - The Instrumentation, Systems, and Automation Society.

Fieldbus Foundation (1998) – *Technical Overview - Foundation fieldbus*, pag 4., Austin, TX.

Fieldbus Foundation (2001). *FF-890 – Foundation Specification – Function Block Application Process – Part 1*, Austin, Texas.

Pagnano, M. A. (2003). AssetView – A Ferramenta de Gerenciamento On Line de Ativos. In: *Controle & Instrumentação*, **80,** 51-55.Valete Editora, São Paulo.

Sean, J. V. (2001). *Foundation Fieldbus High Speed Ethernet Control System.* Fieldbus Inc., Austin.Texas.

Tian, G.Y., Z.X. Zhao and R.W. Baines (1999). A Fieldbus–based intelligent sensor. In *Mechatronics* **10 (2000),** 835-849

Yokogawa, E. C. (2001). *Fieldbus Book – A Tutorial*, Yokogawa Electric Corporation, May 2001.

Copyright © IFAC Information Control Problems in
Manufacturing, Salvador, Brazil, 2004

ELSEVIER
IFAC
PUBLICATIONS
www.elsevier.com/locate/ifac

DEVELOPMENT OF NOVEL RFID TAGS FOR IDENTIFICATION OF METALLIC OBJECTS

Leena Ukkonen, Lauri Sydänheimo, Mikko Keskilammi, Markku Kivikoski

Tampere University of Technology, Institute of Electronics, Rauma Research Unit
Kalliokatu 2, FIN-26100 Rauma, FINLAND
tel. +358 44 5341 507, fax +358 10 2930 262
leena.ukkonen@tut.fi

Abstract: This paper presents novel passive RFID (Radio Frequency Identification) tag designs mountable of metallic objects which can be used in different RFID applications such as supply chain management. The challenges when identifying metallic objects using RFID technology are also considered. The biggest challenge is developing low-profile tag antennas that can be attached to metallic surfaces. In this paper the effects of the vicinity of metal on antenna performance are discussed. Also the performance of antenna types that can exploit metallic surface in their function is presented and analysed.
Copyright © 2004 IFAC
Keywords: automatic recognition, metals, antennas, performance analysis

1. INTRODUCTION

The development of Radio Frequency Identification (RFID) systems has increased recently. RFID can be used for example in identifying objects in warehousing, supply chain management, control and other automation processes. RFID can also be used in developing smart products. In smart products information about the origin of contents or manufacturing process can be saved to an identification tag and can then be read by the consumer with an RFID reader. Information about the origin and manufacturing procedure of canned food can be useful to consumers. In the future RFID will also be used in smart shelf applications, where consumption of different products in the stores can be monitored. When a product is taken out of the shelf it is identified with a RFID reader that is located in the shelf. This way information of the amount of items in the shelf can be monitored. RFID can be utilized also in service logistics and supply chain management. In service logistics the identified objects can be for example metallic motors or other machine parts. In both applications gathering information about different phases of the process is important.

RFID systems consist of radio frequency transponders, i.e. tags, radio frequency transceivers, i.e. readers and a host computer. Tags that consist of an antenna and a microchip are attached to objects. In passive RFID systems, which are considered in this paper, reader unit and tag communicate via backscattering of electromagnetic waves. The reader sends energy to the tag, which then responds by backscattering its identification data to the reader. The tag gets all its energy from the electromagnetic field sent by the reader and therefore there is no internal source of energy in the tag.

For identification of different products a lot of different tag antenna designs have been developed, but one of the biggest challenges nowadays is tagging objects made of metal or other conductive material. These objects can be of varying sizes. Large objects, like the cars of trains and shipping containers, as well as small metallic food cans are required to be identified. Tags that are mountable on metal are needed both in supply chain management and service logistics applications. The vicinity of metal strongly affects the performance of antennas by for instance lowering the radiation efficiency and therefore novel antenna designs are needed for identification of metallic objects. In the past RFID tags mountable on metallic objects have been developed for 125 kHz system frequency. However, since lower frequency systems are based on inductive or capacitive coupling whereas UHF systems use backscattering of electromagnetic waves as coupling method, developing tags to work attached to metal is much more difficult at UHF frequencies.

Studies in different fields of RFID and antenna performance have been carried out at Tampere University of Technology recently. This paper describes the challenges and solutions to resolve the problems occurred in tagging metallic objects.

2. EFFECTS OF THE VICINITY OF METAL ON ANTENNA PERFORMANCE

Theoretically, when an infinite perfect electric ground plane is placed near a short linear antenna the effects can be analytically formulated. In case of a finite conductor plane that is proportional to wavelength and a nonideal antenna a numerical method, for example Finite Element Method, has to be used in antenna analysis. Studies on antenna performance near a conductive surface have been made (Raumonen, *et al.*, 2003) because it is important to know how the antenna performance changes in the presence of metal and these results can also be utilized when designing antennas to work attached to metallic surfaces.

A conductive surface near the antenna affects it performance in a number of ways. In some cases a metal plate near antenna can act as a reflector causing the directivity to increase. Also certain antenna types need a conductive ground plane to function properly. In these cases a metallic surface can be used to improve the performance of the antenna. On the other hand, the wave radiated by the antenna is almost totally reflected by the metallic surface since metal is highly conductive (Cheng, 1993). When electromagnetic wave reflects from metallic surface a 180° phase shift is occurred. This reflected wave cancels the incoming wave and therefore the radiation efficiency of the antenna decreases. These negative effects are strongest when the antenna is very near (for example at a distance of a couple of millimetres or less) the metallic surface (Raumonen, *et al.*, 2003).

In general metallic objects near RFID tags will result in high shut capacitance between tag and a metallic object, low tag voltage and short read range. If generally used label-fabricated RFID tags are directly attached to metal there is no voltage difference between the tag and the metal plate. Therefore the tag is not receiving the power sent by the reader and then will not be identified.

Also the increase in directivity has to be taken into consideration when identifying objects with RFID reader. Since the antenna is not omnidirectional and radiates away from its ground plane it has to be read from certain directions and positions (Ukkonen, *et al.*, 2003).

3. GENERAL ISSUES ABOUT DEVELOPING TAG ANTENNAS

Antenna designs for RFID applications should be small, lightweight, easy to attach to almost every product, cost-effective and capable of working in different environments. Also the system frequency has to be considered when designing tag antennas because it affects the size of the radiating element. The type of the microchip used has to taken into account because there is different joining techniques for different types of microchips. Also the matching impedance of microchips is important, because the antenna structure has to be matched to that impedance to guarantee the maximum possible signal to the microcircuit. The major considerations when choosing the tag antenna design are (Foster and Burberry, 1999)

- The type of antenna
- Its impedance
- RF performance when applied to object
- RF performance when the object has structures around it

Dipole and folded dipole antennas are generally used in RFID tag applications. The impedance of a folded dipole antenna is about four times that of a standard half-wave dipole and therefore folded dipole antenna is easier to match to the impedance of the microchip. With different two-wire and three-wire folded dipole antenna structures impedances that are about 20 times that of a standard dipole can be achieved (Finkenzeller, 2003). Figure 1 shows a folded dipole and a half-wave dipole antennas.

Figure 1. Structures of folded dipole and half-wave dipole antennas

3.1 Attaching the IC Chip to Tag Antennas

The matching impedance of the microchips used in this study was 1200 Ω. The microchips with metallic pads, i.e. straps, were developed by Alien Technology. The size of the microchips was 320 microns square. The width of the metallic pads was 2.5 mm and the total height of the strap was 7 mm. The microchip is electrically connected to the tag antenna via metallic pads. The strap can be contacted to the antenna structure for example by using conductive adhesives.

4. TAG ANTENNA TYPES WORKING ATTACHED TO METAL

When designing tag antennas to work attached to metal it is practical to use metal as a reflector or a ground plane of the antenna. The main idea is to develop antenna structures that, besides the radiating element, also contain other metallic parts in their structure. Therefore suitable antenna types for identification of metallic objects are for example microstrip patch antenna and inverted-F antenna (IFA). Both these antenna types need a metallic ground plane to function. The metallic object on which the tag antennas are mountable can be seen as an extension of the ground plane. All antennas presented here are designed to resonate at 916 MHz, which is the UHF frequency used in RFID in North America. The performance of different antenna types was studied by a series of FEM (Finite Element Method) simulations.

4.1 IFA with Different Substrate Materials

Inverted-F antenna (IFA) is a modification of a quarter-wave monopole antenna. The height of the antenna can be reduced by positioning the radiating element so as parallel with the ground plane while maintaining the resonant length. In Figure 2 the basic structure and dimensions of IFA are presented. The microchip is attached to the feed point between the ground plane and the radiating element (Balanis, 1997).

IFAs need a ground plane to function. When the antenna is designed to be connected to a metallic surface the conductive materials in tagged objects have very little effect on the antenna performance. This is due to using the metallic surface as the ground plane of the antenna. The metallic surface can also act as an extension of the antenna and improve its performance.

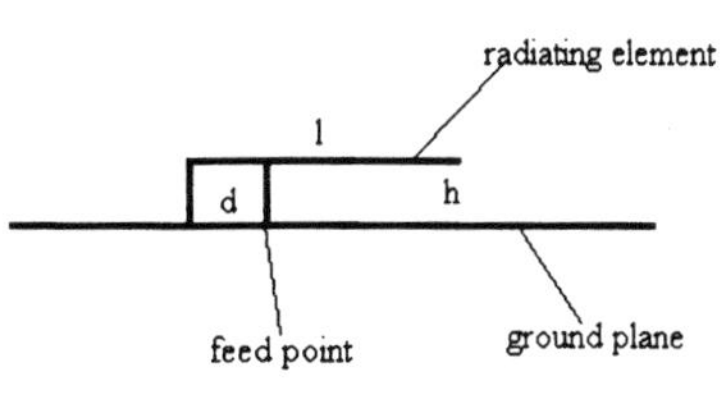

Figure 2. Basic structure and dimensions of IFA

According to earlier studies (Lo and Hwang, 1998; Huang and Wu, 1998) multilayer substrates can be used to improve the antenna parameters of microstrip antennas. Figure 3 shows a typical two-layer structure with substrate and superstrate. A superstrate layer reduces surface waves and so improves the radiation efficiency. The dielectric constant of the superstrate layer is higher than the dielectric constant of the substrate layer. In our study this idea was exploited by adding substrate and superstrate materials to IFA.

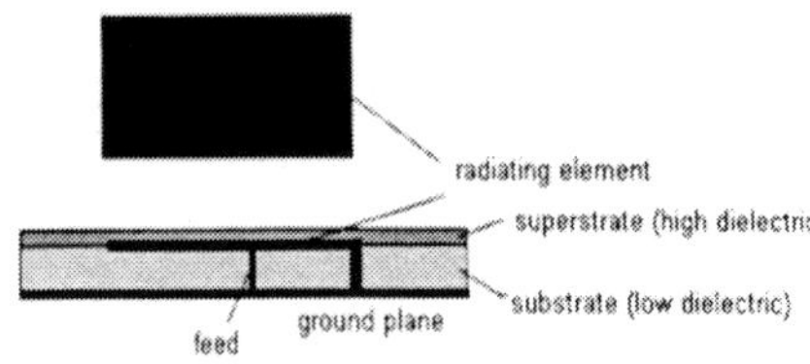

Figure 3. A microstrip patch antenna with substrate and superstrate

In this research, two different substrate materials were used and the antenna was simulated and tested also without any substrate. The design process was started with an IFA with no substrate material. The desired resonant frequency was attained with h = 10 mm, l = 67 mm and the feed point distance d = 7 mm. The matching can be optimised by changing the distance d.

A 10 mm layer of Teflon (dielectric constant ε_r = 2.35) was then added under the antenna element to develop the antenna structure. The substrate material lowered the resonant frequency and therefore the new shortened length was 53 mm while height remained at 10 mm and d at 7 mm. The impedance matching was improved after adding the substrate.

The third step was adding a 3.175 mm thick Rogers TMM10i layer (ε_r = 9.8) above the antenna element and Teflon layer. By using this two-layer structure the antenna element could be shortened again. The impedance matching was further improved compared to first two cases. The new dimensions were l = 34 mm, h = 10 mm and d = 8 mm. The length of the antenna when two substrate layers are used is almost half that of the length without any substrate materials. The height of the antenna element remained at 10 mm in all cases because IFA demands a certain height from the ground plane.

The substrate materials make the antenna structure more robust in applications, but also add weight to the tag. This and the height of the antenna may cause some limitations in objects to which these designs can be attached. On the other hand, wire-type IFA can be easily mounted on different products, because the antenna element can be bent slightly for example around a can and it is relatively cost-effective. In fabricated prototypes the size of the substrate and the ground plane were equal. However, in practical applications both sizes can be diminished to obtain more compact tags. Figure 4 shows a fabricated prototype of IFA without a substrate.

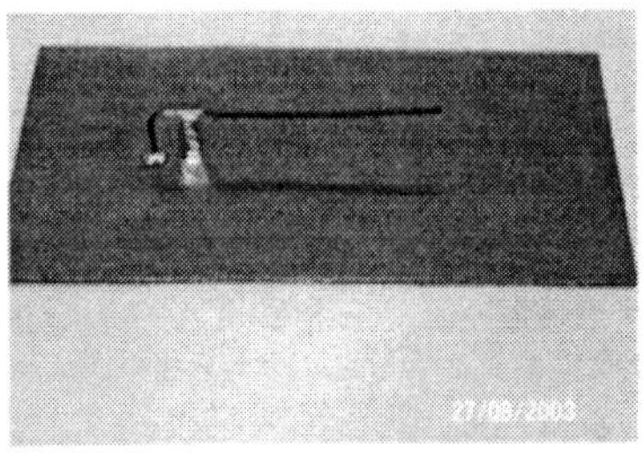

Figure 4. A fabricated prototype of IFA.

4.2 Printed IFA

IFA can also be a printed, microstrip type antenna or it can be made of copper wire, which is placed on top of a substrate. Figure 5 presents a basic structure of printed IFA. The radiating element is connected to the ground plane through the substrate material.

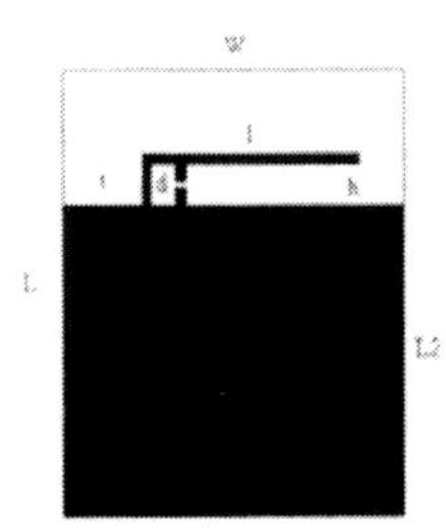

Figure 5. Basic structure of printed IFA

In this study the radiating element is made of copper wire. The diameter of the wire was 1 mm. Only one layer of substrate material does not insulate the antenna structure from metallic surface and therefore two layers of substrate material have to be used. The additional PCB (Printed Circuit Board) layer is placed under the antenna structure presented in Figure 5. The side profile of the two layers is shown in Figure 6.

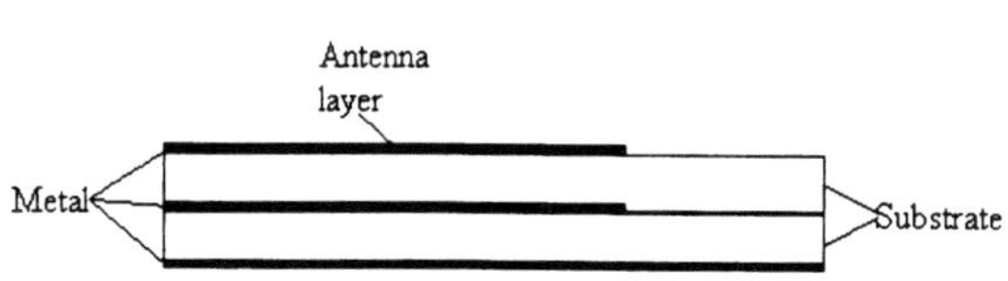

Figure 6. Side profile of the two-layer structure

Rogers 5870 printed circuit board with $\varepsilon_r = 2.35$ and Rogers 3203 with $\varepsilon_r = 3.02$ were used when printed IFA prototypes were fabricated. The physical dimensions of the fabricated model which are shown in Figure 5 were for Rogers 5870 L = 100 mm, W = 80 mm, L2 = 70 mm, length of the radiating element l = 50 mm, distance of the radiating element from the ground plane h = 11 mm, distance of the feed, i.e. strap, d = 4 mm and t = 19.5 mm. When Rogers 3203 was used the dimensions were L = 100 mm, W = 80 mm, L2 = 70 mm, l = 42 mm, h = 11 mm, d = 3.5 mm and t = 14.5 mm. Also in the case of printed IFA the matching can be improved by optimising the distance d. Other dimensions that have an effect on the antenna performance are the length l of the radiating element, height h and the size of the ground plane. When two PCB layers are used the total height of the antenna structure is 6.455 mm for Rogers 5870 and 3.153 mm for Rogers 3203.

Printed IFA can be easily mounted on different containers and its height is lower than the one of regular IFA's. The structure is also robust and relatively easy to fabricate.

4.3 EBG Patch Antenna

Patch antennas are microstrip-type antennas that are typically fabricated on printed circuit board. The basic structure of a patch antenna is shown in Figure 7. The dimensions of a patch antenna are proportional to wavelength and the width and length of the patch can be for example $\lambda/4$ (Balanis, 1997).

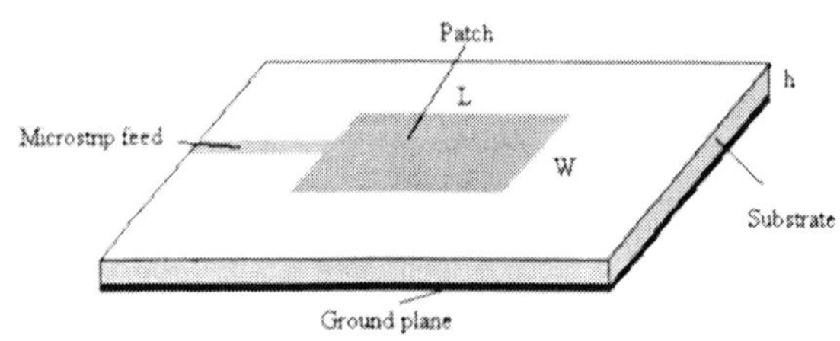

Figure 7. Basic structure of a patch antenna

Different EBG (Electromagnetic Band Gap) structures can be used to improve the radiation parameters of patch antennas. EBG structures are periodic structures that act as a resonator at a certain frequency and forbid the propagation of electromagnetic waves. The size of these structures is proportional to wavelength (Rahmat-Samii and Mosallaei, 2001).

In recent studies the following improvements in the antenna performance have been observed when using EBG structures (Rahmat-Samii and Mosallaei, 2001; Gonzalo *et al.* 1999; Yang and Rahmat-Samii, 2002):

- Increases in gain and radiation efficiency
- Smaller side and back lobes in the radiation pattern
- Increases in bandwidth

When designing antennas to work attached to metal these are the most important improvements.

The EBG structures used in this study were periodically etched holes in the ground plane of the antenna. Two-layer structure was also used with EBG patch antenna similarly as with printed IFA. The patch and EBG ground plane were fabricated on the first PCB layer. The bottom of the second layer was uniformly metallic. The diameter of etched holes varied from 18 mm to 20 mm. Figure 8 shows the three different EBG ground plane structures used in this study.

EBG ground plane 1

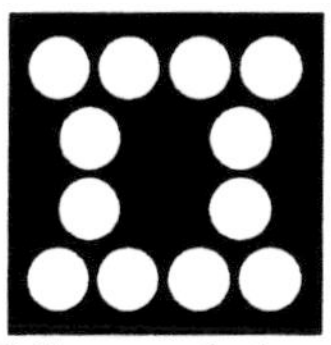

EBG ground plane 2

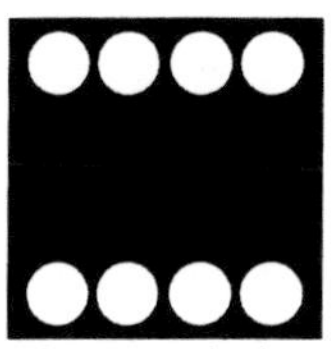

EBG ground plane 3

Figure 8. EBG ground plane structures used with patch antenna

The PCB used when EBG patch prototypes were fabricated was Rogers TMM10i with ε_r = 9.8. The ground plane width and length were 100 mm. The width of the patch was 49 mm and its length was 47 mm. The patch was fed with a 3.5 mm wide microstrip line and the strap was placed as the feed point of that microstrip line. The feed was connected to the EBG ground plane. Total height of the two-layer antenna sturucture was 6.455 mm. By using EBG patch antenna relatively thin and robust antenna structures can be achieved.

5. READ RANGE MEASUREMENTS ATTACHED TO METALLIC OBJECTS

5.1 Comparison of Read Ranges with and without Metal

The read ranges were measured by using Alien Technology 916 MHz reader unit and software. The reader antenna used was linearly polarized. The read ranges of the designs were measured with the original ground plane in air and attached to copper plate. The length of the copper plate was 160 mm and its width was 200 mm.

When IFA was tested it was observed that attaching the tag to metal does not shorten the read ranges. On the contrary, when Teflon substrate is used the read range improves by about 0.4 m attached to metal. Also the amount of reads per second of all prototypes except IFA with no substrate increases when metal is added. The maximum amount of reads per second of these designs is approximately 170, which is fairly good.

The substrate material of IFA has an effect on the variation of the number of reads per second. When no substrate is used the scale of variation increases when 0.8 m read range is exceeded. The number of reads even drops to zero occasionally. When Teflon substrate is used the scale of variation remains stable and there are no drops to zero. Dropping to zero is due to poor energy consumption from the electromagnetic field to the microchip in read ranges near or over the maximum read range. In the case of the two-layer substrate it was also observed that the number of reads per second drops to zero when 0.3 m read range was exceeded. The substrate layers attenuate the incident wave and therefore read range is short compared to other prototypes although the simulated return loss was good for the two-layer substrate case.

The read ranges of printed IFAs attached to metal varied from 0.2 m to 0.8 m. The printed circuit board used affects the read ranges of printed IFA. When thicker Rogers 5870 PCB was used longer read ranges were achieved than with thin Rogers 3203. Occasional drops to zero were also observed when printed IFA was tested. Also the variance of reads per second increases as proportional to read range. Maximum amount of reads per second for printed IFA attached to metal was approximately 160.

Read ranges of the EBG patch antenna are affected by the EBG ground plane used. When a patch antenna with Structure 1 was attached to metal 1.2 m read range was achieved. Different structures work differently attached to metal, which affects the amount of reads per second. Adding the metal plate to EBG patch antenna does not significantly affect the read ranges. Maximum amount of reads per second for EBG patch attached to metal was approximately 160.

5.2 Read Range Measurements Attached to Metallic Cans

Read ranges of EBG patch antennas and printed IFA antennas were also measured attached to a six-pack of metallic cans as shown in Figure 9. When printed IFA was used 0.2 m – 0.5 m read ranges were achieved. The relative shortness of the read range has to be taken into account in applications, but it is sufficient for example for smart shelf applications. When EBG patch antennas were attached to a six-pack of metallic cans 0.8 m read range was achieved.

Figure 9. EBG patch tag attached to six-pack of metallic cans

6. CONCLUSIONS AND FUTURE WORK

This paper presents novel tag designs to be attached to metallic surfaces. Also the challenges when identifying metallic objects with RFID technology are discussed. The biggest challenge is developing low-profile tag antennas that can be attached to metallic surfaces. Metallic surface can be used as a ground plane of an antenna and therefore suitable antenna types are microstrip patch antenna, inverted-F antenna and printed inverted-F antenna. All these antenna types can use the metallic object as an extension of their ground plane and also as a reflector. Different substrate materials can also be used to reduce the size of antenna and improve its performance.

Continuous development of tag antennas that are mountable on metallic objects is going on at Tampere University of Technology. The goal is to further improve the performance of these antenna types so that they can be used in identifying metallic objects with RFID technology in different applications.

7. REFERENCES

Balanis, C. A. (1997). *Antenna Theory, Analysis and Design,* John Wiley & Sons, USA.

Cheng, D. K. (1993) *Fundamentals of Engineering Electromagnetics,* Prentice Hall, USA.

Finkenzeller, K. (2003) *RFID Handbook, 2nd Edition,* John Wiley & Sons, UK.

Foster, P. R. and R. A. Burberry (1999). Antenna Problems in RFID Systems. *RFID Technology (Ref. No. 1999/123), IEE Colloquium on,* pp. 3/1 - 3/5, 25.

Gonzalo, R., P. De Maagt and M. Sorolla (1999). Enhanced Patch-Antenna Performance by Suppressing Surface Waves Using Photonic-Bandgap Substrates. *Microwave Theory and Techniques, IEEE Transactions on,* **Volume: 47,** Issue: 11.

Huang, C-Y. and J-Y. Wu (1998). Compact Microstrip Antenna Loaded with Very High Permittivity Superstrate. *1998 IEEE International Antennas and Propagation Symposium,* **Volume 2**, pp. 680 – 683.

Lo, T. K.-C. and Y. Hwang (1998). Bandwidth Enhancement of PIFA Loaded with Very High Permittivity Material Using FDTD. *1998 IEEE Antennas and Propagation Society International Symposium,* **Volume 2,** pp. 798 – 801.

Raumonen, P., L. Sydänheimo, L. Ukkonen, M. Keskilammi and M. Kivikoski (2003). Folded Dipole Antenna Near Metal Plate. *2003 IEEE International Antennas and Propagation Symposium,* **Volume 1,** pp. 848 – 851.

Rahmat-Samii, Y. and H. Mosallaei (2001). Electromagnetic Band-Gap Structures: Classification, Characterization, and Applications. *11th International Conference on Antennas and Propagation,* **Volume 2,** pp. 560 – 564.

Ukkonen, L., P. Raumonen, M. Keskilammi, L. Sydänheimo, M. Kivikoski (2003). Challenges in the Development of Tag Antennas for Passive RFID of Metallic Objects. *XXVIII Convention on Radio Science & IV Finnish Wireless Communications Workshop*, pp. 267 – 270.

Yang, F. and Y. Rahmat-Samii (2002). Applications of electromagnetic band-gap (EBG) structures in microwave antenna designs. *Microwave and Millimeter Wave Technology, Proceedings on, ICMMT 2002, 3rd International Conference on,* pp. 528 –531.

Copyright © IFAC Information Control Problems in
Manufacturing, Salvador, Brazil, 2004

ELSEVIER
IFAC
PUBLICATIONS
www.elsevier.com/locate/ifac

CONTROL OF ROBOTIZED DISASSEMBLY CELLS IN ELECTR(ON)ICS INDUSTRY

P.Kopacek[1] and B.Kopacek[2]

1) Vienna University of Technology, Vienna Austria
kopacek@ihrt.tuwien.ac.at
2) ECOTRONICS GmbH, Vienna, Austria
bernd.kopacek@ihrt.tuwien.ac.at

Abstract: Fully or semi-automated disassembly will gain more importance in the nearest future especially for electr(on)ic waste. A flexible, modular system for the development of semi-automatized, intelligent disassembly cells including stationary robots and especially a "low cost", hierarchical control structure is presented. The control structure is applied to an industrial disassembly cell for electronic parts on printed boards (PCB's) in use since begin of 2001 and for cellular phones since 2003. *Copyright © 2004 IFAC*

Keywords: Disassembly, Robots, Hierarchical Control, Electronic Scrap.

1. INTRODUCTION

In Europe there is now a clear societal need for take-back and recycling of electronic products (e.g. computers, printers, telephones, household-machines, cellular telephones, ...). "Closing the loop" would serve several (partly interlinked) goals, such as:

- reduction of the amount of materials and environmentally relevant substances going to landfill,
- control of environmentally relevant substances (inorganic) or annihilation of such substances (organic).
- Recycling of components and materials so that they keep maximum value, contributing to sustainability.

The tremendous increasing amount of electro-mechanical and electr(on)ic products to be recycled (and also to be disassembled), makes it necessary to partially automate this disassembly process to increase the efficiency (Hesselbach (1998), Eckert (1998)). The need for such semi-automated solutions can be estimated from the German electrical and electronic equipment market, where more than 2 million tons per year of electronic scrap are expected for the next decade (Kopacek, 1999). Within the EU the electronic waste has already reached staggering 10–12 million tons a year. In some branches there is a tremendous growth. One result of a recent study is that if the current German pace of discarding 3–4 million computers per year continues, disposal costs would alone reach a staggering 200–300 million ECU per year.

Therefore semi- or fully automatized disassembly especially of selected electr(on)ic devices is not only because of the standardization by the European Commission (WEEE Directive) today a hot topic. Today only the toxic components are removed manually and the rest of the materials are shredded and deposed. In Austria as a first implementation of the European WEEE concept was the foundation of the "Multi Life Cycle Center Vienna – MLC^{2}V". One of the tasks of this new centre is to build prototypes of semi automatized disassembly cells for various electr(on)ic products.

Economic disassembly is only possible today by modularity of the robotized cells. The main innovative features of such cells will be: Ability to disassemble different products with only few software modifications of the system, low investment costs, step-by-step investments because of the modularity, the predominant working conditions (toxic vapours ...) could be changed efficiently by the system. Because of the modularity it is easily possible to build up disassembly cells for different product groups.

„Stiff" automatized disassembly in single purpose cells – only for one product – cannot be economically operated today. The number of devices to be collected and concentrated on the place of the disassembly cell is usually to low for a two shift working of the cell. Therefore forming of relevant

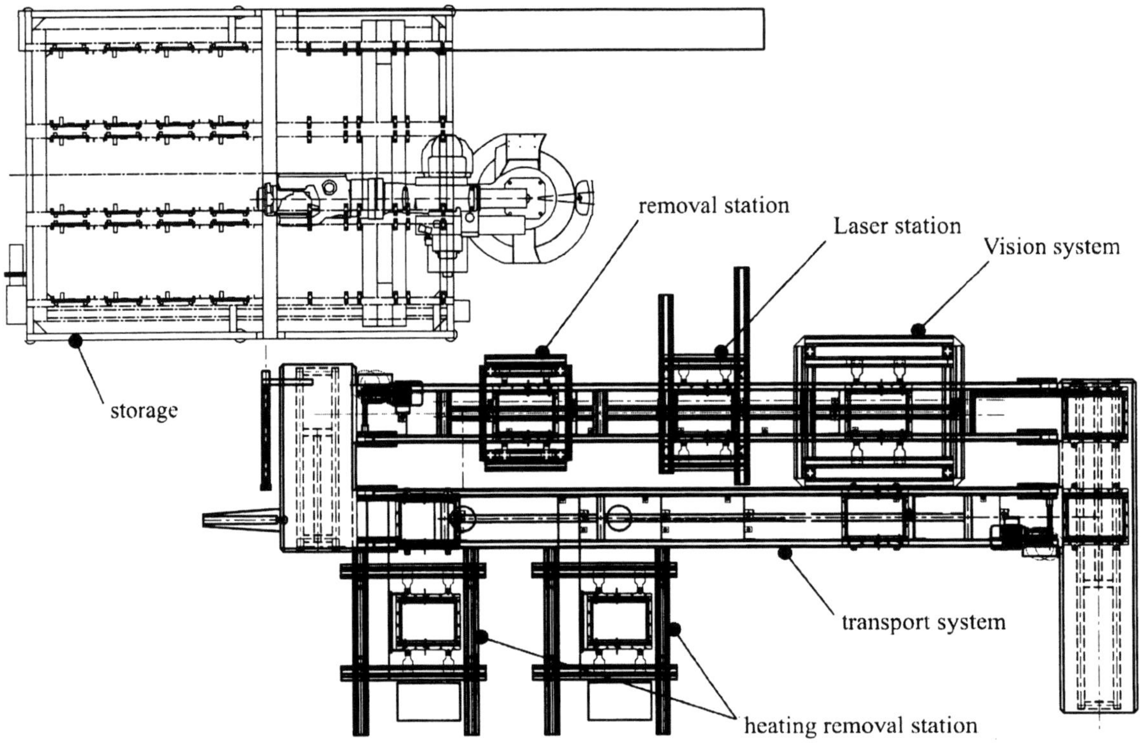

Fig. 1. Layout of the cell

product groups will be the key factor for the whole system. These so called **„disassembly families“** are groups of similar or different products that require nearly the same disassembly operations.

2. EXAMPLES

A disassembly cell for Printed Circuit Boards (PCBs)

The layout of the cell is shown in Fig. 1. The basis of a disassembly cell is a very stiff frame construction developed from commercially available light weight profiles.

The disassembly cell consists of 5 stations:

1. Vision system
2. De-soldering system
3. Removal-station.
4. Infrared heating station
5. Manual feeding and removal station

The vision system has several tasks. It has to recognize the re-useable parts by means of a data base containing the data (kind, production company, assigned, dimensions). It has to detect these parts and to determine the position, the size and the centre of inertia. Furthermore it has to classify the useable parts to be de-soldered or removed from sockets.

The de-soldering station consists of a cross table – two linear axes – controlled to reach every point (center of inertia) on the PCB. The de-soldering process is carried out by laser technology. The de-soldered parts are put on a distinct area outside the laser from which they are removed by the industrial robot and to put into the appropriate magazines.

The third station is the removal station for socketed parts. An industrial robot equipped with special grippers as well as external sensors carries out process. The robot removes these parts and puts them also in the right magazines.

For large, discrete and cheap parts like transistors, condensers, inductors the heating-removal station was developed. Such parts are de-soldered with temperatures between 110 and 380 degree centigrade and they should not be overheated because of their functionality. These parts are usually distributed over the whole PCB. The kernel of this station is a temperature controlled heating device with a power consumption of 3 kW. The control of this station gets information from the vision system and the host computer. In addition to the other stations he receives in addition the temperature limit of heating of each discrete part. According to this information a suction gripper or a computer controlled parallel gripper mounted on a cross-table picks up the parts during the heating process. The parts are then removed in special developed storage devices consisting of 8 plates. After one shift (8 hours) the de-soldered parts will be removed from the station.

In a manual feeding station the PCBs with a maximum size of 300 x 220 mm are attached on special work holding device.

Two of these cells are working since 2001 and 2003 respectively.

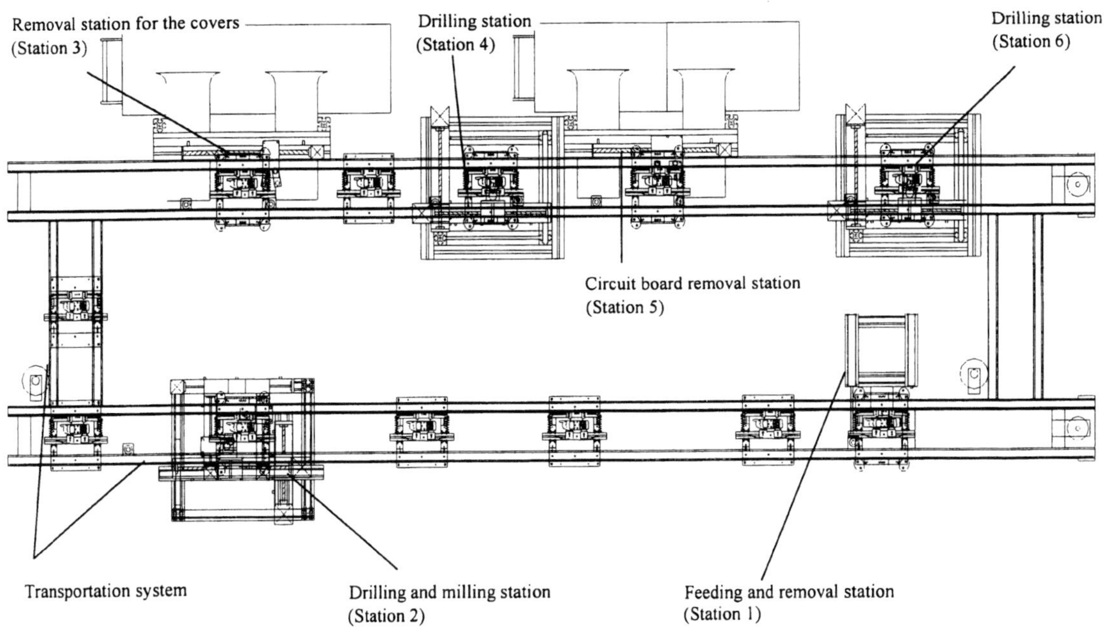

Fig. 2. Layout of the disassembly cell (Kopacek, 2003)

Disassembly cell for mobile phones

Fortunately new mobile phones (2–3 years old) have more and more a standardized construction. The size of the mobile phones will be smaller and smaller – it is only determined by the size of the human fingers to operate the keyboard. Several years ago some producers of mobile phones tried to reduce the number of push buttons on the keyboard. After a very short time they recognized that the users didn't accept this new philosophy. Usually in mobile phones there are lot of re-useable parts available. These are usually some parts e.g. contacts, BGAs, ... from the PCBs. The cover and the housing of the handy is usually a very expensive raw material.

After a detailed analysis of used mobile phones concerning the parts as well as the assembly technology and tests for disassembly with the most frequent mobile phones the following concept for the disassembly cell was created (Fig. 2). It consists on five automatized stations plus a manual feeding and removal station:

- Feeding and removal station
- Drilling and milling station
- Removal station for the covers
- Drilling station
- Circuit board removal station
- Drilling station

The mobile phones were fixed on a pallet in a distinct position. These pallets are moving around on a transportation system. According to the necessary disassembly operations the pallets with the mobile phones to be disassembled are stopped, lifted and fixed in distinct stations.

Before the mobile phone is fixed on a pallet the power supply will be removed and the type of the handy will be recognized by a barcode reader manually. Now the control computer knows exactly the type of the handy. The main dimensions of the handy are stored in a database of the host computer.

In the drilling and milling station (no. 2) the upper part of the handy will be cut off from the lower part and the screws – usually between 4 and 17 – are removed by a simple drilling mechanism. The dust content is removed by air from the pallet.

In the third station – the cover removal station – the cover as well as the keyboard of the handy will be removed by pneumatic sucks. These two parts are separated in two storage devices. In the next station – drilling station; no. 4 – the screws which connect the printed circuit board on the lower part of the housing are removed. In the printed circuit removal station various other parts will be removed from the handy and separated in special storage devices. Because some mobile phones have additional parts connected with the power part of the housing of the handy the remaining screws will be removed in the last drilling station – station 6. Finally the lower part of the handy will be removed in the fixing and removal station.

In developing of this semi-automatized disassembly cell for used mobile phones some previous tests were necessary. For the milling in the drilling and milling station (no. 2) it was necessary to make tests with grinding wheels, with different saws and with milling devices. Finally a milling device was chosen as the most appropriate tool for this task.

Further extensive tests were carried out for the removal of the screws. From the literature there are very high sophisticated, complicated and therefore very expensive and heavy devices known. We found a very simple and very cheap method for the removal of the screws.

For a proper function of the cell a lot of previous tests for solving the following problems were necessary.

1. There are only unsafe information available to recognize from the barcode the exact type of the handy.
2. The mobile phones have different dimensions and different forms.
3. The producers of the mobile phones use usually very different screws with different heads but fortunately with nearly the same diameter.
4. The materials are quite different – therefore it was very difficult to obtain the optimum milling and drilling speed.
5. Some of the mobile phones are not assembly-oriented constructed – e.g. some damping material is glowed on the plastic parts of the mobile phones.
6. It is very difficult to remove some spare parts like microphones, contacts, antennas, LCD displays from the mobile phones.

In former times the construction of the mobile phones was quite different – fortunately in new mobile phones the construction goes towards standard. One cell is working since June 2003 and a second is in construction.

3. HIERARCHICAL CONTROL OF THE DISASSEMBLY CELLS

For this modular, flexible, semi-automated, intelligent, "low cost" disassembly cell a hierarchical control concept is necessary. In such cells binary as well as digital information (data) have to be submitted and handled. Therefore the control hardware consists of a control computer (CC) for data processing and a PLC (programmable logic controllers) for handling of binary information (Fig. 3).

It consists of one CC connected to a database (DB), several PLC's and local control units with an appropriate user interface. In the DB information on the re-usable and therefore parts to be disassembled are stored. This connection is bi-directional: the CC can ask the DB for more or less detailed information about these parts. On then other hand the DB can give additional information probably necessary for recognizing reusable parts. Based on the requests of the CC the DB is able to "learn" – collect and store – additional information (intelligent or self learning DB). The CC supervises the PLC`s and is responsible for logistics in the cell. CC submits binary information like data transmitted to station x; the PLC have to quit this information. Because of the education of the operators of such a cell a user friendly interface is absolutely necessary. Usually a panel with three push bottoms (START, STOP, SECURITY STOP) and several warning devices (lights, sound, ...) might be appropriate. For the system supervisor a computer keyboard, a screen, and a printer are necessary. For very simple

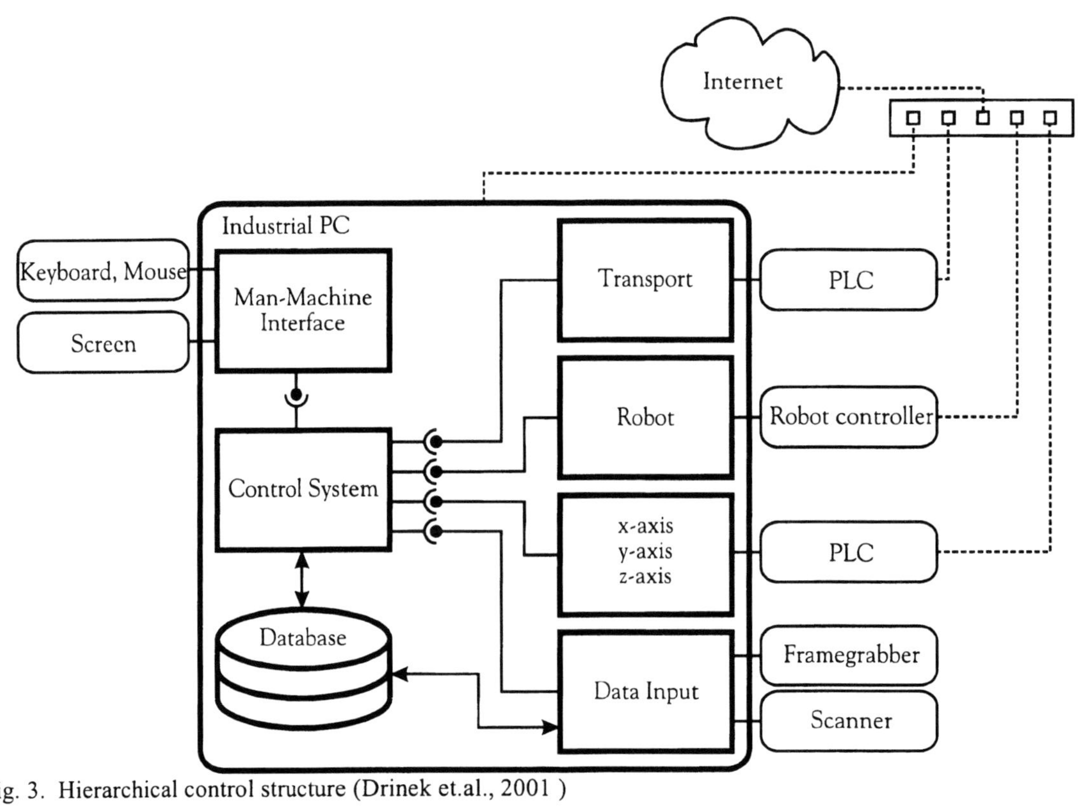

Fig. 3. Hierarchical control structure (Drinek et.al., 2001)

applications the CC can be replaced by an "advanced" PLC.

The PLC's are responsible for the supervision of the cell stations or components. They have to process binary sensor information from the system components and to actuate vice versa components or parts of them. They are connected to distinct components of the cell directly or to their local control units. In the latter case PLC's give only the start commands and are waiting for their execution (Kopacek, 1999).

Local control units are commercially available control devices which are necessary and attached to commercially available components of such a cell (e.g. robots, laser de-soldering, ...). They are connected to the PLC's as well as to the CC. One of the main problems of local control units are the different data interfaces.

An Internet connection between several cells allows the transfer of data for statistics as well as new programs, program upgrades, program parameters. This could be a first step in the direction of so called "e-cells".

But there are also some "classical" control problems to solve. One example is the temperature control in the infrared heating station of the PCB disassembly cell. From the viewpoint of control engineering the dynamic behaviour of this system can be approximated by a first order lag element – gain K and time constant T_1 - with a large dead time (T_t). The time constant T_1 as well as the dead time T_t varies for heating and cooling until a factor of 3 and all the parameters depends from the coordinates on the PCB. Therefore we have to control a direction depended (Kopacek, 1971) distributed parameter system.

CONCLUSION

Disassembly automation especially for electr(on)ic devices is absolutely necessary worldwide in the nearest future because of the dramatically increasing amount of electr(on)ic scrap. Today only some pilot projects in form of fully automated single purpose solutions – and only for one distinct type of one product – are realized. Main problems are the collection and transportation of a sufficient amount of these products to one place in one area. In Austria we developed an industrial robotized disassembly cell for reusable parts on printed circuit boards as well as a disassembly cell for mobile phones. A development cell is currently working at Vienna University of Technology (VUT) and the industrial cells are in use in MLC^2V.

For the control of our disassembly cells a hierarchical structure – several PLCs supervised by a host computer – was choosen. Several control problems have to be solved e.g. logistics, "classical" control ... and some problems are unsolved today.

LITERATURE

Drinek, M., K. Daichendt and H. Zebedin (2001): A new Concept for a fully automatized Disassembly Cell. In: *Proceedings of the 10th International Workshop "Robotics in Alpe Adria Danube Region – RAAD'01"*, Vienna, May 16. – 18. 2001, p.383-388.

Eckerth,G.; T.Bönker and W.Kreis (1998): Intelligent Disassembly Systems for Industrial Reusing and Closed Loop Economy. . *Proceedings of the IFAC Workshop "Intelligent Assembly and Disassembly – IAD'98)*, Pergamon, 1998, p.105-110.

Hesselbach,J. and M. Kühn (1998): Survey of Processes at Disassembly Companies. *Proceedings of the IFAC Workshop "Intelligent Assembly and Disassembly – IAD'98)*, Pergamon, 1998, p.99-104.

Kopacek, P., Kopacek, B. (2003): Robotized Disassembly of Mobile Phones. In: *Preprints of the IFAC Workshop on "Intelligent Assembly and Disassembly (IAD '03)*, Bucharest, Oct. 7- 9, 2003, p.142-144.

Kopacek, B., Kopacek, P. (2001): Semi-Automised Disassembly. In: *Proceedings of the 10th International Workshop on Robotics in Alpe-Adria-Danube Region (RAAD '01)*, Vienna, May 16-18, 2001.

Kopacek,B. and P.Kopacek (1999): *Intelligent Disassembly of Electronic Equipment*. Annual Reviews in Control **23**(1999), p.165-170. Elsevier Publ.Comp. 1999.

Kopacek, P. (1971): *Identification of control systems by pseudorandom binary signals with special emphasis on linear unsymmetric transfer elements*. Dissertation, TU Wien, 1971 (in German).

Copyright © IFAC Information Control Problems in
Manufacturing, Salvador, Brazil, 2004

ELSEVIER
IFAC
PUBLICATIONS
www.elsevier.com/locate/ifac

COOPERATIVE ROBOTS FOR SURVEILLANCE OR TRANSPORTING PARTS IN INDUSTRIES

Carlos M. Soria*, Ricardo Carelli*, Teodiano F. Bastos-Filho, Mário Sarcinelli-Filho****

*Institute of Automatics, National University of San Juan, Argentina, 5400 San Juan, Argentina
**Electrical Engineering Department, Federal University of Espírito Santo, 29060-900 Vitoria-ES, Brazil

Abstract—Surveillance and transport of parts are important tasks associated with mobile robots actuating between cells in a flexible production system or between different sections in an industrial plant. For transporting, depending on the dimensions of the parts, more than one robot is necessary and the part can be shared among several robots. This work presents a methodology for creating a structure containing mobile robots and a vision system, which is used for tasks of surveillance and transport of parts either between cells in a flexible production system or in an industrial plant. *Copyright © 2004 IFAC*

Keywords: computer vision, co-operation, robot control.

1. INTRODUCTION[1]

Surveillance or parts transportation is a current problem in industrial applications. This problem can be resolved using a network of mobile robots, in a cooperative way. The use of cooperative robots is an area of research that involves control, artificial vision, planning, genetic algorithms, neural networks, etc.

When manipulating or transporting large objects, the weight of the object can be shared among several small robots. In addition, a group of robots can accomplish more than one individually by coordinating their actions and by sharing sensors and computing power. Thus, when using several robots it is possible to have different points of view of the object along their paths, and it is possible to exchange sensor information among them.

Teams of robots could be used to perform complex tasks in areas too dangerous or undesirable for humans, such as cleaning up hazardous waste areas, exploring planets, mining in unpopulated areas, participating in search and rescue missions, decommissioning nuclear power plants, and for the surveillance for threats such as biological and chemical warfare weapons.

This work proposes the implementation of a communication protocol between the host PC and the mobile robots, and the development of the software for controlling the robots. A WebCam is used to acquire images of the workspace and a wireless communication is used to link the robots and the PC. The acquired image is processed at the PC, which generates the control signals to be sent to the robots. Figure 1 shows the structure used in this work and Figure 2 shows the whole control system.

[1] This work was supported by CAPES (Brazil) and SETCIP (Argentina)

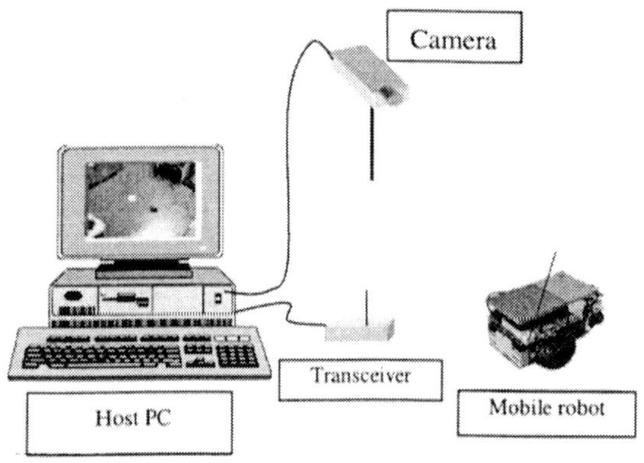

Fig. 1 Structure used in this work.

2. LOW LEVEL VISION

Artificial vision can be defined as a process of obtaining, characterizing and interpreting information of images from a tridimensional world. Such processes can be divided in six main areas: 1) capture, 2) pre-processing, 3) segmentation, 4) description, 5) recognizing and 6) interpretation. Capture is the process where the visual image is obtained. Preprocessing includes techniques such as reduction of noise and enhancing of details. Segmentation is the process that separates an image into objects of our interest. Through process of description, characteristics of the object are obtained (for instance, size and shape), in order to distinguish an object from another. Recognizing is the process that identifies the objects. Finally, interpretation associates a meaning to a group of recognized objects (Gonzales and Woods, 2002).

2.1 Image Acquisition

Visual information is converted to electric signals by a visual sensor. When such signals are sampled spatially and quantified in amplitude, digital images are obtained. The spatial digitization consists of dividing the image in rows and columns as a matrix. The elements of the matrix are called pixel. The value of each pixel is associated with the level of light present in that point of the image. In this work, color image in RGB representation is used, which means that there is three matrices (one for blue, one for green and another for red). Each element of such matrices has a range from 0 to 255. In a binary system, there are 8 bits to represent the level of each pixel at each color, resulting in 24 bits, allowing the recognition of 16 million of colors.

In this work, a WebCam is used as the sensor responsible for capturing the image. The camera has a resolution of 640x480 pixels and a maximum frequency of 15 frames per second.

2.2 Placing the Robot in the Image

There are distinct ways of "placing" the robot. One of the methods is to place it, using as a reference the world coordinate system. It is also possible to place the robot on the camera coordinate system, which was used in this work. With this method, it is possible to use the measurements directly obtained from the image for controlling the robot, not requiring transformation to the world coordinate system. This way, the performance of the system is almost independent of camera calibration.

2.3 Control with Visual Feedback

In order to control the robot movements using an external camera, it is necessary to know its position and orientation at each instant. However, that information should be obtained in a very high frequency, allowing the implementation of a strategy of closed loop control (Santos-Victor, *et al.*, 2002).

The strategy for obtaining the position and orientation of each mobile robot is to find in the image the color tags (blue and green) placed on the top of the robots (Figure 3). The method used to obtain the robot position consists of identifying, in the image, the color blue, generating a binary image and determining the center of mass. The coordinates of the center of mass are associated with the robot position in the image.

In the same way, the color green is identified and the center of mass is obtained. Thus, with that two pairs of coordinates, it is possible to determinate the robot orientation.

2.4 Obtaining the Robot Coordinates

Once the RGB image is acquired and normalized (meaning each element is divided by the sum of the three channels, and each pixel varies between 0 and 1), then a binary image using the blue color as the threshold (color segmentation) is generated (Gonzales and Woods, 2002). Figure 3 shows an example of an image of the robot workspace and the result of the binary image.

From the binary image obtained, the center of mass of the pixels with logic value equal 1 (white color) is calculated. Once the point of the center of mass is calculated, a window around that point is generated, whose size is chosen according to the next estimate robot position. This window allows the calculation of the robot position at each instant of time. Although the camera acquires images from the whole robot workspace, the control algorithm is calculated only at that window, in order to diminish the processing time.

Also another window is generated, of smaller size, in order to calculate the center of mass of the two tags and consequently the robot orientation. In this case, a binary image using red color (the color of the floor) as the threshold is generated. That image is inverted, resulting only in an image with no red color (robot). Thus, a binary image is generated using blue color as the threshold, and an AND operation between the blue binary image and the red binary image is applied.

3. CONTROL SYSTEM

Mobile robots are non holonomic systems, which are characterized for having kinematics geometric restrictions. For being nonlinear and complex systems, the control for a mobile robot is normally based on kinematics models instead of dynamics models.

3.1 Mathematics Model

The mobile robot under study contains two identical parallel wheels, driven by two independent DC motors, and a free wheel at the rear side. It is supposed that mass and inertia of the wheels are neglected and the robot center of mass is located in

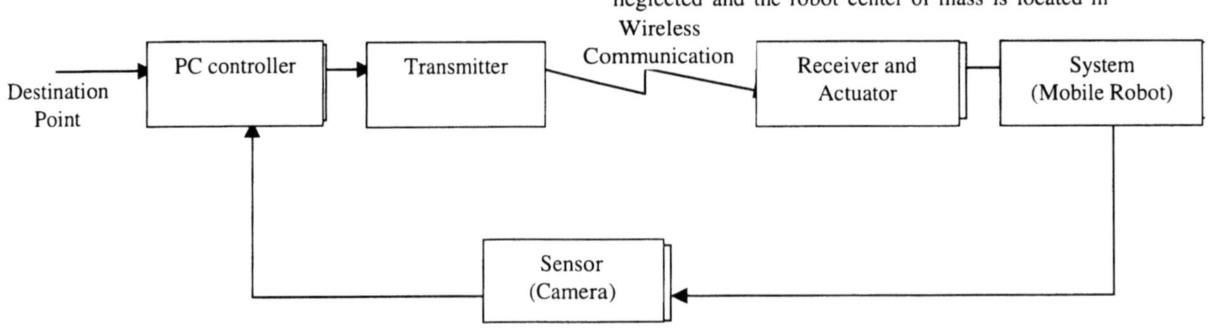

Fig. 2 The whole control system.

Fig. 3 (a) Original image (blue and green tags are on the top of the robot); (b) Binary image obtained from the robot.

the middle of the axle that attaches the two wheels. In addition, the speed of the center of mass is transverse to the two driven wheels.

Let <G> (goal) and <C> (robot) be the frames of inertial references such as shown in Figure 4. It is supposed that the origin Oc of <C> coincides with the center of the axle that attaches the two driven wheels. Let $[x,y]^T$ be the position of Oc in <G>. The robot kinematics and dynamics can be modeled by the following equations

$$\begin{aligned} \dot{x} &= v\cos\theta, \\ \dot{y} &= v\sin\theta, \\ \dot{\theta} &= \omega, \\ m\dot{v} &= F, \\ I\dot{\omega} &= N, \end{aligned} \tag{1}$$

where v and ω are the linear and angular speed, respectively, of <C> (robot) in relation to <G>. The control inputs are the force F along the axis xc and the torque N referred to the vertical axis zc. It can be demonstrate that

$$\begin{aligned} F &= \frac{1}{R}(\tau_1 + \tau_2) \\ N &= \frac{D}{R}(\tau_1 - \tau_2) \end{aligned} \tag{2}$$

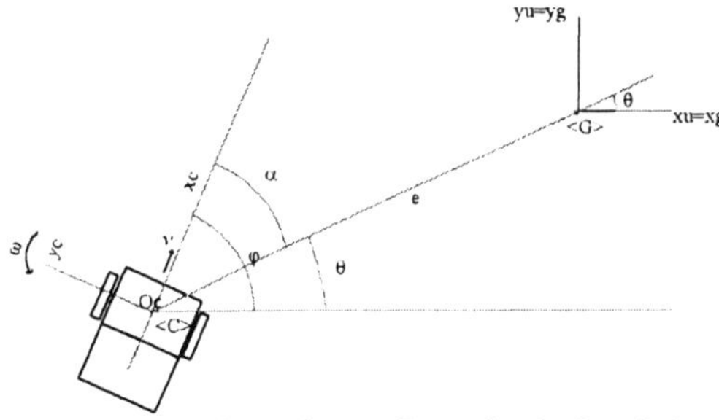

Fig. 4 Inertial reference frames <G> (goal) and <C> (robot).

where R is the radius of the driven wheel, τ_1 τ_2 are the torque of each wheel, $2D$ is the length of the axle that attaches the two wheels, and m and I are the mass and the inertia moment of the mobile robot, respectively.

3.2 Kinematics Model

1) Cartesian Representation

The kinematics equations that involve the Cartesian position of the robot, x and y, and its orientation θ, respect to the frame of inertial reference <G>, such as aforementioned, are

$$\begin{aligned} \dot{x} &= v\cos\theta \\ \dot{y} &= v\sin\theta \\ \dot{\theta} &= \omega \end{aligned} \tag{3}$$

where v is the module of the vector v, and x, y and θ are all the variables measured in relation to the origin of the destination reference frame <G>.

2) Polar Representation

Let the transformation of coordinates

$$\begin{aligned} e &= \sqrt{x^2 + y^2} \\ x &= -e\cos(\theta + \alpha) \\ y &= -e\sin(\theta + \alpha) \\ (\theta + \alpha) &= \tan^{-1}\left(\frac{-y}{-x}\right) \end{aligned} \tag{4}$$

where $\boldsymbol{e}$ is the vector of O_c to O_g, e is the module of $\boldsymbol{e}$, and α is the angle measured of x_c to $\boldsymbol{e}$ (see Figure 4). Differentiating the equations (4) with respect to time, one can obtain the following kinematics equations in the polar coordinate system

$$\begin{aligned} \dot{e} &= -v\cos\alpha \\ \dot{\alpha} &= \frac{\sin\alpha}{e}v - \omega \\ \dot{\theta} &= \omega \end{aligned} \tag{5}$$

The transformation of coordinates is only valid to values distinct of zero of the distance error e, due to when $e=0$ the angle α is not defined (singularity). This will produce a discontinuity in the control law used, which corresponds to the limitation imposed by the theorem of Brockett (1996).

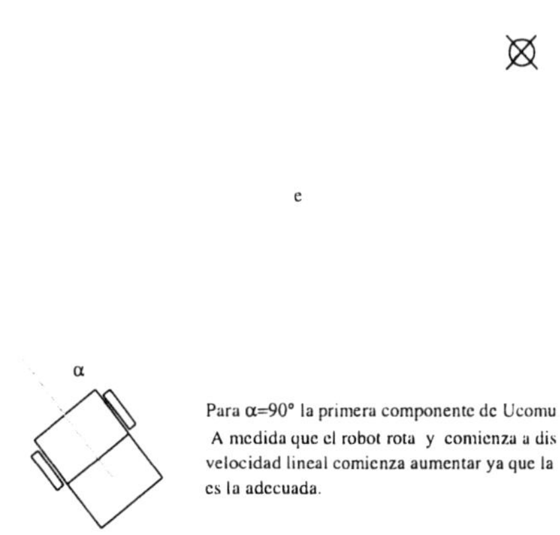

Fig. 5 Performance of the control action Ucom when α<90°.

3.3 Establishing Objectives of Control

The control system was designed to reach a destination point defined in the image. Given an initial position of the robot (x_o,y_o,θ_o) at a distance e distinct of zero of the destination reference frame <G>, the objective of control consists of reaching that destination point, i.e., a final position (x_g,y_g) in the image.

3.4 Controllers of Position

1) Nonlinear Control

The equations of the controller are the following (Carelli, *et. al.*,1999)

$$U_{dif} = \frac{\tau.D}{KR}\{k_a\alpha_1 - k_b\omega_1\}$$

$$U_{com} = \frac{2\tau}{KR}\left(k_c(e_1)e_1\cos\alpha_1 - k_d v_1\right) \qquad (6)$$

$$k_c(e_1) = \frac{k_c}{a+e_1}, k_d > 0$$

where:

U_{dif}:	Differential tension
U_{com}:	Common tension
τ:	Constant of mechanical time of the robot
D:	Distance from the center to the wheels
R:	Radius of the wheels
α_1:	Error of orientation measured between the line that attaches the center of the robot and the destination point, and the main axis of the robot measured in the image.
v_1:	Linear speed in the image.
ω_1:	Angular speed measured in the image.
k_a,k_b,k_c,k_d:	Parameters of the controller.

The control action on the linear speed is composed by two components: the first one regulates the control action U_{com} in order to have a strong control action, if the robot has a good orientation to the goal. Also, if it has a orientation equal or greater than 90 degrees to the goal, the control action on the linear speed is null, allowing only the actuation of the differential control action on the orientation (see Figure 5). This way, when the orientation is being corrected, U_{com} increases, allowing the robot to reach the destination point.

Different experiments were carried out, and the results showed the correct performance of the controller. In case of saturation of the control action,

$$U_{com} = \lim_{e_1\to\infty}\left(\frac{K}{e_1+a}e_1\right) = K = U_{com}\max \qquad (7)$$

where $U_{com}\max$ corresponds to the tension to the motors, in PWM values of saturation. In the same way, in case of cutoff,

$$U_{com} = \lim_{e_1\to e\min}\left(\frac{U_{com}\max}{e\min + a}e\min\right) = U_{com}\min \quad (8)$$

In this case, $e\min$ corresponds to the minimum distance possible to the vision system between the robot position and the goal, representing 10 pixels. $U_{com}\min$ corresponds to the tension in PWM value, when the motors start to move. A practical consideration with respect to the analysis consists of noticing that $e_{1\max}$ of the image corresponds to,

$$e_{1\max} = \sqrt{(h)^2 + (v)^2}$$

where h is the height, in pixel, of the image (480) and v is the width, in pixel, of the image (640).

The second component of U_{com} is related to a factor of damping, diminishing U_{com} when the linear robot speed increases. When the robot starts its movement, this component can be depreciated. However, when the linear speed increases, the control actions U_{com} are smaller.

The parameters of the nonlinear controller are k_a=95, k_b=20, k_c=350, k_d=0.5, a=18. The robot is located on the initial position (60,180) pixel and should get to the destination point X (Figure 6). Figure 7 shows the evolution of the distance to the destination point when the robot moves. Figure 8 shows the control actions applied to the robot motors in order to execute the specified task.

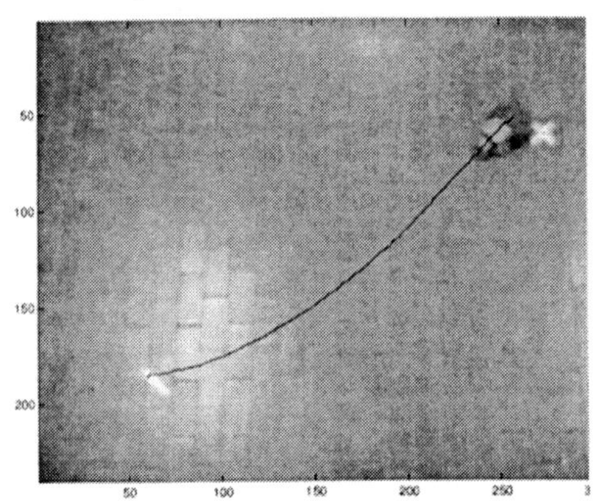

Fig 6 Robot trajectory to reach the destination point.

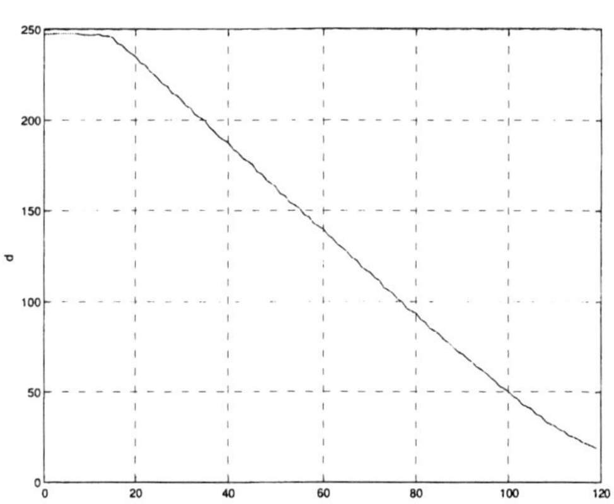

Fig. 7 Distance d to the destination point.

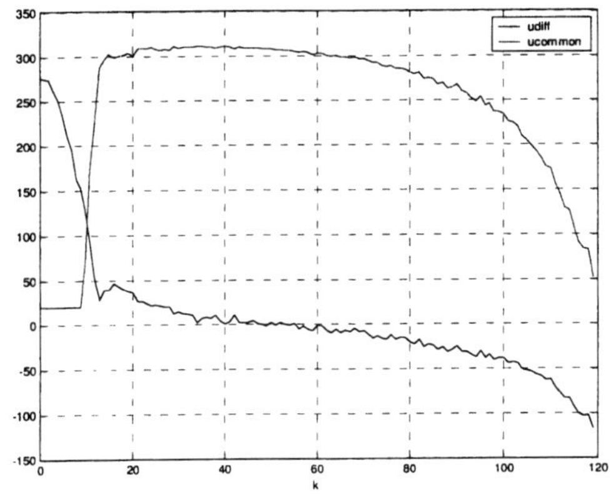

Fig. 8 Differential and common tensions in PWM values.

In order to carry out cooperative tasks, three MindStorm robots were used. One of them is a leader (Fig. 9), located at coordinates (Xc,Yc) and the others robots follow a line formation (Fig. 9a) or triangle formation (Fig. 9b). In both robot formations, there exist a leader and two robots P1 and P2, located at coordinates (Xe1,Ye1) and (Xe2, Ye2), which are calculated by equations (9) and (10).

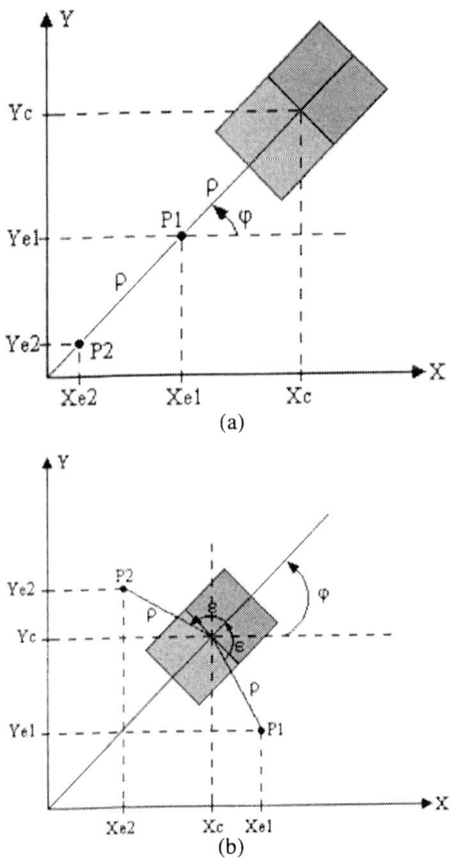

Fig. 9 Robot formation: (a) line; (b) triangle.

$$\begin{aligned} Xe1 &= Xc - \rho \cos\varphi \\ Ye1 &= Yc - \rho \sin\varphi \\ Xe2 &= Xe1 - \rho \cos\varphi \\ Ye2 &= Ye1 - \rho \sin\varphi \end{aligned} \qquad (9)$$

$$\begin{aligned} Xe1 &= Xc + \rho \cos(\varphi - \varepsilon) \\ Ye1 &= Yc + \rho \sin(\varphi - \varepsilon) \\ Xe2 &= Xc + \rho \cos(\varphi + \varepsilon) \\ Ye2 &= Yc + \rho \sin(\varphi + \varepsilon) \end{aligned} \qquad (10)$$

4. CONCLUSIONS

In this work, the development of a system to execute the control of a robot using an artificial vision camera is presented. A control algorithm was implemented in order to allow one robot to reach the goal and three robots follow a line formation and a triangle formation. The next step is to include others three mobile robots (one Pioneer II-DX, one Rug Warrior and one Mind Storm, all of them available at the Electrical Engineering Department in UFES, Brazil) in order to make possible the cooperation among them, allowing the realization of the surveillance and parts transportation in an industrial plant.

REFERENCES

Gonzales, R. C. and R. E. Woods (2002). *Digital Image Processing*. 2nd edition, Addison-Wesley Pub Co., USA.

J. Santos-Victor, R. Carelli, and S. Van der Zwaan, "Nonlinear Visual Control of Remote Cellular Robots," 10th IEEE Mediterranean Conference on Control and Automation, Lisbon, 2002.

Brockett, R. W. (1996). Characteristic Phenomena and Model Problems in Nonlinear Control. In: *Proceedings of the 1996 IFAC World Congress*, pp. 135-140.

Carelli, R., H. Secchi, and V. Mut (1999). Algorithms for Stable Control of Mobile Robots with Obstacle Avoidance. In: *Latin American Applied Research*, pp.191-196.

Copyright © IFAC Information Control Problems in Manufacturing, Salvador, Brazil, 2004

ELSEVIER
IFAC
PUBLICATIONS
www.elsevier.com/locate/ifac

Model Based Monitoring Strategies for Threaded Assembly Systems

Mongkorn Klingajay, Prof. Lakmal D. Seneviratne, Dr. Kaspar Althoefer

Centre of Mechatronics, Division of Engineering, King's College London
London, WC2R 2LS, United Kingdom
{mongkorn.klingajay, lakmal.seneviratne, k.althoefer}@kcl.ac.uk

ABSTRACT: Threaded fastenings are a common assembly method, accounting for over a quarter of all assembly operations. Threaded fastenings are popular because they permit easy disassembly for maintenance, repair, relocation and recycling. A model based approach for monitoring screw fastenings is presented in this paper. The proposed strategy can be integrated with a mechatronics based approach for automating threaded assembly systems. The screw insertion process is typically monitored using the torque vs insertion depth signature signals. For a screw insertion with a given set of parameters, the insertion signature signal will have a unique identity and by comparing the on line signature signals with the ideal signature signal, the integrity of the joint can be predicted. The approach adopted in this study is to use an analytical model developed by the authors to predict ideal insertion signals. However, this model requires various fixed process parameters as input, and it is not always possible to know these parameters in advance with sufficient accuracy. Hence the focus of this paper is the on-line parameter estimation during threaded assembly. A methodology for estimating four unknown parameters of a general self-tapping screw insertion is presented. The approach is based on the Newton Raphson Method (NRM). It is shown that up to four parameters required by the model can be reliably estimated in real time, from on-line torque signature signals. Experimental results are presented to validate the estimation procedure. *Copyright © 2004 IFAC*

Keywords: Threaded assembly automation, Mechatronics, Screw insertions, Parameter estimation.

1. INTRODUCTION

Screw insertion is an assembly operation involving the use of a threaded component to fasten together two or more parts. The threaded component in the form of a screw is driven into a hole in the part, in a manner that allows the helical thread on the screw cylinder to follow a complementary helical groove in the part. The two parts can be clamped together with varying degrees of compressive forces. Threaded components come in many forms and geometries. As they are employed in both critical and the non-critical applications, many standards of performance have been specified by various organizations, to regulate their physical properties, dimensions and thread profiles. The International ISO-Metric screws and the American Unified Standard screws are the common standards in use today. Screw insertions can be carried out either on pre-threaded holes, where the screw is inserted into a pre-tapped hole or using self-tapping screws, where the screw forms a complimentary internal thread in the hole wall as it is driven through [4].

Threaded assemblies are commonly carried out manually using a screwdriver. In contrast to other joining methods, threaded fastenings can be easily disassembled for maintenance, repair, relocation, and recycling. Hence screw fastenings are a very popular assembly method accounting for over 25% of all assembly operations. However, screw insertions are very complex to automate, and hence there is relatively little published research on the topic, with most research on automated assembly focusing on the peg-in-hole assembly problem. Human operators with experience are particularly good at on-line monitoring of the operation, and automated on-line monitoring strategies for the screw fastening process are highly desirable. One such approach is based on the "Torque Vs. Insertion Depth" curve measured in real time; if this signal is within a pre-defined bound of the correct insertion signal, then the insertion is considered to be satisfactory.

The torque signature signal for a correct insertion is usually taught, but can also be predicted analytically, if a sufficiently accurate model is available [3].

2. THE SCREW INSERTION PROCESS

Figures 1 and 2 show a self-tapping screw and the components involved in the insertion process. With self-tapping screw fastenings, the screw is inserted into an unthreaded hole, and the external thread of the screw works as a cutting blade, tapping an internal thread in the hole. Therefore, some problems common to ready-tapped fastenings, such as screw cross threading, do not occur with self-tapping screw fastenings. However, there are a number of possible faults for self-tapping screw fastenings, such as screw jamming and wedging.

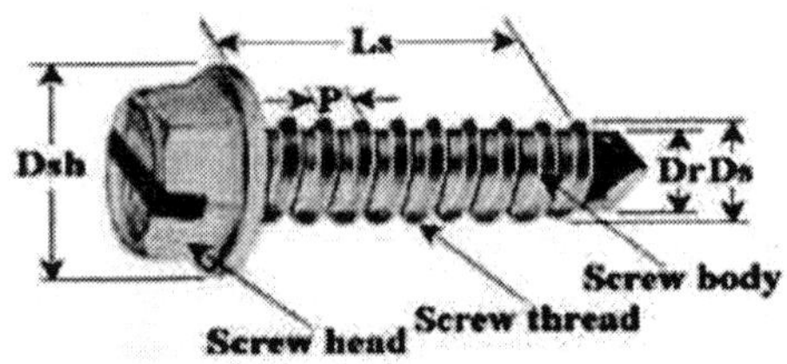

Figure 1. A self-tapping screw.

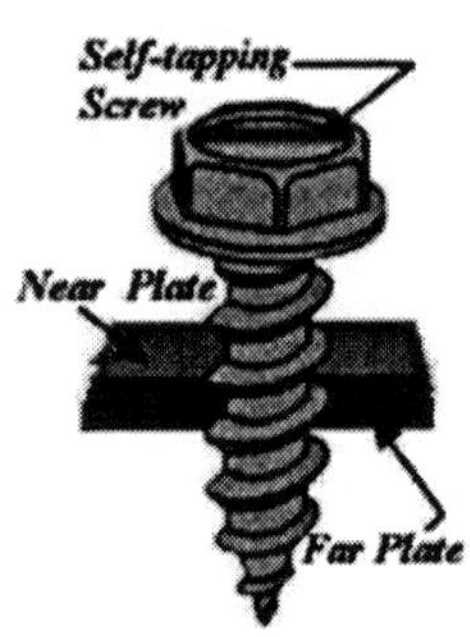

Figure 2. A self-tapping screw insertion.

In [1, 2], the screw insertion process is described by a torque signature signal, Figure 4, which is a two dimensional curve of axial torque Vs insertion depth. The axial torque is necessary to drive the screw through the hole as the insertion develops.

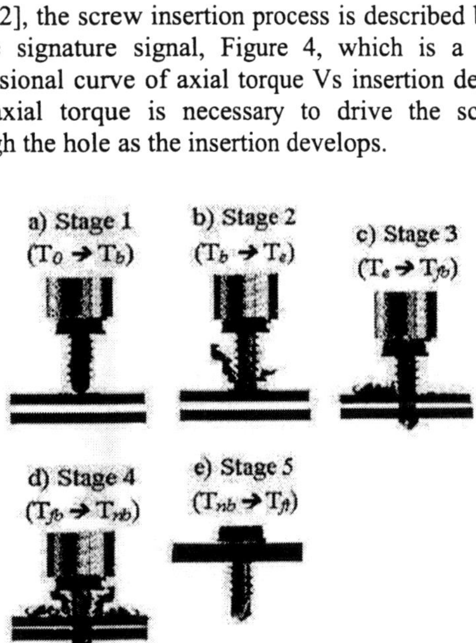

Figure 3. The 5 stages of screw insertion.

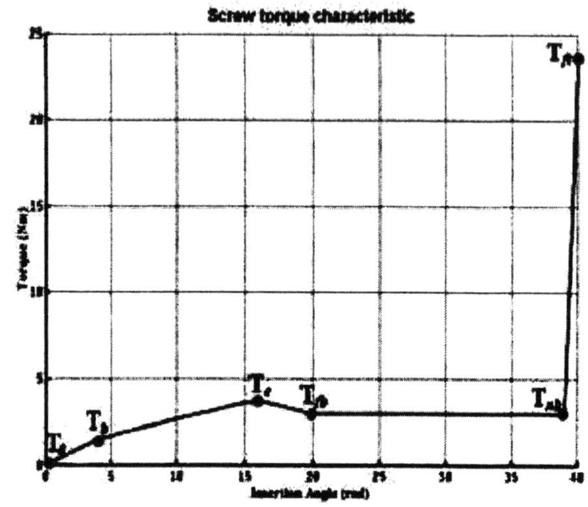

Figure 4. An ideal self-tapping screw insertion signal.

The points T_0, T_b, T_e, T_{fb}, T_{nb} and T_{ft} shown in Figure 3 identify the five main stages during a general self-tapping screw insertion operation [5, 6].

Stage 1 - T_0 - T_b - "Screw Engagement", where the screw first contacts with the hole wall and thread engagement is completed.
Stage 2 – (T_b - T_c) - "Thread Cutting", ends when the screw first breaks through the hole.
Stage 3 – (T_c-T_{fb}) - "Screw Full-Breakthrough" ends when thread forming is completed.
Stage 4 – (T_{fb} -T_{nb}) - "Screw Sliding", where the screw advances without thread cutting until the screw head contacts the top plate.
Stage 5 – (T_{nb}-T_{ft}) - "Screw Tightening", where the final tightening torque is applied.

3. ANALYTICAL MODEL

An analytical model for a general self-tapping screw insertion operation is presented in [1, 4, 5]. The model consists of five equations corresponding to each insertion stage:

$$\tau_1 = \frac{1}{\alpha} R_s A_c \sigma_{UTS} \cos\theta . \phi + \mu R_f K_f \sigma_f \cos\theta(\phi). \quad (1.1)$$

$$\tau_2 = R_s A_c \sigma_{UTS} \cos\theta + 2\mu R_f K_f \sigma_f \cos\theta(\phi - \alpha). \quad (1.2)$$

$$\tau_3 = \frac{1}{\alpha} R_s A_c \sigma_{UTS} \cos\theta + \mu R_f K_f \sigma_f \cos\theta(\phi_b + \phi). \quad (1.3)$$

$$\tau_4 = 2\mu R_f K_f \sigma_f \cos\theta(\phi_b). \quad (1.4)$$

$$\tau_5 = \left[\left(\frac{\mu_h (D_{sh}^3 + D_s^3)}{3(D_{sh}^2 - D_s^2)}\right) + \left(\frac{D_s + D_h}{4}\right)\left(\frac{\pi\mu(D_s + D_h) + 2P}{\pi(D_s + D_h) - 2\mu P}\right)\right]\left[\frac{K_{th} K_{tb}}{K_{th} + K_{th}}\right]. \quad (1.5)$$

Where

$\tau_1, \tau_2, \tau_3, \tau_4, and\, \tau_5$ = Insertion Torque at stages 1-5.
ϕ = Angular rotation of screw.
ϕ_b = Angular position corresponding to screw breakthrough.

More details of the equation and the parameters used in the equations are described in Appendix A.

The model requires a number of parameters, and these can be divided into 3 groups:

a) Screw properties.
b) Plate Properties
c) Material Properties

These parameters can be obtained from databases, manufacturer data sheets, and other published sources.

Table 1 gives the parameters used in this study. The ideal torque Vs insertion depth curve corresponding to the parameters in Table 1 is shown in Figure 6. The accuracy of these parameters depends on a number of factors including manufacturing tolerances. Generally, the cheaper the screw, the larger the parameter uncertainties.

In practice, parameter uncertainties hinder the use of the model in developing automation strategies. This study focuses on identifying system parameters from experimental torque signature signals. Four of the parameters, the friction coefficient (μ), Diameter of hole (D_h), Ultimate tensile stress (σ_{UTS}), and the modulus of elasticity (E), are assumed unknown.

Material Properties	
Material type	Tool Steel
Tap Hole Diameter (D_h)	3.2 mm
Tensile Strength (σ_{UTS})	1100 Mpa
Yield Strength (σ_Y)	450 Mpa
Elastic Modulus (E)	207 Gpa
Coefficient of Friction (μ)	0.238
Plate Properties	
Tap (far) plate thickness (T_2)	6 mm
Tap (near) plate thickness (T_1)	0
Screw Properties	
Screw type	AB6
Screw major diameter (D_s)	3.42 mm
Screw head diameter (D_{sh})	6.52 mm
Screw root diameter (D_r)	2.49 mm
Screw thread pitch (P)	1.19 mm
Screw taper-length (L_t)	2.94 mm
Screw total threaded-length (L_s)	9.67 mm

Table 1. Parameters used in study

4. ESTIMATION STRATEGY

This paper presents the results of an investigation aiming to estimate four unknown parameters [μ, D_h, δ_{UTS}, E], based on the analytical model by Ngemoh et al [1, 4], and the measured torque signature signals, using the Newton Raphson Method (NRM). The NRM gives rapid convergence, and its accuracy approximately doubles with each iteration. Hence this approach allows the development of on-line estimation methods.

Now equations (1.1-1.5) can be written as:

$$\tau = F(V, x, \phi), \tag{3}$$

where ϕ is the screw rotation angle, V is the vector of known parameters, and x is the vector of unknown parameters, representing

$$x(k) = [\,\mu\,(k), D_h\,(k), \sigma_{UTS}\,(k), E\,(k)]^T$$

The NRM works by updating the unknown parameters in order to force an error function $f(x, \phi)$ to approach zero. Equation (3) can be written as:

$$f(x, \phi) = \tau(\phi) - F(V, x, \phi). \tag{4}$$

Where,

$\tau(\phi)$ is the experimental torque signal.

The error function $f(x, \phi)$ is computed at four distinct angles of ϕ, ($f_1(x, \phi_1)$, $f_2(x, \phi_2)$, $f_3(x, \phi_3)$ and $f_4(x, \phi_4)$) providing four independent equations, which can be solved in an iterative manner to find the four unknown parameters.

Let $\bar{x}$ (k) be the kth estimate of the vector of unknown parameters. Vector $\bar{x}$ (k) has four elements as follows,

$$\bar{x}\,(k) = [\,\bar{\mu}\,(k), \overline{D_h}\,(k), \overline{\sigma_{UTS}}\,(k), \bar{E}\,(k)]^T$$

Applying the Newton-Raphson Method (NRM), x can be calculated iteratively:

$$\bar{x}\,(k+1) = \bar{x}\,(k) - J^{-1} \begin{pmatrix} f_1(x, \phi_1) \\ f_2(x, \phi_2) \\ f_3(x, \phi_3) \\ f_4(x, \phi_4) \end{pmatrix} \tag{5}$$

where J is the Jacobean matrix.

Applying Equation (5) iteratively until the error is reduced to a value close to zero will provide an estimate of the unknown parameters.

The equations describing stages 2, 4, and 5 are used for parameter estimation [2]. However, since the model is discontinuous, the torque signals for stage 2 is used to estimate $\bar{\mu}$ and $\overline{D_h}$, the torque signals for stage 4 is used to estimate $\overline{\sigma_{UTS}}$ and the torque signal for stage 5 is used to estimate $\bar{E}$. The above process is repeated iteratively until convergence is reached.

5. TEST RESULTS

Figure 5 presents the estimated μ, D_h, σ_{UTS}, and E values for the case without noise. It is seen that μ, D_h, σ_{UTS}, and E are estimated as 0.238, 2.5 mm, 1100 Mpa, and 207 Gpa. It is noted that the actual values used are μ = 0.238, D_h = 2.5 mm (0.0025 meters), σ_{UTS} = 1100 Mpa (1.1E+9 Meters), and E = 207 Gpa (2.07e+11 meters) see Table 1. It is seen that the results converge to the actual values after 48 iterations, Figure 6.

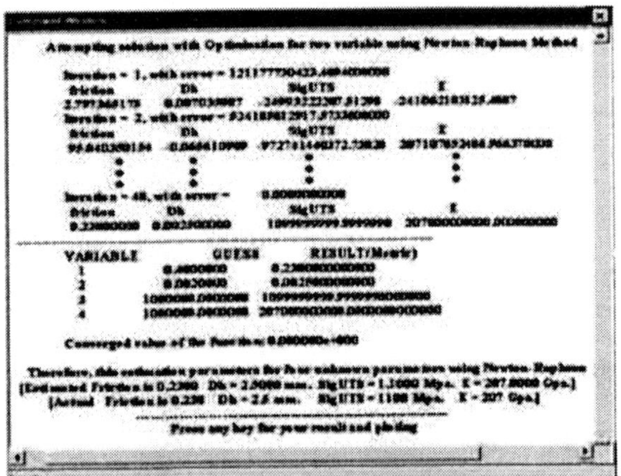

Figure 5. The simulation results

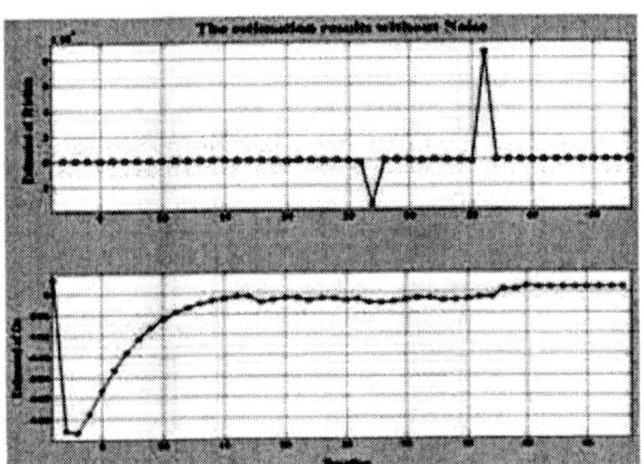

Figure 6(a) : Estimated μ and D_h.

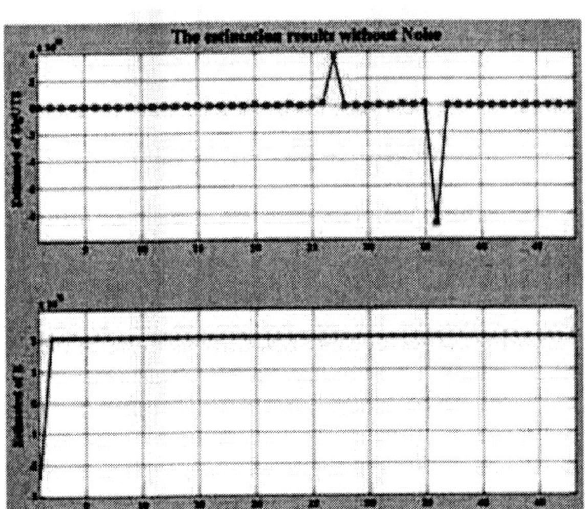

Figure 6(b): Estimated σ_{UTS} and E.
Figure 6. Estimation results without Noise.

The estimated value without Noise for parameters of μ, D_h ,σ_{UTS}, and E have present in Figures 6 (a) and (b). These estimated parameters present with iteration number, showing fast convergence.

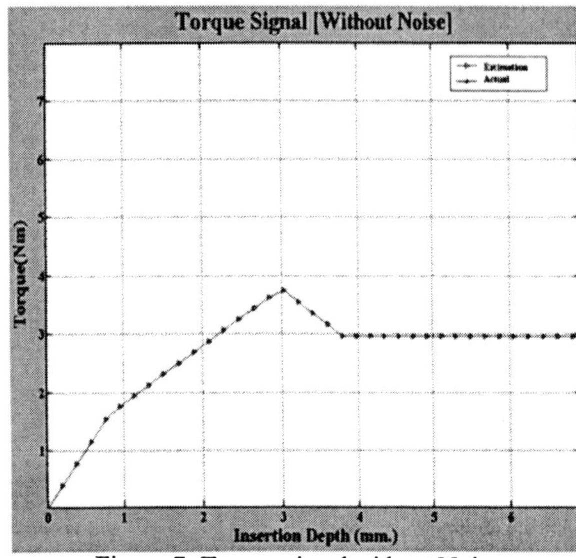

Figure 7. Torque signal without Noise

The torque signature curve predicted with the estimated values is shown in Figure 7, showing close agreement with the actual curve.

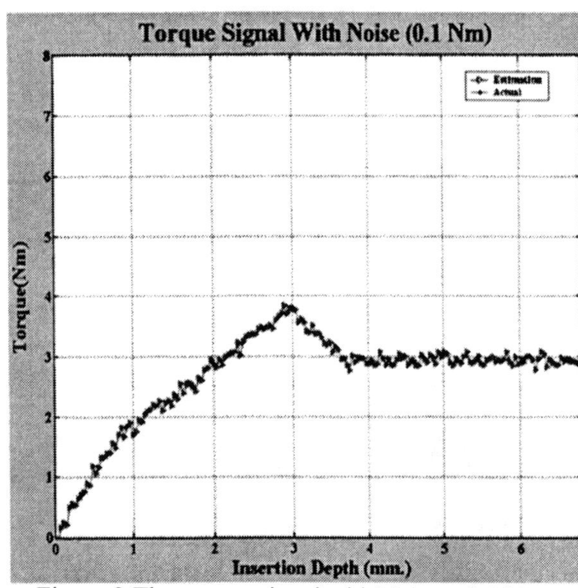

Figure 8 The torque signal with Noise = 0.1 Nm

Figure 8 shows the torque signature signal with a random white noise signal of amplitude 0.1 Nm, superimposed on it, to simulate sensor noise. The accuracy of the estimation algorithm in the presence of noise is next investigated.

The estimated values of μ, D_h, σ_{UTS} and E with noise is shown in Figure 9. Again it is seen that the results converge after 480 iterations, and μ, D_h, σ_{UTS}, and E are estimated as 0.238, 0.0025 meters (2.5 mm.), 1.0999e+9 (1100 Mpa), and 2.07e+11 (207 Gpa) respectively.

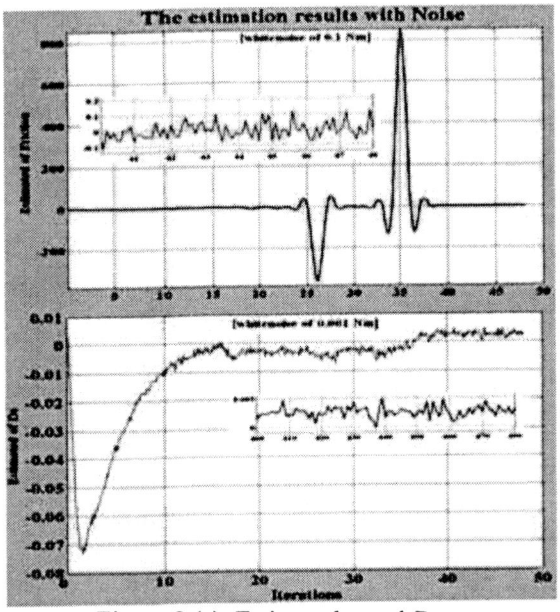

Figure 9 (a): Estimated μ and D_h

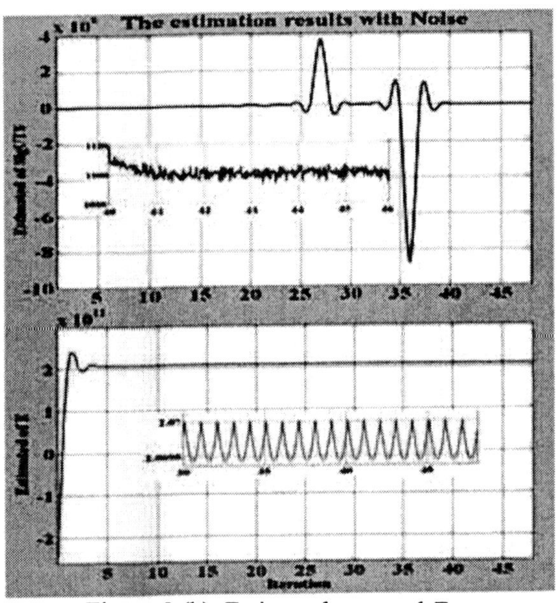

Figure 9 (b): Estimated σ_{UTS} and E.

Figure9. Estimated results μ, D_s, σ_{UTS} and E with Noise

6. CONCLUSIONS

The paper presents a methodology for the on-line estimation of four system parameters during self-tapping threaded fastenings. Test results are presented to validate the estimation methodology. The estimated properties can be used to predict ideal insertion signals and hence develop model based automated monitoring algorithms.

7. ACKNOWLEDGEMENTS

Mongkorn Klingajay is funded by the Thai Government through the Rajamangala Institute of Technology (RIT), Bangkok, Thailand.

8. REFERENCES

[1] L.D. Seneviratne, F.A. Ngemoh and S.W.E. Earles (1992). Theoretical Modeling of Screw Tightening Operations. Proceedings ASME European Joint Conference on Systems, Design and Analysis, Vol.1, pp.189-195.

[2] F.A. Ngemoh (1997). Modelling the Automated Screw Insertion Process. Ph.D.Thesis.University of London.

[3] Visuwan P. (1999). Monitoring Strategies for Self-Tapping Screw Insertion Systems. Ph.D.Thesis. University of London.

[4] L.D. Seneviratne, F.A. Ngemoh, S.W.E. Earles and K. Althoefer (2001). Theoretical modelling of the self-tapping screw fastening process. Proceeding of IMechE, Part C, Journal of Mechanical Engineering Science. 215:135-154.

[5] Klingajay M., L.D. Seneviratne (2002). Friction Estimation in Automated Screw Insertions. Proceedings of the 3rd International Symposium on Robotics and Automation (ISRA2002). Vol. 1, pp.682-688.

[6] Klingajay M., L.D. Seneviratne, Althoefer K. (2002). Parameter estimation during automated screw insertions. Proceedings of IEEE International Conference on Industrial Technology (ICIT'02). Vol. II. pp.1019-1024.

[7] Klingajay M., L. D. Seneviratne, Althoefer K., Y.H. Zweiri (2003). Parameter Estimations for threaded assembly based on the Newton Raphson method. Proceedings of IEEE International Conference on Automation (Automation 2003). Vol.1, pp. 864-869.

[8] Klingajay M., L.D. Seneviratne, N.I.Giannoccaro (2003). Optimization of four estimated parameters for autonomous threaded fastenings. Proceedings of RAAD'03, 12th International Workshop on Robotics in Alpe-Adria-Danube Region.

[9] Klingajay M., Seneviratne L. D., Althoefer K. (2003). Identification of threaded fastening parameters using the Newton Raphson Method. Proceedings of IEEE/RSJ International Conference on Intelligent Robots and Systems (IROS 2003), Las Vegas, USA..

9. APPENDIX A

The detailed analytical model for the 5 stages are given bellowed:

The insertion stage 1

$$\tau_1 = (SigUTS\times((Ds+2\times Dh)/6)\times((0.25\times((Ds-Dh)^2)\times tg(\beta))/\sin(2\times(\mathrm{arct}(((2\times Lt)\times\sqrt{((2\times\pi\times((Ds+2\times Dh)/6))^2+P^2)})/(P\times Ds)))))\times\cos((\mathrm{arct}((P)/(2\times\pi\times((Ds+Dh)/4)))))\times\Phi(i))/(((2\times\pi\times Lt)\times((Ds-Dh)/(P\times Ds))))+\mu\times((Ds+Dh)/4)\times(0.5\times(Ds-Dh)\times\sqrt{((1+(\tan(\beta))^2)\times(((Ds+Dh)/4)^2+(P/(2\times\pi))^2))})\times(0.75\times SigY)\times\cos((\mathrm{arct}((P)/(2\times\pi\times((Ds+Dh)/4)))))\times\Phi(i)$$

The insertion stage 2

$$\tau_2 =((Ds+2\times Dh)/6)\times((0.25\times((Ds-Dh)^2)\times\tan(\beta))/\sin(2\times(\mathrm{arct}(((2\times Lt)\times\surd((2\times\pi\times((Ds+2\times Dh)/6)^2+P^2))/(P\times Ds)))))\times SigUTS\times\cos((\mathrm{arct}((P)/(2\times\pi\times((Ds+Dh)/4)))))+2\times\mu\times((Ds+Dh)/4)\times(0.5\times(Ds-Dh)\times\surd((1+(tg(\beta))^2)\times(((Ds+Dh)/4)^2+(P/(2\times\pi))^2)))\times(0.75\times SigY)\times\cos((\mathrm{arct}((P)/(2\times\pi\times((Ds+Dh)/4)))))\times(\Phi(i)-(((2\times\pi\times Lt)\times((Ds-Dh)/(P\times Ds)))/2))$$

The insertion stage 3

$$\tau_3=(((Ds+2\times Dh)/6)\times((0.25\times((Ds-Dh)^2)\times tg(\beta))/\sin(2\times(\mathrm{arct}(((2\times Lt)\times\surd((2\times\pi\times((Ds+2\times Dh)/6))^2+P^2))/(P\times Ds))))\times SigUTS\times\cos((\mathrm{arct}((P)/(2\times\pi\times((Ds+Dh)/4)))))\times(((2\times\pi\times T2)/(P))-\Phi(i)+((2\times\pi\times Lt)\times((Ds-Dh)/(P\times Ds)))))/(((2\times\pi\times Lt)\times((Ds-Dh)/(P\times Ds))))+\mu\times((Ds+Dh)/4)\times(0.5\times(Ds-Dh)\times\surd((1+(\tan(\beta))^2)\times(((Ds+Dh)/4)^2+(P/(2\times\pi))^2)))\times(0.75\times SigY)\times\cos((\mathrm{arct}((P)/(2\times\pi\times((Ds+Dh)/4)))))\times(((2\times\pi\times T2)/(P))+\Phi(i)-((2\times\pi\times Lt)\times((Ds-Dh)\ (P\times Ds))))$$

The insertion stage 4

$$\tau_4=2\times\mu\times((Ds+Dh)/4)\times(0.5\times(Ds-Dh)\times\surd((1+(tg(\beta))^2)\times(((Ds+Dh)/4^2+(P/(2\times\mathrm{pi}))^2)))\times(0.75\times SigY)\times cos((\mathrm{arct}((P)/(2\times\pi\times((Ds+Dh)/4)))))\times((2\times\pi\times T2)/(P))$$

The insertion stage 5

$$\tau_5=(((\mu\times(Dsh^3+Ds^3))/(3\times(Dsh^2-Ds^3)))+((Ds+Dh)/4)\times((\pi\times\mu\times(Ds+Dh)+2\times P)/(\pi\times(Ds+Dh)-2\times\mu\times P)))\times(((\pi\times E\times(Dsh^2-Ds^2))/(4\times(T1+(\min(((P\times((2\times\pi)\times(Ls\times Ds-T1\times Ds-Lt\times Dh)/(P\times Ds)))/(4\times\pi)),(0.5\times T2))))))\times((((E\times((P-(0.5\times(Ds-Dh)\times(\tan(\beta))))^3)\times E\times((0.5\times(Ds-Dh)\times(tg(\beta)))^3))/((E\times((P-(0.5\times(Ds-Dh)\times(tg(\beta))))^3))+(E\times((0.5\times(Ds-Dh)\times(tg(\beta)))^3))))\times(0.25/(3\times\pi\times((Ds-Dh)^3)))\times(\mathrm{sqrt}(((\pi\times(Ds+Dh))^2)+(4\times P^2))))\times\min(\Phi(i),((2\times\pi\times T2)/(P)))))/(((\pi\times E\times(Dsh^2-Ds^2))/(4\times(T1+(\min(((P\times((2\times\pi)\times(Ls\times Ds-T1\times Ds-Lt\times Dh)/(P\times Ds)))/(4\times\pi)),(0.5\times T2))))))+((((E\times((P-(0.5\times(Ds-Dh)\times(tg(\beta))))^3)\times E\times((0.5\times(Ds-Dh)\times(tg(\beta)))^3))/((E\times((P-(0.5\times(Ds-Dh)\times(tg(\beta))))^3))+(E\times((0.5\times(Ds-Dh)\times(tg(\beta)))^3))))\times(0.25/(3\times\pi\times((Ds-Dh)^3)))\times(\surd(((\pi\times(Ds+Dh))^2)+(4\times P^2))))\times\min(\Phi(i),((2\times\pi\times T2)/(P))))))\times((P\times(\Phi(i)((2\times\pi)\times(Ls\times Ds-T1\times Ds-Lt\times Dh)/(P\times Ds))))/2\times\pi)$$

The parameters used in the analytical model (Equations 1.1 – 1.5) at each stage are described below.

K_{th}= Deflection stiffness coefficient of joint under compression, K_{tb}= Deflection stiffness coefficient of screw and formed threads in combination, A_c= The full-thread cross-sectional area, θ = Thread helix angle, α = Angular position value corresponding to the length of the screw cutting-portion, β = Screw thread crest half-angle, D_h = Tap hole diameter, D_r = Screw root diameter, D_s = Screw major diameter, D_{sh} = Screw head diameter, P = Screw thread pitch, L_t = Screw taper-length, L_s = Screw total threaded-length, T_1 = Thickness of near plate, T_2 = Thickness of far plate, σ_Y = Yield stress, E = Modulus of elasticity, μ = Coefficient of Friction, σ_{UTS} = Ultimate tensile stress of tap-plate.

The equations used in this model between stage 1 and stage 5 have defined to apply for torque signature signal for five stages have explained in each stage below.

$$R_f = \frac{D_s + D_h}{4}$$

$$R_s = \frac{D_s + 2D_h}{6}$$

$$\beta = \frac{\pi}{6}$$

$$K_f = \frac{1}{2}(D_s - D_h)\sqrt{(1+\tan^2\beta)\left(\left(\frac{D_s + D_h}{4}\right)^2 + \left(\frac{P}{2\pi}\right)^2\right)}$$

$$\phi = 2\pi\left(\frac{L_s D_s - t_1 D_s - L_t D_h - t_1 D_s}{pD_s}\right)$$

$$A_c = \frac{1}{4}(D_s - D_h)^2 \tan\beta$$

Copyright © IFAC Information Control Problems in Manufacturing, Salvador, Brazil, 2004

ELSEVIER
IFAC
PUBLICATIONS
www.elsevier.com/locate/ifac

A FLEXIBLE JAVA CLASS LIBRARY FOR SIMULATING AND TELEOPERATING ROBOTS

F. A. Candelas, F. Torres, S. Puente, J. Pomares, V. Segarra, J. Navarrete

Physics, Systems Engineering and Signal Theory Department
University of Alicante, Alicante, Spain

Abstract: This paper presents a Java-3D-based software library for modelling industrial robot arms and programming simulation and tele-operation environments. Its main features are its portability, since it is developed with Java, its simplicity in defining different robots and including them in the final application, as well as its collision-detection management. The library has been applied to the development of a simulation and tele-operation system, focused on the training of Robotics, which is also briefly described here. *Copyright © 2004 IFAC*

Keywords: Java, learning, programming, simulation, robotics, telerobotics.

1. INTRODUCTION

Simulation and tele-operation tools offer many advantages for the training or learning of Robotics, such as flexibility in time-tables and student access to expensive and limited equipment (Gillet et al., 2001; McKee, 2003; Schmid 1988). Many tools have been developed for simulating and/or tele-operating industrial robots; however the majority are focused on specific models or equipment (especially in the case of tele-operation), or are not appropriate for simultaneous use by several users.

This paper presents a software library based on Java and Java 3D that facilitates the object-based programming of robot models, focused on the creation of applications for simulating industrial robot arms. The library also allows tele-operation capabilities (Candelas et al., 2003). Since Java is used as the programming language, the software is very portable, and the applications can be crated as Java applets that the users can execute easily in a computer connected to the Internet.

Java 3D is an API (Application Programming Interface) from Sun's Java (TM) that provides 3D graphics. It is similar to other APIs, such as OpenGL. The main advantages of using Java 3D, in contrast to other APIs for Java, is that Java 3D is a standard extension of Sun's Java and provides good integration with other extensions. Java 3D uses the scene-graph approach, which is a hierarchical approach for describing objects and their relationships among one another (Selman, 2002; Sowizral et al., 2000).

The best sources for obtaining the Java and Java 3D libraries, tools and documentation are The Source for Java Technology (http://java.sun.com/) and The Java 3D Community Site (http://www.j3d.org/).

The paper is structured in the following way. The software library is described in Section 2. Section 3 shows a system for simulating and tele-operating robots developed from the library. Finally, Section 4 offers the main conclusions of this work.

2. THE LIBRARY FOR MODELING ROBOTS

The library presented here is composed of several classes of objects built from Java 3D (Figure 1). These classes are described in the following sections.

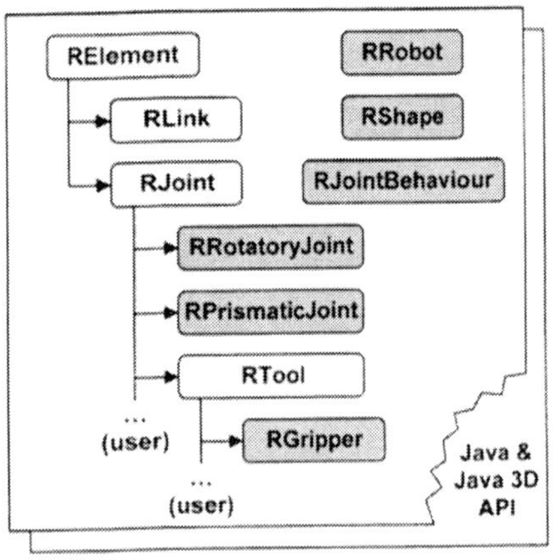

Fig 1. Classes that compose the library.

2.1. The RShape class

An RShape object is a basic object that stores the graphical representation of a robot's components or a scene's element. Its structure, according to Java 3D objects, is shown in Figure 2-a. The node T is a Java 3D transformation-group that allows the definition of the location and orientation of the RShape. S is the Java 3D graphic object to be displayed in a graphical simulation. S can be either a basic Shape3D object of Java 3D, which represents a graphic object to be displayed, or a sub-scene-graph. For the Java 3D to be able to display the graphic object represented by S, it is necessary to add B, which is a Java 3D branch-group that acts as the root of a Java 3D graph. The graphic object of S can be created by programming with Java 3D functions, or can be loaded from a local file with 3D data (WRML; 3D Studio or other formats) or from an URL.

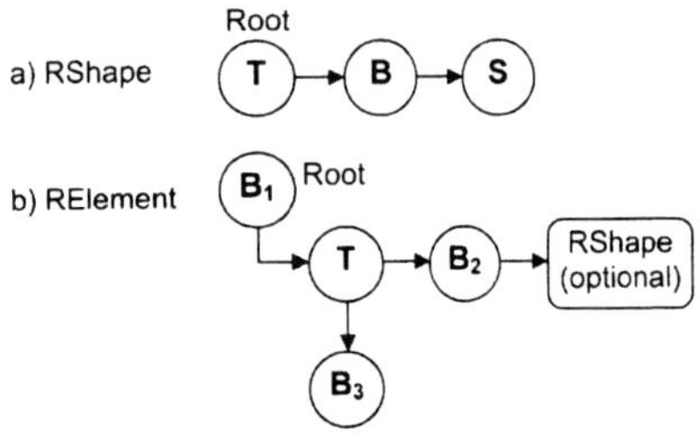

Fig 2. Structure of the RShape and RElement objects.

In addition, the RShape object also defines the method for obtaining the collision-box which is used in collision-detection (see Section 2.7). Basically, a collision-box represents volume that includes all the points of the graphic element. The RShape class can be also used to represent static objects in the simulated scene which can correspond with real objects in the environment of the tele-operated robot, and, thus, the collision-detection also considers these objects.

2.2. The RElement class

RElement defines an abstract class which represents a generic element of a robot. As shown in Figure 1, the RJoint and RLink classes extend the RElement class. As will be shown later on, the Java 3D graphical representation of a robot is a tree composed of linked RJoint and RLink objects.

The structure of a RElement is shown in Figure 2-b. The nodes B_1, B_2 and B_3 are Java 3D branch-groups use for composing a scene-graph structure. B_1 acts as a root node, B_2 is used to link an optional RShape object, and B_3 permits the linking of the RElement to the root-node of another RElement. T is a transformation-group that defines a location and orientation transformation matrix which is applied to all the nodes behind it. The RShape object included in a RElement is used only in the RLink and RTool classes to define their graphical representation. For these two classes, the RElement can also include a material definition. The RJoint-type classes do not require any graphical representation.

The RElement class also defines methods for facilitating its positioning within the 3D scene, moving through the scene-graph or obtaining the current transformation matrix. These methods must be written in derived classes that extend RElement because this is an abstract class.

2.3. The RLink class

The RLink class extends RElement, and represents a physical element of a robot. In this case, the RShape object of the RElement class is used for graphical representation. Besides, an RElement also has methods that permit the access to the collision-box of the RShape.

It should be pointed out that, in an RLink object, there are two transformation nodes (T nodes): one is inherited from the RElement structure and the other is in the RShape object. The first one affects the RLink object and all the other elements behind it in the scene-graph, and is used to situate the represented robot's link and the following links in the 3D space. The second one only affects the graphical representation of the particular RLink object.

2.4. The RJoint class

RJoint is an abstract class that extends RElement, and it represents a robot's generic joint. In contrast to an RLink object, an RJoint does not require any graphic element associated to it. That is, the node B_2 of the

RElement does not point to an RShape object. The T node inherited from the RElement is used in an RJoint to get the robot's joint movement.

An RJoint object includes properties to define the minimum, maximum and home values, as well as four methods to move the joint: *setQ(α)*, *setQF (β)* and *addQ(θ)*, and *setQT(ω, t)*. The first method situates the joint in a value α taking the collision-detection into account. The second one is similar to the first, but it does not consider the collision-detecting. The third one increases the value of the joint according to θ, also considering the collision-detecting. Finally, the fourth method *setQT(ω, t)* allows the programmer to execute a movement in a given time interval, since it performs a movement from the actual joint's value to the new value ω, in t milliseconds. To manage how the joint's values are updated during the movement in the interval, an RJoint uses an RJointBehaviour object, which will be explained in Section 2.6.

As shown before, RJoint is an abstract class which is the base for other more specific classes:

RRotatoryJoint. It represents a rotational movement by defining a rotation axis in the node T of the RElement.

RPrismaticJoint. It represents a movement along a translation axis of the node T of the RElement.

RTool. It represents a tool at the end of a robot. A tool has a movement, which can be considered as one DOF when modelling it. Thus, it is modelled as a joint with this library. This abstract class can be extended to get specific tools, for example, a gripper modelled with RGripper.

2.5. An example of using the classes

Before explaining the rest of the library's classes, an example of modelling a robot with the classes described above is presented next. Figure 3 shows a model that corresponds to a Scorbot ER-IX robot (Intelitek), which has 5 DOF thanks to the rotatory joints q_1 to q_5 plus a tool, as well as 5 links (L_0 to L_4). This robot is not frequently used in industry, but the geometric structure of its links and joints makes this example more illustrative and clear. The model of the robot has been created with the library presented here, as described below.

To create the model, the robot's joints and links must be represented with linked RElment objects, and more specifically, their extensions, RLink and RRotatory objects. Figure 4 shows the scene-graph obtained by linking the first five RElements. This scene-graph obtained is ready to be managed and displayed with Java 3D primitives. Figure 4 also shows the fact that only the RLink objects have RShape objects with their graphic representation.

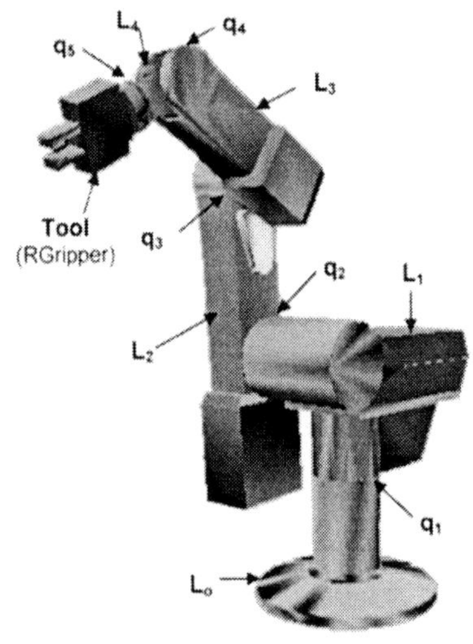

Fig 3. Scorbot ER robot's elements.

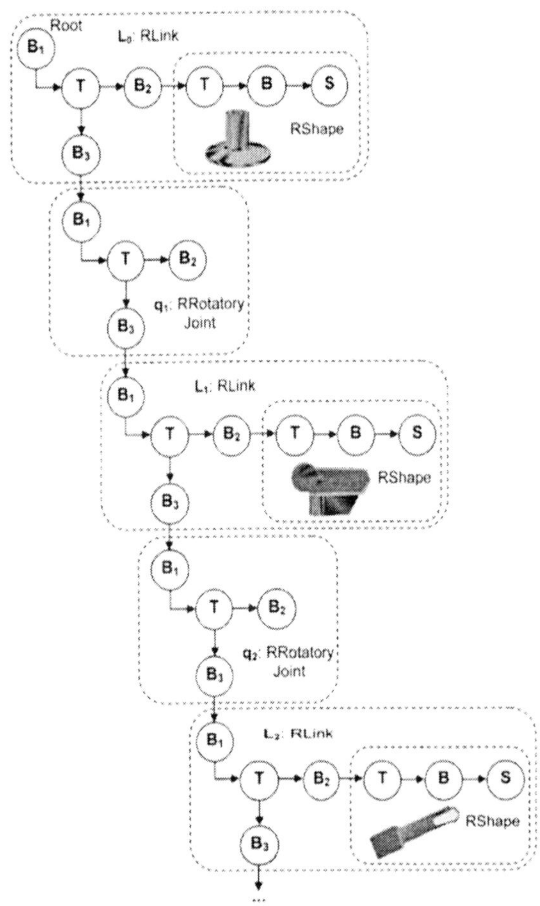

Fig 4. Example of robot description.

Figure 5 shows the basic source-code that creates a Java 3D scene with only the first two links of the previous robot (L_0 and L_1) and a joint (q_1). The program could be completed with a Java even-controller to allow the user to move the robot's joints with the mouse, using the SetQ-style methods of the joint described in Section 2.4.

```
BranchGroup scene;
scene = new BranchGroup();

// Universe initializacion
SimpleUniverse universe;
Universe = new SimpleUniverse(c3d);
initUniverse(universe);

RLink L0,L1;
RRotatoryJoint q1;
L0= new RLink("link0.WRL");
q1= new RRotatoryJoint(
     new Vector3d(0,-1,0));
L1= new RLink("link1.WRL");
L0.setSubRobot(q1);
q1.setSubRobot(L1);
L0.setPosition(0,-0.3,0);
q1.setPosition(0,0.2,0);
q1.setQHome(Math.PI/2);
q1.setVMax(1);
scene.addChild(L0.GetBranchGroup());

// Spot light configuration
SpotLight light;
light = SpotLight();
ConfigureLight(light);
scene.addChild(light);

universe.addBranchGroup(scene);
```

Fig 5. Example of the Java source code.

Although the source-code used to specify a robot with the proposed classes is simple, a class has been created to encapsulate the entire robot representation in a high-level object, as described in the next section.

2.6. The class RJointBehaviour

An RJointBehaviour object is always associated to an RJoint, and is used by the classes that extend the RJoint to perform a movement automatically in the simulation. This class extends the Interpolator and Behaviour classes of Java 3D, and represents an advanced use of the Java 3D API. On the one hand, the Interpolator class basically allows the programmer to obtain samples from 0 to 1. In this case, an object of this class is used to get the sampled values of a joint in a given time interval. And, on the other hand, the Behaviour class is used to define behaviours that are triggered under certain conditions. In the RJointBehaivour class, the behaviour is enabled throughout the time that the movement is been done to get the fastest rendering that the processor permits.

An RJointBehaivior requires the programmer to specify the associated RJoint, an initial value, a final value and a time interval. At the start of the movement, the behaviour is activated. Then, in each triggered event in the behaviour, the object calculates a sample of the joint's value and calls the *setQ(q)* method of the RJoint to position the joint.

The main reason for using this scheme to move the joints is that Java 3D permits the programmer to associate different behaviours to several joints, and to execute them concurrently. Thus, this scheme affords an easy way of moving several joints simultaneously in the rendering.

2.7. The RRobot class

The RRobot class encapsulates the graphic representation and other high-level functions of a robot model in an software object.

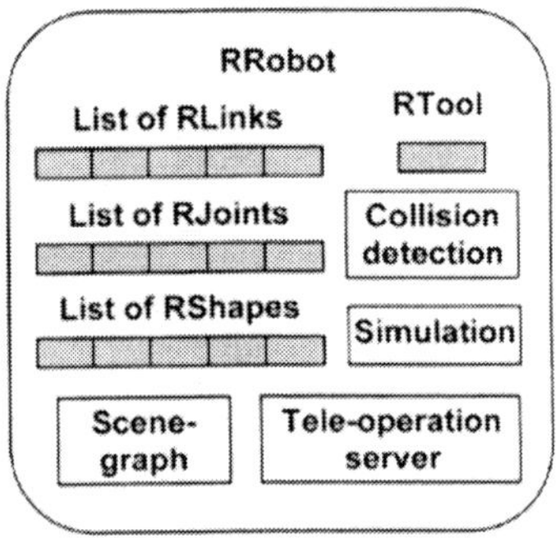

Fig 6. Elements of a robot object.

Figure 6 illustrates the main components of an RRobot object. On the one hand, we have the scene-graph of the RElement-derivated objects, three lists of pointers to the RLinks, RJoints and RShapes, which facilitate the quick access to these components, and one to the RTool. On the other hand, the RRobot is also composed of other modules for performing the simulation, the tele-operation and the collision-detecting tasks.

RRobot is an abstract class that must be extended by defining other derived classes. For example, a new RScorbotER class can be created to model the specific Scorbot ER-IX robot presented before, considering all its characteristics

Inside an RRobot, the simulation module includes the construction and the graphical representation of the robot by means of the Java 3D scene-graph structure, as well as high-level functions to manage the modelled robot. For example, some methods permit the movement of the robot to a desired position, given as, either a vector of the values of joint or as Cartesian co-ordinate within the robot's work-space. Moreover, it is possible to specify whether the joints must be moved sequentially, one by one, or at the same defined time. There are also methods of calculating the direct cinematic transformation from the joint values to the Cartesian co-ordinates of the end tool, as well as the inverse transformation, from the tool co-ordinates to values of joints.

The tele-operation module offers two basic methods that facilitate the use of the library and its simulation capabilities, as a part of a tele-operation system. These methods are called *SendCommand* and *ExecuteCommand*, and they should be implemented in the derived class only if the tele-operation capabilities are desired.

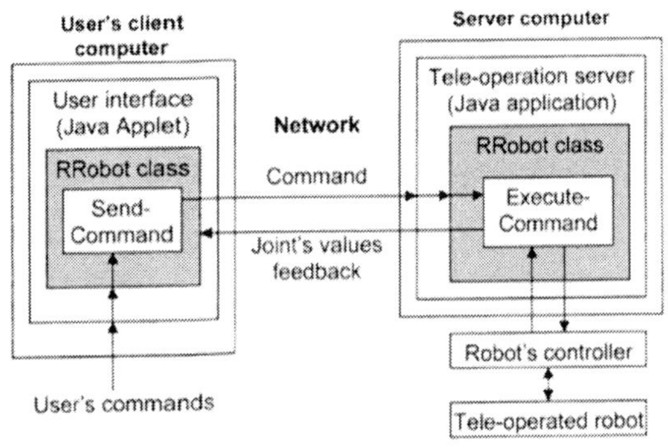

Fig 7. RRobot class in a tele-operation system.

SendCommand permits the programmer to use a class that extends RRobot as part of a user-interface as a client, and it allows the-user interface to send a list of movement commands to a "tele-operation server", as Figure 7 shows. Basically, a command consists of a vector with the new joints' values and another vector with the times associated to the movement. The "tele-operation server" is a Java application which is connected to a real robot through its controller. This application includes the class that extends RRobot, and accepts commands from one or more client. When this server receives a command list, it simulates the commands using the RRobot class simulation capabilities, and, if the simulation is carried out without errors (for example, collisions), the commands are translated to the robot language and sent to the robot's controller by the *ExecuteCommand* method. The simulation guarantees the correct use of the real robot. The server also watches the robot's movements, and sends feedback information about the joints' values to the user's client while the robot is moving, in order to offer the option of updating the simulation with real values. This tele-operation scheme does not implement a control loop of the robot trough the Internet. Instead of it, list of commands are set to the tele-operation server to be finally executed in the robot's controller.

Finally, the collision-detection module has the task of checking the possible collisions, not only between the robot and other objects in the scene, but also between a joint that is moving and the other joints of the robot. The collision-detection is based on the RShape objects, and its collision-boxes. A collision-box is a Bound object of java 3D, which represents a simple 3D-volume, defined by eight 3D-points, that includes all the points of the graphic element. Each time that a movement is done in the simulation, the different collision-boxes are updated and positioned in the 3D space according to the elements that they represent. Next, the intersection of each collision-box with the others is calculated. If some intersection gives a non-empty volume, a collision between objects is indicated, the simulation ends and an error message is given to the application. In this process, the possible collisions between consecutive RLink objects of the robot must be ruled out. The collision-detecting can be carried out quickly, thanks to the fact that the processing of the collision-boxes is a great deal simpler than processing the graphic objects.

3. THE COMPLETE SYSTEM

The library presented here has been used in the developing of a system that allows the simulation and tele-operation of industrial robot arms, called ROBOLAB 2 (www.disclab.ua.es/robolab), which is focused on the training of Robotics for students or operators (Candelas et al, 2003; Puente et al., 2000). This is described briefly in the following sections.

3.1. System's Architecture

Figure 8 shows the elements that compose the system. The robot that can be tele-operated is, in this case, the Scorbot ER-IX. This robot and its controller are the only two pieces of the equipment that require any considerable investment. The "main server" is a PC that affords the web services from the Internet, supplies the Java applet with the user interface and manages the user log-in and account. The "tele-operation server", which includes the object that extends the RRobot class, is in another PC within the private network. It communicates, on the one hand, directly with the robot's controller, and, on the other hand, with the users through the "main server". In addition, there is a "video server" wicht gives the option of a video stream feed-back for tele-operation.

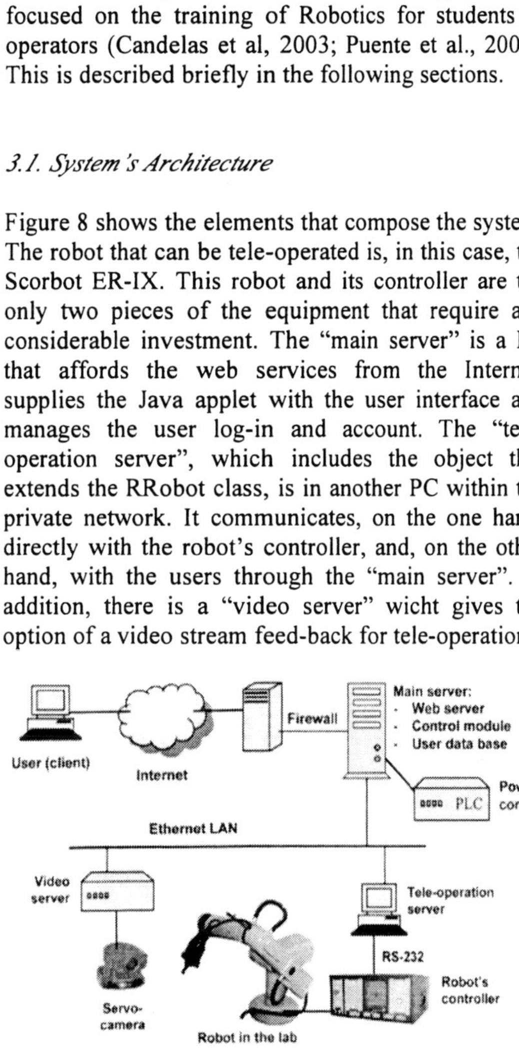

Fig 8. Architecture of ROBOLAB.

The only means that a user requires are a computer connected to the Internet, a web client program and some software components (Java and Java 3D). With these means, the student can access a web page with a Java applet with the user interface and the objects that extend the RRobot class.

3.2. User Interface

The user interface is shown in the Figure 9. It is a Java applet that displays a simulation of an industrial robot arm by using the software library described in this paper. A work-table is also represented in the simulation as a static object in the robot's workspace. The interface also shows the transformation matrix, the values of the joints, and the list of tested commands entered by the user. Not only a keyboard or a mouse, but also a commercial joystick for games with force feedback can be used to define a position or movement for the robot. In the case of a joystick, the collisions detected by the simulation are transmitted to the thought a sensation of resistance by the joystick.

Thanks to the simulation, the students can practice and carry out movements, and store the validated movements in the command list. Once the user has a list of correct commands, he can request the "web server" to remotely execute the sequence with the real robot. To do so, the student must identify himself as an authorized user. The "main server" also manages the access of several users to the robot, guaranteeing their orderly access.

To verify the movement of the real robot, the applet offers the user two different feed-back options. The first one is a compressed video stream, generated by the "video server" and sent to the students' computers while the robot is executing the movement sequence. The second one, is the simulation update with information about the current state of real robot's joints received online from the "tele-operation server", which is based on the RRobot class features (Torres et al., 2001).

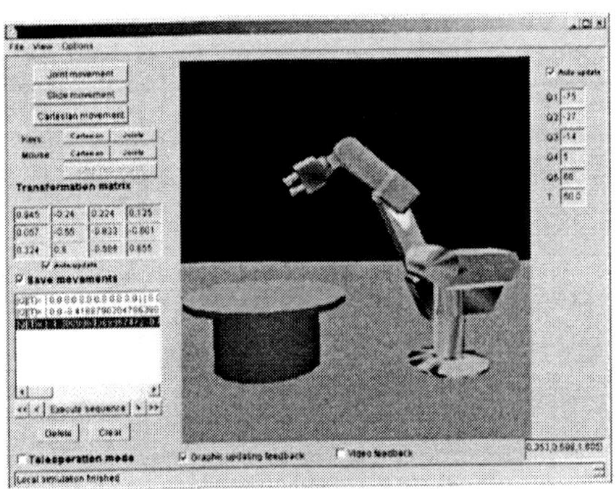

Fig 9. User interface with the Scorbot ER robot.

4. CONCLUSIONS

This paper we have presented, mainly, a new software library that allows the easy programming of robot models for developing Java applications for simulation and tele-operation. The programmer requires some knowledge of Java and Java 3D. The library is very portable since it has been built with Java and Java 3D, and it can be run on many different platforms. Furthermore, the final user only requires basic software to use it, and the end application can even be a Java applet, downloaded from an Internet server.

Another important feature is that the library enables the programmer to easily add new robot models for simulation and even tele-operation, without any changes, either in the interface application or in the system's structure. He only needs to include the new classes that extend the RRobot class and add the new equipment to the system. We have also demonstrated the use of the library for modelling a Scorbot ER-IX robot, as well as a tele-operation system that uses the library with this robot. However, other models for robots, like a PA-10 (Mitsubishi) with 7 DOF or a Cartesian robot with two prismatic joints, have been successfully developed and used for tele-operating.

We are working now in including to the system the capability to recognize basic 3D objects in the real robot's workspace, model these objects, and represent it in the simulation. With these capabilities will be possible tele-manipulate objects with the real robot.

Acknowledgements

We are grateful to the AVCiT of the Generalitat Valenciana regional government for their financial support to the research project CTIDIA/2002/108.

References

Candelas, F. A., Puente, S. T., Torres, F., Ortiz, F. G., Gil and P.; Pomares, J. (2003) A Virtual Laboratory for Teaching Robotics. *International Journal of Engineering Education, Especial Issue "Remote Access/Distance Learning Laboratories"*, vol. 19, pp. 363-370.

Gillet, D., Latchman, H. A., Salzmann, Ch. and Crisalle, O.D. (2001) Hands-On Laboratory Experiments in Flexible and Distance Learning. *J. Engineering Education*, vol. 90, pp. 187-191.

McKee, G. T. (2003) An online robot system for projects in robot intelligence, *International Journal of Engineering Education - Especial Issue Remote Access/Distance Learning Laboratories*, vol. 19, pp. 356-362.

Puente, S.T., Torres, F., Ortiz, F.G. and Candelas, F.A. (2000) Remote Robot Execution through WWW Simulation. *15th Int. Conf. on Pattern Recognition (ICPR 2000)*, vol.4, pp. 503-506.

Schmid, Ch. (1998) Using the World Wide Web for Control Engineering Education. *Journal of Electrical Engineering*, vol. 49, pp. 205-214.

Selman, S. (2002) *Java 3D Programming*. Manning Publications, U.S.A

Sowizral, H., Rushforth, K., and Deering, M. (2000) *The Java 3D API Specification* (2nd edition). Addison-Wesley Pub. Co.

Torres, F., Puente, S.T., Canovas, J., Mangas, J. and Martinez-Larraz, C. (2001) Web teleoperation of robots with simulation feedback. *Telemanipulator and Telepresence Technologies VIII, Proceedings of SPIE*, vol. 4570, pp. 105-112.

Copyright © IFAC Information Control Problems in
Manufacturing, Salvador, Brazil, 2004

ELSEVIER
IFAC
PUBLICATIONS
www.elsevier.com/locate/ifac

SENSORING FOR RETROFITTING OF AN INDUSTRIAL ROBOT [1]

Eduardo José Lima II *
Guilherme Campelo Fortunato Torres *
Carlos Alberto Carvalho Castro *
Alexandre Queiroz Bracarense *
Renato Ventura Bayan Henriques *
Walter Fetter Lages **

* *Federal University of Minas Gerais*
School of Engineering, Mechanical Engineering Department
Av. Antnio Carlos, 6627 - 31270-901 Belo Horizonte, MG, Brazil
ejlima2@hotmail.com,
guilhermefortunato@hotmail.com,
carlosjapao@hotmail.com, bracarense@ufmg.br
bayan@demec.ufmg.br
** *Federal University of Rio Grande do Sul*
School of Engineering, Electrical Engineering Department
Av. Osvaldo Aranha, 103 - 90035-190 Porto Alegre, RS, Brazil
w.fetter@ieee.org

Abstract: Retrofitting of an industrial robot consists on verifying the usability conditions of each component of the robot, replacing obsolete ones, especially electronics and control. This work describes the replacement of the analog sensors (resolvers and tachometers) of a retrofitted ASEA IRB6 robot by digital sensors (incremental optical encoders). The advantages of such replacements are assessed through the use of tachometers and encoders for speed feedback. The control architecture based on a CAN-bus is also presented. *Copyright © 2004 IFAC*

Keywords: Retrofitting, encoders, closed loop control, industrial robots, robot control

1. INTRODUCTION

The Laboratory of Robotics, Welding and Simulation (LRSS) from Federal University of Minas Gerais (UFMG) works currently in a project of retrofitting of an ASEA IRB6 robot, with 5 degrees of freedom, manufactured in 1977 (Figure 1).

Retrofitting of industrial robots consists on verifying the utility of each component of the robot and replacing what is obsolete, especially the electronics and control. The objective of retrofit an old robot is, besides bringing up to date its control technology, to make it available for academic

[1] Authors thank FAPEMIG, FAPERGS, CNPq, CAPES and MANET for financial support.

Fig. 1. ASEA IRB6 robot.

studies, in special studies about robotization of the welding process.

An academic initiative for generating an open standard for robot control is the OROCOS (Open RObot COntrol Software) project (OROCOS, 2002). This project is in its early stages and does not produced a working prototype yet. Also, it is more oriented towards the software architecture of the whole robotic system and does not properly address the supporting hardware architecture.

Open hardware and software architectures for robot control were defined by the PINO project (Yamasaki *et al.*, 2002), but since they were designed with small legged mobile robots in mind, they do not seems adequate for industrial manipulator robots.

There are also commercial robot controllers that can control robots from any manufacturer (URC, 2003). Such a controller eliminates the need to learn different robot programming languages. However, its architecture is not open. Also, its programming language `RobotScript` (Lapham, 1999) can be used for all robots supported by the controller. Nonetheless `RobotScript` is based on `VBScript`, a proprietary language tightly coupled to the Windows operating system. Besides Windows has not been designed with real-time control in mind, it is known for proprietary protocols that change from time to time. Hence it does not appear to be appropriate to serve as a basis for a standard open architecture for robot control.

Fortunately, the mechanics of industrial robots has not changed too much. The main differences from old to newer robots reside in the actuator power drives and controller, including the software. This fact enables us to upgrade old robots to current technology by retrofitting the robot controller.

The first step in robot retrofitting involved the maintenance of the mechanical and electric parts. The robot were disassembled and a verification of all the motors/axis and its electric contacts was made. It was verified that the mechanical structure of the robot was in good state, thus enabling us to reuse it. Power source and motors were in good state as well.

Originally, the ASEA IRB6 used resolvers for position feedback and tachometers for speed feedback to make the closed loop control. Resolvers and tachometers are analog sensors, and threrefore need an analog processing. In order to avoid analog processing, it was chosen to use incremental optical encoders in retrofitting for position feedback. Such sensors are made by light sources, light sensitive devices and a disc in which black and translucent divisions alternate (Jones and Flynn, 1993; Everett, 1995). See Figs. 2 and 3. When the disc turns, the light receivers generate square shaped waves with $\pm 90^{o}$ phase (what allows to determine the turning direction). The frequency of these waves is proportional to the turning speed, as shown in Figure 4.

Thus, using the encoder is possible to get the relative position and speed of the axle by counting signal pulses ocurring in each time interval.

Fig. 2. Incremental optical encoder.

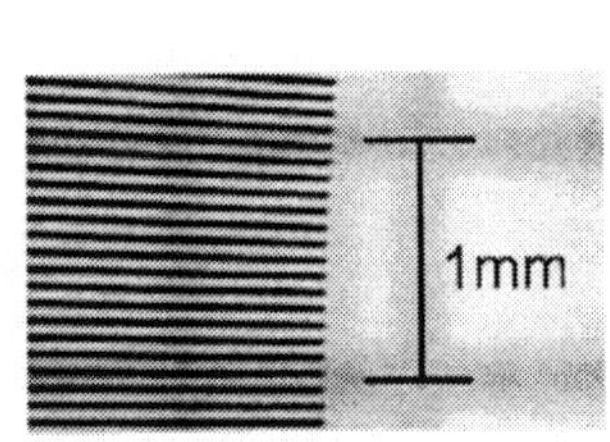

Fig. 3. Detail of an incremental optical encoder.

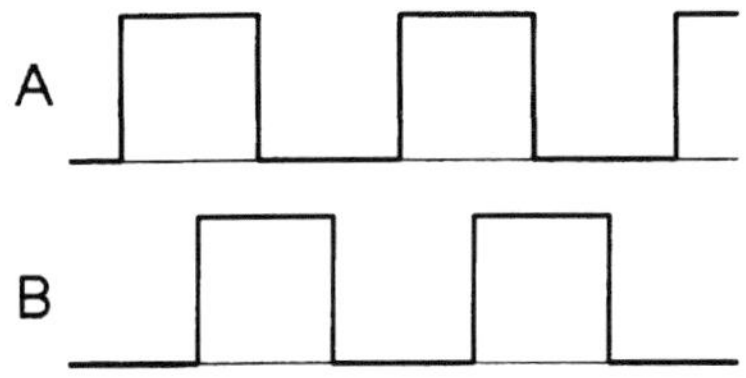

Fig. 4. Signals from channels A e B of an incremental optical encoder.

This work presents an experimental study comparing the use of tachometers and encoders for speed feedback. The hardware implementation of a distributed control system used for robot retrofitting is also described.

2. EXPERIMENTAL PROCEDURE

Before deciding for the replacement of the tachometers by encoders as speed sensors, experiments had been made to verify its accuracy. Basically, the experiment consisted in vary the power supplied to the DC motor and measure its speed using the encoder and the tachometer (Lima II *et al.*, 2003). The speed obtained with the encoder signal was considered in this work as a reference without error. The signals read from the encoder and the tachometer were processed by a digital oscilloscope. Figure 5 shows the experiment basic scheme and Figure 6 shows the adjustable power supply used to drive the motor.

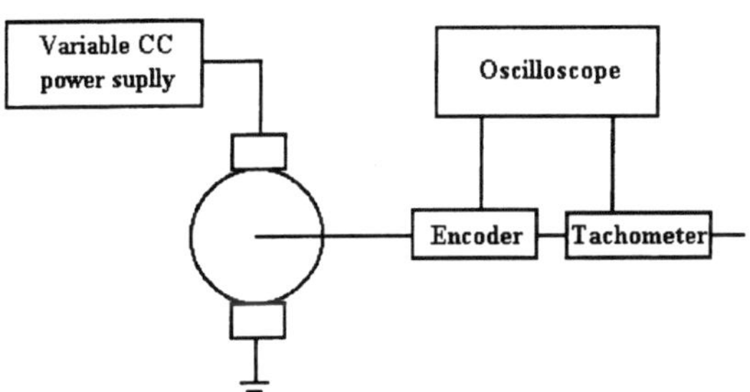

Fig. 5. Experimental setup diagram.

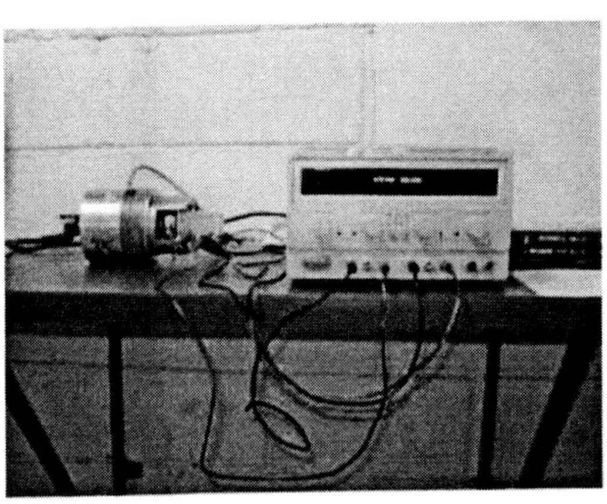

Fig. 6. Experimental setup.

The oscilloscope measures the frequency of the square wave with a precision of $0.1\mu s$ (Figure 7) Since the pulses number (N) of encoder wave is known, the speed of rotation of the motor can be calculated by:

$$\omega = 2\pi \frac{f}{N}$$

where f is the frequency.

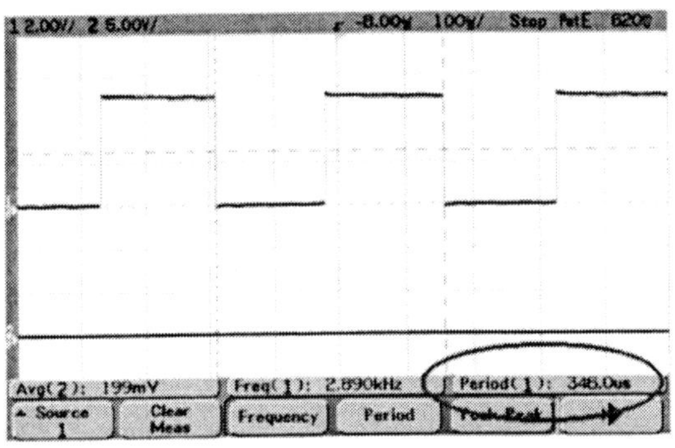

Fig. 7. Period as measured by the digital oscilloscope.

The tachometer voltage was also measured within the use of the oscilloscope. The oscilloscope shows the average voltage with a precision of 1mV (Figure 8).

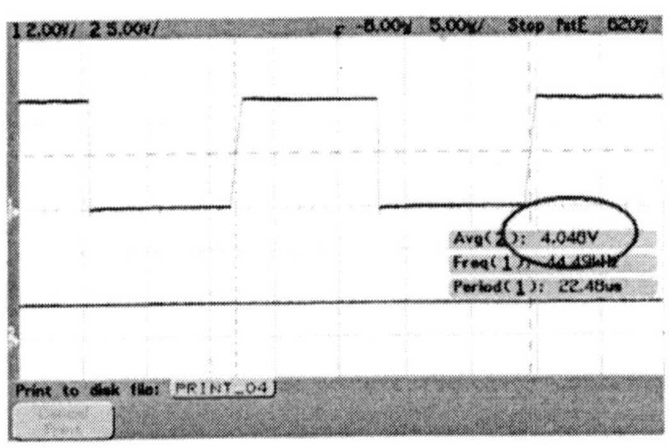

Fig. 8. Average voltage given by the tachometer.

Then, considering the speed ω obtained by the encoder as a referential without error, a linear regression was made between the values of ω and the tachometer voltage (V_{tacho}). Since the motor speed is not exactly a constant with respect to the applied voltage (V_{motor}), it was chosen to use an average of three ω and V_{tacho} measurements for each V_{motor} value.

The coefficients of equation (1) were obtained by linear regression.

$$\omega = a_0 + a_1 V_{tacho} \tag{1}$$

By comparing equation (1) with the characteristic curve of the tachometer:

$$\omega = K V_{tacho}$$

it is expected a value for a_0 close to 0 and a_1 close to K.

The values for a_0 and a_1 were obtained by:

$$a_0 = \frac{\sum x_i \sum x_i y_i - \sum x_i^2 \sum y_i}{(\sum x_i)^2 - N \sum x_i^2}$$

$$a_1 = \frac{\sum x_i \sum y_i - N \sum x_i y_i}{(\sum x_i)^2 - N \sum x_i^2}$$

with x_i the i-th value of ω and y_i the i-th value of V_{tacho}. The correlation coefficient r was obtained by:

$$r = \sqrt{1 - \frac{S_{xy}^2}{S_y^2}}$$

where:

$$S_y^2 = \frac{1}{N-1} \sum_{i=1}^{N} (y_i - \overline{y})^2$$

$$S_{xy}^2 = \frac{1}{N-2} \sum_{i=1}^{N} (y_i - y_{ci})^2$$

where $\overline{y}$ is the average value of V_{tacho} and $y_{ci} = a_0 + a_1 x_i$.

The obtained values were: $a_0 = 6.687$ and $a_1 = 31.706$. Figure 9 shows the comparison between the obtained curve ($\omega = 6.687 + 31.706 V_{tacho}$) and the experimental data. Value of $a_0 = 6.687$ is close to zero, as expected, considering the scale chosen and the ω variation up to 250 rad/s (a_0 is about 2.7% of 250).

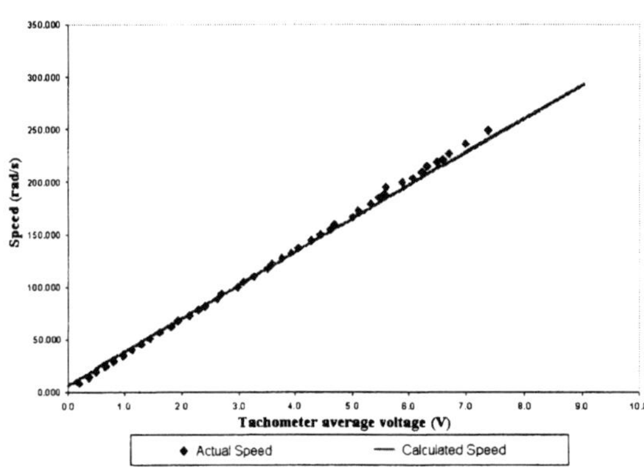

Fig. 9. Linear regression between values of ω and V_{tacho}.

It was found $r = 0.989570$. The factors that make the value of r to be far from 1.000000 are only the tachometer measurement errors, since the motor speed variations are the same measured by the encoder and the tachometer.

Therefore, it can be concluded that besides having the advantage of producing digital signals, that can be directly processed by the digital robot controller, encoders also offer higher precision in measuring speed and better noise immunity when compared to tachometers.

By using the same experimental data, the motor response curve in steady state without load was also obtained. A linear regression relating the voltage applied in the motor (V_{motor}) and the speed computed through the frequency of the encoder signal (ω) was obtained. The resulting values were: $a_0 = -24.731$ and $a_1 = 10.838$, hence:

$$\omega(V_{motor}) = -24.731 + 10.838 V_{motor}$$

Figure 10 shows the related curve. A value of $r = 0.999720$ was obtained, revealing an excellent linearity of the original robot motor, despite being 26 years old.

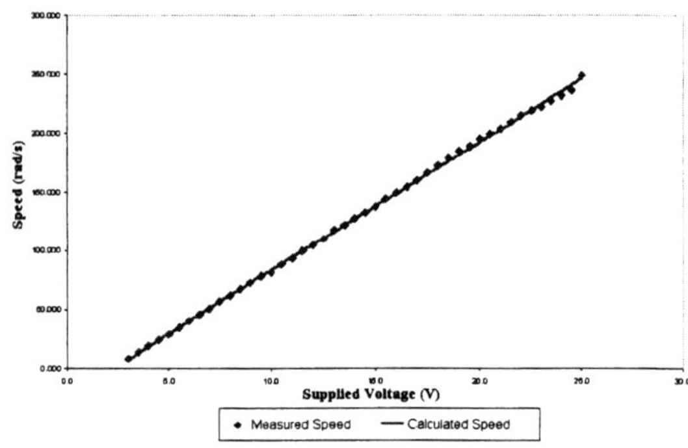

Fig. 10. Steady state response of the motor without load.

Another interesting function is $V_{motor}(\omega)$, which determines the voltage to be applied to the motor to in order to get the desired angular speed in steady state:

$$V_{motor}(\omega) = 2.282 + 0.092\omega \tag{2}$$

Through equation (2) it can be observed that the motor has a dead zone voltage of 2.282V (below this voltage the motor does not move).

3. ENCODERS READING BY ROBOT CONTROLLER

After choosing the encoders and mounting them on the motors, it was necessary to develop a structure to receive and process the signals and send the feedback information to the controller. The original control architecture was analyzed and a new architecture was proposed.

Despite the independent control of each joint, the control was originally centered in only one machine (the robot controller) for each manipulator, bringing the problem of a single point of failure. By considering the possibility of redundant degrees of freedom that the robot can have in its structure (for a given task), a considerable increment in the tolerance for global system failures would be reached by a distributed control. The devices to be arranged in a distributed architecture are those ones responsible for the I/O in each joint (i.e., position sensors, motors and brakes) and, eventually, processors with sufficient performance to implement control strategies.

In order to build a flexible, scalable and lower costing structure, a distributed I/O network based on microcontrollers and CAN-bus was designed (Lages and Bracarense, 2003) (Figure 11). There are also Ethernet connections, so that management tasks and programs can be uploaded to controllers by FTP.

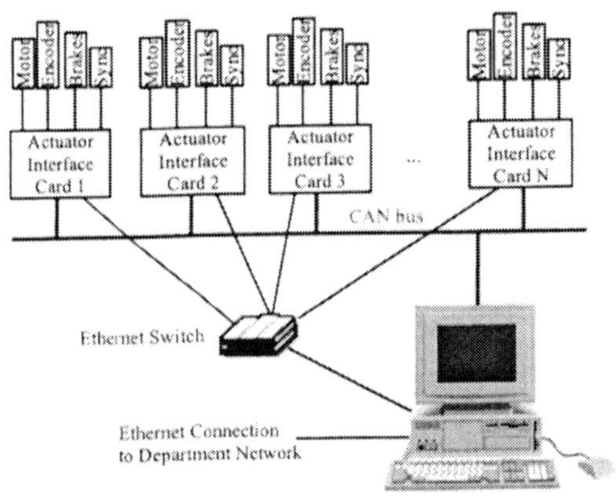

Fig. 11. Distributed I/O architecture over CAN-bus.

What makes CAN-bus an attractive base for the project of flexible real time communication protocols is the possibility of establishing global priorities for the nodes access to the bus, which is made through the following rules: (i) one of the states, 0 or 1, is assigned as dominant and the other as recessive, (ii) in the case of a collision, the node which writes a recessive bit in the bus and reads a dominant bit (i.e., a message with bigger priority than the one which is being transmitted) interrupts the transmission, preventing wasting band with posterior retransmissions of both nodes. This access method to the bus is called CSMA/CA, where CA means Collision Avoidance, although the arbitration mechanism would be better described by "collision solving", because the collision is not prevented, but solved.

The basic idea behind the use of a CAN-bus is simple: each electric device, sensor, actuator, or combination of these in the subsystem is connected to a small computerized module. This represents a total rupture in the concept of centered control connected to modules by a complex wiring system. The CAN protocol is based on simple modules interconnections by a serial data bus, offering a tremendous reduction on electric wiring in the robot.

In the case of ASEA robot, each joint controller can act as a local controller implementing, for instance, a simple PID control law (Fu *et al.*, 1987; Sciavicco and Siciliano, 1995) based on setting points received from the network or as an I/O processor, sending sensors readings and receiving actuators values through the network. When operating in I/O processor mode, the control loop is closed at the controlling PC, enabling the implementation of complex control laws.

Each joint controller is implemented by an custom designed Actuator Interface Card (AIC), implementing PWM amplifiers to drive a motor, an quadrature decoder to interface with optical encoder, a sync switch and brake interfaces. In figure reffig:TINI, it is shown: #1 TINI microcontroller board, which has a microcontroller 8051 with one Java interpreter and a CAN controller on-chip; #2 print circuit board, which has CAN, RS232C, Ethernet and transceivers connectors; #3 Ethernet connector; #4 CAN and RS232 connectors; #5 H bridge.

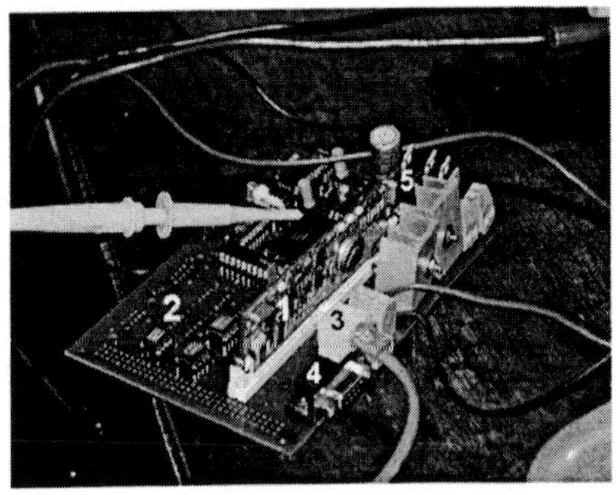

Fig. 12. Actuator interface card (AIC).

The main computer software is based on RTAI (Cloutier *et al.*, 2000; Dozio and Mantegazza, 2003), a real-time variant of the Linux operating system and communicates with the AICs. The control software is extensively based on the use of libraries written in C++, to define objects that represent each robot joint. The motors driving, the brakes releasing and the encoders reading are carried through functions defined in these libraries. Figure 13 shows some program lines which calls these functions.

```
aic.motor.on();
aic.brake.release();
aic.motor=10; //applies 10V to motor
cout<<"Encoder="<<aic.encoder.read();
```

Fig. 13. Examples of program lines which communicates with the AICs.

It is expected, when the project is concluded, an educational controller which may be used in graduate and undergraduate robotic courses. The program will show in real time values of set-points and actual values of position, speed and tool acceleration and each robot joint through graphs. It will allow functioning characteristics learning and other control algorithms development.

Moreover, the controller, with opened architecture, will virtually allow the programming and execution of any robot trajectory, even complex ones, making possible more simplicity in the programming of trajectories.

4. CONCLUSIONS

Comparative tests has shown that encoders produces more accurate speed measurements than tachometers that originally gave information of angular speed to the robot controller.

The experimental data had been also used to obtain the steady state response curve of the original motors without load. The linearity of the curve stated that the motors still are in good functioning condition, enabling their use in the robot retrofitting process.

The distributed control architecture based in a CAN network, and how each node can locally close the control loop or just receive voltage values to be applied in the motors and to inform feedback values to the controller were shown.

The use of an open control architecture makes possible studies in robotics area and development of new control algorithms for the robot and real time welding parameters calculation, specially using covered electrode, by developping advanced control laws that adapt the control system behavior as a function of welding parameters.

REFERENCES

Cloutier, Pierre, Paolo Mantegazza, Steve Papacharalambous, Ian Soanes, Stuart Hughes and Karim Yaghmour (2000). DIAPM-RTAI position paper. In: *Proceedings of the Real-Time Operating Systems Workshop.* Dipartimento di Ingegneria Aerospaziale, Politecnico di Milano. Dipartimento di Ingegneria Aerospaziale, Politecnico di Milano. Milano, Italy. pp. 1–28. http://www.rtai.org.

Dozio, L. and Paulo Mantegazza (2003). Linux real time application interface (RTAI) in low cost high performance motion control. In: *Proceedings Of the Motion Control 2003 Conference.* Associazione Nazionale Italiana per l'Automazione. Milano, Italy.

Everett, H. R. (1995). *Sensors for Mobile Robots: Theory and application.* A K Peters.

Fu, K. S., R. C. Gonzales and C. S. G. Lee (1987). *Robotics: Control, Sensing, Vision, and Intelligence.* McGraw Hill College Div, Inc.

Jones, Joseph L. and Anita M. Flynn (1993). *Mobile Robots: Inspiration to Implementation.* A K Peters.

Lages, Walter Fetter and Alexandre Queiroz Bracarense (2003). Robot retrofitting: A perspective to small and medium size entreprizes. In: *Proceedings of the 3rd Austrian Brazilian Automation Day.* RECOPE/MANET. RECOPE/MANET. São Bernardo do Campo, SP, Brazil. pp. 1–11.

Lapham, John (1999). RobotScript the introduction of a universal robot programming language. *Industrial Robot* **26**, 17. http://www.rwt.com/main/articles/robotscript.html.

Lima II, Eduardo J., Carlos A. C. Castro and Roberto M. Andrade (2003). Levantamento da curva de resposta em regime permanente de um motor de corrente contnua e determinao da constante de um tacmetro. Trabalho da disciplina tcnicas da pesquisa experimental. Universidade Federal de Minas Gerais.

OROCOS (2002). Open robot control software. http://www.orocos.org.

Sciavicco, Lorenzo and Bruno Siciliano (1995). *Modeling and Control of Robot Manipulators.* McGraw Hill, Inc.

URC (2003). Universal robot controller. http://www.rwt.com/main/urc.html.

Yamasaki, F., K. Endo, M. Asada and H. Kitano (2002). Energy-efficient walking for a low-cost humanoid robot, PINO. *AI Magazine* **23**(1), 60–61.

Copyright © IFAC Information Control Problems in
Manufacturing, Salvador, Brazil, 2004

ELSEVIER
IFAC
PUBLICATIONS
www.elsevier.com/locate/ifac

MIXED REALITY WITH HYPER-BONDS - A MEANS FOR REMOTE LABS

F. Wilhelm Bruns
Artec Research Center, University of Bremen
Heinz-H. Erbe
Center of Human-Machine Systems, Technische Universität Berlin

Abstract: The paper presents a learning environment where on-site and remote components merge into a cooperative learning process. The envisaged system allows to work together with complex real and virtual systems, consisting of parts which may be remotely distributed. The learning environment includes a supportive web-database with multimedia learning sequences providing theoretical background information, exercises and help to handle training tasks. Hardware equipment can be connected to the virtual environment with a special bi-directional sensor-actor coupling called Hyper-Bond. The learning environment smoothly integrates equipment and supports full hardware-in-the-loop functionality, allowing to build up real systems as subsystems of complex virtual systems. *Copyright © 2004 IFAC*

Keywords: bond graphs, hyper bonds, e-learning, simulation, remote experiments

1. INTRODUCTION

Engineering education is confronted with the need to develop theoretical integrated with practical learning sequences to fulfill the demands for multi-skilled engineers and also skilled technicians. Tasks and problem solving in complex technical systems requires cognitive and operational knowledge and practical experience about building systems, diagnosis- and maintenance-techniques. However, a significant challenge is that these tasks are essentially characterized by the use of tele-medial systems. Service staff in the professional field need the ability to achieve their aims in (tele) cooperation with others, and they should be able to cooperate in virtual and distributed forms of organization. Concepts concerning pedagogical, technical and organizational aspects to meet these requirements in education and training are in development. Cultural differences and similarities concerning learning and collaboration styles have also to be considered regarding curricula, courseware and teaching methods.

Computers are now used in the classroom and at the job as multimedia tools to provide alternative sources of learning material, to provide interactive learning situations and to provide simulation of systems that cannot for reasons of cost, size or safety be used in reality. The use of the Internet is rapidly increasing and is being seen by some people as the greatest source of knowledge available for learning. The use of simulation tools has a number of benefits to education. The learner is not exposed to the hazards of the real world. The learner is able to explore a range of possible solutions easily and quickly. The learner is able to use the tools that will be available in industry. The cost of simulation tools is significantly less than the real world components and allows more participation and interaction than a limited demonstration. An added benefit is that learners today, enjoy using computer based technology and this enthusiasm fosters the learning process. The question is indeed to what extend real experience can be replaced by learning with simulations? The internet makes it possible for e-learners to have access to remote laboratories. They could change control parameters to study the effect on the performance of a plant equipment. Before doing this they can experiment with the virtual equipment in a computer-simulation to save time when remote experimenting with the real equipment.

Remote laboratory applications are not new. With respect to learning support there are however still many open questions:

- How to synchronize multiple remote actions on one lab-object to support collaboration,
- How to handle feedback from the process: visual, sound, haptics or more general: how to sense all interesting physical phenomena
- How to handle action into the process: electrical, mechanical, thermodynamic or more general: how to generate all interesting physical phenomena
- How to represent learning content, scenarios, prerequisites, tools etc. in a standardized Learning Object Module (LOM) to support a modular, open, easy and flexible use
- How to administrate the learning session running on one original critical equipment, supported by an on-line expert?

Goldberg & Chen (2001) and Song & Goldberg (2003) fruitfully address the problem of collaborative control of robotic cameras to observe a remote laboratory and the distribution of video-streams. On their web-page it is also shown how they face the learning object module problem.

Ferreira et al (2002) and Cooper et al (2002) demonstrate the use of remotely controlled experiments of biochemistry, fundamental physics, automatic visual inspection and digital electronics in the EU-project PEARL (2003). Mueller & Ferreira (2003) report about remote-lab developments in the EU-project MARVEL – a mixed reality learning environment for vocational training in mechatronics.

However, up today, all remote-lab developments strictly separate reality and virtuality, energy and information. One can sense the remote process, view specific parameters, control them by changing parameters and observe the process by video-cameras. The process, as a flow of energy controlled by signals and information is either in reality or completely modeled in virtuality by simulation. Going one step further, we suggest a distributed environment where the process-model of energy and information flow can cross the border between reality and virtuality in an arbitrary bi-directional way. Reality may be the continuation of virtuality or vice versa. This bridging or mixing of reality and virtuality opens up some completely new perspectives not only for learning environments but also for evolutionary systems design and service work. To make this possible, Bruns (1999, 2001) developed a concept of *Complex Objects* being a unit of various closely coupled virtual and real representations and *Hyper-Bonds*, a universal interface type. This name has been chosen because of its relation to bond graph representations. Bond-Graphs being itself a very powerful representation-theory for dynamic systems.

2. BOND-GRAPHS AND HYPER-BONDS

Hyper-Bonds combine the unified abstract systems representation of bond graphs with an implementation of "hyper-connections" between physical phenomena of the computer-external environment and the logical structure of computer-internal representations, a blend of physical systems with their virtual counterparts.
Paynter (1961) introduced the theory of bond graphs as a unifying view on physical phenomena from a continuity of power-flow perspective. Power flows through system components and connections by way that the product of effort and flow is continuous, following typical laws of energy conservation. Effort (e) is the driving force for flow (f) and can be a pressure difference, force and torque, electrical potential difference, temperature difference etc. Flow (f) can be a flow of material, momentum, electric current, entropy. The bond graph theory has been further developed by Karnopp et al (1990). Pairs of effort and flow (e,f) are for example in mechanical systems force (F) and velocity (v), in electrical systems voltage (V) and current (i), in pneumatic/hydraulic systems pressure (P) and volume flow rate (dQ/dt). Figure 1 explains the correlation of pressure p, fluid flow q with force F, velocity v and mass/inertia I, compressibility C of the fluid and friction R in a simple pneumatic equipment. In many automation systems, electro-pneumatic circuits are considered as state automata and the elements can be represented as simple off/on switches like the valve and in/out positioning like the cylinder. These state automata do not require a bond-graph representation, but if one is interested in a more detailed dynamic behavior, then this can be described by graphs like the one in figure 1.

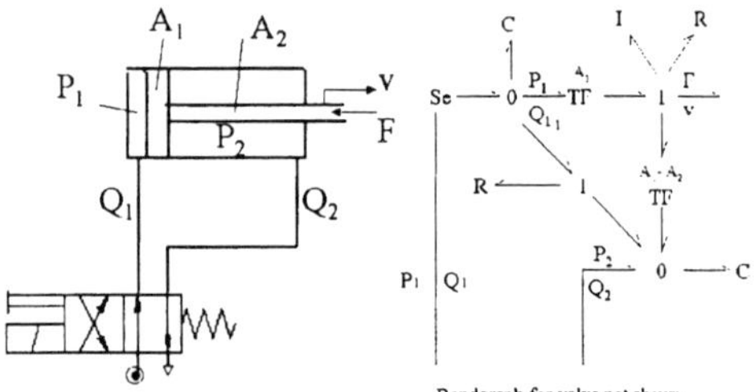

Fig. 1. 4/2 way valve controlling a double-acting cylinder and a bond graph of the cylinder alone

Components (valves, cylinders, etc.) are always connected by bonds having the value pair e and f. Knowing e and f at one connection, resulting from calculation or measurement, allows a cutting of the system in two parts and a separate investigation.
In mechanics one learns the principle of cutting a system at well defined boundaries and replacing the external influences by some observable and measurable relevant variables, reducing the

investigation to the internal dynamics of the rest. In laboratory work this principle is used to construct reproducible experiments, but also mentally it is used to think about systems in hypothesis and mental experiments. Today laboratories, being more and more permeated by computers, a free and easy distribution of a system between reality and virtuality has some advantage. Certain well known aspects of a system can be represented in a formal way by algorithms in the computer, others to be investigated in more detail are represented in reality, but coupled to a dynamic surrounding. This allows completely new forms of easy experimental work and learning. Here Hyper Bonds comes into play.

In order to provide arbitrary boundary conditions, we must have a mechanism to switch between a source of effort and a sink of effort, and to generate and sense phenomena. Figures 2 and 3 show the cutting boundaries between reality and virtuality and its realization with a special sensor/actuator coupling.

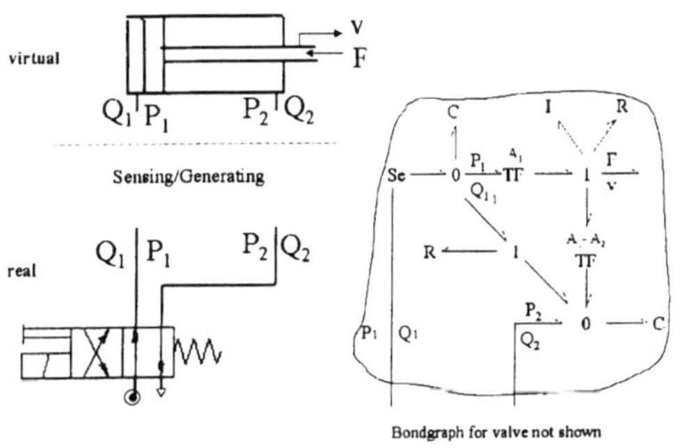

Fig. 2. Cutting of the system of Figure 1 by a sensing/generating mechanism (hyper-bond)

Hyper Bonds are of course also possible for pairs like voltage V and current i, Temperature T and heatflow dQ/dt and mass M and velocity v. Figures 2 and 3 are only simplified examples for use of Hyper-Bonds. The hardware implementation of figure 3 is simple. For every supported physical phenomenon, there must be two sensors (pressure-meter and volume-flow meter for pneumatics) and a controllable source (air-flow and air-pressure). Analog sensor-signals are converted into digital values and then available for the software side of a hyper-bond. The opposite direction requires digital values from the software being converted into analog signals to drive a generating mechanism (force and speed for mechanics).

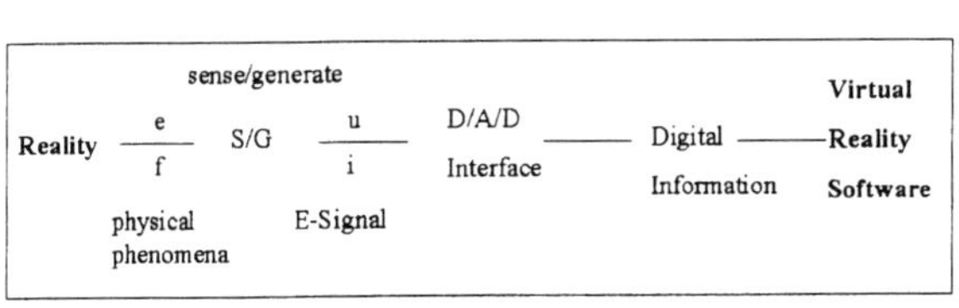

Fig. 3. Hyper-Bond structure

Figure 4 shows a simple connection of a real pneumatic valve, to a pressure source (not seen on the left side) and a hyper-bond below the screen. The valve is under real pressure. If pushed, the tube connecting the valve and the hyper-bond is under pressure and triggers an input signal to generate virtual pressure in the virtual tube-continuation. The simulated cylinder drives out. If the pushbutton is released, the valve opens a pressure release, the real pressure drops and the virtual pressure driven by the back-spring of the cylinder causes an outflow of air through the hyper-bond. This demonstrates the bi-directional character of the hyper-bond.

Fig. 4. Pneumatic equipment: single acting cylinder and valve split into reality and virtuality.

Another simple application can be imagined, having two real double acting cylinders crosswise connected via hyper-bonds through a virtual model. Two persons could then play a pushing-game, figure 5. Or in a more practical application, a concrete remote object can be moved with force-feedback and any computer generated support coming from a virtual model, replacing the one-to-one connection between the two hyper-bonds in figure 5.

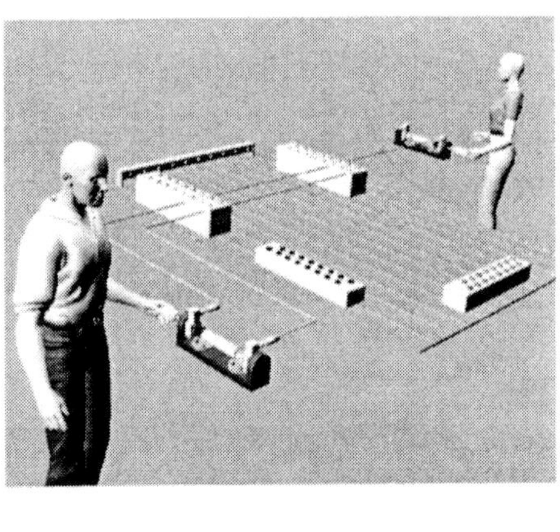

Fig. 5: Two real persons pushing each other by two real cylinders connected via hyper-bonds and virtuality

Hyper-bond is a mechanism based on the translation between physical effort/flow phenomena and digital information like any other analog/digital and digital/analog conversion, however it aims at a unified application oriented solution connecting the physical and its virtual representation and continuation. One of its main feature is, that the modeler does not have to be aware about the direction of energy-flow, hyper-bonds are bi-directional and adapt itself to the environment conditions. Figures 2 and 3 are highly simplified presentations. For a deeper understanding of hybrid bond graphs and how to handle discontinuities, boundary cuts and transfer between power flow, signal flow and logic switches see Mostermann (1997).

3. LEARNING ENVIRONMENTS

The European project "Distributed Real and Virtual Learning Environment for Mechatronics and Tele-service" (DERIVE) used the concept of hyper bonds to describe electro-pneumatics (Bruns et al, 2002) in a unified and didactically expandable way as well as to have a link to several powerful simulation tools supporting bond graph modeling.
DERIVE provides a new learning environment that supports schools for technicians to deliver courses in mechatronics. The support for the learning process will be reflected in a graduation from local real to local virtual to remote virtual to remote real, taking the student from basic knowledge to the full implementation in industry.
The tele-cooperation functionality in the learning environment will allow enterprises to use the training facilities in order to update the knowledge of their employees. With new equipment being more complex and requiring more complex maintenance, the training requirements for workforce and engineers increases. The new environment will allow groups of employees at remote locations to take part at the same training using the same equipment (either simulated or real). This staff will be able to work in a collaborative way to solve problems and explore learning situations. This new kind of interaction will allow the systematic support of skilled workers and engineers. Also, the learning environment is an appropriate tool to realize project orientation in technical training, providing a platform for self-managed and collaborative learning.

A typical learning session could be the following:

Sequence of two Pneumatic Cylinders

Modelling Task:
Build a circuit for a welding operation.
First activate a clamping cylinder with the push of a left button, after this first cylinder has reached its end-position (clamps a work-piece), a second cylinder, the welder is automatically activated. On pressing a right button, the process is terminated and the clamping and welding cylinders move in, releasing work-piece and welding process.
Solve the task by using virtual and real parts as preferred. Criticise the problem definition and suggest improvements.
You can use in virtuality:
Double Acting Cylinder With Magnetic Proximity Switch
Single Acting Cylinders
3/2-way valves with push button, normally closed
3/2-way solenoid valve, normally closed
5/2-way solenoid valve
Sources of pressure and voltage
Hyper-Bond connectors between virtual and real
Tubes and wires
You can use in reality, observable by video camera:
Hyper-Bond connectors between real and virtual
Single Acting Cylinders
3/2-way solenoid valve, normally closed
Tubes and wires

Session:
Virtual solution with double acting cylinder1 upper half, a 5/2 way solenoid valve, a single acting cylinder2 (left lower side) and a dummy cylinder3 (right lower side)

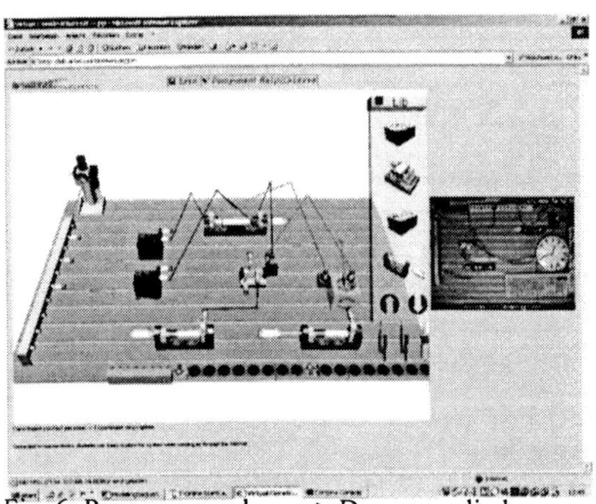

Fig. 6: Pressure supply on -> Dummy cylinder moves out

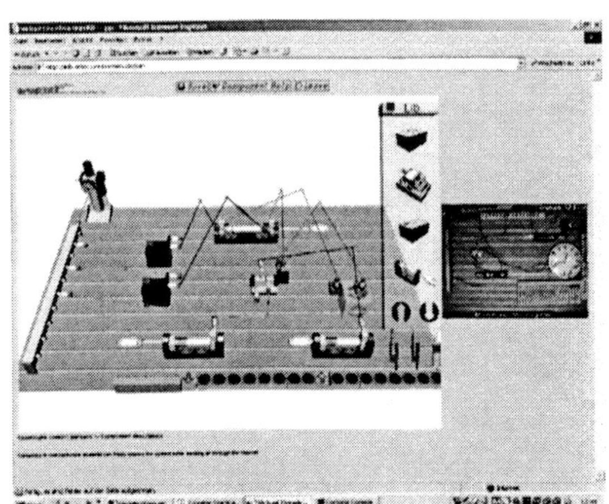

Fig. 7: Virtual circuit in action

In figure 7, if left button is pushed resulting in 1.cylinder out, 2. cylinder out, 3. cylinder in if right button is pushed resulting in 1.cylinder in, 2. cylinder in, 3. cylinder out.
Now export virtual parts into reality!
Connect virtual pressure valve via hyper-bond (first right pneumatic-connector) with a real cylinder seen on video-image upper right, figure 8. Press virtual push-button and see single acting cylinder moving out as long as button is pushed. On release of button, the air flows back from reality into virtuality and cylinder moves in !!! Recognize, that this is a different behavior than the one in virtuality. Why? In figure 9 most virtual parts are exported. Pushing virtual button, highlighting active connectors in virtuality and moving real cylinders

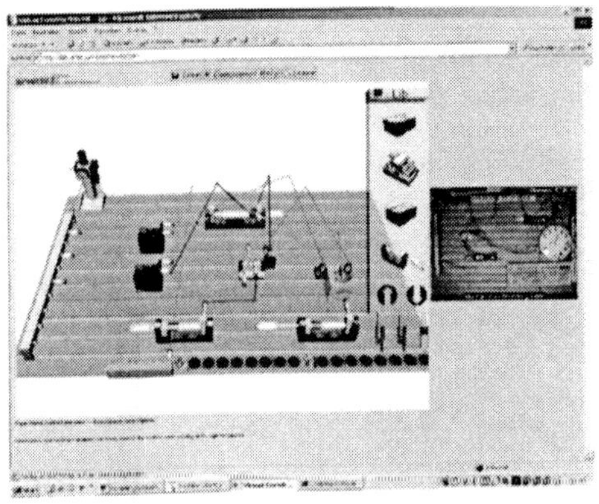

Fig. 8: Exporting pressure

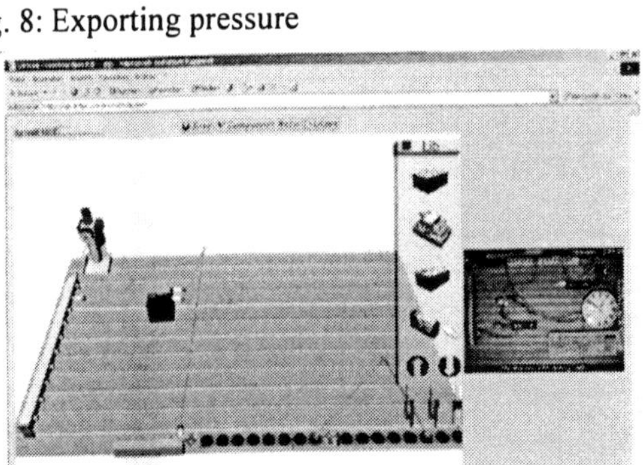

Fig. 9: Most parts exported

4. ADVANTAGES OF MIXED REALITY FOR LEARNING

The project developed a mixed reality human computer interface (Figure 10). To develop adequate technological and pedagogical concepts for e-learning in future technical training, users needs had to be analyzed in depth. The requirements of different user groups (students, teachers, employees) have to be described and consolidated. Acting in an environment, where real world objects and IT-technologies are applied simultaneously, requires new concepts of supporting cooperating local and distributed learning groups. The scientific challenge is to handle physical as well as virtual presence and

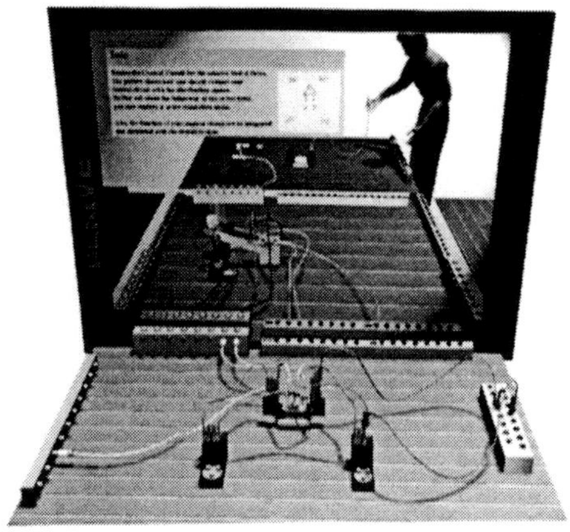

Fig. 10. Real and virtual environment for training in electro-pneumatics to control a Robot-Arm

awareness without confusing side effects for the users. With Hyper-Bonds, a learning environment got realized where on-site and remote components merge into a cooperative learning process. The environment system allows to work together with complex real and virtual systems, consisting of parts which may be distributed remotely. The learning environment includes a supportive web-database with multimedia learning sequences providing theoretical background information, exercises and help to handle training tasks. The learning environment smoothly integrates equipment and supports hardware-in-the-loop functionality, allowing to build up real systems as subsystems of complex virtual systems (Figure 10).
The environment has been evaluated in 3 European vocational schools and colleges and at one industrial site. The results, based on a prototype which was not yet very stable, was already very encouraging for further offering a broad learning-landscape to navigate between concrete and symbolic representations. In a comparative evaluation of

- traditional teaching with blackboard and teacher learner interaction
- support by simulation only (FluidSim-software of FESTO AG, Germany)
- support by real complex system only (Modular Production System/MPS of FESTO AG, Germany)
- support by real and virtual media (DERIVE)

in 20 hours-courses of mechatronics, Grund and Grote (2004) found some hints that students/ apprentices being trained with DERIVE showed better performances in symbol based fault finding.
The following main tendencies are considered important:

- thinking in abstract categories of structure-behavior-function searching for alternative concrete instantiations (e.g. various structures for

one function, or various functions of one structure)
- thinking in categories of information-control-work process and their realization searching for unified system dynamics views (e.g. Petri-Nets and Bond Graphs) for analogous physical phenomena (electricity, pneumatics, force-momentum mechanics ...)
- judging about the adequateness or failure of simulation models versus real systems

Enterprises recognize that employee expertise is a vital and dynamic living treasure. The demand for employee expertise is meaningless unless an organization (enterprise) can develop it in ways that respond to its business needs. Many enterprises rely on off-the-job training (formal learning) without considering its suitability for the learning tasks at hand. On-the-job training (informal learning) has a substantial advantage: it is more close to the problems to be solved and it can be organized in a cooperative way crossing the border between different professions that are involved in a project to fulfill an order of a customer. But on-the-job training is often unplanned and therefore mostly ineffective. With a structure and using well prepared training material usable at the workplace and facilitating advises from outside for cooperative learning makes on-the-job training a powerful tool (Jacobs and Jones, 1995).
Collecting case studies of solved problems in manufacturing and service could help to solve new tasks. If they are released and are available in electronic format, preferably via internet, it is a powerful instrument for learning at workplaces. Via the internet it is also possible to use training material and remote labs offered by training institutions.

5. CONCLUSIONS

A new interface concept, allowing a flexible and user-friendly modeling in a blended real and virtual environment has been introduced. This concept, suitable for an integration into the theory of bond-graphs, allows some learning and working modalities important for training and systems analysis: the stepwise abstraction and concretization of parts of a complex system still within the context of the whole. This feature supports various individual learning styles, systems diagnosis and repair strategies. Further developments and evaluations are undertaken in an EU-project Lab@Future where this interface technology is applied in concepts of future remote laboratories.

REFERENCES

Bruns, F. W. (1999): Complex Construction Kits for Engineering Workspaces. In: Norbert A. Streitz et al (1999): Cooperative Buildings. Lecture Notes in Computer Science 1670. Springer, Berlin, pp 55-68

Bruns, F.W. (2001). *Hyperbonds - Enabling Mixed Reality*. artec paper 82, Bremen ftp://artec-nt.artec.uni-bremen.de /pub/Field1/Publications/ artec-01-Bruns-hyperbonds.pdf

Bruns, F.W. et al (2002). *DERIVE*, Final Report. artec-paper 102, Bremen ftp://artec-nt.artec.uni-bremen.de /pub/Field1/Publications/ artec-02-Bruns-Final_Report_Derive.pdf

Cooper, M., Donnelly, A., Ferreira, J. M. (2002). Remote controlled experiments for teaching over the Internet: A comparison of approaches developed in the PEARL project. In: Proc. Of the 19th ann. Conf. Of the Austral. Soc. Of Computers in Learning in Tertiary Education, Auckland, New Zealand

Ferreira, J. M., Alves, G. Costa, R., Hine, N. (2002). Cooperative learning in a web-connected workbench. In: Lecture Notes on Computer Science 2438, "Groupware: Design, Implementation and Use", Springer, N.Y

Goldberg, K., Chen, B. (2001). Collaborative Control of Robot Motion: Robustness to Error. IEEE/RSJ Intern. Conf. On Robots and Automation

Grund, S., Grote, G. (2004): Specific Learning Effects of a Mixed Reality System in Automation Technology Training. In print

Jacobs, R.L. and M.J. Jones (1995). *Structured On-the-Job Training*. Berrett-Koehler Publisher, San Francisco.

Karnopp, D. C., Margolis, D. L., Rosenberg, R. C.(1990). *System Dynamics – A unified Approach*. John Wiley, New York.

Mueller, D., Feirreira, M. M. (2003): MARVEL: A mixed-reality learning environment for vocational training in mechatronics. In: Proc. Of the Technology Enhanced Learning International Conference, Milan, Italy

Mostermann, P. J. (1997). *Hybrid dynamic systems: A hybrid Bond Graph Modeling Paradigm and its Application in Diagnosis*. Dissertation, Vanderbilt University, Nashiville, Tennessee

Paynter, H. M. (1961). *Analysis and Design of Engineering Systems*. MIT Press, Cambridge, MA.

PEARL (2003). In: http://iet.open.ac.uk/pearl/

Song, D., Goldberg, K. (2003). ShareCam Part I: Interface, System Architecture, and Implementation of a Collaboratively Controlled Robotic Webcam. IEEE/RSJ Intern. Conf. On Robots and Systems

Teleactor (2003). http:///www.teleactor.net/

Lab@Future (2004). http://lab.artec.uni-bremen.de

Copyright © IFAC Information Control Problems in
Manufacturing, Salvador, Brazil, 2004

ELSEVIER
IFAC
PUBLICATIONS
www.elsevier.com/locate/ifac

LEARNING BY TROUBLESHOOTING – A SUITABLE DIDACTICAL SCENARIO FOR ONLINE EXPERIMENTS?

Boris Gauss[*], Andreas Klein[+], Leon Urbas[*], Günter Wozny[+]

[*]*Technische Universität Berlin, Centre of Human-Machine-Systems*
[+] *Technische Universität Berlin, Department of Process Dynamics and Operation*

Web-based technologies enable the conduction of remote experiments with real technical plants or laboratories for engineering education. In this paper, the online control system of a Methanol-water distillation column is presented, which has recently been developed at the Department of Process Dynamics and Operation of the Technische Universität Berlin. A didactical scenario with a troubleshooting task has been created to be realised with the online column. In a formative evaluation study, the scenario has been accepted by the students and has shown a slightly positive effect on learning outcome. The potential of the "learning by troubleshooting" scenario is discussed. *Copyright © 2004 IFAC*

Keywords: Learning Systems, Evaluation, Knowledge Acquisition, On-line control, Laboratory education, Real-time systems.

1. INTRODUCTION

Currently, many institutions for engineering education are making great efforts to provide a web-based access to real experimental facilities to their students (Carnevali & Buttazzo, 2003; Erbe & Bruns, 2003; Zeilmann et al., 2003; Michaud et al., 2001). Such remote laboratories are expected to have a significant pedagogical benefit, especially compared to virtual laboratories, which are based on simulations (Alhalabi et al., 2000). In particular, remote labs are considered as advantageous for being closer to reality, providing real-time, noisy data and a higher degree of freedom for the learners. This is assumed to increase the involvement of the learners, enabling a "thrilling" learning experience and stimulating higher order thinking skills.

In theory, the argument of the advantages of remote laboratories is plausible. Nevertheless, systematic empirical evaluation studies on the actual effects of remote learning scenarios with real facilities on the learners can barely be found in the literature. It is clear that there will be no general pedagogical superiority (or inferiority) of remote labs compared to other learning technologies. Instead, different technologies or media will complement each other. The advantage of a certain technological solution will depend on various factors like the learning task, characteristics of the learners, and the didactical scenario.

This paper presents an attempt to clarify the potential of a specific learning scenario, comprising a troubleshooting task, which has been created to be conducted with a new online distillation column. The scenario is developed for learning chemical process knowledge, in particular understanding the relations between the most important parameters of a Methanol-water distillation process. In an exploratory study with process engineering students, the didactical scenario is systematically evaluated.

The paper is organised as follows: In section 2, the technology and the user interface of the online distillation column are described. Section 3 illustrates the didactical scenario with the troubleshooting task. Section 4 is dedicated to the evaluation study. Section 5 provides the conclusions.

2. THE ONLINE DISTILLATION COLUMN

A laboratory-scaled distillation column has been set up at the Department of Process Dynamics and Operation of the Technische Universität Berlin. This glass column with 30 mm diameter, 20 trays and a height of 3.5 m is fully automated and controlled by the process control system (PCS) ABB *Freelance2000*. It has 21 sensors and 4 actuators. The feed temperature, the reboiler duty, the feed stream and the recycle stream can be controlled via the PCS and also via OPC over the Internet. The column is designed for the separation of a methanol-water-mixture. This is a standard learning example for process and plant technology.

For remote experimentation, the process can be controlled in real-time via internet using a Java applet. Real-time data trend diagrams are displayed for the observation of different sensors like temperature, flow, pressure and concentration. Additionally, a streaming web cam has been installed for the visualisation of the column.

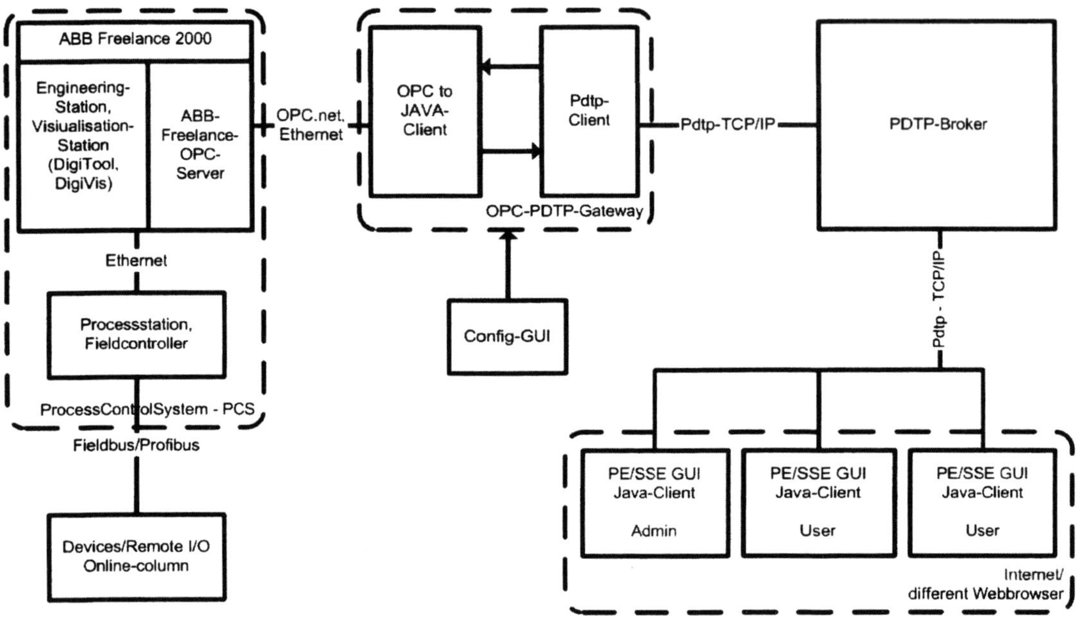

Fig. 1. System architecture

The user (client-side) is connected to the process control system (server-side) via Ethernet network like LAN or wireless-LAN or modem using Java and OPC Technology (fig. 1).

On the client-side, a web browser acts as a container for process visualisation and streaming video. The user works with a proprietary developed java based high fidelity emulation of ABB Freelance2000 PCS called PE/SSE (Proper Educt Small Systems Edition, Urbas, 1999) (fig. 2). The data exchange between the graphical user interface (PE/SSE) and the PCS core is based on an open internet protocol for process data transport (PDTP). The connection to ABB *Freelance2000* via bridge between server and client-side PCS uses modern Ole for Process Control (OPC) technology (Urbas, 1999, Klein et al., 2003).

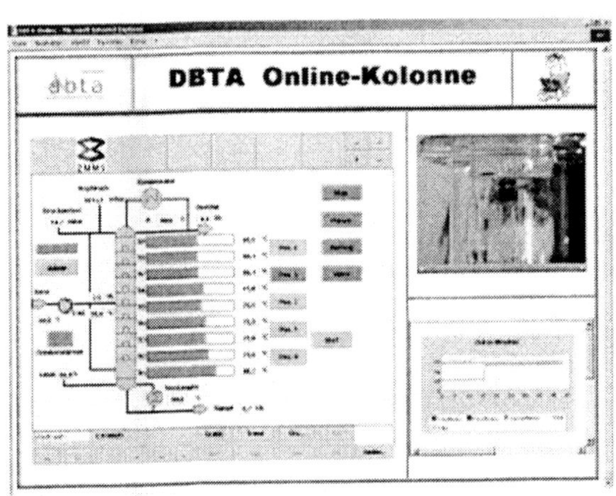

Fig. 2. Online graphical user interface (Java Applet)

3. THE TROUBLESHOOTING SCENARIO

In the troubleshooting scenario, the students play the role of operators who control the Methanol-water distillation plant. The task of the students consists in monitoring the trend diagrams of process parameters provided by the graphical user interface (GUI) of the process control system to detect irregularities in the trend diagrams. Once they have detected an irregularity the students have to identify the malfunction of the system that caused the symptoms visible in the trends. Finally, the students shall initiate the appropriate countermeasure and continue to monitor the process to see whether their countermeasure has been successful. To support the diagnosis task, a malfunctions-symptoms matrix (MSM) is presented to the students. The MSM contains a list of the possible malfunctions and the effects of the malfunctions on the process measurands displayed in the trend diagrams. In the present case, the MSM shows twelve different malfunctions and their effects on eight measurands in the three categories temperature, flow, and pressure. Each malfunction is characterised by a specific pattern of symptoms so that there is exactly one correct solution for each malfunction.

The learning goal of the troubleshooting scenario consists in a deeper understanding of the process, linking theoretical knowledge of thermodynamics, chemistry and physics to the engineering practice. In the role of operators, the students get a new perspective on the theoretical learning contents by a hands-on interaction with the process in a real plant. This is assumed to enhance motivation and understanding. The understanding of the relations between important process parameters provides a basis for the development of advanced modelling skills.

For the creation of "malfunctions", a training console to manipulate process parameters has been developed. The training console is integrated between the OPC server and the process control system, and is activated by a trainer. Therefore, the students can not immediately recognise which process values are changed by the trainer and at what time the changes occur.

4. EVALUATION

4.1. Objectives

The evaluation study presented in the following was a first systematic test of the troubleshooting learning scenario. The main objective of this formative evaluation consisted in gathering information for the optimisation of the learning scenario. In particular, the following questions were addressed:

- Do the students accept the learning by troubleshooting scenario?
- Does troubleshooting enhance the students' knowledge of the process?
- How good do the students perform the troubleshooting task?
- Do learner characteristics affect troubleshooting performance and/or learning outcome?
- Is there a relation between troubleshooting performance and learning outcome?

4.2. Method

The character of the evaluation study was explorative. In view of the fact that the connection between the online GUI and the training console would have required additional efforts with uncertain benefits, the evaluation was carried out using the standard GUI of the PCS ABB *Freelance2000*. The online GUI and the ABB *Freelance2000* GUI provide identical information with very similar navigation, and the only significant difference between both interfaces is the online video stream. While the live video of the process is generally assumed to enhance students' motivation, it provides no information which is necessary to complete the troubleshooting task and could therefore be neglected in the present study.

Framework. Fig. 3 shows the set of variables considered in the evaluation study on the levels of input, processes, and outcomes of the learning process, referring to the heuristic framework for research on e-learning provided by Friedrich et al. (2001).

As input variables, learner characteristics were considered, which are assumed to affect diagnosis as well as learning performance. Prior topic knowledge and motivation generally play a crucial role for learning processes (Unz, 2000). General technical understanding has shown to be a strong predictor for learning outcome in the evaluation of a virtual learning environment for Mechatronics (Grund & Grote, 2002). Velocity of information processing is supposed to be linked to the rapid detection of malfunctions in the trend diagrams.

Materials. To assess topic knowledge, a blank version with four malfunctions of the malfunctions-symptoms-matrix (MSM) was administered. The students had to fill out the MSM with the corresponding symptoms to a given malfunction. For each parameter, the students could choose between four answers: "increasing", "decreasing", "consistent", "do not know". The knowledge test was administered two times, before and after the troubleshooting task, to assess learning performance. Velocity of information processing was measured with the ZVT (number connection test, Oswald & Roth, 1987). General technical understanding was measured with eighteen items of the MTP (Mannheimer Test for physical – technical problem solving, Conrad et al., 1980). The screen of the PCS was recorded during the troubleshooting task to control navigation behaviour. With a video camera, the face (to control line of sight) and verbal comments of the subjects were recorded. Subjects were told to think aloud while completing the troubleshooting task. Afterwards, screen, video and audio recordings were synchronised and transcribed for further analysis of troubleshooting behaviour and performance. An acceptance questionnaire was developed with eleven items and a five point rating scale.

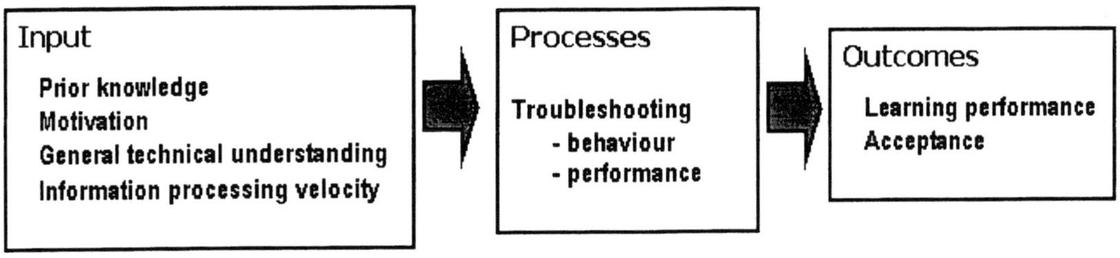

Fig. 3. Methodological framework of the evaluation

Procedure. The evaluation study was carried out in an individual learning scenario with the PCS running on a personal computer, accompanied by two experimenters, a process engineer and a psychologist. After a biographical questionnaire and a short introduction into the PCS by the experimenters, subjects completed the prior MSM knowledge test. In the following, the students attended a 20 minutes trial of the troubleshooting task (as described above in 3). About two minutes after the start, a malfunction was initiated by the experimenters via the training console. As soon as they identified the malfunction, the students had to give a command how to "repair" the malfunction, which was executed by the experimenters via the training console. Subsequently, subjects were told to continue monitoring the process to check whether their command had been the right one. After the first troubleshooting trial, acceptance, general technical understanding and information processing velocity were assessed, followed by a second 20 minutes troubleshooting scenario. Finally, the post MSM knowledge test was administered. Students were paid ten Euros for their participation. Altogether, a session lasted about two hours.

4.3. Results and interpretation

Learner characteristics. Participants were seven undergraduate process engineering students (age: *M*=27.7) in their advanced study period (semester: *M*=11.3) at the TU Berlin. All participants had theoretical background knowledge about distillation columns from the lecture "Process and plant dynamics" (PPD) by Prof. G. Wozny, to which they were attending at that time, but had no practical experience with process control. Their extrinsic motivation for attending the PPD lecture was very high ("I'm attending to the lecture because I have to make the exam": *M*=4.0[1]), intrinsic motivation was moderate ("...because the topic is interesting": *M*=2.7). In the pre MSM knowledge test, students achieved a score of *M*=20.4 (or 63.8% of the maximum 32 points).

Troubleshooting performance. In the first malfunction scenario ("sudden increase of feed stream"), six of the seven participants made the correct diagnosis within the 20 minutes of the trial. In the second scenario ("breakdown of reboiler"), five students were successful. The students correctly identified the first malfunction after *M*=9 min. (SD=5), the second malfunction after *M*=10 min. (SD=4). Only few wrong diagnoses were made (in both scenarios, there were two – different – students who made a total of three wrong diagnoses per scenario[2]). In accordance with the increase in missed diagnoses and time, students also subjectively rated the second malfunction as more difficult ("The troubleshooting task was difficult": *M*=2.6, SD=1.6) than the first malfunction (*M*=1.6, SD=1.4). This difference is statistically significant (Wilcoxon test, Z=-1.8, p<.05). For further analysis, an ordinale "troubleshooting performance" variable was calculated, merging the performance on both malfunctions (see table 1).

Table 1: Troubleshooting performance

	Troubleshooting performance		
	high	medium	low
Both scenarios solved	+	+	-
Only correct diagnoses[2]	+	-	-
n of subjects	3	2	2

Outcomes. Students achieved a post MSM score of *M*=22.6 (70.6% of maximum score). This slight increase in process knowledge of 2.2 points in average was statistically not significant (Wilcoxon test, Z=-1.2, p=.11). Acceptance of the troubleshooting task was high ("Troubleshooting tasks should be part of the study course": *M*=3.6; "Troubleshooting was interesting": *M*=3.1), even though the students considered their learning outcome only as moderate after having completed both scenarios ("Now I can identify malfunctions better than before": *M*=2.0; "Now I know more about the process than before": *M*=1.9).

Relations between variables. There was a significant positive correlation[3] between intrinsic motivation to participate at the evaluation study ("I take part because I want to learn more about troubleshooting") and troubleshooting performance (r=.75, p<.05). The correlation between information processing time (ZVT) and troubleshooting performance was negative (r=-.61, p<.10), indicating that "fast" students performed better. There were no clear relations between learning outcome (difference between post and pre MSM scores) and learner characteristics or troubleshooting performance.

4.4. Discussion

Learning by troubleshooting with the online column found high acceptance among the participants of the evaluation study. Nevertheless, the effect of the troubleshooting scenario on learning outcome remains less clear. On the one hand, this could be attributed to the malfunction-symptoms-matrix (MSM) testing method. Since the behaviour of parameters in complex chemical processes is ambiguous to a certain degree, and the MSM had been administered for the first time in the present study, the validity of the MSM to measure knowledge about the process is still to be verified. However, the slight increase in the MSM score

[1] Unless otherwise noted, values refer to a 5 point rating scale from 0 (="disagree completely") to 4 („agree completely").

[2] Students could correct their diagnosis within the 20 min. if they recognised that it was wrong.

[3] All reported correlations are Spearman-Rho.

corresponds with the students' moderate subjective ratings of their own learning outcome. This supports – at least, roughly – the assumption that the MSM is a valid measure. On the other hand, it is generally often difficult to prove the benefit of experimental learning by "hard facts", since the learning goals can not easily be defined in observable behaviour. Even though the high subjective acceptance of the troubleshooting scenario confirms the argument that learning with experiments on real plants (online as well as "on-site") is a valuable experience for the students (see Introduction), it remains a challenge to define valid, operationalisable criteria for the evaluation of experimental learning scenarios in general and experimental online scenarios in particular.

Taking into account that the students did not have any practical experience with process control, the overall troubleshooting performance was quite satisfactory. Nevertheless, some few participants had problems with the task, and probably would have been frustrated if the troubleshooting had been conducted in a stand-alone online version without direct contact to a tutor. In the evaluation study, the repeated requests to think aloud created a dialogue between student and experimenter. Although the experimenter's questions were open (e.g., "what are you doing now?", "what are you thinking?", "what are you looking at?") and not instructive, the think aloud method created a social situation unlike an individual remote learning scenario. While the experimenters did not answer any questions during the trials, there often was a lively discussion between student and experimenter after the evaluation session had been finished. The students obviously had a need to discuss their experiences with the troubleshooting task. Therefore, an online version of the troubleshooting scenario should provide adequate communication facilities to a tutor and fellow students.

The present results also suggest to look carefully at individual differences of the students, like information processing velocity, which can affect their ability to cope with certain learning scenarios. Especially for individual remote scenarios, the crucial role of intrinsic motivation for the successful completion of learning tasks has to be regarded. The importance of intrinsic motivation in the context of computer-based learning had already been demonstrated in the evaluation of a web-based learning module for steady-state modelling, which had been carried out by some of the authors (Gauss & Urbas, 2003, Gauss et al., 2003).

Besides the results presented in this paper, troubleshooting behaviour will be further studied by a qualitative analysis of the data of the think aloud protocols and the screen recordings gathered in the presented evaluation study. Goal of this analysis is a better understanding of the students' mental models of the process and the identification of typical misconceptions.

In the light of the present findings, it seems necessary to compare the learning by troubleshooting with other didactical scenarios. In particular, it should be examined if troubleshooting, where the students react to a change in the process determined by the system (respectively, the trainer), is different from "classical" experimenting, where the students actively change parameters and study the effects. Such a comparative evaluation will enable a more adequate judgment about the didactical value of learning by troubleshooting. Also, a comparison of "dynamic" troubleshooting at the online column versus "static" troubleshooting with offline material (i.e., screenshots of the process control system) would be very interesting, to investigate whether the higher costs of the online scenario (mainly caused by regular maintenance and support) are justified. In the future, expensive online experiments with real plants will only be preferred to less costly solutions with static material or simulations if their pedagogical benefit is obvious.

5. CONCLUSION

In this paper, a learning by troubleshooting scenario for an online distillation column was presented and evaluated. The results of the evaluation study indicate that this learning scenario was highly accepted by the students. Troubleshooting performance was influenced by intrinsic motivation and information processing velocity of the learners. The effect of troubleshooting on learning outcome was slightly positive. In the future, the didactical potential of the troubleshooting task at the online distillation column should be systematically compared with other learning scenarios.

ACKNOWLEDGEMENT

The work presented in this chapter was funded by the German Federal Ministry of Education and Research within the projects [my:PAT.org] (New Media in Education framework) and MOSES (Notebook University framework), and by the VolkswagenStiftung within the research group MoDyS (Methods of User Modelling in Dynamic Human-Machine-Systems).

REFERENCES

Carnevali, G. & Buttazzo, G. (2003). A virtual laboratory environment for real-time experiments. In *Proceedings of the 5th IFAC International Symposium on Intelligent Components and Instruments for Control Applications (SICICA 2003)*, Aveiro, Portugal, July 9-11, 39-44.

Conrad, W., Baumann, E. & Mohr, V. (1980). *MTP: Mannheimer Test zur Erfassung physikalisch-technischen Problemlösens*. Göttingen: Hogrefe.

Erbe, H.-H. & Bruns, F. W. (2003). Didactical aspects of mechatronics education. In

Proceedings of the 5th IFAC International Symposium on Intelligent Components and Instruments for Control Applications (SICICA 2003), Aveiro, Portugal, July 9-11, 45-50.

Friedrich, H. F., Hron, A. & Hesse, F. W. (2001). A framework for designing and evaluating virtual seminars. *European Journal of Education,* **36 (2)**, 157-174.

Gauss, B. & Urbas, L. (2003) Individual differences in navigation between sharable content objects – an evaluation study of a learning module prototype. *British Journal of Education Technology*, **34 (4)**, 499-509.

Gauss, B., Urbas, L., Hausmanns, C., Zerry, R. & Wozny, G. (2003). Experimental Evaluation of Story-Based Content Structure and Navigation for a Learning Module in SCORM. In: J. Jacko & C. Stephanidis (Eds.), *Human-Computer Interaction. Theory and Practice. Proceedings of the 10th International Conference on Human-Computer Interaction in Crete, Greece, 22-27 June 2003, Vol.1*, 636-640. Mahwah: Erlbaum.

Grund, S. & Grote, G. (2002). *DERIVE final evaluation results*. Retrieved Dec. 9, 2003, from http://www.derive.uni-bremen.de/pdf/D55.pdf

Klein, A.; Zerry, R.; Hausmanns, C.; Urbas, L.; Wozny, G.: *Java based Process control of a distillation column via internet using OPC*; ECCE-4, Granada, Spain, September 21-25, 2003.

Michaud, F., Gentil, S. & Barrault, M. (2001). Expected benefits of web-based learning for engineering education: examples in control engineering. *European Journal of Engineering Education*, **26 (2)**, 151-168.

Oswald, W. D. & Roth, E. (1987). *Der Zahlen-Verbindungs-Test (ZVT)*. Göttingen: Hogrefe.

Unz, D. (2000). Lernen mit Hypertext: Informationssuche und Navigation. Münster: Waxmann.

Urbas, L., *Entwicklung und Realisierung einer Trainings- und Ausbildungsumgebung zur Schulung der Prozessdynamik und des Anlagenbetriebs im Internet, VDI-Fortschrittsberichte Reihe 10 Nr. 614.*

Zeilmann, R. P., Gomes da Silva Jr., J. M., Bazanella, A. S. & Pereira, C. E. (2003). A web-based remote laboratory for control education. In *Proceedings of the 5th IFAC International Symposium on Intelligent Components and Instruments for Control Applications (SICICA 2003)*, Aveiro, Portugal, July 9-11, 51-56.

Copyright © IFAC Information Control Problems in
Manufacturing, Salvador, Brazil, 2004

ELSEVIER
IFAC
PUBLICATIONS
www.elsevier.com/locate/ifac

PROBLEMS, SOLUTIONS AND OPPORTUNITIES IN GLOBAL ENTERPRISES

Carlos Acosta[1], Jorge León[2] and Rene Villalobos[3]

(1) Universidad de las Américas, Puebla, México
(2) Texas A&M University, USA
(3) Arizona State University, USA

Abstract: This study describes important dimensions manufacturing enterprises must account for in working in global markets. Problems encountered in the technical arena as well as in the socio-cultural area are presented. The methodology used was mainly the case-study formulation complemented with strong theory base. The uncovered problems are illustrated through the case-studies in diverse manufacturing companies in North America. Finally some recommendations and solutions are derived. *Copyright © 2004 IFAC.*

Keywords: Manufacturing systems, management systems, plastics industry.

1. INTRODUCTION

Globalization refers to growing economic interdependences among countries, as reflected in increasing cross-border flows of goods and services, capital, and know-how (Govindarajan and Gupta, 2001). Although globalization is not new, the number of companies embarking in global operations, and the economic interchange among countries, continues to have a rapid growth – the World Investment Report (UNCTAD, 2001) reports that, in 1995 there were 40,000 companies with 250,000 affiliates, while in 2000 this increased to more than 60,000 companies having over 800,000 foreign affiliates. Furthermore, the foreign direct investment inflow of developed countries grew about 5 times from $203 billions in 1995 to $1,005 billion in 2000, and that of developing countries doubled within 1995-2000, from $113.3 billion to $240.2 billion, respectively. On the average, the ratio of the number of subsidiaries to a parent company doubles within five years. The increasing trend for globalization is a competitiveness response driven by the forces that are internal and external to the company. Internal forces include the need to reduce costs, increase quality, and rapid need for technology and other resources. External forces include large emerging markets, and global competition.

The opportunity to achieve competitiveness through globalization is augmented today by socio-economical and technological realities. The socio-economical state of the world is more akin to globalization than ever before - more governments are adopting a free-market or open economy that welcome boundary-less commerce through lower tariffs and trade agreements. Technologically the world has become virtually interconnected with exponentially decreasing communications, transportation, and computer costs.

The consideration of global issues in engineering endeavors can result in significant savings and competitiveness for a company. For instance, engineering project implementations can be much faster if the technical team has a good understanding of the local culture, or a product can gain more market share if the product design is tailored according to the country where it will be sold.

In Section 2, a Global Enterprise Model (GEM) is proposed as a foundation for the consideration of globalization factors in technological endeavors. Its use will help engineers to find new opportunities for efficiency, and avoid roadblocks that result from the global nature of a project. In Section 3, real industrial situations are analyzed and presented as casc studies to illustrate the salient dimensions of the model. Finally, in Section 4 concluding remarks are presented.

2. GLOBAL ENTERPRISE MODEL

GEM has 3 interdependent layers: (1) Global Mindset layer, (2) Operational layer, and (3) Societal layer.

2.1. The Global Mindset Layer

Global mindset refers to the knowledge and skills concerning globalization possessed by the people at each level in a company's organization. The knowledge about globalization can range from awareness to detail knowledge about the different cultures, international legal systems, and business environment. As suggested by Govindarajan and Gupta (2001), knowledge and awareness alone are not sufficient to have an effective global mindset, the skills to integrate and exploit this knowledge for competitive advantage is a necessary component.

The global mindset will vary from person to person, and is the challenge of a company to implant the appropriate globalization mentality at different levels and groups in the organization. For instance, when comparing higher management with engineering, it may be more for higher management to know about country laws, politics and international treaties; while it may be more important for engineers to know about the population's skill levels and regional culture.

The global mindset also varies between functions in a company. For instance, a global design group will be aware of the importance of 'localization'. Localization is the process of adapting a product to meet the language, culture and other requirements of a specific target environment or market. Today global enterprises have established large departments dealing with this problem. Siemens, for example, runs "User Interface Design Centers" in Munich, Beijing and Princeton with more than 30 employees. The crucial question is if it is possible to design and produce software as a cultural neutral product and to transfer it into a culturally adapted one by means of localization. If that is possible a design philosophy and technology would be available which could be applied very efficiently. Cultural neutral products (CNP) are not ready for use goods, they are basic technologies or basic products which become usable through user application. CNP require a specification, which abstracts from all culturally shaped, user specific attributes. At the same time it has to guarantee the feasibility of user-related variations of product attributes during the application process (Rauner 2002). As a means to explain the concept of localized globalization, Martinez et al (1997) developed the concept of the "Asian region-oriented machine tool". On the basis of this Makino constructed a machining center (MC) to be available at Singapore and Hong Kong only providing that the operator is China-originated people. This product concept is of "customized-type" or "culture and mindset harmonized type".

2.2. The Operational Layer

The operational layer refers to the value-adding activities, necessary infrastructure, systems and environment associated with the different stages of the life-cycle of a product. GEM's operational layer considers the following dimensions: (1) Markets and competition, (2) Organization, (3) Resources, and (4) Operations.

Global Markets and Competition refers to how global are a company's markets and competition. Why would a company need to globalize if all the current and future markets for its products/services are local? Also, can a company compete efficiently in foreign markets without being present in those markets? If foreign markets are attractive for a company, they are likely to be attractive to other companies as well. Arguably, companies present in a foreign market will have an advantage to those companies that sell remotely in the same market.

Global Organization refers to how a company's organization is global with respect to control and function. In a global organization the control and responsibilities in the organization hierarchy are determined based on efficiency and performance, rather than nationality; e.g., in a vertically global organization, it would be common to find a foreigner in higher management positions. In a global organization, there is significant interaction and collaboration among similar functions and tiers of the organization hierarchy. For instance, R&D teams in different countries will actively exchange recent technological advances, or production facilities in different countries will actively exchange best-practices.

Global Resources refer to capital, workforce, and technology. A company can have all of its sales and organization in one country, and yet be using foreign capitals; e.g. a Korean company, trading in the New York Stock market. The workforce globalization has been promoted by the cost of labor, and skills. A recent example of workforce globalization is the export of engineering functions from developed countries to developing countries with high levels of education but where the salaries are significantly lower. The national origin of the technology can be of importance. For instance, technologies that have military applications in one country may be prohibited from application if other countries. The race to get to market fast has forced companies to identify the needed technology and acquire it regardless of its nationality.

Global Operations refer to all the activities and related systems and infrastructure involved in the design, production and disposal of a good or service. The globalization of the design function can be measured by degree in which the company has design centers in different countries. Global production includes the supply chain, manufacturing, and distribution systems and infrastructure. Factors to consider here include the number of foreign suppliers, in what countries are the manufacturing facilities, warehouses and trans-shipment points located.

2.3. The Societal Layer

The societal layer considers the national characteristics of the countries involved. It includes three dimensions often disregarded by the engineering community; specifically, the nation's cultural, economic and political environments.

Global Culture. Culture is the totality of socially transmitted behavior patterns, beliefs, values, customs, traditions and heritage. Culture is learned, not inherited. Cultural differences among nations make a substantial difference in the way managers and workers behave in organizational settings (Steers et. al, 1992). Using statistical techniques Hofstede (2001) identified cultural dimensions that explain differences between the cultures of different countries; namely: power distance, individualism, masculinity, uncertainty avoidance, and long-term orientation.

Global Economics include the characteristics of each countries' economies, as well ad, special economic treaties between the countries. World trade barriers have fallen substantially in post-war years. Economic Integration has been one of the most significant trends since World War II although almost all integrations have originated regionally (Damanpour, 2000). These international agreements can both facilitate of present significant obstacles to make business in other countries. Examples are the European Union (EU), the North American Free Trade Agreement (NAFTA), Mercado Comun del Cono Sur (MERCOSUR), and the South East Asian Nations (ASEAN). Another important dimension of a national economy is its infrastructure, including the availability, reliability and cost of energy, transportation and communications.

Global Politics include the characteristics of a nation's government and legal system. Important dimensions are: (1) regime type, (2) government volatility, (3) legislation efficiency, and (4) government transparency. A pro free-market government will be more receptive to foreign companies than a controlled market one (Hill, 1994). Political volatility augments the uncertainty associated with foreign investments; for example, both Brazil and Japan are large markets that may be highly attractive in particular industries but the level of political volatility and other country risk has been much higher in Brazil than in Japan (Yip, 1992). The efficiency of the legal system can impact the speed at which operations and transactions are completed. For instance, how fast and effectively can intellectual property be protected will directly affect the establishment of R&D operations in a country. How easy it is to import and export will directly affect materials management decisions. Finally, government ethics and enforcement of international laws must also be considered. For instance, a country with a government that is too lenient with respect to software piracy can discourage companies from participating in an otherwise attractive market.

3. CASE STUDIES

The two case studies in this section were selected to illustrate how some of the components of the GEM model affect the effectiveness of technological projects. Both cases depict how the time to complete the corresponding project could have been significantly shortened if the people involved were better prepared in global issues. Other cases will be presented in the conference presentation but are omitted here due to space problems. The selected cases illustrate how non-technical dimensions can affect an engineering project.

3.1. Autonomous Maintenance and Culture.

A&B is an American company that manufactures family, personal and household products (the company name has being changed and the specific product unspecified for confidentiality reasons). This company has been in Mexico for over 40 years and markets more than 200 brands to nearly five billion consumers in over 140 countries. It has over 100,000 employees in more than 80 countries around the world. Last year profits exceeded 40 billion dollars. In order to further expand its activities A&B purchased a plant in Axcal, Mexico from a Mexican group called PAMEX. The plant has over 500 employees and has a production of a million cases of products per month. A&B also implemented several 'routine' changes in the administration and operation of the plant, some of which created significant opposition due to the cultural background of the Axcalans. More than a year of delays in the successful implementation of AM could have been avoided if A&B's management would have had a better understanding of differences between US and Mexican culture with respect to leadership, teamwork, work ethics, and work force educational level.

Initially A&B hired many workers and managers from other regions in the country creating a high level of resistance to change among the local workers. Americans tend to have a more horizontal relationship between leaders and subordinates; while Mexicans have a high level of subordination to those higher in the company's organizational hierarchy (Hofstede, 1997). This effect was augmented by the Axcalan regional culture that is peculiar even within neighboring regions in Mexico. Axcalans constitute a closed society, distrust foreigners, and are very traditionalists. This distinctiveness has it roots since the pre-colonial times as the people of Axcal were never subordinated by the Aztecs or Mexicans (Rendón, 1996). The initial resistance to implement AM was very high because of A&B's failure to recognize the importance to get a local manager coming from the old company buy-in of the TPM philosophy. The workers were culturally predisposed to follow this local leader. In the second attempt to implement AM, A&B eliminated this resistance by removing this manager from the leadership and improving the incentive program for the workers.

Another cultural issue disregarded by A&B in the initial implementation is the different predisposition that Americans and Mexicans have to work in self-managed teams. In the USA most people leave their parent's home early in their lives, either to go to college or once they get their first jobs after high school graduation. Very differently in Mexico, as in most Latin American cultures, most people live with their families until they get married; even if this is long after they could be financially self sufficient. So while Americans learn to work and trust other people early in their lives, Latin Americans are more used to the security of being surrounded and supported by family and tend to distrust strangers. As a result Americans will be more predisposed to rapidly be effective working in teams, while it may take longer for Mexicans to trust their new team mates. Hofstede's study (1997) also finds significant statistical difference between Americans and Mexicans in terms of how individualist each culture is; the American is much more individualist than the Mexican. According to Hofstede (2001), in the workplace an individualist society 'others' are seen as 'resources', and 'tasks' prevail over 'relationships' making teamwork a natural work method. Interestingly, the importance of relationships for Mexicans rooted in its family-oriented background makes them a very committed and efficient team member once the initial team formation phase is over. It is conjectured that this difference is more obvious in less educated layers of the society, thus the significant resistance of the workers to work in self managed teams in the A&B case. In the second attempt, A&B gave more power to the team leaders, and the reward system directly penalized the team members for not participating in the AM program.

Also American and Mexican cultures are very different with respect to how they perceive rules, procedures and standards. In the work place, Americans dislike rules and prefer less formalization and standardization when compared to Mexicans as suggested by the large difference in the 'uncertainty avoidance' dimension (Hofstede, 2001) between these two countries. In the A&B case, this difference was amplified by several events and management actions such as A&B's purchase of the Axcalan plant, bringing a large number of 'outsiders' into the workforce, and imposing a new work philosophy, created a high level of uncertainty for the Mexican workers. In short, A&B acted in ways which maximized uncertainty in a culture in which uncertainty avoidance is highly salient. As a result it was very difficult to convince the workers about the benefits of the AM program.

Finally, the A&B management failed to recognize the much lower educational level of the workers in PAMEX. Axcal's average education level is one of the lowest in Mexico raising the local's expectations on the company's responsibility to improve the local educational system. The people in the plant showed that there are very enthusiastic and eager to learn but lacked the resources to succeed. In the second attempt A&B opted to renew the workforce with new workers with higher educational level. Some specific modifications recommended to implement AM in Mexico (as compared to USA) include: (1) stronger team leader role and more specific rules, (2) more emphasis in team formation training needed, (3) more technical training needed, (4) monetary reward system should not be out of the question, (5) more training in continuous improvement philosophy, (6) stress timelines, and (7) expect resistance to documentation and report preparation.

3.2. Manufacturing Processes and Global Mindset

Mepo is the fictitious name of a French consortium that produces large automotive plastic parts. In Mexico they have 3 plants and the one related to this case-study is located in the state of Puebla.. Global companies are confronted with the need to know more about the countries in which they have operations. Consequently they need to know more about the culture of the French and the Mexican. Frenchs have a high degree of nationalism. Frenchs have a high degree of centralization, they are bureaucratic and there is a large power distance index between superiors and subordinates. France is high on individualism (Crozier, 1964) and its people have a logical and structured thinking but they are more attentive to the form and aesthetics than they are to the content. Frenchs are polite, friendly and direct in their interpersonal relations. On the other hand Mexicans are used to ask for favors, extra payment and so on. The management style is paternalistic. Mexicans use to say "ahorita," which literally means "right now" really means "I will do it at my personal convenience." Mexicans give a large place to family and friends, being a collectivist society. Mexicans have a centralized type of government, love foreigners and try to solve their problems right away with out any science applied to it.

Mepo has been contracted for the production of front fenders. The fender production uses exact copies of the molds which are currently used for manufacturing this product in France, Turkey, Spain and Brazil. When carrying out the injection tests in Puebla, the initial parts were defective for not meeting geometric specifications. A French technician was sent to the Puebla plant to solve this problem. The specialist arrived but did not take into consideration the Mexican personnel's opinion, rather he decided individually to increase the quantity and pressure of the injected material. This action not only did not correct the problem but also generated excessive burrs and misadjusted the mould. As a direct consequence of this act, production delays and additional expenses were occurred. These dimensional errors generated excessive looseness in assembly with the adjacent parts. A second French expert was called to solve the problem. He made an exhaustive analysis of the problem taking into account the opinion of the workers and Mexican technicians; i.e., that the problem was due to the fact that Puebla, differing from the other MEPO plants, is located 2100 m above sea level. As a result, process parameters were revised such as molding temperature, injection speeds at different percentages of injected plastic mass and retention time of the parts in the mold. To illustrate the importance of these changes, one needs to pay attention to the fact that part shrinkage is actually lower at high molding temperatures. High temperatures lower the viscosity of the polymer allowing more material to be packed into the mold. The effect of this is the same as using higher injection pressures. With changes in these parameters the parts could finally achieve the specified dimensions.

The delays in fixing the dimensional problems using the same molds as in other facilities illustrate the 'individualism' of the French technicians. In global manufacturing, being able to work in teams and listening to all the members of a team and co-workers is crucial to success in production. The logical and structured thinking of the French prevent them to accept opinions from Mexicans in order to solve technical problems. They are more attentive to form (French style) than they are to content, i.e. Mexicans use a very practical way to solve technical problems. The Mexican workers have a fatalistic vision of the world determined by the Catholic religion. They are used to facing problems at work and especially in new projects and they try to solve them with relative tranquility. French people love control. They are not used to coping with the kind of problems with the Mexican people, so they like to find immediate solutions in their own way (French style) as illustrated by the attitude of the first French technician. Also many technology transfer projects coming to Mexico neglect the geographical differences. This is an industrial cultural dimension that was proposed by Moritz (*2003*). In the case study, the barometric pressure, average temperature and moisture were some of the factors affecting the production of sound plastic parts.

CONCLUSIONS AND RECOMMENDATIONS

Successful engineering and technological projects must take into account the effects of globalization. This paper presents a global enterprise model that can be used as a framework for the analysis of an engineering situation from the global perspective. Two case studies are used to illustrate non-engineering aspects of the model.

In the A&B case the failure to consider a very special regional culture caused about a year delays in the implementation of autonomous maintenance, and significant management-worker frictions. Viewing the project using GEM could have helped management avoid or alleviate this situation, especially from the global mindset and cultural dimension perspectives. Also, if management had paid more attention to learn about the skill level of the specific region were the plant was located, instead of assuming that it was the same as in the rest of Mexico, then better hiring strategies could have been implemented and significant training resources could have been saved. The case of A&B was purposely chosen here because it clearly illustrates the effect of culture and global mindset in technical projects.

Localization is found to be a very important concept in the operation of global enterprises. In 2.1 an explanation of it is given regarding design processes, however more in deep detail is needed to take advantage of this concept into manufacturing processes. The importance of it can be seen clearly in the MEPO case-study with the wrong process setup. Here a mixture of technological as well as cultural factors led to severe quality defects. Going a step forward into the MEPO case-study analysis, it was found that, contrary to what a global enterprise needs in organization efficiency (see 2.2), the French management is operating based on nationality of the technician instead of performance. Some conversations held with MEPO supervisors discovered that the solutions to manufacturing problems implemented in the Puebla plant are not communicated to other consortium plants in the world and viceversa. This is contrary to what is proposed in the GEM. Regarding global resources, Mexico is nowadays a country that many product design departments are interested in. Hiring well qualified Mexican design engineers means lowering substantially the product development cost. There are many examples of this in areas such as automotive modules, oil refineries and so on, however the authors found that mold and die design and construction are fields with very little advancement (Kima, 2002). Here the MEPO case-study shows a fertile ground to be seed.

In the race to get to market fast global companies are lacking to evaluate the complete picture. Usually technology and automation is well developed in the headquarters country, however when the technology is transferred to another country many aspects need to be analyzed and adjusted. Culture issues are not collateral side-effects that one may or may not take notice of, depending on the economic situation, but that culture should be at the very core heart of any global development. Technology transfer should be seen from now on as technology cooperation between the manufacturing technology producer and the user.

ACKNOWLEDGEMENT

This research was funded by both CONACYT-Mexico contract 35981 and NSF-USA contract DMI-0116635.

REFERENCES

Crozier, M. (1964). *The Bureaucratic Phenomenon*, University of Chicago Press, Chicago.

Damanpour, F. (2000). NAFTA with a Focus on the U.S. Participation, *Managerial Finance*, **26**, 53-71.

Govindarajan, V., K. Gupta (2001). *The Quest of Global Dominance*, Jossey-Bass, San Fco, CA.

Hill, C.W. (1994), *International Business: Competing in the Global Marketplace*, Irwin, Burr Ridge, IL.

Hofstede, G. (1997) *Culture and Organizations, Software of the Mind*, McGraw Hill, NY.

Hofstede, G. (2001). *Culture Consequences*, Sage, NY.

KIMA S.C: (2002) Investigación de Mercado del Sector Automotriz, Puebla

Martinez I., Galarosa A., Warisawa S, Ito Y (1997) Culture-based Comparative Study of Manufacturing Morphology in Japanese, Taiwanese and Korean Machine Tool Industry, *Proc. of Inter. Conf. on Mfg*, JSME, Tokyo, 205-210.

Moritz E.F. (2003) *Sports, culture and technology – an introductory reader*. Artefact Verlag, Germany

Rauner F. (2002) Localization – The Transfer of Cultural Neutral Technologies into Cultural Applied Products. *IMAC Proc.*, Japan, 26-31.

Rendón, R. (1996) *Breve Historia de Tlaxcala*, Fondo de Cultura Económica, México, DF.

Steers, R., S. Bischoff, and L. Higgins (1992). Cross-cultural management research, *Journal of Management Inquiry*, **1**, 321-330

UNCTAD (2001). *World Investment Report: Promoting Linkages*, United Nations, NY.

Yip G.S. (1992). *Total Global Strategy*, PH, NJ.

Copyright © IFAC Information Control Problems in
Manufacturing, Salvador, Brazil, 2004

ELSEVIER
IFAC
PUBLICATIONS
www.elsevier.com/locate/ifac

A VIRTUAL LABORATORY FOR MOBILE ROBOTICS

L. Enrique Sucar *,[1] **Julieta Noguez** **
Marco A. López-Trinidad ***

* *ITESM Campus Cuernavaca*
Reforma 182-A, Temixco, Morelos, México
esucar@itesm.mx
** *ITESM Campus CCM*
México D.F., México
jnoguez@itesm.mx
*** *ITESM Campus Cuernavaca*
Reforma 182-A, Temixco, Morelos, México
marco.lopez@itesm.mx

Abstract: We are developing a virtual robotics laboratory as a complementary tool for learning basic robotics, in particular for mobile robots. The laboratory includes both types, remote and simulated, and can be used to perform and *observe* the results of simple experiments with different mobile robots. It is based on a generic architecture, which provides a framework for developing virtual laboratories. An intelligent tutor is integrated to the virtual lab, which serves to improve the learning process of the students using the lab. We describe a prototype for the remote and simulated labs, and present some initial experiments in remote robot programming. *Copyright © 2004 IFAC*

Keywords: Virtual Laboratory, Robotics, Intelligent Tutor

1. INTRODUCTION

A *virtual laboratory* provides a way to do experiments with equipment that is not physically present in the same place as the user. There are two types of virtual labs: (i) remote and (ii) simulated. In a remote lab the user interacts with equipment that is physically in another place via sensors and actuators, using a computer and a communication network (i.e. Internet). A simulated lab has computer models of the equipment and its environment, and the user interacts with these models to perform experiments. There are several reasons to build virtual labs:

- To have a wider access to expensive or rare equipment.
- To share equipment that is geographically distributed, for instance among researchers.
- For education or training with equipment that is not easily accessible or which can be damaged by a novice.
- To incorporate other computer tools, such as intelligent tutoring systems.

We are developing a virtual robotics laboratory as a complementary tool for learning basic robotics, in particular for mobile robots. The laboratory includes both types, remote and simulated, and can be used to perform and *observe* the results of simple experiments with different mobile robots. There are two main contributions of this work:

[1] Partially supported by CONACYT, Project No. 40114 and ITESM, *Cátedra de Robótica Móvil*.

(1) It is based on a general architecture for virtual laboratories, such that it can be easily extended to other domains.
(2) It incorporates an intelligent tutor that helps the students using the lab to learn robotics, and it can adapt to the requirements of each student.

Section two reviews related work in general architectures and tutors for virtual labs. Then we present the proposed general architecture, and in section 4 the virtual lab tutor. In section 5 we describe the initial prototypes for the remote and simulated robotics labs. Experimental results in remote robot programming are described in section 6. Finally, we conclude with a summary and a list of our current work.

2. RELATED WORK

Given the advances in computer and communication technologies, in the last years there has been an increased interest in developing virtual laboratories Wagner (1999); Ko et al. (2000). However, most of these efforts are specific to particular domains, so a general methodology for developing virtual laboratories has not emerged. There are few partial models that focus on certain aspects or kinds of virtual laboratories, such as: Levert and Pierre (2000); Licks et al. (2000). This makes the development of a virtual lab a costly and time consuming process. A general framework and standard modules that can be re-utilized, will be an important step to facilitate the expansion of virtual labs to other domains. In this work we propose some initial ideas toward the development of a general architecture for virtual labs.

As with "real" laboratories, a professor or lab assistant that helps the students in the use of the equipment in the lab is required; in particular for teaching laboratories. It is difficult to have a human assistant present each time a student access the virtual lab, so an alternative is to use intelligent tutoring systems. There are some recent developments that combine virtual laboratories with some kind of tutor, such as: TESS and MILAS, Wagner (1999), CMOS, Richard and Gourderes (1999), and GLUCOMAT, Spasic (2000). However, in these systems the tutor is limited, or the virtual lab (usually simulated) is restricted. An important challenge is how to integrate a tutor within a virtual laboratory so that the laboratory is not degraded, and at the same time the tutor can provide effective help to the students. Our approach is to make the tutor as *transparent* as possible to a student using the lab, and limit its intervention to when the experiment is finished, and only in critical cases during the experiment.

3. GENERAL ARCHITECTURE

A block diagram of a general architecture for a virtual laboratory is shown in fig. 1. It includes 3 main parts:

- Guest. A local computer system with which the user interacts with the laboratory. It includes: (a) a graphical user interface, (b) an interface for the remote laboratory, (c) the simulated laboratory, (d) an interface for the simulated laboratory, and (e) a communication interface.
- Host. A remote computer that serves as the interface with the remote laboratory. It has: (a) an interface for controlling the equipment, (b) an interface for observing the experiment, (c) a data base that stores experiments, equipment data, etc., and (d) a communication interface.
- Media. It is the communication media that establishes the link between the Guest and the Host, including the different communication layers required according to the OSI standard.

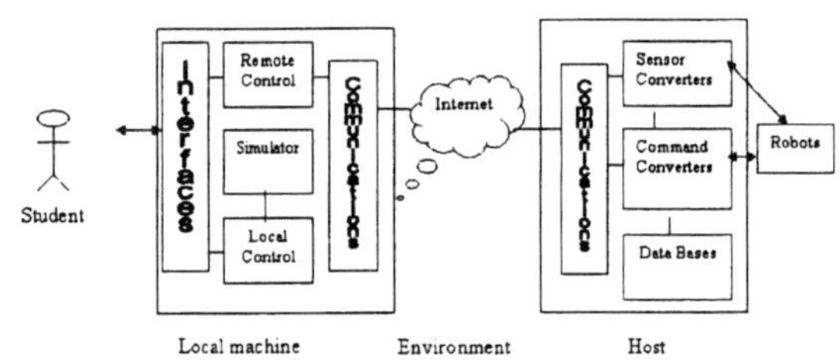

Fig. 1. A general architecture for virtual laboratories.

This architecture provides a general framework for building virtual laboratories in different domains. We expect that several modules developed for the robotics lab could be applied with non or small modifications to other domains, such as the communication interfaces, the interfaces to the virtual and simulated labs, and parts of the user interface. The current implementation considers only one user (Guest) at a time and a single laboratory (Host). So when a user "enters" the laboratory, the access to other users is blocked, and only when the active user "leaves" the lab, other users can enter.

The architecture could be extended for a distributed laboratory in several ways:

- One Guest and one Host: the laboratory (remote or simulated) is distributed among the local (Guest) and remote (Guest) computer systems.
- One Guest with several Hosts: the laboratory (remote or simulated) is distributed among

several Hosts and the user (Guest) access all of them form a single platform.
- Distributed: each system (Participant) can be both, Guest and Host.

Each case implies an increasing level of complexity, in which several aspects have to be considered, such as the synchronization and control mechanisms.

We have developed a prototype for a virtual robotics lab based on the general architecture, that is described in section 5.

4. VIRTUAL LAB TUTOR

Given that the main objective of the virtual lab is to serve as learning tool for students, we have incorporated an intelligent tutoring system (ITS). The inclusion of the tutor extends the general architecture as depicted in figure 2.

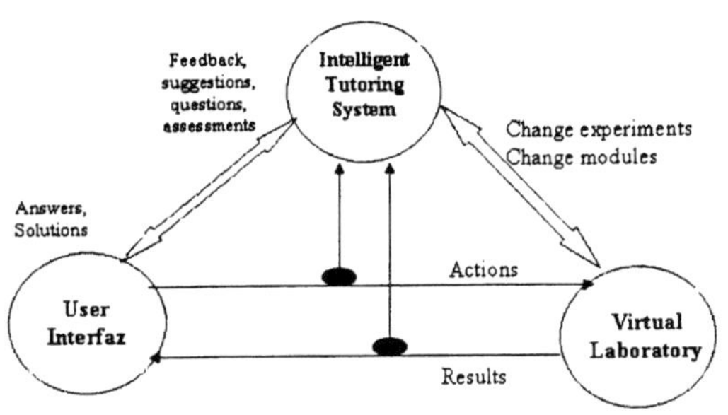

Fig. 2. An intelligent tutoring system in the virtual laboratory

The tutor is basically an *observer* of the experiment, monitoring the user actions and the results as shown in figure 2. In general, it will be transparent to the student in his interaction with the lab, but it could intervene in certain situations. The tutor records the events during each experiment, and maintains a history for each student. Based on the results of the current and previous experiments, it gives advise to the students and can evaluate their performance. It can also suggest experiments to the student, based on her performance and the experiments in the data base. If there are critical errors during an experiment, or if these are repeated in several experiments, it can stop it and point the problems to the student. We consider that providing feedback and personalized help, the student learning process will be improved and she can make a better use of the virtual laboratory. Of course, this hypothesis needs to be validated experimentally.

As most ITS, the virtual lab tutor has 3 main parts: (i) the knowledge base, (ii) the tutor, and (iii) the student model. One of the main differences with other ITS, is that in this case there is not a direct evaluation of the student with questions or problems; the student will be evaluated indirectly based on the results of the experiments. With this information, the tutor has to deduce the state of the student, and decide which actions to take. Given the uncertainty inherent in this task, we are proposing a probabilistic relational student model for the virtual laboratory, Noguez et al. (2003).

4.1 Student Model

Probabilistic relational models provide a new approach to student modeling, that integrates the expressive power of Bayesian networks and of relational models Koller (1999). A probabilistic relational model (PRM) considers the domain to be represented in terms of entities, their properties, and the relations between them. The basic entities in a probabilistic relational models are objects or domain entities. Objects in the domain are partitioned into a set of disjoint classes $X1, ..., Xn$. Each class is associated with a set of attributes $A(Xi)$.

Probabilistic relational models allow to represent each attending student in the same model. Each attribute, $x.a$, of the individual student x, is directly influenced by only a few other parameters related with this student. Each class represents the set of parameters of several students, like in databases, but the model also includes the probabilistic dependences between classes for each student. PRMs allow a compact and natural representation of student models for virtual laboratories. A partial view of the PRM for the students' model is shown in figure 3.

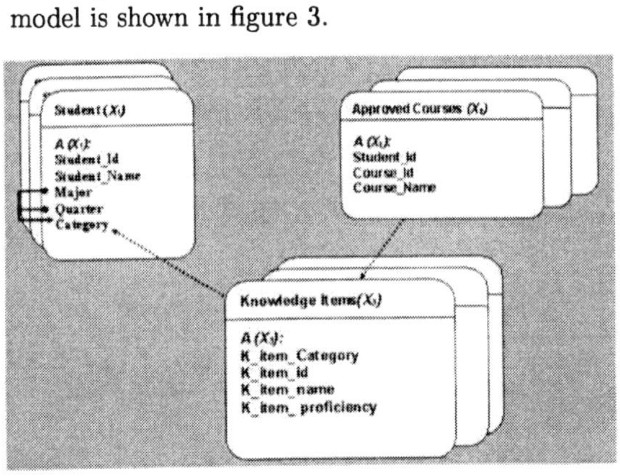

Fig. 3. Part of the PRM for the student model.

From the PRM we can extract a Bayesian network that represents each particular student. A partial view of this network for one student is shown in figure 4. That is, for each student and theme knowledge, the PRM has different probabilities. These values are obtained from evidence in the *Student.Profile*, *Student.Type* and

Student.Background. The *Student.Background* is obtained from experiment performance.

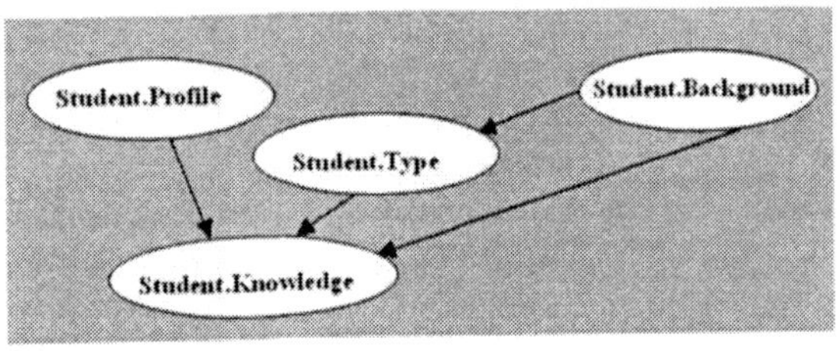

Fig. 4. Conditional dependencies between attributes of the class student in the PRM.

Given the observations on the student performance in a an experiment, these are use to update the information in the student model. So based on this model, the tutor can decide the best way to guide the student in the next experiments. An additional advantage of this representation is that can be used build an initial model of a new student based on the existing ones. By using the student profile, the parameters for new student are a *weighted average* of the similar students in the data base.

To integrate the student model to the virtual lab, we are developing the other components of the ITS, that is the tutor and the knowledge base.

5. PROTOTYPE

We have built an initial prototype of the virtual robotics lab that includes a remote and simulated mode. This prototype is used for performing basic experiments in control and navigation of a mobile robot. The student has to make the robot follow certain trajectory in a simulated or real indoor environment, in one of the following ways:

- Direct manipulation, by given direct navigation commands (advance, turn, etc.) to go from an origin to a goal.
- Programming, via a control program that makes the robot follow a line using light sensors.

In both cases, the user gives high level commands or instructions that are converted to the commands that the real or simulated robot understands. The idea is to use a generic language so different robots could be used in the lab by just changing the interpreter. Following we describe the remote and simulated labs.

5.1 Remote Lab

The remote robotics lab prototype is used to command a mobile robot via Internet, and observe it using a video camera. Based on the general architecture, the remote lab is divided in two main parts, the Guest and the Host. The Guest has 4 modules: (a) the user interface, (b) the command interface, with which the user edits and sends a program to the Host (c) the observer, that displays images of the robot in the environment, and (d) the communication interface. The Host has 5 main modules (a) the navigation module, that verifies and translates the user commands, (b) the security module, that runs a concurrent process that makes sure that the robot does not collide with obstacles using sonars, (c) the drivers to communicate with the robot, (d) the image capture and compression module, and (e) the communication interface.

The interface for the user in the Guest is shown in figure 5. With this interface, the user can command the robot and observe it in its environment. The user can compose a high-level sequence of commands (a program) and send it to the robot via the interface. The host receives the sequence and it translates it to the robot language. Then it sends this program to the robot, and also the security program, and these are executed in the robot. Finally, the user can watch the effects of her program on the robot and its environment, via the *observer* in the lab as a video sequence. In this way the remote lab operates in a *semi-interactive* mode, and can tolerate variable delays in the network. We have performed some initial experiments using the local network in our university, and also from remote locations, which are presented in section 6.

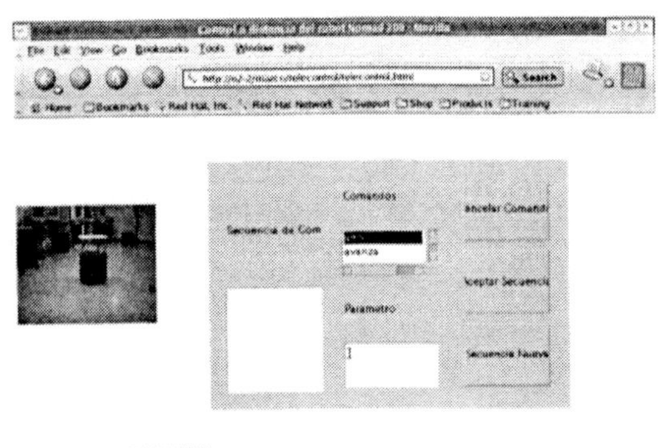

Fig. 5. A view of the remote lab interface.

5.2 Simulated Lab

The simulated lab includes a cinematic model of mobile robot with infrared sensors and a simulated environment (a white line over a black background). The user can program the simulated robot by controlling its motors (differential drive) and using the information from the sensors. The model robot and its environment is depicted using a graphical interface, with which the student can observe the results of the experiment. The main modules for the simulated lab in the Guest machine are: (a) the model, that includes the robot, sensors and the environment, (b) the input module, with which the student selects a configuration for the robot and inputs the control program, (c) the display module, that shows in 3-D the results of the experiment, (d) the monitor, that records the main variables for the robot state during the experiment.

An initial interface for the simulated lab is shown in figure 6. With this interface the user can configure the simulated robot, in particular the number and positions of sensors (infrared). He can also select from several predefined paths that the robot has to follow. Once the user has made a program for the simulated robot, this is executed, and a 2-D view of the robot motion in the virtual environment is depicted. At the end of an experiment, a report on the performance is shown to the student.

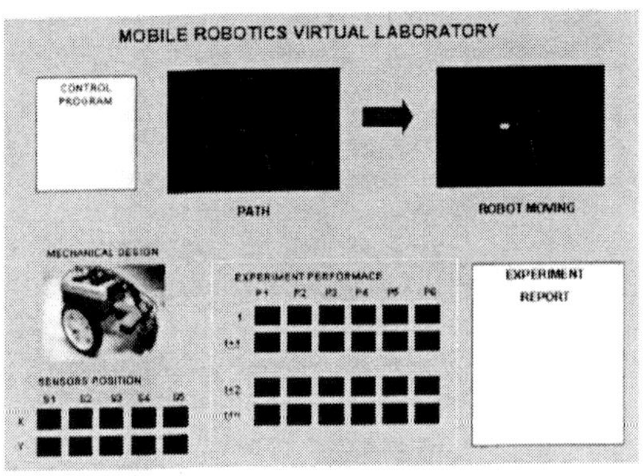

Fig. 6. A view of the simulated lab interface.

We are currently finishing a first prototype for the simulated lab, and integrating the virtual lab tutor, Noguez et al. (2002).

6. EXPERIMENTAL RESULTS

We have done some initial experiments with the remote lab from local and remote (Mexico, Colombia) guests, Sucar and López-Trinidad (2000). For the local experiments, in which the Guest is a computer in the University network, we have measured the delays in both, the program execution and the observation sequence. The experimental setup in this case is the following:

- The Host is a Linux PC that is interconnected to the University Network (Ethernet, 10 MBps).
- The robot is a Scout from Nomadic, that is also connected to the network via a wireless link.
- The Observer is a video camera that has a general view of the laboratory and the robot, and which is connected to other PC in the network.
- The Guest is a user in any machine connected to the University network.

Given this setup, the user makes a program using the Guest interface. The program consists of a sequence of motion commands to the robot, for instance:

```
start
advance 3 m
rotate right 90 degrees
advance 2 m
stop
```

We tested the system from different machines and at different times of the day, and measure the time delays for:

(1) Program execution - the time interval from when a program is submitted in the Guest to when it starts execution in the robot.
(2) Image display - the time interval from when an image is captured in the observer to when it is displayed to the user.

The results are summarized in table 1.

Table 1. Time delays for program execution and image display for the remote lab.

	Average	Standard Deviation
Program execution	51.7 sec.	45.3 sec.
Image display	0.31 sec	0.80 sec.

So in general, the user has to wait aprox. one minute to observe the results of the program in the robot. This includes the delays from the different parts of the process: network delays (Guest–Host and Host–Robot), program translation, and robot start-up. With respect to the image display, the user observes the experiment at about 3 frames per second. There is a wide variation in these delays, mainly due to the different loads in the network, which depend on the time and day of the experiment.

The remote lab can be accessed at this address: http://doc.mor.itesm.mx/~virtualab/telecontrol/telecontrol.html [2].

7. CONCLUSIONS

We have presented a virtual laboratory for learning basic robotics, in particular for mobile robots. The laboratory includes remote and simulated labs, and can be used to perform and *observe* the results of simple experiments with different mobile robots. It is based on a generic architecture,

[2] Send a message to marco.lopez@itesm.mx if you want to do a remote experiment.

which provides a framework for developing virtual laboratories. An ITS is integrated to the virtual lab, which serves to improve the learning process of the students using the lab.

We are currently working on: (i) a formal definition of the general architecture and methodology for virtual laboratories, (ii) a prototype of the simulated lab, (iii) an intelligent tutoring system for virtual labs, and (iv) the definition of a set of experiments for students that will allow us to evaluate the advantages of these tools in a classroom scenario.

REFERENCES

C.C Ko, B.M. Chen, S.H. Chen, V. Ramakrishnan, R. Chen, S.Y. Hu, and Y. Zhuang. A large-scale web-based virtual oscilloscope laboratory experiment. *Engineering Science and Education Journal*, pages 69–76, 2000.

D. Koller. Probabilistic relational models. In *International Workshop on Inductive Logic Programming*, volume 9, pages 3–13, 1999.

C. Levert and S. Pierre. Security, interoperability, and quality of service, aspects in designing a telecomunications platform for virtual laboratories. In *Canadian Conference on Electrical and Computer Engineering*, volume 2, pages 811–815, Canada, 2000.

V. Licks, M. Quiroga, R.Jordan R., and J.S. Correa. Building virtual laboratories-a web based experience. In *International Conference on Engineering Education*, Sao Paulo, Brasil, 2000.

J. Noguez, L. E. Sucar, and M. A. López-Trinidad. Un sistema tutor inteligente para un laboratorio virtual de robtica. In *Workshop on Intelligence Virtual Environments*, pages 27–34, Mexico, 2002. Mexican Internacional Conferencia on Artificial Intelligence.

J. Noguez, L. E. Sucar, and F. Ramos. A probabilistic relational student model for virtual laboratories. In U. Hoppe, F. Verdejo, and J. Kay, editors, *International Conference on Artificial Intelligence in Education*, volume 11, pages 533–534, Sydney, Australia, 2003. IOS Press.

L. Richard and G. Gourderes. An agent-operated simulation-base training system: Presentation of the CMOS project. In S.P. Lajoie and M. Vivet, editors, *International Conference on Artificial Intelligence in Education*, pages 343–351. IOS Press, 1999.

V. Spasic. Intelligent virtual systems in learning. In *Annual EMBS International Conference*, volume 2, pages 1016–1018. IEEE, 2000.

L.E. Sucar and M. López-Trinidad. Laboratorio virtual de robótica móvil. In *Second National Congress on Robotics*, pages 63–68, Toluca, Mexico, 2000. AMRob.

B. Wagner. From computer-based teaching to virtual laboratories in automatic control. In *ASEE/IEEE Frontiers in Education Conference*, volume 3, pages 606–611, San Juan, Puerto Rico, 1999.

Copyright © IFAC Information Control Problems in
Manufacturing, Salvador, Brazil, 2004

ELSEVIER
IFAC
PUBLICATIONS
www.elsevier.com/locate/ifac

REMOTE ELECTRONICS WORKBENCH – TAKING THE LAB HOME

Gustavo Ribeiro Alves[1], António Cardoso[2], José M. Martins Ferreira[2]

[1]*Instituto Superior de Engenharia do Porto*
R. Dr. António Bernardino de Almeida 431, 4200-072 Porto, Portugal
galves@dee.isep.ipp.pt
[2]*Faculdade de Engenharia da Universidade do Porto*
R. Dr. Roberto Frias, 4200-465 Porto, Portugal
{antonio.cardoso, jmf}@fe.up.pt

Abstract: Students usually regard lab classes as the opportunity to practice learnt theory. Some would even pass more time in the lab performing new or the same experiment with slightly different parameters, if allowed to. Benefiting from the experience gained in the PEARL (Practical Experimentation by Accessible Remote Learning) project, the University of Porto has developed a Remote Electronics Workbench (REW) that allows students to carry out real experiments in electronics, from their home computer, through web-based access. The REW includes: interfaces to experimental scenarios; real video feedback from the lab environment; video-conference facilities to enable student-to-student and student-to-tutor dialogue; and registration and booking pages to new and registered users, respectively. It is also a constituting part of a proposal to create a Remote Experimentation Network covering both European and Latin-American countries, under the Alfa II programme. *Copyright © 2004 IFAC*

Keywords: Educational aids, laboratory education, remote experimentation

1. INTRODUCTION

The concept of e-learning is now widely understood by the general educational community at all levels (i.e. at college and university levels), with a broad selection of solutions available both in the market and academia (Landon, 2003; Richards, 2001). E-learning has more recently benefited from the emergence of web-accessible labs providing support for a basic educational component: experimentation. The more recent concept of remote experimentation has been named in several ways (web-based remote-access laboratory, virtual or online laboratory, e-lab, etc.) throughout the last four to five years, in both conference and journal papers. For instance, Ko, *et al.*, (2000; 2001a, b) use the expression "web-based virtual laboratory" in a series of articles that describe different experimental scenarios, developed at the National University of Singapore, that range from the control of relatively simple electronic circuits to more complex dynamic systems. Esche, *et al.*, (2003) refers a series of "remote laboratories" (circa 30), while also referring that "the roots of such systems could be tracked back to a master-slave teleoperator developed at Argonne National Laboratory in 1954". The expressions "remote experimentation" and "remote lab" are used not only in (Esche, *et al.*, 2003) but also in (Johnston, *et al.*, 2001; Cooper, *et al.*, 2002, Callaghan, *et al.*, 2003; Mueller and Ferreira, 2003) and in several other references made by Esche, *et al.*, (2003). The cited examples also provide a fair idea of the overall adoption of this supporting technology in distance education in different engineering fields (chemistry, civil, electronics, mechanical, mechatronics, etc.).
More recently, some authors have referred the pedagogical impact of remote experimentation and discussed ways of improving its benefits in a learning perspective. Terms such as "collaborative" and "cooperative", often associated with actions engaged by students in traditional lab classes, are being transferred into the remote experimentation arena, as seen in (Callaghan, *et al.*, 2003; Ferreira, *et al.*, 2002; Erbe and Bruns, 2003).

The Faculty of Engineering of the University of Porto (FEUP) has been active in the area of remote experimentation since 1999, due to its involvement in several related European R&D projects (Cooper, *et al.*, 2002; Mueller and Ferreira, 2003). This paper describes the implementation of a remote electronics workbench (2004) that re-uses much of the lab infrastructure developed in the past PEARL project, while taking into consideration several functional and pedagogical aspects reported by students during the interim and final validation trials.

2. BACKGROUND: THE PEARL PROJECT

PEARL (2003) was a three-year R&D project that ran from March 2000 till February 2003, with funds provided by the European Commission under the IST (Information Society Technologies) programme. In general terms, the PEARL project aimed to provide access to remote experimentation facilities in selected areas of science and engineering, namely biochemistry, fundamental physics, automatic visual inspection and digital electronics. Following a development phase that took place over the first two years, validation trials were carried out with students in Portugal, England, Scotland and Ireland. Paolo (2002) reported the feedback and comments received during the trials on aspects depicted in table 1. The students' feelings were collected using both a questionnaire and an interview schedule. In total, Paolo analysed almost one hundred questionnaires and several hours of videotaped interviews.

Table 1: Outline of the questionnaire for the user testing involving students

Section	Nature of questions
A: General Computing experience	Designed to gain an insight into the students' past experience of using computers to determine if expertise has an impact on their use of the system or attitudes towards the system.
B: General questions about using the PEARL system	This consists of 22 statements drawn from various usability instruments and covered how easy was it to learn how to operate the system, usefulness of the system and attitudes towards using the system. Each of the statements is responded to on a four-point Likert scale that measures how much the students agrees or disagrees with the statement.
C: Collaboration tools	This section asks students which of the collaboration tools they used and then each tool (video conferencing, audio conferencing and real-time chat) is explored in detail. Students are asked about the quality of the feed, if they experienced any losses in signal and if they encountered any difficulties.
D: Learning with the PEARL system	The students are asked to rank six aims linked to practical work with reference to the activity they performed using the PEARL system. Essentially, the students ask themselves the question "What did I get out of doing this experiment using the PEARL system?"

Some of the main conclusions drawn from the analysis reported by Paolo in (2002) were:

- Real experimentation is not a replacement for simulation, but simulation alone is far less satisfactory than dealing with real devices (it is difficult to develop simulation models sufficiently sophisticated to cope with the unpredictability of real experiments and physical behaviour; students find real experiments more motivating than simulation, leading to better learning effectiveness)
- Synchronous communication facilities are of fundamental importance to enable the benefits of group work, particularly those related to good videoconferencing facilities – however, tutors should take into account that such resources will only be useful if enough bandwidth is available, which requires coordination with on-campus system administrators
- Scientific and pedagogical skills are more than ever required from tutors, who must be aware that script design must lead the students in such a way as to take full advantage of the powerful synergies resulting from distributed teamwork

Having in mind these conclusions, the FEUP team involved in PEARL decided to reuse the remote lab infrastructure developed under the project (after it had finished), for assisting in the lecturing of basic analog electronics concepts. The original idea was to provide students with an opportunity to repeat one or more experiments made on-site, i.e. during a traditional lab class, from their home computer. This idea addressed in particular the first conclusion pointed out by Paolo. The students do have the opportunity to simulate the circuit at home, using shareware such as the student versions of MultiSim or CircuitMaker. However, and as along as they have an Internet connection, students may now access the REW and control the very same circuit used during one of theirs lab class, repeating the experiment with the same or new circuit parameters (e.g. a different capacitor or resistor) and / or with other stimulus (e.g. a rectangular wave instead of a sine wave). Also, this initial motivation (enrich the pedagogical resources available to students) provided us with an excellent opportunity to update some aspects of the remote lab infrastructure.

The hardware part of the remote lab infrastructure, illustrated in figure 1, consists of: an experimental module (typically, a bread board with the circuit under experiment); a PXI system comprising the lab server (i.e. an embedded PC inserted in slot 0) and several test & measurement instruments (function generator, oscilloscope, etc.); and also the web server (a standard desktop). The software part was updated to the latest version of LabVIEW, which enables swift generation of Virtual Instruments (VI) interfaces, remotely controllable through the web. This update allowed us to remove a third-party application from Nacimiento, named AppletVIEW, which was previously used for generating equivalent JAVA applets for the VIs, which enable direct access over the web via any common web browser, without requiring the installation of any plug-ins, as stated by Fidalgo in (2001). However, if the LabVIEW update enabled a serious breakdown in the development time (developing the JAVA applets required some time and effort), the counter part came in the form of the need to install a plug-in at the client side.

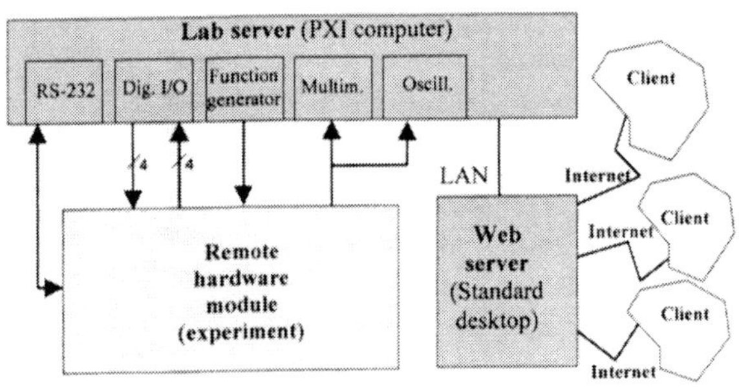

Fig. 1. The hardware component of FEUP's REW.

The 2nd conclusion has also impacted the updating of the synchronous collaboration tools. In PEARL, we used CUSeeMe (2003) for audio / video conference, and although it served fairly as a collaboration tool, several reasons led to its replacement by FlashCom (2003). First, CUSeeMe uses UDP (Uniform Data Protocol), while FlashCom uses TCP/IP (Transport Connection Protocol / Internet Protocol). This difference is crucial in networks where UDP traffic is limited by network administrators, as it is also used for web-based radio broadcast. Second, CUSeeMe faced some problems with the firewall package installed at FEUP. This problem was solved after installing a new service pack, sent by the company responsible for the firewall. Additionally, the CUSeeMe server was installed in Dundee, Scotland. This meant that if two students placed in adjacent rooms, in Porto, Portugal, used CUSeeMe, all the corresponding UDP traffic had to be made via Dundee. Although this would go unnoticed in favourable conditions (enough bandwidth), some of the students reported difficulties in collaborating with each other, due to communication problems. To overcome all these problems, FEUP acquired and locally installed a FlashCom server.

Finally, the last conclusion led to a more careful design of the lab script. Considering that the students had already done the experiment in a traditional lab class, the on-line lab script could be simplified to a great extent, namely by reducing its total length to no more than two screens. This enables the students to cover the all script with just one or two mouse clicks on the side bar, instead of having to scroll up / down, which asks for more motivation, attention, and time to read it. Also the lab script is now presented in the entry page, alongside with the available options, instead of being just one link on the initial HTML page, as happened in PEARL. This caused some students to go directly to the experimental scenario, without even having seen the lab script.

Having clarified the background of the REW, we will now describe its current implementation, in more detail, in the following section.

3. THE REMOTE ELECTRONICS WORKBENCH

3.1 The entry page

Presently, the REW allows students to experiment a simple 2nd-order low-pass filter, namely by remotely controlling its input with a function generator and reading its output with an oscilloscope, both instruments being located at the lab. The access to the experimental scenario is made via an initial HTML page that displays the lab script, illustrated in fig. 2, and the tools selection menu that contains the following options:

- Access to Web cams placed at the lab (fig. 3)
- Access to video-conferencing
- Access to the remote lab:
  - Registration (required)
  - Booking (required for control)
  - Download plug-in (required)
  - Access to the current experiment

The on-line lab script guides the students through the remote experiment explaining: a) what is (are) the objective(s); b) what type of circuit and instruments are used; c) the rationale behind the experiment (e.g. if a certain input is applied or a set of conditions is satisfied, then a certain output or circuit behaviour should be expected); and d) what sort of conclusions or properties should be extracted from the circuit and / or experiment.

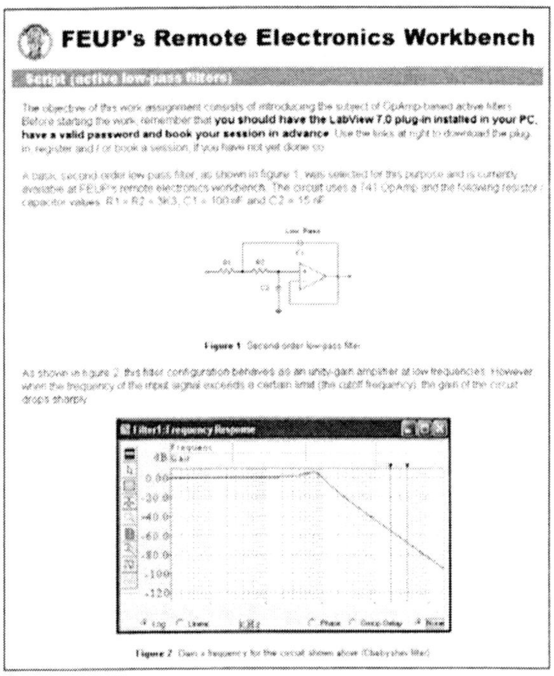

Fig. 2. The lab script contains explicit instructions on what is required and some schemes that help the students understanding the circuit under test.

3.2 The video feedback from the lab

The Web cams provide a live feedback from the lab, which is quite useful for a number of reasons: a) it provides a sense of reality, i.e. the students feel they are accessing something physical, as opposed to running a simulation; b) in certain circuits under test, it provides an image of the current status (LEDs on / off, position of switches, etc.). Both reasons were actually pointed out by students that participated in the first trials conducted under the PEARL project, as expressed by Paolo in (2002).

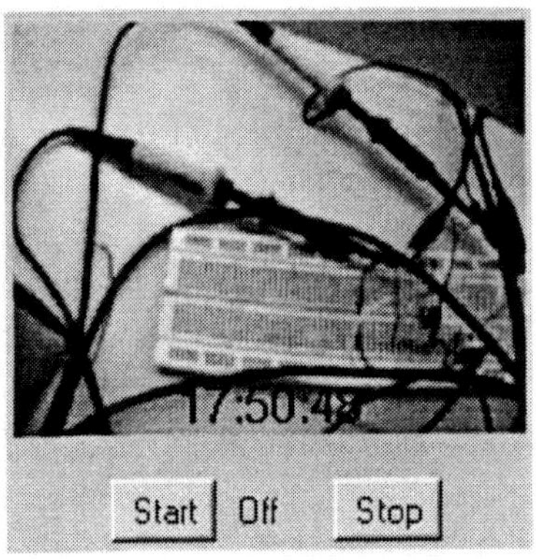

Fig. 3. Image of the circuit under test from one of the Web cams installed at the REW. The interface has start / stop buttons that allow reducing the bandwidth requirements for certain periods, while using the real-time clock as an indicator of the current status.

3.3 Audio / Video communication

As already stated, the audio / video communication is done through FlashCom, a synchronous collaboration tool that enables a student to establish contact with other students, the lab technician, or the tutor. The advantages of collaborative / cooperative work have already been extensively discussed in previous works, so that one may state this type of resource is vital in any remote experimentation environment. Interestingly, some students reported in PEARL a preferential use of the chat mechanism, as opposed to using the audio facilities, due to two reasons: 1) it is more reliable; and 2) when a tutor is present they may send private messages, so as to hide from the tutor any possible misunderstandings or mistakes that may cause a negative impact on him.

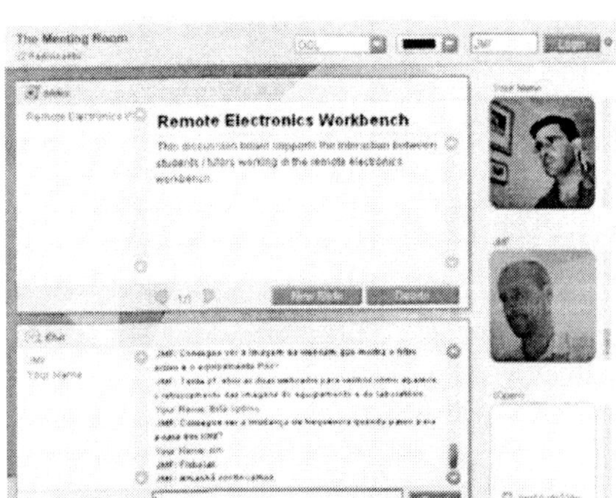

Fig. 4. Student and tutor using the video-conference tool, which contains two or more video frames, one slides window, and one chat window.

3.4 Register / Booking mechanisms

Before being able to access the remote lab (i.e. have a valid login / password), all clients are requested to register themselves with the lab administrator. This may be done through a simple HTML page (link "Registration" on entry page) that asks for some user identification data. After completing the form (some fields are mandatory), the client may either modify or submit it. Submitting the form causes the web server to automatically generate and send to the lab administrator an e-mail message, similar to the one illustrated in figure 5, who may then accept or refuse the registration request, using the two buttons visible at the message bottom. The decision will be reported to the client through an e-mail message, generated automatically by the web server.

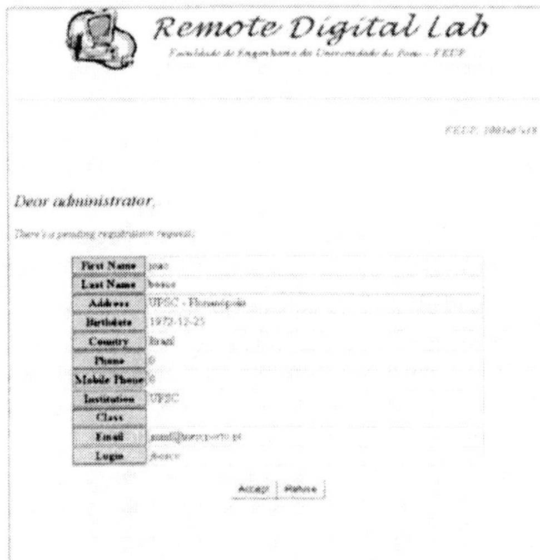

Fig. 5. A pending registration request, as received by the lab administrator.

Our solution for granting access and control to the remote experiment is based on two steps. First, it relies on a booking / scheduling mechanism that requires a valid login / password (obtained through registration, as previously described). The students are presented with a booking page that displays a 7-day x 24-hours map, divided into 1-hour time slots, as illustrated in figure 6. A time slot may be booked by just clicking on it and entering a valid login / password on the small dialogue window prompted by the web server (fig. 6). Only the owner of a booked session or the lab administrator is allowed to free it again, afterwards, using the unbook option. A colour coding scheme indicates if a certain time slot is already booked, is under a booking request, is reserved for lab maintenance, or is free (a protection mechanism prevents booking elapsed time slots, according to the current web server time). Several students may access the lab during the same time slot, by sharing the same login / password (a sort of group key). Regarding control, and while conducting the experiment, only one student is allowed to control the remote instruments, at a given time, as imposed by the functionality of the associated VIs. However, the students are able to pass the control among them, as later explained in section 3.7.

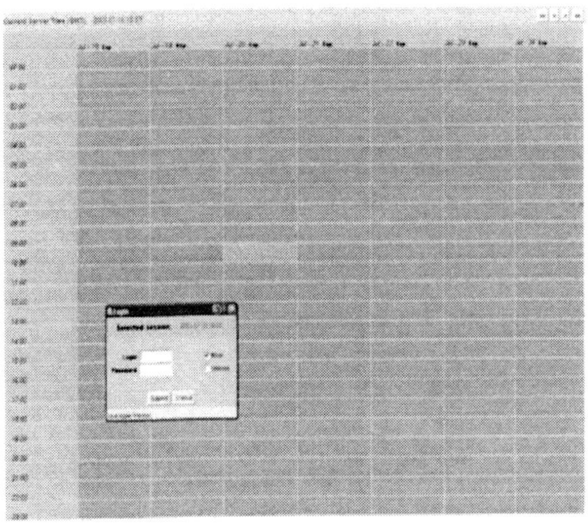

Fig. 6. The booking page while on the process of reserving a 1-hour time slot. The buttons on the top-right corner allow scrolling the visible map (7-day x 24-hours) day-by-day or week-by-week.

3.5 Downloading / installing the plug-in

Before accessing the remote lab, the students are required to download a plug-in (DataSocket Server), provided by National Instruments, in order to be able to control the test and measurement instruments, through the associated VIs. Once downloaded (about 20 Mbytes), the plug-in must be installed at the client computer, which necessarily requires administration privileges. These two details are at times frustrating to the students. For instance, when they are accessing the REW from their home computer for the first time, the size of the plug-in requires a significant time for downloading, even considering a fast connection (ISDN, or even ADSL). On other occasions, the students may not have administration privileges on the computer they are using, which generally happens on computers available at campus.

3.6 Running the experiment

Finally, having previously booked a time slot and having installed the plug-in, the students may then access the remote lab and run the experiment. Each access starts with the download of the VI panels, as illustrated in figure 7, which depicts the situation of using a computer connected to the Internet through a 56K modem. Again, due to the size of the VI panels, this operation may take some minutes, and must be repeated every 1-hour time slot, i.e. even if two contiguous time slots are booked by the same student, the system will close the session on every hour turnover, thus requiring a new download, as illustrated in figure 8.

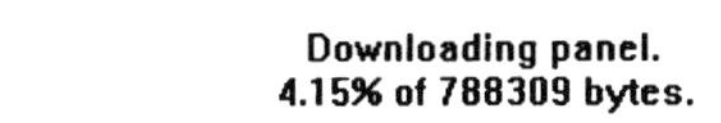

Fig. 7. Downloading the VI panel(s) may take some minutes if the client is connected to the Internet through a low-speed link (e.g. a 56K modem).

Fig. 8. Status bar indicating the current and ending times for a particular remote experiment session.

Apart from this system flaw, and once the VI panel is downloaded, the students gain control over the remote instrument by just clicking on it, as illustrated in figure 9. A request control command is sent to the server that immediately grants it. In working groups, if another student is already controlling the remote instruments, the system returns a warning message. In such occasions, the student may contact his / her group partner(s) - using the collaboration tools, request control, and obtain it as soon as the other student selects the "Release Control of VI" option. LabVIEW also incorporates a timing supervisor mechanism that releases the VI control whenever no actions are detected for a period of 5-minutes (default value). The timer is reset to zero if an action is detected within that period. The students may monitor this timer by selecting the "Show Control Time Remaining" option (shadowed in fig. 9).

The goal of the experiment currently available at the REW is to determine the cut-off frequency of a simple 2nd-order low-pass filter, based on an op-amp with a resistor / capacitor (RC) feedback loop. One process of determining the cut-off frequency is to apply several pure frequencies (sine waves), using the function generator, and check the circuit output with the oscilloscope (channel A connected to the function generator output = filter input; channel B connected to the filter output). For frequencies below the cut-off frequency, the filter output follows the circuit input (circuit gain =1), while for frequencies above the output is attenuated (gain <1). This process requires the student to remotely control both the function generator and the oscilloscope, much in the same way as done locally in the lab class where the experiment is first introduced.

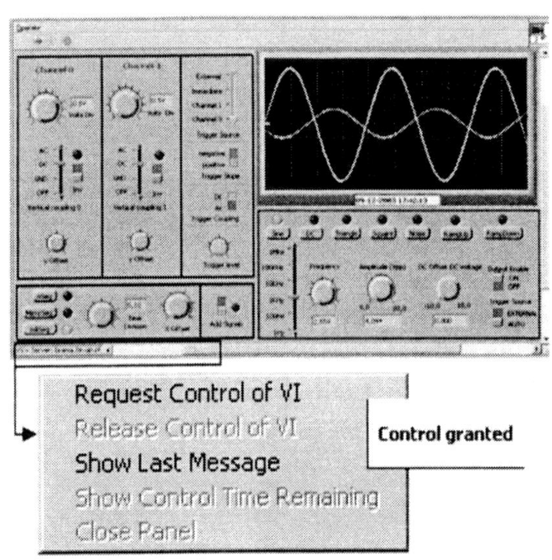

Fig. 9. Remote interfaces of the function generator (bottom-right frame) and of the oscilloscope (remaining frames). The bottom selection menu allows the students to request / release control of the corresponding VI.

4. CONCLUSION AND FUTURE DIRECTIONS

The REW provides students with an opportunity to repeat an experiment made during a traditional class, from their home computer (as long as it is connected to the Internet), thus justifying the expression "taking the lab home". Although only one experiment is currently available (a simple 2nd-order low-pass filter, used in the lecturing of basic analogue concepts – or op-amp basic circuits), the remote lab infrastructure remains the same for any other experiments on electronics, provided they use the test and measurement instruments and resources already available at the lab server. Supporting a new experiment typically requires: a new circuit under test connected to the lab server; developing the VI panels for the experimental scenario (if more instruments are needed – apart from those already controllable through the existing VI panel, or if a different arrangement is preferable); designing the associated lab script; and updating the booking page with information on what experiment is available on each day.

Finally, the REW is also a constituting part of a proposal to create a Remote Experimentation Network (RexNet), covering both European and Latin-American countries, which includes a remote lab for experiments on mechatronics (Erbe and Bruns, 2003) and the RexLab (Alves, 2003), among others. The proposal was elaborated by a consortium comprising 10 institutions and submitted to the 8th-call of the Alfa II programme (Alves, 2004), which is expected to release the evaluation result during the 1st quarter of 2004.

REFERENCES

Alves, R. Gustavo (2004). Remote Experimentation Network - Yielding an Inter-University peer-to-peer e-service. Proposal to the Alfa programme. In: http://www.dee.isep.ipp.pt/~galves/alfa

Alves, João Bosco (2003). Remote Experimentation Laboratory (RexLab). In: http://rexlab.ufsc.br/.

Callaghan, MJ., J. Harkin, C. Peters, TM. McGinnity and LP. Maguire (2003). A Collaborative Environment for Remote Experimentation. In: *Proceedings of the International Conference on Microelectronic Systems Education*, Anaheim, California, USA.

Cooper, Martyn, Alexis Donnelly and José M. Ferreira (2002). Remote controlled experiments for teaching over the Internet: A comparison of approaches developed in the PEARL project. In: *Proceedings of the 19th annual conference of the Australian Society of Computers In Learning In Tertiary Education*, Auckland, New Zealand.

CUSeeMe (2003). QuickNet. In: http://www.cuseeme.com

Erbe, Heinz-Hermann and F. Wilhelm Bruns (2003). Didactical Aspects of Mechatronics Education. In: *Proceedings of the 5th IFAC International Symposium on Intelligent Components and Instruments for Control Applications*, Aveiro, Portugal.

Esche, Sven K., Constantin Chassapis, Jan W. Nazalewicz and Dennis J. Hromin (2002). A scalable System Architecture for Remote Experimentation. In: *Proceedings of the 32nd ASEE/IEEE Frontiers in Education Conference*, Boston, MA, USA.

Ferreira, José M., Gustavo Alves, Ricardo Costa and Nick Hine (2002). Cooperative learning in a web-connected workbench. In: *Lecture Notes on Computer Science 2438*, "Groupware: Design, Implementation and Use", ISBN 3-540-44112-3, Springer, Proceedings of the 8th International Workshop on Groupware, La Serena, Chile.

Fidalgo, André V., Ricardo J. Costa, José M. Ferreira and Gustavo R. Alves (2001). Experimenting the 1149.1 and 1149.4 test infrastructures in a Web-accessible remote Lab (without Plug-ins!). In: *Proceedings of the 16th Conference on Design of Circuits and Integrated Systems*, Porto, Portugal.

Flash Communication Server (2003). Macromedia. In: http://www.macromedia.com/software/flashcom/

Johnston, Murray V., Frederick J. Cox, Gregory J. Forte, Dean C. Nairn, Richard S. Sacher, Anita Z. Schwartz, and Akos Vertes (2001). Remote Experimentation over the Net: Our First Year with MALDI. In: *Analytical Chemistry, bi-monthly Journal of the American Chemical Society*, **Volume 73**, No. 15, pp 440 A-445 A.

Ko, C. C., B. M. Chen, J. Chen, Y. Zhuang and K. C. Tan (2001a). Development of a Web-based laboratory for control experiments on a coupled tank apparatus. In: *IEEE Transactions on Education*, **Volume 44**, No. 1, pp. 76-86.

Ko, C. C., B. M. Chen, S. H. Chen, V. Ramakrishnan, R. Chen, S. Y. Hu and Y. Zhuang (2000). A large scale web-based virtual oscilloscope laboratory experiment. In: *IEE Engineering Science and Education Journal*, **Volume 9**, No. 2, pp. 69-76.

Ko, C. C., B. M. Chen, S. Y. Hu, V. Ramakrishnan, C. D. Cheng, Y. Zhuang and J. Chen (2001b). A web-based virtual laboratory on a frequency modulation experiment. In: *IEEE Transactions on Systems, Man, and Cybernetics, Part C: Applications and Reviews*, **Volume 31**, No. 3, pp. 295-303.

Landon, B. (2003). On-Line Educational Delivery Applications: A Web Tool for Comparative Analysis. In: http://www.edutools.info/index.jsp

Mueller, Dieter and J. M. Martins Ferreira (2003). MARVEL: A mixed-reality learning environment for vocational training in mechatronics. In: *Proceedings of the Technology Enhanced Learning International Conference*, Milan, Italy.

Paolo, Terry Di (2002). Deliverable 9.2 - Plan of the Interim Validation, project PEARL IST-1999-12550. In: http://kmi.open.ac.uk/projects/pearl/

PEARL (2003). In: http://iet.open.ac.uk/pearl/

Remote Electronics Workbench (2004). Faculty of Engineering of the University of Porto. In: http://www.fe.up.pt/~jmf/rew

Richards, P. (2001). MIT to make nearly all course materials available free on the WWW. In: http://web.mit.edu/newsoffice/nr/2001/ocw.html

Copyright © IFAC Information Control Problems in
Manufacturing, Salvador, Brazil, 2004

ELSEVIER
IFAC
PUBLICATIONS
www.elsevier.com/locate/ifac

A WEB-BASED REMOTE MULTIVARIABLE CONTROL EXPERIMENT

J. M. Gomes da Silva Jr., M. A. L. Silveira, C. E. Pereira.

Electrical Engineering Department - DELET
Universidade Federal do Rio Grande do Sul - UFRGS
Av. Osvaldo Aranha 103 - Porto Alegre - Brazil - 90035-190
E-mails: {jmgomes, mlisboa, cpereira}@eletro.ufrgs.br

Abstract: This paper describes the development of a Web-based remote multivariable control experiment on a Foundation Fieldbus network. The control objective is to keep the level in each tank within specified set-points. The plant operation is configurable and programmable over a supervisory application executed along with a Web server. Among other features, the supervisory interface, displayed in the client browser, allows the remote on-line configuration of controller parameters and a real-time graphical visualization of process and manipulated variables. These features allow students to execute open and closed-loop multivariable control experiments. *Copyright © 2004 IFAC*

Keywords: Remote control, Process Control, Fieldbus, Internet, Laboratory education

1. INTRODUCTION

The large dissemination of Internet and world wide web has brought new paradigms and possibilities in technological education. In particular, it allows the remote use of laboratory facilities that can be used to illustrate concepts handled in classroom and serves as a enabling and powerful technology for distance teaching. Through its world wide connectivity, the Internet also allows to have learning materials available to a much larger audience of students, giving them a greater flexibility in terms of defining their own speed and sequence of subjects during learning. (Poindexter and Heck, 1999).

Following this trend, many institutions around the world have been engaged in the development of Web based experimental setting that can be cast into two classes: virtual labs and remote labs.
Virtual labs are associated to the remote iterative simulations of physical devices and systems The simulations are, in general, performed by commercial mathematical software such as MATLAB or LABVIew, which requires specific plug-ins in the client machine, or directly by JAVA applets (see among others (Schmid, 1998), (Robinette and Manseur, 2001) and (Sanchez et al., 2002). The main advantage of the virtual labs approach relies on the reduced cost, since there is no need of expensive experimental equipments and their maintenance (Overstreet and Tzes, 1999). From a pedagogical point of view, virtual labs are useful to illustrate basic theoretical concepts. The students assimilation of these concepts is greatly improved since they can quickly visualize cause-effect relationships by changing simulation parameters. However, it is also necessary that the students realize that models are imprecise representations of the real world. Hence, they should be faced to problems related to the implementation of physical systems such as the presence of noise, physical constraints, time-varying elements and non-linearities, that most of the time are not completely taken into account by simulation models. However, it is well known that academic laboratory experiments are in general expensive and need a continuous maintenance effort. Furthermore, the execution time of the experiment is sometimes prohibitive with respect to the usual classroom time slots. Web Remote labs, where students can interact with real experiments through Internet, appears then as a possible alternative solution to these problems. In this case, expensive laboratory facilities can be made available for a large amount of students, with temporal and location flexibility. Following this new paradigm, in the last few years, many institutions in the world have been invested in the development of Web accessible remote experiments using different approaches and tools. Among many works in this context, we can cite, for instance: (Shen et al., 1999),(Vargas et al, 2000),(Martins Ferreira et al, 2002) regarding the test and characterization of electronic devices and circuits; (Guimarães et al., 2003) where the remote access to a mobile robots infrastructure is provided.

Particularly considering web remote control systems experiments, we can cite (see also references therein): (Johanson, et al 1999), (Batur et al., 2000), (Ramakrishna, et al., 2000), (Ko et al., 2001), (Swamy et al. 2002), (Casini et al. 2003), and (Srinivasagupta and Joseph, 2003).

It can be noticed that most of the control systems available remote labs cited above utilize customized devices and software to make small scale textbook-like experiments remotely available. On the other hand, considering the education on control and automation systems, a key issue is the reduction of the gap between the classical theoretical courses and the real industrial practice. Hence, it is important to allow students to operate with devices, systems and techniques as close as possible to those they will be confronted in advanced industrial settings. Unfortunately, to reproduce in an academic environment a real industrial plant is not an easy task. Industrial equipments are in general expensive (both in terms of acquisition and also in maintenance costs) and usually require a large area. Furthermore, safety constraints should also be taken into account.

All the above mentioned factors restrict the use of real industrial devices in academic laboratories, which in general end up being structured as small scale experiments with little connection to industrial reality. Within this context, making an industrial lab facility available via Web and therefore being accessible to a large number of individuals helps to improve the overall cost-effectiveness of such solution.

Following these ideas, in (Zeilmann, *et al.*, 2003) an infra-structure for web-remote experiments on a Foundation Fieldbus pilot plant was presented. Contrary to most of the above mentioned systems, the implemented pilot plant uses industrial scale devices, which makes it ideal for research and teaching – both distance and in situ - of control systems and industrial automation. The main characteristics of the developed architecture are:

- Independence of devices and industrial protocols, which are running in the pilot plant;
- Clear separation between the Internet communication architecture and the industrial communication protocol in order to avoid that the former interfere in the timing behavior of the second.
- Platform independence on the client side;
- Minimization of requirements on the client side, which only needs a standard Internet browser with Java support to run the experiments;

Considering the same framework and the architecture developed in (Zeilmann, *et al.*, 2003) , the objective of this paper is to describe the implementation of a multivariable control experiment, remotely accessible via internet. With this aim, in Section 2 the physical experiment is presented. In Section 3, the software architecture for the web-based remote access is briefly described. In particular, details concerning the flow of data from experiment to the client and vice-versa are provided. The didactic features and the teaching possibilities of the implemented experiment are discussed in Section 4. Some concluding remarks are given in Section 5.

2. THE PHYSICAL EXPERIMENT

The physical implemented experiment is depicted in Figure 1 and 2. It consists basically of two coupled-tanks. Tank 2 receives a input flow that depends on the level in tank 1 and the opening of the valve 2. On the other hand, part of the output flow of tank 2 is feed to tank 1 by means of an AC pump. The input flow in tank 1 depends also on the opening of the valve 1. A reservoir receives the output flow of tank 2 and it feeds valves 1 and 2 by means of an AC pump. This generates a closed circuit of liquid, which is particularly interesting in a laboratory structure, since it avoids the need for external sources of water.

Figure 1. Pilot Plant

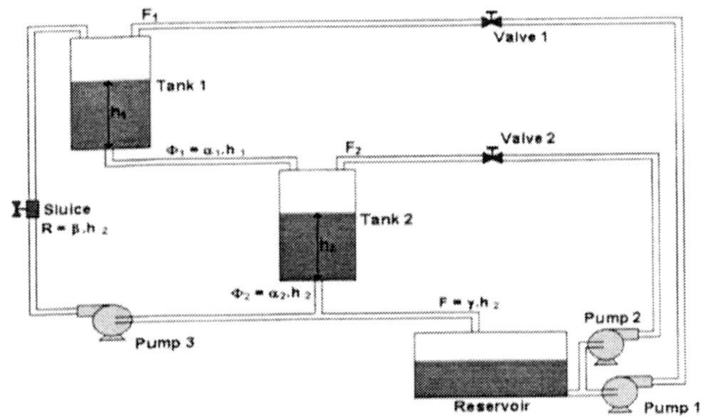

Figure 2. Schematic view of control experiment

The idea is to control the level of both tanks by acting on the proportional valves V1 and V2. In this case, considering a given operating point, the dynamic behavior of the system can be approximated by the following linear multivariable structure:

$$\begin{bmatrix} h_1(s) \\ h_2(s) \end{bmatrix} = \begin{bmatrix} G_{11}(s) & G_{12}(s) \\ G_{21}(s) & G_{22}(s) \end{bmatrix} \begin{bmatrix} v_1(s) \\ v_2(s) \end{bmatrix}$$

where:

- v_1 and v_2 corresponds respectively the opening of valves 1 and 2. They are the manipulated (control) variables.
- h_1 and h_2 are respectively the level in tanks 1 and 2. They are the controlled variables.
- $G_{ij}(s)$ denotes the transfer function from v_i to h_j.

The implemented control strategy is a multi-loop control where the v_1 will be used to control the level in tank 1 and v_2 will be used to control the level in tank 2, as two mono-variable independent loops. In order to minimize the coupling effects, static decouplers can be used (Ogunnaike and Ray, 1994). A block schema of the control structure is shown in Figure 3. $C_1(s)$ and $C_2(s)$ are respectively the controllers of loops 1 and two. K_1 and K_2 are the decoupling gains. Both the sensor levels and the proportional valves are smart devices, which communicate through the Foundation Fieldbus protocol. All the instruments are then connected in a Foundation Fieldbus network control and the control strategy implementation is based on function blocks. In this case we use basically:

- Analog input blocks to perform the unit conversions and sensor signals linearization and make them available for the controllers.
- Analog output blocks to make the controllers outputs available to the proportional valves.
- PID blocks in order to implement the controllers as proportional-integral-derivative controllers.
- Arithmetic blocks in order to implement the decoupling gains.

The system configuration is made by connecting these blocks and setting their parameters. A major feature of these systems is that the function blocks are device independent, that is, a given function can be assigned to run in different physical devices.

Although the controller parameters and the decoupling gains can be modified remotely on-line, as will be seen in the next sections, at this point it is important to emphasize that the control is performed locally, in a distributed way, on the fieldbus network.

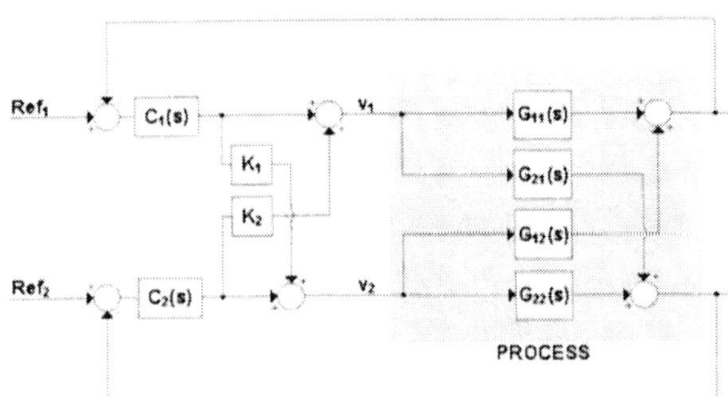

Figure 3. Control Structure

3. SOFTWARE ARCHITECTURE

The implemented software architecture, described in details in (Zeilmann, *et al.*, 2003), is basically divided in three contexts. The experiment context concerns all the software related to the Foundation Fieldbus network and interfaces to the physical experiment. Control algorithms are directly executed on field devices and may be parameterized remotely. This ensures a correct timing behavior due to unavoidable Internet communication delay.. The server context provides synchronization mechanisms for read/write operations of control and process variables, which are stored into a database available both to the Web-server as well as to the Fieldbus-based tools over an OPC interface. Additionally, the server context includes a supervisory system, which manages dynamic Web-pages that will be sent to the remote user as well as live video and audio streams from the physical plant (generated by CCD cameras). The server context also is responsible for session control. Finally, the client context encompasses all software modules on client site. Basically, only a Java-enabled standard browser is necessary and will act as a client to access the supervisory system (the client) . The browser loads the Web page created by the supervisory using the HTTP protocol. The connection between the supervisory system running on the server PC with the Internet browser running at client side (in a student's PC) is established by a Java applet embedded into the html page. By filling in the text fields or navigating the available pages, remote users commands are sent to and processed locally (at server side) by the web-server.

4. REMOTE EXPERIMENTATION

The first step to be followed in order to run an experiment is the user's validation. With this aim, the user should request for a login and a password by filling in an appropriate Web-form. This login and password will be asked and verified by the system at each time the user wants to access the system. Once the login and the user's password are validated, he/she can have access to a scheduling system or, if an experiment has been previously scheduled, the experimental setup can be directly accessed. The scheduling system consists in a Web page where a calendar is available. The user can identify available time slots in the calendar, representing available times for running experiments and select one of them. The idea of this reservation system is to avoid simultaneous access to the experiments (although some experiments can be executed concurrently, most of them requires exclusive access). If a logged-in user tries to access the experiments Web pages at a non scheduled time, access is denied and the scheduling page is automatically loaded.

The time of each session (usually 30 minutes) is monitored by the supervisory system. When the scheduled time is finished, the system automatically logs out the user. A system initialization is then carried out in order to make the experiment available for the next scheduled user. The user can visualize the current time of the experiment by means of a chronometer placed on the right foot of the supervisory Web-page

Figure 5 presents a typical Web page that is displayed in the client browser. The screen is divided into two frames. In the left-side frame users can change control parameters (such as PID and decoupling gains). In this case, an ASP page refreshes the database and the supervisory reads the new values and sends them to the Fieldbus network via the OPC interface. In the main frame, a graphical visualization of the experiment being executed is depicted. It includes (i) real-time graphic visualization of the process (levels) and the control variables (valves opening); (ii) on-line image of the tanks; (iii) a graphical animation of the process; (iv) the current value of the process variables; (v) the current time of the experiment.

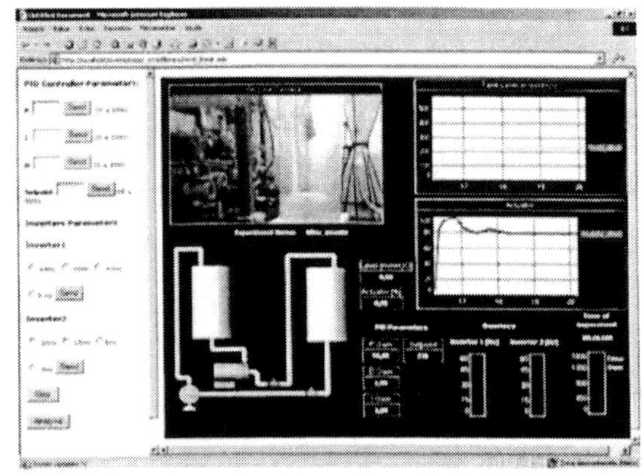

Figure 4. Typical web page

5. TEACHING POSSIBILITIES

The proposed remote experiment can be viewed as an useful support for teaching multivariable control systems. In particular, following topics are handled:

- The modeling of multivariable systems by means of matrix transfer functions.
- The experimental determination of the matrix transfer function .
- The interaction between loops in multivariable systems.
- The multi-loop strategy for controlling multivariable systems.
- The concept of decouplers.
- The design of multivariable PID controllers.
- The design of static decouplers in order to minimize the interaction between loops.

The control system can be operated in manual (open-loop) or automatic mode (closed-loop). The operation mode can be chosen by the user. According to this choice, different screens are loaded in the client browser.

The open-loop screen allows users to set and change (on-line) the opening value of the valves and the speed of the AC pumps. In this screen it is also depicted the behavior of the process variables (levels). In particular, this open-loop mode is useful to determine the different transfer functions by means of the step response method. For this, it suffices to vary the valve opening, while keeping constant the value of the other one, and observe the response of the tank levels.

The closed-loop screen allows users to set and change (on-line) the PID parameters and the decoupling gains. In this screen it is also depicted the behavior of the process and control variables. This mode is useful to evaluate different methods for tuning PID controllers. Also the influence of static decouplers in the behavior of the multivariable control system can be clearly observed.

In both modes, text files containing user experiment data can be generated. The data contained in these files can be used, a posteriori, for more detailed analysis in mathematical dedicated software (such as Matlab, Scilab, etc.).

6. RELATED WORKS

An interesting related web-based level experiment control of a couple tank apparatus is described in (Ko, *et al.*, 2001). The main server runs LabView HTTP server and LabView "G" based programs, which are responsible for the TCP communication with the client, the control implementation, and the data exchange with the plant. Three different control structures can be used: PID, state space and fuzzy. An auxiliary server running MS NetMeeting handles an audio-visual feedback. The data transfer between the client and the server is made by Java applets and ActiveX. Due to the large amount of audio/video data transfer, the experiment is being used only in the internal network of the university (National University of Singapore). NetMeeting is also used in (Swamy et al., 2002). The control algorithm is described by a Matlab-Simulink block diagram, which allows the implementation of any control strategy. Two commercial ECP (Education Product Plants) plants are available: inverted pendulum and torsion disks. No dedicated GUI has been developed. All the commands are executed from standard Matlab windows. This approach does not require a browser, but is restricted to Windows platform, where NetMeeting can be executed. Other examples of remote labs based on Matlab/Simulink are found in (Srinivsgupta and Joseph, 2002), (Casini, et al.., 2003). In (Srinivsgupta and Joseph, 2002), two control experiments are available: a heat exchanger an a mixing tank. The control strategy is also defined by a Simulink block diagram. A block containing as inputs the control signals, and as outputs the measured plant variables, is available. In this case the control algorithm runs in the client machine As pointed in (Overstreet and Tzes, 1999), this can be a problem, since is not possible to ensure the real-time delivery of the process and control variables through an Internet communication. In this paper, the user can elect whether the client or the server will implement and execute the controller. Similarly to (Ko, *et al.*, 2001), LabView is used as front end for data acquisition. The implementation of the control law is not direct. Students should first conceive the control law by using Simulink, design the digital controller, implement it as a C program, and finally create a DLL and execute the code. Images of the controlled system are captured by a National Instruments image acquisition board. In (Casini, et al.., 2003), the user can either select pre-defined control structures (where just parameters and set-points can be set) or create their own Simulink-based controllers. In the last case, the controller is implemented by merging the controller .mdl model with the process Simulink .mdl model. The obtained file is then compiled by the Matlab Real-Time Workshop (RTW) generating an executable file that will exchange data with the user interface. A image system is also incorporated by means of Java applets and specific plug-ins. Four laboratory-scale plants are provided : level control, position control, magnetic levitation and helicopter.

Comparing the proposed work with the above described ones, many common points can be found in the software and hardware architectures . For instance, the role of LabView front end can be compared with the role of the Elipse SCADA System and the use of Java applets, activeX, etc. On the other hand, differently of the previous approaches, in our case the controller runs neither in the client nor in the server, but in the Foundation Fieldbus devices, what impacts positively the control-loops' real-time performance Available control structures are similar PID, lead-lag, gain, and users are limited to set and change, in real time, controller parameters and set-points. Although the use of Matlab-Simulink allows more flexibility in control parameterization, students still have the same "academic" look-and-feel. In our case, the front end screens try to emulate real industrial control room screens. Furthermore, the used devices in our experiment are industrial ones.

7. CONCLUSIONS AND FUTURE WORK

The system is available at the site: http://automation.eletro.ufrgs.br/Plantaweb.htm. The main contribution of this work regarding previously described internet-based experimental facilities is the use of industrial scale equipment and a standardized industrial network protocol. The advantages of such an experimental setting for research and teaching are quite significant and manifold (experiment flexibility, enhanced student motivation, etc). The required bandwidth for having a good image transmission is not always available and sometimes it is difficult to follow the experiment visually. A solution to this problem has been proposed in (Ramakrishnan, *et al., 2000l*) , in which good image transfer performance is achieved by separating the image transfer from data transfer. However, it requires extra hardware and software running in the user's machine.

Due to safety constraints, the presence of a person at the laboratory may still be required in order to handle unexpected situations. This may become a barrier to a 24/7 use of the remote lab.

The presented remote lab capabilities will be expanded and integrated within the context of the Rex-NET (**R**emote **ex**perimentation **Net**work) project, whose goal is to share and spread current competences on Remote Experimentation (and on other forms of e-learning). Project partners are universities from Germany, Portugal, Mexico, and Brazil, By taking advantage of the time difference between participating countries, a broad network of tutors' assistance to remote experimentation sessions will be available.

8 ACKNOWLDEGMENTS

This work has been partly supported by the Brazilian research agencies FINEP, CNPq, and FAPERGS.

REFERENCES

Batur, C., Q. Ma, K. Larson and N. Kettenbauer, (2000). Remote tuning of a PID position controller via internet, *Proceedings of the American Control Conference 2000*, 4403 - 4406, Chicago, USA.

Vargas, H,, Machado, R. Carro, L. (2000) Virtual-Real Lab: An Electronics laboratory using real devices and the Internet In:.*International Conference on M/SET 2000*, Mathematics, Science and Education and Technology. Charlottesville, VA: AACE:

Casini, M., D. Prattichizzo, and A. Vicino, (2003). The Automatic Control Telelab: A User-Friendly Interface for Distance Learning, *IEEE Trans. on Education*, Vol. 46, No. 2, May 2003, 252 – 257.

J. Martins Ferreira, Ricardo J. Costa, Gustavo R. Alves, Martyn Cooper, (2002) "The PEARL Digital Electronics Lab: Full Access to the Workbench via the Web," *13th Annual Conference on Innovations in Education (EAEEIE'02)*, York, England, April 2002

Guimarães, E et al., (2003). REAL: A Virtual Laboratory for Mobile Robot Experiments, *IEEE Trans. on Education*, Vol. 46, No. 1, February 2003, 37 – 42, USA.

Johansson, K., A. Horch and O. Wijk, (1999). Teaching multivariable control using the quadruple tank process, *38th IEEE Conf. on Decision and Control*, 807 – 812, Phoenix USA.

Ko, C. C., B. M. Chen, J. Chen, Y. Zhuang and K. Chen Tan, (2001). Development of a web-based laboratory for control experiments on a coupled tank apparatus, *IEEE Trans. on Education*, Vol. 44, No. 1, February 2001, 76 – 86, USA.

Ogunnaike, B. and W. Ray (1994). Process dynamics, modeling and control. Oxford University Press.

Overstreet, S. J. and A. Tzes, (1999). An Internet-Based Real-Time Control Engineering Laboratory, *IEEE Control Systems Magazine*, October 1999, 19 – 34, USA.

Poindexter, S. and B. Heck, (1999). Using the Web in your Courses: What Can You Do? What Should You Do?, *IEEE Control Systems Magazine*, February 1999, 83 – 92, USA.

Ramakrishnan, V., Y. Zhuang, S.Y. Hu, J.P Chen, K.C Tan, (2000). Development of a web-based control experiment for a coupled tank apparatus, *Proceedings of the American Control Conference 2000*, 4409 – 4413, Chicago, USA.

Robinette, M. F. and R. Manseur, (2001). ROBOT-DRAW, an Internet-based visualization tool for robotics education, *IEEE Trans. On Education* , Vol. 44, No. 1, February 2001, 29 – 34, USA.

Sánchez, J., F. Morilla, S. Dormido, J. Aranda and P. Ruipérez (2002). Virtual and Remote Control Labs Using Java: A Qualitative Approach, *IEEE Control System Mae*, April 2002, 8 – 20, USA.

Schmid, C. (1998). The virtual control lab VCLAB for education on the web, *Proceedings of the American Control Conference 1998*, 1319 - 1325, Philadelphia, USA.

Sebastián, J. M., D. García, and F. M. Sánchez, (2003). Remote-Access Education Based on Image Acquisition and Processing Through the Internet, *IEEE Transactions on Education*, Vol. 46, No. 1, February 2003, 142 – 148, USA.

Shen, H., Zheng Xu, B. Dalager, V. Kristiansen, Ø. Strom, M. S. Shur, T. A. Fjeldly, Jian-Qiang Lü and T. Ytterdal, (1999). Conducting Laboratory Experiments over the Internet, *IEEE Transactions on Education*, Vol. 42, No. 3, August 1999, 180 – 185, USA.

Srinivasagupta, D. and B. Joseph (2002). An Internet-Mediated Process Control Lab, *IEEE Control System Mag*, Feb 2003, 11 – 18, USA.

Swamy, N., O. Kuljaca, and F. L. Lewis, (2002). Internet-Based Educational Control System Lab Using NetMeeting, *IEEE Transactions on Education*, Vol. 45, No. 2, May 2002, 145 – 151, USA.

Zeilmann, R. P., J. M. Gomes da Silva Jr., A. S. Bazanella and C. E. Pereira (2003). Aweb-based remote laboratory for control education. *Proceedings of the 5th IFAC Symposium on Intelligent Components and Instruments for Control Applications (SICICA 2003)*, 51-56, Aveiro, Portugal

Copyright © IFAC Information Control Problems in
Manufacturing, Salvador, Brazil, 2004

ELSEVIER
IFAC
PUBLICATIONS
www.elsevier.com/locate/ifac

SCIENTIFIC COLLABORATION AND KNOWLEDGE SHARING IN THE VIRTUAL MANUFACTURING NETWORK

Olga Nabuco(1), João Maurício Rosário(2), José Reinaldo Silva(3), Khalil Drira(4)

(1) CenPRA – Campinas, São Paulo
(2)LAR-DPM-FEM – UNICAMP, Campinas, São Paulo
(3)EPUSP – São Paulo, São Paulo
(4) LAAS-CNRS – Toulouse, France

Abstract: Most of past research efforts on Collaborative Work Support adopt tool-centric approaches to define cooperation. These approaches are often related to a single category of communication (synchronous or asynchronous) and support only a single type of collaboration (planned or open). In this paper, we describe our view for successful collaboration in Virtual Manet. Our aim is to extend the classical asynchronous computer-supported scientific collaboration by introducing and integrating collaborative work support tools with knowledge management approaches. We are starting by effective requirements analysis, and we will continue by identifying the associated collaboration technologies, and finally defining the Cooperative e-Space integration architecture. The idea can be generalized to encompass the manufacturing business or the collaborative production in virtual enterprises. *Copyright © 2004 IFAC*

Keywords: computer aided work, knowledge engineering, knowledge acquisition.

1. INTRODUCTION

Scientific research activities can be considered a foundation of cooperative work. In a world-wide or country-wide context, scientific communities have been built through remote collaboration for conference planning and face-to-face collaboration during meetings. Joint research projects and research networks are enabling structures for scientific collaboration and constitute the interest of our work. We aim to enforce the existing evolution where communication and publishing technologies are used to increase the collaboration scope by widening and planning face-to-face meetings.

We aim to provide a cooperative electronic space based on software framework for scientific collaboration; integrating different technologies, from synchronous to asynchronous cooperation, coordination and knowledge sharing, providing services as interests maps, electronic scheduling and meetings or topic awareness, for example.

This framework is being elaborated to support the electronic organization of the scientific collaboration within the MANET (Manufacturing Automation Network) research Network. The research environment has problems as a virtual enterprise environment. Consequently it constitutes a good test bed for the proposed collaborative environment.

The paper is structured as follows: Section 2 describes the teams and the Manet organization; Section 3 discusses information sharing requirements that possibly lead to knowledge sharing. Section 4 presents the technology taxonomies that support different categories of collaboration. Section 5 describes the framework components and the general layered integration architecture. Finally, in section 6 the current status.

2. ACTORS OF THE MANET NETWORK

Manufacturing Automation Network - MANET - is a non-profit, multi-institutional and distributed network of researchers, with representatives from many of Brazil's biggest universities and research centers. MANET is one of the networks participating in the Network of Cooperation and Research Program -RECOPE- created by FINEP, an agency linked to the Brazilian Ministry of Science and

Technology. It was created with the aim of making the liaisons of cooperation stronger among researchers. Manet is responsible for aggregating professionals working within manufacturing automation and correlated themes. It involves mainly the participation of electrical, mechanical and computer engineers distributed throughout the Brazilian territory.

Fig. 1. Virtual MANET geographical distribution

Some of Manet's participants share the same research topics, which allows them to be partners and competitors at the same time, just as companies in the same branch. Other members work as complementary labs, such as some companies with their suppliers.

Similarly to a virtual enterprise, where companies are focused on few competence areas, Manet's participants often join their key technical knowledge and assets to fulfil the process of a product creation.

Therefore, Manet can be seen as a metaphor for a distributed manufacturing Research and Development supply chain.

Virtual Manet Project is intended to be a test bed for cooperative tools and techniques. It includes, in the initial phase, four cities, distributed from the Northeast to the South in the Brazilian territory, as shown in Fig. 1. In terms of cooperation and communication, it is similar to a virtual enterprise in the European Community or in the United States.

3. ENABLING DATA, INFORMATION AND KNOWLEDGE SHARING

The idea of associating people in networks is beneficial for both the individual and the community. The benefits can be seen, for instance, when the expertise, the vision, and the skills of different members of the group can be shared by the community.
The research characterizing the relationship in social networks focusing sharing knowledge (Cross et al., 2001) reinforced the hypothesis that critical information received by managers is provided or pointed out by people more than other sources. In other words, the first source of consulting was other people inside a personal network where someone knows, or is acquainted with someone who knows.

Received by the right person, someone who understands its meaning and possibilities and in the adequate delta of time, information can be turned into knowledge. In this case information sharing can lead to knowledge sharing.

Knowledge sharing is a process that embeds the notion of "willingness to share" (Jarvenpaa and Staples, 2000). In this sense, the relation qualities that promote effective knowledge sharing (Cross et al., 2001) are:

Knowledge - knowing what another person knows, even if it is initially inaccurate; other people can only be identified to us if some awareness of their expertise exists. For a group of people to effectively share knowledge they need to know and be aware to each other expertise. Electronic communities can provide access to a group of peers dealing with similar knowledge issues. Through this dynamic process of interchange, knowledge is constantly being regenerated and recontextualized, thus maintaining its relevance to the community (Wasco, 2000).

Access: being able to gain timely access to them. Facts that can influence are closeness of one's relationship, physical proximity, organizational design and collaborative technology. A balance between face-to-face interactions and networked interactions contributes to increase creativity. People that already had face-to-face meetings are also more comfortable in using technology. Advanced communication technology seems most effective when they are used to augment rather than supplant interpersonal contact and experience of working together (Lenders, 2003). Creativity-driven communication requires technology that encourages more spontaneous interactions such as face-to-face meetings, desktop and video conferencing.

Engagement: willingness that a person has in engaging this process. Looking from this perspective the flow coming from knowledge embedded in people through knowledge embedded in a community, can be seen. The knowledge embedded in community perspective views knowledge as collectively owned and maintained by the community (Wasko, 2000). From this perspective, the motivation for knowledge exchange is not self-interest, but care for the community (Wasco, 2000).

Safety: if one feels confident enough to expose one's lack of knowledge to others, the safety quality can promote learning and creativity. Knowledge exchange via open communities reduces the risk of someone having knowledge but being excluded from the exchange, and also reduces the potential information overload on the organization's experts by distributing the obligation to maintain the community to all members.

4. SUPPORTING COORDINATED COMMUNICATION AND COOPERATION FOR KNOWLEDGE SHARING

Manet's partners act in several geographically dispersed universities and research centers. The objective is to offer to people that already have communities of practice, who are already social networked, other environments for sharing information, which, through the special characteristics of this community, can lead to the sharing of knowledge.
An environment that could support Manet' partner activities has to meet several requirements. These requirements were conceived according to a framework structured considering three layers: Cooperation, Coordination and Communication.

Cooperation is the work in process, bringing the specialties of different participants which complement each other, to the network.

Coordination identifies producer and consumer of the information. It has to be medium centralized and used as facilitator, in the sense that it can introduce new members that can be related or identify possibilities of relationship.

Communication means of providing a transparent and well-specified channel where this information flows. The communication enlarges the possibilities of interaction among the participants.

Two groups of supporting tools were envisioned: groupware tools and ontology based tools. The movement is from groupware to knowledge ware in techniques where cooperation, coordination, and communication can always be sensed.
The next two items will bring a small survey on those topics that helped to specify the kind of component tools that fit the requirements.

4.1. Classification of collaboration systems

Groupware is a collaborative technology, which allows people to communicate with each other, co-operate in projects and share information and knowledge.

Different taxonomies classify collaborative systems. Taxonomies that involve time-space: synchronous or asynchronous, same or different space. Task taxonomy such as planning, creativity, intellective, preference, cognitive, conflict, and mixed motive.

Taxonomy according to underlying technology: Keepers – functionality related to storage and access shared data; Coordinators – ordering and synchronization of individual activities that make up the whole process; Communicators – functionality related to unconstrained and explicit communication among the participants; Team-agents – intelligent or semi-intelligent software components.

Taxonomy according to the coordination process (Malone and Crowston, 1994): Managing shared resources (tasks assignment and priorization); managing producer-consumer activity relationship (sequencing pre-requisites); managing simultaneity constraints (synchronizing).

4.2. Categories of collaboration tools

Collaborative systems provide the promise of much increased information sharing within and across organizations. Such systems encourage sharing of ideas in a free-flowing manner as well as in a form of structured repositories. Those systems are used to exchange both knowledge and information (Jarvenpaa and Staples, 2000).

Group file and document handling – from shared view of files/documents to individual editing, document/file management and storing in a central database. Collective authoring and revision of documents; synchronous work on documents. Basic communication capabilities such as e-mail notifications.

Computer conferencing - Space for asynchronous and threaded discussions; real-time text talk and discussions; see and work on documents simultaneously on each other's screen or on a whiteboard; mailing capabilities; audio and video conferencing arc quite common.

Electronic meeting systems - Meeting conduction which can be either regular (same time, same place); synchronous (same time, different place) or asynchronous (different time, different place); meetings participants are notified through e-mail, possibility to chat, conduct real-time discussions, use audio and video conferencing facilities; write or draw in real time; participate in survey-anonymously make group decisions; share documents and files, show and take note of Power Point slides, share live software applications and even work simultaneously on documents and support meeting processes including set-up, maintenance of the agendas and minutes after the meeting.

Electronic Workspace - Provides a common space to coordinate and organize their work. Centered storing for documents and files, capability to work with them, solve problems through discussions, keep to-do lists and address books with info about group contacts, even track project milestones and project interaction. There are different workspaces for different groups and users can be members of several workspaces with each workspace corresponding to each project a user is involved with.

e-Learning tools - these systems use the internet to deliver information, training, educational materials and tools for managing interactions between mentors and students.

4.3. The common language - the role of ontology

Davenport and Prusak stated "People can't share knowledge if they don't speak a common language". Considered a critical part of the development of the semantic web, the ontology approach is used to achieve the presentation, integration of information and to allow software agents to capture information. The role of ontologies is to capture the domain knowledge in a generic way and to provide a commonly agreed understanding of domain, which may be reused and shared within communities or applications.

Four significant methodologies to define ontology design criteria are briefly described below.

Gruber (1993) states that the following characteristics are essential to the construction of ontology:
Clarity - definitions have to be objective, independent of social or computational context. The idea is to solve ambiguities or, in other words, avoid misunderstandings. Definitions have to be stated preferably using logical axioms.
Coherence - "it should sanction inferences that are consistent with definitions". Even defining axioms and informally defined examples should not contradict one another.
Extendibility - an upper ontology should be present in order to anticipate the creation of new terms.
Minimal encoding bias - implementation-dependent codification should be avoided.
Minimal ontological commitment – it means to avoid deep definitions of the intended described domain. The objective is to define only terms that are essential.

Cyc (Reed, 2002) knowledge base was constructed using a three-phase method: first, a manual extraction of pieces of knowledge with common sense of knowledge, and the micro-theories. Followed by common sense knowledge acquisition performed by both specialists and tools. The third phase is mainly performed by tools.

Gruninger and Fox's methodology (Corcho, 2003) proposes the identification of possible scenarios where the ontologies could be used followed by a set of questions, competence questions, to determine the scope of the ontology. The main concepts, the relations and the axioms are then constructed, using first-order logic.

The methodology used by On-To-Knowledge (Schnurr, 2000) includes a first step where the requirements are specified and competency questions are identified. They also investigate the possible use of other ontologies and a first draft is produced. This process is followed by a refinement step, an evaluation process where the requirements are checked against the draft. The ontology is then tested in the application environment and maintenance is started.

5. COMPONENTS AND LAYERS OF THE COOPERATIVE E-SPACE ARCHITECTURE

Virtual Manufacturing Automation Network Cooperative e-Space is designed upon requirements for the use of open software platforms; reusability of applications, and is also intended to be easily repeated by other communities. It can be extended, customized and merged with other components. It is using software components that partners made available and continues in this purpose.

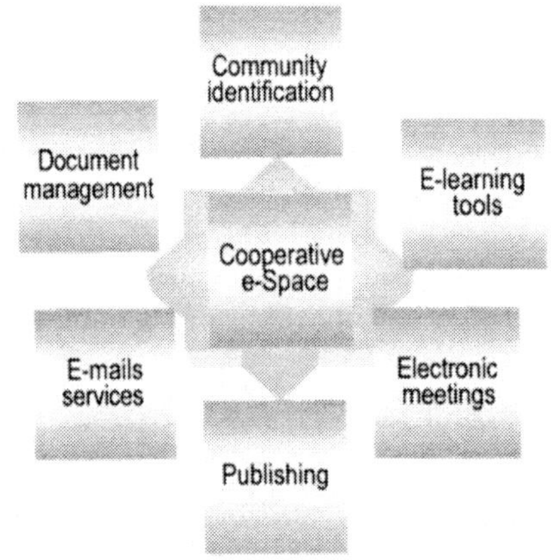

Fig. 2. Components of the cooperative e-Space

The objective is to offer cooperative work support tools, extending new services to the current Manet platform, which make virtual contact, new forums for cultural exchange, possible.

The methodology for constructing the ontology was inspired on the principles presented before but also it aims at providing a dynamic process where its participants can contribute and new concepts can be added.

The three layers identified to support knowledge sharing provide the requirements that have to be satisfied during the construction of the cooperative e-Space

Cooperation – identifies different groups according to ontologies built based on participants' profile. The acquisition knowledge phase of participants' interest is also part of the cooperative process.

Coordination – Tools that perform management conduct this process. The classification process according to different taxonomies and the construction of the relationship among participants and taxonomies is also part of the process. It will be conducted by both human and agent-based tools.

Communication – the representation model represented by RDF (Resource Description Framework, XML (eXtended Markup Language), HTML (Hyper Text Markup Languages), and tools such as VNC (Virtual Network Computing), web servers such as SMTP (Simple Mail Transfer Protocol) server are responsible for making the connection between the participants providing a transparent channel in different time-space situations.

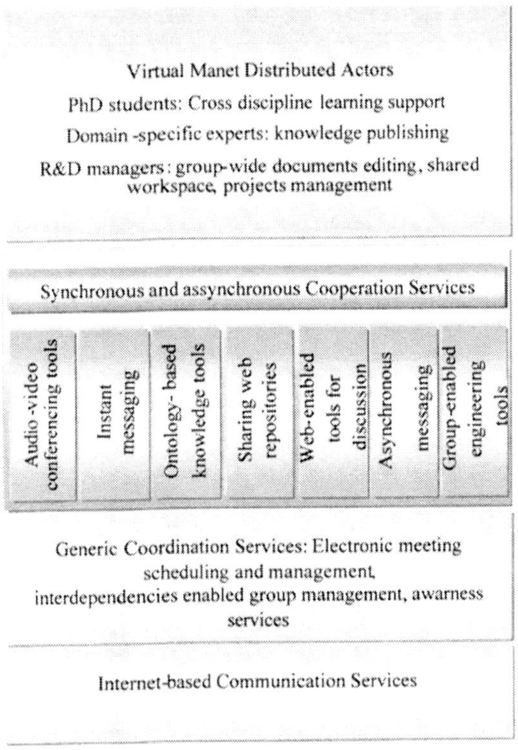

Fig. 3. Virtual Manet's Cooperative e-Space

5.1 Cooperative e-Space approach

The cooperative e-Space shown in Fig 2 is composed of groupware and ontology based tools. Virtual meeting rooms, where members can connect themselves to scheduled meetings, publishing services, where members can have a site for their documents, community identification, where member specialties are classified according to domain of knowledge, etc. Other components will be added to the e-Space facility to increase the possibilities of interaction and relationship. The main endeavor involves the possibility of sharing knowledge electronically.

Fig. 3 shows the components of the e-Space architecture. Components that provide synchronous and asynchronous cooperation services in terms of groupware tools. The components that enclose the knowledge sharing approach are described below using the characteristics pointed by the AKT Model (Shadbolt, 2003).

Knowledge acquisition in such way that it could be turned into a multi-media KA. It will start with keyword recognition through texts (documents as .doc, .pdf, .ps, etc.), using manufacturing domain ontologies to find people who have some form of relationship with each other . The use of distributed agents combined with the ontology domain deals with delivering information to people who are supposed to need it.

Knowledge modeling approach uses ontologies for expressing and structuring conceptual schemes for domains. Other modeling techniques will certainly be developed and will join the library so as to be available.

Knowledge reusability is intended to be available through the ontologies, reusing, merging and extending it. Also the knowledge base already constructed using the ontology can be customized for other users.

The ***knowledge retrieval*** is carried out by identifying the possible pairs in this community. Identifying which participants profile matches each other's keywords that were acquired in a knowledge acquisition process Ontologies that relate profile, keywords and domains are used. Also software agents automates this process freeing the user from the pain of trawling web pages and documents.

Knowledge publishing is related to offering tools that enable sharing documents and co-authoring documents. A possible extension is the integration of the ontology in order to include links to authored documents, projects and markups made on documents.

Knowledge maintenance is related to improving the quality of the information provided by the tools, for instance, people who slightly change their profile, the nature of documents they produce and relate it to new keywords, people, topics and so on. Versioning is also envisioned.

6. CURRENT STATUS

In this paper, we presented our preliminary study aiming at identifying the basic requirements and approaches for information sharing and defining the enabling technologies for communication and collaboration to support the information and knowledge sharing. We claim that this is the key point in any collaborative approach, including in agreements that are made in distributive production. However, to really go into the subject, where experience can be listed, it is necessary to have good tested and collaborative research work, which is one of the best examples, while also being a good contribution in itself.

We presented a layered architecture that integrated different components and services from communication, coordination and cooperation levels. The next step is to integrate and to deploy this architecture. It is planned for the next 6 months and will start by collaboration among CenPRA, University of Campinas and University of Sao Paulo.

REFERENCES

Bafoutsou, Georgia and Mentzas, Gregoris. 2002. Review and functional classification of collaborative systems. *International Journal of Information Management.* **Vol. 22.** Pp. 281-305.

Corcho, Oscar, Fernández-López, Mariano and Gómez-Pérez Asunción, 2003. Methodologies, tools and languages for building ontologies. Where is their meeting point? *Data & Knowledge Engineering*. **Vol. 46**. Pp 41-64.

Cross, Rob, Parker, Andrew, Prusak, Laurence, Borgatti P., Stephen. 2001. Supporting Knowledge Creation and Sharing in Social Networks. *Organizational Dyanmics*. **Vol. 30**. N. 2. Pp 100-120.

Davenport. T. H., Prusak, Laurence. 1998. Working Knowledge. Harvard Business School Press, Boston.

Gennari, J.H., Musen, M.A., Fergerson, R. et al. 2002. The Evolution of Protégé: An Environment for Knowledge-Based Systems Development. *International Journal of Human-Computer Studies*, in press, http://faculty.washington.edu/gennari/#pubs

Gruber, Thomas R., 1993. Toward principles for the design of ontologies used for knowledge sharing. International workshop on formal ontology. March 1993. Padova, Italy.

Gunnlaugsdottir, Johanna. 2003. Seek and you will find, share and you will benefit: organizing knowledge using groupware systems. *International Journal of Information Management*. **Vol. 23**. Pp363-380..

Jarvenpaa, S. L. and Staples, D. S.. 2000. The use of collaborative electronic media for information sharing and explanatory study of determinants. *Journal of Strategic information Systems*. **Vol. 9**. Pp 129-154.

Keller, Richard. 1999 Knowledge Management for Distributed Scientific Project Teams. http://ameslib.arc.nasa.gov/randt/1999/science/space6.html

Leenders, Roger Th. A. J., Engelen Jo and Kratzer Jan. 2003. Virtuality, communication, and new product team creativity: a social network perspective. *Journal of engineering technology and Management JeT-M*. **Vol. 20.** Pp 69-92.

Malone,T.. and Crowston, K. 1994. The interdisciplinary study of coordination.. *ACM Computing Surveys* **Vol.26**. N1.Pp. 87 –119.

Reed, Stephen and Lenat Douglas. 2002.Mapping ontologies into Cyc. http://www.cyc.com/doc/white_papers/mapping-ontologies-into-cyc_v31.pdf

Schnurr, Hans-Peter et al. 2000. On-To-Knowledge Methodology - Baseline Version. http://www.cyc.com/doc/white_papers/mapping-ontologies-into-cyc_v31.pdf

Shadbolt, Nigel and O'Hara Kieron. 2003. AKTuality: an overview of the aims and assumptions of the advance knowledge technologies interdisplinary research collaboration. www.aktors.org. ISBN 0854 327932.

Wasko, M. McLure and Faraj, S., 2000. "It is what one does": why people participate and help others in electronic communities of practice. *Journal of Strategic Information Systems*. **Vol. 9**. Pp155-173.

Copyright © IFAC Information Control Problems in Manufacturing, Salvador, Brazil, 2004

ELSEVIER
IFAC
PUBLICATIONS
www.elsevier.com/locate/ifac

EUROPEAN APPROACH ON INTEROPERABILITY OF ENTERPRISE APPLICATIONS – THEMATIC NETWORK, INTEGRATED PROJECT AND NETWORK OF EXCELLENCE

David Chen[1], Richard Stevens[2], Karsten Schulz[3] and Guy Doumeingts[4]

[1]LAP/GRAI, University Bordeaux 1
351 Cours de la libération, 33405 Talence Cedex, France
Email: chen@lap.u-bordeaux1.fr

[2]Gruppo Formula SPA
Corso IV November 57, 10090 Cascine Vica – Rivoli, Italy
Email: richard.stevens@formula.it

[3]SAP Corporate Research Center Brisbane
SAP Australia Pty Ltd
Email: ka.schulz@sap.com

[4]GRAISOFT
16 cours du Général de Gaulle, 33170 Gradignan, France
Email : gdoumeingts@graisoft.com

Abstract: This paper aims at presenting an overview of a European approach to develop interoperability of enterprise applications and software. The originality of the approach is to address the interoperability problems in an integrated way by bringing together the expertise/knowledge of the three relevant domains: Enterprise Modelling, Architecture & Platform, and Ontology. The paper reviews user requirement, vision and state-of-the-art as well as the gap analysis performed by IDEAS Thematic Network and the resulting roadmaps to develop interoperability. Then the ATHENA Integrated Project (IP) will be outlined and its R&D projects detailed. The INTEROP Network of Excellence (NoE) will be described in the third part of the paper. The complementarities with ATHENA will be discussed and the NoE activities addressed. *Copyright © 2004 IFAC*

Keywords: Interoperability, Enterprise integration, Enterprise engineering, Enterprise application

1. INTRODUCTION

One of the trends in the global market is the increasing collaboration among enterprises during the entire product life cycle. In this context, enterprise systems and applications need to be interoperable in order to achieve seamless business interaction across organisational boundaries and thus realise networked organisations. The European Commission has considered the development of interoperability of enterprise applications and software as a strategic issue for European companies to strengthen their cooperation and gain competitiveness in the market. As a consequence, a thematic network IDEAS (Interoperability Development of Enterprise Applications and Software) was launched. The objective was to elaborate a roadmap to develop interoperability (IST, 2002). Two other main initiatives under the EU Framework Program 6 (FP6) were approved for the period of 2004-2007: ATHENA Integrated Project (IP) and INTEROP Network of Excellence (NoE). ATHENA (Advanced Technologies for Interoperability of Heterogeneous Enterprise Networks and their Applications) consists of a set of projects and will lead to prototypes, technical specifications, guidelines and best practices that form a common European repository of knowledge (Athena, 2003). INTEROP (Interoperability Research for Networked Enterprises Applications and Software) aims at integrating expertise in relevant domains for sustainable structuration of European Research on Interoperability of Enterprise applications (Interop, 2003).

2. IDEAS THEMATIC NETWORK TO ELABORATE ROADMAPS

2.1 Scope and method of the work

In IDEAS, the three main research domains that address interoperability issues were identified as: (1) Enterprise modelling (EM) at the *Business* and *Knowledge* levels dealing with the representation of the inter-networked organisation to establish interoperability requirements; (2) Architecture and platform (A&P) at *Information and Communication Technology (ICT)* level defining the implementation solution to achieve interoperability; and (3) Ontologies (ON) addressing the semantics necessary to assure interoperability.

A state-of-the-art study was performed and user requirements collected. In parallel, visions and business scenarios for the next five-to-ten years were developed. Based on the statements of what are required and what exits to-day, a gap analysis was carried out and activities in research, technology, and standards proposed to reduce the gaps. The resulting roadmaps are a synthesis of a top down approach in the definition of the Vision and a bottom up approach in the State-of-the-Art and User requirements.

2.2 Vision and industrial challenges

As IDEAS intended to articulate the research priorities for interoperability over the ten years following the project, we deemed it important to start with a common understanding of what our ultimate goal would be. Our goal or "Vision" was a target that kept the common effort focused on a common understanding of the challenges ahead of us. We defined that vision with the phrase "By 2010 enterprises will be able to seamlessly interoperate with others." This vision was accompanied with a series of premises that explained and gave concrete substance to the vision itself. To assemble the components that made up this vision we collected input from stakeholders regarding the goals in a wide range of industries. These goals were broken down into technological or organisational components that required effort to bring the company to the desired goal. We examined and mapped these "challenges" to the common vision and found a surprising amount of interdependencies and shared themes. The partners and stakeholders invited to participate quickly converged on the set of premises that together made up the Ideas vision. Some of the key challenges include: the fact that future networked organisations will need self-managing or plug-and-play application solutions to allow their systems and processes to interact with those of their business partners on the fly, or that different platforms will need to seamlessly integrate software architectures, components and services, or that these types of platforms will have to constantly and automatically adapt to new business processes or applications because of the dynamic nature of the business environments. The Vision "challenges" also included concepts tied to users of these systems who, for example, will need to be able to dynamically join working teams and communities through model generated user environments.

The identified industrial challenges highlight the potential development of enterprise knowledge architecture and intelligent infrastructure to support user views, model repositories and knowledge services, in order to achieve cheaper, faster and repeatable solutions delivery, deployment and integration.

The "Vision" supplied by the Ideas Project was an exercise that lead to a strong convergence within the group but it also lid down the scope of the entire project and guided ensuing activities like requirement collection.

2.3 User requirements

IDEAS consortium felt that the future roadmap should highlight research that would have concrete impact on the industrial setting. Defining requirements for evolution of enterprise applications over the following ten years was, however, no easy challenge. Different views from different sectors or organisations could lead to competing requirements. Instead the consortium used a scenario approach. Numerous different potential scenarios were put forward and the impact on applications, data, business models and knowledge were examined and translated into requirements. The results were organised according the quality attributes including security, performance, availability, portability scalability and evolution. Organising the requirement collection in this manner allowed us to map candidate requirements to the same Ideas structure used for other activities like gap analysis. User requirements were therefore easily integrated with the general Ideas framework. In general we can say the requirements showed high need for interoperability between EIS functions not only between business partners in networked organisational scenarios, but even inside the single enterprise. For instance from system engineering scenario, users expressed the need for Enterprise Applications to be broken down into components which were able to be dynamically recomposed to satisfy new functions. In itself this created a subset of requirements like extracting the business logic from applications as business or enterprise models, and using standard and open formats to allow reprocessing in other models or modelling languages. It demonstrated the need for enterprises to manage applications using Enterprise Modelling tools and allow the simulation of application evolution in relation to new Enterprise Models.

Requirements were numerous touching all aspects of enterprise applications from the data structure management (import/export) inside the application itself, to sector-specific standards for product data management or general standards for other data management (process, commercial and knowledge data).

Another of the points that emerged is that ontology aspects must not be information technology dependent. A suitable mechanism for reconciling semantic mismatch, between software applications that need to cooperate, should be developed.

Business requirements were also examined and catalogued. For example it was found that business models should be described (and published) in a standardised form so that companies can compare it with other models in order to choose collaboration partners.

2.4 State-of-the-art

The state-of-the-art (SoA) was carried out around the three themes (EM, A&P, ON). The SoA on EM concluded that there exist many enterprise modelling techniques on the "market". The interoperability between these modelling languages and tools is very low. There is a need to develop a Unified Enterprise Modelling Language (UEML). The SoA on A&P focused on standards that provide architectural specifications relevant to enterprise integration, on e-business frameworks that provide architectural guidelines to support interoperability of inter-enterprise systems and on web services architectures. Most of the standards are still in development and have not reached a mature state. Concerning technology, XML based approaches seem to provide the best hope to achieve data and application interoperability. However, current XML technology still lacks semantic power. Ontologies, tightly connected with the XML solutions, will play a major role in the future platforms for software interoperability. Concerning ontologies, most up-to-date ontology languages are based on the Frame-Slot-Facet paradigm. This is a class-based paradigm very powerful in terms of expressive power but very poor in terms of specificity towards enterprise and business modelling (i.e., semantic adequacy). Existing solutions that exhibit a high semantic adequacy such as MIT Process Handbook and TOVE are too complicate to use in industry.

2.5 Gap analysis and road-mapping

The Road-mapping methodology used is an extension of a combined approach based on some existing ones: MAESTRO and Roadmapper. The methodology consists of: (1) a gap analysis; and (2) a road mapping process. Whilst the gap analysis states missing research, technology, and standards (RTS gaps), the road map process relates gaps, discusses their mutual dependencies, and identifies initiatives and actions to close them. To implement actions to close the gaps, a time horizon and priorities have also been identified. Time horizon is defined as to "when" an activity should be accomplished, whilst priority is defined as to which extent a proposed activity is essential to the overall goal.

The Gap analysis conducted by IDEAS project has identified 34 gaps (Ideas, 2003). They are mapped to the three domains (EM, A&P, and ON). One main observation is that ontologies are becoming an integral and pervasive aspect of enterprise modeling and architecture & platform. It was also observed that there are gaps that require a holistic perspective on EM, A&P, and ON. On this basis it has been conducted a road mapping process that resulted in a fivc horizontal roadmaps in the domains of "EM", "A&P", "EM/ON", "A&P/ON", and "EM/A&P/ON". Each roadmap assigns a time horizon and a priority to each gap previously identified (Ideas, 2003).

The importance of dependencies between RTS gaps was recognised; a Directed Graph (DiGraph) representation used to show such dependencies. This is essential to derive a work plan and coordinate the assignment of resources to actions in order to close gaps strategically and efficiently. The roadmaps elaborated are all horizontal road maps. The gaps identified in the gap analysis process are industry independent and formulate general requirements that any industry must meet to enable the interoperability of enterprise systems. Vertical, industry-specific, road maps are not considered yet. This is because the underlying gaps are currently not reflecting industry specifics due to their general nature. However, the approach taken with gap analysis and road mapping in IDEAS is able to diversify horizontal roadmaps into vertical road maps by introducing an industry attribute to the gaps template.

3. ATHENA INTEGRATED PROJECT (IP)

ATHENA IP aims to enable interoperability by providing reference architectures, methods and infrastructure components. In ATHENA Research

& Development will be executed in synergy and collaboration with Community Building: research will be guided by business requirements defined by a broad range of industrial sectors and integrated into Piloting and Technology Testing as well as Training. ATHENA consists of **three action lines** in which the activities will take place. In Action Line A, the research and development activities will be carried out. Action Line B will take care of the community building whereas Action Line C will host all management activities. Relations between the three action lines are shown figure 1.

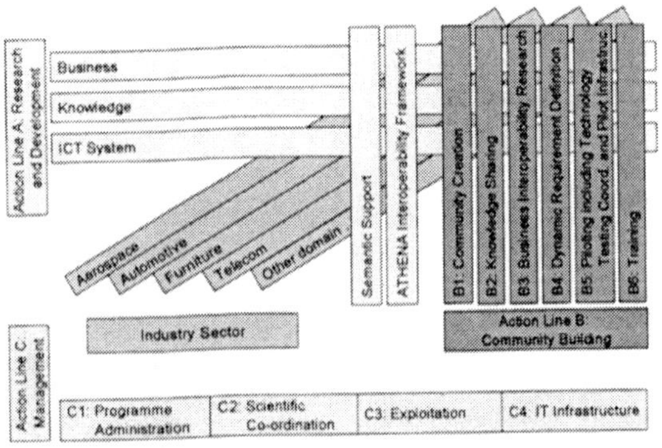

Fig.1 Interaction of ATHENA Action Lines

3.1 Approach

Scenarios play an important role in ATHENA as **sources of requirements** and for the **validation of results**. Scenarios represent the industrial sector-specific implementation of a process and capture industry-specific requirements of a given process. Industry-independent characteristics are described on the process level and are abstracted from the industry-specific scenarios. Scenarios used for validation of results need to be presented with business issues, software and infrastructure, and operational applications used in enterprises. Amongst other criteria, they will be selected based on their general interest to current and future business paradigms, their general (i.e. industry-independent), industry specific and company-specific characteristics, and their coverage of the anticipated ATHENA approach, methods, and tools. These criteria apply for scenarios provided by partners before the start of ATHENA and for future scenarios.

For each selected business scenario, the partners will provide **detailed descriptions**. These descriptions will be analysed and compared in order to identify commonalities and differentiation factors. The description will be based on a common methodology to capture user requirements.

One of the initial tasks (Dynamic Requirements Definition) in Action line B is to provide this methodology to users in order to be able to systematically capture user requirements.

The selection of processes and industries identified in the initial phase will be **extended** during the course of the project. During the first stage, the scenarios of Supply Chain Management, Collaborative Product Development, e-Procurement and Portfolio Management will be investigated. Industrial requirements that are not fully covered by available scenarios will be anticipated through the Community Building Activities in Action Line B. This continuous process of identifying further industries and processes ensures that the quality of scenario selection will be maintained for the whole duration of ATHENA.

3.2 R&D projects

Under Action line A and for the first stage of the IP, six research topics/projects were defined. The definition of these R&D projects was based on the roadmaps elaborated by IDEAS to close some gaps considered as priority.

- *Enterprise Modelling in the Context of Collaborative Enterprises* (A1) aims at developing methodologies for management and modelling of situated processes, flexible resource allocation and assignment and for work management and execution monitoring. For externalising the dynamic dependencies between participants and processes, methodologies of content based routing and collaborative processes will be developed. This project will enable scalable Enterprise Modelling methodologies and infrastructures, repository services and portal server services for "benefit driven Enterprise Modelling involvement".
- *Cross-Organisational Business Processes* (A2) deals with modelling techniques to represent business processes from different organisations on a level that considers the privacy requirements of the involved partners. Such models will have two perspectives: an enterprise modelling perspective that assigns a process to its context in the enterprise, and a formal aspect to perform computational transformations in order to allow for re-use of a process in a cross-organisational environment. Such models need to be enriched through ontological information and need to be executed through IT systems, such as workflow management systems. These systems need to be enabled to operate efficiently in an architectural environment that adapts to particular business scenarios.
- *Knowledge Support and Semantic Mediation Solutions* (A3) is aimed at the development of methods and tools for enterprise knowledge management, with a focus on supporting enterprise and application software interoperability. A key objective is to build a software environment that is able to manage in an integrated fashion the different sorts of knowledge that can be found in an enterprise. Focus is to use formal knowledge, organised in domain ontologies, to annotate the

business processes and the software components in order to reconcile the mismatches that may be encountered in unanticipated cooperation activities.

- *Interoperability Framework and Services for Networked Enterprises* (A4) is concerned with the design and implementation of the infrastructure supporting interoperability in scenarios adopting the Integrated Paradigm (i.e. where there is a standard format for all constituent sub-systems). The project will provide methodologies and develop seamless and configurable inter-connection software components. The resulting toolset will be the basic set of software and engines that would prepare any enterprise in the adoption and exploitation of interoperability support infrastructures.
- *Planned and Customisable Service-Oriented Architectures* (A5) is to develop the understanding, tools and infrastructures required for service-oriented architectures so that they can be achieved more easily through the planning and later customisation of solutions for better application to user scenario requirements. Although the project will consider existing infrastructures, an emphasis will be provided regarding the development of an environment for easier application development that natively provides better customisation.
- *Model-driven and Adaptive Interoperability Architectures* (A6) is to provide new and innovative solutions for the problem of sustaining interoperability through change and evolution, by providing dynamic and adaptive interoperability architecture approaches. The project aims to advance the SoA in this field by applying the principles of model-driven, platform independent architecture specifications, and dynamic and autonomous federated architecture approaches, including the usage of agent technologies.

3.3 Remarks

Though the programme as a whole adopts a holistic perspective on interoperability, during the first stage of the programme, the emphasis of individual projects is more on **individual bricks** rather than on achieving solutions that address interoperability in a holistic manner. However, the individual results of the projects taken together will enable to define later on some projects that provide **integral solutions**.

Four out of six initial research projects (projects A1, A2, A5, and A6) address relatively focused topics. Project A3 is rather **supportive** via its focus on semantic issues; all other projects and the programme itself can benefit from this. Project A4 exploits the results and methodologies reached in the other projects. This project is a first **reconciliation attempt** that will provide an integral solution resulting in the ATHENA Interoperability Framework.

Major impacting results from ATHENA could be tentatively stated as follows:

- A generic, common platform for Networked Organisations, integrated by an open Infrastructure.
- Interoperability services to assist enterprise designers, knowledge workers and system developers in composing and adapting executable solutions, reusing knowledge and software components.
- Specific platforms built to support the research projects, the communities, use-cases, scenarios, test-beds, training and learning facilities, and collaborative experimentation, and

4. INTEROP NETWORK OF EXCELLENCE

4.1 Goals and organisation

Research in the interoperability domain in Europe remains badly structured, fragmented, sometimes overlapping. There is no global vision of the consistency, no co-ordination between European research centres, university laboratories or other bodies. This situation is not only true for research, but also in the training and education areas. Consequently, the primary goal of INTEROP is the *emergence of a lasting European research community on interoperability of enterprise software applications*. The NoE is structured in 4 types of activities: (i) Integrating activities, (ii) Joint research activities, (iii) Spreading excellence activities, and (iv) Management activities.

4.2 Integrating Activities

INTEROP is not a research project but a research integration project. The Integrating Activities are essential to the NoE. Four activities were defined:

- *Knowledge map* (IA1): This activity consists in analysing network's components of knowledge on EM, A&P, ON, in identifying the interlinks between these components, then in elaborating the roadmap for European research on interoperability.
- *Method of work and platform* (IA2): It provides the NoE with appropriate methods for collaborative work, to capitalise the knowledge added by integration and make it open to all, and to develop common platforms and tools.
- *Mobility of researchers* (IA3): Clearly as one of the keys for the de-fragmentation of the research resources in the field, this activity aims at creating suitable scientific, administrative, financial and practical conditions for scientific stays of both researchers and doctoral students in EU states.
- *Method for scientific integration and assessment* (IA4): It manages the performance of the project with implementing performance indicators and makes the performance indicator system alive with disseminating the performance among the partners.

4.3 *Joint Research Activities*

The Joint research activities (JR) aim at developing synergy between the three domains among NoE members. They are expected to generate a set of new concepts, models, methods and tools based on the existing and further developed competences of NoE members. Four activities were defined.

- *Common Enterprise Modelling Framework in distributed environments* (JR1): The target is to integrate the different aspects available in the state-of-the-art (model data, responsibilities, motivation, knowledge, social aspects, etc.) usually fragmented in different domains of sciences and applications to provide a complete reference model of the distributed enterprise. More specifically, the meta-model resulting from UEML project will be maintained and adjusted with respect to IA1.
- *Generation of customised Enterprise Software from Enterprise Modelling* (JR2): The activity is justified by the need for federating the disparate research approaches to develop Enterprise Modelling tools and software in relation with international normalisation. Beyond a necessary state-of-the-art, the purpose is to improve the modelling techniques to allow interoperability between enterprise models and to facilitate the customisation of suitable software.
- *Ontology-based Integration of Enterprise Modelling and Architecture & Platforms* (JR3): As a support for IA1, the purpose is to derive the paradigm of interoperability for enterprise applications from the three research perspectives (EM, A&P, ONTO). The output of this activity will be a complete methodology for putting in practice Interoperability.
- *New architectures and platforms for Interoperability* (JR4): This research area focuses on harmonizing and synthesizing the partner research around new flexible and adaptive architectures for interoperability; such as the model driven approach, service-oriented architecture approaches, peer-2-peer architectures, agent architectures and federated architectures.

4.4 *Spreading Excellence Activities*

To spread excellence beyond the network, disseminate the knowledge and exploit the results, three 'Spreading Excellence' activities (SE) have been designed as follows: (1) *Training by e-learning (SE1)* provides training facilities to spread the knowledge added by INTEROP inside and outside the NoE, during and beyond the EU-funded period, using e-techniques; (2) *Dissemination and Communication (SE2)* is devoted to the external communication and knowledge dissemination to increase awareness on Interoperability; and (3) *Transfer of research to Industry (SE3)* to create an Interoperability Laboratory Network promoting Interoperability services towards SMEs.

4.5 *Remarks*

INTEROP is academic-based and basic research oriented, while ATHENA is industry-driven and applied research oriented. In ATHENA, achieving R&D results is the main objective; while in INTEROP, focus is on restructuring knowledge and researches.

Within JR Activities and for the first stage of the NoE, five workpackages were defined to develop: (i) CEMF-UEML, (ii) Design principles for interoperability, (iii) Method and specification to generate customised Enterprise Software, (iv) Ontology-based Integration of EM and A&P, and (v) Model driven and dynamic, federated enterprise interoperability architectures and interoperability for non-functional aspects.

5. CONCLUSION

In this paper, an overview on European initiatives to develop interoperability of enterprise application and software was presented. The gaps analysis and roadmaps resulted from IDEAS have led to the definition of R&D research projects to carry out by ATHENA. Complementary to ATHENA, dispersed and fragmented knowledge on interoperability and related research activities will be integrated and re-structured by INTEROP. The three initiatives form a coherent and complementary approach that is considered as a strategic development by the European Commission for the years to come.

ACKNOWLEDGEMENT

This paper is partly funded by the EC through Athena IP. The paper does not represent the view of EC, and the authors are solely responsible for the paper's content.

The authors thank and acknowledge the members of the consortiums of IDEAS project (Coordinator: Guy Doumeingts, GRAISOFT), ATHENA IP (Coordinators: Klaus-Dieter Platte and Rainer Ruggaber, SAP) and INTEROP NoE (Coordinator: Jean-Paul Bourrières, University Bordeaux 1).

REFERENCES

ATHENA (2003), Advanced Technologies for Interoperability of Heterogeneous Enterprise Networks and their Applications, FP6-2002-IST-1, Integrated Project Proposal, April 2003.

IDEAS (2003), IDEAS Project Deliverables (WP1-WP7), Public reports, www.ideas-road map.net.

INTEROP (2003), Interoperability Research for Networked Enterprises Applications and Software, NoE Proposal Part B, April 23, 2003.

IST (2002), IST –2001–37368, Thematic Network, IDEAS: Interoperability Development for Enterprise Application and Software – Roadmaps, Description of Work, 13-05-2002

Copyright © IFAC Information Control Problems in Manufacturing, Salvador, Brazil, 2004

ELSEVIER
IFAC
PUBLICATIONS
www.elsevier.com/locate/ifac

A BENCHMARKING SERVICE FOR THE MANUFACTURING CONTROL RESEARCH COMMUNITY

Paul Valckenaers*, Sergio Cavalieri#, Bart Saint Germain*, Paul Verstraete*, Hadeli*, Sergio Terzi$ and Hendrik Van Brussel*,

* *K.U.Leuven – P.M.A.,*
Celestijnenlaan 300B, B-3001 Leuven, Belgium
Paul.Valckenaers@mech.kuleuven.ac.be

#*Università di Bergamo*
Dipartimento di Ingegneria Industriale
Viale Marconi, 5, I - 24044 Dalmine, Italy

$*Politecnico di Milano*
Dipartimento di Ingegneria Gestionale
Piazza Leonardo da Vinci, 32, 20133 – Milano, Italy

Abstract: This paper presents the development – by the IMS Network of Excellence[1] – of a web-based benchmarking service for manufacturing control systems. The paper first discusses the rationale behind this development. Next, the architecture and usage of the benchmarking service is presented and illustrated. Finally, the paper addresses the current status of the benchmarking service. *Copyright © 2004 IFAC*

Keywords: Agents, benchmark examples, control applications, industry automation, performance evaluation, plants, production systems, and production control.

1. INTRODUCTION

A manufacturing control system manages the internal logistics in a production system. It decides about the routings of product instances through the factory, the assignment of workers and raw material or components, and the starting of the processes on semi-finished product instances. Manufacturing control systems do not control the production processes or equipment themselves. Nonetheless, a manufacturing control system has to handle every possible outcome of the activities that it triggers.

Today, manufacturing control developers face harsh challenges posed by the diversity and heterogeneity as well as the changes and disturbances occurring in manufacturing systems and their environment. This makes it equally difficult for researchers to develop control system prototypes, apply them to realistic test cases and compare their results with work by others. In practice, researchers have to restart almost from scratch for every research project or doctoral thesis.

This paper presents an undertaking by the IMS Network of Excellence that aims to remedy this problem.

[1] Cf. www.ims-noe.org

The IMS-NoE network aims to establish a well-coordinated and effective support infrastructure in order to develop co-operation links, coordinate current and future research, facilitate knowledge transfer and dissemination of project results and stimulate the launch of new projects around themes related to IMS.

Within its activities, the network has launched a web-based benchmarking service for manufacturing control systems (possibly extending this service to other research topics in a later phase). This benchmarking service is discussed in this paper. Within the network, a special interest group supports the establishment and elaboration of this web-based service.

This paper first describes the rationale behind the benchmarking service. Next, it describes the service itself (structure, components...). Finally, the current status and the development roadmap are presented.

2. DEVELOPMENT RATIONALE

Communication amongst researchers in the manufacturing control community predominantly uses paper publications as its principal information carrier. This static medium exhibits limitations preventing this research community to join forces adequately concerning (1) the development of realistic test cases, (2) the performing of systematic evaluation campaigns and (3) the generation of benchmarking results suitable for mutual comparison. Typically, research activities lack the resources to move beyond toy test cases and token evaluation campaigns. Moreover, researchers focusing on the evaluation and benchmarking of novel manufacturing control developments face difficulties acquiring, re-implementing and utilizing control systems conceived by other researchers.

In other words, the manufacturing control research community is restarting more or less from scratch whenever investigating a novelty in manufacturing control, and fails to deliver results that are suitable for an in-depth comparison with results from others and the current state of practice. Moreover, the research is unable to test novel ideas on challenges exhibiting realistic size and complexity. The benchmarking service, which is discussed in this manuscript, aims to remedy this situation.

The benchmarking service aims to provide, through web technology, a basis for the development of research prototypes for manufacturing control and the execution of evaluation campaigns. This basis allows researchers to generate performance reports that are suitable for comparison with work by others, and to test their developments on realistic industrial test cases while spending a similar amount of effort as needed to build toy cases from scratch. Furthermore, the benchmarking service permits independent parties to perform the testing and evaluation of manufacturing control systems developed by others.

3. ARCHITECURE AND USAGE

This section presents the main components of the benchmarking service and their relationships. Furthermore, it discusses how the service is used.

3.1 Components of the benchmarking service

The benchmarking service comprises three subsystems. The front-end subsystem is an interactive tool to define the benchmark; this is the Test Bench Assistant or TBA (Terzi, *et al.*, 2003). The second subsystem performs the actual benchmark on an emulated manufacturing system that is connected to a manufacturing control system (Saint Germain, *et al.* 2003b); this is the Test Bench Kernel or TB-kernel. Thirdly, the back-end subsystem processes the results, produces the benchmarking reports, and permits the users to interactively browse the event logs from the second subsystem. In addition, the benchmarking service supports a library of components, which will be growing continuously.

3.2 Test Bench Assistant

The users of the benchmarking service utilize the TBA to define their problem and testing campaign. The TBA produces an XML file that is used by the TB-kernel to bootstrap the emulation and control before this kernel executes the benchmark itself.

The TBA allows its users to define three sets of data. First, the user defines the production system itself. This comprises the processing equipment, storage and transportation, operators, and the topology of these elements. Second, the user defines the product types by specifying the required materials and the process plan for each product type made on the manufacturing system. Third, the user defines the benchmarking scenarios. Such a scenario stochastically defines the workload (order arrivals, order due dates, etc.) and the stochastic behavior of the manufacturing itself (machine breakdowns, processing time variations, processing results, etc.).

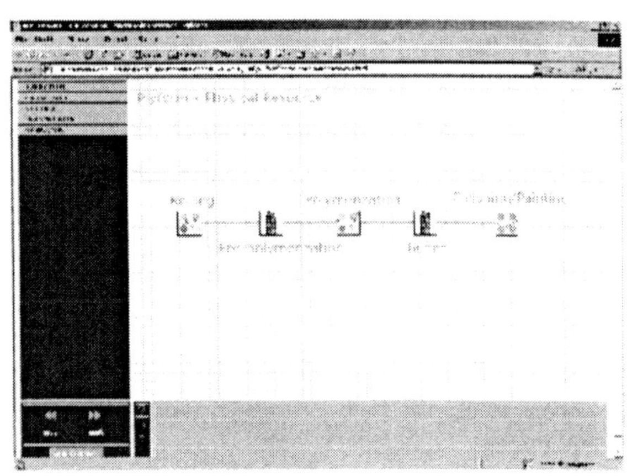

Figure 1 –Screenshot of the current version of TBA

The main purpose of the TBA is to lower the threshold for prospective and novel users of the benchmarking service. The emphasis is therefore on ease-of-use, and not on expressiveness. Figure 1 shows a screenshot of the current implementation. More detailed views can be obtained on the IMS-NoE web site

More advanced applications and users may therefore utilize the TB-kernel directly, thus avoiding limitations of the TBA. Future versions of the TBA intend to enhance the expressiveness of this modeling tool in a layered fashion (advanced buttons) such that inexperienced users are not confused by the enormous diversity and complexity of real-life manufacturing systems.

3.3 Test Bench Kernel

The TB-kernel is the subsystem that performs the actual test and evaluation campaigns. A preprocessor converts the XML files from the TBA into the corresponding input files for the TB-kernel. Advanced users may provide such input files directly to the TB-kernel.

During a boot phase, the input files serve to build an emulation of the factory described by the user. The TBA is unable to generate the code for the emulation of individual manufacturing components. Therefore, the kernel possesses a library of implementations for the emulation of these building blocks. In the current system, these implementations are in Java. When the TB-kernel constructs an emulation of manufacturing components, it retrieves the proper serialized objects in its database, de-serializes the objects and clones them as required. The topology (connections) and parameter settings for these emulation objects are retrieved from the input files and applied to the deserialized objects.

The set of supported building blocks grows as the benchmarking service is applied to an increasing number of manufacturing systems. All building blocks supported through the TBA implement a common interface reflecting the modeling approach of the TBA. Among others, these building blocks accept the parameters that determine their processing capabilities and their stochastic behaviors. More advanced building blocks can be added to the library and have their own methods for retrieving their parameters.

Similarly, the TB-kernel retrieves Java objects corresponding to the product types supported by the manufacturing system. In the current benchmarking service implementation, these objects implement a product agent as defined in the PROSA architecture (Van Brussel, 1998). This agent communicates the set of possible next processing steps for a given product in a given state. The product agent also provides objects containing the state information of a product instance when entering production. Furthermore, a product agent returns an updated state information object when given the current state, the operation performed and the processing result; each piece of information is a Java object in the current implementation. Note that it suffices that the product agent knows the internal details of these information objects. In its most primitive version, such a product agent has a list of processing steps and the state object simply points to the next step that needs executing.

The default object for generating such a product agent takes the process plans generated by the TBA as its parameter. In other words, a single data-driven product agent class handles the full range of plans supported by the TBA. The possibility of providing other product agent classes is provided for advanced applications. For instance, disassembly processes and sheet cutting processes are not supported by the initial version of the TBA; future versions of the TBA will include support for such advanced product agents. Alternatively, process plan objects may be added to accommodate process plan representations from other sources (e.g. from legacy systems). Another category of advanced process plans describes how to perform non-productive tasks such as bringing empty containers to locations where they are required, performing a calibration or executing a maintenance job.

Next, the TB-kernel boots the manufacturing control system. This control system is given references to the product agents and the objects corresponding to factory entries or exits. The control system signals when it has finished its initialization procedures. Such initialization is likely to include the discovery of all reachable manufacturing components from the factory entries and/or exits. The control is also likely to generate control entities corresponding to these manufacturing entities.

With the manufacturing control system in operation, the TB-kernel starts the testing campaign. It processes the input files that define the scenario and informs the control system of order arrivals as specified. Note that executing the scenario includes hiding information from the control system until it would become known in real life.

All emulation entities see time progress in a coherent and consistent manner and generate events as required. TBA-compatible elements generate events as specified. The parameters provided through the XML files specify the stochastic properties for the breakdowns, processing times and processing results. A high-quality random number generator provides a stream of number that is transformed according to the appropriate stochastic distribution. Replications assure that conclusions are not drawn based on a single possible scenario. Likewise the advanced elements see time progress and generate events.

Note that the benchmarking service supports a novel mechanism to ensure that time in emulation and time in the control progresses in a consistent and coherent manner (Saint Germain, *et al.*, 2003b). This mechanism allows the emulation time to progress as fast as possible when the control is inactive whereas the emulation operates in real-time mode when the control is deliberating one or more of its decisions. This imposes a small implementation requirement on the control implementations to be used by the benchmarking service. Control system implementations must indicate to the emulation when they finished processing an event that was reported to them by the emulation, and they have to implement timers through the emulation. Failure to do so will result in an emulation that operates exclusively in real-time mode, which is (very) slow.

Advanced implementations provide these scenario inputs directly instead of through the TBA. Indeed, benchmarking requirements may impose constraints on the scenarios that are to be used. For instance, to obtain measurements for the robustness of manufacturing control systems, the benchmarking service is likely to impose standardized scenarios such that results can be compared. Likewise, the TBA may fail to support scenario attributes needed by advanced applications (e.g. human intervention in the control activities).

During the execution of a scenario, performing an adequate number of replications, the TB-kernel entities create a persistent log of the events that occur in the emulation. This trace is used by the third subsystem to perform the analysis and performance measurement reporting.

3.4 Test Bench Analysis and Reporting

The back-end subsystem provides basic analysis tools for users that simply want to observe and understand what happened in the emulated system during the benchmarking. Secondly, the back-end subsystem will combine the results in more concise reports that are more directly usable for comparisons and communication of the benchmarking results.

The first component of the basic back-end system is a browser for the emulation trace. This system allows the user to visualize what happened. The user will be able to move back and forward in emulation time, jump to a point in time, move in fast forward mode, etc. Initial functionality will be simple and is likely to grow when more experience with the system has been acquired (breakpoints, event filtering...). This part of the back-end subsystem relies on the implementation of the building blocks in the system library to support visualization. In other words, each manufacturing component supported by the library implements the necessary methods to symbolically visualize its state on a 2-dimensional screen. Some minimal requirements for this functionality are likely to emerge as the benchmarking service implementation progresses.

The second type of basic back-end components are simple processes that consume the event log and calculate the basic statistics such as throughput, lead times and due date performance. This set of components will grow, become more accurate and more complete as the benchmarking service operates. The components will calculate the usual range of statistic of these parameters (median, percentiles, average, deviations, worst cases, etc.). The statistic will be calculates over different categories (e.g. lead times for rush orders versus normal orders). These reports will still be quite voluminous and useful for users who want to understand what is happening in a specific system.

These basic components will be subdivided in categories. Indeed, not all measurements are directly related to manufacturing performance. Some statistics only report on the behavior of the system (e.g. equipment utilization rates), others on manufacturing performance (e.g. average and median throughput) and others on control performance (e.g. forecasting horizon time). The benchmarking service aims to reflect state-of-the-art insight in performance measurement, evaluation and interpretation.

In addition to the above basic analysis and reporting components, more advanced benchmarking report generation components will be added. An initial proposal on how to accomplish this is presented in another paper submitted by Sergio Cavalieri to this special session on Benchmarking of INCOM 2004. An important difference to the basic components is that the design of these advanced components implies significant choices on what is relevant and important. Therefore, it is expected that the benchmarking service will become a platform on which multiple design are made available and used such that a solid scientific discussion on the merits and weaknesses of the different proposals emerges.

4. DEVELOPMENT STATUS AND ROADMAP

The envisaged benchmarking service is an ambitious project. Therefore, an incremental approach is used to establish this service. This section first describes which components exist already. Next, it details the manner in which this will evolve into a full-fledged service.

4.1 Current status

Test Bench Assistant. The first version of the TBA currently is operational on the web and can be accessed through the home page of the IMS Network of Excellence (www.ims-noe.org). This interactive and graphical modeling tool allows users to define their benchmarking problems, to store, retrieve and modify these, and to generate the XML files that will drive the TB-kernel. The TBA utilizes an ordinary Internet browser enhanced with a plug-in for visualization in the graphical interface.

Test Bench Kernel. The core components of the TB-kernel are currently running in off-line applications booting from their own input files. Sample factory emulations and manufacturing control systems exist. These are results of research projects on multi-agent manufacturing control (Peeters, 2001; Zamfirescu, 2003; Saint Germain *et al.*, 2003a). Next to the core components, the technology – *Java web start* – to make these off-line applications run remotely on a web server has been selected and validated.

Benchmark Reporting. The implementations of a report-generating subsystem have not yet started. However, elaborated discussions on the content of these reports have taken place. The technology for the graphical interactive event log browser has been tried out already but a comprehensive implementation is not available at this time.

4.2 Roadmap

Test Bench Assistant. A second version of the TBA already is in the pipeline. The emphasis is on enhancing the user-friendliness of the system. For instance, a PDF description of the benchmark specification will be produced next to the current XML version. This helps users to check and communicate their benchmarking specifications. The second version also corrects small errors detected while using the current version, which includes an expansion of its modeling capabilities that do not require a redesign of the current software.

Future versions of the TBA aim to expand its modeling capabilities in more fundamental manners. This work will only commence after experience with applying the benchmarking service to industrial cases provides a comprehensive set of useful features that should be added. Demanding and advanced industrial cases are likely to bypass the current TBA where this is appropriate, thus identifying where upgrading the TBA is desirable.

Consequently, the advanced TBA versions will not be available in a foreseeable time, and priority is given to providing an easy-to-use TBA that lowers the threshold for novel users. More ambitious TBA developments can only be undertaken after the other subsystems of the benchmarking service have been available and used sufficiently to yield the requirements for such a more advanced TBA.

Test Bench Kernel. The TB-kernel is the next subsystem to become operational on the web. The envisaged developments in the immediate future are the connection to the TBA: (1) parsing the XML files to produce the required input files and (2) providing the necessary database with serialized emulation objects, product agents, etc. This system enables to perform benchmarking runs on emulated manufacturing systems designed with the TBA and connected to a simple controller.

In parallel with the above TBA-based developments, the team will provide a number of more challenging industrial test cases. The emulated factories will be very similar to real industrial plants connected to prototype manufacturing controllers that address the industrial control requirements comprehensively (set-ups, empty container transport, tool management, operator scheduling...). The similarity to real industrial plants is of course constrained by the need to account for confidentiality issues. The ability to customize these advanced industrial cases probably will be constrained until an advanced TBA becomes available, except for more skilled users that manage the internal input files of the case studies themselves.

The TB-kernel components generate an event log file for usage by the reporting subsystem. Candidate technologies for these files are XML and serialized java objects.

The main development activity for the TB-kernel is to provide the proper serialized objects for the emulation of factory components, as well as frameworks for implementing the actual manufacturing control systems. It is the intention that the benchmarking service allows researchers to re-use significant parts of a manufacturing control implementation and only provide novel implementations for the parts in which their research contribution is situated. The initial framework will be PROSA-based (Wyns, 1999) and supports stigmergy (Valckenaers, 2001). Later, a choice of frameworks will be made available as research teams provide their technology to the service (e.g. market-based...).

Benchmark reporting. The reporting system initially will provide the graphical browser, which allows users to observe what is happening in the emulated factory. This browser permits to run a movie of the emulate production, jump to a given point in time, go forward and backward, go fast and slow, etc.

Next, the benchmarking service will provide a number of parsers that consume an event trace and calculate a relevant statistic. The collection of supported statistics will expand until all the most relevant ones are available.

Thirdly, the basic statistics will be processed in more advanced reporting and evaluation systems that produce the real benchmarks. It is expected that there will emerge a number of competing solutions aiming to put forward good measures for characteristics like robustness, agility or flexibility. In this respect, the benchmarking service will provide a discussion forum.

Finally, the benchmarking service will provide support to determine what constitutes an adequate testing campaign for the measurement of a given set of system characteristics. In other words, how to make sure that the test runs include the proper excitation of the system being tested.

Overall, the benchmarking service will provide a starting point for developing and testing combinations of manufacturing control systems and factory designs. Developers should be able to pick up intermediate solutions and reusable components allowing to build much more advanced solutions than is feasible today, and they should perform a much more thorough and high-quality evaluation campaign.

5. CONCLUSIONS

This paper presents the motivation, overall design, status, and roadmap of a web-based benchmarking service for manufacturing control.

The service comprises three main components. The TBA provides an interactive user interface to lower the threshold for the prospective users. The TB-kernel performs the specified benchmarks and produces a persistent log file of the events that occurred in the emulated factory connected to the manufacturing control system during the benchmark execution. The back-end provides basic statistics, an interactive and graphical browser for the events logs, and more high-level benchmark reporting subsystems.

The benchmark design is such that it can grow over time and reconciles simple usage for simple cases with advanced usage, by skilled users, for complex and realistic industrial cases. The design makes partial solutions available to researchers who may add their specific research items to such a starting base. Thus, researchers will be able to perform more comprehensive experiments with less effort.

This benchmarking service is work-in-progress and, through its design, it is likely to remain work-in-progress indefinitely, giving increasingly more useful functionality to the manufacturing control community. The next milestone is the development is to make the first version of the TB-kernel available through the IMS-NoE web pages by May 2004 at the IMS Forum in Como (IT).

ACKNOWLEDGEMENTS

This paper presents work funded by the Research Council of the K.U. Leuven – concerted research action on Agents for Coordination and Control – and the 5th framework programme of the European Commission.

REFERENCES

Peeters, P., H. Van Brussel, P. Valckenaers, J. Wyns, L. Bongaerts, M. Kollingbaum, and T. Heikkilae (2001). Pheromone-based emergent shop floor control systems for flexible flow shops. *Artificial Intelligence in Engineering*, **15(4)** 343-352, Elsevier.

Saint Germain, B., P. Valckenaers, H. Van Brussel, Hadeli, O. Bochmann, C. Zamfirescu, and P. Verstraete (2003a). Multi-Agent Manufacturing Control: An Industrial Case Study. *Intelligent Manufacturing Systems 2003,* Proceedings from the 7th IFAC Workshop, Budapest, Hungary, 6-8 April 2003, ISBN 0-08-044289 7, 207–212.

Saint Germain, B., P. Valckenaers, C. Zamfirescu, O. Bochmann, and H. Van Brussel (2003b). Benchmarking of Manufacturing Control Systems in Simulation. *Proceedings of the 3RD International Workshop on Performance Measurement* (IFIP WG5.7 Special Interest Group on Performance Measurement), pp.357-369, IFIP, Bergamo.

Terzi, S., Cavalieri, S., Macchi, M., Garetti, M. and H. Panetto (2003). Development of a reference architecture for benchmarking service of manufacturing systems. *Intelligent Manufacturing Systems 2003,* Proceedings from the 7th IFAC Workshop, Budapest, Hungary, 6-8 April 2003, ISBN 0-08-044289 7, 193–200.

Valckenaers, P., H. Van Brussel, M. Kollingbaum, and O. Bochmann (2001). Multi-agent coordination and control using stigmergy applied to manufacturing control. *Lecture Notes in Artificial Intelligence*, **2086**, 317-334, Springer-Verlag.

Van Brussel, H., J. Wyns, P. Valckenaers, L. Bongaerts and P. Peeters (1998). Reference architecture for holonic manufacturing systems: PROSA. *Computers In Industry* **37**, 255-274.

Wyns J. (1999) *Reference architecture for Holonic Manufacturing Systems – the key to support evolution and reconfiguration*, Ph.D. thesis K.U. Leuven (1999) ISBN 90-5682-164-4

Zamfirescu, C., P. Valckenaers, Hadeli, H. Van Brussel and B. Saint Germain (2003). A case study for modular plant control. *Lecture notes in Artificial Intelligence*, **2744**, 268-279.

Copyright © IFAC Information Control Problems in Manufacturing, Salvador, Brazil, 2004

ELSEVIER
IFAC
PUBLICATIONS
www.elsevier.com/locate/ifac

A UNIFIED ENTERPRISE MODELLING LANGUAGE FOR ENHANCED INTEROPERABILITY OF ENTERPRISE MODELS

Hervé Panetto[1], Giuseppe Berio[2], Khalid Benali[3], Nacer Boudjlida[3], Michaël Petit[4]

[1]*CRAN UMR 7039, University Henri Poincaré Nancy I, F-54506 Vandoeuvre-les-Nancy Cedex*
Herve.Panetto@cran.uhp-nancy.fr
[2]*Dipartimento di Informatica, Università di Torino, Torino, Italy*
berio@di.unito.it
[3]*LORIA UMR 7503, F-54506 Vandoeuvre-les-Nancy Cedex*
{Khalid.Benali, Nacer.Boudjlida}@loria.fr}
[4]*Institut d'Informatique, Facultés Universitaires Notre-Dame de la Paix, Namur, Belgium*
mpe@info.fundp.ac.be

Abstract: There is a serious backwardness in awareness, acceptance and wide use of the Enterprise Modelling (EM) technology in industry because enterprises cannot capitalise from previous modelling efforts. This situation hinders true enterprise integration, interoperability, and enterprise knowledge sharing. A Unified Enterprise Modelling Language, based on meta-modelling of existing EM Languages, would serve as an Interlingua between EM tools providing the business community with a common visual, template based language to be used on top of most commercial enterprise modelling and workflow software tools. *Copyright © 2004 IFAC*

Keywords: enterprise modelling, meta-modelling, interoperability, enterprise system

1. INTRODUCTION

Today's "business trends are clearly towards the need for managing organizational and operational changes within companies in order to face global competition and fluctuating market conditions" (Vernadat, 1996). This situation poses a number of integration (and interoperability) problems that enterprises have to tackle: integration of markets, integration between several development and manufacturing sites, integration between suppliers and manufacturers, integration of design and manufacturing, integration of multi-vendor hardware and software components. Enterprise Integration is therefore, not anymore only a problem of interconnecting physical and software applications but it also requires global business integration, aiming at the use of the existing or new enterprise resources in order to better achieve the overall business objectives. "Things to be integrated and coordinated need to be modelled. Thus, Enterprise Modelling (EM) is clearly a prerequisite for enterprise integration".

According to Vernadat (1996), *enterprise modelling* is the set of activities or processes used to develop the various parts of an enterprise model to address some desired modelling finality. It can also be defined as the art of "externalising" enterprise knowledge, i.e. representing the enterprise in terms of its organisation and operations (e.g. processes, behaviour, activities, information, object and material flows, resources and organisation units, and system infrastructure and architectures). The finality is to make explicit facts and knowledge that add value to the enterprises or can be shared by business applications and users. The prime goal of enterprise modelling is not only to be applied for better enterprise(s) integration but also to support analysis of an enterprise, and more specifically, to represent and understand how the enterprise works, to capitalize acquired knowledge and know-how for later reuse, to design (or redesign) a part of the enterprise, to analyse some aspects of the enterprise (by e.g. economic analysis, organization analysis, qualitative or quantitative analysis,...), to simulate the behaviour of (some part of) the enterprise, to make better decisions about enterprise operations and organization, or to control, coordinate and monitor some parts of the enterprise.

Within the initiative on Computer Integrated Manufacturing (CIM), Enterprise Modelling was born in the United States at the beginning of the 80's and emerged through large CIM projects (e.g. ICAM or CAM-I). In the mid-80's, Europe launched several projects on Enterprise Modelling giving birth to several enterprise modelling languages (including notably GRAI (Doumeingts, *et al.*, 1998) and CIMOSA (AMICE, 1997; Kosanke, *et al.*, 1999)). As a result, in the 90's many commercial tools dealing with EM or business process modelling appeared on the marketplace, e.g. ARIS ToolSet, FirstSTEP, METIS, Enterprise Modeller, KBSI, CimTool, MO^2GO, e-MAGIM and many others.

Because most resources in modern enterprise were computerised systems, the problem of enterprise integration was also been approached with *workflow systems*, each one with its own modelling environment (Action Workflow, COSA, FlowMark, Lotus Notes, Teamware Flow, Ensemble, WorkParty, ...) and, in the late 90's, with EAI (Enterprise Applications Integration) tools. In computerised systems, the situation is currently worse than before. This is mainly due to the "Internet based systems" which, while offering very powerful, low cost and open infrastructures, fall short of integration. In fact, the many related standards (e.g., XML and its customisations) do not directly address the "problem of meaning".

This intensive production of tools has led to a *Tower of Babel situation* in which the many tools, while offering powerful and distinct functionalities, are unable **to interoperate** and can hardly or not at all communicate and exchange models. This is a serious drawback for awareness, acceptance and wide use of the EM technology since enterprises cannot capitalise from previous modelling efforts. This situation hinders true **enterprise integration, interoperability, and sharing enterprise knowledge**.

2. ENTERPRISE INTEGRATION AND ENTERPRISE MODELLING

Enterprise Modelling is an *engineering discipline* closely related to computerised systems. As such, it requires the combined use of Enterprise Modelling Software Environments (EMSE), Enterprise Modelling Languages (EML), and Enterprise Engineering Methodologies (EEM).
According to this point of view, there exists a lot of fragmented approaches to enterprise modelling (including Methodologies, Languages and Tools).
They cover different subsets of the different components of the enterprise engineering world. For instance, the ENV12204 (CEN, 1995) standard of CEN provides an Enterprise Modelling Language, but does not address any of the other components (no tool, methodology or meta-modelling capacity)); GERAM (GERAM, 1997) and ISO/IEC 15288 define methodological guidelines for Enterprise Engineering but without any modelling language; the Workflow Management Coalition (WfMC) provides a modelling language (WPDL) but without methodological support for using this language for modelling processes; the same happens with other EAI tools (such as the so called Integration Brokers).

They also cover different parts of the *enterprise life-cycle*. For instance, ebXML and WfMC focus on software design while approaches like CIMOSA mainly concentrate on enterprise requirements and design. Additionally, some approaches link enterprise models to Enterprise Operation Tools (EOT). These links allow the produced models to be used, for instance, for process *enactment and control* (e.g. in the WfMC approach, WPDL models are used by a workflow engine to control the execution of ongoing work). Models may be also linked to enterprise software applications like ERP (Enterprise Resource Planning) systems (e.g. ARIS and mySAP.com). Some other approaches only aim at modelling for understanding and analysing and they do not provide explicit links to operational systems or to other models in the life-cycle (e.g. ENV12204).

Enterprise modelling approaches may also have very different objectives and needs. As a meaningful example, we may compare the aims of IEM and GRAI. IEM allows representing business processes and it provides specific concepts adapted for assessing quality management procedures, but it cannot directly be employed for an operational implementation of the business processes. In fact, for quality management, it is not necessary to fully define an implemented process. The description has only to be sufficient to enable determining whether the process steps conform to defined quality procedures. This later objective requires, for example, representing outcomes of each process step and their usage as inputs for other process steps. On the other hand, the main objective of GRAI modelling is to define the control system of an enterprise. This requires a very good understanding of the relationships between the business processes at the various control levels (operational, tactical, and strategic). Quality procedures do not guarantee that an enterprise has good performance but only that its products conform to some quality criteria, whereas enterprise control tries guarantee that an enterprise takes into account market data and reflects them in the enterprise internal behaviours.

In enterprise modelling, there seems to be a tendency for approaches combining, in a more or less integrated way, *several sub-languages* or views (see e.g. CIMOSA, GRAI, and ARIS). A combining approach allows taking advantage of the strengths of each of the sub-languages and offers the advantage that the resulting combined method or methodology offers the modeller the capacity to meet more modelling objectives. Models built with the distinct languages have to be related in some way and the *languages have to be integrated.* This need for integrating distinct modelling languages and relating models has also been recognised in domains like *software formal methods* for achieving effective and "practical" solutions to complex problems.

However, the integration of several sub-languages (often called *views*) is currently always performed within a single tool (i.e. in a single approach which creates many overlapping with other existing approaches). No tool (at least in the enterprise modelling domain) currently supports the combination of its own models with models created with a language supported by another tool.

A *completely integrated language* allowing the creation of models combining all needed aspects of the reality is probably unachievable and the supporting tools for that language would be too complex to build. Therefore, the only reasonable approach seems to be to create a modelling environment from "legacy" enterprise modelling tools (and languages) allowing to reuse existing models and to leverage these existing models and tools into an *integrated environment.* This integrated environment should also be complemented with a process for extending, in a controlled way, a set of limited constructs belonging to a *core language.* As we will see in the remainder (section 3.3), a sample of this process has been defined during the UEML project.

3. THE UEML

The UEML project[1] was set up in an attempt to contribute to the solving of the problems of multiple EMLs. The long term objective of UEML is the definition of a core language called Unified Enterprise Modelling Language, which would serve as an *Interlingua* between EM tools. This language will:

- Provide the business community with a common visual, template based language to be used on top of most commercial enterprise modelling and workflow software tools;
- Provide standardised mechanisms for sharing and exchanging enterprise models among projects, overcoming tool dependencies;
- Support the implementation of open and evolutionary enterprise model repositories to leverage enterprise knowledge engineering services and capabilities.

In order to prepare this long term objective, the UEML project was initiated with the objective to create and manage a working group aiming to:

- Create a European Consensus on a core set of modelling constructs and facilitating interoperability in the frame of on-going standardisation efforts in this domain.
- Build a demonstrator portal with services and contents to support and promote, testing, industrial validation, and to collect comments.

The first objective of the project was to analyse the market potential of a UEML, to accurately define the specifications of an embryo of such a language and to demonstrate and disseminate the concepts.

3.1 The need for UEML

Two main efforts related to the definition of a common core language for enterprise modelling are PSL (PSL, 2002) and ENV12204[2] that however do not currently provide a satisfactory answer to critical and also practical problems. PSL is a *logic-based approach* that does not clearly address the problem of the basic *mapping* between a generic EML and PSL itself. This mapping should e.g. state, for instance, that the concept of *Activity* belonging to an EML corresponds to the concept of *Activity* in PSL. Being a *declarative language*, PSL allows discovering *inconsistencies* between distinct models provided in distinct EMLs. However, it neither prevents nor *solves* these inconsistencies. ENV12204 is merely a set of useful concepts for understanding the *domain of enterprise modelling* (or even the set of things that need to be represented by any enterprise modelling language). However, its *syntax* is not well defined and therefore it cannot be used as an *exchange format* between distinct tools. It also does not define mappings between existing EMLs and itself.

The usefulness of a UEML would therefore reside in the availability of a *well-defined syntax* and *well-defined mappings* (possibly *standardised*) between various EMLs and UEML. We believe that the definition of mappings between languages and UEML is important but quite *independent* from the UEML definition itself. Thought, they should be precisely defined and shared (through, for instance, standardisation), they should base on *reasonable hypotheses* and will never be fully (and formally) provable.

Other approaches which attempt to solve the *problems of exchange and interoperability* between computerised systems do not deal clearly with the enterprise modelling area. They can be classified as ways for enabling *business level communication* between distinct *computer-based systems* and therefore as bottom-up approaches. For instance, ebXML, WPDL, EAI techniques are useful for defining communication at *business level* among

[1] UEML IST-TN 2001 34229, www.ueml.org

[2] A new version of ENV 12204, EN ISO 19440 is expected in early 2004.

enterprise software. These approaches are really useful for *programming software layers* such as *middleware* (e.g. CORBA) and *customising software architectures*. This description of purely software aspects can be considered as a kind of *high level programming* that anyway requires *enterprise modelling as a prerequisite*.

However, another prerequisite for the exchange of models (or to make models interoperable) through a common language to be meaningful is to clearly understand the *semantic links* existing between the models themselves. This understanding is fundamental because without it, an exchanged model could be understood in a totally different way by the receiving tool, and thus misinterpreted.

3.2 The need for meta-modelling

In order to understand the links between distinct languages, *meta-modelling* is an important issue. Meta-modelling allows defining the *syntax of a language*[3]. The product of meta-modelling is usually called a *meta-model*.

Meta-models need to be described by using *meta-modelling techniques* (i.e. languages for making meta-models). Approaches like XML (DTDs and Schemas), MOF, Telos, can be used. These techniques are *content-independent* (applicable for the definition of any language). Other meta-modelling techniques are *content-dependent (*and sometimes *domain-specific)*: for instance, XMI is an exchange format based on the meta-model of UML (UML, 2003) in XML designed for enabling exchange of UML models.

Accordingly, a *UEML could be defined as a content-dependent domain-specific meta-model* through a content-independent meta-model. The UEML might just use content-independent meta-modelling techniques[4] as a *way for its definition*.

Currently, several Meta-Modelling Languages (and also tools) exist but none of them are specifically targeted for the definition of EMLs. The reason is that these Meta-Modelling Languages were often developed to design and implement *Information Systems, Knowledge Base Systems and computer-based infrastructures* (environments) allowing to program meta-models.

3.3 The UEML approach

In the UEML project, a UEML meta-model was built on the basis of three existing languages (namely IEM, EEML and GRAI). An illustrative scenario was defined in which models of a common situation are stored in distinct software tools and exchanged. First, models of this scenario where elaborated in the three distinct languages and. the exchange was performed manually by specialists of these languages. More precisely, given a first model in IEM, specialists of GRAI and EEML provided the "semantically equivalent" model in their own languages. Afterward, IEM, EEML and GRAI constructs have been meta-modelled using UML class diagram. This resulted in three so-called "original meta-models". At the same time, the links between the concepts of every original meta-model and their use in the models of the Scenario were defined to illustrate how a unique real-world phenomenon is modelled with the three original meta-models.

With the aim to define a common meta-model for core constructs, we compared and "unified" the three meta-models through an incremental approach. We compared the three meta-models peer-to-peer to find any correspondence between a concept in one meta-model and a concept in another one.

Once the peer-to-peer correspondences (and absence of correspondences) had been defined, a set of common concepts were identified (Table 1) and further elaborated into the first version of the UEML meta-model 1.0 (Fig. 1). This meta-model represents the common concepts underlying the three original EMLs. This meta-model being remarkably different from the three meta-models by the use of an appropriate higher level of abstraction and considering some discrepancies among the three original meta-models, we informally re-defined new correspondences between the UEML meta-model and each original meta-model. Finally, the UEML meta-model and the new correspondences were validated against a subset of the Scenario.

3.4 Defining mappings among EMLs

The clear definition of the meta-models of existing EMLs and of UEML with meta-modelling techniques is necessary but not sufficient to achieve a meaningful exchange of models. The correspondences among constructs between two distinct languages have to be precisely defined by comparing semantics of these constructs. However, this is a difficult task because:

- EMLs are often based on informal semantics, i.e. some natural descriptions of constructs meaning.
- The way in which EMLs are used in specific context and situations may change.

Therefore, as suggested in Sect. 3.1, mappings between languages should rely on reasonable hypotheses should be clearly stated and become the base for building the language, and possibly be standardised further.

[3] A meta-model may also be used to define part or all of the semantics of the language. But this is often not recommended.

[4] It should be noted that the notion of meta-modelling technique is *relative*. In fact, it is often true that a language can be used as a meta-modelling technique for another (sometimes, the same) language.

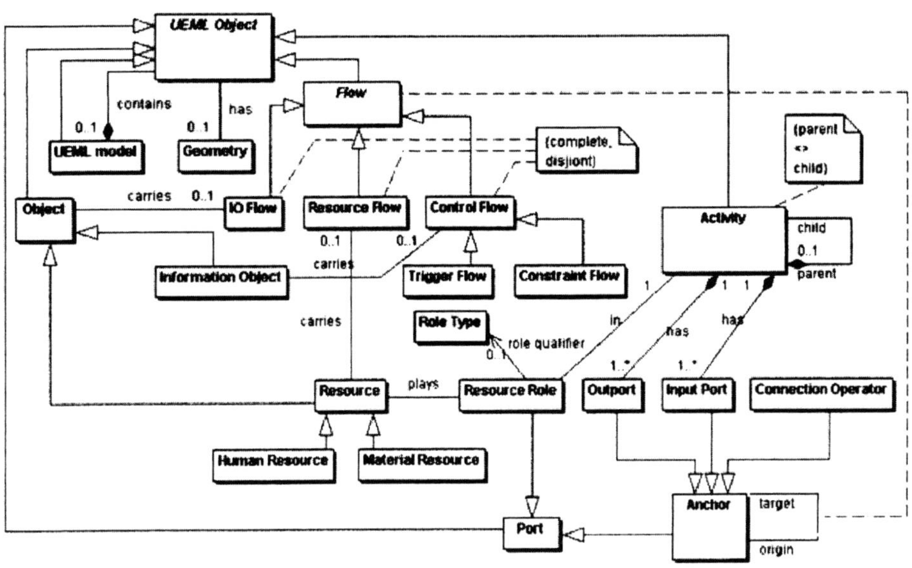

Fig. 1: The UEML 1.0 meta-model

Table 1: Extract of the common concepts of UEML

COMMON CONCEPT	GRAI	IEM	EEML
ACTIVITY	Extended activity	Action state	Task
ROLE	Not explicit	IEM Object state	Role
RESOURCE	Resource	Resource class	Resource
INPUT/OUTPUT FLOW	Input/Flow Ouput/Flow	Successor/ProcessElement	Flow (control flow= false)
CONSTRAINT FLOW	Control/Flow (trigger=false)	No direct	Flow (control flow= false and linked to Role)
CONTROL FLOW	Control/Flow (trigger=true)	ControlSuccessor/ProcessElement	Flow (control flow= true)
RESOURCE FLOW	Resource/Flow (trigger=false)	ResourceSuccessor/Resource State	Role (linked to Task)
CONNECTION OPERATOR	Logical operator	Connection Element State	DecisionPoint (not (Inport or Outport))
PORT	Connector	Port	Decisionpoint (Inport or Outport)

Mappings can be defined, more or less precisely, in various languages. For example, they can be expressed informally in natural language or through the use of a meta-modelling language. From a technical viewpoint, XSLT is a proposal to define transformation of XML documents based on a set of transformation rules expressed on the basis of XML schemas. The advantage of this approach is that software tools are already operational to interpret these mappings and apply them on XML documents. This approach could be considered as a means of implementing correspondences among EMLs, provided that these have XML syntax. Defining relationships at the language level can also be done in an "a priori" manner when new methodologies and methods are under definition. Therefore, a UEML can be a good starting base for placing under control the process of defining new methods and methodologies as well as the rules applied in a specific methodology.

4. CONCLUSION AND OUTLOOK

4.1 Future perspective for enterprise modelling

This analysis of the state of the art in enterprise modelling (Petit, *and al.*, 2002) demonstrates the need to define and develop a UEML approach to solve the current problems faced by workers in the enterprise modelling domain.

But such UEML approach can only be successful and effective under two conditions:

- Providing a global approach of interoperability among enterprise modelling software going further than only providing a common format of exchange;
- Making clear and effective the link between enterprise modelling and Enterprise Applications and Software.

4.2 UEML as a global approach to enterprise modelling

As stated earlier, a common exchange format, if deemed successful, cannot be described independently of mappings to and from existing EMLs. Furthermore, this requires the explicit definition of meta-models of the involved languages and of the mappings among their concepts. However, in order to avoid that UEML, as a common format, becomes yet another language among the large set of existing ones, it requires a larger view of interoperability among EM tools. The UEML language and approach must be flexible to be able to cope with future proprietary emerging languages and with the evolution of existing EMLs. The long term objectives of a

UEML approach would then be to provide the necessary concepts and tools to achieve the following:

- *Interoperability* between already existing supporting *tools as well as newly developed tools*,
- Well-founded integration basis between distinct *enterprise modelling languages*,
- Consistent global models on which also distinct *methodologies* can be integrated,
- Improvement of *existing methodologies* and definition of *new methodologies*.

These objectives pose a number of requirements on the UEML approach:

- The availability of meta-modelling concepts, methods and tools to properly define EMLs (existing ones, new emerging ones, UEML, their extensions and particularisations for specific purposes or applications);
- The availability of concepts, methods and tools to properly define relationships among distinct EMLs and a UEML together with relationships between models created with different EMLs and UEML;
- The concepts and tools to properly define methodologies associated with EMLs, including UEML;
- The specification of an *open architecture* in which all these elements can be implemented to provide an evolutionary multi-language platform for enterprise modelling centred on UEML.

This platform would allow creating *coherent, global and logically centralised* (integrated) models of the enterprise but which may be distributed within different enterprise modelling applications at a *physical level*.

4.3 From enterprise modelling to enterprise systems

We consider that Enterprise Engineering or Enterprise Modelling make sense provided that we are able to link the tools developed in this field with the Enterprise Applications and Software. Enterprise Modelling must be considered as the way to design the *architecture* and the *model* of the enterprise independently from existing or future Enterprise Applications and Software. We see the various levels as shown in the Fig. 2.

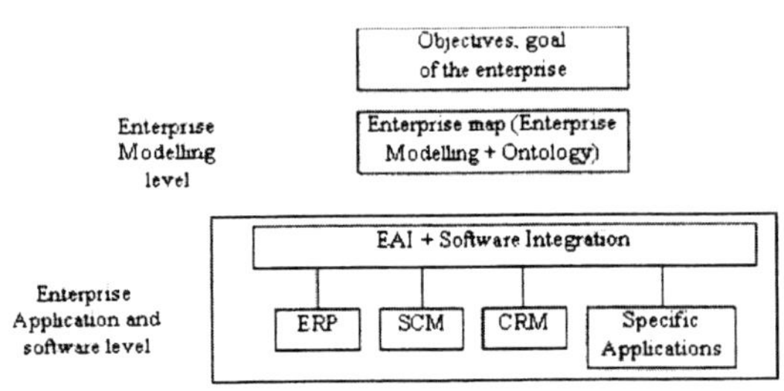

Fig. 2: Enterprise Modelling and Enterprise Application Levels

To our opinion, the future within ten years will be capabilities to generate, from the enterprise modelling level, the specifications allowing to choose or customize Enterprise Applications and Software (EAS) or to derive specifications of software applications for the enterprise models. Moreover, we believe that the EAS such as ERP, SCM or CRM will no longer have a centralised structure. Rather, they will have a modular and distributed structure. Each module will be chosen and customized according to specifications generated from the enterprise models. On the one hand, such a structure might *decrease* the time and cost of EAS development, and *increase* the adequacy of EAS to the needs of the enterprise. This should result in an increased acceptance of EAS by end-users, higher *adaptability* of EAS to the enterprise structure and improved *Return on Investment* of EAS development.

REFERENCES

AMICE. (1993). CIMOSA: Open System Architecture for CIM, *Research Reports ESPRIT*, **1**, Project 688/5288. Springer-Verlag.

Bernus P., Mertins K., and Schmidt G. (Eds) (1998). *Handbook on Architectures of Information Systems*. Springer Verlag, ISBN 3-540 64 453 9.

CEN (1995), CEN/CENELEC ENV 12204. *Advanced Manufacturing Technology, Systems Architecture, Constructs for Enterprise Modelling*, CEN, Brussels.

Doumeingts G., Vallespir B. and Chen D. (1998). Decision modelling GRAI grid, Chapter in: P. Bernus, K. Mertins, G. Schmidt (Eds.) *Handbook on architecture for Information Systems*, Springer-Verlag.

GERAM (1997). Generalized Enterprise Reference Architecture and Methodology Version 1.5, *IFAC-IFIP Task Force on Enterprise Integration*.

Kosanke K., Vernadat F.B., and Zelm M. (Eds.) (1999). CIMOSA: CIM open systems architecture - evolution and application in enterprise engineering and integration. *Computers in Industry, special issue*, **40(2-3)**.

Petit M, *and al.* (2002). *D1.1: State of the Art in Enterprise Modelling*, UEML TN IST 2001 34229, www.ueml.org

PSL (2002). Industrial automation system and integration - *Process specification language: Part 11: PSL-Core*. ISO/CD 18629-11.

UML (2002). *UML 1.5 specification*, OMG.

Vernadat F.B. (1996). *Enterprise modelling and integration: principles and applications*. Chapman & Hall.

ACKNOWLEDGEMENT

The authors would like to thank all the UEML core members for their scientific contribution to this work. This work was funded by the European Commission IST 5th framework programme.

Copyright © IFAC Information Control Problems in
Manufacturing, Salvador, Brazil, 2004

ELSEVIER
IFAC
PUBLICATIONS
www.elsevier.com/locate/ifac

SIMULATION-BASED BENCHMARKING OF PRODUCTION CONTROL APPROACHES FOR COMPLEX MANUFACTURING SYSTEMS

Lars Mönch

Institute of Information Systems
Technical University of Ilmenau
Helmholtzplatz 3
98684 Ilmenau, Germany

Abstract: In this paper, we describe benchmarking efforts for production control approaches applied to complex manufacturing systems. Complex manufacturing systems are characterized by a large number of products, an over time changing product mix, sequence-dependent set-up times, unrelated parallel machines, a mix of different process types and internal and external disturbances. In order to ensure comparison possibilities among different production control approaches, we suggest to use a simulation test-bed and a software architecture that allows for plug in of arbitrary production control software. We discuss different application areas, the advantages and also the limitations of the suggested approach. *Copyright © 2004 IFAC*

Keywords: Complex manufacturing systems, benchmark examples, simulation-based performance evaluation, performance measures, and software architecture.

1. INTRODUCTION

Complex manufacturing systems are characterized by

- a large number of products with an over time changing product mix,
- sequence dependent set-up times,
- unrelated parallel machines,
- a mix of different process types, including batch processes,
- different types of internal and external disturbances,
- re-entrant process flows due to very expensive machinery.

Manufacturing systems with these characteristics can be found in the electronics industry and especially in semiconductor manufacturing.

Currently, it seems that the improvement of operational processes creates the best opportunity to realize the necessary cost reductions. Therefore, the development of efficient planning and control strategies is highly desirable for complex manufacturing systems, especially for the semiconductor wafer fabrication domain (Schömig and Fowler 2000). In the course of the development of new planning and control algorithms, the researchers and developers have to take into account the new opportunities in advanced software technologies.

Usually, it is difficult to predict the performance of new production control strategies in advance on the level of the entire manufacturing system. There are two reasons for this fact. Usually, because of the complexity of the basis manufacturing system and of the basis production process, the new production control approaches influence by design only parts of the manufacturing system. For example, very often only the leading bottleneck machines are considered for more advanced and sophisticated production control strategies. However, in reality, the impact of new production control strategies has to be measured on the entire manufacturing system level. The second reason is the following one. Centralized scheduling heuristics from operations research usually are assessed by a suite of static benchmark problems (cf. for example the OR-Library 2003). In complex manufacturing systems, however, due to the assumed internal disturbances static modelling avoiding the consideration of any stochastic behaviour of the manufacturing systems and of future lot releases is not appropriate.

Therefore, we focus in this paper on the use of discrete-event simulation techniques in order to emulate the complex manufacturing system and its stochastic behaviour.

The paper is organized as follows. In the next section, we discuss several requirements for benchmarking of complex manufacturing systems. Then we describe our benchmarking scheme and discuss related performance measures. In section 3, we describe the used software architecture. In section 4, we discuss a couple of applications of our method and software architecture. In the last section of the paper, we discuss advantages of our approach and discuss also limitations.

2. REQUIREMENT ANALYSIS FOR BENCHMARKING

We are starting with describing requirements for benchmarking of complex manufacturing systems. We distinguish between requirements that arise from a modelling point of view and requirements that take architecture issues into account.

2.1 Modelling Issues

A fixed basis system and a fixed basis process should be considered (cf. Cavalierie et al. 1999) in order to compare different production control approaches. We define the machinery of the shop-floor with its capabilities as basis system. We define the routs and the work load (represented by a work in process (WIP) distribution of the lots) of a certain manufacturing system as the basis process. Both basis system and basis process can be easily represented in a discrete-event simulation model. The simulation model is used for process identification. It is possible to incorporate stochastic factors like machine breakdowns, yield, priority of customer orders and processing times. Because most of the today's commercial discrete-event simulation tools have spreadsheet like interfaces it is possible to represent different simulation scenarios in ASCII files.

Usually it is a huge effort to create and maintain a simulation model for complex manufacturing systems. Verification and validation are also time consuming procedures. Therefore, the models should be available on the web in order to allow a free access for the research community. Using the same models provides the opportunity to better understand research results from the view point of solution quality and run time performance. In the semiconductor manufacturing domain such a test-bed is provided by the MASM test data sets that are available from http://www.eas.asu.edu/~masmlab. The test-bed contains eight simulation models of semiconductor wafer fabrication facilities of different level of complexity. A lot of publications refer to these models and use them for the experimental work presented in the papers (cf. Rose 2002 and Miyashita et al. 2003 amongst others).

Another requirement is the exact description of the treatment of exceptions. If, for example, a production control approach generates schedules for a complex manufacturing system then it is necessary to treat the case of non valid schedules appropriate. It should be described under what conditions a schedule is non valid. An ad hoc reaction and a rescheduling regime have to be documented.

Production control approaches are characterized by the frequency of its application and various parameters of a specific approach. In order to allow for comparison issues a representation of the parameters of the approach is necessary.

2.2 Architecture Issues

There are two possibilities to incorporate production control algorithms into a discrete-event simulation tool. In the first possibility, the production control algorithm basically uses the information and data on the level of the simulation tool. As a result of this strategy, we obtain proprietary source code that is difficult to understand and to maintain. It is rather complicated to implement different production control approaches. However, for comparison purposes the ability of plug in of different production control approaches is highly desirable. Examples for the first strategy are presented by Thiel et al. (1998) in a dispatching and monitoring context. Toba (1999) and Horiguchi et al. (2000) present examples for scheduling-based production control.

The second strategy is much more flexible. We use a blackboard type data layer (Mönch et al. 2003) that acts as a mirror of the production process emulated by the simulation model. In the case of the occurrence of well-defined events the simulation tool is responsible for an update of the corresponding objects of the data layer. The object model of the data layer is much easier to understand and to maintain as the proprietary data structures of a certain simulation tool. If the complex manufacturing system is segmented into different work areas a separate blackboard-type data layer can be assigned to each segment of the manufacturing system. The coupling of a discrete-event simulator and a multi-agent-system for performance assessment without a blackboard is described by Brennan and William (2000).

The architecture has to contain a persistency mechanism that supports object state and performance measure tracing. Using object-oriented databases seems to be appropriate for that purpose because of the highly nested objects from the data layer.

The build in mechanism for job selection and machine load of the discrete-event simulation tool acts as a dispatcher in a natural way.

Building production control applications from scratch is not very common because of available commercial and academic optimization and scheduling libraries, and editors for dispatching rules. Examples are the ILOG solver and scheduler classes (Le Pape 1994), class libraries for genetic algorithms (Pain and Reeves 2002), and the dispatching rule editor from the simulation tool AutoSched AP.

Therefore, an architecture for benchmarking has to taken this fact into account. Because most of the available class libraries are written in the C++ programming language the degree of freedom for choosing an appropriate implementation language for the architecture is reduced.

The suggested architecture allows for a plug in of different production control approaches. The suggested architecture is shown in Figure 1.

The architecture was implemented in the programming language C++ using the simulation tool AutoSched AP 7.1 (Mönch et al. 2003) and in the Java programming language using the simulation tool ParSIMONY (Mönch et al. 2002). The object oriented database POET was used in both implementations. In the first case, we use the subscription/notification mechanism of AutoSched AP in order to update the black-board type data layer. In the latter case, we developed our own event classes in order to update the objects of the data layer.

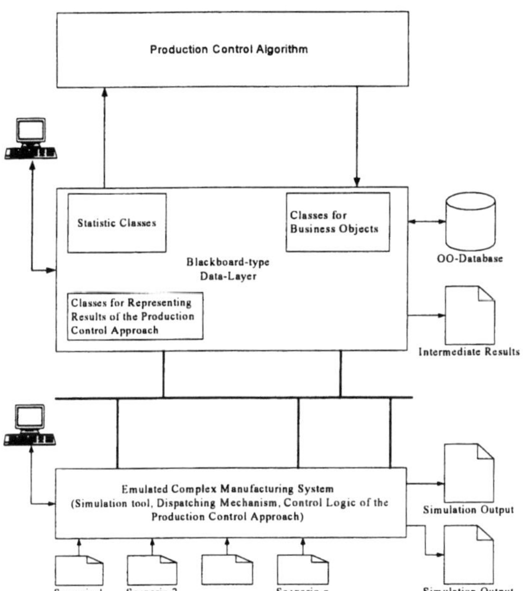

Fig. 1. Architecture for Benchmarking

Comparing our approach to the IMS-NoE (2004) network of excellence benchmarking approach we conclude that the IMS-NoE approach is an open approach that is much more flexible. However, in the situation of complex manufacturing systems where the development of the simulation models is difficult and time consuming our more specialized approach also offers some advantage, because most of the specifics of complex manufacturing systems are included into the data-layer.

3. BENCHMARK SCHEME

3.1 *Overall Methodology*

We consider a fixed production control approach with unknown performance. In order to assess its performance, we suggest the following scheme.

1. Determination of production control approaches used for comparison,
2. Determination and specification of the used performance measures,
3. Specification of the used performance assessment strategy,
4. Description of the hardware and software environment for the benchmark,
5. Description of different scenarios that should be simulated. This includes especially the description of designed experiments.

We distinguish between decentralized and centralized production control approaches. Decentralized approaches work on local data and coordination and cooperation issues become important in order to avoid the limitation of the myopic view of decentralized approaches. Dispatching rules are an example of such type of approaches. Contract-net based approaches from artificial intelligence are other prominent examples for decentralized approaches in manufacturing. Centralized approaches are given by mixed integer programming formulations, constraint satisfaction techniques, and various kinds of local search techniques.

Different possible performance measures are discussed in section 3.2.

The choice of a performance assessment strategy is important because of the stochastic behaviour of the complex manufacturing system. There are two possibilities to assess the performance of a system in simulation modelling.

We are interested in the stationary behaviour of a complex manufacturing system. After the warm-up period, we measure the performance indicators of interest. We have to perform long runs. However, because of the stationary behaviour, a small number of independent runs is enough (cf. Kelton and Law 2000). The performance measure values are taken as the average of the performance measure values for the single runs.

Looking at the transient behaviour of a manufacturing system is the second possibility. Transient behaviour appears during the transition phase from one system state to another one. Product mix changes, ramp up and introduction of new machines leads to transient behaviour of the manufacturing system. Studying a system during the transition phase requires the measurement of the performance measures of interest in a time dependent manner. As opposite to the stationary case, a considerably larger number of runs is required in order to obtain statistically significant results for the performance measures. The length of a single run is determined by the length of the transition phase. The performance measure values are calculated as the average of the performance measure values obtained at a single point of time.

A performance measure assessment strategy is basically given by the decision whether to investigate a stationary manufacturing system or a manufacturing system in the transient phase from a given stationary to another stationary behaviour. Furthermore, the number of replications and the run length are also part of the performance assessment strategy.

In order to compare different production control algorithms from a run time behaviour point of view it is required to fix a hardware and software environment for the benchmark experiments. In the case of distributed production control algorithms it is furthermore required to describe the used load balancing strategy.

Benchmark experiments should organized as different simulation scenarios. A simulation scenario is given by a set of independent parameters, a range of variation for the parameters and a set of dependent variables (basically the performance measure values of interest). A single scenario is represented by a set of ASCII files as shown in Figure 1. By using designed experiments (Montgomery (1997)) and meta modelling techniques like response surface technology it is possible to reduce the number of required simulation runs.

3.2 Performance Measures

We distinguish between direct and indirect performance measures.

Direct performance measures consider quantities that have direct influence on the performance of the manufacturing system. In Table 1, we see examples for direct performance measures.

Table 1 Classes of Direct Performance Measures

Class	Examples
Due-Date-Oriented	Total tardiness
Throughput-Oriented	Throughput of a certain work area
Cycle-Time-Oriented	Average lead-time of a certain product
Load-Oriented	Work in Process

Indirect performance measures do not directly influence the performance of the manufacturing system. They are used to evaluate properties of certain production control algorithms.

Table 2. Classes of Indirect Performance Measures

Class	Examples
Quality of Planes	Average deviation from completion times to planned completion times of a fixed reference plan
Robustness of Planes	Number of required rescheduling activities
Run-Time-Oriented	Run time of a certain production control algorithm

Note, that we mainly consider quantitative performance measures. The consideration of qualitative performance measures like re-configurability or robustness is usually more sophisticated and requires sometimes an over time changing simulation model. Hence this type of measurement is not so well supported.

4. APPLICATIONS

We consider applications in dispatching, scheduling of semiconductor wafer fabrication facilities and multi-agent-based production control of job shops.

4.1 Case Study

We are interested in the behaviour of a dispatching rule applied to a semiconductor manufacturing facility. Therefore, we consider the Mini-Fab-Model from the MASM test-bed. It contains seven machines, sequence dependent set-up times, batch machines and reentrant process flows. We use a variant of the Apparent Tardiness Cost (ATC) dispatching rule that also takes sequence dependent set-up times into account. The corresponding priority index is given by

$$ATCS_j := \frac{w_j}{p_j} exp\left(-\frac{(d_j - p_j - t)^+}{\kappa_1 \bar{p}} \right) exp\left(-\frac{s_j}{\kappa_2 \bar{s}} \right) \quad (1)$$

where we denote by

w_j : weight of job *j*,
p_j : processing time of job *j*,
d_j : due date of job *j*,
t : time for decision-making,
s_j : set-up time of job *j*,
$\bar{p}$: average processing time of the jobs,
$\bar{s}$ average set-up time of the jobs.

Here, we use the notation $x^+ := max(x,0)$ for abbreviation. We choose the job with the highest ATCS index to be processed next if the machine becomes available. Certain variations of the index (1) are possible that take batching into account. Starting from a fixed κ_1 value we are interested in how much we can gain from choosing an appropriate value for the set-up scaling factor κ_2.

For benchmarking purpose, we apply the benchmark scheme described in section 3.1.

1. For comparison we use the FIFO dispatching rule and its variants for batching.
2. We consider two different performance measures. We are interested in the total weighted tardiness of the jobs on an entire manufacturing level. Additionally, we look for the sum of the set-up times.
3. We perform a steady state simulation. Therefore, we run the simulation 180 days including a warm-up time of 15 days. We do five independent replications of each simulation run in order to obtain statistically significant results.

4. The experiments were run on a 2-GHz-Pentium-4-PC. We use the simulation tool AutoSched AP 7.1.
5. We consider two different scenarios that take different work load of the manufacturing system into account. A single scenario in each group is given by choosing a specific value for the set-up scaling factor κ_2.

In Figure 2, we see the results for the different high-load scenarios by means of tardiness and required set-up times. Tardiness and set-up times are represented as percentage of the corresponding FIFO results. The different values for κ_2 are presented at the axis x.

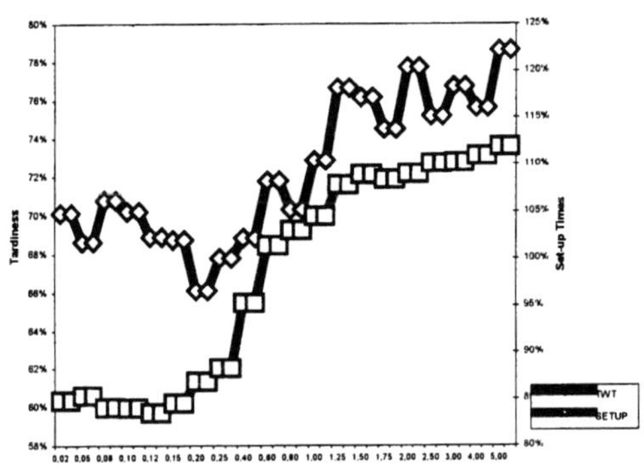

Fig. 2. Performance Values for a High-Load System

In Figure 3, we see the corresponding results for the low-load scenarios.

It can be seen from Figure 2 and Figure 3 that values for the set-up scaling factor κ_2 exist such that both tardiness and set-up times are small. In the case of a system with high load there are on the other hand parameter values for κ_2 such that tardiness is very small and the required set-up times are high. Therefore, the selection of appropriate values for κ_2 should be done in a system-load depending manner.

4.2 Other Work
The performance assessment of certain dispatching heuristics in semiconductor manufacturing based on the described architecture is presented in Mönch and Habenicht (2003). A batch is defined as a set of lots that are processed at the same time on the same machine. The batching heuristics that take future lot arrivals into account are compared against First In First Out (FIFO) dispatching rules. We use the MIMAC 1 simulation model (Fowler and Robinson 1995) from the MASM test-bed. We use the commercial simulation tool AutoSched AP 7.1.

The performance assessment of a production scheduling approach based on the presented architecture is described by Habenicht and Mönch (2002). The production control algorithm is implemented using the ILOG solver libraries. The algorithm is benchmarked against an infinite capacity production scheduling approach. We use the discrete event simulator AutoSched AP 7.1 in this case study. Again, the MIMAC 1 test data set is used in this study.

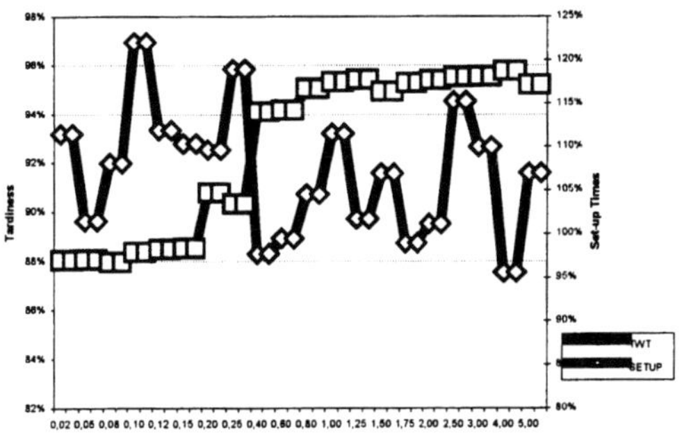

Fig. 3. Performance Values for a Low-Load System

In a third case study (Mönch et al. 2002), we consider a multi-agent-system (MAS) for resource allocation in manufacturing systems. A certain budget is assigned to each single job. On the other hand, a job has to spend a certain amount of his budget in order to be processed by a certain machine at a given point of time. We compare the multi-agent approach against four different dispatching rules. We used the discrete event simulator PARSiMONY and implemented the blackboard-type data layer in the Java programming language. The multi-agent-system was implemented by using the JAFMAS agent framework.

5. ADAVANTAGES AND LIMITATIONS OF THE BENCHMARK SCHEME

5.1 Advantages of the approach
The simulation-based approach has the advantage that we are able to consider dynamic situations. We consider, for example, future lot release information in dynamic situations. Very often, the behaviour and therefore the performance of a certain production control algorithm is different in a static and a dynamic environment.

In real world situations, we have to face with stochastic behaviour of the complex manufacturing system. Examples for stochastic behaviour are machine break downs or changes of job priorities due to changed customer needs. A production control algorithm should take this behaviour into account. The performance of a certain production control algorithm may change under stochastic process conditions compared to deterministic one.

By using a simulation-based approach, we can treat the dynamic and stochastic behaviour of the complex manufacturing system in an appropriate manner.

By using simulation models from a test-bed, it is possible to decrease the data collection, modelling, validation and verification efforts for the model tremendous. On the other hand, the simulation test-bed approach ensures comparability of the results.

5.2 Limitations

The main limitation of the suggested approach is the time consuming behaviour of the simulation runs in connection with the repeated runs of the production control algorithms. Therefore, in the case of complex production systems long simulation runs are often completely impossible in combination with rolling horizon approaches.

A second limitation is given by the design of the software architecture. The objects of the blackboard-type data layer are located in the memory of the computer. In the case of complex manufacturing systems a large number of related objects exist. Hence, the needed memory is also large. The necessary run time for a simulation run increases if the available memory is not enough.

A third major disadvantage comes from the simulation approach itself. Because complex manufacturing systems are stochastic systems, a certain number of independent simulation replications are needed. Hence, the run time of the benchmark experiments increases even more.

6. CONCLUSIONS

In this paper, we discuss benchmarking issues for complex manufacturing systems. We derive requirements for benchmarking from a modelling and also from a software architecture point of view. It turned out that a simulation-based approach has a lot of advantages compared to the common static benchmark schemes from operations research. After discussing a couple of applications of our scheme we show both advantages and disadvantages of the simulation-based approach.

In future work, more research efforts are needed to reduce or even eliminate the limitations of the suggested approach. In order to reduce the run time of the simulation it seems to be worth to consider distributed implementations of the production control algorithms. Another type of effort could be done in reducing the number of independent simulation runs by using more advanced meta modelling techniques.

REFERENCES

Brennan, R. W. and O. William. (2000). A Simulation Test-Bed to Evaluate Multi-Agent Control of Manufacturing Systems. *Proceedings of the 2000 Winter Simulation Conference,* 1747-1756.

Cavalierie, S., L. Bongaerts, M. Macchi, M. Taisch and J. Wyns. (1999). A Benchmark Framework for Manufacturing Control. *Second International Workshop on Intelligent Manufacturing Systems.*

Fowler, J. W. and J. Robinson. (1995). Measurement and improvement of manufacturing capacities (MIMAC): Final report. *Technical Report 95062861A-TR*, SEMATECH, Austin, TX.

Habenicht, I. and L. Mönch. (2002). A Finite-Capacity Beam-Search Algorithm for Production Control in Semiconductor Manufacturing. *Proceedings of the 2002 Winter Simulation Conference,* 1406-1413.

Horiguchi, K., N. Raghavani, R. Uzsoy and S. Venkatheswaran. (2001). Finite-Capacity Production Planning Algorithms for a Semiconductor Wafer Fabrication Facility. *International Journal of Production Research,* Vol. 39, No. 5, 825-842.

IMS-NoE network of excellence (2004). http://www.ims-noe.org/.

Law, A. W. and W. D. Kelton. (2000). *Simulation Modeling and Analysis.* Third edition. New York, McGraw-Hill.

Le Pape, C. (1994). Implementation of Resource Constraints in ILOG SCHEDULE: A Library for the Development of Constraint-Based Scheduling Systems. *Intelligent Systems Engineering,* 3, 55-66.

Miyashita, K., K. Senoh, H. Ozaki and H. Matsuo. (2003). Constant Time Interval Production Planning with Application to WIP Control in Semiconductor Fabrication. *Proceedings of the 2003 Winter Simulation Conference.* 1329-1337.

Mönch, L., M. Stehli, and R. Schulz. (2002). An Agent-based Architecture for Solving Dynamic Ressource Allocation Problems in Manufacturing. *Proceedings of the 14th European Simulation Symposium. Simulation in Industry*, 331–337.

Mönch, L., O. Rose and R. Sturm (2003). Simulation Framework for Performance Assessment of Shop Floor Control Systems. *Simulation – Transactions of the Society for Modelling and Simulation International, Vol. 79, No. 3, 60-67.*

Mönch, L. and I. Habenicht. (2003). Simulation-based Assessment of Batching Heuristics in Semiconductor Manufacturing. *Proceedings of the 2003 Winter Simulation Conference,* 1338-1345.

Montgomery, D. C. (1997*). The Design and Analysis of Experiments.* Fourth Edition, New York, John Wiley and Sons.

OR-Library (2003). http://mscmga.ms.ic.ac.uk/info.-html

Pain, A. R. and C. R. Reeves. (2002). Genetic Algorithm Optimization Software Class Libraries. In S. Voss, David L. Woodruff (eds.) *Optimization Software Class Libraries*, Kluwer Academic Publishers, Boston.

Rose, O. (2002). Some Issues of the Critical Ratio Dipatch Rule in Semiconductor Manufacturing. *Proceedings of the 2002 Winter Simulation Conference. 1401-1405.*

Schömig, A. and J. W. Fowler. (2000). Modeling Semiconductor Manufacturing Operations. *Proceedings of the 9th ASIM Dedicated Conference Simulation in Production and Logistics,* eds.: K. Mertins and M. Rabe, 55-64.

Thiel, M., R. Schulz and P. Gmilkowsky. (1998). Simulation-based Production Control in the Semiconductor Industry. *Proceedings of the 1998 Winter Simulation Conference*, 1029-1033.

Toba, H. (1999). Segment-Based Approach for Real-Time Reactive Rescheduling for Automatic Manufacturing Control. *IEEE Transactions on Semiconductor Manufacturing,* Vol. 13, No.3, 1999, 264-272.

Copyright © IFAC Information Control Problems in
Manufacturing, Salvador, Brazil, 2004

ELSEVIER
IFAC
PUBLICATIONS
www.elsevier.com/locate/ifac

WEB SERVICES: AN INTEROPERABILITY SOLUTION IN EXTENDED/VIRTUAL ENTERPRISES

Nikos Giannopoulos[1], Rajkumar Roy[1], Thierry Divoux[2], Maria-Jose Nunez[3], Amparo Roca de Togores[3], Pedro Malo[4]

[1]*Enterprise Integration, Cranfield University, UK*
{N.Giannopoulos, R.Roy}@cranfield.ac.uk
[2]*Research Centre for Automatic Control, CRAN CNRS, Université Henri Poincaré Nancy I, France*
Thierry.Divoux@cran.uhp-nancy.fr
[3]*AIDIMA, Spain*
{mjnunez, aroca}@aidima.es
[4]*UNINOVA, Intelligent Robotics Centre, Portugal*
pmm@uninova.pt

Abstract: Nowadays, companies form Extended/Virtual Enterprises in order to react better in a fast changing environment. These co-operations require a seamless flow of information between the companies-partners. However, companies today use a wide variety of applications and technologies, which, most of the times are not compatible with each other. This makes interoperability a major issue for a successful Extended/Virtual Enterprise. In this paper, the authors present the Web Services technology and why they are the most prominent technology on solving the interoperability problem. The work described in this paper has been performed in the initial stages of the IMS Smart-fm (IST-2001-52224) project. *Copyright © 2004 IFAC*

Keywords: Web services, extended enterprise, virtual enterprise, interoperability.

1. INTRODUCTION

Following approaches like MRP, JIT, etc., (Jagdev and Browne, 1998), the concept of Extended/Virtual Enterprise (EE/VE) has been proposed as the new organisational form companies have to adopt in order to work more effectively and efficiently in a very competitive environment (Browne and Zhang, 1999). Under the EE/VE concept, companies from different parts of the supply chain (manufacturers, suppliers, etc.) come together and work as one, integrated enterprise, to achieve competitive advantages and offer better level of service to their customers. Advancing from independent companies to an integrated EE/VE, interoperability is a major issue. A reliable IT infrastructure, where there is seamless flow of information, is the backbone of an EE/VE (Slade and Bookma, 2002). However, companies are using a great variety of incompatible applications, technologies, middleware solutions, etc., with not any of these having evolved as a standard (Vawter and Roman, 2001). This makes the exchange of data extremely difficult and proves interoperability as a major issue in an EE/VE.

Recently, the technology of Web Services is being proposed as a solution to this problem. This paper presents the Web Services concept and why this technology could be a solution in the seamless data exchange between different applications. Its structure is as follows: in the second section, the concept of EE/VE is presented. In the third section, an analysis of EE/VE frameworks and infrastructures along with their ability to support interoperability follows. In the fourth section, an assessment of approaches/technologies as a solution to the interoperability issue is performed, whereas in section five, the reader is being introduced to Web Services technology, and why this technology can solve the interoperability problem. At the end, at section 6, the problems and limitations associated with the Web Services technology are being presented.

2. EE/VE AND RELATIVE DEFINITIONS

Literature indicates that there is a terminology confusion surrounding EE/VE, since terms such as EE, VE and Supply Chain Management (SCM) are being used to describe the concept of different companies working together as if they were a single enterprise.

Giannopoulos et al (2003) analyse various definitions of EE, VE and SCM based on their similarities and differences, and they conclude that the frontiers between EE and VE are blur, because EE and VE apart from their differences share many similarities (Browne and Zhang, 1999; Camariha-Matos, et al., 1998).

The similarity between EE and VE concepts is that they both suggest that companies whose expertise lies in different areas should come and work together, and so, by complementing each other, they can exploit more business opportunities, perform better and offer better level of service than each one alone could (Szegheo and Petersen, 2000; Browne and Zhang, 1999).

The major difference between the two concepts is that VE consists of a group of companies that have come together, for a limited time, to pursue a specific target opportunity and it operates only as long as this opportunity it was formed to take advantage of exists (Szegheo and Petersen, 2000; Jagdev and Browne, 1998). EE, on the contrary, requires stability, because it is a group of companies that have formed a permanent partnership, and for this partnership there is no time limit (Browne and Zhang, 1999). VE is based in advanced IT solutions, which have to make all the partner companies IT infrastructures work together seamlessly and efficiently for the time the VE exists (Browne and Zhang, 1999).

An EE is also different from an SCM, because an SCM is designed to accommodate an environment which is highly unlikely to change, or, even if it changes, its changes are easy to predict. The need to cope with unforeseeable changes in a fast changing environment is the reason companies choose to shift from an SCM to an EE/VE (Aerts et al, 2000).

Regardless of their similarities and differences, both EE/VE require a stable and interoperable IT infrastructure to support them. So, for the scope of this paper and only regarding their ICT infrastructure, the terms EE and VE will be used interchangeably.

3. EVALUATION OF EXISTING FRAMEWORKS AND INFRASTRUCTURES FOR EE/VE

There is a large number of frameworks proposed in the literature regarding an EE/VE formation, support and operation, covering from prerequisites and transition phases (Jagdev and Thoben, 2001) to describing and modelling the social environment and the human behaviour of the people involved (Pearson et al, 2002). They are presented, in summary, by Giannopoulos, *et al.* (2003).

However, each company interacts with a fast changing environment and brings along its own cultural, technological and organisational mix (Esterm et al, 2002). So, not all companies could fit in one, single framework; frameworks should rather serve as guides, projecting the potential dangers during the transition from the independent company to EE/VE member status.

In addition to the frameworks, there is a large number of projects worldwide, where interoperability issues are being addressed by proposing common underlying platforms. Typical examples are the E-Construct (www.econstruct.org) and the ECOS (www.funstep.org/ecos) projects.

Giannopoulos et al (2003) present a list of these projects and their proposed infrastructures and analyse them based on their comonalities, differences, and the technology used. By this analysis, it is concluded that two major approaches seem to emerge. The first one is having a central point of contact (portal or web-site) where all enterprises connect they exchange data using a pre-determined communication standard (EDI, XML, etc), whereas the second approach is having all the member companies sharing a common platform for communicating and exchanging data.

This approach follows the 'hub-and-spoke' architecture, with the use of 'adapters' on top of each application and 'message brokers' as the central exchange mechanism. In this architecture, the 'adapter' translates the data in a generic format and sends them to the 'broker', who locates the target application (the application where the data should be sent), performs the appropriate transformations (parses it, duplicates or transforms it) bringing the data in a format the target application can understand and then transmits the data to it (Edwards and Newing, 2000; Slater, 2000).

Nevertheless, neither the portal approach nor the adoption of a platform seems to be the answer to interoperability. When using a portal, if a company wants to connect to it, this company would have to understand the technology, as well as the data representation and categorisation this portal is using, and adjust its technology to the portal's (Borras et al, 2002).

In the case of using the same platform, consider the scenario of a company willing to join an already formed EE/VE. Even if both the EE/VE and the company run on the same platform, it is highly unlikely that they have the same datasets or the same translation and transportation mechanisms for their data, so the new company would have to customise its systems to the format of the EE/VE, something that is not always achievable (Vawter and Roman, 2001).

4. EXISTING TECHNOLOGIES AND EE/VE INTEROPERABILITY

The most used medium for data exchange has been EDI. It is in various standards (Zinner, 1999) and it allows automated information exchange by extracting files out of which it removes any links with other files, and then it translates them in EDI standard format. It uses messages to transfer data from the source to the target application.

Zinner (1999) suggests that with so many alternative EDI standards, it is very difficult for EDI to be broadly adopted. In addition, EDI has so far been adopted by large firms, because it requires costly and complex software to be installed in both the communicated ends, something that not all companies can afford (Stackpole, 2000; Zinner, 1999). The above aforementioned drawbacks of EDI and the growth of Internet as the global e-business basis, gave companies the advantage of using open standards such as Extensible Markup Language (XML) to accommodate their transactions.

Although XML is a cheaper solution than EDI it still does not provide a solution to the interoperability problem. In case a company attempts to access a partner by constructing and sending an XML message, it is highly unlikely the company and the remote partner will be using the same XML schemas, because today more than 500 different XML schemas exist (Whitlock, 2001). XML-based interoperability is also restricted by the creation of tags by various user groups (Cingil et al, 2000), and because it does not support process integration (Duke et al, 1999).

As an answer to the interoperability problem, the Web Services technology is being suggested by various authors and researchers. The concept of Web Services is that each application or process could be decomposed in easy to access, affordable and easy to use reusable components, which are called 'services' (Knorr, 2001). These services can be used by any device and integrate and communicate with any application regardless of the underlying platform, since they can be requested (or published) over the Internet, and brought to work together using XML.

Web Services are based on widely available and established standards and protocols (i.e. XML, HTTP). Since they will be invoked over the Internet a common standard has to be used for formatting the information to be exchanged (Platt, 2000). The most suitable standard is XML, because (a) it is a language that can describe in detail and with consistency any applications content (b) it is designed for document exchange over the Internet (Janssen et al, 2000).

Although XML can be transferred over any transportation protocol (i.e. TCP/IP, SMTP, etc.), HTTP is chosen, since as the Internet's transportation protocol is widely used and it can also pass through firewalls (Platt, 2000). Since the message should be able to be invoked over the internet, there must be a way to define how the request for the specific service should be invoked as well as how the application that will receive this service should process it (Janssen et al, 2000). For the Web Services, messaging protocol is the Simple Object Access Protocol (SOAP), which is the protocol for XML-based messages transportation.

SOAP consists of three parts: the envelope, the encoding rules, and the RPC (Remote Procedure Calls) conventions:

- The envelope contains information about who and how should process it.
- The encoding rules contain information on the data exchange rules
- The RPC conventions contain information on how someone should format his request for a particular service or to respond to the one that it was sent to him.

The other two components of the Web Services technology are the Universal Description, Discovery, and Integration (UDDI) database, and the Web Services Description Language (WSDL). UDDI is the Internet based registry of Web services so services can be located and browsed (McNally, 2003), whereas WSDL describes what a web service does, how to format a request for a Web Service and how to respond to a request.

5. INTEROPERABILITY AND WEB SERVICES

The Web Services concept may sound similar to what both CORBA (Common Object Request Broker Architecture) and DCOM (Distributed Component Object Model) aimed to provide. However, CORBA failed because it is complex and expensive (Bansal and Pal, 2001) and its transportation protocol (IIOP) can not pass firewalls (Faruqui, 2002).

In addition, there are not many CORBA services implemented by the ORB (Object Request Broker) vendors. This means that if a company can not find the CORBA-compatible service it requires, it would have to build one of its own (Ruh et al., 2000). DCOM failed because it can not, like CORBA, work through firewalls too and because it works only in Windows platforms (Platt, 2000; Faruqui, 2002).

Web Services offer an easier and cost effective way for data exchange, since they eliminate the middleware on the 'hub-and-spoke' architecture, they work independently of any underlying platform and even allow legacy systems to integrate and share data using XML. In addition, web services provide the means for locating services (and therefore partners) through the (UDDI) and ways of using these services (through the WDSL) (McNally, 2003), making by this way possible the creation of a process through a number of services. All these make Web Services an easy, quick and cost effective integration medium (Vawter and Roman, 2001; Knorr, 2001).

Borras et al (2002) note that industry benchmarks e-business technologies using a set of criteria, known as the 'perception of risk':

- Acquiring Costs
- Implementation Costs
- Training Costs
- Openness of the solution (proprietary versus established standards, open architecture –easy to integrate and work with).

Using the above criteria and based on the presentation of each of the existing technologies for EE/VE interoperability, table 1 has been derived. However, for the table creation qualitative metrics were used, since it is was not possible to obtain exact values for the associated costs. From Table 1, it can be concluded that Web Services is an easy, quick and cost effective integration medium.

6. PROBLEMS AND LIMITATIONS

From what has been presented so far, Web Services present the potential to be a solution to interoperability problem: They use well established, standard protocols (HTTP, SOAP, etc), they constitute an open platform (since they are based on XML messaging) and they also constitute a quick and cost effective way of integration and data exchange. In other words, they satisfy the 'perception of risk' criteria.

Table 1: Comparison of Interoperability Technologies

Technology	Openness	Costs	Platform Independence	Process Integration	Locating Partners
EDI	Partial	Expensive	YES	NO	NO
XML	Partial	Cheaper than EDI	YES	NO	NO
CORBA	YES	Expensive	YES	Partial	NO
DCOM	YES	Expensive	NO	Partial	NO
Web Services	YES	Cost Effective	YES	YES	YES

Although Web Services are relatively new, there have already been two different approaches emerging, with Microsoft and its .Net platform that runs only under the Windows platform, and Sun, with its Sun ONE J2EE, platform independent, Java specific solution. But companies use a wide variety of platforms and so, the need for all these different platforms to communicate is crucial.

Recently, a group of the biggest IT and consulting companies (BEA Systems, Hewlett-Packard Co., IBM, Accenture, Intel Corp., SAP, etc) formed the Web Services Interoperability (WS-I) Organization (www.ws-i.org), with the aim of providing a framework which will allow engineers to develop mappings between .Net and J2EE.

Nowadays, more than 160 companies have joined the WS-I, including Sun and Oracle. Sun has recently announced that Java will support WS-I, and, in addition, OASIS has joined WS-I in this standardisation procedure (McNally, 2003).

So far Web Services implementations have been mostly kept within the companies that created them until standards for security are set. The problem arises because Web Services allow external users access the company, while, at the same time, traditional security techniques (like SSL, TCL, etc) have been proven inadequate to secure the amount and diversity of the Web Services transactions (Geer, 2003).

New security standards have therefore to be applied. Web Services security is an area with a lot of on-going work. The major initiatives in this domain are SAML, a framework of authentication and authorisation in request/response exchanges based on XML (Naedele, 2003) and WS-SEC (Web Services Security), an initiative by IBM, Microsoft and VeriSign, which works by adding security extensions to SOAP headings.

Other Web Security initiatives are XACML (Extensible Access Control Markup Language), XrML (Extensible Rights Markup Language) and XKMS (XML Key Management Specification). Naedele (2003) presents the associations between all the above aforementioned security initiatives. In April 2002 Microsoft and IBM issued a roadmap on Security in Web Services.

Standardisation work is being done by OASIS Web Services Security Technical Committee (WSS - TC), which issued the first Security Roadmap on April 2002. Work is also being done by W3C Web Services Activity, which composes by the XML Protocol Working Group, the Web Services Description Working group, the Web Services Choreography Working Group and the Semantic Web Services Interest Group.

7. CONCLUSIONS

The aim of this paper was to introduce the concept of Web Services technology and show why Web Services could be the solution to the Interoperability problem today's EE/VE face. For that reason, available frameworks and IT infrastructures suggested for supporting an EE/VE were presented and analysed. The analysis concluded that frameworks serve more as guidelines for a successful EE/VE creation, whereas the currently available IT solutions are not adequate in supporting interoperability.

As a means for solving the interoperability problem, the technology of Web Services was presented. Web Services are gaining more and more popularity over the last years, they are currently being backed up by 160 enterprises (including SAP, Microsoft, Oracle, Sun, etc.) and even though they are still being developed, they represent a very promising technology for an EE/VE, because they provide a powerful mechanism for solving any interoperability problem between different IT infrastructures.

Although so far Web Services implementations have been mostly kept within the companies that created them (until standards for security are set) (McNally, 2003), they have shown their dynamic for becoming the interoperability solution and the integration technology for the future.

Based on the findings presented here, Smart-fm consortium has proceed on initiating further research on the Web Services domain and preparing pilot implementations based on this technology. These pilot demonstrators, which will be available early, next year, will serve as a means for demonstrating to industry the Web Services capabilities. More information can be obtained from the Smart-fm project web site.

REFERENCES

Aerts, A.T.M., Szirbik, N.B., and Goossenaerts, J.B.M., (2000). Flexible Infrastructure for virtual enterprises, *Proceedings of the 4th International Working Conference on the Design of Information Infrastructure Systems for Manufacturing*, November 15-17, Melbourne, Australia, 2000.

Bansal S., and Pal G. (2001). The Web at your (machine's) service, JavaWorld website, http://www.javaworld.com//javaworld/jw-09-2001/ jw-0928-smsservice.html?, 2/8/02

Borras M., Goncalves R., Malo P., Vidagany J., and Steiger-Garcao A. (2002). *ECOS: low-cost e-Business interoperability for SMEs, Challenges and Achievements in E-business and E-work*, IOS Press, 2002

Browne J. and Zhang J. (1999). Extended and virtual enterprises: similarities and Differences, *International Journal of Agile Management Systems*, **1/1**, 1999

Camariha-Matos, L.M. et al. (Eds.) (1998). *Intelligent Systems for Manufacturing: Multi-Agent Systems and Virtual Organization.* Kluwer Academic Publishers, pp. 67-76.

Cingil I., Dogac, A. and Azgin, A. (2000). 'A Broader Approach to Personalization', *Communications of the ACM*, vol. **43**, issue 8

Duke S., Makey, P. and Kiras N. (1999). *Application Integration Management Guide: Strategies and Technologies*, Butler Group Limited, Hull, UK.

Edwards P. and Newing R. (2000). *Application Integration for e-Business,* Business Intelligence 2000, London, UK.

Emmelhainz M. (1993). *EDI: A Total Management Guide*, Van Nostrand Reinhold, New York, USA, 1993.

Esterm W. A., Bowen R. B. and Mills A. C. (2002). Reinventing the wheel, *International Conference on Flexible Automation and Intelligent Manufacturing*, 2002, Dresden, Germany.

Faruqui M. (2002). *SOAP raises the bar for CORBA*, Borland developer network, community.borland.com/article/0,1410,28737, 00.html, 12/08/02.

Geer, D., Taking steps to secure Web Services, Computer, October 2003

Giannopoulos N., Roy R., Panetto H., Ortiz V., Thomas A., Braga H., Sabater M., Amparo Roca A., Lauro L., Lecuona M., Nunez M. J., and Goncalves R. (2003). *State of the art in Extended Enterprises*, Task 1.1.c report, IMS-Smart-fm Project IST-2001-52224.

Jagdev H. S., and Browne J. (1998). The extended enterprise - a context for manufacturing, *Production Planning and Control*, vol. **9**, no3, 1998.

Jagdev H. S., and Thoben K. D. (2001). Anatomy of enterprise collaborations, *Production Planning and Control*, vol **12**, no5, 2001.

Janssen W., Koolwaaij J. and Stefanova M. (2000). XML, hype or hope?, *Telematica Institute*.

Klasell T. and Dudgeon S. (1998). *'Enterprise Application Integration'*, Dain Rauscher Wessels, New York, USA.

Knorr E. (2001). Make way for the web services, *CIO magazine*, 15 October 2001 issue.

Linthicum D. (1999). *Enterprise Application Integration*, Addison-Wesley, Massachusetts, USA, 1999

McNally T. (2003). Software Superglue, *IEE Review*, September 2003 issue.

Naedele, M., Standards for XML and Web Services Security, Computer, April 2003

Pearson D. W., Albert P., Besombes B., Boudarel M. R., Marcon E., and Mnemoi G. (2002). Modelling enterprise networks: a master equation approach, *European journal of operation research*, vol. **138**, 2002.

Patton S. (2001). Web Services in the real world, *CIO magazine*, April 1st, 2001 issue

Platt S. D. (2000). What The Heck Is SOAP, Anyway? *Byte magazine*, 16 August 2000 issue.

Ruh W., Maginnis F. and Brown W. (2000). *Enterprise Application Integration: A Wiley Tech Brief*, Wiley John Wiley & Sons Inc., New York, USA, 2000

Shegzeo O., and Petersen S. A. (2000). Extended Enterprise Engineering: A model based framework, *Concurrent Engineering: Research and Applications*, vol. **8**, no. 1, 2000.

Slade A. J., and Bokma A. F. (2002). Ontologies within Extended Enterprises. *Proceedings of the 35th Hawaii International Conference on System Sciences*, 2002

Slater D. (2000). Middleware Demystified. *CIO magazine*, 15 May 2000 issue.

Stackpole B. (2000). Tag, you are it!, *CIO magazine,* 15 May 2000 issue.

Themistocleous M. (2002). *Evaluating the adoption of Enterprise Application Integration in multinational organizations*, PhD Thesis, Department of Information Systems and Computing, Brunel University, London, 2002.

Vawter C. and Roman Ed. (2001). *J2EE vs. Microsoft .NET*, June 2001, http://www.theserverside.com//resources/article.jsp?l=j2ee-vs-dotnet

Wallace S. (2003). Web Services, Business Week, *European Edition*, 17 March, 2003

Whitlock N. W. (2001). The missing link : Will ebXML bring true global B2B?, http://www-106.ibm.com/developerworks//webservices/library/wsebxml.html, *IBM Developerswork*, January 2000, accessed 20/2, 17:00pm.

Zinner H. (1999). *Diffusion of Information Systems: Barriers and Drivers for Magnifying EDI in Danish Enterprises*, PhD Thesis, Copenhagen Business School, Copenhagen, Demark, 1999.

ACKNOWLEDGEMENT

The authors would like to thank all the Smart-fm members for their scientific contribution to this work. This work was funded by the European Commission IST 5th framework programme.

Copyright © IFAC Information Control Problems in
Manufacturing, Salvador, Brazil, 2004

ELSEVIER
IFAC
PUBLICATIONS
www.elsevier.com/locate/ifac

ONTOLOGY-BASED FRAMEWORK FOR ENHANCED INTEROPERABILITY IN NETWORKED INDUSTRIAL ENVIRONMENTS

Ricardo Jardim-Goncalves, Ricardo Olavo,
Joao Sarraipa and Adolfo Steiger-Garcao

Dep. de Eng. Electrotécnica da Faculdade de Ciências e Tecnologia
da Universidade Nova de Lisboa, UNINOVA, Lisbon Portugal

Abstract: Nowadays, the advent of networked production environments has conducted to increase the number of heterogeneous systems that a manufacturer has to manage. These needs have resulted in major semantic accuracy difficulties when exchanging and integrating information. A normalised representation of the semantics associated with the product and processes, has been identified by the research community as a key to solve the problem. This paper contributes describing a method to develop a harmonized taxonomy in adoption by the furniture industry, prepared using a framework to improve the interoperability of systems within the Product Life Cycle. The proposed platform is based in a standard-based meta-model description assisted by an on-line normalised ontology platform, using state of the art technology for Internet enabled services and data access. The paper reflects major research results from two international projects.
Copyright © 2004 IFAC

Keywords: agile manufacturing, computer integrated manufacturing, enterprise integration, information integration, meta-level knowledge, standards

1 INTRODUCTION

Business success depends on the integration of enterprises' internal processes and relies on collaborations with externals. To support it, networking facilities should be considered to set up an extended and interoperable computing environment (Camarinha-Matos, 1999) (Dsouza, 2001) (Chen, 2000) (Jardim-Goncalves, 2001).

Heterogeneous data models and processes need to be integrated in order to provide interoperability between systems. This situation has been identified not easy to achieve, because typically each application adopts its own data structure and semantics (Clements, 1997) (Lazcano, 2000) (Poole, 2001) (Nagi, 1997).

The adoption of a common standard model, e.g., STEP Application Protocols, could contribute to face this problem. However, these models should not be completely static, to enable dynamic adaptation to the Product Life Cycle Management (PLM) requirements along the time, and within dynamic networking (STEP, 1999) (ISO10303, 2003) (Nayak, 2001) (Umar, 1999) (Jardim-Goncalves, 2002b).

These circumstances conducts to a situation where information is described at meta-level, enabling the foreseen flexibility. On the other hand, this independence requires the definition and agreement of semantics, to describe accurately the meaning of the data to be instantiated by the meta-model, and exchanged between the applications, that should be described using an ontology framework (Radeke, 1999) (Davulcu, 1999) (Jardim-Goncalves, 2002a) (Erwan Berton, 2001).

This ontology would reflect the taxonomy of concepts that will represent the sufficient understanding of the semantics related with the information to exchange. However, when working in worldwide networked environments, collaboration with other enterprises can bring the additional difficulty of natural language. In this case, when the information to be exchanged is described using a native language, requires the other part to understand it. Therefore, the ontology system should be complemented with a multi-language dictionary, where a set of normalized tokens gives the reference to the meanings in many native languages (Heflin, 2000).

A real situation of the described scenario is found in the furniture industry, and has been studied in the scope of the IMS Smart-fm project, where the product data model used is the emerging standard for furniture data exchange, STEP ISO10303 AP236 (AP236, 2002) (funStep, 2003).

This standard specifies an Application Protocol (AP) for exchange of Furniture Product Definition (FPD) data and Interior Design Project (IDP) data among manufacturers, suppliers, retailers, interior designers, and customers. AP236 defines the application model that satisfies the information requirements for furniture product and project data representation and provides the integrated resources necessary to implement it, including representation at meta-level. In its scope, the information exchanged includes catalogues of products and decoration projects with parameterization and graphical specifications.

2 NEEDS FOR A COMMON REFERENCE TAXONOMY

Ontology is the study of the categories of things within a domain. It comes from philosophy and provides a logical framework for research on knowledge representation, including schemas and diagrams to describe classification relationships and definition of concepts (Steven Willmot, 2001).

An ontology reflects a view of a segment of the reality. Two or more communities (e.g., organizations), operating in the same business domain may have different views, leading to different underlying ontologies, and consequently conducting to problems of interoperability when the systems intend to communicate (Uschold, 1996) (Fox, 1997).

The need to have complete and fully integrated systems, that seamlessly communicate and understand each other, requires a complete understanding of the meaning of the data within the domains involved, and that results from a consensus, i.e., harmonization, for a common ontology.

Reaching this harmonization will help the systems to really interoperate, because they will have a unique reference for information exchanged semantics. However, to achieve this common model, it is needed to classify the concepts developing a taxonomy, i.e., the organization of a specific set of information for a particular purpose, described by a hierarchal structure of concepts and definitions (Gruninger, 2002).

That knowledge classification is processed by sub-categorizing them according to essential relevant and cognitive qualities (Leger, 2002). Thus, the taxonomy makes possible the process of classification of data, distributing such categorized information into groups (e.g., classes, families), according to some common relations or affinities.

Foreseeing a common taxonomy, the system's actors must agree in a common classification and definition of its data and services, taking as a basis their particular taxonomies to obtain the harmonized one (Figure 1).

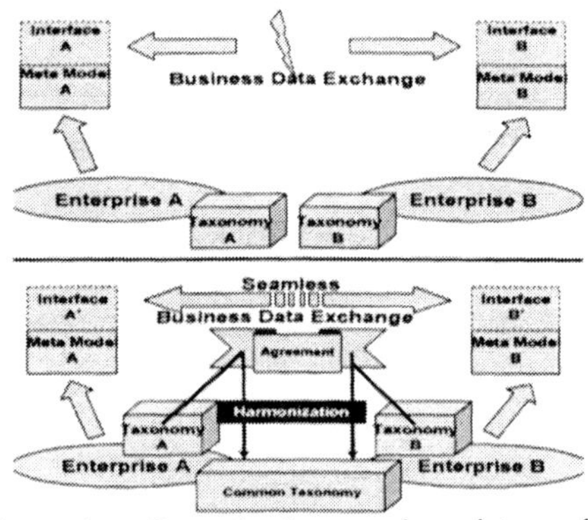

Figure 1 – Scenario for seamless data exchange based on a common harmonised taxonomy

The European project Smart-fm, focusing on improving competitiveness through smart furniture manufacturing, brought to the furniture industry stakeholders a convergence between many classifications of types of products to a common definition that resulted in a common semantics agreement between the different taxonomy owners and users. Figure 2 depicts an example of two taxonomies from Smart-fm, where the "collection" and "series" classes are harmonized in the common class "families".

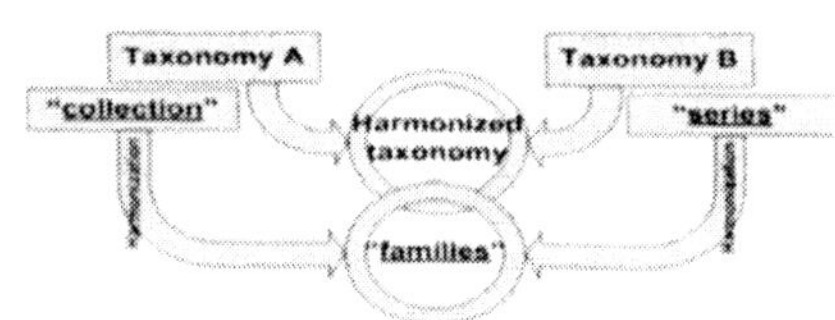

Figure 2 –Example of harmonization in the scope of furniture industry

However, such harmonisation is complex and requires the use of a collaborative working platform allowing users to browse and manage the individual and common taxonomy's structure and data. Because normally the developers are far way from each other, a networked enabled environment should be available to assist in the development, harmonization and updates of the common taxonomy.

3 FRAMEWORK FOR TAXONOMY DEVELOPMENT AND HARMONIZATION

Based on the consolidation of the state-of-the-art in platforms for ontology development and management, it was implemented a framework to assist in the development and harmonisation of taxonomies (Figure 3).

Developed at Stanford University, Protégé-2000 is a well known and widely recognised ontology and knowledge-base editor, and it is an open-source Java tool that provides an extensible architecture for the creation of customized knowledge-based applications (Protege, 2003). The proposed platform uses Java technology, extending the Portégé-2000 editor and integrating a RDBMS, to fulfil the proposed new requirements.

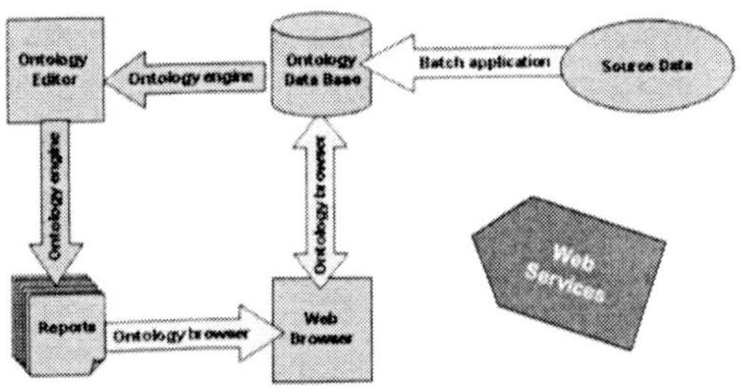

Figure 3 – Architecture for the generator for representation of taxonomies

The *source data* is the entry point of the platform that must be put available in a structured format describing the taxonomy's hierarchic classification, terminology and respective relationships. In a whole, this input represents the ontology described in three complementary layers:

- 1st layer: defines the classes and respective relationships between them, i.e., its hierarchy.

- 2nd layer: defines the terms, and their associations with the classes.

- 3rd layer: describes the classes and terms.

Each level can have more than one class allocated, and each class can have many terms assigned. The same term may exist in different classes, with or without the same definition.

3.1 The framework engine

Web services have been recognised as "XML-enabled standards to allow publishing of applications to other parties across platforms." (Brown, 2003).

Nowadays, the use of web services is expanding rapidly. Over the last year web services have turned into the greatest development craze. They provide a standard means of communication among different software applications in networked environments, independently of the platforms running.

Web services are based on the potential of the combination of XML (Extensible Markup Language), a service-oriented architecture (the Web Services Description Language, or WSDL), registration through a UDDI (Universal Description, Discovery and Integration) repository, and communication encoded using SOAP (the Simple Object Access Protocol) (W3C, 2003).

A web service is a software mechanism identified by a Universal Resource Identifier (URI), whose public interfaces and bindings are defined and described using XML. Its definition can be discovered by any software system, throughout normalised procedures over the Internet.

They work in a way that a service provider hosts a network accessible software module, i.e., an implementation of a web service. The service provider defines the service description for the web service and publishes it to a requestor or service discovery agency.

The service requestor uses a find operation to retrieve the service description locally or from the discovery agency (i.e. a registry or repository) and uses the service description to bind with the service provider and invoke or interact with the web service implementation.

The *Ontology Engine* performs the connection between the *Ontology Data Base* and the *Ontology Editor*. The proposed framework interoperates through the web services technology, performing total integration between the ontology engine server and the ontology engine client.

The *Ontology Engine Server* is composed by an *Ontology Editor* and an *Ontology Interface*. The *Ontology Editor* provides editing and reporting functionalities as a result from client's requests, giving also possibilities to navigate over the data (Figure 4).

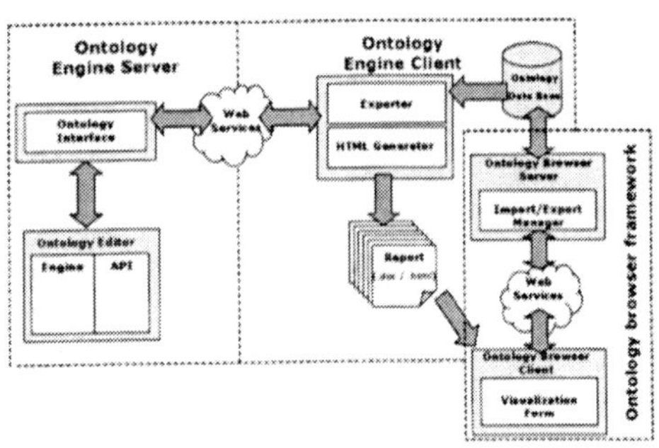

Figure 4 - Ontology Engine framework

The *Ontology Interface* connects both sides of the *Ontology Engine*, i.e., between the *Ontology Editor* and the *Ontology Engine Client*. This interface is characterized as a web services' server.

The communication between the *Data Base* and the *Ontology Editor* is established by the *Ontology Engine Client*. It allows the possibility to execute requests and send information directly to the *Ontology Engine Server*.

Figure 5 presents an extract of the SOAP messages, exchanged between the *Ontology Engine Client* and the *Ontology Engine Server*.

```
<?xml version="1.0" encoding="utf-8"?>
<soap:Envelope      xmlns:xsi="http://www.w3.org/2001/XMLSchema-instance" ... >
<soap:Body>
        <createClass xmlns="http://uninova.pt/ProtegeWriter/">
        <class_name>string</class_name>
        <parent_path>string</parent_path>
        <class_description>string</class_description>
        </createClass>
</soap:Body>
</soap:Envelope>                                                    Request
```

```
<?xml version="1.0" encoding="utf-8"?>
<soap:Envelope
xmlns:xsi="http://www.w3.org/2001/XMLSchema-instance" ... >
<soap:Body>
        <createClassResponse xmlns="http://uninova.pt/ProtegeWriter/">
                <createClassResult>boolean</createClassResult>
        </createClassResponse>
</soap:Body>
</soap:Envelope>                                                    Response
```

Figure 5 – Request/Response SOAP message

3.2 *The framework browser*

The *Ontology Browser* adopts a similar client-server architecture. Hence, this framework is supported by web services technology, the basis for the implementation of the interfaces between the user and the *Ontology Data Base* and the *Visualization Form*.

The implemented browsing services enable the user to access on-line the information in the *Data Base*, and get the *Visualization Form* to be displayed using an Internet browser.

Each time the *Ontology Data Base* is changed, the *Ontology Browser platform* requests an update to the *Ontology Engine Platform* that will generate an update of the reports. Avoiding any multiple access problems issues the platform as an access flag that allows only one user at a time to change the data.

The *Visualization Form* divides the presentation of the ontology's information in two screen regions. The left part shows the classification tree as it is in the data base, including. The right part of the screen shows the descriptions of the taxonomy's classes and terms. The *Ontology Browser* provides capabilities for on-line edition of the taxonomy, where the modifications are made over the *Ontology Data Base*.

The platform allows access to different kind of users. Regular users have complete access to it but without permissions to change data. A more qualified access is provided to registered users that in this case can access to the platform with permission to update and add new data. Typically, users that are responsible for the development and harmonisation of taxonomy apply for qualified access to the platform.

4 METHODOLOGY ADOPTED AND CASE STUDY

The methodology implemented to harmonize the Smart-fm related Ontologies is depicted in Figure 6. This methodology is adapted from (Noy, 2002), that proposes a recognised method to develop a single Ontology, extending it to be appropriate to support the development of harmonized Ontology's structure and contents, from a set of Ontologies identified in a specific domain.

The discussion and agreement tasks were conducted by representatives of the companies were they are in use, and in most of the cases these persons were able to speak only their native language. Therefore, the use of an on-line multi-language dictionary is essential for a common understanding and agreement of the semantics in discussion.

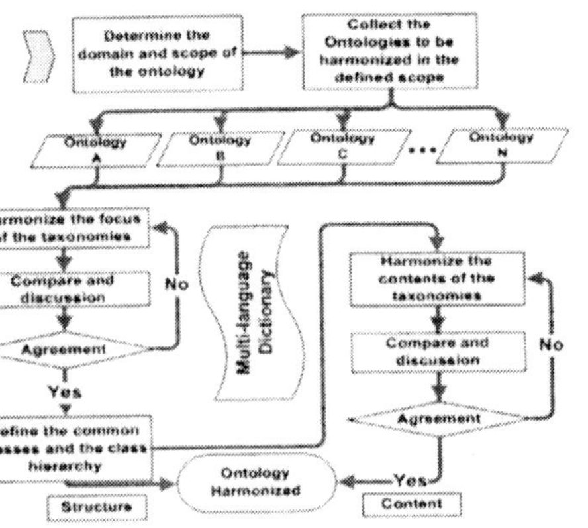

Figure 6 - Extended methodology to support the development of harmonized Ontologies

Next figure depicts part of one of the developments, where can be identified in the Ontology A the *Baby Cots* class from a retailer, represented in the Smart-fm harmonized Ontology as a *Children Beds+Cots*. Further results can be found at www.smart-fm.funstep.org.

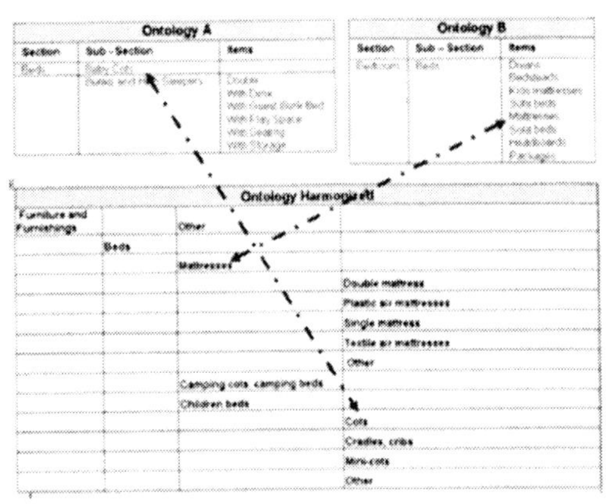

Figure 7 – Snapshot of two Smart-fm source ontologies, and respective harmonization

The harmonization of Ontologies is not an easy task, even when following an established methodology, as it is the case. Especially the harmonization of the focus of the taxonomies and comparing, typically conducts to long discussions before agreement, many times obligating to many iterations.

5 RESULTS OBTAINED AND OUTLOOK

Product manufacturing results from the orchestration of a set of cooperative and interoperable management and production activities. To succeed, information must flow accurately between the systems within and along the Product Life Cycle.

Nowadays, the emergent networking facilities and internet-enabled standards for data exchange have opening new challenges to the manufacturing management, providing new extended ways to communicate. However, interoperability between applications achieves now a new level of complexity, due to the increasing number of heterogeneous systems each application can interoperate.

Typically each application is guided by its own data structure and semantics. To exchange seamless data with others, each application should be able to describe its own information under a common data structure and semantics. Therefore, the adoption of meta-models for the exchange of information can facilitate this task. However, in this case data becomes represented under an open structure, where semantics must be settled in advance between the interlocutors to assure accuracy of the information exchanged.

Harmonization between semantics is thus a requirement to be supported by a common ontology, where a global platform providing services to manage it is of major importance. The services to access the ontology system should be normalised and universally available on-line, using for instance web services technology. This allows an open and wider adoption of the meta-model, contributing for a more accurate networked manufacturing environment.

The framework proposed in this paper was developed by the authors aiming to contribute to solve these requirements. The platform adopts web services technology and can be used on-line, where the user may request an interaction in real-time over the ontology data with the help of an ontology browser.

As well, this platform allows the possibility to provide the information about the ontology in Hyper Text Markup Language (html) format. Also, the user may receive all the data populated in the Data Base in Portable Document Format (pdf) format. These functionalities give to the user popular possibilities to navigate, search and print the ontological information, with the possibility for local complete navigation using hyperlink mechanisms in a totally independent usage from an Internet connection, i.e., offline.

The validation of the research work presented in this paper was done in the range of the furniture industry, with the aim to support interoperability when companies exchange information related with catalogues of furniture products.

In the cases evaluated, the catalogue was represented adopting a subset of the emerging standard ISO10303 AP236, the STEP Application Protocol for furniture product and project data. In the foundations of its reference model, there is the *product_class* entity that can act as the root for the catalogue representation, and its products. Moreover, *product_class* establishes references to *specification_category* entities that describes at catalogue level the many configurations a product can have, e.g., colours, finishing.

Because the many possibilities to represent catalogues of furniture products, and to keep the universality of catalogues for furniture representation, the AP236 presents the description of a catalogue at a meta-level. This provides the flexible mechanism for catalogue representation, providing a standard technique to represent the vast number of possibilities to describe them.

However, when exchanging the catalogue data between applications instantiating such meta-level, semantic problems can seriously take place, resulting in severe misunderstandings of the information exchanged. To assist in this process towards interoperability between the applications, a harmonised ontology must be defined.

The development of such ontology requires a well established methodology, like the one proposed in the paper, together with a distributed access and management by the integrators of the applications in a collaborative networked environment, and includes implementation, discussion and refinement of the ontology components.

When the harmonisation of the ontology is established, and catalogue data is transferred, the systems exchanging the information can access to the platform to clearly understand the semantics of information communicated. For the purpose, the normalised web services could be invoked.

The adoption of web services technology by the platform proposed, offered a complete open framework, enabling that any third party willing to put available similar platforms with the same functionalities and services, or provide further extensions, can do it immediately it through the web services registry mechanism.

The authors developed the research work that resulted in the development and validation of the proposal presented in this paper, in the context of the European project IDEAS (Interoperability Development for Enterprise Application and Software - Roadmaps) international project (IST–2001–37368) (IDEAS, 2003) and IMS project Smart-fm (funStep, 2003).

This work was done focused within one of the PLM stages. In this moment it is in course further research intending to extend these results to cover inter-stage activities. The major challenges already identified are related with PLM inter-stage meta-model definition and the architecture for the representation of the global ontology system.

ACKNOWLEDGEMENTS

The authors would like to thank all the national and international organizations that supported and funded during the last years a significant number of international projects related with the scope of this paper. A special thanks to the Smart-fm and IDEAS partners that somehow contributed for the presented work, and the European Commission that funded these projects.

REFERENCES

AP236 (2002). ISO TC184/SC4, ISO10303 Part 236, Application Protocol: Furniture product and furniture project decoration data

Brown, D. H., Associates, Inc. (2003). http://dhbrown.com

Camarinha-Matos, L. and Afsarmanesh, H. (1999). Tendencies and general requirements for virtual enterprises, IFIP workshop, Chapter 2, Kluwer Academic Publishers, ISBN 0-7923-8639-6

Chen, Q, (2000). Inter-enterprise collaborative business process management. Technical report, HP Labs Palo Alto

Clements, P. (1997). Standard support for the virtual enterprise, Int. Conf. on Enterprise Integration Modeling Technology - ICEIMT'97, Italy

Davulcu, H. (1999). Modeling and analysis of interactions in virtual enterprises. IT for Virtual Enterprises, pages 12–18. IEEE Computer Society

Dsouza, D. (2001). Model-Driven Architecture and Integration: Opportunities and Challenges Version 1.1

Erwan Breton, Jean Bézivin (2001). Using Meta-Model Technologies to Organize Functionalities for Active System Schemes, 5th International Conference on Autonomous Agents, Canada

Fox, M.S. and Gruninger, M. (1997). Ontologies for Enterprise Modelling, Proceedings of ICEMT'97, Italy

funStep (2003). http://www.funstep.org

Gruninger, M. (2002). Using Ontologies to Evaluate Knowledge-based Systems, PerMIS USA

Heflin, Jeff and Hendler, James (2000). Semantic Interoperability on the Web, Proceedings of Extreme Markup Languages. Graphic Communications Association, 2000. pp. 111-120.

IDEAS (2003). D1.1 Part B. Ontology State of the Art. IST-2001-37368, www.ideas-roadmap.net

ISO10303 (2003). www.tc184-sc4.org

Jardim-Gonçalves, R., Steiger-Garção, A. (2001). published in book: Agile Manufacturing: 21st Century Manufacturing Strategy, Chapter 48: "Putting the pieces together" using standards. Elsevier Science Publishers, pp. 735-757, ISBN: 0-08-043567-X

Jardim-Gonçalves, R. and Steiger-Garção, A. (2002a). Implicit multi-level modelling for integration and interoperability in flexible business environments, submitted to Communications of ACM, special issue on Enterprise Components

Jardim-Gonçalves, Ricardo and Steiger-Garção, A. (2002b). Implicit hierarchic meta-modelling - In search of adaptable interoperability for manufacturing and business systems, 5th IEEE/IFIP BASYS2002, Mexico

Lazcano et al. (2000). The wise approach to electronic commerce. International Journal of Computer Systems Science & Engineering, Vol. 15, No. 5.

Léger, Alain, et.al. (2002). OntoWeb, IST-2000-29243, Successful Scenarios for Ontology-based Applications

Nagi, L. (1997). Design and Implementation of a Virtual Information System for Agile Manufacturing, IIE Transactions on Design and Manufacturing, Vol. 29(10), pp. 839-857

Nayak, N., et al (2001). Virtual Enterprises - Building Blocks for Dynamic e-Business, ITVE 2001, IEEE Computer Society, pp 80-87

Noy, Natalya and McGuinness, Deborah (2002). Ontology Development 101: A Guide to Creating Your First Ontology'. Stanford Knowledge Systems Laboratory Technical Report KSL-01-05 and Stanford Medical Informatics Technical Report SMI-2001-0880

Poole, John D. (2001). Model-Driven Architecture: Vision, Standards And Emerging Technologies, ECOOP 2001, Hungary, University of California

Protege (2003). http://protege.stanford.edu/

Radeke, E. et al. (1999). Distributed information management in virtual engineering enterprises by GEN. RIDE-VE'99, pages 36–43. IEEE Computer Society.

STEP (1999). Introducing STEP,ISBN 0-662-64382-8.

Steven Willmott, et.al (2001). Multilingual Agents: Ontologies, Languages and Abstractions, 5th Int. Conference on Autonomous Agents, Canada

Umar, 1999, A framework for analyzing virtual enterprise infrastructure.RIDE-VE'99, pages 4–11. IEEE Computer Society

Uschold, M. and Gruninger, M. (1996). Ontologies: Principles, Methods and Applications Knowledge Engineering Review; Volume 11 Number 2, June 1996

W3C (2003). Web Services Architecture Requirements, www.w3.org/2002/ws/

Copyright © IFAC Information Control Problems in
Manufacturing, Salvador, Brazil, 2004

ELSEVIER
IFAC
PUBLICATIONS
www.elsevier.com/locate/ifac

CREATION OF A KNOWLEDGE ECONOMY FOR INTELLIGENT INDUSTRIAL AUTOMATION: OOONEIDA COMMUNITY IN IMS

F. Auinger[1], R.W. Brennan[2], J. Christensen[3], J. L Martinez Lastra[4], V. Vyatkin[5]

[1]*University for Applied Science Wels, Roseggerstrasse 12, 4600 Wels, Austria*

[2]*Department of Mechanical and Manufacturing Engineering*
University of Calgary, 2500 University Dr. N.W., Calgary, Canada, T2N 1N4

[3] *Rockwell Automation Advanced Technologies, Allen Bradley Drive 1, Mayfield Heights, OH 44121 USA*

[4] *Institute of Production Engineering, Tampere University of Technology, P.O. Box 589 FIN-33101 Tampere, Finland*

[5] *Department of Engineering Sciences, Martin Luther University of Halle-Wittenberg, 06099 Halle, Germany, e-mail: Valeriy.Vyatkin@iw.uni-halle.de*

Abstract: OOONEIDA is a new initiative for enabling decentralized, re-configurable industrial control and automation in discrete manufacturing and continuous process systems. The goal of the OOONEIDA is the creation of the technological infrastructure for a new, open knowledge economy for automation components and automated industrial products. This will be done by further development of the concept of reusable portable software modules (function blocks), and by their application in the time- and cost-effective specification, design, validation, realization and deployment of intelligent mechatronic components in distributed industrial automation and control systems. *Copyright © 2004 IFAC*

Keywords: Industry Automation, Intelligent Manufacturing Systems

1. INTRODUCTION: THE NEW KNOWLEDGE ECONOMY

Efficient handling of embedded intelligence is a key enabler for the agile production facilities that will be indispensable for the sustainable growth and competitiveness of manufacturing industries worldwide.

Open, Object-Oriented kNowledge Economy for Intelligent inDustrial Automation (OOONEIDA) is a new initiative to enable the flexible, open integration and reconfiguration of embedded intelligence in industrial automation systems.

Observing the global market of automation products, one sees the value creation chain as shown in Figure 1.

It is remarkable that there are very similar activities being performed at all levels of the value-adding chain:

- Device vendors design and produce parts of machines (mechatronic components) using basic components such as sensors, actuators, embedded computing devices, system software.

They add their own expertise in devices and partially encapsulate it in software components.

- Machine vendors design functionally complete production machines using the component devices. They also develop software components for control of the machines, communication with other machines and with operators, visualization and diagnostics, etc.
- System integrators fulfill the needs of their customers using machines and components. They encapsulate their integration expertise in the form of software components which make the system run. They also need to integrate the software components obtained from the component vendors into the software of their systems.

Thus, all players in the automation value-adding chain deal with the integration of knowledge and its encapsulation for future use and re-use. As well, all players in the value-adding chain should benefit from the unification and standardization of their components. This is especially true for software where design solutions for different configurations of their products could be used without customization for each vendor requirement.

Knowledge encapsulation would allow for the free substitution of devices of different vendors as required by a particular order. In the case of re-configuration their products would be more flexible.

The goal of OOONEIDA is the creation of the technological infrastructure for a new, open knowledge economy for automation components, products and systems. This new economy will bring tangible benefits for all players in the value creation chain: Device vendors, machine vendors (OEMs), system integrators and industrial enterprises will be able to encapsulate their intellectual property (IP) into reuseable, portable software components (function blocks) and to deploy these components into intelligent devices, machines, systems and automated factories.

2. HISTORY OF THE COMMUNITY FORMATION

Initially the "community" was forming around the R&D activities accompanying the development of new international standards IEC61499 (IEC61499, 2002) and IEC61804 (IEC/TS 61804-1). This work required formulation of the new portable function block concept and numerous feasibility case studies. Some of the first conceptual papers date back to (Schoop and Ferling 1997) and (Neumann and Diedrich 1996).

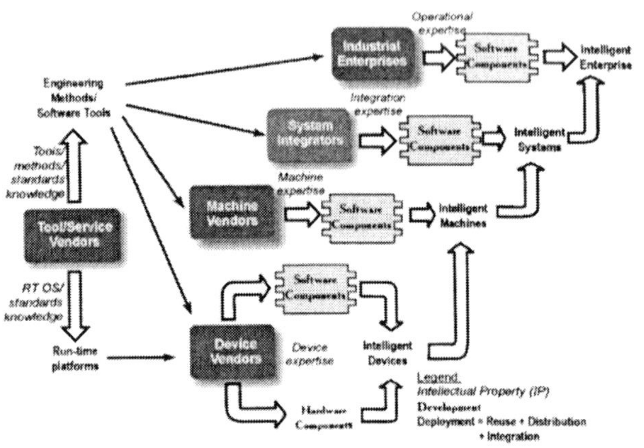

Figure 1. The value creation chain in industrial automation.

In particular, in 1999 the researches of Rockwell Automation (USA), Profactor (Austria) and Softing (Germany) created the first testbed for studying the feasibility of the new function block concept. This work was done in the framework of the IMS project "Holonic Manufacturing Systems" (HMS, 2002). Another testbed was created by Yamatake Electric (Japan) (HOLOBLOC, 2002). Further more extended feasibility studies were accomplished in 2002-2003 in Odo Struger Lab. of the Technical University of Vienna (Austria) and in Martin-Luther University of Halle (Germany) (Vyatkin, 2002).

Currently the "community" includes dozens of research groups and technology transfer centers worldwide, that are enthusiastically conducting the relevant R&D in the directions as follows:

- further development of the architectures and case studies on their application in particular industrial domains, e.g.: (Christensen 2002, Vyatkin and Lastra 2002, Lastra and Tuokko, 2003);

- development of new engineering methodologies and engineering patterns, e.g.: (Christensen 2002);

- portable scalable and reconfigurable hardware/software run-time platforms, e.g.: (Brennan et al. 2002, Rumpl et al. 2002);

- advanced methods of validation of reconfigurable automation systems, e.g.: (Vyatkin 2003, Bonfè and Fantuzzi 2003);

- development of open tool platforms and software tools, e.g.: (Thramboulidis 2002);

3. STRUCTURE OF THE CCI OOONEIDA

The Intelligent Manufacturing Systems (IMS) Program is an international initiative that includes Korea, Japan, Australia, Canada, USA,

Switzerland. The Intelligent Manufacturing Systems (IMS) sponsors the development stages of OOONEIDA as a Community of Common Interest (CCI) as of December 2003. The OOONEIDA CCI is currently being formed with seed funding from the IMS. Industrial and research organizations from four IMS regions (USA, Canada, Switzerland and the EU) are leading the formation of the OOONEIDA CCI. Interest has also been expressed from non-IMS regions including Brazil, New Zealand, Russia, and Singapore. In total, current interested parties include 7 large enterprises with annual turnover of more than $1B each, 5 Small/Medium Enterprises (SMEs), 11 technology transfer centers, 12 academic centers of competence, and 2 government facilities.

The OOONEIDA CCI is administered by the Board of Management that consists of one industrial member from each IMS Region active in the CCI, and is responsible to report progress and financial information to IMS. Working Groups (WGs) are formed of volunteers from the OOONEIDA CCI. At the moment, three Working Groups are already envisioned: WG1 for Automation Knowledge Repository, WG2 for Standardization activities and WG3 for further formation of the community. In particular, WG3 will identify opportunities for research and technology development to support the evolution of the knowledge economy and will formulate proposals for future IMS projects to carry out the required R&D. It is anticipated that the WGs may form Task Forces (TFs) to address specific issues. These TFs will have well-defined scopes of work and schedules of deliverables.

When a prototype repository becomes available, it will be publicly accessible at the website http://www.oooneida.net/. Seed funding from the IMS will decline for a period of three years. Over this period OOONEIDA will gradually become a self-supporting, member-paid organization. It is anticipated that members will want to fund ongoing activities of the CCI in exchange for having their Automation Objects made accessible for sale via a commercial version of the repository.

4. RESEARCH DIRECTIONS

The main research focus of OOONEIDA will be the concept of Automation Object that creates a proper architectural framework for intelligent automation components. The idea of Automation Object stems from the object-oriented software engineering, where an object is a structure encapsulating data and functions. It is envisioned in OOONEIDA that the further development of the Automation Object concept will be based on International Standards, such as IEC 61499, IEC61131, and de-facto industrial standard Unified Modeling Language (UML).

As illustrated in Figure 2, an Automation Object is an abstraction for a mechanical device associated with its *embedded intelligence* i.e. software components of different functional domains. For example, the layout of a component (CAD data) is related to its appearance on its visualization screen (View). The View component can receive data about the dynamic state of the object either from actual process interface or from a simulation model. Control functions can refer directly to the process interface of the object, or to the low-level control functions. The human-machine interface component communicates with both process and controller. It is interesting to note, that there could be internal dependencies between the functions – that means a change in the layout can affect the view, simulation model, and HMI. If these dependencies were described formally it could make possible automatic change of the object's structure in case of customization. Such formal rules could be also considered as a part of the Automation Object concept.

The use of intelligent components and object-oriented design has the potential to bring the benefits of intellectual property **encapsulation** and **reuse** to the design and re-configuration of intelligent mechatronic components, machines and automated production systems.

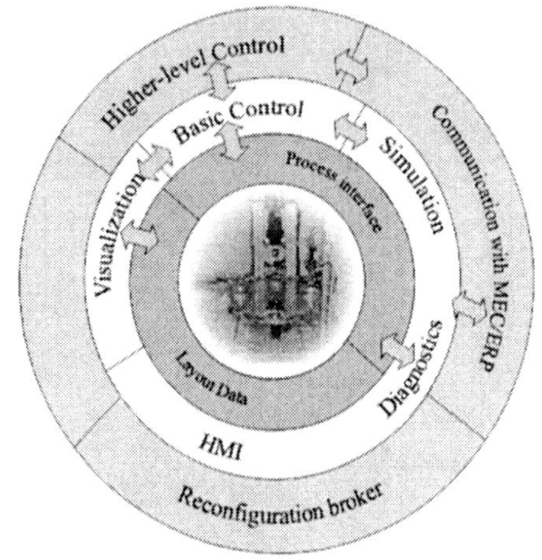

Figure 2. A sketch illustrating the idea of automation object.

Thanks to such parts of the "embedded intelligence" as simulation models, visualization, and human-machine interfaces, a "virtual" automation component can be used even before it is physically implemented. This would open new opportunities for concurrent engineering of automation systems.

OOONEIDA will support a number of key innovations in the automation technologies. One of these is the extensive use of modeling in industrial automation projects. The incorporation of behavioral and analytical models in automation object repositories will bring new opportunities for concurrent engineering and will be indispensable for the agile reconfiguration of production facilities. The vision of OOONEIDA is that the newly built configurations can be formally validated, thus providing significant time and cost reductions during system commissioning and testing, as well as improving the safety and reliability of production systems. Systematic modeling will be enabled through the application of standard modeling notations such as UML, suitably extended and harmonized with the architectural frameworks used in automation. This will simplify the application of advanced control methods in automation.

By targeting the portability, configurability and interoperability issues, OOONEIDA intends to create the foundations at the control and automation level for wide deployment of intelligent production systems, such as Holonic or agent-based manufacturing.

OOONEIDA will help to improve global manufacturing competitiveness through shorter time to market and lower process life cycle costs, and to meet global market-driven requirements for short product life cycles, lower total production volumes, higher product mix, and mass customization.

OOONEIDA will address legacy issues through engineering methodologies and open tools to provide migration paths from current vendor-clustered domains of locally controlled production systems, to the goal of production systems with fully open, distributed control that can be assembled or re-configured in a plug and play manner.

5. REPOSITORY

OOONEIDA members will build and populate a publicly available open repository of Automation Objects based on:

- ongoing, worldwide standardization activities in organizations following in the footsteps of International Electrotechnical Commission (IEC), International Standards Organization (ISO), W3C Consortium and Foundation for Intelligent Physical Agents (FIPA);
- results of research successfully performed in IMS, regional and national projects
- independent development activities of the CCI members.

As illustrated in Figure 3, a key ingredient in supporting the knowledge economy is the role of broker agents to enable all players in the value-adding chain to locate and retrieve the required IP on which to base their development activities, and to deposit results of their activities in searchable form accessible to other players.

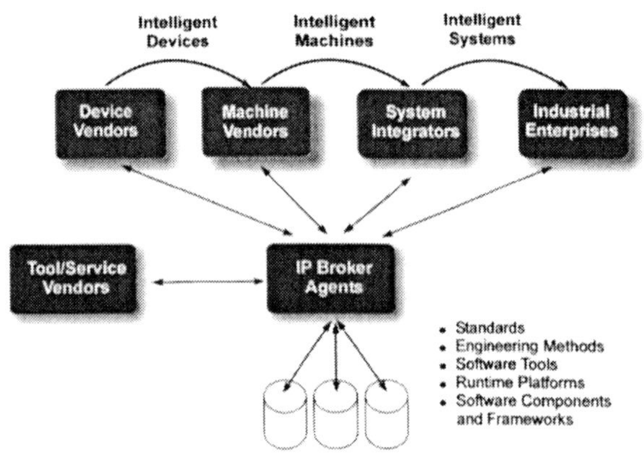

Figure 3. Repository of Automation Intelligence in the automation value-creation chain.

OOONEIDA will demonstrate an architecture-centric approach to industrial embedded systems design, based on open global standards for embedded software and open toolkits. This will enable the time- and cost-effective specification, design, validation, realization and deployment of intelligent mechatronic devices in distributed industrial automation and control systems in both discrete manufacturing and continuous process industries. This architecture will use software agents to enable users to locate, acquire and use the intellectual property needed to perform their value-adding activities and make the results available (possibly for sale) to other players in the knowledge economy.

The IT infrastructure of OOONEIDA can be viewed in more detail as consisting of three domains (Figure 4):

The **open physical domain**, utilizing open, interoperable intelligent mechatronic elements based on the IEC 61499 standard for the use of software modules (function blocks) for industrial process measurement, control and automation.

The **open tools domain**, based on a publicly available, open software tool integration platform such as Eclipse or NetBeans.

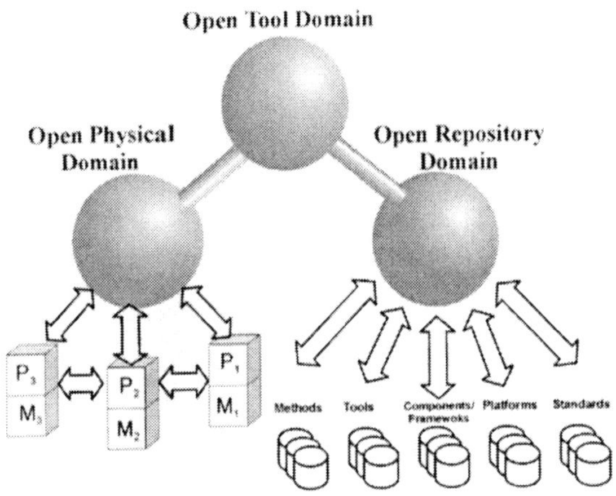

Figure 4. The IT infrastructure of the OOONEIDA initiative.

The **open repository domain**, consisting of distributed repositories for standards, engineering methodologies and their supporting software tools, runtime platform implementations and software components. Systematic indexing and searching of these databases will be supported through appropriate adaptation of a proposed schema for "Automation Objects".

These domains are "open" as defined in part 4 of the IEC 61499 standard; that is, they support **interoperability** of control/automation platforms, software tools and repositories; **portability** of software elements within and between domains; and **configurability** of platforms by tools. Since this infrastructure consists of three open, object oriented domains, it is referred to as the **O^3 infrastructure**.

6. STANDARDIZATION

When the CCI is in full operation, WG2 (Standards) will identify future standardization work necessary to support the OOONEIDA repository and O3 architecture and make proposals to the appropriate international standardization organizations.

7. REFERENCES

OOONEIDA official web site: www.oooneida.info

IEC 65/338/CDV, Committee Draft for Vote, IEC 61499-1, Function blocks - Architecture, International Electrotechnical Commission, Geneva, 2004-01-16; IEC 65/339/CDV, Committee Draft for Vote, IEC 61499-2, Function blocks - Software tool requirements, 2004-01-16.

R. Schoop and H.-D. Ferling: *Fuction Blocks for Distributed Control Systems*, DCCS'97, Proceedings, pp.145--150

IEC 61131-3, Programmable controllers-Programming languages, 2nd Ed., International Electrotechnical Commission, Geneva, 2003.

IEC/TS 61804-1, Function blocks (FB) for process control - Overview of System Aspects, Geneva, 2003-10; IEC/PAS 61804-2, Function blocks (FB) for process control - Specification of FB concept and Electronic Device Description Language (EDDL), Geneva, 2002-10.

R. Lewis, Modeling control systems using IEC 61499, The Institution of Electrical Engineers, 2001

HOLOBLOC™.com - Function Block-Based, Holonic Systems Technology, Access Date: January 2004

IEC TC 65, Working draft – Automation Objects, available at http://www.holobloc.com/stds/iec/sb3/ao/aowd1.zip

Rumpl R., Dutzler C., Zoitl A. und Auinger F., *Platforms for Scaleable Flexible Automation considering concepts of IEC 61499*, In 5th IEEE/IFP International Conference on Information Technology for Balanced Automation Systems in Manufacturing and Services, Cancun, Mexico 2002

Brennan R.W., Zhang X. Xu Y., Norrie D.H., *A reconfigurable concurrent function block model and its implementation in real-time Java*, Integrated computer Aided Engineering, 2002

Vyatkin V. V., Christensen J. H., Martinez Lastra J. L., Auinger F., OONEIDA: An Open, *Object-Oriented kNowledge Economy for Intelligent Distributed Automation*, 1st IEEE International Conference on Industrial Informatics INDIN'03, Proceedings, Banff, Canada, 2003

Vyatkin V. *Intelligent Mechatronic Components: Control System Engineering using an Open Distributed Architecture*, IEEE Conference on Emerging Technologies and Factory Automation (ETFA'03), Proceedings, Lisbon, Portugal, September 2003

K. Thramboulidis: *Development of Distributed Industrial Control Applications: The CORFU Framework*, 4th IEEE International Workshop on Factory Communication Systems, August 2002, Vasteras, Sweden, http://seg.ee.upatras.gr/corfu

Vyatkin V., J. LM Lastra: *Architectural Foundations for Reconfigurable Manufacturing Systems*, 3rd International Symposium on Open Control Systems SoftSympo'03, Helsinki, September, 2003

J.H. Christensen: *IEC 61499 architecture, engineering, methodologies and software tools*, 5th IFIP International Conference on Information Technology for BALANCED AUTOMATION SYSTEMS In Manufacturing and Services, Proceedings, Cancun, Mexico, September, 2002

Marcello Bonfè and Cesare Fantuzzi: "*Design and Verification of Industrial Logic Controllers with UML and Statecharts*", IEEE Conference on Control Application 2003, June 23-35, Istanbul, Turkey

HMS-Holonic Manufacutring Systems Consortium, *Holonic manufacturing systems overview*, Holonic Manufacturing Consortium Web Site, http://hms.ifw.unihannover.de, 2002

Neumann, P.; Diedrich, Ch.: *Distributed Function Block technology - quo vadis?* 1st International PLCopen Conference on Industrial Control Programming ICP ´96, Paris (Frankreich), 30.09.1996 - 03.10.1996. Preprints.

Jose L. Martinez Lastra, Reijo Tuokko: *Co-design architecture for Reconfigurable Assembly Platforms*, IEEE Transactions on Systems, Issue 2, Volume 2, April 2003 ISSN 1109-2777

Vyatkin V., Hanisch H.-M., S. Karras, X.Cai *IEC61499 as an architectural framework to integrate formal models and methods in practical control engineering*, Congress Electric Automation SPS/IPC/Drives, Nuernberg, Germany, November 2002, Proceedings, pp. 310—318

Copyright © IFAC Information Control Problems in
Manufacturing, Salvador, Brazil, 2004

ELSEVIER
IFAC
PUBLICATIONS
www.elsevier.com/locate/ifac

MOBILE ROBOT CONTROL OVER THE INTERNET WITH DELAY COMPENSATION

Vinícius Menezes de Oliveira *
Walter Fetter Lages **

* *Federal University of Rio Grande*
Department of Mathematics
Av. Itália, km 8 - Rio Grande, RS 96201-900
** *Federal University of Rio Grande do Sul*
Electrical Engineering Department
Av. Osvaldo Aranha, 103 - Porto Alegre, RS 90035-190
vinicius@ieee.org, w.fetter@ieee.org

Abstract: This article presents the implementation of a real time controller for mobile robots using an IP network to receive data from sensors and to send data to actuators. In order to compensate for the time delay related to the data transmission it is proposed to estimate the sensors readings that would be taken without delay. The estimation is based on an Extended Kalman Filter. Some experimental results are presented and performance of the control architecture are analyzed. *Copyright ©2004 IFAC*

Keywords: Networked Control, Internet, Mobile Robot Control, Delay Compensation

1. INTRODUCTION

During the past years the Internet growed unexpectedly and by now it is important not only in workplace or scientific field, but even in entertainment of people and to accomplish usual tasks like reading newspapers or buying books. At the same time, the control researchers started to see the Internet as an alternative way of connecting components of complex systems.

There are many ways of using the Internet in control systems. Some works present the teleoperation of robots by using a simple web page, where information from the robot are available to the user and the user can send commands to the robot, thus operating the robot through the Internet (Grange, Fong & Baur 2000).

One of the features of a networked control system is the presence of one or more control loops connected by a communication channel. The replacement of the traditional point to point structure by computer networks brings some advantages: lower costs, reliability, simplicity of installation and maintenance, etc.

The focus of this work is the effective use of conventional computer networks, based on the IP protocol, to implement a distributed control system, transmitting control signals and receiving sensor signals through the Internet. The use of conventional computer networks is a relatively new area of research to the control community, with few articles discussing it, but it is not so new to the industry (Zhang, Branicky & Phillips 2001). Thus, there are various subjects to be studied and analyzed.

This work presents the control of a differential drive mobile robot over the Internet, where the

Extended Kalman Filter (EKF) (Goodwin & Sin 1984) is used to estimate the state transmitted from the robot, in order to compensate for the existing delay in the control network (Alt, da Silva Guerra & Lages 2002), introduced by the network. When a signal is transmitted from a computer to another one a delay can be caused by a highly loaded network or by the existing buffers in communication network. The delay compensation proposed here consists in the estimation of the variables send by the robot by means of an Extended Kalman Filter, based on the system dynamic model.

In the following sections the structure of control over Internet is explained (section 2), as well as its implementation (section 6). The nonlinear model of the mobile robot (section 3) and the derivation of the Extended Kalman Filter equations of estimation and prediction (section 5) are also presented.

2. STRUCTURE OF CONTROL OVER THE INTERNET

Figure 1 details the structure used to control the robot using the Internet. It consists in two independent computers connected to the Internet, one is the controller and the other is the robot on-board computer. The robot sends the sensor readings to the controller, which computes the control law and then sends the control signals to the robot. The robot receives the control signals from the controller and applies them to actuators.

This structure is very suitable to the situation where the mobile robot cannot be equipped with a powerful microprocessor due to restricted power capacity (equipped with batteries). Another situation is a robot with many joints which can be more appropriately controlled by a distributed controller.

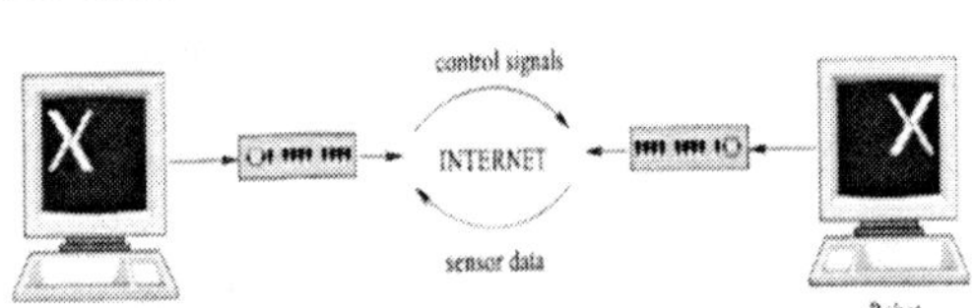

Fig. 1. Structure of control.

The layers in the structure are shown in figure 2. The transport layer adopts the User Datagram Protocol (UDP), instead of the most usual TCP protocol. The TCP protocol is connection oriented and uses error detection and correction methods in order to provide a reliable communication channel between source and destination. The error recovery is based on retransmission of lost packets. The UDP protocol is a datagram oriented protocol. Messages are transmitted in the hope that they arrive at destination, but there is not any verification that they actually reached their destination.

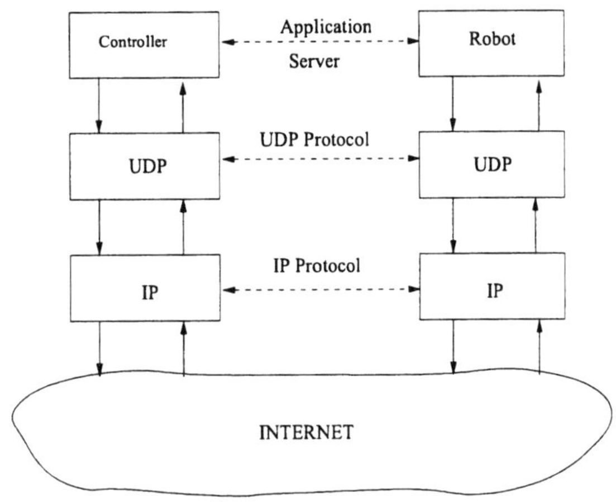

Fig. 2. Layers of the control network.

At a first glance, TCP protocol seems to be more convenient, since it implements a reliable channel. However, TCP does not make any commitment with respect to message timing. In a control network, messages with large network delays are of no use, since the sensor data or control signal they carry would be outdated. Messages discarded by the application protocol due to large delay are also harmful since they used a portion of the system bandwidth which potentially caused delays to other messages. In this context, message retransmission, as performed by TCP, tends to be also harmful since retransmitted messages would suffer larger delay than regular messages, with higher chances to be discarded by the application protocol due to excessive delay. In extreme cases, retransmissions can cause network congestion, which forces messages to be discarded, just to be retransmitted and so on.

On the other hand, in a digital control system, usually sensors are sampled at a constant rate. Therefore, messages with sensor readings are transmitted at the same rate. For the control of the mobile robots typical rates are around 20Hz. At such a rate, whenever a message is received with errors, it could be more convenient to wait for the next message with the sensor readings than retransmit the lost message to receive outdated readings. Also, prior sensor readings can be used to estimate the readings on the lost messages. With these considerations in mind it becomes natural to use the UDP protocol since it uses less bandwidth and is more efficient with respect to timing than TCP protocol. Note, however, that UDP does not provide commitments to respect to timing either. In this work, the problem of delays imposed to messages is dealt with at the application layer. Here the Extended Kalman Filter is used to estimate the signals send by the robot to

Fig. 3. The mobile robot.

the controller. The idea is to estimate the sensor readings that would be taken at the time that the message with sensor readings arrives at controller.

3. MATHEMATICAL MODELING

According to Yamamoto & Yun (1994) and de Wit, Siciliano & Bastin (1997) the configuration dynamic model of a differential drive wheeled mobile robot (figure 3) can be described by the following set of equations:

$$\dot{q} = S(q)\eta \tag{1}$$

$$H\dot{\eta} = F\tau \tag{2}$$

where $q = [x_c \; y_c \; \theta_c \; \phi_r \; \phi_l]^T$ is the generalized coordinates vector, $\eta = [\eta_1 \; \eta_2]^T$ is the velocity vector (η_1 is the linear velocity and η_2 is the angular velocity). The vector $\tau = [\tau_r \; \tau_l^T]$ represents the torques provided by the right (τ_r) and left (τ_l) motors. The matrix $S(q)$ is given by:

$$S(q) = \begin{bmatrix} -\sin(\theta) & 0 \\ \cos(\theta) & 0 \\ 0 & 1 \\ -\frac{1}{r} & -\frac{L}{r} \\ \frac{1}{r} & -\frac{L}{r} \end{bmatrix} \tag{3}$$

and the matrix H is given by:

$$H = \begin{bmatrix} m_b + \frac{2I_\varphi}{r^2} & 0 \\ 0 & m_b + \frac{2L^2 I_\varphi}{r^2} \end{bmatrix} \tag{4}$$

where m_b is the mass of the robot, I_φ and r are, respectively, the inertia and radius of each wheel and L is the distance of each wheel to the center of the robot.

The matrix F is defined as:

$$F = \begin{bmatrix} -\frac{1}{r} & -\frac{L}{r} \\ \frac{1}{r} & -\frac{L}{r} \end{bmatrix} \tag{5}$$

By using the smooth static time-invariant state feedback:

$$\tau = F^* H u \tag{6}$$

where F^* denotes a left inverse of F and u is an auxiliary control input vector, the model can be simplified to

$$\dot{q} = S(q)\eta \tag{7}$$

$$\dot{\eta} = u \tag{8}$$

If the state vector x is defined as as:

$$x \triangleq \left[q^T \; \eta^T\right]^T \tag{9}$$

then the dynamic model of 7 and 8 can be expressed in the affine form:

$$\dot{x} = f(x) + g(x)u \tag{10}$$

with:

$$f(x) = \begin{bmatrix} -\sin(\theta)\eta_1 \\ \cos(\theta)\eta_1 \\ \eta_2 \\ -\frac{\eta_1 + L\eta_2}{r} \\ \frac{\eta_1 - L\eta_2}{r} \\ 0 \\ 0 \end{bmatrix} \tag{11}$$

$$g(x) = \begin{bmatrix} 0_{5\times 2} \\ I_{2\times 2} \end{bmatrix} \tag{12}$$

and the output of the system is given by:

$$y = h(x) = \left[I_{3\times 3} \; 0_{3\times 4}\right] x \tag{13}$$

Now, the system will be described as a discrete system, which is a form more suitable for the derivation of the Extended Kalman Filter (Goodwin & Sin 1984). The discretized dynamic system can be written as:

$$x((k+1)T) = f(x(kT)) + g(x(kT))u(kT) \tag{14}$$

$$y(kT) = h(x(kT)) \tag{15}$$

where T is the sampling period.

For the sake of simplicity, from now on, the constant T will be omitted and the model becomes:

$$x(k+1) = f(x(k)) + g(x(k))u(k) + \omega(k) \quad (16)$$
$$y(k) = h(x(k)) + \nu(k) \quad (17)$$

where $\omega(k)$ is the process noise, representing the modeling errors, and $\nu(k)$ is the measurement noise, representing the sensor noise. It is supposed that $\omega(k)$ and $\nu(k)$ are white noise signals such that:

$$E\left[\omega(k)\omega^T(i)\right] = Q(k)\delta(k-i) \quad (18)$$
$$E\left[\nu(k)\nu^T(i)\right] = R(k)\delta(k-i) \quad (19)$$
$$E\left[\omega(k)\nu^T(i)\right] = 0 \quad (20)$$

where $Q(k)$ and $R(k)$ are the covariance matrices of process and measurement noise, respectively and $\delta(\cdot)$ is the Kroeneker delta function. This assumption is a classical one and implies that process and sensor noise are not correlated.

4. CONTROL LAW

In this work it is used a nonlinear controller based on the fact that the complete model of a differential drive robot is the cascade of a dynamic and a kinematic sub-system (see figure 4). In the kinematics control loop a sliding mode controller is used and in the dynamics control loop uses a neural network. See (de Oliveira & Lages 2003) for details about the development of the control law.

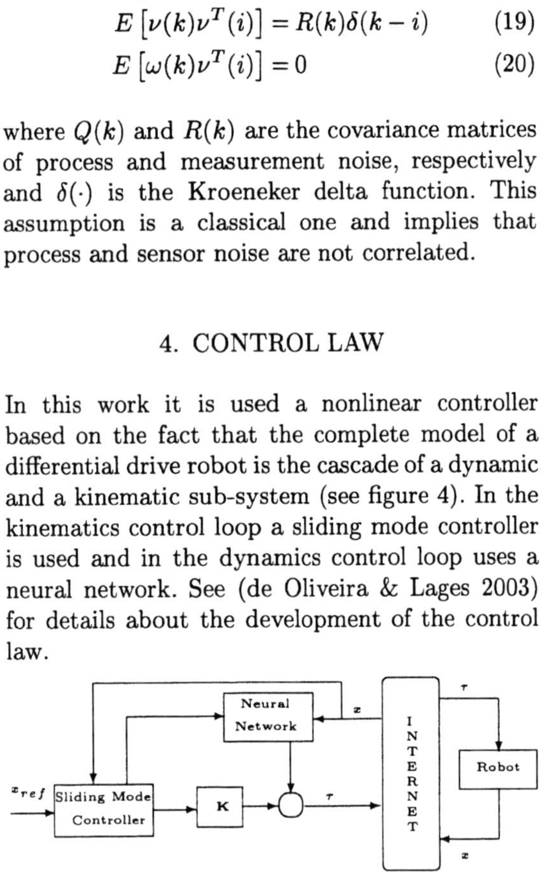

Fig. 4. Control block diagram.

The control law obtained with this control loop is given by the following equation:

$$\tilde{\tau} = \hat{f} + \mathbf{K_4} e_c - \gamma \quad (21)$$

where $\mathbf{K_4}$ is a definite positive gain matrix and $\hat{f}(x)$ is an estimative of the dynamics terms of the robot, given by the neural network.The term γ is present as a matter of robustness. A more detailed demonstration of this controller and its proof can be accessed in (de Oliveira & Lages 2003).

5. EXTENDED KALMAN FILTER

The Kalman Filter (KF) is a linear recursive estimator that calculates the optimal (with respect to the minimum variance) estimate for the estate vector based on observations that are linearly dependent on the unknown state (Kalman 1960). The original proposition intended to deal with linear systems only.

Obviously, the model 10 is nonlinear, therefore its state can not be estimated by using the traditional KF. However, the filter is extended by linearizing the system model around the estimated state vector. In the following it will be shown the linearization of system 16.

The expansion of equations 16 and 17 in Taylor's series around the point $(\hat{x}(k|k), u(k))$ and discarding the high order elements results:

$$x(k+1) \approx f(\hat{x}(k|k)) + g(\hat{x}(k|k))u(k) + \\ + F(k)\left[x(k) - \hat{x}(k|k)\right] + \omega(k) \quad (22)$$
$$y(k) \approx h(\hat{x}(k)) + H(k)[x(k) - \hat{x}(k|k)] + \\ + v(k) \quad (23)$$

where

$$F(k) = \frac{\partial[f(x(k) + g(x(k))u(k))]}{\partial x(k)}\Big|_{x(k)=\hat{x}(k|k)}$$

$$F(k) = \begin{bmatrix} 0 & 0 & -\cos(\theta)\eta_1 & 0 & 0 & -\sin(\theta) & 0 \\ 0 & 0 & -\sin(\theta)\eta_1 & 0 & 0 & \cos(\theta) & 0 \\ 0 & 0 & 0 & 0 & 0 & 0 & 1 \\ 0 & 0 & 0 & 0 & 0 & -\frac{1}{r} & -\frac{L}{r} \\ 0 & 0 & 0 & 0 & 0 & \frac{1}{r} & -\frac{L}{r} \end{bmatrix} \quad (24)$$

$$H(k) = \frac{\partial\left[h(x(k))\right]}{\partial x(k)}\Big|_{x(k)=\hat{x}(k|k)}$$

$$H(k) = \begin{bmatrix} 1 & 0 & 0 & 0 & 0 & 0 & 0 \\ 0 & 1 & 0 & 0 & 0 & 0 & 0 \\ 0 & 0 & 1 & 0 & 0 & 0 & 0 \end{bmatrix} \quad (25)$$

From equations 22 and 23 one can write the expressions for the EKF as (Goodwin & Sin 1984):

$$K(k) = \frac{P(k|k-1)H^T(k)}{H(k)P(k|k-1)H^T(k) + R(k)} \quad (26)$$

$$\hat{x}(k|k) = \hat{x}(k|k-1) + K(k)[y(k) - h(\hat{x}(k|k-1))] \quad (27)$$

and the expressions for prediction are:

$$\hat{x}(k+1|k) = f(\hat{x}(k|k)) + g(\hat{x}(k|k))u(k) \quad (28)$$
$$P(k+1|k) = F(k)P(k|k)F^T(k) + Q(k) \quad (29)$$

Figure 5 shows the block diagram of the control structure used in this work. It presents the Extended Kalman Filter block taking the control signal $u(t)$ from the controller and the signal sent by the robot over the Internet and results in an estimative $\hat{x}(t)$ after the delay of transmission. The nonlinear controller used here consists in a control loop to the kinematics cascaded to a control loop to the dynamics.

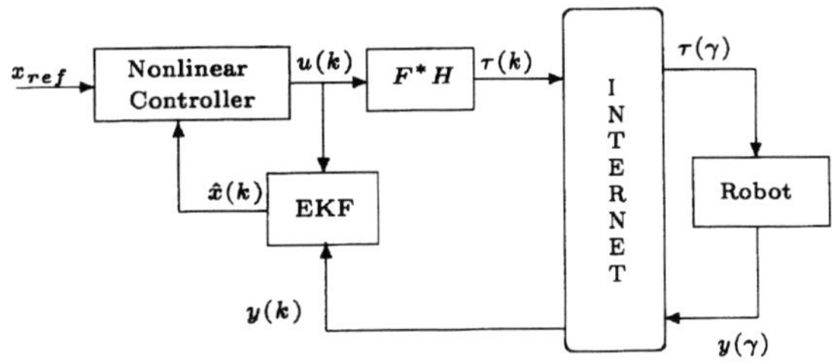

Fig. 5. Block diagram of the control system.

6. IMPLEMENTATION

The software implementation of the control structure is shown in figure 6.

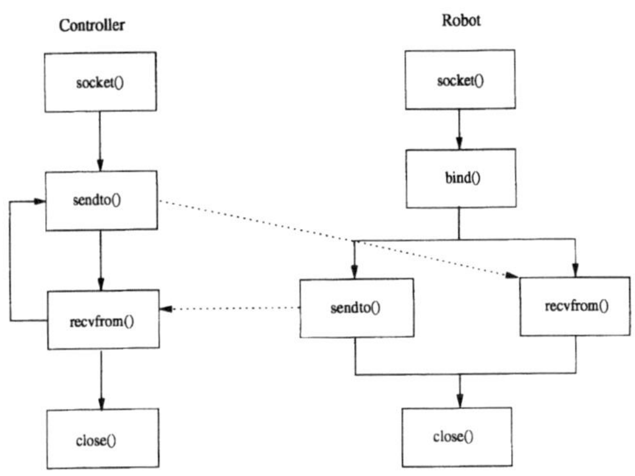

Fig. 6. processes diagram to the control structure.

The part of the robot consists in two threads, one that receives the data from the control task and simulates the robot and the other responsible for reading the sensors data and send back to the control task.

During the initialization of the robot two sockets are created, one socket to the communication between the control task and the thread that receives the control signals and another socket to the communication between the thread that sends the sensor data and the control task. After the creation of the sockets, the task is configured to real time and the tasks are started.

In this structure all the control operations are executed in the control task. When this task is initialized, a socket is created and a local port is defined, a message to turn the robot on is sent and the system continues working.

The program executing in the robot interface node is implemented by three LXRT hard real-time tasks. The first task interacts with the robot hardware to read the joint sensors and send the readings to the controller node at a rate of 100Hz. Another task executes in a free-running (without timing) endless loop. This task waits for commands from controller node and executes them. Note that before the application of commanded voltages to robot actuators, the network delay (measured by time-stamps) is computed and compensated for.

On the controller node side there are also three tasks: Two RTAI LXRT hard real-time tasks and a plain Linux process that implements the user interface. The purpose of the first RTAI task is to keep the time synchronization with the robot interface node, while the other implements the control algorithm.

The control algorithm is implemented by a free-running RTAI task which waits for the sensor readings, computes the delay introduced by the network and compensates for the delay using the robot dynamic model, thus obtaining an updated estimate for the sensor readings. This updated estimate is used to compute the desired voltage levels to be applied to the robot actuators.

7. SIMULATIONS

In this section some simulation results are shown without the implementation of delay compensation.

In order to test the proposed delay compensation method under various network load conditions, a traffic emulator was developed. All network traffic between the robot interface and the controller node is routed through a computer executing the traffic emulator program. The traffic emulator imposes delays to selected packets. These delays are generated accordingly to a statistical network load model resembling real network traffic. Due to the statistical nature of delays imposed to packets a scheduler/dispatcher architecture to implement the traffic emulator. Since random delays are generated, there is an implied packet reordering. By using a scheduler/dispatcher architecture, packet reordering can be easily handled.

Figure 7 shows the trajectory executed by the robot and the reference trajectory and it is possible to observe that the behavior of the robot is not the one expected.

This work intends to include the compensation of the delay that occurs during the transmission of signals over the network. In a first step, it will be considered the compensation of delay during the transmission of the sensor data sent by the robot to the controller using the Extended Kalman Filter. In a second step, the compensation of

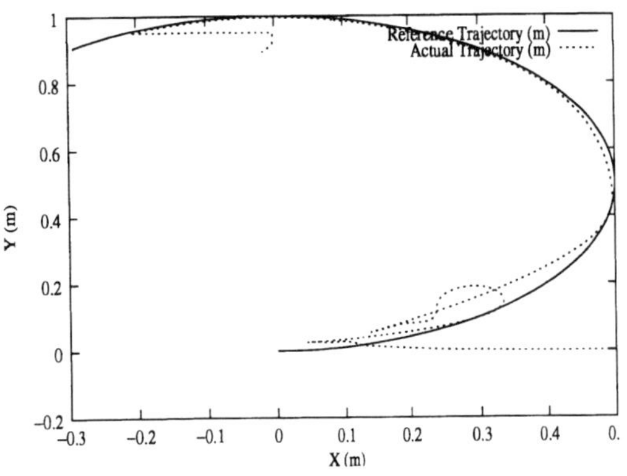

Fig. 7. Trajectory of the robot.

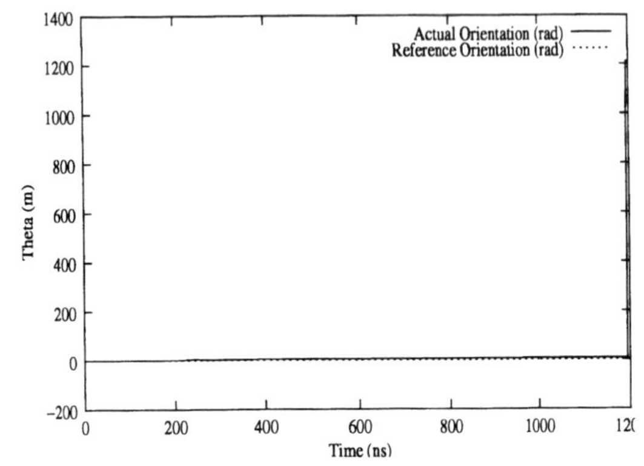

Fig. 10. Orientation of the robot during the motion.

the delay related with the transmission of the control signal to the robot will be implemented. This compensation will be done by computing the control signal for a number of possible delay values and sending all of them to the robot. The robot computer measures the effective delay and selects the most appropriated control value.

The time evolution of each variable of the robot during the execution of the trajectory can be analyzed in figures 8, 9 and 10.

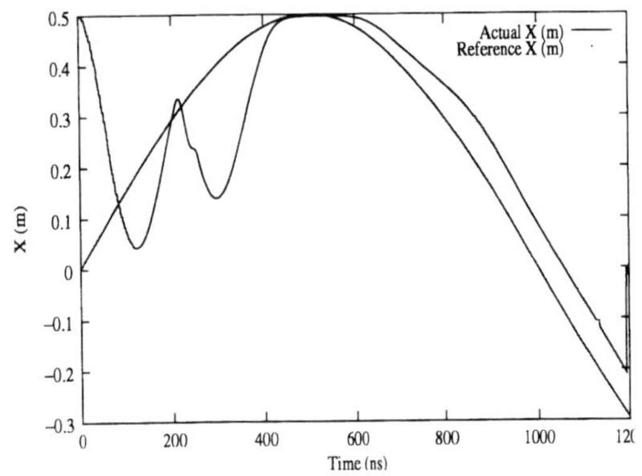

Fig. 8. Position in X axis along the time.

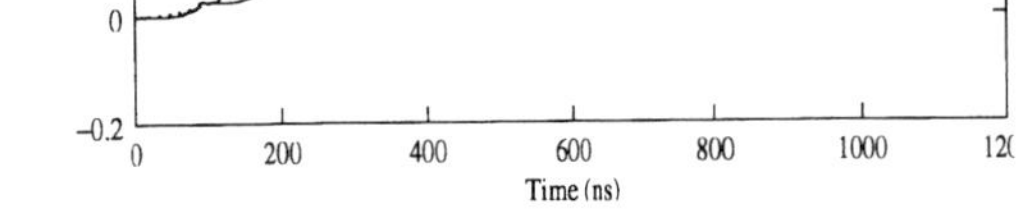

Fig. 9. Position in Y axis along the time.

8. CONCLUSIONS

The networked control of mobile robots over Internet seems to be a possible alternative to remote operation of robots.

Based on the preliminaries results shown in the previous section, we expect to overcome the problems caused by the delay of communication. However, for a more precise evaluation of the entire structure of networked control it is necessary to analyze performance criteria of control systems, such as dumping, overshoot, etc. Hopefully, the final version of this paper will address these issues.

REFERENCES

Alt, G. H., da Silva Guerra, R. & Lages, W. F. (2002). An assessment of real-time robot control over IP network, *Proceedings of the Fourth Real-Time Linux Workshop*, University of Boston, Computer Science Department, Boston, USA, pp. 1–17.

de Oliveira, V. M. & Lages, W. F. (2003). Mobile robot control usin sliding mode and neural network, *Proceedings of the 7th IFAC Symposium on Robot Control.*

de Wit, C. C., Siciliano, B. & Bastin, G. (eds) (1997). *Theory of Robot Control*, 2nd. edn, Springer-Verlag, chapter Modelling and Structural Properties.

Goodwin, G. C. & Sin, K. S. (1984). *Adaptive Filtering, Prediction and Control*, Prentice-Hall Information and System Sciences Series, Prentice-Hall, Inc., Englewood Cliffs, NJ.

Grange, S., Fong, T. & Baur, C. (2000). Effective vehicle teleoperation on the world wide web, *Proceedings of the International Conference on Robotics & Automation*, pp. 2007–2012.

Kalman, R. E. (1960). A new approach to linear filtering and prediction problems, *Transactions of the ASME - Journal of Basic Engineering* (82): 35–45.

Yamamoto, Y. & Yun, X. (1994). Coordinating locomotion and manipulation of a mobile platform, *IEEE Transactions on Robotic and Automation* **39**: 1326–1332.

Zhang, W., Branicky, M. S. & Phillips, S. M. (2001). Stability of networked control systems, *IEEE Control Systems Magazine* **21**(1): 84–99.

Copyright © IFAC Information Control Problems in
Manufacturing, Salvador, Brazil, 2004

ELSEVIER
IFAC
PUBLICATIONS
www.elsevier.com/locate/ifac

A QUALITY-OF-SERVICE (QoS) BASED APPROACH FOR THE COMMUNICATION SUPPORT IN NETWORK-BASED CONTROL SYSTEMS: AN ON-GOING PROJECT

Ricardo Alexandre Reinaldo de Moraes[1] and Francisco Vasques[2]

[1] *Department of Exact and Technological Sciences, UNIPLAC, Planalto Catarinense University Av. Castelo Branco, 170 Lages SC, Brasil*

[2] *Department of Mechanical Engineering, FEUP, University of Porto Rua Dr. Roberto Frias, 4200-465 Porto, Portugal*

e-mail: ricardo@uniplac.net, vasques@fe.up.pt

Abstract:
Due to the increased availability of low cost network technology, the implementation of Network-based Control Systems (NCS) is becoming widely accepted. Current NCS are usually supported by small-dedicated communication networks (known as *fieldbus networks*) operating in access-controlled mediums, which are able to guarantee the delivery of control data within known and bounded delays.
It is however foreseeable that, access-uncontrolled networks (such as packet-based networks) will play an important role in future NCS. In this paper, we present an on-going project, proposing the use of a QoS architecture to implement NCS built upon shared networks with uncontrolled access, *i.e.*, considering communication environments subject to unknown disturbances.
Within this project, several QoS architectures for packet-based networks are looked at carefully, such as the "IntServ" and the "DiffServ" models and the "Multi-Protocol Label Switching" (MPLS) and the "Constraint Based Routing" strategies. Such multiple QoS architectures will be evaluated through the use of an experimental platform enabling the implementation and evaluation of network-based control systems *Copyright © 2004 IFAC*

Keywords: Network-based Control Systems, Quality of Service, Manufacturing Systems.

1. INTRODUCTION

Recently, important technological advances can be observed in the development of devices with sensorial and communication capabilities, able to be ubiquitously incorporated into the physical world. Examples of such devices are the micro-electromechanical based sensors (MEMS-based sensors) used in automobile systems, smart homes, large manufacturing systems, intelligent highways and networked city services, large production systems and their logistics chains. Generally, such devices are characterized by the combination of reduced cost *vs.* large communication and processing capacity.
The advent of ubiquitous, distributed computation, communication, and sensing systems has begun to create an environment in which we have access to enormous amounts of data and the ability to process and communicate that data in ways that were unimagined 20 years ago (Murray *et al.*, 2003). This will have a huge impact in applications where computer systems strongly interact with the environment, enabling the true implementation of the *pervasive distributed computing* paradigm. Within such context, the technological landscape will be highly influenced by state-of-the-art research in the domains of convergence between communications, computing, and control.

Control systems are fundamental elements for the construction of complex computer systems characterized by a close interaction with the exterior environment. Actually, one of the main targets when designing complex computer systems is the optimization of both its performance and reliability in the presence of unexpected disturbances from the outside environment. Such target can be considered similar to that one of a traditional control system, *i.e.* its capability to provide an adequate service in the presence of unexpected disturbances.
The implementation of network-based feedback control systems is becoming widely accepted, due to both economical reasons (inexpensive devices) and the ability to decentralize control functions. Such type of feedback control implementation, where the control loop is closed through a real-time network, is referred as Network-based Control System (NCS).
In a NCS, a timely and robust communication service is fundamental for the adequate behavior of the supported control application. It is well known that, either lost or dropped control messages or unmodeled (and variable) communication delays are difficult to tolerate in current control systems.
Current NCS are usually supported by small-dedicated communication networks (known as fieldbus networks) operating in access-controlled mediums, which are able to guarantee the delivery of control data within known and bounded delays.
It is however foreseeable that, access-uncontrolled networks (such as packet-based networks) will play

an important role in future NCS. When dealing with such timely unpredictable networks, two options can be followed:

a) either a new theory/practice for network-based control systems is developed for systems operating in distributed, asynchronous and packet-based environments;
b) or new methodologies for supporting real-time control communications in such kind of networks are devised.

In this paper, we present the main guidelines of an on-going project, which proposes the use of a QoS architecture to implement NCS built upon shared networks with uncontrolled access, i.e., considering the communication environment subject to unknown disturbances.

Within this project, several QoS architectures for packet-based networks are looked at carefully, such as the "IntServ" and the "DiffServ" architectures and the "Multi-Protocol Label Switching" (MPLS).

Such multiple QoS architectures will be evaluated through the use of an experimental platform, which enables the implementation and evaluation of real NCS. Such platform enables the performance evaluation of both the case of NCS supported by local networks (LANs) or supported by public data communication networks. A special emphasis is given to manufacturing systems that currently use synchronous communication services to support NCS.

This paper is organized as follows. In Section 2 the major challenges concerning the implementation of the pervasive distributed computing paradigm are explained.

In Section 3, several QoS architectures for packet-based networks are looked at carefully, such as the "IntServ" and the "DiffServ" models and the "Multi-Protocol Label Switching" (MPLS).

Finally, in Section 4, an experimental platform intended for the experimental evaluation of multiple QoS architectures is described. Such experimental platform enables the performance evaluation of either the case of NCS supported by local networks (LANs) or the case of NCS involving the use of public data communication networks.

2. NETWORK-BASED COMPUTING SYSTEMS

Carrying out complex computer systems implementing the *pervasive distributed computing* paradigm requires a great research effort on new application domains, mainly in research domains in the intersection between Computer Science, Control Theory, and Communication Systems. The major challenges concerning such implementations have been pointed out (Murray *et al.*, 2003) as being:

a) *Control in distributed, asynchronous, networked environments,* namely for the cases of complex computer systems characterized by a close interaction with the outside environment and interconnected through packet-based communications. This is especially true in applications where one cannot ignore computational and communication constraints in performing control operations.
b) *Building very reliable systems from unreliable components*, able to prevent the global system damage, even in the case of individual component failures. This aspect is especially relevant for the case of control applications, where the system must automatically reconfigure itself in case of individual component failures, so that its performance degrades gradually and not abruptly

Thus, these new systems to be developed must take into consideration the QoS (quality of service) obtained. This model must count on features such as shared resources, high availability, end-to-end requests, and adaptation (follow the environment status and get adapted automatically) (Schmidt *et al.*, 2000; Pedatella *et al.*, 2003). There are many applications of crucial security, where the interruption of a service due to failure can produce important damage or losses. Therefore, fault tolerance becomes a requirement of fundamental importance on these systems, once a fault tolerant system keeps running even with failure to its components.

2.1 Control in Communication Systems vs. Communication in Control Systems

The crossroad between Computer Science, Control Theory, and Communication Systems is precisely one of the research domains that have been identified as one of the most promising I&D areas for future developments in the scope of control systems (Murray *et al.*, 2003)

The implementation of pervasive distributed systems by means of components ubiquitously incorporated into the physical world, presents an unquestionable attractive for large-scale complex systems control. The two main open research domains in the scope of this area of convergence are: the control of large-scale communication systems and the network-based control systems.

The *control of large-scale communication systems* includes topics related to congestion control, routing, information management and power management. The fundamental aspects to consider are related to the large dimension of the systems, its decentralised nature, the difficulty inherent to its stabilization (due to the delays the control actions are subject to) and the uncertainties associated with its peculiar parameters: traffic demand, network topology, transmission channel characteristics, etc.

On the other hand, *network-based control systems* have been assuming a more relevant role in the context of control systems, due to the always-increasing number of systems made up of a large number of components ubiquitously incorporated in the physical world. Due to the continuous evolution of communication systems, as well in economical terms as in working reliability, it can be predicted that, in the near future, even small-located control systems will become network-based. In other words, it will soon be trivial to come across data related to sensors, actuators, diagnose functions, command functions and co-ordination functions being transferred in real time through communication networks.

One of the major advantages of such network-based implementations, where the control functions are distributed throughout multiple computer systems, is the ability to improve the reliability of the global system operation, through the implementation of redundancy.

2.2 Target System

Presently, NCS are still supported by synchronous communication services, which require the use of specific communication protocols in controlled communication environments able to guarantee the information transfer in upper bounded time intervals. The problem will be much more severe if non-controlled communication environments, *i.e.*, when uncontrolled interference between control traffic and other traffics is considered. Thus, innovative methodologies must be considered to deal with this problem.

It is also important to realize that, whatever the communication environment under consideration, its robust operation is bounded to either the weakest individual component of the distributed computing system, or to the possibility of cascade damages that can provoke the system breakdown, in terms of performance and/or running reliability.

In this paper, we propose the use of a QoS architecture to implement a NCS built upon shared packet-based networks with uncontrolled access, i.e., considering communication environments subject to unknown disturbances. Therefore, several QoS architectures for packet-based networks will be looked at carefully, such as the "IntServ" and the "DiffServ" architectures and the "Multi-Protocol Label Switching" (MPLS)" strategies.

Such multiple QoS architectures will be evaluated through the use of an experimental platform enabling the implementation and evaluation of network-based control systems (NCS). This experimental platform enables the performance evaluation of both the cases of NCS supported by local networks (LANs) or involving the use of public packet-based data communication networks. A special emphasis will be given to manufacturing systems that currently use synchronous communication services to support NCS.

3. OPEN PROBLEMS

The implementation of a network-based control system, within which the logical ring among sensors, controllers and actuators is closed through a communication network, requires a real-time communication service guaranteeing that (Martí *et al.*, 2001):

a) the communication delay has an upper bounded and predictable value, when a previously defined set of load assumptions is considered;
b) the jitter associated either to the sampling or the actuation delay, an upper bounded and predictable value, when a previously defined set of load assumptions is considered;
c) the rate of message delivery failure is smaller than an upper bounded value, when a previously defined set of failure assumptions is considered.

Clearly, the considered upper bound for each of these three items must be in accordance with the characteristics of the supported control systems, namely with their control cycle time and their robustness to control data losses. Namely, the maximum jitter associated to the sampling or the actuation delay must be carefully quantified, as the system performance can be strongly influenced by an exceeding value for one of these parameters.

3.1 Real-Time Communication Concepts

Conceptually, a real-time communication service can be provided by any communication system, given its ability to support the implementation of *real-time channels*. Such *real-time channels* concept (Ferrari *et al., 1990*) has been proposed as a mean to provide upper bounded delays to specific connections in conventional packet-switched networks. This concept was later refined and extended in (Kandlur *et al.*, 1994). According to the definition, each packet is assigned a deadline over each link on its route and the packet transmission over a link is scheduled according to its deadline. Using an adequate deadline assignment policy, the network will deliver the packets guaranteeing its timing requirements.

The fundamental problem for a real-time communication service resides on the capability that a specific communication (or routing) protocol will have to support the implementation of *real-time channels*. The following problems will have to be solved for any communication (or routing) protocol under consideration (Zheng and Shin, 1998):

a) The *schedulability problem*: Given a set of real-time channels supported by a communication link, there is the need to verify if all packets travelling through these channels can be delivered before their deadlines.
b) The *minimum guaranteed delay problem*: Given a new real-time channel to be established, calculate the minimum delivery delay upper bound that a link can guarantee, without violating the timing guarantees of the already established real-time channels.

The implementation of *real-time channels* on communication systems, aiming to offer real-time communication services, is a complex problem.

In the following subsections, a non-exhaustive presentation of some of the more relevant protocols that must be explored in the current context is given.

4. PACKET-BASED COMMUNICATION PROTOCOLS TO BE EXPLORED

The demand for services with different quality levels within the Internet brought the development of new service models, within which the application requirements can be specified. However, the service provided today on the Internet is based in the *best effort* paradigm, *i.e.*, it does not guarantee a real-time service to applications, bearing only a compromise of trying to timely deliver the information sent by the emitter to the receiver.

The quality of transport offered to the packet flows is usually known as *Quality of Service* (QoS). The QoS concept in communication networks is a fundamental concept, as it clearly characterizes the provided services and the associated guarantees. For instance, each provided service must be described by a set of parameters and its execution must be in conformity with the service specification. For example, a service with an upper bound of *15*ms for the delay time can deliver information with a *5*ms delay, but never with a *16*ms delay.

An adequate management of the elements involved in the QoS negotiation is required, in order to guarantee a consistent behavior of the communication system. For instance, the addition of a new *router* must not compel a service with a *15*ms delay to be provided with a *16*ms delay. Thus, the adequate management of communication networks with QoS must consider not only the *best effort* paradigm, but also QoS requirements (Granville, 2001).

The use of the TCP/IP architecture to support data communication with timing requirements (deadlines) triggered several research works about QoS issues. The creation by the IETF of several work groups directed to the use of QoS on TCP/IP networks (IETF, 2002) had a consolidation effect on the on-going research works about QoS issues.

Currently, several technologies offer QoS in communication services, pointing out among them: the Integrated Service model (IntServ) and the Differentiated Service model (DiffServ); Traffic Engineering, Multi-Protocol Label Switching (MPLS); and Constraint Based Routing (Wang, 2001).

On the IntServ model, the applications must define the desired route for their packets and must also reserve the required resources before transmitting the data. On the DiffServ model, there is no resource reservation. Packets receive transmission priorities according to pre-defined service classes; packets of the same class receive similar treatment. Traffic engineering main objective is to optimize the network resources. MPLS is a packet routing mechanism within in which packets are labeled; packet routes are defined based on both labels and the constraints imposed by the applications.

To solve the problem of real-time communication, policing traffic mechanisms can also be implemented. Policing mechanisms control and monitor the network flows to guarantee the requested QoS. If in a given moment the network do not provides the required QoS, control measures are taken, such as: packet delay, discard or marking.

Several policing mechanisms haven been proposed. Among the most cited on the literature, there are the *Leaky Bucket* (Butto *et. al., 1991)* and its variants (Berger, 1990), the peak counter, and the window-based mechanisms. Each of these mechanisms use some of the following parameters:

a) the average rate, which defines the transmission average rate (packets per time interval) by which a flow can inject packets or *bytes* into the network;
b) the peak rate, which upper bounds the quantity of packets or *bytes* that can be sent in a short time interval;
c) size of blast, which upper bounds the maximum number of packets or *bytes* that can be sent instantly.

Interoperability among several systems from many suppliers is possible through the use of the TCP/IP protocol. TCP is a protocol from the transport layer of Internet TCP/IP architecture. This protocol is connection-oriented and offers a reliable end-to-end file transfer service. It is responsible for inserting the application messages into the transport datagram, for organizing the arrival of datagrams and also for re-sending lost datagrams. TCP was designed to work using the connectionless network service offered by the IP protocol. Therefore, it does not require a reliable network service, being responsible for the recovery of damaged, lost or doubled data and also for re-ordering data delivered out of order by the network protocol.

Nagle *apud* (Lima and Fonseca, 2002), on the initial phase of Internet, were one of the first to find that routers were vulnerable to congestion collapse. Such phenomenon, despite being rare, can occur when traffic is intense and the network is congested, provoking the discard of packets in the routers and resultant retransmissions, thus increasing round trip time (RTT). When packets are discarded, the resources that were used by them are wasted. Besides, when these packets are retransmitted, new resources are allocated.

Jacobson, who developed a mechanism that forces machines involved on TCP communication to reduce their transmissions of packets if there is congestion, introduced the first solution proposed for congestion problems within TCP/IP networks. This mechanism is composed of two algorithms: *Slow Start* and *Congestion Avoidance*, which are implemented together in practice. Since then, the efficiency of congestion control and prevention mechanisms was improved, with the proposal of two other algorithms: *Fast Retransmit* and *Fast Recovery*. Such mechanisms, composed by the four algorithms (TCP Reno, TCP Vegas, TCP Sack, TCP New Reno), is part of the TCP congestion control (Lima and Fonseca, 2002).

Another approach is proposed in (Liberatore, 2002), where a middleware regulates the network access through a polling scheme. This middleware is built on top of RTP/UDP/IP, which can also provide flexible support for fault tolerance, error recovery, and bandwidth monitoring. This middleware controls every network application, avoiding collision problems and controlling application priorities.

On (Kweon, *et. al.,* 2000) a traffic smoothing solution is introduced, which is implemented between the UDP or TCP layer and the Ethernet MAC layer. This mechanism gives priority to real-

time (RT) packets over packets of applications that do not require real-time communication. Through this mechanism, a reduction of the network collision rate was evidenced, thus guaranteeing RT applications.
(Dias *et al.*, 2003) employ the Lagrangean relaxation to optimize IP network traffic over MPLS technology; in this work, traffic-engineering tasks were modeled as a math-programming problem. The use of Lagrangean relaxation with heuristics was more efficient than the standard routing algorithms.
(Pedattela *et al.* 2003) introduce a Framework that aims to guarantee end-to-end QoS for the Internet through the use of a set of MPLS and DiffServ models and also through the use of a management software based on configuration policies.

5. METHODOLOGY OF ANALYSIS

The implementation of pervasive distributed systems based on the widespread use of components ubiquitously incorporated into the physical world, imposes advanced research work in the domains of convergence between communications, computing, and control.
The proposed work fits completely in the scope of this particular area. Its main objective is to propose and analyze solutions enabling the implementation of network-based control systems supported by access-uncontrolled networks (such as packet-based networks). Within the framework of this project are envisaged both the cases of control systems supported by local networks (LANs), or remote control systems that involve the use of public data communication networks.
The fundamental assertion of this project is that QoS architectures will be the appropriate architectures to support the implementation of NCS, especially when supported by access-uncontrolled networks (i.e., communication environments subject to unknown disturbances).
In order to achieve the targeted objective, a detailed state-of- the-art study is being done on both:

a) Methodologies for support of real-time communication, either through local networks, or public data communication networks.
b) Methodologies for the implementation of network-based control systems.

At the same time, an experimental platform to test network-based control systems is being developed (Figure 1) based on the following components:

a) a 2D table for the swinging of a ball;
b) a 2D position acquisition system for tracking the position of the ball on the table, based on a real-time image processing system;
c) A real-time control and actuation system suited to control the 2D position of a ball.

The interconnection between the acquisition system and the control system will be made through a selectable communication system, which will allow experimental tests to the performance of different communication methodologies, either for local networks (LANs) or for public data communication networks. In order to reproduce the traffic patterns from the several services found on a real system, a disturbance generator interconnected to the system under test will be used. Such disturbance generator will produce real system workloads and, for that, *trace* files will be used containing the actual features of the system in which the system to be controlled will be added to.

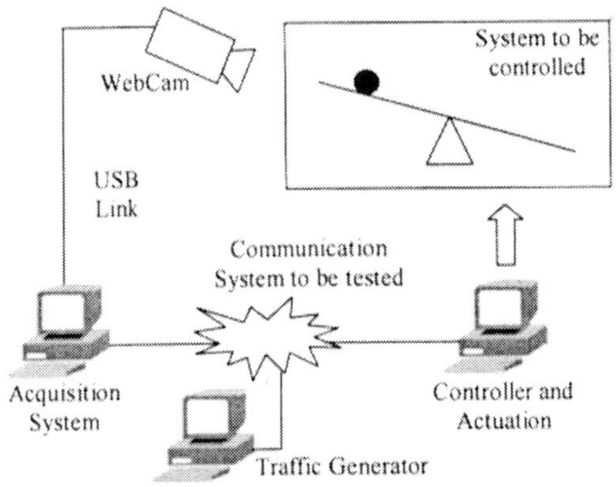

Fig. 1. Experimental platform to test network-based control systems.

According to the mentioned real-time support technologies, the communication system to be tested will be modified. This way, innovative real-time communication mechanisms, appropriate to be implemented on communication protocols traditionally used on either local networks (LANs) or on public data communication networks, will be proposed, tested and analyzed. Thus, with the use of such experimental platform, the impact of different communication methodologies on the quality of service offered to a specific control application will be evaluated. One of the major research targets is the evaluation of which communication parameters have a decisive impact on the control stability.

6. CONCLUSION

The advent of ubiquitous, distributed computation, communication and sensing systems will have a huge impact in applications where computer systems strongly interact with the environment, enabling the true implementation of the *pervasive distributed computing* paradigm.
Control systems are fundamental elements for the construction of complex computer systems characterized by such close interaction with the exterior environment. The implementation of network-based feedback control systems, usually referred as Network-based Control Systems (NCS), will have a strong impact on the implementation of the *pervasive distributed computing* paradigm.
Moreover, considering the innovative domains of application where computer systems strongly interact with the environment, it is our firm believe that there will be strong pressure for the implementation of NCS supported by public data communication networks. However, current technologies are still not able to robustly support control tasks by means of public data communication networks. Thus, there is demand for advanced design methodologies able to implement such kind of NCS.

We argue that there are two kinds of approaches to solve this problem: either new control approaches are developed enabling the consideration of the inherent temporal unpredictability of public data networks; or new communication methodologies are proposed for supporting real-time communications on public data networks.

In this paper, an on going-project was presented (still without experimental results), where it was pointed out that QoS architectures are the appropriate architectures for supporting communication on public data networks, in a temporally predictable way. An experimental platform was also presented, where the selected methodologies for the support of real-time communication will be tested.

REFERENCES

Berger, A. (1990). Performance Analysis of a Rate Control Throttle Where Tokens and Jobs Queue. In: *Proceedings of IEEE INFOCOM*, pp. 30-38.

Butto, M., E. Cavallero and A. Tonietti (1991). Effectiveness of the Leaky Bucket Policing Mechanism in ATM Networks. In: *IEEE JSAC*, pp. 335-342, vol. 9, no. 3.

Dias, A. D., E. Camponogara, J.M. Farines, R. Wilrich, A. Campestrini (2003). Otimização Lagrangeana em Engenharia de Tráfego para redes IP sobre MPLS. In: *21º. Simpósio Brasileiro de Redes de Computadores - SBRC 2003*, pp. 475-490.

Ferrari, D. and D. Verma (1990). A Scheme for Real-Time Channel Establishment in Wide-Area Networks. In: *IEEE Journal on Selected Areas of Communications*, pp. 368-379.

Granville, L.Z. Gerenciamento Integrado de QoS em Redes de Computadores (2001). Tese de Doutorado em Ciência da Computação, Universidade Federal do Rio Grande do SUL – UFRGS. Porto Alegre. Brasil. 159 f.

IETF (2002), INTERNET ENGINEERING TASK FORCE. Disponível em: http://www.ietf.org. Abril 2002.

Kandlur, D., K. Shin and D. Ferrari (1994). Real-Time Communicaiton in Multi-Hop Networks. In: *IEEE Transactions on Parallel and Distributed Systems*, vol. 5, no. 10, pp. 1044-1056.

Kweon, S., K. Shin and G. Workman (2000). Achieving Real-Time Communication over Ethernet with Adaptive Traffic Smoothing. In: *6th IEEE Real Time Technology and Applications Symposium* (RTAS 2000), pp. 90–100.

Liberatore, V. (2002). Scheduling of Network Access for Feedback-based Embedded Systems. In: *Quality of Service over Next Generation Internet*, SPIE ITCom 2002, pp. 73-82.

Lima, M.A.E., N.L.S. Fonseca (2002). Controle de Tráfego Internet. In: *20º. Simpósio Brasileiro de Redes de Computadores - SBRC 2002*, pp. 187-249.

Martí, P., J. Fuertes, G. Fohler and K. Ramamritham (2001). Jitter Compensation for Real-Time Control Systems. In: *Proceedings of the 22nd IEEE Real-Time Systems Symposium (RTSS'01)*, pp. 39-48.

Murray, R., K. Åström, S. Boyd, R. Brockett and G. Stein (2003). Future Directions in Control in an Information-Rich World. *To appear in IEEE Control Systems Magazine*. Disponível em: http://www.cds.caltech.edu/~murray/papers/2003a_mur+03-csm.html

Pedatella, R., E. Madeira, M. Magalhães (2003). Um Framework para Obtenção de QoS fim-a-fim na Internet. In: *21º. Simpósio Brasileiro de Redes de Computadores - SBRC 2003*, pp. 491-506.

Schmidt, D.C. *et al* (2000). Developing Next Generation Distributed Application with QoS – Enabled DPE Middleware, In: *IEEE Communications*, pp. 112-123.

Wang, Z. (2001). *Internet QoS: Architectures and Mechanisms for Quality of Service*. Morgan Kaufmann Publishers.

Zheng, Q. and K. Shin (1998). Fault-Tolerant Real-Time Communication in Distributed Computing Systems. In: *IEEE Transactions on Parallel and Distributed Systems*, vol. 9, no. 5, pp. 470-480.

Copyright © IFAC Information Control Problems in
Manufacturing, Salvador, Brazil, 2004

ELSEVIER
IFAC
PUBLICATIONS
www.elsevier.com/locate/ifac

FUZZY THERMAL CONTROL WITH REMOTE ACCESS FOR BUILDING AUTOMATION

**Adolfo Bauchspiess, Alexandre S. Souza, Antônio A. C. Leite,
Leandro M. A. Ramos, Ênio S. Pereira, Ronaldo J. Santos**

LAVSI –Laboratório de Automação, Visão e Sistemas Inteligentes
ENE - Electrical Engineering Department
UnB - University of Brasília, Brasília-BRAZIL

Abstract: This paper presents the development of a project that aims thermal efficiency in building automation. The internet infrastructure is used to develop a computational system for remote supervision. As building automation involves time-varying, distributed parameter systems we propose the use of fuzzy logic to implement an expert rule based systems that take care of the different contexts of the building management. The implementation of a prototype has been carried out on PIC16F877 microcontrollers. *Copyright © 2004 IFAC*

Keywords: Thermal systems, Fuzzy control, Automation, Microprocessor control.

1. INTRODUCTION

Fuzzy logic, neural networks, genetic algorithms and expert systems are part of a relatively new paradigm known as intelligent systems. These systems were created to supply answers to a great variety of problems which could not be solved by traditional methods, be because of its complexity or the inexistence of a necessary specific tools capable to handle this kind of problem, see (Azevedo, *et al.*, 2000). Fuzzy logic looks for to incorporate the human way of thinking in a control system. The produced controller can be applied to complex dynamic systems.

A lot of work has been published in this field. E.g., Mauris (2002) apply fuzzy symbolic to a comfort sensor, taking care of temperature and humidity to obtain "comfort". Haissig (1999) applied adaptive fuzzy in temperature control for hydronic heating systems. Comparison with PI show better results for the intelligent approach. This work aims to develop tools, which based on the use of intelligent algorithms (fuzzy logic), deals with information generally not used by common building facilities as: external temperature, presence of people in the room, luminous intensity in the control environment among others, processing them and adjusting the air conditioning devices to reduce the energy consumption.

2. THERMAL SYSTEM

Thermal systems are systems where the variables are related with the storage and transport (flow) of heat. In these systems, the heat transference can occur by three distinct forms: conduction, convection and radiation. In practice, one of the three forms is predominant. If considering two forms, in general the radiation can be neglected. Thermal processes are inherently distributed parameters systems that means, its variables vary with the localization or distance. To simplify the problem, we admit that the thermal system can be controlled with sufficient precision using only a few measure points. A step farther, for simulation purposes, would be approximate the system by concentrated parameters (thermal resistances and capacitances).

The prototype thermal system used in this work should represent a typical bureaus working place, Fig 1. Five rooms "working places" have doors and windows that can be opened or closed; it simulates the normal operation of the bureaus. Two heat sources should establish the air temperature in the five rooms.

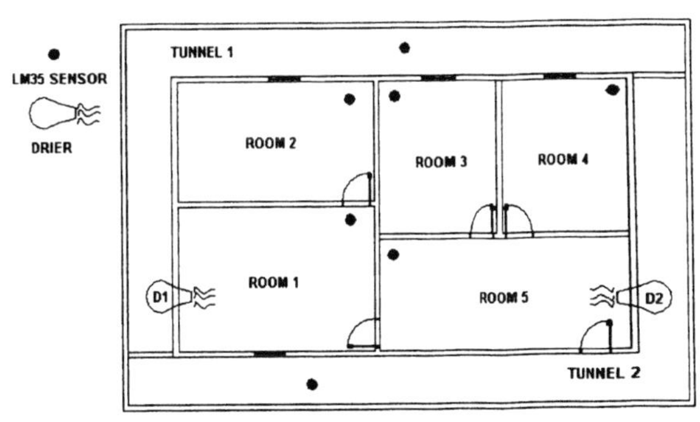

Fig. 1 – Thermal system prototype.

As air heating is much easier to implement than cooling, we used it in our prototype. Heating and cooling are both heat flow problems, with changed sign. So we opted to use heating in our prototype environment. Two 1kW hair driers deliver heat. We expect to obtain preliminary results with heating that can give insights and directions for a later implementation of air conditioning on a real world office environment. The prototype consists of three major areas. The first one is represented by two adjacent rooms (rooms 1 and 2). The larger (room 1) receive heat source D1. The second area comprehends two rooms with half height walls (rooms 3 and 4) and a corridor (room 5), where is the source of heat D2.

These two first areas represent the main part of the prototype, that is, the office. The third area consists of two tunnels (T1 and T2) in which ice bags can be placed to establish different temperatures for inlet D1 and D2. This simulates the effect of the sun on one side and shadow on the other side of a building, which can significantly affect energy efficiency. In each room, corridor and tunnel we installed a temperature sensor to measure the temperatures. A linear, time-invariant analogy of the thermal process can be made by an electrical circuit with capacitors and electrical resistances. In this analogy heat flow is equivalent to electrical current and temperature is equivalent to electrical tension.

The time constant of each room is related with the value of the thermal resistances R of each wall and thermal capacitances C of each room. In a system that involves a capacitance and a resistance in series, the time constant is RC. This time constant can be found experimentally isolating the involved rooms and measuring the time interval to reach 67% of the step response. Calling the heat flow input variable as q and the output temperature variable as T, the model for one room can be represented by the transfer function:

$$\frac{T(s)}{q(s)} = \frac{K}{RCs+1} \quad (1)$$

Where q(s) and T(s) are the heat flow and environment temperature Laplace transforms, respectively, see (Ogata, 1997).

3. MICROCONTROLLER

The microcontroller used in this project is a PIC16F877 which has the following main characteristics: high performance RISC CPU; EEPROM memory with 368 bytes; two pulse width modulation modules; 14 interruptions available; program memory of 8k x14 words; three timers (2x8 bits and 1x16 bits); 8 A/D conversion channels (10 bits); serial communication: synchronous and asynchronous, see (Souza, 2001).

4. MEASUREMENT OF TEMPERATURE

The temperature sensor LM35 used in this project is an integrated circuit that generates an output voltage proportional to the measured temperature, in Celsius degrees. It uses semiconductor properties as measurement principle. It is not necessary to calibrate the sensor. Precision of 0,5°C at 25°C and low current drain, near 60 mA are some advantages of this sensor.

Table 1 - Relevant specifications of the LM35.

Characteristics	Value
Scale factor	10 mV/°C
Operation range	-55 °C to 150 °C
Precision	0,5 °C (at 25 °C)
Current consume	< 60 μA
Supply	4 to 20 V

For the measure of the temperature a ten bit A/D (Analogic/Digital) converter of the microcontroller was used. A successive approximation approach is used, with maximum reference input equal to five volts; see (Souza, 2002). In this conversion type, the two less significant bits are usually discarded. The sensor output signal is amplified to have a range of tension from zero to five volts in the amplifier output, representing a range of temperature from zero to 127.5°C. So, a bit variation in the recorder called ADRESH (0 the 255 in decimal) represents a 0.5°C temperature variation, so, the precision of the sensor is preserved by the A/D conversion.

5. POWER DRIVER

The signal that carries out the control action on the system is a PWM signal with amplitude of five volts. With this amplitude, the signal is not able to supply the necessary energy to the hair drier. It is necessary, then, a power driver which can transform this low power input signal into an output signal with sufficient power. The following circuit was developed:

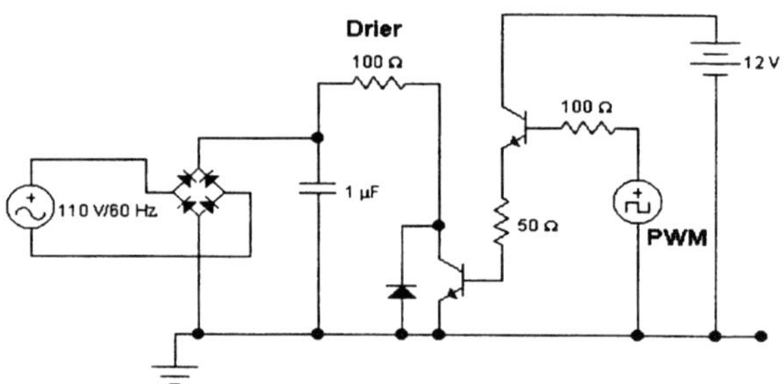

Fig. 2 - Power driver circuit.

The circuit has a first low power transistor BC547, which receives the PWM signal on its base, supplying a tension from 12 volts source to a second high power transistor TIP31C. This second transistor is driven by an 110V AC source, delivering the amount of current required by the hair drier. The two transistors operate in the switched mode; therefore the intensity of the output signal (in the drier) depends on the PWM duty cycle.

6. FUZZY LOGIC

The human creative capacity of inexact reasoning, uncertain or diffuse, contrasts with the computers and machines operation form, driven by binary logic. If these machines lose its "reasoning" restrictions, they would become eventually intelligent, being able to reason by an inexact form. This form of reasoning is known as Fuzzy Logic; see (Wang, 1997).

Fuzzy logic tries to incorporate the human way of thinking in computational systems. When this technique is applied to control loops, it's usual to call it Fuzzy Control. This type of control is generally used with complex dynamic systems.

The Fuzzy technology has an immense practical field, in which it is possible to include human operator's experience (control of processes and industrial plants, computerized controls), making possible decision strategies for complex problems.

Fuzzy systems are relatively easy to understand and use, and the mathematical manipulations of these systems are not so complex as it seems to be in the first sight, however, the fuzzy controller designer needs a deep understanding about doubts and uncertainties that occur in processes and industrial plants, how it affects the usual applications of the modern control theory, making possible the gauging of the relevance and trustworthiness of the gotten results through the application of the Fuzzy logic theory, see (Nascimento, 2000).

A system based on Fuzzy logic has its behaviour schematized for the following set of variables and actions: Input variables, fuzzyficator, inference, base of rules, defuzzyficator and output variables.

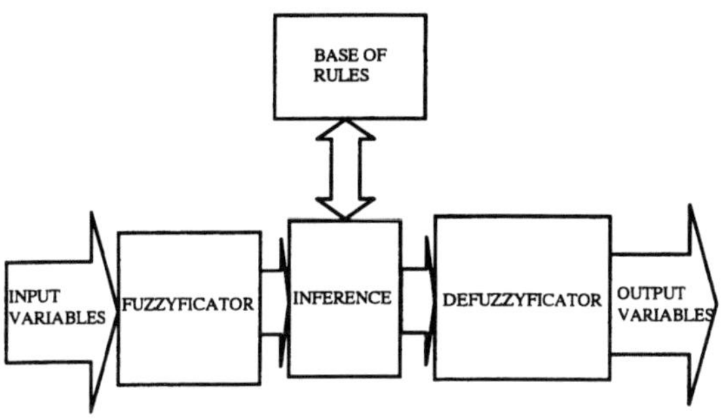

Fig. 3 – Fuzzy system diagram

The fuzzificator is responsible for mapping the numerical inputs in Fuzzy sets (linguistic variables). The fuzzyfication interface is created to receive and treat the numerical data, which came from sensors or computerized input devices, see (Shaw, 1999).

The fuzzification interface use membership functions contained in the knowledge base, converting the input signals into values in the interval [0,1] that are associated to linguistic labels.

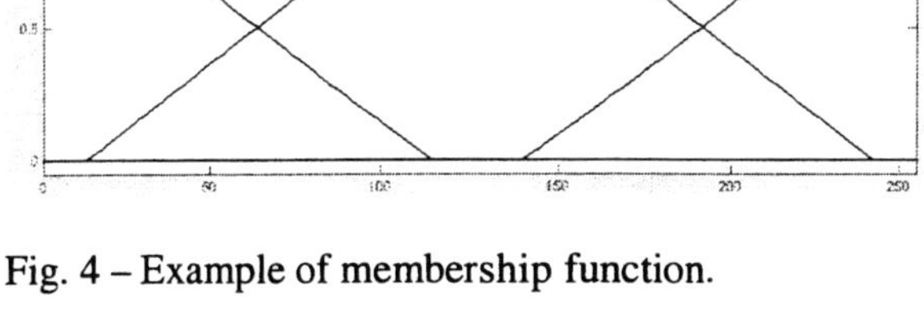

Fig. 4 – Example of membership function.

The inference is made mapping the linguistic input values into linguistic output values using the rules. Fuzzy implication is used to simulate the human decision process, generating control actions, called consequences, from a set of input conditions called antecedents. The knowledge base corresponds to a model of the system to be controlled, consisting in a data base (membership functions) and a base of linguistic fuzzy rules (characterizing the controller targets and the used strategy).

The defuzzificador maps linguistic values in numerical output values. This function is realized by a defuzzification interface, generating a discrete value that can be used in a "real world" control action. As previously said, fuzzy logic emulates the expert operator reaction to process disturbs, as if the operator were operating manually. This form can have a faster reply to disturbances then a conventional PID controller, reducing the overshoot and allowing adjusting the reaction according to the process knowledge.

An interesting potential application of fuzzy logic is the control of air conditioning systems. An example is: A fuzzy controller produces reference signals for the cold and hot water valves, and the humidity control water valve. The control strategy can use different variables, as internal and external temperatures and humidity to determine the operation of the air conditioning system for energy conservation. This is clearly a multivariable controller, with capacity to take care of interdependent variables.

For example, when the temperature increases, the air relative humidity diminishes. This can be explored, through the coordination of the temperature control loop and the humidity subsystem. The humidity controller can actuate before detecting the temperature modification in its sensor. This type of behaviour could improve the quality of this system in terms of predictive behaviour, being able to generate a considerable energy economy, between 10% to 30% (Staefa, 2003). In crisis times in the energy sector (rain shortage, for example) this would represent a considerable amount of saved energy.

In terms of intelligent control on the market we have today fuzzy controlled "window" air conditioning systems. The idea is to use fuzzy logic to implement fuzzy controllers and with them make an intelligent control of temperature and humidity, aiming at to reduce the consumption of the electric energy (economic operation) and providing comfort for the people who work in an office, for example.

For our thermal system, two regions are controlled independently. The first region is represented by rooms 1 and 2 and second for rooms 3, 4 and 5. The tunnels T1 and T2 simulate the external temperature. The same control algorithm is used for both control regions. Such algorithm is recorded individually in distinct microcontrollers PIC16F877A, where each one is responsible for one actuator control (hair drying). The algorithm deals with two input variables: internal and external temperatures and one output variable PWM duty cycle.

First we determined that the system would have three input variables: Reference temperature (to each controlled region), local room temperature and external environment temperature. And one output variable: PWM output (to each controlled hair dryer). All this variables are associated with linguistic variables by pertinence functions. In this project we used 5 membership functions to each input variable and 6 functions for the output variable. We tried to keep the superposition between functions near 50 percent, which seems to be reasonable, as demonstrated by Shaw, 1999.

The rule base was designed with 35 rules, considering only the most relevant process aspects. The target is to minimize energy consumption (each controller, in this case, has a different set of rules). The algorithm has a function that calculates the membership value for a given input signal. Four points are used to define this function: two points represents the minimum points of the trapezoidal membership function, two others, the maximum points of this function. The output value represents the control value for the temperature signal or PWM duty cycle. Moreover, fuzzy functions were implemented: the OR (composition) and AND (aggregation) (inference max-min) and the defuzzyfication function of the MoM type, where the output is calculated by means of the average values that represent the maximum values of pertinence. The algorithm needs ~500 words of program memory, where each word is represented by 14 bits.

One innovative aspect in this project is the use of fuzzy controllers that intercommunicate among them by means of a supervisory computer, which receives the information from all local microcontrollers and manage this. This architecture allows each net device to have its own rule base and consequently fine tune of the air conditioning system. The supervisory computer implements also the interface between the controlled processes and the user (if desired, by a www connection). The key point in energy saving are properly tuned rules that can care adequately with the time-varying, non-linear nature of buildings.

7. ARCHITECTURE FOR REMOTE ACCESS

Remote systems can have one or more customers on-line, monitoring or configuring the same module in any instant, given that it has access granted. Such a system is represented through the next figure. The different types of processes to be interconnected show the necessity to consider a portable remote architecture, or either, compatible with different hardware platforms. This model uses one web server. Obviously, the number of processes is limited by the capacity of the web server, and the net bandwidth.

To implementation the remote system two subsystems have to be implemented. One is the communication between the computer that publishes the pages in the internet and the processes; the other concerns the interaction with remote users.

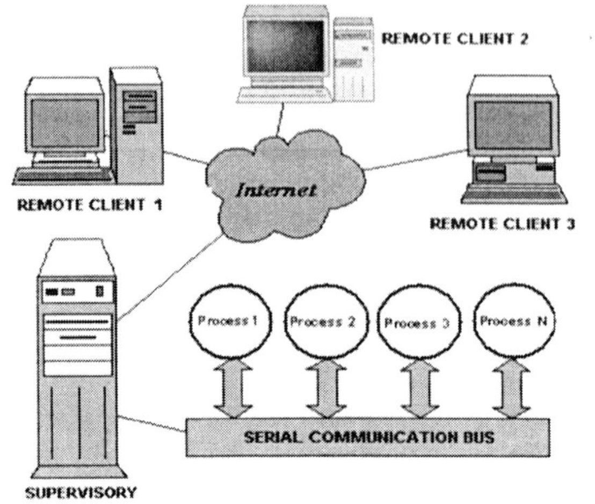

Fig. 5 - Remote supervisory architecture

7.1 Communication between supervisory PC and processes

The standard RS-232 communication interface defines a point-to-point operation; see (Zelenovsky, 2001). However, it is possible to construct a RS-232 bus in which some devices can be connected, and then to share the same set of wires to transmit or receive data.

This implementation is feasible since that having only one master device (supervisory), which can transmit and also make solicitations of data to other devices (PIC microcontrollers).

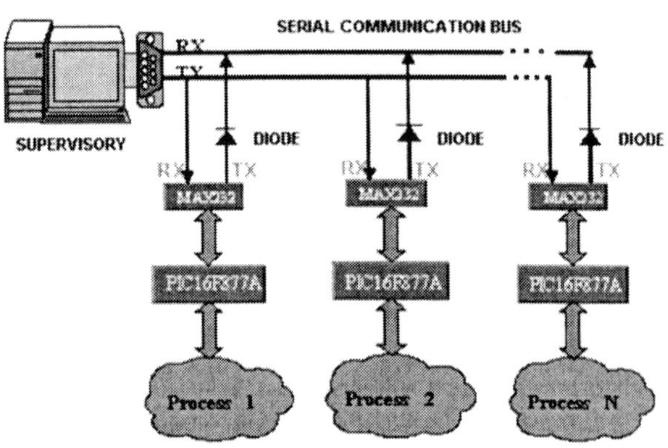

Fig. 6 – Multipoint serial communication.

In the above figure, it is necessary that each slave device has a blocking diode in its transmission line. This electronic element is necessary for the proper operation of the RS-232C standard, which defines the tension of -15V for the transmission lines when these are not active. In this way when a process device starts to transmit data to the supervisory it prevents that the current reaches other slave devices and in this way do not harm the communication drivers between transmitter and receiver.

Based on the above described scheme of communication an electronic circuit was designed to implement the communication between the supervisory PC and four PIC microcontrollers, which are represented by the following schematic:

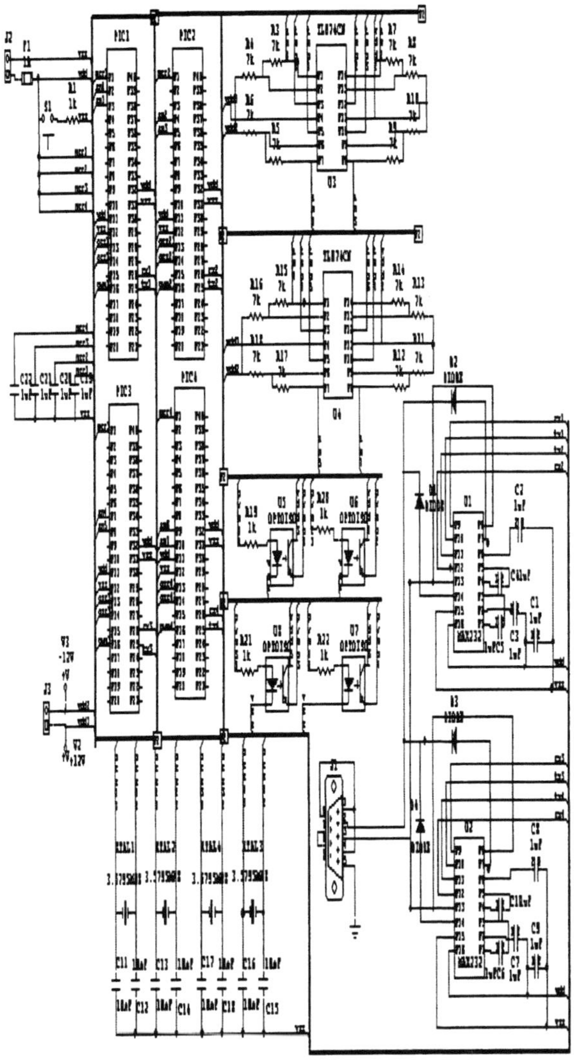

Fig. 7 – Fuzzy control schematic.

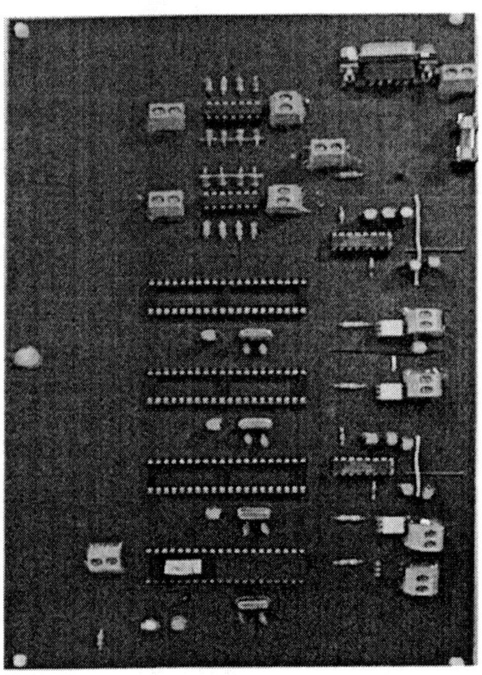

Fig. 8 – Fuzzy control circuit implementation.

Starting from a first design, and after some tests, in order to prevent the typical implementation problems of electronic circuits on protoboards, the circuit in Fig. 7 was assembled on a printed circuit board.

7.2 Supervisory PC and operators Communication

The communication architecture is based on the client/server model and was developed in Java. The operator can access the supervisory system using any operating system that supports a web browser with virtual Java machine. All relatively recent web browsers came with this virtual machine.

The Java language has favourable features for the development of client software as well as the tools necessary to establish communication in real time in a client/server architecture. It is a highly portable language, vastly registered, has mathematical libraries and it is freeware. In general, Java applets (software executed by web browsers) are used in the client machines to establish communication in real time with the Java server software running in the web server (supervisory). From the security point of view the applets offer great advantages because these do not have access permission to the local data (of the client), so, are not able to view/modify any information on the client computer without the initiative of the client himself.

Locally, the web browser is the interface for the system, it downloads the client software in the format of Java applets from the web server and then executes the applet. The web server provides the HTML page and the Java applet. Currently, it is common to find pages developed using languages like PHP (Pre-Hypertext-Processor), ASP (Active Server Page) and JavaScript, where the advantages of these tools are to offer the possibilities of analysis and processing of the data. PHP and ASP execute its scripts in proper web servers, and sending only the results of the requested operation for the client. By this way the necessary processing power on the client computer is reduced.

Java applets are easily updated, a new download lead automatically to read fresh from the web server. Consequently, from the user point or view, no programming and/or installation of software are necessary to update the system.

Through the Java applet the client can configure the thermal process controller, defining its priorities, monitoring the diverse process variables as well as the controller action on the actuator. Based on these ideas the Java applet shown in next figure was build.

Fig. 9 shows the client interface when remotely operating the temperature control of the thermal process prototype. In the below middle a picture of the prototype is depicted. Right below the current temperature in each room, acquired by a LM35 sensor, is shown. A plot of the temperature over time can be chosen checking the correspondent box in the left below. Each curve as a different colour assigned, to simplify the identification of each signal. All the data is collected in a file that can be downloaded by the user, if required.

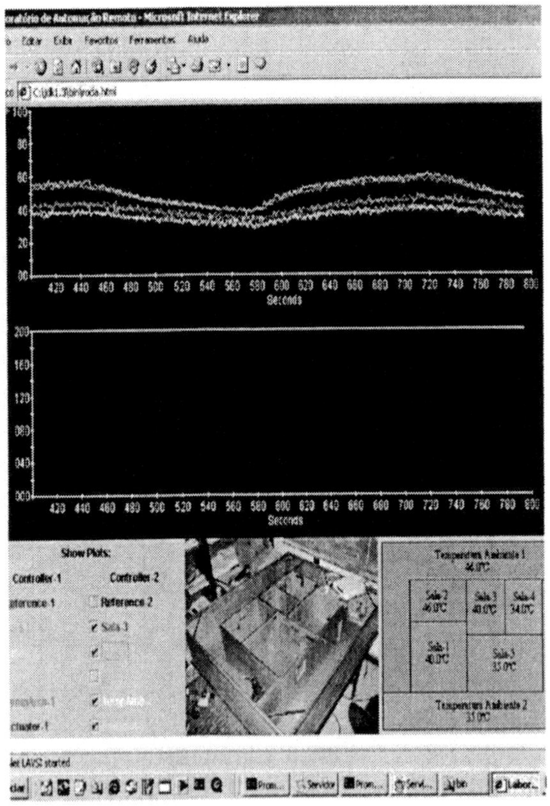

Fig. 9 – Client/server supervisory Java applet.

First experiments with PID control showed that the thermal process is strongly non-linear. For small reference steps (~50°C) we obtained settling times of the order of some minutes. For larger reference steps (~90°C) we obtained settling times of more than three hours. The power of two 1kW hair dryers was clearly insufficient to counterpart the hot loss through the walls of our prototype process.

The fuzzy control of the temperature showed significantly better results than the PID for a fixed set point. Due to specific rules the system converged faster to the set point but, as expected for fuzzy control, some noise was still persistent in steady state. The rule base of the fuzzy control was adjusted empirically until a satisfactory response has been obtained. This is a very important point being investigated. If a practice near approach is to be proposed, it must simplify the acquisition of knowledge necessary to tune each specific building automation. From an expert system is expected to allow the easy commissioning in building automation. Area of each room, disposition and size of the windows, relative position of the room with respect to the sun along a day and along the year, expected use (office, CPD, meeting room etc.), number and position of each air conditioning unit could be used to tune the rules for each temperature control. The presence sensors should "overwrite" the information obtained from the seasonal rules: the normal working day schedule would be *to maintain the comfort temperature (24°C) between 8-12h and 14-18h.* If the presence sensor detects that nobody is in the room for more than 30 min, then an energy saving sate is activated.

8. CONCLUSIONS

In terms of the microcontroller implementation of fuzzy control of thermal processes, it is expected that it offers the necessary hardware/software features for real-time execution of the algorithms. The main considered features were: sufficient memory and necessary functional modules for implementation of the intelligent controller. These until the moment had satisfied the project necessities completely.

The available intelligent controllers on the market have a high cost, therefore, low accessibility. It is expected to develop equipment with minimum costs and good acceptance in building and industrial environment, being able to generate economies of energy for its users, justifying its project, implementation and posterior commercialization. The performance of the intelligent controller must be better, in general terms, of that of conventional PID controllers. This will be observed by the comparison of performance parameters as: settling time and error in permanent regime.

The remote access by the internet must provide control, configuration and monitoring of the building system beyond storage of data for future statistical evaluation and improvements.

REFERENCES

Azevedo, F.M., Brasil, L.M. and de Oliveira, R.C.L. (2000). In: *Redes Neurais com Aplicações em Controle e em Sistemas Especialistas.* Visual Books.

Haissig, Ch. (1999). In: *Adaptive Fuzzy Temperature Control for Hydronic Heating Systems.* IEEE Int. Conference on Control Applications; Hawai 582-588

Mauris, G., and Foulloy, L. (2002). In: *A Fuzzy Symbolic Approach to Formalize Sensory Measurements: An Application to a Comfort Sensor.* IEEE Trans. Instrumentation and Measurement Vol. 51, No.4, Aug.2002, 712-715

Nascimento Jr., C.L and Yoneyama,T. (2000). In: *Inteligência Artificial em Controle e Automação.* Edgard Blücher.

Ogata, K. (1997). In: *Modern Control Engineering.* Prentice Hall, London.

Shaw, Ian S. and Simões, Marcelo Godoy. (1999). In: *Controle e Modelagem Fuzzy.* FAPESP.

Souza, David José and Lavinia, Nicolas C. (2002). In: *Conectando o PIC – Explorando Recursos Avançados,* 1ª Ed. Editora J. J. Carol, São Paulo.

Souza, David José (2001). In: *Desbravando o PIC* (2º Edição). Editora Érica.

Wang,L.X. (1997). In: *A Course in Fuzzy Systems and Control.* Prentice Hall.

Zelenovsky, Ricardo (2001). In: *Processamento de Sinais, Arquitetura e Interfaceamento com Computadores, Microcontroladores.* DC, PUC-RJ, CETUC.

Copyright © IFAC Information Control Problems in
Manufacturing, Salvador, Brazil, 2004

ELSEVIER
IFAC
PUBLICATIONS
www.elsevier.com/locate/ifac

MODELLING AND ANALYSIS OF REAL-TIME DATABASES FOR SENSOR NETWORKS USING COLOURED PETRI NETS

Pedro Fernandes R. Neto *
Maria L. B. Perkusich *** **Angelo Perkusich** **

* *Department of Informatics, State University of Rio Grande do Norte, Mossoro, RN, Brazil*
** *Department of Electrical Engineering, Federal University of Campina Grande, C.P. 10105, 58109-970, Campina Grande, PB, Brasil*
{pedro,perkusic}@dee.ufcg.edu.br
*** *Department of Informatics and Statistics, Catholic University of Pernambuco, Recife, PE, Brazil*
ligia@dei.unicap.br

Abstract: In the last years, the application of sensor networks has been increased. Also, the software development requirements for such systems are getting more complex. The data storage and processing techniques, for these environments, are insufficient for the new requirements. In this paper, we develop and analysis a model for the integration of real-time database technology with an embedded sensor network systems, to tackle such deficiencies. The hierarchical Coloured Petri nets formal method is used to describe the formal model. We detail the analysis of functional and temporal properties related to the sensor network and a real-time database server. *Copyright © 2004 IFAC*

Keywords: Real-time databases, Network sensor, Embedded System, Hierarchical Coloured Petri Nets.

1. INTRODUCTION

In the last years embedded systems Lee (1999, 2002) have been applied in several kinds of computing devices Nierstrasz et al. (2002). The evolution of the technology is making possible to build devices that can execute more complex tasks and it is getting difficult to deal with this increasing complexity without adopting systematic well founded metodologies Lee (1999). Also, the amount of data in such systems has increased as well. Such data must be efficiently managed and stored considering specific non-functional requirements of the application, such as, for example, size and power consumption.

The efficient management of the increasing amount of data in embedded systems is a callenging problem. The techniques for data processing and data storage for smart embedded devices, as is the case of sensor networks Chong and Kumar (2003), are limited and, therefore, they are insufficient in face of the new requirements of the applications, and conventional directly databases cannot be used Tesanovic et al. (2002). Also, the correctness of such systems depends on both, logical consistency of the data and the transactions, and on timing constraints. Real-time databases (RTDB) have been investigated in the last years to allow means to deal with such logical and timing constraints related to embedded systems Tesanovic et al. (2002); Hansson (2002).

As pointed out by Chong and Kumar in Chong and Kumar (2003) sensor networks are a key technology and data management related issues related to them, such as tasking and querying, are hard and challenging problems. Such sensor networks are built using cheap, preprogrammed and smart devices that send data to a central database, where the data can be stored and accessed by applications Yao and Gehrke (2002). This approach has some limitations, such as communication overhead and power consumption, especially in the case of wireless sensor networks. In Chen et al. (2001); Bonnet and Seshadri (2000) it is considered a new architecture, where the environment can be seen as a distributed database system, formed by a server database and each component is a local database.

The objective of this work is to introduce a model for the analysis of an embedded sensor network system together with a real-time database application running on a real-time server. Also, we aim at demonstrating how our RTDB approach based on data warehousing and distributed databases approaches can be used to effectively deal with the inherent complexities related to data management for sensor networks.

In this work we use Hierarchical Coloured Petri Nets (HCPN) Jensen (1992, 1997), and the Design/CPN tool Jensen and al (1999). This tool is used to edit and analyze the HCPN models. Using HCPN we have more compact and organized models because of the use of hierarchal descriptions and complex data type features.

This work is organized as follows. In Section 2 we introduce some concepts related to sensor networks and databases. In Section 3 relevant concepts of real-time databases are discussed. In Section 4, we detail the modeling and timing analysis for a real-time database application for the sensor network previously modeled and analyzed. Finally, in Section 6 we present some conclusions and future works.

2. DATABASES FOR SENSOR NETWORKS

Sensor networks are used to control and to monitor the physical environment and sensor nodes may have different physical sensors and can be used for different application scenarios. In most cases physical resources constraints related to communication, power consumption, computing power, and memory size, must be considered in the design of such systems. In the context of this paper we are mainly concerned with the types of data processing algorithms that can be deployed and how intermediate results can be stored. In such situation, the traditional procedural programming mechanisms can become too complex and limited with respect to flexibility and scalability. As observed in Bonnet and Seshadri (2000), monitoring and control for sensor network are better described in declarative languages. Thus, we adopt a database approach to deal with data management for sensor networks in the context of our research work. Declarative queries are especially suited for sensor networks interactions due to the fact that queries can be formulated independently of the physical structure and organization of the sensors in a specific network Yao and Gehrke (2002, 2003). Users and applications may define queries without knowing details with respect to how data is generated in the network and how the data is processed reply to a query.

In monitoring applications, users and applications typically define three kinds of queries, Bonnet and Seshadri (2000): *historical queries*: these are typically aggregate queries over historical data obtained from the device network; *snapshot queries*: these queries are related to data from the device network at a specific time, and; *long-running queries*: these queries are related to data for a time interval. In order to deal with such queries two approaches can be considered: warehousing and distributed.

For the warehousing approach queries are processed on the data retrieved from the sensors Bonnet et al. (2001). The sensor network is simply used to collect data. The query mechanism is executed in two steps. First, the data is extracted from the devices and stored in a database server. Second, the queries are processed based on the data stored on the server. Such approach is adequate for queries related to historical data. It is important to point out that the sensor network devices are not part of the database. They are accessed using a predefined procedure that accesses the central database. Potential waste of valuable resources for transferring great amounts of data from the server and the difficulties to answer snapshot and long-running queries are the main disadvantages of the warehousing approach. This approach is adopted when the devices of the sensor network have very limited resources. However, with the increasing advances in the technology, more processing and communication resources can be added to the devices, thus allowing the adoption of this approach.

Considering the disadvantages pointed out for the warehousing approach a distributed solution can be adopted. As observed in Bonnet et al. (1999) this approach allows the processing of queries by both, the devices of the network sensor and by a database system, that controls the resources that are used. It is primarily targeted to process snapshot and long-running queries. Also, aggregated queries over historical data can be evaluated directly on the devices instead of in a centralized server Bonnet et al. (2000).

The devices can be seen as components of the database, supporting some functionalities, such as data acquisition, processing and storing, and triggering actions in the physical world.

In our work, we adopt this approach. However, it should be noted that it requires new definitions and techniques, such as mechanisms to deal with the concurrent execution of the transactions and the metadata management.

3. REAL-TIME DATABASE SYSTEMS

Like a conventional database management systems (DBMS), a Real Time-DBMS (RT- DBMS) must process transactions and guarantee that the database consistency is not violated. However, conventional database systems do not emphasize the notion of time constraints or deadlines for transactions and the temporal validity of some data. The performance goal of a system is usually expressed in terms of desired *average* response times, rather than constraints for individual transactions. Thus, when the system makes scheduling decisions, individual real-time constraints are ignored.

For conventional real-time systems individual transaction constraints are the for the designer, but data consistency problems are ignored. Moreover, real-time systems usually deal with simple transactions (called processes), that have simple and predictable data (or resources) requirements. For RT-DBMSs, it is not usually possible to predict data and resources. Therefore, transaction execution times are difficult to predict.

For RT-DBMSs the main goal is to *complete* transaction on time, while maintaining logical and temporal consistency of the data. Thus, for satisfying logical consistency one can use concurrency control techniques, and to satisfy temporal consistency we can use time cognizant transactions processing by tailoring transaction management techniques to explicitly deal with time.

Usually, a RT-DBMS consists of a *controlling system* and a *controlled system* Stankovic et al. (1999). In an automated manufacturing system, the controlling system is composed by the computer and human interfaces, while the controlled system is the manufacturing system (environment), with its robots, assembling stations, parts, and conveyers. The controlling system interacts with its environment based on the data from the environment, such as from various sensors and files (e.g., temperature sensor, part file). The state of the environment, as perceived by the controlling system, must be temporally consistent with the actual state of the environment. Otherwise, the results may be catastrophic. Therefore, timely monitoring the environment as well as timely processing acquired data is necessary. Moreover, the data may be used to derive new data. Thus, the environment history is logged in a temporal databas).

4. MODELING AND ANALYSIS USING COLOURED PETRI NETS

In this Section, we focus on the HCPN model for real-time database application for the sensor network. To help the reader we define a scenario that takes into account two sensors in the sensor network, two update transactions, a query transaction and a server database object. As said before, the sensors acquire the data of the environment, store and transmit the data for the server database object. This consequently reduces the data flow in the network and the energy consumption. The server is updated in order to allow the execution of historical queries. The data on the sensor device are always more accurate than the data on the server. This difference happens due to the delay that exists between the data acquisition and the update execution for the server.

Each sensor is a local database that implements functions for data acquisition and storage. For the model, the local database is modeled as a list. According to the timing constraints imposed to each sensor, it can acquire new data. The temporal parameters for the sensors are defined by the tuple (tl_i, pe_i), where tl_i is the sensor release time, that is, the moment in which all the necessary resources are available and pe_i defines the periodicity for data acquisition.

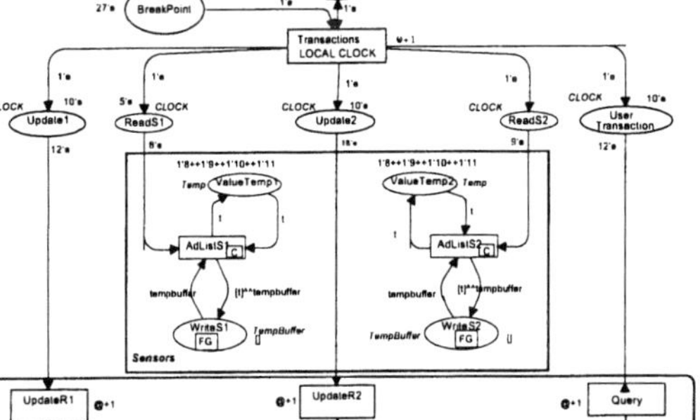

Fig. 1. CPN model for the sensors

The model has local clocks for each sensor that are synchronized by a global clock. In the Figure 1, we show the CPN model for the sensors, the transactions local clocks and the system global clock. From the practical point of the assumption of a global synchronization clock is an important issue, it is out of scope of this paper to discuss such problem, the reader may refer to Li and Rus (2004) for a detailed discussion.

The sensors data acquisition period are modeled by the output arcs inscriptions of the places ReadS1 and ReadS2 that are 6'e and 9'e, respectively, e is the time unit considered in the model. The initial marking for places ReadS1 and ReadS2, models the release time for each transaction. The places ValueTemp1 and ValueTemp2 model the data acquisition and the storage functions modeled by places WriteS1 and WriteS2.

We assume that both query and update transactions are periodic. The first transaction updates the database server, while a query transaction reads data from the server for historical queries. However, the queries can also result in querying directly the sensors to get data for snapshot and long-run queries. It is important to observe that, for the model, we consider the same type data item for the transactions.

For each transaction τ_i the temporal parameters are defined by the quadruple (tl_i, tc_i, pr_i, pe_i), where: tl_i: is the transaction release time, and starting from this instant of time the transaction is ready to be executed; tc_i: defines the transaction computation time; pr_i: defines the maximum transaction execution period, and; pe_i: defines the periodicity of the transaction.

For the server database object, we define three negotiation functions, also called compatibility functions. These functions implement the concurrent control among the conflicting transactions. We also define two methods, the query method and the update method.

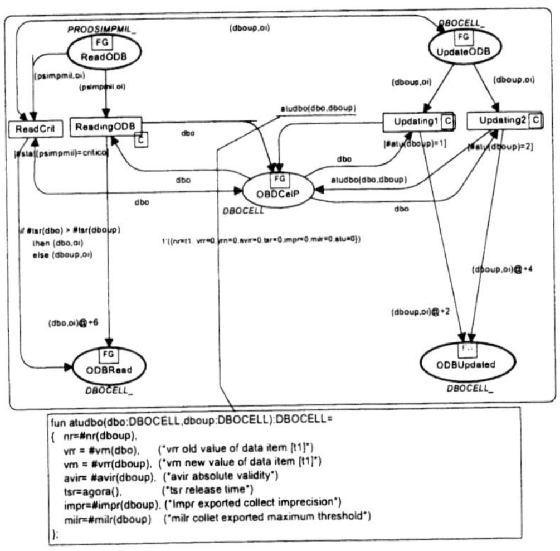

Fig. 2. CPN model for the database server

In the Figure 2, we show model CPN for the server database object. The read method is modelled by the places ReadODB and ODBRead and transitions ReadingODB and ReadCrit. When transition sfsReadingODB fires a data item is read from database, represented by the place OBDCelP. The processing time for the query is illustrated in the input arc inscription of the place ODBRead, and is equal to 6 time units (6'e).

5. TIMING MODEL ANALYSIS FOR THE RTDB APPLICATION

For analysis purposes we consider a situation where the updates and the queries are trying to access the same data item in the database server application. We perform both logical and timing verification for the data and transactions. Thus, we analyze the transactions scheduling, considering the semantic concurrency control, where the correctness criterion is the epsilon serializability considering quality of service.

To analyze the model we consider a situation where two sensors acquire the same type of data. Sensor1: writes periodically in the local database, and the release time is 1 time unit (t.u.) and the period is 6 t.u. We consider that the value of the data item may vary between 8 to 11 units. Sensor2: writes periodically in the local database, and the release time and the period are both is 9 t.u. The value of the data item may vary between 8 to 11 units. Update 1: periodically updates the server database object with respect to the data in Sensor1. The release time is 3 t.u., the computational time is 2 t.u., and the deadline is 12 t.u. and the period is 12 t.u.. Update 2: periodically updates the server database object for sensor2. The release time is 9 t.u., the computational time is 4 t.u., the deadline is 18 t.u. and the period is 18 t.u. Query: periodically queries the real time server database application. The release time is 3 t.u., the computational time is 6 t.u., the deadline is 12 u.t. and the period is 12 t.u.

In Figure 3, we present the message sequence diagrams for the executions of transactions in the sensors. Also, its is shown the execution of transactions as well as the updates in the database server. It is also possible to observe the time properties for each component and when the compatibility function is evaluated due to the transactions concurrent execution. Each arrow represents the transaction execution.

In the time instant 1, as indicated in the line labeled by Sensor Database, a writing for Sensor1 occurred. Each execution is identified by the label Write Operation Sensor1 and the value added. In the line labeled Time Propr. time properties for both sensors and transactions can be observed. Considering the quality of service management functions as detailed in RibeiroNeto et al. (2003), this line represents the specification and monitoring functions. The first execution for Sensor2 is in 9 t.u., as seen in the Figure 3.

In time 3, Query and Update 1 transactions start. As observed in the arrows labels, two conflicting operations tries to access a data record in the Server Database. The first is a query and the second an update (to update the data item t1 to 8).

The transaction Query commits in time instant 9. The transaction Update 1 had to wait the query and then access the server. This can be seen by observing its computational time, where the commit occurred in time instant 13, that should to be in time instant 7. Thus, the operations could not be concurrently executed, or either, the negotiation function (compatibility function) was evaluated to false.

Before the transaction Update 1 commits, transaction Update 2 begins execution. One more time, conflicting operations try to have access to the server at the same time. However, for this concurrent execution, the negotiation function was evaluated to true. The evaluated parameters can be seen in the Figure 3. The line Negotiation with a label with CF evaluate followed by the parameters and its respective execution time values. Thus, transaction Update 2 can execute concurrently.

In time instant 15, one more time conflicting operations try to access the server. At this time we have the situation where an update is executing and a query is invoked, and are concurrently executed. It is important to observe that the data item value $t1$ in the server is 8, however the query considered the value 10 for $t1$, that is the value from the sensor and not the one stored in the database.

In Figure 4, we present the time diagram for the execution of transactions. This diagram was produced using the performance analysis tool for the Design/CPN Lindstrom and Well (1999). In this figure we have a representation of the timing model execution for both transactions and sensors, where the execution are represented by rectangles.

In the vertical, we represent the transactions and the sensors, while that in the horizontal we represent the release time and the periods for both transactions and sensors and computational times and deadline for transactions. For each transaction we consider three states: start, processing and committing. The start of the execution of a transaction is illustrated by a dotted line rectangle. The processing is defined by a filled rectangle and the committing by a continuous line rectangle. For the sensor we only show the state processing.

In the vertical, for time instant 1 we have the execution of transaction Update 1, in time instant 2 we have the execution of transaction Update 2 and in time instant 3 we have the execution of transaction *Query*. The sensors processing in time instant 4 for sensor1 and 5 for sensor2.

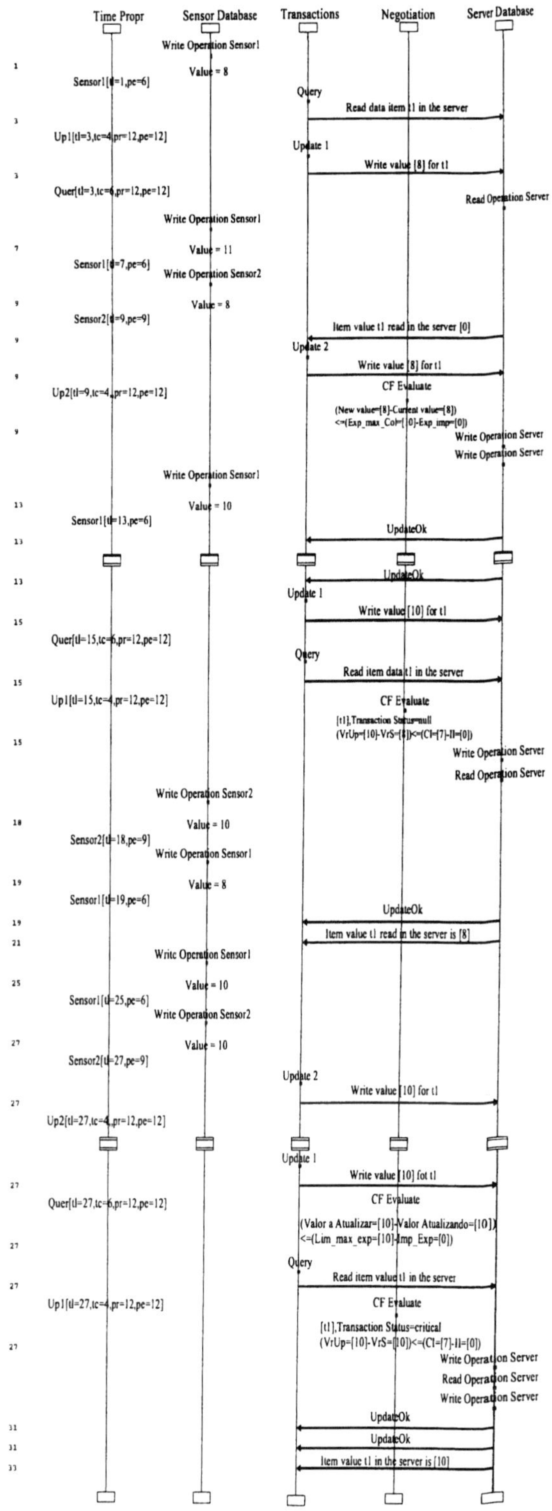

Fig. 3. Message Sequence Diagram

6. CONCLUSION

In this paper we discussed how real-time databases can be applied in the context of sensor networks. We presented the modeling and analysis for a real-time database server for sensor networks using

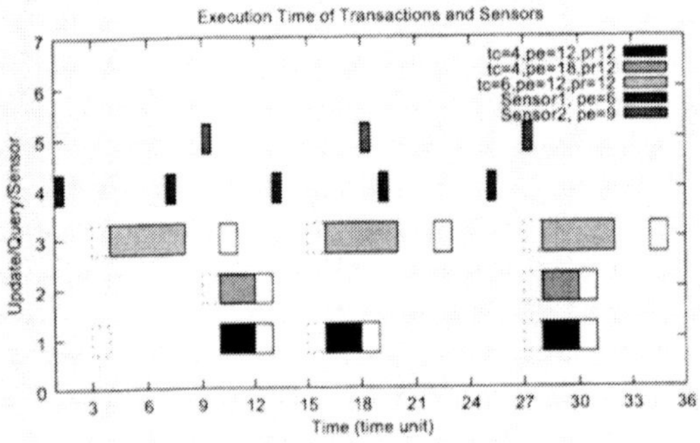

Fig. 4. Time Diagram

Coloured Petri Nets. All the components of a sensor network system have been integrated in the model. The real time server uses a distributed approach for data management.

Currently, there is a software team using the model specified and verified as presented in this work to write down the application code, thus linking the modeling and implementation phases. Also, we are working towards the implementation of a real-time database server and its application to an oil industry sensor network.

REFERENCES

Ph. Bonnet, J. Gehrke, T. Mayr, and P. Seshadri. Query Processing in a Device Database Systems. Technical report ucb/erl no. m99/63, Cornell Technical Report TR99-1775, Outubro 1999.

Ph. Bonnet, J. Gehrke, and P. Seshadri. Querying the Physical World. *IEEE personal Communications. Special Issue Networking the Physical World*, 7(5):10–15, Outubro 2000.

Ph. Bonnet, J. Gehrke, and P. Seshadri. Towards Sensor Database Systems. *2nd International Conference on Mobile Data Management*, pages 3–14, Janeiro 2001. Hong Kong.

Ph. Bonnet and P. Seshadri. Device Database Systems. *Proceedings of the International Conference on Data Engineering ICDE'99*, Maro 2000. San Diego, CA.

Z. Chen, J. Gehrke, and F. Korn. Query Optimization In Compressed Database Systems. *In Proceedings of the 2001 ACM Sigmod International Conference on Management of Data Santa Barbara, California*, Maio 2001.

C.-Y. Chong and S.P. Kumar. Sensor Networks: Evolution, Opportunities, and Chalenges. *Proceedings of IEEE*, pages 1247–1256, August 2003.

J. Hansson. Embedded database systems. Technical report, Dept. of Computer Science, Linkoping University, and Dept. of Computer Engineering, Malardalen University, 2002. Project is funded by CENIIT (Center for Industrial Information Technology at Linkpings Tekniska Hgskola) and is carried out at the Real-Time Systems Laboratory (RTSLAB), Linkping University.

K. Jensen and et. al. *"Design/CPN" 4.0.* Meta Software Corporation and Department of Computer Science, University of Aarhus, Denmark, 1999. On-line version:http://www.daimi.aau.dk/designCPN/.

Kurt Jensen. *Coloured Petri Nets: Basic Concepts, Analysis, Methods and Practical Use.* EACTS – Monographs on Theoretical Computer Science. Springer-Verlag, 1992.

Kurt Jensen. *Coloured Petri Nets: Basic Concepts, Analysis Methods and Practical Use*, volume 2. Springer-Verlag, 1997.

Edward A. Lee. Embedded software - an agenda for research. Technical Report UCB/ERL No. M99/63, University of California, Berkeley, December 1999.

Edward A. Lee. Embedded software. In M. Zelkowitz, editor, *Advances in Computers*, volume 56. Academic Press, 2002.

Qun Li and Daniela Rus, editors. *Global Clock Synchronization in Sensor Networks*, 2004.

B. Lindstrom and L. Well. *"Design/CPN 4.0", Performance Tool Manual.* Department of Computer Science, University of Aarhus, Denmark, 1999. On-line version:http://www.daimi.aau.dk/designCPN/.

Oscar Nierstrasz, Gabriela Arévalo, Stéphane Ducasse, Roel Wuyts, Andrew P. Black, Peter O. Müller, Christian Zeidler, Thomas Genssler, and Reinier van den Born. A component model for field devices. *Lecture Notes in Computer Science*, 2370:200–216, 2002. ISSN 0302-9743.

Pedro Fernandes RibeiroNeto, Angelo Perkusich, and Maria L.B.Perkusich. Real-Time Database Modeling Considering Quality of Service. *5th International Conference on Enterprise Information Systems*, 2003.

J. Stankovic, S. Son, and J. Hansson. Misconceptions about real-time databases. *IEEE Computer*, 32(6):29–36, 1999.

A. Tesanovic, J. Hansson, D. Nystrom, and C. Norstrom. Embedded databases for embedded real-time systems: A component-based approach. Technical report, Dept. of Computer Science, Linkoping University, and Dept. of Computer Engineering, Malardalen University, 2002.

Y. Yao and J. E. Gehrke. The Cougar Approach to In-Network Query Processing in Sensor Networks. *Sigmod Record*, 31(3), Setembro 2002.

Y. Yao and J. E. Gehrke. Query Processing for Sensor Networks. *To appear in the First Biennial Conference on Innovative Data Systems Research (CIDR 2003)*, Janeiro 2003. Asilomar, California.

Copyright © IFAC Information Control Problems in
Manufacturing, Salvador, Brazil, 2004

ELSEVIER
IFAC
PUBLICATIONS
www.elsevier.com/locate/ifac

Applying Petri Nets to Requirements Validation

José Reinaldo Silva
Mechatronics Engineering Dept.
University of São Paulo, Brazil
reinaldo@usp.br

Eston Almança dos Santos
Mechatronics Engineering Dept.
University of São Paulo, Brazil
eston@usp.br

Abstract

Mechatronics and Information Systems can be assumed as complex, and thus have a prohibitive cost for adjustments out of the specification phase. This emphasizes the early development phases as especially critical for the success of projects, besides being very important to guide implementation decisions. We proposed a requirement validation method suitable to a life cycle represented with the Unified Process and addressed to the specification and design of information systems. The key issue of the proposal is the facility to model information flow formally, by the capture of the requirements into Elementary Petri Nets. A software tool is proposed to perform the transformation capturing the dynamic of the processes. Requirements validation is performed in the net, by simulation or by property analysis (suitable to large systems). Finally, we briefly discuss the achievement of a proper (and minimum) set of viewpoints enrolled in information system projects.
Copyright © 2004 IFAC

1. Introduction

There is a growing demand for disciplined methods of development in modern Systems Engineering, specially for automated systems, which includes not only the critical problem of managing the fusion of hardware and software but also the requirements for the software system.
Conceptual and practical schemas appear frequently in the literature, some of which are associated with software development techniques, or connected with specific areas of Engineering, or even with the knowledge acquired in the development of specific artifacts. However, in spite of the consensus that the early phases of the development process are the most important to assure that the target system works properly, still, most of the tools to support system development address the design and implementation phases.

The present paper focus is on the validation of requirements only, supposing that the artifact is related to processes control – meaning the control of business process or industrial processes as introduced in the CIMOSA [1] terminology. In both cases the challenge is to find a systematic transformation between requirements and specifications.
In the informal requirement elicitation it is common to face problems like: i) adjust several viewpoints generated by different agents related with the artifact ii) remove discrepancies, contradictions and iii) validate the acquired requirements. Information Systems Design faces the same problems but requires special attention to the validation of the information flow and control signs (considering the supervisory systems as the same line of artifact).
Petri Nets has being used, as a case tool to the modeling, analysis and documentation of this kind of system and it is also a good framework to express requirements in this area. A simple transformation can be applied to transform Use Cases in an Elementary Petri Net, as we will present in this paper. Thus, Use Cases could be considered a metaphor for requirements informally elicited. Besides, in a first approach, the sequence in which the events are planned can be easily tested by direct simulation or property analysis. The later approach is specially promising to large systems.
Also, Petri Nets are flexible enough to allow that detected anomalies in the requirements specifications could be corrected changing the net directly or suggesting a change in the Use Case. Therefore we could end with a set of isomorphic net models representing different viewpoints of the same information system.
It is not in the scope of this paper to give a complete analysis of generic viewpoints for information systems, but it is not possible to neglect this problem. Therefore, we propose a set of three basic viewpoints: one from the *provider* of the system, one for the *final user* and one for the *developer*. A complete analysis of the formal process of verifying isomorphism among the nets of each viewpoint is also out of the scope of this paper.
The *provider* is the business owner or the controller of the business process, while the *final us*er is the set of agents

or machines that effectively interact with the system without changing it.
An illustrative (and classical) example is presented, based on the last release of the Rational Unified Process (RUP) [2]: an automated bank teller machine. Initial Use Cases were taken from [2] and transformed in a tagged specification, where the tags are specific signals to the Net Translator. The resulting net is sent to a Net simulator where it is possible to choose conveniently an initial marking and following states. Interruptions in the flow of marks can then be inspected as possible mistakes or contradictions in the requirements.

2. Turning Use Cases in a Tagged Representation

Similarly to the development of any artifact, for information systems the requirements represent "what the system does", or, in other words a generic description of its behavior, which is composed of the integrated behavior of subcomponents. We will call these functionalities *business processes*, generalizing the meaning of the term proposed in the terminology of CIMOSA [1].
Thus, a *business process* as used in this work is such that:

- a set of events arranged in a proper schema, that is, a directed sub-graph where there are two distinguished unique elements called start and end, such that all remaining elements belongs to one of the paths connecting these two elements;
- there is a unique *business process* to each functionality;

Each *business process* is composed of a set of events that stands for specific operation in a shop floor, or generically, characteristic operation of the business in focus. We call those operations enterprise activities.
Therefore, a requirement elicitation for an information system is a process that identifies all business processes of the target application (be it a manufacturing site, service in the web, etc.) and its components, the enterprise activities. Each business process can be represented by a net (a sequence of events or operations) as well as the whole system, which is a process composed by several (alternative or integrated) business processes.
As a typical example, let us take the Use Case diagram for an ATM machine [2], already used to illustrate the application of RUP method of development. With this example (for instance the functionality Cash Withdraw) we will show how to transform Use Case in a tagged representation and then in a Petri Net. We used the HPSim [3] simulator to analyze the business processes.
We will show finally that the same approach can be applied to different viewpoints of the same functionality, resulting nets that have isomorphic subnets.

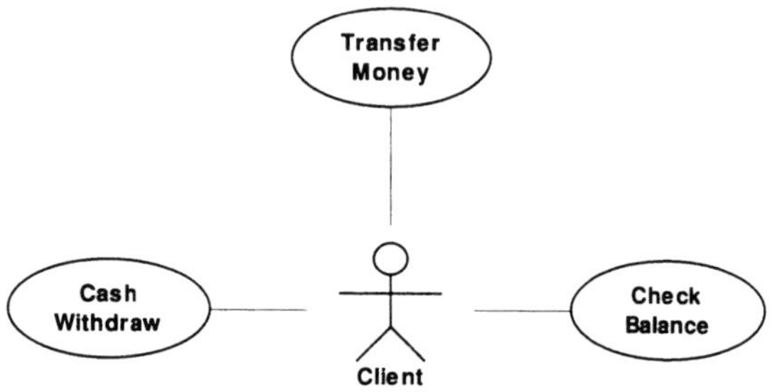

Figure 2-1 Use Cases diagram for ATM

2.1. Use Case for the Cash Withdraw Functionality

Following there are the basic and alternative flows of events for the Cash Withdraw functionality.

2.1.1. Basic Flow of Events

1. Initiate Withdraw – Customer inserts bank's card in the card reader on the ATM machine
2. Verify Bank Card – The ATM reads the account code from the magnetic strip on the bank card and checks if it is an acceptable bank card
3. Enter PIN – The ATM asks for the customer's PIN code (4 digits)
4. Verify PIN – The account code and PIN are verified to determine if the account is valid and if the PIN entered is the correct PIN for the account. For this flow, the account is a valid account and the PIN is the correct PIN associated with this account
5. Select Withdraw – The ATM displays the different alternatives available at this ATM. In this flow, the bank customer always selects "Cash Withdraw"
6. Enter Amount – The ATM asks for the amount to withdraw. For this flow the customer selects a pre-set amount ($10, $20, $50, or $100)
7. Authorization – The ATM initiates the verification process with the Banking System by sending the Card ID, PIN, Amount, and Account information as a transaction. For this flow, the Banking System is online and replies with the authorization to complete the cash withdrawal successfully and updates the account balance accordingly
8. Dispense – The Money is dispensed
9. Receipt – The receipt is printed and dispensed. The ATM also updates the internal log accordingly
10. Return Card – The Bank Card is returned

2.1.2. Alternative Flows of Events

1. Not a valid Card – In Basic Flow Step 2 - Verify Bank Card, if the card is not valid, it is ejected with an appropriate message
2. ATM out of Money – At Basic Flow Step 5 - Select Withdraw, if the ATM is out of money, the "Cash Withdraw" option will not be available
3. Insufficient funds in ATM – At Basic Flow Step 6 - Enter Amount, if the ATM contains insufficient funds to dispense the requested amount, an appropriate message will be displayed, and rejoins the basic flow at Step 6 - Enter Amount.
4. Incorrect PIN – At Basic Flow Step 4 - Verify Account and PIN, the customer has three tries to enter the correct PIN. If an incorrect PIN is entered, the ATM displays the appropriate message and if there are still tries remaining, this flow rejoins Basic Flow at Step 3 - Enter PIN. If, on the final try the entered PIN number is incorrect, the card is retained, ATM returns to Ready State, and this use case terminates
5. No Account – At Basic Flow Step 4 - Verify Account and PIN, if the Banking system returns a code indicating the account could not be found or is not an account which allows withdrawals, the ATM displays the appropriate message and rejoins the Basic Flow at Step 10 - Return Card
6. Insufficient Funds in Account – At Basic Flow Step 6 - Authorization, the Banking system returns a code indicating the account balance is less than the amount entered in Basic Flow Step 6 - Enter Amount, the ATM displays the appropriate message and rejoins the Basic Flow at Step 6 - Enter Amount
7. Daily maximum withdrawal amount reached – At Basic Flow Step 7 - Authorization, the Banking system returns a code indicating that, including this request for withdrawal, the customer has or will have exceeded the maximum amount allowed in a 24 hour period, the ATM displays the appropriate message and rejoins the Basic Flow at Step 6 - Enter Amount
8. Log Error – If at the Basic Flow Step 9 - Receipt, the log cannot be updated, the ATM enters the "secure mode" in which all functions are suspended. An appropriate alarm is sent to the Bank System to indicate the ATM has suspended operation
9. Quit – The customer can, at any time, decide to terminate the transaction (quit). The transaction is stopped and the card ejected
10. "Tilt" – The ATM contains numerous sensors that monitor different functions, such as power, pressure exerted on the various doors and gates, and motion detectors. If at any time a sensor is activated, an alarm signal is sent to the Police and the ATM enters a "secure mode" in which all functions are suspended until the appropriate re-start / re-initialize actions are taken

2.2. The tagged representation

The tagged representation is constructed from the Use Case identifying the start and end points of processes and possible branches [9]. Use the following set of symbols to build the target representation:

Symbol	Description
●	Start of a process (Use Case)
◉	End of a process (Use Case)
→	Start of an event of basic flow of process
↳	Start of an event of alternative flow of process
◇	Start of a conditional event
⤳^{n}	Jump of current iteration to iteration ^{n}
||	Sequence of concurrent events

Table 2-1 Tagged representation of Use Case description

It should be noticed that the construction of the tagged representation does not restrict the elicitation of requirements in Use Case. Thus, the initial process of inception can proceed normally without any constraint. The tagged representation [9], applied to the Cash Withdraw Use Case, is showed in the next subsection.

2.3. Tagged Representation for the Cash Withdraw Functionality

Following there are the basic and alternative flows of events for the tagged representation of the Cash Withdraw functionality.

2.3.1. Basic Flow of Events with Alternative Flows

1 ● Initiate Withdraw – Customer inserts bank's card in the card reader on the ATM machine; (
1.1 ◇ Is the card valid? – Verify Bank Card – The ATM reads the account code from the magnetic strip on the bank's card and checks if it is an acceptable card;

> 1.1.1 ↳ Send message of invalid card – If the card isn't an acceptable card, send an appropriate message. ⤳$^{1.9}$

1.2 → Enter PIN – The ATM asks for the customer's PIN code (4 digits);
1.3 ◇ Is PIN Correct? – Verify PIN – The account code and PIN are verified to determine if the account is valid and if the PIN entered is the correct PIN for the account;

> 1.3.1 ◇ Is the account valid? – Verify account code – The account code is verified;

1.3.1.1 ↳ Send message of invalid account – The Banking system returns a code indicating the account could not be found or is not an account which allows withdrawals. ↝ 1.9

1.3.2 ◇ Is the final try? – Checks the number of tries – The customer has three tries to enter the correct PIN;

1.3.2.1 ↳ Send message of incorrect PIN – The ATM send an appropriate message. ↝ 1.2

1.3.3 ↳ Retain card – on the final try the card is retained, ATM returns to Ready State, and this use case terminates. ↝ 2

1.4 ◇ Have money the ATM? – Select Withdraw – The ATM displays the different alternatives available at this ATM. In this flow, the bank customer always selects "Cash Withdraw";

1.4.1 ↳ Send message ATM out of Money – If the ATM is out of money, the "Cash Withdraw" option will not be available. ↝ 1.4

1.5 ◇ Sufficient funds and does not exceed the daily limit? – Enter amount – The ATM asks for the amount to withdraw. For this flow the customer selects a pre-set amount ($10, $20, $50, or $100);

1.5.1 ◇ Sufficient funds? – Check sufficient funds – Check if the funds are sufficient;

1.5.1.1 ↳ Send message insufficient funds in ATM – The ATM contain insufficient funds to dispense the requested amount. ↝ 1.5

1.5.2 ↳ Send message daily maximum withdrawal amount reached – the Banking system returns a code indicating that, including this request for withdrawal, the customer has or will have exceeded the maximum amount allowed in a 24 hour period. ↝ 1.5

1.6 ◇ Authorized withdraw? – Authorization – The ATM initiates the verification process with the Banking System by sending the Card ID, PIN, Amount, and Account information as a transaction. For this flow, the Banking System is online and replies with the authorization to complete the cash withdrawal successfully and updates the account balance accordingly;

1.6.1 ↳ Send message Insufficient Funds in Account – The Banking system returns a code indicating the account balance is less than the amount entered. ↝ 1.5

1.7 → Dispense – The Money is dispensed;
1.8 → Receipt – The receipt is printed and dispensed. The ATM also updates the internal log accordingly;
1.9 → Return Card – The Bank Card is returned;
1.10 ◇ Finalized Operation? – Updates log – The ATM updates internal log according the operation;

1.10.1 ↳ Cancel Operation – When the log cannot be updated, the ATM enters the "secure mode" in which all functions are suspended. An appropriate alarm is sent to the Bank System to indicate the ATM has suspended operation. ↝ 2

) 2 ⦿ Ends Operation – Use Case ends with the ATM in the Ready State;

2.3.2. Alternative Flows of Events

3 ↳ Cancel Operation – The customer can, at any time, decide to terminate the transaction (quit). The transaction is stopped and the card ejected. ↝ 1.9
4 ↳ Activate Tilt's sensor – The ATM contains numerous sensors that monitor different functions, such as power, pressure exerted on the various doors and gates, and motion detectors. If at any time a sensor is activated, an alarm signal is sent to the Police and the ATM enters a "secure mode" in which all functions are suspended until the appropriate re-start / re-initialize actions are taken. ↝ 2

3. Synthesizing a Petri Net for the Requirements

One of the characteristics of Information Systems is the importance of its dynamics, which is, the integration and synchronizing of its *business processes* components. This is salient in the inception phase as well as in the following phase of elaboration. To validate dynamic, frameworks as Petri Nets [4] have being widely used [5]. Besides, Petri Net is a generic schema, and thus suitable to represent abstract specifications of processes [6].
To translate the tagged representation to Elementary Petri Nets we defined a set of basic net elements to fit the tagged elements. Thus we have the translation specified in BNF (Backus Naur Form) [7] as shown in Table 3-1.
Thus, each tagged element is an operation (enterprise activity) and each Use Case a business process. The set of

all Use Cases functionalities compose the requirement specification for the whole system.

In Table 3-1 each tagged construct is mapped in a defined token, each of which is related to a net element [9]. Table 3-2 shows all these elements in its graphical form, where the connection restrictions were already included, that is, it is always possible to represent an information flow by putting together all net elements.

The general structure is guided by the Use Case, in the sense that the resulting net shows the Basic Flow (BF) of events – if no exception occurs – as the core of the net. The element **BF** defined in Table 3-2 shows the definition of the Basic Flow of events, while the Alternative Flow is defined by the element **AF** in the same table [9].

Conditional events are represented by the element **condition** in Table 3-2.

```
<BF>           ::= <initial> { <event> } <final> { <AF> }
<initial>      ::= <label> "●" <title> "-" <description> ";" "("
<label>        ::= <number> { "." <number> }
<number>       ::= <sequential_number>
<title>        ::= <string>
<description>  ::= <string>
<string>       ::= <any_caracter> { <any_caracter> }
<event>        ::= [ <eventB> | <eventA> ]
<eventB>       ::= <labelB> [ ( "→" <title> "-" <description> ";" ) |

                         ( "◇" <condition> "-" <title> "-" <description> ";" <eventA>
                         ) ] { <eventB> } |

                   "||" "(" <eventB> { <eventB> } ")" "(" <eventB> { <eventB> } ")" "||" {
                   <eventB> }
<labelB>       ::= <label> "." <number>
<condition>    ::= <string>
<eventA>       ::= <labelA> [ ( "↳" <title> "-" <description> <branch> ) |

                         ( "◇" <condition> "-" <title> "-" <description> ";" <eventA>
                         ) ] { <eventA> }

<labelA>       ::= <labelB> "." <number>
<branch>       ::= "~" [ <label> | <labelB> ]
<final>        ::= ")" <label> "◉" <title> "-" <description> ";"
<AF>           ::= <eventUA>
<eventUA>      ::= <label> "↳" <title> "-" <description> <branch> { <eventUA> }

Where:
•  eventB → Event of Basic flow
•  labelB → Event id of Basic flow
•  eventA → Alternative Event of Basic flow
•  labelA → Alternative Event id of Basic flow
•  eventUA → Unconditional Event of Alternative Event
•  labelUA → Unconditional Event id of Alternative Event
```

Table 3-1 Structure definition of the Use Case description

All transformations proposed – from Use Cases to the tagged representation and from these representation to the final representation in Elementary Petri Nets – can be performed by simple algorithms, since they are syntactic and do not include any special knowledge about the target system or artifact. Therefore, a simple case tool can be produced to automate this process. We are developing such tool as an "add on" to be attached to the Rational Suite, since a characteristic of the proposal presented in this paper is to be inside the scope of the RUP (Rational Unified Process).

Applying the mapping of Table 3-2 to the tagged language shown in section 2.3 results in an elementary Petri Net extended to include enabling and inhibitor gates, used to implement the conditional events. The net presented in Fig. 3.1 is the provider's viewpoint of the Cash Withdraw functionality.

Event	BNF notation	Petri net segment
Initial	p ● Title - Description; (	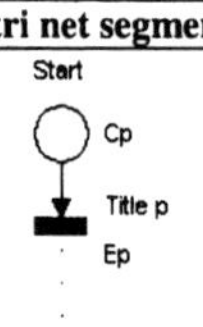
Basic	p → Title - Description;	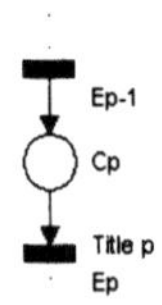
Basic conditional and alternative normal[1]	p◇Condition-Title-Descript.; a ↳ Title-Description ~p'	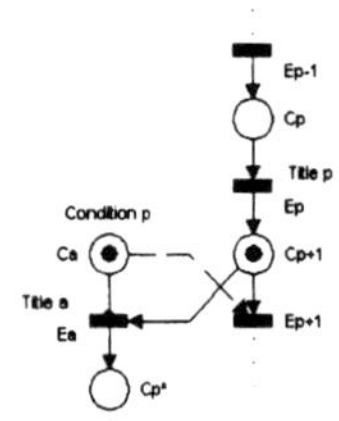
Alternative conditional and normal	a◇Condition-Title-Descript.; a' ↳ Title-Description ~p'	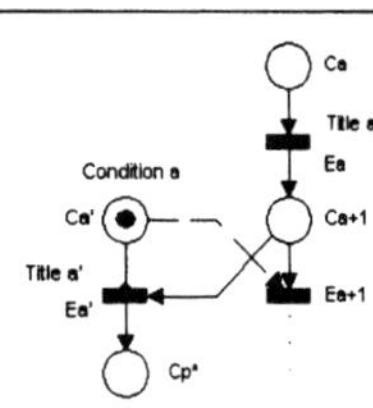
Basic concurrent	|| (p1→ Title - Description; p2→ Title- Description;) (p3→ Title - Description; p4→ Title- Description;) ||	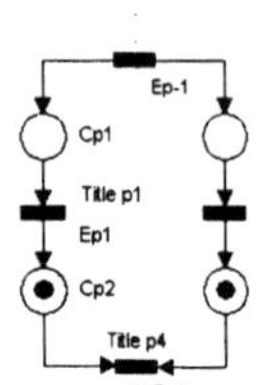
Final	) p ◉ Title - Description;	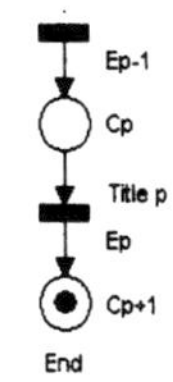
Unconditional Alternative	ai ↳ Title - Description ~p'	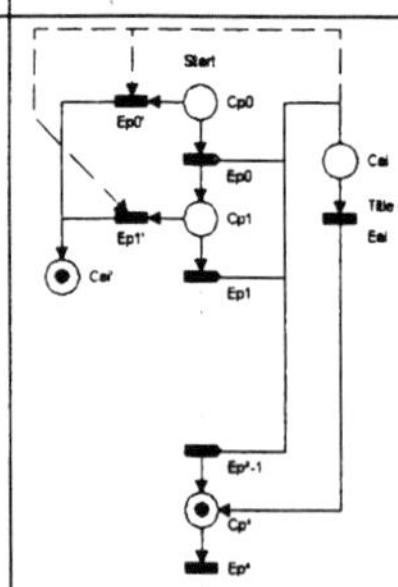

Table 3-2 Petri net segments corresponding to Use Case description

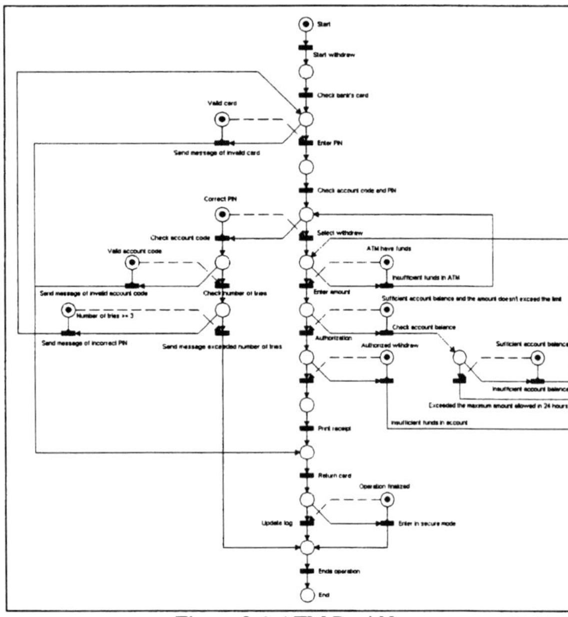

Figure 3-1 ATM Petri Net

4. Analysis and Validation

The net of Fig. 3.1 can be simulated in any token player based in Petri Nets (since it is based in Elementary Nets with extension to include gates). Animation of the token player can be used to validate the process of Cash Withdraw visually. The conditional points are denoted by

[1] Ca represents an external condition, where the arc between the condition Ca and the event Ea is an inhibitor arc, while the arc from Ca to Ep+1 is an enabling arc.

simple places which marking can stand for uncontrollable events such as entering with the wrong password in the ATM section. There is no way to prevent the "value" of this condition and the praxis would be to evaluate the evolution of the process for each possibility.
That means changing the mark in all those places involved in conditional events, what can be done with any reasonable token player tool.
The proposed approach opens the possibility to represent and validate the merging of different viewpoints [10][11]. For instance, Fig. 4-1 shows a viewpoint of the Cash Withdraw not from the developer as we are implicitly doing in the previous sections but from an end user, who is more concerned with the main interaction of obtaining cash.

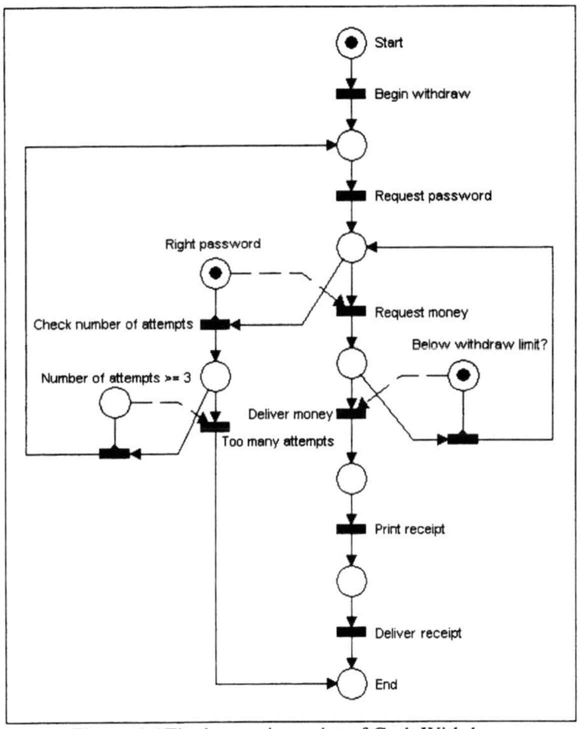

Figure 4-1Final user viewpoint of Cash Withdraw

It could be seen from Fig. 4-1 that the net resulting from the user viewpoint is isomorphic to a sub-net of the Fig 3-1 that represents the developer viewpoint. Thus, the current approach can be used as guide in the merging of viewpoints, by identifying similarities and collapsing the two nets.
Again we are thinking in a direct and informal matching. Automating that with a case tool can be possible even if difficult due to the performance of approximated algorithms to decide about isomorphism between nets.

Combinations of conditions can then be tested and validated as the reaction of the system to (conditional) events that are not controllable but that can be predicted.
The complete system can be validated as a set of business processes (BP). If these processes are all independent the validation is the result of the validation process for each one of the BP components. However, if we have an integrated system, that is, where the BP is dependent, then we will need some integrating places (places that belong to both BP or places that can be collapsed from two or more BP to result in one unique net). Again a token player tool can be used to validate the general behavior.
In both cases the validation process is based on the animation of the token player, that is, in a visual process of following the schematized behavior of the Petri net once an initial state is selected. Thus, the method of validation is suitable for small systems or at least for those where the resulting graph can be easily visualized. This kind of system can also be interpreted, that is, it is not difficult to attribute a semantic meaning for some or all of the places and events, such as "request money", "check the number of attempts", etc. and construct an analogy between the evolution of the schema and a real process. Unfortunately that is not the case for larger systems.
To below larger systems the recommended approach would be to analyze the properties of the nets, principally those directly deducted from its state equation. The state equation is an iterative equation that denotes all reachable states of the net, once defined an initial one. Thus, a generic state can be given by

$$M_i = M_0 + A^T \sum_{0}^{i-1} \sigma_i$$

where A is called the incidence matrix of the net and denotes the existence of arcs between the passive elements (conditions) and the active ones (events).
Several properties can be deduced from this equation including necessary conditions for the existence of deadlock, that in the case of requirement validation would mean an impossible process for the selected value of the uncontrollable conditions.
However, at least theoretically the proposed method above could not admit an analytical or numerical property analysis since the use of internal gates connected to places with persistent tokens makes the incidence matrix inconsistent.
On the other hand there are other possibilities if another kind of net is used. For instance, the GHENeSys (General Hierarchical Enhanced Net System)[8] is an extended object oriented net system where a specific place is used to represent uncontrolled conditions: the pseudobox. Also

this place is not considered as belonging to the core net but as the relationship between the net system and its environment. Thus it is possible to recover a state equation for it even with the presence of gates.
Also, pseudoboxes can be used as collapsed places to connect components in a more generic schema. The same approach presented here could be more suitably represented in the GHENeSys system.

5. Conclusions and further work

We presented a very simple schema for validating requirements, transforming Use Case in Petri Net systems. The approach was inspired in the tedious and informal work that has to be done for information systems with a large set of conditional branches.
In fact this is a work in progress where the early phase of design of discrete dynamic system is supposed to be assisted by computer tools that could help with the dynamic modeling combining RUP method with the Petri Net formalism. We claim this is an approach suited to Manufacturing Information Systems. In the near future we plan to develop this idea further in order to demonstrate this claim formally and present a case study.
A Case tool is being developed as an "add on" to the RequisitePro (Rational Suite) tool as an alternative to validate requirements. However that tool would work fine only to small and medium size systems. The reason for that was already explained in the previous section.
In a further development we will be using an extended net called GHENeSys to reduce the impact of the explosion of states and solve the problem of representing formally the gates used to model conditional branches.

6. Bibliography

[1] Vernadat, F., *Enterprise Modeling and Integration: Principles and Applications*, Chapman & Hall, 1996
[2] Rational Software Corporation. *Rational Unified Process: version 2001A.04.00*. s.l.: s.ed., 2001
[3] Anschuetz, H. "A tool for design and simulation of Petri nets: HPSim version 1.1", s.l.: s.ed., 2001, Available in: <http://home.t-online.de/home/henryk.a/petrinet/e/sim_form.htm>. Accessed in: 13 jul. 2002
[4] Reisig, W., *Introduction to Petri Nets*, Springer Verlag, 1981
[5] Oberweis, A. and Sander, P., "Information system behavior by high level Petri nets", *ACM Transactions on Information Systems*, v.14, n.4, October, 1996
[6] Silva, J.R., "Interactive Design of Integrated Systems", *In Intelligent Systems for Manufacturing: Multi-agent Systems and Virtual Organizations*, Kluwer Academic Pub., p. 567-578, 1998
[7] Naur, Peter (ed.), "Revised report on the algorithmic language ALGOL 60", *Communications of the ACM*, v. 3, n.5, 1960, p. 299-314
[8] del Foyo, P.M.G. and Silva, J.R., "Towards a Unified View of Petri Nets and Object Oriented Modeling", *17th International Congress of Mechanical Engineering*, São Paulo, Brazil, 2003
[9] Santos, E.A., *Systems Requirements Verification by Using Petri Nets*, Master thesis, Escola Politecnica da USP, in Portuguese, December, 2002
[10] Leite, J.C.P., Freeman, P.A., "Requirements validation through viewpoint resolution", *IEEE Transaction on Software Engineering*, v.17, n.12, December, 1991, p. 1253-1269
[11] Kotonya, G. and Somerville, I., *Requirements Engineering*, John Wiley & Sons, 1998
[12] Kotonya, G. and Somerville, I., "Requirements Engineering with viewpoints", *Software Engineering Journal*, v.11, n.1, p.5-18, January, 1996
[13] Vernadat, F., "CIMOSA: Enterprise modeling and enterprise integration using a process-based approach". *In Information Infrastructure Systems for Manufacturing*, H. Yoshikawa and J. Goossenaerts editors, North-Holland, Amsterdam, 1993, pp 65-84 (invited paper).

Copyright © IFAC Information Control Problems in
Manufacturing, Salvador, Brazil, 2004

ELSEVIER
IFAC
PUBLICATIONS
www.elsevier.com/locate/ifac

A SYSTEMATIC AND FORMAL APPROACH TO THE SPECIFICATION OF FLEXIBLE MANUFACTURING SYSTEMS REUSING COLOURED PETRI NETS MODELS

Leandro Dias da Silva [*,1] **Angelo Perkusich** [*]

[*] *Electrical Engineering Department*
Federal University of Campina Grande
P.O. Box 10105
58109-970 Campina Grande, Paraba, Brasil
{leandro,perkusic}@dee.ufcg.edu.br

Abstract: In this work a systematic methodology to reuse Coloured Petri Nets models is described. The methodology is applied to the flexible manufacturing systems domain. A model of a system can be developed reusing parts of previous designs, instead of modeling always from the scratch. The phases of reuse considered in this work are recovery, adaptation, integration, and use verification. We define an automatic integration approach, based on an architectural framework to build new models. The use verification phase is also defined to verify whether the new model do not violate the semantics of the reused models.
Copyright © 2004 IFAC
Keywords: Flexible Manufacturing Systems, Coloured Petri Nets, Reuse

1. INTRODUCTION

The application of formal methods to model complex systems allows automatic analysis and verification (Clarke and Wing, 1996). Petri nets (Murata, 1989) is a formal method that have been widely studied and applied to the formal modeling and analysis of complex concurrent systems (Jensen, 2002; Jensen, 2001). Petri nets models can easily capture structural characteristics as well as behavioral properties such as conflicts, synchronization, non-determinism, and deadlock. Moreover, the mathematical foundation of Petri nets also have a graphical representation, that avoids the use of both possibly ambiguous text or hard to understand mathematical notations. More specifically, Hierarchical Coloured Petri Nets (HCPN) (Jensen, 1992; Jensen, 1997) is a very powerful tool to describe complex models due to the incorporation of complex data types and hierarchy.

In the case of complex systems, such as Flexible Manufacturing Systems (FMS), one possible solution to manage the complexity is to use artifacts that can be reused (Krueger, 1992; McIlroy, 1969). Therefore, in the development process of a new system such artifacts can be used and the design do not always start from the scratch. Reuse techniques may be applied in the design phase of systems (Ribot *et al.*, 1994), then the objects of reuse are models instead of code. In the case that these models are described by means of a formal language, it is possible to investigate the behavior of such models based on formal analysis and simulation (Clarke and Wing, 1996).

The contribution of this work is the introduction of an integration technique, and the application of model checking (Clarke *et al.*, 1999) to verify the integrated model. The relevance of this work is the study and development of techniques and

[1] Partially supported by Capes.

methods to increase the degree of automation in the modeling process, thus reducing the time and coast of the project, and the possibility of human errors. The approach is fully implemented based on HCPN, temporal logic (Emerson, 1990), model checking, and a set of manipulation tools for HCPN models called Design/CPN (Jensen and al, 1999). The major aspects in this paper are related to the study and development of automatic techniques for the integration, and use verification, as well as the working environment of the whole process applied to the FMS domain.

The remaining of this work is organized as follows. In section 2 Colored Petri Nets are presented. A discussion of the reuse of models to build formal models of FMS is presented in Section 3. Finally, in Section 4 are the conclusions and future works suggestions.

2. COLOURED PETRI NETS

Petri nets allow the modeling of properties such as precedence relationships, conflicts, concurrency, synchronization, deadlocks, and resource sharing, among others. Also, the state and action locality characteristic allow the modeling of complex systems using either bottom-up or top-down approaches. Therefore, promoting modularity and reusability, that are important characteristics for the modeling solution introduced in this work.

In the context of this paper the HCPN formalism is used. This formalism incorporates complex data types and hierarchy concepts. An HCPN is a set of non-hierarchical nets (CPN) and each CPN model is then called a CPN page. In the following CPN is informally introduced, the formal definition can be found in (Jensen, 1992).

Hierarchy is defined by the inclusion of substitution transition and fusion places. A substitution transition is a transition that is substituted by a CPN page. The page which the substitution transition belongs to is called superpage and the page represented by that transition is called subpage. The association between subpage and superpage is done by means of sockets and ports. Sockets are all the input and output places of the transition in the superpage. Ports are the places in the subpage associated to the sockets. The ports can be input, output, or input-output. For simulation and state space generation, sockets and ports are glued together and the resulting model is a flat CPN model. The fusion places are physically different places but logically only one place forming a fusion set. Therefore, all the places belonging to a fusion set have always the same marking. Indeed, these two additional mechanisms, substitution transition and fusion places, are only graphical, helping in the organization and visualisation of a CPN model. They favor the modeling of larger and more complex systems by giving the designer the possibility to model by abstraction, composition, or booth.

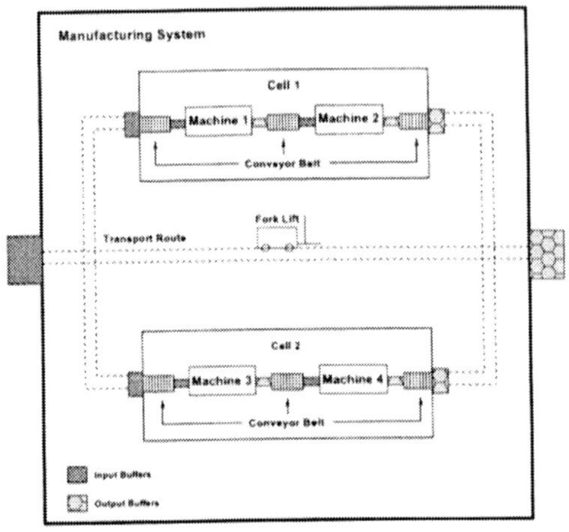

Fig. 1. Flexible manufacturing system

3. REUSE-BASED FLEXIBLE MANUFACTURING SYSTEMS MODELING

A Flexible Manufacturing Systems (FMS) is a production system organized in cells. Each cell can have one or more machines. Also, there is a transport system to move parts from one cell and/or machine to another to produce a product. In Figure 1 a simple FMS is shown.

A framework is necessary for each application domain in order to allow an automatic model integration. In this work the domain is FMS and the framework is realized by a HCPN model. In the framework there are definitions of localization, relationship, and possibly restrictions to be satisfied when building new models based on reuse of models. That is, where the transport agents and production systems are put, how they relate to each other, and how they relate to the framework. Thus, the framework is a generic model for a FMS where it is possible to plug models of transport agents and production systems. Moreover, these models can be changed individually without changing the framework.

In Figure 2 a block diagram for the reuse-based systematic modeling is presented. The repository management activity, recovery and insertion of models, are discussed in details in (Lemos and Perkusich, 2001). The adaptation of recovered models is defined in (Gorgnio and Perkusich, 2002). Therefore, in this work the model integration, and use verification activities are described, as well as the functionality of the reuse process as a whole, unifying all the activities in a systematic modeling method.

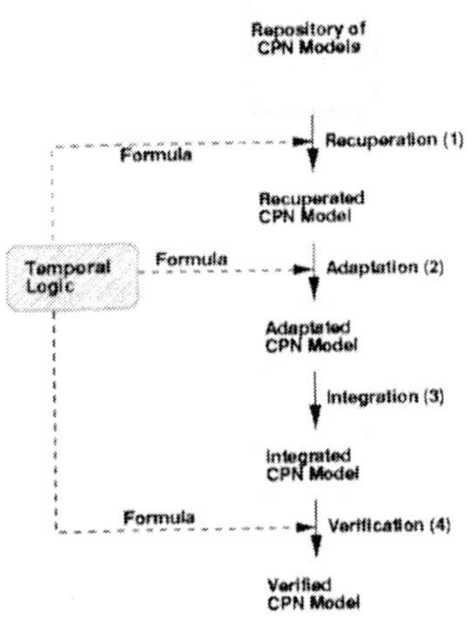

Fig. 2. The phases for reuse-based modeling

3.1 The Framework for the FMS Domain

In Figure 3 the model of the framework for FMS shown. This framework represents the system discussed in Section 3 and illustrated in Figure 1. Using this generic model, it is transparent the way how the entities are modeled. Then, a production system can be defined as a cell structure that has one or more machines, a production sequence and a transport sequence. The production sequence defines the resources needed to have a final product. The transport sequence defines the requests of transport agents needed by an specific production sequence to produce a product. The system can process several different parts or several identical parts at the same time. The description of the places and the transitions in Figure 3 is as follows.

The meaning of the places are:

- SystemIn = System input buffer;
- CellIn = Cell input buffer;
- MachineIn = Machine input buffer;
- MachineOut = Machine output buffer;
- CellOut = Cell output buffer;
- SystemOut = System output buffer;
- MachToMach = Request of transport from the machine output to the machine input, in the same cell;
- MachToCell = Request of transport from the machine output to the cell output;
- CellToCell = Request of transport from the cell output to the cell input;
- CellToOut = Request of transport from the cell output to the system output.

The meaning of the transitions are:

- TranspToCellIn = Transport from the system input to the cell input;
- TranspToMachineIn = Transport from the cell input to the machine input;
- Mach = Machines;
- MTOM = Transport from the machine output to the machine input, in the same cell;
- TranspToCellOut = Transport from the machine output to the cell output;
- CTOC = Transport from the cell output to the cell input;
- CTOO = Transport from the cell output to the system output;
- MachToMachChoice = Choice of transport from the machine output to the machine input, in the same cell;
- MachToCellChoice = Choice of transport from the machine output to the cell output;
- CChoice = Choice of transport from the cell output to the cell input;
- CellToOutChoice = Choice of transport from the cell output to the system output.

The hierarchy mechanisms of coloured Petri nets make possible to define substitution transitions that are substituted by another CPN model. This allow the definition of hierarchies in the framework modeling specific types of resources, or transport agents, keeping the structure of the hierarchical upper level unchanged, characterizing a top-down modeling.

The transition Mach is a substitution transition to the interface of the framework with the models of machines. The transitions TranspToCellIn, TranspToMachineIn, MTOM, TranspToCellOut, CTOC, and CTOO, are substitution transitions to the interface of the framework with the models of transport agents. There are different transitions for the transport system because it is possible to have different transport agents in an FMS.

The production sequence, as well as the transport sequence are specified in the initial marking in place SystemIn. It is possible to generate the transport sequence dynamically using the production sequence, but based on some tests it leads to an overhead of processing, functions, and complexity in the inscriptions, opposing the objectives of this generic proposal. Also the scheduling is compromised with this dynamic definition.

The transitions MachToMachChoice, MachToCellChoice, CChoice, and CellToOutChoice, means a choice to determine what type of transport will be required at this point based on the transport sequence defined in the initial marking of the framework.

3.2 Model Integration

After a model is obtained or recovered it can be integrated in the framework. The integration consists in the selection of a file name with the diagram to be integrated in the model of the framework, the integration environment creation, and the integration to the model. The integration environment consists of the places, transitions and arcs, of the Coloured Petri Net model, as well as the respective inscriptions. While the integration consists in the definition of the hierarchy of the

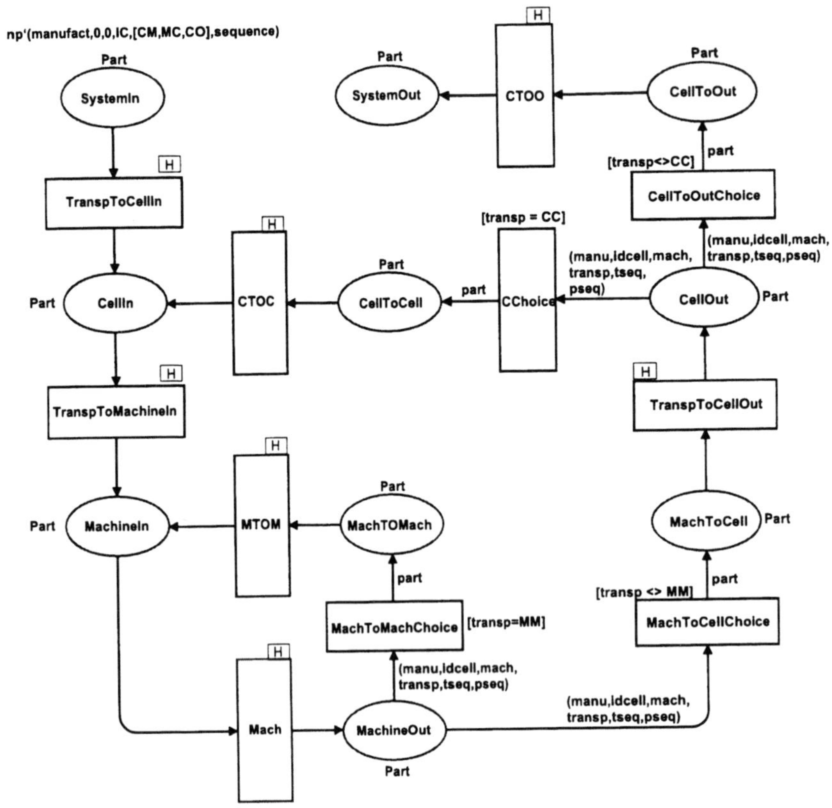

Fig. 3. Framework CPN model

model. The integration, as well as the other activities, are fully implemented using the CPN/ML language (Christensen and Haagh, 1996) using Design/CPN. In the following the steps of the integration activity are listed.

(1) Select the model file
(2) Make places, transitions, and arcs
(3) Define the ports
(4) Define the substitution transition
(5) Associate the sockets to the ports
(6) Adapt the port colors to sockets colors
(7) Insert the declarations of the model into the global declaration node

In Figure 4 it is illustrate how the integration approach is actuality implemented for the Design/CPN tool. First, the designer is asked for the name of the file with the diagram to be integrated in the framework. Then, some functions will be automatically executed to build the integration environment, that is, the places, transitions, arcs, and its respective names, color sets, and inscriptions. The next step is the definition of the input and output ports in the diagram been integrated. After this step, the substitution transition will be defined, and the sockets in the superpage will be associated to its respective ports, previously defined in the subpage. The last step is to select the box with the model declarations, in the model page, to define the ports colors based on the sockets colors, and to append this information in the global declaration node. The file selection needs the user interaction, while all other steps are executed automatically. In Figure 4 just some of the elements and steps of the integration are shown in order to give the reader a more concrete idea of what the model integration is in practice.

For the definition of the algorithm described above some restrictions must be satisfied: (i) unique page name; (ii) the prefix in the color sets names indicates the page name; (iii) The suffix in the place names indicates whether it is a port place or not; (iv) The auxiliary box with model declarations is in dot-dashed line pattern and is unique; (v) Declarations of the port places must be the first ones in the auxiliary box. The first restriction guarantees that the page name for the integrated model is unique, and is a designer responsibility. The model being integrated must have its color set names prefixed with page name. Another restriction is about port places. The input and output places must have names suffixed by IN or OUT, respectively. The last integration restriction is that for the model being integrated there must exist an auxiliary box with *dot-dashed* line pattern. This box must contain all the declarations for this page, including color sets for the ports. Finally, the contents of this box is appended to the global declaration node. Observe that the color sets are renamed to those defined by the user. The resulting model is a HCPN including the model for the components and for the framework.

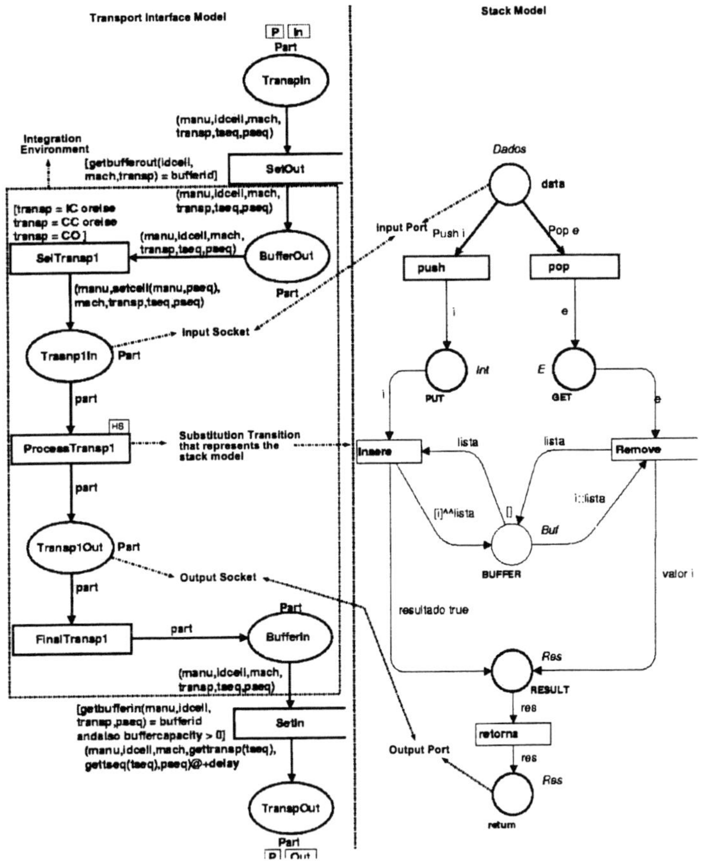

Fig. 4. HCPN model integration

3.3 Use Verification

Besides the activities already discussed we also consider another one named use verification. Such activity is considered because when we model based on reuse it is necessary to guarantee that the semantics of the resulting model satisfies the individual semantics of the reused models.

Therefore, the use verification consists in performing model checking in the framework including the integrated models. The use verification is performed on the entire model and guarantees that the original specifications were preserved, or the components are being correctly used. Such use verification can be seen as a contract verification in the context of components. Also, it is important to note that a new framework is necessary for a new modeling domain. The construction of such frameworks must not violate the specific functionalities of each individual models after the reuse process.

Another argument for the use verification phase is to validate the models developed locally to a specific project in order to make it available to be reused in other subsequent projects. In this case, properties of the model can be specified in temporal logic, and verified using model checking. These properties can be stored in the repository together with the model and it can be used for the use verification phase in subsequent projects.

The steps of the use verification activity are:

(1) Run the occurrence graph tool;
(2) Select the file with the specifications to be verified;
(3) Run model checking.

Initially, the designer must execute the occurrence graph tool in the Design/CPN. The next step is to select the file that has the properties specifications for the model to be verified. These properties are described by temporal logic. The algorithm will execute at this point the model checking in the framework including all the models already integrated. If the properties hold in the resulting model the use verification is successful. In the opposed case the designer will receive one warning that there exists an error in this specific use case. In this case the error is not pointed neither corrected by the procedure of use verification. Therefore it is the designer responsibility to find and to correct the possible errors.

The steps one and two must be done manually by the user, while the third one is automatically executed based on the specifications contained in the file indicated by the designer, that will receive a message saying if the properties are satisfied or not.

Using the example model shown in Figure 4 the following atomic propositions and temporal logic formula can be used to verify if the stack model, after integration, still behaves as a stack model. This model is used to represent a fork lift transport agent, for example.

```
fun PA n = (Mark.Stack'BUFFER 1 n = (1,[4,2]);
fun PB n = (Mark.Stack'RESULT 1 n = (1,4);
val formula = POS(AND(PA,EXIST_NEXT(PB)));
```

The atomic proposition PA specifies the situation where the values 4 and 2 are in the stack buffer. And the atomic proposition PB specifies the situation where the value 4 is in the stack result place. The formula will be evaluated to true if, given that PA is true, it is possible to have a future state where PB is true. That is, the value in the top of the stack will be returned. This simple example is used to illustrate how temporal logic and model checking can be used in the use verification phase. Actually, the formula can be more complicated depending on what the designer needs to prove. And there are several syntactic constructs of the ASK/CTL library that the reader does not need to know, neither we will explain here because space limitations and because it is not the main subject of this work.

4. CONCLUSIONS AND FUTURE WORKS

In this work the description of a reuse based systematic modeling approach using Hierarchical Coloured Petri Nets in the domain of Flexible Manufacturing Systems is presented. All of the reuse activities are fully automatic and considered in this work, namely: recovery, adaptation, integration, and use verification.

For different application domains only the integration step is affected due to the need of defining different integration strategies and an specific framework. Due to the use of framework concept, the designer do not need to care explicitly about the model structure, it is abstracted. Any kind of Flexible Manufacturing Systems can be modeled following the systematic approach described at this work in an easier way reducing time, money, and human errors.

One extension of this work is to diversify the repository. This can be done in several ways. One way is to have several models for a domain and even several domains. Examples of other possible domains are communication systems, integrated circuits, and complex and distributed software. Another way is to put in the repository other kinds of artifacts like code segments, and the other modeling formalisms. Also distributed repositories can be considered.

REFERENCES

Christensen, Soren and Torben Bisgaard Haagh (1996). *Design/CPN Overview of CPN ML Syntax.* 3.0 ed.. University of Aarhus.

Clarke, Edmund M. and Jeannette M. Wing (1996). Formal methods: State of the art and future directions. *ACM Computing Surveys.*

Clarke, Edmund M., Orna Grumberg and Doron A. Peled (1999). *Model Checking.* The MIT Press. Cambridge, Massachusetts.

Emerson, E. Allen (1990). Temporal and modal logic. In: *Handbook of Theoretical Computer Science* (jan Van Leeuwen, Ed.). Vol. B: Formal Models And Semantics. Chap. 16, pp. 995–1072. Elsevier Science.

Gorgnio, K. C. and A. Perkusich (2002). Adaptation of coloured petri nets models of software artifacts for reuse. In: *7th International Conference on Software Reuse, Lecture Notes in Computer Science.* Austin, EUA.

Jensen, K. and et. al (1999). *"Design/CPN" 4.0.* Meta Software Corporation and Department of Computer Science, University of Aarhus, Denmark. On-line version:http://www.daimi.aau.dk/designCPN/.

Jensen, Kurt (1992). *Coloured Petri Nets: Basic Concepts, Analysis, Methods and Practical Use.* EACTS – Monographs on Theoretical Computer Science. Springer-Verlag.

Jensen, Kurt (1997). *Coloured Petri Nets: Basic Concepts, Analysis Methods and Practical Use.* Vol. 2. Springer-Verlag.

Jensen, Kurt, Ed.) (2001). *Third Workshop and Tutorial on Practical Use of Coloured Petri Nets and the CPN Tools, Aarhus, Denmark, August 29-31, 2001.* Vol. PB-554. DAIMI.

Jensen, Kurt, Ed.) (2002). *Fourth Workshop and Tutorial on Practical Use of Coloured Petri Nets and the CPN Tools, Aarhus, Denmark, August 28-30, 2002.* Vol. PB-560. DAIMI.

Krueger, Charles W. (1992). Software reuse. *ACM Computing Surveys* **24**(2), 131–183.

Lemos, A.J.P. and A. Perkusich (2001). Reuse of coloured petri nets software models. In: *Proc. of The Eighth International Conference on Software Engineering and Knowledge Engineering, SEKE'01.* Buenos Aires, Argentina. pp. 145–152.

McIlroy, M. D. (1969). "mass produced" software components. In: *Software Engineering* (P. Naur and B. Randell, Eds.). Scientific Affairs Division, NATO. Brussels. pp. 138–155. Report of a conference sponsored by the NATO Science Co.

Murata, Tadao (1989). Petri nets: Properties, analysis and applications. *Proc. of the IEEE* **77**(4), 541–580.

Ribot, D., B. Blongard and C. Villermain (1994). Development life-cycle with reuse. In: *ACM Symposium on Applied Computing SAC 94.*

Copyright © IFAC Information Control Problems in
Manufacturing, Salvador, Brazil, 2004

ELSEVIER
IFAC
PUBLICATIONS
www.elsevier.com/locate/ifac

A METHODOLOGY PROPOSAL FOR TEACHING AND RESEARCH IN AUTOMATION USING COMPUTER MEDIATED EDUCATION TOOL

Aihara, Cintia Kimie,
Rosário, João Maurício

Universidade Estadual de Campinas – Faculdade de Engenharia Mecânica
cintia@fem.unicamp.br , rosario@fem.unciamp.br

Abstract: The industry automation, with your multidisciplinary feature, involves a large diversity of fundamental importance themes as the communication net developing, dedicated software's, system integration and flexibilization. The industry competition competence and the automated production systems integration has become synonym of flexibility to obtain competitive products and its quality. On this way, the paper shows a didactic platform purpose, which permits the application of methodologies developed for product obtainment. This workbench is important element into the educational process and professional development in industry automation area. It could show the all evolutions of production process using the Computer Mediated Education (EMC). *Copyright © 2004 IFAC*

Keywords: automation, learning control, education, integrated circuit, teaching.

1. INTRODUCTION

The globalization compelled the industries passed for great modifications, with the intention of if becoming more competitive. The modernization of its industrial parks was necessary, aiming at the competitiveness of its products, through the increase of the quality, accessible reduction of costs and prices.

With intention to reach such objective, they had been created a set of techniques, giving itself the AUTOMATION name.

This paper is inserted in this context, having as objective to develop the project and the implantation of a didactic platform, aiming at methodology proposal for the formation and the research in the Integration Automated Industrial Systems area.

2. DEVELOPMENT

The knowledge of automated systems of production is a diversity of independent variables, interlaced the multiple areas, with intensity of modified influences as the moment, being able itself to cite the economy, the general management and the production, the organization of the work; social and psychological aspects as the job, the syndical organization, the feeling of loss of knowledge of the worker, the occupational diseases and the technological side of the automation with "intelligent equipment" as: CAD - Computer Assisted Design, CAM - Manufactures Assisted for Computer, CN - Numerical Command, CNC - Computerized Numerical Command, PLC - Programmable Logical Controller, Robot, FMS - Flexible Manufacture Systems and CIM - Computer Integrated Manufacture. Being about recent subjects, they meet in evolution and they participate of the elaboration of any project for automation in different degrees of relevance. With this, a modern, competitive industrial system, that it aims at productivity and necessary quality to be planned, to use the appropriate technology, with the integration of the parts of form to provide a flexibility. (Vendrameto, 1994)

Knowing the multidisciplinary this area, and the new technology that pierce a knowledge area, development the methodology to teach the automation. (Acar, 1994)

Through one system developed in the Robotic and Integrated Automated Laboratory - LAR, that also uses some elements of the automation and distance education tools as communication element of knowledge transmission.

2.1 Specification book

The specification book of automated production system is the description of behaviour in function of the evolution of the stages that describe the process to be automated. The same it must present a clear description, it needs and without ambiguities, nor omissions, the function of stages to be automated process. With this, the description of the system must be divided in two successive and complementary levels: level 1 and level 2. (Araújo, 1997)

The level 1 - FUNCTIONAL SPECIFICATIONS - describe the operative part – PO - with regard to the command part – PC. The functional specifications allow understanding, which the functions that the automatism must carry through, face to the different situations that can appear. In this level little the form imports that that will carry through one definitive movement, but is important to know where circumstance the displacement must be carried through. It is in this level that foreseen aspects of security for the functioning must be incorporated in the functional specifications, in the ratio where they do not depend directly on the employed technology.

The level 2 - TECHNOLOGICAL AND OPERATIONAL SPECIFICATIONS - a detailing of the conditions of functioning constituent of the process must add to the functional requirements to be automated, through technological and operational specifications. In this level it must have indications on the accurate nature of the transducers and actuators, in the way as these elements will have physically to be inserted in the process that composes the automated system and information concerning its environment of performance. (Bittar, 1993)

These consideration are important for the exploration of the process to automate, considering its repercussions on the economic aspect, that frequently are forgotten during the elaboration tasks of the specification book, therefore are difficult to be express in quantitative way.

2.2 Description Tools

The complexity increasing of the automated production system started the user to find great difficulties in a clear way, concise and not ambiguous, in the functional specifications associates to these systems. This complexity tends to increase still more, with the use of a high number of information of entrances and exits. This form is necessary to describe the system through adequate description tools. (Aihara, 2000)

For the development of automated system production, we must initially describe it clearly, independently of its structure. This description can be global or specific. This language must be the point of view the man, a natural language that expresses in simple way the system specification, and the point of view the control device, a simple description that is easy being interpreted and being executed. This is the biggest difficulty this stage, therefore it is necessary that if it supplies to the designer the information of detailed form, independent the level of the description (Dalbo,1994)

Sometimes the verbal language makes possible the diverse interpretations and ambiguous, the graphical representation starts to be the form of more easy understanding. However, we can find difficulties in finding an accepted graphical representation for all the involved parts in a project, consequently a graphical symbol for each represented function.

2.3 Sequential Function Chart

The IEC defined through IEC 60848 norm - Functional Diagrams for Control Systems - a method representation that combines graphical symbols and literal declarations, that the global description allows in such a way specific, through the necessary relation between the entrances and exits, independently of the technological properties used components or future implementations.

This norm had origin the GRAFCET. Its concepts are based on a set of definitions on which its basic rules in boolean algebra are established (true or false). It is a method of description, formed basically for a set of stages, transitions and the condition associated with the transition and guided linking's. These concepts remain until today, being that the stage represents the partial state of the system, a time that this can be active or inactive, in which the action are carried through. The associated action is only carried through if the stage will be active. The transition indicates the possibility of evolution between the stages and to each transition a condition also known for receptivity exists, representing a decision for the change of state of the system. This means that an effected transition either is necessary that the previous stage to the transition is active and that the logical condition (boolean) associate to the true transition either. The guided linking, indicates the way of evolution in GRAFCET (Georgini, 1999)

With the combination these elements we will have the static representation of the Automated System. Applying the Rules of Evolution, we get the dynamic vision of exactly. This can be verified, in the example Figure 1, where an engine initiates off. When set a button the engine start and only after deactivation the button the engine turned off and thus successively.

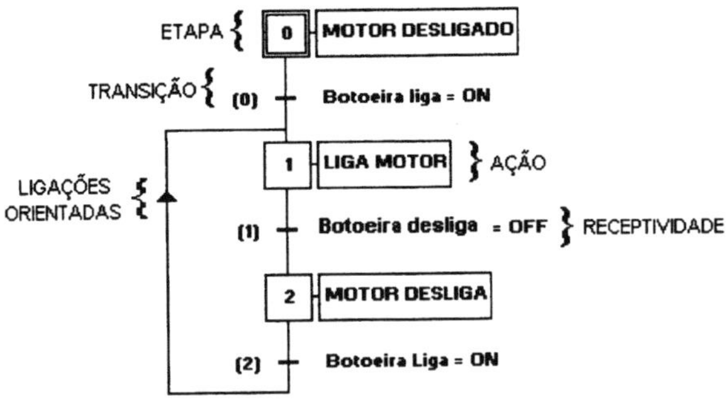

Fig. 1. SFC – Example used SFC

The considered method of representation serves as "communication tool" it enters the different involved technological areas in the development and use of Automated Systems.

2.4 The Didactic Platform

The integration between some automated elements is one of main stages in the implantation of an Automated Production System (SAP). Therefore, the use a consistent language for the description of the system and structure used is important; therefore they guarantee the flexibility and versatility for the system.

The SAP developed as integrant part of the experimental task this paper is idealized in a Didactic Platform for Composed Mixture of Colours with a Part Command, contends PLC's in net and sensors and one Operative Part contend, electric engines, robotic manipulator and pumps.

For the integration these elements, it was used of a consistent language as the SFC therefore, with this, the interpretation easiness and an adequate structure favour the reach of the conclusion objective of integration between all the components.

In this paper, the structure of the tasks and stages to be carried through guarantee the versatility and a bigger flexibility the set, through possibility implementation of new products (colours).

The Platform called SIMTIA - Integrated System for Inks Mixture Automated - it was divided in some work station, being that each work station contain its Command Part and its Operative Part, except the PLC that to property Command Part only and the rotating table that to property Operative Part only. These workstations are:

- Robotic work station for choice of the colours: it has as PC the entrances and exits of the robotic manipulator and microcomputer, and the pumps and the robotic manipulator as PO;
- Robotic work station for laundering and mixture of the ink: it has as PC the entrances and exits of the robotic manipulator and microcomputer and the pump, the mixer and the robotic manipulator as PO;
- Rotative table;
- PLC

The Figure 2, shows the workstation of the Platform and interaction of PC and PO through sensors and actuators, as described previously:

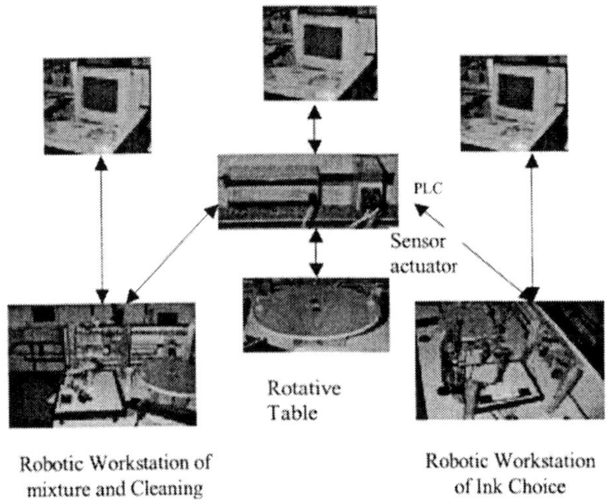

Fig. 2. SIMTIA – Workstation

One complete picture of the system is demonstrated in Figure 3.

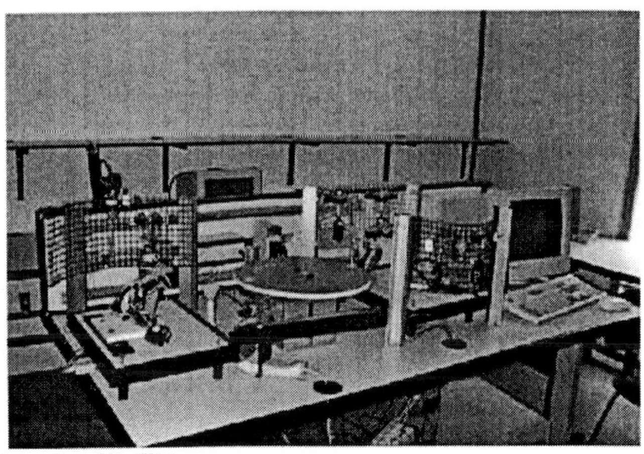

Fig. 3. - SIMTIA

2.5 Education Modulo.

Through the modifications appeared currently in the knowledge area and the learning processes, the distance education (EAD) started to have a great focus as the diffusion of the knowledge

Being the main characteristic of the EAD the establishment of communication for two-ways, which professor and student don't meet together exactly in the surrounding physicist, needing ways that make possible the communication between them (Rocha, 2002), used the TelEduc tool, developed for the Informatics Nucleus in the Education of the UNICAMP.

The Figure 4 show principal site this tool.

Fig. 4. TelEduc Tool – site of entrance the tool

The structure of the course, such as the proper platform, was modulated, when the stage executed to get a level of knowledge.

The modules created are:
Module 0: Introduction the Automated System
Module 1: Description the elements basic of the Automated
Module 2: Structure
Module 3: Unit I: Computer Implementation of the Command Part
Module 4: Part II: Robotic Manipulator Systems
Module 5: Integration
Module 6: SCADA and Communication Networks

For each module theoretical and practical activities are developed that to provide at to supply the necessary minimum knowledge for the understanding the automated production system, vision that the student will have to get the end of the last module only.

Figures 5, 6 and 7, show some screens used for the system, for to put the support materials, readings and activities to the students.

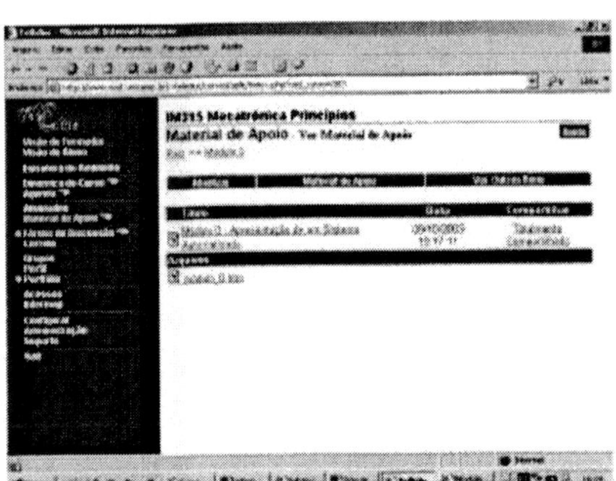

Fig. 5 – TelEduc support material site.

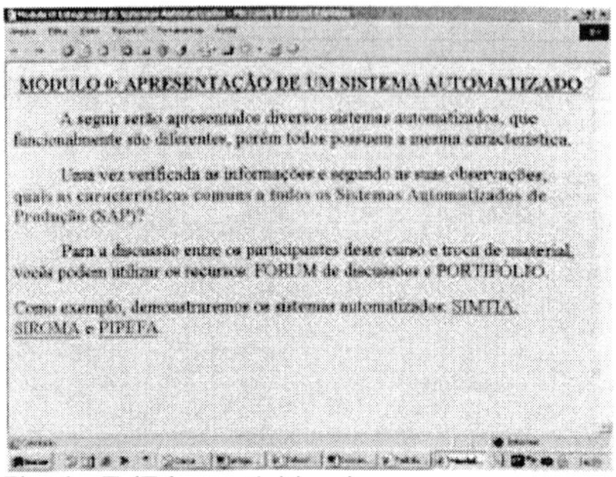

Fig. 6 – TelEduc: activities site.

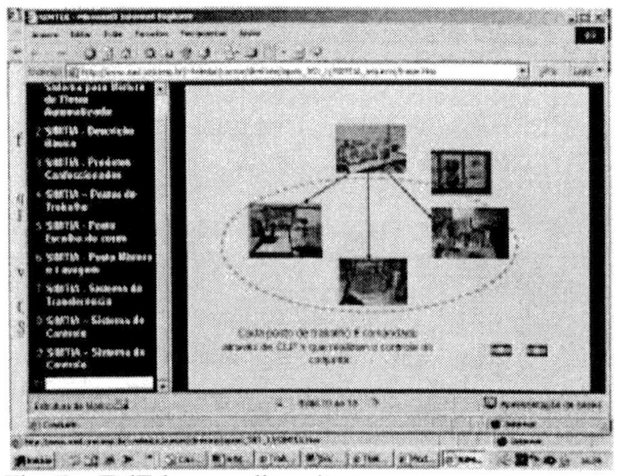

Fig. 7 – TelEduc: reading site.

3. CONCLUSION

This paper, the basic elements of an automated production system had been boarded, involving since its description until its implementation. A methodology for the education this area was also waked up using themselves a tool for distance education, verifying themselves therefore, that more important that to know to implement such system, it is to know to define its tasks and to register them. Amongst the existing forms to describe a system, we used sequential system (SFC - Sequential Function Chart).

All the project and implementation of the Automated System of Production, had been happened through Didactic Platform (SIMTIA), contend a Part Command with PLC's, sensors and microcomputers and an Operative Part with robotic manipulators, rotating table and diverse actuators. The platform was developed of form to allow the inclusion of new technologies.

The complexity of the automated system was possible due to systematization and the structure accepted for the development of the work. The found difficulties of integration, had been in the accomplishment of the communication enter the different forms of programming/software adopted for each element used.

REFERENCES

Acar, M. Mechatronics Engineering Education in the UK. In: *Joint Hungarian British International Mechatronics* Conference, 1994, Budapeste. Proceedings... Budapest: Computational Mechatronics Publ., 1994, p 763- 769.

Aihara, Cintia Kimie, *Projeto e Implementação de uma Plataforma Didática para ensino e pesquisa em automação*: Campinas: Faculdade de Campinas: Faculdade de Engenharia Mecânica, Universidade Estadual de Campinas, 2000, 109 p. Tese (Mestrado)

Araújo, Emerson dos Santos, *Modelagem e Descrição da Parte Comando de um Sistema Automatizado de Produção utilizando o GRAFCET - Aplicado a uma Plataforma Industrial em Automação*. Campinas: Faculdade de Engenharia Mecânica, Universidade Estadual de Campinas, 1997, 91 p. Tese (Mestrado)

Bittar, R. C. S. M. *A Utilização do GRAFCET como Ferramenta na Automação Industrial.* Campinas: Faculdade de Engenharia Mecânica, Universidade Estadual de Campinas, 1993, 105 p. Tese (Mestrado)

Castrucci, Plínio B. L. *Controle Automático Teoria e Projeto.* Editora da Universidade de São Paulo, 1969, 280 p

Dalbó, R. F. Simgraf: *Um Ambiente Computacional para Simulação e Validação de Sistemas Automatizados de Produção Utilizando o GRAFCET.* Campinas: Faculdade de Engenharia Mecânica, Universidade Estadual de Campinas, 1994, 90 p. Tese (Mestrado)

Georgini, João Marcelo, *Elementos para Estruturação e Implementação de Sistemas Automatizados de Produção*. Campinas: Faculdade de Engenharia Mecânica, Universidade Estadual de Campinas, 1999. 209 p. Tese (Mestrado).

IEC - International Eletrotechnical Commission, Genebra. IEC-60848: *Preparation of Function Charts for Control Systems*. Genebra, 1988, 99p.

Kuo, B. C., *Automatic Control Systems* 4th ed. USA: Pratice & Hall, Inc., New Jersey, 1982, 720 p.

Miyagi, P. E., Barretto, M. R. P., Silva, J. R. *Domótica – Controle e Automação* **Volume 2** EDUSP, 1993, 180 p.

Rocha, Heloísa Vieira, *O Ambiente TelEduc para Educação a Distância baseada na Web: Princípios, Funcionalidades e Perspectivas de Desenvolvimento* Campinas: UNICAMP/NIED, 2002 p. 197-205

Santos, José J. Horta, *Automação Industrial* Rio de Janeiro: Livros Técnicos e Científicos, 1979, p. 10 - 23.

Vendrameto, *O. Bases de Conhecimento para a Automação da Manufatura.* São Paulo: Escola Politécnica - Engenharia de Produção, Universidade de São Paulo, 1994. 155 p. Tese (Doutorado).

Williams, D. J. *Manufacturing Systems an introduction to the technologies,* **2nd edition**, Chapman & Hall, London, UK, 1994, 246 p.

AUTHOR INDEX